INTERMEDIATE ALGEBRA

sixth edition

C. L. JOHNSTON
ALDEN T. WILLIS
JEANNE LAZARIS

PWS PUBLISHING COMPANY

I(T)P • An International Thomson Publishing Company

Boston • Albany • Bonn • Cincinnati • Detroit • London • Madrid • Melbourne • Mexico City New York • Paris • San Francisco • Singapore • Tokyo • Toronto • Washington

This book is dedicated to our students, who inspired us to do our best to produce a book worthy of their time.

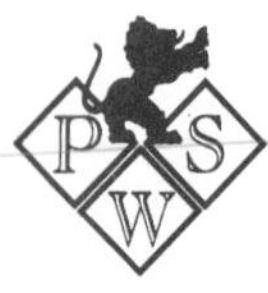

PWS PUBLISHING COMPANY
20 Park Plaza, Boston, MA 02116-4324

Library of Congress Cataloging-in-Publication Data
Johnston, C. L. (Carol Lee).
Intermediate algebra / Johnston, Willis, Lazaris. -- 6th ed.
p. cm.
Includes index.
ISBN 0-534-94470-1
1. Algebra. I. Willis, Alden T. II. Lazaris, Jeanne. III. Title.
QA154.2.J63 1994 94-13965
512.9--dc20 CIP

I(T)P™
International Thomson Publishing
The trademark ITP is used under license.
TI-81 is a registered trademark of Texas Instruments, Inc.

Printed and bound in the United States of America.
01 02 03 04 05 – 10 9 8 7 6 5 4 3

For more information, contact:

PWS Publishing Company
20 Park Plaza
Boston, MA 02116

International Thomson Publishing Europe
Berkshire House I68-I73
High Holborn
London WC1V 7AA
England

Thomas Nelson Australia
102 Dodds Street
South Melbourne, 3205
Victoria, Australia

Nelson Canada
1120 Birchmont Road
Scarborough, Ontario
Canada M1K 5G4

International Thomson Editores
Campos Eliseos 385, Piso 7
Col. Polanco
11560 Mexico D.F., Mexico

International Thomson Publishing GmbH
Königswinterer Strasse 418
53227 Bonn, Germany

International Thomson Publishing Asia
221 Henderson Road
#05-10 Henderson Building
Singapore 0315

International Thomson Publishing Japan
Hirakawacho Kyowa Building, 31
2-2-1 Hirakawacho
Chiyoda-ku, Tokyo 102
Japan

Sponsoring Editor *Susan McCulley Gay*
Editorial Assistant *Hattie Schroeder*
Developmental Editor *Maureen Brooks/Elizabeth Rogerson*
Production Coordinator *Elise S. Kaiser*
Marketing Manager *Marianne C. P. Rutter*
Manufacturing Coordinator *Marcia A. Locke*
Production *Lifland et al., Bookmakers*

Interior/Cover Designer *Elise S. Kaiser*
Interior Illustrator *Scientific Illustrators*
Cover Photo *Jeff Hunter/© The Image Bank*
Compositor *Beacon Graphics Corporation*
Cover Printer *New England Book Components, Inc.*
Text Printer and Binder *R. R. Donnelley/Willard*

CONTENTS

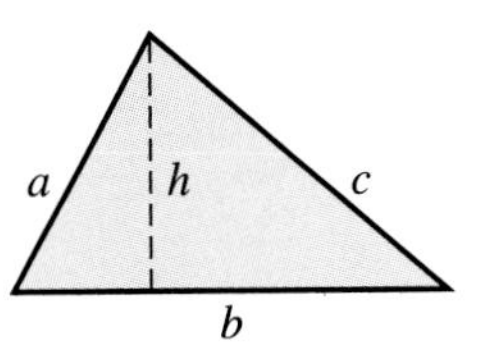
a
h
c
b

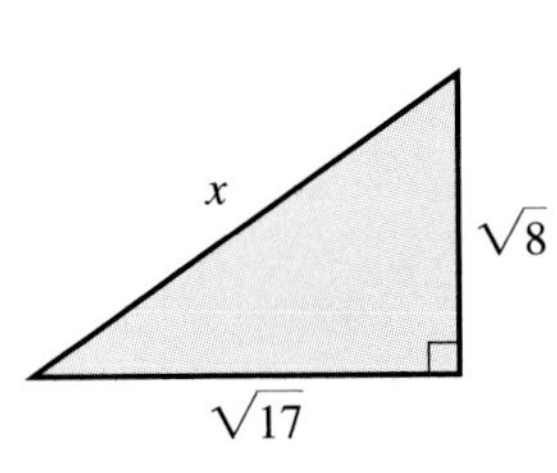
x
√8
√17

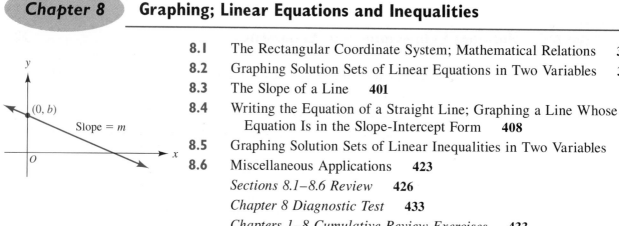

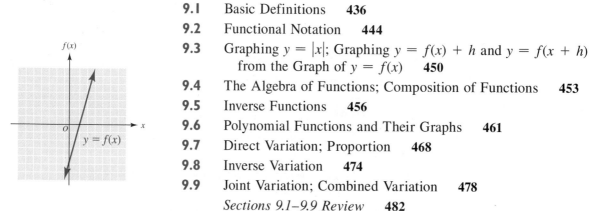

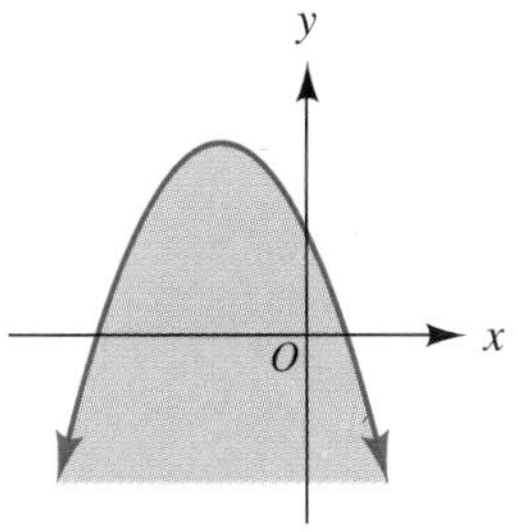

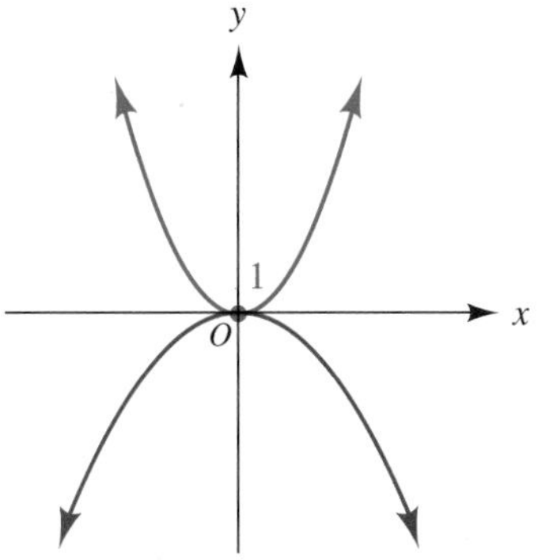

INTERMEDIATE ALGEBRA, Sixth Edition, can be used in an intermediate algebra course in any community college or four-year college, either with a lecture format, in a learning laboratory setting, or for self-study. The main goal of this book is to prepare students for courses in college algebra, statistics, business calculus, pre-calculus, science, or any other subject that has intermediate algebra as a prerequisite.

Changes in the Sixth Edition

This edition incorporates changes that resulted from many helpful comments from users of the first five editions, as well as from the authors' own classroom experience in teaching from the book. The major changes in the Sixth Edition include the following:

1. Writing Problems have been added throughout the book. We believe that writing about mathematics helps students develop a deeper understanding of the subject. Because communication is one of the essential goals of mathematics, we think that students should be encouraged to express mathematical concepts in writing.
2. Critical Thinking and Problem-Solving Exercises have been added at the end of each chapter. These exercises give the student an opportunity to try types of problems not often seen in developmental mathematics books. Techniques used for solving these problems include, but are not limited to, the following: using logic, looking for patterns or relationships, making organized lists, and guessing-and-checking. Some open-ended critical-thinking questions are asked in these exercises, also. Many of the problems are suitable for small-group activities.
3. The objectives for every section are enumerated in an Objectives Checklist found in the printed Test Bank.
4. The use of a graphics calculator is integrated into the text slowly but thoroughly. For example, when absolute value is discussed in the text, the absolute value key on the calculator is introduced. When order of operations is discussed, students are warned about potential problems (especially regarding juxtaposition) that may arise with graphics calculators, and some of the exercises involve using calculators. At the appropriate times, the graphics calculator is used for graphing straight lines, graphing parabolas and other conic sections, graphing exponential and logarithmic functions, solving systems of equations, and evaluating binomial coefficients.
5. At the request of several reviewers, there is no longer a single chapter devoted entirely to solving applied problems. Applied problems now appear more uniformly throughout the text, and more applied problems have been added, including some types of uniform motion problems not previously covered.
6. The number of subsections has been reduced. In some cases, short, closely related subsections were combined; in other cases, subsections were changed to regular sections.

7. New sections on the algebra of functions and composition of functions have been added.

8. Problems (in examples and in exercises) that require the use of the binomial theorem are now marked by a star (★) so that those instructors who wish to do so can postpone or omit discussion of the binomial theorem. At the request of some of the reviewers, more problems have been included in which the student is asked to find just one term of a binomial expansion.

9. Problems have been added in which it is particularly helpful to use a calculator either to factor a quadratic polynomial by completing the square or to solve a quadratic equation by completing the square.

10. Problems have been added in which students are asked to graph a straight line by using the slope and the y-intercept.

11. In the chapter on rational expressions, finding the domain of the variable has been postponed until solving rational equations is discussed, and problems have been added that involve using the correct order of operations with rational expressions.

Features of This Book

The major features of this book include the following:

1. Over 250 Writing Problems in exercise sets throughout the book motivate students to think about mathematical concepts.

2. Over 70 Critical Thinking and Problem-Solving Exercises at the ends of the chapters stimulate creativity in problem solving.

3. The use of the calculator is emphasized throughout the book. The examples and exercises that lend themselves particularly well to the use of a calculator are designated with the corresponding calculator icon: for a scientific calculator, for an RPN calculator, and for a graphics calculator.

4. Two or more approaches to the same problem are often discussed because we believe that students need to learn that there is no single correct method for solving a given problem, and that learning mathematics is more than memorizing a set of rules for solving each kind of problem.

5. For the most part, this book uses a one-step, one-concept-at-a-time approach. That is, major topics are divided into small sections, each with its own examples and often with its own exercises. This approach allows students to master each topic before proceeding confidently to the next section.

6. Many concrete, annotated examples illustrate the general principles covered in each section.

7. In special "Words of Caution" screened boxes, students are warned against common algebraic errors.

8. Interspersed throughout the text and the examples are "Notes" that provide students with additional information regarding the topic or example under discussion. These notes are easily identified with the pointing-finger icon:

9. Visual aids such as shading, color, and annotations guide students through the examples. This new edition makes use of a third color to better differentiate elements in figures and examples.

10. The solutions for many of the examples are explicitly checked in the text, and the importance of checking solutions is emphasized throughout the book.

11. Important concepts and algorithms are enclosed in boxes for easy identification and reference.

12. Examples are given of the problem-solving techniques of making an organized list and of guessing-and-checking.

13. The approach to solving applied problems includes a detailed method for translating English statements into algebraic equations or inequalities and a step-by-step outline of a procedure that can be used in solving many applied problems.

14. A review section with Set I and Set II exercises appears at the end of each chapter, and some chapters also have a mid-chapter review that includes Set I and Set II exercises.

15. The book contains over 7,000 exercises, over 15% of which are new to this series.

    Set I Exercises All the answers for the Set I Exercises are included in the back of the book, along with the **completely worked-out solutions** for nearly all the odd-numbered exercises. Most of the even-numbered Set I Exercises are matched to the odd-numbered exercises. Thus, students can use the solutions of the odd-numbered exercises as examples and study aids for doing the even-numbered Set I exercises.

    Set II Exercises The answers for the Set II exercises are given in the *Annotated Instructor's Edition* and in the printed Test Bank. The odd-numbered exercises of Set II are usually matched to the odd-numbered exercises of Set I.

    Set II Review Exercises The Set II Review Exercises allow space for working problems and for writing answers; the pages can be removed from the book for grading without interrupting the continuity of the text.

16. A Diagnostic Test at the end of each chapter can be used for study and review or as a pretest. Complete solutions to all the problems in these diagnostic tests, together with section references, appear in the answer section at the back of this book.

17. A set of Cumulative Review Exercises is included at the end of each chapter except Chapter 1; the answers are in the answer section at the back of this book.

18. For quick reference, a list of the symbols used in this book appears on the inside front cover, and a list of some algebraic formulas and properties appears on the last book pages and the inside back cover.

Major topics are divided into small manageable sections, as part of the book's one-step, one-concept-at-a-time approach

Boxes enclose important concepts and algorithms, for easy identification and reference

"Notes" to students provide additional information or point out problem-solving hints

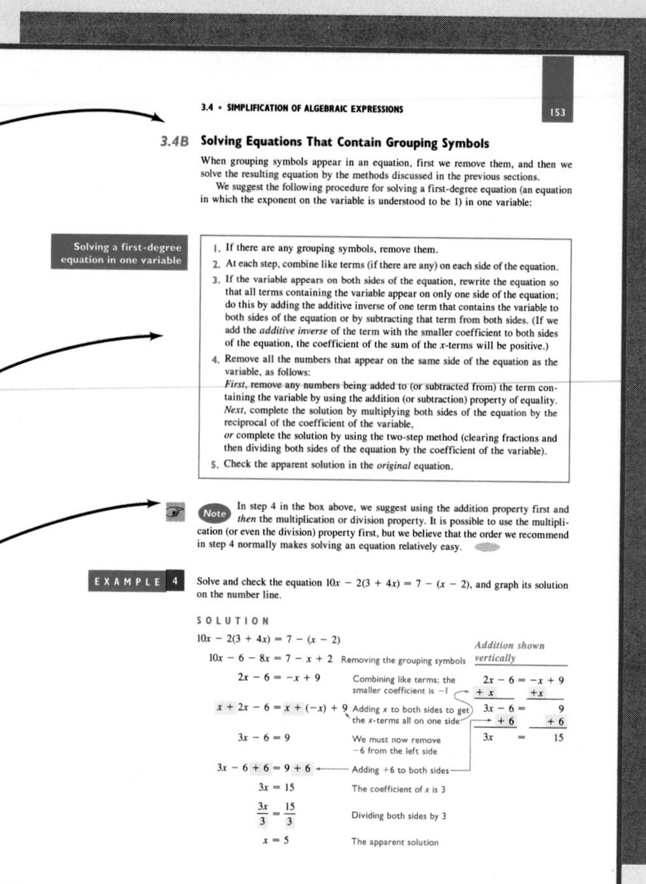

3.4 • SIMPLIFICATION OF ALGEBRAIC EXPRESSIONS 153

3.4B Solving Equations That Contain Grouping Symbols

When grouping symbols appear in an equation, first we remove them, and then we solve the resulting equation by the methods discussed in the previous sections.

We suggest the following procedure for solving a first-degree equation (an equation in which the exponent on the variable is understood to be 1) in one variable:

Solving a first-degree equation in one variable

1. If there are any grouping symbols, remove them.
2. At each step, combine like terms (if there are any) on each side of the equation.
3. If the variable appears on both sides of the equation, rewrite the equation so that all terms containing the variable appear on only one side of the equation; do this by adding the additive inverse of one term that contains the variable to both sides of the equation or by subtracting that term from both sides. (If we add the *additive inverse* of the term with the smaller coefficient to both sides of the equation, the coefficient of the sum of the x-terms will be positive.)
4. Remove all the numbers that appear on the same side of the equation as the variable, as follows:
   First, remove any numbers being added to (or subtracted from) the term containing the variable by using the addition (or subtraction) property of equality.
   Next, complete the solution by multiplying both sides of the equation by the reciprocal of the coefficient of the variable,
   or complete the solution by using the two-step method (clearing fractions and then dividing both sides of the equation by the coefficient of the variable).
5. Check the apparent solution in the *original* equation.

Note In step 4 in the box above, we suggest using the addition property first and *then* the multiplication or division property. It is possible to use the multiplication (or even the division) property first, but we believe that the order we recommend in step 4 normally makes solving an equation relatively easy.

EXAMPLE 4 Solve and check the equation $10x - 2(3 + 4x) = 7 - (x - 2)$, and graph its solution on the number line.

SOLUTION

$10x - 2(3 + 4x) = 7 - (x - 2)$

$10x - 6 - 8x = 7 - x + 2$ Removing the grouping symbols

$2x - 6 = -x + 9$ Combining like terms; the smaller coefficient is -1

$x + 2x - 6 = x + (-x) + 9$ Adding x to both sides to get the x-terms all on one side

$3x - 6 = 9$ We must now remove -6 from the left side

$3x - 6 + 6 = 9 + 6$ Adding $+6$ to both sides

$3x = 15$ The coefficient of x is 3

$\frac{3x}{3} = \frac{15}{3}$ Dividing both sides by 3

$x = 5$ The apparent solution

Addition shown vertically

$$\begin{array}{rcr} 2x - 6 &=& -x + 9 \\ +x & & +x \\ \hline 3x - 6 &=& 9 \\ +6 & & +6 \\ \hline 3x &=& 15 \end{array}$$

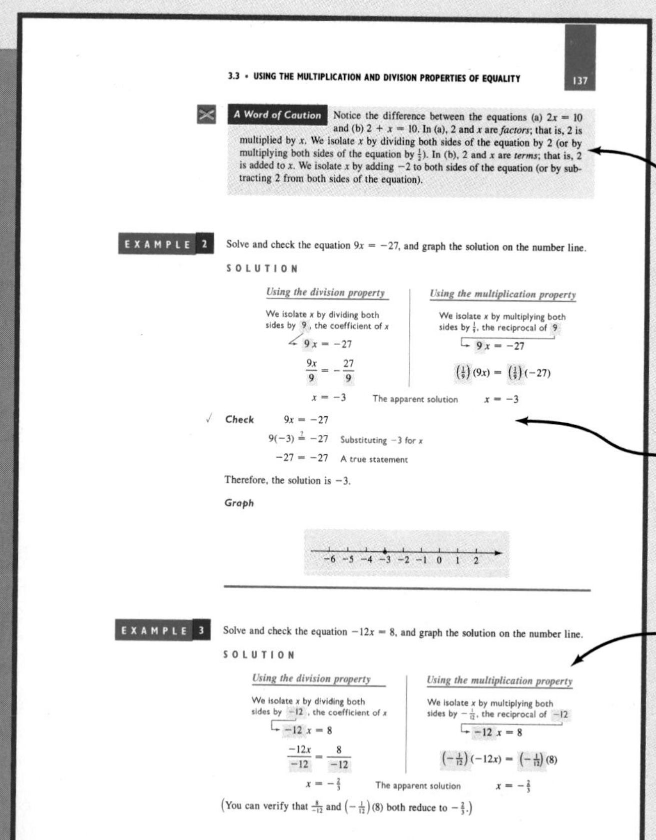

3.3 • USING THE MULTIPLICATION AND DIVISION PROPERTIES OF EQUALITY 137

A Word of Caution Notice the difference between the equations (a) $2x = 10$ and (b) $2 + x = 10$. In (a), 2 and x are *factors*; that is, 2 is multiplied by x. We isolate x by dividing both sides of the equation by 2 (or by multiplying both sides of the equation by $\frac{1}{2}$). In (b), 2 and x are *terms*; that is, 2 is added to x. We isolate x by adding -2 to both sides of the equation (or by subtracting 2 from both sides of the equation).

EXAMPLE 2 Solve and check the equation $9x = -27$, and graph the solution on the number line.

SOLUTION

Using the division property		*Using the multiplication property*
We isolate x by dividing both sides by 9, the coefficient of x		We isolate x by multiplying both sides by $\frac{1}{9}$, the reciprocal of 9
$9x = -27$		$9x = -27$
$\frac{9x}{9} = -\frac{27}{9}$		$\left(\frac{1}{9}\right)(9x) = \left(\frac{1}{9}\right)(-27)$
$x = -3$	The apparent solution	$x = -3$

√ **Check** $9x = -27$

$9(-3) \stackrel{?}{=} -27$ Substituting -3 for x

$-27 = -27$ A true statement

Therefore, the solution is -3.

Graph

EXAMPLE 3 Solve and check the equation $-12x = 8$, and graph the solution on the number line.

SOLUTION

Using the division property		*Using the multiplication property*
We isolate x by dividing both sides by -12, the coefficient of x		We isolate x by multiplying both sides by $-\frac{1}{12}$, the reciprocal of -12
$-12x = 8$		$-12x = 8$
$\frac{-12x}{-12} = \frac{8}{-12}$		$\left(-\frac{1}{12}\right)(-12x) = \left(-\frac{1}{12}\right)(8)$
$x = -\frac{2}{3}$	The apparent solution	$x = -\frac{2}{3}$

(You can verify that $\frac{8}{-12}$ and $\left(-\frac{1}{12}\right)(8)$ both reduce to $-\frac{2}{3}$.)

In special screened boxes labeled "A Word of Caution," students are warned against common errors

The importance of checking solutions is reiterated throughout the book

Two or more approaches to solving a problem are often presented side by side

458 8 • RATIONAL EXPRESSIONS

In Exercises 16–20, set up each problem algebraically, solve, and check. Be sure to state what your variables represent.

16. The sum of two consecutive integers is 33. What are the integers?
17. The sum of two numbers is 5. Their product is −24. What are the numbers?
18. A dealer makes up a 15-lb mixture of oranges. One kind costs 78¢ per pound, and the other costs 99¢ per pound. How many pounds of each kind must be used in order for the mixture to cost 85¢ per pound?
19. Manny has twenty coins with a total value of $1.65. If the coins are all nickels and dimes, how many of each does he have?
20. Ricardo drove at a certain rate for 4 hr. If he had been able to drive 11 mph faster, the trip would have taken 3 hr. How fast did he drive? How far did he drive?

Critical Thinking and Problem-Solving Exercises

1. One day recently, Tad, Ted, and Adam were eating apples. Tad and Adam each ate the same number of apples, and they each ate at least one apple. Ted ate the most apples, and he ate fewer than 10. The *product* of the numbers of apples eaten by all three people was 12. How many apples did each person eat?
2. Julia, Ned, Sean, and Tiffany met for breakfast. Each ordered a different item for breakfast, and each was in a different type of business. The items ordered were pancakes, Belgian waffles, eggs Benedict, and oatmeal. Using the following clues, determine who sat in which chair, what business each person was in, and what each person ordered for lunch.

   The dentist was not sitting in seat 2 or 3.
   The machinist sat opposite the sales associate, who ordered eggs Benedict.
   Julia sat in seat 2.
   Tiffany sat in an odd-numbered seat, opposite the person who ordered pancakes.
   The machinist sat in seat 4 and ate oatmeal.
   Ned is a dentist.
   One person is an optician.

3. Three friends were comparing notes about reading. They discovered that in the last four months, the total number of books that Victor and Nina had read was 16, the total number of books that Nina and Ryan had read was 20, and the total number of books that Victor and Ryan had read was 22. How many books had each person (individually) read?
4. Consider the following algebraic expressions: $x + y$, $x - y$, $y - x$, $y + x$, $-y + x$, $-x + y$, $-x - y$, and $-y - x$.
   a. Name (if there are any) the *pairs* that are *equal* to each other.
   b. Name all the expressions that are the *additive inverses* of $x + y$.
   c. Name all the expressions that are the *additive inverses* of $x - y$.
5. The solution of each of the following problems contains an error. Find and describe the error, and solve each problem correctly.
   a. Simplify $\frac{4}{x} + \frac{2}{x+2}$.
   LCD $= x(x + 2)$

$$\overset{1}{\cancel{x}}(x+2)\frac{4}{\underset{1}{\cancel{x}}} + x\overset{1}{\cancel{(x+2)}}\frac{2}{\underset{1}{\cancel{x+2}}} = 4x + 8 + 2x = 6x + 8$$

   b. Simplify $\frac{1 - \frac{1}{x^2}}{1 - \frac{1}{x}}$.

$$\frac{x^2\left(1 - \frac{1}{x^2}\right)}{x\left(1 - \frac{1}{x}\right)} = \frac{x^2 - 1}{x - 1}$$
$$= \frac{(x+1)\cancel{(x-1)}}{\cancel{x-1}}$$
$$= x + 1$$

NEW!

"Critical Thinking and Problem-Solving Exercises" have been added at the end of every chapter

Large, screened cross-out marks indicate incorrect procedures and anticipate common stumbling blocks for students

All new artwork enhances students' ability to visualize problems and examples

Fully worked out solutions are given, with careful one-step-at-a-time explanation provided and problem-solving points highlighted

Calculator icons designate examples and exercises appropriate for solving with a scientific or graphics calculator

"Writing Problems" have been added throughout the text and are denoted by a pencil icon

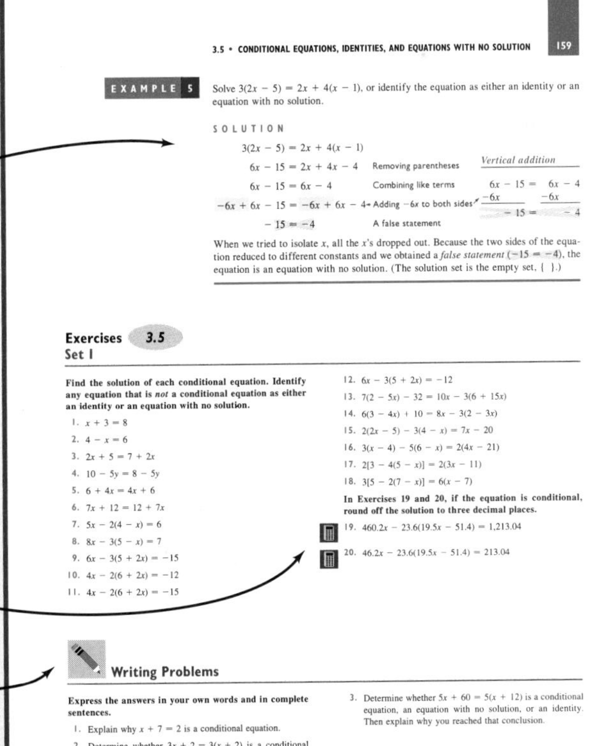

3.5 • CONDITIONAL EQUATIONS, IDENTITIES, AND EQUATIONS WITH NO SOLUTION 159

EXAMPLE 5 Solve $3(2x - 5) = 2x + 4(x - 1)$, or identify the equation as either an identity or an equation with no solution.

SOLUTION

		Vertical addition
$3(2x - 5) = 2x + 4(x - 1)$		
$6x - 15 = 2x + 4x - 4$	Removing parentheses	
$6x - 15 = 6x - 4$	Combining like terms	$6x - 15 = 6x - 4$
$-6x + 6x - 15 = -6x + 6x - 4$	Adding $-6x$ to both sides	$-6x \quad\quad -6x$
$-15 = -4$	A false statement	$-15 = \quad -4$

When we tried to isolate x, all the x's dropped out. Because the two sides of the equation reduced to different constants and we obtained a *false statement* ($-15 = -4$), the equation is an equation with no solution. (The solution set is the empty set, { }.)

Exercises 3.5
Set I

Find the solution of each conditional equation. Identify any equation that is *not* a conditional equation as either an identity or an equation with no solution.

1. $x + 3 = 8$
2. $4 - x = 6$
3. $2x + 5 = 7 + 2x$
4. $10 - 5y = 8 - 5y$
5. $6 + 4x = 4x + 6$
6. $7x + 12 = 12 + 7x$
7. $5x - 2(4 - x) = 6$
8. $8x - 3(5 - x) = 7$
9. $6x - 3(5 + 2x) = -15$
10. $4x - 2(6 + 2x) = -12$
11. $4x - 2(6 + 2x) = -15$
12. $6x - 3(5 + 2x) = -12$
13. $7(2 - 5x) - 32 = 10x - 3(6 + 15x)$
14. $6(3 - 4x) + 10 = 8x - 3(2 - 3x)$
15. $2(2x - 5) - 3(4 - x) = 7x - 20$
16. $3(x - 4) - 5(6 - x) = 2(4x - 21)$
17. $2[3 - 4(5 - x)] = 2(3x - 11)$
18. $3[5 - 2(7 - x)] = 6(x - 7)$

In Exercises 19 and 20, if the equation is conditional, round off the solution to three decimal places.

19. $460.2x - 23.6(19.5x - 51.4) = 1,213.04$
20. $46.2x - 23.6(19.5x - 51.4) = 213.04$

Writing Problems

Express the answers in your own words and in complete sentences.

1. Explain why $x + 7 = 2$ is a conditional equation.
2. Determine whether $3x + 2 = 3(x + 2)$ is a conditional equation, an equation with no solution, or an identity. Then explain why you reached that conclusion.
3. Determine whether $5x + 60 = 5(x + 12)$ is a conditional equation, an equation with no solution, or an identity. Then explain why you reached that conclusion.

Using This Book

As mentioned earlier, *Intermediate Algebra,* Sixth Edition, can be used in three types of instructional programs: lecture, laboratory, and self-study.

The Conventional Lecture Course This book has been class-tested and used successfully in conventional lecture courses by the authors and by many other instructors. It is not a workbook, and therefore it contains enough material to stimulate classroom discussion. Examinations for each chapter are provided in the printed Test Bank, and two different kinds of computer software enable instructors to create their own tests. Tutorial software is available to help students who require extra assistance.

The Learning Laboratory Class This text has also been used successfully in many learning labs. The format of explanation, example, and exercises in each section of the book and the tutorial software make the book easy to use in laboratories. Students can use the diagnostic test at the end of each chapter as a pretest or for review and diagnostic purposes. Because several forms of each chapter test are available in the Test Bank and because test generators are available, a student who does not pass a test can review the material covered on that test and can then take a different form of the test.

Self-Study This book lends itself to self-study because each new topic is short enough for the student to master before continuing, and because more than 700 examples and over 1,700 completely solved exercises show students exactly how to proceed. Students can use the Diagnostic Test at the end of each chapter to determine which parts of that chapter they need to study and can thus concentrate on those areas in which they have weaknesses. The tutorial software extends the usefulness of the new edition in laboratory and self-study settings.

In addition to assigning the writing problems that appear in this text, instructors might also encourage students to keep a journal and/or to do other writing about their feelings about mathematics.

Please note that the even-numbered problems of Set II can be especially challenging to students, as these problems are not necessarily matched to any other problems, there are some problems included for which no examples are given, and there are no answers for these problems in the student edition of the text. (Some instructors assign only these problems as homework to be collected and graded.)

Ancillaries

The following ancillaries are available with this text:

- ***Annotated Instructor's Edition*** includes the complete student text along with answers to most of the exercises printed adjacent to each exercise in a fourth color—purple. Also included are five teaching essays, written by leading mathematics educators, offering ideas to both new and experienced instructors on teaching developmental mathematics.
- ***Test Bank with Answers to Set II Exercises*** includes five different exams for each chapter, two forms for each of three unit exams, and two forms of the final exam. All these exams, which can be easily removed and duplicated for class use, are prepared with adequate workspace; answer keys for these exams, with complete solutions, are provided in this printed manual. The Test Bank also contains the answers to the Set II Exercises and answers to many of the Critical Thinking and Problem-Solving Exercises. An Objectives Checklist is also included, which lists objectives for each section of the text.

- ***Computerized Testing Programs*** are available. **EXPTest** is available in both Windows and DOS versions for the IBM PC and compatibles; **ExamBuilder** is available for the Macintosh. Questions are multiple choice, true-false, and open-ended. Instructors can add to, delete, or change existing questions and produce individual tests. Demos are available.
- ***MathQUEST™ Tutorial Software*** is a text-specific, intuitive tutorial, which runs on Microsoft Windows and Macintosh platforms. It allows students to practice the skills taught in the textbook section by section. The tutorial provides hints at the first wrong answer and a detailed solution at the second wrong answer. Students can also view the detailed solution after working the problem. The format is mainly fill-in-the-blank, with some multiple choice. This diagnostic program keeps track of right and wrong responses and can report on the student's progress to the instructor. A demo is available.
- ***DOS Tutorial Software*** is a text-specific tutoring software system, available for IBM PC and compatibles with DOS. It uses an interactive format to tailor lessons to the specific learning problems of students. The multiple-choice questions help students master mathematical ideas and procedures. Demos are available.
- ***Videotape Series*** is a new series that teaches key topics in the text and features a professional math instructor.

Acknowledgments

We wish to thank the members of the editorial and production staff at PWS Publishing Company for their help with this edition; special thanks go to Susan McCulley Gay, Elise Kaiser, Elizabeth Rogerson, Maureen Brooks, and Judith Mustacchia. Special thanks, too, to Sally Lifland and her staff at Lifland et al., Bookmakers, for their help in the production of this book. In particular we wish to thank Susan Patt for adding to the manuscript the answers that now appear in the *Annotated Instructor's Edition*.

We also give special thanks to Todd Zimmermann for his help in proofreading and in checking the solutions to the problems. We are deeply grateful to Gale Hughes for preparing the *Test Bank with Answers to Set II Exercises* and for preparing the Set II answers for the *Annotated Instructor's Edition*. We would also like to thank the following reviewers for their helpful comments and significant contributions:

Elaine DiPerna
Community College of Allegheny County

Sharon J. Edgmon
Bakersfield College

Doris S. Edwards
Motlow State Community College

James E. Hodge
College of West Virginia

Jeannine Hugill
Highland Community College

Vera Preston
Austin Community College

Ray Stanton
Fresno City College

James A. Wood
Virginia Commonwealth University

Kenneth Word
Central Texas College

To the Student

Mathematics is not a spectator sport! Mathematics is learned by *doing,* not by watching. (The fact that you "understand everything the teacher does on the board" does not mean that you have learned the material!) No textbook can teach mathematics to you, and no instructor can teach mathematics to you; the best that both can do is to help *you* learn.

If you have not been successful in mathematics courses previously and if you really want to succeed now, we strongly recommend that you obtain and study *Mastering Mathematics: How to Be a Great Math Student*, Second Edition, by Richard Manning Smith (Belmont, CA: Wadsworth Publishing Company, 1994). This little book contains *many* fine suggestions for studying, a few of which are listed below:

- Attend all classes. Missing even one class can put you behind in the course by at least two classes. To quote Mr. Smith on this topic:

  > My experience provides irrefutable evidence that you are much more likely to be successful if you attend *all* classes, always arriving on time or a little early. If you are even a few minutes late, or if you miss a class entirely, you risk feeling lost in class when you finally do get there. Not only have you wasted the class time you missed, but you may spend most or all of the first class after your absence trying to catch up. (p. 46)

- Learn as much as you can during class time.
- Sit in the front of the classroom.
- Take complete class notes.
- Ask questions about class notes.
- Solve homework problems on time (*before* the next class).
- Ask questions that deal with course material.
- Organize your notebook.
- Read other textbooks.
- Don't make excuses.

Mastering Mathematics: How to Be a Great Math Student also gives specific suggestions about getting ready for a math course even before the course begins, preparing for tests (including the final exam), coping with a bad teacher, improving your attitude toward math, using class time effectively, avoiding "mental blocks," and so forth.

Two Final Notes (1) If you see a word in this book with which you are not familiar, *look it up in the dictionary.* Learning to use a dictionary effectively is a major part of your education. (2) **It is impossible to overemphasize the value of doing homework!**

It is important for you to realize that, to a very great extent, you can take charge of and have control over your success in mathematics. We believe that studying from *Intermediate Algebra,* Sixth Edition, can help you be successful.

The Real Number System

CHAPTER 1

Algebra is the branch of mathematics that generalizes the operations of arithmetic by using letters to stand for numbers. In intermediate algebra, we review and then develop further the topics that were introduced in elementary algebra. We also discuss a number of topics that were not covered in elementary algebra.

To help you learn the language of algebra, we carefully define the symbols we use. Because we use the equal sign so frequently in algebra, we begin this chapter with a discussion of equality; later in the chapter, we discuss the relations "is less than," "is greater than or equal to," and so on. We also study sets, and we review the properties of real numbers.

1.1 Equality; Sets

Equality

The Equal Sign The symbol $=$ is read is equal to. It is used in a statement when the expression to the left of the equal sign *has the same value or values* or *the same meaning* as the expression to the right of the equal sign. We can say, for example, that $1 + 7 = 3 + 5$, because both $1 + 7$ and $3 + 5$ have the same value, 8.

The Properties of Equality We frequently use the following important properties of equality:

If a, b, and c represent any numbers or expressions, then

$a = a$	The reflexive property of equality
If $a = b$, then $b = a$.	The symmetric property of equality
If $a = b$ and $b = c$, then $a = c$.	The transitive property of equality

The *reflexive property* (which is often used in proofs in geometry) permits us to write a statement such as $3y = 3y$ in proofs.

The *symmetric property* permits us to interchange the two sides of an equation; for example, the symmetric property allows us to rewrite $5 = x$ as $x = 5$. An example from real life might be "If Joy's weight is the same as Ted's weight, then Ted's weight is the same as Joy's weight."

The *transitive property of equality* states that if one quantity equals a second quantity and *if that second quantity* equals a *third* quantity, then the first quantity equals the third quantity. Notice that the two quantities in the *middle* must be identical; for example, if $x = 2t$ and $2t = 3$, then, by the transitive property, $x = 3$. An example from real life might be "If Meg's height is the same as Roberta's height *and* if Roberta's height is the same as Juan's height, then Meg's height is the same as Juan's height." Notice that the two quantities in the *middle* are identical—both are Roberta's height.

The statement "If Meg's height is the same as Roberta's height and if Juan's height is the same as Roberta's height, then Meg's height is the same as Juan's height" is a *true* statement, but it does *not* illustrate the transitive property, because the two quantities in the middle are different from each other—one is Roberta's height, and the other is Juan's height.

The Substitution Principle We also frequently use the **substitution principle**, which is

If $a = b$, then b can be substituted for a in any expression or equation that contains a.

In other words, any number can be substituted for its equal. The statement "If $z = 5y$ and $7 = 5y$, then $z = 7$" illustrates the substitution principle. (It does *not* illustrate the transitive property; "If $z = 5y$ and $5y = 7$, then $z = 7$" would illustrate the transitive property, because the two quantities in the middle are both $5y$.)

The "Unequal to" Symbol The symbol $\neq$ is read is not equal to . It is used in a statement when the expression to the left of the equal sign *does not have exactly the same value or values* or *the same meaning* as the expression to the right of the equal sign.

Sets

A **set** is a collection of numbers, objects, or things.

EXAMPLE 1 Examples of sets:

a. The set of students registered at your college on May 7, 1995, at 8 A.M.
b. The set consisting of the numbers 2, 8, and 12
c. The set of letters in our alphabet; that is, a, b, c, d, and so forth

---

The Elements of a Set The numbers, objects, or things that make up a set are called its **elements** or **members**. A set may contain just a few elements, many elements, or no elements at all.

A set must be *well-defined*. This means that all people must be able to agree that a particular element is in a given set or is not in a given set. For example, a set consisting of the five most beautiful paintings in a museum would *not* be a well-defined set. A painting that is considered beautiful by one person might not be considered beautiful by all other people.

Roster Notation for Representing a Set A roster is a list of the members of a group. To use **roster notation** for representing a set, we list the elements of the set, putting commas between the elements, and we enclose the list within *braces*, { }. For example, roster notation for the set consisting of the numbers 2, 8, and 12 is {2, 8, 12}. We *never* enclose the elements of a set within parentheses, ( ), or brackets, [ ]. Thus, (3, 4, 5) and [7, 8, 9, 10] do *not* denote sets.

EXAMPLE 2 Examples of the elements of sets and roster notation:

a. Set {5, 7, 9} has elements 5, 7, and 9.
b. Set $\{a, f, h, k\}$ has elements a, f, h, and k.

Note It is customary, though not necessary, to arrange numbers and letters in sets in numerical and alphabetical order (to make reading easier) and to represent elements of a set with lowercase letters.

Naming a Set A set is usually named by a capital letter. The expression $A = \{1, 5, 7\}$ can be read "A equals the set whose elements are 1, 5, and 7" or "A is the set whose elements are 1, 5, and 7."

The Set of Digits One important set of numbers is the set of **digits**. This set contains the numerals 0, 1, 2, 3, 4, 5, 6, 7, 8, and 9. These symbols make up our entire number system; *any* number can be written by using some combination of these numerals.

Modified Roster Notation If the number of elements in a set is so large that it is either inconvenient or impossible to list them all, we modify the roster notation. For example, the set of digits could be represented as follows:

$$D = \{0, 1, 2, \ldots, 9\}$$

This is read "D equals the set whose elements are 0, 1, 2, and so on, up to 9." The three dots (called an ellipsis) to the right of the number 2 indicate that the remaining numbers are to be found in the same way we have begun—namely, by adding 1 to each number to find the next number, until we reach 9, the last number in the set. Throughout this chapter, we will use the letter D for naming the set of digits.

The Set of Natural Numbers The numbers 1, 2, 3, 4, 5, 6, and so on, are called **natural numbers** or **counting numbers**. Throughout this text, we will use the letter N for naming this set. Since the set of natural numbers has no largest element, we represent it as follows:

$$N = \{1, 2, 3, 4, \ldots\}$$

This is read "N equals the set whose elements are 1, 2, 3, 4, and so on."

Even Numbers Natural numbers that end in 0, 2, 4, 6, or 8 are called **even natural numbers**.

Odd Numbers Natural numbers that end in 1, 3, 5, 7, or 9 are called **odd natural numbers**.

Consecutive Numbers Numbers that follow one another, in sequence, without interruption are called **consecutive numbers**. For example, 15, 16, 17, and 18 are four consecutive numbers beginning with 15.

Equal Sets Sets A and B are said to be **equal sets**, and we write $A = B$, if every element of A is in B and if every element of B is in A. In other words, both sets must contain exactly the same elements; the elements do not have to be listed in the same order in the two sets.

Unequal Sets Sets E and F are said to be **unequal sets**, and we write $E \neq F$, if there is at least one element of E that is not in F or at least one element of F that is not in E.

EXAMPLE 3 Examples of equal sets and unequal sets:

a. $\{1, 5, 7\} = \{5, 1, 7\}$ Notice that both sets contain exactly the same elements, even though the elements are not listed in the same order.

b. $\{1, 5, 5, 5\} = \{5, 1\}$ Notice that both sets contain exactly the same elements, even though 5 is repeated several times in one of the rosters.

c. $\{0, 7, 11\} \neq \{7, 11\}$ These sets are *not* equal because there is one element in $\{0, 7, 11\}$ that is not in $\{7, 11\}$; that is, the two sets do not contain exactly the same elements.

Set-Builder Notation for Representing a Set In **set-builder notation**, we represent a set by writing a *statement*, or *rule*, that describes the elements of the set. This statement is enclosed within braces, but between the left brace and the statement, we place the x (or whatever variable is being described in the statement) and a vertical bar. The vertical bar is read "such that." The following is an example of set-builder notation:

$$\{x \mid x \text{ is a natural number}\}$$

The statement or rule

It is read "the set of all x such that x is a natural number."

Example 4 demonstrates converting from set-builder notation to roster notation, and Examples 5 and 6 demonstrate converting from roster notation to set-builder notation.

EXAMPLE 4 Write $\{x \mid x \text{ is an even natural number}\}$ in roster notation.

SOLUTION The natural numbers are 1, 2, 3, 4, 5, 6, . . . ; selecting the *even* ones gives the set $\{2, 4, 6, \ldots\}$. Therefore,

$$\{x \mid x \text{ is an even natural number}\} = \{2, 4, 6, \ldots\}$$

EXAMPLE 5 Write {1, 3, 5, ...} in set-builder notation.

SOLUTION We have to find a new way of describing this set, a way that starts "The set of all x such that...." Notice that all of the numbers listed are natural numbers and that they are also all odd numbers. Therefore, we can say

$$\{1, 3, 5, \ldots\} = \{x \mid x \text{ is a natural number and } x \text{ is an odd number}\}$$

or
$$\{1, 3, 5, \ldots\} = \{x \mid x \text{ is an odd natural number}\}$$

EXAMPLE 6 Write {0, 3, 6, 9} in set-builder notation.

SOLUTION We might notice that all of the numbers are digits and that they are all divisible by 3.* We could then say

$$\{0, 3, 6, 9\} = \{x \mid x \text{ is a digit and } x \text{ is divisible by } 3\}$$

or
$$\{0, 3, 6, 9\} = \{x \mid x \text{ is a digit and } x \text{ is 0 or is a multiple of } 3^{\dagger}\}$$

Note It is important to realize that other answers for Examples 5 and 6 could also be correct.

The Symbols ∈ and ∉ If we wish to show that a number, symbol, or object is an element of a given set, we use the symbol ∈, which is read is an element of. Thus, the expression $2 \in A$ is read "2 is an element of A." If we wish to show that a number, symbol, or object is *not* an element of a given set, we use the symbol ∉, which is read is not an element of.

EXAMPLE 7 Examples of using ∈ and ∉:

a. If $A = \{2, 3, 4\}$, we can say $2 \in A$, $5 \notin A$, $3 \in A$, and $4 \in A$.
b. If $F = \{x \mid x \text{ is an even natural number}\}$, then $2 \in F$, $5 \notin F$, $12 \in F$, and $\frac{2}{3} \notin F$.

The Empty Set, or Null Set A set with *no* elements in it is said to be the **empty set** (or *null set*). We use the symbol { } or the symbol ∅ to represent the empty set.

A Word of Caution {∅} is *not* the correct symbol for the empty set. {∅} is a set containing one element (namely, the symbol ∅).

EXAMPLE 8 Examples of empty sets:

a. The set of all people in your math class who are 10 ft tall
b. The set of all the digits greater than 10

Finite Sets If, in counting the elements of a set, the counting comes to an end (or *could* come to an end), the set is called a **finite set**.

---

*When we say that a number is "divisible by n," we mean that when we divide the number by n, the remainder is 0.

†When we say that a number is a "multiple of n," we mean that it is one of the numbers n, $2n$, $3n$, $4n$, $5n$, and so on, where n is any natural number.

EXAMPLE 9 Examples of finite sets:

a. $A = \{5, 9, 10, 13\}$
b. $D =$ The set of digits
c. $S = \{x \mid x \in N$ and x is between 5 and $5{,}000{,}000{,}000\}$
d. $\emptyset = \{\quad\}$

Infinite Sets If, in counting the elements of a set, the counting could never come to an end, the set is called an **infinite set**. A set is infinite if it is not finite.

EXAMPLE 10 Examples of infinite sets:

a. $N = \{1, 2, 3, \ldots\}$ The natural numbers
b. $\{x \mid x$ is an odd natural number$\}$

Subsets Set A is called a **subset** of set B if every element of A is also an element of B. "A is a subset of B" is written $A \subseteq B$.

EXAMPLE 11 Examples of subsets:

a. If $A = \{3, 5\}$ and $B = \{3, 5, 7\}$, then $A \subseteq B$ because every element of A is also an element of B.

A Word of Caution The statement "$3, 5 \subseteq B$" is a *false* statement. A subset must be a *set*, and because there are no braces around 3, 5, "3, 5" does not name a set.

b. If $C = \{10, 7, 5\}$ and $H = \{5, 7, 10\}$, then $C \subseteq H$ because every element of C is also an element of H.

c. If $E = \{4, 7\}$ and $F = \{7, 8, 5\}$, then E is not a subset of F because $4 \in E$ but $4 \notin F$. The symbol for is not a subset of is $\not\subseteq$. Therefore, $E \not\subseteq F$.

Note *Every set is a subset of itself.* Furthermore, mathematicians agree that *the empty set is a subset of every set.*

Problem Solving: Making an Organized List

One widely recognized problem-solving technique is that of **making an organized list**. In an organized list, the items must be listed in some logical, orderly way.

When we are asked to list *all* the subsets of a set that has n elements, we can be sure that we haven't omitted any subsets if we make an *organized* list of the subsets. We might start with the set itself (that set will contain n elements), then list all the subsets that contain $n - 1$ elements, then list all the subsets that contain $n - 2$ elements, and so on, continuing in this manner until we have listed the set with $n - n$, or 0, elements (that set will be the empty set). In Example 12, the given set contains 4 elements; therefore, to make an organized list, we list the set itself, then the subsets that each contain 3 elements, then the subsets that each contain 2 elements, then the subsets that each contain 1 element, and finally the empty set.

EXAMPLE 12 List all the subsets of the set $\{1, 3, 5, 7\}$.

SOLUTION We must be sure to include the set itself *and* the empty set.

The set itself:	$\{1, 3, 5, 7\}$
The subsets with 3 elements:	$\{1, 3, 5\}$, $\{1, 3, 7\}$, $\{1, 5, 7\}$, $\{3, 5, 7\}$
The subsets with 2 elements:	$\{1, 3\}$, $\{1, 5\}$, $\{1, 7\}$, $\{3, 5\}$, $\{3, 7\}$, $\{5, 7\}$
The subsets with 1 element:	$\{1\}$, $\{3\}$, $\{5\}$, $\{7\}$
The subset with 0 elements:	$\{\ \}$

The sixteen subsets are $\{1, 3, 5, 7\}$, $\{1, 3, 5\}$, $\{1, 3, 7\}$, $\{1, 5, 7\}$, $\{3, 5, 7\}$, $\{1, 3\}$, $\{1, 5\}$, $\{1, 7\}$, $\{3, 5\}$, $\{3, 7\}$, $\{5, 7\}$, $\{1\}$, $\{3\}$, $\{5\}$, $\{7\}$, and $\{\ \}$.

A Word of Caution The symbol $\subseteq$ is used to indicate that one set is a *subset* of another set. The symbol $\in$ is used to indicate that a particular *element* is an element of a particular set. Therefore, a statement such as "If $B = \{1, 3, 5\}$, then $3 \subseteq B$" is *false*; a subset must be a *set*, and because there are no braces around 3 in $3 \subseteq B$, "3" does not name a set. The statement "If $B = \{1, 3, 5\}$, then $\{3\} \subseteq B$" *is* true ($\{3\}$ *does* name a set); also, the statement "If $B = \{1, 3, 5\}$, then $3 \in B$" is true.

Exercises 1.1
Set I

In Exercises 1 and 2, N refers to the set of natural numbers, and D refers to the set of digits. Write "true" if the statement is always true; otherwise, write "false."

1. a. The collection consisting of *, &, $, and 5 is a set.
   b. $\{x \mid x \text{ is a natural number}\} = \{1, 2, 3\}$
   c. $\{2, 6, 1, 6\} = \{1, 2, 6\}$
   d. $23 \subseteq N$
   e. $0 \in N$
   f. $\{2\} \in D$
   g. $\{2, 6\} \subseteq N$
   h. $\{\ \} \subseteq N$
   i. $10 \in D$
   j. $0 \in \varnothing$
   k. If $A = \{5, 11, 19\}$, then $11 \in A$.
   l. The collection consisting of all the natural numbers between 0 and 1 is a set.
   m. "If $s = t$ and $s = 5$, then $t = 5$" illustrates the transitive property of equality.
2. a. The collection consisting of a, 3, ^, and # is a set.
   b. $\{0, 3, 6, \ldots\} = \{x \mid x \text{ is a natural number divisible by } 3\}$
   c. $\{2, 2, 7, 7\} = \{7, 2\}$
   d. $2 \in N$
   e. $\{5\} \in N$
   f. $5 \subseteq N$
   g. $\{1, 3, 5\} \subseteq N$
   h. $\{\ \} \subseteq D$
   i. $0 \in D$
   j. $\{3\} \in D$
   k. If $B = \{a, b, c, d, e\}$, then $\{a, b\} \subseteq B$.
   l. The collection consisting of all the digits greater than 12 is a set.
   m. "If $9 = x$, then $x = 9$" illustrates the transitive property of equality.

In Exercises 3 and 4, determine which of the sets are finite and which are infinite.

3. a. the set of even natural numbers
   b. the set of the names of the days of the week
   c. the set of books in the Huntington Library
4. a. the set of natural numbers divisible by 7
   b. the set of the names of the months of the year
   c. the set of odd digits
5. List all the subsets of the set $\{a, b, c\}$.
6. List all the subsets of the set $\{1, 2\}$.
7. If $A = \{3, 5, 10, 11\}$, $B = \{3, 5, 12\}$, and $C = \{5, 3\}$, determine which of the following statements are true and which are false.
   a. $B \subseteq A$
   b. $C \subseteq B$
   c. $C \nsubseteq A$

8. If $E = \{a, b, h, k\}$, $F = \{a, b, c, k\}$, and $G = \{b, k\}$, determine which of the following statements are true and which are false.

   a. $F \not\subseteq E$ b. $G \not\subseteq F$ c. $G \subseteq E$

9. Write each of the given sets in roster notation.

   a. $\{x \mid x \text{ is an even digit}\}$

   b. $\{x \mid x \text{ is a natural number divisible by } 5\}$

10. Write each of the given sets in roster notation.

   a. $\{x \mid x \text{ is an odd natural number}\}$

   b. $\{x \mid x \text{ is a natural number that is a multiple of } 11\}$

11. Write each of the given sets in set-builder notation.

   a. $\{a, b, c, d, e\}$

   b. $\{4, 8, 12\}$

   c. $\{10, 20, 30, \ldots\}$

12. Write each of the given sets in set-builder notation.

   a. $\{0, 2, 4, 6, 8\}$

   b. $\{x, y, z\}$

   c. $\{5, 10, 15, \ldots\}$

Writing Problems

Express the answers in your own words and in complete sentences.

1. Explain why $1, 3 \subseteq \{1, 3, 7\}$ is a false statement.

2. Explain why $\{0\} = \varnothing$ is a false statement.

Exercises 1.1 Set II

1. N refers to the set of natural numbers, and D refers to the set of digits. Write "true" if the statement is always true; otherwise, write "false."

   a. The collection consisting of 3, w, %, and @ is a set.

   b. $\{x \mid x \text{ is a digit}\} = \{0, 1, 2, 9\}$

   c. $\{5, x, 5\} = \{5, x\}$ d. $\{\ \} \subseteq \{1, 3, 5\}$

   e. $\{0\} \in D$ f. $12 \in D$

   g. $0 \notin N$ h. $\{0, 5\} \not\subseteq N$

   i. $0 \notin \{\ \}$ j. $3 \subseteq D$

   k. If $C = \{a, 3, d, 5\}$, then $2 \notin C$.

   l. The collection consisting of all the digits less than 0 is a set.

   m. "If $x = y$ and $y = 1$, then $x = 1$" illustrates the reflexive property of equality.

2. List all the elements of the set $\{2, a, 3\}$.

In Exercises 3 and 4, determine which of the sets are finite and which are infinite.

3. a. the set of even digits

   b. the set of the letters of our alphabet

   c. the set of books in the Library of Congress

4. a. the set of natural numbers divisible by 6

   b. the set of digits that are multiples of 6

   c. $\{1, 2, 3, \ldots, 100\}$

5. List all the subsets of the set $\{*, \$, \&\}$.

6. List all the subsets of the set $\{0, 1, 2, 3\}$.

7. If $A = \{2, 4, 7, 11\}$, $B = \{4, 7, 12\}$, and $C = \{4, 7, 11\}$, determine which of the following statements are true and which are false.

   a. $B \subseteq A$ b. $C \subseteq A$ c. $C \not\subseteq B$

8. If $E = \{1, 2, 3, \ldots, 8\}$, $F = \{2, 3, 5\}$, and $G = \{5\}$, determine which of the following statements are true and which are false.

   a. $F \subseteq E$ b. $G \not\subseteq E$ c. $E \subseteq F$

9. Write each of the given sets in roster notation.

   a. $\{x \mid x \text{ is a digit divisible by } 4\}$

   b. $\{x \mid x \text{ is a natural number that is a multiple of } 8\}$

10. Write each of the given sets in roster notation.

   a. $\{x \mid x \text{ is a natural number that is a multiple of } 6\}$

   b. $\{x \mid x \text{ is a digit divisible by } 5\}$

11. Write each of the given sets in set-builder notation.

   a. $\{0, 3, 6, 9\}$

   b. $\{7, 14, 21, \ldots\}$

   c. $\{a, b, c\}$

12. Write each of the given sets in set-builder notation.

   a. {Sunday, Monday, Tuesday, Wednesday, Thursday, Friday, Saturday}

   b. $\{0, 5\}$

   c. $\{4, 8, 12, \ldots\}$

1.2 Set Union and Set Intersection

Set Union

The **union** of sets A and B, written $A \cup B$, is the set that contains all the elements that are in set A *or* set B *or* both. (The symbol $\cup$ suggests the letter U for *union*.)

EXAMPLE 1 Examples of set union:

a. If $A = \{2, 3, 4\}$ and $B = \{1, 3, 5, 8\}$, then $A \cup B = \{1, 2, 3, 4, 5, 8\}$. (The elements that are in set A *or* set B *or* both are 1, 2, 3, 4, 5, and 8.)
b. If $C = \{3, 4, 8\}$ and $D = \{4, 3, 8\}$, then $C \cup D = \{3, 4, 8\}$.
c. If $G = \{1, 4, 7, 9\}$ and $H = \{2, 12\}$, then $G \cup H = \{1, 2, 4, 7, 9, 12\}$.

Set Intersection

The **intersection** of sets A and B, written $A \cap B$, is the set that contains only those elements that are in *both* A and B. (The symbol $\cap$ suggests the letter A for *and*.)

EXAMPLE 2 Examples of set intersection:

a. If $A = \{2, 3, 4\}$ and $B = \{1, 3, 5, 8\}$, then $A \cap B = \{3\}$. (3 is the only element that is in *both A and B*.)
b. If $C = \{3, 4, 8\}$ and $D = \{4, 3, 8\}$, then $C \cap D = \{3, 4, 8\}$.
c. If $E = \{b, c, g\}$ and $F = \{1, 2, 5, a\}$, then $E \cap F = \{\ \}$. (There are no elements common to E and F.)

To find the union and/or the intersection of more than two sets, we perform operations inside parentheses before other operations (see Example 3).

EXAMPLE 3 If $A = \{1, 4, 7\}$, $B = \{1, 6, 9\}$, and $C = \{4, 7\}$, find $(A \cap B) \cup C$.

SOLUTION Because there are parentheses around $A \cap B$, we first find $A \cap B$: $A \cap B = \{1\}$. *Then* we find the union of $(A \cap B)$ with C:

$$(A \cap B) \cup C = \{1\} \cup \{4, 7\} = \{1, 4, 7\}$$

Therefore,

$$(A \cap B) \cup C = \{1, 4, 7\}$$

Exercises 1.2 Set I

1. Given $A = \{1, 2, 3, 4\}$ and $B = \{2, 4, 5\}$, find the following:

   a. $A \cup B$ b. $A \cap B$ c. $B \cup A$ d. $B \cap A$

2. Given $C = \{2, 5, 6, 12\}$ and $D = \{7, 8, 9\}$, find the following:

   a. $C \cap D$ b. $C \cup D$ c. $D \cap C$ d. $D \cup C$

3. Given $X = \{2, 5, 6, 11\}$, $Y = \{5, 7, 11, 13\}$, and $Z = \{0, 3, 4, 6\}$, find the following:

   a. $X \cap Y$ b. $Y \cup Z$ c. $X \cap Z$
   d. $Y \cap Z$ e. $X \cup Y$ f. $Z \cup Y$
   g. $(X \cap Y) \cap Z$ h. $X \cap (Y \cap Z)$
   i. $(X \cup Y) \cup Z$ j. $X \cup (Y \cup Z)$

4. Given $K = \{a, 4, 7, b\}$, $L = \{m, 4, 6, b\}$, and $M = \{n, 3, 5, t\}$, find the following:

a. $K \cap L$ b. $K \cup L$ c. $K \cup M$
d. $L \cap M$ e. $K \cap M$ f. $L \cup M$
g. $(K \cap L) \cap M$ h. $K \cap (L \cap M)$
i. $(K \cup L) \cup M$ j. $K \cup (L \cup M)$

Writing Problems

Express the answer in your own words and in complete sentences.

1. Explain what the difference is between the union of two sets and the intersection of two sets.

Exercises 1.2 Set II

1. Given $H = \{1, 3, 5, 7\}$ and $K = \{2, 4, 6\}$, find the following:

a. $H \cap K$ b. $H \cup K$ c. $K \cap H$ d. $K \cup H$

2. Given $S = \{1, 5, 10\}$ and $T = \{5\}$, find the following:

a. $S \cap T$ b. $S \cup T$ c. $T \cap S$ d. $T \cup S$

3. Given $A = \{a, 5, 7, b\}$, $B = \{c, 6, 7, b\}$, and $C = \{3, 4, 6\}$, find the following:

a. $A \cap B$ b. $A \cup B$ c. $A \cap C$
d. $A \cup C$ e. $B \cap C$ f. $B \cup C$
g. $(A \cap B) \cap C$ h. $A \cap (B \cap C)$
i. $(A \cup B) \cup C$ j. $A \cup (B \cup C)$

4. Given $E = \{a, b, c\}$, $F = \{c, d, e, f, g\}$, and $G = \{g\}$, find the following:

a. $E \cap F$ b. $F \cap E$ c. $E \cup F$
d. $E \cup G$ e. $E \cap G$ f. $G \cap E$
g. $E \cap (F \cup G)$ h. $(E \cap F) \cup (E \cap G)$
i. $E \cup (F \cap G)$ j. $(E \cup F) \cap (E \cup G)$

1.3 Real Numbers

We have already identified two important sets of numbers—namely, the set of *digits*, $D = \{0, 1, 2, \ldots, 9\}$, and the set of *natural numbers*, $N = \{1, 2, 3, \ldots\}$. We discuss the number line and other important sets of numbers in this section.

The Number Line and the Graph of a Number A **number line** can help us see the relationships among numbers. To construct a number line, we draw a straight line (often a horizontal line), and we select some point on that line and label the point *zero*; then we mark off equally spaced points to the right and to the left of zero. (We put an arrowhead at the right-hand end of the number line, indicating the direction in which numbers get larger; some authors put arrowheads at both ends of the line.) Points to the right of zero represent positive numbers, and points to the left of zero represent negative numbers; zero itself is neither positive nor negative.

We **graph a number** by placing a solid dot *on* the number line *above* that number. In Figure 1, we show a number line with the digits graphed on it.

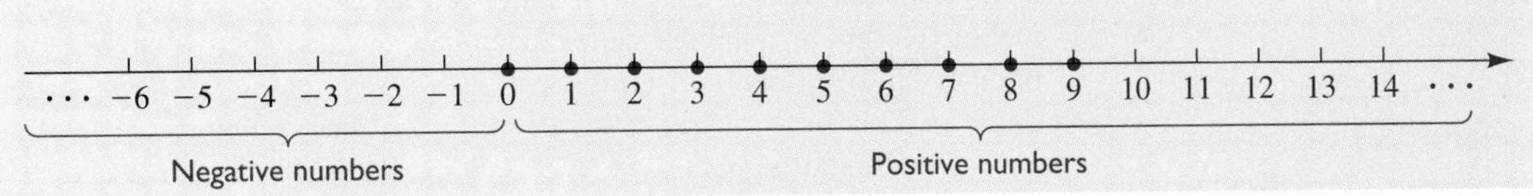

Figure 1

Whole Numbers The set $W = \{0, 1, 2, 3, \ldots\}$ is called the set of **whole numbers**. The set of natural numbers is a subset of the set of whole numbers; that is, $N \subseteq W$.

Integers The set $J = \{\ldots, -3, -2, -1, 0, 1, 2, 3, \ldots\}$ is called the set of **integers**. The set of whole numbers is a subset of the set of integers, and the set of natural numbers is a subset of the set of integers; that is, $W \subseteq J$ and $N \subseteq J$.

Rational Numbers The set of **rational numbers**, Q, is the set of all numbers that can be expressed as quotients of two integers $\left(\text{that is, as fractions of the form } \dfrac{a}{b}\right)$, where the divisor is not zero. In set-builder notation, we have

$$Q = \left\{ \frac{a}{b} \,\middle|\, a, b \in J,\ b \neq 0 \right\}$$

The set of integers is a subset of the set of rational numbers (that is, $J \subseteq Q$), because any integer can be written as a quotient of two integers; for example,

$$5 = \frac{5}{1}, \qquad 17 = \frac{17}{1}, \qquad 0 = \frac{0}{1}, \qquad -7 = \frac{-7}{1}$$

The set of *terminating decimals* (decimals that have only zeros to the right of some specific decimal place) is a subset of the set of rational numbers, because any terminating decimal can be written as a quotient of two integers; for example,

$$0.1 = \frac{1}{10}, \qquad -1.03 = \frac{-103}{100}$$

The set of *nonterminating, repeating decimals** is a subset of the set of rational numbers, because any nonterminating, repeating decimal can be written as a quotient of two integers. A *bar* written above a block of digits indicates that that block repeats. For example,

$$0.666\ldots = 0.\overline{6} = \frac{2}{3}, \qquad -0.272\,727\ldots = -0.\overline{27} = \frac{-3}{11}$$

The set of *mixed numbers* is a subset of the set of rational numbers, because any mixed number can be written as a quotient of two integers; for example,

$$2\frac{1}{3} = \frac{7}{3}, \qquad -5\frac{1}{2} = \frac{-11}{2}$$

It is also true that $N \subseteq Q$ and $W \subseteq Q$; that is, the set of natural numbers and the set of whole numbers are both subsets of the set of rational numbers.

*These repeating decimals repeat a *number* or a *block of numbers.* (In $0.666\ldots$, the 6 repeats, and in $-0.272\,727\ldots$, the block 27 repeats.) In a later chapter, we discuss how to convert $0.272\,727\ldots$ to $\frac{3}{11}$.

Irrational Numbers One very important set of numbers exists that is not a subset of any of the sets of numbers we've discussed so far. When we convert any one of these numbers into its decimal equivalent, we always get a *nonterminating, nonrepeating decimal.* This set is the set of **irrational numbers**, which we will call H. A few examples of irrational numbers are $\sqrt{2}$, $\sqrt{8}$, $\sqrt[3]{55}$, and π.

The digits go on forever, never terminating and never repeating

$\sqrt{2} = 1.414\ 213\ 562\ldots \approx 1.414$ Rounded off to three decimal places

$\approx$ is the symbol for "is approximately equal to"

$\sqrt{8} = 2.828\ 427\ 125\ldots \approx 2.828$ Rounded off to three decimal places

$\sqrt[3]{55} = 3.802\ 952\ 461\ldots \approx 3.803$ Rounded off to three decimal places

$\pi = 3.141\ 592\ 654\ldots \approx 3.1416$ Rounded off to four decimal places

There are infinitely many irrational numbers, each of which becomes a nonterminating, nonrepeating decimal when it is converted to its decimal equivalent.

Irrational numbers cannot be expressed in the form $\frac{a}{b}$, where a and b are integers and $b \neq 0$.

Real Numbers The union of the set of rational numbers and the set of irrational numbers comprises the set of **real numbers**, which we call R. In set notation, $Q \cup H = R$. (We will be concerned only with real numbers until Chapter 7.)

The set of rational numbers is a subset of the set of real numbers, and the set of irrational numbers is a subset of the set of real numbers. *No number can be both a rational number and an irrational number.* That is, $Q \cap H = \{\ \}$.

There is a point on the number line that corresponds to every real number, and there is a real number that corresponds to every point on the number line. Therefore, we say that there is a *one-to-one correspondence* between the set of real numbers and the set of points on the number line, and we often call the number line the *real* number line.

Because of the one-to-one correspondence, we can graph all real numbers on the number line. To graph $\sqrt{5}$, for example, we find (using a calculator) that $\sqrt{5} \approx 2.2$, and so we place a dot on the number line a little to the right of the 2. We graph a few real numbers in Figure 2.

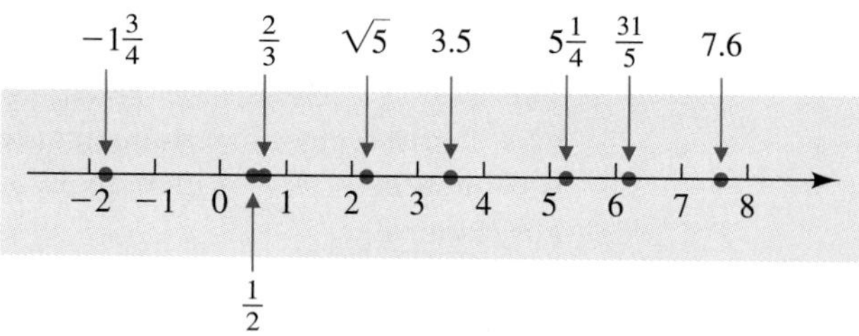

Figure 2

EXAMPLE 1 Consider the set of numbers −4, 2.236 067 97... (never repeats and never terminates), 3, and $-1\frac{3}{4}$. Graph these numbers on the real number line and determine which are (a) natural numbers, (b) integers, (c) rational numbers, (d) irrational numbers, and (e) real numbers.

SOLUTION

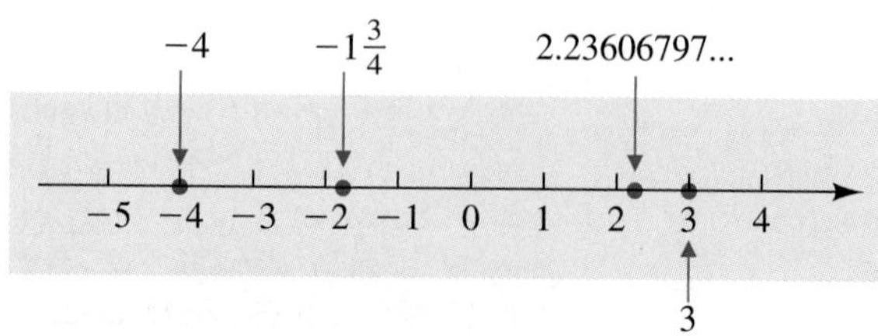

-4 is an integer, a rational number $\left(\text{it can be expressed as } \frac{-4}{1}\right)$,

2.236 067 97... is an irrational number (since it's a nontermina
decimal) and a real number. (Remember, every irrational numbe

3 is a natural number, an integer, a rational number, and a real

$-1\frac{3}{4}$ is a rational number $\left(\text{it can be expressed as } \frac{-7}{4}\right)$ and a real number.

Therefore:

a. There is one natural number in the list: 3.
b. The integers from the list are -4 and 3.
c. The rational numbers are -4, 3, and $-1\frac{3}{4}$.
d. There is one irrational number in the list: 2.236 067 97....
e. The real numbers are -4, 2.236 067 97..., 3, and $-1\frac{3}{4}$.

The relationships among the sets of numbers discussed in this section are shown in Figure 3.

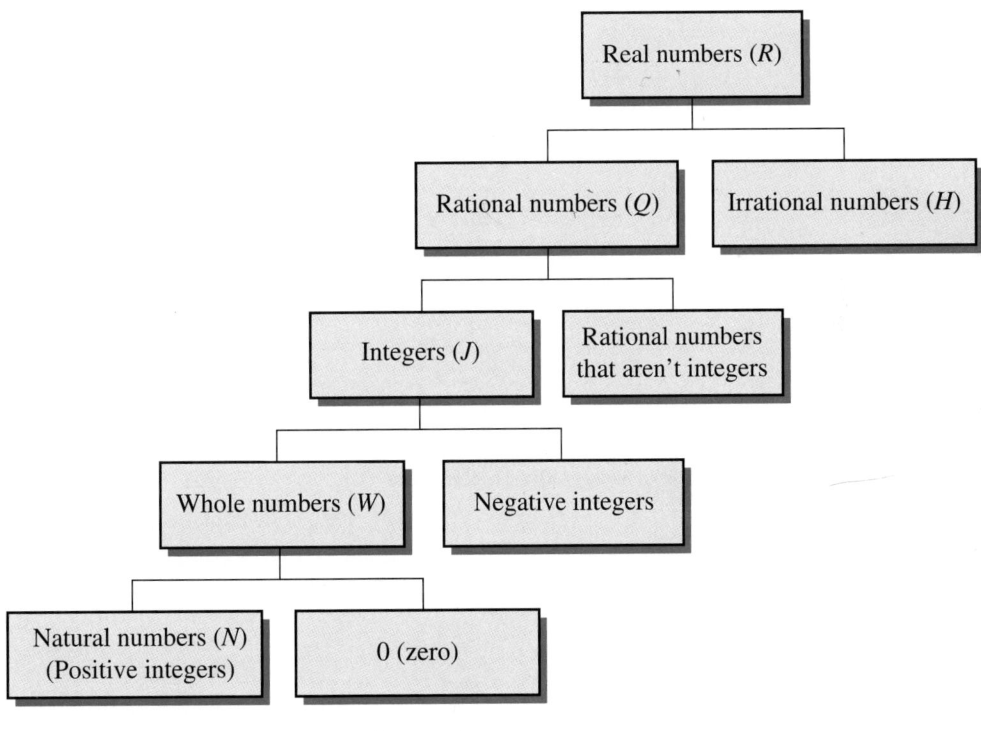

Figure 3

Notice that $N \subseteq W \subseteq J \subseteq Q \subseteq R$, $H \subseteq R$, and $Q \cap H = \{\ \}$.

Note Remember, the symbol $\subseteq$ is used to indicate that one set is a *subset* of another set, whereas the symbol $\in$ is used to indicate that a particular *element* is an element of a particular set. Therefore, if N represents the set of natural numbers, $3 \subseteq N$ is *false*, but $3 \in N$ and $\{3\} \subseteq N$ are true. Similarly, $\{5\} \in N$ is false, but $\{5\} \subseteq N$ and $5 \in N$ are true.

Exercises 1.3
Set I

1. Graph the numbers 7, 2.449 489 734 . . . (never terminates, never repeats), -5, 10, $-\frac{3}{4}$, and $0.\overline{2}$ on a number line, and determine the following:

   a. Which are real numbers?

   b. Which are integers?

   c. Which are natural numbers?

   d. Which are irrational numbers?

   e. Which are rational numbers?

2. Graph the numbers $-2.645\ 751\ 31$ . . . (never terminates, never repeats), -11, $\frac{5}{6}$, $3\frac{1}{3}$, $1.\overline{6}$, and 4 on a number line, and determine the following:

   a. Which are real numbers?

   b. Which are integers?

   c. Which are natural numbers?

   d. Which are irrational numbers?

   e. Which are rational numbers?

In Exercises 3 and 4, replace the question mark with either $\subseteq$ or $\in$ to form a true statement. The letters refer to the sets discussed in this section.

3. a. $\{1, 3\} \ ? \ J$ b. $2 \ ? \ N$ c. $N \ ? \ R$
   d. $\frac{3}{4} \ ? \ Q$ e. $\frac{3}{4} \ ? \ R$ f. $\{3\} \ ? \ R$

4. a. $J \ ? \ Q$ b. $\{2, 3, 4\} \ ? \ R$ c. $\frac{2}{3} \ ? \ R$
   d. $-4 \ ? \ J$ e. $J \ ? \ R$ f. $\{-5\} \ ? \ J$

In Exercises 5 and 6, write "true" if the statement is always true; otherwise, write "false." The letters refer to the sets discussed in this section.

5. a. Every real number is a rational number.

   b. Every integer is a real number.

   c. $R \subseteq W$ d. $3 \in J$

   e. $3 \in H$ f. $H \subseteq R$

   g. 3.316 624 79 . . . (never terminates, never repeats) $\in H$

   h. $-3.872\ 983\ 346$ . . . (never terminates, never repeats) $\in R$

6. a. Every irrational number is a real number.

   b. Every rational number is an integer.

   c. $5 \in H$ d. $5 \in Q$

   e. $Q \subseteq R$ f. $J \subseteq Q$

   g. $0.\overline{13} \in H$

   h. $-4.123\ 105\ 626$ . . . (never terminates, never repeats) $\in H$

Writing Problems

Express the answers in your own words and in complete sentences.

1. Explain why 0.79 is a rational number.
2. Explain what a rational number is.
3. Explain why $-\sqrt{121}$ is a rational number.
4. Explain why 7 is a rational number.

Exercises 1.3
Set II

1. Graph the numbers $-5.196\ 152\ 432$ . . . (never terminates, never repeats), $\frac{5}{7}$, $-2\frac{4}{9}$, $4.\overline{4}$, 0, and 2 on a number line, and determine the following:

   a. Which are real numbers?

   b. Which are integers?

   c. Which are natural numbers?

   d. Which are irrational numbers?

   e. Which are rational numbers?

2. Graph the numbers -3, $0.\overline{12}$, 6, $2\frac{2}{3}$, $-1\frac{1}{2}$, and $-4.216\ 346\ 1$ . . . (never terminates, never repeats) on a number line, and determine the following:

   a. Which are real numbers?

   b. Which are integers?

   c. Which are natural numbers?

   d. Which are irrational numbers?

   e. Which are rational numbers?

In Exercises 3 and 4, replace the question mark with either $\subseteq$ or $\in$ to form a true statement. The letters refer to the sets discussed in this section.

3. a. $\{0, 5\}$? W b. 12 ? W c. $\{12\}$? N
   d. $\frac{8}{9}$? R e. W ? J f. $\{1, 3\}$? D
4. a. 0 ? W b. 12 ? N c. $\{0\}$? W
   d. 5 ? D e. $\{12\}$? R f. $\frac{1}{4}$? R

In Exercises 5 and 6, write "true" if the statement is always true; otherwise write "false." The letters refer to the sets discussed in this section.

5. a. Every whole number is an integer.
   b. Every integer is a rational number.
   c. $W \subseteq H$ d. $7 \in Q$
   e. $7 \in R$ f. $Q \subseteq N$
   g. $-0.\overline{23} \in H$
   h. 3.741 657 387 . . . (never terminates, never repeats) $\in H$
6. a. Every rational number is a real number.
   b. The set of digits is a subset of the set of natural numbers.
   c. $R \subseteq Q$ d. $0 \in H$
   e. $\frac{1}{5} \in H$ f. $H \subseteq Q$
   g. 0.165 132 78 . . . (never terminates, never repeats) $\in H$
   h. 0.165 132 78 . . . (never terminates, never repeats) $\in R$

1.4 Basic Definitions

Constants A **constant** is an object or symbol that does not change its value in a particular problem or discussion. It is usually represented by a number symbol, but it is also often represented by one of the first few letters of the alphabet. Thus, in the expression $4x + 3y - 5$, the constants are 4, 3, and -5. In the expression $ax + by + c$, the constants are understood to be a, b, and c.

Variables In algebra, we often use letters to represent numbers (or sets of numbers), and we call these letters variables. A **variable** is a letter (often one of the last few letters of the alphabet) or symbol that acts as a placeholder for a number that is unknown. In the expression $ax + by + c$, it is understood that the variables are x and y. A variable may assume different values in a particular problem or discussion. An equation or inequality is usually, but not always, *true* for some values of the variables and *false* for other values.

Factors Numbers that are multiplied together to give a product are called the **factors** of that product. Recall from elementary algebra that when two symbols (other than two *numbers*) are written next to each other (this is called **juxtaposition**), as in $7x$, it is understood that they are to be *multiplied* together.

> The number 1 may be considered to be a factor of any number.

EXAMPLE 1 Examples of identifying the factors in an expression:

a. $3 \cdot 5 = 15$ 3 and 5 are factors of 15.
b. $7xyz$ 7, x, y, and z are the factors of $7xyz$.
c. x 1 (understood) and x are the factors of x.

Algebraic Expressions An **algebraic expression** consists of numbers, variables, signs of operation (such as + or −), and signs of grouping. (Not all of these need to be present.) For example, $4x + 3y - 5$ is an algebraic expression.

Terms A **term** of an algebraic expression can consist of one number, one variable (which might be raised to a power), or a *product* of numbers and variables (possibly raised to powers). Thus, the only operations that can be indicated within a *term* are multiplication and powers (or the inverses of these operations: division and roots).

The terms of an algebraic expression are usually the parts of the expression that are separated by addition and subtraction symbols. If a term is preceded by a minus sign, that sign is *part* of the term. *Exception:* An expression *within grouping symbols* is to be considered as a single unit; it is a *term* or it is a *factor of a term*, even if there is more than one term *within* the grouping symbols (see Examples 2b and 2c).

EXAMPLE 2 Examples of identifying the terms in an algebraic expression:

a. $3xy - 5xz + 7y$
It may help to think of this expression as

The first term consists of the factors 3, x, and y
The second term consists of the factors −5, x, and z
The third term consists of the factors 7 and y

$$3xy + (-5xz) + 7y$$

The first term is $3xy$, the second term is $-5xz$, and the third term is $7y$.

b. $3t - 9x(2y + 5z)$
An expression within parentheses is considered as a single unit, and $(2y + 5z)$ is a *factor* of the expression $-9x(2y + 5z)$; therefore, $-9x(2y + 5z)$ consists of *factors* and is considered to be one term.

$$\underbrace{3t}_{\text{First term}} \quad \underbrace{-9x(2y + 5z)}_{\text{Second term}}$$

The factors of the second term are −9, x, and (2y + 5z)

The first term is $3t$, and the second term is $-9x(2y + 5z)$.

c. $\dfrac{2 - x}{xy} + 5(2x - y)$

The fraction bar is a grouping symbol; furthermore, the division $\dfrac{2 - x}{xy}$ can be interpreted as the multiplication $(2 - x) \cdot \dfrac{1}{x} \cdot \dfrac{1}{y}$. Therefore, $\dfrac{2 - x}{xy}$ consists of the factors $(2 - x)$, $\dfrac{1}{x}$, and $\dfrac{1}{y}$ and is a single term.

$$\underbrace{\frac{2 - x}{xy}}_{\text{First term}} + \underbrace{5(2x - y)}_{\text{Second term}}$$

The second term consists of the factors 5 and (2x − y)

The first term is $\dfrac{2 - x}{xy}$, and the second term is $5(2x - y)$.

Coefficients In a term with two factors, the **coefficient** of one factor is the other factor. In a term with more than two factors, the coefficient of each factor is the product of all the other factors in the term.

Numerical Coefficients A **numerical coefficient** is a coefficient that is a *number* rather than a *letter.* When we refer to "the coefficient" of a term, it is understood that we mean the numerical coefficient of that term. If a term has no numerical coefficient written, then the numerical coefficient is understood to be 1.

EXAMPLE 3 Examples of identifying the coefficients of a term:

a. $12xyz$

- xyz is the coefficient of 12
- 12 is the numerical coefficient of xyz, or *the* coefficient of xyz

b. $-xy$

- y is the coefficient of $-x$
- x is the coefficient of $-y$
- -1 is the understood numerical coefficient of xy, since $-xy = (-1)xy$

Exercises 1.4
Set I

In Exercises 1 and 2, (a) list the constants, and (b) list the variables.

1. $xy + 3$
2. $4z - 2y$

In Exercises 3–10, (a) determine the number of terms, (b) write the *second* term, if there is one, and (c) list all of the factors of the *first* term.

3. $E + 5F - 3$
4. $R + 2T - 6$
5. $(R + S) - 2(x + y)$
6. $(A + 2B) - 5(W + V)$
7. $3XYZ + 4$
8. $4ab + 3x$
9. $2A + \dfrac{3B - C}{DE}$
10. $3st + \dfrac{w + z}{xyz}$

In Exercises 11–14, (a) write the numerical coefficient of the first term, and (b) write the variable part of the second term.

11. $2R - 5RT + 3T$
12. $4x - 3xy + z$
13. $-x - y + z$
14. $-a + 2b - c$

Writing Problems

Express the answers in your own words and in complete sentences.

1. Explain what a *factor* is.
2. Explain what a *term* is.

Exercises 1.4
Set II

In Exercises 1 and 2, (a) list the constants, and (b) list the variables.

1. $4X - Z$
2. $3x - 2st$

In Exercises 3–10, (a) determine the number of terms, (b) write the *second* term, if there is one, and (c) list all of the factors of the *first* term.

3. $X + 7Y - 8$
4. $3(x - 5y + 7z)$
5. $(2X - Y) - 4(a + b)$
6. $-4x + \dfrac{1}{6st} - 3z$
7. $7st + 3uv$
8. $x + 2(y + z + 4) - w + s$
9. $3A + \dfrac{2B - 3C}{2AB}$
10. $6 - 3xy + \dfrac{z}{2}$

In Exercises 11–14, (a) write the numerical coefficient of the first term, and (b) write the variable part of the second term.

11. $2X + Y - 3$
12. $9x - 5y + z$
13. $4(X + 2) - 5Y$
14. $z - 3w$

1.5 Inequalities and Absolute Values

Inequalities

"Greater Than" and "Less Than" Symbols The symbol $>$ is read is greater than, and the symbol $<$ is read is less than. These *inequality symbols* are among the symbols that we can use between numbers that are *not* equal to each other. Numbers get larger as we move toward the right on the number line, so $x > y$ if x lies to the right of y on the number line. (This is an informal definition of *greater than*; a more formal definition is that $x > y$ if there exists some *positive* number p such that $x = y + p$. For example, $5 > 3$ because there exists a positive number, 2, such that $5 = 3 + 2$.)

The statements $5 > 3$ and $3 < 5$ give the same information, even though they are read differently. We generalize this fact as follows:

> $a < b$ may always be replaced by $b > a$.
>
> $a > b$ may always be replaced by $b < a$.

Transitive properties hold for *greater than* and for *less than*:

If $x > y$ and $y > z$, then $x > z$.	The transitive property of greater than
If $x < y$ and $y < z$, then $x < z$.	The transitive property of less than

A real-life example of the transitive property of *greater than* might be "If Bill's age is greater than Jane's age and if Jane's age is greater than Ray's age, then Bill's age is greater than Ray's age." (Notice that the two quantities in the *middle* are the same—both are Jane's age.) This example *could* have been worded "If Bill is older than Jane and if Jane is older than Ray, then Bill is older than Ray."

EXAMPLE 1 Examples of using the "greater than" and "less than" symbols:

a. $-1 > -4$ is read "-1 is greater than -4."

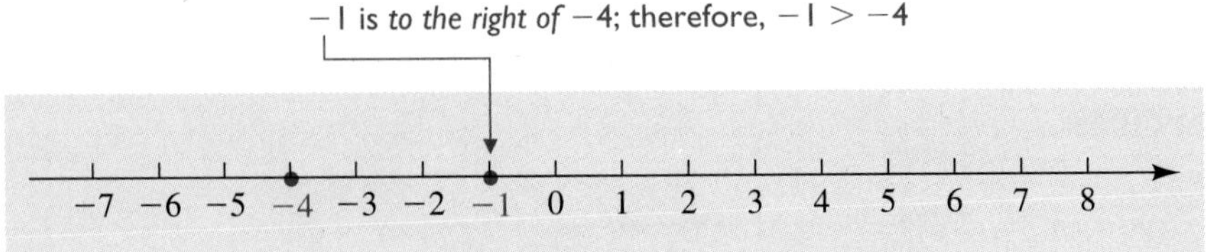

b. $-2 < 1$ is read "-2 is less than 1."

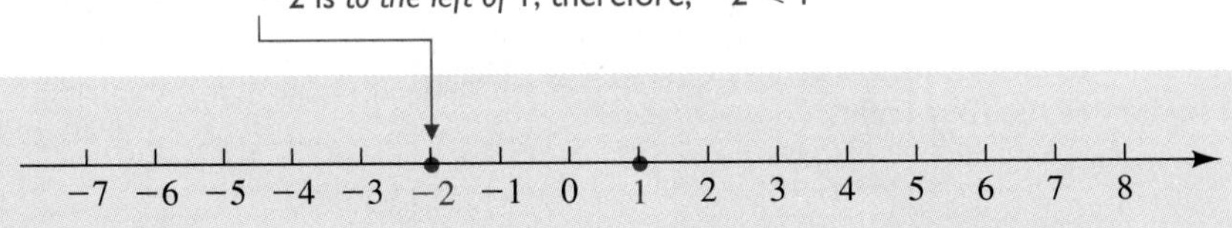

The "Less than or Equal to" Symbol The inequality $a \le b$ is read "a is less than *or* equal to b." This means that if $\left\{\begin{matrix} \text{either } a < b \\ \text{or} \quad a = b \end{matrix}\right\}$ is true, then $a \le b$ is true. For example, $2 \le 3$ is true because $2 < 3$ is true (even though $2 = 3$ is *not* true). Remember, only *one* of the two statements $\left\{\begin{matrix} 2 < 3 \\ 2 = 3 \end{matrix}\right\}$ needs to be true in order for $2 \le 3$ to be true.

The "Greater Than or Equal to" Symbol The inequality $a \ge b$ is read "a is greater than *or* equal to b." This means that if $\left\{\begin{matrix} \text{either } a > b \\ \text{or} \quad a = b \end{matrix}\right\}$ is true, then $a \ge b$ is true. For example, $5 \ge 1$ is true because $5 > 1$ is true (even though $5 = 1$ is *not* true).

Senses of Inequalities If two inequalities are both *less than* (or if one is *less than* and the other is *less than or equal to*), we say that the inequalities have the *same sense.* (Similarly, if both inequalities are *greater than*, we say that they have the same sense.)

Continued Inequalities

A **continued inequality** is a statement such as $a < b < c$ (which can be read "a is less than b, which is less than c"), $a \le b < c$, $a > b \ge c$, and so on. The continued inequality $a < b < c$ really means that $a < b$ *and* $b < c$; furthermore, because of the transitive property of inequalities, it must also be true that $a < c$. Thus, $a < b < c$ is really a combination of three statements: $a < b$, $b < c$, and $a < c$; it also means that b is *between* a and c.

If a continued inequality is to be considered a *valid* continued inequality, both inequality symbols must have the same sense *and* the three inequality statements must all be true. For example, $2 < 5 < 9$ is a valid continued inequality because both symbols are *less than* and the statements $2 < 5$, $5 < 9$, and $2 < 9$ are *all* true.

EXAMPLE 2 Examples of continued inequalities:

a. $4 < 7 < 9$ is read "4 is less than 7, which is less than 9." It can also be read "7 is greater than 4 *and* less than 9." (Remember that $4 < 7$ can be replaced by $7 > 4$.) $4 < 7 < 9$ means that 7 is *between* 4 and 9.

b. $10 > 0 > -3$ is read "10 is greater than 0, which is greater than -3," or "0 is less than 10 and greater than -3." (Remember that $10 > 0$ can be replaced by $0 < 10$.) $10 > 0 > -3$ means that 0 is between -3 and 10.

Note Even though the statement $10 > 0 > -3$ in Example 2b is correct as written, it is customary to write such continued inequalities so that the numbers are in the same order as on the number line—that is, as $-3 < 0 < 10$.

A Word of Caution Notice that $-3 < 0 > -2$ is *not* a valid continued inequality because the inequality symbols are not both *less than* or both *greater than.*

EXAMPLE 3 Examples of determining whether a continued inequality is valid or invalid:

a. $10 < 12 > 5$	The inequality is *invalid* because one symbol is $<$ and the other is $>$.
b. $10 < 15 < 32$	The inequality is *valid* because both symbols have the same sense (both are $<$) *and* it is true that $10 < 15$, $15 < 32$, *and* $10 < 32$.
c. $2 > 5 > 8$	The inequality is *invalid* because $2 > 5$, $5 > 8$, and $2 > 8$ are *false* statements.
d. $9 > 3 < 5$	The inequality is *invalid* because one symbol is $>$ and the other is $<$.

The Slash A slash drawn through a symbol negates that symbol (that is, it puts a "not" in the meaning of the symbol). We've seen this already in such symbols as $\neq$, $\notin$, and $\not\subseteq$. The symbol $\not<$ is read is not less than and is equivalent to the symbol $\geq$. The symbol $\not>$ is read is not greater than and is equivalent to the symbol $\leq$.

EXAMPLE 4 Examples of using the slash:

a. $3 \not< -2$ is read "3 is *not* less than -2." It is equivalent to $3 \geq -2$.
b. $-6 \not> -5$ is read "-6 is *not* greater than -5." It is equivalent to $-6 \leq -5$.

The Additive Inverse of a Number If two numbers are equal numerically but have opposite signs, we say that the numbers are the **additive inverses** of each other. For example, 5 and -5 are the additive inverses of each other, as are $-\sqrt{2}$ and $\sqrt{2}$. In general, we find the additive inverse of a number by changing the sign of the number, so the additive inverse of x is $-x$. (We will define additive inverses more formally in the next section.) The additive inverse of 0 is 0.

The additive inverse of a number is sometimes called the *opposite* of the number or the *negative* of the number.

The Absolute Value of a Number

The symbol for the **absolute value** of a real number x is $|x|$. The formal definition of the absolute value of a real number is

$$|x| = \begin{cases} x & \text{if } x \geq 0 \\ -x & \text{if } x < 0 \end{cases}$$

This definition means that if x is a positive number or zero, then it is its own absolute value. For example, $|6| = 6$. However, if x is negative, the absolute value of x equals the additive inverse of x. For example, if $x = -2$, then the absolute value of x equals the additive inverse of -2, which is 2; in symbols, we have $|-2| = 2$.

We can also think of the absolute value of a number as being the *distance* between that number and zero on the number line, *with no regard for direction* (see Figure 4). The absolute value of a number is never negative.

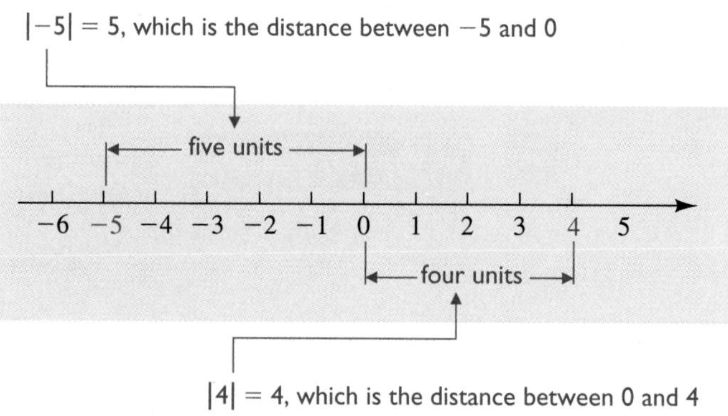

Figure 4

Note A **nonnegative number** is a number that is positive or zero. Thus, we can say that the absolute value of a number is always *nonnegative*.

EXAMPLE 5 Examples of absolute values of real numbers:

a. $|9| = 9$ — Because $9 > 0$, 9 is its own absolute value; furthermore, the distance between 0 and 9 is 9

b. $|-4| = 4$ — The additive inverse of -4 is 4; furthermore, the distance between -4 and 0 is 4

c. $-|-5| = -(+5) = -5$ — Substituting $+5$ for $|-5|$

d. $-|3| = -(+3) = -3$ — Substituting $+3$ for $|3|$

e. $|0| = 0$

Note In c and d, negative signs were located *outside* the absolute value symbols.

Graphics calculators have a key for finding the absolute value of a number. To find $|-5|$, for example, we enter

	Keys	*Display*
On the TI-81™	2nd, x^{-1} (ABS D), (−), 5, ENTER	5
On the Casio fx-8500G™	Shift, x^y (Abs), (−), 5, EXE	5

Exercises 1.5
Set I

In Exercises 1 and 2, determine which symbol, $<$ or $>$, should be used to make each statement true.

1. a. $-15 \,?\, -13$ b. $-15 \,?\, |-13|$
   c. $|-15| \,?\, -13$ d. $|2| \,?\, |-3|$
2. a. $-19 \,?\, -23$ b. $-19 \,?\, |-23|$
   c. $|-19| \,?\, -23$ d. $|-19| \,?\, |-23|$

In Exercises 3 and 4, determine whether each inequality is a valid or an invalid continued inequality. If it is invalid, state *why* the inequality is invalid.

3. a. $3 < 5 < 10$ b. $-2 > 1 > 7$ c. $0 < 4 > 2$
   d. $8 > 3 > -1$ e. $8 > 5 < 9$ f. $0 > 3 > 8$
4. a. $4 > 0 > -2$ b. $-2 < 6 > 5$ c. $3 > 1 < 7$
   d. $-1 < 3 < 6$ e. $4 > 7 > 11$ f. $5 < 0 < -2$

In Exercises 5 and 6, determine whether the expression is true or false.

5. $3 \not< 8$
6. $-4 \not> 5$

In Exercises 7 and 8, evaluate each expression.

7. a. $|7|$ b. $|-5|$ c. $-|-12|$
8. a. $|-34|$ b. $-|5|$ c. $-|-3|$

Writing Problems

Express the answers in your own words and in complete sentences.

1. Explain why $3 < 5 > 4$ is an invalid continued inequality.
2. Explain why $2 > 7 > 9$ is an invalid continued inequality.
3. Explain why $8 > 5 > -9$ is a valid continued inequality.

Exercises 1.5 Set II

In Exercises 1 and 2, determine which symbol, $<$ or $>$, should be used to make each statement true.

1. a. $-18 \; ? \; -12$ b. $-63 \; ? \; |-104|$
   c. $|7| \; ? \; |-8|$ d. $|-3| \; ? \; -5$
2. a. $0 \; ? \; |-5|$ b. $|15| \; ? \; |-3|$
   c. $-8 \; ? \; |-6|$ d. $|10| \; ? \; |-14|$

In Exercises 3 and 4, determine whether the inequality is a valid or an invalid continued inequality. If it is invalid, state *why* the inequality is invalid.

3. a. $-1 < 7 > 5$ b. $0 < 4 < 6$ c. $1 > 3 > 7$
   d. $3 > 2 < 5$ e. $8 < 4 < 2$ f. $8 > 2 > -4$
4. a. $-3 < 5 > 4$ b. $18 > 5 > -6$ c. $2 > 0 < 8$
   d. $-8 < -6 < -2$ e. $-3 > 1 > 5$ f. $8 > 5 > 1$

In Exercises 5 and 6, determine whether the statement is true or false.

5. $-2 \not> 0$ 6. $3 \not< -5$

In Exercises 7 and 8, evaluate each expression.

7. a. $|-15|$ b. $|35|$ c. $-|-22|$
8. a. $-|-5|$ b. $-|13|$ c. $|-2|$

1.6 Operations on Rational Numbers

The properties we discuss in this section apply to operations on all real numbers, both rational and irrational. However, in this section, the examples and exercises involve operations on rational numbers only; we discuss operations involving irrational numbers in a later chapter.

Adding Rational Numbers

The answer to an addition problem is called the **sum**, and the numbers that are added together are called the **terms** or the **addends**.

The Closure Property of Addition One of the properties of the real number system that mathematicians accept as true without proof is the **closure property of addition**, which follows:

The closure property of addition

> If a and b represent any real numbers, then their sum, $a + b$, is a real number.

We assume that you recall from elementary algebra how to add signed rational numbers, but we briefly review the procedure in the following box:

Adding two rational numbers

> 1. When the numbers have the same sign, add their absolute values, and give the sum the sign of both numbers.
> 2. When the numbers have different signs, subtract the smaller absolute value from the larger absolute value, and give the sum the sign of the number with the larger absolute value.

EXAMPLE 1 Examples of adding rational numbers:

a. $-7 + (-11) = -\ 18$ The signs are the same

- The sum of the absolute values: $|-7| + |-11| = 7 + 11 = 18$
- The sign is negative because the given numbers are negative

b. $18.3 + (-32) = -\ 13.7$ The signs are different

- The difference of the absolute values: $|-32| - |18.3| = 32 - 18.3 = 13.7$
- The sign of the number with the larger absolute value (-32)

c. $-\frac{1}{8} + \frac{3}{4} = +\ \frac{5}{8}$ The signs are different

- The difference of the absolute values: $\left|\frac{3}{4}\right| - \left|-\frac{1}{8}\right| = \frac{6}{8} - \frac{1}{8} = \frac{5}{8}$
- The sign of the number with the larger absolute value $\left(\frac{3}{4}\right)$

The Additive Identity Since adding 0 to any real number gives the *identical* number we started with (for example, $7 + 0 = 7$), we call **0** the **additive identity** or the **additive identity element.**

The additive identity property

> The additive identity element is 0.
>
> If a represents any real number, then
>
> $$a + 0 = a \quad \text{and} \quad 0 + a = a$$

Additive Inverses If the sum of two numbers equals the additive identity, 0, we say that the numbers are the **additive inverses** of each other. In the previous section, we mentioned that 5 and -5 are the additive inverses of each other; you can verify that $5 + (-5) = 0$, that $-\frac{1}{2} + \frac{1}{2} = 0$, and so on.

The additive inverse property

> For every real number a, there exists a real number $-a$, called the additive inverse of a, such that
>
> $$a + (-a) = 0 \quad \text{and} \quad -a + a = 0$$

Subtracting Rational Numbers

Subtraction is the *inverse* operation of addition. That is, subtraction "undoes" addition. The answer to a subtraction problem is called the **difference**. The number we're *subtracting* is called the **subtrahend**, and the number we're subtracting *from* is called the **minuend**.

The definition of subtraction

> If a and b represent any real numbers, then
>
> $$a - b = a + (-b)$$

This definition means that we perform a subtraction in two steps. In step 1, we change the subtraction symbol to an addition symbol *and* we change the sign of the number being subtracted. Then, in step 2, we perform the resulting *addition*.

The definition of subtraction leads to the following facts regarding subtraction involving zero:

$a - 0 = a$ This is true because $a - 0 = a + (-0) = a + 0 = a$

$0 - a = -a$ This is true because $0 - a = 0 + (-a) = -a$

A Word of Caution "Subtract a from b" is translated as $b - a$.

EXAMPLE 2 Examples of subtracting rational numbers:

a. Subtract -3 from 0. ["Subtract -3 from 0" translates as $0 - (-3)$.]

$$0 - (-3) = 0 + (+3) = 3$$

b. Subtract 4 from -2. ["Subtract 4 from -2" translates as $(-2) - (+4)$.]

$$(-2) - (+4) = -2 + (-4) = -6$$

c. $2.345 - 11.6 = 2.345 + (-11.6) = -9.255$

d. $\frac{2}{3} - \left(-\frac{2}{5}\right) = \frac{2}{3} + \left(+\frac{2}{5}\right) = \frac{10}{15} + \frac{6}{15} = \frac{16}{15}$, or $1\frac{1}{15}$

Multiplying Rational Numbers

The answer to a multiplication problem is called the **product**, and, as we mentioned earlier, the numbers that are multiplied together are called the **factors** of the product.

The Closure Property of Multiplication Another property of the real number system that mathematicians accept as true without proof is the **closure property of multiplication**, which follows:

The closure property of multiplication

> If a and b represent any real numbers, then their product, ab, is a real number.

We assume that you recall how to multiply signed rational numbers, but we briefly review the procedure in the following box:

Multiplying two nonzero rational numbers

> Multiply the absolute values *and* give the product the correct sign; the sign is *positive* when the numbers have the same sign and *negative* when one number is positive and the other is negative.

The Multiplicative Identity Since multiplying any real number by 1 gives the *identical* number we started with (for example, $0.15 \times 1 = 0.15$), we call **1** the **multiplicative identity** or the **multiplicative identity element**.

The multiplicative identity property

> The multiplicative identity element is 1.
>
> If a represents any real number, then
>
> $$a \cdot 1 = a \quad \text{and} \quad 1 \cdot a = a$$

The Multiplicative Inverse, or Reciprocal, of a Number If the product of two numbers equals the multiplicative identity, 1, we say that the numbers are the **multiplicative inverses**, or **reciprocals**, of each other. For example, 5 and $\frac{1}{5}$ are the multiplicative inverses of each other, as are $-\frac{5}{7}$ and $-\frac{7}{5}$. The multiplicative inverse of 1 is 1. *Zero has no multiplicative inverse.*

The multiplicative inverse property

For every *nonzero* real number a, there exists a real number $\frac{1}{a}$, called the multiplicative inverse, or reciprocal, of a, such that

$$a\left(\frac{1}{a}\right) = 1 \quad \text{and} \quad \left(\frac{1}{a}\right)a = 1$$

Multiplication Involving Zero Multiplying any real number by zero gives a product of zero.

The multiplication property of zero

If a represents any real number, then

$$a \cdot 0 = 0 \quad \text{and} \quad 0 \cdot a = 0$$

EXAMPLE 3

Examples of multiplying rational numbers:

a. $(-0.14)(-10) = +\ 1.40$

The product of the absolute values: $(|-0.14|)(|-10|) = (0.14)(10) = 1.40$

The sign is positive because the given numbers have the same sign

b. $\left(4\frac{1}{2}\right)\left(-1\frac{1}{3}\right) = \left(\frac{9}{2}\right)\left(-\frac{4}{3}\right) = -\ 6$

Converting the mixed numbers to improper fractions

The product of the absolute values: $\left(\left|\frac{9}{2}\right|\right)\left(\left|-\frac{4}{3}\right|\right) = \frac{9}{2} \cdot \frac{4}{3} = \frac{36}{6} = 6$

The sign is negative because one given number is positive and the other is negative

c. $(-7)(0) = 0$ The multiplication property of zero

Dividing Rational Numbers

Division is the *inverse* of multiplication because division "undoes" multiplication. The answer to a division problem is called the **quotient**. The number we are dividing *by* is called the **divisor**, and the number we are dividing *into* is called the **dividend**. If the divisor does not divide exactly into the dividend, the part that is left over is called the **remainder**.

Factors When the remainder is 0, we can say that the divisor and the quotient are both **factors**, or **divisors**, of the dividend.

Because of the inverse relation between division and multiplication, the rules for finding the sign of a quotient are the same as those used for finding the sign of a product.

Dividing one rational number by a nonzero rational number

Divide the absolute value of the dividend by the absolute value of the divisor, *and* give the quotient the correct sign; the sign is *positive* when the numbers have the same sign and *negative* when one number is positive and the other is negative.

To check the answer to a division problem, we verify that the numbers satisfy one of the following conditions:

(Divisor × quotient) + remainder = dividend

or, if the divisor is a *factor* of the dividend,

Divisor × quotient = dividend

Division Involving Zero We next consider division in which zero is the dividend or the divisor or both.

Division of zero by a number other than zero is possible, and the quotient is always 0. For example,

$$0 \div 2 = \tfrac{0}{2} = 0, \quad \text{or} \quad 2\overline{)0}\ \text{(quotient } 0\text{)}, \quad \text{because } 2 \times 0 = 0$$

Division of a nonzero number by zero is impossible. Let's try to divide some nonzero number by zero. For example, let's try $4 \div 0$, or $\frac{4}{0}$. Suppose the quotient is some unknown number we call q. Then $4 \div 0 = q$ means that we must find a number q such that $0 \times q = 4$. But $0 \times q = 4$ is impossible, since zero times any number is zero. Therefore, no answer q exists, and $4 \div 0$ has no answer (or is *undefined*). (It is for this reason that zero has no reciprocal.)

Division of zero by zero cannot be determined. Consider $0 \div 0$, or $\frac{0}{0}$. Suppose that the quotient is 1. Then $0 \div 0 = 1$. This means that 0×1 has to equal zero, and, in fact, it does. This might lead us to assume that the quotient is indeed 1. But now let us suppose that the quotient is zero. (That is, suppose $0 \div 0 = 0$.) This could be true only if 0×0 equals zero, which it does. Furthermore, we could say that $0 \div 0 = 5$, since $0 \times 5 = 0$. The quotient could also be -3 or π or 156, or *any* number! Therefore, we say that $0 \div 0$ cannot be determined.

Because division by zero either is impossible or cannot be determined, we say that division by zero is *undefined*. The important thing to remember about division involving zero is that we cannot divide *by* zero. Division involving zero can be summarized as follows:

Division involving zero

If a represents any real number *except* 0, then

$$\frac{0}{a} = 0$$

$\frac{a}{0}$ is not possible }

$\frac{0}{0}$ cannot be determined } Undefined

EXAMPLE 4

Examples of dividing rational numbers:

a. $(-5) \div (-25) = + \ \frac{1}{5}$

The quotient of the absolute values: $|-5| \div |-25| = 5 \div 25 = \frac{5}{25} = \frac{1}{5}$

The sign is positive because the given numbers have the same sign

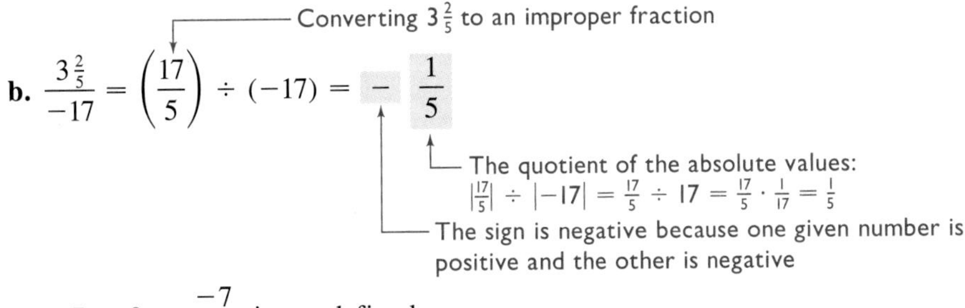
Converting $3\frac{2}{5}$ to an improper fraction
The quotient of the absolute values:
$\left|\frac{17}{5}\right| \div |-17| = \frac{17}{5} \div 17 = \frac{17}{5} \cdot \frac{1}{17} = \frac{1}{5}$
The sign is negative because one given number is positive and the other is negative

b. $\dfrac{3\frac{2}{5}}{-17} = \left(\dfrac{17}{5}\right) \div (-17) = -\dfrac{1}{5}$

c. $-7 \div 0$, or $\dfrac{-7}{0}$, is not defined.

Note The identity properties and the inverse properties are properties of the real number system that mathematicians accept as true without proof.

Using Scientific and Graphics Calculators

Scientific calculators generally use either algebraic logic or Reverse Polish Notation (RPN), and graphics calculators generally use the Equation Operating System (EOS). The type of logic used determines the order in which the keys must be pressed. We give some general guidelines for using calculators with the three types of logic. You should consult the instruction manual for your particular calculator to find out how *your* calculator works (how to turn the calculator on; how to clear the memory; how the "inverse," or "shift," or "second function" keys work; and so on).

To perform an operation that involves just two numbers (an operation such as addition, subtraction, multiplication, or division) on scientific calculators that use algebraic logic or on graphics calculators, we press the keys in the order in which we would read (or write) the problem. For example, to add 2 and 3, we press

Scientific (algebraic) [2] [+] [3] [=]

Graphics [2] [+] [3] [ENTER]

or [2] [+] [3] [EXE]*

However, to perform the same operation on an RPN calculator, we enter both numbers, separating the two numbers with the [ENTER] key, *before* pressing the key for the operation. To add 2 and 3, we press

RPN Scientific [2] [ENTER] [3] [+]

(It is *not* necessary to press [ENTER] after pressing [+].)

Note From now on, we will use the symbol [=] for a calculator with algebraic logic and the symbol [ENTER] for a graphics calculator to represent the key to use for executing a command. If your calculator has an [EXE] key and no [ENTER] key or [=] key, then you press [EXE] when we write [ENTER] or [=].

Negative Numbers and the Calculator Scientific calculators generally have a [+/−] (or [CHS]) key that is used to change the sign of a number, and it is pressed *after* the number is entered. For example, to enter −8, we press [8] [+/−], and the display is [−8] or [−8.0000].

On graphics calculators, the [(−)] key† is used for entering a negative number, and this key is pressed *before* the number is entered. For example, to enter −8, we press [(−)] [8], and the display is [−8].

---

*EXE stands for "execute the command."

†Be sure to use the key that has parentheses around the negative sign; the key *without* parentheses around the negative sign is for *subtraction*.

Note In Examples 5–7, we deliberately work problems that can easily be done *without* a calculator, so that we can mentally check the answers. We suggest that you make up similar problems and work them on your own calculator until you are completely comfortable with it.

EXAMPLE 5 Using a calculator, find $8 - 5$.

SOLUTION

Scientific (algebraic)	8	−	5	=
Graphics	8	−	5	ENTER
RPN Scientific	8	ENTER	5	−

The display is [3] or [3.0000], so $8 - 5 = 3$.

EXAMPLE 6 Using a calculator, find -6×7.

SOLUTION

Scientific (algebraic)	6	+/−	×	7	=
Graphics	(−)	6	×	7	ENTER
RPN Scientific	6	+/−	ENTER	7	×

The display is [−42] or [−42.0000], so $-6 \times 7 = -42$.

EXAMPLE 7 Using a calculator, find $-15 \div (-3)$.

SOLUTION

Scientific (algebraic)	1	5	+/−	÷	3	+/−	=
Graphics	(−)	1	5	÷	(−)	3	ENTER
RPN Scientific	1	5	+/−	ENTER	3	+/−	÷

The display is [5] or [5.0000], so $-15 \div (-3) = 5$.

EXAMPLE 8 Using a calculator, try to find $3 \div 0$.

SOLUTION

Scientific (algebraic)	3	÷	0	=
Graphics	3	÷	0	ENTER
RPN Scientific	3	ENTER	0	÷

The display may be [E], [ERROR 02 MATH], [MA ERROR], or [DIVIDE BY 0]—all error messages.

A Word of Caution On some calculators, the display may be [E 0.], making it easy to overlook the E that stands for "Error" and leading us to believe that the answer is 0. The answer is *not* zero.

In all cases, the error message indicates that $3 \div 0$ has no answer.

Your instructor may or may not want you to use a calculator in solving the problems in these exercises.

Exercises 1.6
Set I

Perform the indicated operations (or write "undefined").

1. $5 + (-9)$
2. $-10 + 7$
3. $-12 + (-7)$
4. $-\frac{7}{8} + \left(-\frac{3}{8}\right)$
5. $-\frac{7}{12} + \left(-\frac{1}{6}\right)$
6. $-\frac{3}{4} + \left(-\frac{3}{8}\right)$
7. $5\frac{1}{4} + \left(-2\frac{1}{2}\right)$
8. $-1\frac{7}{10} + 3\frac{2}{5}$
9. $-13.5 - (-8.06)$
10. $-1.5 - (-1.13)$
11. $2.4 - (-13)$
12. $7.2 - (-89)$
13. $\frac{1}{3} - \left(-\frac{1}{2}\right)$
14. $17\frac{5}{6} - \left(-8\frac{1}{3}\right)$
15. $-5\frac{3}{4} - 2\frac{1}{2}$
16. $-1.56 - 9.7$
17. $16.71 - (-18.9)$
18. $8.9 - (-3.71)$
19. $0 - 3$
20. $0(-3)$
21. $(-26)(-10)$
22. $(-1.1)(-7)$
23. $\frac{3}{8}\left(-\frac{4}{9}\right)$
24. $\dfrac{-150}{10}$
25. $-12 \div 36$
26. $\dfrac{-15}{-6}$
27. $7\frac{1}{2} \div \left(-\frac{1}{2}\right)$
28. $-\frac{5}{4} \div 0$
29. $0 \div \left(-\frac{3}{8}\right)$
30. $0 \div 0$
31. $-\frac{5}{4} \div \left(-\frac{3}{8}\right)$
32. $-25{,}000 \div (-100)$
33. Subtract -3 from -8.
34. Subtract 0.3 from -1.26.
35. Subtract $-\frac{2}{3}$ from 5.
36. $|-8| + (-15)$
37. $|-10| - (-25)$
38. $|-2| - |-9|$
39. $-3|-4|$
40. $|-15| - 4$

Writing Problems

Express the answers in your own words and in complete sentences.

1. Explain why the statement $4 \div 0 = 0$ is false.
2. Explain why the statement $0 \div 8 = 0$ is true.
3. Explain why the statement $0 \div 0 = 0$ is false.
4. Explain why zero does not have a multiplicative inverse.

Exercises 1.6
Set II

Perform the indicated operations (or write "undefined").

1. $-59 + 74$
2. $6 + (-15)$
3. $\frac{5}{8} + \left(-\frac{3}{4}\right)$
4. $\frac{1}{2} + \frac{1}{3}$
5. $-4\frac{1}{2} + 1\frac{3}{4}$
6. $10\frac{3}{8} + \left(-2\frac{3}{4}\right)$
7. $-395.7 + (84.91)$
8. $-1\frac{2}{3} + \left(-3\frac{2}{5}\right)$
9. $281 - 960$
10. $3.62 + (-1.634)$
11. $28.61 - (-37.9)$
12. $-8 - (-3.2)$
13. $-18\frac{3}{10} - 7\frac{2}{5}$
14. $16\frac{2}{3} - \left(-3\frac{7}{12}\right)$
15. $-6.483 - (-2.7)$
16. $3.28 - 7.6$
17. $(-26)(10)$
18. $-5.2 - (7.3)$
19. $0(-13)$
20. $0 - 13$
21. $(-34)(12)$
22. $-6(0)$
23. $\dfrac{3}{7}\left(\dfrac{-14}{15}\right)$
24. $\dfrac{-180}{10}$
25. $-15 \div 60$
26. $-4 \div 12$
27. $\frac{0}{0}$
28. $\frac{2}{3} \div 0$
29. $0 \div 7$
30. $1.6 \div 0$
31. Subtract -1 from -2.
32. Subtract 1.26 from 0.3.
33. $\dfrac{160}{-40}$
34. Subtract -5.2 from -1.63.
35. $|-3| + (-7)$
36. $|-6| - |-10|$
37. $|-5| - (-6)$
38. $4 - |-8|$
39. $|8| + |-18|$
40. $-2 - |-10|$

1.7 The Commutative, Associative, and Distributive Properties of Real Numbers

The Commutative Properties

Addition Is Commutative Reversing the order of the numbers in an addition problem does not change the sum. This property is called the **commutative property of addition.**

The commutative property of addition

> If a and b represent any real numbers, then
> $$a + b = b + a$$

EXAMPLE 1 Verify that the commutative property holds for the sum of 7 and 5.

SOLUTION $7 + 5 = 12$, and $5 + 7 = 12$. Since $7 + 5$ and $5 + 7$ both equal 12, they must equal each other. Therefore, $7 + 5 = 5 + 7$.

Subtraction is not commutative, as a single example will prove.

EXAMPLE 2 Show that subtraction is not commutative by showing that $3 - 2 \neq 2 - 3$.

SOLUTION $3 - 2 = 1$, and $2 - 3 = -1$. Since $1 \neq -1$, $3 - 2$ and $2 - 3$ do not both equal the same number, so $3 - 2 \neq 2 - 3$. By finding one subtraction problem for which the commutative property does *not* hold, we've shown that subtraction is *not* commutative.

Multiplication Is Commutative Reversing the order of the numbers in a multiplication problem does not change the product. This property is called the **commutative property of multiplication.**

The commutative property of multiplication

> If a and b represent any real numbers, then
> $$a \cdot b = b \cdot a$$

EXAMPLE 3 Verify that the commutative property holds for the product of 4 and 5.

SOLUTION $4 \times 5 = 20$, and $5 \times 4 = 20$. Since 4×5 and 5×4 both equal 20, they must equal each other. Therefore, $4 \times 5 = 5 \times 4$.

Division is not commutative, as a single example will prove.

EXAMPLE 4 Show that division is not commutative by showing that $10 \div 5 \neq 5 \div 10$.

SOLUTION $10 \div 5 = 2$, and $5 \div 10 = \frac{1}{2}$. Since $2 \neq \frac{1}{2}$, $10 \div 5$ and $5 \div 10$ do not both equal the same number, so $10 \div 5 \neq 5 \div 10$. By finding one division problem for which the commutative property does *not* hold, we've shown that division is *not* commutative.

The Associative Properties

Addition Is Associative In adding three numbers, we obtain the same sum when we add the first two numbers together first as when we add the last two numbers together first. This property is called the **associative property of addition.**

The associative property of addition

If a, b, and c represent any real numbers, then

$$(a + b) + c = a + (b + c)$$

EXAMPLE 5 Verify that the associative property holds for the sum of 2, 3, and 4.

SOLUTION $(2 + 3) + 4 = 5 + 4 = 9$, and $2 + (3 + 4) = 2 + 7 = 9$. Since $(2 + 3) + 4$ and $2 + (3 + 4)$ both equal 9, they must equal each other. Therefore, $(2 + 3) + 4 = 2 + (3 + 4)$.

Subtraction is not associative, as a single example will prove.

EXAMPLE 6 Show that subtraction is not associative by showing that $(7 - 4) - 8 \neq 7 - (4 - 8)$.

SOLUTION $(7 - 4) - 8 = 3 - 8 = -5$, and $7 - (4 - 8) = 7 - (-4) = 7 + 4 = 11$. Since $-5 \neq 11$, $(7 - 4) - 8$ and $7 - (4 - 8)$ do not both equal the same number, so $(7 - 4) - 8 \neq 7 - (4 - 8)$. By finding one subtraction problem for which the associative property does *not* hold, we've shown that subtraction is *not* associative.

Multiplication Is Associative In multiplying three numbers, we obtain the same product when we multiply the first two numbers together first as when we multiply the last two numbers together first. This property is called the **associative property of multiplication.**

The associative property of multiplication

If a, b, and c represent any real numbers, then

$$(a \cdot b) \cdot c = a \cdot (b \cdot c)$$

EXAMPLE 7 Verify that the associative property holds for the product of 3, 4, and 2.

SOLUTION $(3 \cdot 4) \cdot 2 = 12 \cdot 2 = 24$, and $3 \cdot (4 \cdot 2) = 3 \cdot 8 = 24$. Since $(3 \cdot 4) \cdot 2$ and $3 \cdot (4 \cdot 2)$ both equal 24, they must equal each other. Therefore, $(3 \cdot 4) \cdot 2 = 3 \cdot (4 \cdot 2)$.

Division is not associative, as a single example will prove.

EXAMPLE 8 Show that division is not associative by showing that $(16 \div 4) \div 2 \neq 16 \div (4 \div 2)$.

SOLUTION $(16 \div 4) \div 2 = 4 \div 2 = 2$, and $16 \div (4 \div 2) = 16 \div 2 = 8$. Since $2 \neq 8$, $(16 \div 4) \div 2$ and $16 \div (4 \div 2)$ do not both equal the same number, so $(16 \div 4) \div 2 \neq 16 \div (4 \div 2)$. By finding one division problem for which the associative property does *not* hold, we've shown that division is *not* associative.

How to Determine Whether Commutativity or Associativity Has Been Used In commutativity, the numbers or variables actually exchange places (commute). (Remember, when you go from *home* to *work* and then from *work* to *home*, you are *commuting*.)

$$a + b = b + a \qquad c \cdot d = d \cdot c$$

In associativity, the numbers or variables stay in their original places, but the grouping is changed.

$$a + (b + c) = (a + b) + c \qquad (d \cdot e) \cdot f = d \cdot (e \cdot f)$$

The Distributive Property

The distributive property is one of the most important and most frequently used properties in mathematics. A simple example that may help you understand the distributive property follows: Suppose that Marla has three red roses and seven yellow roses, and suppose that Sam asks her how many roses she would have altogether if he *doubled* the number of roses of each color. Should Marla find the total number of roses she has now and double *that* number, or should she double the number of red roses and double the number of yellow roses and then add those two products? Does it matter? No, since it happens to be true that $2(3 + 7) = (2 \cdot 3) + (2 \cdot 7)$. Therefore, Marla can do the problem either way. It is the distributive property that guarantees that $2(3 + 7) = (2 \cdot 3) + (2 \cdot 7)$.

Multiplication Is Distributive over Addition The **distributive property** can be verified by substituting *any* real numbers for a, b, and c. We may distribute either from the right or from the left.

The distributive property

Multiplication is distributive over addition.

If a, b, and c represent any real numbers, then

$$a(b + c) = (ab) + (ac)$$

$$(b + c)a = (ba) + (ca)$$

Note The distributive property may be extended to include any number of terms inside the parentheses, and the terms may be *subtracted* as well as added. That is,

$$a(b + c - d + e) = (ab) + (ac) - (ad) + (ae)$$

and so forth. 

EXAMPLE 9 Verify that the distributive property holds for $a = 3$, $b = 5$, and $c = 7$.

SOLUTION

$a(b + c)$	$(ab) + (ac)$
$3(5 + 7)$	$(3 \cdot 5) + (3 \cdot 7)$
$= 3 \quad (12)$	$= \quad 15 \quad + \quad 21$
$= \quad 36$	$= \quad 36$

Since $36 = 36$, $a(b + c) = (ab) + (ac)$ if $a = 3$, $b = 5$, and $c = 7$.

A Word of Caution A common error is to think that the distributive property applies to expressions such as $2(3 \cdot 4)$ [that is, to think that $2(3 \cdot 4) = (2 \cdot 3) \cdot (2 \cdot 4)$].

The distributive property applies only when this symbol is an addition or subtraction symbol

$$2(3 \cdot 4) \neq (2 \cdot 3) \cdot (2 \cdot 4)$$
$$2(12) \neq 6 \cdot 8$$
$$24 \neq 48$$

The distributive property permits us to convert addition problems into multiplication problems and multiplication problems into addition problems, which we often need to do when we simplify expressions, when we solve equations, when we find least common multiples or least common denominators, and so on.

The commutative, associative, and distributive properties cannot be proved; they are among the properties that mathematicians accept as true without proof.

EXAMPLE 10 Determine whether each of the following is true or false. If the statement is *true*, name the property that makes the statement true.

SOLUTION

a. $(1.2 + 5) + 2.4 = 1.2 + (5 + 2.4)$ True; associative property of addition

b. $\frac{2}{3}\left(-\frac{5}{8}\right) = -\frac{5}{8}\left(\frac{2}{3}\right)$ True; commutative property of multiplication

c. $-8(1) = -8$ True; multiplicative identity property

d. $0.1 - 5 = 5 - 0.1$ False

e. $17 + (-17) = 0$ True; additive inverse property

f. $3(7 + 9) = (3 \cdot 7) + (3 \cdot 9)$ True; distributive property

g. $2 \div \left(\frac{3}{5}\right) = \left(\frac{3}{5}\right) \div 2$ False

h. $(y \div z) \div x = y \div (z \div x)$ False

i. $4(2 \cdot 3) = (4 \cdot 2) \cdot (4 \cdot 3)$ False

j. $(1 + 2) + 3 = (2 + 1) + 3$ True; commutative property of addition (Notice that the *order* was changed, not the grouping.)

k. $(3 + 7) + 8 = 8 + (7 + 3)$ True; commutative property of addition, used *twice* [That is, $(3 + 7) + 8 = 8 + (3 + 7) = 8 + (7 + 3)$.]

l. $0 = 8(0)$ True; multiplication property of zero

m. $\left(-\frac{1}{8}\right)(-8) = 1$ True; multiplicative inverse property

Exercises 1.7
Set I

In Exercises 1–32, determine whether each statement is true or false. If the statement is *true*, name the property that makes the statement true.

1. $5 + 10 = 10 + 5$
2. $x + y = y + x$
3. $(3)(8 + 2) = (2 + 8)(3)$
4. $(2)(5 + 6) = (6 + 5)(2)$
5. $3(8 + 2) = (3 \cdot 8) + (3 \cdot 2)$
6. $2(5 + 6) = (2 \cdot 5) + (2 \cdot 6)$
7. $6 - (4 - 2) = (6 - 4) - 2$
8. $(12 \div 6) \div 3 = 12 \div (6 \div 3)$

9. $14(7 \cdot 2) = (14 \cdot 7) \cdot (14 \cdot 2)$

10. $25(2 \cdot 6) = (25 \cdot 2) \cdot (25 \cdot 6)$

11. $16 - 4 = 4 - 16$

12. $\left(\frac{2}{3}\right) \div \left(\frac{1}{2}\right) = \left(\frac{1}{2}\right) \div \left(\frac{2}{3}\right)$

13. $6 + (2 \cdot 4) = 6 + (4 \cdot 2)$

14. $(6)(3) + 4 = 4 + (3)(6)$

15. $19 + 0 = 19$

16. $0 + (-5) = -5$

17. $8 + (2 + 4) = (8 + 2) + 4$

18. $(8)(2) - 10 = 10 - (2)(8)$

19. $3(0) = 3$

20. $0(-1) = -1$

21. $(7 + 5)(3) = (7 \cdot 3) + (5 \cdot 3)$

22. $(6 + 2)(5) = (6 \cdot 5) + (2 \cdot 5)$

23. $1(-6) = -6$

24. $5(1) = 5$

25. $4 \div 12 = 12 \div 4$

26. $31 + 2 = 2 + 31$

27. $-8 + 8 = 0$

28. $15 + (-15) = 0$

29. $(6 + 3) + 7 = 7 + (3 + 6)$

30. $8 + (3 + 5) = (5 + 3) + 8$

31. $\left(\frac{1}{7}\right)(7) = 1$

32. $-11 + 11 = 0$

In Exercises 33–44, complete each statement by using the property indicated.

33. Commutative property: $-4(-5) =$ ____________

34. Commutative property: $8(-3) =$ ____________

35. Associative property: $7 + [(-2) + 8] =$ ____________

36. Associative property: $-9 + [12 + (-3)] =$ ____________

37. Distributive property: $-3(2 + 7) =$ ____________

38. Distributive property: $5(-3 + 6) =$ ____________

39. Commutative property: $-7 + 8 =$ ____________

40. Commutative property: $-5 + (-4) =$ ____________

41. Associative property: $9(-4 \cdot 7) =$ ____________

42. Associative property: $9(-3 \cdot 8) =$ ____________

43. Distributive property: $(-8 \cdot 6) + (-8 \cdot 3) =$ ____________

44. Distributive property: $(9 \cdot 6) + (9 \cdot 8) =$ ____________

Writing Problems

Express the answers in your own words and in complete sentences.

1. Explain why $3 + (4 + 9) = 3 + (9 + 4)$ does *not* illustrate the associative property of addition.
2. Explain why $2 \cdot (3 \cdot 7) \neq (2 \cdot 3) \cdot (2 \cdot 7)$.
3. Explain how to determine whether $5 \cdot (3 \cdot 8) = 5 \cdot (8 \cdot 3)$ illustrates the commutative property of multiplication or the associative property of multiplication or both.

Exercises 1.7 Set II

In Exercises 1–32, determine whether each statement is true or false. If the statement is *true*, name the property that makes the statement true.

1. $8 + 3 = 3 + 8$

2. $(6 \cdot 3) + (6 \cdot 8) = 6(3 + 8)$

3. $(7)(2 + 3) = (3 + 2)(7)$

4. $\frac{7}{8} \div 8 = 8 \div \frac{7}{8}$

5. $4(9 + 3) = (4 \cdot 9) + (4 \cdot 3)$

6. $4(2 + 9) = 4(9 + 2)$

7. $4(9 \cdot 3) = (4 \cdot 9) \cdot (4 \cdot 3)$

8. $3(7) = 7(3)$

9. $(7)(8) = (8)(7)$

10. $(6 \cdot 7)5 = (6 \cdot 5) \cdot (7 \cdot 5)$

11. $16 \div 5 = 5 \div 16$
12. $2 - 8 = 8 - 2$
13. $16 - 23 = 23 - 16$
14. $(1 + 4)(5) = (1 \cdot 5) + (4 \cdot 5)$
15. $18 + 0 = 18$
16. $-6 + 6 = 0$
17. $5 + (4 + 2) = (2 + 4) + 5$
18. $\frac{1}{2}(6 + 7) = \left(\frac{1}{2} \cdot 6\right) + \left(\frac{1}{2} \cdot 7\right)$
19. $0(-6) = -6$
20. $1(-6) = -6$
21. $3(5 \cdot 7) = (3 \cdot 5)(7)$
22. $17 \div 1 = 1 \div 17$
23. $(-4)(1) = -4$
24. $-4 + 4 = 0$
25. $(4 + 9)(2) = (4 \cdot 2) + (9 \cdot 2)$
26. $0 \div 5 = 5 \div 0$
27. $13 + (-13) = 0$
28. $13(0) = 0$
29. $(5 + 2) + 6 = 6 + (2 + 5)$
30. $4 + (9 + 2) = (2 + 9) + 4$
31. $(-4)\left(-\frac{1}{4}\right) = 1$
32. $8 + 0 = 8$

In Exercises 33–44, complete each statement by using the property indicated.

33. Commutative property:
$7 + (-2) =$ ________
34. Commutative property:
$-7(-3) =$ ________
35. Associative property:
$4 + [(-6) + 8] =$ ________
36. Associative property:
$6(-5 \cdot 8) =$ ________
37. Distributive property:
$-4(11 + 6) =$ ________
38. Distributive property:
$2(-9 + 7) =$ ________
39. Commutative property:
$-8 + 4 =$ ________
40. Commutative property:
$-7(9) =$ ________
41. Associative property:
$2(-4 \cdot 6) =$ ________
42. Associative property:
$-4 + [5 + (-3)] =$ ________
43. Distributive property:
$(-2 \cdot 5) + (-2 \cdot 7) =$ ________
44. Distributive property:
$(7 \cdot 4) + (7 \cdot 13) =$ ________

1.8 Natural Number Powers of Rational Numbers

Recall from elementary algebra that a shortened notation for a product such as $3 \cdot 3 \cdot 3 \cdot 3$ is 3^4. That is, by definition, the notation 3^4 indicates that 3 is to be used as a factor four times, giving

$$3^4 = 3 \cdot 3 \cdot 3 \cdot 3 = 81$$

In the expression 3^4, 3 is called the *base*, and 4 is called the *exponent*; an exponent must be written as a small number *above* and *to the right of* the base. The entire symbol 3^4 is called an **exponential expression** and is commonly read as "three to the fourth power," "three to the fourth," or "the fourth power of three."

The exponent

$$3^4 = 81$$

The base

> The exponent in an expression always applies *only* to the symbol immediately preceding it. If that symbol is), a right parenthesis, the exponent applies to whatever is inside the parentheses.

A Word of Caution Because an exponent applies only to the immediately preceding symbol, we must distinguish between expressions such as $(-6)^2$ and -6^2; these expressions are *not* the same. For example, in the expression $(-6)^2$, the exponent is immediately to the right of a right parenthesis; therefore, the exponent applies to everything inside the parentheses (that is, to the -6), giving

$$(-6)^2 = (-6)(-6) = 36$$

However, in the expression -6^2, the exponent is immediately to the right of the 6, and so it applies *only* to the 6, giving

$$-6^2 = -(6^2) = -(6 \cdot 6) = -36$$

Since $36 \neq -36$, $(-6)^2 \neq -6^2$.

When the exponent is a 2, as in b^2, we usually read the expression as "b squared" rather than as "b to the second power." If the base is negative, as in $(-4)^2$, we read the expression as "the square of negative 4."

When the exponent is 3, as in b^3, we usually read the expression as "b cubed" rather than as "b to the third power." The expression $(-2)^3$ should be read as "the cube of negative 2."

The Exponent 1 If a represents any real number, then $a^1 = a$. For example, $23^1 = 23$, and $(-0.2)^1 = -0.2$.

Even Powers If a base has an exponent that is an even number, we say that it is an **even power** of the base. For example, 3^2, 5^4, and $(-2)^6$ are even powers.

Odd Powers If a base has an exponent that is an odd number, we say that it is an **odd power** of the base. For example, 3^1, 10^3, and $(-4)^5$ are odd powers.

EXAMPLE 1 Examples of powers of numbers:

a. $2^3 = 2 \cdot 2 \cdot 2 = 8$ — The expression $2^3 = 8$ is read "two cubed equals 8"; the power is an odd power

b. $(-4)^2 = (-4)(-4) = 16$ — The exponent applies to every thing inside the parentheses; the power is an even power

c. $(-0.2)^3 = (-0.2)(-0.2)(-0.2) = -0.008$ — This is an odd power of a negative number; notice that the answer is negative

d. $\left(-\frac{1}{2}\right)^2 = \left(-\frac{1}{2}\right)\left(-\frac{1}{2}\right) = \frac{1}{4}$ — This is an even power of a negative number; notice that the answer is positive

e. $(-1)^4 = (-1)(-1)(-1)(-1) = 1$ — This is an even power of a negative number; notice that the answer is positive

f. $(-1)^7 = (-1)(-1)(-1)(-1)(-1)(-1)(-1) = -1$ — This is an odd power of a negative number; notice that the answer is negative

g. $0^4 = 0 \cdot 0 \cdot 0 \cdot 0 = 0$ — Any nonzero power of zero equals zero

Powers of Negative Numbers An *odd* power of a negative real number is always negative (see Examples 1c and 1f), and an *even* power of a negative real number is always positive (see Examples 1b, 1d, and 1e). In Example 2, we demonstrate finding powers of negative numbers by using these facts.

EXAMPLE 2 Examples of powers of negative numbers:

a. $(-2)^4 = +2^4 = 16$ An even power of a negative number is positive

b. $(-4)^3 = -4^3 = -64$ An odd power of a negative number is negative

---

A Word of Caution When we write an exponential expression, we must *be sure that our exponents look like exponents.* For example, we must be sure that 3^4 doesn't look like 34.

Note We suggest that you memorize the squares of at *least* the first sixteen natural numbers and the cubes of at least the first five natural numbers.

Finding Natural Number Powers with Calculators

Most calculators have an [x^2] key, and on calculators with all three types of logic, this key is pressed *after* the number has been entered (see Example 3). If your calculator has x^2 written *above* some other key, you must press the [INV], [2nd], or [↰] key before pressing the [?] key. (with x^2 written above the [?] key)

To square a negative number with a graphics calculator, we *must* enclose the number in parentheses. To see why, let's try to find $(-1)^2$ by pressing these keys on a graphics calculator: [(−)] [1] [x^2] [ENTER]. The display is [−1], showing that the calculator treats the problem as if it were -1^2. Now let's try pressing [(] [(−)] [1] [)] [x^2] [ENTER]. *Now* the display is [1], showing that we have found $(-1)^2$.

To square a negative number using a (non-graphics) scientific calculator, it is not necessary to enclose the number in parentheses.

EXAMPLE 3 Using a calculator, find $(-1.68)^2$.

SOLUTION

Scientific (algebraic) [1] [.] [6] [8] [+/−] [x^2]

or [1] [.] [6] [8] [+/−] [INV] [?] (x^2 above [?])

Graphics [(] [(−)] [1] [.] [6] [8] [)] [x^2] [ENTER]

RPN Scientific [1] [.] [6] [8] [+/−] [↰] [$\sqrt{x}$] (x^2 above [$\sqrt{x}$])

In all cases, the display is [2.8224], and so $(-1.68)^2 = 2.8224$.

Scientific and graphics calculators have a key for raising to powers other than 2. This key may be labeled [y^x], [x^y], or [∧], and you may have to press the [INV], [2nd], or [↰] key before you press the [y^x] key. (We will use the symbol [y^x] for raising to powers on scientific calculators and the symbol [∧] for raising to powers on graphics calculators.) Examples 4 and 5 demonstrate the use of these keys.

EXAMPLE 4 Using a calculator, find $(-4)^5$.

SOLUTION

Scientific (algebraic) [4] [+/−] [y^x] [5] [=]

or [4] [+/−] [INV] [?] [5] [=] (y^x above the [?] key)

Graphics [(] [(−)] [4] [)] [∧] [5] [ENTER]

RPN Scientific [4] [+/−] [ENTER] [5] [y^x]

The display is [−1024] or [−1024.0000], and so $(-4)^5 = -1{,}024$.

EXAMPLE 5 Using a calculator, find $(-8)^4$.

SOLUTION

Scientific (algebraic) [8] [+/−] [y^x] [4] [=]

or [8] [+/−] [INV] [?] [4] [=] (y^x above the [?] key)

Graphics [(] [(−)] [8] [)] [∧] [4] [ENTER]

RPN Scientific [8] [+/−] [ENTER] [4] [y^x]

The display is [4096] or [4096.0000], and so $(-8)^4 = 4{,}096$.

Note If you have a graphics calculator, you might verify that you do *not* get the correct answer for Example 5 if you do not put parentheses around the −8.

All the exponents in this section are natural numbers. Other exponents will be considered in later chapters.

Exercises 1.8
Set I

Find the value of each of the following expressions. In Exercises 1–16, do *not* use a calculator. (You can *check* your answers by using a calculator.)

1. 4^3 2. 7^2 3. $(-3)^4$ 4. $(-2)^4$
5. -2^4 6. -3^4 7. 0^5 8. 0^6
9. $(-1)^{49}$ 10. $(-1)^{50}$ 11. $\left(\frac{1}{2}\right)^4$ 12. $\left(\frac{7}{8}\right)^2$
13. $(-0.1)^5$ 14. $(0.1)^6$ 15. $\left(\frac{1}{10}\right)^3$ 16. $(-5)^1$

Use a calculator for Exercises 17–24.

17. $(8.7)^4$ 18. $(4.7)^3$ 19. $(11.6)^4$ 20. $(17.3)^2$
21. $(9.2)^3$ 22. $(-1.5)^4$ 23. $(-2.5)^4$ 24. $(-5.3)^3$

Writing Problems

Express the answers in your own words and in complete sentences.

1. Explain why $2^5 \neq 10$.
2. Explain why $(-2)^4 \neq -2^4$.
3. Explain why $(-2)^5 = -2^5$.
4. Explain what $(-2)^5$ means.

Exercises 1.8
Set II

Find the value of each of the following expressions. In Exercises 1–16, do *not* use a calculator. (You can *check* your answers by using a calculator.)

1. 5^3 2. $(-2)^6$ 3. -2^6 4. 0^4
5. $(-15)^2$ 6. $(-3)^3$ 7. $(-1)^{79}$ 8. $(-1)^{64}$
9. $\left(\frac{1}{10}\right)^4$ 10. $(-2)^3$ 11. -2^7 12. $(-1)^{83}$
13. $(-0.1)^3$ 14. $(0.1)^4$ 15. $\left(\frac{1}{10}\right)^5$ 16. $(-4)^1$

Use a calculator for Exercises 17–24.

17. $(31.8)^4$ 18. $\left(-\frac{1}{10}\right)^6$ 19. $\left(-\frac{1}{10}\right)^5$ 20. $(-1.4)^3$
21. $(-1.3)^4$ 22. $(-3.8)^4$ 23. 3^9 24. 18^3

1.9 Roots

Finding Square Roots by Inspection

Just as subtraction is the inverse operation of addition and division is the inverse operation of multiplication, finding roots is an inverse operation of raising to powers. Thus, finding the **square root** of a number is the inverse operation of squaring a number.

Every positive number has both a positive and a negative square root. When we are asked to find a square root of x, we must find a number whose *square* is x. For example, the two square roots of 9 are 3 and -3, since $3^2 = 9$ and $(-3)^2 = 9$.

The Principal Square Root of a Number The *positive* square root of a positive number is called its **principal square root**. The notation for the principal square root of x is $\sqrt{x}$. The symbol $\sqrt{}$ is called the radical sign, and the number under the radical sign is called the radicand.

$\sqrt{x}$, the principal square root of x

If x, the radicand, represents some nonnegative real number and if b represents some nonnegative real number whose square is x (that is, if $x \geq 0$, if $b \geq 0$, and if $b^2 = x$), then

$$\sqrt{x} = b$$

If x, the radicand, represents a *negative* real number, then $\sqrt{x}$ is not real.

Thus, "Find $\sqrt{9}$" means that we must find a *positive* number whose square is 9; therefore, $\sqrt{9} = 3$. (The notation for the *negative* square root of 9 is $-\sqrt{9}$, and $-\sqrt{9} = -3$. Notice that the negative sign is *outside* the radical sign.)

A Word of Caution When the symbol $\sqrt{x}$ is used, it *always* represents the *principal square root* of x. Since the principal square root is always nonnegative, $\sqrt{x}$ is always nonnegative.

If you've memorized the squares of the first sixteen whole numbers, you can find certain square roots "by inspection" (see Example 1).

EXAMPLE 1 Examples of finding square roots by inspection:

a. $\sqrt{36} = 6$ — $\sqrt{36} = 6$ because $6 \geq 0$ and $6^2 = 36$

b. $\sqrt{0} = 0$ — $\sqrt{0} = 0$ because $0 \geq 0$ and $0^2 = 0$

c. $\sqrt{1} = 1$ — $\sqrt{1} = 1$ because $1 \geq 0$ and $1^2 = 1$

d. $-\sqrt{9} = -3$ — $\sqrt{9} = 3$ because $3 \geq 0$ and $3^2 = 9$; therefore, $-\sqrt{9} = -3$

e. $\sqrt{-9}$ is not real. — The radicand is negative.

A Word of Caution Notice that in Example 1d, the negative sign was *outside* the radical sign. Had the problem been $\sqrt{-9}$, we could not have solved it at this time. No *real* number exists whose square is -9, since the square of every real number is always positive or zero. We will discuss square roots of negative numbers in a later chapter.

Finding and Approximating Square Roots with a Calculator

We can find the square root of a number such as 762,129 by using a calculator with a square root key [√] (see Example 2). If the radicand is not the square of a rational number, the number is irrational and we use the calculator to *approximate* the number (see Example 3).

EXAMPLE 2 Using a calculator, find $\sqrt{762{,}129}$.

SOLUTION

Scientific (algebraic) [7] [6] [2] [1] [2] [9] [√]*

or [7] [6] [2] [1] [2] [9] [INV] [?] (√)

Graphics [√] [7] [6] [2] [1] [2] [9] [ENTER]

(You may have to press [2nd] before [√].)

RPN Scientific [7] [6] [2] [1] [2] [9] [√]

The display is [873] or [873.0000], and so $\sqrt{762{,}129} = 873$.

EXAMPLE 3 Using a calculator, approximate $\sqrt{2}$, rounding off the answer to three decimal places.

SOLUTION

Scientific (algebraic) [2] [√]

or [2] [INV] [?] (√)

Graphics [√] [2] [ENTER]

(You may have to press [2nd] before [√].)

RPN Scientific [2] [√]

Since $\sqrt{2}$ is an *irrational* number (there is no rational number whose square is 2), its decimal approximation will not terminate and will not repeat. The display is [1.414213562], and so $\sqrt{2} \approx 1.414$, rounded off to three decimal places.

*On a few scientific calculators, the keystrokes are the same as on a graphics calculator.

Finding Higher Roots

Radicals A **radical** is any indicated root of a number. Some examples of radicals are $\sqrt{9}$, $\sqrt{8}$, $\sqrt[3]{8}$, $\sqrt[3]{55}$, $\sqrt[5]{-32}$, and $\sqrt[4]{17}$. Roots other than square roots are called higher roots; thus, $\sqrt[3]{8}$, $\sqrt[3]{55}$, $\sqrt[5]{-32}$, and $\sqrt[4]{17}$ represent higher roots. The parts of a radical sign are shown below.

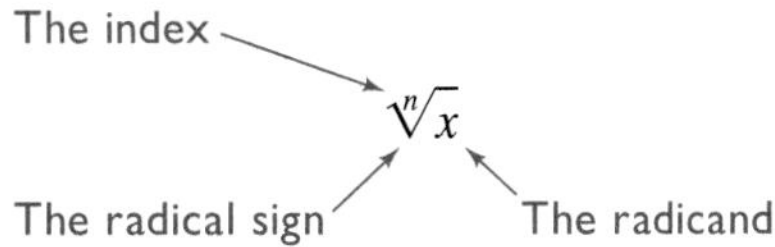

The Meaning of the Symbols Used to Indicate Roots The symbol $\sqrt[3]{x}$ (read "the cube root of x") indicates the *principal cube root* of x. When the index of a radical is 3, we must find some number whose *cube* is x; for example, $\sqrt[3]{8} = 2$ because $2^3 = 8$. You can find selected cube roots "by inspection" if you have memorized the cubes of the first few whole numbers.

The notation $\sqrt[4]{x}$ (read "the fourth root of x") indicates the *principal fourth root* of x. When the index is 4, we must find a number whose *fourth power* is x. The symbol $\sqrt[5]{x}$ indicates the *principal fifth root*, and so on.

When the index is an even number, we call the index an *even index*, and when the index is an odd number, we call it an *odd index*.

Principal Roots When we raise a positive *or* a negative number to an even power, the result is a positive number. Therefore, every even-index radical with a positive radicand has both a positive and a negative real root; the *positive root* is called the **principal root**. (For example, $2^4 = 16$ and $(-2)^4 = 16$; therefore, 16 has two real fourth roots, 2 and -2. The *principal* fourth root, however, is 2; that is, $\sqrt[4]{16} = 2$.)

An even-index radical with a *negative* radicand has *no* real roots. (For example, $\sqrt[4]{-16}$ is not real, because there is no real number whose fourth power is -16.)

When we raise a positive number to an odd power, the result is always positive; therefore, an odd-index radical with a positive radicand always has a positive principal root. (For example, $3^5 = 243$; therefore, the principal fifth root of 243 is 3, or $\sqrt[5]{243} = 3$.)

When we raise a negative number to an odd power, the result is always negative; therefore, an odd-index radical with a negative radicand always has a negative principal root. [For example, $(-2)^7 = -128$; therefore, the principal seventh root of -128 is -2, or $\sqrt[7]{-128} = -2$.]

Principal roots can be summarized as follows:

The principal nth root of x

The symbol $\sqrt[n]{x}$ always represents the principal nth root of x.

If x, the radicand, represents some real number and if b represents some real number whose nth power is x (that is, if $b^n = x$), then

$$\sqrt[n]{x} = b$$

If the radicand is positive, the principal root is positive.

If the radicand is negative and the index is *odd*, the principal root is negative.

If the radicand is negative and the index is *even*, the principal root is *not a real number.*

Problem Solving: Guessing-and-Checking

One of the recognized problem-solving techniques is that of **guessing-and-checking**. We demonstrate this method in Example 4.

EXAMPLE 4 Examples of finding higher roots by inspection or by guessing-and-checking:

a. Find $\sqrt[4]{16}$.

We must find a positive number whose *fourth* power is 16. That is, we must solve $(?)^4 = 16$. If we have memorized that $2^4 = 16$, then this problem is easy to do. If we haven't memorized that fact, then we must use guessing-and-checking. That is, we start with 1 and check to see if $1^4 = 16$. It doesn't. Next, we check to see if $2^4 = 16$. It does. Therefore, $\sqrt[4]{16} = 2$.

b. Find $\sqrt[4]{-10,000}$.

Because the radicand is negative and the index is even, the answer is "not real." (There is no real number whose fourth power is $-10,000$.)

c. Find $-\sqrt[3]{-343}$.

Since the index is odd and the radicand is negative, the principal root of $\sqrt[3]{-343}$ will be negative. We must find a negative number whose cube is -343. We will use guessing-and-checking. We probably know that $(-3)^3 = -27$ and that $(-4)^3 = -64$. Does $(-5)^3 = -343$? No; $(-5)^3 = -125$. Does $(-6)^3 = -343$? No; $(-6)^3 = -216$. Does $(-7)^3 = -343$? Yes! Therefore, $\sqrt[3]{-343} = -7$. But the problem was to find the additive inverse of $\sqrt[3]{-343}$: $-\sqrt[3]{-343}$. We determine that $-\sqrt[3]{-343} = -(-7) = 7$. Therefore, $-\sqrt[3]{-343} = 7$.

We discuss using calculators for finding higher roots in a later chapter.

All roots of nonnegative numbers and all odd roots of negative numbers are *real numbers*; therefore, they can be graphed on the real number line. We graph a few such points in Figure 5. We find decimal approximations for irrational numbers (by using a calculator or by guessing-and-checking) in order to graph them.

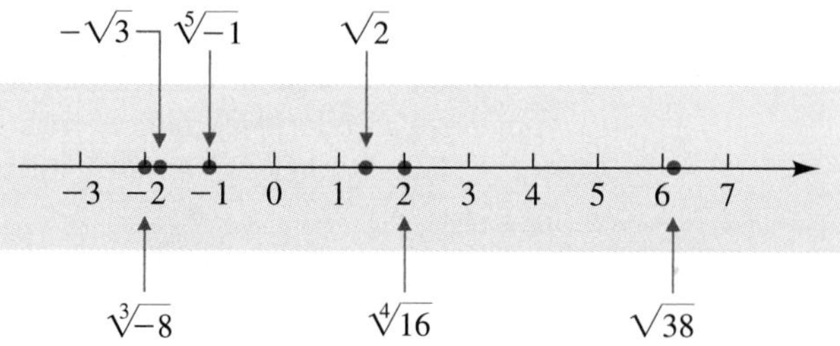

Figure 5

Exercises 1.9
Set I

In Exercises 1–10, find the indicated square roots or write "not real."

1. $-\sqrt{36}$
2. $\sqrt{49}$
3. $-\sqrt{25}$
4. $\sqrt{-64}$
5. $-\sqrt{100}$
6. $-\sqrt{144}$
7. $\sqrt{81}$
8. $\sqrt{121}$
9. $-\sqrt{256}$
10. $-\sqrt{16}$

In Exercises 11–18, use a calculator to find the roots. Round off to three decimal places any answers that are not exact.

11. $\sqrt{209,764}$
12. $\sqrt{389,376}$
13. $\sqrt{12}$
14. $\sqrt{17}$
15. $\sqrt{184}$
16. $\sqrt{191}$
17. $\sqrt{2.8}$
18. $\sqrt{9.38}$

In Exercises 19–30, find the indicated roots or write "not real."

19. $\sqrt[3]{8}$
20. $\sqrt[5]{1}$
21. $\sqrt[4]{-81}$
22. $\sqrt[6]{-64}$
23. $\sqrt[3]{-64}$
24. $\sqrt[3]{-27}$
25. $-\sqrt[5]{32}$
26. $-\sqrt[3]{64}$
27. $\sqrt[3]{-1,000}$
28. $\sqrt[3]{-216}$
29. $-\sqrt[5]{-32}$
30. $-\sqrt[3]{-125}$

Writing Problems

Express the answers in your own words and in complete sentences.

1. Explain why -2 is a square root of 4.

2. Explain why $\sqrt{4} \neq -2$.

Exercises 1.9 Set II

In Exercises 1–10, find the indicated square roots or write "not real."

1. $-\sqrt{169}$
2. $-\sqrt{49}$
3. $\sqrt{225}$
4. $\sqrt{289}$
5. $-\sqrt{400}$
6. $\sqrt{144}$
7. $-\sqrt{900}$
8. $-\sqrt{64}$
9. $\sqrt{100}$
10. $\sqrt{-121}$

In Exercises 11–18, use a calculator to find the roots. Round off to three decimal places any answers that are not exact.

11. $\sqrt{393{,}129}$
12. $\sqrt{395{,}641}$
13. $\sqrt{18}$
14. $\sqrt{23}$
15. $\sqrt{172}$
16. $\sqrt{1.63}$
17. $\sqrt{85.2}$
18. $\sqrt{73.2}$

In Exercises 19–30, find the indicated roots or write "not real."

19. $\sqrt[4]{16}$
20. $-\sqrt[7]{-1}$
21. $\sqrt[4]{-1}$
22. $\sqrt[3]{-125}$
23. $-\sqrt[4]{81}$
24. $-\sqrt[6]{64}$
25. $-\sqrt[3]{216}$
26. $-\sqrt[3]{-8}$
27. $\sqrt[3]{27}$
28. $\sqrt[5]{-243}$
29. $\sqrt[3]{-1}$
30. $\sqrt[4]{10{,}000}$

1.10 Order of Operations

Recall from elementary algebra that we use the following **order of operations** when we need to perform more than one operation:

Order of operations

1. If operations are indicated inside grouping symbols, those operations within the grouping symbols should be performed first. If grouping symbols appear within other grouping symbols, remove the *innermost* grouping symbols first. A fraction bar and the bar of a radical sign are grouping symbols.
2. The evaluation *then* proceeds *in this order*:

   First: Powers and roots are done.

   Next: Multiplication and division are done *in order, from left to right.*

   Last: Addition and subtraction are done *in order, from left to right.*

EXAMPLE 1 Evaluate $8 - 6 - 4 + 7$.

SOLUTION It is important to realize that an expression such as $8 - 6 - 4 + 7$ is evaluated by doing the addition and subtraction *in order, from left to right,* because subtraction is neither associative nor commutative. If the expression is considered as a *sum*—that is, as $8 + (-6) + (-4) + 7$—then the terms can be added in any order, because addition is commutative and associative.

Both methods are correct

Evaluated left to right	*Added in any order*
$8 - 6 - 4 + 7$	$(8) + (-6) + (-4) + (7)$
$= \underbrace{8-6}$ → $2 \quad -4 + 7$	$= \underbrace{(8) + (7)} + \underbrace{(-6) + (-4)}$
$= -2 \quad + 7$	$= 15 \quad + \quad (-10)$
$= 5$	$= 5$

EXAMPLE 2 Evaluate $7 + 3 \cdot 5$.

SOLUTION

$$7 + \underbrace{3 \cdot 5}$$ Multiplication must be done *before* addition

$$= 7 + 15 = 22$$

EXAMPLE 3 Evaluate $16 \div 2 \cdot 4$.

SOLUTION

$$\underbrace{16 \div 2} \cdot 4$$ Division must be done first because it is on the left

$$= 8 \cdot 4$$

$$= 32$$

EXAMPLE 4 Evaluate $(-3)^2 \sqrt[3]{-8} - 2(-6)$.

SOLUTION

$$(-3)^2 \sqrt[3]{-8} - 2(-6)$$ Powers and roots must be done first

$$= (9)(-2) - 2(-6)$$ Multiplication must be done before subtraction

$$= -18 - (-12)$$

$$= -18 + 12 = -6$$

EXAMPLE 5 Evaluate $\dfrac{(-4) + (-2)}{8 - 5}$.

SOLUTION

$$\frac{(-4) + (-2)}{8 - 5}$$ This bar is a grouping symbol for both $(-4) + (-2)$ and $8 - 5$; notice that the bar can be used either above or below the numbers being grouped

$$= \frac{-6}{3} = -2$$

EXAMPLE 6 Evaluate $6 - 4(2 \cdot 3^2 - 12 \div 2)$.

SOLUTION

$6 - 4(2 \cdot 3^2 - 12 \div 2)$	We first evaluate the expression inside the parentheses
$= 6 - 4(2 \cdot 9 - 12 \div 2)$	We raise to powers before multiplying
$= 6 - 4(18 - 6)$	Multiplication and division inside parentheses are done next
$= 6 - 4(12)$	Subtraction inside parentheses is done next
$= 6 - 48$	Multiplication must be done before the final subtraction
$= -42$	

EXAMPLE 7 Evaluate $20 - [5 - (3 - 7)]$.

SOLUTION

$20 - [5 - (3 - 7)]$	When grouping symbols appear within other grouping symbols, we evaluate the *inner* expression first
$= 20 - [5 - (-4)]$	
$= 20 - [5 + 4]$	
$= 20 - 9 = 11$	

Now that we know that multiplication is to be done before addition and/or subtraction, we can restate the distributive property without parentheses around the ab and the ac:

Multiplication is distributive over addition

$$a(b + c) = ab + ac$$

Order of Operations and Scientific Calculators

Most scientific calculators have the correct order of operations "built in." Therefore, they will perform multiplications and divisions before additions and subtractions, they will perform multiplications and divisions from left to right, and so on.

You can tell whether your calculator performs multiplications and divisions before additions and subtractions by trying out simple problems. For example, try the problem $2 + 3 \times 5$ by entering the problem as follows:

Scientific (algebraic) [2] [+] [3] [×] [5] [=]

RPN [2] [ENTER] [3] [ENTER] [5] [×] [+]

If your calculator display is [17] (which *is* the correct answer for $2 + 3 \times 5$), then your calculator does multiplications and divisions before additions and subtractions. If your display is *not* 17, then you must enter the problem differently. Try entering it as $3 \times 5 + 2$ (notice that $2 + 3 \times 5 = 3 \times 5 + 2$ because *addition* is commutative).

You can tell whether your calculator performs multiplications and divisions from left to right by trying the problem $9 \div 3 \times 3$. If you enter

Scientific (algebraic) [9] [÷] [3] [×] [3] [=]

RPN [9] [ENTER] [3] [÷] [3] [×]

and see the display [9] (which *is* the correct answer for $9 \div 3 \times 3$), then your calculator does perform multiplications and divisions from left to right.

To see whether your calculator raises to powers before it multiplies or divides, try the problem 2×3^2. If you enter

Scientific (algebraic) [2] [×] [3] [x^2] [=]

RPN [2] [ENTER] [3] [↰] [$\sqrt{x}$ / x^2] [×]

and see the display [18] (which *is* the correct answer for 2×3^2), then your calculator does raise to powers before it multiplies (or divides).

To see whether your calculator performs additions and subtractions from left to right, try the problems $12 - 6 - 2$ and $15 - 3 + 2$. The answers should be 4 and 14, respectively.

We suggest that you rework the problems in Examples 1–7 with a calculator, being sure to note what keystrokes to use on *your* calculator to obtain the correct answers.

In Examples 8 and 9, we do not show the keystrokes to use for entering the numbers themselves, and in Example 8, we show only the keystrokes to use on a calculator that uses algebraic logic.

EXAMPLE 8 Find $\dfrac{43.6 \times 0.339}{7.42 \times 13.5}$.

SOLUTION Even if your calculator has the correct order of operations built in, you must be *very* careful in entering this problem. Remember that the fraction bar is a grouping symbol; that is, the problem could have been written $(43.6 \times 0.339) \div (7.42 \times 13.5)$. Therefore, if your calculator has parentheses, the problem can be entered as follows:

[(] 43.6 [×] .339 [)] [÷] [(] 7.42 [×] 13.5 [)] [=] — Display: 0.147553159

If your calculator doesn't have parentheses but has a memory key, you can find 7.42×13.5 and *store that answer*, finishing the problem as follows:

43.6 [×] .339 [=] [÷] [RCL] [=] — Display: 0.147553159

The most efficient way to enter the problem (and the way to enter it if your calculator has neither parentheses nor a memory key) is as follows:

43.6 [×] .339 [÷] 7.42 [÷] 13.5 [=] — Display: 0.147553159

(The second [÷]: This is ÷ because 13.5 is in the denominator)

This method works because $\dfrac{43.6 \times 0.339}{7.42 \times 13.5} = \dfrac{43.6}{1} \times \dfrac{0.339}{1} \times \dfrac{1}{7.42} \times \dfrac{1}{13.5}$. The answer is approximately 0.147 553 159.

A Word of Caution You will get an *incorrect* answer to the problem in Example 8 if your calculator has the correct order of operations built in and if you enter the problem as follows:

43.6 [×] .339 [÷] 7.42 [×] 13.5 [=]

The calculator interprets this as $43.6 \times 0.339 \div 7.42 \times 13.5$ and, using the correct order of operations, works from left to right. Therefore, the calculator does the problem as if it were

$$\frac{43.6 \times .339}{7.42} \times 13.5, \quad \text{or} \quad \frac{43.6 \times .339 \times 13.5}{7.42}$$

The calculator display is [26.891563], which is *not* the correct answer for $\dfrac{43.6 \times 0.339}{7.42 \times 13.5}$.

Some calculators have a key for handling common fractions; to use such a key, consult the manual for your calculator. On calculators that use algebraic logic, you can get decimal answers for common fraction problems by using parentheses around the fractions (see Example 9).

EXAMPLE 9 Using a calculator, find $\frac{2}{5} + \frac{2}{3}$.

SOLUTION

Scientific (algebraic) [( ] 2 [÷] 5 [ )] [+] [( ] 2 [÷] 3 [ )] [=]

RPN [2] [ENTER] [5] [÷] [2] [ENTER] [3] [÷] [+]

The calculator display should be [1.06666667]. If you realize that the least common denominator of $\frac{2}{5}$ and $\frac{2}{3}$ is 15, you can find the numerator of the common fraction form of the answer by multiplying the number in the calculator display by 15. If you do this, you see that the common fraction form of the answer is $\frac{16}{15}$, or $1\frac{1}{15}$.

Order of Operations and Graphics Calculators

Graphics calculators do not *always* work from left to right when performing multiplications and divisions. Graphics calculators "understand" juxtaposition (or *implied multiplication*), and at the time of this publication, calculators using the EOS operating system were doing *implied* multiplications *before* multiplications indicated by the × symbol *and before* divisions.

If the problem 2(3), for example, is entered on a graphics calculator as

[2] [( ] [3] [ )] [ENTER]

the calculator will do the multiplication and the display will be [6].

However, if we try to find 6 ÷ 2(3) with a graphics calculator by entering the problem as

[6] [÷] [2] [( ] [3] [ )] [ENTER]

the display is [1], which is an *incorrect* answer. To get the *correct* answer for 6 ÷ 2(3) on a graphics calculator, we must use the × symbol (entering the problem as 6 ÷ 2 × 3) or enter parentheses around 6 ÷ 2, as follows.

[6] [÷] [2] [×] [3] [ENTER] yields [9]

and [( ] [6] [÷] [2] [ )] [( ] [3] [ )] [ENTER] yields [9]

If you have a graphics calculator, be sure to study the section of your manual that explains the order of operations that the calculator uses *and* be sure that you have a good understanding of the *correct* order of operations.

Exercises 1.10 Set I

Evaluate each expression; be sure to perform the operations in the correct order.

1. 16 − 9 − 4 **2.** 18 − 6 − 3 **3.** 12 ÷ 6 ÷ 2

4. 16 ÷ 4 ÷ 2 **5.** 10 ÷ 2(−5) **6.** 15 ÷ 5(−3)

7. $3 \cdot 2^4$ **8.** $5 \cdot 3^2$ **9.** $8 + 6 \cdot 5$

10. $3 + 2 \cdot 4$ **11.** 7 + 5 ÷ 3 **12.** 5 + 3 ÷ 2

13. $10 - 3 \cdot 2$ **14.** $12 - 2 \cdot 4$

15. $10(-15)^2 - 4^3$ **16.** $3(-4)^2 - 2^4$

17. $\frac{1}{2} - 0.02 \times 10^3$ **18.** $\frac{1}{3} - 0.04 \times 10^2$

19. $10^2(5)\sqrt{16}$ **20.** $10^2(3)\sqrt{25}$

21. $2 + 3 \cdot 100 \div 25$ **22.** $3 + 2 \cdot 100 \div 4$

23. 28 + 14/7 **24.** 36 + 18/3

25. $2(2^3 - 5)\sqrt{9}$

26. $5(4^2 - 8)\sqrt{36}$

27. $(-18) \div (-3)(-6)$

28. $(-16) \div (-4)(-2)$

29. $(-10)^3 - 5(10^2)\sqrt[3]{-27}$

30. $(-10)^2 - 5(10^3)\sqrt[3]{-8}$

31. $20 - [5 - (7 - 10)]$

32. $16 - [8 - (2 - 7)]$

33. $\dfrac{7 + (-12)}{8 - 3}$

34. $\dfrac{-14 + (-2)}{9 - 5}$

35. $8 - [5(-2)^3 - \sqrt{16}]$

36. $10 - [3(-3)^2 - \sqrt{25}]$

37. $(3 \cdot 5^2 - 15 \div 3) \div (-7)$

38. $(3 \cdot 4^3 - 72 \div 6) \div (-9)$

39. $15 - \{4 - [2 - 3(6 - 4)]\}$

40. $17 - \{6 - [9 - 2(7 - 2)]\}$

In Exercises 41–54, use a calculator, and show all the digits that your calculator display shows.

41. $\dfrac{63.7 - 14.6}{0.64}$

42. $\dfrac{85.2 - 25.7}{0.32}$

43. $\dfrac{6.79}{3.56 \times 8.623}$

44. $\dfrac{3.45}{1.363 \times 56.8}$

45. $\dfrac{3.79 - 1.062}{45.7 \times 0.245}$

46. $\dfrac{6.78 - 1.479}{0.35 \times 16.2}$

47. $(6.28)^2 + (1.5)^2$

48. $(7.2)^2 + (1.34)^2$

49. $(6.28 + 1.5)^2$

50. $(7.2 + 1.34)^2$

51. $(6.28)^2 + 2(6.28)(1.5) + (1.5)^2$

52. $(7.2)^2 + 2(7.2)(1.34) + (1.34)^2$

53. $\sqrt{(4.6)^2 + (3.7)^2}$

54. $\sqrt{(7.1)^2 + (6.3)^2}$

Writing Problems

Express the answers in your own words and in complete sentences.

1. Find and describe the error in $2 + 3 \cdot 5 = 5 \cdot 5 = 25$.
2. Find and describe the error in $45 \div 9 \cdot 5 = 45 \div 45 = 1$.
3. Find and describe the error in $8 - 4 + 1 = 8 - 5 = 3$.
4. Describe the keystrokes you would use in doing the problem $\dfrac{36.78}{2.35 \times 9.07}$ on a calculator that has the correct order of operations built in.
5. Describe the keystrokes you would use in doing the problem $8 \div 4(2)$ on a graphics calculator.

Exercises 1.10 Set II

Evaluate each expression; be sure to perform the operations in the correct order.

1. $15 - 7 - 2$

2. $3 + 2 \cdot 6$

3. $81 \div 9 \div 3$

4. $12 + 6 \div 6$

5. $18 \div 3(-6)$

6. $8 - 24 \div 8$

7. $2 \cdot 5^2$

8. $8 \cdot 3^2$

9. $9 + 6 \cdot 3$

10. $2(-5) + 6 \div 24$

11. $9 + 7 \div 4$

12. $64 \div 2 \div 2$

13. $17 - 3 \cdot 5$

14. $3(2^3 - 1)\sqrt{64}$

15. $5(-3)^2 - 3^3$

16. $6(4) - 8 \div 24$

17. $\frac{1}{2} - 0.2 \times 10^2$

18. $10^2\sqrt[3]{8} \div 10 \cdot 2$

19. $10^2(4)\sqrt{9}$

20. $75 \div 5^2 \cdot 4 + 12 \cdot 5$

21. $4 + 2 \cdot 100 \div 5$

22. $100 \div 5^2 \cdot 6 + 8 \cdot 75$

23. $24 + 12/6$

24. $10^4 \cdot 4^2 + 100(15)$

25. $4(3^2 - 4)\sqrt{25}$

26. $8^2\sqrt{25} \div 10 \cdot 4$

27. $(-15) \div (-3)(-5)$

28. $10^4 \cdot 2^3 + 10(140)$

29. $(-10)^3 - 4(10^2)\sqrt[3]{-8}$

30. $6 + 3 \cdot 30 \div 5$

31. $18 - [9 - (3 - 8)]$

32. $100 \div 2^2 \cdot 5 + 9 \cdot 25$

33. $\dfrac{8 + (-16)}{12 - 4}$

34. $\frac{3}{5} + 72 \div 9 \cdot 2 - \frac{1}{3}$

35. $(-8)/2 \times (-4)/(-1)$

36. $18 - 11 - 3$

37. $(2 \cdot 3^3 - 63 \div 7) \div (-9)$

38. $10^2\sqrt[3]{27} \div 10 \cdot 3$

39. $20 - \{5 - [9 - 3(6 - 2)]\}$

40. $\frac{2}{3} + 77 \div 11 \cdot 2 - \frac{1}{2}$

In Exercises 41–54, use a calculator, and show all the digits that your calculator display shows.

41. $\dfrac{76.9 - 13.5}{0.25}$

42. $\dfrac{178.3 - 65.8}{0.75}$

43. $\dfrac{5.69}{23.1 \times 14}$

44. $\dfrac{6{,}745}{1{,}145 \times 6.58}$

45. $\dfrac{7.92 - 2.603}{65.3 \times 0.356}$

46. $\dfrac{8.93 + 0.999}{1.45 \times 27}$

47. $(3.17)^2 + (2.4)^2$

48. $(1.3)^2 + (4.67)^2$

49. $(3.17 + 2.4)^2$

50. $(1.3 + 4.67)^2$

51. $(3.17)^2 + 2(3.17)(2.4) + (2.4)^2$

52. $(1.3)^2 + 2(1.3)(4.67) + (4.67)^2$

53. $\sqrt{(7.9)^2 + (5.1)^2}$

54. $\sqrt{(4.9)^2 + (3.7)^2}$

1.11 Prime and Composite Numbers; the Factorization of Natural Numbers; the Least Common Multiple

Prime and Composite Numbers

Prime Numbers A **prime number** is a natural number greater than 1 that has exactly two *unique* natural number factors—namely, itself and 1.

The first ten prime numbers are 2, 3, 5, 7, 11, 13, 17, 19, 23, and 29. There are infinitely many prime numbers, so there is no largest prime number.

Composite Numbers A **composite number** is a natural number greater than 1 that is not prime; it is a natural number that has natural number factors other than itself and 1.

Note One (1) is neither prime nor composite. 

Factoring a Natural Number

Recall that numbers that are multiplied together to give a product are called *factors*. To **factor** a natural number means to rewrite the number, if possible, as a *product* of smaller numbers (its factors). Recall also that when the remainder in a division problem is zero, the divisor and quotient are both *factors* of the dividend. We make use of these facts in *factoring a number.*

The Prime Factorization of a Natural Number The **prime factorization of a natural number** greater than 1 is the indicated product of all the factors of the number *that are themselves prime numbers*. We usually write any repeated factors in exponential form, and we then say that the number is in **prime-factored, exponential form**. (For prime-factored, exponential form, we usually express the product with the *bases* increasing in magnitude from left to right.)

EXAMPLE 1 Find the prime factorization of 18.

SOLUTION

$$\left.\begin{aligned} 18 &= 2 \cdot 9 \\ 18 &= 3 \cdot 6 \end{aligned}\right\}$$ These are not *prime* factorizations because 9 and 6 are not prime numbers

$$\left.\begin{aligned} 18 &= 2 \cdot 9 = \boxed{2 \cdot 3 \cdot 3} \\ 18 &= 3 \cdot 6 = \boxed{3 \cdot 2 \cdot 3} \end{aligned}\right\}$$ The shaded factorizations are prime factorizations because all the factors are prime numbers

Notice that the two ways we factored 18 led to the *same* prime factors, one 2 and two 3's. (See the unique factorization theorem, below.) If we express these factorizations in prime-factored, exponential form, we have $2 \cdot 3^2$ in both cases.

The unique factorization theorem

> A natural number greater than 1 can be expressed as a product of prime numbers in one and only one way, except for the order in which the factors are listed.

In Examples 2 and 3, we show two ways of finding the prime factorization of a number. In the first solution, we make an organized list by dividing first by the *smallest* prime number that divides exactly into the number. In the second solution (the "tree" method), we start with *any* two factors of the number.

EXAMPLE 2 Find the prime factorization of 315.

SOLUTION 1 We first try to divide 315 by the smallest prime number, 2. Two does not divide exactly into 315, and so we try to divide 315 by the next prime number, which is 3. Three *does* divide exactly into 315 and gives a quotient of 105. We again try 3 as a divisor of the quotient, 105. Three *does* divide exactly into 105 and gives a new quotient of 35. We then try to divide *that* quotient, 35, by 3. Three does not divide exactly into 35, and so we try to divide 35 by the next prime number, which is 5. Five *does* divide exactly into 35 and gives a quotient of 7. The process then ends because the quotient, 7, is itself a prime number.

The work of finding the prime factorization of a number can be conveniently arranged by placing the quotients *under* the number we're dividing into, as follows:

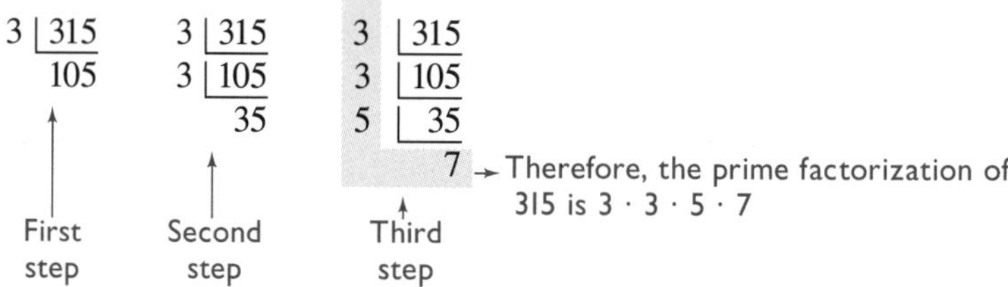

SOLUTION 2 For the tree method, we start with *any* two factors of 315:

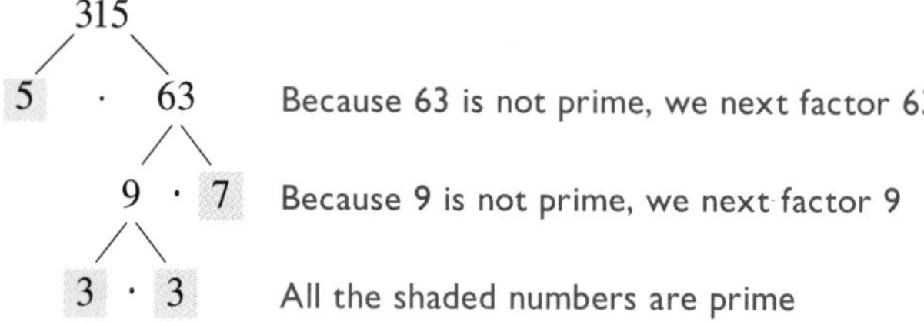

Again, the prime factorization of 315 is $3 \cdot 3 \cdot 5 \cdot 7$. In prime-factored, exponential form, we have $315 = 3^2 \cdot 5 \cdot 7$.

EXAMPLE 3 Find the prime factorization of 48.

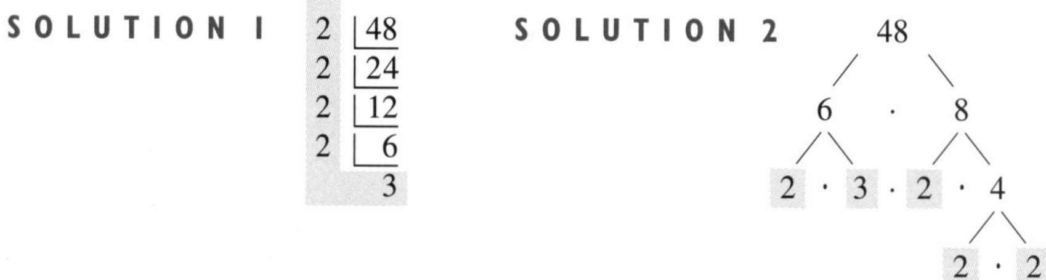

Therefore, the prime factorization of 48 is $2 \cdot 2 \cdot 2 \cdot 2 \cdot 3$. In prime-factored, exponential form, we have $48 = 2^4 \cdot 3$.

When we use guessing-and-checking to find the prime factors of a number, we do not need to try any prime number that has a *square* greater than the number we're trying to factor (see Example 4).

EXAMPLE 4 Find the prime factorization of 97.

SOLUTION

Primes in order of size

2	does not divide 97
3	does not divide 97
5	does not divide 97
7	does not divide 97
11	and larger primes need not be tried because $11^2 = 121$, which is greater than 97

Therefore, 97 is a prime number, and the prime factorization of 97 is simply 97.

Finding All the Integral Factors of a Number Sometimes we need to find *all* the integers, positive and negative, that divide exactly into a given number. When we find all the integral factors, we generally make use of the "plus or minus" symbol, $\pm$. For example, ± 1 is read "plus or minus one," and it means $+1$ *or* -1.

If we're asked to list *all* the factors of some number, we can be sure we haven't left out any factors if we make an *organized list*. We demonstrate this method in Example 5.

EXAMPLE 5 List all the integral factors (or divisors) of 36. Is 36 prime or composite?

SOLUTION We'll first find the *positive* integers that are factors of 36. We start with an outline that contains 1 and 36 and leave some space between these two numbers. We then fill in that outline, writing the work as shown on the right:

$36 \div 1 = 36$ 1, , 36

$36 \div 2 = 18$ 1, 2, , 18, 36

$36 \div 3 = 12$ 1, 2, 3, , 12, 18, 36

$36 \div 4 = 9$ 1, 2, 3, 4, , 9, 12, 18, 36

$36 \div 6 = 6$ We write the 6 only once 1, 2, 3, 4, 6, 9, 12, 18, 36

By checking each positive integer, in turn, to see whether it is a factor of 36 (5 is, of course, *not* a factor of 36) and by "pairing up" the factors in this manner, continuing until we reach the "middle" of the list of factors, we can be sure that we have not omitted any factors. (Because 36 is the square of an integer, when we "pair up" the factors, there is a single, unpaired number left in the center.) We then find *all the integral factors* by placing the symbol $\pm$ in front of each positive factor; thus, the *integral* factors of 36 are ± 1, ± 2, ± 3, ± 4, ± 6, ± 9, ± 12, ± 18, and ± 36. Since 36 has natural number factors other than 1 and 36, it is a composite number.

EXAMPLE 6 List all the integral factors of 31. Is 31 prime or composite?

SOLUTION The integral factors of 31 are ± 1 and ± 31. Since the only natural number factors are 1 and 31, 31 is a prime number.

The Least Common Multiple of Two or More Numbers

We find the *multiples* of a number by multiplying the number by any of the natural numbers. Every number has *infinitely* many multiples. Let's list some multiples of 18 and some multiples of 48. (Multiples of 18 are $18 \cdot 1$, $18 \cdot 2$, $18 \cdot 3$, and so on, and multiples of 48 are $48 \cdot 1$, $48 \cdot 2$, $48 \cdot 3$, and so on.)

Multiples of 18: 18, 36, 54, 72, 90, 108, 126, 144, 162, 180, 198, 216, 234, 252, 270, 288, 306, 324, 342, 360, 378, 396, 414, 432, 450,...

Multiples of 48: 48, 96, 144, 192, 240, 288, 336, 384, 432, 480,...

We see three numbers that are in *both* lists: 144, 288, and 432. These numbers that are *common to both lists* are called *common multiples.* There are infinitely many common multiples of 18 and 48; in addition to 144, 288, and 432, there are also 576, 720, 864, and so on.

The **least common multiple (LCM)** of two numbers is the *smallest* (*least*) number that is a multiple of both numbers; that is, it is the smallest of the common multiples. Therefore, the least common multiple of 18 and 48 is 144. (In Example 7, we use a different method for finding the LCM of 18 and 48.)

We usually do *not* find the LCM of two or more numbers by listing the multiples of all the numbers, as we did above; instead, we use the procedure in the following box:

Finding the LCM of two or more numbers

1. Express each number in prime-factored, exponential form.
2. Write down each *different* base that appears in any of the numbers.
3. Raise each base from step 2 to the highest power to which it occurs in *any* of the factorizations.
4. The LCM is the *product* of the exponential numbers found in step 3.

EXAMPLE 7 Find the LCM of 18 and 48.

SOLUTION $18 = 2 \cdot 3^2$, and $48 = 2^4 \cdot 3$. The bases that appear in one or both numbers are 2 and 3. Therefore, the bases in the LCM are 2 and 3. The *largest* exponent that appears on the 2 (in either of the factorizations) is 4. Therefore, the exponent on the 2 in the LCM must be 4. The *largest* exponent that appears on the 3 in either of the factorizations is 2. Therefore, the exponent on the 3 in the LCM must be 2. The LCM, then, is $2^4 \cdot 3^2 = 144$.

Note 144 is the *smallest* number that is a multiple of both 18 and 48. We could also say that 144 is the smallest number that 18 and 48 both *divide into.*

EXAMPLE 8 Find the LCM of 360, 378, and 108.

SOLUTION $360 = 2^3 \cdot 3^2 \cdot 5$, $378 = 2 \cdot 3^3 \cdot 7$, and $108 = 2^2 \cdot 3^3$. Bases in the LCM must be 2, 3, 5, and 7. The exponent on the 2 must be 3; the exponent on the 3 must be 3; the exponent on the 5 must be 1; the exponent on the 7 must be 1. The LCM, therefore, is $2^3 \cdot 3^3 \cdot 5 \cdot 7 = 7{,}560$. This is the smallest number that 360, 378, and 108 all divide into.

The least common multiple is used in adding and subtracting rational expressions and in solving rational equations.

Exercises 1.11
Set I

In Exercises 1–12, find the prime factorization of each number, and express it in prime-factored, exponential form.

1. 28
2. 30
3. 32
4. 33
5. 43
6. 35
7. 84
8. 75
9. 144
10. 180
11. 156
12. 221

In Exercises 13–22, write "P" if the number is prime or "C" if it is composite. In Exercises 13–18, list *all* the integral factors of each number, and in Exercises 19–22, list the positive factors of each number.

13. 5 14. 8 15. 13

16. 15 17. 12 18. 11

19. 51 20. 42 21. 111

22. 101

In Exercises 23–30, find the LCM of the numbers.

23. 144, 360 24. 84, 35 25. 150, 525

26. 280, 300 27. 78, 130, 156 28. 165, 220, 330

29. 270, 900, 75 30. 140, 105, 98

In Exercises 31 and 32, use guessing-and-checking to solve each problem.

31. List any prime numbers greater than 17 and less than 37 that yield a remainder of 5 when divided by 7.

32. List any prime numbers less than 43 and greater than 13 that yield a remainder of 1 when divided by 6.

Writing Problems

Express the answers in your own words and in complete sentences.

1. Explain what a prime number is.

2. Explain what the least common multiple of two or more numbers is.

3. Explain why the least common multiple of 12 and 18 is *not* 6.

Exercises  1.11 Set II

In Exercises 1–12, find the prime factorization of each number, and express it in prime-factored, exponential form.

1. 21 2. 31 3. 45 4. 87

5. 186 6. 238 7. 19 8. 40

9. 36 10. 27 11. 72 12. 228

In Exercises 13–22, write "P" if the number is prime or "C" if it is composite. In Exercises 13–18, list *all* the integral factors of each number, and in Exercises 19–22, list the positive factors of each number.

13. 21 14. 23 15. 55

16. 41 17. 49 18. 9

19. 44 20. 30 21. 17

22. 87

In Exercises 23–30, find the LCM of the numbers.

23. 87, 58 24. 93, 155 25. 114, 285

26. 435, 580 27. 136, 204, 306 28. 280, 84, 350

29. 85, 102, 170 30. 999, 185

In Exercises 31 and 32, use guessing-and-checking to solve each problem.

31. List any prime numbers greater than 7 and less than 29 that yield a remainder of 3 when divided by 5.

32. List any prime numbers less than 37 and greater than 11 that yield a remainder of 2 when divided by 9.

Sections 1.1–1.11 REVIEW

Equality 1.1

Equal to: The statement $a = b$ is read "a is equal to b" or "a equals b." The equal sign is used when the expression to the left of the equal sign *has the same value or values* as the expression to the right of the equal sign.

The properties of **equality**: If a, b, and c represent any numbers or expressions, then

$a = a$	The reflexive property of equality
If $a = b$, then $b = a$.	The symmetric property of equality
If $a = b$ and $b = c$, then $a = c$.	The transitive property of equality
If $a = b$, then b can be substituted for a in any expression or equation that contains a.	The substitution principle

Not equal to: The statement $m \neq n$ is read "m is not equal to n."

Sets 1.1

A **set** is a collection of symbols, objects, or things.

The **elements** of a set are the objects that make up the set.

$5 \in A$ is read "5 is an element of set A."

$9 \notin B$ is read "9 is not an element of set B."

To use **roster notation** for representing a set, we list its elements within braces.

Modified roster notation—using an ellipsis to represent elements not actually listed—is used when the number of elements in the set is so large that it is not convenient or even possible to list them all.

Set-builder notation can also be used to represent a set.

Equal sets are sets that have exactly the same elements.

The **empty set** is a set that has no elements. Two symbols used for the empty set are $\varnothing$ and $\{\ \}$.

If, in counting the elements of a set, the counting can come to an end, the set is a **finite set**.

If, in counting the elements of a set, the counting will never come to an end, the set is an **infinite set**.

Set A is a **subset** of set B if every element of A is also an element of B:

$A \subseteq B$ is read "A is a subset of B."

$A \not\subseteq B$ is read "A is not a subset of B."

Union and Intersection of Sets 1.2

The **union** of sets A and B, written $A \cup B$, is the set that contains all the elements that are in A *or* B *or* both.

The **intersection** of sets A and B, written $A \cap B$, is the set that contains all the elements that are in *both* A and B.

Sets of Numbers 1.1, 1.3

All the numbers that can be represented by points on the number line are **real numbers**. R represents the set of real numbers. Some subsets of the set of real numbers are

The set of **digits**, $D = \{0, 1, 2, 3, 4, 5, 6, 7, 8, 9\}$

The set of **natural numbers**, $N = \{1, 2, 3, \ldots\}$

The set of **whole numbers**, $W = \{0, 1, 2, \ldots\}$

The set of **integers**, $J = \{\ldots, -3, -2, -1, 0, 1, 2, 3, \ldots\}$

The set of **rational numbers**, $Q = \left\{ \frac{a}{b} \,\middle|\, a, b \in J;\ b \neq 0 \right\}$ (The decimal form of a rational number is always either a terminating decimal or a repeating decimal.)

The set of **irrational numbers**, $H = \{x \mid x \in R, x \notin Q\}$ (The decimal form of an irrational number is always a nonterminating, nonrepeating decimal.)

Definitions 1.4

A **constant** is an object or symbol that does not change its value in a particular problem or discussion.

A **variable** is a letter or symbol that is a placeholder for a number that is unknown.

An **algebraic expression** consists of numbers, letters, signs of operation (such as $+$, $-$, and $\times$), and signs of grouping.

The numbers that are multiplied together to give a product are called the **factors** of that product.

A **term** of an algebraic expression consists of factors only. If a term is preceded by a minus sign, that sign is *part of* the term. An expression within grouping symbols is considered as a single term.

Inequality Symbols 1.5

Greater than: The statement $a > b$ is read "a is greater than b."

Greater than or equal to: The statement $c \geq d$ is read "c is greater than or equal to d."

Not greater than: The statement $e \not> f$ is read "e is not greater than f."

Less than: The statement $g < h$ is read "g is less than h."

Less than or equal to: The statement $i \leq j$ is read "i is less than or equal to j."

Not less than: The statement $k \not< l$ is read "k is not less than l."

Continued Inequalities 1.5

A statement such as $a < b < c$ is a valid **continued inequality** only if both inequalities have the same sense and if the inequalities $a < b$, $b < c$, and $a < c$ are all true.

Absolute Value 1.5

The **absolute value** of a real number x is written $|x|$, where $|x| = \begin{cases} x & \text{if } x \geq 0 \\ -x & \text{if } x < 0 \end{cases}$

Closure for Addition and Multiplication 1.6

If a and b represent any real numbers, then their sum, $a + b$, is a real number.

If a and b represent any real numbers, then their product, ab, is a real number.

Operations on Rational Numbers 1.6

To add rational numbers:

1. When the numbers have the same sign, add their absolute values, and give the sum the sign of both numbers.
2. When the numbers have different signs, subtract the smaller absolute value from the larger absolute value, and give the sum the sign of the number with the larger absolute value.

To subtract one rational number from another:
Change the subtraction symbol to an addition symbol, *and* change the sign of the number being subtracted; add the resulting signed numbers. The following rules apply when one number is zero:

$$a - 0 = a \quad \text{and} \quad 0 - a = -a$$

To multiply two rational numbers:
Multiply the absolute values, *and* attach the correct sign to the product. The sign is *positive* when the numbers have the same sign and *negative* when one number is positive and the other is negative. The following rules apply when one number is zero:

$$a \cdot 0 = 0 \quad \text{and} \quad 0 \cdot a = 0$$ The multiplication property of zero

To divide one rational number by another:
Divide the absolute value of the dividend by the absolute value of the divisor, *and* attach the correct sign to the quotient. The sign is *positive* when the numbers have the same sign and *negative* when one number is positive and the other is negative. The following rules apply when one or both numbers are zero:
If a represents any real number *except* 0,

$$\frac{0}{a} = 0 \qquad \frac{a}{0} \text{ is not possible} \qquad \frac{0}{0} \text{ cannot be determined}$$

Division by zero is undefined

Identity and Inverse Properties 1.6

If a represents any real number,

The **additive identity element** is 0; that is, $a + 0 = 0 + a = a$.

The **multiplicative identity element** is 1; that is, $a \cdot 1 = 1 \cdot a = a$.

The **additive inverse** of a is $-a$; that is, $a + (-a) = -a + a = 0$.

If $a \neq 0$, the **multiplicative inverse** of a is $\frac{1}{a}$; that is, $a\left(\frac{1}{a}\right) = \left(\frac{1}{a}\right)a = 1$.

Commutative Properties 1.7

$$\left.\begin{array}{ll} \textit{Addition}: & a + b = b + a \\ \textit{Multiplication}: & a \cdot b = b \cdot a \end{array}\right\} \quad a, b \in R$$

Subtraction and *division* are not commutative.

Associative Properties 1.7

$$\left.\begin{array}{ll} \textit{Addition}: & (a + b) + c = a + (b + c) \\ \textit{Multiplication}: & (a \cdot b) \cdot c = a \cdot (b \cdot c) \end{array}\right\} \quad a, b, c \in R$$

Subtraction and *division* are not associative.

Distributive Property 1.7

If $a, b, c \in R$,

$$\left.\begin{array}{l} a(b + c) = ab + ac \\ (b + c)a = ba + ca \end{array}\right\}$$ The distributive property can be extended to have any number of terms within the parentheses and to include subtraction as well as addition

Powers of Signed Numbers 1.8

$$(-2)^3 = (-2)(-2)(-2) = -8$$

Exponent: 3; Base: -2

The exponent tells us how many times the base is to be used as a factor

Square Roots 1.9

Every positive number has both a positive and a negative **square root**. The *positive* square root of a positive number is called its **principal square root**. The notation for the principal square root of x is $\sqrt{x}$, and x is called the radicand.

If $x \geq 0$, if $b \geq 0$, and if $b^2 = x$, then $\sqrt{x} = b$.

If $x < 0$, then $\sqrt{x}$ is not real.

Higher Roots 1.9 — Roots other than square roots are called **higher roots**.

Order of Operations 1.10

1. If operations are indicated inside grouping symbols, those operations within the grouping symbols should be performed first. A fraction bar and the bar of a radical sign are grouping symbols.
2. The evaluation then proceeds *in this order*:
   First: Powers and roots are done.
   Next: Multiplication and division are done *in order, from left to right.*
   Last: Addition and subtraction are done *in order, from left to right.*

Prime and Composite Numbers 1.11 — A **prime number** is a natural number greater than 1 that has exactly two unique natural number factors—namely, itself and 1.

A **composite number** is a natural number that has natural number factors other than itself and 1.

Prime Factorization of Natural Numbers 1.11 — The **prime factorization** of a natural number greater than 1 is the indicated product of all the factors of the number that are themselves prime numbers.

Prime-Factored, Exponential Form 1.11 — A number is in **prime-factored, exponential form** if it is written so that all its factors are prime numbers and all repeated factors are expressed in exponential form.

Least Common Multiple (LCM) 1.11 — To find the LCM of two or more numbers, proceed as follows:

1. Express each number in prime-factored, exponential form.
2. Write down each *different* base that appears in any of the numbers.
3. Raise each base to the highest power to which it occurs in *any* of the factorizations.
4. The LCM is the product of all the factors found in step 3.

Sections 1.1–1.11 REVIEW EXERCISES Set I

1. Is the set of natural numbers a finite set or an infinite set?
2. Is the set of digits a finite set or an infinite set?
3. Write the following set in roster notation:
   $\{x \mid x \text{ is a digit less than } 5\}$
4. $P = \{5, x, z\}$. Is $y \in P$?
5. Given $A = \{5, 7, 8\}$, $B = \{2, 5, 7\}$, and $C = \{1, 3, 4, 6\}$, find the following.
   a. $A \cup B$ b. $A \cap B$ c. $C \cap B$ d. $(A \cup B) \cap C$
6. Given the numbers -2, 4.53, $0.\overline{16}$, $\frac{2}{3}$, 2.645 751 3... (never terminates, never repeats), and 0, answer the following questions.
   a. Which are natural numbers?
   b. Which are integers?
   c. Which are rational numbers?
   d. Which are irrational numbers?
   e. Which are real numbers?

In Exercises 7–17, find the value of each expression or write either "undefined" or "not real."

7. $|4|$ 8. $|0|$ 9. $|-10|$ 10. $(-6)^2$
11. -5^2 12. $\frac{0}{5}$ 13. $\frac{7}{0}$ 14. 0^3
15. $6 + 2\sqrt{16} - 2^3$
16. $20 - \{5 - 2[-3(4 - 6) + 2] - 3\}$
17. $6 + 18 \div 6 \div 3$
18. a. Write all the integral factors of 28.
    b. Write the prime factorization of 28.
    c. Write the prime factorization of 168.
    d. Find the least common multiple of 28 and 168.

In Exercises 19–21, if your answer is no, give a reason.

19. Is $-2 < 0 < 8$ a valid continued inequality?
20. Is $1 > 5 > 7$ a valid continued inequality?
21. Is $5 < 8 > 6$ a valid continued inequality?

In Exercises 22–30, determine whether each of the following is true or false. If the statement is *true*, name the property that makes the statement true.

22. $8 - (-15) = -15 - 8$
23. $7 \cdot 1 = 7$
24. $\left(\frac{5}{6} + \frac{1}{8}\right) + \frac{2}{3} = \frac{5}{6} + \left(\frac{1}{8} + \frac{2}{3}\right)$
25. $6 + (-6) = 0$
26. $\left(\frac{1}{2} \div \frac{3}{8}\right) \div \frac{2}{7} = \frac{1}{2} \div \left(\frac{3}{8} \div \frac{2}{7}\right)$
27. $\frac{1}{2} + 0 = \frac{1}{2}$
28. $6 + (2 + 5) = 6 + (5 + 2)$
29. $(-5) \cdot 0 = -5$
30. $\frac{1}{2}\left(\frac{5}{8} + \frac{1}{3}\right) = \frac{1}{2} \cdot \frac{5}{8} + \frac{1}{2} \cdot \frac{1}{3}$

Sections 1.1–1.11 REVIEW EXERCISES Set II

Name ______________________

ANSWERS

1. ______
2. ______
3. ______
4. ______
5. ______
6a. ______
6b. ______
6c. ______
6d. ______
7. ______
8a. ______
8b. ______
8c. ______
8d. ______
8e. ______
9. ______
10. ______
11. ______
12. ______
13. ______
14. ______

1. Is the empty set a subset of the set of integers?

2. Write the following set in roster notation:

$$\{x \mid x \text{ is a whole number less than } 6\}$$

3. Is every irrational number a real number?

4. Given $G = \{1, 4, 8\}$, is $12 \in G$?

5. Is 10 a digit?

6. Given $A = \{2, 4, 6\}$, $B = \{1, 3, 5\}$, and $C = \{3, 5, 7\}$, find the following.

   a. $A \cup B$

   b. $A \cap B$

   c. $C \cap B$

   d. $(A \cap B) \cup C$

7. Is 0 a real number?

8. Given the numbers 6.164 414 0 . . . (never terminates, never repeats), -3, 4.1, $\frac{3}{7}$, 5, and $0.\overline{65}$, determine the following.

   a. Which are natural numbers?

   b. Which are integers?

   c. Which are rational numbers?

   d. Which are irrational numbers?

   e. Which are real numbers?

In Exercises 9–20, find the value of each expression or write either "undefined" or "not real."

9. $|-14|$

10. $-|-23|$

11. $(-11)^2$

12. $(-5)^3$

13. -4^2

14. $\frac{0}{7}$

15. $\frac{0}{0}$

16. $\frac{8}{0}$

17. $10 - 2\sqrt[3]{64} - 3^2$

15. ____________________

16. ____________________

17. ____________________

18. $20 \div 10 \cdot 2$

19. $4 + 16 \div 4 \div 2$

20. $17 + 12 \cdot 2$

18. ____________________

19. ____________________

20. ____________________

21a. ____________________

21b. ____________________

21c. ____________________

21d. ____________________

22. ____________________

23. ____________________

24. ____________________

25. ____________________

26. ____________________

27. ____________________

28. ____________________

29. ____________________

30. ____________________

21. **a.** Write all of the positive integral factors of 132.

b. Write the prime factorization of 132.

c. Write the prime factorization of 110.

d. Find the least common multiple of 132 and 110.

22. Is $7 < 10 > 8$ a valid continued inequality? If your answer is no, give a reason.

In Exercises 23–30, determine whether each of the following is true or false. If the statement is *true*, name the property that makes the statement true.

23. $\frac{4}{5} \cdot \left(\frac{2}{3} \cdot \frac{6}{11}\right) = \left(\frac{4}{5} \cdot \frac{2}{3}\right) \cdot \frac{6}{11}$

24. $\frac{6}{7} \div \frac{5}{2} = \frac{5}{2} \div \frac{6}{7}$

25. $\left(\frac{1}{2} - \frac{1}{8}\right) - \frac{1}{3} = \frac{1}{2} - \left(\frac{1}{8} - \frac{1}{3}\right)$

26. $1 \cdot 17 = 17$

27. $3 \cdot [(-4) \cdot 6] = 3 \cdot [6 \cdot (-4)]$

28. $6(-2 + 6) = 6 \cdot (-2) + 6 \cdot 6$

29. $7 + (4 \cdot 5) = (5 \cdot 4) + 7$

30. $\frac{2}{3} + \frac{1}{6} = \frac{1}{6} + \frac{2}{3}$

Chapter 1 DIAGNOSTIC TEST

The purpose of this test is to see how well you understand sets, the properties of the real number system, and operations on real numbers. We recommend that you work this diagnostic test *before* your instructor tests you on this chapter. Allow yourself about 25 minutes.

Complete solutions for all the problems on this test, together with section references, are given in the answer section at the end of the book. For the problems you do incorrectly, we suggest that you study the sections cited.

In Problems 1–17, write "true" if the statement is always true; otherwise, write "false."

1. $\{2, 7, 7, 2\}$ and $\{2, 7\}$ are equal sets.
2. 0 is a real number.
3. Every irrational number is a real number.
4. $\frac{9}{0} = 0$
5. $(7 \cdot 5) \cdot 2 = 7 \cdot (5 \cdot 2)$ illustrates the commutative property of multiplication.
6. Division is commutative.
7. The set of digits is a subset of the set of natural numbers.
8. $-2 < 7 < 10$ is a valid continued inequality.
9. Every real number is a rational number.
10. x is a factor of the expression $7xy$.
11. "If $x = y$ and $y = 12$, then $x = 12$" illustrates the transitive property of equality.
12. The multiplicative inverse of $-\frac{2}{5}$ is $\frac{5}{2}$.
13. x is a term of the expression $7xy$.
14. $4 < 8 > 5$ is a valid continued inequality.
15. $3 \cdot (7 \cdot 2) = (3 \cdot 7) \cdot (3 \cdot 2)$
16. $\sqrt{-9} = -3$
17. $\sqrt[3]{-64} = -4$

For Problems 18–20, use the sets $A = \{k, w, x, z\}$, $B = \{k, x, y, z\}$, and $C = \{k, x\}$. Determine whether the statement is true or false.

18. $B \subseteq A$ 19. $C \subseteq B$ 20. $k \subseteq C$

For Problems 21–25, use the set $\{2.4, -3, 2.865\,291\,6\ldots$ (never terminates, never repeats), $0, 5, \frac{1}{2}$, and $0.\overline{18}\}$. Determine each of the following:

21. Which numbers from the set are natural numbers?
22. Which numbers from the set are integers?
23. Which numbers from the set are rational numbers?
24. Which numbers from the set are irrational numbers?
25. Which numbers from the set are real numbers?

For Problems 26–29, use the sets $A = \{w, x, z\}$, $B = \{x, y, w\}$, and $C = \{r, s, y\}$. Find each of the following:

26. $A \cup C$ 27. $A \cap C$ 28. $A \cap B$ 29. $A \cup (B \cap C)$

In Problems 30–47, find the value of each expression or write either "undefined" or "not real."

30. $|-17|$ 31. $(-5)^2$ 32. $5 \div (-30)$
33. $-35 - 2$ 34. $\dfrac{-40}{-8}$ 35. $-9(-8)$
36. $-19(0)$ 37. $-27 - (-17)$ 38. $-9 + (-13)$
39. -8^2 40. $-|-3|$ 41. $\sqrt[3]{-27}$
42. $16 \div 4 \cdot 2$ 43. $2\sqrt{9} - 5$ 44. $\sqrt{81}$
45. $\sqrt{-16}$

In Problems 46 and 47, use a calculator; show all the digits that your calculator display shows.

46. $15.3 \div 3(7)$ 47. $\dfrac{(4.56) \times (1.23)}{(2.872)(1.1)}$

48. Find the prime factorization of 78.
49. Find the prime factorization of 65.
50. Find the least common multiple (LCM) of 78 and 65.

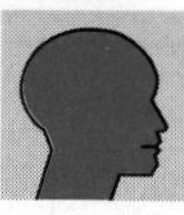

Critical Thinking and Problem-Solving Exercises

Mathematics is much more than a set of rules for solving certain problems; learning mathematics involves analyzing problems, looking for patterns and relationships, examining data, developing models, and so on. In the Critical Thinking and Problem-Solving Exercises that appear at the end of each chapter, you are given the opportunity to devise your own methods of problem solving—in many cases attacking types of problems that are different from any you have seen before. Some are "open-ended" problems—problems that do not have "right" or "wrong" answers. For such problems, analyze the given facts, and draw your own conclusions, which may be different from the conclusions drawn by others.

Problem-solving methods you might try on these exercises include (but are not limited to) guessing-and-checking, making an organized list or a table, looking for patterns, using logic, drawing pictures, making models, simplifying, and working backwards. Your instructor may want you to work in small groups to solve some or all of these problems.

1. By making an organized list, find all the positive factors of 42.

2. Doris is thinking of a certain number. Guess what number Doris is thinking about by using the following four clues.

The number is a three-digit number.

It is less than 130.

The sum of the digits is 7.

One of the digits is a 2.

3. Bruce has six shirts hanging in his closet, one in each of the following six colors: beige, white, brown, yellow, blue, and green. Using these hints, determine the order in which the shirts are hanging.

The white shirt and the yellow shirt are on the two ends.

The beige shirt is to the left of the white shirt and the brown shirt.

The beige shirt is not next to the brown shirt.

The blue shirt and the brown shirt are in the two center positions.

List the shirts—by color—in order from left to right.

4. We know that the reflexive property holds for equality; that is, $a = a$. Does the reflexive property hold for *less than*? Give an example that justifies your answer.

5. We know that the symmetric property holds for equality; that is, if $a = b$, then $b = a$. Does the symmetric property hold for *greater than*? Give an example that justifies your answer.

6. Is the relation "is shorter than" reflexive? Is it symmetric? Is it transitive?

7. A set that contains one element has two subsets (itself and the empty set); a set that contains two elements has four subsets (see the solution for Exercise 6 of Exercises 1.1, Set I); a set that contains three elements has eight subsets (see the solution for Exercise 5 of Exercises 1.1, Set I); a set that contains four elements has sixteen subsets (see Example 12 of Section 1.1). If this pattern continues, how many subsets should a set containing five elements have? How many subsets should a set containing six elements have? How many subsets should a set containing n elements have?

Review of Elementary Topics

CHAPTER

2

In this chapter, we continue our review of elementary algebra. We discuss the properties of integral exponents, scientific notation, and simplifying algebraic expressions. We also discuss evaluating algebraic expressions and using formulas.

2.1 Exponents That Are Positive Integers

Powers of Variables Earlier, we discussed powers of rational numbers. We now discuss powers of variables. Just as 3^4 means $3 \cdot 3 \cdot 3 \cdot 3$, x^4 means $x \cdot x \cdot x \cdot x$. In the exponential expression x^n, the *base* is x, the *exponent* is n, and x^n means that x is to be used as a factor n times. If a variable has no exponent written, then the exponent is understood to be 1; that is, $x = x^1$.

Recall that an exponent always applies *only* to the immediately preceding symbol.

EXAMPLE 1 Examples of the meaning of exponents:

a. $x^2y^3 = xxyyy$

b. $xy^3 = xyyy$ The exponent applies only to the y

c. $(xy)^3 = (xy)(xy)(xy) = xxxyyy$

This exponent applies to whatever is inside the parentheses

d. $y^1 = y$

The Properties of Exponents

We assume that you recall the properties of exponents from elementary algebra; however, we will briefly review them here. (We do not give proofs of the properties of exponents used in this chapter; either they require proof by mathematical induction, a technique not discussed in this book, or they are true by definition.) We list below the first five properties of exponents:

The first five properties of exponents

If $x, y \in R$ and if $m, n \in N$, then the following properties are true:

$x^m \cdot x^n = x^{m+n}$	The product rule
$(x^m)^n = x^{mn}$	The power rule
$(xy)^n = x^n y^n$	The rule for raising a product to a power
$\frac{x^m}{x^n} = x^{m-n}$ if $x \neq 0$	The quotient rule (In this section, $m > n$)
$\left(\frac{x}{y}\right)^n = \frac{x^n}{y^n}$ if $y \neq 0$	The rule for raising a quotient to a power

Note We must add the restrictions $x \neq 0$ and $y \neq 0$ to avoid dividing by zero, which is not defined.

To see that the product rule is reasonable, let's consider the problem $x^3 \cdot x^4$. Because $x^3 = xxx$ and $x^4 = xxxx$,

$$x^3 \cdot x^4 = (xxx)(xxxx) = xxxxxxx = x^7$$

Since x^{3+4} also equals x^7, it must be true that $x^4 \cdot x^3 = x^7$.

We can see that the quotient rule is reasonable if we remember that division is the inverse operation of multiplication. Thus, $\frac{x^7}{x^4} = x^3$, since $x^3 \cdot x^4 = x^7$.

EXAMPLE 2 Examples of using the product rule or of determining that the product rule does not apply:

a. $x^5 \cdot x^9 = x^{5+9} = x^{14}$

When no exponent is written, the exponent is understood to be 1

b. $w \cdot w^7 = w^1 \cdot w^7 = w^{1+7} = w^8$

c. $x^3 \cdot y^2$ The product rule does not apply when the bases are different. x^3y^2 cannot be rewritten in a simpler form.

d. $x^3 + x^2$ The product rule does not apply to addition problems. (We can verify this with numbers; for example, if $x = 2$, we have $2^3 + 2^2 \neq 2^5$, since $2^3 + 2^2 = 8 + 4 = 12$, but $2^5 = 32$.) $x^3 + x^2$ cannot be rewritten in a simpler form.

e. $5^3 \cdot 5^7 = 5^{3+7} = 5^{10}$

A Word of Caution An error commonly made in using the product rule when the bases are constants is to add the exponents and *also* multiply the bases.

Correct method	*Incorrect method*
$10^9 \cdot 10^4 = 10^{9+4}$	$10^9 \cdot 10^4 = (10 \cdot 10)^{9+4} = 100^{13}$
$= 10^{13}$	
$2^3 \cdot 2^2 = 2^{3+2}$	$2^3 \cdot 2^2 = (2 \cdot 2)^{3+2} = 4^5$
$= 2^5 = 32$	$= 1{,}024$

When multiplying powers of the same base, just add the exponents; do *not* multiply the bases as well.

EXAMPLE 3 Examples of using the power rule and the rule for raising a product to a power:

a. $(x^3)^7 = x^{3 \cdot 7} = x^{21}$ Using the power rule

b. $(10^6)^2 = 10^{6 \cdot 2} = 10^{12}$ Using the power rule

c. $(2^a)^b = 2^{a \cdot b} = 2^{ab}$ Using the power rule

d. $(xy)^7 = x^7y^7$ Using the rule for raising a product to a power

e. $(2x)^3 = 2^3x^3$, or $8x^3$ Using the rule for raising a product to a power

A Word of Caution Because an exponent applies only to the immediately preceding symbol, $2x^3 \neq (2x)^3$. In the expression $2x^3$, the exponent applies only to the x, and so $2x^3$ means $2 \cdot x \cdot x \cdot x$. On the other hand, in the expression $(2x)^3$, the exponent applies to everything inside the parentheses because the exponent is immediately to the right of a right parenthesis. Therefore, $(2x)^3$ means $(2x)(2x)(2x)$.

EXAMPLE 4 Examples of using the quotient rule or of determining that the quotient rule does not apply (assume $x \neq 0$ and $y \neq 0$):

a. $\dfrac{x^5}{x^3} = x^{5-3} = x^2$

b. $\dfrac{10^7}{10^3} = 10^{7-3} = 10^4$

c. $\dfrac{y^3}{y} = \dfrac{y^3}{y^1} = y^{3-1} = y^2$

d. $\dfrac{x^5}{y^2}$ The quotient rule does not apply when the bases are different. $\dfrac{x^5}{y^2}$ cannot be rewritten in a simpler form.

e. $z^5 - z^2$ The quotient rule does not apply to subtraction problems. (We can easily verify this if we let $z = 2$; we then have $2^5 - 2^2 \neq 2^3$, since $2^5 - 2^2 = 32 - 4 = 28$, but $2^3 = 8$.) $z^5 - z^2$ cannot be rewritten in a simpler form.

EXAMPLE 5 Examples of raising a quotient to a power (assume $b \neq 0$ and $z \neq 0$):

a. $\left(\dfrac{a}{b}\right)^4 = \dfrac{a^4}{b^4}$ Using the rule for raising a quotient to a power

b. $\left(\dfrac{3}{z}\right)^2 = \dfrac{3^2}{z^2}$, or $\dfrac{9}{z^2}$ Using the rule for raising a quotient to a power

Exercises 2.1
Set I

Use the properties of exponents to simplify each of the following expressions, if possible. Assume that no variable has a value that would make a denominator zero.

1. $10^2 \cdot 10^4$
2. $2^3 \cdot 2^2$
3. $x^2 \cdot x^5$
4. $y^6 \cdot y^3$
5. $x^4 + x^6$
6. $y^3 + y^7$
7. $x^2 \cdot y^7$
8. $m^3 \cdot n^8$
9. $\dfrac{a^8}{a^3}$
10. $\dfrac{x^5}{x^2}$
11. $x^6 - x$
12. $y^4 - y$
13. $(10^3)^2$
14. $(5^2)^4$
15. $(3^a)^b$
16. $(2^m)^n$
17. $(xy)^5$
18. $(uv)^4$
19. $(2x)^6$
20. $(3x)^4$
21. $x^8 \cdot y^5$
22. $s^2 \cdot t^4$
23. $3^{42} \cdot 3^{73}$
24. $7^{31} \cdot 7^{82}$
25. $\dfrac{g^9}{g^4}$
26. $\dfrac{x^7}{x^4}$
27. $\dfrac{x^4}{y^2}$ no
28. $\dfrac{a^5}{b^3}$
29. $\left(\dfrac{x}{y}\right)^3$
30. $\left(\dfrac{u}{v}\right)^7$
31. $\left(\dfrac{3}{x}\right)^4$
32. $\left(\dfrac{x}{5}\right)^2$

Writing Problems

Express the answers in your own words and in complete sentences.

1. Explain why $5^3 \cdot 5^2 \neq 25^5$.
2. Explain why $x^3 + x^4 \neq x^7$.
3. Explain why $(3x)^5 \neq 3x^5$.

Exercises 2.1
Set II

Use the properties of exponents to simplify each of the following expressions, if possible. Assume that no variable has a value that would make a denominator zero.

1. $5^2 \cdot 5^3$
2. $y^3 + y$
3. $X^4 \cdot X^2$
4. $x^4 + x^3$
5. $u^4 + u^7$
6. $4 \cdot 4^3$
7. $a^5 \cdot b^6$
8. $(2v)^3$
9. $\dfrac{P^5}{P^3}$
10. $\dfrac{x^4}{y^3}$
11. $z^7 + z^4$
12. $y^5 - y^2$
13. $(4^2)^3$
14. xx^5
15. $(2^x)^y$
16. xy^5
17. $(ab)^7$
18. $(2^3)^2$
19. $(4x)^2$
20. $(xy)^4$
21. $a^4 \cdot b^2$
22. $3^2 \cdot 3^3$
23. $7^{41} \cdot 7^{75}$
24. $x^2 \cdot y^5$
25. $\dfrac{U^8}{U^7}$
26. $\dfrac{x^6}{y^2}$
27. $\dfrac{r^6}{t^2}$
28. $\dfrac{z^6}{z^2}$
29. $\left(\dfrac{a}{b}\right)^6$
30. $\left(\dfrac{4}{x}\right)^3$
31. $\left(\dfrac{x}{2}\right)^5$
32. $\dfrac{a^{14}}{a^9}$

2.2 The Zero Exponent; Negative Exponents

The Zero Exponent

When we used the quotient rule in the last section, we made sure that the exponent of the dividend (or the numerator) was always greater than the exponent of the divisor (or the denominator). We now consider the case in which the exponents of the dividend and divisor are equal.

We know that $\dfrac{2^7}{2^7} = 1$, because any nonzero number divided by itself is 1. However, if we use the quotient rule in the same problem, we have

$$\frac{2^7}{2^7} = 2^{7-7} = 2^0$$

Since $\dfrac{2^7}{2^7} = 1$, we would *like* 2^0 to equal 1.

Similarly, we know that if $x \neq 0$, $\dfrac{x^4}{x^4} = 1$. However, if we use the quotient rule in the same problem, we have

$$\frac{x^4}{x^4} = x^{4-4} = x^0$$

Therefore, we would *like* x^0 to equal 1.

In fact, mathematicians *define* x^0 to be 1 (if $x \neq 0$).

The zero exponent rule

If $x \in R$ but $x \neq 0$, then

$$x^0 = 1$$

Note If we use the quotient rule on $\dfrac{0}{0}$, we have $\dfrac{0}{0} = \dfrac{0^1}{0^1} = 0^{1-1} = 0^0$. Since $\dfrac{0}{0}$ is undefined, 0^0 must be undefined, also.

EXAMPLE 1 Examples of removing zero exponents (assume that all variables are nonzero):

a. $a^0 = 1$ Using the zero exponent rule
b. $10^0 = 1$ Using the zero exponent rule
c. $(-7)^0 = 1$ The exponent applies to everything inside the parentheses
d. $-7^0 = -(7^0) = -1$ The exponent applies *only* to the 7
e. $(3y)^0 = 1$ The exponent applies to everything inside the parentheses
f. $6x^0 = 6 \cdot x^0 = 6 \cdot 1 = 6$ The exponent applies *only* to the x, not also to the 6

Negative Exponents

Next, we consider using the quotient rule when the exponent of the dividend (or numerator) is *less than* the exponent of the divisor (or denominator).

If we apply the quotient rule to $\frac{2^4}{2^7}$, we have $\frac{2^4}{2^7} = 2^{4-7} = 2^{-3}$, and if we apply the quotient rule to $\frac{3^5}{3^9}$, we have $\frac{3^5}{3^9} = 3^{5-9} = 3^{-4}$. You can verify with a calculator that

$$\frac{2^4}{2^7} = \frac{16}{128} = \frac{1}{8} = \frac{1}{2^3} \quad \text{and} \quad \frac{3^5}{3^9} = \frac{243}{19{,}683} = \frac{1}{81} = \frac{1}{3^4}$$

Therefore, we would *like* 2^{-3} to equal $\frac{1}{2^3}$ and 3^{-4} to equal $\frac{1}{3^4}$.

When mathematicians define x^{-n}, they want all the properties of exponents to hold. In particular, if the product rule and the zero exponent rule are to hold, the following statement must be true:

$$x^n \cdot x^{-n} = x^{n+(-n)} = x^0 = 1 \qquad n + (-n) = 0 \text{ because } n \text{ and } -n \text{ are additive inverses}$$

If the product of x^n and x^{-n} is 1, then x^n and x^{-n} must be the multiplicative inverses of each other; however, it is also true that the multiplicative inverse of x^n is $\frac{1}{x^n}$. So that all the properties we've discussed will be consistent, mathematicians *define* x^{-n} to equal $\frac{1}{x^n}$.

The negative exponent rule: form 1

If $x \in R$ but $x \neq 0$, and if $n \in J$, then

$$x^{-n} = \frac{1}{x^n} \qquad \text{Form 1}$$

Note Suppose we applied the negative exponent rule to the expression 0^{-3}; we would have $0^{-3} = \frac{1}{0^3}$, but $\frac{1}{0^3}$ is undefined. Therefore, we must include the restriction $x \neq 0$ in the negative exponent rule.

The negative exponent rule is valid whether x is a positive number or a negative number, and x^{-n} may be either a positive or a negative number. In fact, if x^n is positive, x^{-n} is also positive, and if x^n is negative, x^{-n} is also negative (see Example 2).

EXAMPLE 2 Examples of using the negative exponent rule to evaluate expressions:

a. $6^{-2} = \frac{1}{6^2}$, or $\frac{1}{36}$ $\quad$ 6^2 and 6^{-2} are both positive

b. $3^{-3} = \frac{1}{3^3}$, or $\frac{1}{27}$ — 3^3 and 3^{-3} are both positive

c. $(-5)^{-3} = \frac{1}{(-5)^3} = \frac{1}{-125}$, or $-\frac{1}{125}$ — $(-5)^3$ and $(-5)^{-3}$ are both negative

d. $(-10)^{-4} = \frac{1}{(-10)^4} = \frac{1}{10^4}$, or $\frac{1}{10{,}000}$ — $(-10)^4$ and $(-10)^{-4}$ are both positive

e. $(-2)^{-5} = \frac{1}{(-2)^5} = \frac{1}{-32}$, or $-\frac{1}{32}$ — $(-2)^5$ and $(-2)^{-5}$ are both negative

---

Note Now that we have defined zero and negative exponents, we no longer need the restriction "$m > n$" in the quotient rule. That is, $\frac{x^m}{x^n} = x^{m-n}$ for all integers m and n.

Your instructor may recommend that you memorize the following properties.

The negative exponent rule: forms 2 and 3

If $x, y \in R$, but $x \neq 0$ and $y \neq 0$, and if $n \in J$, then

$$\frac{1}{x^{-n}} = x^n \qquad \text{Form 2}$$

$$\left(\frac{x}{y}\right)^{-n} = \left(\frac{y}{x}\right)^n \qquad \text{Form 3}$$

In Appendix A, we provide proofs of the two alternative forms of the negative exponent rule. When we cite "the negative exponent rule" in this book, we will be referring to form 1 unless form 2 or form 3 is specified.

EXAMPLE 3 Examples of using all three forms of the negative exponent rule (assume that all variables are nonzero):

a. $x^{-5} = \frac{1}{x^5}$ — Using form 1 of the negative exponent rule

b. $\frac{1}{y^{-2}} = y^2$ — Using form 2 of the negative exponent rule

c. $\left(\frac{a}{b}\right)^{-4} = \left(\frac{b}{a}\right)^4 = \frac{b^4}{a^4}$ — Using form 3 of the negative exponent rule and the rule for raising a quotient to a power

d. $x^{-6a} = \frac{1}{x^{6a}}$ — Using form 1 of the negative exponent rule

---

EXAMPLE 4 Examples of removing negative exponents (assume that all variables are nonzero):

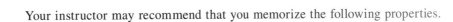

The exponent applies *only* to the x

a. $3x^{-1} = 3 \cdot x^{-1} = \frac{3}{1} \cdot \frac{1}{x} = \frac{3}{x}$ — Using the negative exponent rule: $x^{-1} = \frac{1}{x}$

The exponent applies to everything inside the parentheses

b. $(3x)^{-1} = \frac{1}{(3x)^1} = \frac{1}{3x}$ — Using the negative exponent rule

or $(3x)^{-1} = 3^{-1} \cdot x^{-1} = \frac{1}{3} \cdot \frac{1}{x} = \frac{1}{3x}$ — Using the rule for raising a product to a power and the negative exponent rule

The exponent applies only to the x

c. $6x^{-8} = 6 \cdot x^{-8} = \frac{6}{1} \cdot \frac{1}{x^8} = \frac{6}{x^8}$ — Using the negative exponent rule: $x^{-8} = \frac{1}{x^8}$

d. $(3x)^{-5} = \frac{1}{(3x)^5} = \frac{1}{3^5x^5}$, or $\frac{1}{243x^5}$ — Using the negative exponent rule and the rule for raising a product to a power

or $(3x)^{-5} = 3^{-5} \cdot x^{-5} = \frac{1}{3^5} \cdot \frac{1}{x^5} = \frac{1}{3^5x^5}$ — Using the rule for raising a product to a power and the negative exponent rule

e. $\frac{h^5}{k^{-4}} = \frac{h^5}{1} \cdot \frac{1}{k^{-4}} = \frac{h^5}{1} \cdot \frac{k^4}{1} = h^5k^4$ — Using form 2 of the negative exponent rule on $\frac{1}{k^{-4}}$

f. $\frac{a^{-2}b^4}{c^5d^{-3}} = a^{-2} \cdot b^4 \cdot \frac{1}{c^5} \cdot \frac{1}{d^{-3}}$ — Using forms 1 and 2 of the negative exponent rule

$= \frac{1}{a^2} \cdot \frac{b^4}{1} \cdot \frac{1}{c^5} \cdot \frac{d^3}{1} = \frac{b^4 \cdot d^3}{a^2 \cdot c^5} = \frac{b^4d^3}{a^2c^5}$

Notice that a^{-2} (which was a factor of the numerator) moved to the denominator and its exponent became $+2$, and that d^{-3} (which was a factor of the denominator) moved to the numerator and its exponent became $+3$.

The properties of exponents listed in the previous section also apply when the exponents are negative integers (see Example 5).

EXAMPLE 5 Examples of using the properties of exponents when negative exponents are involved (assume that all variables are nonzero):

a. $y^3 \cdot y^{-2} = y^{3+(-2)} = y^1 = y$ — Using the product rule

b. $b^2 \cdot b^{-6} = b^{2+(-6)} = b^{-4} = \frac{1}{b^4}$ — Using the product rule and the negative exponent rule

c. $\frac{x^4}{x^7} = x^{4-7} = x^{-3} = \frac{1}{x^3}$ — Using the quotient rule and the negative exponent rule

d. $\frac{t^7}{t^{-4}} = t^{7-(-4)} = t^{7+(+4)} = t^{11}$ — Using the quotient rule

e. $(x^3)^{-2} = x^{3(-2)} = x^{-6} = \frac{1}{x^6}$ — Using the power rule and the negative exponent rule

f. $\frac{x^{-3}}{x^2} = x^{-3-2} = x^{-5} = \frac{1}{x^5}$ — Using the quotient rule and the negative exponent rule

or $\frac{x^{-3}}{x^2} = x^{-3} \cdot \frac{1}{x^2} = \frac{1}{x^3} \cdot \frac{1}{x^2} = \frac{1}{x^3 \cdot x^2} = \frac{1}{x^{3+2}} = \frac{1}{x^5}$ — Using the negative exponent rule and the product rule

g. $x^{-4} + x^{-5} = \frac{1}{x^4} + \frac{1}{x^5}$ — Using the negative exponent rule

---

EXAMPLE 6 Examples of evaluating expressions with numerical bases:

a. $10^7 \cdot 10^{-3} = 10^{7+(-3)} = 10^4$, or 10,000 — Using the product rule

b. $5^4 \cdot 5^{-4} = 5^{4+(-4)} = 5^0 = 1$ — Using the product rule and the zero exponent rule

c. $(4^3)^{-1} = 4^{3(-1)} = 4^{-3} = \frac{1}{4^3}$, or $\frac{1}{64}$ Using the power rule and the negative exponent rule

or $(4^3)^{-1} = \frac{1}{(4^3)^1} = \frac{1}{4^{3\cdot 1}} = \frac{1}{4^3}$ Using the negative exponent rule and the power rule

d. $\frac{3^4}{3^{-1}} = 3^{4-(-1)} = 3^{4+(+1)} = 3^5$, or 243 Using the quotient rule

e. $(2^{-3})^2 = 2^{-3(2)} = 2^{-6} = \frac{1}{2^6}$, or $\frac{1}{64}$ Using the power rule and the negative exponent rule

There are times when it is desirable to write an expression with no denominator. We demonstrate the procedure in Example 7.

EXAMPLE 7 Examples of rewriting expressions without denominators (assume that all variables are nonzero):

a. $\frac{m^5}{n^2} = m^5 \cdot \frac{1}{n^2} = m^5 \cdot n^{-2} = m^5n^{-2}$ Using the negative exponent rule, we have $\frac{1}{n^2} = n^{-2}$

b. $\frac{y^3}{z^{-2}} = y^3 \cdot \frac{1}{z^{-2}} = y^3 \cdot z^2 = y^3z^2$ Using form 2 of the negative exponent rule, we have $\frac{1}{z^{-2}} = z^2$

c. $\frac{a}{2c^2} = a \cdot \frac{1}{2} \cdot \frac{1}{c^2} = a \cdot 2^{-1} \cdot c^{-2} = 2^{-1}ac^{-2}$ Using the negative exponent rule

A Word of Caution You may have noticed in the examples that when a *factor* with an exponent is moved from the numerator to the denominator *or* from the denominator to the numerator, the *sign* of its exponent changes.

However, $\frac{x^{-2} + y^{-2}}{x^{-3} + y^{-3}} \neq \frac{x^3 + y^3}{x^2 + y^2}$. x^{-2} and y^{-2} are not *factors* of the numerator, and x^{-3} and y^{-3} are not *factors* of the denominator. We cannot yet simplify $\frac{x^{-2} + y^{-2}}{x^{-3} + y^{-3}}$; we discuss simplifying such expressions in a later chapter.

You need not show all the steps that we have shown in the examples unless your instructor requires you to do so.

Exercises 2.2
Set I

In Exercises 1–40, use the properties of exponents to rewrite each expression with no parentheses, with a base appearing only once, if possible, and with positive exponents or no exponents. Assume that all variables are nonzero.

1. a^{-3}
2. x^{-2}
3. $\frac{1}{z^{-7}}$
4. $\frac{1}{b^{-9}}$
5. $5b^{-7}$
6. $3y^{-2}$
7. $(5b)^{-2}$
8. $(3y)^{-2}$
9. $x^{-3}y^2z^0$
10. $r^3s^{-4}t^0$
11. $xy^{-2}z^{-3}w^0$
12. $z^{-4}bc^{-5}a^0$
13. $\frac{a^3}{b^{-4}}$
14. $\frac{c^4}{d^{-5}}$
15. $\frac{x^{-3}}{y^{-2}}$
16. $\frac{P^{-2}}{Q^{-4}}$

17. $x^{-3} \cdot x^{-4}$ 18. $y^{-1} \cdot y^{-4}$ 19. $(a^3)^{-2}$

20. $(b^2)^{-4}$ 21. $x^8 \cdot x^{-2}$ 22. $a^{-3} \cdot a^5$

23. $(x^{2a})^{-3}$ 24. $(y^{3c})^{-2}$ 25. $\dfrac{y^3}{y^{-2}}$

26. $\dfrac{x^8}{x^{-4}}$ 27. $\dfrac{x^{3a}}{x^{-a}}$ 28. $\dfrac{a^{4x}}{a^{-2x}}$

29. $(3x)^0$ 30. $(2^a)^0$ 31. $5x^0$

32. $2y^0$ 33. $x^{-3} + x^{-5}$ 34. $y^{-2} + y^{-6}$

35. $x^7 - x^{-5}$ 36. $y^{10} - y^{-3}$ 37. $x^0 + y^0$

38. $a^0 + b^0 + c^0$ 39. $(x + y)^0$ 40. $(a + b + c)^0$

In Exercises 41–48, evaluate each expression.

41. $10^5 \cdot 10^{-2}$ 42. $2^4 \cdot 2^{-2}$ 43. $(3^{-2})^{-2}$ 44. $(10^{-1})^{-3}$

45. $(10^0)^5$ 46. $(3^0)^4$ 47. $\dfrac{10^2 \cdot 10^{-1}}{10^{-3}}$ 48. $\dfrac{2^{-3} \cdot 2^2}{2^{-4}}$

In Exercises 49–54, write each expression without a denominator, using negative exponents, if necessary. Assume that all variables are nonzero.

49. $\dfrac{y}{x^3}$ 50. $\dfrac{x}{y^2}$ 51. $\dfrac{x}{a^{-4}}$

52. $\dfrac{m^2}{n^{-3}}$ 53. $\dfrac{x^4y^{-3}}{z^{-2}}$ 54. $\dfrac{a^{-1}b^3}{c^{-4}}$

Writing Problems

Express the answers in your own words and in complete sentences.

1. Explain why $3^{-3} \neq -27$, and find the correct value of 3^{-3}.
2. Explain why $5^0 \neq 0$, and find the correct value of 5^0.
3. Explain why $(x + y)^0 \neq x^0 + y^0$.
4. Explain why $3x^{-1} \neq \dfrac{1}{3x}$.
5. Explain why $5x^0 \neq 1$.

Exercises 2.2 Set II

In Exercises 1–40, use the properties of exponents to rewrite each expression with no parentheses, with a base appearing only once, if possible, and with positive exponents or no exponents. Assume that all variables are nonzero.

1. x^{-4} 2. 16^0 3. $\dfrac{1}{t^{-4}}$

4. x^0b^0 5. $2a^{-4}$ 6. $4x^0$

7. $(2a)^{-2}$ 8. $(-4x)^{-2}$ 9. $x^0y^{-1}z^2$

10. $3x^0y^{-1}$ 11. $x^{-3}yz^{-2}$ 12. $y^3z^{-1}w^0$

13. $\dfrac{a^4}{b^{-3}}$ 14. $\dfrac{x^{-1}y}{z^{-2}}$ 15. $\dfrac{x^{-5}}{y^{-4}}$

16. $\dfrac{a^0b^{-3}c}{d^{-4}}$ 17. $a^{-3} \cdot a^{-5}$ 18. $x^{-3} \cdot x^5$

19. $(a^{-2})^3$ 20. $(-x^2)^{-3}$ 21. x^8x^{-2}

22. $(a^{-3})^2$ 23. $(z^{4b})^{-4}$ 24. $s^{-4} \cdot s^2$

25. $\dfrac{x}{y^{-3}}$ 26. $\dfrac{a^5}{b^2}$ 27. $\dfrac{y^{2b}}{y^{-b}}$

28. $\dfrac{x^4}{x^{-2}}$ 29. $(3R)^0$ 30. $6a^0$

31. $4X^0$ 32. $(3xy)^0$ 33. $a^{-7} + a^{-2}$

34. $x^{-2} - x^3$ 35. $u^4 - u^{-5}$ 36. $3x^{-5}$

37. $a^0 + b^0 + c^0 + d^0$ 38. $x + 5^0$

39. $(a + b + c + d)^0$ 40. a^3b^{-2}

In Exercises 41–48, evaluate each expression.

41. $10^{-3} \cdot 10^5$ 42. -2^{-3} 43. $(2^{-3})^{-2}$ 44. $(10^2)^{-4}$

45. $(7^0)^8$ 46. $(8^0)^2$ 47. $\dfrac{10^{-3} \cdot 10}{10^{-4}}$ 48. $\dfrac{2^{-1}}{2^6 \cdot 2^{-4}}$

In Exercises 49–54, write each expression without a denominator, using negative exponents, if necessary. Assume that all variables are nonzero.

49. $\dfrac{a}{b^4}$ 50. $\dfrac{x^{-2}}{y^3}$ 51. $\dfrac{x}{y^{-3}}$

52. $\dfrac{a^{-4}}{b}$ 53. $\dfrac{x^2y^{-3}}{z^{-5}}$ 54. $\dfrac{a^{-1}b}{c^{-6}}$

2.3 Simplifying Exponential Expressions

An expression with exponents is considered *simplified* when (1) there are no parentheses, (2) each different base appears only once in each separate term, and (3) the exponent on each base is a single natural number. Within each *term* of an expression, it is customary to write any numerical coefficient to the left of the variable(s) and, for ease of reading, to write the variables in alphabetical order.

Throughout this section, we will assume that all variables are nonzero.

EXAMPLE 1

Examples of simplifying exponential expressions:

a. $x^{-2} \cdot x^{7} = x^{-2+7} = x^{5}$

b. $(83x^{5})^{0} = 1$

c. $\dfrac{x^{5}y^{2}}{x^{3}y^{-1}} = x^{5-3}y^{2-(-1)} = x^{2}y^{3}$

d. $(2x)^{-1} = \dfrac{1}{(2x)^{1}} = \dfrac{1}{2x}$ The exponent applies to everything inside the parentheses

e. $2x^{-1} = 2 \cdot x^{-1} = \dfrac{2}{1} \cdot \dfrac{1}{x} = \dfrac{2 \cdot 1}{1 \cdot x} = \dfrac{2}{x}$ The exponent applies only to the x

f. $\left(\dfrac{-3^{-7}x^{9}}{y^{-4}}\right)^{0} = 1$ We need not simplify the expression inside the parentheses; we simply use the fact that $a^{0} = 1$

The power rule, the rule for raising a product to a power, and the rule for raising a quotient to a power can be combined into the general property of exponents, which follows:

The general property of exponents

> If a, b, c, $n \in J$, if x, y, $z \in R$, and if none of the variables has a value that makes a denominator zero, then
>
> $$\left(\frac{x^{a}y^{b}}{z^{c}}\right)^{n} = \frac{x^{an}y^{bn}}{z^{cn}}$$

In applying the general property of exponents, notice the following:

1. x^{a}, y^{b}, and z^{c} are *factors* of, not separate *terms* of, the expression within the parentheses.
2. The exponent of *each* factor within the parentheses must be multiplied by the exponent outside the parentheses.

In Example 2, we show one or two ways of simplifying each expression. In some of the problems, it is also possible to use the properties of exponents in orders *other* than the ones we show. (In most of mathematics, there are many different ways of solving a problem successfully.)

EXAMPLE 2 Examples of using the general property of exponents to simplify expressions:

a. $(x^2y^3)^4 = x^{2\cdot 4}y^{3\cdot 4} = x^8y^{12}$ Notice that x^2 and y^3 are *factors* of the expression inside the parentheses

b. $\left(\dfrac{x^2y^3}{z^4}\right)^7 = \dfrac{x^{2\cdot 7}y^{3\cdot 7}}{z^{4\cdot 7}} = \dfrac{x^{14}y^{21}}{z^{28}}$

The exponent of 2 is understood to be 1, and that 1 must be multiplied by 3

c. $\left(\dfrac{2a^{-3}b^2}{c^5}\right)^3 = \dfrac{2^{1\cdot 3}a^{-3\cdot 3}b^{2\cdot 3}}{c^{5\cdot 3}} = \dfrac{2^3a^{-9}b^6}{c^{15}} = \dfrac{8b^6}{a^9c^{15}}$

If we remove the negative exponent *first*, we have

$$\left(\frac{2a^{-3}b^2}{c^5}\right)^3 = \left(\frac{2^1b^2}{a^3c^5}\right)^3 = \frac{2^{1\cdot 3}b^{2\cdot 3}}{a^{3\cdot 3}c^{5\cdot 3}} = \frac{2^3b^6}{a^9c^{15}}, \text{ or } \frac{8b^6}{a^9c^{15}}$$

d. $\left(\dfrac{3^2c^{-4}}{d^3}\right)^{-1} = \dfrac{3^{2(-1)}c^{(-4)(-1)}}{d^{3(-1)}} = \dfrac{3^{-2}c^4}{d^{-3}} = \dfrac{c^4d^3}{3^2}$, or $\dfrac{c^4d^3}{9}$

If we use form 3 of the negative exponent rule first, we have

$$\left(\frac{3^2c^{-4}}{d^3}\right)^{-1} = \left(\frac{d^3}{3^2c^{-4}}\right)^1 = \frac{d^{3\cdot 1}}{3^{2\cdot 1}c^{(-4)(1)}} = \frac{d^3}{3^2c^{-4}} = \frac{c^4d^3}{3^2}, \text{ or } \frac{c^4d^3}{9}$$

A Word of Caution The general property of exponents *cannot* be used in a problem such as $(x^2 + y^3)^4$. That is,

$$(x^2 + y^3)^4 \neq x^{2\cdot 4} + y^{3\cdot 4}$$

The general property of exponents does not apply, because x^2 and y^3 are *terms* of, not factors of, the expression inside the parentheses. The general property of exponents applies only to *factors* inside the parentheses. We will discuss removing parentheses from expressions such as $(x^2 + y^3)^4$ in a later chapter.

We recommend that you simplify an expression within parentheses *before* you use the general property of exponents to remove the parentheses (especially when the exponent outside the parentheses is positive), although it is *possible* to use the general property first. We show both methods in Example 3.

EXAMPLE 3 Examples of simplifying exponential expressions:

a. $(x^3y^{-1})^5 = \left(\dfrac{x^3}{y^1}\right)^5 = \dfrac{x^{3\cdot 5}}{y^{1\cdot 5}} = \dfrac{x^{15}}{y^5}$ Simplifying inside the parentheses first

or $(x^3y^{-1})^5 = x^{3\cdot 5}y^{(-1)(5)} = x^{15}y^{-5} = \dfrac{x^{15}}{y^5}$ Using the general property of exponents first

b. $(5^0h^{-2})^{-3} = (1h^{-2})^{-3} = (h^{-2})^{-3} = h^{(-2)(-3)} = h^6$ Using the zero exponent property first

or $(5^0h^{-2})^{-3} = 5^{0(-3)}h^{(-2)(-3)} = 5^0h^6 = 1h^6 = h^6$ Using the general property of exponents first

c. $\left(\dfrac{x^5y^4}{x^3y^7}\right)^2 = (x^{5-3}y^{4-7})^2 = (x^2y^{-3})^2 = \left(\dfrac{x^2}{y^3}\right)^2 = \dfrac{x^{2\cdot2}}{y^{3\cdot2}} = \dfrac{x^4}{y^6}$ Simplifying inside the parentheses first

or $\left(\dfrac{x^5y^4}{x^3y^7}\right)^2 = \dfrac{x^{5\cdot2}y^{4\cdot2}}{x^{3\cdot2}y^{7\cdot2}} = \dfrac{x^{10}y^8}{x^6y^{14}} = x^{10-6}y^{8-14} = x^4y^{-6} = \dfrac{x^4}{y^6}$ Using the general property of exponents first

d. $\left(\dfrac{g^{-3}h^3}{g^{-5}h^{-5}}\right)^{-3} = (g^{-3-(-5)}h^{3-(-5)})^{-3} = (g^2h^8)^{-3} = g^{2(-3)}h^{8(-3)} = g^{-6}h^{-24} = \dfrac{1}{g^6h^{24}}$

or $\left(\dfrac{g^{-3}h^3}{g^{-5}h^{-5}}\right)^{-3} = \dfrac{g^{(-3)(-3)}h^{(3)(-3)}}{g^{(-5)(-3)}h^{(-5)(-3)}} = \dfrac{g^9h^{-9}}{g^{15}h^{15}} = g^{9-15}h^{-9-15} = g^{-6}h^{-24} = \dfrac{1}{g^6h^{24}}$

Simplifying Products When Each Factor Contains One Term

When we wish to remove grouping symbols in expressions such as $(3x^2)(7x^3)$ (that is, in expressions in which there is a *single term* inside each set of grouping symbols), the commutative and associative properties of multiplication permit us to rearrange the factors and then multiply, as shown in Example 4.

EXAMPLE 4 Examples of simplifying products when each factor contains only one term:

a. $(3x^2)(7x^3) = 3 \cdot 7 \cdot x^2 \cdot x^3 = 21x^{2+3} = 21x^5$

b. $(2x^3y^2)(4xy^5z)(-3x^2) = 2(4)(-3)x^3y^2xy^5zx^2 = -24x^3xx^2y^2y^5z = -24x^6y^7z$

c. $(-8a^3b^2c^6)(-5ab^3) = (-8)(-5)a^3ab^2b^3c^6 = 40a^4b^5c^6$

You need not show all the steps that we showed in Example 4. You can do all the work mentally, writing only the final answer.

Exercises 2.3

Set I

Write each expression in simplest form.

1. $(5x)^{-3}$
2. $(4y)^{-2}$
3. $7x^{-2}$
4. $3y^{-4}$
5. $\left(\dfrac{3}{x}\right)^3$
6. $\left(\dfrac{7}{y}\right)^2$
7. $\dfrac{8^2}{z}$
8. $\dfrac{2^3}{x}$
9. $(a^2b^3)^2$
10. $(x^4y^5)^3$
11. $(m^{-2}n)^4$
12. $(p^{-3}r)^5$
13. $(x^{-2}y^3)^{-4}$
14. $(w^{-3}z^4)^{-2}$
15. $(10^0k^{-4})^{-2}$
16. $(6^0z^{-5})^{-2}$
17. $(2x^2y^{-4})^3$
18. $(3a^{-1}b^5)^2$
19. $(5m^{-3}n^5)^{-2}$
20. $(8x^8y^{-2})^{-1}$
21. $\left(\dfrac{xy^4}{z^2}\right)^2$
22. $\left(\dfrac{a^3b}{c^2}\right)^3$
23. $\left(\dfrac{M^{-2}}{N^3}\right)^4$
24. $\left(\dfrac{R^5}{S^{-4}}\right)^3$
25. $\left(\dfrac{x^{-5}}{y^4z^{-3}}\right)^{-2}$
26. $\left(\dfrac{a^{-4}}{b^2c^{-5}}\right)^{-3}$
27. $\left(\dfrac{r^7s^8}{r^9s^6}\right)^0$
28. $\left(\dfrac{t^5u^6}{t^8u^7}\right)^0$
29. $\left(\dfrac{3x^2}{y^3}\right)^2$
30. $\left(\dfrac{2a^4}{b^2}\right)^4$
31. $\left(\dfrac{4a^{-2}}{b^3}\right)^{-1}$
32. $\left(\dfrac{m^{-4}}{5n^3}\right)^{-2}$
33. $\left(\dfrac{x^{-1}y^2}{x^4}\right)^{-2}$
34. $\left(\dfrac{u^{-4}v^3}{v^4}\right)^{-3}$
35. $\left(\dfrac{s^{-4}t^{-5}}{s^3t^{-4}}\right)^{-3}$
36. $\left(\dfrac{u^7v^{-4}}{u^{-3}v^{-5}}\right)^{-5}$
37. $(2x^5y^2)(-3x^4)$
38. $(-4a^2b^3)(7b^2)$
39. $(-4x^{-2}y^3)(-2xy^{-1})$
40. $(-5a^4b^{-3})(-2a^{-5}b^4)$
41. $(2s^3t^0u^{-4})(3su^3)(-s)$
42. $(4xy^0z^{-3})(-z)(2x^3z)$

Writing Problems

Express the answer in your own words and in complete sentences.

1. Explain how to use the general property of exponents.

Exercises 2.3 Set II

Write each expression in simplest form.

1. $(2x)^{-4}$ **2.** $2x^{-4}$ **3.** $3y^{-3}$ **4.** $(3y)^{-3}$

5. $\left(\frac{5}{z}\right)^2$ **6.** $\frac{5^2}{z}$ **7.** $\frac{2^4}{y}$ **8.** $\left(\frac{6z}{x}\right)^2$

9. $(x^4y^6)^2$ **10.** $(2xy^5)^5$ **11.** $(x^{-3}y^4)^3$ **12.** $(xy^4)^3$

13. $(a^{-3}b^2)^{-3}$ **14.** $(2x^{-4})^{-4}$ **15.** $(x^0y^{-3})^{-5}$

16. $(3x^{-6}z)^0$ **17.** $(4x^4y^{-2})^4$ **18.** $(3x^4yz^2)^0$

19. $(2x^{-1}y^5)^{-3}$ **20.** $(3xy^2)^{-1}$ **21.** $\left(\frac{a^3b}{c^4}\right)^2$

22. $\left(\frac{5^0x^4}{3^{-1}y^2}\right)^2$ **23.** $\left(\frac{u^{-2}}{v^4}\right)^3$ **24.** $\left(\frac{x^3y^5}{z^4}\right)^{-1}$

25. $\left(\frac{x^{-4}y^2}{z^{-3}}\right)^{-2}$ **26.** $\left(\frac{5s^4t^{-2}}{2x^2}\right)^{-2}$ **27.** $\left(\frac{r^5s^9}{r^2s^{10}}\right)^0$

28. $\left(\frac{ab^3}{b^4}\right)^2$ **29.** $\left(\frac{3a^2}{b^3}\right)^3$ **30.** $\left(\frac{4x^0}{y^3}\right)^2$

31. $\left(\frac{4m^{-3}}{n^5}\right)^{-2}$ **32.** $\left(\frac{u^3v^{-2}}{u^5w^{-3}}\right)^{-1}$ **33.** $\left(\frac{x^3}{x^{-2}y^{-4}}\right)^{-1}$

34. $\left(\frac{g^{-3}}{2h^4}\right)^{-2}$ **35.** $\left(\frac{r^{-3}w^{-4}}{r^6w^{-2}}\right)^{-4}$ **36.** $\left(\frac{p^{-8}q^5}{p^{-6}q^{-4}}\right)^{-3}$

37. $(-3x^2y^4)(8y)$ **38.** $(8x^2y)(-4xy)$

39. $(-4xy^3)(2x^2y^3)$ **40.** $(4^0a^3b^{-2})(3ab^5)$

41. $(5x^0y^{-4}z)(-y)(3y^2z)$ **42.** $(-3x^0y^4)(2xy^{-5})$

2.4 Scientific Notation

Now that we have discussed zero and negative exponents, we can introduce *scientific notation*, a notation used in many sciences and often seen in calculator displays. Scientific notation gives us a shorter way of writing those very large and very small numbers often seen in chemistry and physics, such as Avogadro's number (≈602,000,000,000,000,000,000,000) and Boltzmann's constant (0.000 000 000 000 000 000 138) (see Example 4). We can also use scientific notation to simplify the arithmetic in certain problems that arise in the sciences (see Example 6).

In this text (and in most texts), a positive number written in **scientific notation** must be in the form

$$a \times 10^n, \text{ where } 1 \le a < 10 \text{ and } n \text{ is an integer}$$

For example, 4.32×10^3 is written in scientific notation, because 4.32 is a number that is greater than or equal to 1 but less than 10, and 10^3 is an integral power of 10. Any positive decimal number can be written in scientific notation.

Since a is to be greater than or equal to 1 but less than 10, it must have *exactly one nonzero digit to the left of its decimal point.*

When we convert a number to scientific notation, we do not change its value, because we are using the facts that $10^n \times 10^{-n} = 1$ and that multiplying a number by 1 doesn't change its value. The procedure is demonstrated in Examples 1 and 2.

EXAMPLE 1

Convert 0.0372 to scientific notation.

SOLUTION When 0.0372 is written in scientific notation, the decimal point must be between the 3 and the 7. Since the decimal point must be moved two places to the right, we'll multiply 0.0372 by $10^{2} \times 10^{-2}$.

This product is 1

$$0.0372 = 0.0372 \times \underbrace{10^2 \times 10^{-2}}_{\text{This product is 1}} = (0.0372 \times 10^2) \times 10^{-2} = \underbrace{3.72 \times 10^{-2}}$$

This number is in scientific notation

EXAMPLE 2

Convert 38,500 to scientific notation.

SOLUTION When 38,500 is written in scientific notation, the decimal point must be between the 3 and the 8. Since the decimal point must be moved four places to the left, we'll multiply 38,500 by $10^{-4} \times 10^4$.

This product is 1

$$38{,}500 = 38{,}500 \times \underbrace{10^{-4} \times 10^4}_{\text{This product is 1}} = (38{,}500 \times 10^{-4}) \times 10^4 = \underbrace{3.85 \times 10^4}$$

This number is in scientific notation

In practice, it is not necessary to show the intermediate steps shown in Examples 1 and 2; instead, we can use the following procedure:

Writing a number in scientific notation

1. Finding a:
   a. Replace the decimal point with a caret ($_\wedge$). (If there is no decimal point, put the caret immediately to the right of the last digit of the number.)
   b. Place a decimal point in the number so that there is exactly one nonzero digit to the left of the decimal point. (The number just written is a.)
2. Finding the correct power of 10:
   a. The number of digits separating the caret and the decimal point in step 1 gives the absolute value of the exponent of 10. If the decimal point and the caret coincide (lie one on top of the other), the exponent is zero.
   b. The sign of the exponent of 10 is positive if the caret is to the right of the decimal point and negative if the caret is to the left of the decimal point.

The number in scientific notation is the product of the two numbers found in steps 1 and 2. (We drop the caret in the final answer.)

This procedure implies that if the number to be converted to scientific notation is greater than or equal to 10, the exponent of the 10 will be positive; if the number is greater than zero but less than 1, the exponent of the 10 will be negative. If the number is greater than or equal to 1 but less than 10, the exponent of the 10 will be zero.

EXAMPLE 3 Examples of writing decimal numbers in scientific notation:

	Decimal notation	*Finding a*	*Scientific notation*
a.	2,450	$2.450_\wedge \times 10^?$	2.45×10^3
b.	2.45	$2\overset{\wedge}{.}45 \times 10^?$	2.45×10^0
c.	0.0245	$0_\wedge 02.45 \times 10^?$	2.45×10^{-2}
d.	92,900,000	$9.290\ 000\ 0_\wedge \times 10^?$	9.29×10^7
e.	0.000 561 8	$0_\wedge 0005.618 \times 10^?$	5.618×10^{-4}

a, the number between 1 and 10 — The power of 10

Notice that in parts c and e, the caret is to the *left* of the decimal point and the exponent on the 10 is *negative*; in parts a and d, the caret is to the *right* of the decimal point and the exponent on the 10 is *positive.* In part b, the decimal point and the caret coincide, and the exponent on the 10 is zero.

EXAMPLE 4 Express (a) Avogadro's number and (b) Boltzmann's constant in scientific notation.

SOLUTION

a. Avogadro's number is approximately 602,000,000,000,000,000,000,000.

The caret is to the right of the decimal point; there are 23 digits between the caret and the decimal point

$$602{,}000{,}000{,}000{,}000{,}000{,}000{,}000 = 6.\overbrace{020\,000\,000\,000\,000\,000\,000\,00}_\wedge \times 10^{23}$$

$$= 6.02 \times 10^{23}$$

b. Boltzmann's constant is 0.000 000 000 000 000 138.

The caret is to the left of the decimal point; there are 16 digits between the caret and the decimal point

$$0.000\,000\,000\,000\,000\,138 = 0_\wedge\overbrace{000\,000\,000\,000\,0001}.38 \times 10^{-16}$$

$$= 1.38 \times 10^{-16}$$

To convert from scientific notation to decimal notation, we simply multiply by the power of 10. For example,

$$4.32 \times 10^3 = 4{,}320$$

$$2.3 \times 10^{-3} = 0.0023$$

It is sometimes necessary to convert a number such as 732.4×10^3 to scientific notation or to decimal notation. Example 5 demonstrates how to do this.

EXAMPLE 5 Convert 732.4×10^3 to scientific notation and then to decimal notation.

SOLUTION Notice that $732.4 = 7.324 \times 10^2$.

$$732.4 \times 10^3 = (7.324 \times 10^2) \times 10^3 = 7.324 \times 10^5 \quad \text{In scientific notation}$$

$$732.4 \times 10^3 = 732{,}400 \quad \text{In decimal notation}$$

EXAMPLE 6 Use scientific notation in solving this problem: $\dfrac{30{,}000{,}000 \times 0.0005}{0.000\,000\,6 \times 80{,}000}$.

SOLUTION

$$\frac{30{,}000{,}000 \times 0.000\,5}{0.000\,000\,6 \times 80{,}000} = \frac{(3 \times 10^7) \times (5 \times 10^{-4})}{(6 \times 10^{-7}) \times (8 \times 10^4)}$$ Writing each factor in scientific notation

$$= \frac{(\overset{1}{\cancel{3}} \times 5) \times (10^7 \times 10^{-4})}{(\underset{2}{\cancel{6}} \times 8) \times (10^{-7} \times 10^4)}$$ Collecting the powers of 10

$$= \frac{5}{16} \times \frac{10^3}{10^{-3}}$$ Simplifying what's inside the parentheses *and* rewriting the expression as a product of two fractions

$$= 0.3125 \times (10^3 \times 10^3)$$ Writing 5/16 in decimal form and writing $1/10^{-3}$ as 10^3

Converting 0.3125 to scientific notation

$$= (3.125 \times 10^{-1}) \times 10^6$$

$$= 3.125 \times 10^5$$ Combining powers of 10

$$= 312{,}500$$ Decimal notation

Calculators and Scientific Notation

Scientific calculators usually express very large or very small answers in scientific notation (see Examples 7 and 8). On the calculator, however, numbers in scientific notation are displayed in a different (and possibly misleading) way. The calculator display [2.45 ^{04}] does *not* mean 2.45 to the fourth power. It means 2.45×10^4. (The calculator displays [2.45 04] and [2.45 E 4] also mean 2.45×10^4.)

EXAMPLE 7 Use a calculator to find 600,000 × 300,000.

SOLUTION The display may be [1.8 11], [1.8 ^{11}], or [1.8 E 11]. These displays all mean 1.8×10^{11}, *not* 1.8^{11}. Observe that whereas $1.8 \times 10^{11} =$ 180,000,000,000, 1.8^{11} is approximately 642.684 100 7.

EXAMPLE 8 Use a calculator to find 0.000 06 ÷ 500.

SOLUTION The display may be [1.2 −07], [1.2 $^{-07}$], or [1.2 E −07]. These displays all mean $1.2 \times 10^{-7} = 0.000\,000\,12$.

Some calculators will give the *wrong answer* if very large (or very small) numbers are entered *unless* the numbers are entered in scientific notation (see Example 9). The keystrokes used to indicate operations vary, of course, with the brand of calculator. To indicate in Example 9 that we're *entering* a number in scientific notation, we will use the symbol [EE] for scientific and graphics calculators and [E] for RPN calculators. (That key on your calculator might be marked [EXP].) You must consult your calculator manual for details about *your* calculator. In Example 9, we do not show the keystrokes to use for entering the numbers themselves.

EXAMPLE 9 Using a calculator, find 4,300,000,000,000 × 0.000 000 053, entering both numbers in scientific notation.

SOLUTION Converting the numbers to scientific notation, we have

$$4{,}300{,}000{,}000{,}000 = 4.300\,000\,000\,000_{\wedge} \times 10^{12} = 4.3 \times 10^{12}$$

$$.000\,000\,053 = {}_{\wedge}000\,000\,05.3 \times 10^{-8} = 5.3 \times 10^{-8}$$

Scientific (algebraic) 4.3 [EE] 12 [×] 5.3 [EE] 8 [+/−] [=]

Graphics 4.3 [EE] 12 [×] 5.3 [EE] [(−)] [8] [ENTER]

RPN Scientific 4.3 [E] 12 [ENTER] 5.3 [E] 8 [+/−] [×]

The display is [227900] or [227900.0000], and so

$$4{,}300{,}000{,}000{,}000 \times 0.000\,000\,053 = 227{,}900$$

A Word of Caution In Example 9, watch your calculator display carefully if you *start* entering the problem as follows:

[4] [3] [0] [0] [0] [0] [0] [0] [0] [0] [0] [0] [0]

Your calculator may *ignore* the last few zeros; if it does, you will probably get an incorrect answer.

Note: The even-numbered applied problems at the end of these exercise sets are not "matched" to the odd-numbered problems.

Exercises 2.4 Set I

In Exercises 1–12, write each number in scientific notation.

1. 28.56 **2.** 375.4

3. 0.061 84 **4.** 0.003 056

5. 78,000 **6.** 1,400

7. 0.2006 **8.** 0.000 095

9. 0.362×10^{-2} **10.** 0.6314×10^{-3}

11. 245.2×10^{-5} **12.** 31.7×10^{-4}

In Exercises 13–16, perform the calculations by first converting all the factors to scientific notation.

13. $\dfrac{0.000\,06 \times 800{,}000{,}000}{50{,}000{,}000 \times 0.0003}$

14. $\dfrac{65{,}000{,}000{,}000 \times 0.0007}{0.0013 \times 1{,}400{,}000}$

15. $\dfrac{0.000\,006\,3 \times 5{,}500{,}000}{350{,}000 \times 0.000\,033}$

16. $\dfrac{55{,}000{,}000 \times 0.001\,17}{0.000\,003 \times 15{,}000}$

In Exercises 17–20, perform the indicated operations with a calculator, and express each answer in scientific notation.

17. 560,000 × 23,000 **18.** 0.000 06 ÷ 20,000

19. $\sqrt{0.000\,002\,56}$ **20.** $\sqrt{0.000\,000\,81}$

In Exercises 21–24, use a scientific calculator.

21. By definition (in chemistry and physics), 1 mole of any substance contains approximately 6.02×10^{23} molecules. How many molecules will 600 moles of oxygen contain? (Express the answer in scientific notation.)

22. The indebtedness of one of the developing nations is $120,000,000,000. If the annual (simple) interest rate is 7.8%, what is the interest on this debt for one year?

23. If the spacecraft Voyager traveled 4,400,000,000 mi in 12 years, what was its average speed in miles per hour? (Use "rate = distance/time." Assume that each year has 365 days. Round off the answer to the nearest mile per hour.)

24. The speed of light is about 186,000 mi per second. How many miles does light travel in one day? (Use "rate × time = distance." Express the answer in scientific notation and in decimal notation.)

Writing Problems

Express the answers in your own words and in complete sentences.

1. Explain why scientific notation is useful.
2. Explain why 345.67×10^4 is not correctly written in scientific notation.
3. Explain what the calculator display [3.12 07] means.

Exercises 2.4 Set II

In Exercises 1–12, write each number in scientific notation.

1. 50.48
2. 0.0878
3. 4,500.9
4. 0.000 505
5. 289.3
6. 2,478,000
7. 0.006 12
8. 0.000 01
9. 63.7×10^4
10. 0.0357×10^{-5}
11. 0.492×10^{-3}
12. 0.251×10^4

In Exercises 13–16, perform the calculations by first converting all the factors to scientific notation.

13. $\dfrac{0.000\,28 \times 40{,}000{,}000}{5{,}000{,}000 \times 0.0007}$

14. $\dfrac{510{,}000{,}000 \times 0.000\,05}{0.000\,34 \times 3{,}000}$

15. $\dfrac{0.000\,024 \times 5{,}500{,}000}{320{,}000 \times 0.000\,033}$

16. $\dfrac{144}{0.000\,06 \times 800 \times 0.000\,15}$

In Exercises 17–20, perform the indicated operations with a calculator, and express each answer in scientific notation.

17. $340{,}000 \times 680{,}000$
18. $0.000\,25 \div 500$
19. $\sqrt{0.000\,002\,89}$
20. $\sqrt{0.000\,000\,01}$

In Exercises 21–24, use a scientific calculator.

21. The indebtedness of one of the developing nations is \$52,000,000,000. If the annual (simple) interest rate is 7.4%, what is the interest on this debt for two years?
22. The planet Neptune is about 2,700,000,000 mi from the earth. What is this distance in kilometers? (1 mi $\approx$ 1.61 km.)
23. A unit used in measuring the length of light waves is the Ångström. One micron is a millionth of a meter, and one Ångström is one ten-thousandth of a micron. One Ångström is what part of a meter? (Express the answer in scientific notation.)
24. The speed of light is about 186,000 mi per second. How long will it take light to reach us from the planet Neptune? (Use "time = distance/rate." Assume that Neptune is about 2,700,000,000 mi from the earth. Round off the answer to the nearest hour.)

2.5 Evaluating and Substituting in Algebraic Expressions

Evaluating Algebraic Expressions

The process of substituting a numerical value for a variable in an algebraic expression (and then simplifying) is called *evaluating* the expression. If the number being substituted is a *negative* number, it is almost always necessary to enclose the number within parentheses (see Examples 1 and 2).

EXAMPLE 1 Find the value of $3x$ if $x = -2$.

SOLUTION If $x = -2$, $3x = 3(-2) = -6$.

Note In Example 1, writing $3 - 2$ is *incorrect*; $3x$ means that we are to find 3 *times x*, and $3 - 2$ means that we are to *subtract* 2 from 3. The correct notation is $3(-2)$.

EXAMPLE 2 Find the value of x^2 if $x = -4$.

SOLUTION If $x = -4$, $x^2 = (-4)^2 = 16$.

Note In Example 2, writing -4^2 is *incorrect*, since -4^2 means $-(4)^2 = -(16)$. We want -4 to be squared; therefore, the correct notation is $(-4)^2$.

EXAMPLE 3 Find the value of $3x^2 - 5y^3$ if $x = -4$ and $y = -2$.

SOLUTION

$$
\begin{aligned}
& 3x^2 - 5y^3 \\
&= 3(-4)^2 - 5(-2)^3 \quad \text{Substituting} \\
&= 3(16) - 5(-8) \\
&= 48 + 40 \\
&= 88
\end{aligned}
$$

EXAMPLE 4 Find the value of $2a - [b - (3x - 4y)]$ if $a = -3$, $b = 4$, $x = -5$, and $y = -2$.

SOLUTION

$$
\begin{aligned}
& 2a - [b - (3x - 4y)] \\
&= 2(-3) - \{(4) - [3(-5) - 4(-2)]\} \quad \text{Substituting} \\
&= 2(-3) - \{4 - [-15 + 8]\} \\
&= 2(-3) - \{4 - [-7]\} \\
&= 2(-3) - \{4 + [+7]\} \\
&= 2(-3) - \{11\} \\
&= -6 - 11 \\
&= -17
\end{aligned}
$$

EXAMPLE 5 Evaluate $b - \sqrt{b^2 - 4ac}$ when $a = 3$, $b = -7$, and $c = 2$.

SOLUTION

This bar is a grouping symbol for $b^2 - 4ac$

$$
\begin{aligned}
& b - \sqrt{b^2 - 4ac} \\
&= (-7) - \sqrt{(-7)^2 - 4(3)(2)} \\
&= (-7) - \sqrt{49 - 24} \\
&= (-7) - \sqrt{25} \\
&= (-7) - 5 = -12
\end{aligned}
$$

A Word of Caution In Example 5, *two* errors are made if we write

$$b - \sqrt{b^2 - 4ac} = -7 - \sqrt{-7^2 - 4(3)(2)} = -7 - \sqrt{49 - 24}$$

It is incorrect to omit the parentheses around -7 here, and $-7^2 = -49$, not 49.

Substituting in Algebraic Expressions

It is sometimes desirable to substitute one expression containing variables for another. If the expression being substituted contains more than one term, *we must be sure to enclose that expression within parentheses* (see Example 7).

EXAMPLE 6 Substitute x for $2a^2 - 3a$ in the expression $(2a^2 - 3a)^2 + 5(2a^2 - 3a) - 15$.

SOLUTION When we let $x = 2a^2 - 3a$ in the expression $(2a^2 - 3a)^2 + 5(2a^2 - 3a) - 15$, we get $x^2 + 5x - 15$.

EXAMPLE 7 Substitute $x^3 - 7x$ for b in the expression $2b^2 - 5$.

SOLUTION When we let $b = x^3 - 7x$ in the expression $2b^2 - 5$, we get $2(x^3 - 7x)^2 - 5$.* Notice that it is necessary to put $x^3 - 7x$ *inside parentheses* when we make the substitution.

Exercises 2.5
Set I

In Exercises 1–18, evaluate each expression, given that $a = \frac{1}{3}$, $b = -5$, $c = -1$, $x = 5$, $y = -6$, $D = 0$, $E = -1$, $F = 5$, $G = -15$, $H = -4$, and $J = 2$.

1. $2y^2 + 3x$
2. $c^3 - y$
3. $b - 12a^2$
4. $y - 24a^2$
5. $b^2 - 4xy$
6. $y^2 - 5cx$
7. $(x + y)^2$
8. $(b + c)^2$
9. $x^2 + 2xy + y^2$
10. $b^2 + 2bc + c^2$
11. $x^2 + y^2$
12. $b^2 + c^2$
13. $\dfrac{3D}{F + G}$
14. $\dfrac{5D}{G + H}$
15. $2E - [F - (D - 5G)]$
16. $3G - [D - (F - 2H)]$
17. $-E - \sqrt{E^2 - 4HF}$
18. $-E + \sqrt{E^2 - 4GJ}$

In Exercises 19–22, make the indicated substitutions.

19. Substitute b for $x^3 - 7x$ in the expression $5(x^3 - 7x)^2 - 4(x^3 - 7x) + 8$.
20. Substitute c for $a^5 + a$ in the expression $4(a^5 + a) + 3$.
21. Substitute $x^2 - 4x$ for a in the expression $2a^2 - 3a + 7$. (Do not attempt to simplify the result.)
22. Substitute $y^4 + 2$ for b in the expression $b^2 - 2b$. (Do not attempt to simplify the result.)

Writing Problems

Express the answer in your own words and in complete sentences.

1. Describe what is wrong with the following statement:

   If we substitute $x + 3$ for b in the expression $b^2 - 4$, we get $x + 3^2 - 4$.

*We have not yet discussed simplifying expressions such as this.

Exercises 2.5
Set II

In Exercises 1–18, evaluate each expression, given that $a = \frac{3}{4}$, $b = 5$, $c = -4$, $d = 3$, $e = -2$, and $f = -8$.

1. $4c^2 + 5e$
2. $c^2 + e$
3. $b - 14a^2$
4. $(b + c)^2$
5. $c^2 - 4bc$
6. $b^2 + c^2$
7. $(b + e)^2$
8. $b^2 + 2bc + c^2$
9. $b^2 + 2be + e^2$
10. $(f + e)^2$
11. $b^2 + e^2$
12. $f^2 + 2ef + e^2$
13. $\dfrac{5c}{e + f}$
14. $f^2 + e^2$
15. $2b - [f - (c - b)]$
16. $c - \{b - (d - f) - e\}$
17. $-e + \sqrt{e^2 - 4df}$
18. $(b + c + d)^2$

In Exercises 19–22, make the indicated substitutions.

19. Substitute d for $x^2 + 2x$ in the expression $3(x^2 + 2x)^2 + 4(x^2 + 2x) - 5$.
20. Substitute a for $x^3 - x$ in the expression $7(x^3 - x)^2 - (x^3 - x) - 3$.
21. Substitute $z^2 + 3z - 5$ for c in the expression $3c^4 - 2c + 4$. (Do not attempt to simplify the result.)
22. Substitute $a^2 + 5a$ for x in the expression $-2x^3 - 4x^2 + 3x - 2$. (Do not attempt to simplify the result.)

2.6 Using Formulas

One reason for studying algebra is that algebra allows us to use *formulas*. Students encounter formulas in many college courses, as well as in real-life situations. In the examples and exercises in this section, we list the subject areas in which the given formulas are used.

A formula is used, or evaluated, the same way we evaluate any expression that contains numbers and variables.

EXAMPLE 1 Given the formula $A = \frac{1}{2}h(B + b)$, find A when $h = 5$, $B = 3$, and $b = 7$.

SOLUTION

$$A = \tfrac{1}{2}h(B + b) \qquad \text{Geometry}$$
$$A = \tfrac{1}{2}(5)(3 + 7)$$
$$A = \tfrac{1}{2}(5)(10) = 25$$

EXAMPLE 2  Given the formula $T = \pi\sqrt{\dfrac{L}{g}}$, find T when $\pi \approx 3.14$, $L = 96$, and $g = 32$. Round off the answer to two decimal places.

SOLUTION

$$T = \pi\sqrt{\frac{L}{g}} \qquad \text{Physics}$$
$$T \approx (3.14)\sqrt{\frac{96}{32}} = (3.14)\sqrt{3} \approx (3.14)(1.732)$$
$$T \approx 5.44 \qquad \text{Rounded off to 2 decimal places}$$

Formulas will be discussed further in a later chapter.

Exercises 2.6 Set I

Use each formula, substituting the given values for the variables. Answers that are not exact should be rounded off to two decimal places.

For Exercises 1 and 2, find q, using this formula from nursing: $q = \frac{DQ}{H}$.

1. $D = 5, H = 30, Q = 420$
2. $D = 25, H = 90, Q = 450$

For Exercises 3 and 4, find A, using this formula from business: $A = P(1 + rt)$.

3. $P = 500, r = 0.09, t = 2.5$
4. $P = 400, r = 0.07, t = 3.5$

For Exercises 5 and 6, find A, using this formula from business: $A = P(1 + i)^n$.

5. $P = 600, i = 0.085, n = 2$
6. $P = 700, i = 0.075, n = 2$

For Exercises 7 and 8, find C, using this formula from chemistry: $C = \frac{5}{9}(F - 32)$.

7. $F = -10$
8. $F = -7$

For Exercises 9 and 10, find s, using this formula from physics: $s = \frac{1}{2}gt^2$.

9. $g = 32, t = 8\frac{1}{2}$
10. $g = 32, t = 4\frac{3}{4}$

For Exercises 11 and 12, find Z, using this formula from physics: $Z = \frac{Rr}{R + r}$.

11. $R = 22, r = 8$
12. $R = 55, r = 25$

For Exercises 13 and 14, find S, using this formula from geometry: $S = 2\pi r^2 + 2\pi rh$. (Use $\pi \approx 3.14$.)

13. $r = 3, h = 20$
14. $r = 6, h = 10$

Writing Problems

Express the answer in your own words and in complete sentences.

1. Explain how to use a formula.

Exercises 2.6 Set II

Use each formula, substituting the given values for the variables. Answers that are not exact should be rounded off to two decimal places.

For Exercises 1 and 2, find q, using this formula from nursing: $q = \frac{DQ}{H}$.

1. $D = 15, H = 80, Q = 320$
2. $D = 20, H = 24, Q = 300$

For Exercises 3 and 4, find A, using this formula from business: $A = P(1 + rt)$.

3. $P = 450, r = 0.08, t = 2.5$
4. $P = 1{,}000, r = 0.06, t = 5.5$

For Exercises 5 and 6, find A, using this formula from business: $A = P(1 + i)^n$.

5. $P = 900, i = 0.095, n = 2$
6. $P = 400, i = 0.055, n = 3$

For Exercises 7 and 8, find C, using this formula from chemistry: $C = \frac{5}{9}(F - 32)$.

7. $F = 95$
8. $F = 14$

For Exercises 9 and 10, find s, using this formula from physics: $s = \frac{1}{2}gt^2$.

9. $g = 32, t = 5\frac{1}{2}$
10. $g = 32, t = 6\frac{1}{4}$

For Exercises 11 and 12, find Z, using this formula from physics: $Z = \frac{Rr}{R + r}$.

11. $R = 150, r = 25$
12. $R = 80, r = 35$

For Exercises 13 and 14, find S, using this formula from geometry: $S = 2\pi r^2 + 2\pi rh$. (Use $\pi \approx 3.14$.)

13. $r = 3.6, h = 5.1$
14. $r = 6.5, h = 1.2$

2.7 Simplifying Square Roots: An Introduction

Perfect Squares We can call an expression a **perfect square** if any numerical coefficient is the square of an integer and if there are even exponents on *all* the variables. For example, $9x^2$ is a perfect square, as is $100a^6b^{10}$; however, $7y^4$ is *not* a perfect square, because 7 is not the square of an integer, and $25xy^6$ is *not* a perfect square, because the exponent on the variable x is not an even number.

We have already discussed the method of finding square roots of nonnegative numbers. In this section, we find square roots of algebraic expressions containing variables when the radicands are perfect squares that are also products.

Note In this section, we will assume that all variables represent nonnegative numbers; *because* we assume this, we will be able to write $\sqrt{x^2} = x$. In a later chapter, we explain why we must write $\sqrt{x^2} = |x|$ if there's a possibility that x might be negative.

When we simplify the square roots *given in this section*, the radical signs will always be removed completely, because we will be simplifying square roots in which the radicand is a perfect square. To simplify such square roots, we can use the procedure in the following box:

Finding the principal square root of a perfect square

1. Find the principal square root of the numerical coefficient by inspection or by guessing-and-checking.
2. Find the square root of each variable factor by dividing its exponent by 2 and dropping the radical sign.

Note The justification for this procedure is given in a later chapter, where we consider simplifying square roots in which the radicands are *not* perfect squares.

EXAMPLE 1

Examples of finding principal square roots:

a. $\sqrt{9x^2} = 3x$ Because $(3x)^2 = 9x^2$

b. $\sqrt{25x^2} = 5x$ Because $(5x)^2 = 25x^2$

c. $\sqrt{100a^6b^{10}} = 10a^{6/2}b^{10/2} = 10a^3b^5$ Because $(10a^3b^5)^2 = 100a^6b^{10}$

Exercises 2.7
Set I

Find the principal square root of each expression. Assume that all variables represent nonnegative numbers.

1. $\sqrt{4x^2}$
2. $\sqrt{9y^2}$
3. $\sqrt{m^4n^2}$
4. $\sqrt{u^{10}v^6}$
5. $\sqrt{25a^4b^2}$
6. $\sqrt{100b^4c^2}$
7. $\sqrt{x^{10}y^4}$
8. $\sqrt{x^{12}y^8}$
9. $\sqrt{100a^{10}y^2}$
10. $\sqrt{121a^{24}b^4}$
11. $\sqrt{81m^8n^{16}}$
12. $\sqrt{49c^{18}d^{10}}$

Writing Problems

Express the answer in your own words and in complete sentences.

1. Explain why $\sqrt{4x^2y^4} = 2xy^2$.

Exercises 2.7 Set II

Find the principal square root of each expression. Assume that all variables represent nonnegative numbers.

1. $\sqrt{100a^8}$
2. $\sqrt{49b^6}$
3. $\sqrt{36e^8f^2}$
4. $\sqrt{81h^{12}k^{14}}$
5. $\sqrt{9a^4b^2c^6}$
6. $\sqrt{144x^8y^2z^6}$
7. $\sqrt{121a^4b^8}$
8. $\sqrt{144x^6y^2}$
9. $\sqrt{169a^8b^6c^4}$
10. $\sqrt{4s^6t^6}$
11. $\sqrt{16u^2v^8}$
12. $\sqrt{x^{12}y^{16}}$

2.8 Removing Grouping Symbols

The common grouping symbols are parentheses ( ), brackets [ ], braces { }, and a bar —— (generally used with fractions and radicals).

An algebraic expression is not considered simplified unless all grouping symbols have been removed. In order to remove grouping symbols successfully, you must be able to distinguish between addition problems, subtraction problems, and multiplication problems. The problem $2z + (4x - y)$ is an *addition* problem, whereas the problem $2z(4x - y)$ is a *multiplication* problem. The problem $(5x + 3y)(-2z)$ is a *multiplication* problem, whereas the problem $(5x + 3y) - (2z)$ is a *subtraction* problem.

Removing Grouping Symbols in Addition Problems

If there are grouping symbols in an *addition* problem, we can often simply drop the grouping symbols (see Examples 1a and 1b); however, if the first term inside the grouping symbols has a *written* sign, we must also drop any addition symbol in front of the grouping symbols (see Example 1c).

EXAMPLE 1 Remove the grouping symbols.

a. $(3x - 5) + 6y$

SOLUTION This is an *addition* problem; we simply drop the parentheses:

$$(3x - 5) + 6y = 3x - 5 + 6y$$

b. $5z + (4y + 7)$

SOLUTION This is an *addition* problem; we simply drop the parentheses:

$$5z + (4y + 7) = 5z + 4y + 7$$

c. $5z + (-3y + 7)$

SOLUTION This is an *addition* problem; because the first term inside the parentheses has a *written* sign, we drop the parentheses *and* the addition symbol in front of them:

$$5z + (-3y + 7) = 5z - 3y + 7$$

Removing Grouping Symbols in Multiplication Problems

To remove grouping symbols in a multiplication problem *when there is a single term* inside each set of grouping symbols, we use the method discussed near the end of Section 2.3; for example, $(3x^2)(5xy^3) = 3 \cdot 5 \cdot x^2xy^3 = 15x^3y^3$.

To remove grouping symbols in a multiplication problem when at least one of the factors contains *more than* one term, we use the distributive property:

$$a(b + c) = ab + ac$$

The distributive property also has the forms $(b + c)a = ba + ca$ and $a(b - c) = ab - ac$.

To use the distributive property to remove grouping symbols, we multiply *each term* inside the grouping symbols by the factor that is outside them.

EXAMPLE 2 Examples of using the distributive property to remove parentheses:

a. $4x(x^2 - 2xy + y^2)$

$= 4x[x^2 + (-2xy) + y^2]$

$= (4x)(x^2) + (4x)(-2xy) + (4x)(y^2)$

$= 4x^3 - 8x^2y + 4xy^2$

or

$4x(x^2 - 2xy + y^2)$

$= (4x)(x^2) - (4x)(2xy) + (4x)(y^2)$

$= 4x^3 - 8x^2y + 4xy^2$

b. $(-2x^2 + xy^2)(-3xy)$

$= (-2x^2)(-3xy) + (xy^2)(-3xy)$

$= 6x^3y - 3x^2y^3$

Removing Grouping Symbols in Subtraction Problems

The Additive Inverse of an Algebraic Expression We can find the **additive inverse of an algebraic expression** by multiplying that expression by -1. For example, the additive inverse of $12x^3$ is $-1(12x^3)$, or $-12x^3$; the additive inverse of $(8y + 2z - 5w)$ is

$$-1(8y + 2z - 5w) = (-1)(8y) + (-1)(2z) + (-1)(-5w) = -8y - 2z + 5w$$

To remove grouping symbols in a subtraction problem, we use the definition of subtraction:

$$a - b = a + (-b)$$

That is, we *add* the additive inverse of the expression being subtracted (the subtrahend) to the expression we're subtracting from (the minuend); we actually change the subtraction problem to an addition problem. See Example 3.

EXAMPLE 3 Examples of removing the grouping symbols in subtraction problems:

2z is being subtracted

Adding −2z, the additive inverse of 2z, to the minuend

a. $(5x + 3y) - (2z) = (5x + 3y) + (-2z) = 5x + 3y - 2z$

Adding the additive inverse of the subtrahend to the minuend

b. $3x - (8y + 2z - 5w) = 3x + (-1)(8y + 2z - 5w)$

$= 3x + (-8y - 2z + 5w)$

$= 3x - 8y - 2z + 5w$

Adding the additive inverse of the subtrahend to the minuend

c. $2x - (-8 - 6z + w) = 2x + (-1)(-8 - 6z + w)$

$= 2x + (8 + 6z - w)$

$= 2x + 8 + 6z - w$

It is not necessary to show all the steps that we showed in Example 3.

A Word of Caution Be sure to distinguish between problems such as $(5x + 3y) - (2z)$ and $(5x + 3y)(-2z)$. In $(5x + 3y) - (2z)$, the $2z$ is being *subtracted from* $5x + 3y$, whereas $(5x + 3y)(-2z)$ is a *multiplication problem.*

To remove grouping symbols that are preceded by a negative sign, we can insert a 1 *between* the negative sign and the left grouping symbol and then use the distributive property. (Recall that multiplying an expression by 1 does not change the value of the expression, and inserting a 1 between the negative sign and the left grouping symbol is equivalent to multiplying the expression by 1.) For example,

$$-(2x + 3y - z) = -1(2x + 3y - z) = -2x - 3y + z$$

If grouping symbols occur *within* other grouping symbols, we follow the rules for order of operations and remove the *innermost* grouping symbols first (see Example 4).

EXAMPLE 4 Remove the grouping symbols in $3 + 2[a - 5(x - 4y)]$.

SOLUTION

Changing the subtractions to additions

$3 + 2[a - 5(x - 4y)] = 3 + 2\{a + (-5)[x + (-4y)]\}$

$= 3 + 2\{a + (-5x) + (20y)\}$ Using the distributive property on $(-5)[x + (-4y)]$

$= 3 + 2a - 10x + 40y$ Using the distributive property on $2\{a + (-5x) + (20y)\}$

A Word of Caution The following are some common errors made in removing grouping symbols:

Correct method	*Common error*	
$-(x - 2y) = -x + 2y$	$-(x - 2y) = -x - 2y$	$-1(-2y) = +2y$, not $-2y$
$6(y - 3) = 6y - 18$	$6(y - 3) = 6y - 3$	-3 was not multiplied by the 6
$3 + 2(x + y) = 3 + 2x + 2y$	$3 + 2(x + y) = 5(x + y)$	Multiplication must be done *before* addition

Exercises 2.8 Set I

Remove the grouping symbols in each expression.

1. $10 + (4x - y)$
2. $8 + (3a - b)$
3. $(2x + 7y) + (3z - 6)$
4. $(5z - 3w) + (4x + 2y)$
5. $3a(6 + x)$
6. $5b(7 + y)$
7. $(x - 5)(-4)$
8. $(y - 2)(-5)$
9. $-3(x - 2y + 2)$
10. $-2(x - 3y + 4)$
11. $x(xy - 3)$
12. $a(ab - 4)$
13. $3a(ab - 2a^2)$
14. $4x(3x - 2y^2)$
15. $(3x^3 - 2x^2y + y^3)(-2xy)$
16. $(4z^3 - z^2y - y^3)(-2yz)$
17. $(-2ab)(3a^2b)(6abc^3)$
18. $(5x^2y)(-2xy^3)(3xyz^2)$
19. $4xy^2(3x^3y^2 - 2x^2y^3 + 5xy^4)$
20. $-2x^2(5x^4y - 2x^3y^2 - 3xy^3)$
21. $(3mn^2)(-2m^2n)(5m^2 - n^2)$
22. $(6a^2b)(-3ab^2)(2a^2 - b^2)$
23. $(2x^2)(-4xy^2z)(3xy - 2xz + 5yz)$
24. $(7a^2b)(3a - 2b - c)(ab^2)$
25. $(7x + 3y) - 2x^2$
26. $(3a - b) - 4c$
27. $-(3x - 2y)$
28. $-(7z + 2w)$
29. $7 - (-4R - S)$
30. $9 - (-3m - n)$
31. $6 - 2(a - 3b)$
32. $12 - 3(2R - S)$
33. $3 - 2x(x - 4y)$
34. $2 - 5x(2x - 3y)$
35. $-(x - y) + (2 - a)$
36. $-(a - b) + (x - 3)$
37. $(a - b)(2) - 6$
38. $(x - y)(3) - 5$
39. $x - [a + (y - b)]$
40. $y - [m + (x - n)]$
41. $5 - 3[a - 4(2x - y)]$
42. $7 - 5[x - 3(2a - b)]$
43. $2 - [a - (b - c)]$
44. $5 - [x - (y + z)]$
45. $9 - 2[-3a - 4(2x - y)]$
46. $P - \{x - [y - (4 - z)]\}$

Writing Problems

Express the answers in your own words and in complete sentences.

1. Explain the difference between the expressions $(3x^2 - 4y)(-2x^2)$ and $(3x^2 - 4y) - (2x^2)$.
2. Explain the difference between the expressions $(3x^2 - 4y)(-2x^2)$ and $(3x^2)(-4y)(-2x^2)$.
3. Explain why $4x - (3y + 4z) \neq 4x - 3y + 4z$.
4. Explain why $3x(2x - 5) \neq 6x^2 - 5$.

Exercises 2.8 Set II

Remove the grouping symbols in each expression.

1. $5 + (2x - y)$
2. $7 - (a - 3b)$
3. $(9s + 12t) + (6u - v)$
4. $(7 - 8w) - (4t + 9u)$
5. $2R(R + S)$
6. $7x(x - 5)$
7. $(x - 7)(-3)$
8. $(x - 7) - y$
9. $-4(x - 5y + 3)$
10. $(a + 7b - c) - 3$
11. $u(uv - 5)$
12. $(a + 7b - c)(-3)$
13. $4xy^2z(3x^2 - xy)$
14. $2a(a^2 + 3b - c^3)$
15. $(2x^3 - 5xy + y^2)(-3xy)$
16. $(3x^3y - 4x + 2z) - 3y$
17. $(5xy^2z^2)(-3yz)(2xz^2)$
18. $(-2xy^3)(4xz)(-2yz^4)$
19. $2x^2y(-4xy^2z + 3xz - y^2z)$
20. $3a^2b(-2ab^2 + 8ab - 1)$
21. $(4xy^2)(-2xy)(5x^2 - xy)$
22. $9x^2y(2y^2)(3x^3 - y^2)$
23. $4u^2v(-3uv^2)(u - 3v - w)$
24. $2x^3z(-3xz)(8x^2 + 2z^2)$

25. $(8c^3 - 3bc) - b^2$
26. $(8c - 3bc)(-b^2)$
27. $-(12r - 7v)$
28. $-(13 + 5s)$
29. $4 - (-3a - b)$
30. $8 - 5(z - w)$
31. $7 - 2(6R - S)$
32. $2 - 2(3x + y)$
33. $4 - 6x(3x - 2y)$
34. $(9 - z) - x$
35. $-(c - d) + (a - b)$
36. $(a + b) - (c - d)$
37. $(c - d)(5) - 4$
38. $(a + b)(3) - 2$
39. $z - [a + (b - c)]$
40. $s - \{t - (u - v) - w\}$
41. $6 - 4[x - 3(a - 2b)]$
42. $5 - 2(3x - [y - a])$
43. $3 - [x - (y - z)]$
44. $(x + y)(2) - 3z$
45. $a - \{b - [c - (2 - d)]\}$
46. $x - \{y - [z + w]\}$

2.9 Combining Like Terms; Simplifying Algebraic Expressions

A *term* with exponents is considered simplified when each different base appears only once and when the exponent on each base is a single *positive* integer (this means that we cannot leave any negative or zero exponents in the term). It is customary to write any numerical coefficient first and to list the variables in each term in alphabetical order.

Like Terms **Like terms** are terms that, *in simplified form* (with the variables listed in alphabetical order), would have *identical variable parts*; only the numerical coefficients can be different. Also, terms consisting only of *numbers* are like terms with each other.

EXAMPLE 1 Examples of like terms:

a. $5x^2$, $8xx$, $\frac{3}{4}x^2$, and $2.3x^2$ are like terms. They are called *x^2-terms*. (Note that $8xx = 8x^2$.)

b. 9, -2, $\frac{1}{3}$, and 1.7 are like terms. They are called *constant terms*.

c. $34x^2y$, $8yx^2$, xyx, $2.8x^2y$, and $\frac{1}{5}x^2y$ are like terms. ($yx^2 = x^2y$ and $xyx = x^2y$; therefore, the terms are like terms)

Unlike Terms **Unlike terms** are terms that, *in simplified form* (with the variables listed in alphabetical order), would *not* have identical variable parts.

EXAMPLE 2 Examples of unlike terms:

a. 3 and $3x$ are unlike terms. One is a constant term, and the other is an x-term

b. $5x^2$ and $5x$ are unlike terms. The variable parts, x^2 and x, are different

c. $7y$ and $7Y$ are unlike terms. It is understood that the lowercase y and the capital Y represent different numbers

d. $6x$ and $6y$ are unlike terms. The variable parts, x and y, are different

Combining Like Terms

Combining like terms is rewriting a sum or difference of two or more *like terms* as a single term. For example, $3x$ and $5x$ are like terms, and, according to the distributive property, $(3 + 5)x = 3x + 5x$; therefore, it must also be true that $3x + 5x = (3 + 5)x = 8x$. When we rewrite $3x + 5x$ as $8x$, we are *combining like terms.*

In combining like terms, we generally think of all problems as *addition problems*, even when some of the terms are preceded by negative signs. For example, we interpret $7x^2y - 15x^2y$ as $7x^2y + (-15x^2y)$; then, using the distributive property "in reverse," we have a sum of $[7 + (-15)]x^2y$, or $-8x^2y$.

EXAMPLE 3 Examples of using the distributive property to combine like terms:

$x^2y = 1x^2y$

a. $5x^2y - 8x^2y + x^2y = [5 + (-8) + 1]x^2y = -2x^2y$

$ba = ab$; therefore, the terms are all like terms

b. $4ab + ba - 6ab = [4 + 1 + (-6)]ab = -1ab = -ab$

We *cannot* combine unlike terms. For example, we cannot write $5y + 7x$ as a single term ($5y$ and $7x$ are *unlike terms*), nor can we write $19 + 6x$ as a single term (19 and $6x$ are *unlike terms*).

A Word of Caution A common error is to think that

$$19 + 6x = 25x$$

This is not correct, as we can see if we substitute any number except 1 for x. If we let $x = 2$, we have $19 + 6(2) = 19 + 12 = 31$, but $25(2) = 50$. Since $31 \neq 50$, we have verified that $19 + 6x \neq 25x$.

Note Although it is not possible to *combine* (add) unlike terms, it *is* possible to *multiply* unlike terms together; for example, $19 + 6x$ cannot be simplified, but $(19)(6x) = 114x$. Similarly, $3x - 3x^2$ cannot be simplified, but $(3x)(-3x^2) = -9x^3$.

In practice, we usually do not show the step in which we use the distributive property; instead, we follow the procedure in the following box:

Combining (adding) like terms

> Add the numerical coefficients *of the like terms*; the number so obtained is the numerical coefficient of the sum. The variable part of the sum is the same as the variable part of any *one* of the like terms.

EXAMPLE 4 Examples of combining like terms:

a. $5z^3 - 8z^3 - 12z^3 + z^3 = -14z^3$

b. $7s^2t^3 - 12s^2t^3 + s^2t^3 = -4s^2t^3$

When we combine like terms, we often change the grouping and/or the order in which the terms appear. (We call this process *collecting* like terms.) The commutative and associative properties of addition guarantee that when we do this, the sum remains unchanged.

EXAMPLE 5 Examples of collecting and combining like terms:

a. $12a - 7b - 9a + 4b$

$= 12a + (-9a) + (-7b) + 4b$ — Collecting like terms, using the commutative and associative properties of addition

$= 3a + (-3b)$ — Combining like terms

$= 3a - 3b$

b. $9 + 7x - 2y - 11x - 4y$
$= 9 + 7x + (-11x) + (-2y) + (-4y)$ Collecting like terms
$= 9 + (-4x) + (-6y)$ Combining like terms
$= 9 - 4x - 6y$

Note It is not necessary to show any of the intermediate steps that we showed in Example 5. You can simply write the final answer.

Simplifying Algebraic Expressions

Simplifying an algebraic expression

1. Remove all grouping symbols.
2. Express each *term* in simplest form. (In each term, each different base should appear only once, the exponent on each base should be a single *positive* integer, and the numerical coefficient should be written first. It is customary to list the variables in alphabetical order.)
3. Simplify all radicals.*
4. Collect and combine like terms.

*In this section, all radical signs will be completely removed, since we will be simplifying square roots of perfect squares.

Note We will add to this list in later chapters.

EXAMPLE 6 Examples of simplifying algebraic expressions:

a. $x(x^2 + xy + y^2) - y(x^2 + xy + y^2)$
$= x^3 + x^2y + xy^2 - x^2y - xy^2 - y^3$ Using the distributive property
$= x^3 + x^2y - x^2y + xy^2 - xy^2 - y^3$ Collecting like terms
$= x^3 + 0 + 0 - y^3$ Combining like terms
$= x^3 - y^3$

b. $\sqrt{16} - \{8[-5(3x - 2) + 13] - 11x\}$
$= 4 - \{8[-15x + 10 + 13] - 11x\}$ Using the distributive property inside the brackets
$= 4 - \{8[-15x + 23] - 11x\}$ Combining like terms inside the brackets
$= 4 - \{-120x + 184 - 11x\}$ Using the distributive property inside the braces
$= 4 - \{-131x + 184\}$ Combining like terms inside the braces
$= 4 + 131x - 184$ Removing grouping symbols
$= 131x - 180$ Combining like terms

c. $\sqrt{9x^2} + x^{-1}x^2 - 4x^0(12x) - (3x)^0$ (Assume $x \geq 0$)
$= 3x + x^1 - 4(1)(12x) - 1$
$= 3x + x - 48x - 1$ Simplifying each term
$= -44x - 1$ Combining like terms

In Example 7, we need to combine like terms in order to use the product rule and the quotient rule.

EXAMPLE 7 Examples of simplifying algebraic expressions (assume $z \neq 0$):

a. $2^{3x} \cdot 2^{5x} = 2^{3x+5x} = 2^{8x}$ **b.** $\left(\dfrac{z^{2n}z^{-5n}}{z^{3n}}\right)^2 = (z^{2n+(-5n)-3n})^2 = (z^{-6n})^2 = z^{-12n} = \dfrac{1}{z^{12n}}$

EXAMPLE 8 In $-3(a - 2b) + 2(-a - 3b)$, substitute $x + y$ for a and $x - y$ for b and then simplify.

SOLUTION

$-3(a - 2b) + 2(-a - 3b)$	
$= -3([x + y] - 2[x - y]) + 2(-[x + y] - 3[x - y])$	Substituting
$= -3(x + y - 2x + 2y) + 2(-x - y - 3x + 3y)$	Removing the inner grouping symbols
$= -3(-x + 3y) + 2(-4x + 2y)$	Combining like terms inside the parentheses
$= 3x - 9y - 8x + 4y$	Using the distributive property
$= -5x - 5y$	Combining like terms

Exercises 2.9
Set I

In Exercises 1–58, express each algebraic expression in simplest form. Assume that all variables represent positive numbers.

1. $5x - 8x + x$
2. $3a - 5a + a$
3. $8x^2y - 2x^2y$
4. $10ab^2 - 3ab^2$
5. $6xy^2 + 8x^2y$
6. $5a^2b - 4ab^2$
7. $2xy - 5yx + xy$
8. $8mn - 7nm + 3mn$
9. $5xyz^2 - 2(xyz)^2$
10. $3(abc)^3 - 4abc^3$
11. $5xyz^2 - xyz^2 - 4xyz^2$
12. $7a^2bc - a^2bc - 5a^2bc$
13. $7x^2y - 2xy^2 - 4x^2y$
14. $4xy^2 - 5x^2y - 2y^2x$
15. $3ab - a + b - ab$
16. $5xy - x - y + xy$
17. $2x^3 - 2x^2 + 3x - 5x$
18. $5y^2 - 3y^3 + 2y - 4y$
19. $4x - 3y + 7 - 2x + 4 - 6y$
20. $3b - 5a - 9 - 2a + 4 - 5b$
21. $a^2b - 5ab + 7ab^2 - 3a^2b + 4ab$
22. $xy^2 + y - 5x^2y + 3xy^2 + x^2y$
23. $2h(3h^2 - k) - k(h - 3k^3)$
24. $4x(2y^2 - 3x) - x(2x - 3y^2)$
25. $(3x - 4) - 5x$
26. $(5x - 7) - 8x$
27. $(3x - 4)(-5x)$
28. $(5x - 7)(-8x)$
29. $2 + 3x$
30. $5 + 8y$
31. $3x - [5y - (2x - 4y)]$
32. $2x - [7y - (3x - 2y)]$
33. $-10[-2(3x - 5) + 17] - 4x$
34. $-20[-3(2x - 4) + 20] - 5x$
35. $8 - 2(x - [y - 3x])$
36. $9 - 4(u - [t - 2u])$
37. $2x(4 + 5x) - \sqrt{16x^2}$
38. $5y(2 + 5y) - \sqrt{36y^2}$
39. $(3u - v) - \{2u - (10 - v) - 20\} - \sqrt{64v^2}$
40. $(5x - y) - \{3x - (8 - y) - 15\} - \sqrt{49x^2}$
41. $50 - \{-2t - [5t - (6 - 2t)]\} + 7^0$
42. $24 - \{-4x - [2x - (3 - 5x)]\} + 3^0$
43. $100v - 3\{-4[-2(-4 - v) - 5v]\}$
44. $60z - 4\{-3[-4(-2 - z) - 3z]\}$
45. $w^2(w^2 - 4) + 4(w^2 - 4)$
46. $x^2(x^2 - 9) + 9(x^2 - 9)$
47. $3x(5 \cdot 4x^2)(2x^3)$
48. $2y(2 \cdot 3y^3)(5y^4)$
49. $5X^{-4}X^6 + 3X^0$
50. $3Y^3Y^{-1} + 2Y^0$
51. $13^{3x} \cdot 13^{6x}$
52. $7^{9m} \cdot 7^{4m}$
53. $5^x \cdot 5^{3x}$
54. $6^{5x} \cdot 6^{8x}$
55. $\dfrac{3^{9x}}{3^{6x}}$
56. $\dfrac{5^{7t}}{5^{3t}}$
57. $\left(\dfrac{x^{3n}x^{-4n}}{x^{5n}}\right)^3$
58. $\left(\dfrac{w^{5m}w^{-2m}}{w^{6m}}\right)^2$

In Exercises 59–62, substitute $x + 2y$ for a and $3x - y$ for b and simplify.

59. $3a - 5b$ 60. $4a - 2b$ 61. $2(3a - b)$ 62. $3(5b - a)$

Writing Problems

Express the answers in your own words and in complete sentences.

1. Explain why $4 + 3x \neq 7x$.
2. Explain why $3R$ and $3r$ are not like terms.
3. Explain why $-5x^2y$ and $3xy^2$ are not like terms.
4. Explain why $4ab$ and $6ba$ are like terms.
5. Explain why $-8xyx$, $3yxx$, and $-12x^2y$ are like terms.
6. Explain what must be done to *simplify* an algebraic expression.

Exercises 2.9
Set II

In Exercises 1–58, express each algebraic expression in simplest form. Assume that all variables represent positive numbers.

1. $4y + y - 10y$
2. $10a - 16a + a$
3. $a^2b - 3a^2b$
4. $2x^3y - 9yx^3$
5. $4a^2b + 6ab^2$
6. $12x^3y + 8xy^3$
7. $5uv - 2vu + uv$
8. $ab + 3ab - 18ba$
9. $3(stu)^2 - 5stu^2$
10. $2 + 4xy$
11. $8ab^2c - ab^2c - 4ab^2c$
12. $6fg^2 - 2(fg)^2$
13. $5ab^2c - 7a^2bc - 2ab^2c$
14. $17x(yz)^2 - 12xy^2z^2$
15. $6st + s - t - st$
16. $8 - 2xy - 6yx$
17. $5R^3 - 2R + 3R^2 - R$
18. $9x + 2y - 3z$
19. $6x - 6 + 4y - 3 + 2x - 7y$
20. $4 + 2a - 3b - 7 - 6b - 2a$
21. $xy - 2xy^2 - 5x^2y - 3xy + 2x^2y - 4xy^2$
22. $st^2 - 3s + 2t - s^2t - 5s + 3t$
23. $3x(2x^2 - y) - y(x - 4y^2)$
24. $5a^2(4a - ab + b^2)$
25. $(4s - 7) - 2s$
26. $(2x + 3y) - 9x$
27. $(4y - 7)(-2y)$
28. $(2x + 3y)(-9x)$
29. $4 + 5z$
30. $(8 + 3x)y$
31. $2s - [4t - (3s - 5t)]$
32. $(8 + 3x) + y$
33. $-5[-2(2x - 4) + 15] - 2x$
34. $\sqrt{121x^6z^4} + 3x^3z^2$
35. $6 - 4(y - [x - 3y])$
36. $8 + 2(x - [y - 4])$
37. $2z(3 + 7z) - \sqrt{16z^2}$
38. $3x + 7y - 2z$
39. $(2x - y) - \{5x - (6 - y) - 12\} - \sqrt{121y^2}$
40. $16 - 4\{6 - 2(x - 3y) + x\}$
41. $36 - \{-2x - [6x - (5 - 2x)]\} + 3x^0$
42. $(3x + 14y - 3z^2[3z + x])^0$
43. $50x - 5\{-2[-3(-3 - x) - 2x]\}$
44. $x^0 + y^0 + z^0 + 8^0$
45. $y^2(y^2 - 16) + 16(y^2 - 16)$
46. $(x + y + z + 8)^0$
47. $4x(3 \cdot 5x^2)(6x^4)$
48. $3x(2x^2)(-16xy^3)$
49. $4z^{-7}z^{10} + 7z^0$
50. $-3x^{-3}x^5 - 18x^0$
51. $11^{3z} \cdot 11^{9z}$
52. $3^{6x} \cdot 3^{2x}$
53. $17^x \cdot 17^{5x}$
54. $2^{3y} \cdot 2^{-6y}$
55. $\dfrac{5^{4t}}{5^{8t}}$
56. $\dfrac{3^{-t}}{3^{4t}}$
57. $\left(\dfrac{x^{4n}x^{-6n}}{x^{3n}}\right)^4$
58. $\left(\dfrac{x^{2n-1}y^{3n}}{x^n y^{2n+2}}\right)^2$

In Exercises 59–62, substitute $x + 3y$ for a and $2x - y$ for b and simplify.

59. $7a - 2b$
60. $12a - b$
61. $5(2a - b)$
62. $b - 3a$

Sections 2.1–2.9 REVIEW

The Properties of Exponents If $x, y \in R$ and if $m, n \in N$, then the following properties are true:

2.1	$x^m \cdot x^n = x^{m+n}$	The product rule
	$(x^m)^n = x^{mn}$	The power rule
	$(xy)^n = x^n y^n$	The rule for raising a product to a power
	$\dfrac{x^m}{x^n} = x^{m-n}$ if $x \neq 0$	The quotient rule
	$\left(\dfrac{x}{y}\right)^n = \dfrac{x^n}{y^n}$ if $y \neq 0$	The rule for raising a quotient to a power
2.2	$x^0 = 1$ if $x \neq 0$	The zero exponent rule
	$x^{-n} = \dfrac{1}{x^n}$ if $x \neq 0$ and $n \in J$	The negative exponent rule, form 1
	$\dfrac{1}{x^{-n}} = x^n$ if $x \neq 0$ and $n \in J$	The negative exponent rule, form 2
	$\left(\dfrac{x}{y}\right)^{-n} = \left(\dfrac{y}{x}\right)^n$ if $x \neq 0$ and $y \neq 0$ and if $n \in J$	The negative exponent rule, form 3
2.3	$\left(\dfrac{x^a y^b}{z^c}\right)^n = \dfrac{x^{an}y^{bn}}{z^{cn}}$ if $a, b, c, n \in J$, if $x, y, z \in R$, and if none of the variables has a value that makes a denominator zero	The general property of exponents

Scientific Notation 2.4

A number written in **scientific notation** is written in the form

$$a \times 10^n, \text{ where } 1 \le a < 10 \text{ and } n \text{ is an integer}$$

Evaluating Algebraic Expressions and Using Formulas 2.5, 2.6

1. First replace each variable by its numerical value.
2. Then carry out all operations in the correct order.

Simplifying Square Roots of Perfect Squares 2.7

1. Find the principal square root of the numerical coefficient by inspection or by guessing-and-checking.
2. Find the square root of each variable factor by dividing its exponent by 2 and dropping the radical sign.

Removing Grouping Symbols 2.8

In addition problems: Drop the grouping symbols; if the first term inside the grouping symbols has a *written* sign, also drop any addition symbol in front of the grouping symbols.

In multiplication problems: When at least one of the factors contains more than one term, use the distributive property: $a(b + c) = ab + ac$.

In subtraction problems: Add the additive inverse of the expression being subtracted to the expression being subtracted from.

To find the additive inverse of an algebraic expression, we can multiply that expression by -1.

To remove grouping symbols that are preceded by a negative sign, we can insert a 1 *between* the negative sign and the left grouping symbol and then use the distributive property.

If grouping symbols occur *within* other grouping symbols, we remove the *innermost* grouping symbols first.

Simplifying Algebraic Expressions 2.9

1. Remove all grouping symbols.
2. Express each *term* in simplest form. (In each term, each different base should appear only once, the exponent on each base should be a single *positive* integer, and the numerical coefficient should be written first.)
3. Simplify all radicals.
4. Collect and combine like terms.

Sections 2.1–2.9 REVIEW EXERCISES Set I

In Exercises 1–18, simplify each expression, if possible, using only positive exponents (or no exponents) in the answers. Assume that no variable has a value that would make a denominator zero.

1. $x^3 \cdot x^5$
2. $x^4 + x^2$
3. $(N^2)^3$
4. $s^6 - s^2$
5. $\dfrac{a^5}{a^2}$
6. $\dfrac{x^6}{y^4}$
7. $\left(\dfrac{2a}{b^2}\right)^3$
8. x^4y^{-2}
9. $\left(\dfrac{x^{-4}y}{x^{-2}}\right)^{-1}$
10. $s^0 + t^0$
11. $(s + t)^0$
12. xy^4
13. $3c^3d^2(c - 4d)$
14. $8 - 2(3x - y)$
15. $(-10x^2y^3)(-8x^3)(-xy^2z^4)$
16. $(4x^3 + 2x - y)(-2x)$
17. $5 - 2[3 - 5(x - y) + 4x - 6]$
18. $(4x^3 + 2x - y) - 2x$
19. Write 148.6 in scientific notation.
20. Write 3.17×10^{-3} in decimal notation.

In Exercises 21–24, solve each problem by using a calculator and the given formula.

21. $A = P(1 + rt)$
    Find A when $P = 550$, $r = 0.09$, $t = 2.5$.
22. $C = \frac{5}{9}(F - 32)$
    Find C when $F = 104$.
23. $S = R\left[\dfrac{(1 + i)^n - 1}{i}\right]$
    Find S when $R = 750$, $i = 0.09$, $n = 3$.
24. $S = \dfrac{a(1 - r^n)}{1 - r}$
    Find S when $a = 7.5$, $r = 2$, $n = 6$.

ANSWERS

1. ______
2. ______
3. ______
4. ______
5. ______
6. ______
7. ______
8. ______
9. ______
10. ______
11. ______
12. ______
13. ______
14. ______
15. ______
16. ______

____

Name

In Exercises 1–18, simplify each expression, if possible, using only positive exponents (or no exponents) in the answers. Assume that no variable has a value that would make a denominator zero.

1. $y^6 + y^5$

2. $x^4 \cdot x^8$

3. $(z^4)^3$

4. $\dfrac{a^5}{b^2}$

5. $\dfrac{y^6}{y^4}$

6. $s^6 - s^2$

7. $\left(\dfrac{x}{2b^2}\right)^3$

8. $\left(\dfrac{x^{-4}y}{x^{-2}}\right)^{-2}$

9. x^4x^{-2}

10. x^0y^0

11. $(x + y)^0$

12. $x^0 + y^0$

13. $5 - 2(x - 6y)$

14. $(3x^2 - 2) - 3x$

15. $(5ab^2)(-2a^2)(-3a^3b)$

16. $3xy^2(4x^2 - 5y^3)$

17. $2x(4x^2 + 2xy + y^2) - y(4x^2 + 2xy + y^2)$

18. $20 - 3\{x - 2[5x - 3(x - y) - 3y] + 4x\}$

17. ______

18. ______

19. ______

20. ______

21. ______

22. ______

23. ______

24. ______

In Exercises 19–21, solve each problem by using a calculator and the given formula.

19. $I = Prt$ Find I when $P = 1{,}250$, $r = 0.08$, $t = 3.5$.

20. $F = \frac{9}{5}C + 32$ Find F when $C = 25.5$.

21. $A = P(1 + i)^n$ Find A when $P = 950$, $i = 0.09$, $n = 4$.
(Round off the answer to two decimal places.)

22. Write 0.000 538 in scientific notation.

23. Write 1.452×10^5 in decimal notation.

24. Use scientific notation in solving this problem:

$$\frac{700{,}000{,}000 \times 0.000\,09}{0.000\,000\,3 \times 40{,}000}$$

Chapter 2 DIAGNOSTIC TEST

The purpose of this test is to see how well you understand the properties of integral exponents, scientific notation, using formulas, and simplifying and evaluating algebraic expressions. We recommend that you work this diagnostic test *before* your instructor tests you on this chapter. Allow yourself about 50 minutes.

Complete solutions for all the problems on this test, together with section references, are given in the answer section at the end of the book. For the problems you do incorrectly, we suggest that you study the sections cited.

In Problems 1–20, simplify each expression. Assume that all variables represent nonnegative numbers and that no variable has a value that would make a denominator zero.

1. $2^5 \cdot 2^7$
2. $x^2 \cdot x^{-5}$
3. $(N^2)^4$
4. $\left(\dfrac{2X^3}{Y}\right)^2$
5. $\left(\dfrac{xy^{-2}}{y^{-3}}\right)^2$
6. $\dfrac{1}{a^{-3}}$
7. $(15x)^0$
8. $15x^0$
9. $5z^{-2}$
10. $(5z)^{-2}$
11. $(-2)^0$
12. $(3^{-2})^{-1}$
13. $10^{-3} \cdot 10^5$
14. $\dfrac{2^{-4}}{2^{-7}}$
15. $7x - 2(5 - x) + \sqrt{81x^2}$
16. $(3x^2y - 2x) - 3x$
17. $(3x^2y - 2x)(-3x)$
18. $(3x^2y)(-2x)(-3x)$
19. $6x(2xy^2 - 3x^3) - 3x^2(2y^2 - 6x^2)$
20. $7x - 2\{6 - 3[8 - 2(x - 3) - 2(6 - x)]\}$
21. a. Express 81,300,000 in scientific notation.
    b. Express 0.000 000 000 38 in scientific notation.
22. Using a scientific calculator, find (81,300,000) (0.000 000 000 38).
23. Evaluate $\dfrac{-b - \sqrt{b^2 - 4ac}}{2a}$ if $a = 3$, $b = -1$, and $c = -2$.
24. Given the formula $S = \dfrac{a(1 - r^n)}{1 - r}$, find S when $a = -8$, $r = 3$, and $n = 2$.

25. Use a scientific calculator and the formula $I = Prt$ to find the (simple) interest for 3 years on a debt of \$3,400,000,000 if the annual interest rate is $7\frac{1}{2}$%.

Chapters 1–2 CUMULATIVE REVIEW EXERCISES

In Exercises 1–15, write "true" if the statement is always true; otherwise, write "false."

1. 0 is a real number.
2. Every rational number is a real number.
3. If $A = \{3, 6, 9\}$ and $B = \{4, 8, 12, 16, 20\}$, then $A \cap B = \{\ \}$.
4. If $A = \{3, 6, 9\}$ and $B = \{4, 8, 12, 16, 20\}$, then $A \cup B = \{\ \}$.
5. Subtraction is commutative.
6. $7x^0 = 1$
7. $5xy$ is a factor of the expression $3 + 5xy$.
8. Division is not associative.
9. $3 \cdot (4 \cdot 9) = (3 \cdot 4) \cdot (3 \cdot 9)$ because of the distributive property.
10. The additive inverse of -3 is $-\frac{1}{3}$.
11. $(x + y)^0 = x^0 + y^0$
12. The multiplicative inverse of 6 is -6.
13. $(8 \times 6) \times 2 = 8 \times (6 \times 2)$ because multiplication is associative.
14. $3 < 9 > 2$ is a valid continued inequality.
15. The set of integers is a subset of the set of rational numbers.

In Exercises 16–31, evaluate each expression or write either "not defined" or "not real."

16. $|-24|$
17. -3^4
18. $(-3)^4$
19. 3^{-4}
20. -3^{-4}
21. $(-3)^{-4}$
22. $(-2)^{-5}$
23. $45 \div 0$
24. $2^3 \cdot 2^4$
25. $3^2 \cdot 3^{-6}$
26. $\sqrt{49}$
27. $\sqrt[5]{-32}$
28. $\dfrac{5^7}{5^4}$
29. $\frac{0}{0}$
30. $0 \div 16$
31. $\dfrac{1}{4^{-2}}$

In Exercises 32–39, simplify each expression. Assume that $a \neq 0$, $b \neq 0$, and $d \neq 0$.

32. $(a + b + c)^0$
33. $a^{-5} \cdot a^3$
34. $2 + 3(x - 3y + z)$
35. $(8xy^2)(-4x)(-2x)$
36. $(8xy^2 - 4x)(-2x)$
37. $(8xy^2 - 4x) - (2x)$
38. $d^{-5} + d^{-3}$
39. $b^{-4} \cdot b^{-6}$
40. Evaluate $\dfrac{-b + \sqrt{b^2 - 4ac}}{2a}$ if $a = 5$, $b = -3$, and $c = -2$.

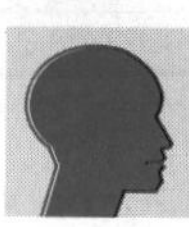

Critical Thinking and Problem-Solving Exercises

1. Gloria was thinking about numbers one day, and she made the following observations:

$$1 = 1^2$$
$$1 + 3 = 2^2$$
$$1 + 3 + 5 = 3^2$$
$$1 + 3 + 5 + 7 = 4^2$$

Verify that her observations are correct, and *without doing the actual addition*, predict the answer for $1 + 3 + 5 + 7 + 9$. Then check your prediction by doing the addition.

2. How many different three-digit numbers are there that are less than 300, that have an odd digit in the units place, and that have a 4, 6, or 8 in the tens place?

3. Jean, Joan, Josie, John, Jim, and Juan are sitting in the classroom, in front of their professor, in the arrangement shown in the figure. Using the following clues, decide who is sitting where.

   Jean is between Joan and Jim.

   Joan is next to the professor.

   John is next to Jean, and he is closer to the professor than Josie is.

4. Bob and Barbara are saving money. Barbara adds \$300 to her savings account every month, and Bob adds \$100 to his savings account every month. Right now, Barbara has \$300 in her account, and Bob has \$900 in his. How long will it be before Barbara has exactly twice as much in her account as Bob has in his account?

5. Esperanza has to decide between two part-time jobs; she can work 10 hr per week as a student-worker in the math department, earning \$4.35 per hour, or she can work at a local fast-food chain for 12 hr per week, earning \$4.25 per hour. What are some of the things she might consider as she decides between the two jobs? Which job would you take if you were Esperanza? Why? What other information might be useful in making your decision?

6. Suppose your instructor asked you to explain to a student who had missed class *why* $x^{-n} = \frac{1}{x^n}$. What would your explanation be?

First-Degree Equations and Inequalities in One Variable

CHAPTER 3

Most problems in algebra are solved by using equations or inequalities. In this chapter, we show how to solve certain equations and inequalities that contain only one variable. We discuss other types of equations and inequalities in later chapters.

3.1 First-Degree Equations, Their Solutions, and Their Graphs

An **equation** is a *statement* that two quantities are equal. It may be a true statement, a false statement, or a statement that is sometimes true and sometimes false.

First-Degree Equations in One Variable

A **first-degree equation in one variable** is, as its name implies, an equation in which there is only one variable and the highest power of that variable is the first power; for example, $3x + 8 = 17$ is a first-degree equation in one variable.

The Domain of the Variable The set of all the numbers that can be used in place of a variable is called the **domain** of the variable. For example, in the expression $1/x$, x cannot be 0 because we cannot allow division by 0. We would say, therefore, that 0 is not in the domain of the variable or that the domain is the set of all real numbers except 0.

In some applications, the solution set must be restricted to some subset of the set of real numbers. For example, if x were to represent the number of dimes in a collection of coins, it wouldn't make sense to say that we had 1.72 dimes. In this case, we would need to restrict the domain of the variable to the set of whole numbers.

In this chapter, if the domain of the variable is not mentioned, the domain is understood to be the set of real numbers, R.

Solutions of Equations

To **solve** an equation means to find values of the variable that make the equation (the *statement*) true. Therefore, a **solution** of an equation is a number from the domain of the variable that, when substituted for the variable, makes the two sides of the equation equal (in other words, it makes the statement true). A solution of an equation is also called a **root** or a **zero** of the equation.

Kinds of Equations

Conditional Equations A **conditional equation** is a statement that is true for certain values of the variable (the solutions) but is not true for others. For example, $3x + 8 = 17$ is a conditional equation; the statement is true if 3 is substituted for x but is *not* true if any other number is substituted for x. When we solve a first-degree conditional equation in one variable, our final equation should be of the form $x = k$, where k is some number. When the equation is in this form, we say that we have *isolated* the variable—that is, we have the variable by itself on one side of the equal sign and a constant on the other side of the equal sign.

Identities An **identity** is a statement that is true when *any* value from the domain is substituted for the variable. For example, $2(x + 3) = 2x + 6$ is an identity; the statement is true if we substitute 7 for x, 0 for x, -23 for x, or any other real number for x.

Equations with No Solution An **equation with no solution** is a statement that is false for all values of the variable. For example, $x + 1 = x + 2$ is such an equation; no values of x will make the statement true. Such equations can be called *contradictions* or *defective equations*.

Solution Sets of Equations

The **solution set** of an equation is the set of all the numbers that are solutions of that equation. For example, the solution set of the equation $3x + 8 = 17$ is $\{3\}$, since 3 *is* a solution of the equation and it is the *only* value of x that makes $3x + 8 = 17$ true. (Notice that for the solution *set*, 3 is written within braces.) The solution set of an identity is the set of all the numbers in the domain of the variable; for example, the solution set of $2(x + 3) = 2x + 6$ is the set of all real numbers. The solution set of an equation that has *no* solution is the empty set; for example, the solution set of $x + 1 = x + 2$ is $\{\ \}$.

Equivalent Equations

Equations that have exactly the same solution set are called **equivalent equations**.

EXAMPLE 1 Examples of solutions, solution sets, and equivalent equations:

a. As we mentioned above, 3 *is* a solution of $3x + 8 = 17$, and the *solution set* of the equation $3x + 8 = 17$ is $\{3\}$.

b. The solution set of $3x = 9$ is also $\{3\}$, because $3(3) = 9$ (that is, 3 *is* a solution of $3x = 9$) and 3 is the *only* value of x that makes $3x = 9$ true.

c. The solution set of $x = 3$ is *also* $\{3\}$, since $3 = 3$ is true (that is, 3 *is* a solution of $x = 3$) and 3 is the *only* value of x that makes $x = 3$ true.

The equations $3x + 8 = 17$, $3x = 9$, and $x = 3$ are *equivalent equations* because they all have exactly the same solution set.

Solving First-Degree Equations

When we **solve a first-degree equation**, we want to *isolate the variable*. That is, we want to find an equivalent equation that is of the form $x = k$. We use the following principles of equality in writing equations equivalent to the given equation:

The addition property of equality

For all $a, b, c \in R$,

$$\text{if} \quad a = b, \quad \text{then} \quad a + c = b + c$$

In words: If the same number is added to both sides of an equation, the new equation is equivalent to the original equation.

The subtraction property of equality

For all $a, b, c \in R$,

$$\text{if} \quad a = b, \quad \text{then} \quad a - c = b - c$$

In words: If the same number is subtracted from both sides of an equation, the new equation is equivalent to the original equation.

The multiplication property of equality

For all $a, b, c \in R$, but $c \neq 0$,

$$\text{if} \quad a = b, \quad \text{then} \quad ac = bc$$

In words: If both sides of an equation are multiplied by the same nonzero number,* the new equation is equivalent to the original equation.

*If we multiply both sides of an equation by zero, we always get the equation $0 = 0$, which is usually not equivalent to the original equation.

The division property of equality

For all $a, b, c \in R$, but $c \neq 0$,

$$\text{if} \quad a = b, \quad \text{then} \quad \frac{a}{c} = \frac{b}{c}$$

In words: If both sides of an equation are divided by the same nonzero number,* the new equation is equivalent to the original equation.

*We cannot divide both sides of an equation by zero, since division by zero is not permitted.

The symmetric property of equality

For all $a, b \in R$,

$$\text{if} \quad a = b, \quad \text{then} \quad b = a$$

In words: If the two sides of an equation are interchanged, the new equation is equivalent to the original equation.

Note The symmetric property, which was introduced earlier, permits us to rewrite an equation such as $4 = x$ as $x = 4$. (The equation $x = 4$ is in the *form* $x = k$.)

Note It is possible to solve all first-degree equations without using either the subtraction property or the division property of equality. (This is so because subtraction is the inverse operation of addition and division is the inverse operation of multiplication.) We do not show the use of the subtraction property in this book, but some instructors do use it.

When we use the principles of equality listed above to solve a first-degree equation in one variable, three outcomes are possible:

1. If the equation can be reduced to the form $x = k$, where k is some number, the equation is a *conditional equation* (see Example 2). If there are restrictions on the domain, the equation may have *no solution* even though it reduces to the form $x = k$ (see Example 3).
2. If the two sides of the equation reduce to the same constant so that we obtain a true statement (for example, $0 = 0$), the equation is an *identity* (see Example 4).
3. If the two sides of the equation reduce to unequal constants so that we obtain a false statement (for example, $2 = 5$), the equation is an *equation with no solution* (see Example 5).

Although there is no single correct way to solve an equation, you may find the following suggestions helpful. (The first five steps can be done in any order.) Remember: Our goal is to *isolate the variable*, if possible.

Solving a first-degree equation in one variable

Always write each new equation *under* the previous equation.

1. Remove denominators, if there are any, by multiplying both sides of the equation by the least common multiple (the LCM) of all the denominators.
2. Remove all grouping symbols.
3. Combine like terms on each side of the equal sign.
4. Move all the terms that contain the variable to one side of the equal sign (usually the *left* side) and all the constants to the other side by adding the appropriate terms to both sides of the equation.
5. Divide both sides of the equation by the coefficient of the variable.
6. Determine whether the equation is a conditional equation, an identity, or an equation with no solution.
7. If the equation is a conditional equation, check the solution.

To check the solution of an equation, perform the following steps:

1. Determine whether the *apparent* solution (the number found by following steps 1–5 in the box above) is in the domain of the variable. If it is not, the apparent solution is *not* a solution. If the apparent solution *is* in the domain, continue with the following steps.
2. Replace the variable in the given equation with the apparent solution.
3. Perform the indicated operations on both sides of the equal sign.
4. If the resulting numbers on both sides of the equal sign are the same, the solution checks.

EXAMPLE 2 Find the solution set of $8x - 3[2 - (x + 4)] = 4(x - 2)$, and graph the solution on the number line.

SOLUTION

$8x - 3[2 - (x + 4)] = 4(x - 2)$	
$8x - 3[2 - x - 4] = 4x - 8$	Removing the parentheses
$8x - 3[-x - 2] = 4x - 8$	Combining like terms inside the brackets
$8x + 3x + 6 = 4x - 8$	Removing the brackets
$11x + 6 = 4x - 8$	Combining like terms on the left side
$11x + 6 - 4x - 6 = 4x - 8 - 4x - 6$	Adding $-4x - 6$ to both sides to get the x-term on one side and the constant on the other
$7x = -14$	Combining like terms on each side
$x = -2$	Dividing both sides by 7

The apparent solution is -2. Since the equation reduces to the form $x = k$, the equation is a conditional equation.

Check The domain is the set of all real numbers, so the apparent solution, -2, is in the domain.

$$8x - 3[2 - (x + 4)] = 4(x - 2)$$
$$8(-2) - 3[2 - (\{-2\} + 4)] \stackrel{?}{=} 4(\{-2\} - 2)$$
$$-16 - 3[2 - (2)] \stackrel{?}{=} 4(-4)$$
$$-16 - 3[0] \stackrel{?}{=} -16$$
$$-16 = -16 \quad \text{True}$$

The solution checks. Therefore, $x = -2$ is the condition necessary to make the two sides of the equation equal to each other.

The solution set is $\{-2\}$, and the *graph* of the solution (or of the solution set) is

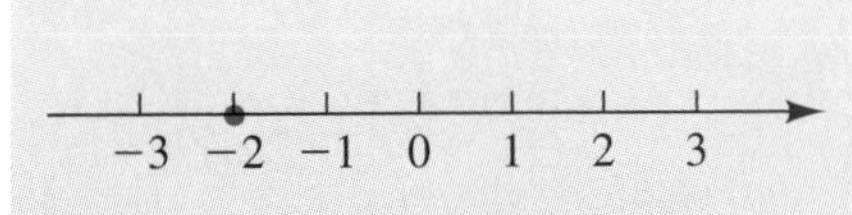

A Word of Caution In solving the equation $11x + 6 = 4x - 8$, it is *incorrect* to write*

$$11x + 6 = 4x - 8 = 7x = -14 = x = -2$$

An equal sign here implies that $4x - 8 = 7x$, which is incorrect

An equal sign here implies that $-14 = -2$, which is incorrect

The separate equations must be written *under* each other.

*It is *possible* to use the symbol $\Rightarrow$ for *implies* between the separate equations, writing $11x + 6 = 4x - 8 \Rightarrow 7x = -14 \Rightarrow x = -2$; this is read "$11x + 6 = 4x - 8$ *implies* $7x = -14$, which *implies* $x = -2$." However, we do not recommend writing the equations this way.

In Example 3, the domain is restricted by the statement $x \in J$. In this case, we attempt to solve the equation in the usual way, but if the apparent solution is not an *integer*, we must reject the solution.

EXAMPLE 3 Find $\{x \mid 2(3x + 5) = 14, x \in J\}$.

SOLUTION The domain is the set of integers; that is, we are interested only in *integral* solutions to the equation.

$$2(3x + 5) = 14$$
$$6x + 10 = 14$$
$$6x = 4$$
$$x = \tfrac{4}{6} = \tfrac{2}{3}$$

The only apparent solution, $\frac{2}{3}$, is not in the domain; therefore, the equation has no solution, and $\{x \mid 2(3x + 5) = 14, x \in J\} = \{\ \}$. If we had been asked to graph the solution set, there would be no points to graph.

EXAMPLE 4 Find the solution set of $\dfrac{x+3}{6} - \dfrac{2x-3}{9} = \dfrac{5}{6} - \dfrac{x}{18}$, and graph the solution set on the number line.

SOLUTION The domain is the set of all real numbers.

$$\frac{x+3}{6} - \frac{2x-3}{9} = \frac{5}{6} - \frac{x}{18}$$

$$\frac{18}{1} \cdot \left(\frac{x+3}{6} - \frac{2x-3}{9}\right) = \frac{18}{1} \cdot \left(\frac{5}{6} - \frac{x}{18}\right)$$ Multiplying both sides by 18, the LCM of 6, 9, and 18

$$\frac{18}{1} \cdot \frac{x+3}{6} - \frac{18}{1} \cdot \frac{2x-3}{9} = \frac{18}{1} \cdot \frac{5}{6} - \frac{18}{1} \cdot \frac{x}{18}$$ Using the distributive property

$$3(x + 3) - 2(2x - 3) = 15 - x$$ Simplifying

$$3x + 9 - 4x + 6 = 15 - x$$ Using the distributive property

$$-x + 15 = 15 - x$$ Combining like terms

$$-x + 15 + x - 15 = 15 - x + x - 15$$ Adding $x - 15$ to both sides

$$0 = 0$$ A *true* statement

(If we had added just $+x$ to both sides, our final statement would have been $15 = 15$, which is also a true statement.) *Notice that no variable appears in the last step*; we were *not* able to get the equation into the form $x = k$ because x no longer appears in the equation. Since the two sides of the equation reduced to the *same* constant (that is, since we were left with the *true* statement $0 = 0$), this equation is an identity. *Any* real number will make the two sides of the equation equal.

Check We show the check for two real numbers selected at random; we'll try -1 and 7.

If we substitute -1 for x, we have

$$\frac{-1+3}{6} - \frac{2(-1)-3}{9} \stackrel{?}{=} \frac{5}{6} - \frac{(-1)}{18}$$

$$\frac{8}{9} = \frac{8}{9}$$

Both sides simplify to $\frac{8}{9}$

If we substitute 7 for x, we have

$$\frac{7+3}{6} - \frac{2(7)-3}{9} \stackrel{?}{=} \frac{5}{6} - \frac{(7)}{18}$$

$$\frac{4}{9} = \frac{4}{9}$$ Both sides simplify to $\frac{4}{9}$

As our answer, we state that the solution set is $\{x \mid x \in R\}$—the set of all real numbers—and that the given equation is an *identity*. The graph of the solution set of the equation is

EXAMPLE 5 Find the solution set of $4(x-1) - 3(4-x) = 7x - 13$.

SOLUTION The domain is the set of all real numbers.

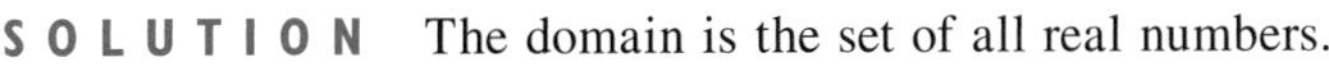

$$4(x-1) - 3(4-x) = 7x - 13$$

$$4x - 4 - 12 + 3x = 7x - 13$$

$$7x - 16 = 7x - 13$$

$$7x - 16 - 7x = 7x - 13 - 7x$$ Adding $-7x$ to both sides

$$-16 = -13$$ A *false* statement

Notice that no variable appears in the last step; we were *not* able to get the equation into the form $x = k$ because x no longer appears in the equation. Since the two sides of the equation reduced to *different* constants (that is, we were left with the *false* statement $-16 = -13$), this equation is an equation with *no solution*. *No* real number can be found that will make the two sides of the equation equal.

As our answer, we state that the solution set is the empty set, { }, and that the equation has *no solution*. If we had been asked to graph the solution set, there would be no points to graph.

Exercises 3.1
Set I

In each exercise, find the solution set. Identify any equation that is not a conditional equation as either an identity or an equation with no solution. For Exercises 1–10, graph the solution set of each conditional equation on the real number line.

1. $4x + 3(4 + 3x) = -1$
2. $5x + 2(4 + x) = -6$
3. $7y - 2(5 + 4y) = 8$
4. $5y - 3(4 + 2y) = 3$
5. $4x + 12 = 2(6 + 2x)$
6. $6x + 15 = 3(5 + 2x)$
7. $3[5 - 2(5 - z)] = 2(3z + 7)$
8. $5[4 - 3(2 - z)] = 3(5z + 6)$
9. $\frac{x}{3} - \frac{x}{6} = 18$
10. $\frac{x}{4} - \frac{x}{8} = 16$
11. $\frac{y+3}{8} - \frac{3}{4} = \frac{y+6}{10}$
12. $\frac{y+7}{12} - \frac{5}{6} = \frac{y+4}{9}$
13. $\{z \mid 5z - 3(2 + 3z) = 6, z \in N\}$
14. $\{z \mid 8z + 6 = 2(7z + 9), z \in N\}$
15. $\{x \mid 7x - 2(5 + 4x) = 8, x \in J\}$
16. $\{x \mid 7x + 15 = 3(3x + 5), x \in J\}$
17. $\{x \mid 2(3x - 6) - 3(5x + 4) = 5(7x - 8), x \in J\}$

18. $\{z \mid 4(7z - 9) - 7(4z + 3) = 6(9z - 10), z \in J\}$

19. $\dfrac{2(y-3)}{5} - \dfrac{3(y+2)}{2} = \dfrac{7}{10}$

20. $\dfrac{5(x-4)}{6} - \dfrac{2(x+4)}{9} = \dfrac{5}{18}$

In Exercises 21 and 22, use a calculator, and round off each solution to two decimal places.

21. $6.23x + 2.5(3.08 - 8.2x) = -14.7$

22. $9.84 - 4.6x = 5.17(9.01 - 8.23x)$

Writing Problems

Express the answers in your own words and in complete sentences.

1. Explain what the solution of an equation is. (Don't explain how to find it.)

2. Describe what you would do *first* in solving the equation $\dfrac{2(x-3)}{3} - \dfrac{3(x+2)}{4} = \dfrac{5}{6}$, and explain why that would be your first step.

3. Explain why $3x + 7 = 4$ and $x + 8 = 7$ are equivalent equations.

Exercises 3.1 Set II

In each exercise, find the solution set. Identify any equation that is not a conditional equation as either an identity or an equation with no solution. For Exercises 1–10, graph the solution set of each conditional equation on the real number line.

1. $6x + 3(2 + 2x) = -6$
2. $7x + 5(2 + x) = -2$
3. $8y - 3(2 + 3y) = 5$
4. $8y + 4(6 - 2y) = -2$
5. $8x + 12 = 4(2x + 3)$
6. $14x - 3(4 - x) = 2(x + 9)$
7. $3[15 - (5 - 3x)] = 9(x + 18)$
8. $3 + 5(x - 4) = 9x + 1$
9. $\dfrac{z}{9} - \dfrac{z}{2} = \dfrac{7}{3}$
10. $\dfrac{x}{5} - \dfrac{x}{10} = 8$
11. $\dfrac{x+1}{4} - \dfrac{5}{8} = \dfrac{x-1}{8}$
12. $\dfrac{x+3}{5} - \dfrac{x}{2} = \dfrac{5-3x}{10}$
13. $\{x \mid 4x + 5(-4 - 5x) = 22, x \in N\}$
14. $\{x \mid 3x - 4(2 - x) = 6, x \in N\}$
15. $\{y \mid 10 - 7y = 4(11 - 6y), y \in J\}$
16. $\{y \mid -37 - 9(y + 1) = 4(2y - 3), y \in J\}$
17. $\{x \mid 6(3x - 5) = 3[4(1 - x) - 7], x \in J\}$
18. $\{x \mid 16 - 8(2x - 2) = 5(5x - 10), x \in N\}$
19. $\dfrac{2(x-3)}{3} - \dfrac{3(x+2)}{4} = \dfrac{5}{6}$
20. $\dfrac{3(x-4)}{4} - \dfrac{5(x+2)}{6} = \dfrac{7}{12}$

In Exercises 21 and 22, use a calculator, and round off each solution to two decimal places.

21. $7.02(5.3x - 4.28) = 11.6 - 2.94x$
22. $7.23 - 6.1x = 3.2(1.08 - 5.3x)$

3.2 Simple First-Degree Inequalities, Their Solutions, and Their Graphs

A conditional inequality is an inequality that is true for some values of the variable and false for other values; for example, $x + 3 \geq 5$ and $7(3 - 2x) < 4(x - 3)$ are conditional inequalities. In this section, we discuss solving conditional first-degree inequalities that contain *one* of the symbols $<$, $>$, $\leq$, or $\geq$; such inequalities are often called *simple* (or *singular*) inequalities. (*Continued* inequalities and *combined* inequalities, which we will study in the next two sections, contain at least *two* such symbols.)

Solutions and Solution Sets of Conditional Inequalities

A **solution of a conditional inequality** is any number that, when substituted for the variable, makes the inequality a true statement.

The **solution set** of an inequality is the set of all numbers that are solutions of the inequality. Whereas the solution set of a first-degree conditional *equation* normally has just one element in it, the solution set of an *inequality* usually contains infinitely many elements.

If an inequality is in one of the forms $x > k$, $x \geq k$, $x < k$, or $x \leq k$, it is easy to see what the solution set is. For example, the solution set of the inequality $x \geq 3$ is the set of *all* the numbers that are greater than or equal to 3.

The Graph of the Solution Set of an Inequality in One Variable

The solution set of an inequality in one variable can be graphed on the number line.

To graph $x \geq k$, we first locate k. We put a *left bracket* at k, indicating that k (the endpoint) *and points to its right* are included in the graph; then we shade the number line from k toward the *right*, adding an arrowhead (pointing toward the right) to indicate that the graph continues indefinitely toward the right. For example, the graph of $x \geq -2$ is as follows:

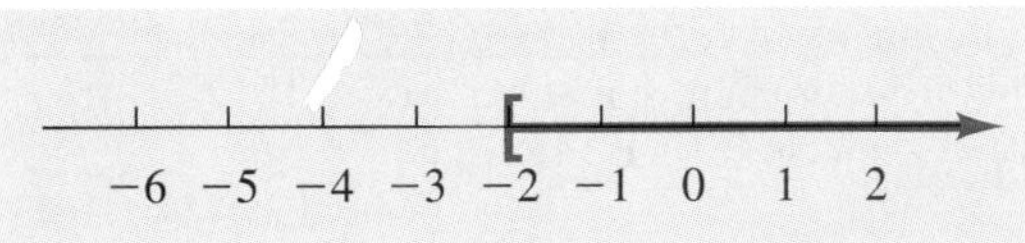

We graph $x > k$ in a similar way, except that we put a *left parenthesis* at k, indicating that the endpoint, k, is *not* included in the graph but that all points to the right of k *are* included. For example, the graph of $x > 3$ is as follows:

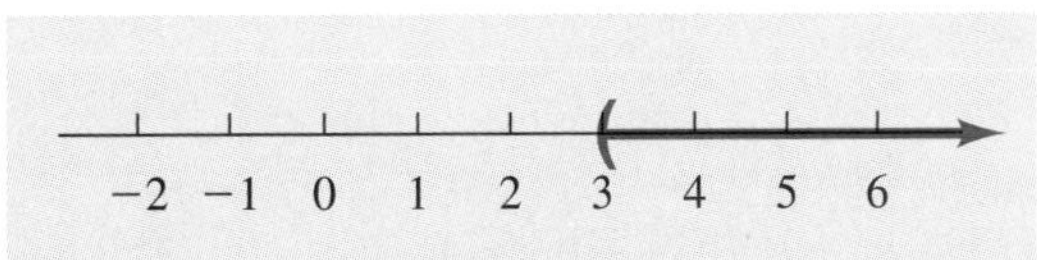

(This notation means that points such as 3.2, 3.01, and 3.0005 *are* included in the graph.)

To graph $x \leq k$, we first locate k. We put a *right bracket* at k, indicating that k (the endpoint) *and points to its left* are included in the graph; then we shade the number line from k toward the *left*, adding an arrowhead (pointing toward the left) to indicate that the graph continues indefinitely toward the left. For example, the graph of $x \leq 1$ is as follows:

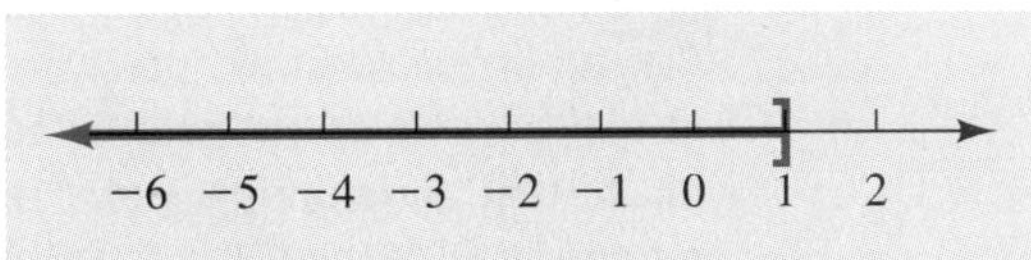

We graph $x < k$ in a similar fashion, except that we put a *right parenthesis* at k, indicating that the endpoint k is *not* included in the graph but that points to the left of k *are* included. For example, the graph of $x < -4$ is as follows:

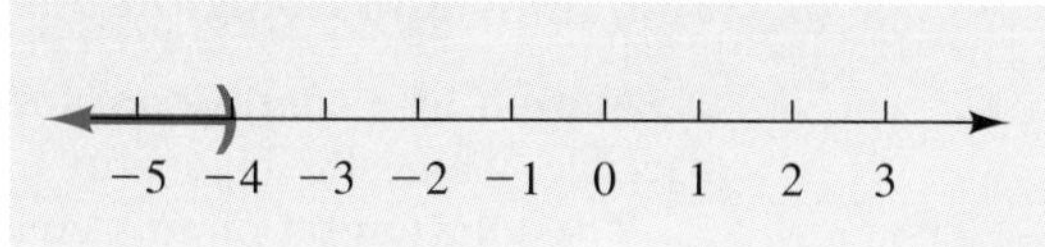

(This notation means that points such as -4.3, -4.02, and -4.0003 *are* included in the graph.)

Infinity Symbols In mathematics, we sometimes make use of the symbols for **positive infinity**, $+\infty$, and **negative infinity**, $-\infty$. The infinity symbols are *not* symbols for real numbers, and they cannot be treated as such. We consider $+\infty$ to be larger than any real number and $-\infty$ to be smaller than any real number.

Unbounded Intervals and Interval Notation

An **unbounded interval** is an interval that has no "end"; that is, it is a set of numbers in which the numbers increase (or decrease) without bound. For example, the interval that contains all the real numbers that are greater than 3 is an unbounded interval. (There is no largest number in the interval.)

Interval notation gives us a compact way of describing intervals. The following table compares set-builder notation, interval notation, and graphs for unbounded sets:

Set-builder notation	*Interval notation*	*Graph*
$\{x \mid x > k\}$	$(k, +\infty)$ The left parenthesis indicates that k is *not* included in the solution set	k
$\{x \mid x \geq k\}$	$[k, +\infty)$ The left bracket indicates that k *is* included in the solution set	k
$\{x \mid x < k\}$	$(-\infty, k)$ The right parenthesis indicates that k is *not* included in the solution set	k
$\{x \mid x \leq k\}$	$(-\infty, k]$ The right bracket indicates that k *is* included in the solution set	k
$\{x \mid x \in R\}$	$(-\infty, +\infty)$ This is interval notation for the set of *all* real numbers	

Notice that the use of brackets and parentheses in *interval notation* corresponds to the use of brackets and parentheses in *graphing* inequalities; that is, we use a *bracket* to indicate that an endpoint *is included* in the interval, and we use a *parenthesis* to indicate that an endpoint is *not* included in the interval. We *always* use a parenthesis, never a bracket, next to the infinity symbol.

In interval notation, the smaller number or symbol must always be *to the left of the larger number or symbol*. In particular, $-\infty$ must always be to the *left* of any number or of $+\infty$, and $+\infty$ must always be to the *right* of any number or of $-\infty$.

If an interval does not include either endpoint, we call it an **open interval**; if an interval includes *both* endpoints, we call it a **closed interval** (we'll see closed intervals in the next section); and if an interval includes one endpoint but not the other, we call it a **half-open interval**. Therefore, we generally read $(3, +\infty)$ as the open interval from 3 to infinity, and $(-\infty, 2]$ as the half-open interval from negative infinity to 2.

EXAMPLE 1 Examples of converting from set-builder notation to interval notation:

a. $\{x \mid x \geq 5\}$ is equivalent to $[5, +\infty)$.
b. $\{x \mid x < -4\}$ is equivalent to $(-\infty, -4)$.
c. $\{x \mid x > -2\}$ is equivalent to $(-2, +\infty)$.
d. $\{x \mid x \leq 0\}$ is equivalent to $(-\infty, 0]$.

Writing an Algebraic Statement to Describe a Graph

When we write an algebraic statement that describes a graph, we can use set-builder notation or, sometimes, interval notation. (We *cannot* use interval notation to describe a graph if the graph consists of separate *points*). In this section, we write algebraic statements to describe graphs of unbounded intervals. In the next section, we will write algebraic statements to describe other graphs.

EXAMPLE 2 Use set-builder notation *and* interval notation to describe the set that is graphed.

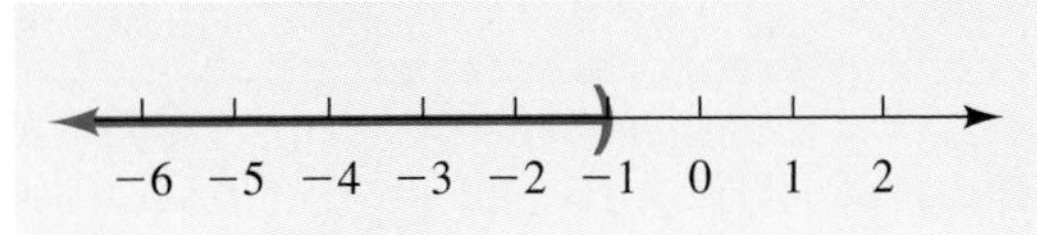

SOLUTION The set graphed is the set of all real numbers that are less than -1. In set-builder notation, we write this set as $\{x \mid x < -1\}$.

The graph extends infinitely far to the left, so, for interval notation, we write the symbol $-\infty$ *to the left of* -1. Because the right endpoint of the interval, -1, is *not* included in the interval, we use a *right parenthesis* after -1. Therefore, in interval notation, the set is $(-\infty, -1)$.

The Sense of an Inequality

Recall that the sense of an inequality refers to whether the inequality symbol is a *greater than* symbol or a *less than* symbol.

$\left.\begin{matrix} a > b \\ c > d \end{matrix}\right\}$ ← The same sense (both are $>$)

$\left.\begin{matrix} a < b \\ c > d \end{matrix}\right\}$ ← Opposite senses (one is $>$, one is $<$)

$\left.\begin{matrix} a \geq b \\ c \leq d \end{matrix}\right\}$ ← Opposite senses

$\left.\begin{matrix} a \leq b \\ c \leq d \end{matrix}\right\}$ ← The same sense

The Properties of Inequalities

In the last section, we solved equations by adding the same number to both sides of the equation, multiplying both sides of the equation by the same number, and so on. In Example 3, we examine the effect of adding the same number to both sides of an *inequality*, multiplying both sides of an *inequality* by the same number, and so forth.

EXAMPLE 3 Determine whether the sense of the inequality $8 < 12$ changes when we perform various arithmetic operations on both sides of the inequality (note that $8 < 12$ is a *true* statement).

a. Does the sense of the inequality change if we add +5 to both sides?

SOLUTION

$8 < 12$

$8 + 5 \ ? \ 12 + 5$

$13 \ ? \ 17$ — 13 is less than 17

$13 < 17$ — The sense of the inequality is *unchanged* when we add the same number to both sides

The senses are the same.

b. Does the sense of the inequality change if we subtract 9 from both sides?

SOLUTION

$8 < 12$

$8 - 9 \ ? \ 12 - 9$

$-1 \ ? \ 3$ — -1 is less than 3

$-1 < 3$ — The sense of the inequality is *unchanged* when we subtract the same number from both sides

The senses are the same.

c. Does the sense of the inequality change if we multiply both sides by 3 (a *positive* number)?

SOLUTION

$8 < 12$

$8(3) \ ? \ 12(3)$

$24 \ ? \ 36$ — 24 is less than 36

$24 < 36$ — The sense of the inequality is *unchanged* when we multiply both sides by the same *positive* number

The senses are the same.

d. Does the sense of the inequality change if we divide both sides by 2 (a *positive* number)?

SOLUTION

$8 < 12$

$\frac{8}{2} \ ? \ \frac{12}{2}$

$4 \ ? \ 6$ — 4 is less than 6

$4 < 6$ — The sense of the inequality is *unchanged* when we divide both sides by the same *positive* number

The senses are the same.

e. Does the sense of the inequality change if we multiply both sides by -3 (a *negative* number)?

SOLUTION

$8 < 12$

$8(-3) \ ? \ 12(-3)$

$-24 \ ? \ -36$ — *-24 is greater than -36*

$-24 > -36$ — The sense of the inequality *changes* when we multiply both sides by the same *negative* number

The senses are *opposite*.

f. Does the sense of the inequality change if we divide both sides by -4 (a *negative* number)?

SOLUTION

$$8 < 12$$

$$\frac{8}{-4} \; ? \; \frac{12}{-4}$$

$$-2 \; ? \; -3 \qquad -2 \text{ is greater than } -3$$

$$-2 > -3 \qquad \text{The sense of the inequality changes when we divide both sides by the same negative number}$$

The senses are *opposite*.

We see that when we multiply or divide both sides of an inequality by a *negative number*, the *sense* of the inequality changes. When we solve simple *inequalities* that contain $>$, $\geq$, $<$, or $\leq$, we can use any of the following properties of inequalities:

The addition and subtraction properties of inequalities

If an inequality contains the symbol $>$, $<$, $\geq$, or $\leq$, the sense of the inequality is unchanged if the same number is added to or subtracted from both sides of the inequality. For example, for all $a, b, c \in R$,

$$\text{if } a < b, \text{ then } a + c < b + c \text{ and } a - c < b - c$$

The senses are the same

The multiplication and division properties of inequalities

If an inequality contains the symbol $>$, $<$, $\geq$, or $\leq$, the sense of the inequality is unchanged if both sides of the inequality are multiplied or divided by the same *positive* number. For example, for all $a, b, c \in R$,

$$\text{if } a < b \text{ and if } c > 0, \text{ then } ac < bc \text{ and } \frac{a}{c} < \frac{b}{c}$$

However, the sense of the inequality must be *changed* if both sides of the inequality are multiplied or divided by the same *negative* number. For example, for all $a, b, c \in R$,

The senses are opposite

$$\text{if } a < b \text{ and if } c < 0, \text{ then } ac > bc \text{ and } \frac{a}{c} > \frac{b}{c}$$

The senses are opposite

Solving a Simple First-Degree Inequality

When we solve a simple first-degree inequality, we must find *all* the values of the variable that satisfy the inequality. Therefore, we want our final inequality to be in one of the forms $x < k$, $x > k$, $x \leq k$, or $x \geq k$, where k is some real number. We can use the following procedure:

Solving an inequality that contains $<$, $\leq$, $>$, or $\geq$

We proceed as if we were solving an equation, except that we must *change the sense of the inequality if we multiply or divide both sides of the inequality by a negative number.*

Then, if necessary, we use the facts that $a < b$ can be rewritten as $b > a$, $a \geq b$ can be rewritten as $b \leq a$, and so on, to get the inequality into one of the forms $x < k$, $x > k$, $x \leq k$, or $x \geq k$.

EXAMPLE 4 Find the solution set of $3x - 2(2x - 7) \leq 2(3 + x) - 4$. Describe the solution set in set-builder notation *and* in interval notation, and graph the solution set.

SOLUTION The domain is the set of all real numbers.

$$3x - 2(2x - 7) \leq 2(3 + x) - 4$$
$$3x - 4x + 14 \leq 6 + 2x - 4$$
$$-x + 14 \leq 2 + 2x$$
$$-x + 14 + x - 2 \leq 2 + 2x + x - 2$$
Adding $x - 2$ to both sides to get the x-term on one side and the constant on the other
$$12 \leq 3x$$
$$\frac{12}{3} \leq \frac{3x}{3}$$
The sense is not changed if we divide both sides by a positive number
$$4 \leq x$$
$$x \geq 4$$
Remember: $4 \leq x$ can be replaced by $x \geq 4$

It is also possible to solve $-x + 14 \leq 2 + 2x$ as follows:

$$-x + 14 \leq 2 + 2x$$
$$-x + 14 - 2x - 14 \leq 2 + 2x - 2x - 14$$
Adding $-2x - 14$ to both sides to get the x-term on one side and the constant on the other
$$-3x \leq -12$$
$$\frac{-3x}{-3} \geq \frac{-12}{-3}$$
The sense *must* be changed when we divide both sides by a negative number
$$x \geq 4$$

In set-builder notation, the solution set is $\{x \mid x \geq 4\}$, and in interval notation, the solution set is $[4, +\infty)$.

For the graph of the solution set, we put a *left bracket* at the 4, indicating that 4 is included in the solution set, and we shade the number line to the *right* of 4, adding an arrowhead that points toward the right:

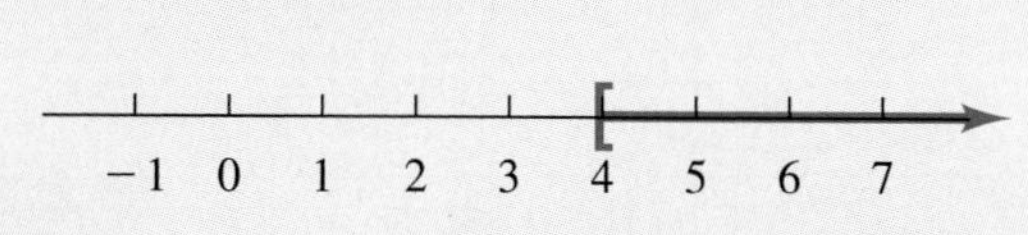

A Word of Caution We solve an inequality by a method similar to the method used for solving an equation. For this reason, a common error is to substitute an *equal sign* for the *inequality symbol.*

$$-x + 14 \leq 2 + 2x$$
$$-x + 14 - 2x - 14 = 2 + 2x - 2x - 14$$
$$-3x = -12$$
$$\frac{-3x}{-3} = \frac{-12}{-3}$$
$$x = 4$$

(The inequality $-x + 14 \leq 2 + 2x$ was correctly solved as part of Example 4.) Remember that when we add the same number to both sides of an inequality, we get an *inequality.* We do *not* get an equation.

EXAMPLE 5

Solve $\frac{x+2}{15} > \frac{x+3}{5} - \frac{1}{3}$. Describe the solution set in set-builder notation *and* in interval notation, and graph the solution set.

SOLUTION The domain is the set of all real numbers. The LCM of 15, 5, and 3 is 15.

$$\frac{x+2}{15} > \frac{x+3}{5} - \frac{1}{3}$$

$$\frac{15}{1} \cdot \left(\frac{x+2}{15}\right) > \frac{15}{1} \cdot \left(\frac{x+3}{5} - \frac{1}{3}\right)$$ Multiplying both sides by *positive* 15 does not change the sense of the inequality

$$x + 2 > 3(x + 3) - 5$$

$$x + 2 > 3x + 9 - 5$$

$$x + 2 > 3x + 4$$

$$x + 2 - x - 4 > 3x + 4 - x - 4$$

$$-2 > 2x$$

$$\frac{-2}{2} > \frac{2x}{2}$$ ← The sense is *un*changed

$$-1 > x$$

$$x < -1$$ ← Equivalent statements

It is also possible to solve $x + 2 > 3x + 4$ as follows:

$$x + 2 > 3x + 4$$

$$x + 2 - 3x - 2 > 3x + 4 - 3x - 2$$

$$-2x > 2$$

$$\frac{-2x}{-2} < \frac{2}{-2}$$ ← The sense must be changed when we divide both sides by -2

$$x < -1$$

In set-builder notation, the solution set is $\{x \mid x < -1\}$, and in interval notation, the solution set is $(-\infty, -1)$.

For the graph of the solution set, we put a *right parenthesis* at the -1, indicating that -1 is not included in the solution set, and we shade the number line to the *left* of -1, adding an arrowhead that points toward the left:

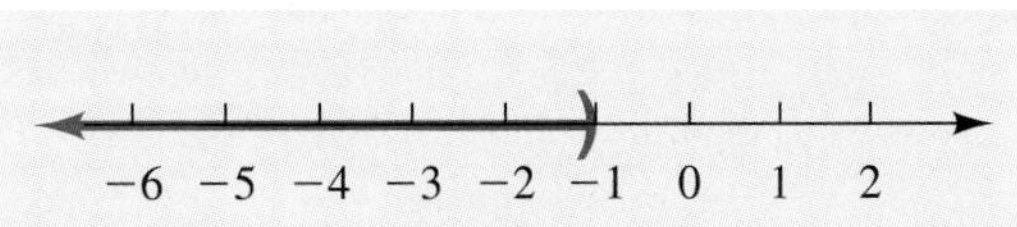

EXAMPLE 6

Convert $\{x \mid 4(3x - 5) < 10, x \in N\}$ to roster notation, and graph the solution set on the real number line.

SOLUTION The domain is the set of *natural numbers*. We first solve the inequality $4(3x - 5) < 10$, and from the solution set of that inequality, we select just those numbers that are natural numbers.

$$4(3x - 5) < 10$$
$$12x - 20 < 10$$
$$12x < 30 \quad \text{Adding 20 to both sides}$$
$$x < \tfrac{30}{12} \quad \text{Dividing both sides by 12}$$
$$x < \tfrac{5}{2}$$

The *natural numbers* that are less than $\frac{5}{2}$, or $2\frac{1}{2}$, are 1 and 2. Therefore,

$$\{x \mid 4(3x - 5) < 10, x \in N\} = \{1, 2\}$$

The graph is

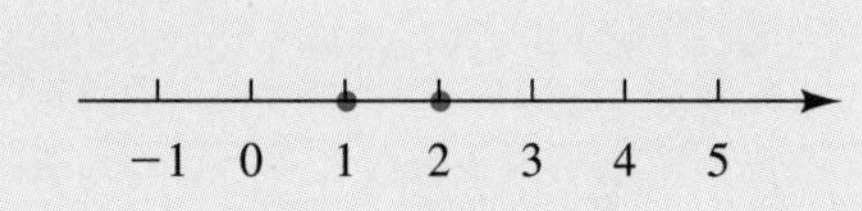

Note We *cannot* write the solution set for the problem in Example 6 in interval notation, since the solution set consists of separate numbers, not an interval.

Exercises 3.2
Set I

In Exercises 1–4, graph each solution set, and convert the description of each set from set-builder notation to interval notation.

1. $\{x \mid x \geq 5\}$
2. $\{x \mid x \leq 2\}$
3. $\{x \mid x < -3\}$
4. $\{x \mid x > -1\}$

In Exercises 5–8, write, in set-builder notation *and* in interval notation, the algebraic statement that describes the set of numbers graphed.

5.

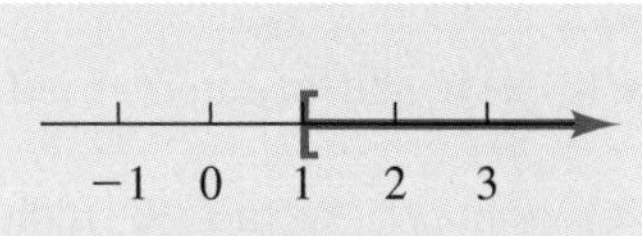

6.

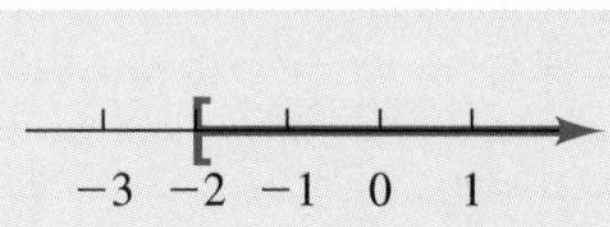

7.

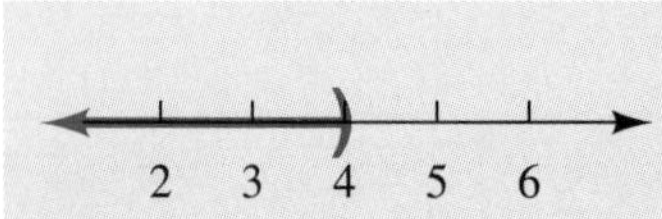

8.

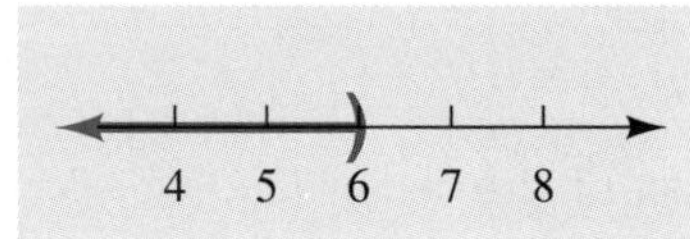

In Exercises 9–24, solve each inequality, describing the solution set in set-builder notation *and* in interval notation; graph each solution set.

9. $3x - 1 < 11$
10. $7x - 12 < 30$
11. $17 \geq 2x - 9$
12. $33 \geq 5 - 4x$
13. $2y - 16 > 17 + 5y$
14. $6y + 7 > 4y - 3$
15. $4z - 22 < 6(z - 7)$
16. $8(a - 3) > 15a - 10$
17. $9(2 - 5m) - 4 \geq 13m + 8(3 - 7m)$
18. $18k - 3(8 - 4k) \leq 7(2 - 5k) + 27$
19. $10 - 5x > 2[3 - 5(x - 4)]$
20. $3[2 + 4(y + 5)] < 30 + 6y$
21. $\dfrac{z}{3} > 7 - \dfrac{z}{4}$
22. $\dfrac{t}{5} - 8 > -\dfrac{t}{3}$
23. $\dfrac{1}{3} + \dfrac{w + 2}{5} \geq \dfrac{w - 5}{3}$
24. $\dfrac{u - 2}{3} - \dfrac{u + 2}{4} \geq -\dfrac{2}{3}$

In Exercises 25 and 26, solve each inequality. Use a calculator, and round off each answer to three decimal places, writing the answer in set-builder notation.

25. $14.73(2.65x - 11.08) - 22.51x \geq 13.94x(40.27)$
26. $1.065 - 9.801x \leq 5.216x - 2.740(9.102 - 7.641x)$

In Exercises 27–30, convert to roster notation, and graph the solution set on the real number line.

27. $\{x|x + 3 < 10, x \in N\}$
28. $\{x|x + 5 < 8, x \in N\}$
29. $\{x|2(x + 3) \le 11, x \in N\}$
30. $\{x|3(x + 1) \le 17, x \in N\}$

Writing Problems

Express the answers in your own words and in complete sentences.

1. Explain the main differences between solving an equation and solving an inequality.
2. Explain why 3 is not a solution of $5x < 15$.
3. Describe the graph of the solution set of $x > 6$.
4. Describe what you would do *first* in solving the inequality $7 - 2x > 3 + x$, and explain why that would be your first step.
5. Describe what you would do *first* in solving the inequality $\frac{1}{2} + \frac{x + 9}{5} \le \frac{x + 1}{2}$, and explain why that would be your first step.

Exercises 3.2 Set II

In Exercises 1–4, graph each solution set, and convert the description of each set from set-builder notation to interval notation.

1. $\{x|x \le -7\}$
2. $\{x|x < 5\}$
3. $\{x|x > 6\}$
4. $\{x|x \ge 99\}$

In Exercises 5–8, write, in set-builder notation *and* in interval notation, the algebraic statement that describes the set of numbers graphed.

5.
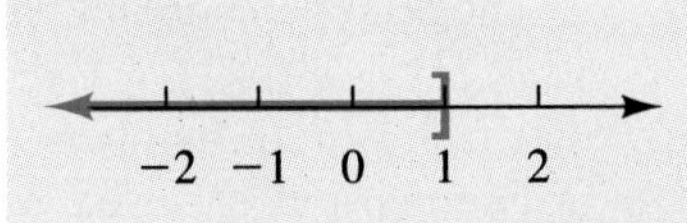

6.
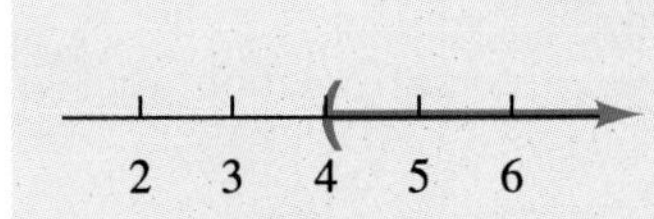

7.
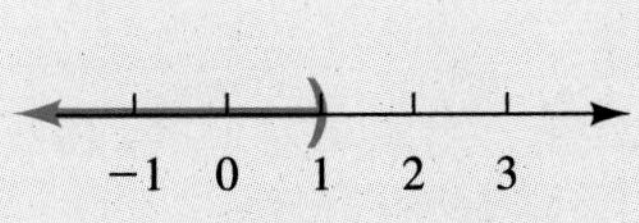

8.
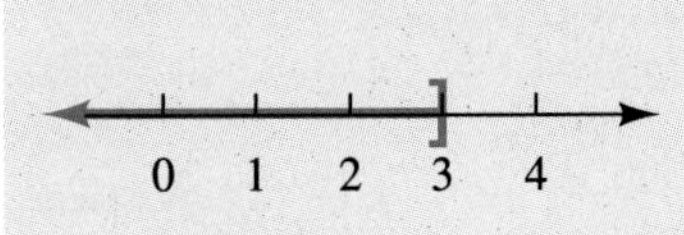

In Exercises 9–24, solve each inequality, describing the solution set in set-builder notation *and* in interval notation; graph each solution set.

9. $-3 \le x + 4$
10. $3x + 2 \ge 8$
11. $18 - 7y > -3$
12. $5 - 3y \le 8$
13. $11z - 7 < 5z - 13$
14. $3x - 8 \le 7 - 2x$
15. $3(2 + 3x) \ge 5x - 6$
16. $4(x - 1) \le 7x + 2$
17. $6(10 - 3t) + 25 \ge 4t - 5(3 - 2t)$
18. $6x - 4(3 - 2x) > 5(3 - 4x) + 7$
19. $6z < 2 - 4[2 - 3(z - 5)]$
20. $28 - 7x \ge [6 - 2(x - 1)]$
21. $\frac{w}{3} > 12 - \frac{w}{6}$
22. $\frac{x}{2} \le 5 - \frac{x}{3}$
23. $\frac{1}{2} + \frac{u + 9}{5} \ge \frac{u + 1}{2}$
24. $\frac{x - 4}{2} - \frac{x + 4}{4} < -\frac{3}{2}$

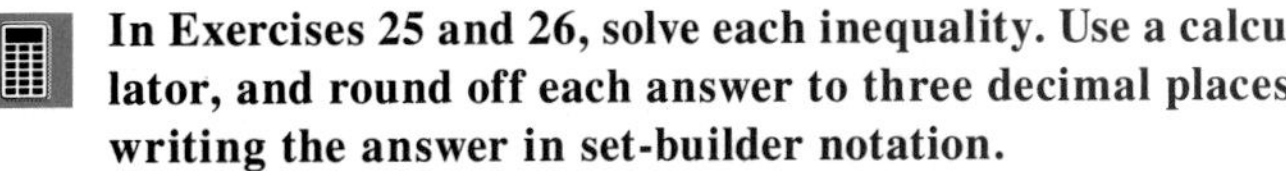

In Exercises 25 and 26, solve each inequality. Use a calculator, and round off each answer to three decimal places, writing the answer in set-builder notation.

25. $54.7x - 48.2(20.5 - 37.6x) \le 81.9(60.3x - 19.1) + 97.4$
26. $3.7 - 1.06x < 8.62 - 1.4(6.2 - 3.2x)$

In Exercises 27–30, convert to roster notation, and graph the solution set on the real number line.

27. $\{x|x + 2 < 8, x \in N\}$
28. $\{x|x - 1 < 4, x \in N\}$
29. $\{x|4(x + 4) \le 42, x \in N\}$
30. $\{x|3(x + 5) < 27, x \in N\}$

3.3 Combined Inequalities, Their Solutions, and Their Graphs

Combined Inequalities

Combined inequalities result when we connect two or more simple inequalities with the words *or* or *and* or when we find the union or the intersection of the solution sets of two or more simple inequalities. The word *or* indicates set union, whereas the word *and* indicates set intersection.

A *continued inequality* is one kind of combined inequality, because the statement $a < x < b$ is equivalent to the compound statement $a < x$ *and* $x < b$ (that is, $a < x < b$ is equivalent to the intersection of sets $\{x \mid x > a\}$ and $\{x \mid x < b\}$). If $a < b$, then $a < x < b$ is a valid continued inequality, and so is $b > x > a$.

EXAMPLE 1 Examples of combined inequalities:

a. $x > 3$ or $x \le -1$ means $\{x \mid x > 3\} \cup \{x \mid x \le -1\}$.
b. $x < -1$ and $x \ge -4$ means $\{x \mid x < -1\} \cap \{x \mid x \ge -4\}$.
c. $-3 \le x \le 4$ means $\{x \mid x \ge -3\} \cap \{x \mid x \le 4\}$.

Before we discuss *solving* combined inequalities, we should discuss bounded intervals.

Bounded Intervals and Interval Notation A **bounded interval** is an interval that *does* have "ends." The set $\{x \mid -3 < x < 4\}$, for example, is bounded; no number in the set is less than -3 (we say that -3 is the *lower bound*), and no number in the set is greater than 4 (we say that 4 is the *upper bound*). The set consists of all the real numbers *between* -3 and 4. The set $\{x \mid 2 \le x \le 5\}$ is bounded; its lower bound is 2, and its upper bound is 5.

As we mentioned in the last section, an *open interval* contains neither endpoint; $\{x \mid -3 < x < 4\}$ is an open interval. A *closed interval* contains both endpoints; $\{x \mid 2 \le x \le 5\}$ is a closed interval. *Half-open intervals* contain one endpoint but not the other; $\{x \mid -1 \le x < 5\}$ is a half-open interval.

To describe a bounded set in interval notation, we write a *pair* of numbers (with the smaller number *always* on the left) separated by a comma and enclosed within brackets and/or parentheses; a *bracket* indicates that an endpoint is included, whereas a *parenthesis* indicates that an endpoint is not included.

The following table compares set-builder notation, interval notation, and graphs for bounded sets.

Set-builder notation	*Interval notation*	*Graph*
$\{x \mid a \le x \le b\}$	$[a, b]$ A closed interval The brackets indicate that both a and b are included in the solution set	a b
$\{x \mid a < x < b\}$	(a, b) An open interval The parentheses indicate that neither a nor b is included in the solution set	a b
$\{x \mid a \le x < b\}$	$[a, b)$ A half-open interval	a b
$\{x \mid a < x \le b\}$	$(a, b]$ A half-open interval	a b

Notice that each graph is an *unbroken portion* of the number line. When we describe a graph algebraically, we can write a single statement, using a *continued inequality* or *interval notation*, if the graph occupies an *unbroken portion of the line* (see Example 2). If the graph occupies two separate portions of the number line, we must write the algebraic description as *two separate statements*, connecting the two separate statements with the symbol for set union or with the word *or* (see Example 3).

EXAMPLE 2 Write the algebraic statement that describes the set of numbers that is graphed, using set-builder notation and using interval notation.

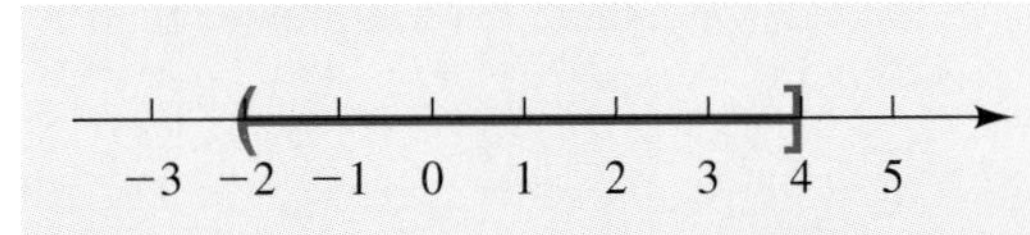

SOLUTION The graph is an unbroken portion of the number line. In set-builder notation, the set is $\{x \mid -2 < x \leq 4\}$. It *can* also be written as $\{x \mid x > -2\} \cap \{x \mid x \leq 4\}$, since the graph includes those numbers that are greater than -2 *and also* less than or equal to 4.

In interval notation, the set is $(-2, 4]$.

EXAMPLE 3 Write the algebraic statement that describes the set of numbers that is graphed, using set-builder notation and using interval notation.

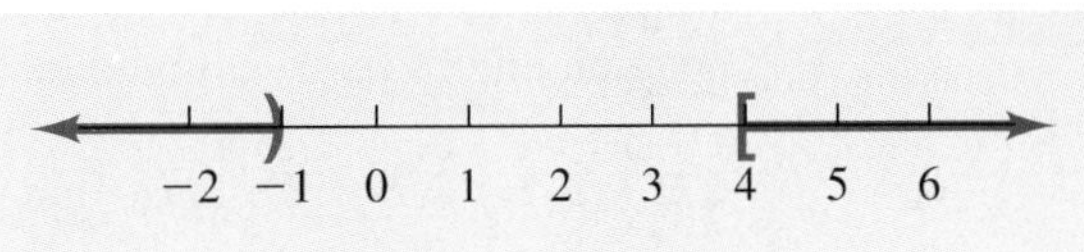

SOLUTION The graph occupies two separate portions of the number line, so we must write the algebraic description as two separate statements. In set-builder notation, the set is $\{x \mid x < -1\} \cup \{x \mid x \geq 4\}$. (The set can also be written as $\{x \mid x < -1$ *or* $x \geq 4\}$.)

In interval notation, the set is $(-\infty, -1) \cup [4, +\infty)$.

A Word of Caution A common error is to write the algebraic statement for the graph of Example 3 as $\{x \mid -1 > x \geq 4\}$, which is incorrect. The statement $-1 > x \geq 4$ is an *invalid* inequality, because $-1 \geq 4$ is a false statement.

Another common error is to write the algebraic statement for the graph of Example 3 as $\{x \mid -1 < x \geq 4\}$, and still another error is to write $\{x \mid -1 > x \leq 4\}$. Both of these are incorrect, because both inequality symbols in a continued inequality must have the *same sense*.

Solving Combined Inequalities

The solution set of a combined inequality that contains the word *or* is the *union* of the solution sets of the two simple inequalities (see Example 4). (We use the techniques learned in the last section to find the solution set of each simple inequality.)

EXAMPLE 4 Find and graph the solution set of $x - 1 > 2$ *or* $x - 1 \leq -2$.

SOLUTION We first solve each inequality separately:

$$x - 1 > 2 \quad \text{or} \quad x - 1 \leq -2$$

$$x > 3 \quad \text{or} \quad x \leq -1$$

The solution set is $\{x \mid x > 3 \text{ or } x \leq -1\}$, which can also be written $\{x \mid x > 3\} \cup \{x \mid x \leq -1\}$ or, in interval notation, $(-\infty, -1] \cup (3, +\infty)$. We must graph all the numbers that are greater than 3 or less than or equal to -1.

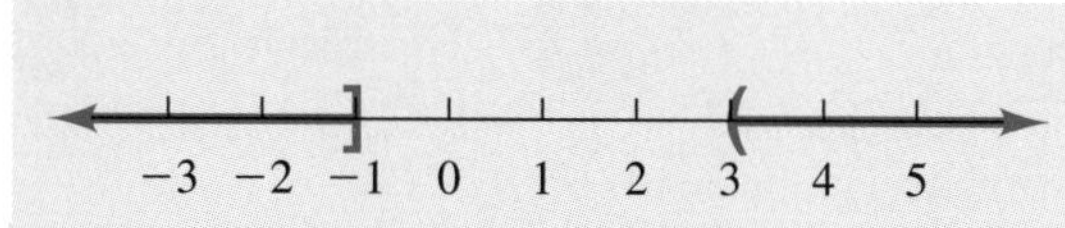

The solution set of a combined inequality that contains the word *and* is the *intersection* of the solution sets of the two simple inequalities. We often find the intersection by graphing the two simple inequalities separately and finding where the two graphs overlap (see Example 5).

EXAMPLE 5 Graph the solution set of $x > -2$ *and* $x < 4$. Describe the solution set in set-builder notation and in interval notation.

SOLUTION We must find the set of all the numbers that are greater than -2 *and at the same time* less than 4.

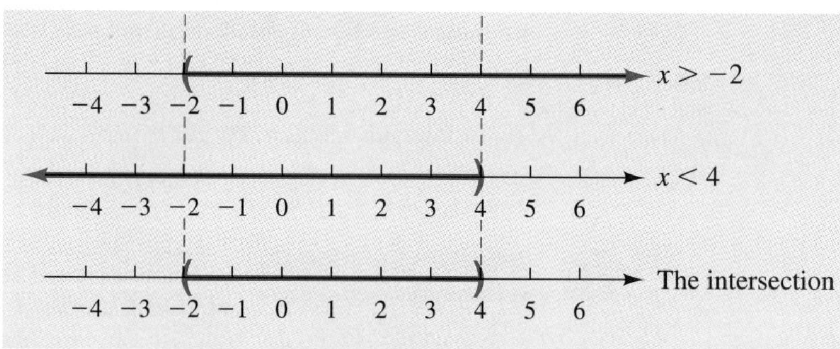

The last graph is the graph of the solution set of $x > -2$ and $x < 4$.

Because the final answer occupies a single portion of the number line, we can describe the set with a single statement—a continued inequality; because -2 is *not* in $x > -2$ and 4 is *not* in $x < 4$, we cannot include -2 and 4 in the final answer. In set-builder notation, the solution is $\{x \mid -2 < x < 4\}$, and in interval notation, it is $(-2, 4)$.

If $a > b$, then $a < x < b$ is an *invalid* continued inequality, and the solution set is the empty set (see Example 6).

EXAMPLE 6 Graph the set $\{x \mid 7 < x < 2\}$.

SOLUTION The statement $7 < x < 2$ is an *invalid* continued inequality because $7 < 2$ is *false*. The solution set is $\{\ \}$, and the graph will have no points graphed on it.

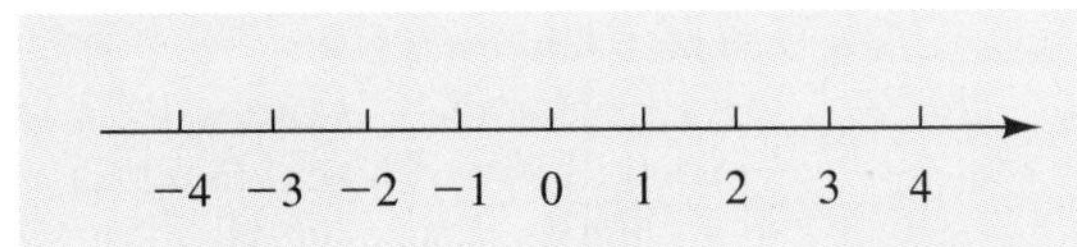

Note We cannot write the empty set in interval notation.

EXAMPLE 7 Solve $x > 1$ *or* $x < 4$ graphically, writing the solution set in set-builder notation and in interval notation.

SOLUTION We graph each inequality separately; then we include in our final answer any number that is in one graph *or* the other *or* both.

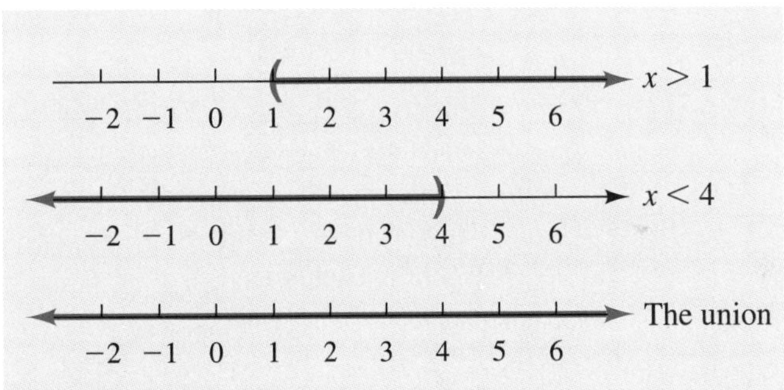

The last graph is the graph of the solution set of $x > 1$ or $x < 4$.

The solution set is the set of *all* real numbers. In set-builder notation, the solution set is $\{x \mid x \in R\}$, and in interval notation, it is $(-\infty, +\infty)$.

Solving Continued Inequalities

When we solve a continued inequality, we want to rewrite the inequality so that x is all by itself between the two inequality symbols. That is, we want our answer to be in the form $a < x < b$ (if $a < b$) or $a > x > b$ (if $a > b$).

EXAMPLE 8 Find and graph the solution set of $2 < x + 5 \leq 9$.

SOLUTION Notice first of all that $2 \leq 9$ is a true statement, so the continued inequality is valid. Also notice that no restrictions were put on the domain of the variable.

Since $\{x \mid 2 < x + 5 \leq 9\} = \{x \mid x + 5 > 2\} \cap \{x \mid x + 5 \leq 9\}$, we could first find the solution set of each of the inequalities separately and then find the intersection of the two solution sets. However, the continued inequality may be more conveniently solved as follows:

$$2 < x + 5 \le 9$$

$$2 - 5 < x + 5 - 5 \le 9 - 5 \quad \text{Adding } -5 \text{ to all three parts of the inequality}$$

$$-3 < x \le 4$$

The solution set is $\{x \mid -3 < x \le 4\}$, or, in interval notation, $(-3, 4]$. The graph must not include the -3 (we will have a left parenthesis at -3), but it should include the 4 (we will have a right bracket at 4), and we will shade the portion of the line between -3 and 4.

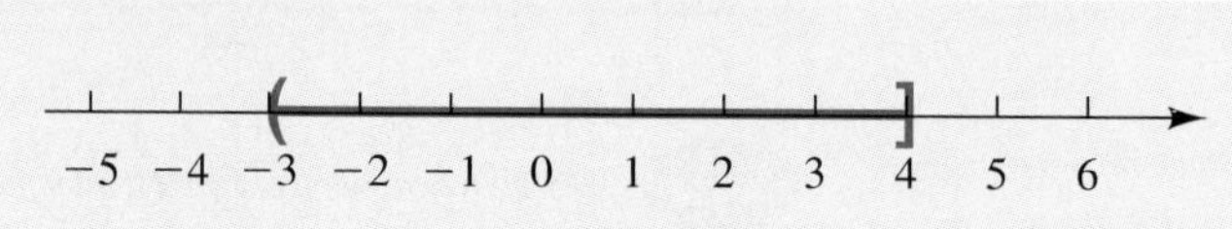

EXAMPLE 9

Solve $-7 \le 2x + 1 \le 5$, writing the solution set in set-builder notation and in interval notation, and graph the solution set.

SOLUTION The inequality is a valid continued inequality, since $-7 \le 5$ is a true statement.

$$-7 \le 2x + 1 \le 5$$

$$-7 - 1 \le 2x + 1 - 1 \le 5 - 1 \quad \text{Adding } -1 \text{ to all three parts of the inequality}$$

$$-8 \le 2x \le 4$$

$$-4 \le x \le 2$$

Dividing all three parts of the inequality by *positive* 2 does not change the senses of the inequalities

The solution set is $\{x \mid -4 \le x \le 2\}$, or, in interval notation, $[-4, 2]$, and the graph is as follows:

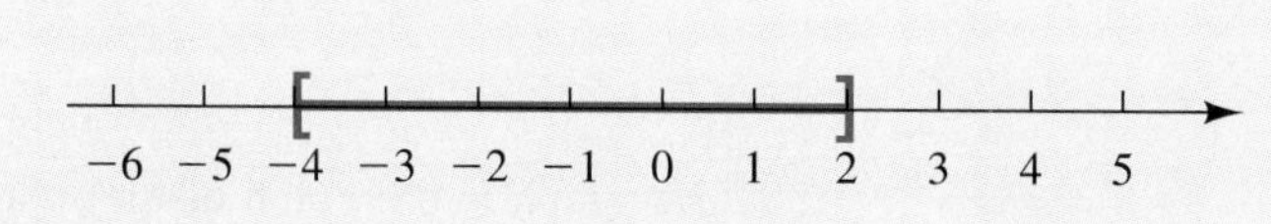

EXAMPLE 10

Write $\{x \mid -7 \le 2x + 1 \le 5,\ x \in N\}$ in roster notation, and graph the set.

SOLUTION The inequality is the same inequality as in Example 9, but now a restriction has been put on the domain. The only *natural numbers* that satisfy the inequality $-4 \le x \le 2$ are 1 and 2; therefore,

$$\{x \mid -7 \le 2x + 1 \le 5, x \in N\} = \{1, 2\}$$

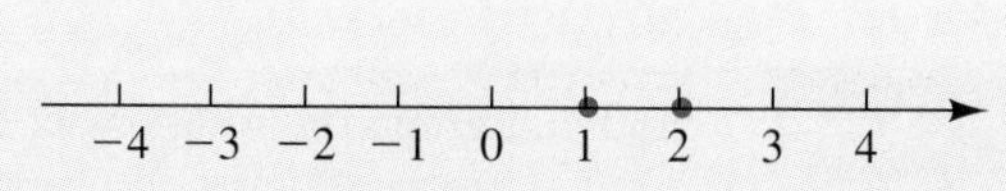

Note We *cannot* use interval notation for the result in Example 10, since the solution set does not include *all real numbers between* 1 and 2.

Exercises 3.3
Set I

In Exercises 1–8, write, in set-builder notation and in interval notation, the algebraic statement that describes the set that is graphed.

1. 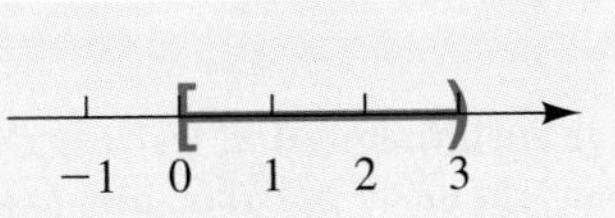

2. 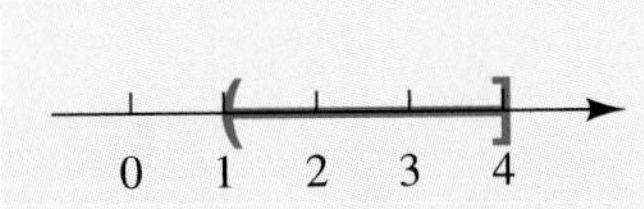

3. 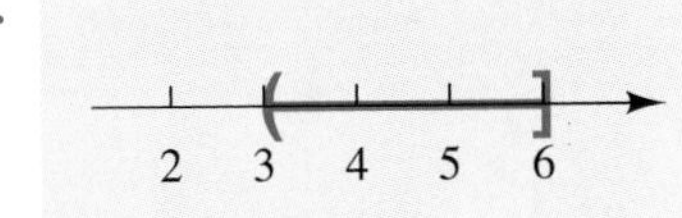

4. 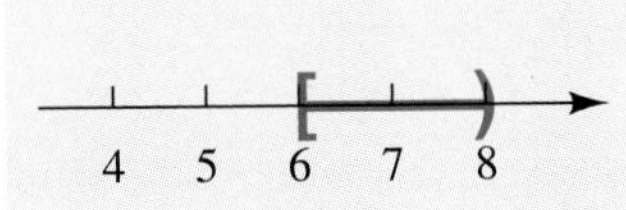

5. 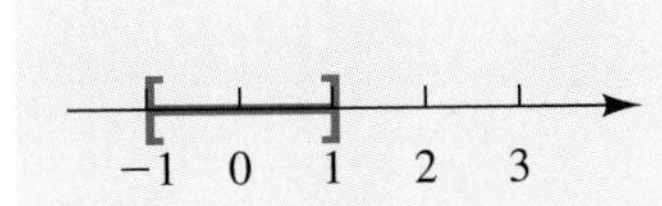

6. 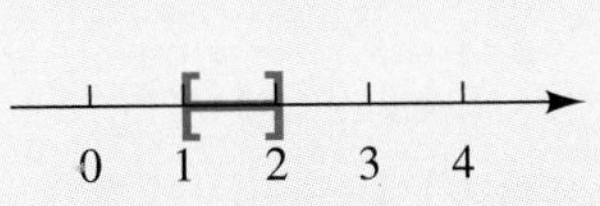

7. 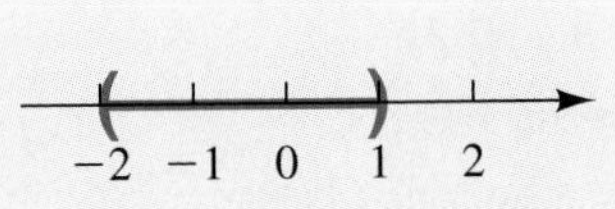

8. 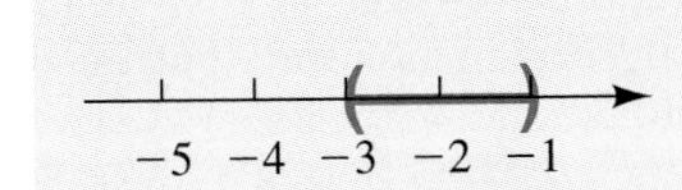

In Exercises 9–22, solve each inequality, writing the solution in interval notation *whenever possible*, and graph the solution set, unless it is { }.

9. $5 > x - 2 \geq 3$
10. $7 > x - 3 \geq 4$
11. $-5 \geq x - 3 \geq 2$
12. $-3 \geq x - 2 \geq 4$
13. $-4 < 3x - 1 \leq 7$
14. $-6 < 4x - 2 \leq 5$
15. $x - 1 > 3$ or $x - 1 < -3$
16. $x - 2 > 5$ or $x - 2 < -5$
17. $2x + 1 \geq 3$ or $2x + 1 \leq -3$
18. $3x - 2 \geq 5$ or $3x - 2 \leq -5$
19. $x > 4$ and $x \geq 2$
20. $x < 3$ and $x < -1$
21. $x > 4$ or $x \geq 2$
22. $x < 3$ or $x < -1$

In Exercises 23–28, rewrite each inequality in roster notation and graph it.

23. $\{x \mid -5 \leq x - 3 \leq 2, x \in N\}$
24. $\{x \mid -3 \leq x - 2 \leq 4, x \in N\}$
25. $\{x \mid 4 \geq x - 3 > -5, x \in J\}$
26. $\{x \mid 6 \geq x - 2 > -4, x \in J\}$
27. $\{x \mid -3 \leq 2x + 1 \leq 7, x \in N\}$
28. $\{x \mid -5 \leq 2x + 3 \leq 5, x \in N\}$

Writing Problems

Express the answers in your own words and in complete sentences.

1. Explain the meaning of $-3 \leq x < 7$.
2. Describe how you would graph $-3 \leq x < 7$.
3. Describe what you would do *first* in solving $2 < 3x + 4 \leq 5$, and explain why that would be your first step.
4. Explain why $8 > 2x + 5 \geq 4$ is a valid continued inequality.
5. Describe how you would find the solution set of $x > 4$ *and* $x \leq 9$.
6. Describe how you would find the solution set of $x > 4$ *or* $x \leq 9$.

Exercises 3.3
Set II

In Exercises 1–8, write, in set-builder notation and in interval notation, the algebraic statement that describes the set that is graphed.

1.
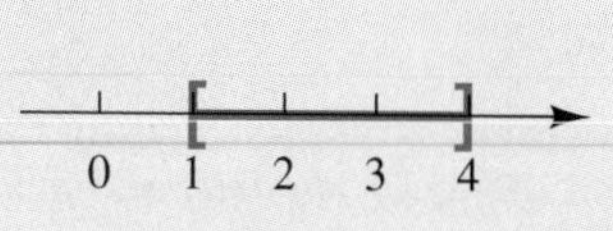

2.
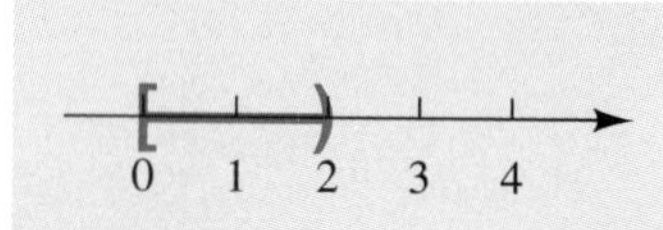

3.
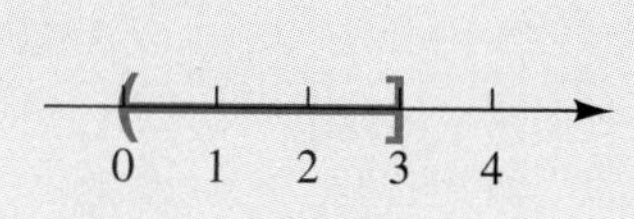

4.
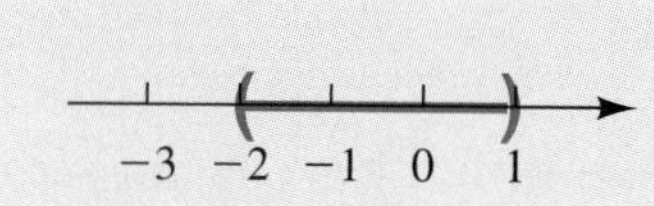

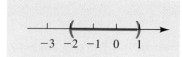

5.
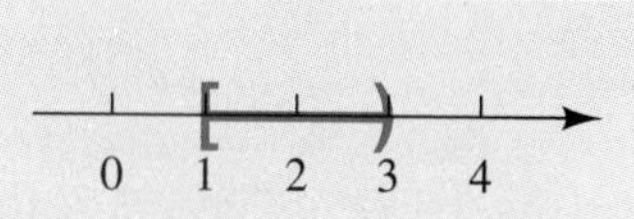

6.
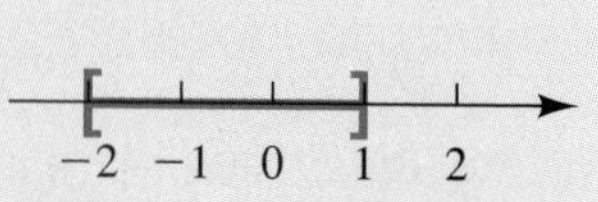

7.
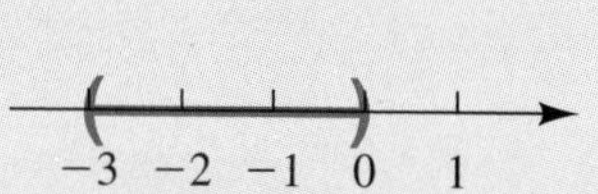

8.
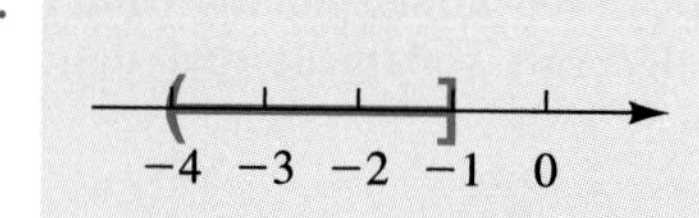

In Exercises 9–22, solve each inequality, writing the solution in interval notation *whenever possible*, and graph the solution set, unless it is { }.

9. $8 > x - 1 \geq 2$
10. $-3 < x + 2 < 5$
11. $-6 \geq x - 4 \geq 3$
12. $7 > x + 3 \geq -1$
13. $-5 < 3x - 2 \leq 4$
14. $0 \leq 2x + 4 < 8$
15. $x - 3 > 4$ or $x - 3 < -4$
16. $x + 5 > 1$ or $x + 5 < -1$
17. $4x - 1 \geq 2$ or $4x - 1 \leq -2$
18. $3x + 2 > 5$ or $3x + 2 < -5$
19. $x < 6$ and $x < -2$
20. $x > 3$ and $x < 6$
21. $x < 6$ or $x < -2$
22. $x > 3$ or $x < 6$

In Exercises 23–28, rewrite each inequality in roster notation and graph it.

23. $\{x \mid -4 \geq x - 2 > 1, x \in N\}$
24. $\{x \mid 2 \leq x + 4 < 8, x \in N\}$
25. $\{x \mid 5 \geq x - 2 > -3, x \in J\}$
26. $\{x \mid -2 \leq x - 1 < 4, x \in J\}$
27. $\{x \mid -2 \leq 2x + 4 \leq 10, x \in N\}$
28. $\{x \mid -6 < 2x - 3 \leq 5, x \in J\}$

3.4 Absolute Value Equations and Inequalities, Their Solutions, and Their Graphs

Recall that the definition of the absolute value of x is

$$|x| = \begin{cases} x & \text{if } x \geq 0 \\ -x & \text{if } x < 0 \end{cases}$$

In this section, we consider solving conditional equations and inequalities that contain absolute value symbols, where the expression within the absolute value symbols is usually not just a simple x.

Solving Equations of the Form $|X| = k$

We first consider solving equations of the form $|X| = k$, where X represents some algebraic expression and where $k > 0$.

Recall that $|x|$ can be interpreted as the distance between x and 0 on the number line. As we can see from the graph in Figure 1, if $k > 0$, there are *two* numbers that are k units from 0; therefore, the equation $|x| = k$ has *two* solutions, k and $-k$.

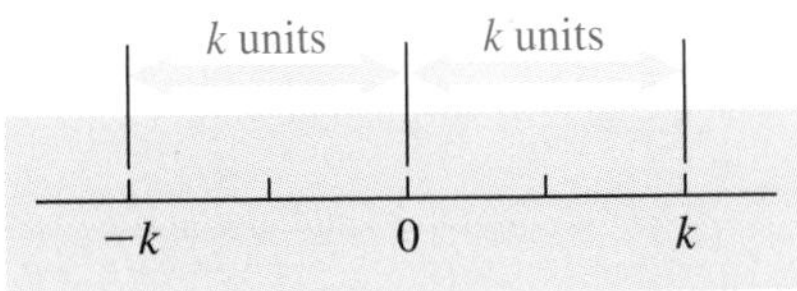

Figure 1

We generalize this idea in Property 1 (which is proved in Appendix A).

Property 1 for absolute values: solving an equation of the form $|X| = k$

For all positive real numbers k,

$$\text{if} \quad |X| = k, \quad \text{then} \quad X = k \quad \text{or} \quad X = -k$$

where X represents any algebraic expression.

Note It is also possible to solve an equation of the form $|X| = k$ by using the definition of absolute value, which states that $|X| = X$ or $|X| = -X$. In this case, we write "If $|X| = k$, then $X = k$ or $-X = k$." Then $-X = k$ can be rewritten as $X = -k$, and we have the results given in Property 1.

Note If $k = 0$, then $|X| = k$ has just one solution (we find the solution by solving the equation $X = 0$). If $k < 0$, then $|X| = k$ has *no* solution, since $|X|$ is always positive and, thus, can never equal a negative number.

Property 1 permits us to rewrite an equation that contains absolute value symbols as two separate equations containing *no* absolute value symbols; we then solve each resulting equation separately (see Example 1).

EXAMPLE 1 Solve $\left|\frac{3-2x}{5}\right| = 2$, and graph its solution set.

SOLUTION Because the given equation is of the form $|X| = k$, we can use Property 1. According to Property 1,

if $$\left|\frac{3-2x}{5}\right| = 2$$

then $$\frac{3-2x}{5} = 2 \quad \text{or} \quad \frac{3-2x}{5} = -2$$

$$3 - 2x = 10 \quad \text{or} \quad 3 - 2x = -10$$

$$-2x = 7 \quad \text{or} \quad -2x = -13$$

$$x = -\tfrac{7}{2} \quad \text{or} \quad x = \tfrac{13}{2}$$

√ **Check for** $-\frac{7}{2}$ $\left|\frac{3-2\left(-\frac{7}{2}\right)}{5}\right| = \left|\frac{3+7}{5}\right| = \left|\frac{10}{5}\right| = |2| = 2$

√ **Check for** $\frac{13}{2}$ $\left|\frac{3-2\left(\frac{13}{2}\right)}{5}\right| = \left|\frac{3-13}{5}\right| = \left|\frac{-10}{5}\right| = |-2| = 2$

Therefore, the solution set is $\left\{-3\frac{1}{2}, 6\frac{1}{2}\right\}$. The graph of the solution set is

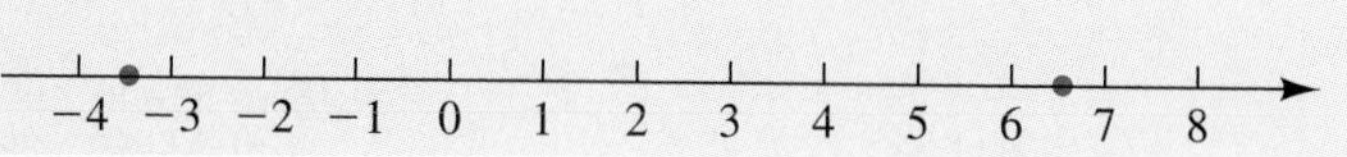

EXAMPLE 2 Solve $|x| = -3$.

SOLUTION We know that $|x| \geq 0$, and no number that is greater than or equal to 0 can equal -3. Therefore, there is *no* solution.

Solving Inequalities of the Forms $|X| < k$ and $|X| \leq k$

In Example 3, we first find the solution set of an inequality of the form $|X| \leq k$ by using the problem-solving technique of guessing-and-checking.

EXAMPLE 3 Find the solution set of $|x| \leq 3$.

SOLUTION The statement $|x| \leq 3$ is equivalent to "$|x| = 3$ *or* $|x| < 3$." Solving $|x| = 3$, we find the solutions 3 and -3. We graph these points (they *are* in the solution set) below, and we see that they separate the number line into three portions: the portion to the left of -3, the portion between -3 and 3, and the portion to the right of 3.

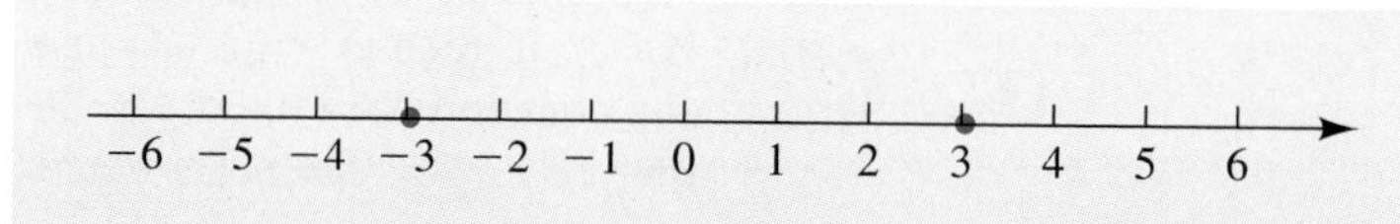

Now we try one number in each interval:

Does -5 make the statement $|x| \leq 3$ true? No. $|-5| = 5$, and $5 \not\leq 3$.

Does -2 make the statement $|x| \leq 3$ true? Yes! $|-2| = 2$, and $2 \leq 3$.

Does 4 make the statement $|x| \leq 3$ true? No. $|4| = 4$, and $4 \not\leq 3$.

You might try *any* other numbers between -3 and 3; they will all satisfy $|x| \leq 3$. In the graph of the solution set, we use brackets at -3 and 3, since they *are* in the solution set, and we shade the line between those points (see the graph below).

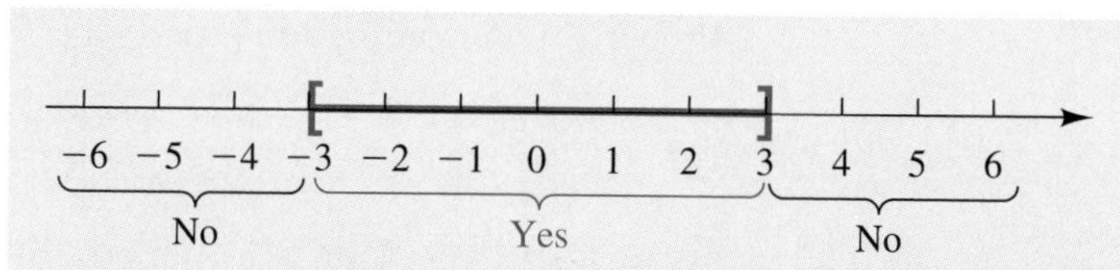

We can also solve $|x| < 3$ by reasoning as follows: $|x|$ can be interpreted as being the distance between x and 0 on the number line, so to solve the inequality $|x| < 3$, we need to find all the numbers that are *less than* 3 units from 0. This means that all the numbers *between* -3 and 3 will be in the solution set.

The solution set of $|x| \leq 3$ is $\{x \mid -3 \leq x \leq 3\}$, or, in interval notation, $[-3, 3]$.

In general, because $|x|$ can be interpreted as the distance between x and 0 on the number line, to solve the inequality $|x| < k$ (where $k > 0$) means to find all the numbers that are *less than* k units from 0. As we can see from the graph in Figure 2, all the numbers *between* $-k$ *and* k will be in the solution set.

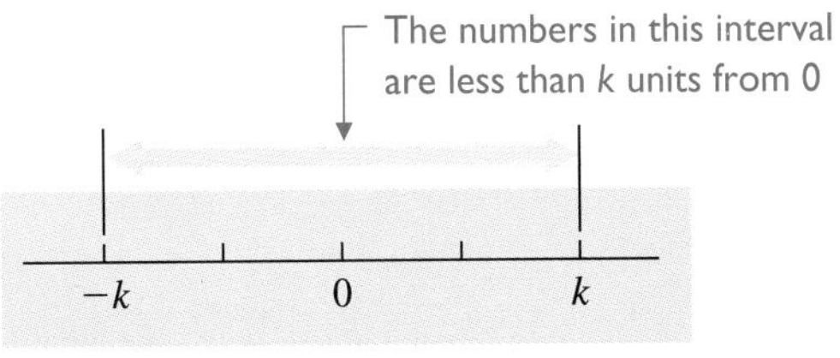

Figure 2

Because the graph occupies a single portion of the number line, we can write the solution set in set-builder notation as $\{x \mid -k < x < k\}$. We generalize this idea in Property 2, which is proved in Appendix A.

Property 2 for absolute values: solving inequalities of the forms $|X| < k$ and $|X| \le k$

For all positive real numbers k,

$$\text{if} \quad |X| < k, \quad \text{then} \quad -k < X < k$$

$$\text{if} \quad |X| \le k, \quad \text{then} \quad -k \le X \le k$$

where X represents any algebraic expression. The graph of the solution set will *always* be an uninterrupted portion of the number line.

Note If $k < 0$, then there is no solution for $|X| < k$ or for $|X| \le k$. For example, $|x| < -2$ has no solution, and neither does $|x| \le -2$, since $|x|$ is always greater than or equal to 0 and cannot be less than or equal to -2.

If $k = 0$, then the inequality $|X| \le k$ becomes $|X| \le 0$, and this is equivalent to the *equation* $X = 0$, which can be solved. If $k = 0$, the inequality $|X| < k$ becomes $|X| < 0$, and there is *no solution* for $|X| < 0$. Therefore, the solution set of $|x| \le 0$ is $\{0\}$, and the solution set of $|x| < 0$ is $\{\ \}$.

In Examples 4 and 5, we use Property 2 for solving conditional inequalities that are of the forms $|X| < k$ and $|X| \le k$.

EXAMPLE 4 Solve $|2x - 3| < 5$, and graph its solution set.

SOLUTION Because the inequality is of the form $|X| < k$, we can use Property 2 to rewrite the inequality *with no absolute value symbols.*

$$|2x - 3| < 5$$

$$-5 < 2x - 3 < 5 \quad \text{Using Property 2}$$

$$-2 < 2x < 8 \quad \text{Adding 3 to all three parts}$$

$$-1 < x < 4 \quad \text{Dividing all three parts by 2}$$

The solution set is $\{x \mid -1 < x < 4\}$, or, in interval notation, $(-1, 4)$. The graph is as follows:

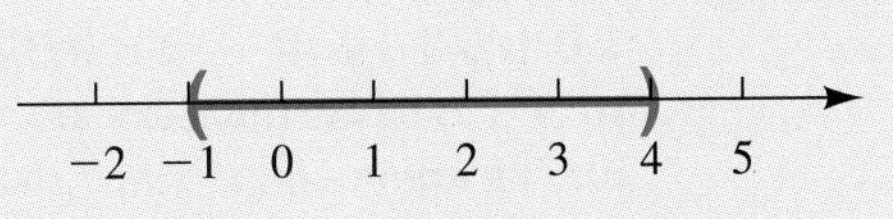

EXAMPLE 5 Solve $\left|\frac{4-3x}{2}\right| \le 6$, and graph its solution set.

SOLUTION Because the inequality is of the form $|X| \le k$, we can use Property 2 to rewrite the inequality with no absolute value symbols.

$$\left|\frac{4-3x}{2}\right| \le 6$$

$$-6 \le \frac{4-3x}{2} \le 6 \quad \text{Using Property 2}$$

$$-12 \le 4 - 3x \le 12 \quad \text{Multiplying all three parts by 2}$$

$$-16 \le -3x \le 8 \quad \text{Adding } -4 \text{ to all three parts}$$

$$\frac{-16}{-3} \ge \frac{-3x}{-3} \ge \frac{8}{-3} \quad \text{When we divide all three parts by } -3 \text{, we must } \textit{change the senses} \text{ of the inequalities}$$

$$\tfrac{16}{3} \ge x \ge -\tfrac{8}{3}$$

As we mentioned earlier, it is customary (though not essential) to write a continued inequality so that the smaller number is on the left. If we do this, we find that the solution set is $\{x \mid -\frac{8}{3} \le x \le \frac{16}{3}\}$, or, in interval notation, $\left[-\frac{8}{3}, \frac{16}{3}\right]$. The graph of the solution set is as follows:

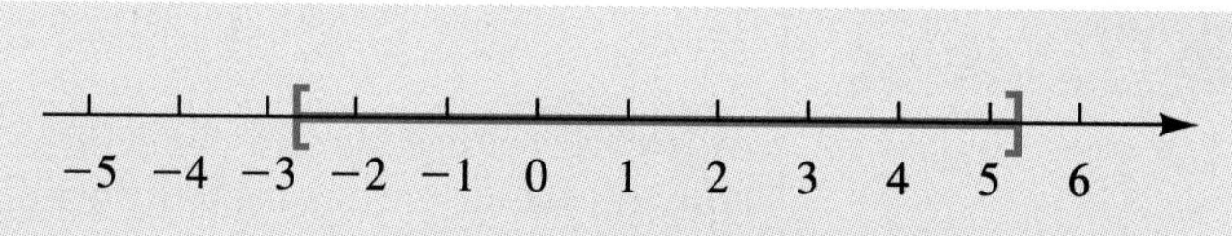

A Word of Caution In Example 5, it is acceptable to write the solution set as $\{x \mid \frac{16}{3} \ge x \ge -\frac{8}{3}\}$. However, in *interval* notation, the solution *cannot* be written as $[\frac{16}{3}, -\frac{8}{3}]$. In interval notation, the smaller number *must* be written on the left.

EXAMPLE 6 Find the solution set of $|x + 3| < -5$.

SOLUTION Since $|x + 3|$ is always positive or zero, it can't be less than -5. Therefore, the solution set is the empty set, { }.

Solving Inequalities of the Forms $|X| > k$ and $|X| \ge k$

In Example 7, we first find the solution set of an inequality of the form $|X| > k$ by using the problem-solving technique of guessing-and-checking. To use this technique, we must temporarily substitute an equal sign for the inequality symbol, in order to find the points that separate the number line into intervals; then we test one number in each interval.

EXAMPLE 7 Find the solution set of $|x| > 3$.

SOLUTION Substituting an equal sign for the greater than symbol, we have $|x| = 3$, and its solutions are 3 and -3. We graph these points on the following number line as *hollow dots* (since they are *not* in the solution set), and we see that they separate the number line into three portions: the portion to the left of -3, the portion between -3 and 3, and the portion to the right of 3.

We try one number in each interval:

Does -6 make the statement $|x| > 3$ true? Yes! $|-6| = 6$, and $6 > 3$.

Does -1 make the statement $|x| > 3$ true? No. $|-1| = 1$, and $1 \not> 3$.

Does 5 make the statement $|x| > 3$ true? Yes! $|5| = 5$, and $5 > 3$.

You might try *any* other numbers less than -3 or greater than 3; they will all satisfy $|x| > 3$. In the graph of the solution set, we must use a *right parenthesis* at -3 (indicating that -3 is not included in the graph but that points to its left *are* included) and a *left parenthesis* at 3 (indicating that 3 is not included in the graph but that points to its right *are* included). Then we shade the number line to the *left* of -3 and to the *right* of 3 (see the graph below).

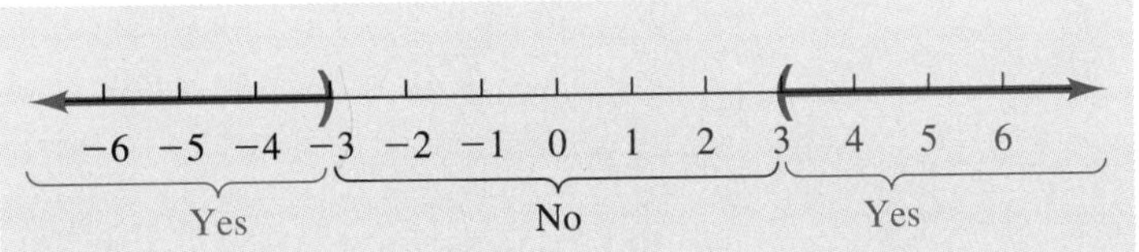

We can also solve $|x| > 3$ by reasoning as follows: To solve $|x| > 3$, we need to find all the numbers that are *more than* 3 units from 0. This means that all the numbers *less than* -3 or *greater than* 3 will be in the solution set.

The graph of the solution set occupies two separate portions of the number line. Therefore, we must write the solution set as two separate statements. The solution set of $|x| > 3$ is $\{x \mid x < -3\} \cup \{x \mid x > 3\}$, or, in interval notation, $(-\infty, -3) \cup (3, +\infty)$.

In general, because $|x|$ can be interpreted as the distance between x and 0 on the number line, to solve the inequality $|x| > k$ (where $k > 0$) means to find all the numbers that are *more than* k units from 0. As we can see from the graph in Figure 3, all the numbers *less than* $-k$ *or more than* $+k$ will be in the solution set.

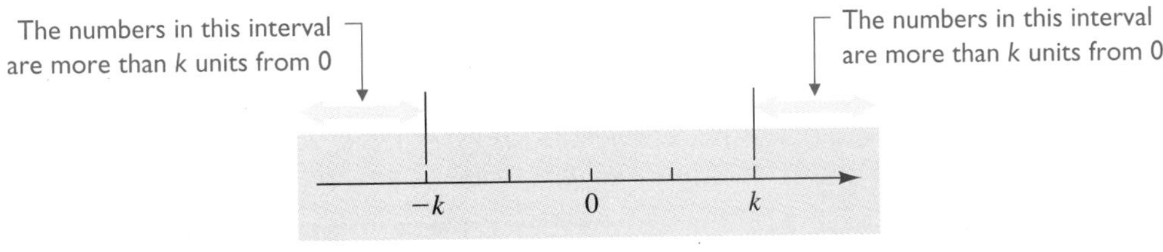

Figure 3

Because the graph occupies two separate portions of the number line, we write the solution set in set-builder notation as $\{x \mid x < -k\} \cup \{x \mid x > k\}$. We generalize this idea in Property 3, which is proved in Appendix A.

Property 3 for absolute values: solving inequalities of the forms $|X| > k$ and $|X| \geq k$

For all positive real numbers k,

$$\text{if} \quad |X| > k, \quad \text{then} \quad X > k \quad \text{or} \quad X < -k$$

$$\text{if} \quad |X| \geq k, \quad \text{then} \quad X \geq k \quad \text{or} \quad X \leq -k$$

where X represents any algebraic expression. The graph of the solution set will *always* be along two separate portions of the number line.

Note If $k < 0$, then the solution set of the inequality $|X| > k$ is the set of *all* real numbers, as is the solution set of $|X| \geq k$. For example, $|x| > -2$ is true for *all* values of x, since the absolute value of any real number is greater than or equal to zero and zero is, of course, greater than -2.

If $k = 0$, then the solution set of the inequality $|X| \geq k$ is R, the set of all real numbers; that is, $|X| \geq 0$ is true for all real numbers. If $k = 0$, then the solution set of $|X| > k$ is all X *except* 0.

A Word of Caution Notice in Property 3 that when we rewrite $|X| > k$ without absolute value symbols, the two inequalities have *opposite senses*. That is, if $|X| > k$, then $X > k$ or $X < -k$; *the senses are opposite*— X must be *greater than positive* k or *less than negative* k. Similarly, if $|X| \geq k$, then $X \geq k$ or $X \leq -k$; *the senses are opposite.*

In Examples 8 and 9, we use Property 3 for solving conditional inequalities that are of the forms $|X| > k$ and $|X| \geq k$.

EXAMPLE 8 Solve $|5 - 2x| \geq 3$, writing the solution set in set-builder notation and in interval notation, and graph the solution set.

SOLUTION Because the inequality is of the form $|X| \geq k$, we can use Property 3 to rewrite the inequality *with no absolute value symbols.*

$$|5 - 2x| \geq 3$$

It is *incorrect* to use the word *and* here

$5 - 2x \geq 3$	or	$5 - 2x \leq -3$	Using Property 3
$-2x \geq -2$	or	$-2x \leq -8$	Adding -5 to both sides of both inequalities
$\dfrac{-2x}{-2} \leq \dfrac{-2}{-2}$	or	$\dfrac{-2x}{-2} \geq \dfrac{-8}{-2}$	When we divide both sides of both inequalities by -2, we must *change the senses* of the inequalities
$x \leq 1$	or	$x \geq 4$	

The solution set is $\{x \mid x \leq 1\} \cup \{x \mid x \geq 4\}$, or, in interval notation, $(-\infty, 1] \cup [4, +\infty)$. The graph of the solution set is

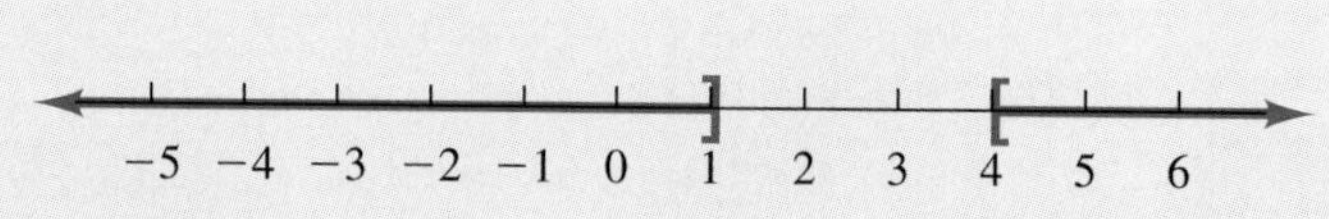

A Word of Caution A common error in Example 8 is to *incorrectly* write the answer as $1 \leq x \geq 4$; this is an *invalid* continued inequality, since the senses are different. (Furthermore, $1 \leq x$ is not equivalent to $x \leq 1$.) The solution set *must* be written as *two separate statements.*

EXAMPLE 9 Solve $\left|3 - \frac{x}{2}\right| > 5$, and graph its solution set.

SOLUTION Because the inequality is of the form $|X| > k$, we can use Property 3 to rewrite the inequality with no absolute value symbols.

$$\left|3 - \frac{x}{2}\right| > 5$$

It is *incorrect* to use the word *and* here

$3 - \frac{x}{2} > 5$	or	$3 - \frac{x}{2} < -5$	Using Property 3
$2\left(3 - \frac{x}{2}\right) > 2(5)$	or	$2\left(3 - \frac{x}{2}\right) < 2(-5)$	Multiplying both sides of both inequalities by 2
$6 - x > 10$	or	$6 - x < -10$	
$-x > 4$	or	$-x < -16$	Adding -6 to both sides of both inequalities
$\frac{-x}{-1} < \frac{4}{-1}$	or	$\frac{-x}{-1} > \frac{-16}{-1}$	When we divide both sides of both inequalities by -1, we must *change the senses* of the inequalities
$x < -4$	or	$x > 16$	

The solution set is $\{x \mid x < -4\} \cup \{x \mid x > 16\}$, or, in interval notation, $(-\infty, -4) \cup (16, +\infty)$. The graph of the solution set is

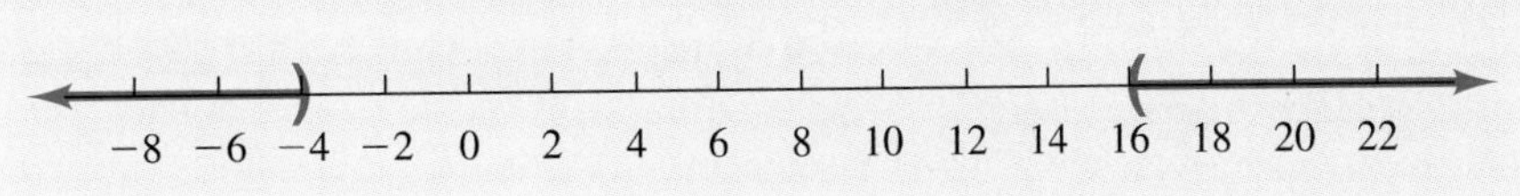

Exercises 3.4
Set I

Solve each of the following equations or inequalities, and graph the solution set, unless it is { }.

1. $|x| = 3$
2. $|x| = 5$
3. $|3x| = 12$
4. $|2x| = 10$
5. $|x| < 2$
6. $|x| < 7$
7. $|4x| < 12$
8. $|3x| < 9$
9. $|5x| \le 25$
10. $|2x| \le 2$
11. $|x| > 2$
12. $|x| > 3$
13. $|3x| \ge -3$
14. $|4x| \ge -5$
15. $|x + 2| = 5$
16. $|x + 3| = 3$
17. $|x - 3| < 2$
18. $|x - 4| < 1$
19. $|x + 4| \le -3$
20. $|x + 2| \le -4$
21. $|x + 1| > 3$
22. $|x - 2| > 4$
23. $|x + 5| \ge 2$
24. $|x + 4| \ge 1$
25. $|3x + 4| = 3$
26. $|4x + 3| = 5$
27. $|2x - 3| < 4$
28. $|3x - 1| < 5$
29. $|3x - 5| \ge 6$
30. $|4x - 1| \ge 3$
31. $|1 - 2x| \le 5$
32. $|2 - 3x| < 4$
33. $|2 - 3x| > 4$
34. $|5 - 2x| \ge 6$
35. $\left|\frac{5x + 2}{3}\right| \ge 2$
36. $\left|\frac{4x + 3}{5}\right| \ge 3$
37. $\left|\frac{3x - 4}{5}\right| < 1$
38. $\left|\frac{5x - 1}{2}\right| < 2$
39. $\left|\frac{1 - x}{2}\right| = 3$
40. $\left|\frac{3 - x}{2}\right| = 2$
41. $\left|\frac{5 - x}{2}\right| \le 2$
42. $\left|\frac{4 - x}{3}\right| \le 1$
43. $\left|3 - \frac{x}{2}\right| > 2$
44. $\left|4 - \frac{x}{3}\right| > 1$
45. $\left|4 - \frac{x}{2}\right| < 2$
46. $\left|5 - \frac{x}{3}\right| < 2$

Writing Problems

Express the answers in your own words and in complete sentences.

1. Explain why $|x + 5| = -3$ has no solution.
2. Explain why $|3x + 8| < -8$ has no solution.
3. Explain why *every real number* is a solution of $|5x - 8| > -1$.
4. Explain why $|5x - 8| \neq 5x + 8$.

Exercises 3.4 Set II

Solve each of the following equations or inequalities, and graph the solution set, unless it is { }.

1. $|x| = 4$
2. $|x| = -3$
3. $|4x| = 16$
4. $|x| = 0$
5. $|x| < 4$
6. $|x| < -1$
7. $|3x| < 12$
8. $|2x| \leq 4$
9. $|3x| \leq 15$
10. $|8x| < 12$
11. $|x| > 1$
12. $|x| > 0$
13. $|2x| \geq -6$
14. $|5x| < 15$
15. $|x + 7| = 5$
16. $|x - 3| = 2$
17. $|x - 5| < 3$
18. $|x + 1| \leq 1$
19. $|x + 3| \leq -2$
20. $|x - 2| < 0$
21. $|x + 3| > 5$
22. $|x + 2| \geq -3$
23. $|x + 5| \geq 2$
24. $|x - 1| > 1$
25. $|2x + 3| = 5$
26. $|3x - 1| = 2$
27. $|3x - 4| < 7$
28. $|5x + 3| \leq -8$
29. $|5x - 3| \geq 4$
30. $|4x + 2| > 2$
31. $|3 - 4x| \leq 7$
32. $|4 - 2x| < 1$
33. $|2 - 4x| \geq 5$
34. $|7 - 3x| > 0$
35. $\left|\frac{2x + 5}{5}\right| \geq 3$
36. $\left|\frac{5x - 2}{3}\right| \leq 2$
37. $\left|\frac{3x - 4}{2}\right| < 1$
38. $\left|\frac{4x + 7}{3}\right| \geq 1$
39. $\left|\frac{3 - x}{4}\right| = 5$
40. $\left|\frac{4 - x}{3}\right| = 2$
41. $\left|\frac{3 - x}{2}\right| \leq 5$
42. $\left|\frac{6 - 2x}{3}\right| \leq 4$
43. $\left|5 - \frac{x}{2}\right| > 1$
44. $\left|7 - \frac{x}{3}\right| > 5$
45. $\left|4 - \frac{x}{3}\right| < 1$
46. $\left|8 - \frac{x}{2}\right| \leq 2$

3.5 Applications: An Introduction

The main reason for studying algebra is to equip yourself with the tools necessary to solve problems, and most problems are expressed in words. The skills learned in this section can be applied to solving mathematical problems encountered in many fields of learning as well as in real-life situations.

We do not use algebra in the problem-solving techniques of guessing-and-checking and making an organized list. In this section, however, we *do* use algebra in problem solving.

Although we can't give you a definite set of rules that will enable you to solve all problems that are expressed in words, we do suggest a procedure that should help you get started. In later chapters, we will discuss several different types of applied problems (money problems, mixture problems, motion problems, and so forth) in separate sections. However, the general *method* of attacking applied problems is the same for *all* types of problems, and it is this *method* on which you should concentrate.

In the following suggestions for solving problems that are expressed in words, we use the notation "Step 1," "Step 2," and so forth, for the steps you will be *writing*.

Suggestions for solving applied problems

Read First, read the problem very carefully. *Be sure you understand the problem.* Read it several times, if necessary.

Think Determine what *type* of problem it is (money problem, motion problem, and so forth), if possible. Determine what is unknown. What is being asked for is often found in the last sentence of the problem, which may begin with "What is the . . . ?" or "Find the . . . ." Is enough information given so that you *can* solve the problem? Do you need a special formula? What operation(s) must be used? What is the domain of the variable? What kind of number is reasonable as the answer? What kind of number might have to be rejected as the answer?

Sketch Draw a sketch *with labels*, if that might be appropriate and helpful.

Step 1. Represent one unknown number by a variable, and *declare its meaning* in a sentence of the form "Let $x =$ . . . ." Then reread the problem to see how you can represent any other unknown numbers in terms of the *same* variable, declaring their meaning in sentences that begin "Then . . . ."

Reread Reread the sentences, breaking them up into key words and phrases.

Step 2. Translate each word or phrase into an algebraic expression, and arrange these expressions into an equation or inequality that represents the facts given in the problem. (We call this equation or inequality a *mathematical model* of the problem.)

Step 3. Solve the equation or inequality.

Step 4. Solve for *all* the unknowns asked for in the problem.

Step 5. Check the solution(s) *in the word statement.*

Step 6. State the results clearly.

The following list of key expressions and their corresponding algebraic operations can help you translate English sentences into an algebraic equation or inequality:

+	−	×	÷	=
the sum of	minus	times	divided by	is
added to	decreased by	the product of	the quotient of	equals
increased by	less than	multiplied by	per	is equal to
plus	subtracted from	by		
more than	the difference of	of (in fraction and percent problems)		
all	less			
total	exceeds			

Note Because subtraction is not commutative, care must be taken to get the numbers in an applied problem involving subtraction in the correct order. For example, whereas the statements "m minus n," "m decreased by n," "the difference of m and n," and "m less n" are translated as $m - n$, the expressions "m subtracted from n" and "m less than n" are translated as $n - m$.

EXAMPLE 1 Three times an unknown number, decreased by 5, is 13. What is the unknown number?

SOLUTION

Step 1. Let $x =$ the unknown number

Reread Three | times | an unknown number, | decreased by | 5, | is | 13

Step 2. $3 \quad \cdot \quad x \quad - \quad 5 \quad = \quad 13$

Step 3.
$$3x - 5 = 13$$
$$3x = 18 \quad \text{Adding 5 to both sides}$$

Step 4.
$$x = 6 \quad \text{Dividing both sides by 3}$$

√ **Step 5.** *Check* Three | times | an unknown number, | decreased by | 5, | is | 13

$3 \quad \cdot \quad (6) \quad - \quad 5 \quad = \quad 13$

$$3(6) - 5 \stackrel{?}{=} 13 \quad \text{The "unknown number" is replaced by 6}$$
$$18 - 5 \stackrel{?}{=} 13$$
$$13 = 13$$

Step 6. Therefore, the unknown number is 6.

Note To check an applied problem, we must check the solution in the *word statement*. Any error that may have been made in writing the equation will not be discovered if we simply substitute the solution into the equation.

EXAMPLE 2 Find three consecutive integers such that the sum of the first two is 23 less than 3 times the third.

SOLUTION

Think The domain is the set of integers.

Step 1. Let $x =$ the first integer
Then $x + 1 =$ the second consecutive integer
and $x + 2 =$ the third consecutive integer

Reread Sum of first two integers | is | 23 less than | 3 times the third

Step 2.
$$x + (x + 1) = 3(x + 2) - 23$$
On the right side, $23 - 3(x + 2)$ would be *incorrect*

Step 3.
$$x + x + 1 = 3x + 6 - 23$$
$$2x + 1 = 3x - 17$$
$$18 = x \quad \text{Adding } -2x + 17 \text{ to both sides}$$

Step 4. $x = 18, \quad x + 1 = 19, \quad x + 2 = 20$

√ **Step 5.** *Check* The sum of 18 and 19 is 37, and 3 times 20 is 60; 37 is 23 less than 60. The problem checks.

Step 6. Therefore, the integers are 18, 19, and 20.

Note Adding 2 to any odd integer gives the *next* odd integer; that is, $5 + 2 = 7$, $-17 + 2 = -15$, and so forth. Therefore, if x is an odd integer, the next odd integer is $x + 2$, and the one after that is $x + 4$ (see Example 3). Also, if x is an even integer, the next even integer is $x + 2$, and the one after that is $x + 4$ (see Example 4).

EXAMPLE 3 Find three consecutive odd integers whose sum is 72.

SOLUTION

Think The domain is the set of odd integers.

Step 1. Let x = the first odd integer
Then $x + 2$ = the second consecutive odd integer
and $x + 4$ = the third consecutive odd integer

Reread Sum of the integers is 72

Step 2. $x + (x + 2) + (x + 4) = 72$

Step 3. $3x + 6 = 72$ Combining like terms

$3x = 66$ Adding -6 to both sides

Step 4. $x = 22$ Dividing both sides by 3

Steps 5 and 6. Since 22 is not an odd integer, it is not in the domain. Therefore, there is no solution for the problem.

When a problem deals with *inequalities* rather than with quantities that are *equal* to each other, you will almost always find more than one correct answer (see Example 4).

EXAMPLE 4 Find four consecutive even integers whose sum is between 21 and 45.

SOLUTION

Think The domain is the set of even integers.

Step 1. Let x = the first even integer
Then $x + 2$ = the second consecutive even integer
and $x + 4$ = the third consecutive even integer
and $x + 6$ = the fourth consecutive even integer

Because the sum of the integers is to be *between* 21 and 45, the algebraic statement can be written as a continued inequality.

Step 2. $21 < x + (x + 2) + (x + 4) + (x + 6) < 45$

Step 3. $21 < 4x + 12 < 45$ Combining like terms

$9 < 4x < 33$ Adding -12 to all three parts

$\frac{9}{4} < x < \frac{33}{4}$ Dividing all three parts by 4

Step 4. The even integers between $\frac{9}{4}$ and $\frac{33}{4}$ are 4, 6, and 8; therefore, we will have *three* sets of answers:

$x = 4$	$x = 6$	$x = 8$
$x + 2 = 6$	$x + 2 = 8$	$x + 2 = 10$
$x + 4 = 8$	$x + 4 = 10$	$x + 4 = 12$
$x + 6 = 10$	$x + 6 = 12$	$x + 6 = 14$

Step 5. *Check* $4 + 6 + 8 + 10 = 28$, $6 + 8 + 10 + 12 = 36$, and $8 + 10 + 12 + 14 = 44$. All these sums of four consecutive even integers are between 21 and 45.

Step 6. Therefore, four consecutive even integers whose sum is between 21 and 45 are 4, 6, 8, and 10; 6, 8, 10, and 12; *and* 8, 10, 12, and 14.

Exercises 3.5
Set I

In *all* exercises, set up the problem algebraically. Be sure to state what your variables represent.

In Exercises 1–6, solve for the unknown number, and check the solution.

1. Seven more than twice an unknown number is 23.
2. Eleven more than 3 times an unknown number is 38.
3. Four times an unknown number, decreased by 7, is 25.
4. Five times an unknown number, decreased by 6, is 49.
5. When an unknown number is decreased by 7, the difference is half the unknown number.
6. When an unknown number is decreased by 12, the difference is half the unknown number.

In Exercises 7–18, solve for the unknowns, and check the solutions.

7. A 12-cm length of string is to be cut into two pieces. The first piece is to be 3 times as long as the second piece. Find the length of each piece.
8. A 12-m length of rope is to be cut into two pieces. The first piece is to be twice as long as the second piece. Find the length of each piece.
9. The sum of three consecutive integers is 19. Find the integers.
10. The sum of three consecutive integers is 40. Find the integers.
11. A 42-cm piece of wire is to be cut so that the first piece is 8 cm longer than the second piece. Find the length of each piece.
12. A 50-m piece of hose is to be cut so that the first piece is 6 m longer than the second piece. Find the length of each piece.
13. The sum of the first two of three consecutive integers less the third integer is 10. What are the integers?
14. The sum of the first two of three consecutive integers less the third integer is 17. What are the integers?
15. Find three consecutive odd integers such that 3 times the sum of the last two odd integers is 40 more than 5 times the first integer.
16. Find three consecutive odd integers such that 5 times the sum of the last two odd integers is 60 more than 5 times the first integer.
17. David bought 4 more cans of corn than cans of peas and 3 times as many cans of green beans as cans of peas. If he bought 24 cans of these three vegetables altogether, how many cans of each kind did he buy?
18. John bought 6 more cans of peaches than cans of pears and 3 times as many cans of cherries as cans of pears. If he bought 21 cans of these three fruits altogether, how many cans of each kind did he buy?

In Exercises 19–22, find the unknown number, and check the solution.

19. Three times the sum of 8 and an unknown number is equal to twice the sum of the unknown number and 7.
20. Four times the sum of 5 and an unknown number is equal to 3 times the sum of the unknown number and 9.
21. When twice the sum of 5 and an unknown number is subtracted from 8 times the unknown number, the result is equal to 4 times the sum of 8 and twice the unknown number.
22. When 7 times the sum of 2 and an unknown number is subtracted from 4 times the sum of 3 and twice the unknown number, the result is equal to 0.

In Exercises 23–26, find *all possible solutions* for each problem.

23. Find three consecutive even integers whose sum is between 21 and 45.
24. Find three consecutive even integers whose sum is between 81 and 93.
25. In a certain mathematics class, a student needs between 560 and 640 points in order to receive a C. The final exam is worth 200 points. If Clarke has 396 points just before the final, what range of scores on the final exam will give him a C for the course?
26. In a certain English class, a student needs between 720 and 810 points in order to receive a B. The final exam is worth 200 points. If Cathy has 584 points just before the final, what range of scores on the final exam will give her a B for the course?

Exercises 3.5
Set II

In *all* exercises, set up the problem algebraically. Be sure to state what your variables represent.

In Exercises 1–6, solve for the unknown number, and check the solution.

1. Nine more than 4 times an unknown number is 33.

2. Seven plus an unknown number is equal to 17 decreased by the unknown number.

3. Five times an unknown number, decreased by 8, is 12.

4. Four plus an unknown number is equal to 20 decreased by the unknown number.

5. When an unknown number is decreased by 12, the difference is one-third the unknown number.

6. Two plus an unknown number is equal to 8 decreased by the unknown number.

In Exercises 7–18, solve for the unknowns, and check the solutions.

7. A 36-cm length of cord is to be cut into two pieces. The first piece is to be 5 times as long as the second piece. Find the length of each piece.

8. A 32-cm piece of string is to be cut so that the first piece is one-third the length of the second piece. Find the length of each piece.

9. The sum of three consecutive integers is 73. Find the integers.

10. The sum of three consecutive integers is 84. What are the integers?

11. A 52-m piece of rope is to be cut so that the first piece is 14 m longer than the second piece. Find the length of each piece.

12. A piece of pipe 60 cm long is to be cut so that one piece is 34 cm shorter than the other piece. Find the length of each piece.

13. The sum of the first two of three consecutive integers less the third integer is 20. What are the integers?

14. The sum of three consecutive even integers is -54. What are the integers?

15. Find three consecutive odd integers such that 4 times the sum of the last two odd integers is 48 more than 7 times the first integer.

16. The sum of the first two of three consecutive integers less the third integer is 85. What are the integers?

17. Tom bought 4 more cans of tomato soup than cans of split pea soup and 3 times as many cans of vegetable soup as cans of split pea soup. If he bought 29 cans of these three soups altogether, how many cans of each kind did he buy?

18. The total receipts for a concert were $7,120. Some tickets were $8.50 each, and the rest were $9.50 each. If 800 tickets were sold altogether, how many of each kind of ticket were sold?

In Exercises 19–22, solve for the unknown number, and check the solution.

19. Four times the sum of 6 and an unknown number is equal to 3 times the sum of the unknown number and 10.

20. When an unknown number is subtracted from 42, the difference is one-sixth the unknown number.

21. When 3 times the sum of 4 and an unknown number is subtracted from 8 times the unknown number, the result is equal to 4 times the sum of 3 and twice the unknown number.

22. When an unknown number is subtracted from 16, the difference is one-third the unknown number.

In Exercises 23–26, find *all possible solutions* for each problem.

23. Find three consecutive even integers whose sum is between 39 and 51.

24. Find four consecutive odd integers whose sum is between -5 and 20.

25. In a certain history class, a student needs between 600 and 675 points in order to receive a B. The final exam is worth 150 points. If Manuel has 467 points just before the final, what range of scores on the final exam will give him a B for the course?

26. In a certain physics class, a student needs at least 450 points in order to receive an A. The final exam is worth 100 points. If Cindy has 367 points just before the final, what range of scores on the final exam will give her an A for the course?

Sections 3.1–3.5 REVIEW

Types of Equations
3.1

A **conditional equation** is an equation whose two sides are equal only when certain numbers (called *solutions*) are substituted for the variable.

An **identity** is an equation whose two sides are equal no matter what permissible number is substituted for the variable.

No solution exists for an equation whose two sides are unequal no matter what permissible number is substituted for the variable.

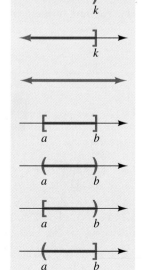

Solving a First-Degree Equation in One Variable
3.1

1. Remove denominators by multiplying both sides of the equation by the least common multiple (LCM) of all the denominators.
2. Remove all grouping symbols.
3. Combine like terms on each side of the equal sign.
4. Move all the terms that contain the variable to one side of the equal sign and all constants to the other side.
5. Divide both sides of the equation by the coefficient of the variable.
6. Determine whether the equation is a conditional equation, an identity, or an equation with no solution.
7. If the equation is a conditional equation, check the solution.

Interval Notation and Graphs
3.2, 3.3

A bracket indicates that an endpoint is included; a parenthesis indicates that an endpoint is not included.

Set-builder notation	*Interval notation*	*Type of interval*	*Graph*
$\{x \mid x > k\}$	$(k, +\infty)$	Open	k
$\{x \mid x \geq k\}$	$[k, +\infty)$	Half-open	k
$\{x \mid x < k\}$	$(-\infty, k)$	Open	k
$\{x \mid x \leq k\}$	$(-\infty, k]$	Half-open	k
$\{x \mid x \in R\}$	$(-\infty, +\infty)$	Open	
$\{x \mid a \leq x \leq b\}$	$[a, b]$	Closed	a b
$\{x \mid a < x < b\}$	(a, b)	Open	a b
$\{x \mid a \leq x < b\}$	$[a, b)$	Half-open	a b
$\{x \mid a < x \leq b\}$	$(a, b]$	Half-open	a b

Solving a First-Degree Inequality in One Variable 3.2

We proceed in the same way we solve equations, except that the *sense* must be changed when we multiply or divide both sides by a negative number.

Solving Combined Inequalities 3.3

The solution set of a **combined inequality** that contains the word *or* is the *union* of the solution sets of the two simple inequalities.

The solution set of a combined inequality that contains the word *and* is the *intersection* of the solution sets of the two simple inequalities.

Solving Continued Inequalities 3.3

A **continued inequality** can be solved by using any of the properties of inequalities on *all three parts* of the continued inequality.

Solving an Equation with Absolute Value Symbols 3.4

An equation that contains absolute value symbols is equivalent to *two* equations without absolute value symbols.

If $k > 0$ and if $|X| = k$, then $X = k$ or $X = -k$ Property 1

Solving an Inequality with Absolute Value Symbols 3.4

If $k > 0$ and if $|X| < k$, then $-k < X < k$ Property 2

If $k > 0$ and if $|X| \le k$, then $-k \le X \le k$ Property 2

If $k > 0$ and if $|X| > k$, then $X > k$ or $X < -k$ Property 3

If $k > 0$ and if $|X| \ge k$, then $X \ge k$ or $X \le -k$ Property 3

Solving Applied Problems 3.5

Read First, read the problem very carefully. *Be sure you understand the problem.* Read it several times, if necessary.

Think Determine what *type* of problem it is, if possible. Determine what is unknown. Is enough information given so that you *can* solve the problem? Do you need a special formula? What operation(s) must be used? What is the domain of the variable? What kind of number is reasonable as the answer? What kind of number might have to be rejected as the answer?

Sketch Draw a sketch *with labels*, if that might be appropriate and helpful.

Step 1. Represent one unknown number by a variable, and *declare its meaning* in a sentence of the form "Let $x = \ldots$." Then reread the problem to see how you can represent any other unknown numbers in terms of the *same* variable, declaring their meaning in sentences that begin "Then . . . ."

Reread Reread the sentences, breaking them up into key words and phrases.

Step 2. Translate each word or phrase into an algebraic expression, and arrange these expressions into an equation or inequality that represents the facts given in the problem.

Step 3. Solve the equation or inequality.

Step 4. Solve for *all* the unknowns asked for in the problem.

Step 5. Check the solution(s) *in the word statement.*

Step 6. State the results clearly.

Sections 3.1–3.5 REVIEW EXERCISES Set I

In Exercises 1–5, find the solution set of each equation. Identify any equation that is not a conditional equation as either an identity or an equation with no solution.

1. $7 - 2(M - 4) = 5$
2. $5[-13z - 8(4 - 2z) + 20] = 15z - 17$
3. $\dfrac{3(x + 3)}{4} - \dfrac{2(x - 3)}{3} = 1$
4. $\dfrac{x - 1}{4} + \dfrac{x - 8}{2} = x - 4 - \dfrac{x + 1}{4}$
5. $2[-7y - 3(5 - 4y) + 10] = 10y - 12$

In Exercises 6–10, find the solution set of each inequality.

6. $10 - 3(x + 2) \geq 9 - 2(4 - 3x)$

7. $\frac{3z}{5} - \frac{2z}{3} < \frac{1}{2}$

8. $\frac{2w}{3} - \frac{5w}{6} < \frac{7}{12}$

9. $\frac{2(x + 6)}{10} + \frac{3x}{20} < 3$

10. $\frac{5(x - 3)}{7} - \frac{3x}{2} < 5$

In Exercises 11–23, find each set or solution set and graph it. *Whenever possible*, express the answer in interval notation.

11. $\{x \mid 3x + 2 = 11, x \in J\}$

12. $|3x| = 9$

13. $|x| \leq -3$

14. $|x| \geq 2$

15. $|6 - 2x| = 10$

16. $3|x - 2| \leq 6$

17. $2|x - 3| \leq 4$

18. $\{x \mid -3 < 2x - 1 < 8\}$

19. $\{x \mid -2 < x + 1 < 0, x \in J\}$

20. $\left\{x \,\middle|\, \frac{2(x - 1)}{6} + \frac{x}{9} = 1, x \in J\right\}$

21. $\left\{x \,\middle|\, \frac{3(x - 1)}{4} + \frac{x}{8} = 1, x \in J\right\}$

22. $\left|2 - \frac{x}{3}\right| \leq 1$

23. $\left|\frac{2x - 4}{3}\right| \geq 2$

In Exercises 24–26, write, in set-builder notation and in interval notation, the algebraic statement that describes each graph.

24.

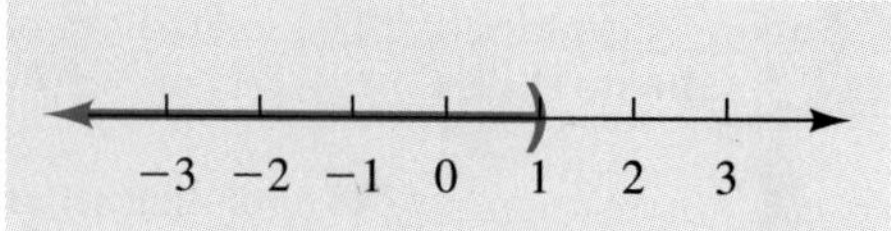

25.

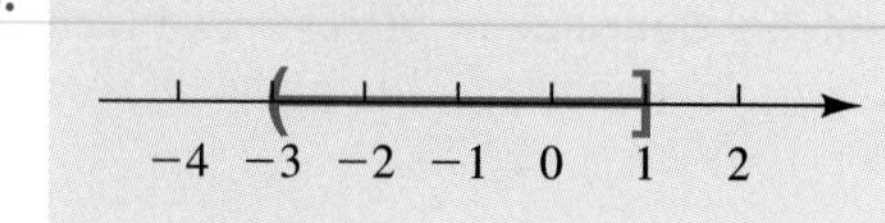

26.

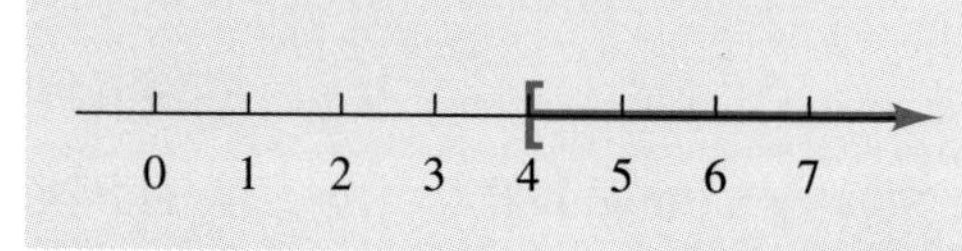

In Exercises 27 and 28, set up each problem algebraically, solve, and check. Be sure to state what your variables represent.

27. When 5 times an unknown number is subtracted from 28 and the result is divided by 3, the quotient is 3 times the unknown number. Find the unknown number.

28. The sum of three consecutive integers is 23. What are the integers?

Name ______________________

ANSWERS

1. ______
2. ______
3. ______
4. ______
5. ______
6. ______
7. ______
8. ______
9. ______
10. ______
11. ______
12. ______
13. ______
14. ______

In Exercises 1–5, find the solution set of each equation. Identify any equation that is not a conditional equation as either an identity or an equation with no solution.

1. $13 = 8 - 5(2a + 13)$

2. $5(t - 3) + 1 = 3t - 2(7 - t)$

3. $3[-8x - 5(3 - 4x) + 10] = 36x + 2$

4. $\dfrac{x + 3}{4} = \dfrac{x - 2}{3} + \dfrac{1}{4}$

5. $\dfrac{3(x - 3)}{4} + 1 = \dfrac{8(x + 2)}{7}$

In Exercises 6–10, find the solution set of each inequality.

6. $2(x - 4) - 5 \geq 7 + 3(2x - 1)$

7. $15 - 4(m - 6) \leq 2(2 - 5m) + 8$

8. $\dfrac{2w}{4} - \dfrac{7}{9} < \dfrac{5w}{6}$

9. $\dfrac{5x}{3} < 11 - \dfrac{7(x - 9)}{12}$

10. $\dfrac{x}{3} - \dfrac{x + 2}{5} < 1$

In Exercises 11–23, find each set or solution set. *Whenever possible*, express the answer in interval notation.

11. $|8x| = 12$

12. $\{x \mid 13 + 2x = 1, x \in N\}$

13. $|11 - 7x| = 17$

14. $|x| > -5$

15. $\{x \mid -2 < x + 3 < 1, x \in J\}$

16. $\left\{x \,\middle|\, \frac{13}{15} - \frac{x}{5} = \frac{2(8 - x)}{15}, x \in J\right\}$

17. $\left|4 - \frac{x}{2}\right| \le 2$

18. $8|2 - x| < 40$

19. $\left|\frac{3x + 2}{4}\right| > 2$

20. $\{x \mid 5 < 3x - 1 < 2\}$

21. $\left|5 - \frac{x}{3}\right| \ge 1$

22. $\{x \mid -3 \le 2x - 1 \le 7, x \in N\}$

23. $|x - 4| = -3$

In Exercises 24–26, write, in set-builder notation and in interval notation, the algebraic statement that describes each graph.

24.

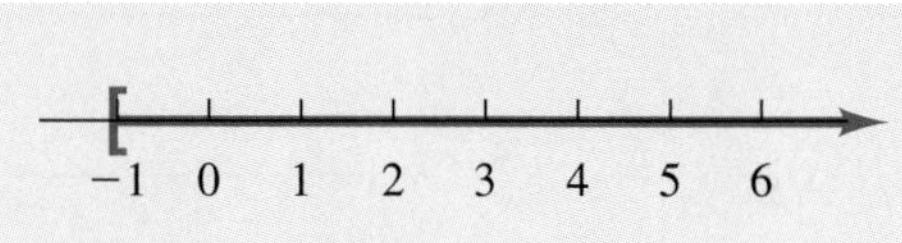

25.

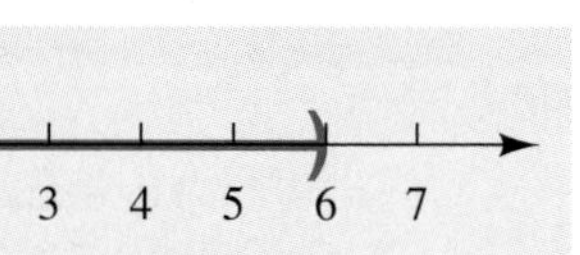

26.

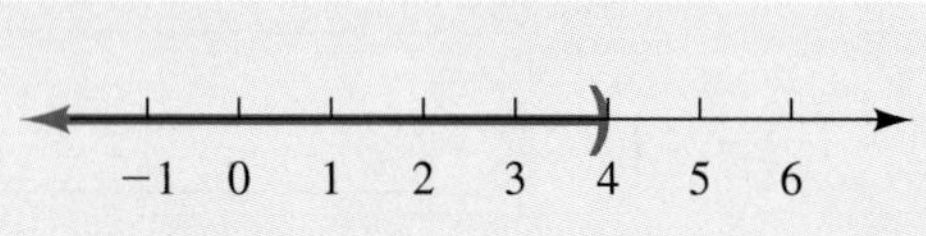

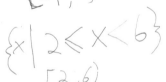

In Exercises 27 and 28, set up each problem algebraically, solve, and check. Be sure to state what your variables represent.

27. When 7 times an unknown number is subtracted from 120 and the result is divided by 2, the quotient is 46. Find the unknown number.

28. The sum of three consecutive integers is −222. Find the integers.

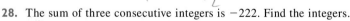

15. ______
16. ______
17. ______
18. ______
19. ______
20. ______
21. ______
22. ______
23. ______
24. ______
______
25. ______
______
26. ______
______
27. ______
28. ______
______
______

Chapter 3 DIAGNOSTIC TEST

The purpose of this test is to see how well you understand first-degree equations and inequalities in one variable and their applications. We recommend that you work this diagnostic test *before* your instructor tests you on this chapter. Allow yourself about 50 minutes.

Complete solutions for all the problems on this test, together with section references, are given in the answer section at the end of the book. For the problems you do incorrectly, we suggest that you study the sections cited.

In Problems 1–4, find the solution set of each equation. Identify any equation that is not a conditional equation as either an identity or an equation with no solution.

1. $12(3x - 5) = 8[5 - 2(x + 4)]$
2. $2[7x - 4(1 + 3x)] = 5(3 - 2x) - 23$
3. $\frac{x}{6} - \frac{x + 2}{4} = \frac{1}{3}$
4. $3(x - 6) = 5(1 + 2x) - 7(x - 4)$

In Problems 5 and 6, convert the description of each set from set-builder notation to interval notation, and graph each solution set.

5. $\{x | x < -3\}$ 6. $\{x | x \geq -1\}$

In Problems 7 and 8, write, in set-builder notation *and* in interval notation, the algebraic statement that describes the set of numbers graphed.

7.

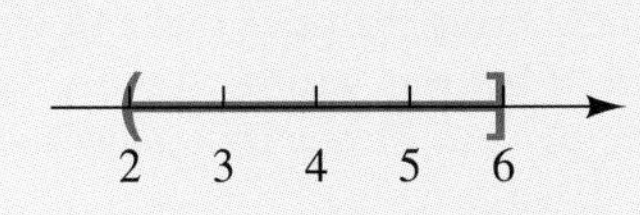

8.

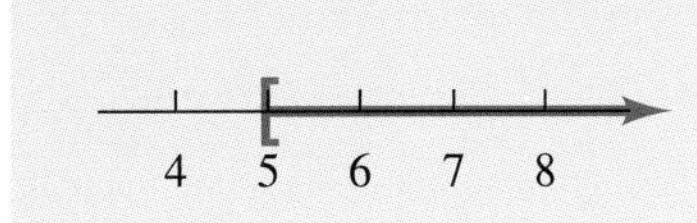

In Problems 9–18, find each set or solution set, and graph it. *Whenever possible*, express the answer in interval notation.

9. $5w + 2 \leq 10 - w$
10. $13h - 4(2 + 3h) \geq 0$
11. $\{x | -3 < x + 1 < 5, x \in J\}$
12. $\{x | 4 \geq 3x + 7 > -2, x \in R\}$
13. $\left\{x \,\middle|\, \frac{5(x - 2)}{3} + \frac{x}{4} \leq 12, x \in R\right\}$
14. $\left\{x \,\middle|\, \left|\frac{2x + 3}{5}\right| = 1\right\}$
15. $|3x - 1| > 2$
16. $|7 - 3x| \geq 6$
17. $\left|\frac{5x + 1}{2}\right| \leq 7$
18. $\{x \| 2x - 5| < 11\}$

In Problems 19 and 20, set up each problem algebraically, solve, and check. Be sure to state what your variables represent.

19. When 23 is added to [illegible] times an unknown number, the sum is 31. Find the unknown number.
20. The sum of two consecutive integers is 55. Find the integers.

Chapters 1–3 CUMULATIVE REVIEW EXERCISES

In Exercises 1–30, find the value of each expression or write either "undefined" or "not real."

1. $(-14) - (-22)$ 2. $(-12)(-4)$ 3. $(-1)^7$
4. $(-6) + (-8)$ 5. $(-2)^4$ 6. $\sqrt{64}$
7. $\frac{9}{0}$ 8. $\frac{0}{-7}$ 9. $\frac{|-20|}{-5}$
10. $(-18) - (7)$ 11. $\sqrt[3]{-64}$ 12. -6^2
13. $(35) \div (-7)$ 14. $(-10)(0)(8)$ 15. $\sqrt[9]{-1}$
16. $10^{-2} \cdot 10^5$ 17. $(4^0)^3$ 18. $(2^{-3})^{-1}$
19. $-3 - 2^2 \cdot 6$ 20. $7 - [4 - (13 - 5)]$
21. $(-11) + 15$ 22. $(14)(-2)$
23. $|0|$ 24. $\sqrt{81}$
25. $\sqrt[4]{81}$ 26. $3 + 2 \cdot 5$
27. $\frac{0}{0}$ 28. $16 \div (-2)^2 - \frac{7 - 1}{2}$
29. $0 \div 15$ 30. $36 \div 18 \times 2$
31. Write the prime factorization of 78.
32. Find all the prime numbers between 11 and 31 that yield a remainder of 3 when divided by 4.
33. What is the additive identity element?
34. Complete this statement by using the associative property:

$$3 + (2 + 7) = \_\_\_\_\_\_\_\_\_\_$$

In Exercises 35–38, simplify the expression.

35. $y - 2(x - y) - 3(1 - y) - \sqrt{4y^2}$

36. $2x(x^2 + 1) - x(x^2 + 3x - 2)$

37. $(5x)^2(3x^2)^3$

38. $(2x^3 - 4y^2) - 5x^3$

In Exercises 39–46, find the solution set of each equation or inequality. Identify any *equation* that is not a conditional equation as either an identity or an equation with no solution.

39. $8x - 4(2 + 3x) = 12$

40. $2[-5y - 6(y - 7)] < 6 + 4y$

41. $6(3x - 5) + 7 = 9(3 + 2x) - 1$

42. $5x - 7 \leq 8x + 4$

43. $-5 < x + 4 \leq 3$

44. $8\{4 + 3(x - 2)\} = 3(8x + 3) - 25$

45. $|2x + 3| > 1$

46. $|x - 4| \leq 3$

In Exercises 47 and 48, set up each problem algebraically, solve, and check. Be sure to state what your variables represent.

47. Lupe is twice as old as Juana. In 9 yr, Juana will be five-sevenths as old as Lupe is then. How old is Juana now?

48. Three court reporters are planning to buy some equipment (a computer, the necessary software, and a laser printer) to use in their work. They plan to share the cost equally. If they allow one more person to join their group and to share the cost equally with them, the cost for each of the original three reporters will be reduced by $1,125. What is the total cost of the equipment?

Critical Thinking and Problem-Solving Exercises

1. Bella is saving for a CD player. She was able to save $40 toward the CD player in January, $43 in February (her income was gradually increasing), $46 in March, and $49 in April. If she continues in this way, how much will Bella save in May? How much will she save in June? If the CD player that she wants costs $400, when will she be able to afford it? Explain how you arrived at your answer.

2. Eddie entered a florist shop to buy some flowers. He saw 4 more pink carnations than orange poppies, and the sum of the number of pink carnations and orange poppies was equal to the total number of white gardenias. There were twice as many white gardenias as there were yellow daisies. There were one-third as many white gardenias as there were red roses, and there were 48 red roses. Altogether, how many flowers did Eddie see?

3. Five-year-old Derek was visiting his grandmother's farm, and he decided to practice counting by counting the *animals* in one of the fields and also the *feet* of those animals. He counted 8 animals and 20 feet; some animals were cows and some were chickens. How many cows were in the field? How many chickens?

4. Toshiko is thinking of a number. Using the following clues, guess what the number is. The number is a two-digit number that is less than 80, and it is prime. The sum of the digits is 10, and the units digit is less than 6.

5. A merchant marked up the price of an item that cost him $120 by 20%. Because the item hadn't sold after a long period of time, he marked the selling price down by 20%, thinking that that would bring the price down to what he had paid for the item. What was wrong with his reasoning? What would the percent of decrease have to be to bring the price back down to $120?

6. Make up an applied problem that leads to a first-degree equation or inequality in one variable, similar to any of the problems in the last section of this chapter. Then solve and check the problem. (If you're working in groups, let other students in your group solve *your* problem, and *you* solve *their* problems.)

Polynomials

CHAPTER 4

In this chapter, we study in detail a particular type of algebraic expression called a *polynomial.* We will see how to add, subtract, multiply, divide, and simplify polynomials; we will also see how to raise a polynomial with two terms to any natural number power. Much of the work in algebra involves operations with polynomials.

4.1 Basic Definitions

Polynomials

A **polynomial in one variable** is an algebraic expression that, in simplified form, contains only terms that can be expressed in the form ax^n, where a represents any real number (that is, $a \in R$), x represents any variable, and n represents any *nonnegative* integer. The exponent on the variable *cannot* be a negative integer, but it *can* be zero. Therefore, the expression -8, for example, is a polynomial, because $-8 = -8x^0$, so -8 *can* be expressed in the form ax^n. Similarly, $3x^3 - 4x^2 + 2x - 8$ is a polynomial, since every *term* can be expressed in the form ax^n.

A polynomial with only one term is called a **monomial**, a polynomial with two terms is called a **binomial**, and a polynomial with three terms is called a **trinomial**. There are no commonly used special names for polynomials with more than three terms.

EXAMPLE 1 Examples of algebraic expressions that are polynomials in one variable:

a. $4z^4 - 2z^2$ This polynomial is a binomial in z

b. $7x^2 - 5x + 2$ This polynomial is a trinomial

c. 5 This is a monomial; it can be expressed in the form $5x^0$, since $5x^0 = 5 \cdot 1 = 5$

If an algebraic expression in simplest form contains terms with variables in a denominator or under a radical sign or if any variable contains an exponent that is not an integer,* then it is *not* a polynomial. (If there is a negative exponent on a variable that is in a numerator, the expression is *not* in simplest form, and, in simplest form, the expression would have a variable in the denominator; thus, it is not a polynomial.) See Example 2.

EXAMPLE 2 Examples of algebraic expressions that are *not* polynomials:

a. $\dfrac{2}{x - 5}$ This expression is not a polynomial because there is a variable in the denominator

b. $\sqrt{x - 5}$ This expression is not a polynomial because there is a variable under a radical sign

c. $4x^{-1}$ This expression is not a polynomial because it has a negative exponent on a variable; in simplest form, $4x^{-1}$ becomes $\dfrac{4}{x}$, and then it has a variable in the denominator

*Exponents that are not integers are discussed in a later chapter.

An algebraic expression containing two variables is a **polynomial in two variables** if, in its simplified form, (1) there are no variables in denominators, (2) there are no variables under radical signs, and (3) the exponents on all variables are nonnegative integers.

EXAMPLE 3 Examples of algebraic expressions that are polynomials in two variables:

a. $x^2y\sqrt{5} + \frac{1}{2}xy^2$ — This polynomial is a binomial; note that *constants* can be under radical signs and in denominators

b. $4u^2v^2 - 7uv + 6$ — This polynomial is a trinomial in u and v

c. $(x - y)^2 - 2(x - y) - 8$

The Degree of a Term of a Polynomial

To find the **degree of any term** of a polynomial, we first write the polynomial in simplest form. Then, if the polynomial contains only one variable, the degree of any of its terms equals the exponent *on the variable* in that term; if the polynomial contains more than one variable, the degree of any of its terms equals the *sum* of the exponents *on the variables* in that term.

EXAMPLE 4 Examples of finding the degree of a term:

a. 5^2x^3 Third degree — We consider only exponents on *variables*

b. $6x^2y$ Third degree because $6x^2y = 6x^2y^1$ ($2 + 1 = 3$)

c. 14 Zero degree because $14 = 14x^0$

We see from Example 4c that the degree of the constant term 14 is zero. Actually, the degree of a (nonzero) constant term is *always* zero.

The Degree of a Polynomial

The **degree of a polynomial** is defined as the degree of its highest-degree term. Therefore, to find the degree of a polynomial, we first find the degree of each of its terms. The *largest* of these numbers is the degree of the polynomial.*

EXAMPLE 5 Examples of finding the degree of a polynomial:

a. $9x^3 - 7x + 5$

- 5: 0 degree term
- $-7x$: 1st degree term
- $9x^3$: 3rd degree term ← Highest-degree term

Therefore, $9x^3 - 7x + 5$ is a third-degree polynomial.

b. $14uv^3 - 11u^5v + 8$

- 8: 0 degree term
- $-11u^5v$: 6th degree term ← Highest-degree term
- $14uv^3$: 4th degree term

Therefore, $14uv^3 - 11u^5v + 8$ is a sixth-degree polynomial.

---

*Mathematicians define the zero polynomial, 0, as having *no* degree.

The Leading Coefficient The **leading coefficient** of a polynomial is the numerical coefficient of its highest-degree term.

Descending and Ascending Powers

In a polynomial, if the exponents on one variable get *smaller* as we read the terms from left to right, we say that the polynomial is arranged in **descending powers** of that variable. If the exponents on one variable get *larger* as we read the terms from left to right, we say that the polynomial is arranged in **ascending powers** of that variable.

EXAMPLE 6 Arrange $5 - 2x^2 + 4x$ in descending powers of x, and name the leading coefficient.

SOLUTION $-2x^2 + 4x + 5$. The leading coefficient is -2.

A polynomial with more than one variable can be arranged in descending powers of *any one* of its variables.

EXAMPLE 7 Arrange $3x^3y - 5xy + 2x^2y^2 - 10$ first in descending powers of x, then in descending powers of y, and then in ascending powers of x.

SOLUTION

$3x^3y + 2x^2y^2 - 5xy - 10$ In descending powers of x

$2x^2y^2 + 3x^3y - 5xy - 10$ In descending powers of y

Since y is to the same power in both terms, we write the higher-degree term first

$-10 - 5xy + 2x^2y^2 + 3x^3y$ In ascending powers of x

Polynomial Equations

A **polynomial equation** is an equation that has a polynomial on each side of the equal sign (the polynomial on one side of the equal sign can be the zero polynomial, 0). The **degree of a polynomial equation** is the same as the degree of the *highest-degree* term on either side of the equal sign. We have already solved first-degree polynomial equations.

EXAMPLE 8 Examples of polynomial equations:

a. $5x - 3 = 3x + 4$ This equation is a 1st degree polynomial equation in one variable

b. $2x^2 - 4x + 7 = 0$ This equation is a 2nd degree polynomial equation in one variable; it is also called a *quadratic equation*

Exercises 4.1
Set I

In Exercises 1–16, if the expression is a polynomial, find its degree. If it is *not* a polynomial, write "not a polynomial."

1. $2x^2 + \frac{1}{3}x$
2. $\frac{1}{9}y^2 - 5$
3. $x^{-2} + 5x^{-1} + 4$
4. $y^{-3} + y^{-2} + 6$
5. 10
6. 20
7. $\dfrac{4}{x} + 7x - 3$
8. $\dfrac{3}{x} - 2x + 5$
9. $x\sqrt{5} + 6$
10. $y\sqrt{2} - 5$
11. $\sqrt{x + 4}$
12. $\sqrt{z - 3}$
13. $\dfrac{1}{2y^2 - 5y} - 3y$
14. $\dfrac{1}{2x^2 + 4x} + 5x$
15. $x^3y^3 - 3^7x^2y + 3^4xy^2 - y^3$
16. $2^7y^2z^3 - 3yz^5 + 6z^4$

In Exercises 17–20, write each polynomial in descending powers of the indicated variable, and find the leading coefficient.

17. $7x^3 - 4x - 5 + 8x^5$ Powers of x

18. $10 - 3y^5 + 4y^2 - 2y^3$ Powers of y

19. $3x^2y + 8x^3 + y^3 - y^5$ Powers of y

20. $6y^3 + 7x^2 - 4y^2 + y$ Powers of y

Writing Problems

Express the answers in your own words and in complete sentences.

1. Explain why $\sqrt{x^2 - 4}$ is not a polynomial.

2. Explain why $2^5x^3y^4$ is a seventh-degree polynomial.

Exercises 4.1 Set II

In Exercises 1–16, if the expression is a polynomial, find its degree. If it is *not* a polynomial, write "not a polynomial."

1. $4x^3 - \frac{1}{2}x$
2. $x\sqrt{3} - 5$
3. $x^{-3} - 3x^2 + 3$
4. $4x^3 + 2x^2z^2 - 6$
5. 7
6. $\sqrt{8 - 3x}$
7. $\dfrac{3}{x} - 4x + 2$
8. $\frac{3}{4}x^5 - 2x^3y^3 + 3$
9. $y\sqrt{3} + 2$
10. $\dfrac{3}{x^3 + 2x^2 - 5x + 1}$
11. $\sqrt{x - 2}$
12. $3^5x + \sqrt{5}$
13. $5z + \dfrac{1}{3x^3 + 7x}$
14. $y^{-3} - 2y^{-2} + 3y$
15. $x^4y^3 + 2^8x^3y^2 - 3xy^5$
16. x

In Exercises 17–20, write each polynomial in descending powers of the indicated variable, and find the leading coefficient.

17. $12b^2 - 14b^4 + 8 - 7b$ Powers of b

18. $8 - 2x + 13x^2y + 2x^4$ Powers of x

19. $3xy + 4y^5 - 3y^2 - 3$ Powers of y

20. $4st^2 - 9s^2t + 3s^2t^3 - 5t^6$ Powers of t

4.2 Adding and Subtracting Polynomials

Adding Polynomials

Because polynomials *are* algebraic expressions, we have already added many polynomials. Polynomials are added horizontally by removing the grouping symbols and combining like terms. It is often helpful, as you're adding, to underline all like terms with the same kind of marking.

EXAMPLE 1 Example of adding polynomials:

$$(5x^3y^2 - 3x^2y^2 + 4xy^3) + (4x^2y^2 - 2xy^2) + (-7x^3y^2 + 6xy^2 - 3xy^3)$$

$$= 5x^3y^2 - 3x^2y^2 + 4xy^3 + 4x^2y^2 - 2xy^2 - 7x^3y^2 + 6xy^2 - 3xy^3$$

$$= -2x^3y^2 + x^2y^2 + xy^3 + 4xy^2$$

Vertical addition is sometimes used. In this case, it is important to align all like terms vertically.

Adding polynomials vertically

1. Arrange the polynomials under one another so that like terms are in the same vertical column.
2. Find the sum of the terms in each vertical column by adding the numerical coefficients.

EXAMPLE 2

Add $(3x^2 + 2x - 1) + (2x + 5) + (4x^3 + 7x^2 - 6)$.

SOLUTION

$$\begin{array}{rrrr} & 3x^2 & + 2x & - 1 \\ & & 2x & + 5 \\ 4x^3 & + 7x^2 & & - 6 \\ \hline 4x^3 & + 10x^2 & + 4x & - 2 \end{array}$$

Subtracting Polynomials

Polynomials can be subtracted horizontally by removing the grouping symbols and combining like terms. When subtraction problems are written in words, remember that "subtract m from n" means $n - m$, *not* $m - n$.

EXAMPLE 3

Subtract $(-4x^2y + 10xy^2 + 9xy - 7)$ from $(11x^2y - 8xy^2 + 7xy + 2)$.

SOLUTION

$$(11x^2y - 8xy^2 + 7xy + 2) - (-4x^2y + 10xy^2 + 9xy - 7)$$

$$= 11x^2y - 8xy^2 + 7xy + 2 + 4x^2y - 10xy^2 - 9xy + 7$$ Adding the additive inverse of the subtrahend to the minuend

$$= 15x^2y - 18xy^2 - 2xy + 9$$

It is necessary to know how to subtract polynomials vertically in order to be able to use long division in dividing one polynomial by another.

Subtracting polynomials vertically

1. Write the polynomial being subtracted *under* the polynomial it is being subtracted from. Write like terms in the same vertical column.
2. Change the sign of each term in the polynomial being subtracted.*
3. Find the sum of the *resulting* terms in each vertical column by adding the numerical coefficients.

*Your instructor may allow or require you to *show* the sign changes. See the alternative method in Example 4.

EXAMPLE 4

Subtract $(4x^2y^2 + 7x^2y - 2xy + 9)$ from $(6x^2y^2 - 2x^2y + 5xy + 8)$ vertically.

SOLUTION

Signs changed mentally

$$\begin{array}{r} 6x^2y^2 - 2x^2y + 5xy + 8 \\ 4x^2y^2 + 7x^2y - 2xy + 9 \\ \hline 2x^2y^2 - 9x^2y + 7xy - 1 \end{array}$$

Alternative method: Sign changes shown

$$\begin{array}{r} 6x^2y^2 - 2x^2y + 5xy + 8 \\ \ominus \qquad \ominus \qquad \oplus \qquad \ominus \\ 4x^2y^2 + 7x^2y - 2xy + 9 \\ \hline 2x^2y^2 - 9x^2y + 7xy - 1 \end{array}$$

Exercises 4.2
Set I

In Exercises 1–4, perform the indicated operations.

1. $(-3x^4 - 2x^3 + 5) + (2x^4 + x^3 - 7x - 12)$
2. $(-5y^3 + 3y^2 - 3y) + (3y^3 - y + 4)$
3. $(7 - 8v^3 + 9v^2 + 4v) - (9v^3 + 6 - 8v^2 + 4v)$
4. $(3x + 7x^3 - x^2 - 5) - (3x^2 - 6x^3 + 2 - 5x)$

In Exercises 5 and 6, add the polynomials.

5. $\begin{array}{r} 4x^3 + 7x^2 - 5x + 4 \\ \underline{2x^3 - 5x^2 + 5x - 6} \end{array}$

6. $\begin{array}{r} 3y^4 - 2y^3 + 4y + 10 \\ \underline{-5y^4 + 2y^3 + 4y - 6} \end{array}$

7. Subtract $(6 + 3x^5 - 4x^2)$ from $(4x^3 + 6 + x)$.
8. Subtract $(7 - 4x^4 + 3x^3)$ from $(x^3 + 7 - 3x)$.
9. Subtract $(3y^4 + 2y^3 - 5y)$ from $(2y^4 + 4y^3 + 8)$.
10. Subtract $(2y^3 - 3y^2 + 4)$ from $(7y^3 + 5y^2 - 5y)$.

In Exercises 11–14, subtract the lower polynomial from the one above it.

11. $\begin{array}{r} -3x^4 - 2x^3 \quad\quad + 4x - 3 \\ \underline{2x^4 + 5x^3 - x^2 \quad\quad - 1} \end{array}$

12. $\begin{array}{r} -2x^4 \quad\quad + 3x^2 - x + 1 \\ \underline{- x^3 - 2x^2 + x - 5} \end{array}$

13. $\begin{array}{r} 5x^3 - 3x^2 + 7x - 17 \\ \underline{-2x^3 - 7x^2 \quad\quad - 6} \end{array}$

14. $\begin{array}{r} 2x^3 + 7x^2 - 3x - 12 \\ \underline{-5x^3 + 9x^2 - 2x \quad\quad} \end{array}$

15. Subtract $(-3m^2n^2 + 2mn - 7)$ from the sum of $(6m^2n^2 - 8mn + 9)$ and $(-10m^2n^2 + 18mn - 11)$.
16. Subtract $(-9u^2v + 8uv^2 - 16)$ from the sum of $(7u^2v - 5uv^2 + 14)$ and $(11u^2v + 17uv^2 - 13)$.
17. Subtract the sum of $(x^3y + 3xy^2 - 4)$ and $(2x^3y - xy^2 + 5)$ from the sum of $(5 + xy^2 + x^3y)$ and $(-6 - 3xy^2 + 4x^3y)$.
18. Subtract the sum of $(2m^2n - 4mn^2 + 6)$ and $(-3m^2n + 5mn^2 - 4)$ from the sum of $(5 + m^2n - mn^2)$ and $(3 + 4m^2n + 2mn^2)$.

In Exercises 19 and 20, perform the indicated operations.

19. $(8.586x^2 - 9.030x + 6.976) - [1.946x^2 - 41.45x - (7.468 - 3.914x^2)]$

20. $(24.21 - 35.28x - 73.92x^2) - [82.04x - 53.29x^2 - (64.34 - 19.43x^2)]$

Writing Problems

Express the answers in your own words and in complete sentences.

1. Explain why, if we subtract $-3x^2 + 2x - 5$ from $2x^2 - 4x - 1$, we do *not* get $-5x^2 + 6x - 4$.
2. Explain why, if we subtract $-3x^2 + 2x - 5$ from $2x^2 - 4x - 1$, we do *not* get $-x^2 - 2x - 6$.

Exercises 4.2
Set II

In Exercises 1–4, perform the indicated operations.

1. $(-3x^5 + 2x^3 - 4x + 1) + (-4x^5 + 2x^2 - 4x - 6)$
2. $(-5x^4 - 2x^3 + 6x^2 - 3) - (-2x^4 + 4x^3 + 3x + 4)$
3. $(4x^5 + 3x^4 - 2x^2 + 5) - (-x^4 + x^3 - x^2 + x)$
4. $(-8y^3 - y^2 + 7y - 6) - (-8y^3 - y^2 + 7y - 6)$

In Exercises 5 and 6, add the polynomials.

5. $\begin{array}{r} 13h^3 - 8h^2 + 16h - 14 \\ \underline{6h^3 + 5h^2 - 18h + 9} \end{array}$

6. $\begin{array}{r} 3x^3 - 2x^2 + 5x - 3 \\ \underline{-7x^3 \quad\quad - 9x - 4} \end{array}$

7. Subtract $(22 + 8y^2 - 14y)$ from $(11y^2 - 5y - 12y^3)$.
8. Subtract $(5x^3 - 3x - x^4)$ from $(-3x^2 - x^4 + 7 - x)$.
9. Subtract $(w - 12w^3 - 15 - 18w^2)$ from $(18w^3 + 5 - 9w)$.
10. Subtract $(z^3 - 3z^2 + 2 - z)$ from $(5z - z^3 + z^2)$.

In Exercises 11–14, subtract the lower polynomial from the one above it.

11. $\begin{array}{r} 2x^3 - 5x^2 \qquad - 3 \\ \underline{5x^3 + 2x^2 - 3x + 7} \end{array}$

12. $\begin{array}{r} -4y^4 + 6y^3 + 2y^2 \qquad - 6 \\ \underline{3y^4 \qquad + 6y^2 - y + 5} \end{array}$

13. $\begin{array}{r} 7z^3 - 6z^2 + 4z \qquad \\ \underline{-4z^3 - 8z^2 + 8z - 2} \end{array}$

14. $\begin{array}{r} -2x^3 + x^2 \qquad\qquad \\ \underline{-6x^3 - 3x^2 - 5x + 4} \end{array}$

15. Subtract $(5xy - 12 + 3xy^2)$ from the sum of $(7x^2y + 4xy^2 - 5)$ and $(8xy^2 - 7x^2y + xy)$.

16. Subtract the sum of $(2x^3 - 5x + 4)$ and $(-5x^3 + 2x^2 - 2)$ from $(x^3 - 2x^2 - 3x)$.

17. Subtract the sum of $(4m^3n^3 - 10mn)$ and $(-10m^2n^2 - 15mn)$ from the sum of $(5m^3n^3 - 8m^2n^2 + 20mn)$ and $(-14m^2n^2 + 5mn)$.

18. Given the polynomials $(10a^3 - 8a + 12)$, $(11a^2 + 9a - 14)$, and $(-6a^3 + 17a)$, subtract the sum of the first two from the sum of the last two.

In Exercises 19 and 20, perform the indicated operations.

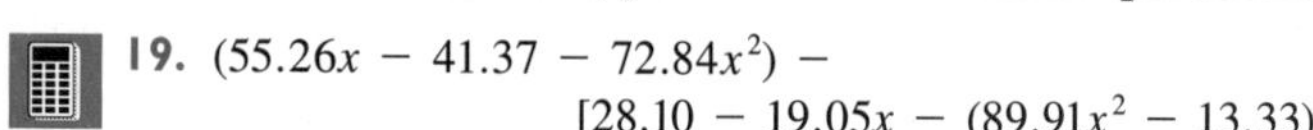

19. $(55.26x - 41.37 - 72.84x^2) - [28.10 - 19.05x - (89.91x^2 - 13.33)]$

20. $(23.1x^2 - 3.4x + 2) - [3.05x - 4.6x^2 - (5.13x^2 + 8.1)]$

4.3 Multiplying Polynomials

We have already multiplied some polynomials together. For example, when we found that $(3x^2)(7x^3) = 21x^5$ in an earlier chapter, we were really multiplying a monomial by a monomial. Also, when we found that $4x(x^2 - 2xy + y^2) = 4x^3 - 8x^2y + 4xy^2$, we were multiplying a polynomial by a monomial. In this section, we discuss multiplying two polynomials when both have two or more terms.

Multiplying Two Binomials

We often need to find the product of two binomials, and so it is convenient to be able to find their product by inspection (that is, by doing the multiplication mentally and writing only the final answer). First, however, we show the step-by-step method for multiplying two binomials. When we multiply two binomials, it is necessary to use the distributive property *more than once* (see Example 1).

EXAMPLE 1 Find the product $(2x + 3)(5y + 7)$.

SOLUTION We first distribute $5y + 7$ over $2x + 3$ (that is, we treat $5y + 7$ as if it were a single number).

Step 1. $(2x + 3)(5y + 7) = (2x)(5y + 7) + (3)(5y + 7)$ Using the distributive property

Using the distributive property again

Step 2. $= (2x)(5y) + (2x)(7) + (3)(5y) + (3)(7)$

Step 3. $= 10xy + 14x + 15y + 21$

Let's agree on some terminology so that we can more easily discuss finding products of two binomials by inspection. Consider the product from Example 1 again: Because $2x + 3$ and $5y + 7$ are binomials, we call the $2x$ and the $5y$ the *First terms* of the binomials and the 3 and the 7 the *Last terms* of the binomials.

The *F irst terms* of the binomials; the *product* of these terms is $(2x)(5y)$, or $10xy$

$$(2x + 3)(5y + 7)$$

The *L ast terms* of the binomials; the *product* of these terms is $(3)(7)$, or 21

Because the $2x$ and the 7 are the "outside" terms, when we multiply them together we call their product the *Outer product.* Because the 3 and the $5y$ are the "inside" terms, we call their product the *Inner product.*

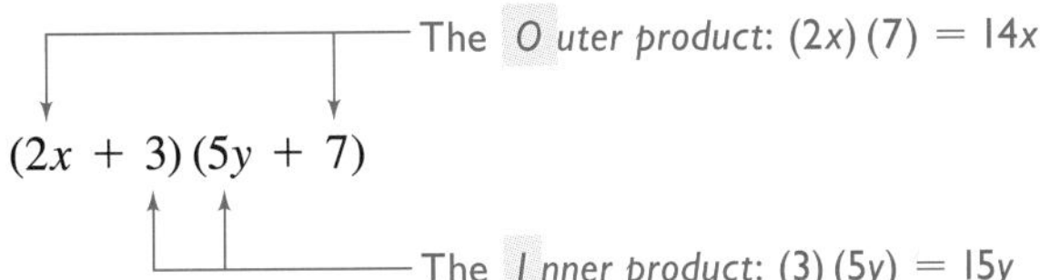

Observe from step 3 of Example 1 that in the answer for $(2x + 3)(5y + 7)$, the first term was $10xy$, the product of the two *F* irst terms; the second term was $14x$, the *O* uter product; the third term was $15y$, the *I* nner product; and the last term was 21, the product of the two *L* ast terms.

The FOIL Method In general, the product of two binomials is a sum of four terms. The first term is the product of the two *F* irst terms, the second term is the *O* uter product, the third term is the *I* nner product, and the fourth term is the product of the two *L* ast terms. We do not *have* to write the terms in this order, but if we do, we can use the letters F, O, I, L as a memory device and say that we're using the **FOIL method** to find the product of two binomials by inspection. You must remember, however, that the four products so found are *terms* and must be added together. It might be even better to think of the procedure as the $F + O + I + L$ method. We use the FOIL method in Example 2.

EXAMPLE 2 Find $(3x^2 + 2)(x + 5)$, and simplify the result.

SOLUTION Using the FOIL method, we have

First	+	Outer	+	Inner	+	Last
$(3x^2 + 2)(x + 5)$		$(3x^2 + 2)(x + 5)$		$(3x^2 + 2)(x + 5)$		$(3x^2 + 2)(x + 5)$
$3x^3$	+	$15x^2$	+	$2x$	+	10

The product, then, is $3x^3 + 15x^2 + 2x + 10$.

In Examples 1 and 2, the inner and outer products were *not* like terms, so they could not be combined.

When the FOIL method is used in a multiplication problem that is of the form $(ax + by)(cx + dy)$, the inner and outer products will always be *like terms* and they must be combined. The multiplication can be done very quickly by using the procedure given in the following box:

Multiplying $(ax + by)(cx + dy)$

1. The *first term* of the product is the product of the first terms of the binomials.
2. The *middle term* of the product is the *sum* of the inner and outer products.
3. The *last term* of the product is the product of the last terms of the binomials.

EXAMPLE 3 Find $(3x + 2y)(4x - 5y)$.

SOLUTION

$$(3x + 2y)(4x - 5y) = 12x^2 - 7xy - 10y^2$$

$12x^2$: The product of the first terms

$-10y^2$: The product of the last terms

$-7xy$: The sum of the inner and outer products

We recommend that you learn to multiply two binomials by inspection so that you can write the answer without having to write any intermediate steps.

Multiplying Two Polynomials When Both Contain Two or More Terms

When we multiply two polynomials that both contain two or more terms, we must apply the distributive property more than once. The problems can be set up horizontally or vertically. If the work is done vertically, we must be sure to write only like terms in the same vertical column when we write the partial products (see Example 4, Solution 2).

EXAMPLE 4 Find the product $(x^2 - 3x + 2)(x - 5)$.

SOLUTION 1

$$\begin{aligned}(x^2 - 3x + 2)(x - 5) &= (x^2 - 3x + 2)\,x + (x^2 - 3x + 2)(-5)\\ &= x^3 - 3x^2 + 2x - 5x^2 + 15x - 10\\ &= x^3 - 8x^2 + 17x - 10\end{aligned}$$

We call $x^3 - 3x^2 + 2x$ and $-5x^2 + 15x - 10$ the *partial products*

SOLUTION 2 When the multiplication is done vertically, we can multiply either from right to left or from left to right. When one of the polynomials contains more terms than the other, it is usually easier to begin by writing the polynomial with more terms *above* the other one.

Multiplying from right to left

$$\begin{array}{r} x^2 - 3x + 2\\ x - 5\\ \hline -5x^2 + 15x - 10\\ x^3 - 3x^2 + 2x \phantom{{}-10}\\ \hline x^3 - 8x^2 + 17x - 10\end{array}$$

$-5x^2 + 15x - 10$ ← Partial product $(x^2 - 3x + 2)(-5)$

$x^3 - 3x^2 + 2x$ ← Partial product $(x^2 - 3x + 2)(x)$

Multiplying from left to right

$$\begin{array}{l} x^2 - 3x + 2\\ x - 5\\ \hline x^3 - 3x^2 + 2x\\ - 5x^2 + 15x - 10\\ \hline x^3 - 8x^2 + 17x - 10\end{array}$$

We *add* the partial products—the polynomials that are shaded. (Notice that like terms *in the partial products* are in the same vertical column.)

Missing Terms Consider the polynomial $5x^3 + 7x + 2$. Because there is *no* x^2-term written, we say that there is a **missing term.** We can write $5x^3 + 7x + 2$ as $5x^3 + 0 + 7x + 2$ or, since $0x^2 = 0$, as $5x^3 + 0x^2 + 7x + 2$. Therefore, the coefficient of x^2 in $5x^3 + 7x + 2$ is understood to be 0; when we rewrite the polynomial as $5x^3 + 0x^2 + 7x + 2$, we call $0x^2$ a *placeholder.*

A Word of Caution It is *incorrect* to write $5x^3 + 7x + 2 = 5x^3 + x^2 + 7x + 2$; the coefficient of x^2 in $5x^3 + x^2 + 7x + 2$ is understood to be 1, not zero. Thus, $5x^3 + 7x + 2 \neq 5x^3 + x^2 + 7x + 2$.

When we multiply polynomials vertically and when we divide polynomials, it is usually desirable to use a placeholder for any missing terms; for example, we can write $2m^4 - 3m^2 + 2m - 5$ as $2m^4 + 0m^3 - 3m^2 + 2m - 5$, where $0m^3$ is a placeholder (see Example 5).

EXAMPLE 5 Find the product $(2m + 2m^4 - 5 - 3m^2)(2 + m^2 - 3m)$.

SOLUTION Multiplication is simplified by first arranging the polynomials in descending powers of m.

Note that $0m^3$ was written as a placeholder

$$\begin{array}{rrrrrrr} & & 2m^4 & +\,0m^3 & -\,3m^2 & +\,2m & -\,5 \\ & & & & m^2 & -\,3m & +\,2 \\ \hline & & 4m^4 & +\,0m^3 & -\,6m^2 & +\,4m & -\,10 \\ & -\,6m^5 & +\,0m^4 & +\,9m^3 & -\,6m^2 & +\,15m & \\ 2m^6 & +\,0m^5 & -\,3m^4 & +\,2m^3 & -\,5m^2 & & \\ \hline 2m^6 & -\,6m^5 & +\,m^4 & +\,11m^3 & -\,17m^2 & +\,19m & -\,10 \end{array}$$

Like terms are aligned in the partial products

EXAMPLE 6 Find $(3a^2b - 6ab^2)(2ab - 5)$.

SOLUTION

$$\begin{array}{rrrr} & & 3a^2b & -\,6ab^2 \\ & & 2ab & -\,5 \\ \hline & & -\,15a^2b & +\,30ab^2 \\ 6a^3b^2 & -\,12a^2b^3 & & \\ \hline 6a^3b^2 & -\,12a^2b^3 & -\,15a^2b & +\,30ab^2 \end{array}$$

Notice that the second line is moved over far enough so that we won't have unlike terms in the same vertical column

Exercises 4.3
Set I

Find each product.

1. $(x + 5)(x + 4)$
2. $(y + 3)(y + 7)$
3. $(y + 8)(y - 9)$
4. $(z + 10)(z - 3)$
5. $(3x + 4)(2x - 5)$
6. $(2y + 5)(4y - 3)$
7. $(4x + 5y)(4x - 5y)$
8. $(s - 2t)(s + 2t)$
9. $(2x - 3y)(5x - y)$
10. $(6u - v)(3u - 4v)$
11. $(7z - 2)(8w + 3)$
12. $(9z - 1)(4x + 5)$
13. $(8x - 9y)(8x + 9y)$
14. $(7w + 2x)(7w - 2x)$
15. $(2s^2 + 5)(s + 1)$
16. $(3u^2 + 2)(u + 1)$
17. $(2x - y)(2x - y)$
18. $(5z - w)(5z - w)$
19. $(2x + 3)(3y - 4)$
20. $(4u - 2)(3v - 5)$
21. $(7x^2 - 3)(7x^2 - 3)$
22. $(4x^2 - 5)(4x^2 - 5)$
23. $(3x - 4)(2x^2 + 5)$
24. $(5y - 2)(3y^2 + 1)$
25. $3x(2x - 4)(x - 1)$
26. $7x(x + 3)(2x - 5)$
27. $(2h - 3)(4h^2 - 5h + 7)$
28. $(5k - 6)(2k^2 + 7k - 3)$
29. $(4 + a^4 + 3a^2 - 2a)(a + 3)$
30. $(3b - 5 + b^4 - 2b^3)(b - 5)$
31. $(4 - 3z^3 + z^2 - 5z)(4 - z)$
32. $(3 + 2v^2 - v^3 + 4v)(2 - v)$
33. $(3u^2 - u + 5)(2u^2 + 4u - 1)$
34. $(2w^2 + w - 7)(5w^2 - 3w - 1)$
35. $-2xy(-3x^2y + xy^2 - 4y^3)$
36. $-3xy(-4xy^2 - x^2y + 3x^3)$
37. $(a^3 - 3a^2b + 3ab^2 - b^3)(-5a^2b)$
38. $(m^3 - 3m^2p + 3mp^2 - p^3)(-4mp^2)$
39. $(x^2 + 2x + 3)^2$
40. $(z^2 - 3z - 4)^2$
41. $[(x + y)(x^2 - xy + y^2)][(x - y)(x^2 + xy + y^2)]$
42. $[(a - 1)(a^2 + a + 1)][(a + 1)(a^2 - a + 1)]$

Writing Problems

Express the answers in your own words and in complete sentences.

1. Explain why $(2x - 4)(x - 1) \neq 2x^2 + 4$.

2. Describe what your first step would be in working the problem $(x - 3x^3 + 2 + 4x^2)(5 - x + 4x^2)$, and explain why that would be your first step.

Exercises 4.3 Set II

Find each product.

1. $(z + 2)(z + 5)$
2. $(x - 9)(x + 2)$
3. $(z + 4)(z - 5)$
4. $(y - 6)(y - 4)$
5. $(6x + 2)(2x - 6)$
6. $(3x^2 + 3)(x - 1)$
7. $(3x + 4y)(3x - 4y)$
8. $(x + 8y)(x + 8y)$
9. $(3x - y)(5x - 7y)$
10. $(2x - 3)(5 - y)$
11. $(8x + 3)(3y - 2)$
12. $(4x + 5)(2x^2 + 7)$
13. $(6y - 5z)(6y + 5z)$
14. $(7x - y)(7x - y)$
15. $(8x^2 + 7)(x + 1)$
16. $(3x^3 - 4)(2x^3 + 3)$
17. $(3x - y)(3x - y)$
18. $(4x^2 - 1)(4x^2 + 1)$
19. $(2x - 1)(y - 4)$
20. $(2x - 1)(x - 4)$
21. $(2x^2 - 6)(2x^2 - 6)$
22. $(x - 6)(2x - 3)$
23. $(7x - 1)(2x^2 + 3)$
24. $(x + 1)(x + 1)$
25. $5x(3x + 2)(2x - 4)$
26. $3x^2(4x - 1)(x + 2)$
27. $(4a - 3)(6a^2 - 8a + 15)$
28. $(3x - 5)(4x^2 - 5x + 2)$
29. $(8 + 2z^4 - z^2 - 9z)(z - 4)$
30. $(3x + 1)(x - 5 - 2x^3 + x^2)$
31. $(6 - k)(4k - 7 + k^4 - 5k^3)$
32. $(3x + 2y - 1)(3x + 2y + 1)$
33. $(5h^2 - h + 8)(h^2 + 3h - 2)$
34. $(2x^2 + 3x - 4)(2x^2 + 3x - 4)$
35. $-4yz(6y^3 - y^2z + 3yz^2)$
36. $(5xy)(-2y^2)(3x^2y)$
37. $(2e^3 - 4e^2f + 7ef^2 - f^3)(-5ef^2)$
38. $(2x - y - 1)(2x + y + 1)$
39. $(3p^2 - 5p + 4)^2$
40. $(x + 1)(x^4 - x^3 + x^2 - x + 1)$
41. $[(2a - b)(4a^2 + 2ab + b^2)][(2a + b)(4a^2 - 2ab + b^2)]$
42. $[(x + 3y)(x - 3y)][(x^2 - 3xy + 9y^2)(x^2 + 3xy + 9y^2)]$

4.4 Special Products

A few products occur so frequently that it is helpful to learn special formulas (or methods) for finding them.

The Product of the Sum and Difference of Two Terms

Because $a + b$ is a *sum* of two terms and $a - b$ is a *difference* of two terms, we call the product $(a + b)(a - b)$ the **product of the sum and difference of two terms.** The formula for finding such a product is given in Property 1.

PROPERTY 1 FOR POLYNOMIALS
The product of the sum and difference of two terms

$$(a + b)(a - b) = a^2 - b^2$$

In words: The product of the sum and difference of two terms equals the square of the first term of one of the binomials minus the square of the second term of that binomial.

PROOF OF PROPERTY 1

$$(a + b)(a - b) = a^2 - ab + ba - b^2 = a^2 - b^2$$ ■

EXAMPLE 1 Examples of using Property 1:

a. $(2x + 3y)(2x - 3y) = (2x)^2 - (3y)^2 = 4x^2 - 9y^2$
b. $(13s - 5t)(13s + 5t) = (13s)^2 - (5t)^2 = 169s^2 - 25t^2$

In Example 2, we are able to use Property 1 to find a product of two *trinomials*, because it happens to be possible to insert grouping symbols and express the product in the *form* $(a + b)(a - b)$. (Not all trinomials can be converted to this form.)

EXAMPLE 2 Simplify $(x + y + 2)(x + y - 2)$.

SOLUTION Because the first *two* terms of both trinomials are identical, we can insert grouping symbols around the identical terms; we then have a product of the *form* $(a + b)(a - b)$.

$(x + y + 2)(x + y - 2) = [(x + y) + 2][(x + y) - 2]$ ← This is of the form $(a + b)(a - b)$

$= (x + y)^2 - 2^2$ — Using Property 1

$= (x + y)(x + y) - 4$

$= x^2 + 2xy + y^2 - 4$ — Using the FOIL method on $(x + y)(x + y)$

Squaring a Binomial

Property 2 gives the formulas for **squaring a binomial**. We *strongly* urge you to learn these formulas and to learn how and when to apply them.

PROPERTY 2 FOR POLYNOMIALS
The square of a binomial

$$(a + b)^2 = a^2 + 2ab + b^2 \quad \text{Form 1}$$

$$(a - b)^2 = a^2 - 2ab + b^2 \quad \text{Form 2}$$

In words: The square of a binomial has *three* terms.

1. The *first term* of the result is the square of the first term of the binomial.
2. The *middle term* of the result is twice the product of the two terms of the binomial.
3. The *last term* of the result is the square of the last term of the binomial.

PROOF OF FORM 1 OF PROPERTY 2

$ba = ab$, so ba and ab are like terms

$$(a + b)^2 = (a + b)(a + b) = a^2 + ab + ba + b^2 = a^2 + 2ab + b^2$$ ■

PROOF OF FORM 2 OF PROPERTY 2

$-ba = -ab$, so $-ba$ and $-ab$ are like terms

$$(a - b)^2 = (a - b)(a - b) = a^2 - ab - ba + b^2 = a^2 - 2ab + b^2$$ ■

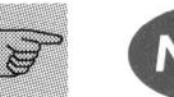

Note Form 2 of Property 2 is not really needed; we *can* treat $(a - b)^2$ as if it were $[a + (-b)]^2$ and use Form 1 (see Example 3c).

EXAMPLE 3 Examples of squaring binomials:

a. $(5m + 2n^2)^2 = (5m)^2 + 2(5m)(2n^2) + (2n^2)^2$ — Using form 1 of Property 2

$= 25m^2 + 20mn^2 + 4n^4$

b. $(15x - 4y)^2 = (15x)^2 - 2(15x)(4y) + (4y)^2$ — Using form 2 of Property 2

$= 225x^2 - 120xy + 16y^2$

c. $(15x - 4y)^2 = [15x + (-4y)]^2 = (15x)^2 + 2(15x)(-4y) + (-4y)^2$ — Using form 1 of Property 2

$= 225x^2 - 120xy + 16y^2$

EXAMPLE 4 Simplify $[3 - (x + y)]^2$.

SOLUTION This expression is of the form $(a - b)^2$, so we can use form 2 of Property 2. (We could also treat the given expression as $\{3 + [-(x + y)]\}^2$ and use form 1.)

$$[3 - (x + y)]^2 = 3^2 - 2(3)(x + y) + (x + y)^2$$
$$= 9 - 6(x + y) + (x + y)^2$$
$$= 9 - 6x - 6y + x^2 + 2xy + y^2$$

Using form 1 of Property 2 on $(x + y)^2$

A Word of Caution A common error is to confuse squaring a *product* of two *factors* with squaring a *sum* of two *terms*. It *is* true that

$$(ab)^2 = a^2b^2$$

Here, a and b are *factors*

However, $(a + b)^2$ cannot be found by simply adding the squares of a and b.

Correct method	*Incorrect method*
$(a + b)^2 = a^2 + 2ab + b^2$	$(a + b)^2 = a^2 + b^2$

($2ab$): This is the term that is often left out

We can verify by substituting numbers for a and b that $(a + b)^2 = a^2 + 2ab + b^2$ and that $(a + b)^2 \neq a^2 + b^2$. For example, if we let $a = 3$ and $b = 4$, we have

$(a + b)^2$	$a^2 + 2ab + b^2$	$a^2 + b^2$
$= (3 + 4)^2$	$= 3^2 + 2(3)(4) + 4^2$	$= 3^2 + 4^2$
$= (7)^2$	$= 9 + 24 + 16$	$= 9 + 16$
$= 49$	$= 49$	$= 25$

We see that $(a + b)^2 = a^2 + 2ab + b^2$ and that $(a + b)^2 \neq a^2 + b^2$.

Exercises 4.4
Set I

Simplify each expression.

1. $(2u + 5v)(2u - 5v)$
2. $(3m - 7n)(3m + 7n)$
3. $(2x^2 - 9)(2x^2 + 9)$
4. $(10y^2 - 3)(10y^2 + 3)$
5. $(x^5 - y^6)(x^5 + y^6)$
6. $(a^7 + b^4)(a^7 - b^4)$
7. $(7mn + 2rs)(7mn - 2rs)$
8. $(8hk + 5ef)(8hk - 5ef)$
9. $(12x^4y^3 + u^7v)(12x^4y^3 - u^7v)$
10. $(11a^5b^2 + 9c^3d^6)(11a^5b^2 - 9c^3d^6)$
11. $(a + b + 2)(a + b - 2)$
12. $(x + y + 5)(x + y - 5)$
13. $(x^2 + y + 5)(x^2 + y - 5)$
14. $(u^2 - v + 7)(u^2 - v - 7)$
15. $7x(x - 1)(x + 1)$
16. $3y(2y - 5)(2y + 5)$
17. $(2x + 3)^2$
18. $(6x + 5)^2$
19. $(5x - 3)^2$
20. $(9x - 6)^2$
21. $(7x - 10y)^2$
22. $(4u - 9v)^2$
23. $(7x^2 + 9y)^2$
24. $(8s + 11t^3)^2$
25. $(15u - v^2)^2$
26. $(16w - t^2)^2$
27. $[s + (t + u)]^2$
28. $[v + (x + s)]^2$
29. $([u + v] + 7)^2$
30. $([s + t] + 4)^2$
31. $([2x - y] - 3)^2$
32. $([3z - w] - 7)^2$
33. $(x + [u + v])^2$
34. $(s + [y + 2])^2$
35. $(y - [x - 2])^2$
36. $(x - [z - 5])^2$
37. $5(x - 3)^2$
38. $7(x + 1)^2$

Writing Problems

Express the answers in your own words and in complete sentences.

1. Describe the method you would use in working the problem $(3x^2 + 5y)(3x^2 - 5y)$, and explain why you would use that method.
2. Explain why $(2x + 5)^2 \neq 4x^2 + 25$.

Exercises 4.4
Set II

Simplify each expression.

1. $(6x - 5)(6x + 5)$
2. $(4 + 3z)(4 - 3z)$
3. $(7u - 8v)(7u + 8v)$
4. $(x - 8y)(x + 8y)$
5. $(x^6 - y^5)(x^6 + y^5)$
6. $(x^2 + y^3)(x^2 - y^3)$
7. $(2xy + 3uv)(2xy - 3uv)$
8. $(3abc - ef)(3abc + ef)$
9. $(13z^2w^3 + 5v^5u^7)(13z^2w^3 - 5v^5u^7)$
10. $(16x^3y^2 - 11z^4)(16x^3y^2 + 11z^4)$
11. $(s + t - 1)(s + t + 1)$
12. $(2x - y - 3)(2x + y + 3)$
13. $(x^2 + y + 5)(x^2 + y - 5)$
14. $(x^2 - y + 5)(x^2 + y - 5)$
15. $3z(5z - 1)(5z + 1)$
16. $4ab^2(a - 2)(a + 2)$
17. $(3x + 4)^2$
18. $(6x - 5y)^2$
19. $(5x - 2)^2$
20. $(8yz + 2)^2$
21. $(6x - 7y)^2$
22. $(3 - [x - y])^2$
23. $(8r^2 + 6s)^2$
24. $(9x + 12y^3)^2$
25. $(13p - q^2)^2$
26. $(11t^2 - u^3)^2$
27. $[w + (x + y)]^2$
28. $[(x + y) + z]^2$
29. $([x + y] + 5)^2$
30. $([x + y] + [a + b])^2$
31. $([7x - y] - 3)^2$
32. $(4x + [a - b])^2$
33. $(s + [x + 3])^2$
34. $3x(x - 1)^2$
35. $(z - [y - 3])^2$
36. $(3x + 4)^2x^2$
37. $8(a + 5)^2$
38. $3x(-4x)^2(2x^3)$

4.5 The Binomial Theorem

Before we discuss raising a binomial to a power, we introduce two new symbols.

Factorial Notation

Mathematicians define $n!$ (read *n* factorial) as follows:

$$0! = 1$$

$$1! = 1$$

$$n! = n(n-1)(n-2)\cdots(3)(2)(1), \text{ where } n \text{ is an integer greater than } 1$$

That is, $n!$ is a product of n factors; the *first* factor is n, the *second* factor is $n - 1$, the *third* factor is $n - 2$, and so on (with each succeeding factor decreasing by 1), until we reach the *last* factor, which is 1. For example, $3! = 3 \cdot 2 \cdot 1 = 6$ (the *first* factor is 3, the *second* factor is $3 - 1$, or 2, and the *last* factor is $3 - 2$, or 1).

EXAMPLE 1 Evaluate 5!.

SOLUTION The first factor is 5, the second factor is 4, the third factor is 3, the fourth factor is 2, and the *last* factor is 1. Therefore,

$$5! = 5 \cdot 4 \cdot 3 \cdot 2 \cdot 1 = 120$$

Most scientific calculators have a factorial key labeled [!], and it is pressed *after* the number; that is, to find 6!, press [6] [!]. You may or may not have to press [2nd], [SHIFT], or [INV] before pressing [!] and [=] or [ENTER] after pressing [!]. The display is 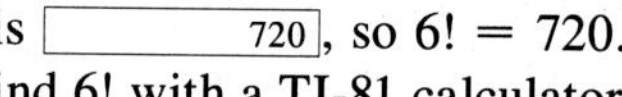 [720], so 6! = 720.

We proceed as follows to find 6! with a TI-81 calculator. We first press [6] and then [MATH]. This brings up one of the MATH *menus*. We press [▼] four times so that the *fifth* entry, 5:! , is highlighted, and we press [ENTER] to *select* that entry. Pressing [ENTER] *again* gives the value of 6!, which is 720.

Binomial Coefficients

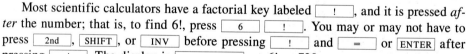

When we raise a binomial to a power, the result is a polynomial, and the coefficients of that polynomial are called **binomial coefficients**. For example, $(a + b)^2 = 1\,a^2 + 2\,ab + 1\,b^2$, and the binomial coefficients are 1, 2, and 1. The symbol we will use for the binomial coefficient is $\binom{n}{r}$.* (We will use this notation when we are raising a binomial to the nth power, and, as we shall soon see, we will let $r = 0$, then 1, then 2, and so on, up to n.)

---

*An alternative notation for the binomial coefficient is ${}_nC_r$; that is, ${}_nC_r = \binom{n}{r}$. In probability theory (which will not be discussed in this book), ${}_nC_r$ is the symbol for the combination of n things taken r at a time; for this reason, $\binom{n}{r}$ is usually read as "combination n, taken r at a time."

Mathematicians define the binomial coefficient, $\binom{n}{r}$, as follows:

$$\binom{n}{0} = 1$$

$$\binom{n}{r} = \frac{n(n-1)(n-2)\cdots(n-r+1)}{r!},$$ where n and r are *positive* integers and $r \le n$

Notice that, according to the definition, the denominator is $r!$ and the numerator consists of several factors; the *first* factor is n, and each succeeding factor decreases by 1 until we reach the *last* factor, which is $n - r + 1$. (If $r = 1$, the numerator will contain just *one* factor, since, if $r = 1$, $n - r + 1 = n - 1 + 1 = n$.)

To use the definition of $\binom{n}{r}$, we must identify n (it is the *upper* number) and r (it is the *lower* number), and, to determine what the *last* factor of the numerator should be, we must calculate $n - r + 1$ (see Example 2).

EXAMPLE 2 Evaluate $\binom{8}{5}$.

SOLUTION The lower number is 5, so $r = 5$, and the denominator will be 5 factorial (5!). The upper number is 8, so $n = 8$, and the *first* factor in the numerator will be 8. Then $n - r + 1$ (the *last* factor in the numerator) equals $8 - 5 + 1$, or 4; therefore, the *last* factor in the numerator will be 4. That is, the factors in the numerator will start with 8 and will decrease by 1 until we reach 4, the last factor. (Then we reduce the resulting fraction).

The first factor of the numerator is 8

The last factor of the numerator is 4

$$\binom{8}{5} = \frac{8 \cdot 7 \cdot 6 \cdot 5 \cdot 4}{5!} = \frac{8 \cdot 7 \cdot \overset{1}{\cancel{6}} \cdot \overset{1}{\cancel{5 \cdot 4}}}{\underset{1}{\cancel{5 \cdot 4}} \cdot \underset{1}{\cancel{3 \cdot 2}} \cdot 1} = 56$$

The denominator is 5!, since r is 5

EXAMPLE 3 Evaluate $\binom{6}{2}$.

SOLUTION The lower number is 2, so $r = 2$, and the denominator will be 2!. The upper number is 6, so $n = 6$; $n - r + 1 = 6 - 2 + 1 = 5$, so the first factor of the numerator will be 6 and the last factor will be 5.

$$\binom{6}{2} = \frac{6 \cdot 5}{2!} = \frac{\overset{3}{\cancel{6}} \cdot 5}{\underset{1}{\cancel{2}} \cdot 1} = 15$$

EXAMPLE 4 Evaluate $\binom{6}{3}$.

SOLUTION The lower number is 3, so $r = 3$, and the denominator will be 3!. The upper number is 6, so $n = 6$; $n - r + 1 = 6 - 3 + 1 = 4$, so the first factor of the numerator will be 6 and the last factor will be 4.

$$\binom{6}{3} = \frac{6 \cdot 5 \cdot 4}{3!} = \frac{\overset{1}{\cancel{6}} \cdot 5 \cdot 4}{\underset{1}{\cancel{3 \cdot 2}} \cdot 1} = 20$$

EXAMPLE 5 Evaluate $\binom{4}{0}$.

SOLUTION By definition, $\binom{4}{0} = 1$.

A number of theorems about the binomial coefficient can be proved. Some of them are shown below.

$$\binom{n}{n} = 1; \quad \binom{n}{1} = n; \quad \binom{n}{n-1} = n; \quad \binom{n}{r} = \binom{n}{n-r}; \quad \binom{n}{r} = \frac{n!}{r!\,(n-r)!}$$

You may find the theorem $\binom{n}{r} = \dfrac{n!}{r!\,(n-r)!}$ easier to remember and to use than the definition of $\binom{n}{r}$.

EXAMPLE 6 Examples of evaluating $\binom{n}{r}$ using the formula $\binom{n}{r} = \dfrac{n!}{r!\,(n-r)!}$:

a. $\binom{5}{2}$ $\quad$ n is 5 and r is 2; $\binom{5}{2} = \dfrac{5!}{2!\,(5-2)!} = \dfrac{5!}{2!\,(3!)} = \dfrac{5 \cdot 4 \cdot 3 \cdot 2 \cdot 1}{(2 \cdot 1)(3 \cdot 2 \cdot 1)} = 10$

b. $\binom{4}{3}$ $\quad$ n is 4 and r is 3; $\binom{4}{3} = \dfrac{4!}{3!\,(4-3)!} = \dfrac{4!}{3!\,(1!)} = \dfrac{4 \cdot 3 \cdot 2 \cdot 1}{(3 \cdot 2 \cdot 1)(1)} = 4$

c. $\binom{8}{3}$ $\quad$ n is 8 and r is 3; $\binom{8}{3} = \dfrac{8!}{3!\,(8-3)!} = \dfrac{8!}{3!\,(5!)}$

$$= \frac{8 \cdot 7 \cdot 6 \cdot 5 \cdot 4 \cdot 3 \cdot 2 \cdot 1}{(3 \cdot 2 \cdot 1)(5 \cdot 4 \cdot 3 \cdot 2 \cdot 1)} = 56$$

Note Because $5! = 5 \cdot 4 \cdot 3!$ (Can you verify this?), in Example 6a, we could have divided both numerator and denominator by 3!, as follows:

$$\frac{5!}{2!\,(3!)} = \frac{5 \cdot 4 \cdot 3!}{2 \cdot 1(3!)} = 10$$

Because $4! = 4 \cdot 3!$ and $8! = 8 \cdot 7 \cdot 6 \cdot 5!$, the work in Examples 6b and 6c could have been shown as follows:

$$\frac{4!}{3!(1!)} = \frac{4 \cdot 3!}{3!(1)} = 4 \quad \text{and} \quad \frac{8!}{3!(5!)} = \frac{8 \cdot 7 \cdot 6 \cdot 5!}{3 \cdot 2 \cdot (5!)} = 56$$

Note In Example 6b, we found that $\binom{4}{3} = 4$. Since $3 = 4 - 1$, $\binom{4}{3}$ fits the pattern $\binom{4}{4-1}$, and we have verified that $\binom{n}{n-1} = n$ if $n = 4$.

We found in Example 6c that $\binom{8}{3} = 56$, and we found in Example 2 that $\binom{8}{5} = 56$. Since $\binom{8}{3}$ and $\binom{8}{5}$ both equal 56, $\binom{8}{3} = \binom{8}{5}$. Substituting $\binom{8}{8-3}$ for $\binom{8}{5}$, we have $\binom{8}{3} = \binom{8}{8-3}$, and we have verified that $\binom{n}{r} = \binom{n}{n-r}$ for $n = 8$ and $r = 3$.

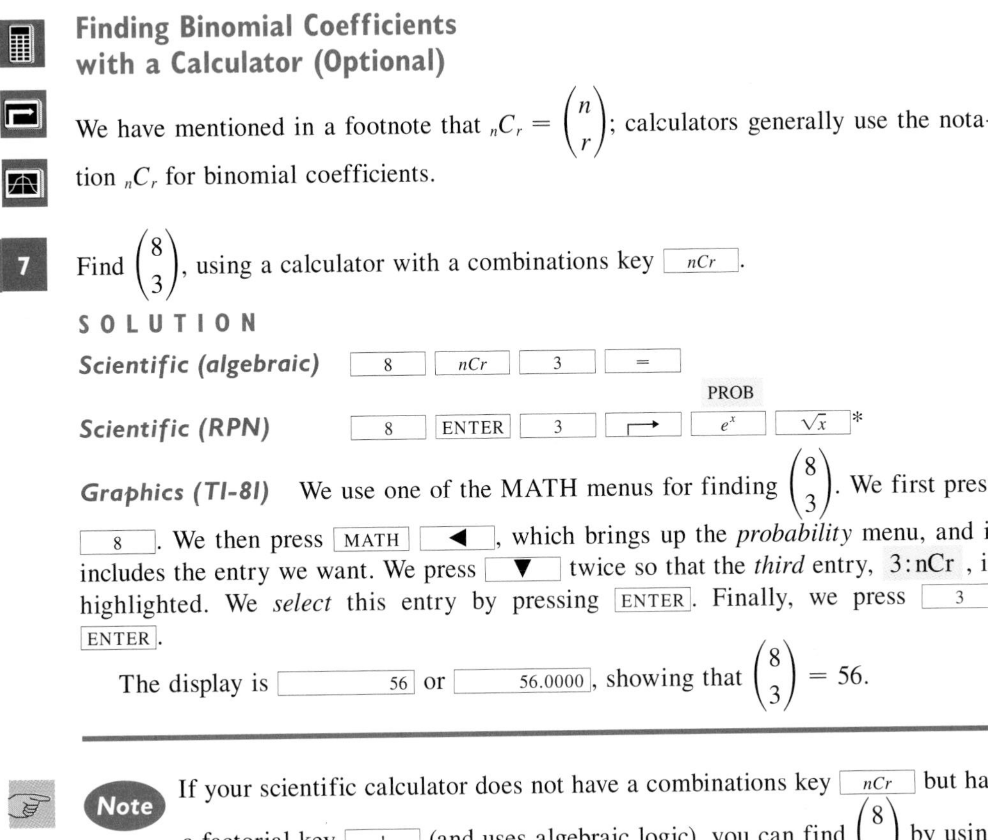

Finding Binomial Coefficients with a Calculator (Optional)

We have mentioned in a footnote that ${}_nC_r = \binom{n}{r}$; calculators generally use the notation ${}_nC_r$ for binomial coefficients.

EXAMPLE 7 Find $\binom{8}{3}$, using a calculator with a combinations key [nCr].

SOLUTION

Scientific (algebraic) [8] [nCr] [3] [=]

Scientific (RPN) [8] [ENTER] [3] [↱] [PROB e^x] [$\sqrt{x}$]*

Graphics (TI-81) We use one of the MATH menus for finding $\binom{8}{3}$. We first press [8]. We then press [MATH] [◀], which brings up the *probability* menu, and it includes the entry we want. We press [▼] twice so that the *third* entry, 3:nCr , is highlighted. We *select* this entry by pressing [ENTER]. Finally, we press [3] [ENTER].

The display is [56] or [56.0000], showing that $\binom{8}{3} = 56$.

Note If your scientific calculator does not have a combinations key [nCr] but has a factorial key [!] (and uses algebraic logic), you can find $\binom{8}{3}$ by using the fact that $\binom{8}{3} = \dfrac{8!}{3!\,(5!)}$ and entering the problem as follows:

[8] [!] [÷] [3] [!] [÷] [5] [!] [=]

Powers of Binomials

Powers of binomials occur so frequently that it is convenient to have a shortened method for raising a binomial to any nonnegative, integral power. We know that $(a + b)^0 = 1$, $(a + b)^1 = a + b$, and $(a + b)^2 = a^2 + 2ab + b^2$. Let's find $(a + b)^3$ and $(a + b)^4$.

$$\begin{aligned}
(a + b)^3 &= (a + b)^2(a + b) \\
&= (a^2 + 2ab + b^2)(a + b) \\
&= (a^2 + 2ab + b^2)(a) + (a^2 + 2ab + b^2)(b) \\
&= a^3 + 2a^2b + ab^2 + a^2b + 2ab^2 + b^3 \\
&= a^3 + 3a^2b + 3ab^2 + b^3 \\
(a + b)^4 &= \{(a + b)^2\}^2 \\
&= (a^2 + 2ab + b^2)^2
\end{aligned}$$

Let's finish the multiplication vertically:

$$\begin{array}{rrrrr}
 & & a^2 + 2ab + b^2 & & \\
 & & a^2 + 2ab + b^2 & & \\
\hline
 & & a^2b^2 & + 2ab^3 & + b^4 \\
 & 2a^3b & + 4a^2b^2 & + 2ab^3 & \\
a^4 & + 2a^3b & + a^2b^2 & & \\
\hline
a^4 & + 4a^3b & + 6a^2b^2 & + 4ab^3 & + b^4
\end{array}$$

*The RPN calculator we use does not actually have a combinations *key*. Instead, we access the *probability menu* by pressing [↱] [PROB e^x]. Then the [$\sqrt{x}$] key is directly *below* "Cn, r" on the calculator display, and so pressing [$\sqrt{x}$] enables us to find the desired binomial coefficient.

Now let's consider the following binomial expansions. [We suggest that you verify our results for $(a + b)^5$ and $(a + b)^6$ by further multiplication.]

$$(a + b)^0 = 1$$
$$(a + b)^1 = 1a + 1b$$
$$(a + b)^2 = 1a^2 + 2ab + 1b^2$$
$$(a + b)^3 = 1a^3 + 3a^2b + 3ab^2 + 1b^3$$
$$(a + b)^4 = 1a^4 + 4a^3b + 6a^2b^2 + 4ab^3 + 1b^4$$
$$(a + b)^5 = 1a^5 + 5a^4b + 10a^3b^2 + 10a^2b^3 + 5ab^4 + 1b^5$$
$$(a + b)^6 = 1a^6 + 6a^5b + 15a^4b^2 + 20a^3b^3 + 15a^2b^4 + 6ab^5 + 1b^6$$

A careful examination of the above expansions shows that $(a + b)^n$ is always a polynomial in a and b, that there are always $n + 1$ terms, and that the degree of each term is n (that is, in each term, the sum of the exponents on the variables is n).

Let's look carefully at the way in which $(a + b)^6$ *could* have been written:

The exponents on the *a*'s are 6 in the first term, 5 in the second term, 4 in the third term, 3 in the fourth term, and so on; that is, the exponents on the *a*'s *decrease by 1* as we read the terms from left to right

The exponents on the *b*'s are 0 in the first term, 1 in the second term, 2 in the third term, and so on; that is, the exponents on the *b*'s *increase by 1* as we read the terms from left to right

$$(a + b)^6 = 1a^6b^0 + 6a^5b^1 + 15a^4b^2 + 20a^3b^3 + 15a^2b^4 + 6a^1b^5 + 1a^0b^6$$

These are equal

These are both *n*

These are both 1

The coefficients of the first and last terms are 1, the coefficients of the second and next-to-last terms are 6 (n is 6 in this case), the coefficients of the third and third-from-last terms are 15, and the coefficient of the middle term is 20. Because of the "matching" of the coefficients of the first and last terms, the second and next-to-last terms, and so on, we often say that the coefficients are *symmetrical*.

Now let's look at some binomial coefficients. By definition, $\binom{6}{0} = 1$. Also, according to the theorems given earlier, $\binom{6}{6} = 1$, $\binom{6}{1} = 6$, and $\binom{6}{5} = 6$. We saw in Example 3 that $\binom{6}{2} = 15$, and, since $\binom{n}{r} = \binom{n}{n-r}$, $\binom{6}{4} = \binom{6}{6-4}$, or $\binom{6}{4} = \binom{6}{2} = 15$. We saw in Example 4 that $\binom{6}{3} = 20$. Therefore, the entire expansion could be written as

$$(a + b)^6$$
$$= \binom{6}{0}a^6b^0 + \binom{6}{1}a^5b^1 + \binom{6}{2}a^4b^2 + \binom{6}{3}a^3b^3 + \binom{6}{4}a^2b^4 + \binom{6}{5}a^1b^5 + \binom{6}{6}a^0b^6$$

There are seven terms, and every term is a sixth-degree term; that is, the *sum* of the exponents on the variables in any term is always 6. Each *term* is of the form $\binom{n}{r}a^{n-r}b^r$, where r is 0 in the first term, 1 in the second term, 2 in the third term, and so on; finally, in the *last* (or seventh) term, r is 6. Because of this pattern, we say that the $(r + 1)$st term is $\binom{n}{r}a^{n-r}b^r$.

The Binomial Theorem

The **binomial theorem**, which follows, can be used for raising any binomial to a power. It can also be used to find any *term* of a binomial expansion (see Examples 14 and 15). The proof of the binomial theorem requires the use of mathematical induction and is, therefore, beyond the scope of this text.

The binomial theorem

If n is any positive integer, then

$$(a+b)^n = \binom{n}{0}a^nb^0 + \binom{n}{1}a^{n-1}b^1 + \binom{n}{2}a^{n-2}b^2 + \cdots + \binom{n}{r}a^{n-r}b^r + \cdots + \binom{n}{n-2}a^2b^{n-2} + \binom{n}{n-1}a^1b^{n-1} + \binom{n}{n}a^0b^n$$

where the $(r+1)$st term is $\binom{n}{r}a^{n-r}b^r$. That is, r is 0 in the first term, 1 in the second term, 2 in the third term, and so on; in the *last* term, r equals n.

Note Observe that the first term in the binomial expansion simplifies to a^n and the second term to $na^{n-1}b$; also observe that the next-to-last term simplifies to nab^{n-1} and the last term to b^n.

In using the binomial theorem, we can write 1 in place of $\binom{n}{0}$ and $\binom{n}{n}$, and we can write n in place of $\binom{n}{1}$ and $\binom{n}{n-1}$. Also, because the coefficients are symmetrical, we need to calculate only the first half of the coefficients; the last half are symmetrical with respect to the first half.

One way of using the binomial theorem is given in the following box:

Using the binomial theorem

To expand $(a+b)^n$, where n is a positive integer, first make a blank outline with $(n+1)$ terms:

$$\underbrace{(\ )a^{\square}b^{\square} + (\ )a^{\square}b^{\square} + (\ )a^{\square}b^{\square} + (\ )a^{\square}b^{\square} + \cdots + (\ )a^{\square}b^{\square}}_{(n+1)\text{ terms}}$$

1. a. Fill in the coefficients.
   b. Fill in the powers of a and b, as shown.

The exponent of a is n in the first term, and the exponents of a *decrease* by 1 in each succeeding term

The exponent of b is 0 in the first term, and the exponents of b *increase* by 1 in each succeeding term

$$(1)a^nb^0 + (n)a^{n-1}b^1 + \binom{n}{2}a^{n-2}b^2 + \cdots + \binom{n}{n-2}a^2b^{n-2} + (n)a^1b^{n-1} + (1)(a)^0(b)^n$$

2. Rewrite the entire expression with each term simplified. The result is the expansion of $(a+b)^n$.

EXAMPLE 8 Use the binomial theorem to simplify $(a + b)^8$.

SOLUTION There will be nine terms. The blank outline is

$$(a+b)^8 = (\ )a^{\square}b^{\square} + (\ )a^{\square}b^{\square} + (\ )a^{\square}b^{\square} + (\ )a^{\square}b^{\square} + (\ )a^{\square}b^{\square} + (\ )a^{\square}b^{\square} + (\ )a^{\square}b^{\square} + (\ )a^{\square}b^{\square} + (\ )a^{\square}b^{\square}$$ Nine terms

Step 1. $(a+b)^8 = 1a^8b^0 + 8a^7b^1 + \binom{8}{2}a^6b^2 + \binom{8}{3}a^5b^3 + \binom{8}{4}a^4b^4 + \binom{8}{5}a^3b^5 + \binom{8}{6}a^2b^6 + 8a^1b^7 + 1a^0b^8$

Step 2. We found in Example 6c that $\binom{8}{3} = 56$; $\binom{8}{5}$ also equals 56, because $\binom{8}{5} = \binom{8}{8-3} = \binom{8}{3}$. We must calculate $\binom{8}{2}$ and $\binom{8}{4}$.

For $\binom{8}{2}$, n is 8 and r is 2; therefore,

$$\binom{8}{2} = \frac{8!}{2!(8-2)!} = \frac{8!}{2!(6!)} = \frac{8\cdot 7\cdot 6!}{2\cdot 1(6!)} = 28$$

Then $\binom{8}{6}$ also equals 28, since $\binom{8}{6} = \binom{8}{8-2} = \binom{8}{2}$.

For $\binom{8}{4}$, n is 8 and r is 4; therefore,

$$\binom{8}{4} = \frac{8!}{4!(8-4)!} = \frac{8!}{4!(4!)} = \frac{8\cdot 7\cdot 6\cdot 5\cdot 4!}{4\cdot 3\cdot 2\cdot 1(4!)} = 70$$

Therefore, $(a+b)^8 = a^8 + 8a^7b + 28a^6b^2 + 56a^5b^3 + 70a^4b^4 + 56a^3b^5 + 28a^2b^6 + 8ab^7 + b^8$

Note The coefficients for raising a binomial to a power can be found especially easily on the TI-81 graphics calculator. To find the coefficients for raising a binomial to the eighth power, for example, we can first find $\binom{8}{0}$ by pressing

[8] [MATH] [◄] [▼] [▼] [ENTER] [0] [ENTER]

The display is 1 (which is what we would expect). Pressing [▲] duplicates the row [8 nCr 0] of the display, and then pressing [◄] [1] enables us to replace the 0 with a 1; pressing [ENTER], then, tells us that $\binom{8}{1} = 8$. Pressing [▲] *again* duplicates the row [8 nCr 1] of the display, and then pressing [◄] [2] enables us to replace the 1 with a 2; pressing [ENTER] tells us that $\binom{8}{2} = 28$. Continuing in this fashion (pressing [▲], then [◄], then another number, and then [ENTER]), we can easily find all the coefficients for raising a binomial to a power.

It is also possible to determine the numerical coefficients for expanding a binomial by memorizing the following pattern of coefficients:

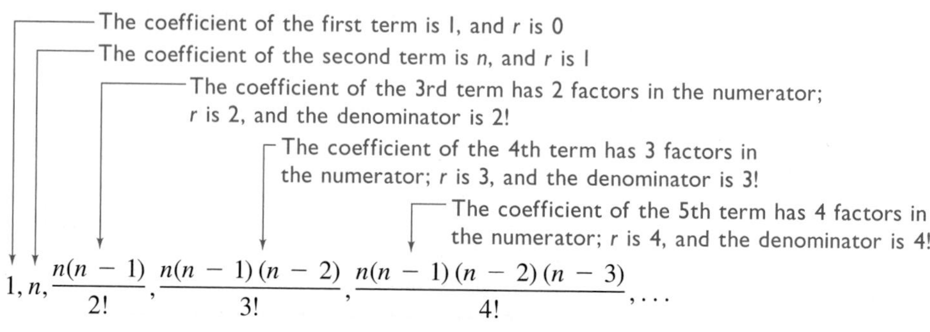

The coefficient of the first term is 1, and r is 0
The coefficient of the second term is n, and r is 1
The coefficient of the 3rd term has 2 factors in the numerator; r is 2, and the denominator is 2!
The coefficient of the 4th term has 3 factors in the numerator; r is 3, and the denominator is 3!
The coefficient of the 5th term has 4 factors in the numerator; r is 4, and the denominator is 4!

$$1, n, \frac{n(n-1)}{2!}, \frac{n(n-1)(n-2)}{3!}, \frac{n(n-1)(n-2)(n-3)}{4!}, \ldots$$

Still another method of determining the numerical coefficients for raising a binomial to a power is to use Pascal's triangle, which is discussed below.

Pascal's Triangle

If binomial expansions are written for $n = 0$, $n = 1$, $n = 2$, $n = 3$, and $n = 4$, we have

$$(a + b)^0 = 1$$
$$(a + b)^1 = 1a + 1b$$
$$(a + b)^2 = 1a^2 + 2ab + 1b^2$$
$$(a + b)^3 = 1a^3 + 3a^2b + 3ab^2 + 1b^3$$
$$(a + b)^4 = 1a^4 + 4a^3b + 6a^2b^2 + 4ab^3 + 1b^4$$
$$\vdots$$

If we delete the variables and the plus signs from the right sides of the equations, we get a triangular array of numbers known as **Pascal's triangle**:

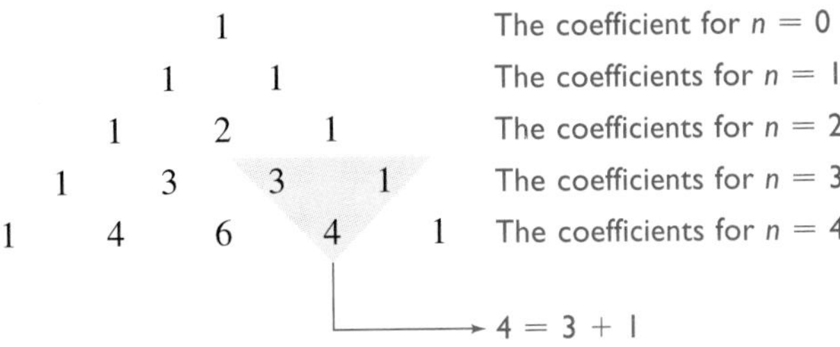

A close examination of the numbers in Pascal's triangle reveals that

1. The first and last numbers in any row are 1.
2. Any other number in Pascal's triangle is the sum of the two closest numbers in the row above it (see the shaded triangle).

As we stated earlier, Pascal's triangle can be used to find the coefficients in a binomial expansion. The triangle we have shown can be extended to any size by using the two rules given above. Try extending the triangle by one more row (that is, until the second number is a 5); if you do this correctly, you will get the coefficients for $(a + b)^5$ (see Example 9).

EXAMPLE 9 Using Pascal's triangle, find $(s - t)^5$.

SOLUTION We rewrite $(s - t)^5$ as $[s + (-t)]^5$; then we use the facts that

$$(-t)^0 = 1, \quad (-t)^1 = -t, \quad (-t)^2 = t^2, \quad (-t)^3 = -t^3, \quad (-t)^4 = t^4, \quad (-t)^5 = -t^5$$

We repeat here the row of Pascal's triangle for $n = 4$ (that is, the row for raising a binomial to the fourth power); then we find the next row:

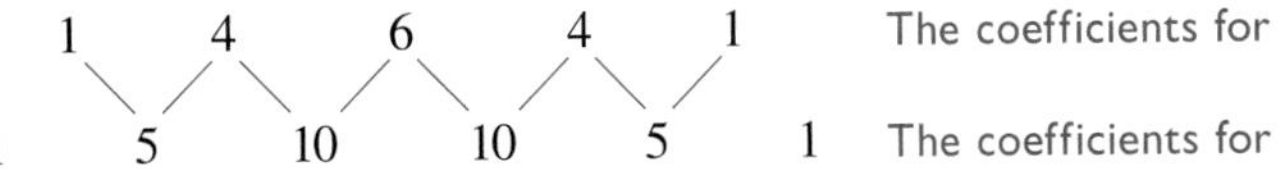

Therefore,

$$\begin{aligned}[s+(-t)]^5 &= 1\ s^5(-t)^0 + 5\ s^4(-t)^1 + 10\ s^3(-t)^2 + 10\ s^2(-t)^3 + 5\ s^1(-t)^4 + 1\ s^0(-t)^5 \\ &= s^5 - 5s^4t + 10s^3t^2 - 10s^2t^3 + 5st^4 - t^5\end{aligned}$$

Notice in Example 9 that we were raising a *difference* of two terms to a power and that the signs of the terms of the expansion *alternated*. (See also Examples 12 and 13.) The signs in the final answer will always alternate when we raise a binomial *difference* to a power.

We recommend that you simplify $(a + b)^8$ by using Pascal's triangle and verify that you get the same result that we got in Example 8.

The terms of the binomials being expanded often consist of something other than single variables (see Example 10).

EXAMPLE 10 Examples of binomials with terms that consist of more than a single variable:

a. $(3x^2 + 5y)^3$ In this binomial, $a = 3x^2$ and $b = 5y$

b. $(2e^3 - f^2)^5$ In this binomial, $a = 2e^3$ and $b = -f^2$

c. $\left(\frac{x}{2} - 4\right)^4$ In this binomial, $a = \frac{x}{2}$ and $b = -4$

EXAMPLE 11 Expand $(3x^2 + 5y)^3$, using any of the methods discussed for finding the coefficients.

SOLUTION

Step 1. The coefficients are 1, 3, 3, and 1:

$$1\,(3x^2)^3(5y)^0 + 3\,(3x^2)^2(5y)^1 + 3\,(3x^2)^1(5y)^2 + 1\,(3x^2)^0(5y)^3$$

Step 2.
$$\begin{aligned}(3x^2 + 5y)^3 &= 3^3x^6 + 3(3^2x^4)(5y) + 3(3x^2)(5^2y^2) + 5^3y^3 \\ &= 27x^6 + 27(5)x^4y + 9(25)x^2y^2 + 125y^3 \\ &= 27x^6 + 135x^4y + 225x^2y^2 + 125y^3\end{aligned}$$

EXAMPLE 12 Expand $(2e^3 - f^2)^5$.

SOLUTION

Step 1. The coefficients are 1, 5, 10, 10, 5, and 1.

$$\begin{aligned}&1\,(2e^3)^5(-f^2)^0 + 5\,(2e^3)^4(-f^2)^1 + 10\,(2e^3)^3(-f^2)^2 \\ &\quad + 10\,(2e^3)^2(-f^2)^3 + 5\,(2e^3)^1(-f^2)^4 + 1\,(2e^3)^0(-f^2)^5\end{aligned}$$

Step 2.
$$\begin{aligned}(2e^3 - f^2)^5 &= 2^5e^{15} + 5(2^4e^{12})(-f^2) + 10(2^3e^9)f^4 \\ &\quad + 10(2^2e^6)(-f^6) + 5(2e^3)f^8 + (-f^{10}) \\ &= 32e^{15} - 5(16)e^{12}f^2 + 10(8)e^9f^4 - 10(4)e^6f^6 + 10e^3f^8 - f^{10} \\ &= 32e^{15} - 80e^{12}f^2 + 80e^9f^4 - 40e^6f^6 + 10e^3f^8 - f^{10}\end{aligned}$$

EXAMPLE 13 Simplify $\left(\frac{x}{2} - 4\right)^4$.

SOLUTION

Step 1. The coefficients are 1, 4, 6, 4, and 1.

$$1\left(\frac{x}{2}\right)^4(-4)^0 + 4\left(\frac{x}{2}\right)^3(-4)^1 + 6\left(\frac{x}{2}\right)^2(-4)^2 + 4\left(\frac{x}{2}\right)^1(-4)^3 + 1\left(\frac{x}{2}\right)^0(-4)^4$$

Step 2. $$\left(\frac{x}{2} - 4\right)^4 = \frac{x^4}{2^4} - \frac{16x^3}{2^3} + \frac{6(16)x^2}{2^2} - \frac{4(64)x}{2} + 256$$

$$= \frac{x^4}{16} - 2x^3 + 24x^2 - 128x + 256$$

In Examples 14 and 15, we demonstrate how to find just one term of an expansion.

EXAMPLE 14 Find and simplify the fifth term of $(x + y)^9$.

SOLUTION Recall that the formula for the $(r + 1)$st term of the binomial expansion of $(a + b)^n$ is $\binom{n}{r}a^{n-r}b^r$. For $(x + y)^9$, $n = 9$, and for the *fifth* term, $r + 1 = 5$, so $r = 4$. (Remember: for the *first term*, $r = 0$; for the *second term*, $r = 1$; and so forth. Therefore, it is reasonable that for the *fifth term*, r should be 4.) Substituting into the formula for the $(r + 1)$st term, we have

$$\binom{n}{r}a^{n-r}b^r = \binom{9}{4}x^{9-4}y^4 = \frac{9!}{4!\,(9-4)!}x^5y^4 = \frac{9 \cdot 8 \cdot 7 \cdot 6 \cdot 5!}{4 \cdot 3 \cdot 2 \cdot 1 \cdot (5!)}x^5y^4 = 126x^5y^4$$

Therefore, the fifth term of $(x + y)^9$ is $126x^5y^4$.

EXAMPLE 15 Find and simplify the sixth term of $(2x - 3y^2)^8$.

SOLUTION For $(2x - 3y^2)^8$, $n = 8$, and for the *sixth* term, $r + 1 = 6$, so $r = 5$. Substituting into the formula for the $(r + 1)$st term, we have

$$\binom{n}{r}a^{n-r}b^r = \binom{8}{5}(2x)^{8-5}(-3y^2)^5 = \frac{8!}{5!\,(8-5)!}(2x)^3(-3y^2)^5$$

$$= \frac{8 \cdot 7 \cdot 6 \cdot 5!}{(5!) \cdot 3 \cdot 2 \cdot 1}(8x^3)(-3^5y^{10}) = 56(8)(-243)x^3y^{10} = -108{,}864x^3y^{10}$$

Therefore, the sixth term of $(2x - 3y^2)^8$ is $-108{,}864x^3y^{10}$.

In Example 16, we demonstrate finding just *some* of the terms of an expansion.

EXAMPLE 16 Write the first five terms of the expansion of $(x + y^2)^{12}$.

SOLUTION

Step 1. $$1(x)^{12}(y^2)^0 + 12(x)^{11}(y^2)^1 + \frac{12!}{(2!)\,(10!)}(x)^{10}(y^2)^2$$

$$+ \frac{12!}{(3!)\,(9!)}(x)^9(y^2)^3 + \frac{12!}{(4!)\,(8!)}(x)^8(y^2)^4 + \cdots$$

Step 2. $$(x + y^2)^{12} = x^{12} + 12x^{11}y^2 + 66x^{10}y^4 + 220x^9y^6 + 495x^8y^8 + \cdots$$

In this section, we discussed only the binomial expansion for powers that are *nonnegative* integers. Other numbers can be used for powers of binomials, but binomial expansions using such numbers are not discussed in this book; they are studied in higher-level courses.

Exercises 4.5
Set I

In Exercises 1–6, evaluate each expression.

1. $4!$
2. $6!$
3. $\binom{3}{2}$
4. $\binom{7}{3}$
5. $\binom{6}{4}$
6. $\binom{4}{0}$

In Exercises 7–22, simplify each expression.

7. $(x + y)^5$
8. $(r + s)^4$
9. $(x - 2)^5$
10. $(y - 3)^4$
11. $(3r + s)^6$
12. $(2x + y)^6$
13. $(x + y^2)^4$
14. $(u + v^2)^5$
15. $\left(2x - \frac{1}{2}\right)^5$
16. $\left(3x - \frac{1}{3}\right)^4$
17. $\left(\frac{1}{3}x + \frac{3}{2}\right)^4$
18. $\left(\frac{1}{5}x + \frac{5}{2}\right)^4$
19. $(4x^2 - 3y^2)^5$
20. $(3x^2 - 2y^3)^5$
21. $(x + x^{-1})^4$
22. $(x^{-1} - x)^4$
23. Find the sixth term of $(a + b)^9$.
24. Find the fourth term of $(a + b)^7$.
25. Find the eighth term of $(3x^2 - y)^{10}$.
26. Find the fifth term of $(2x^2 - 3w)^9$.
27. Write the first four terms of the expansion of $(x + 2y^2)^{10}$.
28. Write the first four terms of the expansion of $(x + 3y^2)^8$.
29. Write the first four terms of the expansion of $(x - 3y^2)^{10}$.
30. Write the first four terms of the expansion of $(x - 2y^3)^{11}$.

Writing Problems

Express the answers in your own words and in complete sentences.

1. Describe the method you would use in working the problem $(x + 2y)^3$, and explain why you would use that method.
2. Describe the method you would use in working the problem $(3x - y)^7$, and explain why you would use that method.
3. Explain how to find the correct exponents to use for the x's in expanding $(x + y)^{10}$.
4. Explain how to find the correct exponents to use for the y's in expanding $(x + y)^{10}$.

Exercises 4.5
Set II

In Exercises 1–6, evaluate each expression.

1. $7!$
2. $2!$
3. $\binom{5}{4}$
4. $\binom{6}{6}$
5. $\binom{9}{5}$
6. $\binom{7}{4}$

In Exercises 7–22, simplify each expression.

7. $(s + t)^6$
8. $(x - y)^6$
9. $(x + 4)^4$
10. $(2x - 3y)^3$
11. $(4x - y)^5$
12. $(s - t)^5$
13. $(x + y^2)^3$
14. $(x - y)^5$
15. $\left(2x - \frac{1}{2}\right)^6$
16. $(y + y^{-1})^5$
17. $\left(\frac{1}{4}x + \frac{3}{2}\right)^4$
18. $(x + x^{-1})^6$
19. $(3s^2 - 3t^3)^5$
20. $(x^{-1} - x)^3$
21. $(3x + 2y^{-1})^4$
22. $(y^{-1} - y)^7$
23. Find the fourth term of $(a + b)^9$.
24. Find the third term of $(a + b)^{10}$.

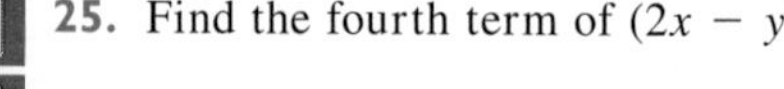

25. Find the fourth term of $(2x - y^2)^{11}$.

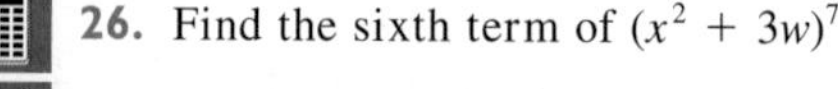

26. Find the sixth term of $(x^2 + 3w)^7$.

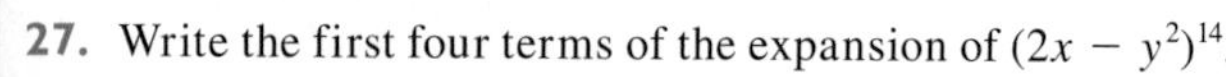

27. Write the first four terms of the expansion of $(2x - y^2)^{14}$.

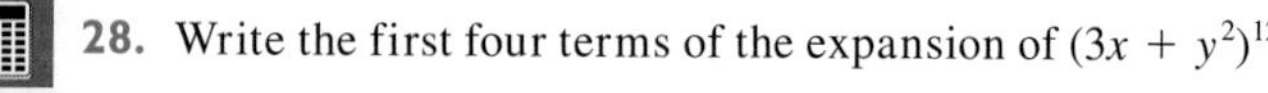

28. Write the first four terms of the expansion of $(3x + y^2)^{12}$.

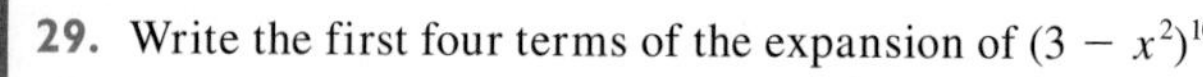

29. Write the first four terms of the expansion of $(3 - x^2)^{10}$.
30. Write the first four terms of the expansion of $(2h - k^2)^{11}$.

4.6 Dividing Polynomials

Dividing a Monomial by a Monomial

To divide a monomial by a monomial, we use the quotient rule; for example,

$$\frac{x^5}{x^3} = x^{5-3} = x^2$$

If the monomials contain numerical coefficients, we reduce the resulting fraction to lowest terms; for example,

$$\frac{12x^6}{8x^2} = \frac{3}{2}x^{6-2} = \frac{3}{2}x^4$$

Dividing a Polynomial by a Monomial

Next, we consider problems in which the *divisor* is a monomial but the dividend is a polynomial containing more than one term. We can convert a division problem that is of the form $\frac{a+b}{c}$ to the form $\frac{a}{c} + \frac{b}{c}$ $\left(\text{and we can convert } \frac{a-b}{c} \text{ to the form } \frac{a}{c} - \frac{b}{c}\right)$.

PROOF THAT $\frac{a+b}{c} = \frac{a}{c} + \frac{b}{c}$

$$\frac{a+b}{c} = (a+b)\left(\frac{1}{c}\right)$$ Writing the division problem as a multiplication problem

$$= a\left(\frac{1}{c}\right) + b\left(\frac{1}{c}\right)$$ Using the distributive property

$$= \frac{a}{c} + \frac{b}{c} \quad \blacksquare$$

$\left(\text{The proof that } \frac{a-b}{c} = \frac{a}{c} - \frac{b}{c} \text{ is similar.}\right)$

Dividing a polynomial by a monomial

Divide *each* term of the polynomial by the monomial, and simplify each term. The quotient is the sum of the simplified terms.

EXAMPLE 1

Examples of dividing a polynomial by a monomial:

a. $$\frac{4x^3 - 6x^2}{2x} = \frac{4x^3}{2x} - \frac{6x^2}{2x} = 2x^2 - 3x$$

b. $$\frac{4x^2y - 8xy^2 + 12xy}{-4xy} = \frac{4x^2y}{-4xy} + \frac{-8xy^2}{-4xy} + \frac{12xy}{-4xy} = -x + 2y - 3$$

A Word of Caution A common error in a problem such as $\frac{x^5 - y^3}{x^2}$ is to divide only x^5 by x^2, writing $\frac{x^5 - y^3}{x^2} = x^3 - y^3$. *This is incorrect.* We must either leave the given expression unchanged or write

$$\frac{x^5 - y^3}{x^2} = \frac{x^5}{x^2} - \frac{y^3}{x^2} = x^3 - \frac{y^3}{x^2}$$

Similarly, $\frac{4+3}{4} \neq 1 + 3$, and $\frac{x^2 - y^2}{x^2} \neq 1 - y^2$, and $\frac{x^2 - y^2}{x^2} \neq -y^2$.

Dividing a Polynomial by a Polynomial with More Than One Term

The method used to divide one polynomial (the dividend) by a polynomial (the divisor) with two or more terms is similar to the method used to divide one whole number by another (using long division) in arithmetic. The long-division procedure can be summarized as follows. (Reading Hint: Read the steps in the summary, and *as you are doing so*, look also at the steps in Example 2 to see that the authors are following the step-by-step method of the summary.)

Dividing one polynomial by another

1. Arrange the divisor and the dividend in descending powers of one variable. In the *dividend*, use placeholders (or leave spaces) for any missing terms.
2. Find the first term of the quotient by dividing the first term of the dividend by the first term of the divisor.
3. Multiply the *entire* divisor by the first term of the quotient. Place this product under the dividend, aligning like terms.
4. Subtract the product found in step 3 from the dividend, bringing down at least one term of the dividend. This difference is the *remainder.* If the degree of the remainder is not less than the degree of the divisor, continue with steps 5 through 8.
5. Find the next term of the quotient by dividing the first term of the remainder by the first term of the divisor.
6. Multiply the entire divisor by the term found in step 5, placing this product under the remainder found in step 4 and aligning like terms.
7. Subtract the product found in step 6 from the polynomial above it, bringing down at least one more term of the dividend.
8. Repeat steps 5 through 7 until the remainder is 0 *or* until the degree of the remainder is less than the degree of the divisor. If there is a nonzero remainder, it should be written as the numerator of a fraction whose denominator is the *divisor* of the division problem. This fraction (which *must* be preceded by either a plus sign or a minus sign) is written as the last term of the final answer.
9. Check the answer by verifying that divisor $\times$ quotient $+$ remainder $=$ dividend.

Note In steps 4 and 7, it is acceptable to "bring down" *all* the remaining terms of the dividend (after subtracting) rather than just one term.

A Word of Caution Most errors in division problems are *subtraction* errors.

EXAMPLE 2 Divide $(27x + 19x^2 + 6x^3 + 10)$ by $(5 + 3x)$.

SOLUTION

Step 1. Arranging the terms of the divisor and the dividend in descending powers of x, we have

$$3x + 5\overline{)6x^3 + 19x^2 + 27x + 10}$$

Step 2. $3x + 5\overline{)6x^3 + 19x^2 + 27x + 10}$, quotient $2x^2$ ← The first term of the quotient is $\frac{6x^3}{3x} = 2x^2$

Step 3.

$$\begin{array}{r} 2x^2 \\ 3x + 5\overline{)6x^3 + 19x^2 + 27x + 10} \\ 6x^3 + 10x^2 \end{array}$$

← This is $2x^2(3x + 5)$; like terms are aligned

Step 4.

$$\begin{array}{r} 2x^2 \\ 3x + 5\overline{)6x^3 + 19x^2 + 27x + 10} \\ \underline{6x^3 + 10x^2} \\ 9x^2 + 27x \end{array}$$

Remainder → $9x^2 + 27x$ ← We subtract and bring down the next term

The division must be continued because the degree of the remainder is not less than the degree of the divisor.

Step 5.

$$\begin{array}{r} 2x^2 + 3x \\ 3x + 5\overline{)6x^3 + 19x^2 + 27x + 10} \\ \underline{6x^3 + 10x^2} \\ 9x^2 + 27x \end{array}$$

← The second term of the quotient is $\frac{9x^2}{3x} = 3x$

Step 6.

$$\begin{array}{r} 2x^2 + 3x \\ 3x + 5\overline{)6x^3 + 19x^2 + 27x + 10} \\ \underline{6x^3 + 10x^2} \\ 9x^2 + 27x \\ 9x^2 + 15x \end{array}$$

← This is $3x(3x + 5)$; like terms are aligned

Step 7.

$$\begin{array}{r} 2x^2 + 3x \\ 3x + 5\overline{)6x^3 + 19x^2 + 27x + 10} \\ \underline{6x^3 + 10x^2} \\ 9x^2 + 27x \\ \underline{9x^2 + 15x} \\ 12x + 10 \end{array}$$

Remainder → $12x + 10$ ← We subtract and bring down the next term

The degree of the remainder is *still* not less than the degree of the divisor. Therefore, we must repeat steps 5, 6, and 7.

Step 8.

$$\begin{array}{r} 2x^2 + 3x + 4 \\ 3x + 5\overline{)6x^3 + 19x^2 + 27x + 10} \\ \underline{6x^3 + 10x^2} \\ 9x^2 + 27x \\ \underline{9x^2 + 15x} \\ 12x + 10 \\ \underline{12x + 20} \\ -10 \end{array}$$

← The third term of the quotient is $\frac{12x}{3x} = 4$

$12x + 20$ ← This is $4(3x + 5)$

Remainder → -10

The division is finished because the degree of the remainder is less than the degree of the divisor.

Step 9. ***Check*** $(3x + 5)(2x^2 + 3x + 4) + (-10) = 6x^3 + 19x^2 + 27x + 10$

Answer $2x^2 + 3x + 4 - \dfrac{10}{3x + 5}$, or $2x^2 + 3x + 4 + \dfrac{-10}{3x + 5}$

EXAMPLE 3 Divide $(x^3 - 3 + x^4 - 5x)$ by $(x^2 - 1 - x)$.

SOLUTION We arrange the terms of the dividend and the divisor in descending powers of x. (Notice that an x^2-term is missing from the dividend, so we write $0x^2$ as a placeholder.)

$$
\begin{array}{r}
x^2 + 2x + 3 \\
x^2 - x - 1\overline{)x^4 + x^3 + 0x^2 - 5x - 3} \\
\underline{x^4 - x^3 - x^2} \qquad\qquad \\
2x^3 + x^2 - 5x \qquad \\
\underline{2x^3 - 2x^2 - 2x} \qquad \\
3x^2 - 3x - 3 \\
\underline{3x^2 - 3x - 3} \\
0
\end{array}
$$

√ **Check** $(x^2 + 2x + 3)(x^2 - x - 1) = x^4 + x^3 - 5x - 3$

Answer $x^2 + 2x + 3$

Factors Recall that when the final remainder is 0, we can say that the *divisor* is a *factor* of the dividend and that the *quotient* is a *factor* of the dividend. Therefore, in Example 3, we can say that both $x^2 + 2x + 3$ and $x^2 - x - 1$ are factors of $x^4 + x^3 - 5x - 3$.

If we "lose" *more* than one term when we subtract, then we must bring down more than one term from the dividend (see Example 4).

EXAMPLE 4 Solve $(6x^3 + 4x^4 - 1 + 2x) \div (3x - 1 + 2x^2)$.

SOLUTION Arranging both divisor and dividend in descending powers of x, we have

$$
\begin{array}{r}
2x^2 \qquad\quad + 1 \\
2x^2 + 3x - 1\overline{)4x^4 + 6x^3 + 0x^2 + 2x - 1} \\
\underline{4x^4 + 6x^3 - 2x^2} \qquad\qquad \\
2x^2 + 2x - 1 \\
\underline{2x^2 + 3x - 1} \\
-x \qquad
\end{array}
$$

We "lost" two terms here

Subtracting and bringing down the next *two terms*

√ **Check** $(2x^2 + 1)(2x^2 + 3x - 1) - x = 4x^4 + 6x^3 + 2x - 1$

Answer $2x^2 + 1 + \dfrac{-x}{2x^2 + 3x - 1}$, or $2x^2 + 1 - \dfrac{x}{2x^2 + 3x - 1}$

EXAMPLE 5 Solve $(17ab^2 + 12a^3 - 10b^3 - 11a^2b) \div (3a - 2b)$.

SOLUTION

$$
\begin{array}{r}
4a^2 - ab + 5b^2 \\
3a - 2b\overline{)12a^3 - 11a^2b + 17ab^2 - 10b^3} \\
\underline{12a^3 - 8a^2b} \qquad\qquad\qquad \\
-3a^2b + 17ab^2 \qquad \\
\underline{-3a^2b + 2ab^2} \qquad \\
15ab^2 - 10b^3 \\
\underline{15ab^2 - 10b^3} \\
0
\end{array}
$$

Arranging both divisor and dividend in descending powers of a

You might verify that if we had expressed the divisor and the dividend in descending powers of b, the answer would have been $5b^2 - ab + 4a^2$, which equals $4a^2 - ab + 5b^2$.

√ **Check** $(4a^2 - ab + 5b^2)(3a - 2b) = 12a^3 - 11a^2b + 17ab^2 - 10b^3$

Answer $4a^2 - ab + 5b^2$

Since the remainder is 0, we can say that both the divisor and the quotient are factors of the dividend.

EXAMPLE 6 Solve $(1 - x^5) \div (1 - x)$.

SOLUTION Arranging both divisor and dividend in descending powers of x, we have

$$\begin{array}{r|l} & x^4 + x^3 + x^2 + x + 1 \\ \hline -x + 1 & -x^5 + 0x^4 + 0x^3 + 0x^2 + 0x + 1 \\ & \underline{-x^5 + x^4} \\ & -x^4 + 0x^3 \\ & \underline{-x^4 + x^3} \\ & -x^3 + 0x^2 \\ & \underline{-x^3 + x^2} \\ & -x^2 + 0x \\ & \underline{-x^2 + x} \\ & -x + 1 \\ & \underline{-x + 1} \\ & 0 \end{array}$$

√ **Check** $(1 - x)(x^4 + x^3 + x^2 + x + 1) = -x^5 + 1 = 1 - x^5$

Answer $x^4 + x^3 + x^2 + x + 1$

Notice that both $x^4 + x^3 + x^2 + x + 1$ and $1 - x$ are factors of $1 - x^5$.

Exercises 4.6
Set I

Perform the indicated divisions.

1. $\dfrac{18x^5 - 24x^4 - 12x^3}{6x^2}$
2. $\dfrac{16y^4 - 36y^3 + 20y^2}{-4y^2}$
3. $\dfrac{55a^4b^3 - 33ab^2}{-11ab}$
4. $\dfrac{26m^2n^4 - 52m^3n}{-13mn}$
5. $\dfrac{-15x^2y^2z^2 - 30xyz}{-5xyz}$
6. $\dfrac{-24a^2b^2c^2 - 16abc}{-8abc}$
7. $(13x^3y^2 - 26x^5y^3 + 39x^4y^6) \div (13x^2y^2)$
8. $(21m^2n^3 - 35m^3n^2 - 14m^3n^5) \div (7m^2n^2)$
9. $(x^2 + 10x - 5) \div (x + 7)$
10. $(x^2 + 9x - 5) \div (x + 4)$
11. $(x^2 + 3x - 10) \div (x + 7)$
12. $(x^2 - x - 72) \div (x - 9)$
13. $(x^2 - 2x - 5) \div (x - 3)$
14. $(x^2 - 5x - 2) \div (x - 6)$
15. $(6z^3 - 13z^2 - 4z + 15) \div (3z - 5)$
16. $(6y^3 + 7y^2 - 11y - 12) \div (2y + 3)$
17. $(8x - 4x^3 + 10) \div (2 - x)$
18. $(12x - 15 - x^3) \div (3 - x)$
19. $(28x - x^3 - 10) \div (5 - x)$
20. $(19y - y^3 + 4) \div (4 - y)$
21. $(x^4 - 7x + 6) \div (x - 2)$
22. $(x^4 + 6x + 3) \div (x - 1)$
23. $(2x^4 + 7x^2 + 5) \div (x^2 + 2)$
24. $(3x^4 + 11x^2 + 1) \div (x^2 + 3)$

25. $(v^4 - 3v^3 - 8v^2 - 9v - 5) \div (v + 1)$

26. $(w^4 - 2w^3 - 12w^2 - 13w - 10) \div (w + 2)$

27. $(17xy^2 + 12x^3 - 10y^3 - 11x^2y) \div (3x - 2y)$

28. $(13x^2y - 11xy^2 + 10x^3 - 12y^3) \div (2x + 3y)$

29. $(x + 8x^2 + 4x^4 - 3x^3 + 2) \div (x + 1 + 4x^2)$

30. $(11x^2 + 10 + 2x^4 - x^3 + 3x) \div (x + 2x^2 + 2)$

31. $(2x^5 - 10x^3 + 3x^2 - 15) \div (3 + 2x^3)$

32. $(3x^5 - 6x^3 + x^2 - 2) \div (1 + 3x^3)$

33. $(3x^4 + 14x^3 + 2x^2 + 3x + 2) \div (3x^2 - x + 1)$

34. $(2x^4 + 13x^3 - 10x^2 + 9x - 2) \div (2x^2 - x + 1)$

Writing Problems

Express the answers in your own words and in complete sentences.

1. Explain why $\dfrac{6 + 3x^2}{6} \neq 1 + 3x^2$.

2. Explain why $\dfrac{7x + 8}{8} \neq 7x + 1$.

3. Describe what you would do *first* in working the problem $(3x - 4x^3 + 2x^2 + 8) \div (4 + x)$, and explain why you would do that first.

4. Describe what you would do *first* in working the problem $(x^3 - x + 8) \div (x + 2)$, and explain why you would do that first.

5. Explain how to check a division problem in algebra.

Exercises 4.6 Set II

Perform the indicated divisions.

1. $\dfrac{24h^6 - 56h^4 - 40h^3}{8h^2}$

2. $\dfrac{5x^4 - 40x^3 + 10x^2}{-5x^2}$

3. $\dfrac{60e^5f^3 - 84e^2f^4}{12ef^2}$

4. $\dfrac{7y^5 - 35y^4 + 14y^3}{7y^2}$

5. $\dfrac{-45r^3s^2t^4 + 63r^2s^3t^3}{-9rs^2t}$

6. $\dfrac{42u^4 - 56u^2 + 28u^6}{-14u^2}$

7. $(-30m^4n^2 + 60m^2n^3 - 45m^3n^4) \div (15m^2n^2)$

8. $(32x^3y^2z - 64xy^2z + 48xy^3z^2) \div (-16xy^2z)$

9. $(x^2 - 3x - 11) \div (x + 3)$

10. $(x^4 - 1) \div (x - 1)$

11. $(x^2 + x - 28) \div (x - 5)$

12. $(y^2 + 2y + 1) \div (y + 1)$

13. $(z^2 - 7z - 20) \div (z + 2)$

14. $(x^2 - 81y^2) \div (x + 9y)$

15. $(3x^4 + 2x^3 - 6x^2 - x + 5) \div (3x + 2)$

16. $(x^5 - 1) \div (x^2 + x - 1)$

17. $(8x - x^3 + 25) \div (4 - x)$

18. $(x^3 + x^2 + x + 3) \div (x + 1)$

19. $(37x - 6 - x^3) \div (6 - x)$

20. $(x^3 - 1) \div (x - 1)$

21. $(x^4 - 29x + 4) \div (x - 3)$

22. $(4x + x^3 + 4 + x^2) \div (x + 1)$

23. $(2x^4 + 7x^2 + 3) \div (x^2 + 1)$

24. $(2x^2 + 6 + 3x + x^3) \div (x + 2)$

25. $(z^4 - 2z^3 - 4z^2 - 10z - 9) \div (z + 1)$

26. $(2x^4 + 3x^3 + 2x^2 - x + 3) \div (2x + 3)$

27. $(23x^2y + 16xy^2 + 6x^3 + 3y^3) \div (3x + y)$

28. $(14 + 20w^3 - 29w - 23w^2) \div (5w - 2)$

29. $(2x + 3x^4 - 2x^3 + 3 + 9x^2) \div (x + 3x^2 + 1)$

30. $(2x^4 + x^3 + 4x^2 - 4x - 5) \div (2x + 1)$

31. $(3x^5 - 12x^3 + 2x^2 - 8) \div (2 + 3x^3)$

32. $(15x^3 - 5xy^2 + 12y^3 - 23x^2y) \div (3x - 4y)$

33. $(2x^4 + 7x^3 + 10x^2 + x - 4) \div (2x^2 + x - 1)$

34. $(6x^4 + x^3 - 2x^2 - 3x - 2) \div (3x^2 - x - 2)$

4.7 Synthetic Division (Optional)

Synthetic division may be used when we divide a polynomial by a *first-degree* binomial. In using synthetic division, we write the coefficients of the quotient *below* the coefficients of the dividend.

Synthetic Division and Long Division

Before we show how synthetic division is done, we will show how synthetic division is derived from long division.

EXAMPLE 1 Divide $(19x + 2x^3 - 5 - 11x^2)$ by $(x - 3)$, using *long division.*

SOLUTION

$$\begin{array}{r} 2x^2 - 5x + 4 \quad \text{R } 7 \\ x - 3\overline{)2x^3 - 11x^2 + 19x - 5} \\ \underline{2x^3 - 6x^2 \qquad\qquad} \\ -5x^2 + 19x \qquad \\ \underline{-5x^2 + 15x \qquad} \\ 4x - 5 \\ \underline{4x - 12} \\ 7 \end{array}$$

We note that when we divide any polynomial by a binomial of the form $x + a$ or $x - a$, (1) the degree of the quotient is *always one less than* the degree of the dividend, (2) the quotient is in descending powers of the variable, and (3) the leading coefficient of the quotient always equals the leading coefficient of the dividend. Therefore, in order to shorten the division process, we can *temporarily* omit the variables. (We will put them back in later.) When we omit the variables, we must insert zeros for any missing terms. If we leave out the variables and the plus signs from the problem given in Example 1, we have

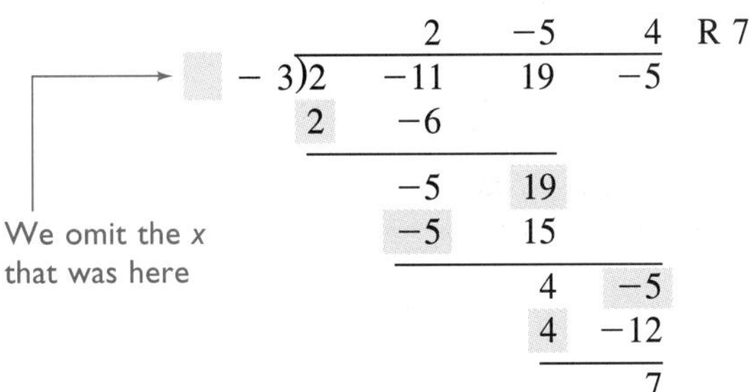

The shaded numbers are not needed, because they are duplicates of the numbers directly above them; deleting the shaded numbers gives

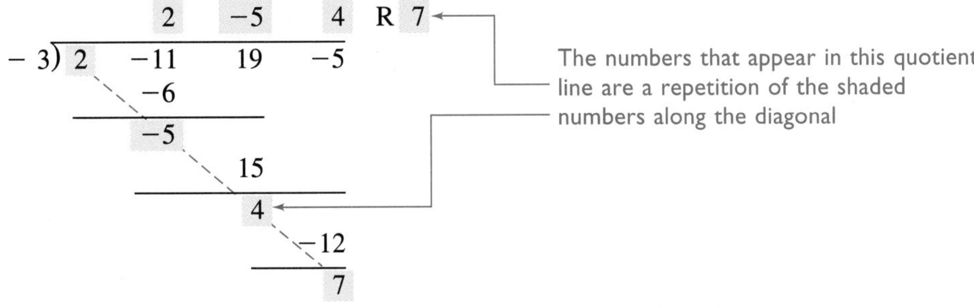

Some of the numbers still appear twice. The 2 *should* appear twice, because it is the leading coefficient of the dividend (and must, therefore, also be the leading coefficient of the quotient). However, the -5, the 4, and the 7 do *not* need to appear twice; because we want to write the coefficients of the quotient *below* the coefficients of the dividend, we delete -5, 4, and 7 from the quotient line, close up the empty spaces, and move 2 (the *first* number from the quotient line) so that it is to the left of and aligns with the remaining -5, 4, and 7. We then have the array of numbers shown below:

$$\begin{array}{r|rrrr} -3 & 2 & -11 & 19 & -5 \\ & & -6 & 15 & -12 \\ \hline & 2 & -5 & 4 & 7 \end{array}$$

Now notice that if we *subtract* the second row from the top row, we get the bottom row

This 2 was moved from the quotient line

Also notice that if we multiply each of the first three numbers in the bottom row by -3, we get the numbers in the second row. Changing the -3 to a $+3$ would permit us to *add* instead of *subtract*.

The Synthetic Division Process

We now demonstrate the synthetic division procedure for the division problem in Example 1. To get ready to use synthetic division on the problem $(19x + 2x^3 - 5 - 11x^2) \div (x - 3)$, we arrange the numbers as follows.

$$\begin{array}{r|rrrr} 3 & 2 & -11 & 19 & -5 \\ \hline \end{array}$$

The coefficients of the dividend, arranged in descending powers of the variable

Notice the sign change (we are dividing by $x - 3$, but we use $+3$ here so that we can *add*, rather than subtract)

We next bring down the leading coefficient, 2, from the dividend:

$$\begin{array}{r|rrrr} 3 & 2 & -11 & 19 & -5 \\ \hline & 2 & & & \end{array}$$

Bringing down the leading coefficient

We now multiply and add, as follows:

$$\begin{array}{r|rrrr} 3 & 2 & -11 & 19 & -5 \\ & & 6 & & \\ \hline & 2 & & & \end{array}$$

Multiply

The *product* of 3 and 2 is written under the -11

Add

$$\begin{array}{r|rrrr} 3 & 2 & -11 & 19 & -5 \\ & & 6 & & \\ \hline & 2 & -5 & & \end{array}$$

The *sum* of -11 and 6 is written under the 6

$$\begin{array}{r|rrrr} 3 & 2 & -11 & 19 & -5 \\ & & 6 & -15 & \\ \hline & 2 & -5 & & \end{array}$$

The *product* of 3 and -5 is written under the 19

Multiply

Add

$$\begin{array}{r|rrrr} 3 & 2 & -11 & 19 & -5 \\ & & 6 & -15 & \\ \hline & 2 & -5 & 4 & \end{array}$$

The *sum* of 19 and -15 is written under the -15

3	2	−11	19	−5
		6	−15	12
	2	−5	4	

Multiply. The *product* of 3 and 4 is written under the −5

Add

3	2	−11	19	−5
		6	−15	12
	2	−5	4	7

The *sum* of −5 and 12 is written under the 12

7 is the remainder

These are the coefficients of the quotient; the quotient must be *2nd degree*, since the dividend is *3rd degree*

The last number in the bottom row is the *remainder*. The other numbers in the last row are the coefficients of the quotient. We now supply the missing variables, remembering that the *degree* of the quotient is *one less than* the degree of the dividend. The quotient is (apparently) $2x^2 - 5x + 4$, with a remainder of 7.

√ ***Check*** $(x - 3)(2x^2 - 5x + 4) + 7 = 2x^3 - 11x^2 + 19x - 5$

Therefore, $(19x + 2x^3 - 5 - 11x^2) \div (x - 3) = 2x^2 - 5x + 4 + \dfrac{7}{x - 3}$.

EXAMPLE 2 Divide $2x^4 - 3x^2 + 5$ by $x - 1$.

SOLUTION

1	2	0	−3	0	5
		2	2	−1	−1
	2	2	−1	−1	4
	$2x^3$	$+2x^2$	$-1x$	−1	R 4

Notice the sign change (we're dividing by $x - 1$)

The dividend is 4th degree; zeros are inserted for all missing terms

The quotient must be 3rd degree

√ ***Check*** $(x - 1)(2x^3 + 2x^2 - x - 1) + 4 = 2x^4 - 3x^2 + 5$

Therefore, $\dfrac{2x^4 - 3x^2 + 5}{x - 1} = 2x^3 + 2x^2 - x - 1 + \dfrac{4}{x - 1}$.

EXAMPLE 3 Find $\dfrac{x^5 + 32}{x + 2}$.

SOLUTION

−2	1	0	0	0	0	32
		−2	4	−8	16	−32
	1	−2	4	−8	16	0
	$1x^4$	$-2x^3$	$+4x^2$	$-8x$	+16	

Notice the sign change (we're dividing by $x + 2$)

The dividend is 5th degree

The quotient must be 4th degree

√ ***Check*** $(x + 2)(x^4 - 2x^3 + 4x^2 - 8x + 16) = x^5 + 32$

Therefore, $\dfrac{x^5 + 32}{x + 2} = x^4 - 2x^3 + 4x^2 - 8x + 16$. Since the remainder is 0, we can say that $x + 2$ and $x^4 - 2x^3 + 4x^2 - 8x + 16$ are *factors* of $x^5 + 32$.

EXAMPLE 4 Divide $8x^4 - 6x^3 - 4x^2 + 7x - 3$ by $x - \frac{3}{4}$.

SOLUTION

$$\begin{array}{r|rrrrr}
\frac{3}{4} & 8 & -6 & -4 & 7 & -3 \\
 & & 6 & 0 & -3 & 3 \\
\hline
 & 8 & 0 & -4 & 4 & \fbox{0} \\
 & 8x^3 + & 0x^2 & -4x & +4 &
\end{array}$$

The dividend is 4th degree

The quotient must be 3rd degree

✓ *Check* $\left(x - \frac{3}{4}\right)(8x^3 - 4x + 4) = 8x^4 - 6x^3 - 4x^2 + 7x - 3$

Therefore, $(8x^4 - 6x^3 - 4x^2 + 7x - 3) \div \left(x - \frac{3}{4}\right) = 8x^3 - 4x + 4$.

Synthetic division *can* be used when the divisor is of the form $ax + b$ or $ax - b$; however, in this book we use it only when the divisor is of the form $x + a$ or $x - a$, where a is an integer, a common fraction, or a decimal. (Synthetic division *cannot* be used if the divisor is not a first-degree binomial.)

Synthetic division is a very useful tool; we will use it later in *factoring* and in *graphing* polynomial functions.

Exercises 4.7

Set I

Use synthetic division to perform the divisions.

1. $(x^2 + 2x - 18) \div (x - 3)$
2. $(x^2 + 4x - 10) \div (x - 2)$
3. $(x^3 + 3x^2 - 5x + 6) \div (x + 4)$
4. $(x^3 + 6x^2 + 4x - 7) \div (x + 5)$
5. $(x^4 + 6x^3 - x - 4) \div (x + 6)$
6. $(2x^4 + 5x^3 + 10x - 2) \div (x + 3)$
7. $\dfrac{x^4 - 16}{x - 2}$
8. $\dfrac{x^7 - 1}{x - 1}$
9. $\dfrac{x^6 - 3x^5 - 2x^2 + 3x + 5}{x - 3}$
10. $\dfrac{x^6 - 3x^4 - 7x - 2}{x - 2}$
11. $(3x^4 - x^3 + 9x^2 - 1) \div \left(x - \frac{1}{3}\right)$
12. $(3x^4 - 4x^3 - x^2 - x - 2) \div \left(x + \frac{2}{3}\right)$
13. $\dfrac{x^5 + x^4 - 45x^3 - 45x^2 + 324x + 324}{x + 3}$
14. $\dfrac{x^5 + x^4 - 45x^3 - 45x^2 + 324x + 324}{x + 6}$
15. $\dfrac{4x^4 - 45x^2 + 3x + 100}{x - 2}$
16. $\dfrac{9x^4 - 13x^2 + 2x + 6}{x - 1}$

In Exercises 17–20, use a calculator.

17. $(2.6x^3 + 1.8x - 6.4) \div (x - 1.5)$
18. $(3.8x^3 - 1.4x^2 - 23.9) \div (x - 2.5)$
19. $(2.7x^3 - 1.6x + 3.289) \div (x - 1.2)$
20. $(3x^3 + 1.2x^2 - 1.5) \div (x - 1.6)$

Writing Problems

Express the answer in your own words and in complete sentences.

1. Explain why synthetic division cannot be used in the problem

$$(4x^3 - 3x^2 + 2x - 4) \div (x^2 - 1)$$

Exercises 4.7 Set II

Use synthetic division to perform the divisions.

1. $(x^2 + 7x + 10) \div (x + 3)$
2. $(x^2 + 6x + 12) \div (x - 2)$
3. $(x^3 - 8x^2 + 11x - 14) \div (x - 6)$
4. $(x^4 + 5x^3 - x^2 - 4x + 7) \div (x + 5)$
5. $(2x^4 - 10x^3 - 6x + 15) \div (x - 5)$
6. $(2x^3 + 3x^2 + x - 15) \div \left(x - \frac{3}{2}\right)$
7. $\dfrac{x^4 - 81}{x - 3}$
8. $(x^5 - 32) \div (x - 2)$
9. $(x^6 - 4x^5 - 12x^3 + 48x^2 - 15x + 43) \div (x - 4)$
10. $(x^5 - 4x^4 + 4x^3 + 4x^2 - 11x + 8) \div (x - 2)$
11. $(8x^4 - 2x^3 - 6x^2 + 4x + 3) \div \left(x + \frac{3}{4}\right)$
12. $(2x^4 + 15x^3 + 6x^2 - 8x - 7) \div (x + 7)$
13. $\dfrac{x^5 + x^4 - 45x^3 - 45x^2 + 324x + 324}{x - 6}$
14. $(x^5 + 1) \div (x + 1)$
15. $\dfrac{4x^4 - 13x^2 + 2x + 6}{x - 2}$
16. $(2x^3 + 5x^2 + 5x - 4) \div \left(x - \frac{1}{2}\right)$

In Exercises 17–20, use a calculator.

17. $(1.4x^3 - 2.6x^2 + 56.3) \div (x + 3.5)$
18. $(3x^3 + 8x^2 + 6x - 3) \div \left(x - \frac{1}{3}\right)$
19. $(2.4x^3 + 1.32x^2 - 3.6) \div (x - 1.1)$
20. $(x^5 - 5x^4 + 10x^3 - 10x^2 + 5x - 1) \div (x - 1)$

4.8 Applications: Dry Mixture Problems

The equations used to solve many applied problems, including the ones in this section and the next, are first-degree polynomial equations. Such equations are solved by using the techniques we have already discussed.

In this section, we concentrate on solving those mixture problems in which the *value* or *cost* of the ingredients or items to be mixed is important. We will refer to these problems as **dry mixture problems**, although the mixtures can be of dry ingredients, liquid ingredients, coins, tickets, amounts of interest earned, and so on. Because we are concerned with the *value* of the ingredients or items, these mixture problems (which are often especially important to merchants) can also be regarded as *money* problems.

A chart is often helpful in solving dry mixture problems. The relationships used in filling in the chart we show are as follows, where the *columns* go up and down (they are *vertical*) and the *rows* go from side to side (they are *horizontal*).

Three important relationships used in solving dry mixture problems

1. $\begin{pmatrix} \textit{Unit cost} \\ \text{of one} \\ \text{ingredient} \end{pmatrix} \times \begin{pmatrix} \textit{amount} \text{ of} \\ \text{that} \\ \text{ingredient} \end{pmatrix} = \begin{pmatrix} \textit{total cost} \\ \text{of that} \\ \text{ingredient} \end{pmatrix}$ We use this relationship *down the columns* of the chart

2. $\begin{pmatrix} \textit{Amount} \text{ of} \\ \text{ingredient A} \end{pmatrix} + \begin{pmatrix} \textit{amount} \text{ of} \\ \text{ingredient B} \end{pmatrix} = \begin{pmatrix} \textit{total amount} \\ \text{of mixture} \end{pmatrix}$ We use this relationship *in row 2* of the chart

3. $\begin{pmatrix} \textit{Total value} \text{ of} \\ \text{ingredient A} \end{pmatrix} + \begin{pmatrix} \textit{total value} \text{ of} \\ \text{ingredient B} \end{pmatrix} = \begin{pmatrix} \textit{total value} \\ \text{of mixture} \end{pmatrix}$ We use this relationship *in the last row* of the chart

The *last row* of the chart gives us the equation to use.

We find the *total cost* (or *total value*) of any item by *multiplying* the *unit cost* (the cost per pound, per ounce, per gram, or whatever) of that item by the *amount* of that item. For example, if a certain coffee costs \$3.40 per pound, the value of 15 lb of that coffee is $\left(\frac{\$3.40}{1\ \cancel{\text{lb}}}\right)(15\ \cancel{\text{lb}}) = \$51.00$. If a 50-lb mixture of nuts is to be worth \$3.54 per pound, the total value of that mixture is $\left(\frac{\$3.54}{1\ \cancel{\text{lb}}}\right)(50\ \cancel{\text{lb}}) = \$177.00$. Similarly, the total value of 18 nickels is $\left(\frac{5¢}{1\ \cancel{\text{nickel}}}\right)(18\ \cancel{\text{nickels}}) = 90¢$. This is the relationship we use down the *columns* of the chart.

The basic chart is as follows:

	Ingredient A	Ingredient B	Mixture
Unit cost			
Amount	+	=	
Total cost or value	+	=	

← This square is sometimes empty (Mixture, Unit cost)

← This row gives us the equation (Total cost or value)

It is not *necessary* to use a chart for solving mixture problems; the filled-in chart is simply an organized summary of the facts given in the problem.

EXAMPLE 1 A wholesaler needs to make up a 50-lb mixture of two kinds of coffee, one worth \$3.40 per pound and the other worth \$3.60 per pound. How many pounds of each kind of coffee must be used if the mixture is to be worth \$3.54 per pound?

SOLUTION

Step 1. Let $x =$ the number of pounds of the less expensive coffee
Then $50 - x =$ the number of pounds of the more expensive coffee

A Word of Caution

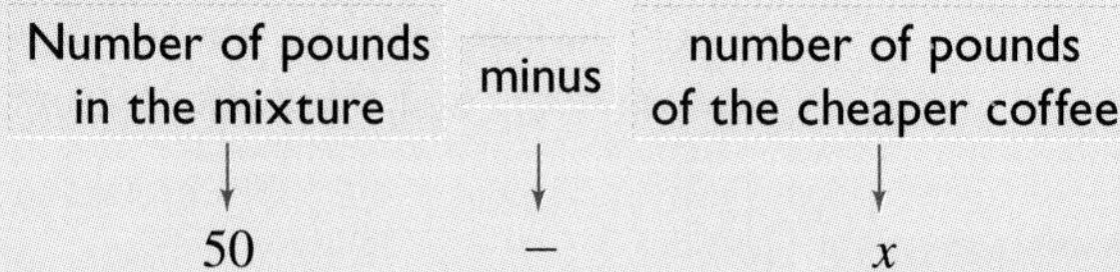

$$50 \quad - \quad x$$

equals the number of pounds left over (the number of pounds of the more expensive coffee). A common error is to write $x - 50$ or $x + 50$ as the amount of the more expensive coffee; however, in neither case would there be 50 lb in the mixture:

	Amount of ingredient A	+	amount of ingredient B	=	total amount of mixture	
Trying $x - 50$:	x	+	$(x - 50)$	=	$2x - 50$,	not 50
Trying $x + 50$:	x	+	$(x + 50)$	=	$2x + 50$,	not 50
Using $50 - x$:	x	+	$(50 - x)$	=	50	

Chart

	Less expensive coffee		More expensive coffee		Mixture
Unit cost	\$3.40		\$3.60		\$3.54
Amount	x	+	$50 - x$	=	50
Total value	$\$3.40x$	+	$\$3.60(50 - x)$	=	\$3.54(50)

Reread

Notice that [Amount of less expensive coffee] + [amount of more expensive coffee] = [amount of mixture]

$$x + (50 - x) = 50$$ Second row of chart

and that [Total value of less expensive coffee] + [total value of more expensive coffee] = [total value of mixture]

Step 2. $3.40x + 3.60(50 - x) = 3.54(50)$ Last row of chart

Step 3. $340x + 360(50 - x) = 354(50)$ Multiplying both sides by 100 to clear decimals

$$340x + 18{,}000 - 360x = 17{,}700$$

$$-20x = -300$$

Step 4. $x = 15$ Number of lb of less expensive coffee

$50 - x = 35$ Number of lb of more expensive coffee

√ **Step 5.** ***Check*** Total value of less expensive coffee is 15(\$3.40) = \$51.00
Total value of more expensive coffee is 35(\$3.60) = +126.00
\$177.00

Total value of mixture is 50(\$3.54) = \$177.00

Step 6. Therefore, the wholesaler should use 15 lb of the \$3.40 coffee and 35 lb of the \$3.60 coffee.

EXAMPLE 2 Bill has \$3.85 in nickels, dimes, and quarters. If he has twice as many nickels as quarters and 34 coins altogether, how many of each kind of coin does he have?

SOLUTION

Step 1. Let $x =$ the number of quarters
Then $2x =$ the number of nickels ← There are twice as many nickels as quarters
and $34 - [x + 2x] =$ the number of dimes

Note There are 34 coins in all, so when we subtract $[x + 2x]$ (the *sum* of the number of quarters and the number of nickels) from 34, the amount left over gives the number of dimes.

Chart

	Quarters		Nickels		Dimes		Mixture
Value of each coin	\$0.25		\$0.05		\$0.10		
Number of coins	x	+	$2x$	+	$34 - 3x$	=	34
Total value	$\$0.25x$	+	$\$0.05(2x)$	+	$\$0.10(34 - 3x)$	=	\$3.85

Step 2. $0.25x + 0.05(2x) + 0.10(34 - 3x) = 3.85$ Bottom row of chart

Step 3. $25x + 5(2x) + 10(34 - 3x) = 385$ Multiplying both sides by 100

$25x + 10x + 340 - 30x = 385$ Simplifying

$5x = 45$

Step 4. $x = 9$ The number of quarters

$2x = 18$ The number of nickels

$34 - 3x = 7$ The number of dimes

Step 5. ***Check***

9 quarters: \$2.25
18 nickels: 0.90
7 dimes: 0.70
34 coins \$3.85

There are twice as many nickels as quarters, and there are 34 coins altogether

The total value is \$3.85

Step 6. Therefore, Bill has 9 quarters, 18 nickels, and 7 dimes.

Note For some of the problems in the following exercises, you will need to remember that the formula for the perimeter, P, of a rectangle of length l and width w is $P = 2l + 2w$.

For other problems in the exercises, you'll need to find the amount of interest earned. We find the total amount of simple interest earned by using the formula $I = Prt$, where I is the amount of interest earned, P is the amount invested (the principal), r is the annual interest rate (usually expressed in decimal form), and t is the number of years. For example, if \$4,000 is invested for one year at 3% simple interest, the amount of interest earned is (\$4,000) (0.03) (1) = \$120.

The problems in the following exercises are not all dry mixture problems, but they *do* all result in first-degree polynomial equations.

Exercises 4.8
Set I

In each exercise, set up the problem algebraically, solve, and check. Be sure to state what your variables represent.

1. A dealer needs to make up a 100-lb mixture of Colombian coffee worth \$3.90 per pound and Brazilian coffee worth \$3.60 per pound. How many pounds of each kind must be used in order for the mixture to be worth \$3.72 per pound?

2. A dealer wants to make up a 50-lb mixture of cashews and peanuts. If the cashews are worth \$7.60 per pound and the peanuts are worth \$2.60 per pound, how many pounds of each kind of nut must be used in order for the mixture to be worth \$4.20 per pound?

3. A 10-lb mixture of almonds and walnuts is worth \$39.72. If walnuts are worth \$4.50 per pound and almonds are worth \$3.00 per pound, how many pounds of each kind of nut are there in the mixture?

4. Mrs. Diederich paid \$15.81 for a 6-lb mixture of granola and dried apple chunks. If the granola cost \$2.10 per pound and the dried apple chunks cost \$4.24 per pound, how many pounds of granola did she buy? How many pounds of dried apple chunks?

5. Margaretha is putting crown-molding around the walls in her living room. The room is 3 ft longer than it is wide. If the crown-molding costs \$79.20 (without tax) and if the cost per foot of the molding is \$1.20, how wide is Margaretha's living room?

6. Jorge fenced a small dog-run for his dog, Fifi. The fence cost \$227.70 (without tax), and the fencing was \$3.45 per meter. If the run is 20 m longer than it is wide, how wide is the dog-run?

7. Amanda counted the nickels and dimes in her purse and found that she had 17 coins in all. If the total value of the coins was \$1.15, how many of each kind of coin did she have?

8. Heather counted the dimes and quarters in her little sister's bank and found that there were 17 coins in all. If the total value of the coins was \$3.35, how many of each kind of coin were in the bank?

9. Elaine has \$3.20 in nickels, dimes, and quarters. If she has 7 more dimes than quarters and 3 times as many nickels as quarters, how many of each kind of coin does she have?

10. Trisha has \$4.15 in nickels, dimes, and quarters. If she has 5 more nickels than dimes and twice as many quarters as dimes, how many of each kind of coin does she have?

11. Mrs. Robinson plans to mix 15 lb of English toffee worth \$6.60 per pound with peanut brittle worth \$4.20 per pound. How many pounds of peanut brittle must she use to make a mixture worth \$5.10 per pound?

12. Mrs. Reid needs to mix 24 lb of macadamia nuts worth \$8.30 per pound with peanuts that are worth \$3.30 per pound. How many pounds of peanuts must she use to make a mixture worth \$6.30 per pound?

13. George invested some money at 5% simple (annual) interest and some money at 4% simple (annual) interest. If the total amount invested was \$7,000 and if, at the end of one year, the amount of interest earned was \$306.50, find the amount invested at each rate.

14. In one year, Henrietta earned \$383.60 on \$8,000 that she had invested at simple interest. If some of the money earned 4% interest and the rest earned 5% interest, how much was invested at each rate?

15. Joyce spent \$108.50 for 23 tickets to a movie. Some were for children and some for adults. If each child's ticket cost \$3.50 and each adult's ticket cost \$5.25, how many of each kind of ticket did she buy?

16. Jerry spent \$9.18 for 45 stamps. Some were 18¢ stamps and some were 22¢ stamps. How many of each kind did he buy?

Exercises 4.8
Set II

In each exercise, set up the problem algebraically, solve, and check. Be sure to state what your variables represent.

1. A merchant wants a 100-lb mixture of Delicious and Spartan apples to be worth 96¢ per pound. If the Delicious apples are worth 99¢ per pound and the Spartan apples are worth 89¢ per pound, how many pounds of each kind should be used in the mixture?

2. Lori, who owns a delicatessen, made up a 40-pt mixture of peaches and pears for a fruit salad. Peaches were worth 71¢ per pint, and pears were worth 79¢ per pint. If the 40-pt mixture was worth \$29.76, how many pints of each kind of fruit were used in the mixture?

3. Mrs. Koontz paid \$36.85 for a 10-lb mixture of Brand A and Brand B coffees. If Brand A cost \$3.85 per pound and Brand B cost \$3.60 per pound, how many pounds of each brand did she buy?

4. A 5-lb mixture of caramels and nougats costs \$11.27. If nougats are worth \$2.30 per pound and caramels are worth \$2.20 per pound, how many pounds of each kind are there?

5. Henry is putting a wallpaper border around the top of his bedroom walls. The bedroom is 4 ft longer than it is wide. If the wallpaper border costs 95¢ per foot and if the total cost (without tax) is \$45.60, how long is Henry's bedroom?

6. Claire bought a sandwich, a piece of pie, and a salad. The pie cost \$1.20 more than the salad, and the sandwich cost as much as the pie and the salad together. What was the cost of the salad if the total cost for the lunch (without tax) was \$6.20?

7. Michelle has 14 coins with a total value of \$1.85. If all the coins are dimes and quarters, how many of each kind of coin does she have?

8. Wing has 18 coins consisting of nickels, dimes, and quarters. If he has 5 more dimes than quarters and if the total value of the coins is \$1.90, how many of each kind of coin does he have?

9. Michael has \$2.25 in nickels, dimes, and quarters. If he has 3 fewer dimes than quarters and as many nickels as the sum of the dimes and quarters, how many of each kind of coin does he have?

10. A 50-lb mixture of Delicious and Granny Smith apples is worth \$47.70. If the Delicious apples are worth 99¢ per pound and the Granny Smith apples are worth 89¢ per pound, how many pounds of each kind are there?

11. Mrs. Curtis plans to mix 16 lb of Brand A coffee worth \$3.50 per pound with Brand B coffee worth \$3.60 per pound. If she wants the mixture to be worth \$3.56 per pound, how many pounds of Brand B should she use?

12. Jennifer's piggy bank contains nickels, dimes, and quarters. She has 4 more dimes than quarters and 3 times as many nickels as dimes. The total value of the coins is \$5.00. How many of each kind of coin does she have?

13. In one year, Michael earned \$375.50 on \$8,000 that he had invested at simple interest. If some of the money earned 4% interest and the rest earned 5% interest, how much was invested at each rate?

14. Melissa bought shampoo, hand lotion, and nail polish. She bought one more bottle of hand lotion than of nail polish and twice as many bottles of shampoo as of hand lotion. If shampoo was \$3.89 per bottle, hand lotion was \$3.99 per bottle, and nail polish was \$2.49 per bottle and if the total cost (without tax) of all the items Melissa bought was \$40.29, how many bottles of nail polish did she buy?

15. A 50-lb mixture of Delicious and Golden Delicious apples is worth \$53.30. If the Delicious apples are worth \$1.29 per pound and the Golden Delicious apples are worth 89¢ per pound, how many pounds of each kind are there in the mixture?

16. A grocer plans to make up a 10-lb mixture of dried tropical fruit and raisins. The tropical fruit is worth \$1.58 per pound, and the raisins are worth \$1.38 per pound. How many pounds of each should be used if the mixture is to be worth \$1.49 per pound?

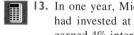

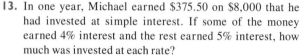

4.9 Applications: Solution Mixture Problems

Another type of mixture problem, usually found in pharmacy, chemistry, and nursing, involves the mixing of liquids. Such problems are called *solution mixture problems* (or *solution problems*) because a mixture of two or more liquids is, under certain conditions, a solution.* Solution mixture problems are similar to dry mixture problems, except that in solution mixture problems, it is the *strength* of the solutions that is important, rather than the cost of the solutions. The problems in this section lead to first-degree polynomial equations.

Recall (from arithmetic and/or elementary algebra) that a 60% solution of alcohol is a mixture that is 60% pure alcohol and 40% water. Recall also that to find the number of milliliters of pure alcohol in, for example, 80 mL of a 60% solution of alcohol, we convert 60% to its decimal form (0.6) and then multiply 0.6 by 80 mL. (We find, then, that there is 48 mL of pure alcohol in 80 mL of a 60% solution.)

An alloy of metals is often a solution, too. To find the amount of pure tin in, for example, 10 oz of a lead-tin alloy that is 18% tin, we convert 18% to 0.18 and then multiply 0.18 by 10 oz. (This tells us that there is 1.8 oz of pure tin in 10 oz of the alloy.)

If we add pure alcohol to an alcohol solution to get a *stronger* solution, we are adding a 100% solution of alcohol, and if we add water to an alcohol solution to get a *weaker* solution, we are adding a 0% solution of alcohol.

Solution mixture problems can be solved by using a chart similar to the chart used for dry mixture problems. We use the following relationships in solving solution mixture problems:

Three relationships used in solving solution mixture problems

1. $$\begin{pmatrix}\text{Strength}\\ \text{of one}\\ \text{solution}\end{pmatrix} \times \begin{pmatrix}\text{amount}\\ \text{of that}\\ \text{solution}\end{pmatrix} = \begin{pmatrix}\text{total amount of}\\ \text{the given substance}\\ \text{in that solution}\end{pmatrix}$$ *We use this relationship down the columns of the chart*

2. $$\begin{pmatrix}\text{Amount of}\\ \text{solution A}\end{pmatrix} + \begin{pmatrix}\text{amount of}\\ \text{solution B}\end{pmatrix} = \begin{pmatrix}\text{amount of}\\ \text{mixture}\end{pmatrix}$$ *We use this relationship in row 2 of the chart*

3. $$\begin{pmatrix}\text{Total amount of}\\ \text{given substance}\\ \text{in solution A}\end{pmatrix} + \begin{pmatrix}\text{total amount of}\\ \text{given substance}\\ \text{in solution B}\end{pmatrix} = \begin{pmatrix}\text{total amount of}\\ \text{given substance}\\ \text{in the mixture}\end{pmatrix}$$ *We use this relationship in the last row of the chart*

*A solution is a homogeneous mixture of two or more substances.

We show the strength being expressed in decimal form, but it could also be expressed in common fraction form. The chart is as follows:

Chart

	Solution A	Solution B	Mixture
Strength (in decimal form)			
Amount of solution	+	=	
Total amount of pure substance	+	=	

← This row gives us the equation

EXAMPLE 1 How many milliliters (mL) of water must be added to 60 mL of a 5% glycerin solution to reduce it to a 2% solution?

SOLUTION

Step 1. Let x = the number of mL of water to be added

Water is 0% glycerin

Chart

	5% solution		Water		Mixture
Strength (in decimal form)	0.05		0.0		0.02
Amount of solution	60	+	x	=	$60 + x$
Total amount of pure glycerin	0.05(60)	+	$0.0x$	=	$0.02(60 + x)$

Reread	Amount of glycerin in 5% solution	+	amount of glycerin in water added	=	amount of glycerin in 2% solution
Step 2.	0.05(60)	+	0	=	$0.02(60 + x)$

Step 3. $5(60) + 0 = 2(60 + x)$ Multiplying both sides by 100 to clear decimals

$$300 = 120 + 2x$$

$$180 = 2x$$

Step 4. $x = 90$ The number of mL of water

Step 5. *Check* In 60 mL of a 5% glycerin solution, there is (60 mL) (0.05), or 3 mL, of glycerin. The total volume of the mixture is 90 mL + 60 mL, or 150 mL. If there were 3 mL of glycerin in 150 mL of solution (and if no more glycerin were added), the solution would be a $\left(\frac{3 \text{ mL}}{150 \text{ mL}} \times 100\right)\%$, or 2%, solution, as required.

Step 6. Therefore, 90 mL of water must be added to reduce the 5% solution to a 2% solution.

EXAMPLE 2 How many liters (L) of a 20% alcohol solution must be added to 3 L of a 90% alcohol solution to obtain an 80% solution?

SOLUTION

Step 1. Let x = the number of L of the 20% solution

Chart

	20% solution		90% solution		Mixture
Strength (in decimal form)	0.20		0.90		0.80
Amount of solution	x	+	3	=	$x + 3$
Total amount of pure alcohol	$0.20x$	+	0.90(3)	=	$0.80(x + 3)$

Reread Amount of alcohol in 20% solution + amount of alcohol in 90% solution = amount of alcohol in 80% solution

Step 2. $0.20x + 0.90(3) = 0.80(x + 3)$

Step 3. $2x + 9(3) = 8(x + 3)$ Multiplying both sides by 10 to clear decimals

$$2x + 27 = 8x + 24$$

$$3 = 6x$$

Step 4. $x = \frac{1}{2}$ The number of L of the 20% solution

Step 5. *Check* Three liters of a 90% alcohol solution contains (3 L) (0.90), or 2.7 L, of pure alcohol. One-half liter of a 20% alcohol solution contains $\left(\frac{1}{2}\text{ L}\right)(0.20)$, or 0.1 L, of pure alcohol. Therefore, there is 2.8 L of pure alcohol in the mixture. The total volume of the mixture is 3.5 L, and the strength of the mixture is $\left(\frac{2.8\text{ L}}{3.5\text{ L}} \times 100\right)\%$, or 80%, as was desired.

Step 6. Therefore, $\frac{1}{2}$ L of the 20% alcohol solution must be added.

EXAMPLE 3 A pharmacist needs 4 L of a 25% solution of boric acid. If she has to mix a 50% solution and a 10% solution to obtain the correct strength, how much of each must she use?

SOLUTION

Step 1. Let x = the number of L of the 50% solution
Then $4 - x$ = the number of L of the 10% solution

Chart

	50% solution		10% solution		Mixture
Strength (in decimal form)	0.50		0.10		0.25
Amount of solution	x	+	$4 - x$	=	4
Total amount of pure boric acid	$0.50x$	+	$0.10(4 - x)$	=	0.25(4)

Step 2. $0.50x + 0.10(4 - x) = 0.25(4)$

Step 3. $50x + 10(4 - x) = 25(4)$ Multiplying both sides by 100

$50x + 40 - 10x = 100$ Simplifying

$40x = 60$

Step 4. $x = 1.5$ The number of L of the 50% solution

$4 - x = 2.5$ The number of L of the 10% solution

√ **Step 5.** *Check* There is 0.50(1.5 L), or 0.75 L, of pure boric acid in 1.5 L of a 50% solution. There is 0.10(2.5 L), or 0.25 L, of pure boric acid in 2.5 L of a 10% solution. Therefore, there is (0.75 L + 0.25 L), or 1.0 L, of pure boric acid going *into* the mixture. The total amount of solution in the mixture is 4 L, and the *strength* of the mixture will be $\frac{1.0 \text{ L}}{4 \text{ L}}$, or 0.25, or 25%, as was desired.

Step 6. Therefore, the pharmacist must use 1.5 L of the 50% solution and 2.5 L of the 10% solution.

Exercises 4.9 Set I

In each exercise, set up the problem algebraically, solve, and check. Be sure to state what your variables represent.

1. How many milliliters of water must be added to 500 mL of a 40% solution of hydrochloric acid to reduce it to a 25% solution?
2. How many liters of water must be added to 10 L of a 30% solution of antifreeze to reduce it to a 20% solution?
3. How many liters of pure alcohol must be added to 10 L of a 20% solution of alcohol to make a 50% solution?
4. How many milliliters of pure alcohol must be added to 1,000 mL of a 10% solution of alcohol to make a 40% solution?
5. How many cubic centimeters (cc) of a 20% solution of sulfuric acid must be mixed with 100 cc of a 50% solution to make a 25% solution of sulfuric acid?
6. How many pints of a 2% solution of disinfectant must be mixed with 5 pt of a 12% solution to make a 4% solution of disinfectant?
7. If 100 gal of a 75% glycerin solution is made up by combining a 30% glycerin solution and a 90% glycerin solution, how much of each solution must be used?
8. Two copper alloys, one containing 65% copper and the other 20% copper, are to be mixed to produce an alloy that is 35% copper. How much of the 65% alloy should be used to obtain 1,200 g of an alloy that is 35% copper?
9. A chemist has two solutions of hydrochloric acid. One is a 40% solution, and the other is a 90% solution. How many liters of each should be mixed to get 10 L of a 50% solution?
10. If whole milk contains 3.25% butterfat and nonfat milk contains 0.5% butterfat, how much whole milk should be added to 1,000 mL of nonfat milk to obtain milk with 2% butterfat?

Exercises 4.9 Set II

In each exercise, set up the problem algebraically, solve, and check. Be sure to state what your variables represent.

1. How many cubic centimeters (cc) of water must be added to 600 cc of a 20% solution of potassium chloride to reduce it to a 5% solution?
2. How many milliliters of a 36% solution of hydrochloric acid should be added to 1,000 mL of a 60% solution to make a 56% solution?
3. How many liters of pure alcohol must be added to 200 L of a 5% solution to make a 62% solution?
4. A 30% solution of antifreeze is to be mixed with an 80% solution of antifreeze to make 1,000 mL (1 L) of a 40% solution. How many milliliters of each should be used?

5. How many cubic centimeters (cc) of a 25% solution of nitric acid must be mixed with 100 cc of a 5% solution to make a 21% solution?

6. A 30% acetic acid solution must be mixed with a 5% acetic acid solution to obtain 100 L of a 25% acetic acid solution. How many liters each of the 30% solution and the 5% solution must be used?

7. A chemist has two solutions of hydrochloric acid, one a 40% solution and the other a 90% solution. How many liters of each should be mixed to get 20 L of a 75% solution?

8. If 1,500 cc of a 10% dextrose solution is made up by combining a 20% solution with a 5% solution, how much of each solution must be used?

9. A certain fertilizer has a 20% nitrogen content. Another fertilizer is 40% nitrogen. How much of each should be mixed together to get a 100-kg mixture with a 32% nitrogen content?

10. How many liters of a 24% acetic acid solution must be mixed with 3 L of a 32% solution to make a 28.8% solution?

Sections 4.1–4.9 REVIEW

Polynomials 4.1

A **polynomial** in one variable is an algebraic expression that, in simplified form, contains only terms that can be expressed in the form ax^n, where a represents any real number, x represents any variable, and n represents any *nonnegative* integer. An algebraic expression with two variables is a polynomial if, in its simplified form, (1) there are no variables in denominators, (2) there are no variables under radical signs, and (3) the exponents on all variables are nonnegative integers.

A **monomial** is a polynomial with only one term.

A **binomial** is a polynomial with two terms.

A **trinomial** is a polynomial with three terms.

Degree of a Polynomial 4.1

The **degree of a term** of a polynomial is the sum of the exponents of its variables. The **degree of a polynomial** is the same as that of its highest-degree term.

Adding Polynomials 4.2

Add (or combine) *like* terms.

Subtracting Polynomials 4.2

Change the subtraction symbol to an addition symbol and change the sign of each term of the polynomial being subtracted; then add the resulting polynomials.

Multiplying Two Binomials 4.3

To multiply two binomials, the FOIL method can be used.

Multiplying a Polynomial by a Polynomial 4.3

Multiply the first polynomial by *each term* of the second polynomial; then add the resulting products.

Special Products 4.4

$(a + b)(a - b) = a^2 - b^2$ The product of the sum and difference of two terms

$(a + b)^2 = a^2 + 2ab + b^2$ The square of a binomial, form 1

$(a - b)^2 = a^2 - 2ab + b^2$ The square of a binomial, form 2

Factorial Notation *4.5

$0! = 1$

$1! = 1$

$n! = n(n - 1)(n - 2) \cdot \cdots \cdot 3 \cdot 2 \cdot 1$, where n is an integer greater than 1

* Section 4.5 may have been postponed.

The Binomial Coefficients ★4.5

$\binom{n}{0} = 1$

$\binom{n}{r} = \frac{n(n-1)(n-2)\cdot \cdots \cdot (n-r+1)}{r!} = \frac{n!}{r!(n-r)!}$, where n and r are *positive* integers and $r \le n$

Binomial Theorem ★4.5

$$(a+b)^n = \binom{n}{0}a^nb^0 + \binom{n}{1}a^{n-1}b^1 + \binom{n}{2}a^{n-2}b^2 + \cdots + \binom{n}{r}a^{n-r}b^r + \cdots + \binom{n}{n-2}a^2b^{n-2} + \binom{n}{n-1}a^1b^{n-1} + \binom{n}{n}a^0b^n$$

where n is any positive integer and where the $(r+1)$st term is $\binom{n}{r}a^{n-r}b^r$

Dividing a Polynomial by a Monomial 4.6

Divide *each term* of the polynomial by the monomial; then add the resulting quotients.

Dividing a Polynomial by a Polynomial 4.6

To divide a polynomial by a polynomial, we use a method similar to that of long division of whole numbers in arithmetic.

Synthetic Division *4.7

Synthetic division can be used to divide a polynomial by a first-degree binomial.

Dry Mixture Problems 4.8

In dry mixture problems, we are concerned with the *cost* involved when two or more "dry" ingredients are mixed.

Solution Mixture Problems 4.9

In solution mixture problems, we are concerned with the *strengths* involved when two or more wet ingredients (or metals) are mixed.

Sections 4.1–4.9 REVIEW EXERCISES Set I

In Exercises 1 and 2, if the expression is a polynomial, find its degree. Otherwise, write "not a polynomial."

1. $7x^2 - \frac{3}{5}xy^2 + 2$

2. $\frac{16}{3z - w} - 2w^2$

★ 3. Evaluate 6!.

★ 4. Evaluate $\binom{7}{4}$.

In Exercises 5 and 6, perform the indicated operations, and simplify the results.

5. $(13x - 6x^3 + 14 - 15x^2) + (-17 - 23x^2 + 4x^3 + 11x)$

6. $(5x^2y + 3xy^2 - 4 + y^2) - (8 - 4x^2y + 2xy^2 - y^2)$

In Exercises 7 and 8, subtract the lower polynomial from the one above it.

7. $\begin{array}{r} x^3 \qquad\qquad - 4x + 2 \\ \underline{3x^3 + 2x^2 + x - 5} \end{array}$

8. $\begin{array}{r} x^4 \qquad + x^2 \qquad + 1 \\ \underline{x^3 \qquad\quad + x \qquad} \end{array}$

9. Subtract $(3k^2 - 5k - 6)$ from the sum of $(2k^3 - 7k + 11)$ and $(4k^3 + k^2 - 9k)$.

In Exercises 10–25, perform the indicated operations.

10. $(5a^2b + 3a - 2c)(-3ab^2)$

11. $(5x - 4y)(7x - 2y)$

★ Section 4.5 may have been postponed; in the remainder of this book, problems from this section will be marked with a black star (★).

*Section 4.7 was optional.

★ 12. $(x^2 + 3)^4$

13. $(4x^2 - 5x + 1)(2x^2 + x - 3)$

14. $4x(x^2 - y^2)(x^2 + y^2)$

★ 15. $\left(2x^2 + \frac{1}{2}\right)^5$

16. $(3 + x + y)^2$

17. $(a - 5)^2$

18. $(z + 3)(z^2 - 3z + 9)$

19. $\dfrac{-15a^2b^3 + 20a^4b^2 - 10ab}{-5ab}$

20. $(6x^2 - 9x + 10) \div (2x - 3)$

21. $(10a^2 + 23ab - 5b^2) \div (5a - b)$

22. $(3x^4 - 2x^3 + 2x^2 + 2x - 15) \div (3x^2 - 2x + 5)$

23. $\dfrac{x^4 - 81}{x + 3}$

In Exercises 24 and 25, use synthetic division if Section 4.7 was studied.

24. $(3x^5 + 7x^4 - 4x^2 + 4) \div (x + 2)$

25. $(5x^5 - 2x^4 + 10x^3 - 4x^2 + 2) \div \left(x - \frac{2}{5}\right)$

In Exercises 26 and 27, set up each problem algebraically, solve, and check. Be sure to state what your variables represent.

26. How many cubic centimeters of water must be added to 10 cc of a 17% solution of a disinfectant to reduce it to a 0.2% solution?

27. A dealer makes up a 15-lb mixture of different candies worth \$2.20 and \$2.60 per pound. How many pounds of each kind of candy must be used for the mixture to be worth \$2.36 per pound?

Name

ANSWERS

1.
2.
3.
4.
5.
6.
7.
8.
9.
10.
11.
12.
13.
14.
15.
16.
17.

In Exercises 1 and 2, if the expression is a polynomial, find its degree. Otherwise, write "not a polynomial."

1. $\frac{1}{2}a^2b^3 - 2^3ab^2 + 5$

2. $9m^3 - \dfrac{8}{m - 5n}$

★3. Evaluate 4!.

★4. Evaluate $\dbinom{7}{2}$.

In Exercises 5 and 6, perform the indicated operations, and simplify the results.

5. $(7z - 20 + 11z^3 - 9z^2) + (-16z^3 + 15 - 3z^2 + 12z)$

6. $(18ef^2 - 14 - 6e^2f + 4f) - (17 + 13f - 14e^2f + 10ef^2)$

In Exercises 7 and 8, subtract the lower polynomial from the one above it.

7.
$$\begin{array}{r} x^3 - 2x^2 + 1 \\ 5x^3 + 2x^2 - x + 3 \\ \hline \end{array}$$

8.
$$\begin{array}{r} x^4 + x^3 + 1 \\ x^2 + x \\ \hline \end{array}$$

9. Subtract $(13h^3 - 2h^2 + 7)$ from the sum of $(12h^2 - 5h - 18)$ and $(9h^3 + 10h - 6)$.

In Exercises 10–25, perform the indicated operations.

10. $-4ab^2(7ab^2 - 5b + 4c)$

11. $(y^2 - 3y + 5)(2y - 3)$

12. $(3n^2 - 6n + 2)(n^2 + 4n - 5)$

13. $(w - 2x - 5)^2$

14. $3a(a^2 + 5b)(a^2 - 5b)$

15. $(2x - 3y)^2$

★16. $(x^2 + 1)^6$

★17. $(x^2 + x^{-1})^4$

18. $(2x - 3)(4x^2 + 6x + 9)$

19. $\dfrac{x^6 - 64}{x - 2}$

20. $\dfrac{24m^4n^3 - 30m^2n^2 + 18m^4n}{-6m^2n}$

21. $(21c^2 - 29c - 18) \div (3c - 5)$

22. $(12a^2 - 7ab - 10b^2) \div (4a - 5b)$

23. $(20k^4 - 8k^3 - 39k^2 + 6k - 9) \div (5k^2 - 2k - 6)$

In Exercises 24 and 25, use synthetic division if Section 4.7 was studied.

24. $(4x^5 - 16x^3 - 6x + 3) \div (x - 2)$

25. $(12x^5 + x^4 - 6x^3 + 8x + 11) \div \left(x + \frac{3}{4}\right)$

18. ______________

19. ______________

20. ______________

21. ______________

22. ______________

23. ______________

24. ______________

25. ______________

26. ______________

____

27. ______________

In Exercises 26 and 27, set up the problem algebraically, solve, and check. Be sure to state what your variables represent.

26. Dimitri has 19 coins in her pocket with a total value of \$3.10. If all the coins are dimes and quarters, how many of each does she have?

27. How many cubic centimeters of a 50% phenol solution must be added to 400 cc of a 5% solution to make it a 10% solution?

Chapter 4 DIAGNOSTIC TEST

The purpose of this test is to see how well you understand operations with polynomials and solving mixture problems. We recommend that you work this diagnostic test *before* your instructor tests you on this chapter. Allow yourself about 50 minutes.

Complete solutions for all the problems on this test, together with section references, are given in the answer section at the end of the book. For the problems you do incorrectly, we suggest that you study the sections cited.

1. For the polynomial $2x^2y^3 - 5^2x^2y^2 - \frac{1}{3}xy$, find
   a. The degree of the third term
   b. The degree of the polynomial

2. Add:
$$\begin{array}{rrrr} -4x^3 & -\ 3x^2 & & +\ 5 \\ & 2x^2 & +\ 6x & -\ 10 \\ 3x^3 & & -\ 2x & +\ 8 \\ \hline \end{array}$$

3. Subtract the lower polynomial from the polynomial above it:
$$\begin{array}{l} 3x^4 - x^3 \qquad\quad + x - 2 \\ \underline{5x^4 + x^3 + x^2 - 3x - 5} \end{array}$$

4. Subtract $(8 - z + 4z^2)$ from $(-3z^2 - 6z + 8)$.

5. Simplify $(3ab^2 - 5ab) - (a^3 + 2ab) + (4ab - 7ab^2)$.

In Problems 6–15, perform the indicated operations.

6. $-3ab(6a^2 - 2ab^2 + 5b)$

7. $\dfrac{9z^3w + 6z^2w^2 - 12zw^3}{3zw}$

8. $(x - 2)(x^2 + 2x + 4)$

9. $(2x^4 + 3)(2x^4 - 3)$

10. $(5m - 2)(3m + 4)$

11. $(3R^2 - 5)^2$

12. $(12y^2 - 4y + 1) \div (3y + 2)$

13. Use synthetic division if you studied Section 4.7: $(2x^4 + 3x^3 - 7x^2 - 5) \div (x + 3)$.

14. $(m^2 - 2m + 5)^2$

15. $(6x^4 - 3x^3 + 2x^2 + 5x - 7) \div (2x^2 - x + 4)$

★ 16. a. Write the *meaning* of 7!. (Do not evaluate it.)
   b. Evaluate $\binom{7}{5}$.

★ 17. Simplify $(2x + 1)^5$.

In Problems 18–20, set up each problem algebraically, solve, and check. Be sure to state what your variables represent.

18. Linda has 14 coins with a total value of \$2.15. If all the coins are dimes and quarters, how many of each kind of coin does she have?

19. A grocer needs to make up a 60-lb mixture of cashews and peanuts. If the cashews are worth \$7.40 per pound and the peanuts are worth \$2.80 per pound, how many pounds of each kind of nut must be used for the mixture to be worth \$4.18 per pound?

20. How many cubic centimeters of water must be added to 600 cc of a 20% solution of potassium chloride to reduce it to a 15% solution?

Chapters 1–4 CUMULATIVE REVIEW EXERCISES

1. Given the numbers -2, $\frac{1}{2}$, 4.5, 1.4142136... (never terminates, never repeats), 10, 0, and $0.\overline{234}$, answer the following questions.
   a. Which are natural numbers?
   b. Which are real numbers?
   c. Which are rational numbers?
   d. Which are integers?
   e. Which are irrational numbers?

2. Simplify $\left(\dfrac{x^2y^{-1}}{y^{-4}}\right)^{-1}$.

3. Find the LCM of 108 and 360.

4. Simplify $5a - \{9 - 2(3 - a) - 3a\}$.

In Exercises 5–9, find the solution set of each equation or inequality.

5. $4(x - 3) = 4 - (x + 6)$

6. $1 - x \le 2(x - 4)$

7. $|5 - x| = 6$

8. $|3x - 5| > 7$

9. $|2x - 1| \le 4$

In Exercises 10–12, perform the indicated operations, and simplify the results.

10. $(3x^4 + 7x^2 - 2x)(x + 2)$

★ 11. $(x - 1)^6$

12. $(x^3 + 2x^2 - 13x - 16) \div (x + 4)$

13. Solve $-3 < 2x + 1 < 5$.

In Exercises 14–17, set up each problem algebraically, solve, and check. Be sure to state what your variables represent.

14. If raisins are worth \$1.38 per pound and granola is worth \$2.10 per pound, how many pounds of raisins must be mixed with 15 lb of granola for the mixture to be worth \$37.71?

15. Consuelo invested \$13,500, putting part of the money into an account with an annual interest rate of 7.2% and the rest into an account with an annual interest rate of 7.4%. If she earned \$986.60 in interest for the first year, how much did she invest at each rate?

16. Patricia is filling an 18-cu.-ft planter with a mixture of potting soil and perlite. She wants to use twice as many cubic feet of potting soil as perlite. The potting soil comes in $1\frac{1}{2}$-cu.-ft bags, and the perlite comes in 2-cu.-ft bags. How many bags of potting soil should she buy? How many bags of perlite?

17. Mrs. Rice invested part of \$23,000 at 10% interest and the remainder at 8%. Her total yearly income from these two investments is \$2,170. How much is invested at each rate?

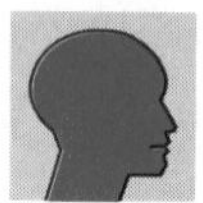

Critical Thinking and Problem-Solving Exercises

1. Helen is looking at a rectangle that is 18 cm long and 24 cm wide, and she wants to construct a *square* that has the same *perimeter* as this rectangle. What would the *area* of the square be?

2. Ed wants to use three-digit code-numbers for some articles he is manufacturing. How many different three-digit code-numbers can he make up if he uses just the digits 1, 2, 3, 4, and 5, assuming that the *order* in which the digits are used makes a difference and that no digit can be repeated (that is, 231 would be considered different from 123, and 112 would not be permitted)?

3. Jamilah was arranging a display of sports equipment in her sporting goods store. She placed the basketballs between the footballs and the baseballs. She put the tennis balls to the right of the golf balls. She put the volleyballs on the extreme left end of the shelf, right next to the footballs. She put the golf balls between the baseballs and the tennis balls. List (from left to right) the order in which the balls were displayed.

4. Five people are attending a meeting. If each person shakes hands with each other person exactly once, how many handshakes will occur? How many handshakes will occur if twelve people attend the meeting?

5. José arranged three coins into a small triangle. Then he formed a second (equilateral) triangle by adding one more row of coins, using three additional coins in the third row, so that there were six coins used altogether. He formed a third triangle by adding another row, using a total of ten coins. If he keeps on in this way, how many coins will he add for the last row of the fourth triangle, and how many coins will be used altogether? How many coins will be in the last row of the eighth triangle?

★6. Prove that $\dfrac{n!}{(n-1)!} = n$.

★7. Prove that $\dbinom{n}{n-r} = \dbinom{n}{r}$.

★8. Prove that $\dbinom{n}{n} = 1$.

Factoring

CHAPTER 5

This chapter deals with factoring polynomials. We review the methods of factoring learned in elementary algebra, but we also discuss additional types of factoring. Then we solve equations whose solution involves factoring, and we discuss solving applied problems that lead to such equations. It is essential that the techniques of factoring be mastered, because factoring is used extensively in work with algebraic fractions, in graphing, and in solving equations.

5.1 The Greatest Common Factor (GCF)

Factoring an algebraic expression means rewriting it (if possible) as a single term that is a *product of prime factors*. That is, when we *factor* an algebraic expression, we are "*undoing*" multiplication. Earlier, we used the distributive property to rewrite a product of factors as a sum of terms—that is, we rewrote $a(b + c)$ as $ab + ac$. We now use the distributive property to rewrite a sum of terms as a product of factors (that is, as a single term) whenever possible. When we use the distributive property to get

$$ab + ac = a(b + c)$$

we say that we are factoring out the greatest common factor. Notice that the right side of this equation has only *one term*.

Finding the Greatest Common Factor (GCF) of an Algebraic Expression

Earlier, we discussed finding all the factors of a number by making an organized list. Using that method, let's list the positive factors of 12, 18, and 30.

The factors of 12: 1, 2, 3, 4, 6, 12

The factors of 18: 1, 2, 3, 6, 9, 18

The factors of 30: 1, 2, 3, 5, 6, 10, 15, 30

We see that 1, 2, 3, and 6 are *common* to all three lists; these are the *common factors* of 12, 18, and 30. The *largest* of the common factors is 6, and so we call 6 the *greatest common factor (GCF)*; this means that 6 is the largest number that *divides exactly* into *all* the numbers 12, 18, and 30.

The **greatest common factor (GCF)** of two or more terms of an algebraic expression is the *largest* expression that is a factor of all the terms. In practice, we usually don't find the GCF by listing all the factors of the terms; rather, we use the following procedure:

Finding the greatest common factor (GCF)

1. Write each term in simplest form, writing any numerical coefficients in prime-factored, exponential form.
2. Write down each base (numerical or variable) *that is common to all the terms.*
3. Raise each of the bases from step 2 to the *lowest* power to which it occurs in any of the terms.
4. The *greatest common factor* (GCF) is the *product* of all the expressions found in step 3. It may be positive or negative.

EXAMPLE 1

Find the GCF for the terms of $-35xy^3z^3 + 21x^3y^3z^3 - 28x^2y^2z^4 + 7xy^2z^3$.

SOLUTION

$$-35xy^3z^3 + 21x^3y^3z^3 - 28x^2y^2z^4 + 7xy^2z^3$$
$$= -5 \cdot 7xy^3z^3 + 3 \cdot 7x^3y^3z^3 - 2^2 \cdot 7x^2y^2z^4 + 7xy^2z^3$$

The bases common to all four terms are 7, x, y, and z. The smallest exponent on the 7 and on the x is an understood 1, the smallest exponent on the y is 2, and the smallest exponent on the z is 3. Therefore, the GCF (the *largest* expression that divides exactly into all four terms) is $7xy^2z^3$ or $-7xy^2z^3$.

Factoring Out the Greatest Common Factor

To rewrite the *sum of terms* $ab + ac$ as the *product of factors* $a(b + c)$—that is, to "undo" the multiplication—we can follow the suggestions in the following box. They are based on the fact that we're asking ourselves a question such as "If $-35xy^3z^3 + 21x^3y^3z^3 - 28x^2y^2z^4 + 7xy^2z^3$ had been the *answer to a multiplication problem*, what would the original factors have been?"

Factoring an expression with a common factor

1. Combine like terms, if there are any.
2. Find the GCF of all the terms. It will often, but not always, be a monomial.
3. Find the *polynomial factor** by dividing each term of the polynomial being factored by the GCF. The polynomial factor will always have as many terms as the expression in step 1. It should contain only *integer* coefficients.
4. Rewrite the expression as the product of the factors found in steps 2 and 3.
5. Check the result by using the distributive property to remove the parentheses; the resulting product should be the polynomial from step 1.

*We call this factor the polynomial factor because it will be a polynomial and will always contain more than one term.

Note If an expression has been factored, *it contains only one term*. However, even if an expression contains only one term, it still may not be factored completely. Any expression inside parentheses should always be examined carefully to see whether it can be factored further.

When we use the procedure described in the box above, we are factoring *over the integers*; that is, all the constants in the new polynomials are *integers*. In this chapter, when we say to factor a polynomial, we mean to factor *over the integers*.

Factorization of polynomials over the integers is *unique*; that is, a polynomial can be completely factored over the integers in one and only one way (except for the order in which the factors are written and the signs of the factors).

EXAMPLE 2

Factor $-35xy^3z^3 + 21x^3y^3z^3 - 28x^2y^2z^4 + 7xy^2z^3$.

SOLUTION We found in Example 1 that the GCF is $7xy^2z^3$ or $-7xy^2z^3$. To find the polynomial factor, we divide each term of the given polynomial by the GCF. If we use $7xy^2z^3$ as the GCF, the polynomial factor is

$$\frac{-35xy^3z^3}{7xy^2z^3} + \frac{21x^3y^3z^3}{7xy^2z^3} - \frac{28x^2y^2z^4}{7xy^2z^3} + \frac{7xy^2z^3}{7xy^2z^3} = -5y + 3x^2y - 4xz + 1$$

Therefore, $-35xy^3z^3 + 21x^3y^3z^3 - 28x^2y^2z^4 + 7xy^2z^3$

$= 7xy^2z^3(-5y + 3x^2y - 4xz + 1)$ ← Factored form

√ *Check* $7xy^2z^3(-5y + 3x^2y - 4xz + 1)$

$= -35xy^3z^3 + 21x^3y^3z^3 - 28x^2y^2z^4 + 7xy^2z^3$

If we use $-7xy^2z^3$ as the GCF, the polynomial factor is

$$\frac{-35xy^3z^3}{-7xy^2z^3} + \frac{21x^3y^3z^3}{-7xy^2z^3} - \frac{28x^2y^2z^4}{-7xy^2z^3} + \frac{7xy^2z^3}{-7xy^2z^3} = 5y - 3x^2y + 4xz - 1$$

Therefore, $-35xy^3z^3 + 21x^3y^3z^3 - 28x^2y^2z^4 + 7xy^2z^3$

$= -7xy^2z^3(5y - 3x^2y + 4xz - 1)$ ← Factored form

√ *Check* $-7xy^2z^3(5y - 3x^2y + 4xz - 1)$

$= -35xy^3z^3 + 21x^3y^3z^3 - 28x^2y^2z^4 + 7xy^2z^3$

The answers $7xy^2z^3(-5y + 3x^2y - 4xz + 1)$ and $-7xy^2z^3(5y - 3x^2y + 4xz - 1)$ each contain just one term; in each answer, the expression within the parentheses has no common factor left in it. Both answers are correct and acceptable.

Prime (or Irreducible) Polynomials

A polynomial is said to be **prime**, or **irreducible**, over the integers if it cannot be expressed as a product of polynomials of lower degree such that all the constants in the new polynomials are integers.

EXAMPLE 3 Factor $3x - 7y$.

SOLUTION Although 3 and x are factors of the first term of $3x - 7y$, they are not factors of the second term. In fact, $3x$ and $-7y$ have no common integral factors other than ± 1. Although it is true that

$$3x - 7y = 3\left(x - \tfrac{7}{3}y\right)$$

and that

$$3x - 7y = 7\left(\tfrac{3}{7}x - y\right)$$

the parentheses in both cases now contain constants that are *not* integers. We must conclude that $3x - 7y$ is *not factorable* over the integers. It is a prime polynomial.

Sometimes an expression has a GCF that is not a monomial. Such an expression can still be factored using the same rules as above. The type of factoring seen in Examples 4 and 5 is used in *factoring by grouping* (covered in the next section) and is also used extensively in calculus.

EXAMPLE 4 Factor $a(x + y) + b(x + y)$.

SOLUTION This expression contains two terms and thus is *not* in factored form. The common factor, $x + y$, is not a monomial; it is a binomial. Nevertheless, $x + y$ is the GCF. We factor as follows:

First term Second term

$a(x + y) + b(x + y)$ $(x + y)$ is a common factor of the two terms

$= (x + y)(a + b)$ Factoring out the GCF

This term is $\dfrac{b(x + y)}{(x + y)} = b$

This term is $\dfrac{a(x + y)}{(x + y)} = a$

√ *Check* $(x + y)(a + b) = (a + b)(x + y) = a(x + y) + b(x + y)$

Therefore, $a(x + y) + b(x + y)$ factors into $(x + y)(a + b)$. The answer, $(x + y)(a + b)$, contains just one term, so it is in factored form.

EXAMPLE 5 Factor $72x^2(m^2 + n)^2 + 36(m^2 + n)$.

SOLUTION The GCF is $2^2 \cdot 3^2(m^2 + n)^1$. Therefore,

This exponent means that there are *two* factors of $(m^2 + n)$

$$72x^2(m^2 + n)^2 + 36(m^2 + n) = 36(m^2 + n)(2x^2[m^2 + n] + 1)$$

When we remove the inner grouping symbols, we have

$$72x^2(m^2 + n)^2 + 36(m^2 + n) = 36(m^2 + n)(2x^2m^2 + 2x^2n + 1)$$

√ *Check*

$$\begin{aligned} 36(m^2 + n)(2x^2m^2 + 2x^2n + 1) &= 36(m^2 + n)(2x^2[m^2 + n] + 1) \\ &= 72x^2(m^2 + n)^2 + 36(m^2 + n) \end{aligned}$$

Therefore, $72x^2(m^2 + n)^2 + 36(m^2 + n)$ factors into $36(m^2 + n)(2x^2m^2 + 2x^2n + 1)$.

Exercises 5.1
Set I

Factor each of the following expressions, or write "not factorable."

1. $54x^3yz^4 - 72xy^3$
2. $225ab^5c - 105a^3b^2c^6$
3. $16x^3 - 8x^2 + 4x$
4. $27a^4 - 9a^2 + 3a$
5. $3x^3 - 4y^2 + 5z$
6. $7z^3 + 3y^2 - 4x$
7. $6my + 15mz - 5n - 4n$
8. $4nx + 8ny + 16z - 4z$
9. $-35r^7s^5t^4 - 55r^8s^9u^4 + 40p^8r^9s^8$
10. $-120a^8b^7c^5 + 40a^4c^3d^9 - 80a^5c^5$
11. $10x^2 - 21y + 11z^3$
12. $15a^3 + 14b^5 - 13c$
13. $-24x^8y^3 - 12x^7y^4 + 48x^5y^5 + 60x^4y^6$
14. $64y^9z^5 + 48y^8z^6 - 16y^7z^7 - 80y^4z^8$
15. $m(a + b) + n(a + b)$
16. $3a(a - 2b) + 2(a - 2b)$
17. $x(y + 1) - (y + 1)$
18. $2e(3e - f) - 3(3e - f)$
19. $5(x - y) - (x - y)^2$
20. $4(a + b) - (a + b)^2$
21. $8x(y^2 + 3z)^2 - 6x^4(y^2 + 3z)$
22. $12a^3(b - 2c^5)^3 - 15a^2(b - 2c^5)^2$
23. $5(x + y)^3(a + b)^5 + 15(x + y)^2(a + b)^6$
24. $14(s + t)^4(u + v)^7 + 7(s + t)^5(u + v)^6$

Writing Problems

Express the answers in your own words and in complete sentences.

1. Explain what the GCF of two or more terms is. (Do *not* explain how to find it.)
2. Explain why the GCF of $6x^3y - 12x^2y^2$ is not $6xy$.
3. Explain what a prime polynomial is.

Exercises 5.1 Set II

Factor each of the following expressions, or write "not factorable."

1. $168s^4 - 126st^3u$
2. $-5x^3 + 10x^5y - 15x^3y^2$
3. $25y^3 - 15y^2 + 5y$
4. $17 + 34x^2 - 85xy$
5. $5a^2 - 2b + 3c$
6. $3x(x + y) - 6x^2(x + y)^2$
7. $7xy + 3yx - 5z + z$
8. $7xy + 49xy^2 - 14x^2y$
9. $-36x^3y^7 - 18x^4y^5 + 27x^5y^3 - 9x^6y^2$
10. $5x^4 - 7y^3 + 3z$
11. $25s^2 - 28t^3 + 9u$
12. $6a(b - c)^3 - 9a^2(b - c)^2$
13. $10a^2b^4 - 15a^3c^3 - 25a^5c^7$
14. $3a^3 + 17b - 8c^2$
15. $a(a + b) + 3b(a + b)$
16. $15x^4y^3 - 18x^5y^6 + 27y^4$
17. $3x(2y - 5) - 2(2y - 5)$
18. $3x^3(y + 3z)^2 - 6x^4(y + 3z)^3$
19. $6(a - b)^2 - (a - b)$
20. $12(x + y)^3 - 3(x + y)^4$
21. $6s^4(t^3 + 2u)^3 - 9s^3(t^3 + 2u)^2$
22. $24x^3(a - 5b)^5 - 32x^4(a - 5b)^4$
23. $21(a + b)^4(c + d)^7 + 3(a + b)^5(c + d)^6$
24. $54s^4(u^2 - v^3)^3 - 36s^3(u^2 - v^3)^4$

5.2 Factoring by Grouping

Earlier, we worked a few problems in which the product of two factors was a polynomial that contained four terms. In this section, we "undo" such problems. That is, we start with a polynomial that contains four (or more) terms, assume that it is the product of two factors, and attempt to find the factors that, when multiplied together, would give such a product.

We begin by considering those four-term polynomials that can be factored by *grouping* the four terms into two groups of two terms each. In all the problems in this section, when we factor by grouping there will be a *common factor* after we've grouped the terms and factored each group. (In later sections of this chapter, we will consider other methods of factoring by grouping.) To "undo" the multiplication, we can use the suggestions in the following box. They are based on the fact that we're asking ourselves a question such as "If $ax + ay + bx + by$ had been the *answer to a multiplication problem*, what would the original factors have been?" (See Example 1.) The suggestions in the box assume that any factor common to *all four* terms has already been factored out.

Factoring an expression of four terms by grouping two and two

1. If possible, arrange the four terms into two groups of two terms each so that each group of two terms is factorable.
2. Factor each *group*. There will now be two terms. *The expression will not yet be factored.*
3. Factor the two-term expression resulting from step 2, if possible. If the two terms resulting from step 2 do not have a common factor, try a different arrangement of the original four terms.

EXAMPLE 1 Factor $ax + ay + bx + by$.

SOLUTION

The GCF = a — The GCF = b

$$ax + ay + bx + by$$

$$= a(x + y) + b(x + y)$$ ← This expression is not yet in factored form because it has two terms; the GCF is $(x + y)$

$$= (x + y)(a + b)$$ Factoring out $(x + y)$, the GCF

√ **Check** $(x + y)(a + b) = (x + y)a + (x + y)b = ax + ay + bx + by$

Therefore, $ax + ay + bx + by$ factors into $(x + y)(a + b)$.

It is often possible to group terms differently and still be able to factor the expression (see Examples 2 and 4). *The same factors are obtained no matter what grouping is used*, because factorization over the integers is unique.

EXAMPLE 2 Factor $ab - b + a - 1$.

SOLUTION 1

The GCF = b

$$ab - b + a - 1$$ — $a - 1 = 1(a - 1)$

$$= b(a - 1) + 1(a - 1)$$ ← This is not yet factored; the GCF is $(a - 1)$

$$= (a - 1)(b + 1)$$ Factoring out $(a - 1)$

SOLUTION 2 If we rearrange the terms, we have

The GCF = a — The GCF = -1

$$ab + a - b - 1$$

$$= a(b + 1) - 1(b + 1)$$ ← Notice that when we factor -1 from $-b - 1$, we must change the sign of each term that goes *into* the parentheses

$$= (b + 1)(a - 1)$$

√ **Check** $(a - 1)(b + 1) = ab - b + a - 1$

Therefore, $ab - b + a - 1$ factors into $(a - 1)(b + 1)$ or $(b + 1)(a - 1)$.

EXAMPLE 3 Factor $2x^2 - 6xy + 3x - 9y$.

SOLUTION

The GCF = $2x$ — The GCF = 3

$$2x^2 - 6xy + 3x - 9y$$

$$= 2x(x - 3y) + 3(x - 3y)$$ $(x - 3y)$ is the GCF

$$= (x - 3y)(2x + 3)$$ Factoring out $(x - 3y)$

√ **Check** $(x - 3y)(2x + 3) = 2x^2 - 6xy + 3x - 9y$

Therefore, $2x^2 - 6xy + 3x - 9y$ factors into $(x - 3y)(2x + 3)$.

If we need to factor an expression that has six terms, we can try either three groups of two terms each *or* two groups of three terms each.

EXAMPLE 4 Factor $ax^2 - ax + 5a + bx^2 - bx + 5b$.

SOLUTION 1 If we try two groups of three terms each, we have

$$ax^2 - ax + 5a + bx^2 - bx + 5b$$
$$= a(x^2 - x + 5) + b(x^2 - x + 5) \quad \text{There is a common } \textit{trinomial} \text{ factor}$$
$$= (x^2 - x + 5)(a + b)$$

SOLUTION 2 If we try three groups of two terms each, we rearrange the terms and have

$$ax^2 + bx^2 - ax - bx + 5a + 5b$$
$$= x^2(a + b) - x(a + b) + 5(a + b) \quad (a + b) \text{ is the GCF}$$
$$= (a + b)(x^2 - x + 5)$$

√ **Check** $(x^2 - x + 5)(a + b) = (a + b)(x^2 - x + 5)$
$$= ax^2 + bx^2 - ax - bx + 5a + 5b$$
$$= ax^2 - ax + 5a + bx^2 - bx + 5b$$

Therefore, $ax^2 - ax + 5a + bx^2 - bx + 5b$ factors into $(x^2 - x + 5)(a + b)$.

A Word of Caution An expression is *not* factored until it has been written as a single term that is a product of factors. For an illustration, consider Example 1 again:

$$ax + ay + bx + by \quad \text{(Example 1)}$$
$$= \underbrace{a(x + y)}_{\text{First term}} + \underbrace{b(x + y)}_{\text{Second term}} \quad \text{This expression is } \textit{not} \text{ in factored form because it has } \textit{two} \text{ terms}$$
$$= \underbrace{(x + y)(a + b)}_{\text{A single term}} \quad \text{The } \textit{factored form} \text{ of } ax + ay + bx + by$$

A Word of Caution In Example 1, even though the final answer is correct, *two* errors are made if we write

$$ax + ay + bx + by \quad \text{There } \textit{must} \text{ be a plus sign here}$$
$$= a(x + y)\ b(x + y) \quad ax + ay + bx + by \neq a(x + y)\ b(x + y)$$
$$= (x + y)(a + b) \quad a(x + y)\ b(x + y) \neq (x + y)(a + b)$$

Exercises 5.2
Set I

Factor each expression, or write "not factorable."

1. $mx - nx - my + ny$
2. $ah - ak - bh + bk$
3. $xy + x - y - 1$
4. $ad - d + a - 1$
5. $3a^2 - 6ab + 2a - 4b$
6. $2h^2 - 6hk + 5h - 15k$
7. $6e^2 - 2ef - 9e + 3f$
8. $8m^2 - 4mn - 6m + 3n$
9. $x^3 + 3x^2 - 2x - 6$
10. $a^3 - a^2 - 2a + 2$
11. $b^3 + 4b^2 + 5b - 20$
12. $6x^3 + 3x^2 - 2x + 1$
13. $2a^3 + 8a^2 - 3a - 12$
14. $5y^3 - 10y^2 + 2y - 4$
15. $acm + bcm + acn + bcn$
16. $cku + ckv + dku + dkv$
17. $a^2x + 2ax + 5x + a^2y + 2ay + 5y$
18. $ax^2 + 3ax + 7a + bx^2 + 3bx + 7b$
19. $s^2x - sx + 4x + s^2y - sy + 4y$
20. $at^2 - at + 3a + bt^2 - bt + 3b$
21. $ax^2 + ax + a - x^2 - x - 1$
22. $xy^2 + xy + 2x - y^2 - y - 2$

Writing Problems

Express the answer in your own words and in complete sentences.

1. Find and describe the *two* errors in the following:

$$\begin{aligned} &3x^3 + 5x^2 + 6x + 10 \\ &= x^2(3x + 5)\ 2(3x + 5) \\ &= (3x + 5)(x^2 + 2) \end{aligned}$$

Exercises 5.2
Set II

Factor each expression, or write "not factorable."

1. $hw - kw - hz + kz$
2. $3a - 5b + 6ac - 10bc$
3. $xy - x + y - 1$
4. $3x^3 - 7x^2 + 3x - 7$
5. $12m^2 - 6mn + 14m - 7n$
6. $a^3m + 4a^2m - a^2n - 4an$
7. $6x^2 - 3xy - 4x + 2y$
8. $6x + 3y - 12ax - 6ay$
9. $x^3 + x^2 - 2x - 2$
10. $7xy - 3z - 14axy + 6az$
11. $4a^3 + 8a^2 - a + 2$
12. $8x^2 - 11xy - 16x + 22y$
13. $3t^3 - 12t^2 + 2t - 8$
14. $ab - ac - bd - cd$
15. $acx + acy + bcx + bcy$
16. $x^4 + x^3 + x^2 + x$
17. $ax^2 + 3ax + 8a + bx^2 + 3bx + 8b$
18. $x^3 + x^2a + x^2 + ax + 4x + 4a$
19. $a^2x - ax + 4x + a^2y - ay + 4y$
20. $2x^3 - 2ax^2 + x^2 - ax + x - a$
21. $b^2y + 2by + 3y - b^2 - 2b - 3$
22. $a^2x + 3ax + 4x - a^2 - 3a - 4$

5.3 Factoring a Difference of Two Squares

Any polynomial that can be expressed in the form $a^2 - b^2$ is called a **difference of two squares**. Because $(a + b)(a - b) = a^2 - b^2$ (we proved this earlier), it must be true that $a^2 - b^2$ *factors into* $(a + b)(a - b)$.

Factoring a difference of two squares

1. Find the principal square root of each of the terms by inspection.
2. The two binomial factors are the sum of and the difference of the square roots found in step 1.

In symbols: $a^2 - b^2 = (a + b)(a - b)$

EXAMPLE 1 Factor $49a^6b^2 - 81c^4d^8$.

SOLUTION

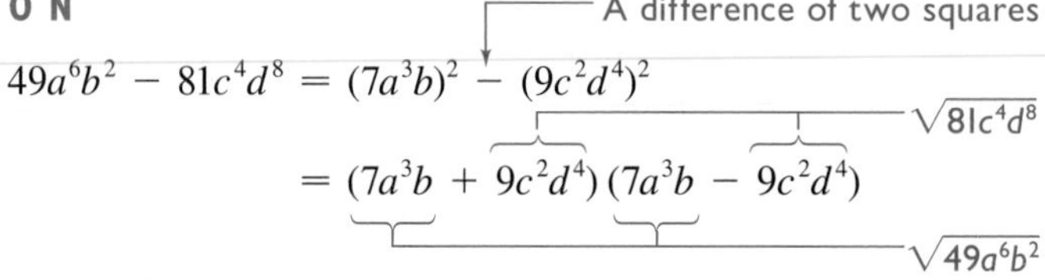

$$49a^6b^2 - 81c^4d^8 = (7a^3b)^2 - (9c^2d^4)^2$$
$$= (7a^3b + 9c^2d^4)(7a^3b - 9c^2d^4)$$

✓ **Check** $(7a^3b + 9c^2d^4)(7a^3b - 9c^2d^4) = 49a^6b^2 - 81c^4d^8$

Therefore, $49a^6b^2 - 81c^4d^8$ factors into $(7a^3b + 9c^2d^4)(7a^3b - 9c^2d^4)$.

A Word of Caution A *sum* of two squares is *not factorable* over the integers. (*Exception*: If the exponents on the variables are multiples of 4 or 6, the polynomial *may* be factorable by using methods of factoring covered later in this chapter.)

It is clear that there is no common factor for $a^2 + b^2$. We suggest that you verify that $a^2 + b^2$ is not factorable by assuming that it factors into a product of two binomials and attempting to find those binomials that, multiplied together, give $a^2 + b^2$. You will be unsuccessful. You might try this *again* after you've studied the next section.

We consider an expression to be *completely factored* when all its factors are prime. By this, we mean that no more factoring can be done by any method.

EXAMPLE 2 Factor $a^4 - b^4$.

SOLUTION $a^4 - b^4$ is a difference of two squares.

$$a^4 - b^4 = (a^2 + b^2)(a^2 - b^2) = (a^2 + b^2)(a + b)(a - b)$$

(Arrow to $(a^2 - b^2)$: This factor is not prime, so it must be factored)

(Arrow to $(a^2 + b^2)$: This factor is a sum of two squares and cannot be factored)

✓ **Check** $(a^2 + b^2)(a + b)(a - b) = (a^2 + b^2)(a^2 - b^2) = a^4 - b^4$

Therefore, $a^4 - b^4$ factors into $(a^2 + b^2)(a + b)(a - b)$.

A Word of Caution When there is a common factor, find and remove the GCF before trying any other factoring.

EXAMPLE 3 Factor $27x^4 - 12y^2$.

SOLUTION The GCF is 3.

$$27x^4 - 12y^2 = 3(9x^4 - 4y^2)$$ ← This expression has been factored, but this factor is not prime, so it must be factored

$$= 3(3x^2 + 2y)(3x^2 - 2y)$$ ← Factored form

(The checking will not be shown.)

Sometimes the quantities that are squares are not monomials. If the entire expression is of the form $a^2 - b^2$, we can still factor it (see Example 4).

EXAMPLE 4 Factor $(x - y)^2 - (a + b)^2$.

SOLUTION

$(x - y)^2 - (a + b)^2$ — A difference of two squares

$= [(x - y) + (a + b)][(x - y) - (a + b)]$ — $\sqrt{(x - y)^2}$; $\sqrt{(a + b)^2}$

$= (x - y + a + b)(x - y - a - b)$ — Removing the inner grouping symbols

✓ **Check** $(x - y + a + b)(x - y - a - b)$

$= ([x - y] + [a + b])([x - y] - [a + b])$

$= [x - y]^2 - [a + b]^2$

Therefore, $(x - y)^2 - (a + b)^2$ factors into $(x - y + a + b)(x - y - a - b)$.

EXAMPLE 5 Factor $a^2 - b^2 + a - b$.

SOLUTION If we group the terms into two groups of two terms each, we have

$a^2 - b^2 + a - b$ — A difference of two squares

$= (a + b)(a - b) + (a - b)$ — $a - b$ is a common binomial factor

$= (a + b)(a - b) + 1(a - b)$ — Writing 1 as the coefficient of $(a - b)$

$= (a - b)([a + b] + 1)$ — Factoring out $(a - b)$, the GCF

$= (a - b)(a + b + 1)$

✓ **Check** $(a - b)(a + b + 1) = a^2 - b^2 + a - b$

Therefore, $a^2 - b^2 + a - b$ factors into $(a - b)(a + b + 1)$.

Exercises 5.3
Set I

Factor each polynomial completely, or write "not factorable."

1. $2x^2 - 8y^2$
2. $3x^2 - 27y^2$
3. $98u^4 - 72v^4$
4. $243m^6 - 300n^4$
5. $x^4 - y^4$
6. $a^4 - 16$
7. $4c^2 + 1$
8. $16d^2 + 1$
9. $4h^4k^4 - 1$
10. $9x^4 - 1$
11. $25a^4b^2 - a^2b^4$
12. $x^2y^4 - 100x^4y^2$
13. $16x^2 + 8x$
14. $25y^2 + 5y$
15. $9x^2 + 7y$
16. $36a^2 + 5b$
17. $(x + y)^2 - 4$
18. $(a + b)^2 - 9$
19. $(a + b)x^2 - (a + b)y^2$
20. $(x + y)a^2 - (x + y)b^2$
21. $9x^2 + y^2$
22. $25x^2 + y^2$
23. $x^2 - 9y^2 + x - 3y$
24. $x^2 - y^2 + x - y$
25. $x + 2y + x^2 - 4y^2$
26. $a + b + a^2 - b^2$

Writing Problems

Express the answers in your own words and in complete sentences.

1. Describe what your first step would be in factoring $s^2 + s - t^2 - t$, and explain why you would do that first.
2. Verify that $6m^2 - 54n^2 = (2m + 6n)(3m - 9n)$; then explain why the completely factored form of $6m^2 - 54n^2$ is not $(2m + 6n)(3m - 9n)$.

Exercises 5.3 Set II

Factor completely, or write "not factorable."

1. $6m^2 - 54n^4$
2. $3x^2 - 27x$
3. $50a^2 - 8b^2$
4. $4x^2 - 9$
5. $x^4 - 81$
6. $x^2 + 16$
7. $16y^4z^4 + 1$
8. $1 - 25a^2b^2$
9. $16y^4z^4 - 1$
10. $81x^2 - y^2$
11. $3b^2x^4 - 12b^2y^2$
12. $y^4 - 256$
13. $4x^2 + 8x$
14. $28a^2b^2 - 7a^2$
15. $25x^2 + 3y$
16. $49a^2 - 7a$
17. $(s + t)^2 - 16$
18. $(s + t)a^2 - (s + t)b^2$
19. $(s + t)x^2 - (s + t)y^2$
20. $(a - b)^2 - 1$
21. $4 + x^2$
22. $9x^2 - 3x$
23. $s^2 - t^2 + s - t$
24. $4x + 4y + x^2 - y^2$
25. $4x + y + 16x^2 - y^2$
26. $49x^2 - a^2 + 14x - 2a$

5.4 Factoring the General Trinomial

A trinomial that can be expressed in the form $ax^2 + bx + c$ is often called the **general trinomial**. Earlier, we worked many multiplication problems in which the product of two binomials was a trinomial. In this section, we "undo" such problems. That is, we start with a trinomial, *assume* that it is a product of two binomial factors, and attempt to find those factors. (Although many trinomials are not factorable, many others *will* factor into the product of two binomials.)

5.4A Factoring a Trinomial of the Form $x^2 + bx + c$

The easiest type of trinomial to factor is one that can be expressed in the form $x^2 + bx + c$ (a second-degree trinomial in which the leading coefficient is 1), and that is the type we now consider.

Careful examination of the following four multiplication problems leads us to the rule of signs that follows.

The same sign

$(x + 3)(x + 2) = x^2 + 5x + 6$

The same sign

$(x - 3)(x - 2) = x^2 - 5x + 6$

A plus sign here means that the signs of the last terms of the two binomials were *the same*

Different signs

$(x + 3)(x - 2) = x^2 + x - 6$

Different signs

$(x - 3)(x + 2) = x^2 - x - 6$

A minus sign here means that the signs of the last terms of the two binomials were *different*

The rule of signs for factoring trinomials that can be expressed in the form $ax^2 + bx + c$

Arrange the trinomial in descending powers of one variable.

1. If $a > 0$ and if the sign of the last term of the trinomial is a plus sign, the signs of the last terms of the two binomials will be the same:
   a. If the sign of the middle term of the trinomial is a plus sign, the signs of the last terms of the binomials will both be positive.
   b. If the sign of the middle term of the trinomial is a minus sign, the signs of the last terms of the binomials will both be negative.
2. If the sign of the last term of the trinomial is a minus sign, the signs of the last terms of the two binomials will be different (one +, one −); the signs will be inserted *as part of* the last terms of the binomials.

The method of factoring a trinomial that can be expressed in the form $x^2 + bx + c$ is summarized as follows:

Factoring a trinomial that can be expressed in the form $x^2 + bx + c$

(It is assumed that like terms have been combined.)

1. Factor out the GCF, if there is one.
2. Arrange the trinomial in descending powers of the variable.
3. Make a blank outline. (We *assume* that the trinomial will factor into the product of two binomials.) If the rule of signs indicates that the signs will both be positive or both be negative, you can insert those signs.
4. The first term inside each set of parentheses will be the square root of the first term of the trinomial. If the third term of the trinomial contains a variable, the square root of *that* variable must be a *factor* of the second term of each binomial.
5. Find the last term of each binomial:
   a. List (mentally, at least) *all* pairs of integral factors of the coefficient of the last term of the trinomial.
   b. Select the particular pair of factors that has a sum equal to the coefficient of the middle term of the trinomial. If no such pair of factors exists, the trinomial is *not factorable*.
6. Check the result by multiplying the binomials together; the resulting product should be the trinomial from step 2. Be sure that neither factor can be factored further.

Note *Alternative Step 5:* It is also possible to consider only the *positive* factors of the coefficient of the last term of the trinomial and to insert the signs *last*. In this case, when the rule of signs tells us that the signs are going to be the same, we select those factors whose *sum* equals the absolute value of the coefficient of the middle term of the trinomial, and when the rule of signs tells us that the signs are going to be *different*, we select those factors whose *difference* equals the absolute value of the coefficient of the middle term of the trinomial. Finally, we insert the signs so that we obtain the correct middle term.

EXAMPLE 1 Factor $z^2 + 8zw + 12w^2$.

SOLUTION

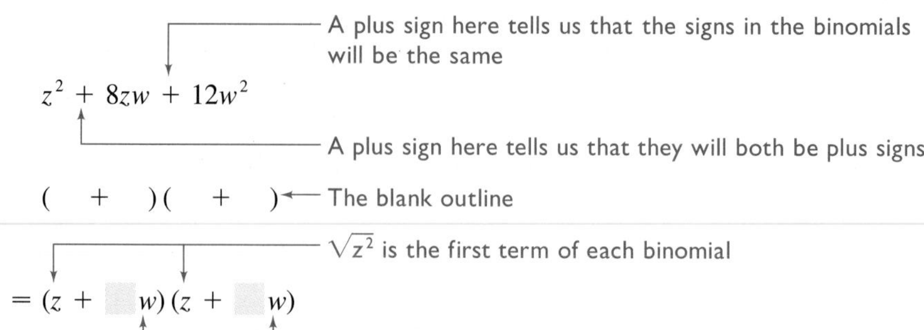

A plus sign here tells us that the signs in the binomials will be the same
$z^2 + 8zw + 12w^2$
A plus sign here tells us that they will both be plus signs
(+)(+)
The blank outline
$\sqrt{z^2}$ is the first term of each binomial
$= (z + \quad w)(z + \quad w)$
$\sqrt{w^2}$ is a *factor* of the second term of each binomial

The pairs of integral factors of $+12$ are $(1)(12)$, $(-1)(-12)$, $(2)(6)$, $(-2)(-6)$, $(3)(4)$, and $(-3)(-4)$. Because the coefficient of the middle term of the trinomial is 8, we must select the pair whose sum is $+8$—the pair 2 and 6. We then have

$$z^2 + 8zw + 12w^2 = (z + 2w)(z + 6w) \text{ or } (z + 6w)(z + 2w)$$

√ **Check** $(z + 2w)(z + 6w) = (z + 6w)(z + 2w) = z^2 + 8zw + 12w^2$

Therefore, $z^2 + 8zw + 12w^2$ factors into $(z + 2w)(z + 6w)$ or $(z + 6w)(z + 2w)$.

A Word of Caution *Both sets of parentheses are essential.* Note:

$$(z + 2w)z + 6w = z^2 + 2zw + 6w \neq z^2 + 8zw + 12w^2$$

$$z + 2w(z + 6w) = z + 2zw + 12w^2 \neq z^2 + 8zw + 12w^2$$

$$z + 2w \cdot z + 6w = z + 2zw + 6w \neq z^2 + 8zw + 12w^2$$

EXAMPLE 2 Factor $w^2 + 8w - 20$.

SOLUTION

$$w^2 + 8w - 20$$ A minus sign here tells us that the signs in the binomials will be *different*

We must find two integers whose product is -20 and whose sum is $+8$. The integers are $+10$ and -2. Therefore,

$$w^2 + 8w - 20 = (w - 2)(w + 10) \quad or \quad (w + 10)(w - 2)$$ The order of the factors is unimportant

√ **Check** $(w - 2)(w + 10) = (w + 10)(w - 2) = w^2 + 8w - 20$

Therefore, $w^2 + 8w - 20$ factors into $(w - 2)(w + 10)$ or $(w + 10)(w - 2)$.

EXAMPLE 3 Factor $x^2 + 3x - 5$.

SOLUTION We must find two integers whose product is -5 and whose sum is 3. The only pairs of integers whose product is -5 are $(-1)(5)$ and $(1)(-5)$. The sum of neither pair is $+3$. Therefore, this trinomial is *not factorable* over the integers; it is a prime polynomial.

EXAMPLE 4 Factor $2ax^2 - 14axy - 60ay^2$.

SOLUTION The three terms have a common factor, and the GCF is $2a$. When we factor out the GCF, we have

$$2ax^2 - 14axy - 60ay^2 = 2a(x^2 - 7xy - 30y^2)$$

The trinomial has been factored, but not completely. We must now factor $x^2 - 7xy - 30y^2$. To do this, we must find two integers whose product is -30 and whose sum is -7. The only such pair is (3) and (-10). Therefore,

$$2ax^2 - 14axy - 60ay^2 = 2a(x + 3y)(x - 10y)$$

Check $2a(x + 3y)(x - 10y) = 2a(x^2 - 7xy - 30y^2) = 2ax^2 - 14axy - 60ay^2$

Therefore, $2ax^2 - 14axy - 60ay^2$ factors into $2a(x + 3y)(x - 10y)$.

Sometimes the terms of the trinomial are not monomials (see Example 5).

EXAMPLE 5 Factor $(x - y)^2 - 2(x - y) - 8$.

SOLUTION This problem can be solved by using a substitution. If we let $(x - y) = a$, we have

$$(x - y)^2 - 2(x - y) - 8 = a^2 - 2a - 8$$
$$= (a + 2)\cdot(a - 4)$$

The pair of factors whose product is -8 and whose sum is -2 is the pair $(+2)$ and (-4)

Substituting back, we get

$$(x - y)^2 - 2(x - y) - 8 = [(x - y) + 2][(x - y) - 4]$$
$$= (x - y + 2)(x - y - 4)$$

Therefore, $(x - y)^2 - 2(x - y) - 8$ factors into $(x - y + 2)(x - y - 4)$.

Exercises 5.4A

Set I

Factor each polynomial completely, or write "not factorable."

1. $t^2 + 7t - 30$
2. $m^2 - 13m - 30$
3. $m^2 + 13m + 12$
4. $x^2 + 15x + 14$
5. $x^2 + 7x + 1$
6. $x^2 + 5x + 1$
7. $x^2 - 7x + 18$
8. $x^2 - 3x + 18$
9. $x^2 + 4x + 6x$
10. $y^2 + 7y + 3y$
11. $u^4 - 15u^2 + 14$
12. $y^4 - 16y^2 + 15$
13. $u^2 + 12u - 64$
14. $v^2 - 30v - 64$
15. $3x + x^2 + 2$
16. $7a + a^2 + 10$
17. $x^4 - 6x^3 + 2x^2$
18. $y^6 - 2y^4 + 2y^2$
19. $x^2 + xy - 2y^2$
20. $x^2 + 6xy + 9y^2$
21. $(a + b)^2 + 6(a + b) + 8$
22. $(m + n)^2 + 9(m + n) + 8$
23. $(x + y)^2 - 13(x + y) - 30$
24. $(x + y)^2 - 10(x + y) - 24$

Writing Problems

Express the answers in your own words and in complete sentences.

1. Explain why $x^2 + 8x + 12$ factors into $(x + 2)(x + 6)$.
2. Explain why $x^2 + 3x + 3$ is not factorable.
3. Explain why $x^2 + 5x + 6$ does not factor into $(x + 1)(x + 6)$.
4. Explain why $x^2 - 5x - 6$ does not factor into $(x - 2)(x - 3)$.

Exercises 5.4A Set II

Factor each polynomial completely, or write "not factorable."

1. $z^2 - 9z + 20$
2. $x^3 - 6x^2 - 16x$
3. $u^2 + 11u + 10$
4. $x^2 - 12x + 35$
5. $x^2 + 4x + 1$
6. $x^2 + x - 56$
7. $x^2 + 9x + 10$
8. $x^2 + 6x + 6$
9. $x^2 + 5x + 3x$
10. $x^2 + 12x - 13$
11. $x^4 + 2x^2 - 3$
12. $x^4 + 2x^3 - 3x^2$
13. $f^2 - 7f - 18$
14. $x^2 + 8x + 7x$
15. $10v + v^2 + 16$
16. $15 - 8x + x^2$
17. $z^4 + 3z^3 + 5z^2$
18. $x^4 - 21 - 4x^2$
19. $a^2 + 4ab + 3b^2$
20. $x^2 + 5x - 24$
21. $(a + b)^2 + 9(a + b) + 14$
22. $(x + y)^2 - (x + y) - 12$
23. $(x + y)^2 - 2(x + y) - 15$
24. $(a - b)^2 - 6(a - b) + 5$

5.4B Factoring a Trinomial of the Form $ax^2 + bx + c$

We show two methods of factoring a trinomial that can be expressed in the form $ax^2 + bx + c$ (the general trinomial). In the first method, we use guessing-and-checking, and we use factoring by grouping in the second method, the master product method. (When we try to factor the general trinomial, we *assume* that it will factor into a product of two binomials; we may decide later that the trinomial does *not* factor.)

The Guessing-and-Checking Method

The following procedure can be used for factoring any trinomial that can be expressed in the form $ax^2 + bx + c$:

Factoring a trinomial that can be expressed in the form $ax^2 + bx + c$

1. Factor out the GCF, if there is one.
2. Arrange the trinomial in descending (or ascending) powers of one variable. (If the leading coefficient is negative, either arrange the trinomial in ascending powers or factor out -1 before proceeding.)
3. a. If the leading coefficient is 1, factor by using the techniques given in the preceding section.

   b. If the leading coefficient is *not* 1, proceed with steps 4–7.
4. Make a blank outline, and fill in the *variable parts* of each binomial. (In deciding what the variable parts of the terms of the binomials should be, use the method given in the preceding section.) If the rule of signs indicates that the signs of the last terms of the binomials will both be positive or both be negative, you may insert those signs in the outline.
5. List (mentally, at least) all pairs of factors of the coefficient of the first term of the trinomial and all pairs of factors of the coefficient of the last term of the trinomial.
6. By guessing-and-checking, select the pairs of factors (if they exist) from step 5 that make the sum of the inner and outer products of the binomials equal to the middle term of the trinomial. If no such pairs exist, the trinomial is *not factorable*.
7. Check the result by multiplying the binomials together. The product should be the trinomial from step 2. Check to be sure that neither factor can be factored further.

Hint If the original trinomial does not contain a common factor, then neither *binomial factor* can contain a common factor. This fact can help eliminate some combinations from consideration (see Examples 6 and 8).

EXAMPLE 6

Factor $6x^2 - 89xy - 15y^2$.

SOLUTION The partially filled-in outline

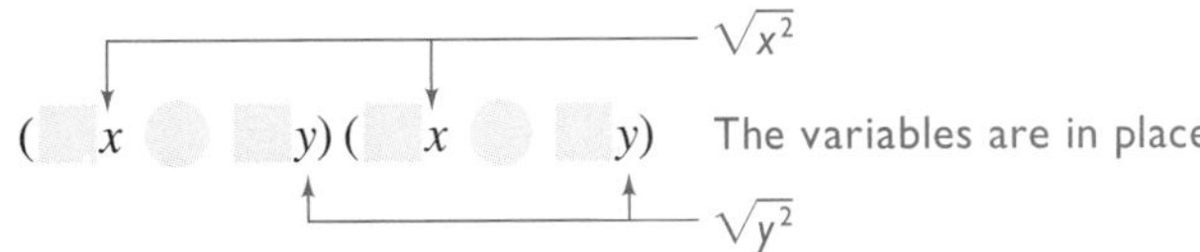

There is no common factor for all three terms. The pairs of positive factors of 6 are (2)(3) and (1)(6). The pairs of factors of -15 are $(3)(-5)$, $(-3)(5)$, $(15)(-1)$, and $(-15)(1)$. Thus, the *possible* factorizations are

$(2x - 3y)(3x + 5y)$	$(2x + 3y)(3x - 5y)$
$^*(2x - 5y)(3x + 3y)$	$^*(2x + 5y)(3x - 3y)$
$^*(2x - y)(3x + 15y)$	$^*(2x + y)(3x - 15y)$
$(2x - 15y)(3x + y)$	$(2x + 15y)(3x - y)$
$^*(6x - 3y)(x + 5y)$	$^*(6x + 3y)(x - 5y)$
$(6x - 5y)(x + 3y)$	$(6x + 5y)(x - 3y)$
$^*(6x - 15y)(x + y)$	$^*(6x + 15y)(x - y)$
$(6x - y)(x + 15y)$	$(6x + y)(x - 15y)$

This is the *only* combination in which the sum of the inner and outer products is $-89xy$

Check $(6x + y)(x - 15y) = 6x^2 - 89xy - 15y^2$

Therefore, $6x^2 - 89xy - 15y^2$ factors into $(6x + y)(x - 15y)$.

The starred combinations all contain a common factor of 3 in one of the binomials. Since the original trinomial had no common factor, these combinations need not be given any consideration.

EXAMPLE 7

Factor $3x^2 - x + 7$.

SOLUTION Possible factorizations:

$$(3x - 7)(x - 1)$$
$$(3x - 1)(x - 7)$$

In both of these factorizations, the product of the two first terms is $3x^2$ and the product of the two last terms is $+7$. However, in neither one is the sum of the inner and outer products $-x$. Therefore, $3x^2 - x + 7$ is *not factorable*.

EXAMPLE 8

Factor $12 + 7x - 10x^2$.

SOLUTION If we arranged this trinomial in *descending* powers of x, the leading coefficient would be negative, so we leave it in *ascending* powers. Note that 12, 7, and 10 have no common factor. The pairs of positive factors of 12 are (3)(4), (2)(6), and (1)(12). The pairs of factors of -10 are $(2)(-5)$, $(-2)(5)$, $(1)(-10)$, and $(-1)(10)$.

This time, rather than write *all* possible combinations of binomials, we will stop when we find a combination that gives us $12 + 7x - 10x^2$. In each starred combination, one of the binomials contains a common factor of 2; therefore, these combinations need not be considered.

$$^*(3 + 5x)(4 - 2x)$$
$$(3 + 2x)(4 - 5x)$$
$$^*(3 - 5x)(4 + 2x)$$
$$(3 - 2x)(4 + 5x)$$ ← The sum of the inner and outer products is $+7x$

√ **Check** $(3 - 2x)(4 + 5x) = 12 + 7x - 10x^2$

Therefore, $12 + 7x - 10x^2$ factors into $(3 - 2x)(4 + 5x)$.

Note We could have factored out -1 from the original trinomial before factoring. The answer would have been $(-1)(2x - 3)(5x + 4)$, which is equivalent to $(3 - 2x)(4 + 5x)$.

EXAMPLE 9

Factor $15x^2 - 25xy - 10y^2$.

SOLUTION

$5(3x^2 - 5xy - 2y^2)$ Factoring out 5, the GCF

$= 5(x - 2y)(3x + 1y)$ The outer product is $1xy$, and the inner product is $-6xy$, so the sum of the inner and outer products is $-5xy$

√ **Check** $5(x - 2y)(3x + y) = 5(3x^2 - 5xy - 2y^2) = 15x^2 - 25xy - 10y^2$

Therefore, $15x^2 - 25xy - 10y^2$ factors into $5(x - 2y)(3x + y)$.

A Word of Caution If the 5 isn't factored out first in Example 9, we will get

$15x^2 - 25xy - 10y^2 = (5x - 10y)(3x + y)$ Not factored completely

or $15x^2 - 25xy - 10y^2 = (x - 2y)(15x + 5y)$ Not factored completely

Neither of these answers is *completely* factored.

EXAMPLE 10

Factor $5(2y - z)^2 + 12(2y - z) + 4$.

SOLUTION We can use a substitution here. Let $(2y - z) = a$. Then

$$5(2y - z)^2 + 12(2y - z) + 4 = 5a^2 + 12a + 4 = (1a + 2)(5a + 2)$$

The outer product is $2a$, and the inner product is $10a$, so the sum of the inner and outer products is $+12a$

Therefore, $5a^2 + 12a + 4 = (a + 2)(5a + 2)$. Substituting back $2y - z$ for a, we have

$$5(2y - z)^2 + 12(2y - z) + 4 = [(2y - z) + 2][5(2y - z) + 2]$$
$$= (2y - z + 2)(10y - 5z + 2)$$

√ **Check** $([2y - z] + 2)(5[2y - z] + 2) = 5(2y - z)^2 + 12(2y - z) + 4$

Therefore, $5(2y - z)^2 + 12(2y - z) + 4$ factors into $(2y - z + 2)(10y - 5z + 2)$.

EXAMPLE 11

Factor $2x^2 + 5xy - 3y^2 + 8x - 4y$.

SOLUTION Since this expression contains more than three terms, we will have to factor by grouping. We'll try grouping the first three terms and the last two terms (if this doesn't work, we'll try a different grouping).

$$2x^2 + 5xy - 3y^2 + 8x - 4y$$

Factoring each group, we have

$$(2x - y)(x + 3y) + 4(2x - y)$$

Notice that we still have two terms. Therefore, the expression has not yet been factored. However, we now have a common *binomial* factor—namely, $2x - y$. When we factor this out, we have

$$(2x - y)[(x + 3y) + 4]$$

✓ **Check** $(2x - y)(x + 3y + 4) = 2x^2 + 5xy - 3y^2 + 8x - 4y$

Therefore, $2x^2 + 5xy - 3y^2 + 8x - 4y$ factors into $(2x - y)(x + 3y + 4)$.

The Master Product Method (Optional)

The **master product method**, which makes use of factoring by grouping, is a method of factoring trinomials that eliminates some of the guesswork. It can also be used to determine whether or not a trinomial is factorable. The procedure described in the following box assumes that any common factors have been factored out.

Factoring a trinomial of the form $ax^2 + bxy + cy^2$ by the master product method

Arrange the trinomial in the form $ax^2 + bxy + cy^2$.

1. Find the master product (MP) by multiplying the first and last coefficients of the trinomial being factored (MP $= a \cdot c$).
2. List the pairs of factors of the master product.
3. Choose the pair of factors whose sum equals the coefficient of the middle term (b). If there is no such pair of factors, then the trinomial is *not factorable*.
4. Rewrite the given trinomial, replacing the middle term with a sum of two terms whose coefficients are the pair of factors found in step 3.
5. Factor the expression from step 4 by grouping.

Check the answer by multiplying the binomial factors to see whether their product is the given trinomial. Check to be sure that neither factor can be factored further.

EXAMPLE 12

Factor $5x + 2x^2 + 3$.

SOLUTION $2x^2 + 5x + 3$ Arranging in descending powers of x

$$\text{MP} = (2)(+3) = +6$$

$6 = (+1)(+6) = (-1)(-6)$ We must select the pair of factors of 6 whose sum equals 5

$= (+2)(+3) = (-2)(-3)$

$(+2) + (+3) = 5$ ← The middle coefficient

$2x^2 + 5x + 3 = 2x^2 + 2x + 3x + 3$ ← Writing $5x$ as $2x + 3x$; we now factor by grouping

$= 2x(x + 1) + 3(x + 1)$ The GCF is $(x + 1)$

$= (x + 1)(2x + 3)$

✓ **Check** $(x + 1)(2x + 3) = 2x^2 + 5x + 3$

Therefore, $5x + 2x^2 + 3$ factors into $(x + 1)(2x + 3)$.

EXAMPLE 13 Factor $3m^2 - 2mn - 8n^2$.

SOLUTION $\text{MP} = (3)(-8) = -24$

$$
\begin{aligned}
-24 &= (-1)(24) = (1)(-24) \\
&= (-2)(12) = (2)(-12) \\
&= (-3)(8) = (3)(-8) \\
&= (-4)(6) = (4)(-6)
\end{aligned}
$$

We must select the pair of factors of -24 whose sum is -2

$(+4) + (-6) = -2$ ← The middle coefficient

$$
\begin{aligned}
3m^2 - 2mn - 8n^2 &= 3m^2 + 4mn - 6mn - 8n^2 \\
&= m(3m + 4n) - 2n(3m + 4n) \\
&= (3m + 4n)(m - 2n)
\end{aligned}
$$

Writing $-2mn$ as $+4mn - 6mn$; we now factor by grouping

The GCF is $(3m + 4n)$

√ *Check* $(3m + 4n)(m - 2n) = 3m^2 - 2mn - 8n^2$

Therefore, $3m^2 - 2mn - 8n^2$ factors into $(3m + 4n)(m - 2n)$.

EXAMPLE 14 Factor $12x^2 - 5x + 10$.

SOLUTION $\text{MP} = 12(10) = 120$

$$
\begin{aligned}
120 &= 1(120) = (-1)(-120) \\
&= 2(60) = (-2)(-60) \\
&= 3(40) = (-3)(-40) \\
&= 4(30) = (-4)(-30) \\
&= 5(24) = (-5)(-24) \\
&= 6(20) = (-6)(-20) \\
&= 8(15) = (-8)(-15) \\
&= 10(12) = (-10)(-12)
\end{aligned}
$$

None of the sums of these pairs is -5. Therefore, the trinomial is not factorable.

Factoring Perfect Square Trinomials

We learned earlier that

$$(a + b)^2 = a^2 + 2ab + b^2$$

and

$$(a - b)^2 = a^2 - 2ab + b^2$$

Trinomials that fit the patterns $a^2 + 2ab + b^2$ and $a^2 - 2ab + b^2$ are called **perfect square trinomials**.

$$a^2 \pm 2ab + b^2$$

The first and last terms are perfect squares

$2(a)(b)$, where $a = \sqrt{a^2}$ and $b = \sqrt{b^2}$

The middle term is twice the product of the square roots of the first and last terms

Therefore, $a^2 \pm 2ab + b^2 = (\sqrt{a^2} \pm \sqrt{b^2})^2 = (a \pm b)^2$.

Factoring $a^2 \pm 2ab + b^2$ (a perfect square trinomial)

1. Find a, the square root of the first term, and b, the square root of the last term.
2. Substitute into one of the following formulas:

$$a^2 + 2ab + b^2 = (a + b)^2$$

$$a^2 - 2ab + b^2 = (a - b)^2$$

Note Perfect square trinomials can *also* be factored by guessing-and-checking or by the master product method.

EXAMPLE 15 Factor $4x^2 - 12x + 9$.

SOLUTION $4x^2 - 12x + 9$ *might* be a perfect square trinomial, since $4x^2 = (2x)^2$ and $9 = 3^2$. Because the middle term equals $-2(2x)(3)$ (the middle term fits the pattern $-2ab$), the expression *is* a perfect square trinomial.

$\sqrt{4x^2} = 2x$

$\sqrt{9} = 3$

$$4x^2 - 12x + 9 = (2x - 3)^2$$

√ ***Check*** $(2x - 3)^2 = (2x)^2 - 2(2x)(3) + 3^2 = 4x^2 - 12x + 9$

Therefore, $4x^2 - 12x + 9$ factors into $(2x - 3)^2$.

EXAMPLE 16 Factor $36x^4 - 13x^2 + 1$.

SOLUTION $36x^4 - 13x^2 + 1$ *might* be a perfect square trinomial, since $36x^4 = (6x^2)^2$ and $1 = 1^2$. We check: Does $(6x^2 - 1)^2$ equal $36x^4 - 13x^2 + 1$? No, $(6x^2 - 1)^2 = 36x^4 - 12x^2 + 1$. However, the expression still might be factorable. We try

$$(9x^2 - 1)(4x^2 - 1) \stackrel{?}{=} 36x^4 - 13x^2 + 1$$

This *does* check, and the expression has been factored. However, neither factor is prime (each factor is a difference of two squares), so we must factor again:

$$36x^4 - 13x^2 + 1 = (9x^2 - 1)(4x^2 - 1) = (3x - 1)(3x + 1)(2x - 1)(2x + 1)$$

√ ***Check*** $(3x - 1)(3x + 1)(2x - 1)(2x + 1) = (9x^2 - 1)(4x^2 - 1) = 36x^4 - 13x^2 + 1$

Therefore, $36x^4 - 13x^2 + 1$ factors into $(3x - 1)(3x + 1)(2x - 1)(2x + 1)$.

More on Factoring by Grouping

When we discussed factoring by grouping previously, we always had a common factor after factoring each group. We can *also* complete the factoring after we've grouped if we have a difference of two squares (see Example 17) or a factorable trinomial (see Example 18).

EXAMPLE 17 Factor $x^2 + 4xy + 4y^2 - 9$.

SOLUTION If we grouped the first two terms and the last two terms, we would *not* have a common binomial factor. (Try it!) However, the first *three* terms form a perfect square trinomial; that is, $x^2 + 4xy + 4y^2 = (x + 2y)^2$. Therefore, we'll try grouping the first three terms and leaving the last term by itself.

$$(x^2 + 4xy + 4y^2) - (9)$$

$$= (x + 2y)^2 - (3)^2 \leftarrow \text{This is a difference of two squares}$$

$$= ([x + 2y] + 3)([x + 2y] - 3)$$

$$= (x + 2y + 3)(x + 2y - 3)$$

√ ***Check*** $(x + 2y + 3)(x + 2y - 3) = (x + 2y)^2 - 3^2 = x^2 + 4xy + 4y^2 - 9$

Therefore, $x^2 + 4xy + 4y^2 - 9$ factors into $(x + 2y + 3)(x + 2y - 3)$.

The type of factoring seen in Example 17 is used later when we factor by completing the square.

EXAMPLE 18 Factor $5x^2 - 30xy + 45y^2 - 14x + 42y - 3$.

SOLUTION Since the polynomial has more than three terms, we must factor by grouping. Let's look at the first three terms only:

$$5x^2 - 30xy + 45y^2 = 5(x^2 - 6xy + 9y^2) = 5(x - 3y)^2$$

That result leads us to believe that we may be able to make a substitution that will give us a factorable trinomial. Let's try the following grouping:

$$5x^2 - 30xy + 45y^2 \quad - \quad 14x + 42y \quad - \quad 3$$

$$= 5(x - 3y)^2 - 14(x - 3y) - 3$$

Double check: If we removed the parentheses, would we get back the original polynomial? Yes. Continuing, we let $a = x - 3y$. We then have

$$5a^2 - 14a - 3 = (5a + 1)(a - 3)$$

Substituting back, we get

$$5x^2 - 30xy + 45y^2 - 14x + 42y - 3$$

$$= (5[x - 3y] + 1)([x - 3y] - 3)$$

$$= (5x - 15y + 1)(x - 3y - 3)$$

Checking, which is not shown, will verify that $5x^2 - 30xy + 45y^2 - 14x + 42y - 3$ factors into $(5x - 15y + 1)(x - 3y - 3)$.

Exercises 5.4B
Set I

Factor each polynomial completely, or write "not factorable."

1. $5x^2 + 9x + 4$
2. $5x^2 + 12x + 4$
3. $7 - 22b + 3b^2$
4. $7 - 10u + 3u^2$
5. $x + 3x^2 + 1$
6. $x + 11x^2 + 1$
7. $3n^2 + 14n - 5$
8. $3n^2 + 2n - 5$
9. $3t^2 - 17tz - 6z^2$
10. $3x^2 - 7xy - 6y^2$
11. $-7 + 4x^2 + x$
12. $-13 + 6x^2 + x$
13. $8 + 12z - 8z^2$
14. $9 + 21z - 18z^2$

15. $1 + 4a^2 + 4a$
16. $1 + 9b^2 + 6b$
17. $2x^2 - 18$
18. $5y^2 - 80$
19. $4 + 7h^2 - 11h$
20. $4 + 7h^2 - 16h$
21. $24xy + 72x^2 + 2y^2$
22. $16xy + 32x^2 + 2y^2$
23. $x^2 - 7xy + 49y^2$
24. $x^2 - 5xy + 25y^2$
25. $2x^2y + 8xy^2 + 8y^3$
26. $3x^3 + 6x^2y + 3xy^2$
27. $6e^4 - 7e^2 - 20$
28. $10f^4 - 29f^2 - 21$
29. $12x^4 - 75x^2y^2$
30. $36x^4 - 16x^2y^2$
31. $a^4 + 2a^2b^2 + b^4$
32. $x^4 + 6x^2y^2 + 9y^4$
33. $2(a + b)^2 + 7(a + b) + 3$
34. $3(a - b)^2 + 7(a - b) + 2$
35. $4(x - y)^2 - 8(x - y) - 5$
36. $4(x + y)^2 - 4(x + y) - 3$
37. $5x^2 + 10xy + 5y^2 - 21x - 21y + 4$
38. $5x^2 - 10xy + 5y^2 - 12x + 12y + 4$
39. $(2x - y)^2 - (3a + b)^2$
40. $(2x + 3y)^2 - (a - b)^2$
41. $x^2 + 10xy + 25y^2 - 9$
42. $x^2 + 8xy + 16y^2 - 25$
43. $a^2 - 4x^2 - 4xy - y^2$
44. $x^2 - 9a^2 - 6ab - b^2$
45. $4x^4 - 13x^2 + 9$
46. $9x^4 - 13x^2 + 4$
47. $3x^2 - 7xy - 6y^2 - x + 3y$
48. $3t^2 - 17tz - 6z^2 - 3t - z$
49. $3n^2 + 2mn - 5m^2 + 3n + 5m$
50. $3n^2 + 14mn - 5m^2 + 3n - m$

Writing Problems

Express the answers in your own words and in complete sentences.

1. Describe what your first step would be in factoring $17x - 12 + 5x^2$, and explain why you would do that first.
2. Describe what your first step would be in factoring $-2 + 9x - 4x^2$, and explain why you would do that first.

Exercises 5.4B Set II

Factor each polynomial completely, or write "not factorable."

1. $3x^2 + 7x + 4$
2. $7x^2 + 13x - 2$
3. $2 - 9x + 4x^2$
4. $3x^2 - 5x - 8$
5. $x + 7x^2 + 1$
6. $3x^2 + 6x - 6$
7. $5k^2 + 2k - 7$
8. $13a^2 - 18a + 5$
9. $3w^2 + 7wx - 6x^2$
10. $2x^2 - 2x - 12$
11. $-5 + 8x^2 + x$
12. $29x - 5 + 6x^2$
13. $8 - 14v - 4v^2$
14. $64c^2 - 16c + 1$
15. $1 + 16m^2 + 8m$
16. $5x^2 + 5x + 1$
17. $3z^2 - 75$
18. $5k + 3k^2 - 2$
19. $6 + 7x^2 - 23x$
20. $21x^2 + 22x - 8$
21. $128x^2 + 32xy + 2y^2$
22. $2y^2 - 72$
23. $x^2 - 2xy + 4y^2$
24. $11x^2 + 32xy - 3y^2$
25. $18mn^2 - 24m^2n + 8m^3$
26. $5x^2 + 7xy + 3x$
27. $2e^4 + 11e^2 - 6$
28. $5x^4 - 17x^2 + 6$
29. $36x^4 - 81x^2y^2$
30. $-24 - 6x + 45x^2$
31. $w^4 + 8w^2y^2 + 16y^4$
32. $17u - 12 + 5u^2$
33. $3(a - b)^2 + 16(a - b) + 5$
34. $4x^2 + 8xy + 4y^2 + 11x + 11y + 7$
35. $6(x + y)^2 - 7(x + y) - 3$
36. $6a^2 + 12ab + 6b^2 + 11a + 11b + 5$
37. $5a^2 + 10ab + 5b^2 - 16a - 16b + 3$
38. $4a^2 - 8ab + 4b^2 + 9a - 9b + 5$
39. $(3x + y)^2 - (a - 2b)^2$
40. $-8 + 14v + 4v^2$
41. $x^2 + 6xy + 9y^2 - 16$
42. $5z^3 + 40z^2 + 15z$
43. $y^2 - 25x^2 - 10xz - z^2$
44. $8x^3 + 24x^2 + 8x$
45. $9x^4 - 37x^2 + 4$
46. $7y^3 + 28y^2 + 14y$
47. $3w^2 + 7wx - 6x^2 - w - 3x$
48. $5(a + b) - 2 + 12(a + b)^2$
49. $5k^2 + 2km - 7m^2 + 5k + 7m$
50. $-8 + 10(x - y) + 12(x - y)^2$

Sections 5.1–5.4 REVIEW

Greatest Common Factor 5.1

The **greatest common factor (GCF)** of two or more terms of an algebraic expression is the largest expression that is a factor of all the terms. If a polynomial contains a common factor, find the greatest common factor and factor it out. Then try to factor the polynomial factor.

Factoring by Grouping 5.2

A polynomial with more than three terms may be factorable by grouping.

Factoring a Difference of Two Squares 5.3

A **difference of two squares** is always factorable:

$$a^2 - b^2 = (a + b)(a - b)$$

A *sum* of two squares is not factorable unless the exponents on the variables are multiples of 4 or of 6.

Factoring Trinomials 5.4A

See the suggestions in the box in Section 5.4A for trying to factor a trinomial that can be expressed in the form $x^2 + bx + c$.

5.4B

Use **guessing-and-checking** or the **master product method** to try to factor a trinomial that can be expressed in the form $ax^2 + bx + c$.

Factoring Perfect Square Trinomials 5.4B

A **perfect square trinomial** is always factorable:

$$a^2 + 2ab + b^2 = (a + b)^2 \quad \text{and} \quad a^2 - 2ab + b^2 = (a - b)^2$$

Check to see if any factor can be factored further.

Check the solution.

Sections 5.1–5.4 REVIEW EXERCISES Set I

Factor each polynomial completely, or write "not factorable."

1. $65x^2y^3 - 39xy^4 - 13xy$
2. $3xyz - 5a + 3b$
3. $3x^3 + 9x^2 - 12x$
4. $x^3y^6 + 5$
5. $x^2 + 13x + 40$
6. $x^2 + x - 20$
7. $8 + x^2 + 8x$
8. $x^2 - 11x + 18$
9. $x^2 - 256$
10. $x^2 - 14x + 13$
11. $3x^2 - x - 14$
12. $2x^2 - 11x - 40$
13. $3xy + 12x + 2y + 8$
14. $x^3 + 5x^2 - x - 5$
15. $5ag - 7x + 10cd - 14$
16. $x^2 - x + 12$
17. $8x^3 - 2x$
18. $x^2 + 9$
19. $4x^2 + 11x - 3$
20. $11a - 10 + 8a^2$

Name

ANSWERS

1.

2.

3.

4.

5.

6.

7.

8.

9.

10.

11.

12.

13.

14.

15.

16.

Factor each polynomial completely, or write "not factorable."

1. $14ac + 56ab - 7a$
2. $x^2 - 144$
3. $x^3 - 7x^2 + x - 4$
4. $x^2 - 14x + 48$
5. $7x^2 + 5x - 2$
6. $x^2 + 16$
7. $5x^2 + 11x - 12$
8. $6 - 15x + 6x^2$
9. $2xy + 5y + 8x + 20$
10. $4x^2 - y^2 + 2x - y$
11. $6x + 5x^2 - 8$
12. $x^3 + 8x^2 - 4x - 32$
13. $2 + 15x + 7x^2$
14. $3x^3 - x^2 + 3x - 1$
15. $5x^2 + x + 1$
16. $8x^2 + 30x - 8$

17. $3x^2y^2 + 9xy^3 - 21xy^2$

18. $x^3 - x^2 - 8x + 8$

17. ______________________

18. ______________________

19. ______________________

19. $11x^2 + 31x - 6$

20. $5ax - 10a + 3bx - 6b - x + 2$

20. ______________________

5.5 Factoring a Sum or Difference of Two Cubes

A **sum of two cubes** is a binomial that can be expressed in the form $a^3 + b^3$. For example, $8x^3 + 27y^3$ is a sum of two cubes, because it can be expressed as $(2x)^3 + (3y)^3$, which is in the *form* $a^3 + b^3$. A **difference of two cubes** is a binomial that can be expressed in the form $a^3 - b^3$.

The formulas for factoring $a^3 + b^3$ and $a^3 - b^3$ are based on two special products. Consider the products $(a + b)(a^2 - ab + b^2)$ and $(a - b)(a^2 + ab + b^2)$.

$$\begin{array}{rrrr} & a^2 & -ab & +b^2 \\ & & a & +b \\ \hline & a^2b & -ab^2 & +b^3 \\ a^3 & -a^2b & +ab^2 & \\ \hline a^3 & & & +b^3 \end{array} \qquad \begin{array}{rrrr} & a^2 & +ab & +b^2 \\ & & a & -b \\ \hline & -a^2b & -ab^2 & -b^3 \\ a^3 & +a^2b & +ab^2 & \\ \hline a^3 & & & -b^3 \end{array}$$

Since $(a + b)(a^2 - ab + b^2) = a^3 + b^3$, $a^3 + b^3$ factors as follows:

$$a^3 + b^3 = (a + b)(a^2 - ab + b^2)$$

A sum of two cubes

The same sign

Also, since $(a - b)(a^2 + ab + b^2) = a^3 - b^3$, $a^3 - b^3$ factors as follows:

$$a^3 - b^3 = (a - b)(a^2 + ab + b^2)$$

A difference of two cubes

The same sign

Notice that each product contains exactly one negative sign. When we're factoring a *difference* of two cubes, the negative sign is in the *binomial factor.* When we're factoring a *sum* of two cubes, the negative sign is the sign of the *second term* of the *trinomial factor.* The sign of the *third* term of the trinomial is positive whether we're factoring a sum of two cubes or a difference of two cubes.

A Word of Caution $a^3 + b^3 \neq (a + b)^3$

Remember that

$$\begin{aligned} (a + b)^3 &= (a + b)(a + b)(a + b) \\ &= a^3 + 3a^2b + 3ab^2 + b^3 \end{aligned}$$

whereas

$$a^3 + b^3 = (a + b)(a^2 - ab + b^2)$$

In order to use the new factoring formulas, you must be able to find the cube root of a term. The cube root of the numerical coefficient is usually found by inspection. The cube root of a *variable* can be found by dividing the exponent on the variable by 3; for example, $\sqrt[3]{x^3} = x^{3/3} = x^1$ because $(x^1)^3 = x^3$.

EXAMPLE 1 Find $\sqrt[3]{27a^6b^3}$.

SOLUTION

$$\sqrt[3]{27a^6b^3} = 3\ a^2\ b$$

$\sqrt[3]{27}$, found by inspection

$b^{3/3} = b^1$

$a^{6/3} = a^2$

The formulas are summarized as follows:

Factoring $a^3 + b^3$ or $a^3 - b^3$ (a sum or difference of two cubes)

1. Find a, the cube root of the first term, and b, the cube root of the second term. (If necessary, express numerical coefficients in prime-factored, exponential form in order to find the cube roots.)
2. Substitute into one of the following formulas:

$$a^3 + b^3 = (a + b)(a^2 - ab + b^2)$$
$$a^3 - b^3 = (a - b)(a^2 + ab + b^2)$$

A Word of Caution A common error is to think that the middle term of the trinomial factor is $2ab$ instead of ab.

$$a^3 + b^3 = (a + b)(a^2 - ab + b^2)$$

NOT $2ab$

Note The expression $a^2 - ab + b^2$ does *not* factor further. You might verify this by attempting to find two binomials whose product is $a^2 - ab + b^2$; you might also verify that $a^2 + ab + b^2$ does not factor further.

EXAMPLE 2 Factor $x^3 + 8$.

SOLUTION

$\sqrt[3]{x^3} = x^{3/3} = x^1$

$\sqrt[3]{8} = 2$

$$x^3 + 8 = (\ x\ )^3 + (\ 2\ )^3$$

Always the same sign

Always +

$$= (\quad + \quad)(\quad - \quad + \quad) \qquad \text{The blank outline}$$

Always opposite signs

$$= (x + 2)[(x)^2 - (x)(2) + (2)^2]$$
$$= (x + 2)(x^2 - 2x + 4)$$

√ **Check** $(x + 2)(x^2 - 2x + 4) = x^3 - 2x^2 + 4x + 2x^2 - 4x + 8 = x^3 + 8$

Therefore, $x^3 + 8$ factors into $(x + 2)(x^2 - 2x + 4)$.

EXAMPLE 3 Factor $27x^3 - 64y^3$.

SOLUTION

$$27x^3 - 64y^3 = (3x)^3 - (4y)^3$$

Always the same sign — Always +

$$= (\quad - \quad)(\quad + \quad + \quad)$$ The blank outline

Always opposite signs

$$= (3x - 4y)[(3x)^2 + (3x)(4y) + (4y)^2]$$

$$= (3x - 4y)(9x^2 + 12xy + 16y^2)$$ ← Factored form

(Checking will verify that the factoring was correct.)

EXAMPLE 4 Factor $(x + 2)^3 - (y - 2z)^3$.

SOLUTION Let $(x + 2) = a$ and $(y - 2z) = b$. Then

$$(x + 2)^3 - (y - 2z)^3 = a^3 - b^3 = (a - b)(a^2 + ab + b^2)$$

Since $a = x + 2$ and $b = y - 2z$, substituting back gives

$$(x + 2)^3 - (y - 2z)^3$$
$$= [(x + 2) - (y - 2z)][(x + 2)^2 + (x + 2)(y - 2z) + (y - 2z)^2]$$
$$= [x + 2 - y + 2z][x^2 + 4x + 4 + xy + 2y - 2xz - 4z + y^2 - 4yz + 4z^2]$$

↑ Factored form

(Checking will verify that the factoring was correct.)

EXAMPLE 5 Factor $64v^6 + w^6$.

SOLUTION

$$64v^6 + w^6$$ ← If we view this as a sum of two *squares*—that is, as $(8v^3)^2 + (w^3)^2$—we cannot factor it at all

$$= (4v^2)^3 + (w^2)^3$$ It *is* factorable, however, as we see if we view it as a sum of two *cubes*—that is, as $(4v^2)^3 + (w^2)^3$

$$= (4v^2 + w^2)([4v^2]^2 - [4v^2][w^2] + [w^2]^2)$$
$$= (4v^2 + w^2)(16v^4 - 4v^2w^2 + w^4)$$

Note Let's consider factoring $x^6 - y^6$. If we view $x^6 - y^6$ as a difference of two *cubes*, we have

$$x^6 - y^6 = (x^2)^3 - (y^2)^3 = (x^2 - y^2)(x^4 + x^2y^2 + y^4)$$

This, however, has not been factored completely, because $x^4 + x^2y^2 + y^4$ factors into $(x^2 - xy + y^2)(x^2 + xy + y^2)$. [You might verify that $(x^2 - xy + y^2)(x^2 + xy + y^2) = (x^4 + x^2y^2 + y^4)$.] We will see in the next section how to factor expressions such as $x^4 + x^2y^2 + y^4$.

If we view $x^6 - y^6$ as a difference of two *squares*, we have

$$x^6 - y^6 = (x^3)^2 - (y^3)^2 = (x^3 + y^3)(x^3 - y^3)$$ One factor is a sum of two cubes, and the other is a difference of two cubes

$$= (x + y)(x^2 - xy + y^2)(x - y)(x^2 + xy + y^2)$$

which *has* been completely factored.

Therefore, whenever there is a choice between treating an expression as a difference of two squares and treating it as a difference of two cubes, we recommend treating it as a difference of two squares. (Then each resulting *factor* must be factored further.)

Exercises 5.5 Set I

Factor each polynomial completely, or write "not factorable."

1. $x^3 - 8$
2. $x^3 - 27$
3. $64 + a^3$
4. $8 + b^3$
5. $125 - x^3$
6. $1 - a^3$
7. $8x^3 - 2x$
8. $27x^3 - 3x$
9. $c^3 - 27a^3b^3$
10. $c^3 - 64a^3b^3$
11. $8x^3y^6 + 27$
12. $64a^6b^3 + 125$
13. $125x^6y^4 - 1$
14. $64a^6b^3 - 9$
15. $a^4 + ab^3$
16. $x^3y + y^4$
17. $81 - 3x^3$
18. $40 - 5b^3$
19. $(x + y)^3 + 1$
20. $1 + (x - y)^3$
21. $64x^3 - y^6$
22. $125w^3 - v^6$
23. $4a^3b^3 + 108c^6$
24. $5x^3y^6 + 40z^9$
25. $x^6 - 729$
26. $y^6 - 64$
27. $(x + 1)^3 - (y - z)^3$
28. $(x - y)^3 - (a + b)^3$

Writing Problems

Express the answers in your own words and in complete sentences.

1. Explain *why* $a^3 + b^3$ factors into $(a + b)(a^2 - ab + b^2)$.
2. Describe what your first step would be in factoring $5 + 40x^3$, and explain why you would do that first.

Exercises 5.5 Set II

Factor each polynomial completely, or write "not factorable."

1. $a^3 - 125$
2. $b^3 + 1$
3. $27 + c^3$
4. $16 + 2x^3$
5. $8 - b^3$
6. $x^3y^3 + 64$
7. $64a^3 - 4a$
8. $8a^3 + 27$
9. $x^3 - 64y^3z^3$
10. $x^3 - xy^2$
11. $8a^6b^3 + 1$
12. $250 - 2c^3$
13. $8x^9y^5 - 27$
14. $x^3 + 4x$
15. $x^4y^3 + x$
16. $a^4 + 8a$
17. $16 - 2m^3$
18. $3x^3 - 81$
19. $(a + b)^3 - (x - y)^3$
20. $(x + y)^3 + (u + v)^3$
21. $27a^3 - b^6$
22. $(s - 2)^3 + (t - 3v)^3$
23. $2x^3y^3 + 16z^6$
24. $x^6 + 64y^6$
25. $z^6 - 1$
26. $6x^4 + 48x$
27. $(a - 2)^3 - (b + c)^3$
28. $(a + x)^3 - 8(b - z)^3$

5.6 Factoring by Completing the Square of a Polynomial (Optional)

Factoring Trinomials of the Form $x^2 + bx + c$ by Completing the Square

Trinomials of the form $x^2 + bx + c$, where c is a large number, can conveniently be factored by using a calculator and the technique of completing the square.

Before we try to complete the square of a polynomial, let's examine some factoring problems:

$$x^2 + 6x + 9 = (x + 3)^2$$

$$x^2 - 10x + 25 = (x - 5)^2$$

$$x^2 - 14x + 49 = (x - 7)^2$$

What if the constants were missing? That is, what if we had

$$x^2 + 6x + ? = (x + ?)^2$$

$$x^2 - 10x + ? = (x - ?)^2 \quad \text{See Example 1}$$

$$x^2 - 14x + ? = (x - ?)^2 \quad \text{See Example 1}$$

and we needed to determine the numbers that should replace the question marks? What number needs to be added to $x^2 + 6x + \square$ in order to make the expression the square of some binomial? A little experimenting and some thought will show that if we find $\frac{1}{2}$ of 6 (that is, half of the coefficient of x) and square that number, we'll have

$$x^2 + 6x + \boxed{9} = (x + \boxed{3})^2 \quad \tfrac{1}{2}(6) = 3\text{, and } 3^2 = 9$$

When we find the constant to add to $x^2 + 6x$ (for example) to make the expression a *perfect square trinomial* (that is, the *square* of a binomial), we say that we are **completing the square of the polynomial**.

EXAMPLE 1 Examples of finding the numbers for completing the square:

a. $x^2 - 10x + ? = (x - ?)^2$

$$x^2 - 10x + 25 = (x - 5)^2 \quad \text{Half of } -10 \text{ is } -5\text{, and } [-5]^2 = 25$$

b. $x^2 - 14x + ? = (x - ?)^2$

$$x^2 - 14x + 49 = (x - 7)^2 \quad \text{Half of } -14 \text{ is } -7\text{, and } [-7]^2 = 49$$

In Example 2, we demonstrate the method of factoring a trinomial of the form $x^2 + bx + c$ by completing the square. Keep in mind that whatever we add to the polynomial (to complete the square) *we must also subtract from the polynomial* so that we don't change the *value* of the original expression.

EXAMPLE 2 Factor $x^2 + 4x - 780$ by completing the square, or write "not factorable."

SOLUTION We rewrite the trinomial to be factored, leaving spaces between the last two terms, and then we determine what number would have to be added to the first two terms to "complete the square."

$x^2 + 4x \qquad - 780$	Half of 4 is 2, and $2^2 = 4$
$= x^2 + 4x + 4 - 4 - 780$	Adding 4 *and also subtracting* 4 leaves the value unchanged
$= (x^2 + 4x + 4) - 784$	Using a calculator, we find that $\sqrt{784} = 28$
$= (x + 2)^2 - 28^2$	This is now a difference of two squares
$= [(x + 2) + 28][(x + 2) - 28]$	
$= (x + 30)(x - 26)$	

Check $(x + 30)(x - 26) = x^2 + 4x - 780$

Therefore, $x^2 + 4x - 780$ factors into $(x + 30)(x - 26)$.

Note We could factor $x^2 + 4x - 780$ by guessing-and-checking, but it might take some time to find two numbers whose product is 780 and whose difference is 4.

In Example 2, we were able to factor $(x + 2)^2 - 28^2$ because it is a difference of two squares. If we do *not* have a difference of two squares after we have completed the square, then the given polynomial is not factorable (see Examples 3 and 4).

EXAMPLE 3 Factor $x^2 - 24x - 480$ by completing the square, or write "not factorable."

SOLUTION

$x^2 - 24x \qquad - 480$	Half of -24 is -12, and $(-12)^2 = 144$
$= x^2 - 24x + 144 - 144 - 480$	Adding 144 *and also subtracting* 144 leaves the value unchanged
$= (x^2 - 24x + 144) - 624$	
$= (x - 12)^2 - 624$	624 is not a perfect square

Since $(x - 12)^2 - 624$ is *not* a difference of two squares, $x^2 - 24x - 480$ is *not factorable.*

EXAMPLE 4 Factor $x^2 + 10x + 386$ by completing the square, or write "not factorable."

SOLUTION

$x^2 + 10x \qquad + 386$	Half of 10 is 5, and $5^2 = 25$
$= x^2 + 10x + 25 - 25 + 386$	Adding 25 *and also subtracting* 25 leaves the value unchanged
$= (x^2 + 10x + 25) + 361$	Using a calculator, we find that $\sqrt{361} = 19$
$= (x + 5)^2 + 19^2$	This is a *sum* of two squares, not a difference of two squares

Since $(x + 5)^2 + 19^2$ is *not* a difference of two squares, $x^2 + 10x + 386$ is *not factorable.*

Factoring Fourth-Degree Trinomials and Binomials by Completing the Square

Some fourth-degree trinomials and binomials *may* be factored into a product of two *trinomials* by completing the square, although the procedure is a little different from the one demonstrated in Examples 2–4. A fourth-degree binomial or trinomial cannot be factored by using this new procedure unless it satisfies the following conditions:

1. The first and last terms of the expression to be factored must be perfect squares.
2. The degrees of any variable factors in the first and last terms must be multiples of *four.*

If these conditions are satisfied, we then try to find a *positive perfect square term* to add to the given binomial or trinomial to make it a perfect square trinomial. Of course, *when a term is added, the same term must also be subtracted* so that the value of the original expression is unchanged. We will then have an expression that is a difference of two squares, and a difference of two squares is *always* factorable. If you need to, you can use steps 1–5 below to help you form this difference of two squares.

A complete procedure for factoring a polynomial that is of the form $ax^4 + bx^2y^2 + cy^4$, where a and c are perfect squares, follows:

Factoring a polynomial of the form $ax^4 + bx^2y^2 + cy^4$

Note: a and c must be squares of natural numbers.

1. Form the binomials $\sqrt{ax^4} + \sqrt{cy^4}$ and $\sqrt{ax^4} - \sqrt{cy^4}$.
2. Square one of the binomials found in step 1.
3. Subtract the polynomial being factored *from* the trinomial found in step 2.
4. If the difference found in step 3 is a positive term that is a perfect square, *add it to* and *subtract it from* the polynomial being factored, and continue with step 5. If it is not, repeat steps 2 and 3, using the other binomial from step 1. If the difference found in step 3 is still not a positive term that is a perfect square, the polynomial cannot be factored over the integers.
5. Regroup the terms to form a difference of two squares.
6. Factor the difference of two squares.
7. Determine whether any factor can be factored further.
8. Check the result.

Note The polynomials found in steps 1, 2, and 3 do *not* equal the polynomial that is being factored, and steps 1, 2, and 3 need not be shown.

EXAMPLE 5 Factor $x^4 + x^2 + 1$.

SOLUTION

Step 1. The binomials are $\sqrt{x^4} + \sqrt{1} = x^2 + 1$ and $\sqrt{x^4} - \sqrt{1} = x^2 - 1$.
Step 2. We'll try $x^2 + 1$:

$$(x^2 + 1)^2 = x^4 + 2x^2 + 1$$

The difference is positive and is a perfect square

Step 3. $(x^4 + 2x^2 + 1) - (x^4 + x^2 + 1) = x^2$

$x^2 - x^2 = 0$

Step 4. $x^4 + x^2 + 1 = x^4 + x^2 + 1 + x^2 - x^2$

Step 5. $= (x^4 + x^2 + 1 + x^2) - x^2$

$= (x^4 + 2x^2 + 1) - x^2$

$= (x^2 + 1)^2 - x^2$ A difference of two squares

Step 6. $= [(x^2 + 1) + x][(x^2 + 1) - x]$

Step 7. $= (x^2 + x + 1)(x^2 - x + 1)$ Both factors are prime

√ **Step 8.** ***Check*** $(x^2 + x + 1)(x^2 - x + 1)$

$= x^2(x^2 - x + 1) + x(x^2 - x + 1) + 1(x^2 - x + 1)$

$= x^4 - x^3 + x^2 + x^3 - x^2 + x + x^2 - x + 1$

$= x^4 + x^2 + 1$

Therefore, $x^4 + x^2 + 1$ factors into $(x^2 + x + 1)(x^2 - x + 1)$.

EXAMPLE 6 Factor $h^4 - 14h^2k^2 + 25k^4$.

SOLUTION

Step 1. The binomials are $\sqrt{h^4} + \sqrt{25k^4} = h^2 + 5k^2$ and $\sqrt{h^4} - \sqrt{25k^4} = h^2 - 5k^2$.
Step 2. We'll try $h^2 + 5k^2$:

$$(h^2 + 5k^2)^2 = h^4 + 10h^2k^2 + 25k^4$$

Step 3. $(h^4 + 10h^2k^2 + 25k^4) - (h^4 - 14h^2k^2 + 25k^4) = 24h^2k^2$

$24h^2k^2$ is not a perfect square. We will repeat steps 2 and 3, using the binomial $h^2 - 5k^2$.

Step 2. $(h^2 - 5k^2)^2 = h^4 - 10h^2k^2 + 25k^4$

Step 3. $(h^4 - 10h^2k^2 + 25k^4) - (h^4 - 14h^2k^2 + 25k^4) = 4h^2k^2$

Step 4. $4h^2k^2$ is positive and is a perfect square. $4h^2k^2 - 4h^2k^2 = 0$

$h^4 - 14h^2k^2 + 25k^4 = h^4 - 14h^2k^2 + 25k^4 + 4h^2k^2 - 4h^2k^2$

Step 5. $= (h^4 - 14h^2k^2 + 25k^4 + 4h^2k^2) - 4h^2k^2$

$= (h^4 - 10h^2k^2 + 25k^4) - 4h^2k^2$

$= (h^2 - 5k^2)^2 - (2hk)^2$

Step 6. $= [(h^2 - 5k^2) + 2hk][(h^2 - 5k^2) - 2hk]$

Step 7. $= (h^2 + 2hk - 5k^2)(h^2 - 2hk - 5k^2)$

Both factors are prime.

Step 8. Checking, which is not shown, will confirm that $h^4 - 14h^2k^2 + 25k^4$ factors into $(h^2 + 2hk - 5k^2)(h^2 - 2hk - 5k^2)$.

EXAMPLE 7 Factor $a^4 + 4$.

SOLUTION

Step 1. The binomials are $\sqrt{a^4} + \sqrt{4} = a^2 + 2$ and $\sqrt{a^4} - \sqrt{4} = a^2 - 2$.
Step 2. We'll try $a^2 + 2$:

$$(a^2 + 2)^2 = a^4 + 4a^2 + 4$$

Step 3. $(a^4 + 4a^2 + 4) - (a^4 + 4) = 4a^2$ ← A positive, perfect square

$4a^2 - 4a^2 = 0$

Step 4. $a^4 + 4 = a^4 + 4 + 4a^2 - 4a^2$

Step 5. $= (a^4 + 4a^2 + 4) - 4a^2$

$= (a^2 + 2)^2 - (2a)^2$

Step 6. $= [(a^2 + 2) + 2a][(a^2 + 2) - 2a]$

Step 7. $= (a^2 + 2a + 2)(a^2 - 2a + 2)$

Both factors are prime.

Step 8. Checking, which is not shown, will verify that $a^4 + 4$ factors into $(a^2 + 2a + 2)(a^2 - 2a + 2)$.

EXAMPLE 8 Factor $h^4 - 6h^2k^2 + 25k^4$.

SOLUTION

Step 1. The binomials are $\sqrt{h^4} + \sqrt{25k^4} = h^2 + 5k^2$ and $\sqrt{h^4} - \sqrt{25k^4} = h^2 - 5k^2$.
Step 2. We'll try $h^2 - 5k^2$:

$$(h^2 - 5k^2)^2 = h^4 - 10h^2k^2 + 25k^4$$

Step 3. $(h^4 - 10h^2k^2 + 25k^4) - (h^4 - 6h^2k^2 + 25k^4) = -4h^2k^2$

The difference is negative, not positive. We repeat steps 2 and 3, trying the binomial $h^2 + 5k^2$.

Step 2. $(h^2 + 5k^2)^2 = h^4 + 10h^2k^2 + 25k^4$

Step 3. $(h^4 + 10h^2k^2 + 25k^4) - (h^4 - 6h^2k^2 + 25k^4) = 16h^2k^2$

$16h^2k^2 - 16h^2k^2 = 0$

Step 4. $h^4 - 6h^2k^2 + 25k^4 = h^4 - 6h^2k^2 + 25k^4 + 16h^2k^2 - 16h^2k^2$

Step 5. $= (h^4 - 6h^2k^2 + 25k^4 + 16h^2k^2) - 16h^2k^2$

$= (h^4 + 10h^2k^2 + 25k^4) - 16h^2k^2$

$= (h^2 + 5k^2)^2 - (4hk)^2$

Step 6. $= [(h^2 + 5k^2) + 4hk][(h^2 + 5k^2) - 4hk]$

Step 7. $= (h^2 + 4hk + 5k^2)(h^2 - 4hk + 5k^2)$

Both factors are prime.

Step 8. Checking, which is not shown, will confirm that $h^4 - 6h^2k^2 + 25k^4$ factors into $(h^2 + 4hk + 5k^2)(h^2 - 4hk + 5k^2)$.

EXAMPLE 9 Factor $x^4 - 2x^2 + 9$.

SOLUTION

Step 1. The binomials are $\sqrt{x^4} + \sqrt{9} = x^2 + 3$ and $\sqrt{x^4} - \sqrt{9} = x^2 - 3$.
Step 2. We'll try $x^2 + 3$:

$$(x^2 + 3)^2 = x^4 + 6x^2 + 9$$

Step 3. $(x^4 + 6x^2 + 9) - (x^4 - 2x^2 + 9) = 8x^2$

$8x^2$ is not a perfect square. We repeat steps 2 and 3, using the binomial $x^2 - 3$.

Step 2. $(x^2 - 3)^2 = x^4 - 6x^2 + 9$

Step 3. $(x^4 - 6x^2 + 9) - (x^4 - 2x^2 + 9) = -4x^2$

The difference is negative, not positive. We have tried both binomials from step 1, and neither one gave us a difference that was a positive, perfect square. Therefore, $x^4 - 2x^2 + 9$ is *not factorable* over the integers.

Exercises 5.6 Set I

In Exercises 1–4, find the two missing numbers in each equation.

1. $x^2 + 12x + ? = (x + ?)^2$
2. $x^2 + 18x + ? = (x + ?)^2$
3. $x^2 + 8x + ? = (x + ?)^2$
4. $x^2 + 22x + ? = (x + ?)^2$

In Exercises 5–34, factor each polynomial completely, or write "not factorable."

5. $x^2 - 14x - 480$
6. $x^2 - 12x - 640$
7. $x^2 + 6x - 1{,}015$
8. $x^2 + 8x - 1{,}073$
9. $x^2 - 2x - 532$
10. $x^2 - 4x - 628$
11. $x^2 + 16x + 793$
12. $x^2 + 6x + 685$
13. $x^4 + 3x^2 + 4$
14. $x^4 + 5x^2 + 9$
15. $4m^4 + 3m^2 + 1$
16. $9u^4 + 5u^2 + 1$
17. $64a^4 + b^4$
18. $a^4 + 4b^4$
19. $x^4 - 3x^2 + 9$
20. $x^4 - x^2 + 16$
21. $a^4 - 17a^2b^2 + 16b^4$
22. $a^4 - 37a^2 + 36$
23. $x^2 + x + 1$
24. $x^2 + 5x + 9$
25. $a^4 - 3a^2b^2 + 9b^4$
26. $a^4 - 15a^2b^2 + 25b^4$
27. $a^4 + 9$
28. $x^4 + 25$
29. $50x^4 - 12x^2y^2 + 2y^4$
30. $32x^4 - 2x^2y^2 + 2y^4$
31. $8m^4n + 2n^5$
32. $3m^5 + 12mn^4$
33. $50x^4y + 32x^2y^3 + 8y^5$
34. $48x^5 + 21x^3y^2 + 3xy^4$

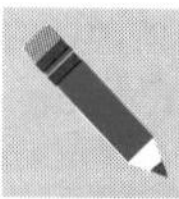

Writing Problems

Express the answers in your own words and in complete sentences.

1. Explain why factoring by completing the square won't work for factoring $x^2 + x + 1$.
2. Explain why factoring by completing the square won't work for factoring $x^4 + x^2 + 5$.

Exercises 5.6 Set II

In Exercises 1–4, find the two missing numbers in each equation.

1. $x^2 + 18x + ? = (x + ?)^2$
2. $x^2 + 5x + ? = (x + ?)^2$
3. $x^2 + 20x + ? = (x + ?)^2$
4. $x^2 - 30x + ? = (x - ?)^2$

In Exercises 5–34, factor each polynomial completely, or write "not factorable."

5. $x^2 - 10x - 551$
6. $x^2 - 8x - 768$
7. $x^2 + 12x - 1{,}120$
8. $x^2 + 14x - 912$
9. $x^2 - 12x - 790$
10. $x^2 - 8x - 1{,}029$
11. $x^2 + 4x + 845$
12. $x^2 + 10x + 986$
13. $u^4 + 4u^2 + 16$
14. $x^4 + 7x^2 + 16$
15. $y^4 - 9y^2 + 16$
16. $16m^4 + 8m^2 + 1$
17. $4a^4 + 1$
18. $x^4 + 10x^2 + 49$
19. $x^4 - 21x^2 + 4$
20. $4x^4 - 76x^2 + 9$
21. $x^4 - 10x^2 + 9$
22. $x^4 - 3x^2 + 1$
23. $x^2 + 3x + 4$
24. $2x^4 - 38x^2 + 18$
25. $z^4 + z^2w^2 + 25w^4$
26. $x^4 - 14x^2 + 25$
27. $y^4 + 1$
28. $9x^4 - 7x^2y^2 + y^4$
29. $2x^8y - 6x^4y + 18y$
30. $x^2 + 7x + 16$
31. $2a^5 + 128a$
32. $x^4 - 17x^2y^2 + 64y^4$
33. $27x^4 + 6x^2y^2 + 3y^4$
34. $8x^4 - 80x^2 + 2$

5.7 Factoring by Using Synthetic Division (Optional)

Subscripts In mathematics, we sometimes use **subscripts**—small numbers written to the *right* of a letter and *below* the normal line of type—to distinguish between constants. For example, r_1, read r sub-one, and r_2, read r sub-two, represent two *different* constants.

Consider the factored polynomial $(x - r_1)(x - r_2)(x - r_3)$. If these factors are multiplied together, the *constant* term in their product must be $r_1r_2r_3$, the product of the constant terms in each factor. For example,

$$(x - 5)(x + 2)(x - 3) = x^3 - 6x^2 - x + 30$$

$$(-5)(+2)(-3) = 30$$

If a polynomial *does* have a factor of the form $x - r$ or $x + r$, then r must be a factor of the constant term of the polynomial. We know that $x - 7$, for example, cannot be a factor of $x^3 - 6x^2 - x + 30$, because 7 is not a factor of 30.

Recall that if the remainder in a division problem is zero, both the *divisor* and the *quotient* are *factors* of the dividend. Factoring by using synthetic division makes use of this fact. The procedure is summarized below. In this procedure, it is assumed that any common factors have been factored out and that the polynomial has been arranged in descending powers of the variable.

Factoring a polynomial with a leading coefficient of 1 by using synthetic division

1. Find all the integral factors, positive and negative, of the constant term of the polynomial.
2. Begin dividing the polynomial synthetically by each of the factors found in step 1. Stop when a remainder of zero is obtained.
3. If a remainder of zero is obtained for r, then both $(x - r)$ *and* the quotient are factors of the polynomial.
4. Apply *any* method of factoring to see whether the quotient can be factored further. If synthetic division is used again, *it is important to try again any value of r that gave a remainder of zero in step 3.*
5. Check the solution.

EXAMPLE 1 Factor $x^3 - x^2 + 2x - 8$.

SOLUTION

Step 1. The integral factors of the constant term 8 are $\pm 1, \pm 2, \pm 4, \pm 8$.
Step 2. We will divide synthetically by ± 1, ± 2, ± 4, and ± 8 until a remainder of zero is obtained.

$$1x^3 - 1x^2 + 2x - 8$$

Dividing by 1:

$$\begin{array}{r|rrrr} 1 & 1 & -1 & 2 & -8 \\ & & 1 & 0 & 2 \\ \hline & 1 & 0 & 2 & \boxed{-6} \end{array}$$

The remainder is *not* zero; therefore, $(x - 1)$ is not a factor

Dividing by −1:

$$\begin{array}{r|rrrr} -1 & 1 & -1 & 2 & -8 \\ & & -1 & 2 & -4 \\ \hline & 1 & -2 & 4 & \boxed{-12} \end{array}$$

The remainder is *not* zero; therefore, $(x + 1)$ is not a factor

Dividing by 2:

$$\begin{array}{r|rrrr} 2 & 1 & -1 & 2 & -8 \\ & & 2 & 2 & 8 \\ \hline & 1 & 1 & 4 & \boxed{0} \end{array}$$

The remainder *is* zero; therefore, $(x - 2)$ is a factor

$$1x^2 + 1x + 4$$ ← The quotient is another factor

Step 3. $x^3 - x^2 + 2x - 8 = (x - 2)(x^2 + x + 4)$
Step 4. Applying methods of factoring trinomials, we see that the quotient $x^2 + x + 4$ cannot be factored further, so it is an *irreducible* quadratic factor.

√ **Step 5.** ***Check***

$$\begin{aligned} &(x - 2)(x^2 + x + 4) \\ &= x^3 + x^2 + 4x - 2x^2 - 2x - 8 \\ &= x^3 - x^2 + 2x - 8 \end{aligned}$$

Therefore, $x^3 - x^2 + 2x - 8$ factors into $(x - 2)(x^2 + x + 4)$.

Note Remember: If we divide synthetically by $x - 2$, we change the sign to $+2$. Here, we're working backwards. Therefore, because the remainder was 0 when we divided by $+2$, the corresponding *factor* is $x - 2$.

EXAMPLE 2 Factor $x^5 - 3x^4 + 2x^3 - x^2 + 3x - 2$.

SOLUTION

Step 1. The integral factors of the constant term 2 are $\pm 1, \pm 2$.
Step 2. We will divide synthetically by ± 1 and ± 2 until a remainder of zero is obtained.

$$\begin{array}{r|rrrrrr} 1 & 1 & -3 & 2 & -1 & 3 & -2 \\ & & 1 & -2 & 0 & -1 & 2 \\ \hline & 1 & -2 & 0 & -1 & 2 & \boxed{0} \end{array}$$

The remainder *is* zero; therefore, $(x - 1)$ is a factor

$$1x^4 - 2x^3 - 1x + 2$$ ← The quotient is another factor

Step 3. $x^5 - 3x^4 + 2x^3 - x^2 + 3x - 2 = (x - 1)(x^4 - 2x^3 - x + 2)$
Step 4. To see whether the quotient $x^4 - 2x^3 - x + 2$ can be factored further, we'll use synthetic division again, even though this particular quotient can be factored by grouping.

Step 1. Factors of the constant term 2 are ±1 and ±2. *It is important to try dividing by 1 again.*

Step 2.

$$\begin{array}{r|rrrrr} 1 & 1 & -2 & 0 & -1 & 2 \\ & & 1 & -1 & -1 & -2 \\ \hline & 1 & -1 & -1 & -2 & 0 \end{array}$$

0 ← The remainder *is* zero; therefore, $(x - 1)$ is a factor

$1x^3 - 1x^2 - 1x - 2$ ← The quotient is another factor

Step 3. $x^5 - 3x^4 + 2x^3 - x^2 + 3x - 2 = (x - 1)(x^4 - 2x^3 - x + 2)$
$= (x - 1)(x - 1)(x^3 - x^2 - x - 2)$

Step 4. To see whether the quotient $x^3 - x^2 - x - 2$ can be factored further, we'll use synthetic division *again*.

Step 1. Factors of the constant term -2 are ±1 and ±2.

Step 2.

$$\begin{array}{r|rrrr} 1 & 1 & -1 & -1 & -2 \\ & & 1 & 0 & -1 \\ \hline & 1 & 0 & -1 & -3 \end{array}$$

-3 ← The remainder is *not* zero

$(x - 1)$ *is not* a factor of $x^3 - x^2 - x - 2$.
We'll try $+2$:

$$\begin{array}{r|rrrr} 2 & 1 & -1 & -1 & -2 \\ & & 2 & 2 & 2 \\ \hline & 1 & 1 & 1 & 0 \end{array}$$

0 ← The remainder *is* zero; therefore, $(x - 2)$ is a factor

The *quotient* is $1x^2 + 1x + 1$. You can verify that it will not factor further.

Therefore, $x^5 - 3x^4 + 2x^3 - x^2 + 3x - 2$
$= (x - 1)(x^4 - 2x^3 - x + 2)$
$= (x - 1)(x - 1)(x^3 - x^2 - x - 2)$
$= (x - 1)^2(x - 2)(x^2 + x + 1)$ ← Factored form

√ **Step 5.** ***Check*** $(x - 1)^2(x - 2)(x^2 + x + 1)$
$= (x^2 - 2x + 1)(x - 2)(x^2 + x + 1)$
$= (x^3 - 4x^2 + 5x - 2)(x^2 + x + 1)$
$= x^5 - 3x^4 + 2x^3 - x^2 + 3x - 2$

Therefore, $x^5 - 3x^4 + 2x^3 - x^2 + 3x - 2$ factors into $(x - 1)^2(x - 2)(x^2 + x + 1)$.

EXAMPLE 3 Factor $x^5 + 1$.

SOLUTION The integral factors of the constant term $+1$ are ±1.

$1x^5 + 0x^4 + 0x^3 + 0x^2 + 0x + 1$

$$\begin{array}{r|rrrrrr} 1 & 1 & 0 & 0 & 0 & 0 & 1 \\ & & 1 & 1 & 1 & 1 & 1 \\ \hline & 1 & 1 & 1 & 1 & 1 & 2 \end{array}$$

2 ← The remainder is *not* zero; therefore, $(x - 1)$ is not a factor

$$\begin{array}{r|rrrrr|r} -1 & 1 & 0 & 0 & 0 & 0 & 1 \\ & & -1 & 1 & -1 & 1 & -1 \\ \hline & 1 & -1 & 1 & -1 & 1 & \boxed{0} \end{array}$$

← The remainder *is* zero; therefore, $(x + 1)$ is a factor

$1x^4 - 1x^3 + 1x^2 - 1x + 1$ ← The quotient is another factor

To see whether the quotient can be factored further, we'll use synthetic division again.

$$1x^4 - 1x^3 + 1x^2 - 1x + 1$$

± 1 are the only factors of the constant term 1; since we already know that $(x - 1)$ is not a factor, the only factor we need to try is $(x + 1)$

$$\begin{array}{r|rrrr|r} -1 & 1 & -1 & 1 & -1 & 1 \\ & & -1 & 2 & -3 & 4 \\ \hline & 1 & -2 & 3 & -4 & \boxed{5} \end{array}$$

← The remainder is *not* zero

Therefore, $x^4 - x^3 + x^2 - x + 1$ is a prime polynomial, and

$$x^5 + 1 = (x + 1)(x^4 - x^3 + x^2 - x + 1)$$

√ **Check** $(x + 1)(x^4 - x^3 + x^2 - x + 1)$

$$= x^5 - \cancel{x^4} + \cancel{x^3} - \cancel{x^2} + \cancel{x} + \cancel{x^4} - \cancel{x^3} + \cancel{x^2} - \cancel{x} + 1 = x^5 + 1$$

Therefore, $x^5 + 1$ factors into $(x + 1)(x^4 - x^3 + x^2 - x + 1)$.

It is shown in higher-level mathematics courses that if the leading coefficient of a polynomial is *not* 1, then the polynomial may have factors of the form $\left(x - \dfrac{p}{q}\right)$, where p is a factor of the *constant term* of the polynomial and q is a factor of the *leading coefficient* of the polynomial. For example, to factor $2x^4 + x^3 - x + 1$ by using synthetic division, we might need to try $\pm\frac{1}{2}$, using factors of the constant term, 1, in the numerator and factors of the leading coefficient, 2, in the denominator (see Example 4). In practice, we usually try ± 1 (factors of the constant term 1) *before* trying any fractions.

EXAMPLE 4

Factor $2x^4 + x^3 - x + 1$.

SOLUTION The integral factors of 1 are ± 1. Because $2x^4 + x^3 - x + 1 = 2(x^4 + \frac{1}{2}x^3 - \frac{1}{2}x + \frac{1}{2})$, we may also need to try $\pm\frac{1}{2}$, but we will try the integers first.

$$2x^4 + x^3 - x + 1 = 2x^4 + x^3 + 0x^2 - x + 1$$

$$\begin{array}{r|rrrr|r} 1 & 2 & 1 & 0 & -1 & 1 \\ & & 2 & 3 & 3 & 2 \\ \hline & 2 & 3 & 3 & 2 & \boxed{3} \end{array}$$

← The remainder is not zero

$$\begin{array}{r|rrrr|r} -1 & 2 & 1 & 0 & -1 & 1 \\ & & -2 & 1 & -1 & 2 \\ \hline & 2 & -1 & 1 & -2 & \boxed{3} \end{array}$$

← The remainder is not zero

$$\begin{array}{r|rrrr|r} \frac{1}{2} & 2 & 1 & 0 & -1 & 1 \\ & & 1 & 1 & \frac{1}{2} & -\frac{1}{4} \\ \hline & 2 & 2 & 1 & -\frac{1}{2} & \boxed{\frac{3}{4}} \end{array}$$

← The remainder is not zero

$$\begin{array}{r|rrrr|r} -\frac{1}{2} & 2 & 1 & 0 & -1 & 1 \\ & & -1 & 0 & 0 & \frac{1}{2} \\ \hline & 2 & 0 & 0 & -1 & \boxed{\frac{3}{2}} \end{array}$$

← The remainder is not zero

The polynomial is not factorable; it is prime, or irreducible.

EXAMPLE 5 Factor $4x^3 - 12x^2 - x + 3$ by using synthetic division.

SOLUTION (Actually, $4x^3 - 12x^2 - x + 3$ is easily factored by grouping.) To factor $4x^3 - 12x^2 - x + 3$ by using synthetic division, we may need to try ± 1 and ± 3 (these are all the integral factors of the constant term), and we may *also* need to try $\pm\frac{1}{2}$, $\pm\frac{3}{2}$, $\pm\frac{1}{4}$, and $\pm\frac{3}{4}$. Notice that the *numerators* are all 1's and 3's (factors of the constant term, 3) and the denominators are all 2's and 4's (factors of the leading coefficient, 4). It is not necessary to write $\pm\frac{1}{1}$ and $\pm\frac{3}{1}$ as numbers to try, since they equal ± 1 and ± 3, respectively. Also, it is not necessary to write, for example, $\frac{\pm 1}{\pm 2}$; these are combined into $\pm\frac{1}{2}$.

In practice, we usually try the integers first, but for demonstration purposes, we will try $\frac{1}{2}$ first.

$$\begin{array}{r|rrrr} \frac{1}{2} & 4 & -12 & -1 & 3 \\ & & 2 & -5 & -3 \\ \hline & 4 & -10 & -6 & 0 \end{array}$$

0 ← The remainder *is* zero; therefore, $\left(x - \frac{1}{2}\right)$ is a factor

$4x^2 - 10x - 6$ ← The quotient is another factor

We now factor the quotient:

$$4x^2 - 10x - 6 = 2(2x^2 - 5x - 3) = 2(2x + 1)(x - 3)$$

If we then write that $4x^3 - 12x^2 - x + 3$ factors into $\left(x - \frac{1}{2}\right)(2)(2x + 1)(x - 3)$, we haven't factored *over the integers*; however, if we multiply the first two factors together, we have $\left(x - \frac{1}{2}\right)(2) = (2x - 1)$. Now the constants are all integers. Checking will verify that $4x^3 - 12x^2 - x + 3$ factors into $(2x - 1)(2x + 1)(x - 3)$.

Exercises 5.7
Set I

Factor each polynomial completely, or write "not factorable."

1. $x^3 + x^2 + x - 3$
2. $x^3 + x^2 - 5x - 2$
3. $x^3 - 3x^2 - 4x + 12$
4. $x^3 - 2x^2 - 5x + 6$
5. $2x^3 - 8x^2 + 2x + 12$
6. $2x^3 + 12x^2 + 22x + 12$
7. $6x^3 - 13x^2 + 4$
8. $x^3 - 7x - 6$
9. $x^4 - 3x^2 + 4x + 4$
10. $x^4 - x^3 - 5x^2 + 5x + 6$
11. $x^4 + 2x^3 - 3x^2 - 8x - 4$
12. $x^4 + 4x^3 + 3x^2 - 4x - 4$
13. $3x^4 - 4x^3 - 1$
14. $2x^4 + 3x^3 + 2x^2 + x + 1$
15. $x^4 - 4x^3 - 7x^2 + 34x - 24$
16. $x^4 + 6x^3 + 3x^2 - 26x - 24$
17. $6x^3 + x^2 - 11x - 6$
18. $6x^3 - 19x^2 + x + 6$

Writing Problems

Express the answers in your own words and in complete sentences.

1. Explain why $x + 2$ cannot be a factor of $x^5 - 4x^4 + 3x - 5$.
2. Explain why $x + \frac{2}{3}$ *might* be a factor of $3x^5 - 4x^4 + 3x - 2$.

Exercises 5.7
Set II

Factor each polynomial completely, or write "not factorable."

1. $x^3 + 2x^2 + 2x - 5$
2. $x^3 + 7x^2 + 11x + 2$
3. $x^3 + 2x^2 - 5x - 6$
4. $x^3 - 11x^2 - 25x + 3$
5. $2x^3 - 12x^2 + 22x - 12$
6. $x^3 - 2x^2 + 2x - 1$
7. $x^3 - 7x + 6$
8. $x^4 + x^3 - 3x^2 - 4x - 4$
9. $x^4 + 3x^3 + 3x^2 + x - 2$
10. $2x^3 + 8x^2 + 8x$
11. $x^4 + 6x^3 + 8x^2 - 6x - 9$
12. $2x^3 + x^2 + x + 1$
13. $2x^4 + 2x^3 + 3x^2 + x - 1$
14. $x^3 - 3x^2 - 8x - 10$
15. $x^4 + 2x^3 - 13x^2 - 38x - 24$
16. $x^4 + 3x^3 - 3x^2 - 7x + 6$
17. $6x^3 + 7x^2 - 7x - 6$
18. $x^7 + 1$

5.8 Selecting the Appropriate Method of Factoring

With so many different kinds of factoring to choose from, you might be confused as to which method of factoring to try first. The following is a procedure you can use to select the correct method of factoring a particular algebraic expression.

First, check for a common factor, no matter how many terms the expression has. If there is a common factor, find the greatest common factor and factor it out.

If the polynomial to be factored contains two terms:

1. Is it a *difference* of two *squares*? (Section 5.3)
2. Is it a *sum* of two *squares*? If so, it is *not factorable*, unless its degree is a multiple of four and it can be factored by completing the square or unless it is *also* a sum of two cubes.
3. Is it a *sum* or *difference* of two *cubes*? (Section 5.5)
4. Can it be factored by completing the square? (Section 5.6)
5. Can it be factored by using synthetic division? (Section 5.7)

If the polynomial to be factored contains three terms:

1. Can it be expressed in the form $x^2 + bx + c$? (Section 5.4A)
2. Are the first and last terms perfect squares? If so, is it a perfect square trinomial? (Section 5.4B)
3. Can it be expressed in the form $ax^2 + bx + c$? (Section 5.4B)
4. Can it be factored by completing the square? (Section 5.6)
5. Can it be factored by using synthetic division? (Section 5.7)

If the polynomial to be factored contains four or more terms:

1. Can it be factored by grouping? (Section 5.2)
2. Can it be factored by using synthetic division? (Section 5.7)

Check to see whether any factor can be factored further. When the expression is *completely factored*, the same factors are obtained no matter what method is used.

Check the result by multiplying the factors together.

EXAMPLE 1 Examples of selecting the method of factoring:

a. $6x^2y - 12xy + 4y$ — A common factor; the GCF is $2y$
$= 2y(3x^2 - 6x + 2)$ — $3x^2 - 6x + 2$ is not factorable

b. $3x^3 - 27xy^2$ — A common factor; the GCF is $3x$
$= 3x(x^2 - 9y^2)$ — $x^2 - 9y^2$ is a difference of two squares
$= 3x(x + 3y)(x - 3y)$

c. $2ac - 3ad + 10bc - 15bd$ — Factor by grouping
$= a(2c - 3d) + 5b(2c - 3d)$
$= (2c - 3d)(a + 5b)$

d. $2a^3b + 16b$ — A common factor; the GCF is $2b$
$= 2b(a^3 + 8)$ — $a^3 + 8$ is a sum of two cubes
$= 2b(a + 2)(a^2 - 2a + 4)$

e. $6x^2 + 9x - 10$ — The master product method shows that this is not factorable (can you verify this?)

f. $12x^2 - 13x - 4$ — A trinomial of the form $ax^2 + bx + c$
$= (3x - 4)(4x + 1)$

g. $2xy^3 - 14xy^2 + 24xy$ — A common factor; the GCF is $2xy$
$= 2xy(y^2 - 7y + 12)$ — $y^2 - 7y + 12$ is a trinomial of the form $x^2 + bx + c$
$= 2xy(y - 3)(y - 4)$

h. $16x^2 - 24xy + 9y^2$ — A perfect square trinomial
$= (4x - 3y)^2$

i. $x^2 - y^2 + 2x - 2y$ — Factor by grouping
$= (x + y)(x - y) + 2(x - y)$
$= (x - y)(x + y + 2)$

Exercises 5.8
Set I

Factor each polynomial completely, or write "not factorable."

1. $12e^2 + 13e - 35$
2. $30f^2 + 17f - 21$
3. $6ac - 6bd + 6bc - 6ad$
4. $10cy - 6cz + 5dy - 3dz$
5. $2xy^3 - 4xy^2 - 30xy$
6. $3yz^3 - 6yz^2 - 24yz$
7. $3x^3 + 24h^3$
8. $54f^3 - 2g^3$
9. $9e^2 - 30ef + 25f^2$
10. $16m^2 + 56mp + 49p^2$
11. $x^3 + 3x^2 - 4x - 12$
12. $a^3 - 2a^2 - 9a + 18$
13. $a^2 - b^2 - a + b$
14. $x^2 - y^2 - x - y$
15. $x^2 + x - 10$
16. $x^2 + x + 14$
17. $x^3 - 8y^3 + x^2 - 4y^2$
18. $a^3 - b^3 + a^2 - b^2$
19. $10x^2 + 2x - 21$
20. $10x^2 + 10x - 21$
21. $x^2 - 4xy + 4y^2 - 5x + 10y + 6$
22. $x^2 - 6xy + 9y^2 - 8x + 24y + 15$
23. $x^2 - 6xy + 9y^2 - 25$
24. $a^2 - 8ab + 16b^2 - 1$

Writing Problems

Express the answers in your own words and in complete sentences.

1. Describe the method you would try first in factoring $4x^2 + 4x + 1 - 9y^2$.
2. Describe the method you would try first in factoring $2x^4 - 3x^3 - x^2 + 3x - 1$.

Exercises 5.8 Set II

Factor each polynomial completely, or write "not factorable."

1. $5x^2 + 12x + 4$
2. $2a^2mn - 18b^2mn$
3. $4ac + 4bc - 8ad - 8bd$
4. $15x^2 - 13x - 2$
5. $6x^3y + 4x^2y - 10xy$
6. $3x^2 - 18x + 15$
7. $4x^3 + 32y^3$
8. $4x^2 + 4x + 1 - y^2$
9. $16x^2 - 24xy + 9y^2$
10. $x^2 + 100$
11. $b + b^2 - 16b - 16$
12. $1 + 4a^2$
13. $x^2 - 4y^2 - x + 2y$
14. $4x^2 - y^2 + 6x + 3y$
15. $x^2 + x + 3$
16. $27 + 12x^2 + 36x$
17. $x^3 + y^3 + x^2 - y^2$
18. $2x^3 - x^2 - 18x + 9$
19. $10x^2 + 28x - 21$
20. $6x^2 + 7xy - 10y^2$
21. $x^2 - 2xy + y^2 - 10x + 10y + 21$
22. $9x^2 + 6xy + y^2 - z^2 - 2z - 1$
23. $x^2 - 10xy + 25y^2 - 16$
24. $25x^2 - 10xy + y^2 + 35x - 7y + 12$

5.9 Solving Equations by Factoring

We are now ready to solve quadratic (second-degree) and higher-degree equations. One method of solving such equations is based on the zero-factor property, which is stated without proof.

The zero-factor property

> If $a, b \in R$ and if $ab = 0$, then $a = 0$ or $b = 0$.
>
> *In words*: If the product of two factors is zero, then one or both factors must be zero.

The zero-factor property can be extended to include more than two factors; that is, if a product of any number of factors is zero, then at least one of the factors must be zero.

Many higher-degree equations can be solved by factoring. The method is summarized below.

Solving an equation by factoring

1. Move all nonzero terms to one side of the equal sign by adding the same expression to both sides. *Only zero must remain on the other side.* Then arrange the polynomial in descending powers.
2. Factor the nonzero polynomial completely.
3. Applying the zero-factor property, set each factor equal to zero.* (Any factors that are nonzero constants *cannot* equal zero; see Example 4.)
4. Solve each resulting first-degree equation.
5. Check all apparent solutions in the original equation.

*If any of the factors that contain variables are *not* first-degree binomials, we cannot solve the equation at this time.

EXAMPLE 1 Find the solution set of $4x^2 - 16x = 0$.

SOLUTION The polynomial must be factored first.

$$4x^2 - 16x = 0$$

$$4x(x - 4) = 0 \qquad \text{This is of the form } ab = 0$$

$$4x = 0 \quad \Big| \quad x - 4 = 0 \qquad \text{Applying the zero-factor property}$$

$$x = 0 \quad \Big| \quad x = 4 \qquad \text{Solving each resulting equation}$$

√ ***Check for*** $x = 0$ $\quad 4(0^2) - 16(0) = 0$

√ ***Check for*** $x = 4$ $\quad 4(4^2) - 16(4) = 64 - 64 = 0$

The solution set is $\{0, 4\}$.

EXAMPLE 2 Find the solution set of $x^2 - x = 6$.

SOLUTION

$$x^2 - x = 6$$

$$x^2 - x - 6 = 0 \qquad \text{Adding } -6 \text{ to both sides}$$

$$(x + 2)(x - 3) = 0 \qquad \text{Factoring}$$

$$x + 2 = 0 \quad \Big| \quad x - 3 = 0 \qquad \text{Applying the zero-factor property}$$

$$x = -2 \quad \Big| \quad x = 3$$

√ ***Check for*** $x = -2$ $\quad (-2)^2 - (-2) = 4 + 2 = 6$

√ ***Check for*** $x = 3$ $\quad (3)^2 - (3) = 9 - 3 = 6$

The solution set is $\{-2, 3\}$.

Sometimes, it is convenient to get all terms to the right side of the equation, leaving only zero on the left side (see Example 3).

EXAMPLE 3 Find the solution set of $4 - x = 3x^2$.

SOLUTION We first add $(-4 + x)$ to both sides and arrange the terms in descending powers.

$$4 - x = 3x^2$$

$$0 = 3x^2 + x - 4 \qquad \text{Adding } (-4 + x) \text{ to both sides}$$

$$0 = (x - 1)(3x + 4) \qquad \text{Factoring}$$

$$x - 1 = 0 \quad \Big| \quad 3x + 4 = 0 \qquad \text{Applying the zero-factor property}$$

$$x = 1 \quad \Big| \quad 3x = -4$$

$$\Big| \quad x = -\tfrac{4}{3}$$

(The check will not be shown.) The solution set is $\left\{1, -\frac{4}{3}\right\}$.

A Word of Caution The product must equal *zero*, or no conclusions can be drawn about the factors. Suppose

$$(x-1)(x-3) = 8 \quad \longleftarrow \text{No conclusion can be drawn because the product } \neq 0$$

A common error is to think that if $(x-1)(x-3) = 8$, then $x - 1 = 8$ *or* $x - 3 = 8$. Both these assumptions are incorrect. Consider the following:

If $x - 1 = 8$, then $x = 9$.

Check for $x = 9$

$$(x-1)(x-3) = 8$$
$$(9-1)(9-3) \stackrel{?}{=} 8$$
$$8 \cdot 6 \stackrel{?}{=} 8$$
$$48 \neq 8$$

If $x - 3 = 8$, then $x = 11$.

Check for $x = 11$

$$(x-1)(x-3) = 8$$
$$(11-1)(11-3) \stackrel{?}{=} 8$$
$$10 \cdot 8 \stackrel{?}{=} 8$$
$$80 \neq 8$$

If the product of two numbers is 8, we *don't* know what the factors are; one factor could be 2 and the other 4, one factor could be 16 and the other $\frac{1}{2}$, one factor could be -1 and the other -8, and so forth.

The correct solution is

$$(x-1)(x-3) = 8$$
$$x^2 - 4x + 3 = 8$$
$$x^2 - 4x - 5 = 0 \qquad \text{Adding } -8 \text{ to both sides}$$
$$(x-5)(x+1) = 0 \qquad \text{Factoring}$$
$$x - 5 = 0 \quad \bigg| \quad x + 1 = 0 \qquad \text{Applying the zero-factor property}$$
$$x = 5 \quad \bigg| \quad x = -1$$

(The check will not be shown.) The solution set is $\{5, -1\}$.

★ **EXAMPLE 4** Find the solution set of $(x+2)^3 = x^3 + 56$.

SOLUTION We first remove the grouping symbols, using the binomial theorem.

$$(x+2)^3 = x^3 + 56$$
$$x^3 + 3x^2(2) + 3x(2)^2 + (2)^3 = x^3 + 56 \qquad \text{Using the binomial theorem on the left side}$$
$$x^3 + 6x^2 + 12x + 8 - x^3 - 56 = 0 \qquad \text{Moving all terms to the left side}$$
$$6x^2 + 12x - 48 = 0$$
$$6(x^2 + 2x - 8) = 0 \qquad \text{Factoring out 6, the GCF}$$
$$6(x+4)(x-2) = 0 \qquad \text{Continuing the factoring}$$
$$6 \neq 0 \quad \bigg| \quad x + 4 = 0 \quad \bigg| \quad x - 2 = 0 \qquad \text{Applying the zero-factor property}$$
$$x = -4 \quad \bigg| \quad x = 2$$

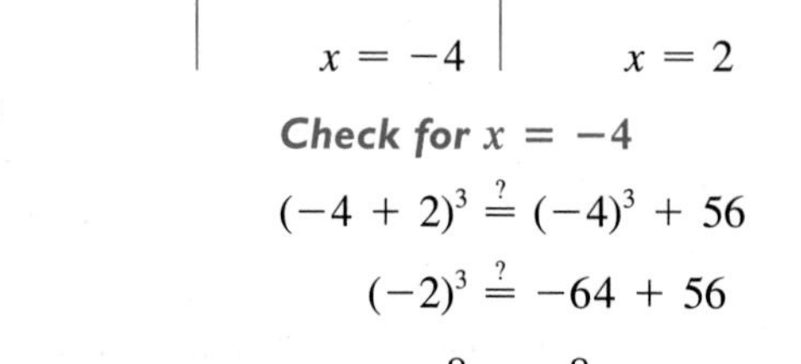

√ **Check for** $x = -4$

$$(-4+2)^3 \stackrel{?}{=} (-4)^3 + 56$$
$$(-2)^3 \stackrel{?}{=} -64 + 56$$
$$-8 = -8$$

Check for $x = 2$

$$(2+2)^3 \stackrel{?}{=} (2)^3 + 56$$
$$(4)^3 \stackrel{?}{=} 8 + 56$$
$$64 = 64$$

The solution set is $\{-4, 2\}$.

Exercises 5.9
Set I

Find the solution set of each of the following equations.

1. $3x(x - 4) = 0$
2. $5x(x + 6) = 0$
3. $4x^2 = 12x$
4. $6a^2 = 9a$
5. $x(x - 4) = 12$
6. $x(x - 2) = 15$
7. $2x^3 + x^2 = 3x$
8. $4x^3 = 10x - 18x^2$
9. $2x^2 - 15 = -7x$
10. $3x^2 - 10 = -13x$
11. $4x^2 + 9 = 12x$
12. $25x^2 + 4 = 20x$
13. $21x^2 + 60x = 18x^3$
14. $68x^2 = 30x^3 + 30x$
15. $4x(2x - 1)(3x + 7) = 0$
16. $5x(4x - 3)(7x - 6) = 0$
17. $x^3 + 3x^2 - 4x = 12$
18. $x^3 + x^2 - 9x = 9$
19. $x^4 - 10x^2 + 9 = 0$
20. $x^4 - 13x^2 + 36 = 0$

★ 21. $(x + 3)^3 = x^3 + 63$

★ 22. $(x + 1)^3 = x^3 + 37$

★ 23. $(x + 3)^4 = x^4 + 108x + 81$

★ 24. $(x + 2)^4 = x^4 + 32x + 16$

Writing Problems

Express the answers in your own words and in complete sentences.

1. Explain why the solution of the equation $x^2 - x - 6 = 0$ is not $(x + 2)(x - 3)$.
2. Describe what is wrong with the following statement: If $xy = 4$, then $x = 4$ or $y = 4$.
3. Describe what is wrong with the following statement: If $(x - 3)(x - 4) = 12$, then $x - 3 = 12$ or $x - 4 = 12$.

Exercises 5.9
Set II

Find the solution set of each of the following equations.

1. $4x(x - 2) = 0$
2. $16x^2 = 5x$
3. $x(3x - 1) = 2$
4. $x^2 + 9 = 6x$
5. $x(x + 1) = 12$
6. $(x + 2)^3 = x^3 + 8$
7. $6x^3 + 10x^2 = 4x$
8. $4x^3 + 10x^2 = 6x$
9. $3x^2 - 2 = x$
10. $4x^4 - 5x^2 + 1 = 0$
11. $6x^2 + 12 = 17x$
12. $2x^3 + 3x^2 = 2x + 3$
13. $30x - 3x^2 = 9x^3$
14. $x^3 - 5x^2 - 4x + 20 = 0$
15. $3x(3x - 1)(5x + 7) = 0$
16. $4x^3 = 20x^2 + 9x - 45$
17. $z^3 - 2z^2 + 2 = z$
18. $14x^4 + 41x^3 = 3x^2$
19. $x^4 - 5x^2 + 4 = 0$

★ 20. $(x - 1)^4 = x^4 - 2x + 1$

★ 21. $(x + 2)^3 = x^3 + 26$

★ 22. $(x - 2)^3 = x^3 - 2$

★ 23. $(x + 1)^4 = x^4 + 4x + 1$

24. $7x^2 + 4 = 29x$

5.10 Applied Problems That Involve Factoring

Many applications, including many problems dealing with geometric figures, lead to equations that can be solved by factoring. Several formulas relating to geometric figures are given in Figure 1; you are urged to memorize any of these that you don't already know.

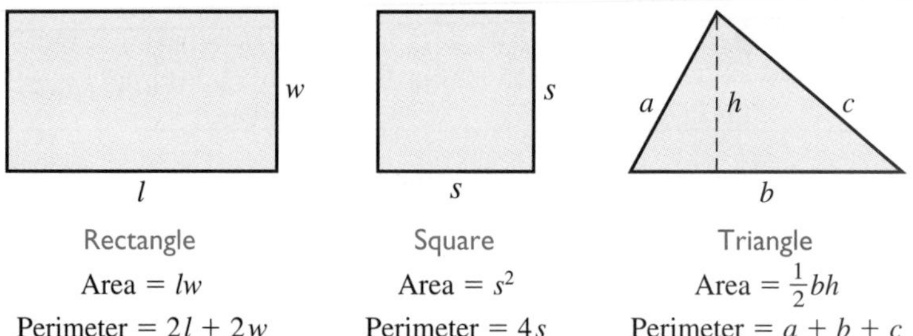

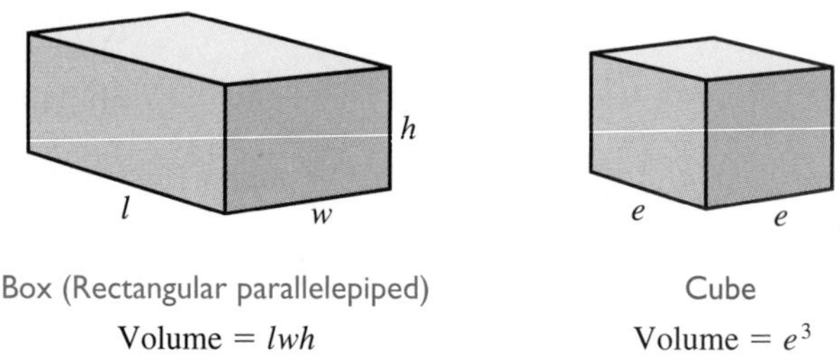

Figure 1

Note We assume that you recall that the *perimeter* of a plane figure (a two-dimensional figure) is the distance *around* the figure, and we assume that you have some idea of what we mean by the *area* of a plane figure and the *volume* of a three-dimensional figure. The *domain* in such problems is the set of *positive* real numbers, since the length of a side of a geometric figure cannot be negative (or zero).

In solving problems that involve geometric figures, it is often helpful to sketch *and label* the figures.

EXAMPLE 1 The width of a rectangle is 5 cm less than its length. Its area is 10 more (numerically) than its perimeter. What are the dimensions of the rectangle?

SOLUTION

Think The domain is the set of positive real numbers, since the length of a rectangle cannot be negative or zero.

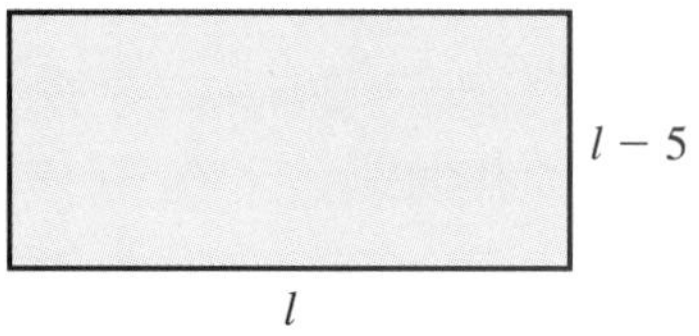

Step 1. Let l = the length (in cm)
Then $l - 5$ = the width (in cm)

(Rectangle sketch: length l, width $l - 5$)

Think Area $= lw = i(l - 5)$
Perimeter $= 2l + 2w = 2l + 2(l - 5)$

Reread Its area | is | 10 | more than | its perimeter

Step 2. $l(l - 5) \quad = \quad 10 \quad + \quad 2l + 2(l - 5)$

Step 3.

$$l^2 - 5l = 10 + 2l + 2l - 10$$
$$l^2 - 9l = 0$$
$$l(l - 9) = 0$$

$~~l = 0~~$ | $l - 9 = 0$ Applying the zero-factor property

Step 4. Not in the domain | $l = 9$ The length (in cm)

$l - 5 = 4$ The width (in cm)

Step 5. ***Check*** The area is (9 cm) (4 cm) = 36 sq. cm.
The perimeter is 2(9 cm) + 2(4 cm) = 18 cm + 8 cm = 26 cm.
36 is 10 more than 26.

Step 6. Therefore, the rectangle has a length of 9 cm and a width of 4 cm.

EXAMPLE 2

The base of a triangle is 5 cm more than its altitude. If the area is 33 sq. cm, find the altitude and the base of the triangle.

SOLUTION

Think The domain is the set of positive real numbers, since neither the altitude nor the base of a triangle can be negative or zero.

Step 1. Let $h =$ the altitude of the triangle (in cm)
Then $5 + h =$ the base of the triangle (in cm)

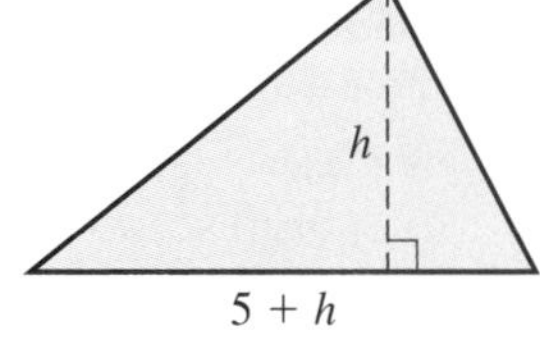

Think The area of a triangle $= \frac{1}{2}bh$, or $\frac{1}{2}hb$

Reread

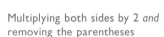

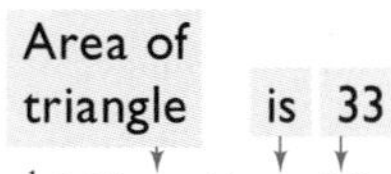

Step 2. $\frac{1}{2}h(5 + h) = 33$

Step 3. $5h + h^2 = 66$ Multiplying both sides by 2 *and* removing the parentheses

$$h^2 + 5h - 66 = 0$$

$$(h - 6)(h + 11) = 0$$

	$h - 6 = 0$	$h + 11 = 0$	Applying the zero-factor property
Step 4.	$h = 6$	~~$h = -11$~~	-11 is not in the domain, since an altitude can't be negative
	$5 + h = 11$		

√ **Step 5.** *Check* The area is $\frac{1}{2}(6 \text{ cm})(11 \text{ cm}) = 33$ sq. cm.

Step 6. Therefore, the altitude is 6 cm, and the base is 11 cm.

★ **EXAMPLE 3**

One edge of a cube is 2 cm longer than an edge of a second cube. If the volume of the larger cube is 56 cc (cubic centimeters) more than the volume of the smaller cube, find the length of an edge of each cube.

SOLUTION

Think The domain is the set of all positive real numbers.

Step 1. Let $x =$ the length of an edge of the smaller cube (in cm)
Then $x + 2 =$ the length of an edge of the larger cube (in cm)

Think The volume of a cube is e^3.

Reread

Volume of larger cube	is	56	more than	volume of smaller cube
$(x + 2)^3$	$=$	56	$+$	x^3

Step 2. $(x + 2)^3 = 56 + x^3$

Step 3. $x^3 + 6x^2 + 12x + 8 = 56 + x^3$ Using the binomial theorem to expand $(x + 2)^3$

$$6x^2 + 12x - 48 = 0$$

$$6(x^2 + 2x - 8) = 0$$

$$6(x - 2)(x + 4) = 0$$

	$6 \neq 0$	$x - 2 = 0$	$x + 4 = 0$	Applying the zero-factor property
Step 4.		$x = 2$	~~$x = -4$~~	-4 is not in the domain
		$x + 2 = 4$		

√ **Step 5.** *Check* The volume of the smaller cube is $(2 \text{ cm})^3 = 8 \text{ cm}^3$, or 8 cc.
The volume of the larger cube is $(4 \text{ cm})^3 = 64 \text{ cm}^3$, or 64 cc.
$64 \text{ cc} = 56 \text{ cc} + 8 \text{ cc}$

Step 6. Therefore, the length of an edge of the smaller cube is 2 cm, and the length of an edge of the larger cube is 4 cm.

Exercises 5.10
Set I

Set up each problem algebraically, solve, and check. Be sure to state what your variables represent.

1. Find three consecutive even integers such that the product of the first two is 38 more than the third number.
2. Find three consecutive odd integers such that the product of the first two is 52 more than the third number.
3. The height of a rectangular box equals the length of an edge of a certain cube. The width of the box is 3 in. more than its height, and the length of the box is 4 times its height. The volume of the box is 8 times the volume of the cube. Find the *dimensions* of the box, and find the *volume* of the cube.
4. The height of a rectangular box equals the length of an edge of a certain cube. The width of the box is twice its height, and the length of the box is 10 cm more than its height. The volume of the box is 6 times the volume of the cube. Find the *dimensions* of the box, and find the *volume* of the cube.
5. The width of a rectangle is 5 m less than its length; its area is 46 more (numerically) than its perimeter. What are the dimensions of the rectangle?
6. The width of a rectangle is 7 m less than its length; its area is 4 more (numerically) than its perimeter. What are the dimensions of the rectangle?
7. The base of a triangle is 7 cm more than its altitude. If the area is 39 sq. cm, find the altitude and the base of the triangle.
8. The base of a triangle is 4 m more than its altitude. If the area is 48 sq. m, find the altitude and the base of the triangle.
9. One square has a side 6 cm shorter than the side of a second square. The area of the larger square is 9 times as great as the area of the smaller square. Find the length of a side of each square.
10. One square has a side 4 cm shorter than the side of a second square. The area of the larger square is 4 times as great as the area of the smaller square. Find the length of a side of each square.
11. ★ One cube has an edge 3 cm longer than an edge of a second cube. If the volume of the larger cube is 63 cc more than the volume of the smaller one, find the length of an edge of each cube.
12. ★ One cube has an edge 1 cm longer than an edge of a second cube. If the volume of the larger cube is 37 cc more than the volume of the smaller one, find the length of an edge of each cube.
13. A box with no top is to be formed from a rectangular sheet of metal by cutting 2-in. squares from the corners and folding up the sides. The length of the box is to be 3 in. more than its width, and its volume is to be 80 cu. in. (See the figure.)
    a. Find the dimensions of the original sheet of metal.
    b. Find the dimensions of the box.

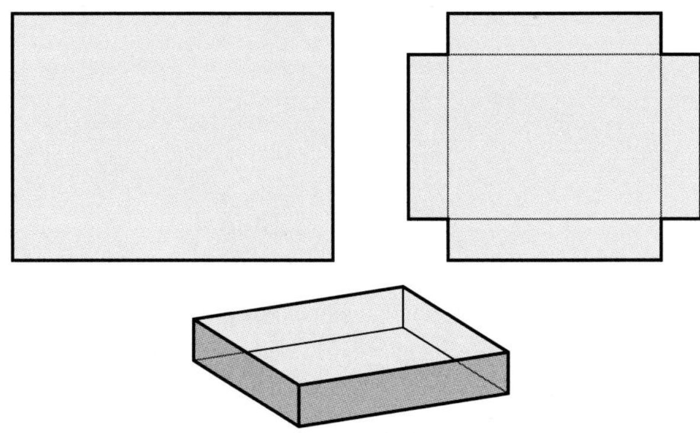

14. A box with no top is to be formed from a rectangular sheet of metal by cutting 3-in. squares from the corners and folding up the sides. The length of the box is to be 3 in. more than its width, and its volume is to be 30 cu. in.
    a. Find the dimensions of the original sheet of metal.
    b. Find the dimensions of the box.
15. Thirty-five square yards of carpet is laid in a rectangular room. The length of the room is 3 yd less than two times its width. Find the dimensions of the room.
16. Fifty-four square yards of carpet is laid in a rectangular room. The length of the room is 3 yd less than two times its width. Find the dimensions of the room.
17. David is working on a square needlepoint picture, and the yarn necessary for completing *one square inch* of the picture costs 30¢. If the picture were 3 in. longer and 2 in. narrower, the yarn would cost \$45. What are the dimensions of the picture?
18. Chang's family room is square. If the room were 1 yd narrower and 1 yd longer, the total cost of covering it with carpeting that costs \$24 per square yard, installed, would be \$1,920. What are the dimensions of the room?

Exercises 5.10
Set II

Set up each problem algebraically, solve, and check. Be sure to state what your variables represent.

1. Find three consecutive even integers such that the product of the first two is 16 more than the third number.
2. Find three consecutive odd integers such that the product of the first and third is 13 more than 16 times the second.

3. The width of a rectangular box equals the length of an edge of a certain cube. The height of the box is 2 ft more than its width, and the length of the box is 3 times its width. The volume of the box is 6 times the volume of the cube. Find the *dimensions* of the box, and find the *volume* of the cube.

4. A 2-cm-wide mat surrounds a picture. The area of the picture itself is 286 sq. cm. If the length of the outside of the mat is twice its width, what are the dimensions of the outside of the mat? What are the dimensions of the picture?

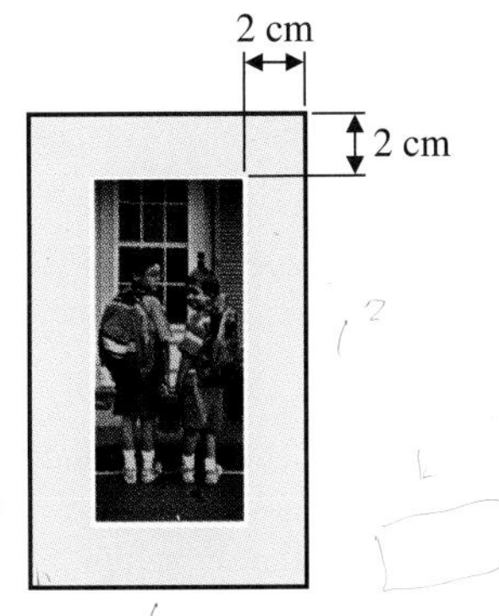

5. The width of a rectangle is 6 yd less than its length; its area is 23 more (numerically) than its perimeter. What are the dimensions of the rectangle?

6. Patricia is 3 years older than Sandra. The product of their ages is 154. How old is each?

7. The base of a triangle is 6 cm more than its altitude. If the area is 20 sq. cm, find the altitude and the base of the triangle.

8. Find three consecutive odd integers such that the product of the first and third is 9 more than 12 times the second.

9. One square has a side 2 m longer than the side of a second square. The area of the larger square is 16 times the area of the smaller square. Find the length of a side of each square.

10. The length of a rectangular box is twice its width, and its height is 2 cm less than its width. The volume of the box is 150 cc. Find the dimensions of the box.

★ 11. One cube has an edge 2 cm longer than an edge of a second cube. If the volume of the larger cube is 26 cc more than the volume of the smaller one, find the length of an edge of each cube.

12. Find three consecutive odd integers such that the product of the first and third is 11 more than 14 times the second.

13. A box with no top is to be formed from a square sheet of metal by cutting 3-in. squares from the corners and folding up the sides. The box is to have a volume of 12 cu. in.

 a. Find the dimensions of the sheet of metal.

 b. Find the dimensions of the box.

★ 14. One cube has an edge 2 cm longer than an edge of a second cube. If the volume of the larger cube is 152 cc more than the volume of the smaller one, find the length of an edge of each cube.

15. Forty square yards of carpet is laid in a rectangular room. The length of the room is 2 yd less than three times its width. Find the dimensions of the room.

16. The base of a triangle is 5 m more than its altitude, and its area is 12 sq. m. Find its base and its altitude.

17. Mr. Christ's kitchen is square, and he plans to have the floor covered with linoleum that costs \$18 per square yard, installed. If the kitchen were 1 yd narrower and 2 yd longer, the cost of the linoleum would be \$324. What are the dimensions of the kitchen?

18. A box with no top is to be formed from a rectangular piece of cardboard by cutting squares from the corners and folding up the sides. The width of the box is to be 2 in. less than the length, the height is to be 3 in., and the volume is to be 105 cu in.

 a. What size squares should be cut from the corners?

 b. What are the dimensions of the box?

 c. What are the dimensions of the original piece of cardboard?

Sections 5.5–5.10 REVIEW

Factoring a Sum or Difference of Two Cubes 5.5

A **sum of two cubes** is always factorable: $a^3 + b^3 = (a + b)(a^2 - ab + b^2)$

A **difference of two cubes** is always factorable: $a^3 - b^3 = (a - b)(a^2 + ab + b^2)$

Factoring by Completing the Square 5.6

Some binomials and some trinomials can be factored by **completing the square**.

Factoring by Using Synthetic Division 5.7

Some polynomials can be factored by **using synthetic division**.

Selecting the Method of Factoring 5.8

If there is a common factor, find the greatest common factor and factor it out. Then try to factor the polynomial factor.

If the polynomial to be factored contains two terms:

Is it a difference of two squares? (Section 5.3)
Is it a sum or difference of two cubes? (Section 5.5)
Can it be factored by completing the square? (Section 5.6)
Can it be factored by using synthetic division? (Section 5.7)

If the polynomial to be factored contains three terms:
Can it be expressed in the form $x^2 + bx + c$? (Section 5.4A)
Can it be expressed in the form $ax^2 + bx + c$? (Section 5.4B)
Is it a perfect square trinomial? (Section 5.4B)
Can it be factored by completing the square? (Section 5.6)
Can it be factored by using synthetic division? (Section 5.7)
If the polynomial to be factored contains four or more terms:
Can it be factored by grouping? (Section 5.2)
Can it be factored by using synthetic division? (Section 5.7)
Check to see whether any factor can be factored further.
Check the result by multiplying the factors together.

Solving an Equation by Factoring 5.9

1. Move all nonzero terms to one side of the equation by adding the same expression to both sides. *Only zero must remain on the other side.* Then arrange the polynomial in descending powers.
2. Factor the nonzero polynomial completely.
3. Applying the zero-factor property, set each factor equal to zero.
4. Solve each resulting first-degree equation.
5. Check the apparent solutions in the original equation.

Applied Problems 5.10

Many applied problems lead to equations that can be solved by factoring.

Applied Problems Involving Geometric Figures 5.10

In solving applied problems involving geometric figures, it may help to sketch *and label* the figures as an aid in writing the equation.

Sections 5.5–5.10 REVIEW EXERCISES Set I

In Exercises 1–29, factor each polynomial completely, or write "not factorable."

1. $15u^2v - 3uv$
2. $3n^2 + 16n + 5$
3. $4x^2y - 8xy^2 + 4xy$
4. $5x^2 + 11x + 2$
5. $17u - 12 + 5u^2$
6. $4 + 21x + 5x^2$
7. $6u^3v^2 - 9uv^3 - 12uv$
8. $15a^2 + 15ab - 30b^2$
9. $81 - 9m^2$
10. $x^2 + 25$
11. $10x^2 - xy - 24y^2$
12. $28x^2 - 13xy - 6y^2$
13. $8a^3 - 27b^3$
14. $64h^3 - 125k^3$
15. $1 + 4y + 4y^2$
16. $4 + 12x + 9x^2$
17. $x^3 + 8y^3$
18. $27x^3 + y^3$
19. $x^2 + x - y - y^2$
20. $x^2 - x + y - y^2$
21. $8a^2 - 8ab + 2b^2$
22. $18h^2 - 12hk + 2k^2$
23. $x^3 - 4x^2 - 4x + 16$
24. $x^2 + x + 9$
25. $x^3 - 2x^2 - 9x + 18$
26. $x^2 + 8x + 16 - 25y^2$
27. $(2x + 3y)^2 + (2x + 3y) - 6$
28. $a^2 + 6ab + 9b^2 - 7a - 21b + 12$
29. $x^2 + 4xy + 4y^2 - 5x - 10y + 6$

In Exercises 30–38, find the solution set of each equation.

30. $5z^2 - 12z + 7 = 0$
31. $x^2 = 3(6 + x)$
32. $x^3 = 36x^2$
33. $6x^2 = 13x + 5$
34. $x^3 - 2x^2 - 9x + 18 = 0$
*35. $(x + 3)^3 = x^3 + 27$
36. $4 + 5x^2 + 12x = 0$
*37. $(x + 3)^4 = x^4 + 12x^3 + 81$
*38. $(x + 2)^4 = x^4 + 8x^3 + 16$

In Exercises 39–44, set up each problem algebraically, solve, and check. Be sure to state what your variables represent.

39. Find three consecutive odd integers such that the product of the first and third is 5 more than 8 times the second.
40. The height of a room is 8 ft, and the room is 5 ft longer than it is wide. If the volume of the room is 2,400 cu. ft, find the width and the length of the room.
41. The cost of the carpeting for a family room is \$520. If the carpeting costs \$26 per sq. yd, installed, and if the room is 1 yd longer than it is wide, find the dimensions of the room.
42. The length of a rectangle is 7 more than its width, and its area is 40 more (numerically) than its perimeter. What are the dimensions of the rectangle?
*43. One cube has an edge 4 cm longer than an edge of a second cube. If the volume of the larger cube is 316 cc more than the volume of the smaller one, find the length of an edge of each cube.
44. One square has a side 3 cm longer than the side of a second square. The area of the larger square is 16 times the area of the smaller square. Find the length of a side of each square.

Sections 5.5–5.10 REVIEW EXERCISES Set II

ANSWERS

Name

1. ______
2. ______
3. ______
4. ______
5. ______
6. ______
7. ______
8. ______
9. ______
10. ______
11. ______
12. ______
13. ______
14. ______
15. ______
16. ______

In Exercises 1–29, factor each polynomial completely, or write "not factorable."

1. $2a^2 - 9a + 10$
2. $a^2 - 2ab - 15b^2$
3. $5x^2 + x - 1$
4. $3x^2 + 13x - 10$
5. $6x^2 + xy - 15y^2$
6. $5a^2 + 13ab - 6b^2$
7. $2ac + bc - 3bd - 6ad$
8. $7x^3 + 40x^2 - 12x$
9. $64x^3 + 27y^3$
10. $6x^2 + 29x + 20$
11. $x^2 + 100$
12. $3x^3 - 12xy^2$
13. $x^2 + x + 17$
14. $4y^2 - 5y - 6$
15. $5a^4 - 40ab^3$
16. $x^3 + x^2 - 9x - 9$

17. $x^2 - 10xy + 25y^2 - 9$

18. $3x^3 - 81$

19. $3n^2 - 8n - 5$

20. $x^3 - 3x^2 - 4x + 12$

21. $y^2 + 36$

22. $6x^2 + x - 2$

23. $x^2 + x + 15$

24. $y^2 + 10y + 25 - 4x^2$

25. $(5x - 2)^2 + 8(5x - 2)y + 15y^2$

26. $a^3 + a^2b - b^2a - b^3$

27. $x^2 + 6xy + 9y^2 - 2x - 6y - 24$

28. $(3x - 2y)^2 - (3x - 2y) - 6$

29. $9x^2 + 12xy + 4y^2 - 4a^2 - 4ab - b^2$

17. ____________________

18. ____________________

19. ____________________

20. ____________________

21. ____________________

22. ____________________

23. ____________________

24. ____________________

25. ____________________

26. ____________________

27. ____________________

28. ____________________

29. ____________________

30. ____________________

31. ____________________

In Exercises 30–38, find the solution set of each equation.

30. $x(x - 2) = 15$

31. $x^3 = 16x^2$

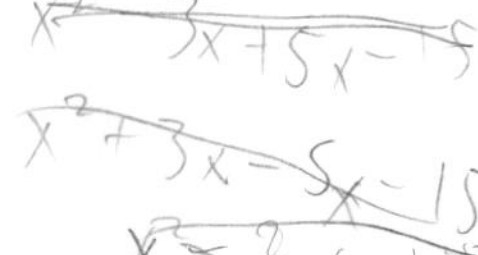

32. $x^2 = 4(5 + 2x)$

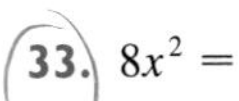

33. $8x^2 = 10x + 12$

32. ______________________

33. ______________________

34. ______________________

35. ______________________

36. ______________________

37. ______________________

38. ______________________

39. ______________________

34. $x^3 = 9x^2$

35. $x^3 + 4x^2 - 25x = 100$

$x^3 + 4x^2 - 25x - 100 = 0$
$x^2(x+4) - 25(x+4)$
$(x^2 - 25)(x+4)$
$(x+5)(x-5)(x+4)$
$\pm 5, -4$

36. $x^3 - 5x^2 - 4x + 20 = 0$

★37. $(x + 4)^3 = x^3 + 64$

★38. $(x - 2)^3 = x^3 - 8$

In Exercises 39–44, set up each problem algebraically, solve, and check. Be sure to state what your variables represent.

39. Find three consecutive odd integers such that the product of the first and third is 4 less than 25 times the second.

40. The length of a rectangle is 8 in. less than its width, and its area is 560 sq. in. Find the length and the width of the rectangle.

40. ______________

____

41. ______________

____

42. ______________

43. ______________

44. ______________

★**41.** One cube has an edge 1 cm longer than an edge of a second cube. If the volume of the larger cube is 91 cc more than the volume of the smaller one, find the length of an edge of each cube.

42. Find three consecutive odd integers such that the product of the first and third is 7 more than 10 times the second.

43. The area of a square is 3 times its perimeter (numerically). What is the length of its side?

44. The height of a rectangular box is 2 cm less than the width, and its length is 7 times its width. If the volume is 525 cc, find the dimensions of the box.

Chapter 5 DIAGNOSTIC TEST

The purpose of this test is to see how well you understand factoring and solving equations by factoring. We recommend that you work this diagnostic test *before* your instructor tests you on this chapter. Allow yourself about 50 minutes.

Complete solutions for all the problems on this test, together with section references, are given in the answer section at the end of the book. For the problems you do incorrectly, we suggest that you study the sections cited.

In Problems 1–12, factor each polynomial completely, or write "not factorable."

1. $4x - 16x^3$
2. $43 + 7x^2 + 6$
3. $7x^2 + 23x + 6$
4. $x^2 + 81$
5. $2x^3 + 4x^2 + 16x$
6. $6x^2 - 5x - 6$
7. $y^3 - 1$
8. $3ac + 6bc - 5ad - 10bd$
9. $4x^2 + 4x + 1 - y^2$
10. $y^2 - 4$
11. $4z^2 + z + 1$
12. $8x^3 + y^3$

In Problems 13–18, find the solution set of each equation.

13. $x^2 = 16$
14. $z^2 + 7z = 0$
15. $2x^2 + x - 15 = 0$
16. $8y^2 = 4y$
17. $2x^3 + 3x = 7x^2$

★18. $(x + 1)^3 = x^3 + 1$

In Problems 19 and 20, set up each problem algebraically, solve, and check. Be sure to state what your variables represent.

19. Find three consecutive even integers such that the product of the first two is 68 more than the third number.
20. The base of a triangle is 8 cm more than its altitude. If the area of the triangle is 64 sq. cm, find its altitude and its base.

Chapters 1–5 CUMULATIVE REVIEW EXERCISES

1. Evaluate $18 \div 2\sqrt{9} - 4^2 \cdot 3$.
2. Given the formula $h = 48t - 16t^2$, find h when $t = 2$.
3. Find the solution set of $2(x - 3) - 5 = 6 - 3(x + 4)$.
4. Find the solution set of $-6 < 4x - 2 < 10$.
5. Find the solution set of $|2x - 3| \geq 7$.
6. State which of the following are real numbers: $\frac{2}{3}$, 4.7958315 . . . , 0.
7. Find the solution set of $2x^2 = 9x + 5$.

In Exercises 8 and 9, factor each expression completely, or write "not factorable."

8. $3x^3 + 81$
9. $x^3 + 5x^2 - x - 5$

In Exercises 10–14, perform the indicated operations and simplify the result.

10. $(2x^2 - 3x + 4) - (x^2 - 5x - 6) + (3 - x + 5x^2)$
11. $(a - 4)(a^2 - 2a + 5)$
12. $(3x + 5)^2$

★13. $(2x - 1)^5$

14. $(x^3 - 2x^2 - 6x + 8) \div (x + 2)$

In Exercises 15–19, fill in each blank with the word or phrase that fits best.

15. Numbers that are being multiplied together are called ________.
16. In a division problem, the number we are dividing *into* is called the ________.
17. If every element of set A is also in set B, we say that A is a ________ of B.
18. If the sum of two numbers is 0, we call the numbers the ______________ of each other.
19. A number that can be expressed in the form $\frac{a}{b}$, where $a \in J$ and $b \in J$, but $b \neq 0$, is called a ________ number.

In Exercises 20–24, set up each problem algebraically, solve, and check. Be sure to state what your variables represent.

20. The sum of three consecutive odd integers is 58. What are the integers?
21. How many quarts of pure antifreeze must be added to 10 qt of a 20% solution of antifreeze to make a 50% solution?
22. Lyndsey is three times as old as Nathan. Two years ago, Lyndsey was seven times as old as Nathan. How old is Nathan now?
23. Yang has 27 coins. He has 4 times as many nickels as quarters. The product of the number of quarters and twice the number of dimes equals the number of nickels. How many coins of each kind does he have?
24. Michael and Kay are going to carpet their living room. The carpeting they have chosen is \$27 per square yard, and the total cost (without tax) of the carpeting is \$810. If their living room is 1 yd longer than it is wide, what is the width of their living room?

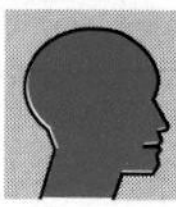

Critical Thinking and Problem-Solving Exercises

1. Discover the patterns and fill in the blanks below with a name that continues each pattern. (Many answers are possible.)

   Dick, Mick, Nick, __________

   Joe, Kate, Lorna, __________

   Pauline, Olivia, Nancy, __________

2. Melissa, Wen Lie, and Jiang are friends. One of them is a supervisor, one is a professor of English, and one is an engineer. Their last names are Smith, Jones, and White. By using the following clues, can you match the first name, last name, and profession of each of these three women?

   Ms. White is a supervisor.

   Ms. Smith and the English professor dined together.

   Jiang is an engineer.

   Melissa is not a supervisor.

3. A survey of 80 students was taken recently, and the following results were reported:

   57 enjoy reading biographies.

   50 enjoy reading western novels.

   52 enjoy reading mysteries.

   26 enjoy reading western novels and mysteries.

   31 enjoy reading western novels and biographies.

   34 enjoy reading mysteries and biographies.

   11 enjoy reading all three types of books.

   1 student does not enjoy reading any of these three types of books.

   How many students enjoy *only* western novels?

4. A container is in the shape of a right circular cylinder, and its radius is 5 cm. At present, it is one-fourth full of water. If 300 cc of water is added, it will be one-third full. What is the volume of the container?

5. If Janice's son, who is in kindergarten, writes the numbers 1 through 100, how many 7's will he be writing altogether?

6. In a large county in 1991, 1.6 million people attended the county fair, and the earnings were $4.2 million. In 1992, the county fair officials raised the prices for admission by 20%, charging $10 for adults and $5 for children. The economy was very poor in 1992, and only 1.2 million people attended the fair; the earnings decreased to $850,000. In 1993, the fair officials lowered the prices to $8 for adults and $4 for children. Do you think the decision to lower the prices for 1993 was the right one? Why? What factors other than the high admission prices might have affected attendance?

7. Suppose you need to construct a rectangle whose perimeter *has* to be 36 cm and whose length and width have to be in whole centimeters, not in fractional parts of centimeters. Make an organized list of possible lengths and widths for the rectangle, and include the *area* for each rectangle. Which dimensions give the largest area?

8. Make up an applied problem that results in a second-degree equation that can be solved by factoring. (If you're working in groups, you can then solve each other's problems.)

Rational Expressions and Equations

CHAPTER

6

In this chapter, we define rational expressions and discuss how to perform operations on them and how to solve equations and applied problems that involve them. A knowledge of the different methods of factoring is *essential* to our work with rational expressions.

6.1 Rational Expressions

A **rational expression** (also sometimes called a **fraction** or an **algebraic fraction**) is an algebraic expression of the form $\dfrac{P}{Q}$, where P and Q are polynomials and $Q \neq 0$.* We call P the **numerator** and Q the **denominator** of the rational expression. We also call P and Q the **terms** of the rational expression.

$$\frac{P}{Q}$$

The numerator

The terms of the rational expression

The fraction bar

The denominator (it cannot be zero)

Note In Sections 6.1–6.6, whenever a rational expression is written, it is understood that none of the variables has a value that would make a denominator zero.

A Word of Caution If you are accustomed to writing fractions in arithmetic with a *slanted bar* (/), you will probably find it to your advantage to break the habit, because it is incorrect to write an answer that should be $\dfrac{x-5}{x+3}$, for example, as $x - 5/x + 3$. Notice that $x - 5/x + 3 = x - \dfrac{5}{x} + 3$ (division must be done before addition and subtraction). Then, since $\dfrac{x-5}{x+3} \neq x - \dfrac{5}{x} + 3$, $\dfrac{x-5}{x+3} \neq x - 5/x + 3$. If a slanted bar *is* used for rational expressions, any numerator or denominator that contains more than one term *must* be enclosed within parentheses.

If a slanted bar is used, parentheses are necessary for denominators with more than one *factor,* also. In this case, a common error is to write an answer that should be $\dfrac{1}{2x}$ as $1/2x$. Note, however, that $1/2x = \dfrac{1}{2}x$ (the division must be done first because it's on the *left*). Then, since $\dfrac{1}{2x} \neq \dfrac{1}{2}x$, $\dfrac{1}{2x} \neq 1/2x$.

Equivalent Rational Expressions and the Fundamental Property of Rational Expressions

Just as equivalent fractions in arithmetic are fractions that have the same value, **equivalent rational expressions** are rational expressions that have the same value. The **fundamental property of rational expressions,** which follows, allows us to reduce rational expressions and also to "build" rational expressions in order to add and subtract them.

*We usually refer to expressions in which the numerator and denominator are natural numbers as *fractions* and to expressions in which the numerator and denominator are polynomials (other than natural numbers) as *rational expressions.*

The fundamental property of rational expressions

If P, Q, and C represent polynomials and if $Q \neq 0$ and $C \neq 0$, then

$$\frac{P \cdot C}{Q \cdot C} = \frac{P}{Q}$$

The fundamental property of rational expressions, therefore, permits us to do the following:

1. Multiply both numerator and denominator by the same nonzero number; that is, if $Q \neq 0$ and $C \neq 0$, then $\frac{P}{Q} = \frac{P \cdot C}{Q \cdot C}$.
2. Divide both numerator and denominator by the same nonzero number; that is, if $Q \neq 0$ and $C \neq 0$, then $\frac{P \cdot C}{Q \cdot C} = \frac{P}{Q}$. This can be expressed another way: If $Q \neq 0$ and $D \neq 0$, then $\frac{P \div D}{Q \div D} = \frac{P}{Q}$.

A Word of Caution When we use the fundamental property of rational expressions and multiply (or divide) both numerator and denominator by the same expression, we do not change the value of the original expression. However, we do *not* get a rational expression equivalent to the one we started with if we *add* the same expression to or *subtract* the same expression from both the numerator and the denominator. For example, although $\frac{2}{3} = \frac{2 \cdot 4}{3 \cdot 4}$, it is *not* true that $\frac{2}{3} = \frac{2 + 4}{3 + 4}$; and although $\frac{6}{9} = \frac{6 \div 3}{9 \div 3}$, it is *not* true that $\frac{6}{9} = \frac{6 - 3}{9 - 3}$.

EXAMPLE 1 Determine whether each pair of rational expressions is equivalent.

a. $\frac{x + 3}{x + 6}, \frac{2(x + 3)}{2(x + 6)}$

SOLUTION Yes; the second rational expression can be obtained from the first by multiplying both numerator and denominator by 2.

b. $\frac{3 + x}{5 + x}, \frac{3}{5}$

SOLUTION No; the second rational expression cannot be obtained from the first by multiplying or dividing both numerator and denominator by the same expression.

Exercises 6.1

Set I

Determine whether each pair of rational expressions is equivalent.

1. $\frac{3}{6y}, \frac{1}{2y}$
2. $\frac{8}{4x}, \frac{2}{x}$
3. $\frac{x}{5}, \frac{x + 5}{10}$
4. $\frac{3}{z}, \frac{9}{z + 6}$
5. $\frac{2 + x}{3 - y}, \frac{8(2 + x)}{8(3 - y)}$
6. $\frac{4 - x}{y + 2}, \frac{3(4 - x)}{3(y + 2)}$
7. $\frac{x - y}{y - x}, \frac{y - x}{x - y}$
8. $\frac{z - 5}{5 - z}, \frac{5 - z}{z - 5}$

Writing Problems

Express the answers in your own words and in complete sentences.

1. Describe one way of determining whether two rational expressions are equivalent.
2. Find and describe the error in the following:

$$\frac{x-1}{x+2} = x - 1/x + 2$$

Exercises 6.1 Set II

Determine whether each pair of rational expressions is equivalent.

1. $\frac{12}{3y}, \frac{4}{y}$
2. $\frac{12}{3+y}, \frac{4}{y}$
3. $\frac{z}{2}, \frac{z+6}{8}$
4. $\frac{6+3y}{8+4y}, \frac{3}{4}$
5. $\frac{9-y}{x+2}, \frac{2(9-y)}{2(x+2)}$
6. $\frac{y+2}{y+4}, \frac{1}{2}$
7. $\frac{s-t}{t-s}, \frac{t-s}{s-t}$
8. $\frac{x-y}{x+y}, \frac{y-x}{-x-y}$

6.2 Reducing Rational Expressions to Lowest Terms

A rational expression is in *lowest terms* if the greatest common factor (GCF) of its numerator and denominator is 1. In this section and in the remainder of the book, it is understood that all rational expressions are to be reduced to lowest terms unless otherwise indicated.

One way to reduce a rational expression to lowest terms is as follows:

Reducing a rational expression to lowest terms

1. Factor the numerator and denominator completely.
2. Using the fundamental property of rational expressions, divide both numerator and denominator by all common factors.

If the numerator and denominator of a rational expression do not have a common factor (other than 1), then the rational expression is already in lowest terms, and we cannot reduce it further.

Recall that

$$-(a-b) = -1(a-b) = -a+b = b-a$$

Therefore, $b - a$ can always be substituted for $-(a - b)$, and $-(a - b)$ can always be substituted for $b - a$. We use this property in Example 1c.

EXAMPLE 1 Examples of reducing rational expressions to lowest terms or of determining that expressions cannot be reduced:

a. $\frac{4x^2y}{2xy} = \frac{\overset{2x}{\cancel{4x^2y}}}{\underset{1}{\cancel{2xy}}} = 2x$ The slashes indicate that we're dividing both numerator and denominator by the common factors 2, x, and y

b. $\frac{3x^2-5xy-2y^2}{6x^3y+2x^2y^2} = \frac{(x-2y)\overset{1}{\cancel{(3x+y)}}}{2x^2y\underset{1}{\cancel{(3x+y)}}} = \frac{x-2y}{2x^2y}$ The slashes indicate that we're dividing both numerator and denominator by $(3x + y)$

c. $\dfrac{2b^2 + ab - 3a^2}{4a^2 - 9ab + 5b^2}$

$$= \frac{(b-a)(2b+3a)}{(a-b)(4a-5b)} = \frac{\overset{1}{\cancel{(b-a)}}(2b+3a)}{-1\underset{1}{\cancel{(b-a)}}(4a-5b)} = \frac{(2b+3a)}{(-1)(4a-5b)} = \frac{2b+3a}{5b-4a}$$

Substituting $-1(b - a)$ for $(a - b)$

Substituting $5b - 4a$ for $-1(4a - 5b)$

Note It is also possible to do this problem by multiplying both numerator and denominator by -1, so other forms of the answer are possible; among these are

$$\frac{-(3a+2b)}{4a-5b}, \quad \frac{-3a-2b}{4a-5b}, \text{ and } \frac{2b+3a}{-(4a-5b)}.$$

A difference of two cubes

d. $$\frac{x^3 - 8}{x^3 - 2x^2 + 4x - 8} = \frac{\overset{1}{\cancel{(x-2)}}(x^2+2x+4)}{\underset{1}{\cancel{(x-2)}}(x^2+4)} = \frac{x^2+2x+4}{x^2+4}$$

The denominator can be factored by grouping

e. $\dfrac{x+3}{x+6}$ This rational expression cannot be reduced, since neither x nor 3 is a *factor* of the numerator or of the denominator

A Word of Caution A common error in Example 1e is to write

$$\frac{\overset{1}{\cancel{x}} + \overset{1}{\cancel{3}}}{\underset{1}{\cancel{x}} + \underset{2}{\cancel{6}}} = \frac{1+1}{1+2} = \frac{2}{3} \quad \text{or} \quad \frac{\cancel{x} + \overset{1}{\cancel{3}}}{\cancel{x} + \underset{2}{\cancel{6}}} = \frac{1}{2}$$

Both of these are incorrect because we can only divide by common *factors*. This is made *very* clear if we substitute 5 for x:

$$\frac{5+3}{5+6} = \frac{8}{11} \neq \frac{2}{3} \quad \text{and} \quad \frac{8}{11} \neq \frac{1}{2}$$

When a rational expression contains only one term in the denominator but several terms in the numerator, it can be reduced as a single rational expression or it can be rewritten as a sum of several expressions with each resulting expression then being reduced (see Example 2).

EXAMPLE 2 Examples of reducing rational expressions to lowest terms:

x is *not* a factor of the numerator

a. $\dfrac{x+y}{x}$

This rational expression cannot be reduced. ← This is an acceptable answer

If we write $\dfrac{x+y}{x}$ as a *sum* of separate rational expressions, we have the following:

$$\frac{x+y}{x} = \frac{x}{x} + \frac{y}{x} = 1 + \frac{y}{x}$$ ← This is also an acceptable answer

This is an acceptable answer

b. $\dfrac{7pqr - 21p^2r + 28qr^3}{7pqr} = \dfrac{7r(pq - 3p^2 + 4qr^2)}{7pqr} = \dfrac{pq - 3p^2 + 4qr^2}{pq}$

The same problem can also be done as follows:

$$\frac{7pqr - 21p^2r + 28qr^3}{7pqr} = \frac{7pqr}{7pqr} - \frac{21p^2r}{7pqr} + \frac{28qr^3}{7pqr} = 1 - \frac{3p}{q} + \frac{4r^2}{p}$$

This is also an acceptable answer

Notice that

$$\frac{pq - 3p^2 + 4qr^2}{pq} = \frac{pq}{pq} - \frac{3p^2}{pq} + \frac{4qr^2}{pq} = 1 - \frac{3p}{q} + \frac{4r^2}{p}$$

Exercises 6.2
Set I

Reduce each rational expression to lowest terms, or write "cannot be reduced."

1. $\dfrac{12m^3n}{4mn}$
2. $\dfrac{-6hk^4}{24hk}$
3. $\dfrac{15a^4b^3c^2}{-35ab^5c}$
4. $\dfrac{40e^2f^2g}{16e^4fg^3}$
5. $\dfrac{40x - 8x^2}{5x^2 + 10x}$
6. $\dfrac{16y - 16y^3}{24y^2 - 24y}$
7. $\dfrac{c^2 - 4}{4}$
8. $\dfrac{9 + d^2}{9}$
9. $\dfrac{24w^2x^3 - 16wx^4}{18w^3x - 12w^2x^2}$
10. $\dfrac{18c^5d + 45c^4d^2}{12c^2d^3 + 30cd^4}$
11. $\dfrac{x^2 - 16}{x^2 - 9x + 20}$
12. $\dfrac{x^2 - 2x - 15}{x^2 - 9}$
13. $\dfrac{2x^2 - xy - y^2}{y^2 - 4xy + 3x^2}$
14. $\dfrac{2k^2 + 3hk - 5h^2}{3h^2 - 7hk + 4k^2}$
15. $\dfrac{2x^2 - 3x - 9}{12 - 7x + x^2}$
16. $\dfrac{15 + 7y - 2y^2}{4y^2 - 21y + 5}$
17. $\dfrac{2y^2 + xy - 6x^2}{3x^2 + xy - 2y^2}$
18. $\dfrac{10y^2 + 11xy - 6x^2}{4x^2 - 4xy - 15y^2}$
19. $\dfrac{2x^2 - 9x - 5}{2x^2 + 5x + 3}$
20. $\dfrac{3x^2 - 11x - 4}{3x^2 + 5x + 2}$
21. $\dfrac{a^3 - 1}{1 - a^2}$
22. $\dfrac{x^3 + y^3}{y^2 - x^2}$
23. $\dfrac{x^2 + 4}{x^2 + 4x + 4}$
24. $\dfrac{9 + y^2}{9 + 6y + y^2}$
25. $\dfrac{13x^3y^2 - 26xy^3 + 39xy}{13x^2y^2}$
26. $\dfrac{21m^2n^3 - 35m^3n^2 - 14mn}{7m^2n^2}$
27. $\dfrac{6a^2bc^2 - 4ab^2c^2 + 12bc}{6abc}$
28. $\dfrac{8a^3b^2c - 4a^2bc - 10ac}{4abc}$

Writing Problems

Express the answers in your own words and in complete sentences.

1. Explain how to reduce a rational expression to lowest terms.
2. Find and describe any errors in the following:

$$\frac{\overset{x}{\cancel{2x^{2}}} - \overset{5}{\cancel{15}}}{\underset{1}{\cancel{2x}} - \underset{1}{\cancel{3}}} = x + 5$$

Exercises 6.2
Set II

Reduce each rational expression to lowest terms, or write "cannot be reduced."

1. $\dfrac{15c^2d^5}{12c^2d^3}$

2. $\dfrac{24x^4y^2}{48x^6y^3}$

3. $\dfrac{14e^5f^2g^3}{-42e^3f^4g}$

4. $\dfrac{x^2+9}{x^2-9}$

5. $\dfrac{32m^2-24m^3}{80m^3-60m^4}$

6. $\dfrac{x^2+x-6}{x^2+6x+9}$

7. $\dfrac{6}{6-x}$

8. $\dfrac{2x^2-7x-15}{2x^2+7x-15}$

9. $\dfrac{12a^4b-30a^3b^2}{18a^2b^2-45ab^3}$

10. $\dfrac{x-4}{x^2-16}$

11. $\dfrac{h^2-2h-24}{h^2-36}$

12. $\dfrac{5x^2-11x+2}{5x^2+7x+2}$

13. $\dfrac{24x^2+14xy-5y^2}{3y^2-14xy+8x^2}$

14. $\dfrac{8x^3+1}{8x^3+4x^2+2x+1}$

15. $\dfrac{6x^2+5x-4}{4-3x-10x^2}$

16. $\dfrac{x+4}{3x^2+9x-12}$

17. $\dfrac{21m^2-mn-2n^2}{4n^2-9nm-9m^2}$

18. $\dfrac{2x^3-x^2-8x+4}{2-3x-2x^2}$

19. $\dfrac{5x^2-4x-1}{5x^2+7x+2}$

20. $\dfrac{6x^2-x-2}{8x-10x^2-3x^3}$

21. $\dfrac{4x^2-w^2}{w^3-8x^3}$

22. $\dfrac{5-x}{9x^2-39x-30}$

23. $\dfrac{x^2+16}{x^2+8x+16}$

24. $\dfrac{x^4+4}{x^4-4}$

25. $\dfrac{-30m^4n^2+60m^2n^3-45m^3n}{15m^3n^2}$

26. $\dfrac{15xyz^3-12xz^2+10y^2z}{6y^2z}$

27. $\dfrac{24abc+18b^2c-8bc}{12ac}$

28. $\dfrac{32x^3yz-12xy^2+48xy^2z}{-16xy^2z}$

6.3 Multiplying and Dividing Rational Expressions

Multiplying Rational Expressions

Multiplication of rational expressions is defined as follows:

The definition of multiplication of rational expressions

If P, Q, R, and S represent polynomials and if $Q \neq 0$ and $S \neq 0$, then

$$\frac{P}{Q} \cdot \frac{R}{S} = \frac{P \cdot R}{Q \cdot S}$$

In practice, however, we can often reduce the resulting rational expression. Therefore, we give the following suggestions for multiplying rational expressions:

Multiplying rational expressions

1. Factor any numerators or denominators that contain more than one term.
2. Divide the numerators and denominators by all factors common to both. (The common factors can be, but do not have to be, in the same rational expression.)
3. The numerator of the product is the product of all the factors remaining in the numerators, and the denominator of the product is the product of all the factors remaining in the denominators. If the only factor remaining in the numerator is 1, that 1 must be *written.*

We can write the product as a single rational expression either *after* we divide by the common factors (see Examples 1a, 1c, and 1d) or *before* we divide by the common factors (see Example 1b).

EXAMPLE 1 Examples of multiplying rational expressions:

a. $\frac{a}{3} \cdot \frac{1}{a^2} = \frac{\cancel{a}^{1}}{3} \cdot \frac{1}{\cancel{a^2}_{a}} = \frac{1}{3a}$

b. $\frac{2y^3}{3x^2} \cdot \frac{12x}{5y^2} = \frac{2\cancel{y^3}^{y} \cdot \cancel{12x}^{4}}{\cancel{3x^2}_{x} \cdot 5\cancel{y^2}_{1}} = \frac{8y}{5x}$

c. $\frac{10xy^3}{x^2 - y^2} \cdot \frac{2x^2 + xy - y^2}{15x^2y} = \frac{\cancel{10xy^3}^{2y^2}}{\cancel{(x + y)}_{1}(x - y)} \cdot \frac{\cancel{(x + y)}^{1}(2x - y)}{\cancel{15x^2y}_{3x}}$

$= \frac{2y^2(2x - y)}{3x(x - y)}$

d. $\frac{a^3 - b^3}{a^2 - b^2 - a + b} \cdot \frac{8a^2b}{4a^3 + 4a^2b + 4ab^2}$

A difference of two cubes ($a^3 - b^3$)

Factoring by grouping: $(a + b)(a - b) - 1(a - b) = (a - b)(a + b - 1)$

$= \frac{\cancel{(a - b)}^{1}\cancel{(a^2 + ab + b^2)}^{1}}{\cancel{(a - b)}_{1}(a + b - 1)} \cdot \frac{\cancel{8}^{2}aab}{\cancel{4}_{1}a\cancel{(a^2 + ab + b^2)}_{1}}$

$= \frac{2ab}{a + b - 1}$

Now that we have discussed multiplication of rational expressions, we can give an alternative method for reducing rational expressions, based on the definition of multiplication and the identity property of multiplication. We can reduce the rational expression $\frac{P \cdot C}{Q \cdot C}$ as follows (assuming $Q \neq 0$ and $C \neq 0$):

$$\frac{P \cdot C}{Q \cdot C} = \frac{P}{Q} \cdot \frac{C}{C} = \frac{P}{Q} \cdot 1 = \frac{P}{Q}$$

Using the definition of multiplication (first step); $\frac{C}{C} = 1$ (second step); $\frac{P}{Q} \cdot 1 = \frac{P}{Q}$ (third step)

Dividing Rational Expressions

Recall that two numbers are the *multiplicative inverses* (or *reciprocals*) of each other if their product is 1. This implies that the multiplicative inverse of $\frac{a}{b}$ is $\frac{b}{a}$, since $\frac{a}{b} \cdot \frac{b}{a} = 1$. We use these facts in *dividing* rational expressions.

The rule for dividing rational expressions is as follows:

Dividing rational expressions

> Multiply the dividend by the multiplicative inverse (the reciprocal) of the divisor.
>
> $$\frac{P}{Q} \div \frac{S}{T} = \frac{P}{Q} \cdot \frac{T}{S}$$
>
> ($\frac{P}{Q}$: The dividend; $\frac{S}{T}$: The divisor)
>
> where P, Q, S, and T represent polynomials and $Q \neq 0$, $S \neq 0$, and $T \neq 0$.

This method works (in algebra *and* in arithmetic) because a division problem is equivalent to a rational expression $\left(\text{that is, } P \div Q = \frac{P}{Q}\right)$ and because of the multiplicative identity and multiplicative inverse properties. That is,

Writing the division problem as a single rational expression

$\frac{T}{S}$ is the multiplicative inverse of the divisor

$$\frac{P}{Q} \div \frac{S}{T} = \frac{\frac{P}{Q}}{\frac{S}{T}} = \frac{\frac{P}{Q}}{\frac{S}{T}} \cdot \frac{\frac{T}{S}}{\frac{T}{S}} = \frac{\frac{P}{Q} \cdot \frac{T}{S}}{\frac{S}{T} \cdot \frac{T}{S}} = \frac{\frac{P}{Q} \cdot \frac{T}{S}}{1} = \frac{P}{Q} \cdot \frac{T}{S}$$

The value of this rational expression is 1, and multiplying a number by 1 does not change its value

$\frac{S}{T} \cdot \frac{T}{S} = 1$

Therefore, $\frac{P}{Q} \div \frac{S}{T} = \frac{P}{Q} \cdot \frac{T}{S}$.

EXAMPLE 2 Examples of dividing rational expressions:

a. $$\frac{3y^3 - 3y^2}{16y^5 + 8y^4} \div \frac{y^2 + 2y - 3}{4y + 12} = \frac{3y^3 - 3y^2}{16y^5 + 8y^4} \cdot \frac{4y + 12}{y^2 + 2y - 3}$$

$$= \frac{3y^2(y - 1)}{8y^4(2y + 1)} \cdot \frac{4(y + 3)}{(y - 1)(y + 3)} = \frac{3}{2y^2(2y + 1)}$$

b. $$\frac{x^2 + x - 2}{x + 2} \div \frac{x^2 + 2x - 3}{x + 3} = \frac{x^2 + x - 2}{x + 2} \cdot \frac{x + 3}{x^2 + 2x - 3}$$

$$= \frac{(x + 2)(x - 1)}{x + 2} \cdot \frac{x + 3}{(x + 3)(x - 1)} = 1$$

c. $$\frac{y^2 - x^2}{4xy - 2y^2} \div \frac{2x - 2y}{2x^2 + xy - y^2} = \frac{y^2 - x^2}{4xy - 2y^2} \cdot \frac{2x^2 + xy - y^2}{2x - 2y}$$

$$= \frac{(y + x)(y - x)}{2y(2x - y)} \cdot \frac{(x + y)(2x - y)}{2(x - y)}$$

$$= \frac{(x + y)^2 (y - x)}{4y(x - y)} = \frac{(x + y)^2 (-1)(x - y)}{4y(x - y)}$$

$$= \frac{(-1)(x + y)^2}{4y} = -\frac{(x + y)^2}{4y}$$

Order of Operations When Multiplication and Division Are Involved

The usual order of operations applies when we're multiplying and dividing rational expressions. If division is interpreted as *and changed to* multiplication, then the multiplication can be done in any order (see Example 3).

EXAMPLE 3 Perform the indicated operations.

a. $\frac{3x}{5y} \div \frac{7x^2}{6y^3} \div \frac{4y}{9z}$

SOLUTION 1 Working from left to right, we have

$$\frac{3x}{5y} \div \frac{7x^2}{6y^3} \div \frac{4y}{9z} = \left(\frac{3x}{5y} \div \frac{7x^2}{6y^3}\right) \div \frac{4y}{9z} = \left(\frac{3x}{5y} \cdot \frac{6y^3}{7x^2}\right) \div \frac{4y}{9z} = \frac{18y^2}{35x} \cdot \frac{9z}{4y} = \frac{81yz}{70x}$$

SOLUTION 2 If we change all divisions to multiplications,

$$\frac{3x}{5y} \div \frac{7x^2}{6y^3} \div \frac{4y}{9z} = \frac{3x}{5y} \cdot \frac{6y^3}{7x^2} \cdot \frac{9z}{4y} = \frac{81yz}{70x}$$

b. $\frac{7x^2}{3y} \div \frac{5x}{2y^3} \cdot \frac{3y}{4x}$

SOLUTION 1 Working from left to right, we have

$$\frac{7x^2}{3y} \div \frac{5x}{2y^3} \cdot \frac{3y}{4x} = \left(\frac{7x^2}{3y} \div \frac{5x}{2y^3}\right) \cdot \frac{3y}{4x} = \left(\frac{7x^2}{3y} \cdot \frac{2y^3}{5x}\right) \cdot \frac{3y}{4x} = \frac{14xy^2}{15} \cdot \frac{3y}{4x} = \frac{7y^3}{10}$$

SOLUTION 2 If we change the division to multiplication,

$$\frac{7x^2}{3y} \div \frac{5x}{2y^3} \cdot \frac{3y}{4x} = \frac{7x^2}{3y} \cdot \frac{2y^3}{5x} \cdot \frac{3y}{4x} = \frac{7y^3}{10}$$

The Three Signs of a Rational Expression

Every rational expression has three signs associated with it, even if those signs are not written: the sign of the rational expression itself, the sign of the numerator, and the sign of the denominator. Consider the rational expression $\frac{5}{8}$:

The sign of the numerator

The sign of the rational expression → $+\frac{+5}{+8}$

The sign of the denominator

Recall that $\frac{-1}{-1} = 1$ and that $(-1)(-1) = 1$, and remember that the multiplicative identity property guarantees that when we multiply a number by 1, the value of the number is unchanged. Therefore, we can multiply a rational expression by $\frac{-1}{-1}$ or by $(-1)(-1)$ without changing its value. Recall, too, that $\frac{-1}{1} = -1$ and that $\frac{1}{-1} = -1$. Therefore, in the expression $(-1)(-1)$, we can replace the second factor with $\frac{-1}{1}$ or with $\frac{1}{-1}$, so $(-1)\left(\frac{-1}{1}\right) = 1$ and $(-1)\left(\frac{1}{-1}\right) = 1$.

Let's see what happens to the three signs of the rational expression $\frac{5}{8}$ as we multiply it by $\frac{-1}{-1}$, by $(-1)\left(\frac{-1}{1}\right)$, and by $(-1)\left(\frac{1}{-1}\right)$, all of which equal 1.

$$+\frac{+5}{+8} = +\left(\frac{-1}{-1}\right)\left(\frac{5}{8}\right) = +\frac{(-1)(5)}{(-1)(8)} = +\frac{-5}{-8}$$ Multiplying by $\frac{-1}{-1}$ leaves the value unchanged

The sign of the numerator and the sign of the denominator are different from those in $+\frac{+5}{+8}$; that is, *two* of the three signs are different

$$+\frac{+5}{+8} = +(-1)\left(\frac{-1}{1}\right)\left(\frac{5}{8}\right) = -\frac{(-1)(5)}{(1)(8)} = -\frac{-5}{+8}$$ Multiplying by $(-1)\left(\frac{-1}{1}\right)$ leaves the value unchanged

The sign of the rational expression and the sign of the numerator are different from those in $+\frac{+5}{+8}$; that is, *two* of the three signs are different

$$+\frac{+5}{+8} = +(-1)\left(\frac{1}{-1}\right)\left(\frac{5}{8}\right) = -\frac{(1)(5)}{(-1)(8)} = -\frac{+5}{-8}$$ Multiplying by $(-1)\left(\frac{1}{-1}\right)$ leaves the value unchanged

The sign of the rational expression and the sign of the denominator are different from those in $+\frac{+5}{+8}$; that is, *two* of the three signs are different

Therefore, $\frac{5}{8} = \frac{-5}{-8} = -\frac{-5}{8} = -\frac{5}{-8}$.

You might also verify, by multiplying $\frac{-3}{5}$ by $\frac{-1}{-1}$, by $(-1)\left(\frac{-1}{1}\right)$, and by $(-1)\left(\frac{1}{-1}\right)$, that $\frac{-3}{5} = \frac{3}{-5} = -\frac{3}{5} = -\frac{-3}{-5}$.

We see, then, that changing two of the three signs of a rational expression does not change the value of that rational expression.

The rule of signs for rational expressions

> If any *two* of the three signs of a rational expression are changed, the value of the rational expression is unchanged; that is, the new rational expression is equivalent to the original one.

This rule of signs is sometimes helpful when we perform addition, multiplication, and so forth, on rational expressions and also when we reduce rational expressions.

Note The rule of signs for rational expressions implies that when a rational expression contains a negative factor, the negative sign of that factor can be written in front of the numerator *or* in front of the rational expression *or* in front of the denominator. For example, $\frac{-2}{3x} = -\frac{2}{3x} = \frac{2}{-3x}$. We can even think of the rule of signs as permitting us to *move* a negative sign from the numerator to in front of the rational expression, from the numerator to the denominator, and so on. (However, most authors and instructors prefer that the negative sign *not* be in the *denominator.*)

When the numerator and denominator each have just one term, many instructors believe that the rational expression should not contain more than one negative sign. For example, many instructors insist that $\frac{-5x}{-y}$ be rewritten as $\frac{5x}{y}$, and that $-\frac{-4z}{-3w}$ be rewritten as $-\frac{4z}{3w}$ or as $\frac{-4z}{3w}$.

EXAMPLE 4 Find the missing term in each expression.

a. $-\dfrac{-5}{xy} = \dfrac{?}{xy}$

SOLUTION Because the signs of the *denominators* are the same in both rational expressions (they are both understood to be +) and the signs of the *rational expressions* are different, the signs of the *numerators* must be different. Therefore,

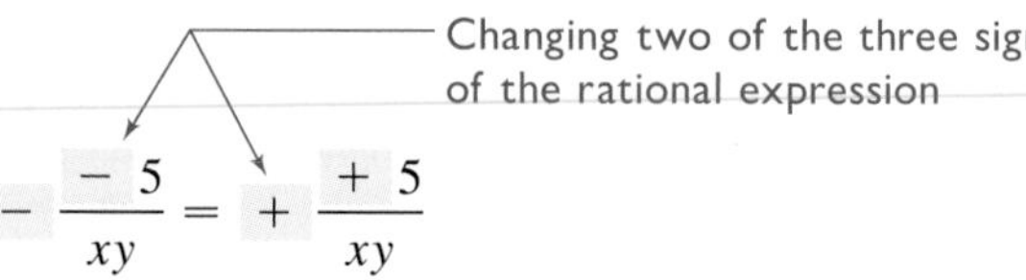

$$-\frac{-5}{xy} = +\frac{+5}{xy}$$

The missing term is +5, or 5.

b. $\dfrac{8}{-x} = -\dfrac{8}{?}$

SOLUTION Because the signs of the *numerators* are the same in both rational expressions (they are both understood to be +) and the signs of the *rational expressions* are different, the signs of the *denominators* must be different. Therefore,

Changing two of the three signs of the rational expression

$$+\frac{8}{-x} = -\frac{8}{+x}$$

The missing term is $+x$, or x.

In Example 5, we use the fact that $b - a$ can always be substituted for $-(a - b)$.

EXAMPLE 5 For each of the following, find the missing term.

a. $-\dfrac{1}{2 - x} = \dfrac{1}{?}$

SOLUTION The signs of the numerators are the same (both are understood to be +), and the signs of the rational expressions are different. Therefore, the signs of the denominators must also be different.

Changing two of the three signs of the rational expression

$$-\frac{1}{+(2 - x)} = +\frac{1}{-(2 - x)} = \frac{1}{x - 2}$$

Substituting $x - 2$ for $-(2 - x)$

Therefore, the missing term is $x - 2$.

b. $\dfrac{y - 5}{3} = -\dfrac{?}{3}$

SOLUTION The signs of the denominators are both understood to be +, and the signs of the rational expressions are different. Therefore, the signs of the numerators must be different.

Changing two of the three signs of the rational expression

Substituting $5 - y$ for $-(y - 5)$

$$+\frac{+(y - 5)}{3} = -\frac{-(y - 5)}{3} = \frac{5 - y}{3}$$

The missing term is $5 - y$.

c. $\dfrac{x - y}{(a - b)(c + d)} = \dfrac{?}{(b - a)(c + d)}$

SOLUTION The signs of the denominators are different because $b - a = -(a - b)$. If we multiply both numerator and denominator of the first rational expression by -1, we will have the denominator that we want in the second rational expression. (Multiplying both numerator and denominator by -1 gives us what we want in this case because the signs of the rational expressions are the *same*.)

Multiplying both numerator and denominator by -1

Substituting $y - x$ for $-(x - y)$

$$\frac{x - y}{(a - b)(c + d)} = \frac{-1\,(x - y)}{-1\,(a - b)(c + d)} = \frac{y - x}{(b - a)(c + d)}$$

Substituting $b - a$ for $-(a - b)$

The missing term is $y - x$.

Exercises 6.3
Set I

In Exercises 1–28, perform the indicated operations.

1. $\dfrac{27x^4y^3}{22x^5yz} \cdot \dfrac{55x^2z^2}{9y^3z}$

2. $\dfrac{13b^2c^4}{42a^4b^3} \cdot \dfrac{35a^3bc^2}{39ac^5}$

3. $\dfrac{mn^3}{18n^2} \div \dfrac{5m^4}{24m^3n}$

4. $\dfrac{27k^3}{h^5k} \div \dfrac{15hk^2}{-4h^4}$

5. $\dfrac{15u - 6u^2}{10u^2} \cdot \dfrac{15u^3}{35 - 14u}$

6. $\dfrac{-22v^2}{63v + 84} \cdot \dfrac{42v^3 + 56v^2}{55v^3}$

7. $\dfrac{-15c^4}{40c^3 - 24c^2} \div \dfrac{35c}{35c^2 - 21c}$

8. $\dfrac{40d - 30d^2}{d^2} \div \dfrac{24d^2 - 18d^3}{12d^3}$

9. $\dfrac{d^2e^2 - d^3e}{12e^2d} \div \dfrac{d^2e^2 - de^3}{3e^2d + 3e^3}$

10. $\dfrac{9m^2n + 3mn^2}{16mn^2} \div \dfrac{2mn^2 - m^2n}{10mn^2 - 20n^3}$

11. $\dfrac{w^2 - 2w - 8}{6w - 24} \cdot \dfrac{5w^2}{w^2 - 3w - 10}$

12. $\dfrac{-15k^3}{8k + 32} \div \dfrac{15k^2 - 5k^3}{k^2 + k - 12}$

13. $\dfrac{4a^2 + 8ab + 4b^2}{a^2 - b^2} \div \dfrac{6ab + 6b^2}{b - a}$

14. $\dfrac{u^2 - v^2}{7u^2 - 14uv + 7v^2} \div \dfrac{2u^2 + 2uv}{14v - 14u}$

15. $\dfrac{4 - 2a}{2a + 2} \div \dfrac{2a^3 - 16}{a^2 + 2a + 1}$

16. $\dfrac{18 + 6a}{4 - 2a} \div \dfrac{2a^3 + 54}{4a - 8}$

17. $\dfrac{x^3 + y^3}{2x - 2y} \div \dfrac{x^2 - xy + y^2}{x^2 - y^2}$

18. $\dfrac{x^3 - y^3}{3x + 3y} \div \dfrac{x^2 + xy + y^2}{x^2 - y^2}$

19. $\dfrac{e^2 + 10ef + 25f^2}{e^2 - 25f^2} \cdot \dfrac{3e - 3f}{f - e} \div \dfrac{e + 5f}{5f - e}$

20. $\dfrac{3x - y}{y + x} \div \dfrac{y^2 + yx - 12x^2}{4x^2 + 5xy + y^2} \cdot \dfrac{10x^2 - 8xy}{8xy - 10x^2}$

21. $\dfrac{(x + y)^2 + x + y}{(x - y)^2 - x + y} \cdot \dfrac{x^2 - 2xy + y^2}{x^2 + 2xy + y^2} \div \dfrac{x - y}{x + y}$

22. $\dfrac{ac + bc + ad + bd}{ac - ad - bc + bd} \cdot \dfrac{c^2 - 2cd + d^2}{a^2 + 2ab + b^2} \div \dfrac{c + d}{a - b}$

23. $\dfrac{7x^2}{15y^3} \div \dfrac{14y}{5x} \div \dfrac{8x^2}{3y}$

24. $\dfrac{10s^3}{9t^2} \div \dfrac{25t}{19s^2} \div \dfrac{38s^4}{3t^2}$

25. $\dfrac{11x^3}{7xy} \div \dfrac{22y}{3x} \cdot \dfrac{14xy^3}{9y^2}$

26. $\dfrac{12a^3b}{5b^4} \div \dfrac{4b^3}{6a} \cdot \dfrac{8ab^3}{48ab}$

27. $\dfrac{x - 5}{x + 3y} \div \dfrac{x - 3y}{x^2 - 9y^2} \cdot \dfrac{x}{xy - 5y}$

28. $\dfrac{x - 3}{x + 4y} \div \dfrac{x - 4y}{x^2 - 16y^2} \cdot \dfrac{y}{x^2 - 3x}$

In Exercises 29–38, find the missing term.

29. $-\frac{5}{8} = \frac{5}{?}$

30. $-\frac{6}{-k} = \frac{?}{k}$

31. $\frac{-x}{5} = \frac{x}{?}$

32. $\frac{6}{-k} = \frac{?}{k}$

33. $\frac{8 - y}{4y - 7} = \frac{y - 8}{?}$

34. $\frac{w - 2}{5 - w} = \frac{?}{w - 5}$

35. $\frac{u - v}{a - b} = \frac{v - u}{?}$

36. $\frac{2 - x}{y - 5} = \frac{?}{5 - y}$

37. $\frac{a - b}{(3a + 2b)(a - 5b)} = \frac{?}{(3a + 2b)(5b - a)}$

38. $\frac{(e + 4f)(7e - 3f)}{2e - f} = \frac{(e + 4f)(3f - 7e)}{?}$

Writing Problems

Express the answer in your own words and in complete sentences.

1. Using the correct terminology, explain how to divide one rational expression by another.

Exercises 6.3 Set II

In Exercises 1–28, perform the indicated operations.

1. $\frac{11m^4p^2}{30mn^2} \cdot \frac{21n^5p}{22m^3np^4}$

2. $\frac{65x^5y^3}{78x^2y^4} \cdot \frac{4y^2}{x^4}$

3. $\frac{5m^2n^3}{18mn^2} \div \frac{15n^3}{36m^2n}$

4. $\frac{x + 3}{6} \cdot \frac{1}{3x + 9}$

5. $\frac{7b}{24a^2b^2} \cdot \frac{32a^2b^3}{7 - 21a}$

6. $\frac{34a^3b^3}{7ab^2} \cdot \frac{21a^3b^2}{17a^2b}$

7. $\frac{-7f^2}{40f + 16f^2} \div \frac{28f^3}{30f^3 + 12f^4}$

8. $\frac{12e^2 - 18e}{25e^4} \cdot \frac{15e^2}{18e^2 - 27e}$

9. $\frac{6x^2y - 2xy^2}{9xy + 18y^2} \div \frac{4x^2y^2 - 12x^3y}{3x^2 + 6xy}$

10. $\frac{9x^2 - 30x + 25}{9x^2 - 25} \div \frac{3x^3 + 3x^2}{3x^2 - 2x - 5}$

11. $\frac{6v^2 - 36v}{25 - 5v} \cdot \frac{v^2 - 11v + 30}{-12v^2}$

12. $\frac{x^3 + 27}{x^2 - 9} \div \frac{x^2 - 3x + 9}{x^2 + 9}$

13. $\frac{5h^2k - h^3}{9hk + 18k^2} \div \frac{h^2 - 7hk + 10k^2}{3h^2 - 12k^2}$

14. $\frac{x^2 - 2x - 3}{x^4 - 10x^2 + 9} \cdot \frac{x - 1}{x + 3}$

15. $\frac{32 + 4b^3}{b^2 - b - 6} \div \frac{b^2 - 2b + 4}{15b - 5b^2}$

16. $\frac{2a^2 - 7a + 6}{a^2 + a - 6} \cdot \frac{2a^2 + 13a + 15}{15a - 7a^2 - 2a^3}$

17. $\frac{5y^2 + 15yz + 45z^2}{9z^2 + 3zy} \div \frac{2y^3 - 54z^3}{9z^2 - y^2}$

18. $\frac{27x^3 + 8}{9x^2 - 4} \div (9x^2 - 6x + 4)$

19. $\frac{6uv - 6v^2 + 6v}{10uv} \cdot \frac{(u + v) - (u + v)^2}{(u - v)^2 + (u - v)} \div \frac{3v^2 + 3vu}{5uv - 5u^2}$

20. $\frac{x^2 + 6x + 9 - y^2}{5x - 5y + 15} \cdot \frac{15x - 15y}{9x^2 - 9y^2} \cdot \frac{1}{6 + 2x + 2y}$

21. $\frac{10nm - 15n^2}{3n + 2m} \div \frac{9mn - 6m^2}{4m^2 + 12mn + 9n^2} \div \frac{4m^2 - 9n^2}{27n - 18m}$

22. $\frac{xs - ys + xt - yt}{sx + tx + sy + ty} \cdot \frac{sx + sy - ty - xt}{xs - ys - tx + yt}$

23. $\frac{9s^2}{7t^3} \div \frac{15t}{13s^2} \div \frac{26s}{9t}$

24. $\frac{15m^3}{4n^2} \div \frac{12n}{17m^3} \cdot \frac{3m^4}{34n^2}$

25. $\frac{8u^3}{9uv} \div \frac{6v}{5u^4} \cdot \frac{3uv^3}{16v^2}$

26. $\frac{6a^3b}{19b^4} \div \frac{18b^3}{33a^4} \div \frac{11ab^3}{38ab}$

27. $\frac{x - 7}{x - 2y} \div \frac{x + 2y}{x^2 - 4y^2} \cdot \frac{x}{xy - 7y}$

28. $\frac{4s - 5}{4s + 5t} \div \frac{4s - 5t}{16s^2 - 25t^2} \cdot \frac{t}{8st - 10t}$

In Exercises 29–38, find the missing term.

29. $-\frac{2}{7} = \frac{2}{?}$

30. $\frac{-8}{x} = -\frac{?}{x}$

31. $\frac{-14}{3m} = \frac{14}{?}$

32. $\frac{4 - x}{x + 3} = -\frac{?}{x + 3}$

33. $\frac{5 - x}{3y - 2} = \frac{x - 5}{?}$

34. $\frac{?}{x + 5} = \frac{x + 10}{-(x + 5)}$

35. $\frac{x - y}{c - d} = \frac{?}{d - c}$

36. $\frac{y - 2}{?} = \frac{2 - y}{y + 7}$

37. $\frac{5t - 2u}{(t - 4u)(3t + u)} = \frac{?}{(4u - t)(3t + u)}$

38. $\frac{(8 - x)(8 + x)}{?} = \frac{(x - 8)(x + 8)}{(3 + x)(5 + x)}$

6.4 Finding the Least Common Denominator (LCD)

We ordinarily use the **least common denominator (LCD)** when we add or subtract rational expressions that have different denominators. The LCD is the least common multiple (LCM) of all the denominators. (We discussed finding the LCM of *numbers* in an earlier chapter.) The LCD is the *smallest* expression that is a multiple of all the denominators; it is the smallest expression that all the denominators divide into *exactly* (that is, without a remainder).

It's easy to misunderstand the concept of the least common denominator because of its name. The word *least* makes many people think that the expression should be small, which is not correct. It is better to think of the LCD as being *at least as large* as the largest of the individual denominators.

To find the LCD of two or more rational expressions, we can reduce all the rational expressions to lowest terms and then follow the procedure described in the following box:

Finding the LCD of two or more rational expressions

1. Factor each denominator completely, if possible, expressing any repeated factors in exponential form.
2. Write down each different base that appears in any denominator.
3. Raise each base to the highest power to which it occurs in *any* denominator.
4. The LCD is the product of all the expressions found in step 3.

EXAMPLE 1

Find the LCD for $\frac{2}{x} + \frac{x}{x + 2}$.

SOLUTION

Step 1. The denominators cannot be factored.
Step 2. x, $(x + 2)$ All the different bases
Step 3. x^1, $(x + 2)^1$ The highest powers of the bases
Step 4. The LCD is $x(x + 2)$.

A Word of Caution In Example 1, you might think that the LCD was $(x + 2)$ because the other denominator, x, is a *part* of $(x + 2)$. However, x is a *term* of $x + 2$, *not* a factor of it, and so we need *both* x and $x + 2$ as *factors* of the LCD. Making a substitution will confirm this. Suppose that, in Example 1, $x = 3$. Then $\frac{2}{x} + \frac{x}{x + 2} = \frac{2}{3} + \frac{3}{5}$, and the LCD for $\frac{2}{3} + \frac{3}{5}$ is clearly $3 \cdot 5$, or 15, *not* 5.

EXAMPLE 2 Find the LCD for $\frac{2x-3}{x^2+10x+25} - \frac{5}{4x^2+20x} + \frac{4x-3}{x^2+2x-15}$.

SOLUTION

Step 1. $x^2 + 10x + 25 = (x+5)^2$
$4x^2 + 20x = 2^2 \cdot x \cdot (x+5)$
$x^2 + 2x - 15 = (x+5)\ (x-3)$
} The denominators in factored form

Step 2. $2,\ x,\ (x+5),\ (x-3)$ All the different bases

Step 3. $2^2,\ x^1,\ (x+5)^2,\ (x-3)^1$ The highest powers of the bases

Step 4. The LCD is $4x(x+5)^2(x-3)$.

EXAMPLE 3 Find the LCD for $\frac{x+2}{x^3+8} + \frac{5x}{x^2-2x+4} - \frac{3x^2}{x^3-2x^2}$.

SOLUTION It happens that two of these rational expressions are reducible:

$$\frac{x+2}{x^3+8} = \frac{\overset{1}{\cancel{x+2}}}{(\underset{1}{\cancel{x+2}})(x^2-2x+4)} = \frac{1}{x^2-2x+4}$$

and

$$\frac{3x^2}{x^3-2x^2} = \frac{3\overset{1}{\cancel{x^2}}}{\underset{1}{\cancel{x^2}}(x-2)} = \frac{3}{x-2}$$

Now the problem is reduced to $\frac{1}{x^2-2x+4} + \frac{5x}{x^2-2x+4} - \frac{3}{x-2}$, and so the denominators are $x^2 - 2x + 4$, $x^2 - 2x + 4$, and $x - 2$.

The LCD is $(x-2)^1(x^2-2x+4)^1$, or $(x-2)(x^2-2x+4)$.

Exercises 6.4

Set I

Find the LCD in each exercise. ***Do not add the rational expressions.***

1. $\frac{9}{25a^3} + \frac{7}{15a}$
2. $\frac{13}{18b^2} + \frac{11}{12b^4}$
3. $\frac{49}{60hk^3} + \frac{71}{90h^2k^4}$
4. $\frac{44}{42x^2y^2} - \frac{45}{49x^3y}$
5. $\frac{11}{2w-10} - \frac{15}{4w}$
6. $\frac{27}{2m^2} + \frac{19}{8m-48}$
7. $\frac{15b}{9b^2-c^2} + \frac{12c}{(3b-c)^2}$
8. $\frac{14e}{(5f-2e)^2} + \frac{27f}{4e^2-25f^2}$
9. $\frac{5}{2g^3} - \frac{3g-9}{g^2-6g+9} + \frac{12g}{4g^2-12g}$
10. $\frac{5y-30}{y^2-12y+36} + \frac{7}{9y^2} - \frac{15y}{3y^2-18y}$
11. $\frac{2x-5}{2x^2-16x+32} + \frac{4x+7}{x^2+x-20}$
12. $\frac{8k-1}{5k^2-30k+45} + \frac{3k-4}{k^2+4k-21}$
13. $\frac{35}{3e^2} - \frac{2e}{e^2-9} - \frac{13}{4e-12}$
14. $\frac{3}{8u^3} - \frac{5u-1}{6u^2+18u} - \frac{6u+7}{u^2-5u-24}$
15. $\frac{x^2+1}{12x^3+24x^2} - \frac{4x+3}{x^2-4x+4} - \frac{1}{x^2-4}$
16. $\frac{2y+5}{y^2+6y+9} + \frac{7y}{y^2-9} - \frac{11}{8y^2-24y}$

Writing Problems

Express the answer in your own words and in complete sentences.

1. Explain what the least common denominator of two or more denominators is.

Exercises 6.4
Set II

Find the LCD in each exercise. ***Do not add the rational expressions.***

1. $\dfrac{17}{50c} + \dfrac{23}{40c^2}$

2. $\dfrac{83}{12x^2} + \dfrac{5}{18x^4}$

3. $\dfrac{35}{24m^2n} + \dfrac{25}{63mn^3}$

4. $\dfrac{5}{28b^3c^2} - \dfrac{8}{49bc^5}$

5. $\dfrac{50}{9z - 63} - \dfrac{16}{3z}$

6. $\dfrac{18}{5 - 25x} + \dfrac{3}{10 - 50x}$

7. $\dfrac{6uv}{9v^2 - 16u^2} + \dfrac{12u^2v}{(4u - 3v)^2}$

8. $\dfrac{13x}{x^2 - 4} - \dfrac{15x^2}{x^2 + 4}$

9. $\dfrac{3}{25h} - \dfrac{20h^3}{5h^4 - 25h^3} + \dfrac{4h - 20}{h^2 - 10h + 25}$

10. $\dfrac{9}{4x^2 - 9} - \dfrac{7x}{6x + 9} + \dfrac{5}{2x^3 - x^2 - 3x}$

11. $\dfrac{5a - 3}{3a^2 - 30a + 72} + \dfrac{7a + 2}{a^2 - 12a + 36}$

12. $\dfrac{3x - 7}{15x^2 - 55x - 20} + \dfrac{5x - 20}{6x^2 + 20x + 6}$

13. $\dfrac{15}{4m^2} - \dfrac{11m}{2m^2 - 50} + \dfrac{2m - 5}{m^2 - 2m - 35}$

14. $\dfrac{5}{6x^2 + 30x} - \dfrac{7x}{2x^2 - 50} + \dfrac{8x + 40}{3x^2 - 9x - 30}$

15. $\dfrac{17}{2z^2 - 98} - \dfrac{9z + 11}{4z^3 - 28z^2} - \dfrac{13z}{z^2 - 14z + 49}$

16. $\dfrac{x - 5}{2x^3 + 7x^2 - 15x} - \dfrac{4x^2 + 9}{4x^3 - 12x^2 + 9x} + \dfrac{9}{4x^4 - 9x^2}$

6.5 Adding and Subtracting Rational Expressions

Adding Rational Expressions That Have the Same Denominator

Rational expressions that have the same denominator can be added by using the following procedure:

Adding rational expressions that have the same denominator

1. The numerator of the sum is the sum of the numerators, and the denominator of the sum is the same as any *one* of the denominators. That is, if P, Q, and R represent polynomials and if $Q \neq 0$, then

$$\frac{P}{Q} + \frac{R}{Q} = \frac{P + R}{Q}$$

2. Reduce the resulting rational expression to lowest terms whenever possible.

PROOF First recall that $\dfrac{P}{Q}$ is equivalent to $P\left(\dfrac{1}{Q}\right)$ and that $\dfrac{R}{Q}$ is equivalent to $R\left(\dfrac{1}{Q}\right)$. Therefore,

Factoring out $\dfrac{1}{Q}$

$$\frac{P}{Q} + \frac{R}{Q} = P\left(\frac{1}{Q}\right) + R\left(\frac{1}{Q}\right) = \frac{1}{Q}(P + R) = \frac{1}{Q} \cdot \frac{P + R}{1} = \frac{P + R}{Q}$$

■

EXAMPLE 1 Examples of adding rational expressions that have the same denominator:

Factoring numerator and denominator

a. $\dfrac{4}{6+3x} + \dfrac{2x}{6+3x} = \dfrac{4+2x}{6+3x} = \dfrac{2(2+x)}{3(2+x)} = \dfrac{2\overset{1}{\cancel{(2+x)}}}{3\underset{1}{\cancel{(2+x)}}} = \dfrac{2}{3}$

Factoring -3 from $15 - 3d$

b. $\dfrac{15}{d-5} + \dfrac{-3d}{d-5} = \dfrac{15+(-3d)}{d-5} = \dfrac{15-3d}{d-5} = \dfrac{-3(d-5)}{d-5} = \dfrac{-3\overset{1}{\cancel{(d-5)}}}{\underset{1}{\cancel{d-5}}} = -3$

Subtracting Rational Expressions That Have the Same Denominator

Any subtraction problem can be changed to an addition problem by using the definition of subtraction, $a - b = a + (-b)$, even when the problem involves rational expressions. Subtraction of rational expressions that have the same denominator can also be done by using the following procedure:

Subtracting rational expressions that have the same denominator

1. The numerator of the difference is the difference of the numerators, and the denominator of the difference is the same as any *one* of the denominators. That is, if P, Q, and R represent polynomials and if $Q \neq 0$, then

$$\frac{P}{Q} - \frac{R}{Q} = \frac{P-R}{Q}$$

2. Reduce the resulting rational expression to lowest terms whenever possible.

EXAMPLE 2 Examples of subtracting rational expressions that have the same denominator:

a. $\dfrac{3}{4a} - \dfrac{5}{4a} = \dfrac{3}{4a} + \left(-\dfrac{5}{4a}\right) = \dfrac{3}{4a} + \left(\dfrac{-5}{4a}\right) = \dfrac{3+(-5)}{4a} = \dfrac{-2}{4a} = -\dfrac{1}{2a}$ Changing subtraction to addition

or $\dfrac{3}{4a} - \dfrac{5}{4a} = \dfrac{3-5}{4a} = \dfrac{-2}{4a} = -\dfrac{1}{2a}$ Using the procedure in the box

b. $\dfrac{7}{x-2} - \dfrac{4}{x-2} = \dfrac{7}{x-2} + \left(-\dfrac{4}{x-2}\right) = \dfrac{7}{x-2} + \left(\dfrac{-4}{x-2}\right) = \dfrac{7+(-4)}{x-2} = \dfrac{3}{x-2}$

or $\dfrac{7}{x-2} - \dfrac{4}{x-2} = \dfrac{7-4}{x-2} = \dfrac{3}{x-2}$ Using the procedure in the box

(We show the second method only in part c.)

c. $\dfrac{4x}{2x-y} - \dfrac{2y}{2x-y} = \dfrac{4x-2y}{2x-y} = \dfrac{2(2x-y)}{2x-y} = \dfrac{2\overset{1}{\cancel{(2x-y)}}}{\underset{1}{\cancel{2x-y}}} = 2$

A Word of Caution In Example 2a, $-1/2a$ is an *incorrect* answer, and in Example 2b, $3/x - 2$ is an *incorrect* answer. That is, $-1/2a = -\dfrac{1}{2}a$, *not* $-\dfrac{1}{2a}$, and $3/x - 2 = \dfrac{3}{x} - 2$, *not* $\dfrac{3}{x-2}$.

A Word of Caution A common error in adding or subtracting rational expressions is to multiply the rational expressions by the same number. This is *incorrect* (unless that number is 1).

Correct method	*Incorrect method*
$\dfrac{6}{5a}+\dfrac{2}{5a}=\dfrac{8}{5a}$	$\dfrac{6}{5a}+\dfrac{2}{5a}=(5a)\left(\dfrac{6}{5a}\right)+(5a)\left(\dfrac{2}{5a}\right)=8$

In a subtraction problem, if the numerator of the rational expression being subtracted contains more than one term, we *must* put parentheses around that numerator when we rewrite the problem as a single rational expression (see Example 3).

EXAMPLE 3 Subtract $\dfrac{3}{x+5}-\dfrac{x+2}{x+5}$.

SOLUTION

Subtracting the *entire* numerator $(x+2)$

$$\frac{3}{x+5}-\frac{x+2}{x+5}=\frac{3-(x+2)}{x+5}=\frac{3-x-2}{x+5}=\frac{1-x}{x+5}$$

$\left(\dfrac{-x+1}{x+5}\right.$ is also a correct answer.$\left.\right)$

Adding and Subtracting Rational Expressions When the Denominators Are Additive Inverses

In an addition problem, when two denominators are not identical *but are the additive inverses of each other,* we can make the denominators equal by multiplying both numerator and denominator of one of the rational expressions by −1. (We must make the denominators equal, because we can add rational expressions only when the *denominators are the same.*)

A Word of Caution Whenever we change the sign of a numerator or denominator that contains more than one term, *we must put parentheses around that numerator or denominator.*

$$\frac{a+b}{c-d}=\frac{-a+b}{-c-d} \quad \text{Incorrect}$$

Rather,

$$\frac{a+b}{c-d}=\frac{-(a+b)}{-(c-d)}=\frac{-a-b}{d-c}$$

EXAMPLE 4 Add $\dfrac{9}{x-2}+\dfrac{5+x}{2-x}$.

SOLUTION The rational expressions do *not* have the same denominator; however, the denominators are the additive inverses of each other. Therefore, we will multiply both numerator and denominator of $\dfrac{5+x}{2-x}$ by −1; when we do this, we *must* put parentheses around the numerator and the denominator.

Multiplying both numerator and denominator by −1

$$\frac{9}{x-2}+\frac{5+x}{2-x}=\frac{9}{x-2}+\frac{-1\,(5+x)}{-1\,(2-x)}=\frac{9}{x-2}+\frac{-5-x}{x-2}=\frac{9-5-x}{x-2}=\frac{4-x}{x-2}$$

Replacing $-1(2-x)$ with $x-2$

Note Other acceptable answers for Example 4 include $\frac{x-4}{2-x}$ and $-\frac{x-4}{x-2}$.

In a subtraction problem in which the denominators are the additive inverses of each other, changing the sign of the rational expression being subtracted *and* the sign of its denominator converts the problem to an addition problem (see Example 5). (Recall that we can change any *two* of the three signs of a rational expression without changing the *value* of the expression.)

EXAMPLE 5 Subtract $\frac{9}{y-6} - \frac{3}{6-y}$.

SOLUTION Changing the sign of the rational expression being subtracted and the sign of its denominator

$$\frac{9}{y-6} - \frac{3}{(6-y)} = \frac{9}{y-6} + \frac{3}{-(6-y)} = \frac{9}{y-6} + \frac{3}{y-6} = \frac{12}{y-6}$$

A Word of Caution In Example 5, $12/y - 6$ is an *incorrect* answer, because $12/y - 6 = \frac{12}{y} - 6$, *not* $\frac{12}{y-6}$.

Adding and Subtracting Rational Expressions with Different Denominators

When denominators are not identical and are *not* the additive inverses of each other, we must convert the rational expressions to rational expressions with the same denominator; this conversion is often called "building fractions." We can build fractions by using the identity property of multiplication—that is, by multiplying by a rational expression whose numerator and denominator are equal. This produces a rational expression equivalent to the original one, since $\frac{x}{x} = 1$, $\frac{x+2}{x+2} = 1$, and so forth. We can also build fractions by using the fundamental property of rational expressions. The two methods are compared below.

Using the identity property of multiplication	*Using the fundamental property of rational expressions*
$\frac{x}{x+2} = \frac{x}{x+2} \cdot \frac{x}{x} = \frac{x^2}{x(x+2)}$	$\frac{x}{x+2} = \frac{x \cdot x}{(x+2) \cdot x} = \frac{x^2}{x(x+2)}$
$\frac{2}{x} = \frac{2}{x} \cdot \frac{x+2}{x+2} = \frac{2x+4}{x(x+2)}$	$\frac{2}{x} = \frac{2 \cdot (x+2)}{x \cdot (x+2)} = \frac{2x+4}{x(x+2)}$

We can use the procedure in the following box for adding and subtracting rational expressions that have different denominators. It is usually helpful to reduce the given rational expressions to lowest terms *before* using this procedure.

Adding or subtracting rational expressions that do not have the same denominator

1. Find the LCD.
2. Convert each rational expression to an equivalent expression that has the LCD as its denominator by using either the identity property of multiplication or the fundamental property of rational expressions. (Dividing the LCD by the denominator will give you the factor to multiply that numerator and denominator by.)
3. Add or subtract the resulting rational expressions.
4. Reduce the sum or difference to lowest terms whenever possible.

A Word of Caution A common error is to *reduce* rational expressions just after they have been converted to equivalent expressions with the LCD as the denominator. The addition then cannot be done, because the rational expressions no longer have the same denominator; we simply get an addition problem identical to the one we started with. The following, for example, is *not* the way to add rational expressions:

This is the problem we started with

$$\frac{2}{x} + \frac{x}{x+2} = \frac{2(x+2)}{x(x+2)} + \frac{x(x)}{(x+2)(x)} = \frac{2\cancel{(x+2)}}{x\cancel{(x+2)}} + \frac{x\cancel{(x)}}{(x+2)\cancel{(x)}} = \frac{2}{x} + \frac{x}{x+2}$$

We can't perform this addition because the denominators are different

EXAMPLE 6 Add $\dfrac{7}{18x^2y} + \dfrac{5}{8xy^4}$.

SOLUTION

Step 1. $18 = 2 \cdot 3^2$; $8 = 2^3$; the LCD is $2^3 \cdot 3^2x^2y^4$, or $72x^2y^4$.

$$\frac{\text{LCD}}{\text{denominator}} = \frac{72x^2y^4}{18x^2y} = 4y^3$$

Step 2. $\dfrac{7}{18x^2y} = \dfrac{7\ (4y^3)}{18x^2y\ (4y^3)} = \dfrac{28y^3}{72x^2y^4}$

$\dfrac{5}{8xy^4} = \dfrac{5\ (9x)}{8xy^4\ (9x)} = \dfrac{45x}{72x^2y^4}$

$$\frac{\text{LCD}}{\text{denominator}} = \frac{72x^2y^4}{8xy^4} = 9x$$

Step 3. $\dfrac{7}{18x^2y} + \dfrac{5}{8xy^4} = \dfrac{28y^3}{72x^2y^4} + \dfrac{45x}{72x^2y^4} = \dfrac{28y^3 + 45x}{72x^2y^4}$

Step 4. $\dfrac{28y^3 + 45x}{72x^2y^4}$ cannot be reduced.

EXAMPLE 7 Subtract $3 - \dfrac{2a}{a+2}$.

SOLUTION

Step 1. The LCD is $a + 2$.

Notice the parentheses

Step 2. $3 = \dfrac{3}{1} \cdot \dfrac{a+2}{a+2} = \dfrac{3(a+2)}{1(a+2)} = \dfrac{3a+6}{a+2}$

Step 3. $3 - \dfrac{2a}{a+2} = \dfrac{3a+6}{a+2} - \dfrac{2a}{a+2} = \dfrac{3a+6-2a}{a+2} = \dfrac{a+6}{a+2}$

Step 4. $\dfrac{a+6}{a+2}$ cannot be reduced.

In step 2 of Example 8, we will need to remove the grouping symbols in the *numerators* so that we can combine like terms in step 3; it is not *necessary* to remove the grouping symbols in the denominators, although it is acceptable to do so (and your instructor may want you to do so).

EXAMPLE 8 Subtract $\frac{z+1}{z+2} - \frac{z-1}{z-2}$.

SOLUTION

Step 1. The LCD is $(z+2)(z-2)$.

Notice the parentheses

Step 2. $$\frac{z+1}{z+2} = \frac{z+1}{z+2} \cdot \frac{z-2}{z-2} = \frac{(z+1)(z-2)}{(z+2)(z-2)} = \frac{z^2-z-2}{(z+2)(z-2)}$$

$$\frac{z-1}{z-2} = \frac{z-1}{z-2} \cdot \frac{z+2}{z+2} = \frac{(z-1)(z+2)}{(z-2)(z+2)} = \frac{z^2+z-2}{(z-2)(z+2)}$$

Step 3. $$\frac{z+1}{z+2} - \frac{z-1}{z-2} = \frac{z^2-z-2}{(z+2)(z-2)} - \frac{z^2+z-2}{(z-2)(z+2)}$$

$$= \frac{z^2-z-2-(z^2+z-2)}{(z+2)(z-2)}$$ The parentheses are essential

$$= \frac{z^2-z-2-z^2-z+2}{(z+2)(z-2)}$$

$$= \frac{-2z}{(z+2)(z-2)} = -\frac{2z}{z^2-4}$$ Changing the sign of the rational expression and the sign of the numerator

Step 4. $-\frac{2z}{z^2-4}$ cannot be reduced.

EXAMPLE 9 Add $\frac{3x+9}{x^2+7x+10} + \frac{14}{x^2+3x-10}$.

SOLUTION

Step 1. $x^2+7x+10 = (x+2)(x+5)$
$x^2+3x-10 = (x+5)(x-2)$
The LCD is $(x+2)(x+5)(x-2)$.

You can verify that $\frac{3x+9}{x^2+7x+10}$ cannot be reduced

Step 2. $$\frac{3x+9}{x^2+7x+10} = \frac{(3x+9)(x-2)}{(x+2)(x+5)(x-2)}$$

$$\frac{14}{x^2+3x-10} = \frac{14(x+2)}{(x+5)(x-2)(x+2)}$$

Step 3. $$\frac{3x+9}{x^2+7x+10} + \frac{14}{x^2+3x-10}$$

$$= \frac{(3x+9)(x-2)}{(x+2)(x+5)(x-2)} + \frac{14(x+2)}{(x-2)(x+5)(x+2)}$$

$$= \frac{3x^2+3x-18+14x+28}{(x+2)(x+5)(x-2)}$$

$$= \frac{3x^2+17x+10}{(x+2)(x+5)(x-2)}$$

Step 4. $$= \frac{(3x+2)\overset{1}{\cancel{(x+5)}}}{(x+2)\underset{1}{\cancel{(x+5)}}(x-2)}$$ Factoring the numerator and then dividing both numerator and denominator by $(x+5)$

$$= \frac{3x+2}{(x+2)(x-2)}$$

When several operations are indicated, we must be sure to follow the correct order of operations (see Example 10).

EXAMPLE 10 Perform the indicated operations: $\frac{x+3}{x+2}+\frac{1}{x-2}\div\frac{x+3}{x^2-4}$.

SOLUTION The division must be done *before* the addition.

$$\frac{x+3}{x+2}+\frac{1}{x-2}\div\frac{x+3}{x^2-4}=\frac{x+3}{x+2}+\frac{1}{x-2}\cdot\frac{x^2-4}{x+3}$$ Changing the division to multiplication

$$=\frac{x+3}{x+2}+\frac{1}{\cancel{x-2}}\cdot\frac{(x+2)\cancel{(x-2)}}{x+3}$$ Factoring x^2-4

$$=\frac{x+3}{x+2}+\frac{x+2}{x+3}$$ Simplifying; now the LCD is $(x+2)(x+3)$

$$=\frac{(x+3)(x+3)}{(x+2)(x+3)}+\frac{(x+2)(x+2)}{(x+3)(x+2)}$$ Using the fundamental property of rational expression

$$=\frac{x^2+6x+9+x^2+4x+4}{(x+3)(x+2)}$$ Adding the rational expressions

$$=\frac{2x^2+10x+13}{(x+3)(x+2)}$$

A Word of Caution In the addition in Example 10, it would be *incorrect,* in adding $\frac{x+3}{x+2}+\frac{x+2}{x+3}$, to divide the numerator of one rational expression and the denominator of the *other* one by $x+3$ (or by $x+2$). That is, the following is incorrect:

$$\frac{x+3}{x+2}+\frac{x+2}{x+3}=\frac{\cancel{x+3}}{\cancel{x+2}}+\frac{\cancel{x+2}}{\cancel{x+3}}$$

Such dividing can be done *only* in multiplication problems.

Exercises 6.5

Set I

Perform the indicated operations.

1. $\frac{5a}{a+2}+\frac{10}{a+2}$
2. $\frac{6b}{b-3}-\frac{18}{b-3}$
3. $\frac{8m}{2m-3n}-\frac{12n}{2m-3n}$
4. $\frac{21k}{4h+3k}+\frac{28h}{4h+3k}$
5. $\frac{x-3}{5x+7}-\frac{6x+3}{5x+7}$
6. $\frac{y+6}{8y+3}-\frac{5y+4}{8y+3}$
7. $\frac{6z-5}{3z-5}-\frac{2z-4}{3z-5}$
8. $\frac{2t-7}{7t-3}-\frac{6t-5}{7t-3}$
9. $\frac{-15w}{1-5w}-\frac{3}{5w-1}$
10. $\frac{-35}{6w-7}-\frac{30w}{7-6w}$
11. $\frac{7z}{8z-4}+\frac{6-5z}{4-8z}$
12. $\frac{8x}{6x-5}+\frac{10-4x}{5-6x}$
13. $\frac{12x-31}{12x-28}-\frac{18x-39}{28-12x}$
14. $\frac{13-30w}{15-10w}-\frac{10w+17}{10w-15}$
15. $\frac{9}{25a^3}+\frac{7}{15a}$
16. $\frac{13}{18b^2}+\frac{11}{12b^4}$
17. $\frac{49}{60h^2k^2}-\frac{71}{90hk^4}$
18. $\frac{44}{42x^2y^2}-\frac{45}{49x^3y}$
19. $\frac{5}{t}+\frac{2t}{t-4}$
20. $\frac{6r}{r-8}-\frac{11}{r}$

21. $\dfrac{3k}{8k-4}-\dfrac{7}{6k}$

22. $\dfrac{2}{9j}+\dfrac{4j}{18j+12}$

23. $x^2-\dfrac{3}{x}+\dfrac{5}{x-3}$

24. $y^2-\dfrac{2}{y}+\dfrac{3}{y-5}$

25. $\dfrac{2a+3b}{b}+\dfrac{b}{2a-3b}$

26. $\dfrac{3x-5y}{y}+\dfrac{y}{3x+5y}$

27. $\dfrac{2}{a+3}-\dfrac{4}{a-1}$

28. $\dfrac{5}{b-2}-\dfrac{3}{b+4}$

29. $\dfrac{x+2}{x-3}-\dfrac{x+3}{x-2}$

30. $\dfrac{x-4}{x+6}-\dfrac{x-6}{x+4}$

31. $\dfrac{x+2}{x^2+x-2}+\dfrac{3}{x^2-1}$

32. $\dfrac{5}{x^2-4}+\dfrac{x+1}{x^2-x-2}$

33. $\dfrac{2x}{x-3}-\dfrac{2x}{x+3}+\dfrac{36}{x^2-9}$

34. $\dfrac{m}{m+6}-\dfrac{m}{m-6}-\dfrac{72}{m^2-36}$

35. $\dfrac{x-2}{x^2+4x+4}-\dfrac{x+1}{x^2-4}$

36. $\dfrac{x-2}{x^2-1}-\dfrac{x+1}{x^2-2x+1}$

37. $\dfrac{4}{x^2+2x+4}+\dfrac{x-2}{x+2}$

38. $\dfrac{x+3}{x-9}+\dfrac{3}{x^2-3x+9}$

39. $\dfrac{5}{2g^3}-\dfrac{3g-9}{g^2-6g+9}+\dfrac{12g}{4g^2-12g}$

40. $\dfrac{5y-30}{y^2-12y+36}+\dfrac{7}{9y^2}-\dfrac{15y}{3y^2-18y}$

41. $\dfrac{2x-5}{2x^2-16x+32}+\dfrac{4x+7}{x^2+x-20}$

42. $\dfrac{8k-1}{5k^2-30k+45}+\dfrac{3k-4}{k^2+4k-21}$

43. $\dfrac{35}{3e^2}-\dfrac{2e}{e^2-9}-\dfrac{3}{4e-12}$

44. $\dfrac{3}{8u^3}-\dfrac{5u-1}{6u^2+18u}-\dfrac{6u+7}{u^2-5u-24}$

45. $\dfrac{x^2+1}{12x^3+24x^2}-\dfrac{4x+3}{x^2-4x+4}-\dfrac{1}{x^2-4}$

46. $\dfrac{2y+5}{y^2+6y+9}+\dfrac{7y}{y^2-9}-\dfrac{11}{8y^2-24y}$

47. $\dfrac{7}{3y^3-12y^2+48y}+\dfrac{y^2+4}{y^3+64}-\dfrac{y}{y^2+8y+16}$

48. $\dfrac{x^2+9}{x^3+27}+\dfrac{5}{2x^3-6x^2+18x}-\dfrac{x}{x^2+6x+9}$

49. $\dfrac{x-1}{x^3+x^2-9x-9}-\dfrac{x+3}{x^3-3x^2-x+3}$

50. $\dfrac{x+2}{x^3-2x^2-x+2}-\dfrac{x-1}{x^3+x^2-4x-4}$

51. $\dfrac{x+6}{x-5}+\dfrac{1}{x+4}\div\dfrac{x+6}{x^2-x-20}$

52. $\dfrac{2x-3}{x+4}+\dfrac{1}{x-1}\div\dfrac{2x-3}{x^2+3x-4}$

53. $\dfrac{3}{8x^2}-\dfrac{x+3}{4x^2-8x}\div\dfrac{x^2+2x-3}{x^2-2x}$

54. $\dfrac{8}{9x^2}-\dfrac{x+5}{3x^2-12x}\div\dfrac{x^2+8x+15}{x^2-4x}$

55. $\dfrac{x+3}{x-5}-\dfrac{x-2}{x^2-x-12}\div\dfrac{x+7}{x^2+3x-28}$

56. $\dfrac{x+4}{x-2}-\dfrac{x-6}{x^2+x-12}\div\dfrac{x+5}{x^2+2x-15}$

Writing Problems

Express the answers in your own words and in complete sentences.

1. Describe how to build $\dfrac{x}{x-4}$ to a rational expression with a denominator of $x^2-7x+12$.
2. Explain why multiplying a rational expression by $\dfrac{x-3}{x-3}$ does not change its value.
3. Explain why $\dfrac{5x}{x+3}-\dfrac{2x+1}{x+3}\neq\dfrac{3x+1}{x+3}$.
4. Find and describe the error(s) in the following:
$$\frac{x+2}{x-4}+\frac{x-4}{x+2}=1+1=2$$
5. Find and describe the error(s) in the following:
$$\frac{x+3}{x-5}+\frac{x-2}{x+3}=\frac{x+3+x-2}{x-5+x+3}=\frac{2x+1}{2x-2}$$
6. Find and describe the error(s) in the following:
$$\frac{x+3}{2x-7}-\frac{x+5}{2x-7}=\frac{x+3-x+5}{2x-7}=\frac{8}{2x-7}$$
7. Find and describe the error(s) in the following:
$$\frac{x+3}{x-5}+\frac{x+2}{x+1}$$
$$=(x-5)(x+1)\left(\frac{x+3}{x-5}\right)+(x-5)(x+1)\left(\frac{x+2}{x+1}\right)$$
$$=x^2+4x+3+x^2-3x-10=2x^2+x-7$$

Exercises 6.5
Set II

Perform the indicated operations.

1. $\frac{24}{8+c} + \frac{3c}{8+c}$
2. $\frac{35}{2-x} + \frac{15}{2-x}$
3. $\frac{30c}{4b-5c} - \frac{24b}{4b-5c}$
4. $\frac{16y}{4y+5} + \frac{20}{4y+5}$
5. $\frac{x-6}{2x+9} - \frac{4x+7}{2x+9}$
6. $\frac{s+3}{5s+9} - \frac{6s-1}{5s+9}$
7. $\frac{7u-11}{3u-13} - \frac{4u-9}{3u-13}$
8. $\frac{5t-12}{10t-7} - \frac{t-1}{10t-7}$
9. $\frac{-35t}{3-7t} - \frac{15}{7t-3}$
10. $\frac{x+3}{3x-5} - \frac{2x+4}{5-3x}$
11. $\frac{5u}{6u-15} + \frac{30-7u}{15-6u}$
12. $\frac{5+a}{2a-5} + \frac{3a-2}{5-2a}$
13. $\frac{12x-5}{18x-6} - \frac{15x-4}{6-18x}$
14. $\frac{16-3x}{2x-7} + \frac{3x-5}{7-2x}$
15. $\frac{17}{50c} + \frac{23}{40c^2}$
16. $\frac{18}{25x^2} + \frac{7}{30x^3}$
17. $\frac{35}{27m^2n} - \frac{25}{63mn^3}$
18. $\frac{13}{15x^3y} - \frac{41}{40xy^3}$
19. $\frac{6}{e} + \frac{3e}{5-e}$
20. $\frac{8}{x+5} + x$
21. $\frac{9}{12d} - \frac{4}{18-27d}$
22. $\frac{6}{3x+6} - \frac{5}{9x+18}$
23. $\frac{5}{n} - n - \frac{8}{n-6}$
24. $\frac{3-x}{x^2} - \frac{x}{4x+x^2} - \frac{1}{x+4}$
25. $\frac{k}{h-k} + \frac{k+h}{k}$
26. $\frac{c-d}{c} - \frac{d}{d-c}$
27. $\frac{7}{x-5} - \frac{13}{x+4}$
28. $\frac{8}{4+a} + \frac{8}{2+a}$
29. $\frac{z-3}{z-6} - \frac{z+6}{z+3}$
30. $\frac{x+5}{x+1} - \frac{x-1}{x-5}$
31. $\frac{16-4a}{a^2+a-12} + \frac{3}{a^2-16}$
32. $\frac{2x+4}{x^2+2x-8} + \frac{3}{5x^2-7x-6}$
33. $\frac{3c}{c+5} - \frac{3c}{c-5} + \frac{150}{c^2-25}$
34. $\frac{5x}{x+3} - \frac{5x}{x-3} + \frac{90}{x^2-9}$
35. $\frac{m+1}{m^2-9} - \frac{m-2}{m^2+4m-21}$
36. $x - \frac{x}{2x^2+x-3} + \frac{4}{3x^2-x-2}$
37. $\frac{u-4}{u+8} + \frac{8}{u^2+4u+16}$
38. $\frac{x}{x^3-8} - \frac{2}{x^2-4} + \frac{5}{x+2}$
39. $\frac{3}{25h} - \frac{20h^3}{5h^4-25h^3} + \frac{4h-20}{h^2-10h+25}$
40. $\frac{8-x}{4-x} - \frac{x}{x-4} + \frac{4x}{3x^2-48}$
41. $\frac{5a-3}{3a^2-30a+72} + \frac{7a+2}{a^2-12a+36}$
42. $\frac{x+3}{6x^2-7x-6} - \frac{3x-2}{2x^2+5x-3}$
43. $\frac{15}{4m^2} - \frac{11m}{2m^2-50} + \frac{2m-5}{m^2-2m-35}$
44. $\frac{8}{3x^3+6x^2-9x} - \frac{4}{9x^3-9x^2} - \frac{x+2}{5x^2+15x}$
45. $\frac{17}{2z^2-98} - \frac{9z+11}{4z^3-28z^2} - \frac{13z}{z^2-14z+49}$
46. $\frac{5}{4x^2+12x+9} - \frac{8}{4x^2-9} - \frac{1}{6x^2+9x}$
47. $\frac{x^2+1}{x^3+1} + \frac{3}{5x^3-5x^2+5x} - \frac{x}{x^2+2x+1}$
48. $\frac{x}{8x+4} - \frac{3}{8x^3+1} + \frac{5}{8x^2-4x+2}$
49. $\frac{x+5}{x^3-5x^2-x+5} - \frac{x+1}{x^3-x^2-25x+25}$
50. $\frac{x-4}{x^3-5x^2-4x+20} - \frac{x+5}{x^3-2x^2-25x+50}$
51. $\frac{x+7}{x-2} + \frac{1}{x-3} \div \frac{x+7}{x^2-5x+6}$
52. $\frac{3x-5}{x-4} - (x+1) \div \frac{3x^2-2x-5}{x-4}$
53. $\frac{5}{6x^2} - \frac{x+3}{5x^2-15x} \div \frac{x^2+7x+10}{x^2+2x-15}$
54. $\frac{3x-1}{2x^3} - \frac{3x-1}{3x^2-7x+2} \cdot \frac{x^2-3x+2}{4x^2-3x-1}$
55. $\frac{x-5}{x+3} - \frac{x+2}{x^2+7x+12} \div \frac{3x^2+6x}{x^2+3x-4}$
56. $\frac{x^2+1}{x^2-1} - \frac{2x^2+x-1}{x-1} \div \frac{2x^2-5x+2}{x-2}$

6.6 Simplifying Complex Fractions

If the numerator or denominator (or both) of an expression is itself a rational expression, the expression is called a **complex fraction** (or **compound fraction**). The following are examples of complex fractions:

$$\frac{\frac{2}{x}}{3}, \quad \frac{a}{\frac{1}{c}}, \quad \frac{\frac{3}{z}}{\frac{5}{z}}, \quad \frac{\frac{3}{x}-\frac{2}{y}}{\frac{5}{x}+\frac{3}{y}}$$

$$\frac{\frac{1}{x}+\frac{3}{y}}{\frac{5}{x}-\frac{2}{y}}$$

← The primary numerator
← The main fraction bar
← The primary denominator

$$\frac{\frac{1}{x}+\frac{3}{y}}{\frac{5}{x}-\frac{2}{y}}$$

Secondary rational expressions (top: $\frac{1}{x}$, $\frac{3}{y}$; bottom: $\frac{5}{x}$, $\frac{2}{y}$)

Simplifying complex fractions

> *Method 1:* Multiply the complex fraction by a rational expression in which both the numerator and the denominator equal the LCD of the secondary fractions (the value of this rational expression will be 1); then simplify the results.
>
> *Method 2:* First, rewrite the primary numerator and the primary denominator of the complex fraction so that each is a single rational expression; then divide the simplified numerator by the simplified denominator.

In some of the examples (and in some of the exercises), the solution can be found more easily by Method 1 than by Method 2, and in other examples and exercises, the opposite is true. Before either method is used, we recommend inserting a denominator of 1 whenever there is no denominator in a secondary term, thereby writing *all* the secondary terms as rational expressions.

EXAMPLE 1 Simplify $\dfrac{\frac{2}{x}-\frac{3}{x^2}}{5+\frac{1}{x}}$.

SOLUTION 1 Using Method 1, we have

$$\frac{\frac{2}{x}-\frac{3}{x^2}}{\frac{5}{1}+\frac{1}{x}}$$

The LCD of the secondary denominators x, x^2, 1, and x is x^2

$$\frac{x^2}{x^2}\cdot\frac{\frac{2}{x}-\frac{3}{x^2}}{5+\frac{1}{x}} = \frac{\frac{x^2}{1}\cdot\frac{2}{x}-\frac{x^2}{1}\cdot\frac{3}{x^2}}{\frac{x^2}{1}\cdot\frac{5}{1}+\frac{x^2}{1}\cdot\frac{1}{x}} = \frac{2x-3}{5x^2+x}$$

This method clears all secondary denominators, "collapsing" the complex fraction into an ordinary rational expression.

SOLUTION 2 Using Method 2, we have

$$\frac{\frac{2}{x}-\frac{3}{x^2}}{\frac{5}{1}+\frac{1}{x}} = \left(\frac{2}{x}-\frac{3}{x^2}\right) \div \left(\frac{5}{1}+\frac{1}{x}\right)$$

$$= \left(\frac{2}{x}\cdot\frac{x}{x}-\frac{3}{x^2}\right) \div \left(\frac{5}{1}\cdot\frac{x}{x}+\frac{1}{x}\right)$$

$$= \left(\frac{2x}{x^2}-\frac{3}{x^2}\right) \div \left(\frac{5x}{x}+\frac{1}{x}\right) = \frac{2x-3}{x^2} \div \frac{5x+1}{x}$$

$$= \frac{2x-3}{\cancel{x^2}_{\,x}} \cdot \frac{\cancel{x}^{\,1}}{5x+1} = \frac{2x-3}{5x^2+x}$$

This method converts the complex fraction into a division problem.

We show only Method 1 for Example 2. We recommend that you also simplify the given complex fraction by using Method 2.

EXAMPLE 2 Simplify $\dfrac{x-\frac{4}{x}}{x-\frac{2}{x+1}}$.

SOLUTION Using Method 1, we have

$$\frac{\frac{x}{1}-\frac{4}{x}}{\frac{x}{1}-\frac{2}{x+1}}$$

The LCD of the secondary denominators is $x(x + 1)$

$$\frac{x(x+1)}{x(x+1)} \cdot \frac{\frac{x}{1}-\frac{4}{x}}{\frac{x}{1}-\frac{2}{x+1}} = \frac{\frac{x(x+1)}{1}\cdot\frac{x}{1}-\frac{x(x+1)}{1}\cdot\frac{4}{x}}{\frac{x(x+1)}{1}\cdot\frac{x}{1}-\frac{x(x+1)}{1}\cdot\frac{2}{(x+1)}}$$

$$= \frac{x^2(x+1)-4(x+1)}{x^2(x+1)-2x} = \frac{(x+1)(x^2-4)}{x^3+x^2-2x}$$

$$= \frac{(x+1)\cancel{(x+2)}^{\,1}(x-2)}{x(x-1)\cancel{(x+2)}_{\,1}} = \frac{(x+1)(x-2)}{x(x-1)}$$

Sometimes expressions with negative exponents on the variables become complex fractions when the negative exponents are removed (see Example 3).

EXAMPLE 3 Simplify $\dfrac{16x^{-2} - y^{-2}}{4x^{-1} - y^{-1}}$.

SOLUTION

$$\frac{16x^{-2} - y^{-2}}{4x^{-1} - y^{-1}} = \frac{\dfrac{16}{x^2} - \dfrac{1}{y^2}}{\dfrac{4}{x} - \dfrac{1}{y}}$$

Rewriting with positive exponents

The LCD of the secondary denominators is x^2y^2

$$= \frac{x^2y^2}{x^2y^2} \cdot \frac{\dfrac{16}{x^2} - \dfrac{1}{y^2}}{\dfrac{4}{x} - \dfrac{1}{y}}$$

Using Method I

$$= \frac{\dfrac{x^2y^2}{1} \cdot \dfrac{16}{x^2} - \dfrac{x^2y^2}{1} \cdot \dfrac{1}{y^2}}{\dfrac{x^2y^2}{1} \cdot \dfrac{4}{x} - \dfrac{x^2y^2}{1} \cdot \dfrac{1}{y}}$$

$$= \frac{16y^2 - x^2}{4xy^2 - x^2y} = \frac{\overset{1}{\cancel{(4y - x)}}(4y + x)}{xy\underset{1}{\cancel{(4y - x)}}}$$

$$= \frac{4y + x}{xy}$$

EXAMPLE 4 Simplify $\dfrac{3}{x + \dfrac{2}{x + \dfrac{1}{5x}}}$.

SOLUTION This type of complex fraction is sometimes called a *continued fraction*. The primary denominator itself contains a complex fraction; we will simplify this complex fraction first:

$$\frac{3}{x + \dfrac{2}{x + \dfrac{1}{5x}}} = \frac{3}{x + \dfrac{5x}{5x} \cdot \dfrac{\dfrac{2}{1}}{\dfrac{x}{1} + \dfrac{1}{5x}}}$$

Using Method I to simplify $\dfrac{2}{x + \dfrac{1}{5x}}$

$$= \frac{3}{x + \dfrac{10x}{5x^2 + 1}}$$

The LCD of the secondary denominators is now $5x^2 + 1$

$$= \frac{5x^2 + 1}{5x^2 + 1} \cdot \frac{\dfrac{3}{1}}{\dfrac{x}{1} + \dfrac{10x}{5x^2 + 1}}$$

Using Method I to simplify the main complex fraction

$$= \frac{(5x^2 + 1)(3)}{(5x^2 + 1)\left(\dfrac{x}{1} + \dfrac{10x}{5x^2 + 1}\right)} = \frac{15x^2 + 3}{5x^3 + x + 10x} = \frac{15x^2 + 3}{5x^3 + 11x}$$

Exercises 6.6
Set I

Simplify each complex fraction.

1. $\dfrac{\dfrac{21m^3n}{14mn^2}}{\dfrac{20m^2n^2}{8mn^3}}$

2. $\dfrac{\dfrac{10a^2b}{12a^4b^3}}{\dfrac{5ab^2}{16a^2b^3}}$

3. $\dfrac{\dfrac{15h-6}{18h}}{\dfrac{30h^2-12h}{8h}}$

4. $\dfrac{\dfrac{9k^4}{20k^2-35k^3}}{\dfrac{12k}{16-28k}}$

5. $\dfrac{\dfrac{c}{d}+2}{\dfrac{c^2}{d^2}-4}$

6. $\dfrac{\dfrac{x^2}{y^2}-1}{\dfrac{x}{y}-1}$

7. $\dfrac{a+2-\dfrac{9}{a+2}}{a+1+\dfrac{a-7}{a+2}}$

8. $\dfrac{x-3+\dfrac{x-3}{x+2}}{x+4-\dfrac{4x+23}{x+2}}$

9. $\dfrac{\dfrac{x+y}{y}+\dfrac{y}{x-y}}{\dfrac{y}{x-y}}$

10. $\dfrac{\dfrac{a-b}{a}-\dfrac{a}{a+b}}{\dfrac{b^2}{a+b}}$

11. $\dfrac{\dfrac{x}{x+1}+\dfrac{4}{x}}{\dfrac{x}{x+1}-2}$

12. $\dfrac{\dfrac{4x}{4x+1}+\dfrac{1}{x}}{\dfrac{2}{4x+1}+2}$

13. $\dfrac{\dfrac{x+4}{x}-\dfrac{3}{x-1}}{x+1+\dfrac{2x+1}{x-1}}$

14. $\dfrac{\dfrac{2x-8}{x^2-6x}+\dfrac{x}{x-6}}{x-\dfrac{16}{x}}$

15. $\dfrac{4x^{-2}-y^{-2}}{2x^{-1}+y^{-1}}$

16. $\dfrac{x^{-2}-9y^{-2}}{x^{-1}-3y^{-1}}$

17. $\dfrac{\dfrac{x-2}{x+2}-\dfrac{x+2}{x-2}}{\dfrac{x-2}{x+2}+\dfrac{x+2}{x-2}}$

18. $\dfrac{\dfrac{m+3}{m-3}+\dfrac{m-3}{m+3}}{\dfrac{m+3}{m-3}-\dfrac{m-3}{m+3}}$

19. $\dfrac{\dfrac{2x+y}{x}+\dfrac{3x+y}{y-x}}{\dfrac{x+y}{y}+\dfrac{2(x+y)}{x-y}}$

20. $\dfrac{\dfrac{a+b}{b}+\dfrac{2(a+b)}{a-b}}{\dfrac{2a-b}{a}-\dfrac{5b-a}{b-a}}$

21. $\dfrac{1}{x+\dfrac{1}{x+\dfrac{1}{x+x}}}$

22. $\dfrac{2}{y+\dfrac{2}{y+\dfrac{2}{y+y}}}$

23. $\dfrac{x+\dfrac{1}{2+\dfrac{x}{3}}}{x-\dfrac{3}{4+\dfrac{x}{2}}}$

24. $\dfrac{x+\dfrac{5}{1+\dfrac{x}{2}}}{x-\dfrac{4}{2+\dfrac{x}{3}}}$

Writing Problems

Express the answers in your own words and in complete sentences.

1. Describe the method you would use in simplifying the complex fraction $\dfrac{\dfrac{x-2}{x+3}}{\dfrac{x-2}{x+1}}$, and explain why you would use that method.

2. Describe the method you would use in simplifying the complex fraction $\dfrac{\dfrac{2}{x^2}-\dfrac{1}{x}}{\dfrac{2}{x^2}+\dfrac{3}{x}}$, and explain why you would use that method.

Exercises 6.6
Set II

Simplify each complex fraction.

1. $\dfrac{\dfrac{15e^3f^3}{36e^2f^4}}{\dfrac{5e^2f}{33e^3f^2}}$

2. $\dfrac{\dfrac{8x^2}{y^3}}{\dfrac{4x^3}{5y^4}}$

3. $\dfrac{\dfrac{30h^2-40hk}{2hk^2}}{\dfrac{54hk^2-72k^3}{6k^3}}$

4. $\dfrac{\dfrac{8x-16}{5x^2-5x-60}}{\dfrac{4x-8}{5x+15}}$

5. $\dfrac{\dfrac{9a^2}{5b^2}-5}{\dfrac{3a}{b}+5}$

6. $\dfrac{\dfrac{x^2}{y^2}-4}{\dfrac{x+2y}{y^3}}$

7. $\dfrac{z + 5 + \dfrac{z + 5}{z - 3}}{z - 4 - \dfrac{2}{z - 3}}$

8. $\dfrac{x - 3 - \dfrac{16}{x + 3}}{x - 4 - \dfrac{9}{x + 4}}$

9. $\dfrac{\dfrac{5m}{n - m}}{\dfrac{m + n}{n} + \dfrac{n}{m - n}}$

10. $\dfrac{\dfrac{a + 2b}{a} - \dfrac{3b}{a - 2b}}{a - b - \dfrac{6b^2}{a - 2b}}$

11. $\dfrac{\dfrac{1}{w} + \dfrac{9w}{6w + 1}}{\dfrac{5}{6w + 1} + 5}$

12. $\dfrac{\dfrac{1}{x} - \dfrac{3}{x + 2}}{\dfrac{5}{x + 2} - \dfrac{1}{x^2}}$

13. $\dfrac{2x - 5 - \dfrac{3}{x}}{\dfrac{1}{x} + \dfrac{x}{x - 12}}$

14. $\dfrac{3x + \dfrac{9x}{x - 1}}{x + 8 + \dfrac{18}{x - 1}}$

15. $\dfrac{5c^{-1} + 2d^{-1}}{25c^{-2} - 4d^{-2}}$

16. $\dfrac{25a^{-2} - b^{-2}}{5a^{-1} + b^{-1}}$

17. $\dfrac{\dfrac{2t - 1}{2t + 1} - \dfrac{2t + 1}{2t - 1}}{\dfrac{2t - 1}{2t + 1} + \dfrac{2t + 1}{2t - 1}}$

18. $\dfrac{\dfrac{x + 1}{x - 3} - \dfrac{x - 2}{x + 2}}{\dfrac{x + 3}{x + 2} - \dfrac{x + 3}{x - 3}}$

19. $\dfrac{\dfrac{u - 3v}{u} + \dfrac{3u - 4v}{u - v}}{\dfrac{2u - 7v}{v} + \dfrac{2u - v}{u - v}}$

20. $\dfrac{\dfrac{x}{x^2 - 1} + \dfrac{3x + 3}{x - 1}}{\dfrac{2x - 1}{x - 1} + \dfrac{x}{1 - x}}$

21. $\dfrac{1}{2z + \dfrac{1}{2z + \dfrac{1}{2z}}}$

22. $\dfrac{5}{1 + \dfrac{1}{x + \dfrac{1}{x}}}$

23. $\dfrac{x - \dfrac{1}{2 + \dfrac{x}{3}}}{x + \dfrac{1}{1 + \dfrac{x}{5}}}$

24. $\dfrac{\dfrac{1}{3 + \dfrac{x}{2}} + x}{x - \dfrac{1}{2 + \dfrac{x}{3}}}$

Sections 6.1–6.6 REVIEW

The Fundamental Property of Rational Expressions 6.1

If P, Q, and C represent polynomials and if $Q \neq 0$ and $C \neq 0$, then

$$\frac{P \cdot C}{Q \cdot C} = \frac{P}{Q}$$

Reducing a Rational Expression to Lowest Terms 6.2

1. Factor the numerator and denominator completely.
2. Divide both numerator and denominator by all common factors.

Multiplying Rational Expressions 6.3

If P, Q, R, and S represent polynomials and if $Q \neq 0$ and $S \neq 0$, then

$$\frac{P}{Q} \cdot \frac{R}{S} = \frac{P \cdot R}{Q \cdot S}$$

1. Factor any numerators or denominators that contain more than one term.
2. Divide the numerators and denominators by all factors common to both. (The common factors can be, but do not have to be, in the same rational expression.)
3. The numerator of the product is the product of all the factors remaining in the numerators, and the denominator of the product is the product of all the factors remaining in the denominators. If the only factor remaining in the numerator is 1, that 1 must be *written.*

Dividing Rational Expressions 6.3

Multiply the dividend by the multiplicative inverse of the divisor:

$$\frac{P}{Q} \div \frac{S}{T} = \frac{P}{Q} \cdot \frac{T}{S}$$

where P, Q, S, and T represent polynomials and $Q \neq 0$, $S \neq 0$, and $T \neq 0$.

The Rule of Signs for Rational Expressions 6.3

If any *two* of the three signs of a rational expression are changed, the value of the rational expression is unchanged.

Finding the LCD 6.4

1. Factor each denominator completely, if possible, expressing any repeated factors in exponential form.
2. Write down each different base that appears in any denominator.
3. Raise each base to the highest power to which it occurs in *any* denominator.
4. The LCD is the product of all the expressions found in step 3.

Adding Rational Expressions That Have the Same Denominator 6.5

1. The numerator of the sum is the sum of the numerators, and the denominator of the sum is the same as any *one* of the denominators. That is, if P, Q, and R represent polynomials and if $Q \neq 0$, then

$$\frac{P}{Q} + \frac{R}{Q} = \frac{P + R}{Q}$$

2. Reduce the resulting rational expression to lowest terms whenever possible.

Subtracting Rational Expressions That Have the Same Denominator 6.5

1. The numerator of the difference is the difference of the numerators, and the denominator of the difference is the same as any one of the denominators. That is, if P, Q, and R represent polynomials and if $Q \neq 0$, then

$$\frac{P}{Q} - \frac{R}{Q} = \frac{P - R}{Q}$$

2. Reduce the resulting rational expression to lowest terms whenever possible.

Adding or Subtracting Rational Expressions That Have Different Denominators 6.5

It is often helpful to reduce the rational expressions to lowest terms before using step 1 of the following procedure.

1. Find the LCD.
2. Convert each rational expression to an equivalent expression that has the LCD as its denominator.
3. Add or subtract the resulting rational expressions.
4. Reduce the sum or difference to lowest terms whenever possible.

Simplifying Complex Fractions 6.6

Method 1: Multiply the complex fraction by a rational expression in which both the numerator and the denominator equal the LCD of the secondary fractions (the value of this rational expression will be 1); then simplify the results.

Method 2: First, rewrite the primary numerator and the primary denominator of the complex fraction so that each is a single rational expression; then divide the simplified numerator by the simplified denominator.

Sections 6.1–6.6 REVIEW EXERCISES Set I

In Exercises 1 and 2, determine whether each pair of rational expressions is equivalent.

1. $\dfrac{2x + 6}{10}, \dfrac{x + 3}{5}$

2. $\dfrac{5x}{y}, \dfrac{5x + 1}{y + 1}$

In Exercises 3 and 4, find each missing term.

3. $\dfrac{9}{x - 3} = \dfrac{?}{3 - x}$

4. $-\dfrac{11}{2 - x} = \dfrac{11}{?}$

In Exercises 5–8, reduce each rational expression to lowest terms, or write "cannot be reduced."

5. $\dfrac{4z^3 + 4z^2 - 24z}{2z^2 + 4z - 6}$

6. $\dfrac{6k^3 - 12k^2 - 18k}{3k^2 + 3k - 36}$

7. $\dfrac{a^3 - 27b^3}{a^2 - 3ab + 2a - 6b}$

8. $\dfrac{2x + 1}{8x^3 + 1}$

In Exercises 9–22, perform the indicated operations; reduce all answers to lowest terms.

9. $\dfrac{-35mn^2p^2}{14m^3p^3} \cdot \dfrac{13m^4n}{52n^3p}$

10. $\dfrac{10b^2c}{6ab^4} \div \dfrac{15abc^2}{-12ac^3}$

11. $\dfrac{z^2 + 3z + 2}{z^2 + 2z + 1} \div \dfrac{z^2 + 2z - 3}{z^2 - 1}$

12. $\dfrac{a^2 - a - 2}{a^2 - 4a + 4} \cdot \dfrac{a^2 - 4}{a^2 + 3a + 2}$

13. $\dfrac{x^3 + y^3}{3x^2 - 3xy + 3y^2} \div \dfrac{x^2 - y^2}{x^2 + xy - 2y^2} \cdot \dfrac{15x^2y}{5x^2y + 10xy^2}$

14. $\dfrac{a^2 - 4b^2}{a^2 + 2ab + b^2} \cdot \dfrac{a^2 + 2ab + 4b^2}{a + 2b} \div \dfrac{a^3 - 8b^3}{2a^2 + 4ab + 2b^2}$

15. $\dfrac{20y - 7}{6y - 8} + \dfrac{17 + 2y}{8 - 6y}$

16. $\dfrac{23 - 22z}{30z - 24} + \dfrac{38z - 25}{24 - 30z}$

17. $\dfrac{11}{30e^3f} - \dfrac{7}{45e^2f^2}$

18. $\dfrac{3}{28u^2v^2} - \dfrac{7}{40u^4v}$

19. $\dfrac{a + 1}{a^2 - a - 2} - \dfrac{a - 2}{a^2 + a - 6}$

20. $\dfrac{x + 2}{x^2 + x - 2} - \dfrac{x - 3}{x^2 - x - 6}$

21. $\dfrac{15x}{5x^2 + 20x} - \dfrac{7}{3x^2} - \dfrac{3x + 12}{x^2 + 8x + 16}$

22. $\dfrac{4y - 20}{y^2 - 10y + 25} - \dfrac{5}{11y^3} + \dfrac{24y}{30y - 6y^2}$

In Exercises 23–26, simplify each complex fraction.

23. $\dfrac{\dfrac{x}{y + 1} + 2}{\dfrac{x}{y + 1} - 2}$

24. $\dfrac{3 - \dfrac{a}{b + 2}}{2 + \dfrac{a}{b + 2}}$

25. $\dfrac{8R^{-3} + T^{-3}}{4R^{-2} - T^{-2}}$

26. $\dfrac{m^{-2} - 16n^{-2}}{m^{-3} - 64n^{-3}}$

Sections 6.1–6.6 REVIEW EXERCISES Set II

Name

ANSWERS

1.

2.

3.

4.

5.

6.

7.

8.

9.

10.

11.

12.

13.

In Exercises 1 and 2, determine whether each pair of rational expressions is equivalent.

1. $\dfrac{x}{8y}, \dfrac{x+3}{8y+3}$

2. $\dfrac{2x-3}{y+2}, \dfrac{8x-12}{4y+8}$

In Exercises 3 and 4, find each missing term.

3. $-\dfrac{13}{6-7x} = \dfrac{13}{?}$

4. $\dfrac{17}{x-11} = \dfrac{?}{11-x}$

In Exercises 5–8, reduce each rational expression to lowest terms, or write "cannot be reduced."

5. $\dfrac{x^2+5x}{x^3-2x^2-35x}$

6. $\dfrac{2x^2+3x-20}{x^3+64}$

7. $\dfrac{m^3+8n^3}{m^2-6n+2mn-3m}$

8. $\dfrac{6a^2+11ab-2b^2}{ac+2bc-ad-2bd}$

In Exercises 9–22, perform the indicated operations; reduce all answers to lowest terms.

9. $\dfrac{55f^5g^3}{-8e^2f^2} \cdot \dfrac{4e^3g}{15ef^2g^2}$

10. $\dfrac{39a^3b^5}{5ac^2} \cdot \dfrac{30c^3d^2}{65a^2b}$

11. $\dfrac{a^2+4a-21}{6a^3+42a^2} \div \dfrac{3a^2-27}{36a^3}$

12. $\dfrac{2x^2-11x-21}{2x^2+11x+12} \div \dfrac{3x^2+13x+4}{3x^3-20x^2-7x}$

13. $\dfrac{4h^2+8hk+16k^2}{15hk^2+30k^3} \div \dfrac{3h^3-24k^3}{18kh^3} \div \dfrac{16h^2-8hk}{5h^2-20k^2}$

14. $\dfrac{7x^2 + 19x - 6}{2x^2 - 11x + 15} \cdot \dfrac{5x^2 - 16x + 3}{x^3 + 2x^2 - 3x} \cdot \dfrac{2x^2 - 7x + 5}{70x^2 - 34x + 4}$

15. $\dfrac{18 - 27u}{15 - 25u} + \dfrac{23u - 12}{25u - 15}$

16. $\dfrac{4x + 2}{3x - 5} + \dfrac{6x - 1}{5 - 3x}$

17. $\dfrac{7}{20a^3b^2} - \dfrac{13}{30ab^4}$

18. $\dfrac{5}{8x^2y} - \dfrac{3}{14xy^3}$

19. $\dfrac{3w - 15}{w^2 - 3w - 10} - \dfrac{5w + 30}{w^2 + w - 30}$

20. $\dfrac{x + 1}{x^2 - x - 6} - \dfrac{x - 3}{x^2 + 3x + 2}$

21. $\dfrac{12y^2 + 18y}{4y^2 - 9} + \dfrac{18y - 12y^2}{4y^2 - 12y + 9} - \dfrac{9}{8y^2}$

22. $\dfrac{7}{5y^2} - \dfrac{y + 1}{5y^2 - 9y - 2} - \dfrac{2y - 1}{2y^2 - y - 6}$

14. ______________

15. ______________

16. ______________

17. ______________

18. ______________

19. ______________

20. ______________

21. ______________

22. ______________

23. ______________

24. ______________

25. ______________

26. ______________

In Exercises 23–26, simplify each complex fraction.

23. $\dfrac{\dfrac{3}{w} + \dfrac{4w}{4w + 3}}{\dfrac{6}{4w + 3} + 2}$

24. $\dfrac{3 + \dfrac{7}{x - 1}}{\dfrac{5}{x + 2} + 1}$

25. $\dfrac{8y^{-3} - 27x^{-3}}{4y^{-2} - 9x^{-2}}$

26. $\dfrac{64x^{-3} + 1}{16x^{-2} - 1}$

6.7 Solving Rational Equations

A **rational equation** is an equation that contains one or more rational expressions.

The Domain of the Variable

Before we discuss solving rational equations, we must discuss finding the domain of the variable when rational expressions are involved. We learned earlier that the *domain of a variable* is the set of all the numbers that can be used in place of that variable, and we learned that we cannot divide by zero. Consequently, the domain of the variable in a rational expression is the set of all real numbers, *except that* we must exclude any values of the variable that would make the denominator equal zero.

We find the values to exclude by *setting the denominators equal to zero* and solving the resulting equations (see Example 1). If there are no variables in the denominator, no values of the variable need be excluded.

EXAMPLE 1 Examples of finding the domain of a rational expression:

a. $\frac{x}{3}$ There are no variables in the denominator; therefore, there are no values to exclude, so the domain is the set of all real numbers.

b. $\frac{5}{x}$ $x = 0$ Setting the denominator equal to zero

Therefore, the domain is the set of all real numbers except 0.

c. $\frac{2x - 5}{x - 1}$ $x - 1 = 0$ Setting the denominator equal to zero

$x = 1$

Therefore, the domain is the set of all real numbers except 1.

d. $\frac{x^2 + 2}{x^2 - 3x - 4}$ $x^2 - 3x - 4 = 0$ Setting the denominator equal to zero

$(x - 4)(x + 1) = 0$

$x - 4 = 0 \quad | \quad x + 1 = 0$

$x = 4 \quad | \quad x = -1$

Therefore, the domain is the set of all real numbers except 4 and -1.

e. $\frac{2}{3}$ There are no variables in the denominator; therefore, the domain is the set of all real numbers.

Solving Rational Equations

The technique for solving rational equations differs greatly from the techniques for performing addition and subtraction on rational expressions.

We usually solve a rational equation by first "clearing fractions"—that is, by multiplying both sides of the equation by the least common multiple of *all* the denominators. This procedure, however, can lead to an equation that is *not equivalent* to the original equation. (Recall that *equivalent equations* are equations with identical solution sets.)

Equivalent equations are obtained when

1. The same number or expression is added to or subtracted from both sides of the given equation.
2. Both sides of the given equation are multiplied or divided by the same *nonzero* constant.

Nonequivalent equations may be obtained when

1. Both sides of the given equation are *multiplied* by an expression containing the variable. In this case, we may obtain an *apparent* solution to the equation that is *not in the domain of the variable.* Such a solution is an **extraneous root** and must be rejected. (Any value of the variable that, when checked, gives a false statement, such as $3 = 0$, is also an extraneous root. We will see roots of this kind later, when we solve radical equations.)
2. Both sides of the given equation are *divided* by an expression containing the variable. In this case, roots of the given equation may be lost. For example, the equation $(x - 5)(x + 4) = 0$ has the solution set $\{5, -4\}$. If, however, we divide both sides of that equation by $x - 5$, we obtain the equation $x + 4 = 0$, and *its* solution set is $\{-4\}$. Since the two equations do not have the same solution set, they are *not equivalent.* Because we do not get an equation equivalent to the original equation (and we often *lose* a solution of the equation), we *should not divide both sides of an equation by an expression containing the variable.*

When we solve a rational equation, if any of the denominators contain a variable, the LCD will also contain a variable. Because we are not permitted to multiply both sides of an equation by zero, we must *exclude* any values of the variable that make the LCD (the expression we're multiplying by) zero; otherwise, we would be multiplying both sides of the equation by zero.

After we have "cleared fractions," we have an equation that can be solved by the methods learned in earlier sections of this book. Then, after we have solved the new equation, some (or all) of the *apparent* solutions to the new equation *may* not be in the domain of the variable—these solutions are extraneous roots, and they must be rejected. (If we forget to reject an apparent solution that was *not* in the domain and instead check it in the original equation, we will always get zeros in denominators, which will indicate that that solution must be rejected.)

We can solve rational equations by using the addition, subtraction, multiplication, and division properties of equality, as outlined below.

Solving a rational equation

1. Find the domain of the variable, and find the LCD of all the denominators.
2. Remove denominators by multiplying *both sides of the equation* by the LCD.
3. a. Remove all grouping symbols.
   b. Collect and combine like terms on each side of the equal sign.

First-degree equations	*Second-degree equations*
4. a. Move all terms that contain the variable to one side of the equal sign, and move all other terms to the other side. **b.** Divide both sides of the equation by the coefficient of the variable, or multiply both sides of the equation by the reciprocal of that coefficient.	**4. a.** Move *all* nonzero terms to one side of the equal sign. *Only zero must remain on the other side.* Then arrange the terms in descending powers. **b.** Factor the nonzero polynomial.* **c.** Set each factor equal to zero, and solve each resulting equation.

5. Reject any apparent solutions that are not in the domain.
6. To find any errors, check any other apparent solutions in the original equation.

*If the polynomial cannot be factored, the equation cannot be solved at this time.

EXAMPLE 2 Find the solution set of $\frac{3}{x+1} - \frac{2}{x} = \frac{5}{2x}$.

SOLUTION

Step 1. The domain of the variable is the set of all real numbers except -1 and 0. The LCD is $2x(x + 1)$.

Step 2.
$$2x(x+1)\left(\frac{3}{x+1} - \frac{2}{x}\right) = 2x(x+1)\left(\frac{5}{2x}\right)$$

$$\frac{2x(x+1)}{1} \cdot \frac{3}{(x+1)} - \frac{2x(x+1)}{1} \cdot \frac{2}{x} = \frac{2x(x+1)}{1} \cdot \frac{5}{2x}$$
Using the distributive property on the left side

Step 3.
$$6x - 4(x+1) = 5(x+1)$$
$$6x - 4x - 4 = 5x + 5$$
$$2x - 4 = 5x + 5$$

Step 4.
$$-9 = 3x$$
$$x = -3$$

Step 5 does not apply, since -3 *is* in the domain of the variable.

Step 6. *Check*
$$\frac{3}{x+1} - \frac{2}{x} = \frac{5}{2x}$$
$$\frac{3}{-3+1} - \frac{2}{-3} \stackrel{?}{=} \frac{5}{2(-3)}$$
$$\frac{3}{-2} - \frac{2}{-3} \stackrel{?}{=} \frac{5}{-6}$$
$$-\frac{9}{6} + \frac{4}{6} \stackrel{?}{=} -\frac{5}{6}$$
$$-\frac{5}{6} = -\frac{5}{6}$$ A true statement

Therefore, the solution is -3, and the solution set is $\{-3\}$.

EXAMPLE 3 Find the solution set of $\frac{x}{x-3} = \frac{3}{x-3} + 4$.

SOLUTION

Step 1. The domain is the set of all real numbers except 3. The LCD is $x - 3$.

Step 2.
$$(x-3)\left[\frac{x}{x-3}\right] = (x-3)\left[\frac{3}{x-3} + 4\right]$$

$$\frac{(x-3)}{1} \cdot \frac{x}{(x-3)} = \frac{(x-3)}{1} \cdot \frac{3}{(x-3)} + \frac{(x-3)}{1} \cdot \frac{4}{1}$$
Using the distributive property on the right side

Step 3.
$$x = 3 + 4(x-3)$$
$$x = 3 + 4x - 12$$
$$x = 4x - 9$$

Step 4.
$$9 = 3x$$

Step 5.
$$3 = x$$
Since 3 is not in the domain, this equation has *no solution*

An extraneous root

If we try to check 3 in the original equation, we obtain expressions that are not defined:

$$\frac{x}{x-3} = \frac{3}{x-3} + 4$$

$$\frac{3}{3-3} \stackrel{?}{=} \frac{3}{3-3} + 4$$

$$\frac{3}{0} \stackrel{?}{=} \frac{3}{0} + 4$$

Not defined

Therefore, 3 is not a root, or solution, of the given equation. This equation has no roots. The solution set of the equation is { }.

EXAMPLE 4 Find the solution set of $\frac{x+2}{x-2} = \frac{14}{x+1}$.

SOLUTION

Step 1. The domain is the set of all real numbers except 2 and −1. The LCD is $(x-2)(x+1)$.

Step 2. $$\overset{1}{\cancel{(x-2)}}(x+1)\left(\frac{x+2}{\underset{1}{\cancel{x-2}}}\right) = (x-2)\overset{1}{\cancel{(x+1)}}\left(\frac{14}{\underset{1}{\cancel{x+1}}}\right)$$

$$(x+2)(x+1) = 14(x-2)$$

Step 3. $x^2 + 3x + 2 = 14x - 28$ Because there is a second-degree term, we will move all nonzero terms to the left side

Step 4. $x^2 - 11x + 30 = 0$

$(x-5)(x-6) = 0$ Factoring the polynomial

$x - 5 = 0 \quad | \quad x - 6 = 0$ Setting each factor equal to zero

$x = 5 \quad | \quad x = 6$

Step 5 does not apply, because 5 and 6 *are* in the domain of the variable.

Step 6. ***Check for*** $x = 5$

$$\frac{x+2}{x-2} = \frac{14}{x+1}$$

$$\frac{5+2}{5-2} \stackrel{?}{=} \frac{14}{5+1}$$

$$\frac{7}{3} \stackrel{?}{=} \frac{14}{6}$$

$$\frac{7}{3} = \frac{7}{3}$$

Check for $x = 6$

$$\frac{x+2}{x-2} = \frac{14}{x+1}$$

$$\frac{6+2}{6-2} \stackrel{?}{=} \frac{14}{6+1}$$

$$\frac{8}{4} \stackrel{?}{=} \frac{14}{7}$$

$$2 = 2$$

The solution set is {5, 6}.

EXAMPLE 5 Find the solution set of $\frac{2}{x} + \frac{3}{x^2} = 1$.

SOLUTION

Step 1. The domain is the set of all real numbers except zero. The LCD is x^2.

Step 2. $x^2\left[\frac{2}{x}+\frac{3}{x^2}\right]=x^2\,[1]$

$$\frac{\overset{x}{\cancel{x^2}}}{1}\left(\frac{2}{\underset{1}{\cancel{x}}}\right)+\frac{\overset{1}{\cancel{x^2}}}{1}\left(\frac{3}{\underset{1}{\cancel{x^2}}}\right)=\frac{x^2}{1}\left(\frac{1}{1}\right)$$ Using the distributive property on the left side

Step 3. $2x+3=x^2$ ← A second-degree term

Step 4. $0=x^2-2x-3$

$$0=(x-3)(x+1)$$

$x-3=0$	$x+1=0$
$x=3$	$x=-1$

Step 5 does not apply, since 3 and -1 *are* in the domain of the variable.

√ **Step 6.**

Check for $x=3$	***Check for*** $x=-1$
$\frac{2}{x}+\frac{3}{x^2}=1$	$\frac{2}{x}+\frac{3}{x^2}=1$
$\frac{2}{3}+\frac{3}{3^2}\stackrel{?}{=}1$	$\frac{2}{-1}+\frac{3}{(-1)^2}\stackrel{?}{=}1$
$\frac{2}{3}+\frac{1}{3}\stackrel{?}{=}1$	$-2+3\stackrel{?}{=}1$
$1=1$	$1=1$

The solution set is $\{3, -1\}$.

A Word of Caution When we multiply both sides of an *equation* by the LCD, the denominators are removed completely. When we add or subtract rational expressions, the denominators can*not* be removed. A common error is to confuse solving an *equation,* such as $\frac{2}{x}+\frac{3}{x^2}=1$, with working an *addition* (or subtraction) *problem,* such as $\frac{2}{x}+\frac{3}{x^2}$.

The equation

Both sides are multiplied by the LCD to remove the denominators:

$$\frac{2}{x}+\frac{3}{x^2}=1 \quad \text{The LCD} = x^2$$

$$\frac{x^2}{1}\left(\frac{2}{x}+\frac{3}{x^2}\right)=\frac{x^2}{1}\cdot\frac{1}{1}$$

$$2x+3=x^2$$

In Example 5, we solved this equation by factoring. In this case, the result is two numbers, -1 and 3, that are solutions of the given equation.

The addition problem

Each rational expression is converted to an equivalent rational expression with the LCD as its denominator:

$$\frac{2}{x} + \frac{3}{x^2} \quad \text{The LCD} = x^2$$

This equals 1

$$= \frac{2}{x} \cdot \frac{x}{x} + \frac{3}{x^2}$$

$$= \frac{2x}{x^2} + \frac{3}{x^2} = \frac{2x + 3}{x^2}$$

In this case, the result is a rational expression that represents the sum of the given rational expressions.

The usual mistake is to multiply both terms of the *sum* by the LCD:

$$\frac{2}{x} + \frac{3}{x^2} = \frac{x^2}{1} \cdot \frac{2}{x} + \frac{x^2}{1} \cdot \frac{3}{x^2} = 2x + 3$$

Multiplying the *terms* of an *expression* by an expression that does not equal 1 *changes* the value of the original expression.

EXAMPLE 6 Find the solution set of $\dfrac{x}{x-3} - \dfrac{3x}{x^2 - x - 6} = \dfrac{4x^2 - 4x - 18}{x^2 - x - 6}$.

SOLUTION

Step 1. Since $x^2 - x - 6 = (x + 2)(x - 3)$, the domain is the set of all real numbers except -2 and 3. The LCD is $(x + 2)(x - 3)$.

Step 2. $$(x+2)(x-3)\left[\frac{x}{x-3} - \frac{3x}{x^2 - x - 6}\right] = \overset{1}{\cancel{(x+2)(x-3)}}\left[\frac{4x^2 - 4x - 18}{\underset{1}{\cancel{x^2 - x - 6}}}\right]$$

$$(x+2)\overset{1}{\cancel{(x-3)}}\left[\frac{x}{\underset{1}{\cancel{x-3}}}\right] - \overset{1}{\cancel{(x+2)}}\overset{1}{\cancel{(x-3)}}\left[\frac{3x}{\underset{1}{\cancel{(x+2)}}\underset{1}{\cancel{(x-3)}}}\right] = 4x^2 - 4x - 18$$

Step 3.
$$x(x + 2) - 3x = 4x^2 - 4x - 18$$
$$x^2 + 2x - 3x = 4x^2 - 4x - 18$$
$$x^2 - x = 4x^2 - 4x - 18$$

Step 4.
$$0 = 3x^2 - 3x - 18$$
$$0 = 3(x^2 - x - 6)$$
$$0 = 3(x + 2)(x - 3)$$

$x + 2 = 0$	$x - 3 = 0$
$x = -2$	$x = 3$

Step 5. Neither -2 nor 3 is in the domain of the variable. Therefore, there is no solution. The solution set is { }.

EXAMPLE 7 Find the solution set of $\frac{5}{x-7} - \frac{1}{2x} = \frac{9x+7}{2x^2-14x}$.

SOLUTION

Step 1. Since $2x^2 - 14x = 2x(x-7)$, the domain is the set of all real numbers except 0 and 7. The LCD is $2x(x-7)$.

Step 2. $$2x(x-7)\left[\frac{5}{x-7} - \frac{1}{2x}\right] = 2x(x-7)\left[\frac{9x+7}{2x^2-14x}\right]$$

$$\overset{1}{2x(\cancel{x-7})}\left(\frac{5}{\underset{1}{\cancel{x-7}}}\right) - \overset{1}{2x}(x-7)\left(\frac{1}{\underset{1}{\cancel{2x}}}\right) = \overset{1}{\cancel{2x(x-7)}}\left[\frac{9x+7}{\underset{1}{\cancel{2x^2-14x}}}\right]$$

Step 3. $$10x - (x-7) = 9x + 7$$
$$10x - x + 7 = 9x + 7$$
$$9x + 7 = 9x + 7$$

Step 4. $$7 = 7$$

This is an identity; however, the solution set is *not* the set of all real numbers, since 0 and 7 are not in the domain of the variable. The solution set, therefore, is the set of all real numbers *except* 0 and 7.

Exercises 6.7
Set I

In Exercises 1–8, find the domain of the variable.

1. $\frac{10-7y}{y+4}$
2. $\frac{3z+2}{5-z}$
3. $\frac{5x}{9}$
4. $\frac{7}{2x}$
5. $\frac{a^2+1}{a^2-25}$
6. $\frac{h^2+5}{h^2-h-6}$
7. $\frac{4c+3}{c^4-13c^2+36}$
8. $\frac{2x-5}{x^3-5x^2-9x+45}$

In Exercises 9–32, find the solution set of each equation.

9. $\frac{2}{k-5} - \frac{5}{k} = \frac{3}{4k}$
10. $\frac{4}{h} - \frac{6}{h-7} = \frac{2}{3h}$
11. $\frac{x}{x-2} = \frac{2}{x-2} + 5$
12. $\frac{x}{x+5} = 4 - \frac{5}{x+5}$
13. $\frac{12m}{2m-3} = 6 + \frac{18}{2m-3}$
14. $\frac{40w}{5w+6} = 8 - \frac{48}{5w+6}$
15. $\frac{2y}{7y+5} = \frac{1}{3y}$
16. $\frac{2b}{3-4b} = \frac{1}{2b}$
17. $\frac{3e-5}{4e} = \frac{e}{2e+3}$
18. $\frac{2x-1}{3x} = \frac{3}{2x+7}$
19. $\frac{1}{2} - \frac{1}{x} = \frac{4}{x^2}$
20. $\frac{7}{4} - \frac{17}{4x} = \frac{3}{x^2}$
21. $\frac{4}{x+1} = \frac{3}{x} + \frac{1}{15}$
22. $\frac{1}{x+1} = \frac{3}{x} + \frac{1}{2}$
23. $\frac{6}{x+4} = \frac{5}{x+3} + \frac{4}{x}$
24. $\frac{7}{x+5} = \frac{3}{x-1} - \frac{4}{x}$
25. $\frac{6}{x^2-9} + \frac{1}{5} = \frac{1}{x-3}$
26. $\frac{6-x}{x^2-4} - 2 = \frac{x}{x+2}$
27. $\frac{x+2}{x-2} - \frac{x-2}{x+2} = \frac{16}{x^2-4}$
28. $\frac{x-5}{x+5} - \frac{x+5}{x-5} = \frac{100}{x^2-25}$
29. $\frac{1}{2x^2-11x+15} + \frac{x-1}{2x^2+x-15} = \frac{-4}{x^2-9}$
30. $\frac{1}{3x^2-10x+8} + \frac{x+8}{3x^2+2x-8} = \frac{6}{x^2-4}$
31. $\frac{8}{x^3+64} + \frac{3}{x^2-16} = \frac{-1}{x^2-4x+16}$
32. $\frac{12}{x^3+27} + \frac{1}{x^2-9} = \frac{1}{x^2-3x+9}$

Writing Problems

Express the answers in your own words and in complete sentences.

1. Verify that the steps shown below are the correct steps for solving the equation:

$$\frac{15}{3-x} = 5 + \frac{5x}{3-x}$$

$$(3-x)\left(\frac{15}{3-x}\right) = (3-x)\left(5 + \frac{5x}{3-x}\right)$$

$$15 = 15 - 5x + 5x$$

$$15 = 15 \quad \text{True; the equation is an identity}$$

Explain why the solution set for the equation is *not* the set of all real numbers.

2. Explain the differences between the methods for adding rational expressions and for solving a rational equation.

Exercises 6.7 Set II

In Exercises 1–8, find the domain of the variable.

1. $\frac{4y+5}{2y-9}$ 2. $\frac{x+2}{5}$ 3. $\frac{11}{6d}$

4. $\frac{3-x}{x+2}$ 5. $\frac{2a^2+13}{a^2+3a-28}$ 6. $\frac{3x^2-13x+12}{9x^2-24x+16}$

7. $\frac{u^2+10u+1}{u^3-16u+2u^2-32}$ 8. $\frac{5}{v^3+3v^2-v-3}$

In Exercises 9–32, find the solution set of each equation.

9. $\frac{7}{2t-3} - \frac{4}{5t} = \frac{3}{t}$ 10. $\frac{6}{x} = \frac{7}{x-3} - \frac{1}{2}$

11. $\frac{8}{6+5e} = 3 - \frac{15e}{6+5e}$ 12. $5 = \frac{3}{x-2} - \frac{8}{2-x}$

13. $\frac{36}{4-3d} = 9 + \frac{27d}{4-3d}$ 14. $\frac{13}{3-x} = 8 - \frac{5x}{3-x}$

15. $\frac{4a}{33-5a} = \frac{2}{a}$ 16. $\frac{9}{x} = \frac{24-3x}{5}$

17. $\frac{3}{1-2x} + \frac{5}{2-x} = 2$ 18. $\frac{x}{8} = \frac{5}{6} - \frac{1}{x-1}$

19. $\frac{1}{2} - \frac{1}{6x} = \frac{7}{3x^2}$ 20. $\frac{1}{5x} + \frac{1}{x^2} = \frac{2}{4x+5}$

21. $\frac{4}{3x-1} = \frac{2}{x} + 1$ 22. $\frac{3}{1-2x} = \frac{2x+1}{x-2}$

23. $\frac{11}{2-3x} = \frac{6}{x+6} + \frac{3}{x}$ 24. $\frac{x-5}{3+2x} = \frac{2x-1}{x-9}$

25. $\frac{x+25}{x^2-25} = 1 - \frac{12x}{x-5}$ 26. $\frac{x+4}{x^2-16} + 1 = \frac{-7}{x-4}$

27. $\frac{x-4}{x+4} - \frac{x+4}{x-4} = \frac{64}{x^2-16}$

28. $\frac{4}{x^2-x-6} - \frac{2}{x^2-2x-3} = \frac{-1}{x+1}$

29. $\frac{1}{4x^2+11x-3} + \frac{24x-1}{4x^2-13x+3} = \frac{6}{x^2-9}$

30. $\frac{6}{x^2+x-6} - \frac{5}{x^2-x-2} = \frac{-1}{x+1}$

31. $\frac{12}{x^3+8} - \frac{1}{x^2-4} = \frac{1}{x^2-2x+4}$

32. $\frac{14}{x^3+1} + \frac{4}{x^2-1} = \frac{7}{x^2-x+1}$

6.8 Solving Literal Equations

Literal equations are equations that contain more than one variable. Such equations are also called *equations in two or more variables*. **Formulas** are literal equations that have applications in real-life situations. Sometimes literal equations have already been solved for one of the variables.

EXAMPLE 1 Examples of literal equations:

a. $A = P(1 + rt)$ This is an equation in four variables; it is also a formula, because it has applications in business. It has already been solved for A; you might be asked to solve it for P, for r, or for t.

b. $\dfrac{4ab}{d} = 15$ This is an equation in three variables; it is not a formula. It has not been solved for any of its variables; you might be asked to solve it for a, for b, or for d.

When we solve a literal equation for one of its variables, the solution will contain the other variables as well as constants. We must isolate the variable we are solving for; that is, *that variable must appear only once, all by itself, on one side of the equal sign.* All other variables and all constants must be on the other side of the equal sign. The suggestions in the following box are based on the addition, subtraction, multiplication, and division properties of equality.

Solving a literal equation

1. Remove denominators (if there are any) by multiplying both sides of the equation by the LCD.
2. Remove any grouping symbols, and on each side of the equal sign, collect and combine any like terms.
3. Move all the terms that contain the variable you are solving for to one side of the equal sign, and move all other terms to the other side.
4. Factor out the variable you are solving for (if it appears in more than one term).
5. Divide both sides of the equation by the coefficient of the variable you are solving for (or multiply both sides of the equation by the multiplicative inverse of that coefficient).

EXAMPLE 2 Solve $A = P(1 + rt)$ for t.

SOLUTION In the final equation, t must appear only once, by itself, on one side of the equal sign. All other variables and all constants must be on the other side of the equal sign. (Steps 1 and 4 do not apply.)

$$A = P(1 + rt)$$

Step 2. $A = P + Prt$ Using the distributive property

Step 3. $A - P = Prt$ Moving all the terms with the variable we are solving for (t) to one side and all other terms to the other side

Step 5. $\dfrac{A - P}{Pr} = \dfrac{\overset{1}{\cancel{Pr}}t}{\underset{1}{\cancel{Pr}}}$ Dividing both sides by Pr, the coefficient of t

$$t = \frac{A - P}{Pr}$$ This equation has been solved for t

EXAMPLE 3 Solve $\frac{1}{F} = \frac{1}{G} + \frac{1}{H}$ for G.

SOLUTION The LCD is FGH.

Step 1. $FGH\left(\frac{1}{F}\right) = FGH\left(\frac{1}{G} + \frac{1}{H}\right)$ — Removing denominators by multiplying both sides by the LCD

Step 2. $\frac{FGH}{1}\left(\frac{1}{F}\right) = \frac{FGH}{1}\left(\frac{1}{G}\right) + \frac{FGH}{1}\left(\frac{1}{H}\right)$ — Using the distributive property on the right side

$$GH = FH + FG$$

Step 3. $GH - FG = FH$ — Moving all terms that contain G to one side and all other terms to the other side

Step 4. $G(H - F) = FH$ — Factoring G from the left side

Step 5. $\frac{G(H - F)}{(H - F)} = \frac{FH}{H - F}$ — Dividing both sides by $H - F$, the coefficient of G

$$G = \frac{FH}{H - F}$$ — This equation has been solved for G

EXAMPLE 4 Solve $I = \frac{nE}{R + nr}$ for n.

SOLUTION

Step 1. $(R + nr)\, I = \overset{1}{\cancel{(R + nr)}}\,\frac{nE}{\underset{1}{\cancel{R + nr}}}$ — Multiplying both sides by the LCD

Step 2. $IR + Inr = nE$

Step 3. $IR = nE - Inr$ — Moving all terms that contain n to one side and all other terms to the other side

Step 4. $IR = n(E - Ir)$ — Factoring n from the right side

Step 5. $\frac{IR}{E - Ir} = \frac{n(E - Ir)}{E - Ir}$ — Dividing both sides by $(E - Ir)$, the coefficient of n

$$n = \frac{IR}{E - Ir}$$ — This equation has been solved for n

Exercises 6.8
Set I

Solve each equation for the variable listed after the semicolon.

1. $2(3x - y) = xy - 12;\ y$
2. $2(3x - y) = xy - 12;\ x$
3. $z = \frac{x - m}{s};\ m$
4. $s^2 = \frac{N - n}{N - 1};\ N$
5. $\frac{2x}{5yz} = z + x;\ x$
6. $\frac{xy}{z} = x + y;\ y$
7. $C = \frac{5}{9}(F - 32);\ F$
8. $A = \frac{h}{2}(B + b);\ B$
9. $s = c\left(1 + \frac{a}{c}\right);\ c$
10. $Z = \frac{Rr}{R + r};\ R$
11. $A = P(1 + rt);\ r$
12. $S = \frac{1}{2}g(2t - 1);\ t$
13. $v^2 = \frac{2}{r} - \frac{1}{a};\ a$
14. $\frac{1}{p} = 1 + \frac{1}{s};\ s$
15. $S = \frac{a}{1 - r};\ r$
16. $I = \frac{E}{R + r};\ R$
17. $\frac{1}{F} = \frac{1}{u} + \frac{1}{v};\ F$
18. $\frac{1}{c} = \frac{1}{a} + \frac{1}{b};\ b$
19. $L = a + (n - 1)d;\ n$
20. $A = 2\pi rh + 2\pi r^2;\ h$
21. $C = \frac{a}{1 + \frac{a}{\pi A}};\ a$
22. $R = \frac{V + v}{1 + \frac{vV}{c^2}};\ v$

Writing Problems

Express the answer in your own words and in complete sentences.

1. Suppose that someone has solved a literal equation for y and gives this answer: $y = 3x - 2y$. Explain why this is not an acceptable final answer.

Exercises 6.8 Set II

Solve each equation for the variable listed after the semicolon.

1. $5(2x + y) = 2xy + 25; x$
2. $\dfrac{mn}{m + n} = 1; m$
3. $3ab - 5 = 4(2a + b); b$
4. $\dfrac{x - y}{x} = x + y; y$
5. $\dfrac{1}{f} = \dfrac{1}{a} + \dfrac{1}{b}; a$
6. $A = \dfrac{h}{2}(B + b); b$
7. $A = \dfrac{h}{2}(B + b); h$
8. $\dfrac{1}{f} = \dfrac{1}{a} + \dfrac{1}{b}; f$
9. $a = b\left[1 + \dfrac{c}{b}\right]; b$
10. $PV = nRT; V$
11. $\dfrac{2}{15ab - 8a^2} = \dfrac{7}{20a^2 - 12ab}; b$
12. $x = \dfrac{y}{z + 4}; z$
13. $v^2 = \dfrac{2}{r} - \dfrac{1}{a}; r$
14. $\dfrac{2}{x} = \dfrac{y}{3} + \dfrac{1}{y}; x$
15. $I = \dfrac{E}{R + r}; r$
16. $A = P + Prt; P$
17. $\dfrac{1}{R} = \dfrac{1}{r_1} + \dfrac{1}{r_2}; r_1$
18. $y = mx + b; x$
19. $C = \dfrac{a}{1 + \dfrac{a}{\pi A}}; A$
20. $\dfrac{x}{a} + \dfrac{y}{b} = 1; b$
21. $E = \dfrac{k(1 - A)\pi R^2}{r^2}; A$
22. $S = \dfrac{n}{2}(A + L); A$

6.9 Ratio Problems

A **ratio** is used to compare the sizes of two or more *like* quantities. (When we say that we compare "like quantities," we mean that we compare inches with inches, pounds with pounds, cubic feet with cubic feet, and so forth.)

The ratio of a to b can be written as $a:b$ or as the division problem $\dfrac{a}{b}$.

The ratio of b to a can be written as $b:a$ or as $\dfrac{b}{a}$.

We call a and b the *terms* of the ratio; the terms of a ratio may be any kind of number, except that the divisor (or denominator) cannot be zero.

The notation $a:b:c$ (read "the ratio a to b to c") is used when we're comparing more than two quantities. For example, if the lengths of the sides of a triangle are 14 ft, 21 ft, and 35 ft, we can divide *all three numbers* by the largest number that divides exactly into all of them (7) and say that the lengths are in the ratio 2:3:5. (It is also acceptable to say that the lengths are in the ratio 14:21:35.)

The key to solving ratio problems is to use the given ratio to help represent the unknown numbers, as shown in the examples.

Representing the unknowns in a ratio problem

1. Multiply each term of the ratio by x.
2. Let the resulting products represent the unknown numbers.

EXAMPLE 1 Two numbers are in the ratio of 3 to 5. Their sum is -80. Find the numbers. (The same problem could have been worded as "Divide -80 into two parts whose ratio is 3:5" or "Separate -80 into two parts whose ratio is 3:5.")

SOLUTION

Think $3:5$ The ratio
$3x:5x$ Multiplying each term of the ratio by x

Let the resulting products, $3x$ and $5x$, represent the unknown numbers.

Step 1. Let $3x$ = one number

Let $5x$ = the other number

Reread The sum is -80

Step 2. $3x + 5x = -80$

Step 3. $8x = -80$

$x = -10$

Although we have solved for x, we *are not finished yet.* Now we must replace x with -10 in the expressions $3x$ and $5x$, since the unknown numbers are $3x$ and $5x$.

Step 4. Therefore, $3x = 3(-10) = -30$ One number

$5x = 5(-10) = -50$ The other number

Step 5. ***Check*** The ratio of -30 to -50 is $\frac{-30}{-50}$, or $\frac{3}{5}$, or 3 to 5. The sum of -30 and -50 is -80.

Step 6. Therefore, one number is -30, and the other is -50.

Some problems relating to rectangles, squares, and triangles are ratio problems (see Examples 2 and 3).

EXAMPLE 2 The three sides of a triangle are in the ratio 2:3:4. The perimeter is 63 ft. Find the lengths of the three sides.

SOLUTION

Think $2:3:4$ The ratio
$2x:3x:4x$ Multiplying each term of the ratio by x

Step 1. Let $2x$ = the number of ft in the first side

Let $3x$ = the number of ft in the second side

Let $4x$ = the number of ft in the third side

Reread The perimeter is 63

Step 2. $2x + 3x + 4x = 63$

Step 3. $9x = 63$

$x = 7$

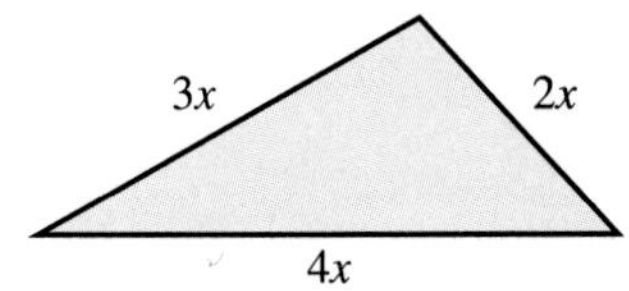

Step 4. Therefore, $2x = 2(7) = 14$ The number of ft in the first side

$3x = 3(7) = 21$ The number of ft in the second side

$4x = 4(7) = 28$ The number of ft in the third side

√ **Step 5.** ***Check*** Because $14 \div 7 = \mathbf{2}$, $21 \div 7 = \mathbf{3}$, and $28 \div 7 = \mathbf{4}$, the lengths 14 ft, 21 ft, and 28 ft are in the ratio 2:3:4. The perimeter is 14 ft + 21 ft + 28 ft, or 63 ft.

Step 6. Therefore, the lengths of the three sides are 14 ft, 21 ft, and 28 ft.

EXAMPLE 3

The length and width of a rectangle are in the ratio 5:3. The perimeter is to be less than 48. What values can the length of the rectangle have?

SOLUTION

Think 5 :3 The ratio

$5x:3x$ Multiplying each term of the ratio by x

Step 1. Let $3x$ = the width

Let $5x$ = the length

Think The perimeter is $2(3x) + 2(5x)$; it must be greater than 0 *and* less than 48.

Step 2. $0 < 2(3x) + 2(5x) < 48$

Step 3. $0 < 6x + 10x < 48$

$0 < 16x < 48$

$0 < x < 3$ Dividing all three parts by 16

Step 4. $0 < 5x < 15$ The *length* is $5x$

√ **Step 5.** ***Check*** We show checks for lengths of 10 and 8 (two values for the length chosen arbitrarily between 0 and 15). If the length were 10, the width would have to be 6 and the perimeter would be 32. If the length were 8, the width would have to be $\frac{24}{5}$ and the perimeter would be $\frac{128}{5}$, or 25.6. (You might check the problem using several other values between 0 and 15 for the length of the rectangle.)

Step 6. Therefore, the length must be greater than 0 *and* less than 15.

Exercises 6.9
Set I

In each exercise, set up the problem algebraically and solve. Be sure to state what your variables represent. In Exercises 1–12, check your solutions.

1. Two numbers are in the ratio of 4 to 5. Their sum is 81. Find the numbers.

2. Two numbers are in the ratio of 8 to 3. Their sum is 77. Find the numbers.

3. The three sides of a triangle are in the ratio 3:4:5, and the perimeter is 108 m. Find the lengths of the three sides.

4. The three sides of a triangle are in the ratio 4:5:6, and the perimeter is 120 in. Find the lengths of the three sides.

5. Fifty-four hours of a student's week are spent in study, class, and work. The times spent in these activities are in the ratio 4:2:3. How many hours are spent in each activity?

6. Forty-eight hours of a student's week are spent in study, class, and work. The times spent in these activities are in the ratio 4:2:6. How many hours are spent in each activity?

7. The length and width of a rectangle are in the ratio of 7 to 6. The perimeter is 78 ft. Find the length and width.

8. The length and width of a rectangle are in the ratio of 9 to 5. The perimeter is 196 cm. Find the length and width.

9. The length and width of a rectangle are in the ratio of 3 to 2. If the area of the rectangle is 150 sq. m, find its length and its width.

10. The length and width of a rectangle are in the ratio of 4 to 3. If the area of the rectangle is 192 sq. in., find its width and its length.

11. Divide (separate) 143 into two parts whose ratio is 7:6.

12. Divide (separate) 221 into two parts whose ratio is 8:5.

13. The three sides of a triangle are in the ratio 4:5:6. The perimeter is to be less than 60. What is the range of values that the *shortest* side can have?

14. The three sides of a triangle are in the ratio 8:9:10. The perimeter is to be less than 81. What is the range of values that the *shortest* side can have?

Exercises 6.9 Set II

In each exercise, set up the problem algebraically and solve. Be sure to state what your variables represent. In Exercises 1–12, check your solutions.

1. Two numbers are in the ratio of 2 to 7. Their sum is 99. Find the numbers.

2. Three numbers are in the ratio 4:7:3. Their sum is 56. Find the numbers.

3. The sides of a triangle are in the ratio 5:6:7, and the perimeter is 72 ft. Find the lengths of the sides.

4. The ratio of pineapple juice to orange juice to lemon juice in a punch recipe is 11 to 6 to 1. If 36 L of punch is to be made (with only these three juices), how much pineapple juice should be used?

5. The formula for a particular shade of green paint calls for mixing 3 parts of blue paint with 1 part of yellow paint. Find the number of liters of blue paint and the number of liters of yellow paint needed to make 14 L of the desired shade of green paint.

6. In Mrs. Aguilar's bread recipe, flour and water are to be mixed in the ratio of 3 parts water to 5 parts flour. How many parts of flour are needed to make a mixture of 64 parts altogether of water and flour?

7. The length and width of a rectangle are in the ratio of 8 to 3, and the perimeter is 88 m. Find the length and the width of the rectangle.

8. Nick needs to mix gasoline and oil in the ratio of 50 to 1 for his motorcycle. If he needs to mix 382.5 oz altogether, how much gasoline should he use? How much oil?

9. The length and width of a rectangle are in the ratio of 5 to 3. If the area of the rectangle is 735 sq. yd, find its length and its width.

10. The base and altitude of a triangle are in the ratio 3:2. If the area of the triangle is 48 sq. cm, find its base and its altitude.

11. Divide (separate) 207 into two parts whose ratio is 5:4.

12. Todd has nickels, dimes, and quarters in the ratio 5:8:7. If he has 140 of these coins altogether, how many of each kind does he have?

13. The three sides of a triangle are in the ratio 4:5:8. The perimeter is to be less than 85. What is the range of values that the *shortest* side can have?

14. The three sides of a triangle are in the ratio 3:5:7. The perimeter is to be greater than 60. What is the range of values that the *longest* side can have?

6.10 Motion Problems

Motion problems (sometimes called *rate-time-distance* problems) are often seen in real life; they are concerned with one or more *moving* objects. The formula relating *distance* traveled d, *rate* of travel (or speed) r, and *time* of travel t is

$$d = r \cdot t \qquad \text{or} \qquad r \cdot t = d$$

For example, if you're driving your car at an average rate of 50 miles per hour (mph), then

In 2 hr, you travel a distance of 100 mi, because $50\frac{\text{mi}}{\cancel{\text{hr}}}(2\ \cancel{\text{hr}}) = 100$ mi;

In 3 hr, you travel a distance of 150 mi, because $50\frac{\text{mi}}{\cancel{\text{hr}}}(3\ \cancel{\text{hr}}) = 150$ mi;

and so on.

When motion problems involve two *different* rates, students often find a chart helpful for organizing the given information; the chart we use for motion problems is a little different from the chart we use for mixture problems.

The chart method for solving distance-rate-time problems

Read Read the problem carefully. *Be sure you understand the problem.*

Think If the problem is a distance-rate-time problem involving two different rates, a chart can be used.

Step 1. Represent one unknown number by a variable, and declare its meaning in a sentence of the form "Let $x = \ldots$." Express other unknowns in terms of x, if possible.

Chart Make a blank chart, as follows:

	r ·	t =	d
One object			
Other object			

Fill in Fill in two of the boxes in each row, using the variable and the given information.

Reread Fill in the remaining box in each row, using the formula $r \cdot t = d$.

Step 2. Are the distances equal? If so, the equation is obtained by setting the two d-values from the chart equal to each other. Do the distances differ by some constant? If so, write the equation using this information. Are the *times* equal, or do you know something about the *sum* or *difference* of the times? If so, write the equation using this information.

Step 3. Solve the equation.

Step 4. Solve for *all* the unknowns asked for in the problem.

Step 5. Check the solution(s) *in the word statement.*

Step 6. State the results clearly.

EXAMPLE 1

A car loaded with campers left Los Angeles, traveling toward Lake Havasu, at 8:00 A.M. A second car full of campers left Los Angeles at 8:30 A.M. and traveled 10 mph faster over the same road. If the second car overtook the first at 10:30 A.M., what was the average speed of each car?

SOLUTION

Think This is a distance-rate-time problem (the formula is $r \cdot t = d$).

Step 1. Let x = the rate of the slower car in mph
Then $x + 10$ = the rate of the faster car in mph

Partially Filled in Chart

	r	$\cdot$ t =	d
Slower car	$x\dfrac{\text{mi}}{\text{hr}}$	$\dfrac{5}{2}$ hr	
Faster car	$(x + 10)\dfrac{\text{mi}}{\text{hr}}$	2 hr	

Slower car left at 8:00 A.M. and was overtaken at 10:30 A.M.; therefore, time was $\frac{5}{2}$ hr

Faster car left at 8:30 A.M. and overtook first car at 10:30 A.M.; therefore, time was 2 hr

Use the formula $r \cdot t = d$ to fill in these blanks as follows:

Completed Chart

	r	$\cdot$ t =	d	$r \cdot t = d$
Slower car	$x\dfrac{\text{mi}}{\text{hr}}$	$\dfrac{5}{2}$ hr	$x\left(\dfrac{5}{2}\right)$ mi	$x\left(\dfrac{5}{2}\right) = d$
Faster car	$(x + 10)\dfrac{\text{mi}}{\text{hr}}$	2 hr	$(x + 10)(2)$ mi	$(x + 10)(2) = d$

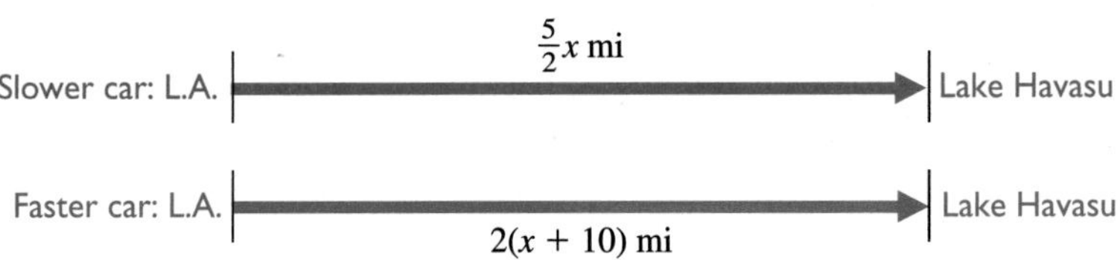

Since both cars started at the same place, they had each driven the same distance when the second car overtook the first; therefore, we set the two *distances* equal to each other.

Step 2. $\frac{5}{2}x = 2(x + 10)$ We *drop* the units

Step 3. $5x = 4(x + 10)$ Multiplying both sides by 2

$5x = 4x + 40$

Step 4. $x = 40$ The rate of the slower car in mph

$x + 10 = 40 + 10 = 50$ The rate of the faster car in mph

√ **Step 5.** ***Check*** The first car traveled $\left(2\frac{1}{2}\ \cancel{\text{hr}}\right)\left(40\frac{\text{mi}}{\cancel{\text{hr}}}\right) = 100$ mi.

The second car traveled $(2\ \cancel{\text{hr}})\left(50\frac{\text{mi}}{\cancel{\text{hr}}}\right) = 100$ mi.

Step 6. Therefore, the average speed of the slower car was 40 mph, and the average speed of the faster car was 50 mph.

EXAMPLE 2

Joe drove from his home to Anchorage to pick up his wife, Linda. He drove at 45 mph. Linda drove at 54 mph on the return trip. If the total driving time for the round trip was 11 hr, how far from their home is Anchorage?

SOLUTION

Step 1. Let $x =$ the number of hr for the trip to Anchorage

Then $11 - x =$ the number of hr for the return trip

Chart

	r ·	t =	d
Trip to Anchorage	$45\frac{\text{mi}}{\text{hr}}$	x hr	$45x$ mi
Return trip	$54\frac{\text{mi}}{\text{hr}}$	$(11 - x)$ hr	$54(11 - x)$ mi

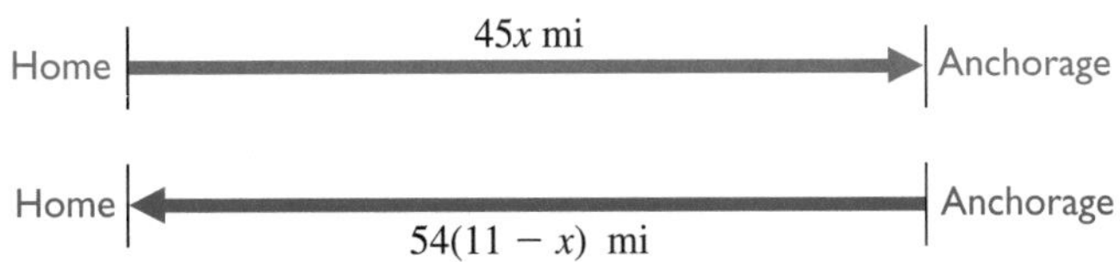

We now use the fact that the distances are equal.

Step 2. $45x = 54(11 - x)$ We drop the units

Step 3. $45x = 594 - 54x$

$99x = 594$

Step 4. $x = 6$ The number of hr for the trip to Anchorage

$45x = 270$ The number of mi Joe drove

√ **Step 5.** *Check* Hours for the return trip: $11 - 6 = 5$

Miles Linda drove: $\left(54\frac{\text{mi}}{\cancel{\text{hr}}}\right)(5\ \cancel{\text{hr}}) = 270$ mi

Step 6. Therefore, Joe and Linda live 270 mi from Anchorage.

EXAMPLE 3 Jill drove for 270 mi at a constant rate. If her rate had been 9 mph faster, the trip would have taken 1 hr less. Find Jill's actual rate.

SOLUTION We solve the formula $r \cdot t = d$ for t: $t = \frac{d}{r}$.

Step 1. Let x = Jill's actual rate (the slower rate) in mph
Then $x + 9$ = the faster rate in mph

Also, $\frac{270}{x}$ = Jill's time (in hr) at the slower rate $\left(\text{using } t = \frac{d}{r}\right)$

and $\frac{270}{x + 9}$ = the time (in hr) if she goes faster $\left(\text{using } t = \frac{d}{r}\right)$

Chart

	r ·	t =	d
Actual trip	x	$\frac{270}{x}$	270
Faster trip	$x + 9$	$\frac{270}{x + 9}$	270

We now use the fact that

Number of hr at slower rate	equals	number of hr at faster rate	plus	1 hr

Step 2. $$\frac{270}{x} = \frac{270}{x+9} + 1$$

Step 3. $$\overset{1}{\cancel{x}}(x+9)\left(\frac{270}{\underset{1}{\cancel{x}}}\right) = x(x+9)\left(\frac{270}{x+9} + 1\right)$$ Multiplying both sides by $x(x+9)$, the LCD

$$270(x+9) = x\overset{1}{\cancel{(x+9)}}\left(\frac{270}{\underset{1}{\cancel{x+9}}}\right) + x(x+9)(1)$$ Using the distributive property

$$270x + 2{,}430 = 270x + x^2 + 9x$$ This is a quadratic equation

$$270x + 2{,}430 = 279x + x^2$$

$$0 = x^2 + 9x - 2{,}430$$

$$0 = (x+54)(x-45)$$ Using a calculator to help find the factors of 2,430

Setting each factor equal to zero

$$x + 54 = 0 \qquad\qquad x - 45 = 0$$

Step 4. $$x = -54 \qquad\qquad x = 45$$

($x = -54$: Reject)

√ **Step 5.** ***Check*** The time for Jill to drive 270 mi at a rate of $45\ \frac{\text{mi}}{\text{hr}}$ is $\frac{270\text{ mi}}{45\ \frac{\text{mi}}{\text{hr}}} = 6\text{ hr}$.

The time for her to drive 270 mi going $9\ \frac{\text{mi}}{\text{hr}}$ faster—that is, at a rate of $54\ \frac{\text{mi}}{\text{hr}}$—is $\frac{270\text{ mi}}{54\ \frac{\text{mi}}{\text{hr}}} = 5\text{ hr}$, 1 hr less than at the slower rate.

Step 6. Therefore, Jill's rate is $45\ \frac{\text{mi}}{\text{hr}}$, or 45 mph.

Some motion problems involve a boat running in moving water. In this case, if the water is moving at w miles per hour and the speed of the boat in still water is b miles per hour, when the boat is going *upstream* (*against* the current, or against the movement of the water), the actual speed of the boat will be $(b - w)$ miles per hour; when the boat is going *downstream* (*with* the current), the actual speed will be $(b + w)$ miles per hour (see Example 4).

Similarly, if the wind is blowing at w miles per hour and an airplane is flying with a speed that would be s miles per hour in still air, when the plane flies *with* the wind, its actual speed will be $(s + w)$ miles per hour; when it's flying *against* the wind, its actual speed will be $(s - w)$ miles per hour.

EXAMPLE 4 A boat cruised downstream for 4 hr before heading back. After traveling upstream for 5 hr, it was still 16 mi short of its starting point. If the speed of the river was 4 mph, find the speed of the boat in still water.

SOLUTION

Think This is a distance-rate-time problem; the formula is $r \cdot t = d$.

Step 1. Let x = the speed (in mph) of the boat in still water
Then $x + 4$ = the speed (in mph) of the boat going downstream
and $x - 4$ = the speed (in mph) of the boat going upstream

Chart

	r ·	t =	d
Downstream	$(x + 4)\dfrac{\text{mi}}{\text{hr}}$	4 hr	$(x + 4)(4)$ mi
Upstream	$(x - 4)\dfrac{\text{mi}}{\text{hr}}$	5 hr	$(x - 4)(5)$ mi

$r \cdot t = d$
$(x + 4)(4) = d$

$r \cdot t = d$
$(x - 4)(5) = d$

We now use the fact that there is a 16 mi difference between the distances, as shown in the diagram below:

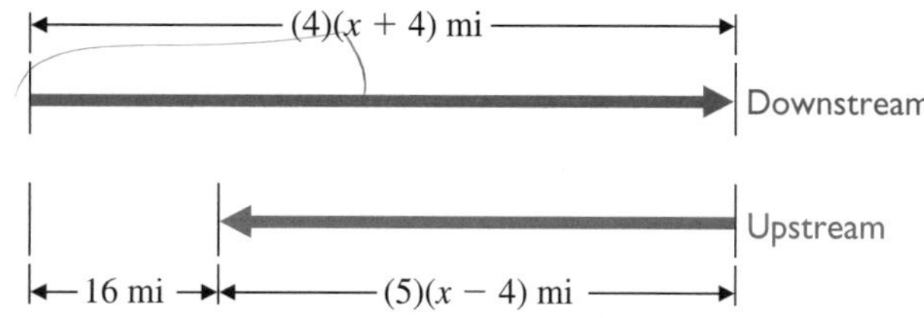

Reread Number of mi downstream equals 16 mi plus number of mi upstream

Step 2. $(x + 4)(4) = 16 + (x - 4)(5)$ Dropping the units

Step 3.
$$4x + 16 = 16 + 5x - 20$$
$$4x + 16 = 5x - 4$$
$$20 = x$$

Step 4. $x = 20$ The speed (in mph) of the boat in still water

Step 5. ***Check*** The speed downstream was (20 + 4) mph, or 24 mph, so the distance downstream was $\left(24 \dfrac{\text{mi}}{\cancel{\text{hr}}}\right)(4\ \cancel{\text{hr}})$, or 96 mi. The speed upstream was (20 − 4) mph, or 16 mph, so the distance upstream was $\left(16 \dfrac{\text{mi}}{\cancel{\text{hr}}}\right)(5\ \cancel{\text{hr}})$, or 80 mi. The distance downstream was 16 mi more than the distance upstream.

Step 6. Therefore, the speed of the boat in still water was 20 mph.

Exercises 6.10
Set I

In each exercise, set up the problem algebraically, solve, and check. Be sure to state what your variables represent.

1. The Malone family left San Diego by car at 7 A.M., driving toward San Francisco. Their neighbors, the King family, left in their car at 8 A.M., also driving toward San Francisco. By traveling 9 mph faster, the Kings overtook the Malones at 1 P.M.
   a. Find the average speed of each car.
   b. Find the total distance traveled by each car before they met.

2. The Duran family left Ames, Iowa, by car at 6 A.M., driving toward Yellowstone National Park. Their neighbors, the Silva family, left in their car at 8 A.M., also driving toward Yellowstone. By traveling 10 mph faster, the Silvas overtook the Durans at 4 P.M.
   a. Find the average speed of each car.
   b. Find the total distance traveled before they met.

3. Eric hiked from his camp to a lake in the mountains and returned to camp later in the day. He walked at a rate of 2 mph going to the lake and 5 mph coming back. The trip to the lake took 3 hr longer than the trip back.

   a. How long did it take him to hike to the lake?

   b. How far is it from his camp to the lake?

4. Lee hiked from her camp up to an observation tower in the mountains and returned to camp later in the day. She walked up at the rate of 2 mph and jogged back at the rate of 6 mph. The trip to the tower took 2 hr longer than the return trip.

   a. How long did it take her to hike to the tower?

   b. How far is it from her camp to the tower?

5. Fran and Ron live 54 mi apart. Both left their homes at 7 A.M. by bicycle, riding toward each other. They met at 10 A.M. If Ron's average speed was four-fifths of Fran's, how fast did each cycle?

6. Tran and Atour live 60 mi apart. Both left their homes at 10 A.M. by bicycle, riding toward each other. They met at 2 P.M. If Atour's average speed was two-thirds of Tran's, how fast did each cycle?

7. Colin paddled a kayak downstream for 3 hr. After having lunch, he paddled upstream for 5 hr. At that time, he was still 6 mi short of his starting point. If the speed of the river was 2 mph, how fast did Colin's kayak move in still water? How far downstream did he travel?

8. Jessica paddled a kayak downstream for 4 hr. After having lunch, she paddled upstream for 6 hr. At that time, she was still 4 mi short of her starting point. If the speed of the river was 1 mph, how fast did Jessica's kayak move in still water? How far downstream did she travel?

9. Mr. Zaleva flew his private plane from his office to his company's storage facility, bucking a 28-mph head wind all the way. He flew home the same day, with the same wind at his back. The round trip took 2.5 hr of flying time. If the speed of the plane would have been 140 mph in still air, how far is the storage facility from his office?

10. Mrs. Summers drove her motorboat upstream a certain distance while pulling her son Brian on a water ski. She returned to the starting point pulling her other son Derek. The round trip took 25 min of skiing time. On both legs of the trip, the speedometer read 30 mph. If the speed of the current was 6 mph, how far upstream did she travel?

11. It takes one plane $\frac{1}{2}$ hr longer than another to fly a certain distance. Find that distance if one plane flies at 500 mph and the other at 400 mph.

12. It took Barbara 10 min longer to ride her bicycle a certain distance than it took Ed to cover that same distance. Find the distance if Barbara's speed was 9 mph and Ed's was 12 mph.

13. A cyclist traveled for 90 mi at a constant rate. If his rate had been 3 mph faster, the trip would have taken 1 hr less. Find the actual rate of the cyclist.

14. A skiboat traveled (in still water) for 120 mi at a constant rate. If the boat had gone 10 mph slower, the same trip would have taken 2 hr longer. Find the actual rate of the boat.

15. Two airplanes left Miami at the same time, one heading due east at 210 mph and the other heading due west at 190 mph. In how many hours were the planes 1,400 mi apart? (Assume that the planes were unaffected by wind.)

16. Two boats left the tip of an island at the same time, one heading due south at 24 mph and the other heading due north at 28 mph. In how many hours were the boats 117 mi apart?

Exercises 6.10 Set II

In each exercise, set up the problem algebraically, solve, and check. Be sure to state what your variables represent.

1. The Dent family left Chino by car at 6 A.M., driving toward Lake Mojave. Their neighbor, Mr. Scott, left in his car at 7 A.M., also driving toward Lake Mojave. By traveling 10 mph faster, Mr. Scott overtook the Dents at noon.

   a. Find the average speed of each car.

   b. Find the total distance traveled before they met.

2. Bill and Andrew live 34 mi apart. Both left their homes at 9 A.M., walking toward each other. They met at 1 P.M. If Andrew's average speed was $\frac{1}{2}$ mph faster than Bill's, how fast did each walk? How far from Bill's house did they meet?

3. Lori hiked from her camp to a waterfall and returned to camp later in the day. She walked at a rate of $1\frac{1}{2}$ mph going to the waterfall and at a rate of 3 mph coming back. The trip to the waterfall took 1 hr longer than the trip back.

a. How long did it take her to hike to the waterfall?

b. How far is it from her camp to the waterfall?

4. David jogged from his home to a park at the rate of 7 mph. He later walked home at the rate of 5 mph. If the trip home took 0.6 hr longer than the trip to the park, how long did it take David to get from his home to the park? How far is the park from his home?

5. Matthew and Lucas live 63 mi apart. Both left their homes at 8 A.M. by bicycle, riding toward each other. They met at 11 A.M. If Lucas's average speed was three-fourths of Matthew's, how fast did each cycle?

6. Anthony and Mark left a marina at the same time, traveling in the same direction. The speed of Mark's boat was $\frac{7}{8}$ the speed of Anthony's boat. In 3 hr, Anthony was 12 mi ahead of Mark. How fast was each going?

7. In their houseboat, the Powitzsky family motored upstream for 4 hr. After lunch, they motored downstream for 2 hr. At that time, they were still 12 mi away from the marina where they had begun. If the speed of the houseboat in still water was 15 mph, what was the speed of the river? How far upstream did the Powitzskys travel?

8. Plane A can travel a distance of 630 mi in $2\frac{1}{2}$ hr. The speed of plane B is $\frac{6}{7}$ that of plane A. If both planes leave an airport at the same time and fly in opposite directions, how long will it be before they are 1,872 mi apart?

9. Ms. Lee flew her private plane from her home to a nearby recreation area, bucking a 25-mph head wind all the way. She flew back home the same day, with the same wind at her back. The round trip took 8 hr. If the speed of the plane would have been 100 mph in still air, how far is her home from the recreation area?

10. Two trains were traveling in the same direction, with the first train going 6 mph slower than the second train. The first train left a depot at 4 A.M. The second left the same depot at 5 A.M. and passed the first train at 10 A.M. How fast was each train going?

11. It takes Mina 10 min longer than Susan to jog eight laps around the school soccer field. Find the distance around the soccer field if Mina's speed is 3 mph and Susan's is 4 mph.

12. The rate of Tom's boat in still water is 30 mph. If it can travel 132 mi *with* the current in the same time that it can travel 108 mi *against* the current, find the speed of the current.

13. A jogger traveled for 48 mi at a constant rate. If she had jogged 2 mph slower, it would have taken her 2 hr longer to cover the same distance. Find her actual rate.

14. San Francisco and Susanville are 297 mi apart. Rebecca left San Francisco at noon driving toward Susanville; she averaged 54 mph. Susan left Susanville at the same time, driving toward San Francisco; she averaged 45 mph. At what time did they meet? How many miles did Susan drive before they met?

15. Two planes left Houston at noon, one flying due east at 204 mph and the other heading due west at 192 mph. In how many hours were the planes 693 mi apart?

16. A cyclist traveled for 80 mi at a constant rate. If his rate had been 6 mph slower, the same trip would have taken 3 hr longer. Find the actual rate of the cyclist.

6.11 Other Applied Problems That Lead to Rational Equations

All the applied problems that we have solved in previous chapters can lead to equations that contain rational expressions; there will be some problems of these kinds in the exercises at the end of this section.

In this section, however, we also introduce a new application that involves rational expressions; these problems are often called *work problems*. One basic relationship used in solving work problems is the following:

	Rate $\times$ time $=$ amount of work done
In symbols:	$r \cdot t = w$

For example, suppose José can assemble one radio in 3 days.

$$\text{José's } rate = \frac{1 \text{ radio}}{3 \text{ days}} = \frac{1}{3} \text{ radio per day}$$

If his working *time* is 5 days, then the amount of work done is

Rate · time = amount of work done

$$\frac{1}{3}\frac{\text{radio}}{\cancel{\text{day}}} \cdot 5 \cancel{\text{days}} = \frac{5}{3} \text{ radios assembled}$$

The other basic relationship used to solve work problems is the following:

$$\begin{pmatrix}\text{Amount of work A}\\ \text{does in}\\ \text{time } x\end{pmatrix} + \begin{pmatrix}\text{amount of work B}\\ \text{does in}\\ \text{time } x\end{pmatrix} = \begin{pmatrix}\text{amount of work done}\\ \text{together in}\\ \text{time } x\end{pmatrix}$$

EXAMPLE 1 Albert can paint a house in 6 days. Ben can paint the same house in 8 days. How long would it take Albert and Ben to paint the house if they worked together?

SOLUTION

Think

$$\text{Albert's rate} = \frac{1 \text{ house}}{6 \text{ days}} = \frac{1}{6}\frac{\text{house}}{\text{day}}, \text{ or } \frac{1}{6} \text{ house per day}$$

$$\text{Ben's rate} = \frac{1 \text{ house}}{8 \text{ days}} = \frac{1}{8}\frac{\text{house}}{\text{day}}, \text{ or } \frac{1}{8} \text{ house per day}$$

Step 1. Let x = the number of days for Albert and Ben together to paint the house

Reread

Albert's rate · Albert's time = amount Albert paints

$$\frac{1}{6}\frac{\text{house}}{\cancel{\text{day}}} \cdot x \cancel{\text{days}} = \frac{x}{6} \text{ house}$$

Ben's rate · Ben's time = amount Ben paints

$$\frac{1}{8}\frac{\text{house}}{\cancel{\text{day}}} \cdot x \cancel{\text{days}} = \frac{x}{8} \text{ house}$$

Therefore, Amount Albert paints in x days + amount Ben paints in x days = amount painted together in x days

Step 2.

$$\frac{x}{6} \text{ house} + \frac{x}{8} \text{ house} = 1 \text{ house}$$

← One house painted

$$\frac{x}{6} + \frac{x}{8} = 1$$

We drop the units

Step 3.

$$\frac{\overset{4}{24}}{1} \cdot \frac{x}{\underset{1}{\cancel{6}}} + \frac{\overset{3}{24}}{1} \cdot \frac{x}{\underset{1}{\cancel{8}}} = \frac{24}{1} \cdot \frac{1}{1}$$

The LCD is 24

$$4x + 3x = 24$$

$$7x = 24$$

Step 4. $$x = \frac{24}{7} = 3\tfrac{3}{7}$$

√ **Step 5.** *Check* $$\left(\frac{1}{\cancel{6}_{1}}\ \frac{\text{house}}{\cancel{\text{day}}}\right)\left(\frac{\overset{4}{\cancel{24}}}{7}\ \cancel{\text{days}}\right) + \left(\frac{1}{\cancel{8}_{1}}\ \frac{\text{house}}{\cancel{\text{day}}}\right)\left(\frac{\overset{3}{\cancel{24}}}{7}\ \cancel{\text{days}}\right)$$

$$= \tfrac{4}{7}\text{ house} + \tfrac{3}{7}\text{ house} = 1\text{ house}$$

Step 6. Therefore, it would take Albert and Ben $3\frac{3}{7}$ days to paint the house if they worked together.

EXAMPLE 2

Machine A can make 200 brackets in 6 hr. How long does it take machine B to make 100 brackets if the two machines working together can make 200 brackets in 5 hr?

SOLUTION

Step 1. Let x = the number of hr for machine B to make 100 brackets

Think $$\text{A's rate} = \frac{200\text{ brackets}}{6\text{ hr}} = \frac{100}{3}\ \frac{\text{brackets}}{\text{hr}}, \text{ or } \frac{100}{3}\text{ brackets per hr}$$

$$\text{B's rate} = \frac{100\text{ brackets}}{x\text{ hr}} = \frac{100}{x}\ \frac{\text{brackets}}{\text{hr}}, \text{ or } \frac{100}{x}\text{ brackets per hr}$$

The two machines working together can make 200 brackets if each machine runs for 5 hr. Therefore,

Reread Number of brackets A makes in 5 hr + number of brackets B makes in 5 hr = number of brackets made together in 5 hr

Step 2. $$\frac{100}{3}\ \frac{\text{brackets}}{\cancel{\text{hr}}}(5\ \cancel{\text{hr}}) + \frac{100}{x}\ \frac{\text{brackets}}{\cancel{\text{hr}}}(5\ \cancel{\text{hr}}) = 200\text{ brackets}$$

$$\frac{500}{3} + \frac{500}{x} = 200$$ Dropping the units

Step 3. $$\frac{\overset{1}{\cancel{3}}x}{1}\cdot\frac{500}{\cancel{3}_{1}} + \frac{3\overset{1}{\cancel{x}}}{1}\cdot\frac{500}{\cancel{x}_{1}} = \frac{3x}{1}\cdot\frac{200}{1}$$ The LCD is $3x$

$$500x + 1{,}500 = 600x$$

$$1{,}500 = 100x$$

Step 4. $$x = 15$$

√ **Step 5.** *Check* $$\left(\frac{100}{3}\ \frac{\text{brackets}}{\cancel{\text{hr}}}\right)(5\ \cancel{\text{hr}}) + \left(\frac{100}{\cancel{15}_{3}}\ \frac{\text{brackets}}{\cancel{\text{hr}}}\right)(\overset{1}{\cancel{5}}\ \cancel{\text{hr}})$$

$$= \frac{500}{3}\text{ brackets} + \frac{100}{3}\text{ brackets} = \frac{600}{3}\text{ brackets} = 200\text{ brackets}$$

Step 6. Therefore, it takes machine B 15 hr to make 100 brackets.

EXAMPLE 3 It takes pipe 1 12 min to fill a particular tank. It takes pipe 2 only 8 min to fill the same tank. Pipe 3, which is a drain pipe, takes 6 min to *empty* the same tank. How long does it take to fill the tank when the valves of all three pipes are open?

SOLUTION

Think The rate of pipe 1 $= \dfrac{1 \text{ tank}}{12 \text{ min}} = \dfrac{1}{12}\dfrac{\text{tank}}{\text{min}}$, or $\dfrac{1}{12}$ tank per min

The rate of pipe 2 $= \dfrac{1 \text{ tank}}{8 \text{ min}} = \dfrac{1}{8}\dfrac{\text{tank}}{\text{min}}$, or $\dfrac{1}{8}$ tank per min

The rate of pipe 3 $= \dfrac{1 \text{ tank}}{6 \text{ min}} = \dfrac{1}{6}\dfrac{\text{tank}}{\text{min}}$, or $\dfrac{1}{6}$ tank per min

Step 1. Let $x =$ the number of min for all three pipes together to fill the tank.

Reread Amount 1 does in time x + amount 2 does in time x − amount 3 does in time x = 1 full tank

Minus because pipe 3 *empties* the tank

Step 2. $\left(\dfrac{1}{12}\dfrac{\text{tank}}{\cancel{\text{min}}}\right)(x\ \cancel{\text{min}}) + \left(\dfrac{1}{8}\dfrac{\text{tank}}{\cancel{\text{min}}}\right)(x\ \cancel{\text{min}}) - \left(\dfrac{1}{6}\dfrac{\text{tank}}{\cancel{\text{min}}}\right)(x\ \cancel{\text{min}}) = 1 \text{ tank}$

$$\frac{x}{12} + \frac{x}{8} - \frac{x}{6} = 1 \qquad \text{Dropping the units}$$

Step 3. $\dfrac{\overset{2}{\cancel{24}}}{1} \cdot \dfrac{x}{\underset{1}{\cancel{12}}} + \dfrac{\overset{3}{\cancel{24}}}{1} \cdot \dfrac{x}{\underset{1}{\cancel{8}}} - \dfrac{\overset{4}{\cancel{24}}}{1} \cdot \dfrac{x}{\underset{1}{\cancel{6}}} = \dfrac{24}{1} \cdot \dfrac{1}{1}$ The LCD is 24

$$2x + 3x - 4x = 24$$

Step 4. $x = 24$

Step 5. ***Check*** $\left(\dfrac{1}{\underset{1}{\cancel{12}}}\dfrac{\text{tank}}{\cancel{\text{min}}}\right)(\overset{2}{\cancel{24}}\ \cancel{\text{min}}) + \left(\dfrac{1}{\underset{1}{\cancel{8}}}\dfrac{\text{tank}}{\cancel{\text{min}}}\right)(\overset{3}{\cancel{24}}\ \cancel{\text{min}}) - \left(\dfrac{1}{\underset{1}{\cancel{6}}}\dfrac{\text{tank}}{\cancel{\text{min}}}\right)(\overset{4}{\cancel{24}}\ \cancel{\text{min}})$

$= 2$ tanks $+ 3$ tanks $- 4$ tanks $= 1$ tank

Step 6. Therefore, it takes 24 min to fill the tank when the valves of all three pipes are open.

Some applied problems involve the individual digits of a number. Consider the number 59. In that number, 5 is the *tens digit* and 9 is the *units digit*. The number itself equals 5(10) + 9, or the tens digit times 10, *plus* the units digit. A three-digit number equals the hundreds digit times 100, *plus* the tens digit times 10, *plus* the units digit.

EXAMPLE 4 The units digit of a two-digit number is three times the tens digit. The product of the digits divided by the sum of the digits is $\frac{3}{2}$. Find the number.

SOLUTION

Step 1. Let $x =$ the tens digit
Then $3x =$ the units digit

Reread The product of the digits divided by the sum of the digits is $\frac{3}{2}$

Step 2. $\dfrac{x(3x)}{x + 3x} = \dfrac{3}{2}$

$$\frac{3x^2}{4x} = \frac{3}{2}$$

Step 3. $\dfrac{4x}{1} \cdot \dfrac{3x^2}{4x} = \dfrac{4x}{1} \cdot \dfrac{3}{2}$ The LCD is $4x$

$$3x^2 = 6x$$

$$3x^2 - 6x = 0$$

$$3x(x - 2) = 0$$

$3x = 0$ | $x - 2 = 0$

Step 4. $x = 0$ | $x = 2$ The tens digit

Reject ⟶ $3x = 0$ | $3x = 6$ The units digit

(The number would be 00.) The number is apparently 26.

√ **Step 5.** ***Check*** The units digit is three times the tens digit; the product of the digits is $2 \cdot 6$, or 12, and the sum of the digits is $2 + 6$, or 8. Then $\frac{12}{8} = \frac{3}{2}$.

Step 6. Therefore, the number is 26.

Exercises 6.11
Set I

Set up each problem algebraically, solve, and check. Be sure to state what your variables represent.

1. Henry can paint a house in 5 days, and Teri can paint the same house in 4 days. If Henry and Teri work together, how many days will it take them to paint the house?

2. Jill can paint a house in 6 days, and Karla can paint it in 9 days. How many days will it take Jill and Karla to paint the house if they work together?

3. Trisha can type 100 pages of manuscript in 3 hr. How long does it take David to type 80 pages if he and Trisha working together can type a 500-page manuscript in 10 hr?

4. Carlos can wrap 200 newspapers in 55 min. How long does it take Heather to wrap 150 papers if she and Carlos working together can wrap 400 papers in 1 hr?

5. It takes machine A 36 hr to do a job that machine B does in 24 hr. If machine A runs for 12 hr before machine B is turned on, how long will it take both machines running together to finish the job?

6. It takes machine A 18 hr to do a job that machine B does in 15 hr. If machine A runs for 3 hr before machine B is turned on, how long will it take both machines running together to finish the job?

7. Two numbers differ by 8. One-fourth the larger is 1 more than one-third the smaller. Find the numbers.

8. Two numbers differ by 6. One-fifth the smaller exceeds one-half the larger by 3. Find the numbers.

9. An automobile radiator contains 14 qt of a 45% solution of antifreeze. How much of the solution must be drained out and replaced by pure antifreeze to make a 50% solution?

10. An automobile radiator contains 12 qt of a 30% solution of antifreeze. How much of the solution must be drained out and replaced with pure antifreeze to make a 50% solution?

11. The tens digit of a two-digit number is 1 more than its units digit. The product of the digits divided by the sum of the digits is $\frac{6}{5}$. Find the number.

12. The units digit of a two-digit number is 1 more than the tens digit. The product of the digits divided by the sum of the digits is $\frac{12}{7}$. Find the number.

13. It takes pipe 2 1 hr longer than pipe 1 to fill a particular tank. Pipe 3, a drain pipe, can drain the tank in 2 hr. If it takes 3 hr to fill the tank when the valves of all three pipes are open, how long does it take pipe 1 to fill the tank alone?

14. It takes pipe 2 2 hr longer than pipe 1 to fill a particular tank. Pipe 3, a drain pipe, can drain the tank in 4 hr. If it takes 2 hr to fill the tank when the valves of all three pipes are open, how long does it take pipe 1 to fill the tank alone?

15. Ruth can proofread 230 pages of a deposition in 4 hr. How long does it take Sandra to proofread 60 pages if, when they work together, they can proofread 525 pages in 6 hr?

16. Alice can crochet 3 sweaters in 48 hr. How long does it take Willa to crochet 2 sweaters if together they can crochet 15 sweaters in 112 hr?

17. In a film-processing lab, machine A can process 5,700 ft of film in 60 min. How long does it take machine B to process 4,300 ft of film if the two machines running together can process 15,500 ft of film in 50 min?

18. In a film-processing lab, machine C can process 3,225 ft of film in 15 min. How long does it take machine D to process 12,750 ft of film if the two machines running together can process 28,800 ft of film in 45 min?

Exercises 6.11 Set II

Set up each problem algebraically, solve, and check. Be sure to state what your variables represent.

1. Merwin can wax all his floors in 4 hr. His wife June can do the same job in 3 hr. How long will it take them to wax the floors if they work together?

2. The denominator of a fraction exceeds the numerator by 6. The value of the fraction is $\frac{3}{5}$. Find the fraction.

3. It takes Kevin 8 hr to clean the yard and mow the lawn. He and his brother Jason can do the same job in 3 hr when they work together. How long does it take Jason to do the job when he works alone?

4. Machine A can make 400 bushings in 8 hr. How long does it take machine B to make 50 bushings if the two machines working together can make 800 bushings in 12 hr?

5. It takes machine A 20 hr to do a job that machine B does in 28 hr. If machine A runs for 8 hr before machine B is turned on, how long will it take both machines running together to finish the job?

6. The sum of a fraction and its reciprocal is $\frac{73}{24}$. Find the fraction and its reciprocal.

7. Two numbers differ by 22. One-eighth the larger exceeds one-sixth the smaller by 2. Find the numbers.

8. Sherma can type 15 words per minute faster than Karen. If Sherma types 3,850 words in the same time that Karen types 3,100 words, find the rate of each.

9. One-fifth of the 10% antifreeze solution in a car radiator is drained and replaced by an antifreeze solution of unknown concentration. If the resulting mixture is 25% antifreeze, what was the unknown concentration?

10. The units digit of a two-digit number is 2 more than the tens digit. The product of the digits divided by the sum of the digits is $\frac{3}{4}$. Find the number.

11. The tens digit of a two-digit number is 4 more than its units digit. The product of the digits divided by the sum of the digits is $\frac{3}{2}$. Find the number.

12. The speed of a plane in still air is 480 mph. If it can fly 3,030 miles *with* the wind in the same time that it can fly 2,730 miles *against* the same wind, what is the speed of the wind?

13. A refinery tank has one fill pipe and two drain pipes. Pipe 1 can fill the tank in 3 hr. It takes pipe 2 6 hr longer than pipe 3 to drain the tank. If it takes 12 hr to fill the tank when the valves of all three pipes are open, how long does it take pipe 3 to drain the tank alone?

14. It takes Matt 30 min longer than Rebecca to walk a certain distance. Find that distance if Matt's speed is 3 mph and Rebecca's is 4 mph.

15. Lori can knit 4 scarves in 24 hr. How long does it take Anita to knit one scarf if together they can knit 21 scarves in 72 hr?

16. The speed of the current in a river is 6 mph. If a boat can travel 222 mi *with* the current in the same time it can travel 150 mi *against* the current, what is the speed of the boat in still water?

17. In a film-processing lab, machine A can process 4,275 ft of film in 45 min. How long does it take machine B to process 10,625 ft of film if the two machines running together can process 41,600 ft of film in 80 min?

18. The tens digit of a two-digit number is 2 more than the units digit. If the product of the digits divided by the sum of the digits is $\frac{12}{5}$, find the number.

Sections 6.7–6.11 REVIEW

The Domain of a Rational Expression 6.7 The domain of the variable in a rational expression is the set of all real numbers *except* those that would make the denominator equal zero.

Solving a Rational Equation 6.7

1. Find the domain of the variable, and find the LCD of all the denominators.
2. Remove denominators by multiplying *both sides of the equation* by the LCD.
3. a. Remove all grouping symbols.
   b. Collect and combine like terms on each side of the equal sign.

First-degree equations

4. a. Move all terms that contain the variable to one side of the equal sign, and move all other terms to the other side.
   b. Divide both sides of the equation by the coefficient of the variable, or multiply both sides of the equation by the reciprocal of that coefficient.

Second-degree equations

4. a. Move *all* nonzero terms to one side of the equal sign. *Only zero must remain on the other side.* Then arrange the terms in descending powers.
   b. Factor the nonzero polynomial.
   c. Set each factor equal to zero, and solve each resulting equation.

5. Reject any apparent solutions that are not in the domain.
6. Check any other apparent solutions in the original equation.

Literal Equations 6.8 Literal equations are equations that contain more than one variable.

Solving a Literal Equation 6.8 Proceed in the same way used to solve an equation with one variable. The solution will contain the other variables as well as constants.

Ratios 6.9 A **ratio** is used to compare the sizes of two or more *like* quantities. "The ratio of a to b" can be written as $a:b$ or as the division problem $\frac{a}{b}$.

Motion Problems 6.10 The formula for solving motion problems is $r \cdot t = d$, where r is the rate, t is the time, and d is the distance.

Work Problems 6.11 The formula for solving work problems is $r \cdot t = w$, where r is the rate, t is the time, and w is the amount of work done.

Sections 6.7–6.11 REVIEW EXERCISES Set I

In Exercises 1 and 2, find the domain of the variable.

1. $\dfrac{15}{20 - 45a^2}$
2. $\dfrac{7m - 10}{6m^2 - 21m - 90}$

In Exercises 3–10, find the solution set of each equation.

3. $\dfrac{5a - 4}{6a} = \dfrac{-2}{3a + 10}$
4. $\dfrac{3b}{5b - 4} = \dfrac{7 - 2b}{3}$
5. $\dfrac{3}{x} - \dfrac{8}{x^2} = \dfrac{1}{4}$
6. $\dfrac{4}{x^2} - \dfrac{3}{x} = \dfrac{5}{2}$
7. $\dfrac{9}{x + 1} = \dfrac{4}{x} + \dfrac{1}{x - 1}$
8. $\dfrac{4}{3x - 1} - \dfrac{2}{x + 1} = \dfrac{1}{x}$
9. $\dfrac{x}{x + 5} = 8 - \dfrac{5}{x + 5}$
10. $\dfrac{3}{x - 3} - \dfrac{1}{3x} = \dfrac{8x + 3}{3x^2 - 9x}$

In Exercises 11–14, solve each equation for the variable listed after the semicolon.

11. $5(x - 2y) = 14 + 3(2x - y)$; y
12. $2(5 - 2x) = 22 - 2(y - 5x)$; x
13. $R = \dfrac{R_1R_2}{R_1 + R_2}$; R_1
14. $F = \dfrac{1}{\dfrac{1}{u} + \dfrac{1}{v}}$; v

In Exercises 15–20, set up each problem algebraically, solve, and check. Be sure to state what your variables represent.

15. The sides of a triangle are in the ratio 3:7:8. The perimeter is to be less than 36. What is the range of values that the *shortest* side can have?

16. The moving walkway at an airport moves at the rate of 5 ft per sec. When Karla walks on the walkway that is going in the same direction that she is walking, she gets to the end of the walkway in 50 sec. However, when she really wants a challenge, she likes to walk on the walkway that is going in the *opposite* direction. When she does this (walking at the same rate as before), it takes her 550 sec to reach the end of the walkway. How fast does she walk when she is *not* on the walkway at all?

17. Using tractor A, a man can cultivate his corn in 20 hr. With tractor B, the same corn can be cultivated in 15 hr. After one man cultivates with tractor A for 5 hr, he is joined by a second man using tractor B. How long will it take both men and tractors working together to finish the job?

18. Section gang A can lay a rail in 16 min. Section gang B can lay a rail in 12 min. After gang A has been working 4 min, section gang B joins them; how long will it take both gangs working together to finish laying one rail?

19. The denominator of a fraction is 10 more than its numerator. If 1 is added to both numerator and denominator, the value of the new fraction is $\frac{2}{3}$. Find the original fraction.

20. Pipe 1 can fill a tank in 4 hr; pipe 2 can fill it in 6 hr. When the valves of pipes 1, 2, and 3 are open, the tank is filled in 5 hr. How long does it take pipe 3, a drain pipe, to drain the tank by itself?

Name ____________________

ANSWERS

1. ________
2. ________
3. ________
4. ________
5. ________
6. ________
7. ________
8. ________
9. ________
10. ________
11. ________
12. ________
13. ________
14. ________

In Exercises 1 and 2, find the domain of the variable.

1. $\dfrac{11y - 6}{6y^2 + 8y - 64}$

2. $\dfrac{3x + 9}{2x^2 + 9x - 5}$

In Exercises 3–10, find the solution set of each equation.

3. $\dfrac{4u + 1}{5u} = \dfrac{20}{3u + 6}$

4. $\dfrac{2x + 1}{4x + 2} = \dfrac{3x - 2}{2}$

5. $\dfrac{17}{6x} + \dfrac{5}{2x^2} = \dfrac{2}{3}$

6. $\dfrac{3}{x} - \dfrac{1}{x^2} = \dfrac{11}{16}$

7. $\dfrac{9}{2x - 1} = \dfrac{4}{x} - \dfrac{1}{x - 3}$

8. $\dfrac{6}{6 + x} - \dfrac{1}{x} = \dfrac{13}{4}$

9. $\dfrac{4}{x - 4} - \dfrac{1}{2x} = \dfrac{7x + 4}{2x^2 - 8x}$

10. $\dfrac{x}{x - 8} = 7 + \dfrac{8}{x - 8}$

In Exercises 11–14, solve each equation for the variable listed after the semicolon.

11. $4(2y + x) = xy + 32;\ x$

12. $ab = \dfrac{c}{a} - \dfrac{b}{a};\ b$

13. $\dfrac{1}{p} = 1 + \dfrac{1}{s};\ p$

14. $\dfrac{a}{b - 3} = \dfrac{1}{a} + \dfrac{1}{4};\ b$

In Exercises 15–20, set up each problem algebraically, solve, and check. Be sure to state what your variables represent.

15. ________________

16. ________________

17. ________________

18. ________________

19. ________________

20. ________________

15. The ratio of the length of a rectangle to its width is 7 to 3. The perimeter is to be less than or equal to 40. What is the range of values that the width can have?

16. One car leaves Boston at 8 A.M., traveling 40 mph. One and a half hours later, a second car starts from the same point, traveling 45 mph along the same route. How long will it take the second car to overtake the first car?

17. A tank is filled through pipes 1 and 2 and drained through pipe 3. Pipe 1 can fill the tank in 5 hr, and pipe 2 can fill it in 3 hr. When the valves of pipes 1, 2, and 3 are open, the tank is filled in 6 hr. How long does it take pipe 3 to drain the tank by itself?

18. Machine A can produce 25 sprocket wheels in 30 min. How many sprocket wheels can machine B produce in 1 hr if both machines working together can produce 315 sprocket wheels in $3\frac{1}{2}$ hours?

19. The numerator of a fraction is 32 less than its denominator. If the numerator and denominator are each increased by 3, the new fraction has a value of $\frac{1}{2}$. Find the original fraction.

20. Machine A can produce 25 sprocket wheels in 30 min. How long does it take machine B to produce 112 sprocket wheels if both machines working together can produce 477 sprocket wheels in $4\frac{1}{2}$ hr?

Chapter 6 DIAGNOSTIC TEST

The purpose of this test is to see how well you understand operations on rational expressions and solving rational equations. We recommend that you work this diagnostic test *before* your instructor tests you on this chapter. Allow yourself about 50 minutes.

Complete solutions for all the problems on this test, together with section references, are given in the answer section at the end of the book. For the problems you do incorrectly, we suggest that you study the sections cited.

1. Are $\dfrac{3x + 4}{2y + 4}$ and $\dfrac{3x}{2y}$ equivalent rational expressions?

In Problems 2 and 3, reduce each rational expression to lowest terms, or write "cannot be reduced."

2. $\dfrac{f^2 + 5f + 6}{f^2 - 9}$
3. $\dfrac{x^4 - 2x^3 + 5x^2 - 10x}{x^3 - 8}$

In Problems 4–8, perform the indicated operations.

4. $\dfrac{x^2 - 2x - 24}{x^2 - 36} \cdot \dfrac{x^2 + 7x + 6}{x^2 + x - 12} \div (x^3 + 1)$
5. $\dfrac{3m + 3n}{m^3 - n^3} \div \dfrac{m^2 - n^2}{m^2 + mn + n^2}$
6. $\dfrac{20a + 27b}{12a - 20b} + \dfrac{44a - 13b}{20b - 12a}$
7. $\dfrac{x}{x + 4} - \dfrac{x}{x - 4} - \dfrac{32}{x^2 - 16}$
8. $\dfrac{3}{x^2 + x - 6} - \dfrac{2}{x^2 - 4} - \dfrac{3}{x^2 + 5x + 6}$

In Problems 9 and 10, find each missing term.

9. $\dfrac{4}{-h} = \dfrac{-4}{?}$
10. $-\dfrac{3}{k - 2} = \dfrac{3}{?}$
11. Simplify $\dfrac{6 - \dfrac{4}{w}}{\dfrac{3w}{w - 2} + \dfrac{1}{w}}$.

In Problems 12 and 13, find the domain of the variable.

12. $\dfrac{2x + 3}{x^2 - 4x}$
13. $\dfrac{y + 2}{3y^2 - y - 10}$

In Problems 14–16, find the solution set of each equation.

14. $\dfrac{2}{3a + 5} - \dfrac{6}{a - 2} = 3$
15. $\dfrac{x}{x + 7} = 3 - \dfrac{7}{x + 7}$
16. $\dfrac{3}{2z} + \dfrac{3}{z^2} = -\dfrac{1}{6}$
17. Solve the equation $I = \dfrac{E}{R + r}$ for r.

In Problems 18–20, set up each problem algebraically, solve, and check. Be sure to state what your variables represent.

18. The length and width of a rectangle are in the ratio of 5 to 4. The perimeter is 90 ft. Find the length and the width.
19. Roy hiked from his camp to a mountain lake one day and returned the next day. He walked up to the lake at a rate of 3 mph and returned to camp at the rate of 5 mph. The hike to the lake took 2 hr longer than the return trip.
    a. How long did it take Roy to hike to the lake?
    b. How far was the lake from Roy's camp?
20. Rachelle can make 24 bushings on a lathe in 8 hr. How long does it take Melissa to make 8 bushings if she and Rachelle can make 14 bushings in 4 hr when they work together?

Chapters 1–6 CUMULATIVE REVIEW EXERCISES

1. Simplify $\left(\dfrac{x^4y^{-3}}{x^{-2}}\right)^{-2}$.
2. Solve $\dfrac{x - 2}{4} = \dfrac{x + 3}{2} + \dfrac{1}{4}$.
3. Solve $9 - 2(m - 6) \le 3(4 - m) + 5$.
4. Is 0 a rational number?
5. Is 17 a real number?

In Exercises 6–11, perform the indicated operations, and simplify the results.

6. $(2x - 5)(x^2 - 4x + 3)$
7. $(8a^2 + 6a + 1) \div (2a + 3)$
8. $(x - 2)^6$
9. $\dfrac{x + 3}{2x^2 + 3x - 2} \cdot \dfrac{2x^2 - 9x + 4}{3x^2 + 10x + 3} \div \dfrac{x - 4}{x + 2}$
10. $\dfrac{3x}{2x^2 + 15x + 7} + \dfrac{4}{2x^2 - 5x - 3}$
11. $\dfrac{a - 3}{a^3 + 27} - \dfrac{1}{a^2 - 9}$

In Exercises 12–15, factor completely, or write "not factorable."

12. $10x^2 + 3xy - 18y^2$

13. $x^4 - x^2 - 12$

14. $x^3 - 8$

15. $6ax + 3bx - 2ay - by$

In Exercises 16–22, set up algebraically, solve, and check.

16. The length of a rectangle is 7 in. more than its width; its area is 60 sq. in. What are the dimensions of the rectangle?

17. Mrs. Kishinami spent $9.70 for 50 stamps. If she bought only 18¢ and 20¢ stamps, how many of each kind did she buy?

18. Debbie invested $1,400. Part of the money earned 6.25% per year, and the rest earned 5.75%. At the end of one year, the interest paid was $83.50. How much did Debbie place in each account?

19. A boater is running her boat's engine so that in still water she would be going 28 mph; the speed of the current is 4 mph. How far up the river (against the current) can she go if she *must* be back at her starting point in $3\frac{1}{2}$ hr?

20. Mr. Liao must mix a 2% and a 10% solution of disinfectant in order to obtain 12 L of a 5% solution. How much of the 2% solution should he use? How much of the 10% solution?

21. Alberto left Ft. Lauderdale at 3 A.M., traveling at 45 mph. Two hours later, Carlos left from the same point, traveling along the same route at 55 mph. How long did it take Carlos to overtake Alberto?

22. Forty-five square yards of carpet is laid in a rectangular room. The length is 1 yd less than twice the width. Find the dimensions of the room.

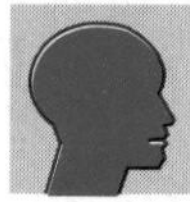

Critical Thinking and Problem-Solving Exercises

1. Six players are competing in a ping-pong tournament. If, in the first round, each player plays the other players exactly once, how many games will be played altogether in the first round?

2. Predict what the next two numbers should be for the following list.

   7, 5, 10, 8, 13, 11, 16, __, __

3. Kristy stores her accessories in seven numbered, colored boxes on a shelf in her room. The accessories are scarves, pins, belts, rings, necklaces, gloves, and hats. Using the figure and the following clues, determine the color of each box and the contents of each box:

   The rings are at the top of the stack.

   The peach-colored box is immediately above the hats on the left side of the stack.

   Box 3 is orchid.

   The white box of gloves is on the left side of the shelf, and the yellow box is on the right and one level higher than the gloves.

   The gray box is immediately below the belts.

   The aqua box, which has a number less than 6, is to the immediate left of the belts.

   The box on the bottom right holds the necklaces, and the other bottom box is white and does not hold the pins.

   The yellow box is directly below the scarves.

   The box that contains the belts has a number that is 4 more than that of the blue box.

Figure for Exercise 3

4. Find and describe the error in reasoning in the following:

$$\left(\frac{x}{y}+\frac{y}{x}\right) \div \left(\frac{x}{y}-\frac{y}{x}\right) = \left(\frac{x}{y}+\frac{y}{x}\right) \cdot \left(\frac{y}{x}-\frac{x}{y}\right)$$

Exponents and Radicals

CHAPTER 7

In this chapter, we extend the properties of exponents to include rational exponents (that is, exponents that are not integers), and we define the relation between exponents and radicals. Radicals are discussed in detail in this chapter, and *complex numbers* are introduced.

7.1 Rational Exponents

Let's review the eight properties of integral exponents.

The properties of exponents

$x^m \cdot x^n = x^{m+n}$	The product rule
$(x^m)^n = x^{mn}$	The power rule
$(xy)^n = x^n y^n$	The rule for raising a product to a power
$\dfrac{x^m}{x^n} = x^{m-n}$ if $x \neq 0$	The quotient rule
$\left(\dfrac{x}{y}\right)^n = \dfrac{x^n}{y^n}$ if $y \neq 0$	The rule for raising a quotient to a power
$x^0 = 1$ if $x \neq 0$	The zero exponent rule
$x^{-n} = \dfrac{1}{x^n}$ if $x \neq 0$	The negative exponent rule, form I
$\left(\dfrac{x^a y^b}{z^c}\right)^n = \dfrac{x^{an} y^{bn}}{z^{cn}}$	The general property of exponents

In this section, we define additional properties of exponents that make it possible to work with exponents that are rational numbers that are not integers.

Before we define such exponents, let's review the definitions and properties of radicals. Recall that in the expression $\sqrt[n]{p}$, the symbol $\sqrt[n]{p}$ represents the number whose nth power is p; p is called the *radicand*, and n is called the *index* (plural: *indices*). (It is understood that n represents a natural number that is greater than 1. Also, when no index is written, the index is understood to be 2 and the radical represents a *square root*.)

Principal roots

The symbol $\sqrt[n]{p}$ always represents the *principal* nth root of p.

If the *radicand is positive*, the principal root is positive.

If the *radicand is negative*:

1. When the index is odd, the principal root is negative.
2. When the index is even, the principal root *is not a real number.**

*Numbers that are not real numbers are discussed later in this chapter.

Basic Definitions

We now want to give some meaning to expressions in which the *exponent* is a rational number that is not an integer—that is, to expressions such as $4^{1/2}$, $8^{-2/3}$, and $x^{3/5}$.

We want to define rational exponents so that all of the properties of exponents will be valid even with these new (rational) exponents. For example, to find $(4^{1/2})^2$ (that is, to *square* $4^{1/2}$), we want to be able to apply the power rule and *multiply* the exponents, even though one of the exponents is not an integer. If we *do* multiply the exponents in the expression $(4^{1/2})^2$, we have

$$(4^{1/2})^2 = 4^{(1/2)(2)} = 4^1 = 4$$

But $(\sqrt{4})^2$ also equals 4. To see this, recall that $\sqrt{4} = 2$; therefore, $(\sqrt{4})^2 = 2^2 = 4$. Since $(\sqrt{4})^2 = 4$ and since 4 is the result we got when we squared $4^{1/2}$, perhaps we should *define* $4^{1/2}$ to equal $\sqrt{4}$.

Suppose we do define $4^{1/2}$ to equal $\sqrt{4}$. Will we get the correct result if we apply the product rule and *add* the exponents in order to perform the multiplication $4^{1/2} \cdot 4^{1/2}$? If we add the exponents, we have

$$4^{1/2} \cdot 4^{1/2} = 4^{1/2+1/2} = 4^1 = 4$$

But $\sqrt{4} \cdot \sqrt{4}$ also equals 4, since $\sqrt{4} \cdot \sqrt{4} = 2 \cdot 2 = 4$. Therefore, it does seem reasonable to define $4^{1/2}$ to equal $\sqrt{4}$.

Note The discussion given above is not a proof. *Definitions cannot be proved.* The discussion was included to help you understand *why* we define rational exponents the way we do.

The rational exponent $1/n$ is defined as follows:

The exponent $1/n$

$x^{1/n} = \sqrt[n]{x}$ if $x \geq 0$ when n is an even natural number

$x^{1/n} = \sqrt[n]{x}$ for all x when n is an odd natural number greater than 1

$x^{1/n}$ is not real if $x < 0$ when n is an even natural number.*

*An even root of a negative number is not real; therefore, $x^{1/n}$ is not real if $x < 0$ when n is even.

We say that $x^{1/n}$ is in **exponential form** and that $\sqrt[n]{x}$ is in **radical form**.

EXAMPLE 1 Examples of converting from exponential form to radical form:

a. $a^{1/4} = \sqrt[4]{a}$ if $a \geq 0$, but $a^{1/4}$ is not real if $a < 0$

b. $b^{1/7} = \sqrt[7]{b}$ for all b

c. $3^{1/3} = \sqrt[3]{3}$

d. $5^{1/2} = \sqrt{5}$

e. $(-4)^{1/3} = \sqrt[3]{-4}$

EXAMPLE 2 Examples of converting from radical form to exponential form:

a. $\sqrt[4]{z} = z^{1/4}$ if $z \geq 0$

b. $\sqrt[5]{y} = y^{1/5}$

c. $\sqrt[3]{-10} = (-10)^{1/3}$

d. $\sqrt{14} = 14^{1/2}$

A Word of Caution When we write rational exponents, we must be sure that the exponents *look like exponents*! That is, we must be sure that $2^{\frac{1}{4}}$ doesn't look like $2\frac{1}{4}$ and that $2^{1/4}$ doesn't look like 21/4.

EXAMPLE 3 Examples of evaluating expressions that contain rational exponents:

a. $27^{1/3} = \sqrt[3]{27} = 3$ **b.** $100^{1/2} = \sqrt{100} = 10$

c. $(-8)^{1/3} = \sqrt[3]{-8} = -2$ **d.** $32^{1/5} = \sqrt[5]{32} = 2$

e. $(-9)^{1/2}$ is not real Notice that 2, the denominator of the exponent, is even, and $-9 < 0$

The exponent applies *only* to the 16

f. $-16^{1/4} = -(16^{1/4}) = -\sqrt[4]{16} = -2$

The rational exponent m/n is defined as follows:

The exponent m/n

$x^{m/n} = \sqrt[n]{x^m} = (\sqrt[n]{x})^m$ if $x \geq 0$ when n is an even natural number

$x^{m/n} = \sqrt[n]{x^m} = (\sqrt[n]{x})^m$ for all x when n is an odd natural number greater than 1

$x^{m/n}$ is not real if m/n is in lowest terms and if $x < 0$ when n is an even natural number.

EXAMPLE 4 Examples of converting from exponential form to radical form:

a. $d^{4/5} = \sqrt[5]{d^4}$ or $(\sqrt[5]{d})^4$ **b.** $a^{2/3} = \sqrt[3]{a^2}$ or $(\sqrt[3]{a})^2$

c. $y^{3/4} = \sqrt[4]{y^3}$ if $y \geq 0$, but $y^{3/4}$ is not real if $y < 0$

EXAMPLE 5 Examples of converting from radical form to exponential form:

a. $\sqrt[5]{x^3} = x^{3/5}$ **b.** $(\sqrt[5]{x})^3 = x^{3/5}$

c. $\sqrt[6]{z^5} = z^{5/6}$ if $z \geq 0$, but $\sqrt[6]{z^5}$ is not real if $z < 0$

d. $(\sqrt[6]{z})^5 = z^{5/6}$ if $z \geq 0$, but $(\sqrt[6]{z})^5$ is not real if $z < 0$

EXAMPLE 6 Examples of evaluating expressions that contain rational exponents:

a. $8^{2/3} = (\sqrt[3]{8})^2 = 2^2 = 4$, or $8^{2/3} = (\sqrt[3]{8^2}) = \sqrt[3]{64} = 4$

b. $(-32)^{2/5} = (\sqrt[5]{-32})^2 = (-2)^2 = 4$

c. $(-4)^{3/4}$ is not real since the base (-4) is negative and 4, the denominator of the rational exponent, is an even number.

The exponent applies *only* to the 64

d. $-64^{5/6} = -(64^{5/6}) = -(\sqrt[6]{64})^5 = -(2^5) = -32$

Note The problem in Example 6b *could* be done by writing $(-32)^{2/5} = \sqrt[5]{(-32)^2} = \sqrt[5]{1{,}024} = 4$; however, it's *easier* to do the problem as shown in Example 6b. Similarly, without using a calculator, it would be *very* difficult to do the problem in Example 6d by writing $-(64^{5/6})$ as $-\sqrt[6]{64^5} = -\sqrt[6]{1{,}073{,}741{,}824} = -32$.

Calculators and Rational Exponents

EXAMPLE 7 Examples of using a calculator to evaluate expressions that contain rational exponents:

a. $3^{1/2}$

Scientific (Algebraic)

[3] [y^x] [(] [1] [÷] [2] [)] [=]

or, since $1/2 = .5$, [3] [y^x] [.] [5] [=]

Scientific (RPN) [3] [ENTER] [.] [5] [y^x]

Graphics [3] [∧] [.] [5] [ENTER]

The display is [1.732050808]; therefore, $3^{1/2} \approx 1.732\ 050\ 808$.

Note We can also find $3^{1/2}$ by using the square root key, since, by definition, $3^{1/2} = \sqrt{3}$. You might verify that $\sqrt{3} \approx 1.732\ 050\ 808$.

b. $(-64)^{1/3}$

Scientific (Algebraic)

[6] [4] [+/−] [y^x] [(] [1] [÷] [3] [)] [=]*

The calculator display may be [−4], but on some scientific calculators there will be an error message. The error message appears because some calculators do not allow the base to be negative when the [y^x] key is used.

Scientific (RPN) [6] [4] [+/−] [ENTER] [3] [↰] [y^x]* ($\sqrt[x]{y}$)

Graphics

[(−)] [6] [4] [∧] [(] [1] [÷] [3] [)] [ENTER]

The calculator display is [−4]; it *is* true that $(-64)^{1/3} = -4$.

c. $(-8)^{2/3}$

Scientific (Algebraic)

[(] [8] [+/−] [)] [y^x] [(] [2] [÷] [3] [)] [=]

Graphics

[(] [(−)] [8] [)] [∧] [(] [2] [÷] [3] [)] [ENTER]

The display (probably) shows an error message. Does this mean that $(-8)^{2/3}$ is not real? No! As we mentioned, some calculators do not allow the base to be negative when the [y^x] key (or the [∧] key) is used, and some of those that allow the base to be negative when the numerator of the exponent is 1 will give an error message when the numerator of the exponent is *not* 1. Let's evaluate $(-8)^{2/3}$ by converting it to radical form:

$$(-8)^{2/3} = (\sqrt[3]{-8})^2 = (-2)^2 = 4, \quad \text{or} \quad (-8)^{2/3} = \sqrt[3]{(-8)^2} = \sqrt[3]{64} = 4$$

We see that there *is* a real answer! It's 4.

Using the fact that $(-8)^{2/3} = ((-8)^{1/3})^2$, we can do the problem on *some* algebraic scientific calculators as follows:

[(] [(] [8] [+/−] [)] [y^x] [(] [1] [÷] [3] [)] [)] [x^2]

And on a graphics calculator, we can proceed as follows:

[(] [(−)] [8] [)] [∧] [(] [1] [÷] [3] [)] [∧] [2] [ENTER]

The calculator display is [4].

---

*See the Note at the end of this section.

A Word of Caution Examples 7b and 7c show that you must be very careful about checking the *reasonableness* of *all* answers, especially of calculator answers. You must make every effort to understand the order of operations that *your* calculator uses, and you must not recklessly accept all the results that you see in a calculator display.

Note On some scientific calculators, cube roots, fourth roots, and so on, can be found by using the inverse of the [y^x] key. For example, to find $\sqrt[3]{64}$, we can press [6] [4] [INV] [y^x] [3] [=].* The display is [4]. (We press [3] after [INV] [y^x] because we're finding a cube root; to find a fourth root, we would press [4] after [INV] [y^x]; and so on.) On some calculators, this procedure does *not* work for finding roots of negative numbers.

We're actually using the inverse of the [y^x] key on RPN calculators when we use the keystrokes shown in Example 7b. To find a fourth root on an RPN calculator, we would press [4] before pressing the left arrow key; to find a fifth root, we would press [5] before pressing the left arrow key; and so forth.

Exercises 7.1 Set I

In Exercises 1–10, convert each radical to exponential form; assume that all variables represent nonnegative numbers.

1. $\sqrt{5}$ 2. $\sqrt{7}$ 3. $\sqrt[3]{z}$ 4. $\sqrt[4]{x}$
5. $\sqrt[4]{x^3}$ 6. $\sqrt[5]{y^2}$ 7. $(\sqrt[3]{x^2})^2$ 8. $(\sqrt[4]{x^3})^2$
9. $\sqrt[n]{x^{2n}}$ 10. $\sqrt[n]{y^{5n}}$

In Exercises 11–16, convert each expression to radical form; assume that all variables represent nonnegative numbers.

11. $7^{1/2}$ 12. $5^{1/3}$ 13. $a^{3/5}$ 14. $b^{2/3}$
15. $x^{m/n}$ 16. $x^{(a+b)/a}$

In Exercises 17–28, evaluate each expression *without* using a calculator. (The answer might be "not real.") Then, if your instructor permits, work the exercises again *with* a calculator.

17. $8^{1/3}$ 18. $27^{1/3}$ 19. $(-27)^{2/3}$
20. $(-8)^{2/3}$ 21. $4^{3/2}$ 22. $9^{3/2}$
23. $(-16)^{3/4}$ 24. $(-1)^{1/4}$ 25. $-4^{1/2}$
26. $-121^{1/2}$ 27. $-64^{5/6}$ 28. $-81^{3/4}$

Writing Problems

Express the answers in your own words and in complete sentences.

1. Find and describe the error in the following:
$$16^{1/2} = \frac{16}{2} = 8$$
Also, evaluate $16^{1/2}$ correctly.
2. The definition of the exponent m/n states that $x^{m/n} = \sqrt[n]{x^m} = (\sqrt[n]{x})^m$ if $x \geq 0$ when n is an even natural number. Explain why the restriction "... if $x \geq 0$ when n is an even natural number" is necessary.

Exercises 7.1 Set II

In Exercises 1–10, convert each radical to exponential form; assume that all variables represent nonnegative numbers.

1. $\sqrt{17}$ 2. $\sqrt{31}$ 3. $\sqrt[3]{y}$ 4. $\sqrt[5]{x^2}$
5. $\sqrt[4]{x^7}$ 6. $\sqrt[3]{z^4}$ 7. $(\sqrt[3]{x})^2$ 8. $(\sqrt{y})^3$
9. $\sqrt[n]{x^{3n}}$ 10. $\sqrt[n]{x^{7n}}$

In Exercises 11–16, convert each expression to radical form; assume that the variables represent nonnegative numbers.

11. $13^{1/3}$ 12. $7^{1/4}$ 13. $x^{4/7}$ 14. $c^{2/5}$

*On some scientific calculators, the keystrokes are [3] [INV] [y^x] [6] [4] [=].

15. $y^{a/b}$ 16. $z^{(a-b)/b}$

In Exercises 17–28, evaluate each expression *without* using a calculator. (The answer might be "not real.") Then, if your instructor permits, work the exercises again *with* a calculator.

17. $64^{2/3}$ 18. $16^{3/4}$ 19. $8^{2/3}$ 20. $(-1)^{1/5}$ 21. $27^{4/3}$ 22. $1^{7/8}$ 23. $25^{3/2}$ 24. $(-64)^{2/3}$ 25. $-25^{1/2}$ 26. $-81^{1/4}$ 27. $-256^{3/4}$ 28. $-1^{7/8}$

7.2 Using the Properties of Exponents with Rational Exponents

All of the properties of exponents reviewed at the beginning of this section can be used when the exponents are *any* rational numbers, as long as the base is nonnegative whenever the denominator of the exponent is an even number.

EXAMPLE 1

Examples of using the properties of exponents to simplify expressions with rational exponents (assume that all variables represent nonnegative numbers):

a. $a^{1/2}a^{-1/4} = a^{\frac{1}{2}+\left(-\frac{1}{4}\right)} = a^{\frac{2}{4}-\frac{1}{4}} = a^{1/4}$ Using the product rule

b. $x^{-1/3} = \dfrac{1}{x^{1/3}}$ Using the negative exponent rule

c. $(z^{-1/2})^4 = z^{-\frac{1}{2}\cdot 4} = z^{-2} = \dfrac{1}{z^2}$ Using the power rule and the negative exponent rule

d. $\dfrac{y^{2/3}}{y^{1/2}} = y^{\frac{2}{3}-\frac{1}{2}} = y^{\frac{4}{6}-\frac{3}{6}} = y^{1/6}$ Using the quotient rule

e. $x^{-1/2}xx^{2/5} = x^{-\frac{1}{2}+1+\frac{2}{5}}$ Using the product rule

$= x^{-\frac{5}{10}+\frac{10}{10}+\frac{4}{10}} = x^{9/10}$

f. $\dfrac{b^{3/2}b^{-2/3}}{b^{5/6}} = b^{\frac{3}{2}-\frac{2}{3}-\frac{5}{6}}$ Using the product rule and the quotient rule

$= b^{\frac{9}{6}-\frac{4}{6}-\frac{5}{6}} = b^0 = 1$ Using the zero exponent rule

g. $(9h^{-2/5}k^{4/3})^{-3/2} = (3^2h^{-2/5}k^{4/3})^{-3/2}$

$= 3^{2\left(-\frac{3}{2}\right)}h^{-\frac{2}{5}\left(-\frac{3}{2}\right)}k^{\frac{4}{3}\left(-\frac{3}{2}\right)}$ Using the general property of exponents

$= 3^{-3}h^{3/5}k^{-2} = \dfrac{h^{3/5}}{27k^2}$ Using the negative exponent rule

h. $\left(\dfrac{xy^{-2/3}}{z^{-4}}\right)^{-3/4} = \left(\dfrac{xz^4}{y^{2/3}}\right)^{-3/4} = \left(\dfrac{y^{2/3}}{xz^4}\right)^{3/4}$ Using the negative exponent rule, forms 1 and 3

$= \dfrac{y^{\frac{2}{3}\cdot\frac{3}{4}}}{x^{\frac{3}{4}}z^{\frac{4}{1}\cdot\frac{3}{4}}} = \dfrac{y^{1/2}}{x^{3/4}z^3}$ Using the general property of exponents

i. $\left(\dfrac{c^{5/2}d^{-3}}{c^{-1/2}}\right)^{2/3} = \dfrac{c^{\frac{5}{2}\cdot\frac{2}{3}}d^{-\frac{3}{1}\cdot\frac{2}{3}}}{c^{-\frac{1}{2}\cdot\frac{2}{3}}}$ Using the general property of exponents

$= \dfrac{c^{5/3}d^{-2}}{c^{-1/3}}$ We will use the quotient rule on $\frac{c^{5/3}}{c^{-1/3}}$, and $\frac{5}{3}-\left(-\frac{1}{3}\right)=\frac{5}{3}+\frac{1}{3}=\frac{6}{3}=2$; we will use the negative exponent rule on d^{-2}

$= \dfrac{c^{\frac{5}{3}-\left(-\frac{1}{3}\right)}}{d^2}$

$= \dfrac{c^2}{d^2}$

EXAMPLE 2 Examples of evaluating expressions:

a. $16^{-1/4} = \dfrac{1}{16^{1/4}} = \dfrac{1}{\sqrt[4]{16}} = \dfrac{1}{2}$

b. $(-64)^{-1/3} = \dfrac{1}{(-64)^{1/3}} = \dfrac{1}{\sqrt[3]{-64}} = \dfrac{1}{-4} = -\dfrac{1}{4}$

c. $16^{-3/4}(-27)^{2/3} = \dfrac{1}{16^{3/4}}(\sqrt[3]{-27})^2 = \dfrac{1}{(\sqrt[4]{16})^3}(-3)^2 = \dfrac{1}{2^3}(9) = \dfrac{9}{8}$

Exercises 7.2
Set I

In Exercises 1–30, perform the indicated operations. Express the answers in exponential form with positive exponents only; assume that all variables represent nonnegative numbers.

1. $x^{1/2}x^{3/2}$
2. $y^{5/4}y^{3/4}$
3. $a^{3/4}a^{-1/2}$
4. $b^{5/6}b^{-1/3}$
5. $z^{-1/2}z^{2/3}$
6. $N^{-1/3}N^{3/4}$
7. $(H^{3/4})^2$
8. $(s^{5/6})^3$
9. $(x^{-3/4})^{1/3}$
10. $(y^{-2/3})^{1/2}$
11. $\dfrac{a^{3/4}}{a^{1/2}}$
12. $\dfrac{b^{2/3}}{b^{1/6}}$
13. $\dfrac{x^{1/2}}{x^{-1/3}}$
14. $\dfrac{z^{1/4}}{z^{-1/3}}$
15. $x^{2/3}xx^{-1/2}$
16. $x^{3/4}xx^{-1/3}$
17. $(x^{-1/2})^3(x^{2/3})^2$
18. $(x^{3/4})^2(x^{-2/3})^3$
19. $\dfrac{u^{1/2}v^{-2/3}}{u^{-1/4}v^{-1}}$
20. $\dfrac{u^{2/3}v^{-3/5}}{u^{-1/3}v^{-1}}$
21. $(16x^{-2/5}y^{4/9})^{3/2}$
22. $(8x^{9/8}y^{-3/4})^{2/3}$
23. $\left(\dfrac{a^6d^0}{b^{-9}c^3}\right)^{7/3}$
24. $\left(\dfrac{y^{10}z^{-5}}{x^0w^{-15}}\right)^{7/5}$
25. $\left(\dfrac{x^3y^0z^{-1}}{32x^{-1}z^2}\right)^{2/5}$
26. $\left(\dfrac{ab^{-1}c^0}{8a^{-4}b^4}\right)^{2/3}$
27. $\left(\dfrac{x^{-1}y^{2/3}}{z^{-5}}\right)^{-3/5}$
28. $\left(\dfrac{a^{3/4}b^{-1}}{c^{-2}}\right)^{-2/3}$
29. $\left(\dfrac{9x^{-2/3}y^{2/9}}{x^{-2}}\right)^{-3/2}$
30. $\left(\dfrac{27R^{3/5}S^{-5/2}}{S^{-4}}\right)^{-2/3}$

In Exercises 31–36, evaluate each expression.

31. $8^{-2/3}$
32. $27^{-2/3}$
33. $4^{1/2}(9)^{-3/2}$
34. $8^{1/3}(16)^{-3/2}$
35. $100^{-1/2}(-8)^{1/3}$
36. $(10^4)^{-1/2}(-27)^{1/3}$

Writing Problems

Express the answer in your own words and in complete sentences.

1. Describe what your first step would be in simplifying $\left(\dfrac{3^{-4}x^{3/4}y^{-1/4}}{z^{-1/2}}\right)^{-4/3}$, and explain why that would be your first step.

Exercises 7.2
Set II

In Exercises 1–30, perform the indicated operations. Express the answers in exponential form with positive exponents only; assume that all variables represent nonnegative numbers.

1. $a^{1/4}a^{3/4}$
2. $b^{3/5}b^{1/4}$
3. $x^{5/6}x^{-1/3}$
4. $c^{2/3}c^{-1/4}$
5. $z^{-1/5}z^{2/3}$
6. $y^{7/8}y^{-1/3}$
7. $(x^{3/5})^2$
8. $(a^{2/5})^{1/3}$
9. $(x^{-3/7})^{1/3}$
10. $(a^{-1/4})^{-8/9}$
11. $\dfrac{x^{7/8}}{x^{1/4}}$
12. $\dfrac{y^{5/6}}{y^{1/3}}$
13. $\dfrac{p^{3/5}}{p^{-1/2}}$
14. $\dfrac{a^{-2/5}}{a^{-7/10}}$
15. $z^{2/5}z^{-3/5}z^{7/10}$

16. $x^{-1/3}x^{2/5}x^{3/4}$ 17. $(x^{-1/3})^4(x^{2/3})^5$ 18. $(a^{-3/4})^3(a^{5/6})^2$

19. $\dfrac{x^{1/3}y^{-4/5}}{x^{-1/4}y^{-1}}$ 20. $\dfrac{a^{-2/5}b^{2/3}}{a^{1/2}b^{1/2}}$ 21. $(36R^{-4/3}S^{2/5})^{1/2}$

22. $(49x^{-1/5}y^{3/4})^{1/2}$ 23. $\left(\dfrac{s^3t^0}{u^{-6}v^9}\right)^{4/3}$ 24. $\left(\dfrac{a^0b^4}{c^8d^{-12}}\right)^{1/4}$

25. $\left(\dfrac{y^{2/3}z^{-2/3}}{4z^{-3}}\right)^{3/2}$ 26. $\left(\dfrac{8a^{-12}}{b^9c^3}\right)^{2/3}$ 27. $\left(\dfrac{R^{-3/2}S^{1/2}}{8R^{-2}}\right)^{2/3}$

28. $\left(\dfrac{16x^8y^4}{x^{12}}\right)^{3/4}$ 29. $\left(\dfrac{x^2y^{-4}z^6}{32x^{-2}y^0z}\right)^{3/5}$ 30. $\left(\dfrac{64a^6b^8}{a^{18}b^2}\right)^{5/6}$

In Exercises 31–36, evaluate each expression.

31. $64^{-2/3}$ 32. $16^{-3/4}$ 33. $9^{1/2}(25)^{3/2}$

34. $(10^6)^{-1/2}(125)^{1/3}$ 35. $\dfrac{(10^4)^{1/2}(10^{-2})^{3/2}}{(10^5)^{-1/5}}$ 36. $\dfrac{(2^4)^{1/4}(2^{-8})^{1/2}}{(2^{-6})^{-1/3}}$

7.3 Simplifying Radicals That Do Not Involve Division

The Principal Square Root of x^2

When we simplified square roots earlier, we assumed that any variables represented *nonnegative* numbers. If we assume that $x \geq 0$, we can write $\sqrt{x^2} = x$. But what if x is negative?

Recall that the square root symbol $\sqrt{\ }$ always represents a *positive* number (or zero), and recall that $|x|$ always represents a positive number or zero. Therefore, if $x < 0$, we can write $\sqrt{x^2} = |x|$; both sides of the equation represent the same positive number. For example, suppose x is -5. Then $\sqrt{(-5)^2} = \sqrt{25} = 5$, and $|-5| = 5$. Therefore, $\sqrt{(-5)^2} = |-5|$. (Notice that *without* the absolute value symbols, we would have $\sqrt{(-5)^2} = -5$, *which is false*.)

For all real numbers x,	$\sqrt{x^2} = \|x\|$

EXAMPLE 1 Examples of finding square roots:

a. $\sqrt{w^6} = \sqrt{(w^3)^2} = |w^3|$ Because w^3 is negative if w is negative and because $\sqrt{w^6}$ is positive, we need and must retain the absolute value symbols

b. $\sqrt{z^{12}} = \sqrt{(z^6)^2} = |z^6| = z^6$ Because z^6 is positive for all values of z, we do not need to retain the absolute value symbols

c. $\sqrt{(a-b)^2} = |a-b|$ Even if a and b are both *known* to be positive, $a - b$ can be negative; therefore, we need to retain the absolute value symbols

What happens if, in problems such as the ones in Example 1, we convert from radical form to exponential form? For example, let's consider $\sqrt{x^6}$. If we convert $\sqrt{x^6}$ to exponential form, we have $\sqrt{x^6} = x^{6/2} = x^3$ if $x \geq 0$; if $x < 0$, x^3 is negative, so absolute value symbols are necessary, and we have $\sqrt{x^6} = |x^{6/2}| = |x^3|$. Notice that in both cases, we divide the exponent 6 by 2 (the understood index of the radical) to find the exponent of the answer. (Because the answer contains an exponent that is an odd number, absolute value symbols are necessary unless we *know* that x is nonnegative.)

Now let's consider $\sqrt{y^8}$. If we convert $\sqrt{y^8}$ to exponential form, we have $\sqrt{y^8} = |y^{8/2}| = |y^4|$; y^4 is positive for *all* y, so absolute value symbols are not really necessary, and we can write $\sqrt{y^8} = y^4$.

We can generalize these results as follows:

> To simplify a square root when the radicand contains just one factor that is a perfect square, divide the exponent on the variable by 2 and drop the radical sign. Place absolute value symbols around the result if the result contains an *odd* exponent.

The justification for the procedure given in the box above is as follows:

$$\sqrt{x^{2n}} = |(x^{2n})^{1/2}| = |x^{2n/2}| = |x^n|$$

The absolute value symbols in the answer are necessary if n is an odd number.

The Principal nth Root of x^n

Recall that the radical sign always stands for the principal root of a number. Also recall that if the radicand is positive, the principal root is positive; if the radicand is negative, the principal root is negative when the index is odd. Therefore, the following properties can be used:

> If n is an odd natural number greater than 1, $\sqrt[n]{x^n} = x$ for all x.
>
> If n is an even natural number, $\sqrt[n]{x^n} = |x|$ for all x.

EXAMPLE 2 Examples of finding principal nth roots of nth powers:

a. $\sqrt[4]{x^4} = |x|$ — Absolute value symbols are necessary because the index is even

b. $\sqrt[4]{2^4} = |2| = 2$ — We place absolute value symbols around 2 temporarily, but then $|2| = 2$

c. $\sqrt[4]{(-3)^4} = |-3| = 3$ — $\sqrt[4]{(-3)^4}$ is a *positive* number; absolute value symbols are necessary around -3; then $|-3| = 3$

d. $\sqrt[3]{y^3} = y$ — If y is negative, y^3 is negative, and so $\sqrt[3]{y^3}$ should also be negative; absolute value symbols would be *incorrect*

e. $\sqrt[5]{(-2)^5} = -2$ — $\sqrt[5]{(-2)^5}$ is a *negative* number; absolute value symbols would be *incorrect*

Perfect Powers When the exponent on a factor of a radicand is exactly divisible by the index of the radical, we call that factor a **perfect power**. For example, in the expression $\sqrt[3]{x^9}$, we call x^9 a perfect power because the exponent, 9, is exactly divisible by the index, 3. In the expression $\sqrt[5]{t^{20}}$, we call t^{20} a perfect power because the exponent, 20, is exactly divisible by the index, 5. In the expression $\sqrt{z^8}$, we call z^8 a perfect power because the exponent, 8, is exactly divisible by the index, which is understood to be 2.

If we convert $\sqrt[3]{x^9}$ to exponential form, we have $\sqrt[3]{x^9} = x^{9/3} = x^3$. If we convert $\sqrt[5]{t^{20}}$ to exponential form, we have $\sqrt[5]{t^{20}} = t^{20/5} = t^4$. We generalize these results as follows:

> To simplify a radical when the radicand contains just one factor that is a perfect power, divide the exponent by the index of the radical and drop the radical sign. If the index is an *even* number, (1) the radicand must be nonnegative, and (2) it may be necessary to place absolute value symbols around the result.

Because the principal root is negative if the radicand is negative when the index is odd, cube roots of negative numbers will be negative (you might verify that $\sqrt[3]{-8} = -\sqrt[3]{8}$), fifth roots of negative numbers will be negative (you can verify that $\sqrt[5]{-32} = -\sqrt[5]{32}$), and so forth. This suggests that a negative sign can be "moved across" the radical sign if the index of the radical is an odd number.

A Word of Caution $\sqrt{-9} \neq -\sqrt{9}$, and $\sqrt[4]{-16} \neq -\sqrt[4]{16}$. Remember: $-\sqrt{9} = -3$, but $\sqrt{-9}$ is not a real number; and $-\sqrt[4]{16} = -2$, but $\sqrt[4]{-16}$ is not a real number. (Numbers such as $\sqrt{-9}$ will be discussed later in the chapter.)

EXAMPLE 3 Examples of simplifying radicals when the radicand is a perfect power:

a. $\sqrt[4]{z^{12}} = |z^{12/4}| = |z^3|$ Absolute value symbols are necessary because the index was even and the exponent of the answer is odd

b. $\sqrt[7]{x^{14}} = x^{14/7} = x^2$ Absolute value symbols are not needed because the index was an odd number

c. $\sqrt[7]{x^{21}} = x^{21/7} = x^3$ It would be *incorrect* to use absolute value symbols because the index was an odd number

The cube root of a negative number is negative

d. $\sqrt[3]{-b^6} = -\sqrt[3]{b^6} = -b^{6/3} = -b^2$

Notice that $-b^2$ is a *negative* number. Let's verify for $b = 2$ that $\sqrt[3]{-b^6}$ and $-\sqrt[3]{b^6}$ both equal $-b^2$:

$$\sqrt[3]{-2^6} = \sqrt[3]{-64} = -4 = -2^2 \quad \text{and} \quad -\sqrt[3]{2^6} = -\sqrt[3]{64} = -4 = -2^2$$

Simplifying Radicals When the Radicand Contains More Than One Factor

A radicand often consists of two of more factors. In this case, the following basic property of radicals can be used to simplify the radical:

The product rule for radicals

$$\sqrt[n]{ab} = \sqrt[n]{a}\sqrt[n]{b}$$

if $\sqrt[n]{a}$ and $\sqrt[n]{b}$ are both real numbers.*

*We discuss the case in which $\sqrt[n]{a}$ and $\sqrt[n]{b}$ are not necessarily real later in this chapter.

PROOF The product rule for radicals is easily proved if we convert the radicals to exponential form (for simplicity in the proof, we assume that $a \geq 0$ and $b \geq 0$):

$$\sqrt[n]{ab} = (ab)^{1/n} = a^{1/n}b^{1/n} = \sqrt[n]{a}\sqrt[n]{b}$$ ■

It is *also* true that $\sqrt[n]{a}\sqrt[n]{b} = \sqrt[n]{ab}$.

In this section, we will use the product rule for radicals for *simplifying* radicals; later we will use it for *multiplying* radicals.

EXAMPLE 4 Examples verifying that the product rule for radicals is true:

a. $\sqrt{16 \cdot 25} \stackrel{?}{=} \sqrt{16}\sqrt{25}$

$\sqrt{400} \stackrel{?}{=} 4 \cdot 5$

$20 = 20$

b. $\sqrt[3]{8}\sqrt[3]{27} \stackrel{?}{=} \sqrt[3]{8 \cdot 27}$

$2 \cdot 3 \stackrel{?}{=} \sqrt[3]{216}$

$6 = 6$

Example 5 shows how the product rule for radicals can be used for simplifying a radical.

EXAMPLE 5

Examples of using the product rule for radicals to simplify radicals:

a. $\sqrt{9x^6y^8} = \sqrt{9}\sqrt{x^6}\sqrt{y^8} = 3|x^3|y^4$ Absolute value symbols around x^3 are essential; the answer can also be written as $3|x^3y^4|$ or as $|3x^3y^4|$

The seventh root of a negative number is negative

b. $\sqrt[7]{-2^7y^{14}} = -\sqrt[7]{2^7y^{14}} = -\sqrt[7]{2^7}\sqrt[7]{y^{14}} = -2y^2$

Note We now have an alternative way of approaching a problem such as $\sqrt[3]{-b^6}$. We can think of it this way:

$$\sqrt[3]{-b^6} = \sqrt[3]{(-1)b^6} = \sqrt[3]{(-1)}\sqrt[3]{b^6} = -1(b^2) = -b^2$$

In the examples we've done so far, all the factors of the radicand were perfect powers, and so there were no radical signs left in the answers. This will not always be the case. In Example 6, we simplify radicals in which some (or all) of the factors of the radicand are not perfect powers; in such problems, there will always be a radical sign in the answer. It is customary to write all rational factors to the left of the radical sign.

EXAMPLE 6

Examples of simplifying radicals when some factors of the radicand are not perfect powers:

a. $\sqrt{50} = \sqrt{25 \cdot 2} = \sqrt{25}\sqrt{2} = 5\sqrt{2}$ **b.** $\sqrt[3]{24} = \sqrt[3]{8 \cdot 3} = \sqrt[3]{8}\sqrt[3]{3} = 2\sqrt[3]{3}$

c. $\sqrt[4]{32} = \sqrt[4]{16 \cdot 2} = \sqrt[4]{16}\sqrt[4]{2} = 2\sqrt[4]{2}$

Simplest Radical Form

A number of conditions must be satisfied in order for a radical to be in **simplest radical form**. We will explore these conditions in the balance of this section and in the following six sections.

The *first* condition for simplest radical form is that the radicand must be positive. (That is, $\sqrt[3]{-5}$ is not in simplest radical form; it must be rewritten as $-\sqrt[3]{5}$.)

The *second* condition is that if the radicand were to be expressed in prime-factored, exponential form, none of the factors could have an exponent that was greater than or equal to the index of the radical.

In Example 6, we changed radicals to simplest radical form by inspection; that is, we used guessing-and-checking to break the radicand up into factors that were perfect powers. A more formal procedure for simplifying radicals (when division is not involved) is as follows:

Simplifying a radical that does not involve division

1. Express the radicand in prime-factored, exponential form.
2. If the exponent of any factor of the radicand is greater than the index, break up that factor into the product of a perfect power and a nonperfect power *whose exponent is less than the index.*
3. Use the product rule for radicals to rewrite the radical as a product of radicals.
4. Remove perfect powers from the radical by dividing their exponents by the index; place absolute value symbols around the result if necessary. All remaining exponents should be less than the index; factors with such exponents remain under the radical sign.
5. The simplified radical is the *product* of all the factors from step 4. It is customary to write all rational factors to the left of any irrational ones.
6. It may be possible to reduce the order of the radical (see Section 7.5).

EXAMPLE 7

Examples of simplifying radicals:

a. $\sqrt{27} = \sqrt{3^3}$ — 3^3 is the prime-factored, exponential form of 27

$= \sqrt{3^2 \cdot 3}$ — 3^2 is a perfect power; the other exponent (an understood 1) is smaller than the index

$= \sqrt{3^2}\sqrt{3}$ — Using the product rule for radicals

$= 3\sqrt{3}$ — This is in simplest radical form

b. $\sqrt{48} = \sqrt{2^4 \cdot 3}$ — $2^4 \cdot 3$ is the prime-factored, exponential form of 48; 2^4 is a perfect power; the exponent on the 3 (an understood 1) is smaller than the index

$= \sqrt{2^4}\sqrt{3}$ — Using the product rule for radicals

$= 4\sqrt{3}$ — This is in simplest radical form

c. $\sqrt[3]{54} = \sqrt[3]{2 \cdot 3^3}$ — $2 \cdot 3^3$ is the prime-factored, exponential form of 54

$= \sqrt[3]{2}\sqrt[3]{3^3}$ — Using the product rule for radicals

$= \sqrt[3]{2}\,(3)$ — Simplifying $\sqrt[3]{3^3}$

$= 3\sqrt[3]{2}$ — Writing the rational factor to the left of the irrational factor

d. $\sqrt{360} = \sqrt{2^3 \cdot 3^2 \cdot 5}$ — $2^3 \cdot 3^2 \cdot 5$ is the prime-factored, exponential form of 360

$= \sqrt{2^2 \cdot 2 \cdot 3^2 \cdot 5}$ — The shaded factors (those whose exponents are understood 1's) must remain under the radical sign

$= \sqrt{2^2 \cdot 3^2 \cdot 2 \cdot 5}$ — Rearranging the factors

$= \sqrt{2^2}\sqrt{3^2}\sqrt{2 \cdot 5}$ — Using the product rule for radicals

$= 2 \cdot 3\sqrt{10}$, or $6\sqrt{10}$ — This is in simplest radical form

In part e, assume that $y \geq 0$.

e. $\sqrt{12x^4y^3} = \sqrt{2^2 \cdot 3 \cdot x^4 \cdot y^2 \cdot y}$ — The shaded factors must remain under the radical sign

$= \sqrt{2^2 \cdot x^4 \cdot y^2 \cdot 3 \cdot y}$ — Rearranging the factors

$= \sqrt{2^2}\sqrt{x^4}\sqrt{y^2}\sqrt{3y}$ — Using the product rule for radicals

$= 2x^2y\sqrt{3y}$ — We don't need absolute value symbols since x^2 is always positive and $y \geq 0$

f. $\sqrt[3]{-16a^5b^7} = -\sqrt[3]{2^4a^5b^7}$ — The cube root of a negative number is negative

$= -\sqrt[3]{2^3 \cdot 2 \cdot a^3 \cdot a^2 \cdot b^6 \cdot b}$ — The shaded factors must remain under the radical sign

$= -\sqrt[3]{2^3 \cdot a^3 \cdot b^6 \cdot 2 \cdot a^2 \cdot b}$ — Rearranging the factors

$= -\sqrt[3]{2^3}\sqrt[3]{a^3}\sqrt[3]{b^6}\sqrt[3]{2a^2b}$ — Using the product rule for radicals

$= -2 \cdot a \cdot b^2\sqrt[3]{2a^2b}$

$= -2ab^2\sqrt[3]{2a^2b}$

In part g, assume that $h \geq 0$ and $k \geq 0$.

g. $\sqrt[4]{96h^{11}k^3m^4} = \sqrt[4]{2^5 \cdot 3 \cdot h^{11}k^3m^4}$

$= \sqrt[4]{2^4 \cdot 2 \cdot 3 \cdot h^8 \cdot h^3 \cdot k^3 \cdot m^4}$ — The shaded factors must remain under the radical sign

$= \sqrt[4]{2^4 \cdot h^8 \cdot m^4 \cdot 2 \cdot 3 \cdot h^3 \cdot k^3}$ — Rearranging the factors

$= \sqrt[4]{2^4}\sqrt[4]{h^8}\sqrt[4]{m^4}\sqrt[4]{2 \cdot 3 \cdot h^3 \cdot k^3}$ — Using the product rule for radicals

$= 2 \cdot h^2 \cdot |m|\sqrt[4]{2 \cdot 3 \cdot h^3 \cdot k^3}$

$= 2h^2|m|\sqrt[4]{6h^3k^3}$ — Absolute value symbols around m are essential

In part h, assume that $y \geq 0$ and $z \geq 0$.

h. $5xy^2\sqrt{28x^4y^9z^3} = 5xy^2\sqrt{2^2 \cdot 7 \cdot x^4 \cdot y^8 \cdot y \cdot z^2 \cdot z}$

$= 5xy^2\sqrt{2^2 \cdot x^4 \cdot y^8 \cdot z^2 \cdot 7 \cdot y \cdot z}$ — Rearranging the factors

$= 5xy^2\sqrt{2^2}\sqrt{x^4}\sqrt{y^8}\sqrt{z^2}\sqrt{7yz}$ — Using the product rule for radicals

$= 5xy^2(2x^2y^4z)\sqrt{7yz}$

$= 10x^3y^6z\sqrt{7yz}$

  It is not necessary to show all the steps that we have shown in Example 7.

Remember: You do not *have* to simplify a radical by writing the radicand in prime-factored, exponential form if, by inspection, you see that the radicand contains a factor that is a perfect power (see Example 6).

Exercises 7.3 Set I

Simplify each radical.

1. $\sqrt{4x^2}$
2. $\sqrt{9y^2}$
3. $\sqrt[3]{8x^3}$
4. $\sqrt[3]{27y^3}$
5. $\sqrt[4]{16x^4y^8}$
6. $\sqrt[4]{81u^8v^4}$
7. $\sqrt{2^5}$
8. $\sqrt{3^3}$
9. $\sqrt{(-2)^2}$
10. $\sqrt{(-3)^2}$
11. $\sqrt[3]{-3^5}$
12. $\sqrt[3]{-2^8}$
13. $\sqrt[4]{32}$
14. $\sqrt[4]{48}$
15. $\sqrt[5]{-x^7}$
16. $\sqrt[5]{-z^8}$
17. $\sqrt{8a^4b^2}$
18. $\sqrt{20m^8u^2}$
19. $3m\sqrt{18m^{12}n^6}$
20. $2h\sqrt{50h^8k^6}$
21. $5\sqrt[3]{-24a^5b^2}$
22. $6\sqrt[3]{-54c^4d}$
23. $\sqrt[5]{64m^{11}p^{15}u}$
24. $\sqrt[5]{128u^4v^{10}w^{16}}$
25. $\sqrt[3]{8(a+b)^3}$
26. $\sqrt[3]{27(x-y)^6}$
27. $\sqrt{4(x-y)^2}$
28. $\sqrt{9(3z-w)^2}$

Writing Problems

Express the answers in your own words and in complete sentences.

1. Find and describe the error in the following: $\sqrt{4x^2} = 2x$.
2. Explain why absolute value symbols are not needed in this statement: $\sqrt{16x^4} = 4x^2$
3. Find and describe the error in the following: $\sqrt[3]{x^3} = |x|$
4. Find and describe the error in the following:
$$3\sqrt{5} = \sqrt{3 \cdot 5} = \sqrt{15}$$
5. Find and describe the error in the following:
$$\sqrt{50} = \sqrt{2 \cdot 5^2} = 2\sqrt{5}$$

Exercises 7.3 Set II

Simplify each radical.

1. $\sqrt{25x^2}$
2. $\sqrt{64a^4}$
3. $\sqrt[6]{64y^6}$
4. $\sqrt[5]{32b^8}$
5. $\sqrt[4]{256a^8b^{12}}$
6. $\sqrt[4]{81x^4y^{12}}$
7. $\sqrt{2^7}$
8. $\sqrt[3]{5^4}$
9. $\sqrt{(-7)^2}$
10. $\sqrt[5]{-3^5}$
11. $\sqrt[3]{-2^5}$
12. $\sqrt{(-3)^4}$
13. $\sqrt{40}$
14. $\sqrt[5]{128}$
15. $\sqrt[5]{-x^7}$
16. $\sqrt[3]{a^8}$
17. $\sqrt{75x^8y^4}$
18. $\sqrt{48a^6b^8c^2}$
19. $\sqrt[3]{27xy^8z^5}$
20. $\sqrt[3]{-16x^3y^2}$
21. $3\sqrt{32x^6y^{10}}$
22. $\sqrt[3]{-27ab^2c^3}$
23. $\sqrt[7]{256x^9y^{15}}$
24. $\sqrt{21x^4y^6z^{10}}$
25. $\sqrt[3]{(a+b)^6}$
26. $-8\sqrt[3]{-8s^2t^7}$
27. $\sqrt{16(u-5v)^2}$
28. $\sqrt[4]{16(2x-3y)^4}$

7.4 Simplifying Radicals That Involve Division

The following property can be used in simplifying radicals that involve division:

The quotient rule for radicals

$$\sqrt[n]{\frac{a}{b}} = \frac{\sqrt[n]{a}}{\sqrt[n]{b}}$$

if $\sqrt[n]{a}$ and $\sqrt[n]{b}$ are both real numbers and if $b \neq 0$.

PROOF (For simplicity in the proof, we assume that $a \geq 0$ and $b > 0$.) If we convert the radicals to exponential form, we have

$$\sqrt[n]{\frac{a}{b}} = \left(\frac{a}{b}\right)^{1/n} = \frac{a^{1/n}}{b^{1/n}} = \frac{\sqrt[n]{a}}{\sqrt[n]{b}} \qquad \blacksquare$$

It is also true that $\dfrac{\sqrt[n]{a}}{\sqrt[n]{b}} = \sqrt[n]{\dfrac{a}{b}}$.

The quotient rule for radicals can be used in simplifying radicals in which the radicand is a rational expression (see Example 1), and later we will use it in *dividing* radicals.

A *third* condition that must be satisfied in order for a radical to be in *simplest radical form* is that the radicand cannot be a rational expression.

EXAMPLE 1

Examples of using the quotient rule for radicals to simplify radicals in which the radicand is a rational expression (assume that $y \neq 0$ and $k \neq 0$):

a. $\sqrt{\dfrac{25}{36}} = \dfrac{\sqrt{25}}{\sqrt{36}} = \dfrac{5}{6}$

b. $\sqrt[3]{\dfrac{8}{27}} = \dfrac{\sqrt[3]{8}}{\sqrt[3]{27}} = \dfrac{2}{3}$

c. $\sqrt{\dfrac{x^4}{y^6}} = \dfrac{\sqrt{x^4}}{\sqrt{y^6}} = \dfrac{x^2}{|y^3|}$

d. $\sqrt[5]{\dfrac{32x^{10}}{y^{15}}} = \dfrac{\sqrt[5]{2^5x^{10}}}{\sqrt[5]{y^{15}}} = \dfrac{2x^2}{y^3}$

e. $\sqrt{\dfrac{3x^2}{4}} = \dfrac{\sqrt{3x^2}}{\sqrt{4}} = \dfrac{|x|\sqrt{3}}{2}$

f. $\sqrt{\dfrac{50h^2}{2k^4}} = \sqrt{\dfrac{25h^2}{k^4}} = \dfrac{\sqrt{25h^2}}{\sqrt{k^4}} = \dfrac{5|h|}{k^2}$ (Reducing $\frac{50}{2}$ to 25)

Rationalizing the Denominator When the Radicand Is a Rational Expression

In Example 1, the denominators were all perfect powers (in Example 1f, the denominator became a perfect power after we reduced the rational expression). This will not always be the case.

If the radicand is a rational expression in which the denominator is *not* a perfect power, we must *build* the radicand in such a way that we force the denominator to become a perfect power (see Example 2). This procedure causes the denominator to become a rational number, and so it is called **rationalizing the denominator.**

EXAMPLE 2

Examples of rationalizing the denominator when the radicand is a fraction or rational expression:

a. $\sqrt{\dfrac{3}{5}} = \sqrt{\dfrac{3}{5} \cdot \dfrac{5}{5}}$ Multiplying (under the radical sign) by $\frac{5}{5}$ makes the denominator 5^2, a perfect power

$= \sqrt{\dfrac{3 \cdot 5}{5^2}}$

$= \dfrac{\sqrt{15}}{\sqrt{5^2}}$ Using the quotient rule for radicals

$= \dfrac{\sqrt{15}}{5}$ The denominator is now a rational number

b. $\sqrt[3]{\frac{7}{3^2}} = \sqrt[3]{\frac{7}{3^2} \cdot \frac{3}{3}}$ Multiplying (under the radical sign) by $\frac{3}{3}$ makes the denominator 3^3, a perfect power

$= \sqrt[3]{\frac{7 \cdot 3}{3^3}}$

$= \frac{\sqrt[3]{21}}{\sqrt[3]{3^3}}$ Using the quotient rule for radicals

$= \frac{\sqrt[3]{21}}{3}$ The denominator is now a rational number

c. $\sqrt[5]{\frac{1}{7x^2}} = \sqrt[5]{\frac{1}{7x^2} \cdot \frac{7^4x^3}{7^4x^3}}$ Multiplying (under the radical sign) by $\frac{7^4x^3}{7^4x^3}$ makes the denominator 7^5x^5, a perfect power

$= \sqrt[5]{\frac{7^4x^3}{7^5x^5}}$

$= \frac{\sqrt[5]{7^4x^3}}{\sqrt[5]{7^5x^5}}$ Using the quotient rule for radicals

$= \frac{\sqrt[5]{2{,}401x^3}}{7x}$ Using a calculator to find 7^4; the denominator is now a rational number

In the problems in Example 3, we combine rationalizing the denominator with what we learned in the last section.

EXAMPLE 3 Examples of simplifying radicals (assume that no variable has a value that would make a denominator zero):

a. $\sqrt{\frac{4}{11}} = \sqrt{\frac{4}{11} \cdot \frac{11}{11}}$ Multiplying (under the radical sign) by $\frac{11}{11}$ makes the denominator 11^2, a perfect square

$= \sqrt{\frac{2^2 \cdot 11}{11^2}}$

$= \frac{\sqrt{2^2} \cdot \sqrt{11}}{\sqrt{11^2}}$ Using the product rule for radicals and the quotient rule for radicals

$= \frac{2\sqrt{11}}{11}$ This is in simplest radical form

b. $\sqrt{\frac{x^6}{8}} = \sqrt{\frac{x^6}{2^3} \cdot \frac{2}{2}}$ Multiplying (under the radical sign) by $\frac{2}{2}$ makes the denominator 2^4, or 16, a perfect square

$= \sqrt{\frac{x^6 \cdot 2}{16}}$

$= \frac{\sqrt{x^6}\sqrt{2}}{\sqrt{16}}$ Using the product rule for radicals and the quotient rule for radicals

$= \frac{|x^3|\sqrt{2}}{4}$ This is in simplest radical form

Note In part b, we *could* multiply under the radical sign by $\frac{8}{8}$, but then the numbers we would have to work with would be unnecessarily large.

c. $\sqrt[3]{\dfrac{a^4b^3}{-2c}} = -\sqrt[3]{\dfrac{a^4b^3}{2c} \cdot \dfrac{2^2c^2}{2^2c^2}}$ Multiplying (under the radical sign) by $\dfrac{2^2c^2}{2^2c^2}$ makes the denominator 2^3c^3, a perfect power

$= -\sqrt[3]{\dfrac{a^3 \cdot a \cdot b^3 \cdot 2^2 \cdot c^2}{2^3 \cdot c^3}}$

$= -\sqrt[3]{\dfrac{a^3 \cdot b^3 \cdot 2^2 \cdot a \cdot c^2}{2^3 \cdot c^3}}$ Rearranging the factors

$= -\dfrac{ab\sqrt[3]{4ac^2}}{2c}$ Using the product rule for radicals and the quotient rule for radicals

d. $\sqrt[5]{\dfrac{24x^2y^6}{64x^4y}} = \sqrt[5]{\dfrac{3y^5}{8x^2}}$ Reducing the radicand to lowest terms

$= \sqrt[5]{\dfrac{3y^5}{2^3x^2} \cdot \dfrac{2^2x^3}{2^2x^3}}$ Multiplying (under the radical sign) by $\dfrac{2^2x^3}{2^2x^3}$ makes the denominator 2^5x^5, a perfect power

$= \dfrac{\sqrt[5]{y^5 \cdot 2^2 \cdot 3 \cdot x^3}}{\sqrt[5]{2^5x^5}}$ Rearranging the factors

$= \dfrac{y\sqrt[5]{12x^3}}{2x}$ The denominator is now a rational number

In part e, assume that $x > 0$ and $y > 0$.

e. $\dfrac{8y}{x^4}\sqrt{\dfrac{5x^5y^2}{2xy^3}} = \dfrac{8y}{x^4}\sqrt{\dfrac{5x^4}{2y}}$ Reducing the radicand to lowest terms

$= \dfrac{8y}{x^4}\sqrt{\dfrac{5x^4}{2y} \cdot \dfrac{2y}{2y}}$ Multiplying (under the radical sign) by $\dfrac{2y}{2y}$ makes the denominator 2^2y^2, a perfect square

$= \dfrac{8y}{x^4}\sqrt{\dfrac{x^4 \cdot 5 \cdot 2 \cdot y}{2^2 \cdot y^2}}$ Rearranging the factors

$= \dfrac{\overset{4}{\cancel{8y}}}{\underset{x^2}{\cancel{x^4}}} \cdot \dfrac{\overset{1}{\cancel{x^2}}\sqrt{10y}}{\underset{1}{\cancel{2y}}}$

$= \dfrac{4\sqrt{10y}}{x^2}$, or $\dfrac{4}{x^2}\sqrt{10y}$

Rationalizing the Denominator When the Denominator Contains a Radical Sign

A *fourth* condition that must be satisfied if a radical is to be in *simplest radical form* is that no denominator can contain a radical. If a denominator contains a radical sign, it may be necessary to build the expression so that the denominator becomes a perfect power and thus a rational number. This procedure is also called rationalizing the denominator.

When we rationalize the denominator in problems in which the denominator has one term and contains a radical sign (see Example 4), we use the product rule for radicals to rewrite a product of two radicals as a single radical; for example,

$$\sqrt{7}\sqrt{7} = \sqrt{7 \cdot 7} = \sqrt{7^2} = 7 \quad \text{and} \quad \sqrt[3]{5^2}\sqrt[3]{5} = \sqrt[3]{5^2 \cdot 5} = \sqrt[3]{5^3} = 5$$

EXAMPLE 4 Examples of rationalizing the denominator when it contains a radical sign:

a. $\frac{2}{\sqrt{5}} = \frac{2}{\sqrt{5}} \cdot \frac{\sqrt{5}}{\sqrt{5}}$ Multiplying by $\frac{\sqrt{5}}{\sqrt{5}}$ makes the denominator $\sqrt{5}\sqrt{5}$, which equals 5, a rational number

$= \frac{2\sqrt{5}}{\sqrt{5}\sqrt{5}}$ $\sqrt{5}\sqrt{5} = (\sqrt{5})^2 = 5$

$= \frac{2\sqrt{5}}{5}$ The denominator is now a rational number

b. $\frac{6}{\sqrt[3]{2}} = \frac{6}{\sqrt[3]{2}} \cdot \frac{\sqrt[3]{2^2}}{\sqrt[3]{2^2}}$ Multiplying by $\frac{\sqrt[3]{2^2}}{\sqrt[3]{2^2}}$ makes the denominator $\sqrt[3]{2^3}$, which equals 2, a rational number

$= \frac{6\sqrt[3]{2^2}}{\sqrt[3]{2}\sqrt[3]{2^2}}$ Using the product rule for radicals: $\sqrt[3]{2}\sqrt[3]{2^2} = \sqrt[3]{2 \cdot 2^2} = \sqrt[3]{2^3} = 2$

$= \frac{\overset{3}{\cancel{6}}\sqrt[3]{4}}{\underset{1}{\cancel{2}}}$

$= 3\sqrt[3]{4}$

In part c, assume that $y > 0$.

c. $\frac{3xy}{\sqrt[4]{y}} = \frac{3xy}{\sqrt[4]{y}} \cdot \frac{\sqrt[4]{y^3}}{\sqrt[4]{y^3}}$ Multiplying by $\frac{\sqrt[4]{y^3}}{\sqrt[4]{y^3}}$ makes the denominator $\sqrt[4]{y^4}$, which equals y, a rational number

$= \frac{3xy\sqrt[4]{y^3}}{\sqrt[4]{y}\sqrt[4]{y^3}}$ Using the product rule for radicals: $\sqrt[4]{y}\sqrt[4]{y^3} = \sqrt[4]{y \cdot y^3} = \sqrt[4]{y^4} = y$

$= \frac{3x\overset{1}{\cancel{y}}\sqrt[4]{y^3}}{\underset{1}{\cancel{y}}}$

$= 3x\sqrt[4]{y^3}$

A Word of Caution In the answer for Example 4c, be sure that the index is written so that it appears *inside* the "arm" of the radical symbol. Otherwise, the expression $3x\sqrt[4]{y^3}$ may look like $3x^4\sqrt{y^3}$. You may even want to write the expression as $3x \cdot \sqrt[4]{y^3}$ or as $3x(\sqrt[4]{y^3})$.

We can see that to rationalize the denominator (whether because the denominator contains a radical sign or because the radicand is a rational expression), if the index of the radical is an understood 2, we want the exponents of all the factors in the denominator to be multiples of 2; if the index of the radical is 3, we want the exponents of all the factors in the denominator to be multiples of 3; if the index of the radical is 4, we want the exponents of all the factors in the denominator to be multiples of 4; and so forth.

Note Before handheld calculators were available, we rationalized denominators to make it easier to find decimal approximations of irrational numbers. Tables of square roots told us that $\sqrt{2} \approx 1.414$, but if we needed a decimal approximation for $\frac{1}{\sqrt{2}}$, we had to either divide 1 by 1.414 (without a calculator, of course) or rationalize the denominator, getting $\frac{\sqrt{2}}{2}$, and then divide 1.414 by 2 (a *much* easier calculation to perform). The approximation of $\frac{1}{\sqrt{2}}$ is 0.707.

Exercises 7.4
Set I

Simplify each radical; assume that all variables in denominators are nonzero.

1. $\sqrt{\dfrac{16}{25}}$ 2. $\sqrt{\dfrac{64}{100}}$ 3. $\sqrt[3]{\dfrac{-27}{64}}$ 4. $\sqrt[3]{\dfrac{-8}{125}}$

5. $\sqrt[4]{\dfrac{a^4b^8}{16c^0}}$ 6. $\sqrt[4]{\dfrac{c^8d^{12}}{81e^0}}$ 7. $\sqrt{\dfrac{4x^3y}{xy^3}}$ 8. $\sqrt{\dfrac{x^5y}{9xy^3}}$

9. $\dfrac{10}{\sqrt{5}}$ 10. $\dfrac{14}{\sqrt{2}}$ 11. $\dfrac{5}{\sqrt{3}}$ 12. $\dfrac{8}{\sqrt{7}}$

13. $\dfrac{9}{\sqrt[3]{3}}$ 14. $\dfrac{10}{\sqrt[3]{5}}$ 15. $\dfrac{8}{\sqrt[5]{4}}$ 16. $\dfrac{5}{\sqrt[4]{3}}$

17. $\sqrt[3]{\dfrac{m^5}{-3}}$ 18. $\sqrt[3]{\dfrac{k^4}{-5}}$ 19. $\dfrac{n}{2m}\sqrt[3]{\dfrac{8m^2n}{2n^3}}$

20. $\dfrac{x}{3y}\sqrt[3]{\dfrac{18xy^2}{2x^3}}$ 21. $\sqrt[4]{\dfrac{3m^7}{4m^3p^2}}$ 22. $\sqrt[4]{\dfrac{5a^9}{2a^5b^2}}$

23. $\sqrt[5]{\dfrac{15x^4y^7}{24x^6y^2}}$ 24. $\sqrt[5]{\dfrac{30mp^8}{48m^4p^3}}$ 25. $\dfrac{4x^2}{5y^3}\sqrt[3]{\dfrac{3x^2y^3}{8x^5y}}$

26. $\dfrac{3x}{4y^2}\sqrt[3]{\dfrac{2xy^5}{3x^3y}}$ 27. $\dfrac{2x^3}{y}\sqrt[3]{\dfrac{5y^0z}{16x^7z^8}}$ 28. $\dfrac{5c}{b}\sqrt[3]{\dfrac{b^4cd^2}{25a^0c^5}}$

Writing Problems

Express the answers in your own words and in complete sentences.

1. Explain how you would rationalize the denominator for $\dfrac{5}{\sqrt[7]{2^3}}$.

2. Without using a calculator, find a decimal approximation, correct to three decimal places, for $\dfrac{3}{\sqrt{7}}$ *without* rationalizing the denominator. Use $\sqrt{7} \approx 2.646$. Then rationalize the denominator, and work the problem again. Write a short paragraph comparing the two methods.

Exercises 7.4
Set II

Simplify each radical; assume that all variables in denominators are nonzero.

1. $\sqrt{\dfrac{36}{81}}$ 2. $\sqrt{\dfrac{100}{49}}$ 3. $\sqrt[3]{\dfrac{-1}{27}}$ 4. $\sqrt[4]{\dfrac{16}{81}}$

5. $\sqrt{\dfrac{x^6y^0}{v^8}}$ 6. $\sqrt[4]{\dfrac{5^0}{a^4b^8}}$ 7. $\sqrt[5]{\dfrac{x^{15}y^{10}}{-243}}$ 8. $\sqrt[3]{\dfrac{6^0st^3}{s^3t^2}}$

9. $\dfrac{18}{\sqrt{3}}$ 10. $\dfrac{6}{\sqrt{15}}$ 11. $\dfrac{3}{\sqrt[5]{8}}$ 12. $\dfrac{16}{\sqrt[3]{12}}$

13. $\dfrac{8}{\sqrt[7]{4}}$ 14. $\dfrac{1}{\sqrt[3]{16}}$ 15. $\dfrac{2}{\sqrt[3]{3}}$ 16. $\dfrac{3}{\sqrt[3]{81}}$

17. $\sqrt[3]{\dfrac{x^7}{-2}}$ 18. $\sqrt[5]{\dfrac{-y^3}{4}}$ 19. $\dfrac{4}{a}\sqrt[3]{\dfrac{3a^2b}{4b^3}}$

20. $\dfrac{x}{3}\sqrt[3]{\dfrac{75x^8}{3x}}$ 21. $\sqrt[3]{\dfrac{8a^4b^5}{81ab^2}}$ 22. $\sqrt[3]{\dfrac{48x^3y^5}{8x^6y^2}}$

23. $\sqrt[7]{\dfrac{2xy^5}{12x^2y^2}}$ 24. $\sqrt[3]{\dfrac{y^5zw^2}{4x^0z^5}}$ 25. $\dfrac{3x^4}{2y^5}\sqrt[3]{\dfrac{96xy^4}{2x^2y}}$

26. $\dfrac{5x}{2y^4}\sqrt[3]{\dfrac{8xy^6}{5x^3y}}$ 27. $\dfrac{7xz}{y}\sqrt[3]{\dfrac{3x}{4^0yz^2}}$ 28. $\dfrac{s^3}{t^2}\sqrt[3]{\dfrac{2st^4}{8s^2t^5}}$

7.5 Reducing the Order of a Radical

In this section, we will assume that all variables represent nonnegative numbers. If we convert $\sqrt[4]{x^2}$ to exponential form, we have $x^{2/4}$. But since $\frac{2}{4} = \frac{1}{2}$, $x^{2/4} = x^{1/2}$. Also, by definition, $x^{1/2} = \sqrt{x}$. Therefore, $\sqrt[4]{x^2} = \sqrt{x}$. Rewriting a radical as an *equivalent* radical with a *smaller index* is called **reducing the order** of the radical.

Whenever there is some number that is a factor of the index of the radical and *also* a factor of every exponent under the radical, the order of the radical can be reduced. The easiest way to do this is to convert the radical to exponential form and reduce the rational exponent; then convert the expression back to radical form.

EXAMPLE 1

Examples of reducing the order of radicals:

a. $\sqrt[4]{x^2} = x^{2/4} = x^{1/2} = \sqrt{x}$

b. $\sqrt[6]{8b^3} = (2^3b^3)^{1/6} = ((2b)^3)^{1/6} = (2b)^{3\cdot\frac{1}{6}} = (2b)^{1/2} = \sqrt{2b}$

c. $\sqrt[8]{16x^4y^4z^4} = (2^4x^4y^4z^4)^{1/8} = ((2xyz)^4)^{1/8} = (2xyz)^{4\cdot\frac{1}{8}} = (2xyz)^{1/2} = \sqrt{2xyz}$

A *fifth* condition that must be satisfied if a radical is to be in simplest radical form is that the *order of the radical* must be as small as possible.

In Example 2, we combine reducing the order of the radical with the procedures from the previous two sections.

EXAMPLE 2

Express the radicals in simplest radical form (assume that all variables represent positive numbers).

a. $\sqrt[8]{16x^{12}y^4z^{20}}$

SOLUTION $\sqrt[8]{16x^{12}y^4z^{20}} = (2^4x^{12}y^4z^{20})^{1/8} = ((2x^3yz^5)^4)^{1/8} = (2x^3yz^5)^{4\cdot\frac{1}{8}}$
$= (2x^3yz^5)^{1/2} = \sqrt{2x^3yz^5} = xz^2\sqrt{2xyz}$

b. $\sqrt[6]{\dfrac{8y^9}{x^3}}$

SOLUTION

$$\sqrt[6]{\frac{8y^9}{x^3}} = \sqrt[6]{\frac{2^3y^9}{x^3}} = \left(\frac{2^3y^9}{x^3}\right)^{1/6} = \left(\left(\frac{2y^3}{x}\right)^3\right)^{1/6} = \left(\frac{2y^3}{x}\right)^{3\cdot\frac{1}{6}} = \left(\frac{2y^3}{x}\right)^{1/2}$$

$$= \sqrt{\frac{2y^3}{x}} = \sqrt{\frac{2y^2y}{x}\cdot\frac{x}{x}} = \frac{y\sqrt{2xy}}{x}, \text{ or } \frac{y}{x}\sqrt{2xy}$$

Let's summarize the rules we've discussed so far for expressing a radical in *simplest radical form*:

Conditions that must be met for a single radical to be in simplest form

All of the following must be true:

1. The radicand is a positive number.
2. If the radicand were expressed in prime-factored, exponential form, no exponent would be greater than or equal to the index of the radical.
3. The radicand is not a rational expression.
4. There are no radical signs in the denominator.
5. The order of the radical cannot be reduced.

We will add to this list in later sections.

Exercises 7.5
Set I

Express each radical in simplest radical form; assume that all variables represent nonnegative numbers.

1. $\sqrt[6]{x^3}$ 2. $\sqrt[6]{x^2}$ 3. $\sqrt[8]{a^6}$ 4. $\sqrt[8]{a^2}$
5. $\sqrt[6]{27b^3}$ 6. $\sqrt[6]{4b^4}$ 7. $\sqrt[4]{49a^2}$ 8. $\sqrt[4]{144x^2}$
9. $\sqrt[8]{81x^4y^0z^{12}}$ 10. $\sqrt[6]{27x^9y^3z^0}$ 11. $\sqrt[6]{256x^8y^4z^{10}}$
12. $\sqrt[4]{64x^8y^2z^6}$ 13. $\sqrt[6]{\dfrac{x^3}{27}}$ 14. $\sqrt[6]{\dfrac{x^2}{4}}$
15. $\dfrac{1}{\sqrt[6]{a^3}}$, $a \neq 0$ 16. $\dfrac{1}{\sqrt[4]{x^2}}$, $x \neq 0$

Writing Problems

Express the answers in your own words and in complete sentences.

1. Describe what your first step would be in simplifying $\sqrt[6]{x^2y^4}$, and explain why you would do that first.
2. Explain why $\sqrt[4]{16x^2y^6}$ is not in simplest radical form.

Exercises 7.5
Set II

Express each radical in simplest radical form; assume that all variables represent nonnegative numbers.

1. $\sqrt[6]{x^4}$ 2. $\sqrt[10]{y^8}$ 3. $\sqrt[6]{z^9}$ 4. $\sqrt[8]{a^{10}}$
5. $\sqrt[6]{81x^4}$ 6. $\sqrt[8]{16y^6}$ 7. $\sqrt[4]{25y^2}$ 8. $\sqrt[6]{64x^6y^9}$
9. $\sqrt[6]{16x^4y^0z^8w^{12}}$ 10. $\sqrt[4]{81a^6b^8c^2}$ 11. $\sqrt[6]{81a^6b^8c^{10}}$
12. $\sqrt[6]{5^0x^3y^9z^6}$ 13. $\sqrt[6]{\dfrac{a^3}{64}}$ 14. $\sqrt[6]{\dfrac{16x}{x^5}}$, $x \neq 0$
15. $\dfrac{1}{\sqrt[8]{y^4}}$, $y \neq 0$ 16. $\dfrac{3}{\sqrt[4]{a^6}}$, $a \neq 0$

7.6 Combining Radicals

Like Radicals **Like radicals** are radicals that have the *same index* and the *same radicand.* The *coefficients* of the radicals can be different.

EXAMPLE 1 Examples of like radicals:

a. $3\sqrt{5}, 2\sqrt{5}, -7\sqrt{5}$ Index = 2; radicand = 5
b. $2\sqrt[3]{x}, -9\sqrt[3]{x}, 11\sqrt[3]{x}$ Index = 3; radicand = x
c. $\frac{2}{3}\sqrt[4]{5ab}, -\frac{1}{2}\sqrt[4]{5ab}$ Index = 4; radicand = $5ab$

Unlike Radicals **Unlike radicals** are radicals that have different indices or different radicands or both.

EXAMPLE 2 Examples of unlike radicals:

a. $\sqrt{7}, \sqrt{5}$ Different radicands

b. $\sqrt[3]{x}, \sqrt{x}$ Different indices

c. $\sqrt[5]{2y}, \sqrt[3]{2}$ Different indices and different radicands

Combining Like Radicals

We can **combine like radicals**, as we can combine any like terms, by using the distributive property.

EXAMPLE 3 Examples of combining like radicals:

a. $5\sqrt{2} + 3\sqrt{2} = (5 + 3)\sqrt{2} = 8\sqrt{2}$

b. $6\sqrt[3]{x} - 4\sqrt[3]{x} = (6 - 4)\sqrt[3]{x} = 2\sqrt[3]{x}$

c. $2x\sqrt[4]{5x} - x\sqrt[4]{5x} = (2x - x)\sqrt[4]{5x} = x\sqrt[4]{5x}$

d. $\frac{2}{3}\sqrt[5]{4xy^2} - 6\sqrt[5]{4xy^2} = \left(\frac{2}{3} - 6\right)\sqrt[5]{4xy^2} = -\frac{16}{3}\sqrt[5]{4xy^2}$

We can see from Example 3 that the coefficient of the answer is the sum of the coefficients of the like radicals, and the radical part of the answer is the same as the radical part of any *one* of the terms. In other words, we can treat $5\sqrt{2} + 3\sqrt{2}$ as if it were $5x + 3x$, omitting the middle step shown in Example 3. In Example 3a, for instance, we can simply write $5\sqrt{2} + 3\sqrt{2} = 8\sqrt{2}$.

A Word of Caution A common error is to think that $8\sqrt{2} = \sqrt{8 \cdot 2}$. This is not correct! A factor that is *outside* a radical cannot be "moved across" the radical sign. (Actually, $8\sqrt{2} = \sqrt{8^2 \cdot 2}$, but $\sqrt{8^2 \cdot 2}$ is not in simplest radical form.)

A Word of Caution Another common error is to think that $\sqrt{a} + \sqrt{a} = \sqrt{a + a}$. We can see that this is not correct by letting $a = 4$:

$$\sqrt{4} + \sqrt{4} = 2 + 2 = 4 \qquad \Big| \qquad \sqrt{4 + 4} = \sqrt{8}$$

Since $4 \neq \sqrt{8}$, $\sqrt{4} + \sqrt{4} \neq \sqrt{4 + 4}$. Therefore, $\sqrt{a} + \sqrt{a} \neq \sqrt{a + a}$.

In general, $\sqrt{a} + \sqrt{b} \neq \sqrt{a + b}$. To verify this, we can let $a = 9$ and $b = 16$. Then

$$\sqrt{9} + \sqrt{16} = 3 + 4 = 7, \quad \text{but} \quad \sqrt{9 + 16} = \sqrt{25} = 5$$

Therefore, $\sqrt{9} + \sqrt{16} \neq \sqrt{9 + 16}$.

Combining Unlike Radicals

When two or more *unlike* radicals are connected with addition or subtraction symbols, we *usually* cannot express the sum or difference with just one radical sign. However, some terms that are not like radicals *become* like radicals after they have been expressed in simplest radical form; in this case, any like radicals can then be combined.

Combining unlike radicals

1. Simplify the radical in each term.
2. Then combine any like radicals.

A radical expression with two or more terms is not in *simplest radical form* unless each term is in simplest radical form *and* all like terms have been combined. This condition must be added to the list of rules in the previous section for expressing a radical in simplest form.

EXAMPLE 4

Rewrite each expression in simplest radical form; assume that all variables represent positive numbers.

a. $\sqrt{8} + \sqrt{18}$

SOLUTION $\sqrt{8} + \sqrt{18} = \sqrt{4 \cdot 2} + \sqrt{9 \cdot 2} = 2\sqrt{2} + 3\sqrt{2} = 5\sqrt{2}$

b. $\sqrt{12} + \sqrt{50}$

SOLUTION $\sqrt{12} + \sqrt{50} = \sqrt{4 \cdot 3} + \sqrt{25 \cdot 2} = 2\sqrt{3} + 5\sqrt{2}$

This expression cannot be simplified further; the radicals are not like radicals. We could find a decimal *approximation* of the sum by using a calculator.

c. $\sqrt[4]{4} + \sqrt{2}$

SOLUTION $\sqrt[4]{4}$ is not in simplest radical form; the order of the radical can be reduced: $\sqrt[4]{4} = \sqrt[4]{2^2} = 2^{2/4} = 2^{1/2} = \sqrt{2}$. Then

$$\sqrt[4]{4} + \sqrt{2} = \sqrt{2} + \sqrt{2} = (1 + 1)\sqrt{2} = 2\sqrt{2}$$

d. $\sqrt{12x} - \sqrt{27x} + 5\sqrt{3x}$

SOLUTION
$$\begin{aligned}\sqrt{12x} - \sqrt{27x} + 5\sqrt{3x} &= \sqrt{4 \cdot 3 \cdot x} - \sqrt{9 \cdot 3 \cdot x} + 5\sqrt{3x}\\ &= 2\sqrt{3x} - 3\sqrt{3x} + 5\sqrt{3x}\\ &= (2 - 3 + 5)\sqrt{3x} = 4\sqrt{3x}\end{aligned}$$

e. $\sqrt[3]{24a^2} - 5\sqrt[3]{3a^5} + 2\sqrt[3]{81a^2}$

SOLUTION
$$\begin{aligned}&\sqrt[3]{24a^2} - 5\sqrt[3]{3a^5} + 2\sqrt[3]{81a^2}\\ &= \sqrt[3]{2^3 \cdot 3 \cdot a^2} - 5\sqrt[3]{3 \cdot a^3 \cdot a^2} + 2\sqrt[3]{3^3 \cdot 3 \cdot a^2}\\ &= 2\sqrt[3]{3a^2} - 5a\sqrt[3]{3a^2} + 2 \cdot 3\sqrt[3]{3a^2}\\ &= (2 - 5a + 6)\sqrt[3]{3a^2} = (8 - 5a)\sqrt[3]{3a^2}\end{aligned}$$

f. $2\sqrt{\frac{1}{2}} - 6\sqrt{\frac{1}{8}} - 10\sqrt{\frac{4}{5}}$

SOLUTION
$$\begin{aligned}&2\sqrt{\tfrac{1}{2}} - 6\sqrt{\tfrac{1}{8}} - 10\sqrt{\tfrac{4}{5}}\\ &= 2\sqrt{\frac{1}{2}\cdot\frac{2}{2}} - 6\sqrt{\frac{1}{2^3}\cdot\frac{2}{2}} - 10\sqrt{\frac{4}{5}\cdot\frac{5}{5}} \quad \text{Rationalizing the denominators}\\ &= 2\left(\sqrt{\frac{2}{4}}\right) - 6\left(\sqrt{\frac{2}{2^4}}\right) - 10\left(\sqrt{\frac{4 \cdot 5}{25}}\right)\\ &= \overset{1}{\cancel{2}}\left(\frac{\sqrt{2}}{\underset{1}{\cancel{2}}}\right) - \overset{3}{\cancel{6}}\left(\frac{\sqrt{2}}{\underset{2}{\cancel{4}}}\right) - \overset{2}{\cancel{10}}\left(\frac{2\sqrt{5}}{\underset{1}{\cancel{5}}}\right)\\ &= \sqrt{2} - \tfrac{3}{2}\sqrt{2} - 4\sqrt{5}\\ &= \left(1 - \tfrac{3}{2}\right)\sqrt{2} - 4\sqrt{5} = -\tfrac{1}{2}\sqrt{2} - 4\sqrt{5}\end{aligned}$$

g. $5\sqrt[4]{\dfrac{3x}{8}}+\dfrac{2x}{3}\sqrt[4]{\dfrac{1}{6^3x^3}}$

SOLUTION $5\sqrt[4]{\dfrac{3x}{8}}+\dfrac{2x}{3}\sqrt[4]{\dfrac{1}{6^3x^3}}$

$$=5\sqrt[4]{\frac{3x}{2^3}\cdot\frac{2}{2}}+\frac{2x}{3}\sqrt[4]{\frac{1}{6^3x^3}\cdot\frac{6x}{6x}}$$ Rationalizing the denominators

$$=5\left(\sqrt[4]{\frac{6x}{2^4}}\right)+\frac{2x}{3}\left(\sqrt[4]{\frac{6x}{6^4x^4}}\right)$$

$$=5\left(\frac{\sqrt[4]{6x}}{2}\right)+\frac{\overset{1}{\cancel{2x}}}{3}\left(\frac{\sqrt[4]{6x}}{\underset{3}{\cancel{6x}}}\right)$$

$$=\tfrac{5}{2}\sqrt[4]{6x}+\tfrac{1}{9}\sqrt[4]{6x}$$

$$=\left(\tfrac{5}{2}+\tfrac{1}{9}\right)\sqrt[4]{6x}=\left(\tfrac{45}{18}+\tfrac{2}{18}\right)\sqrt[4]{6x}=\tfrac{47}{18}\sqrt[4]{6x}$$

h. $\dfrac{-5+\sqrt{25-13}}{2}$

SOLUTION We must subtract *before* simplifying the square root

$$\frac{-5+\sqrt{25-13}}{2}=\frac{-5+\sqrt{12}}{2}=\frac{-5+2\sqrt{3}}{2}$$

We do *not* have like radicals in the numerator

Simplest form

Exercises 7.6
Set I

In each exercise, write the given expression in simplest radical form; assume that all variables represent positive numbers.

1. $8\sqrt{2}+3\sqrt{2}$
2. $12\sqrt{5}+6\sqrt{5}$
3. $3\sqrt{6}+\sqrt{6}$
4. $\sqrt{7}+5\sqrt{7}$
5. $\sqrt{15}+\sqrt{10}$
6. $\sqrt{2}+\sqrt{14}$
7. $3\sqrt{5}-\sqrt{5}$
8. $4\sqrt{3}-\sqrt{3}$
9. $\sqrt{9}+\sqrt{12}$
10. $\sqrt{45}+\sqrt{25}$
11. $\sqrt{12}-\sqrt{8}$
12. $\sqrt{28}-\sqrt{3}$
13. $\sqrt{18}+\sqrt[4]{4}$
14. $\sqrt{3}+\sqrt[4]{9}$
15. $5\sqrt[3]{xy}+2\sqrt[3]{xy}$
16. $7\sqrt[4]{ab}+3\sqrt[4]{ab}$
17. $2\sqrt{50}-\sqrt{32}$
18. $3\sqrt{24}-\sqrt{54}$
19. $3\sqrt{32x}-\sqrt{8x}$
20. $4\sqrt{27y}-3\sqrt{12y}$
21. $\sqrt{125M}+\sqrt{20M}-\sqrt{45M}$
22. $\sqrt{75P}-\sqrt{48P}+\sqrt{27P}$
23. $\sqrt[3]{27x}+\frac{1}{2}\sqrt[3]{8x}$
24. $\frac{3}{4}\sqrt[3]{64a}+\sqrt[3]{27a}$
25. $\sqrt[3]{a^4}+2a\sqrt[3]{8a}$
26. $H\sqrt[3]{8H^2}+3\sqrt[3]{H^5}$
27. $\sqrt[5]{x^2y^6}+\sqrt[5]{x^7y}$
28. $\sqrt[5]{a^3b^8}+\sqrt[5]{a^8b^3}$
29. $3\sqrt{\frac{1}{6}}+\sqrt{12}-5\sqrt{\frac{3}{2}}$
30. $3\sqrt{\frac{5}{2}}+\sqrt{20}-5\sqrt{\frac{1}{10}}$
31. $10\sqrt{\dfrac{5b}{4}}-\dfrac{3b}{2}\sqrt{\dfrac{4}{5b}}$
32. $12\sqrt[3]{\dfrac{x^3}{16}}+x\sqrt[3]{\dfrac{1}{2}}$
33. $2k\sqrt[4]{\dfrac{3}{8k}}-\dfrac{1}{k}\sqrt[4]{\dfrac{2k^3}{27}}+5k^2\sqrt[4]{\dfrac{6}{k^2}}$
34. $6\sqrt[3]{\dfrac{a^4}{54}}+2a\sqrt[3]{\dfrac{a}{2}}$
35. $\sqrt{4x^2+4x+1}$
36. $\sqrt{16a^2+8a+1}$
37. $\dfrac{-10-\sqrt{100-48}}{6}$
38. $\dfrac{13+\sqrt{169-44}}{2}$
39. $\dfrac{5-\sqrt{25-12}}{6}$
40. $\dfrac{-9+\sqrt{81-64}}{16}$
41. $\dfrac{7+\sqrt{49+8}}{4}$
42. $\dfrac{11+\sqrt{121+28}}{14}$

Writing Problems

Express the answers in your own words and in complete sentences.

1. Explain why $\sqrt{75} + \sqrt{27}$ is not in simplest radical form.
2. Explain why $\sqrt[6]{2^3} + \sqrt{2}$ is not in simplest radical form.
3. Make up an example (different from the one in the text) using numbers to show that $\sqrt{a} + \sqrt{b} \neq \sqrt{a + b}$.

Exercises 7.6 Set II

In each exercise, write the given expression in simplest radical form; assume that all variables represent positive numbers.

1. $5\sqrt{7} + 10\sqrt{7}$
2. $8\sqrt{5} + 2\sqrt{5}$
3. $9\sqrt{11} + \sqrt{11}$
4. $\sqrt{6} + \sqrt{6}$
5. $\sqrt{7} + \sqrt{2}$
6. $10\sqrt{3} + \sqrt{3}$
7. $5\sqrt{13} - 3\sqrt{13}$
8. $\sqrt{6} - \sqrt{2}$
9. $\sqrt{49} + \sqrt{28}$
10. $\sqrt{7} + \sqrt{18}$
11. $5\sqrt{50} - \sqrt{2}$
12. $3\sqrt{16} - \sqrt{2}$
13. $\sqrt{5} + \sqrt[4]{25}$
14. $\sqrt{50} + \sqrt{18}$
15. $8\sqrt[5]{5a} + \sqrt[5]{5a}$
16. $3\sqrt[3]{6x^2} - 5\sqrt[3]{6x^2}$
17. $7\sqrt{98} - 3\sqrt{50}$
18. $2\sqrt[3]{54a^2} + 4\sqrt[3]{16a^2}$
19. $8\sqrt{75x^5} + 2x\sqrt{3x^3}$
20. $3\sqrt{25} + 2\sqrt{72} - \sqrt{16}$
21. $7\sqrt{14x} - 2\sqrt{5x} + \sqrt{3x}$
22. $4\sqrt{27a^5} - a\sqrt{12a^3}$
23. $\sqrt[3]{64x} + \frac{1}{5}\sqrt[3]{8x}$
24. $\frac{1}{3}x\sqrt[4]{4x^2} + x\sqrt{8x}$
25. $\sqrt[3]{x^7} - 5x^2\sqrt[3]{x}$
26. $\sqrt{x^5y^7} + 2xy\sqrt{16xy^3} - \sqrt{xy^3}$
27. $\sqrt[6]{a^6b^3} + 6a\sqrt{a^2b^3}$
28. $\sqrt[3]{8x^5} + x\sqrt[6]{x^4} - \sqrt[3]{x}$
29. $5\sqrt{\frac{1}{8}} - 7\sqrt{\frac{1}{18}} - 5\sqrt{\frac{1}{50}}$
30. $8\sqrt[3]{\frac{a^3}{32}} + a\sqrt[3]{54}$
31. $12\sqrt{\frac{7z}{9}} - \frac{4z}{5}\sqrt{\frac{25}{7z}}$
32. $12\sqrt[3]{\frac{x^5}{24}} + 6x\sqrt[3]{\frac{x^2}{3}}$
33. $3h^2\sqrt[4]{\frac{4}{27h}} + 4h^2\sqrt[4]{\frac{5}{8}} - 2h^2\sqrt[4]{\frac{12}{h}}$
34. $\sqrt{\frac{x^3}{3y^3}} + xy\sqrt{\frac{x}{3y^5}} - \frac{y^2}{3}\sqrt{\frac{3x^3}{y^7}}$
35. $\sqrt{a^2 + 6a + 9}$
36. $\sqrt{25x^2 - 10x + 1}$
37. $\dfrac{8 - \sqrt{64 - 16}}{4}$
38. $\dfrac{-13 + \sqrt{169 - 52}}{2}$
39. $\dfrac{-3 - \sqrt{9 + 20}}{10}$
40. $\dfrac{12 + \sqrt{144 - 48}}{8}$
41. $\dfrac{3 + \sqrt{9 + 12}}{6}$
42. $\dfrac{13 + \sqrt{169 + 24}}{8}$

7.7 Multiplying Radicals When the Indices Are the Same

To multiply two or more radicals when the *indices* are the same, we use the product rule for radicals "in reverse"; that is, we use the fact that $\sqrt[n]{a}\sqrt[n]{b} = \sqrt[n]{ab}$ if $\sqrt[n]{a}$ and $\sqrt[n]{b}$ are both real. (We did some multiplication like this in an earlier section.)

Multiplying radicals when the indices are the same

1. Use the product rule for radicals: $\sqrt[n]{a}\sqrt[n]{b} = \sqrt[n]{ab}$ if $\sqrt[n]{a}$ and $\sqrt[n]{b}$ are both real.
2. Simplify the resulting radical.

EXAMPLE 1

Examples of multiplying radicals when the indices are the same (assume that all variables represent nonnegative numbers):

a. $\sqrt{3y}\sqrt{12y^3} = \sqrt{3y \cdot 12y^3} = \sqrt{36y^4} = 6y^2$

b. $2\sqrt[3]{2x^2}(3\sqrt[3]{4x}) = 2 \cdot 3\sqrt[3]{8x^3} = 6 \cdot 2x = 12x$

c. $\sqrt[4]{8ab^3}\sqrt[4]{4a^3b^2} = \sqrt[4]{2^3ab^3 \cdot 2^2a^3b^2} = \sqrt[4]{2^5a^4b^5}$

$= \sqrt[4]{2^4 \cdot 2 \cdot a^4 \cdot b^4 \cdot b} = 2ab\sqrt[4]{2b}$

d. $(4\sqrt{3})^2 = (4\sqrt{3})(4\sqrt{3}) = 16\sqrt{3 \cdot 3} = 16 \cdot 3 = 48$

A final condition to be added to our list for simplest radical form is that no term of the expression can contain more than one radical sign.

Conditions that must be met for an algebraic expression to be in simplest radical form

All of the following must be true:

1. All radicands are positive numbers.
2. If the radicands were expressed in prime-factored, exponential form, no exponent in any radicand would be greater than or equal to the index of that radical.
3. No radicand is a rational expression.
4. No denominator contains a radical.
5. In each term, the order of the radical cannot be reduced.
6. All like radicals have been combined.
7. No *term* of the expression contains more than one radical sign.

If we need to multiply an algebraic expression that contains two or more terms with radicals by another algebraic expression, we use the distributive property (see Example 2).

EXAMPLE 2

Find $\sqrt{8x}(\sqrt{8x} - 3\sqrt{2})$ and simplify. Assume that $x > 0$.

SOLUTION

$$\begin{aligned}\sqrt{8x}(\sqrt{8x} - 3\sqrt{2}) &= \sqrt{8x}(\sqrt{8x}) - \sqrt{8x}(3\sqrt{2}) \\ &= \sqrt{64x^2} - 3\sqrt{16x} \\ &= 8x - 3 \cdot 4\sqrt{x} \\ &= 8x - 12\sqrt{x}\end{aligned}$$

If we need to multiply two algebraic expressions containing two terms each, we can use the FOIL method, even when the expressions are not polynomials (see Example 3).

EXAMPLE 3 Find $(\sqrt{3} - 5)(\sqrt{3} + 2)$ and simplify.

SOLUTION Using the FOIL method, we have

$$\begin{aligned}(\sqrt{3} - 5)(\sqrt{3} + 2) &= \sqrt{3} \cdot \sqrt{3} + 2\sqrt{3} - 5\sqrt{3} - 5 \cdot 2 \\ &= 3 - 3\sqrt{3} - 10 \\ &= -7 - 3\sqrt{3}\end{aligned}$$

In Example 4, Solution 1, we use the rule for squaring a binomial because we are squaring a quantity that contains two terms. In Solution 2, we write the expression $(3\sqrt{5} - 4)$ next to itself and use the FOIL method.

EXAMPLE 4 Find $(3\sqrt{5} - 4)^2$ and simplify.

SOLUTION 1 We square this as we would square a binomial

$$\begin{aligned}(3\sqrt{5} - 4)^2 &= (3\sqrt{5})^2 - 2(3\sqrt{5})(4) + 4^2 \\ &= 9 \cdot 5 - 24\sqrt{5} + 16 \\ &= 45 - 24\sqrt{5} + 16 \\ &= 61 - 24\sqrt{5}\end{aligned}$$

SOLUTION 2

$$\begin{aligned}(3\sqrt{5} - 4)(3\sqrt{5} - 4) &= (3\sqrt{5})(3\sqrt{5}) + (3\sqrt{5})(-4) + (-4)(3\sqrt{5}) + (-4)(-4) \\ &= 45 - 12\sqrt{5} - 12\sqrt{5} + 16 \\ &= 61 - 24\sqrt{5}\end{aligned}$$

In Examples 5 and 6, we use the binomial theorem. It is also possible to do the problem in Example 5 by squaring $(\sqrt[4]{x} - 1)$ and then squaring that answer again, and it is also possible to do the problem in Example 6 by squaring $(x^{1/3} + y^{1/3})$ and then multiplying that answer by $(x^{1/3} + y^{1/3})$.

★ **EXAMPLE 5** Find $(\sqrt[4]{x} - 1)^4$ and simplify the result; assume that $x \geq 0$.

SOLUTION We use the binomial theorem here, even though $\sqrt[4]{x} - 1$ is not a binomial. (Do you know why $\sqrt[4]{x} - 1$ is not a binomial?) The coefficients are 1, 4, 6, 4, and 1.

$$(\sqrt[4]{x} - 1)^4 = 1(\sqrt[4]{x})^4 - 4(\sqrt[4]{x})^3(1) + 6(\sqrt[4]{x})^2(1)^2 - 4(\sqrt[4]{x})^1(1)^3 + 1(1)^4$$

$(\sqrt[4]{x})^2 = \sqrt[4]{x^2} = \sqrt{x}$

$$= x - 4\sqrt[4]{x^3} + 6\sqrt{x} - 4\sqrt[4]{x} + 1$$

★ **EXAMPLE 6** Find $(x^{1/3} + y^{1/3})^3$ and simplify the result.

SOLUTION We use the binomial theorem, even though $x^{1/3} + y^{1/3}$ is not a binomial. (Do you know why $x^{1/3} + y^{1/3}$ is not a binomial?) The coefficients are 1, 3, 3, and 1.

$$\begin{aligned}(x^{1/3} + y^{1/3})^3 &= 1(x^{1/3})^3 + 3(x^{1/3})^2(y^{1/3}) + 3(x^{1/3})(y^{1/3})^2 + 1(y^{1/3})^3 \\ &= x + 3x^{2/3}y^{1/3} + 3x^{1/3}y^{2/3} + y\end{aligned}$$

★ **EXAMPLE 7** Find the value of $x^4 + 1$ if $x = \sqrt{3} + 1$.

SOLUTION If $x = \sqrt{3} + 1$, $x^4 + 1 = (\sqrt{3} + 1)^4 + 1$. We'll use the binomial theorem in simplifying $(\sqrt{3} + 1)^4$. The coefficients are 1, 4, 6, 4, and 1.

$$\begin{aligned}(\sqrt{3} + 1)^4 + 1 &= 1(\sqrt{3})^4 + 4(\sqrt{3})^3(1) + 6(\sqrt{3})^2(1)^2 + 4(\sqrt{3})(1)^3 + 1(1)^4 + 1\\ &= 9 + 4(3\sqrt{3}) + 6(3) + 4\sqrt{3} + 1 + 1\\ &= 9 + 12\sqrt{3} + 18 + 4\sqrt{3} + 1 + 1\\ &= 29 + 16\sqrt{3}\end{aligned}$$

Exercises 7.7
Set I

In Exercises 1–38, simplify each expression; assume that all variables represent nonnegative numbers.

1. $\sqrt{3}\sqrt{3}$
2. $\sqrt{7}\sqrt{7}$
3. $\sqrt[3]{3}\ \sqrt[3]{9}$
4. $\sqrt[3]{4}\sqrt[3]{16}$
5. $\sqrt[4]{9}\sqrt[4]{9}$
6. $\sqrt[4]{25}\sqrt[4]{25}$
7. $\sqrt{5ab^2}\sqrt{20ab}$
8. $\sqrt{3x^2y}\sqrt{27xy}$
9. $3\sqrt[5]{2a^3b}(2\sqrt[5]{16a^2b})$
10. $4\sqrt[5]{4cb^4}(5\sqrt[5]{8c^2b})$
11. $(5\sqrt{7})^2$
12. $(4\sqrt{6})^2$
13. $\sqrt{2}(\sqrt{2} + 1)$
14. $\sqrt{3}(\sqrt{3} + 1)$
15. $\sqrt{x}(\sqrt{x} - 3)$
16. $\sqrt{y}(4 - \sqrt{y})$
17. $\sqrt{3}(2\sqrt{3} + 1)$
18. $\sqrt{5}(3\sqrt{5} + 1)$
19. $\sqrt{3x}(\sqrt{3x} - 4\sqrt{12})$
20. $\sqrt{5a}(\sqrt{10} + 3\sqrt{5a})$
21. $(\sqrt{7} + 2)(\sqrt{7} + 3)$
22. $(\sqrt{3} + 2)(\sqrt{3} + 4)$
23. $(5 + \sqrt{3})(5 - \sqrt{3})$
24. $(\sqrt{5} + \sqrt{3})(\sqrt{5} - \sqrt{3})$
25. $(2\sqrt{3} - 5)^2$
26. $(5\sqrt{2} - 3)^2$
27. $2\sqrt{7x^3y^3}(5\sqrt{3xy})(2\sqrt{7x^3y})$
28. $3\sqrt{5xy}(4\sqrt{2x^5y^3})(5\sqrt{5x^3y^5})$
29. $(3\sqrt{2x} + 5)^2$
30. $(4\sqrt{3x} - 2)^2$
31. $(\sqrt{2x} + 3)(\sqrt{2x} - 3)$
32. $(\sqrt{5x} + 7)(\sqrt{5x} - 7)$
33. $(\sqrt{xy} - 6\sqrt{y})^2$
34. $(\sqrt{ab} + 2\sqrt{a})^2$
35. ★ $(\sqrt[5]{x} + 1)^5$
36. ★ $(1 + \sqrt[3]{y})^3$
37. ★ $(x^{1/4} + 1)^4$
38. $(x^{1/2} + y^{1/2})^2$
39. ★ Find the value of $x^4 - 1$ if $x = 1 - \sqrt{2}$.
40. ★ Find the value of $x^3 + 1$ if $x = 1 + \sqrt{2}$.

Writing Problems

Express the answers in your own words and in complete sentences.

1. Find and describe the error in the following:

$$(\sqrt[5]{x} + 1)^5 = x + 1$$

2. Find and describe the error in the following:

$$(x^{1/6} + y^{1/6})^6 = x + y$$

Exercises 7.7
Set II

In Exercises 1–38, simplify each expression; assume that all variables represent nonnegative numbers.

1. $\sqrt{13}\sqrt{13}$
2. $\sqrt{2}\sqrt{32}$
3. $\sqrt[3]{5}\ \sqrt[3]{25}$
4. $\sqrt[5]{48}\sqrt[5]{4}$
5. $\sqrt[4]{49}\sqrt[4]{49}$
6. $\sqrt[4]{2}\sqrt[4]{8}$
7. $\sqrt[5]{4x^2y^3}\sqrt[5]{8x^3y^2}$
8. $\sqrt[3]{4x^2y^2}\sqrt[3]{2x^2y}$
9. $5\sqrt[5]{x^8y^6}(4\sqrt[5]{x^2y^4})$
10. $6\sqrt[3]{24a^4b^2}(5a\sqrt[3]{3ab^2})$
11. $(3\sqrt{2})^2$
12. $(2\sqrt{5})^3$

13. $\sqrt{5}(\sqrt{5}+1)$
14. $\sqrt{8}(\sqrt{2}-4)$
15. $\sqrt{z}(\sqrt{z}-5)$
16. $\sqrt{5x}(\sqrt{5x}-2)$
17. $\sqrt{7}(3\sqrt{7}+2)$
18. $\sqrt{27}(4\sqrt{3}+5)$
19. $\sqrt{2y}(10\sqrt{10y}+2\sqrt{6})$
20. $\sqrt{3a}(4\sqrt{3a}+2)$
21. $(\sqrt{11}+2)(\sqrt{11}+5)$
22. $(\sqrt{7}+3)(\sqrt{7}-3)$
23. $(\sqrt{6}-\sqrt{3})(\sqrt{6}+\sqrt{3})$
24. $(2-\sqrt{6})(2+\sqrt{6})$
25. $(2\sqrt{5}-4)^2$
26. $(\sqrt{5}+\sqrt{3})^2$
27. $3\sqrt{5x^4y^3}(2\sqrt{15xy^5})(5\sqrt{2x^2y})$
28. $8\sqrt[4]{6a^3b^5}(3\sqrt[4]{2ab^3})(2\sqrt[4]{8a^2b^4})$
29. $(5\sqrt{3x}-4)^2$
30. $(4\sqrt{5}-2a)^2$
31. $(\sqrt{7x}+3)(\sqrt{7x}-3)$
32. $(8-\sqrt{7x})(8+\sqrt{7x})$
33. $(\sqrt{x}+3\sqrt{y})^2$
34. $(3\sqrt{a}-\sqrt{2b})^2$
★35. $(\sqrt[4]{x}+2)^4$
★36. $(\sqrt[4]{x}-\sqrt[4]{y})^4$
★37. $(x^{1/3}-1)^3$
★38. $(x^{1/5}+2)^5$
39. Find the value of y^2-2y-5 if $y=1-\sqrt{6}$.
★40. Find the value of x^4+1 if $x=1+\sqrt{2}$.

7.8 Dividing Radicals When the Indices Are the Same

To divide radicals when the *indices* are the same, we use the quotient rule for radicals "in reverse"; that is, we use the fact that $\dfrac{\sqrt[n]{a}}{\sqrt[n]{b}}=\sqrt[n]{\dfrac{a}{b}}$ if $\sqrt[n]{a}$ and $\sqrt[n]{b}$ are both real and if $b \neq 0$.

Dividing radicals when the indices are the same

1. Use the quotient rule for radicals: $\dfrac{\sqrt[n]{a}}{\sqrt[n]{b}}=\sqrt[n]{\dfrac{a}{b}}$ if $\sqrt[n]{a}$ and $\sqrt[n]{b}$ are both real and if $b \neq 0$.
2. Simplify the resulting radical.

EXAMPLE 1 Examples of dividing radicals (assume that all variables represent positive numbers):

a. $\dfrac{12\sqrt[3]{a^5}}{2\sqrt[3]{a^2}}=\dfrac{6}{1}\sqrt[3]{\dfrac{a^5}{a^2}}=6\sqrt[3]{a^3}=6a$

b. $\dfrac{5\sqrt[4]{28xy^6}}{10\sqrt[4]{7xy}}=\dfrac{1}{2}\sqrt[4]{\dfrac{28xy^6}{7xy}}=\dfrac{1}{2}\sqrt[4]{4y^5}=\dfrac{1}{2}\sqrt[4]{4y^4y}=\dfrac{y}{2}\sqrt[4]{4y}$, or $\dfrac{y\sqrt[4]{4y}}{2}$

c. $\dfrac{\sqrt{5x}}{\sqrt{10x^2}}=\sqrt{\dfrac{5x}{10x^2}}=\sqrt{\dfrac{1}{2x}}=\sqrt{\dfrac{1}{2x}\cdot\dfrac{2x}{2x}}=\dfrac{\sqrt{2x}}{2x}$, or $\dfrac{1}{2x}\sqrt{2x}$

When the dividend contains more than one term but the divisor contains only one term, we divide *each term* of the dividend by the divisor (see Example 2). (This procedure is similar to the procedure we used earlier when we divided a polynomial by a monomial.) It may be necessary to rationalize denominators.

EXAMPLE 2 Examples of dividing radicals when the dividend (or numerator) contains more than one term:

a. $\dfrac{3\sqrt{14} - \sqrt{8}}{\sqrt{2}} = \dfrac{3\sqrt{14}}{\sqrt{2}} - \dfrac{\sqrt{8}}{\sqrt{2}} = 3\sqrt{\dfrac{14}{2}} - \sqrt{\dfrac{8}{2}} = 3\sqrt{7} - \sqrt{4} = 3\sqrt{7} - 2$

b. $\dfrac{3\sqrt[5]{64a} - 9\sqrt[5]{4a^3}}{3\sqrt[5]{2a}} = \dfrac{3\sqrt[5]{64a}}{3\sqrt[5]{2a}} - \dfrac{9\sqrt[5]{4a^3}}{3\sqrt[5]{2a}} = \sqrt[5]{\dfrac{64a}{2a}} - 3\sqrt[5]{\dfrac{4a^3}{2a}}$

$= \sqrt[5]{32} - 3\sqrt[5]{2a^2} = 2 - 3\sqrt[5]{2a^2}$

Rationalizing the denominator; $\dfrac{\sqrt{5}}{\sqrt{5}} = 1$

c. $\dfrac{4 - \sqrt{5}}{\sqrt{5}} = \dfrac{4}{\sqrt{5}} - \dfrac{\sqrt{5}}{\sqrt{5}} = \dfrac{4}{\sqrt{5}} \cdot \dfrac{\sqrt{5}}{\sqrt{5}} - 1 = \dfrac{4\sqrt{5}}{5} - 1$

If we rationalize the denominator *first*, we have

$$\frac{4 - \sqrt{5}}{\sqrt{5}} = \frac{(4 - \sqrt{5})}{\sqrt{5}} \cdot \frac{\sqrt{5}}{\sqrt{5}} = \frac{4\sqrt{5} - \sqrt{5}\sqrt{5}}{5} = \frac{4\sqrt{5} - 5}{5}$$

which is also correct.

Rationalizing a Denominator That Contains Square Roots and Has Two Terms

What if we need to rationalize a denominator when the denominator contains square roots and has two terms? Suppose, for example, that we need to rationalize the denominator of $\dfrac{5}{2 - \sqrt{2}}$. We might try to rationalize the denominator by multiplying by $\dfrac{\sqrt{2}}{\sqrt{2}}$; we would have

$$\frac{5}{2 - \sqrt{2}} \cdot \frac{\sqrt{2}}{\sqrt{2}} = \frac{5\sqrt{2}}{(2 - \sqrt{2})\sqrt{2}} = \frac{5\sqrt{2}}{2\sqrt{2} - 2}$$

We see that we didn't get rid of the radical in the denominator at all. (You might multiply $\dfrac{5\sqrt{2}}{2\sqrt{2} - 2}$ by $\dfrac{\sqrt{2}}{\sqrt{2}}$; you'll see that there is *still* a radical in the denominator.)

What if we tried to rationalize the denominator of $\dfrac{5}{2 - \sqrt{2}}$ by multiplying by $\dfrac{2 - \sqrt{2}}{2 - \sqrt{2}}$? We would have

$$\frac{5}{2 - \sqrt{2}} \cdot \frac{2 - \sqrt{2}}{2 - \sqrt{2}} = \frac{5(2 - \sqrt{2})}{4 - 4\sqrt{2} + 2}, \quad \text{or} \quad \frac{5(2 - \sqrt{2})}{6 - 4\sqrt{2}}$$

There is *still* a radical in the denominator.

In order to rationalize the denominator in expressions in which the denominator contains square roots and has two terms, we need a new definition.

The conjugate of an algebraic expression

> The conjugate of the algebraic expression $a + b$ is $a - b$.

EXAMPLE 3 Examples of conjugates:

a. The conjugate of $\sqrt{3} + 1$ is $\sqrt{3} - 1$.

b. The conjugate of $\sqrt{5} - \sqrt{3}$ is $\sqrt{5} + \sqrt{3}$.

c. The conjugate of $-7 + \sqrt{2}$ is $-7 - \sqrt{2}$.

d. The conjugate of $-\sqrt{11} - \sqrt{13}$ is $-\sqrt{11} + \sqrt{13}$.

e. The conjugate of $\sqrt{x} - \sqrt{y}$ is $\sqrt{x} + \sqrt{y}$.

If an algebraic expression contains *square roots* and has two terms, then the product of that expression and its conjugate will always be a rational number. Consider, for example, the product of $\sqrt{3} + 1$ and its conjugate:

$$(\sqrt{3} + 1)(\sqrt{3} - 1) = (\sqrt{3})^2 - (1)^2 = 3 - 1 = 2 \quad \text{A rational number}$$

Because of this fact, the following procedure should be used when a denominator (or divisor) with two terms contains square roots.

Rationalizing a denominator that contains square roots and has two terms

Multiply both numerator and denominator by the conjugate of the denominator.

Note Multiplying both numerator and denominator by the same number is equivalent to multiplying the expression by 1.

In problems like the one in Example 4a, we recommend that you leave the numerator in factored form until you see whether the expression reduces.

EXAMPLE 4 Rationalize each denominator and simplify.

a. $\dfrac{2}{\sqrt{3} + 1}$

SOLUTION Multiplying both numerator and denominator by $\sqrt{3} - 1$, the conjugate of the denominator $\sqrt{3} + 1$

$$\frac{2}{\sqrt{3} + 1} = \frac{2(\sqrt{3} - 1)}{(\sqrt{3} + 1)(\sqrt{3} - 1)} = \frac{2(\sqrt{3} - 1)}{3 - 1} = \frac{2(\sqrt{3} - 1)}{2} = \sqrt{3} - 1$$

b. $\dfrac{\sqrt{5} + \sqrt{3}}{\sqrt{5} - \sqrt{3}}$

SOLUTION Multiplying both numerator and denominator by $\sqrt{5} + \sqrt{3}$, the conjugate of the denominator $\sqrt{5} - \sqrt{3}$

$$\frac{\sqrt{5} + \sqrt{3}}{\sqrt{5} - \sqrt{3}} = \frac{(\sqrt{5} + \sqrt{3})(\sqrt{5} + \sqrt{3})}{(\sqrt{5} - \sqrt{3})(\sqrt{5} + \sqrt{3})} = \frac{5 + 2\sqrt{5}\sqrt{3} + 3}{5 - 3} = \frac{8 + 2\sqrt{15}}{2}$$

$$= \frac{8}{2} + \frac{2\sqrt{15}}{2} = 4 + \sqrt{15}$$

Exercises 7.8
Set I

Simplify each expression; assume that all variables represent positive numbers.

1. $\dfrac{\sqrt{32}}{\sqrt{2}}$
2. $\dfrac{\sqrt{98}}{\sqrt{2}}$
3. $\dfrac{\sqrt[3]{5}}{\sqrt[3]{4}}$
4. $\dfrac{\sqrt[3]{7}}{\sqrt[3]{2}}$
5. $\dfrac{12\sqrt[4]{15x}}{4\sqrt[4]{5x}}$
6. $\dfrac{15\sqrt[4]{18y}}{5\sqrt[4]{3y}}$
7. $\dfrac{\sqrt[5]{128z^7}}{\sqrt[5]{2z}}$
8. $\dfrac{\sqrt[5]{3^7b^8}}{\sqrt[5]{3b^2}}$
9. $\dfrac{\sqrt{72x^3y^2}}{\sqrt{2xy^2}}$
10. $\dfrac{\sqrt{27x^2y^3}}{\sqrt{3x^2y}}$
11. $\dfrac{\sqrt{20} + 5\sqrt{10}}{\sqrt{5}}$
12. $\dfrac{2\sqrt{6} + \sqrt{18}}{\sqrt{6}}$
13. $\dfrac{\sqrt{75} - \sqrt{6}}{\sqrt{3}}$
14. $\dfrac{\sqrt{98} + \sqrt{21}}{\sqrt{7}}$
15. $\dfrac{2 - \sqrt{7}}{\sqrt{7}}$
16. $\dfrac{\sqrt{13} - 4}{\sqrt{13}}$
17. $\dfrac{8\sqrt[4]{16x} + 6\sqrt[4]{8x^4}}{2\sqrt[4]{2x}}$
18. $\dfrac{6\sqrt[5]{81x^4} + 9\sqrt[5]{27x^7}}{3\sqrt[5]{3x^3}}$
19. $\dfrac{6\sqrt[4]{2^5m^2}}{2\sqrt[4]{2m^3}}$
20. $\dfrac{7\sqrt[4]{3^5H^3}}{14\sqrt[4]{3H^4}}$
21. $\dfrac{4\sqrt[3]{8x} + 6\sqrt[3]{32x^4}}{2\sqrt[3]{4x}}$
22. $\dfrac{10\sqrt[3]{81a^7} + 15\sqrt[3]{6a}}{5\sqrt[3]{3a}}$
23. $\dfrac{6}{\sqrt{3} - 1}$
24. $\dfrac{10}{\sqrt{3} + 1}$
25. $\dfrac{\sqrt{2}}{\sqrt{3} + \sqrt{2}}$
26. $\dfrac{\sqrt{7}}{\sqrt{7} - \sqrt{2}}$
27. $\dfrac{\sqrt{7} + \sqrt{3}}{\sqrt{7} - \sqrt{3}}$
28. $\dfrac{\sqrt{11} - \sqrt{5}}{\sqrt{11} + \sqrt{5}}$
29. $\dfrac{4\sqrt{3} - \sqrt{2}}{4\sqrt{3} + \sqrt{2}}$
30. $\dfrac{\sqrt{x + 1} - \sqrt{x}}{\sqrt{x + 1} + \sqrt{x}}$
31. $\sqrt{\dfrac{a^2 + 2a - 3}{a^2 + 4a + 3}}, a > 1$
32. $\sqrt{\dfrac{m^2 - m - 2}{m^2 - 3m + 2}}, m > 2$

Writing Problems

Express the answers in your own words and in complete sentences.

1. Explain how to rationalize the denominator of $\dfrac{5}{\sqrt{5} - 2}$.
2. Explain how to rationalize the denominator of $\dfrac{5}{\sqrt{5x} - 3}$.

Exercises 7.8
Set II

Simplify each expression; assume that all variables represent positive numbers.

1. $\dfrac{\sqrt{72}}{\sqrt{2}}$
2. $\dfrac{\sqrt{75}}{\sqrt{3}}$
3. $\dfrac{\sqrt[5]{3}}{\sqrt[5]{4}}$
4. $\dfrac{\sqrt[3]{x^8y}}{\sqrt[3]{5xy^2}}$
5. $\dfrac{30\sqrt[4]{64x}}{6\sqrt[4]{4x}}$
6. $\dfrac{35\sqrt[5]{96x^5}}{7\sqrt[5]{6x^3}}$
7. $\dfrac{\sqrt{x^4y}}{\sqrt{5y}}$
8. $\dfrac{\sqrt{m^6n}}{\sqrt{3n}}$
9. $\dfrac{\sqrt{300a^5b^2}}{\sqrt{3ab^2}}$
10. $\dfrac{\sqrt[4]{2xy^3}}{\sqrt[4]{27x^2y}}$
11. $\dfrac{\sqrt{34} + 2\sqrt{6}}{\sqrt{2}}$
12. $\dfrac{3\sqrt{27} - \sqrt{2}}{\sqrt{3}}$
13. $\dfrac{\sqrt{45} - \sqrt{10}}{\sqrt{5}}$
14. $\dfrac{\sqrt{11} + \sqrt{55}}{\sqrt{11}}$
15. $\dfrac{8 - \sqrt{6}}{\sqrt{6}}$
16. $\dfrac{\sqrt{15} - 3}{\sqrt{15}}$
17. $\dfrac{10\sqrt[4]{12x} + 15\sqrt[4]{18x^3}}{5\sqrt[4]{6x}}$
18. $\dfrac{8\sqrt[5]{32x^2} + 12\sqrt[5]{2x^6}}{4\sqrt[5]{2x^2}}$
19. $\dfrac{6\sqrt[3]{4B^2}}{9\sqrt[3]{8B^4}}$
20. $\dfrac{5\sqrt[3]{3K^3}}{15\sqrt[3]{9K^5}}$
21. $\dfrac{30\sqrt[4]{32a^6} - 6\sqrt[4]{24a^2}}{3\sqrt[4]{2a^2}}$
22. $\dfrac{8\sqrt[3]{2x} - 4\sqrt[3]{3x^4}}{2\sqrt[3]{5x^2}}$
23. $\dfrac{8}{\sqrt{5} - 1}$
24. $\dfrac{3}{\sqrt{6} + 1}$
25. $\dfrac{\sqrt{7}}{\sqrt{3} + \sqrt{7}}$
26. $\dfrac{21}{3\sqrt{5} + 2\sqrt{6}}$
27. $\dfrac{\sqrt{5} + \sqrt{2}}{\sqrt{5} - \sqrt{2}}$

28. $\dfrac{3\sqrt{2}+\sqrt{6}}{3\sqrt{2}-\sqrt{6}}$

29. $\dfrac{5\sqrt{7}+\sqrt{3}}{5\sqrt{7}-\sqrt{3}}$

30. $\dfrac{x-4}{\sqrt{x}+2}$

31. $\sqrt{\dfrac{a^2+2ab+b^2}{a^2-b^2}},\ a>b$

32. $\sqrt{\dfrac{x^2-y^2}{x^2-2xy+y^2}},\ x>y$

7.9 Multiplying and Dividing Radicals When the Indices Are Not the Same

It is possible to multiply and divide radicals when the indices are not the same. (In this section, we will assume that all variables represent positive numbers.)

Multiplying and dividing radicals when all radicands can be expressed as powers of the same base

1. Write any numerical factors of the radicands in prime-factored, exponential form.
2. Convert the radicals to exponential form.
3. Perform the indicated multiplication or division.
4. Convert back to radical form.
5. Express the answer in simplest radical form.

EXAMPLE 1

Examples of multiplying and dividing radicals with different indices:

a. $\sqrt{x}\sqrt[3]{x} = x^{1/2}x^{1/3} = x^{3/6}x^{2/6} = x^{5/6} = \sqrt[6]{x^5}$

b. $\sqrt{2}\sqrt[3]{-32} = \sqrt{2}(-\sqrt[3]{2^5}) = 2^{1/2}(-2^{5/3}) = -2^{1/2}\cdot 2^{5/3} = -2^{1/2+5/3}$

$= -2^{3/6+10/6} = -2^{13/6} = -\sqrt[6]{2^{13}} = -\sqrt[6]{2^{12}\cdot 2}$

$= -2^2\sqrt[6]{2} = -4\sqrt[6]{2}$

c. $\sqrt[4]{a^3}\sqrt{a}\sqrt[3]{a^2} = a^{3/4}a^{1/2}a^{2/3} = a^{9/12}a^{6/12}a^{8/12} = a^{23/12} = \sqrt[12]{a^{23}}$

$= \sqrt[12]{a^{12}\cdot a^{11}} = a\sqrt[12]{a^{11}}$

d. $\dfrac{\sqrt[3]{-d^2}}{\sqrt[4]{d^3}} = \dfrac{-\sqrt[3]{d^2}}{\sqrt[4]{d^3}} = -\dfrac{d^{2/3}}{d^{3/4}} = -d^{2/3-3/4} = -d^{8/12-9/12} = -d^{-1/12}$

$= -\dfrac{1}{d^{1/12}} = -\dfrac{1}{\sqrt[12]{d}} = -\dfrac{1}{\sqrt[12]{d}}\cdot\dfrac{\sqrt[12]{d^{11}}}{\sqrt[12]{d^{11}}} = -\dfrac{\sqrt[12]{d^{11}}}{\sqrt[12]{d^{12}}} = -\dfrac{\sqrt[12]{d^{11}}}{d}$

Rationalizing the denominator

If all the radicands *cannot* be expressed as powers of the *same* base, we can multiply and divide radicals with different indices as follows:

Multiplying and dividing radicals with different indices when all the radicands cannot be expressed as powers of the same base

1. Convert the radicals to exponential form.
2. Find the LCD of the denominators of the exponents, and build all the exponents so that they have this denominator.
3. Convert from exponential form back to radical form.
4. Perform the multiplication or division, expressing the answer in simplest radical form.

EXAMPLE 2 Examples of multiplying radicals with different indices:

a. $\sqrt{x}\sqrt[3]{y} = x^{1/2}y^{1/3} = x^{3/6}y^{2/6} = \sqrt[6]{x^3}\sqrt[6]{y^2} = \sqrt[6]{x^3y^2}$

b. $\sqrt[4]{x^3}\sqrt{t} = x^{3/4}t^{1/2} = x^{3/4}t^{2/4} = \sqrt[4]{x^3}\sqrt[4]{t^2} = \sqrt[4]{x^3t^2}$

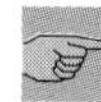

Note It is also possible to do the problems in Example 2 as follows:

a. $\sqrt{x}\sqrt[3]{y} = x^{1/2}y^{1/3} = x^{3/6}y^{2/6} = (x^3y^2)^{1/6} = \sqrt[6]{x^3y^2}$

b. $\sqrt[4]{x^3}\sqrt{t} = x^{3/4}t^{1/2} = x^{3/4}t^{2/4} = (x^3t^2)^{1/4} = \sqrt[4]{x^3t^2}$

Exercises 7.9 Set I

In each exercise, perform the indicated operation. Express the answer in simplest radical form. Assume that all variables represent positive numbers.

1. $\sqrt{a}\sqrt[4]{a}$
2. $\sqrt{b}\sqrt[3]{b}$
3. $\sqrt{8}\sqrt[3]{16}$
4. $\sqrt{27}\sqrt[3]{81}$
5. $\sqrt[3]{x^2}\sqrt[4]{x^3}\sqrt{x}$
6. $\sqrt{y}\sqrt[3]{y}\sqrt[4]{y^3}$
7. $\sqrt[3]{-8z^2}\sqrt[3]{-z}\sqrt[4]{16z^3}$
8. $\sqrt[3]{-27w}\sqrt[3]{-w^2}\sqrt[4]{16w^3}$
9. $\dfrac{\sqrt[4]{G^3}}{\sqrt[3]{G^2}}$
10. $\dfrac{\sqrt[5]{H^4}}{\sqrt{H}}$
11. $\dfrac{\sqrt[3]{-x^2}}{\sqrt[6]{x^5}}$
12. $\dfrac{\sqrt[6]{y^3}}{\sqrt[3]{-y^2}}$
13. $\sqrt[4]{x}\sqrt{y}$
14. $\sqrt[8]{s}\sqrt[4]{t}$
15. $\sqrt[6]{h}\sqrt[4]{x^3}$
16. $\sqrt[5]{w^2}\sqrt{t}$

Writing Problems

Express the answer in your own words and in complete sentences.

1. Describe what your first step would be in simplifying $\sqrt{2}\sqrt[4]{2}\sqrt[3]{2}$, and explain why you would do that first.

Exercises 7.9 Set II

In each exercise, perform the indicated operation. Express the answer in simplest radical form. Assume that all variables represent positive numbers.

1. $\sqrt[3]{a^2}\sqrt[6]{a^2}$
2. $\sqrt[2]{x^3}\sqrt[5]{x^4}$
3. $\sqrt{2}\sqrt[6]{8}$
4. $\sqrt[5]{3}\sqrt[5]{81}$
5. $\sqrt{x}\sqrt[4]{x^3}\sqrt[8]{x^6}$
6. $\sqrt[3]{a}\sqrt[5]{a^2}\sqrt[6]{a^4}$
7. $\sqrt[3]{-5x^2}\sqrt[4]{x^5}\sqrt[3]{-25x^3}$
8. $\sqrt[3]{x}\sqrt[5]{x^3}\sqrt[4]{x^6}$
9. $\dfrac{\sqrt[4]{a^3}}{\sqrt[3]{a^5}}$
10. $\dfrac{\sqrt[3]{x^2}}{\sqrt[5]{x^6}}$
11. $\dfrac{\sqrt[3]{-x^4}}{\sqrt{x^3}}$
12. $\dfrac{\sqrt[4]{32x^3}}{\sqrt{2x}}$
13. $\sqrt[6]{x}\sqrt{y}$
14. $\sqrt[3]{m}\sqrt[6]{n}$
15. $\sqrt[8]{w^3}\sqrt{z}$
16. $\sqrt[4]{f^3}\sqrt[8]{g^3}$

Sections 7.1–7.9 REVIEW

Relationship Between Exponents and Radicals 7.1

$x^{1/n} = \sqrt[n]{x}$ if $x \geq 0$ when n is an even natural number

$x^{1/n} = \sqrt[n]{x}$ for all x when n is an odd natural number greater than 1

$x^{1/n}$ is not real if $x < 0$ when n is an even natural number.

$x^{m/n} = \sqrt[n]{x^m} = (\sqrt[n]{x})^m$ if $x \geq 0$ when n is an even natural number

$x^{m/n} = \sqrt[n]{x^m} = (\sqrt[n]{x})^m$ for all x when n is an odd natural number greater than 1

$x^{m/n}$ is not real if m/n is in lowest terms and if $x < 0$ when n is an even natural number.

Properties of Exponents 7.2 — All the properties of exponents given in an earlier chapter are valid, even when some or all of the exponents are rational numbers that are not integers.

Principal Roots 7.3 — $\sqrt{x^2} = |x|$ for all real numbers x

$\sqrt[n]{x^n} = x$ for all x if n is an odd natural number greater than 1

$\sqrt[n]{x^n} = |x|$ for all x if n is an even natural number

Simplifying Radicals 7.3 — The product rule for radicals:

$$\sqrt[n]{ab} = \sqrt[n]{a}\sqrt[n]{b} \text{ if } \sqrt[n]{a} \text{ and } \sqrt[n]{b} \text{ are both real numbers}$$

To simplify a radical that does not involve division:

1. Express the radicand in prime-factored, exponential form.
2. If the exponent of any factor of the radicand is greater than the index, break up that factor into the product of a perfect power and a nonperfect power *whose exponent is less than the index.*
3. Use the product rule for radicals to rewrite the radical as a product of radicals.
4. Remove perfect powers from the radical by dividing their exponents by the index; place absolute value symbols around the result if necessary. All remaining exponents should be less than the index; factors with such exponents remain under the radical sign.
5. The simplified radical is the *product* of all the factors from step 4. It is customary to write all rational factors to the left of any irrational ones.

7.5 6. It may be possible to reduce the order of the radical.

7.4 The quotient rule for radicals:

$$\sqrt[n]{\frac{a}{b}} = \frac{\sqrt[n]{a}}{\sqrt[n]{b}} \text{ if } \sqrt[n]{a} \text{ and } \sqrt[n]{b} \text{ are both real numbers and } b \neq 0$$

Rationalizing the Denominator 7.4 — *If the denominator has a single term*: Multiply by a rational expression in which numerator and denominator are equal; select this rational expression so that multiplying by it will make the exponent of every factor of the denominator exactly divisible by the index of the radical.

7.8 *If the denominator contains square roots and has two terms*: Multiply both numerator and denominator by the *conjugate* of the denominator.

Adding and Subtracting Radicals 7.6 — *Like radicals*: The coefficient of the sum is the sum of the coefficients. The radical part of the sum is the same as the radical part of any *one* of the terms.

Unlike radicals:

1. Simplify the radical in each term.
2. Then combine any like radicals.

Multiplying Radicals 7.7, 7.9 — Use the property $\sqrt[n]{a}\sqrt[n]{b} = \sqrt[n]{ab}$ if $\sqrt[n]{a}$ and $\sqrt[n]{b}$ are both real; then simplify. If the indices are not the same, convert to exponential form.

Dividing Radicals 7.8, 7.9 — Use the property $\frac{\sqrt[n]{a}}{\sqrt[n]{b}} = \sqrt[n]{\frac{a}{b}}$ if $\sqrt[n]{a}$ and $\sqrt[n]{b}$ are both real and if $b \neq 0$; then simplify. If the indices are not the same, convert to exponential form; rationalize the denominator if necessary.

Simplified Form of a Radical Expression — An algebraic expression is in simplest radical form if

7.3 1. All radicands are positive numbers.

7.3 2. If the radicands were expressed in prime-factored, exponential form, no exponent in any radicand would be greater than or equal to the index of that radical.

7.4 3. No radicand is a rational expression.

7.4 4. No denominator contains a radical.

7.5 5. In each term, the order of the radical cannot be reduced.

7.6 6. All like radicals have been combined.

7.7 7. No *term* of the expression contains more than one radical sign.

Sections 7.1–7.9 REVIEW EXERCISES Set I

In Exercises 1–3, convert each expression to radical form. Assume that $a \geq 0$ and $y \geq 0$.

1. $a^{3/4}$
2. $(3y)^{3/4}$
3. $(2x^2)^{2/5}$

In Exercises 4–6, convert each radical to exponential form. Assume that $b \geq 0$.

4. $\sqrt[4]{b^3}$
5. $\sqrt[5]{8x^4}$
6. $\sqrt[5]{27x^3}$

In Exercises 7 and 8, evaluate each expression. You might check the answers by using a calculator.

7. $(-64)^{2/3}$
8. $(-27)^{2/3}$

In Exercises 9 and 10, perform the indicated operations. Express the answers in exponential form, using positive exponents. Assume that $R \geq 0$, $a > 0$, and $b > 0$.

9. $(P^{2/3}R^{3/4})^{2/3}$
10. $\left(\dfrac{27a^{1/3}b^{2/3}}{3b^{-1}}\right)^{-3/2}$

In Exercises 11–16, simplify each expression. Assume that $y > 0$.

11. $\sqrt[3]{32}$
12. $\sqrt[3]{-125x^3}$
13. $\sqrt[4]{16y^8}$
14. $\frac{1}{3}\sqrt[5]{3^5m^6p}$
15. $\sqrt{\dfrac{20x^3}{5y}}$
16. $\dfrac{15}{\sqrt{3y}}$

In Exercises 17–30, perform the indicated operations, and write the answers in simplest radical form. Assume that all variables represent positive numbers.

17. $\sqrt[3]{-2x}\sqrt[3]{-4x^4}$
18. $(5\sqrt{3x})^2$
19. $\dfrac{\sqrt{3}+\sqrt{7}}{\sqrt{3}-\sqrt{7}}$
20. $\dfrac{8}{\sqrt{5}+1}$
21. $(\sqrt{2}+3)(4\sqrt{2}-1)$
22. $(\sqrt{13}+\sqrt{3})^2$
23. $\sqrt[3]{16x^5}+x\sqrt[3]{54x^2}$
24. $(\sqrt{13}-2)(\sqrt{13}+2)$
25. $\sqrt[3]{-z^2}\sqrt[4]{z^2}$
★26. $(\sqrt[4]{2}-x)^4$
27. $\dfrac{\sqrt[4]{G^2}}{\sqrt[5]{G}}$
28. $3\sqrt[4]{\dfrac{2ab}{32a^3}}-\dfrac{1}{2}\sqrt[4]{\dfrac{b}{a^2}}$
29. $\sqrt{t}\sqrt[3]{f^2}$
★30. $(2^{1/5}+1)^5$

Sections 7.1–7.9 REVIEW EXERCISES Set II

ANSWERS

1. ______
2. ______
3. ______
4. ______
5. ______
6. ______
7. ______
8. ______
9. ______
10. ______
11. ______
12. ______
13. ______
14. ______
15. ______
16. ______

____

Name

In Exercises 1–3, convert each expression to radical form.

1. $R^{4/5}$
2. $(4x^2)^{3/5}$
3. $B^{2/3}$

In Exercises 4–6, convert each radical to exponential form. Assume that $x \geq 0$.

4. $\sqrt[3]{8y^2}$
5. $(\sqrt[4]{x^3})^2$
6. $\sqrt[5]{a^2}$

In Exercises 7 and 8, evaluate each expression. You might check the answers by using a calculator.

7. $(-32)^{2/5}$
8. $(-125)^{2/3}$

In Exercises 9 and 10, perform the indicated operations. Express the answers in exponential form, using positive exponents. Assume that $x > 0$ and $y \geq 0$.

9. $(x^{3/2}y^{3/4})^{2/3}$
10. $\left(\dfrac{27x^{-1}y^{2/3}}{3x^{1/3}}\right)^{1/2}$

In Exercises 11–16, simplify each expression. Assume that $b > 0$ and $a \geq 0$.

11. $\sqrt[3]{81}$
12. $\sqrt[5]{32x^{10}}$
13. $\sqrt[3]{36x^4}$
14. $3b\sqrt[4]{\dfrac{a^3}{27b}}$
15. $\frac{1}{2}\sqrt[3]{16m^4n^7}$
16. $\dfrac{b}{3}\sqrt[3]{\dfrac{-27a}{b^2}}$

In Exercises 17–30, perform the indicated operations, and write the answers in simplest radical form. Assume that all variables represent positive numbers.

17. $\sqrt[3]{-3a^2}\sqrt[3]{-9a^2}$

18. $\dfrac{3+\sqrt{5}}{1-\sqrt{5}}$

19. $(4+\sqrt{2x})^2$

20. $z\sqrt[3]{32z}+\sqrt[3]{4z^4}$

21. $\dfrac{3\sqrt{2}-1}{2\sqrt{2}+3}$

22. $2\sqrt[4]{\dfrac{xy}{81x^3}}-\dfrac{1}{3}\sqrt[4]{\dfrac{y^2}{x^2y}}$

23. $(\sqrt{7z}-\sqrt{3})(\sqrt{7z}+\sqrt{3})$

★24. $(\sqrt[4]{x}+2)^4$

25. $(2\sqrt{3z})^3$

26. $\sqrt[3]{-M^2}\sqrt[6]{M^4}$

27. $\dfrac{\sqrt[6]{H^4}}{\sqrt[5]{H^3}}$

28. $(x-\sqrt{3})^2$

29. $\sqrt[4]{x^3}\sqrt{y}$

★30. $(5^{1/3}-1)^3$

17. ______________

18. ______________

19. ______________

20. ______________

21. ______________

22. ______________

23. ______________

24. ______________

25. ______________

26. ______________

27. ______________

28. ______________

29. ______________

30. ______________

7.10 Solving Radical Equations

Radical Equations A **radical equation** is an equation in which the variable appears in a radicand. Recall that even roots of negative numbers are not real numbers. Therefore, if the index of the radical is an *even* number, the domain of the variable is the set of all real numbers that will make the radicand greater than or equal to zero. If the index is an *odd* number, the domain is the set of all real numbers.

EXAMPLE 1 Find the domain for each of these radical equations (*do not solve the equations*).

a. $\sqrt{x+2} = 3$

SOLUTION The index is understood to be 2, an even number.

$$x + 2 \geq 0 \quad \text{Because the index is even, the radicand must be } \geq 0$$

$$x \geq -2$$

The domain is $\{x \mid x \geq -2\}$.

b. $\sqrt[3]{x-1} = 4$

SOLUTION Because the index of the radical is an odd number, the domain is the set of all real numbers.

c. $\sqrt{4x+5} - \sqrt{x-1} = \sqrt{14-x}$

SOLUTION Because the indices are all understood to be 2, all the radicands must be greater than or equal to zero. We must first find the domain of each term:

$$4x + 5 \geq 0 \quad \textit{and} \quad x - 1 \geq 0 \quad \textit{and} \quad 14 - x \geq 0$$

$$x \geq -\tfrac{5}{4} \quad \textit{and} \quad x \geq 1 \quad \textit{and} \quad x \leq 14$$

Then the domain of the *equation* will be the *intersection* of these domains. We'll find the intersection graphically:

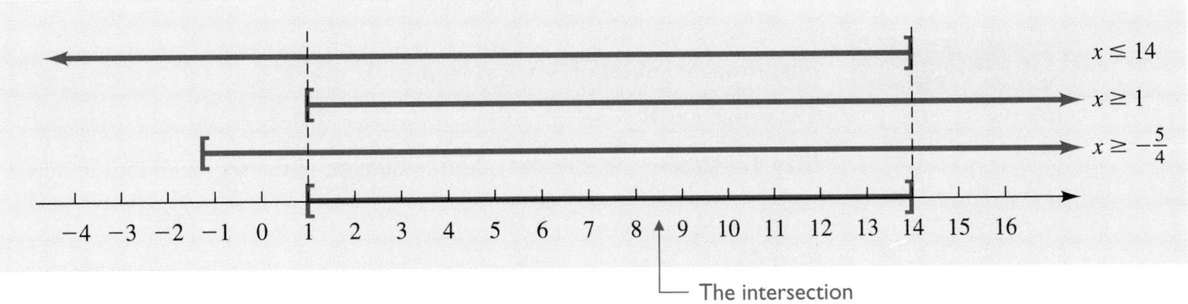

The numbers that satisfy *all three* conditions are the real numbers between 1 and 14 (*including* 1 and 14). Therefore, the domain is $\{x \mid 1 \leq x \leq 14\}$.

Extraneous Roots In order to solve a radical equation, we must eliminate the radical signs by raising both sides of the equation to some power. When we do this, the new equation may *not* be equivalent to the original equation. (Recall that *equivalent equations* are equations that have exactly the same solution set.)

If we square both sides of the equation $x = 3$, we get $x^2 = 9$. Let's consider the solution sets of these equations. The solution set of $x = 3$ is $\{3\}$, but the solution set of $x^2 = 9$ is $\{3, -3\}$. Since the two equations do not have exactly the same solution set, they are *not* equivalent equations; -3 is not a solution of $x = 3$—it is an **extraneous root** that was introduced when we squared both sides of $x = 3$.

Whenever we raise both sides of an equation to the same power, we may introduce extraneous roots; therefore, all apparent solutions *must* be checked in the *original* equation.

Solving a Radical Equation

Solving a radical equation

1. Find the domain of the variable.
2. Arrange the terms so that one term with a radical is by itself on one side of the equal sign.
3. Raise each side of the equation to a power equal to the index of the radical.
4. Simplify each side of the equation.

   If a radical still remains, repeat steps 2, 3, and 4.
5. Solve the resulting equation.
6. Check apparent solutions in the original equation, because there may be extraneous roots.

EXAMPLE 2 Find the solution set of $\sqrt{x+2} = 3$.

SOLUTION

Step 1. Because the index is even, the radicand must be nonnegative, and so the domain is $\{x \mid x \geq -2\}$.

Step 2 does not apply.

$$\sqrt{x+2} = 3$$

Step 3. $(\sqrt{x+2})^2 = 3^2$ Squaring both sides

Step 4. $x + 2 = 9$

Step 5. $x = 7$ 7 is in the domain

√ **Step 6.** ***Check*** $\sqrt{7+2} = \sqrt{9} = 3$

The solution set is $\{7\}$.

EXAMPLE 3 Find the solution set of $\sqrt[3]{x-1} + 4 = 0$.

SOLUTION

Step 1. Because the index is odd, the domain is the set of all real numbers.

$$\sqrt[3]{x-1} + 4 = 0$$

Step 2. $\sqrt[3]{x-1} = -4$ Getting the radical by itself on one side

Step 3. $(\sqrt[3]{x-1})^3 = (-4)^3$ Cubing both sides

Step 4. $x - 1 = -64$

Step 5. $x = -63$ -63 is in the domain

√ **Step 6.** ***Check*** $\sqrt[3]{x-1} + 4 = 0$

$$\sqrt[3]{-63-1} + 4 \stackrel{?}{=} 0$$

$$\sqrt[3]{-64} + 4 \stackrel{?}{=} 0$$

$$-4 + 4 \stackrel{?}{=} 0$$

$$0 = 0$$

The solution set is $\{-63\}$.

EXAMPLE 4 Find the solution set of $\sqrt[4]{2x+1} = -2$.

SOLUTION

Step 1. Because the index is even, the radicand must be nonnegative; that is, $2x + 1 \ge 0$, or $x \ge -\frac{1}{2}$.

We know that the left side of the equation, $\sqrt[4]{2x+1}$, must be *positive* (since the principal root of a radical with an even index must be positive), and no positive number can equal -2. Therefore, this equation does not have a solution that is a real number.

If we did not notice that there could not be a real solution and used the procedure outlined in the preceding box, we would obtain the same result (with quite a bit more work), as shown below.

Step 2 does not apply.

$$\sqrt[4]{2x+1} = -2$$

Step 3. $(\sqrt[4]{2x+1})^4 = (-2)^4$ Raising both sides to the fourth power

Step 4. $2x + 1 = 16$

$2x = 15$

Step 5. $x = \frac{15}{2}$ $\frac{15}{2}$ is in the domain

√ **Step 6.** ***Check*** $\sqrt[4]{2x+1} = -2$

$$\sqrt[4]{2\left(\tfrac{15}{2}\right) + 1} \stackrel{?}{=} -2$$

$$\sqrt[4]{15+1} \stackrel{?}{=} -2$$

$$\sqrt[4]{16} \stackrel{?}{=} -2$$

The principal root ⟶ $2 \ne -2$

Therefore, $\frac{15}{2}$ is not a solution. The solution set is { }.

A Word of Caution If one side of the equation contains more than one term, squaring each term is *not* the same as squaring both sides of the equation. That is, if $a = b + c$, $a^2 \ne b^2 + c^2$; rather, $a^2 = (b + c)^2$, and $(b + c)^2 = b^2 + 2bc + c^2$. The squaring of $(b + c)$ will *always* result in a trinomial.

EXAMPLE 5 Find the solution set of $\sqrt{2x+1} = x - 1$.

SOLUTION

Step 1. The domain is $\{x \mid x \ge -\frac{1}{2}\}$.

$$\sqrt{2x+1} = x - 1$$

Step 3. $(\sqrt{2x+1})^2 = (x-1)^2$ Squaring both sides

When squaring $(x - 1)$, do not forget this middle term

Step 4. $2x + 1 = x^2 - 2x + 1$

Step 5. $0 = x^2 - 4x$

$0 = x(x - 4)$

$x = 0 \quad | \quad x - 4 = 0$

$x = 4$ Both apparent solutions are in the domain

√ **Step 6.** ***Check for*** $x = 0$

$$\sqrt{2x+1} = x - 1$$
$$\sqrt{2(0)+1} \stackrel{?}{=} (0) - 1$$
$$\sqrt{1} \stackrel{?}{=} -1$$
$$1 \neq -1$$

The symbol $\sqrt{1}$ represents the *principal* square root of 1, which is 1 (*not* -1)

Check for $x = 4$

$$\sqrt{2x+1} = x - 1$$
$$\sqrt{2(4)+1} \stackrel{?}{=} (4) - 1$$
$$\sqrt{9} \stackrel{?}{=} 3$$
$$3 = 3$$

Therefore, 0 *is not a solution*, whereas 4 *is* a solution. The solution set is $\{4\}$.

EXAMPLE 6 Find the solution set of $\sqrt{4x+5} - \sqrt{x-1} = \sqrt{14-x}$.

SOLUTION

Step 1. The domain is $\{x \mid 1 \le x \le 14\}$ (see Example 1c).

$$\sqrt{4x+5} - \sqrt{x-1} = \sqrt{14-x}$$ One radical is by itself on one side of the equal sign

Step 3. $$(\sqrt{4x+5} - \sqrt{x-1})^2 = (\sqrt{14-x})^2$$ Squaring both sides

When squaring, don't forget this middle term

Step 4. $$4x + 5 - 2\sqrt{(4x+5)(x-1)} + x - 1 = 14 - x$$ Simplifying; a radical remains

$$5x + 4 - 2\sqrt{(4x+5)(x-1)} = 14 - x$$ Combining like terms

Step 2. $$-2\sqrt{(4x+5)(x-1)} = -6x + 10$$ Adding $-5x - 4$ to both sides

$$\sqrt{(4x+5)(x-1)} = 3x - 5$$ Dividing both sides by -2

Step 3. $$(\sqrt{4x^2 + x - 5})^2 = (3x-5)^2$$ Squaring both sides again

Step 4. $$4x^2 + x - 5 = 9x^2 - 30x + 25$$

Step 5. $$0 = 5x^2 - 31x + 30$$ Getting all terms on one side

$$0 = (5x-6)(x-5)$$ Factoring

$$5x - 6 = 0 \qquad x - 5 = 0$$
$$x = \tfrac{6}{5} \qquad x = 5$$ Both apparent solutions are in the domain

√ **Step 6.** ***Check for*** $x = \frac{6}{5}$

$$\sqrt{4x+5} - \sqrt{x-1} = \sqrt{14-x}$$
$$\sqrt{4\left(\tfrac{6}{5}\right)+5} - \sqrt{\tfrac{6}{5}-1} \stackrel{?}{=} \sqrt{14-\tfrac{6}{5}}$$
$$\sqrt{\tfrac{24}{5}+\tfrac{25}{5}} - \sqrt{\tfrac{6}{5}-\tfrac{5}{5}} \stackrel{?}{=} \sqrt{\tfrac{70}{5}-\tfrac{6}{5}}$$
$$\sqrt{\tfrac{49}{5}} - \sqrt{\tfrac{1}{5}} \stackrel{?}{=} \sqrt{\tfrac{64}{5}}$$
$$\frac{7}{\sqrt{5}} - \frac{1}{\sqrt{5}} \stackrel{?}{=} \frac{8}{\sqrt{5}}$$
$$\frac{6}{\sqrt{5}} \neq \frac{8}{\sqrt{5}}$$

Check for $x = 5$

$$\sqrt{4x+5} - \sqrt{x-1} = \sqrt{14-x}$$
$$\sqrt{4(5)+5} - \sqrt{5-1} \stackrel{?}{=} \sqrt{14-5}$$
$$\sqrt{20+5} - \sqrt{4} \stackrel{?}{=} \sqrt{9}$$
$$\sqrt{25} - 2 \stackrel{?}{=} 3$$
$$5 - 2 \stackrel{?}{=} 3$$
$$3 = 3$$

Therefore, $\frac{6}{5}$ is *not* a solution, whereas 5 *is* a solution; the solution set is $\{5\}$.

Solving Equations That Contain Rational Exponents

When we need to solve an equation that contains rational exponents, *if the numerators of the exponents are odd numbers*, we can solve the equation by raising both sides of the equation to the same power (see Example 7). In a later chapter, we will consider solving equations that contain rational exponents whose numerators are even numbers.

EXAMPLE 7 Find the solution set of $x^{-1/4} = 2$.

SOLUTION The power to which we raise both sides must be the multiplicative inverse of $-1/4$ so that the *product of the two exponents* will be 1. Therefore, we must raise both sides to the negative fourth power.

Raising both sides to the negative fourth power

$$(x^{-1/4})^{-4} = 2^{-4}$$

$$x = 2^{-4} = \frac{1}{2^4} = \frac{1}{16}$$

(You might verify that we get the same result if we convert the equation to radical form and solve.)

√ ***Check*** $\left(\frac{1}{16}\right)^{-1/4} = (16)^{1/4} = \sqrt[4]{16} = 2$

The solution set is $\left\{\frac{1}{16}\right\}$.

Applications Involving Radical Equations

There are many applications involving radical equations in the sciences, business, engineering, and so forth.

EXAMPLE 8 The formula from electricity that gives the relationship among current, power, and resistance is $I = \sqrt{\frac{P}{R}}$, where I, the current, is measured in amps (amperes); P, the power, is measured in watts; and R, the resistance, is measured in ohms. Using this formula, find P, the amount of power consumed, if an appliance has a resistance of 5 ohms and draws 10 amps of current.

SOLUTION We're told that the resistance is 5 ohms, so $R = 5$, and we're told that the current is 10 amps, so $I = 10$. We must find the power, P.

$$I = \sqrt{\frac{P}{R}}$$

$$10 = \sqrt{\frac{P}{5}}$$

$$(10)^2 = \left(\sqrt{\frac{P}{5}}\right)^2$$

$$100 = \frac{P}{5}$$

$$P = 500$$

The check, which is not shown, will confirm that 500 watts of power is consumed.

Exercises 7.10 Set I

In Exercises 1–24, find the solution set of each equation.

1. $\sqrt{3x+1}=5$
2. $\sqrt{7x+8}=6$
3. $\sqrt{x+1}=\sqrt{2x-7}$
4. $\sqrt{3x-2}=\sqrt{x+4}$
5. $\sqrt[4]{4x-11}-1=0$
6. $\sqrt[4]{3x+1}-2=0$
7. $\sqrt{4x-1}=2x$
8. $\sqrt{6x-1}=3x$
9. $\sqrt{x+7}=2x-1$
10. $\sqrt{x+5}=3x-9$
11. $\sqrt{8x}+3=\sqrt{8x+33}$
12. $\sqrt{2x}+1=\sqrt{2x-3}$
13. $\sqrt{3x+1}-\sqrt{x+4}=1$
14. $\sqrt{2x+1}-\sqrt{x}=1$
15. $\sqrt{3x+4}-\sqrt{2x-4}=2$
16. $\sqrt{2x-1}+2=\sqrt{3x+10}$
17. $\sqrt[3]{2x+3}-2=0$
18. $\sqrt[3]{4x-3}-3=0$
19. $\sqrt{4u+1}-\sqrt{u-2}=\sqrt{u+3}$
20. $\sqrt{2-v}+\sqrt{v+3}=\sqrt{7+2v}$
21. $x^{1/2}=5$
22. $x^{1/3}=3$
23. $2x^{-5/3}=64$
24. $5x^{-3/2}=40$

In Exercises 25 and 26, solve each problem for the required unknown, using the formula $I=\sqrt{\dfrac{P}{R}}$, as given in Example 8.

25. Find the amount of power, P, consumed if an appliance has a resistance of 16 ohms and draws 5 amps of current.
26. Find the amount of power, P, consumed if an appliance has a resistance of 9 ohms and draws 8 amps of current.

In Exercises 27 and 28, use this formula from statistics: $\sigma=\sqrt{npq}$, where σ (*sigma*) is the standard deviation, n is the number of trials, p is the probability of success, and q is the probability of failure.

27. Find the probability of success, p, if the standard deviation is $3\frac{1}{3}$, the probability of failure is $\frac{2}{3}$, and the number of trials is 50.
28. Find the probability of success, p, if the standard deviation is 12, the probability of failure is $\frac{3}{5}$, and the number of trials is 600.

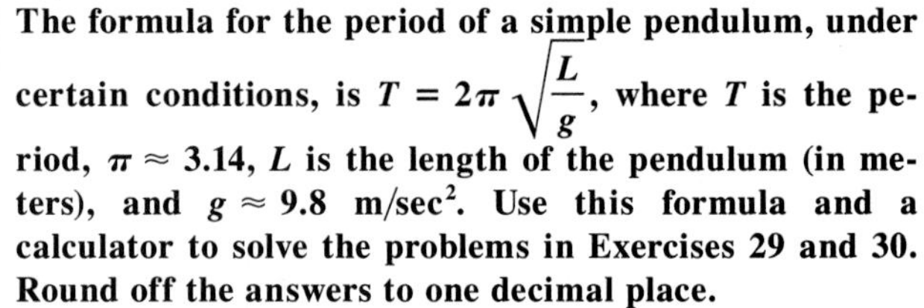

The formula for the period of a simple pendulum, under certain conditions, is $T=2\pi\sqrt{\dfrac{L}{g}}$, where T is the period, $\pi\approx 3.14$, L is the length of the pendulum (in meters), and $g\approx 9.8$ m/sec^2. Use this formula and a calculator to solve the problems in Exercises 29 and 30. Round off the answers to one decimal place.

29. Find the length of a simple pendulum if the period is 1.32 sec.
30. Find the length of a simple pendulum if the period is 1.04 sec.

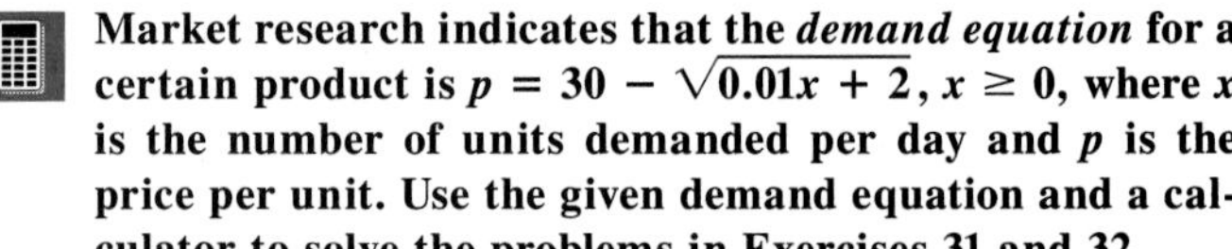

Market research indicates that the *demand equation* for a certain product is $p=30-\sqrt{0.01x+2}$, $x\geq 0$, where x is the number of units demanded per day and p is the price per unit. Use the given demand equation and a calculator to solve the problems in Exercises 31 and 32.

31. Find the demand, x, if $p=\$21.50$.
32. Find the demand, x, if $p=\$24.50$.

Writing Problems

Express the answers in your own words and in complete sentences.

1. Describe what your first step would be in solving the equation $\sqrt{x-3}+5=x$, and explain why you would do that first.
2. Explain why, when we square both sides of $\sqrt{x-3}=x+2$, we *don't* get $x-3=x^2+4$.
3. Without attempting to solve the equation, explain how you can tell that the equation $\sqrt{x-2}=-3$ has no solution.
4. Explain why $x=7$ and $x^2=49$ are not equivalent equations.
5. Describe what your first step would be in solving the equation $x^{1/4}-2=0$, and explain why you would do that first.

Exercises 7.10 Set II

In Exercises 1–24, find the solution set of each equation.

1. $\sqrt{3x-2}=x$
2. $x=\sqrt{10-3x}$
3. $\sqrt{3x-2}=\sqrt{5x+4}$
4. $\sqrt{5x-6}=x$
5. $\sqrt[5]{7x+4}-2=0$
6. $\sqrt{x+4}-\sqrt{2}=\sqrt{x-6}$

7. $x = \sqrt{12x - 36}$

8. $\sqrt{3x + 1} = \sqrt{1 - x}$

9. $\sqrt{x - 3} = x - 5$

10. $\sqrt{4x + 5} + 5 = 2x$

11. $\sqrt{4x} + 1 = \sqrt{4x + 17}$

12. $\sqrt{3x} - 1 = \sqrt{3x - 17}$

13. $\sqrt{3x + 4} - \sqrt{x + 5} = 1$

14. $\sqrt{x + 6} = \sqrt{x - 2} + 2$

15. $\sqrt{v + 7} - \sqrt{v - 2} = 5$

16. $\sqrt{x} = \sqrt{x + 16} - 2$

17. $\sqrt[3]{x - 1} = 2$

18. $\sqrt{2 + x} = 1$

19. $\sqrt[5]{3x - 14} - 1 = 0$

20. $\sqrt{2x - 9} - \sqrt{4x} + 3 = 0$

21. $x^{1/4} = 1$

22. $x^{1/5} = 2$

23. $x^{3/5} = 8$

24. $\sqrt{4v + 1} = \sqrt{v + 4} + \sqrt{v - 3}$

In Exercises 25 and 26, solve each problem for the required unknown, using the formula $I = \sqrt{\frac{P}{R}}$, as given in Example 8.

25. Find the amount of power, P, consumed if an appliance has a resistance of 25 ohms and draws 4 amps of current.

26. Find R, the resistance, for an electrical system that consumes 600 watts and draws 5 amps of current.

In Exercises 27 and 28, use this formula from statistics: $\sigma = \sqrt{npq}$, where σ (*sigma*) is the standard deviation, n is the number of trials, p is the probability of success, and q is the probability of failure.

27. Find the probability of success, p, if the standard deviation is 8, the probability of failure is $\frac{4}{5}$, and the number of trials is 400.

28. Find the probability of success, p, if the standard deviation is 20, the probability of failure is $\frac{5}{6}$, and the number of trials is 2,880.

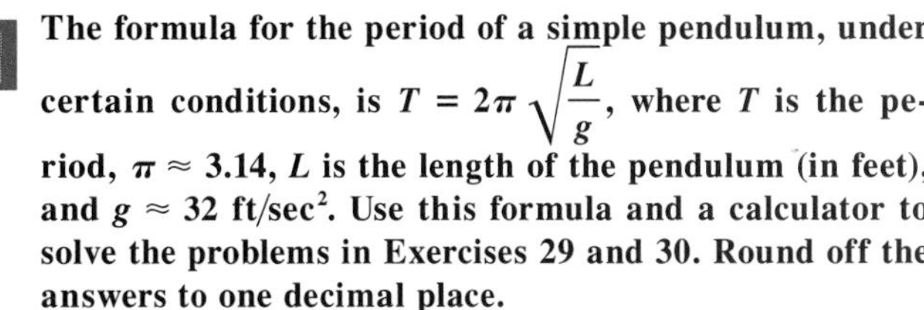

The formula for the period of a simple pendulum, under certain conditions, is $T = 2\pi\sqrt{\frac{L}{g}}$, where T is the period, $\pi \approx 3.14$, L is the length of the pendulum (in feet), and $g \approx 32$ ft/sec^2. Use this formula and a calculator to solve the problems in Exercises 29 and 30. Round off the answers to one decimal place.

29. Find the length of a simple pendulum if the period is 1.38 sec.

30. Find the length of a simple pendulum if the period is 1.06 sec.

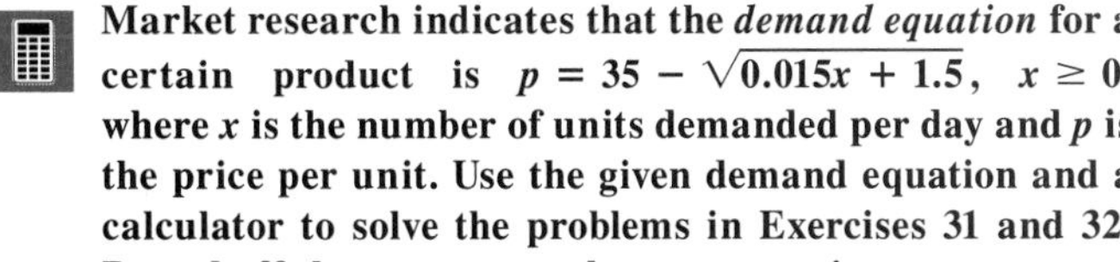

Market research indicates that the *demand equation* for a certain product is $p = 35 - \sqrt{0.015x + 1.5}$, $x \geq 0$, where x is the number of units demanded per day and p is the price per unit. Use the given demand equation and a calculator to solve the problems in Exercises 31 and 32. Round off the answers to the nearest unit.

31. Find the demand, x, if $p = \$26.83$.

32. Find the demand, x, if $p = \$31.54$.

7.11 Using the Pythagorean Theorem

Right Triangles A triangle that has one *right angle* (a 90° angle) is called a **right triangle**. The longest side of a right triangle is called its **hypotenuse**, and the other two sides are called the **legs**. A **diagonal** of a rectangle divides the rectangle into two right triangles. The parts of a right triangle are shown in Figure 1a, and a rectangle with one of its diagonals is shown in Figure 1b.

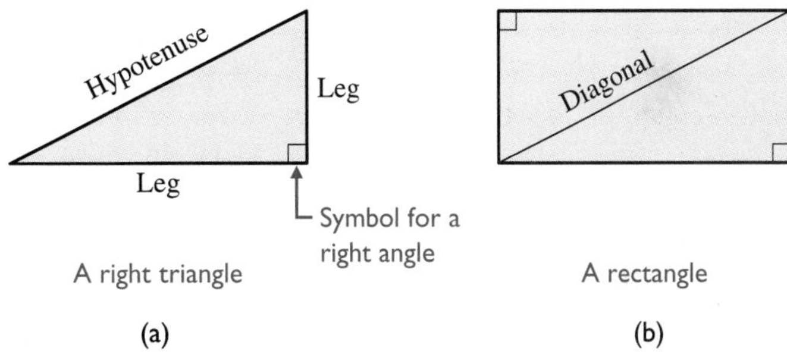

Figure 1

The Pythagorean Theorem

The following theorem is proved in geometry:

The Pythagorean theorem

The square of the hypotenuse of a right triangle is equal to the sum of the squares of the two legs:

$$c^2 = a^2 + b^2$$

where a and b represent the lengths of the legs and c represents the length of the hypotenuse.

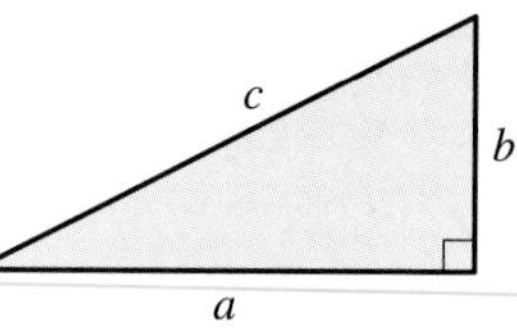

Note The Pythagorean theorem applies only to *right triangles*.

The Pythagorean theorem can be used to find one side of a right triangle when the other two sides are known.

EXAMPLE 1

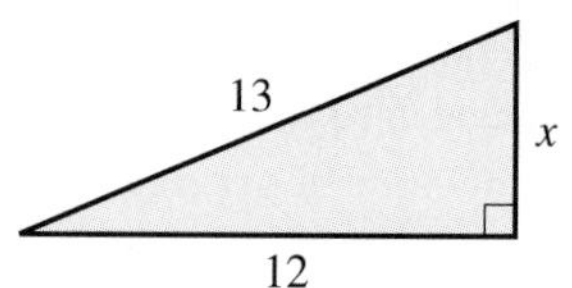

Find x, using the Pythagorean theorem.

SOLUTION The domain is the set of all *positive* numbers, since the length of a side of a triangle must be positive.

$$c^2 = a^2 + b^2 \quad \text{The Pythagorean theorem}$$
$$13^2 = 12^2 + x^2$$
$$169 = 144 + x^2$$
$$0 = x^2 - 25 \quad \text{Adding } -169 \text{ to both sides}$$
$$0 = (x + 5)(x - 5)$$
$$x + 5 = 0 \quad \Big| \quad x - 5 = 0$$
$$x = -5 \quad \Big| \quad x = 5$$

5 and -5 are both solutions of the original equation; however, -5 is not in the domain of the variable. Therefore, the only solution is 5.

In using the Pythagorean theorem, it is generally helpful to draw and label a sketch of the figure.

EXAMPLE 2

Find the length of the hypotenuse of a right triangle with legs that are $\sqrt{17}$ and $\sqrt{8}$ units long.

SOLUTION The domain is the set of all positive numbers. Let $x =$ the length of the hypotenuse.

$$c^2 = a^2 + b^2 \quad \text{The Pythagorean theorem}$$
$$x^2 = (\sqrt{8})^2 + (\sqrt{17})^2$$
$$x^2 = 8 + 17$$
$$x^2 = 25$$
$$x^2 - 25 = 0$$
$$(x + 5)(x - 5) = 0$$

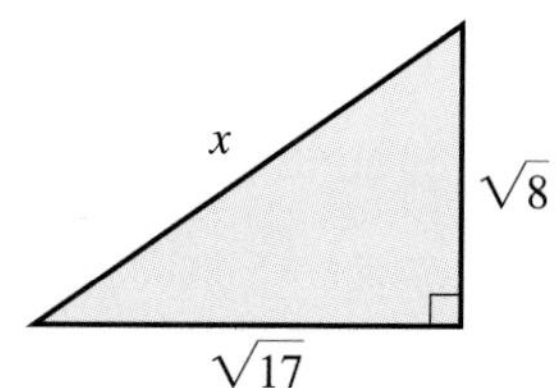

$$
\begin{array}{l|l}
x + 5 = 0 & x - 5 = 0 \\
\quad x = -5 & \quad x = 5
\end{array}
$$

Both -5 and 5 are solutions of the original equation. However, because -5 is not in the domain of the variable, the only solution for the problem is 5. Therefore, the length of the hypotenuse is 5 units.

Solving Equations of the Form $x^2 = p$ If $p \geq 0$, we can use the following rule for solving an equation of the form $x^2 = p$.

> If $x^2 = p$ and if $p \geq 0$, then
> $$x = \sqrt{p} \quad \text{or} \quad x = -\sqrt{p}$$

We'll see in a later chapter that there are two solutions for every quadratic (*second-degree*) equation, although sometimes the two solutions are identical and sometimes they are not real; notice that there are two possible solutions of the quadratic equation $x^2 = p$. We can easily verify that $\sqrt{p}$ and $-\sqrt{p}$ are solutions of $x^2 = p$. If $p \geq 0$, then

$$\text{if } x = \sqrt{p}, \qquad x^2 = (\sqrt{p})^2 = \sqrt{p}\sqrt{p} = p$$
$$\text{if } x = -\sqrt{p}, \qquad x^2 = (-\sqrt{p})^2 = (-\sqrt{p})(-\sqrt{p}) = p$$

We often combine the two solutions and write $x = \pm\sqrt{p}$.

We use this method of solving an equation in Example 3.

EXAMPLE 3 Find x, using the Pythagorean theorem.

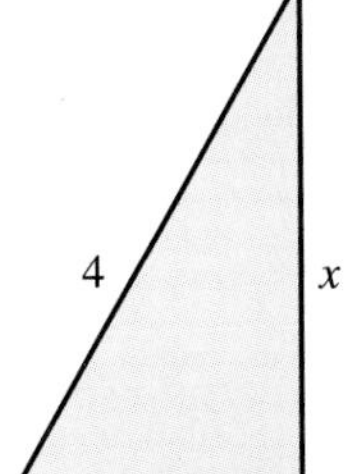

SOLUTION The domain is the set of positive numbers.

$$c^2 = a^2 + b^2 \quad \text{The Pythagorean theorem}$$
$$4^2 = 2^2 + x^2$$
$$16 = 4 + x^2$$
$$x^2 = 12$$

$$
\begin{array}{l|l}
x = \sqrt{12} & x = -\sqrt{12} \\
x = 2\sqrt{3} & x = -2\sqrt{3}
\end{array}
$$

Because $-2\sqrt{3}$ is not in the domain, $2\sqrt{3}$ is the only solution.

EXAMPLE 4 The length of a rectangle is 2 cm more than its width. If the length of its diagonal is 10 cm, find the dimensions of the rectangle.

SOLUTION

Think The domain is the set of all positive numbers.

Step 1. Let $x =$ the width (in cm)
Then $x + 2 =$ the length (in cm)

Step 2. $(10)^2 = (x)^2 + (x + 2)^2$ Using the Pythagorean theorem

Step 3. $100 = x^2 + x^2 + 4x + 4$

$0 = 2x^2 + 4x - 96$

$0 = x^2 + 2x - 48$ Dividing both sides by 2

$0 = (x + 8)(x - 6)$

	$x + 8 = 0$	$x - 6 = 0$	
Step 4.	$x = -8$	$x = 6$	The width (in cm)
	Not in the domain	$x + 2 = 8$	The length (in cm)

√ **Step 5.** ***Check*** $6^2 + 8^2 = 36 + 64 = 100$, and $10^2 = 100$

Step 6. Therefore, the width of the rectangle is 6 cm, and the length is 8 cm.

There are many applications in the real world that require the use of the Pythagorean theorem (see Example 5).

EXAMPLE 5 Two airplanes left Denver at the same time. One plane was flying due east, and it was flying 50 mph faster than the other plane, which was flying due south. After 2 hr, the planes were 500 mi apart. What was the speed of the slower plane? (We assume that there was *no* wind blowing.)

SOLUTION

Step 1. Let $x =$ the speed of the slower plane (in mph)

Then $x + 50 =$ the speed of the faster plane (in mph)

Think After 2 hr, the slower plane had flown

$$\left(x \frac{\text{mi}}{\cancel{\text{hr}}}\right)(2\ \cancel{\text{hr}}) = 2x \text{ mi}$$

and the faster plane had flown

$$\left([x + 50] \frac{\text{mi}}{\cancel{\text{hr}}}\right)(2\ \cancel{\text{hr}}) = 2(x + 50) \text{ mi}$$

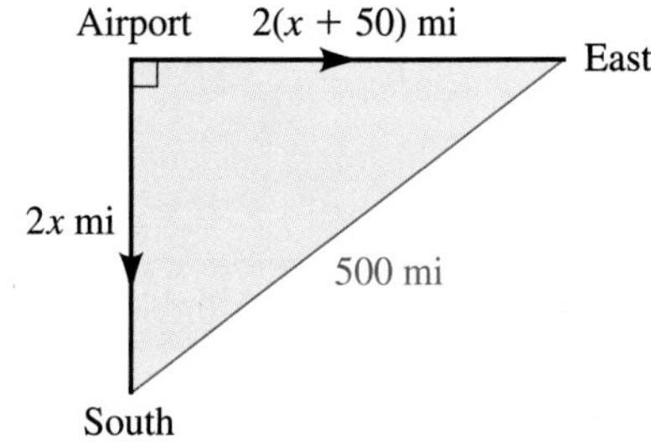

Step 2. $(2x)^2 + [2(x + 50)]^2 = 500^2$ Using the Pythagorean theorem

Step 3. $4x^2 + (2x + 100)^2 = 250{,}000$

$4x^2 + 4x^2 + 400x + 10{,}000 = 250{,}000$

$8x^2 + 400x - 240{,}000 = 0$

$x^2 + 50x - 30{,}000 = 0$ Dividing both sides by 8

$(x - 150)(x + 200) = 0$

	$x - 150 = 0$	$x + 200 = 0$
Step 4.	$x = 150$	$x = -200$
	$x + 50 = 200$	We reject this solution, since speed cannot be negative

√ **Step 5.** ***Check*** The faster plane was flying 50 mph faster than the slower plane. The slower plane flew $\left(150 \frac{\text{mi}}{\cancel{\text{hr}}}\right)(2\ \cancel{\text{hr}}) = 300$ mi, and the faster plane flew $\left(200 \frac{\text{mi}}{\cancel{\text{hr}}}\right)(2\ \cancel{\text{hr}}) = 400$ mi. The sum of the squares of the distances is $300^2 + 400^2$, or $90{,}000 + 160{,}000 = 250{,}000$, which is 500^2.

Step 6. Therefore, the slower plane was flying at 150 mph.

Exercises 7.11
Set I

In Exercises 1–6, use the Pythagorean theorem to find x in each figure.

1.
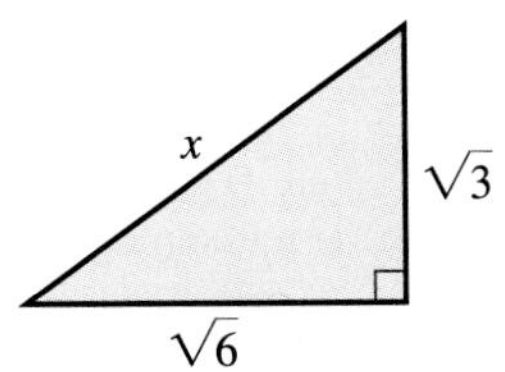

2.
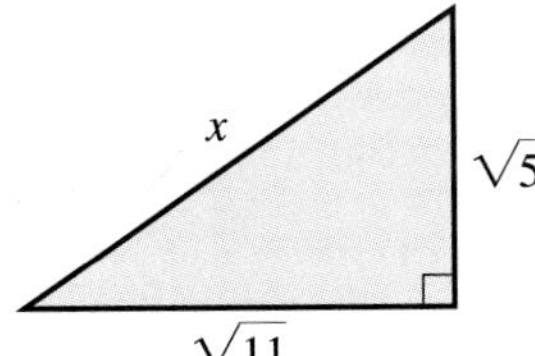

3.
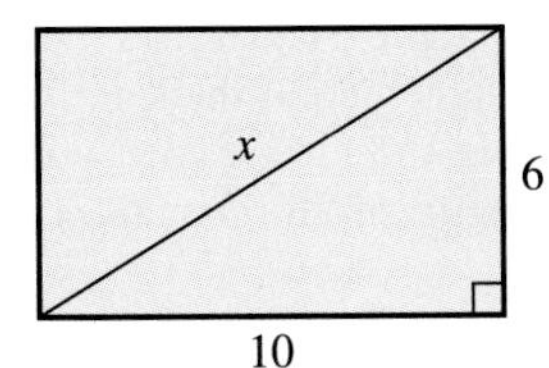

4.
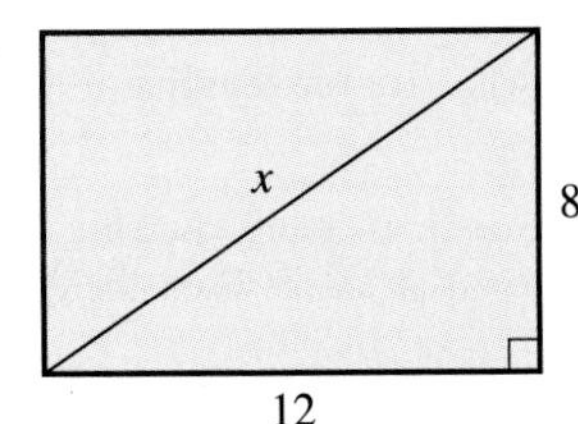

5.
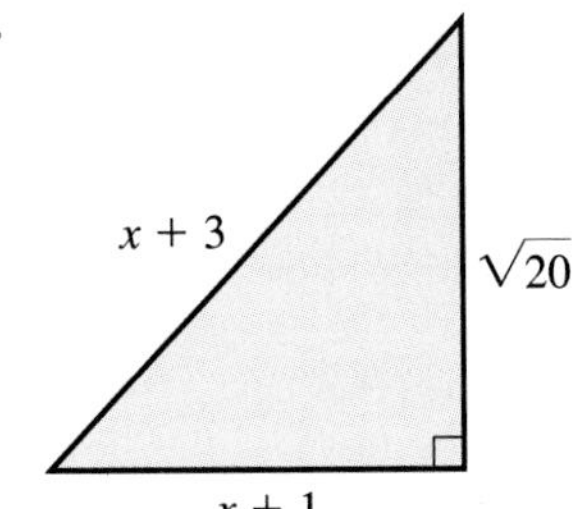

6.
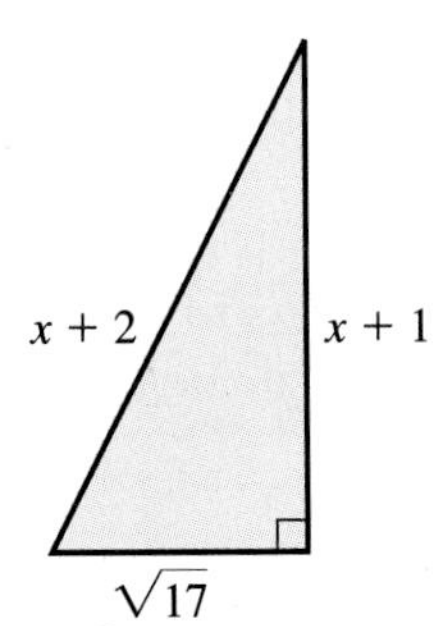

In Exercises 7–16, set up each problem algebraically, solve, and check.

7. The length of each side of a square is 4 in. Find the length of a diagonal of the square.

8. The length of each side of a square is 3 m. Find the length of a diagonal of the square.

9. The length of a rectangle is 24 cm, and the length of its diagonal is 25 cm. Find the width of the rectangle.

10. The length of a rectangle is 40 ft, and the length of its diagonal is 41 ft. Find the width of the rectangle.

11. The length of one leg of a right triangle is 4 cm less than twice the length of the other leg. If the length of the hypotenuse is 10 cm, what are the lengths of the two legs?

12. The length of one leg of a right triangle is 4 m more than twice the length of the other leg. If the length of the hypotenuse is $\sqrt{61}$ m, what are the lengths of the two legs?

13. Jaime and Hy left a corner at the same time. Jaime was jogging due west; Hy was walking due north. Jaime's speed was 7 mph faster than Hy's. After 1 hr, the distance between the two men was 13 mi. How fast was Hy walking?

14. Rob and Julie started from the same point at the same time, walking at right angles to each other. Julie was walking 1 mph faster than Rob. After 3 hr, they were 15 mi apart. How fast was Julie walking?

15. The length of a diagonal of a square is 3.87 ft. Find the length of each side of the square. Round off the answer to two decimal places.

16. The length of a diagonal of a square is 4.24 m. Find the length of each side of the square. Round off the answer to two decimal places.

Writing Problems

Express the answers in your own words and in complete sentences.

1. Find and describe the error in the following statement:

   If $a^2 + b^2 = c^2$, then $a + b = c$.

2. Make up an example using numbers to show that if $a^2 + b^2 = c^2$, it need not be true that $a + b = c$.

Exercises 7.11
Set II

In Exercises 1–6, use the Pythagorean theorem to find x in each figure.

1.

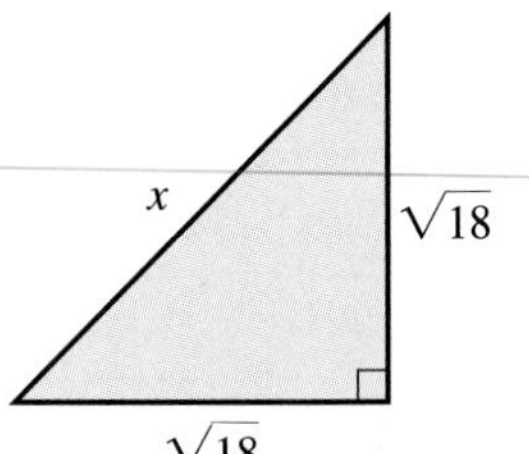

2.

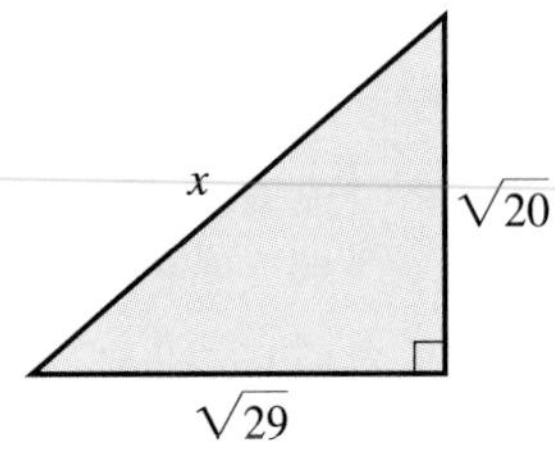

3.

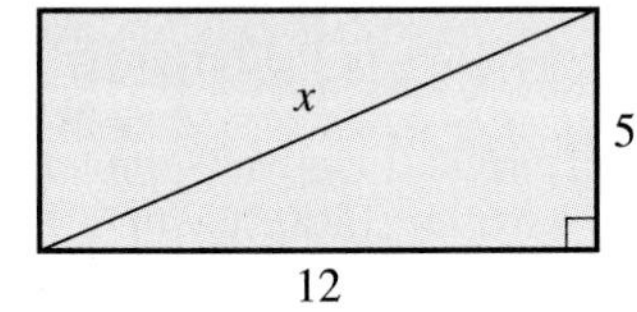

4.

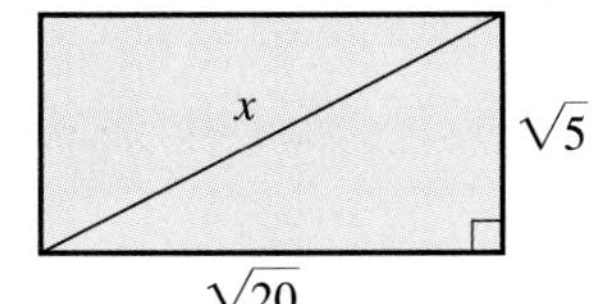

5.

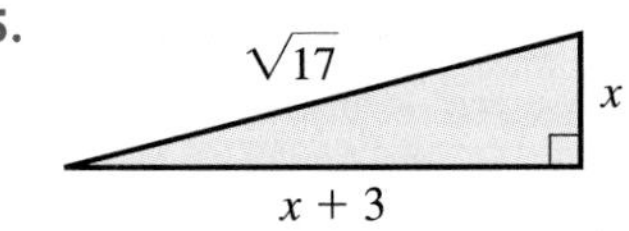

6.

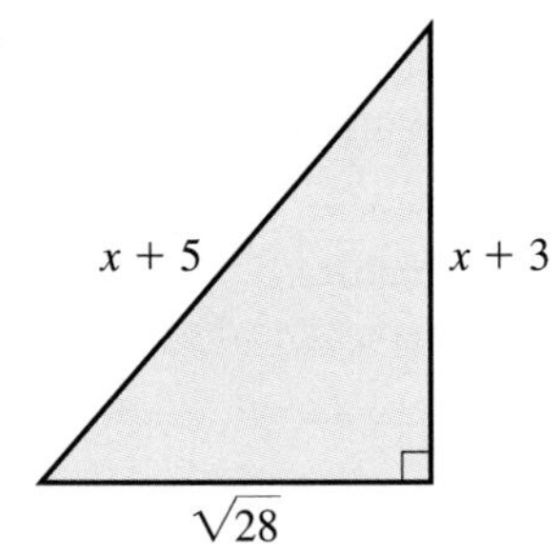

In Exercises 7–16, set up each problem algebraically, solve, and check.

7. The length of each side of a square is 5 m. Find the length of a diagonal of the square.

8. The width of a rectangle is 8 ft, and its length is 11 ft. Find the length of a diagonal of the rectangle.

9. The length of a rectangle is 15 cm, and the length of its diagonal is 17 cm. Find the width of the rectangle.

10. The length of one leg of a right triangle is 2 yd more than twice the length of the other leg. If the length of the hypotenuse is 2 yd less than three times the length of the shorter leg, find the lengths of all three sides of the triangle.

11. The length of one leg of a right triangle is 3 km more than twice the length of the other leg. If the length of the hypotenuse is $\sqrt{137}$ km, what are the lengths of the two legs?

12. The length of a rectangle is 3 in. more than its width, and the length of its diagonal is 15 in. Find the dimensions of the rectangle.

13. Tran and Francisco left a corner at the same time. Francisco was running due south, and Tran was walking due west. Francisco's speed was 7 mph faster than Tran's. After 1 hr, the distance between the two men was 17 mi. How fast was Francisco running?

14. Two cars left an intersection at the same time, one going north and the other going east. The car going east was traveling 10 mph faster than the other car. After 30 min, the cars were 25 mi apart. How fast was the slower car going?

15. The length of a diagonal of a square is 18.3 cm. Find the length of each side of the square. Round off the answer to two decimal places.

16. The length of a diagonal of a square is 1.414 ft. Find the length of each side of the square. Round off the answer to the nearest unit.

7.12 Complex Numbers: Basic Definitions

All the numbers discussed so far in this book have been *real* numbers. Recall that the set of real numbers is the union of the set of rational numbers with the set of irrational numbers. In this section, we discuss a new kind of number that is *not* a real number but is essential in many applications of mathematics.

In order for us to be able to solve equations such as $x^2 = -4$, we must invent, or *define*, a new set of numbers that will *not* be real numbers.

The basis for our new set of numbers is the number i, which is defined as follows:

$$i^2 = -1$$

or

$$i = \sqrt{-1}$$

Note $\sqrt{-1}$ is not in simplest form, since the radicand is negative; simplest form for $\sqrt{-1}$ is i. We will see in a later section that i^2 is not in simplest form either.

Pure Imaginary Numbers

A **pure imaginary number** is a number that can be written as bi, where b is a real number and $i = \sqrt{-1}$. If b is an irrational number, the i is usually written first, as in ib.

EXAMPLE 1 Examples of pure imaginary numbers:

$$-5i, \quad -2i\sqrt{3}, \quad -\tfrac{2}{5}i, \quad 0.63i$$

A Word of Caution A common error is to think that $i = -1$. *This is not true!* $i = \sqrt{-1}$, *not* -1.

Another common error is to write a number that should be $\sqrt{5}i$ as $\sqrt{5i}$; the i should *not* be under the radical sign. (Numbers such as $\sqrt{5i}$ are discussed in higher-level courses.) If $\sqrt{5}i$ is written as $i\sqrt{5}$, it is *very* clear that i is not under the radical sign.

Complex Numbers

A **complex number** is a number that can be expressed in the form $a + bi$, where a and b are real numbers and $i = \sqrt{-1}$.

The real part → a + bi ← The imaginary part

$$a + bi$$

EXAMPLE 2 Examples of complex numbers:

$$2 - 3i, \quad \sqrt{5} - \tfrac{1}{2}i, \quad -0.7 + 0.4i, \quad 4 + 0i$$

Because the set of complex numbers is a completely new set, we need definitions for equality, addition, subtraction, and so forth.

We should emphasize that the *ordering* property does not hold for complex numbers; that is, the relations "less than" and "greater than" are *not defined* for the set of complex numbers.

Equality of complex numbers is defined as follows:

Definition of equality of two complex numbers

If $a + bi = c + di$, then

$a = c$ (The real parts are equal)

and $b = d$ (The coefficients of i are equal)

Furthermore, by definition,

$$0i = 0$$

When the *imaginary* part is zero, the complex number is a *real number.*

$$2 + 0i = 2$$

b is zero (points to 0); A real number (points to 2)

The set of real numbers is a subset of the set of complex numbers.

When the *real* part is zero, the complex number is a *pure imaginary number.*

$$0 + 3i = 3i$$

a is zero (points to 0); A pure imaginary number (points to $3i$)

The set of pure imaginary numbers is a subset of the set of complex numbers.

Note Some authors call the number $a + bi$ an *imaginary* number (but not a *pure* imaginary number) if $b \neq 0$. We call such a number a complex number.

The product rule for radicals implies that for *square roots*, $\sqrt{ab} = \sqrt{a}\sqrt{b}$ only if a and b are both positive. Actually, it is *also* true that $\sqrt{ab} = \sqrt{a}\sqrt{b}$ if a and b have *opposite signs*; for example, $\sqrt{9(-1)} = \sqrt{9}\sqrt{-1}$. We use this fact in simplifying square roots (see Example 3).

EXAMPLE 3 Examples of writing numbers in the form $a + bi$:

Using the product rule for radicals

a. $\sqrt{-9} = \sqrt{9(-1)} = \sqrt{9}\sqrt{-1} = 3i = 0 + 3i$

b. $\sqrt{-17} = \sqrt{17(-1)} = \sqrt{17}\sqrt{-1} = \sqrt{17}i = 0 + i\sqrt{17}$

c. $2 - \sqrt{-25} = 2 - \sqrt{25(-1)} = 2 - \sqrt{25}\sqrt{-1} = 2 - 5i$

A Word of Caution The product rule for radicals does *not* apply when n is even and *both* a and b are negative. For example,

$\sqrt{(-4)(-9)} = \sqrt{36} = 6$ | $\sqrt{-4}\sqrt{-9} = 2i \cdot 3i = 6i^2 = 6(-1) = -6$

Since $6 \neq -6$, $\sqrt{(-4)(-9)} \neq \sqrt{-4}\sqrt{-9}$. In general, $\sqrt[n]{ab} \neq \sqrt[n]{a}\sqrt[n]{b}$ if *both* a and b are negative when n is even.

We can use the definition of equality of complex numbers in solving equations involving complex numbers. We simply set the two *real* parts equal to each other and solve that equation; then we set the coefficient of i from one side of the equation equal to the coefficient of i from the other side and solve *that* equation (see Example 4).

EXAMPLE 4 Examples of solving equations involving complex numbers:

a. If $x - 3i = 5 + yi$, then

	$x = 5$	Setting the real parts equal to each other
and	$-3 = y$	Setting the coefficients of i equal to each other
or	$y = -3$	

b. If $3x + 7yi = 10 - 2i$, then

	$3x = 10$	Setting the real parts equal to each other
or	$x = \frac{10}{3}$	
and	$7y = -2$	Setting the coefficients of i equal to each other
or	$y = -\frac{2}{7}$	

Exercises 7.12
Set I

In Exercises 1–12, convert each expression to the form $a + bi$. Express all radicals in simplest radical form.

1. $3 + \sqrt{-16}$
2. $4 - \sqrt{-25}$
3. $\sqrt{-64}$
4. $\sqrt{-100}$
5. $5 + \sqrt{-32}$
6. $6 + \sqrt{-18}$
7. $\sqrt{-36} + \sqrt{4}$
8. $\sqrt{9} - \sqrt{-25}$
9. $2i - \sqrt{9}$
10. $3i - \sqrt{16}$
11. 14
12. -7

In Exercises 13–20, solve for x and y.

13. $3 - 4i = x + 2yi$
14. $3x + 5i = 6 + yi$
15. $5x - 3i = 6 - 7yi$
16. $-3 - yi = 2x + 3i$
17. $x\sqrt{3} - yi = 2 + i\sqrt{2}$
18. $3 + yi\sqrt{5} = x\sqrt{8} - i$
19. $\frac{3}{4}x - \frac{1}{3}yi = \frac{3}{5}x + \frac{1}{2}yi$
20. $\frac{2}{3}x - yi = \frac{1}{2}x + 3yi$

Writing Problems

Express the answer in your own words and in complete sentences.

1. Find and describe the error in the following statement: $\sqrt{-4} = \sqrt{2}i$

Exercises 7.12
Set II

In Exercises 1–12, convert each expression to the form $a + bi$. Express all radicals in simplest radical form.

1. $5 + \sqrt{-49}$
2. $\sqrt{-4} - 6$
3. $\sqrt{-81}$
4. -16
5. $3 + \sqrt{-8}$
6. $\sqrt{7} + \sqrt{-36}$
7. $\sqrt{-100} + \sqrt{16}$
8. 7
9. $5i - \sqrt{7}$
10. $\sqrt{8} - \sqrt{-18}$
11. $\sqrt{50}$
12. $\sqrt{-27}$

In Exercises 13–20, solve for x and y.

13. $5 - yi = x + 4i$
14. $8x - 6i = 12 + yi$
15. $3x + 7yi = 2 + 3i$
16. $2x - 3yi = 5 + 2i$
17. $4 + yi\sqrt{2} = x\sqrt{5} - 3i$
18. $x\sqrt{7} + 2i = \sqrt{2} + 4yi$
19. $\frac{1}{3}x - \frac{3}{4}i = \frac{1}{2} + \frac{1}{5}yi$
20. $\frac{3}{5}x - \frac{1}{2}yi = 3 - 4i$

7.13 Adding and Subtracting Complex Numbers

The commutative, associative, and distributive properties hold for complex numbers. Therefore, we can add complex numbers by treating i as if it were a variable and performing the operations as usual. Addition and subtraction are formally defined as follows:

Adding and subtracting complex numbers

$$(a + bi) + (c + di) = (a + c) + (b + d)i$$
$$(a + bi) - (c + di) = (a - c) + (b - d)i$$

EXAMPLE 1 Examples of adding complex numbers:

a. $(2 + 3i) + (-4 + 5i) = [2 + (-4)] + (3 + 5)i = -2 + 8i$

b. $7 + (-5 + 3i) = (7 + 0i) + (-5 + 3i) = [7 + (-5)] + (0 + 3)i = 2 + 3i$

c. $(8 + 7i) + (-5i) + (-13 + 4i) = [8 + (-13)] + [7 + (-5) + 4]i = -5 + 6i$

d. $(-7 + 4i) + (6 - 3i) = (-7 + 6) + [4 + (-3)]i = -1 + i$

Note Your instructor may not require you to show the intermediate steps that we have shown.

EXAMPLE 2 Examples of subtracting complex numbers:

a. $(-5 + 2i) - (6 - 2i) = (-5 - 6) + [2 - (-2)]i = -11 + 4i$

or $(-5 + 2i) - (6 - 2i) = -5 + 2i - 6 + 2i = -11 + 4i$

b. $-2 - (-9 - 4i) = [-2 - (-9)] + [0 - (-4)]i = 7 + 4i$

or $-2 - (-9 - 4i) = -2 + 9 + 4i = 7 + 4i$

c. $(7 - i) - (6 - 10i) - (-4) = [7 - 6 - (-4)] + [(-1) - (-10)]i = 5 + 9i$

or $(7 - i) - (6 - 10i) - (-4) = 7 - i - 6 + 10i + 4 = 5 + 9i$

When complex numbers are written in radical form (with a negative radicand), we must rewrite them in the form $a + bi$ before we perform any operation on them.

EXAMPLE 3 Simplify $(5 + \sqrt{-4}) - (-7 - \sqrt{-9}) + \sqrt{-16}$.

SOLUTION

$$\begin{aligned}(5 + \sqrt{-4}) - (-7 - \sqrt{-9}) + \sqrt{-16} &= (5 + 2i) - (-7 - 3i) + (4i)\\ &= [5 - (-7)] + [2 - (-3) + 4]i\\ &= 12 + 9i\end{aligned}$$

A Word of Caution In Example 3, remember that $\sqrt{-4} \neq -\sqrt{4}$ and that $-\sqrt{-9} \neq +\sqrt{+9}$.

Exercises 7.13
Set I

In Exercises 1–10, perform the indicated operations; write the answers in the form $a + bi$.

1. $(4 + 3i) + (5 - i)$ **2.** $(6 - 2i) + (-3 + 5i)$

3. $(7 - 4i) - (5 + 2i)$ **4.** $(8 - 3i) - (4 + i)$

5. $(2 + i) + (3i) - (2 - 4i)$

6. $(3 - i) + (2i) - (-3 + 5i)$

7. $(2 + 3i) - (x + yi)$

8. $(x - i) - (7 + yi)$

9. $(9 + \sqrt{-16}) + (2 + \sqrt{-25}) + (6 - \sqrt{-64})$

10. $(13 - \sqrt{-36}) - (10 - \sqrt{-49}) + (8 + \sqrt{-4})$

In Exercises 11–14, solve for x and y.

11. $(4 + 3i) - (5 - i) = (3x + 2yi) + (2x - 3yi)$

12. $(3 - 2i) - (x + i) = (2x - yi) - (x + 2yi)$

13. $(2 - 5i) - (5 + 3i) = (3x + 2yi) - (5x + 3yi)$

14. $(4x - 3yi) - (7x + 2yi) = (7 - 2i) - (3 + 4i)$

Writing Problems

1. Show, step by step, how the commutative, associative, and distributive properties are used when we write

$$(2 + 3i) + (-4 + 5i) = [2 + (-4)] + (3 + 5)i$$

Exercises 7.13 Set II

In Exercises 1–10, perform the indicated operations; write the answers in the form $a + bi$.

1. $(8 - 5i) + (7 + 2i)$
2. $(5 - 2i) - (4) - (3 + 4i)$
3. $(9 - 12i) - (7 + i)$
4. $(7i) - (2 + i) + (-3 - 5i)$
5. $(7 + 3i) - (2i) - (2 - 5i)$
6. $(8 + 3i) - (2 - i) + (-9i)$
7. $(3 - xi) - (y - 5i)$
8. $(3x - 4i) + (x - yi) - (3 + yi)$
9. $(3 + \sqrt{-9}) - (4 - \sqrt{-81}) + (2 - \sqrt{-1})$
10. $(5 - \sqrt{-16}) + (\sqrt{-4} - 5) - (3 - \sqrt{-25})$

In Exercises 11–14, solve for x and y.

11. $(5 - i) + (-3 + 2i) = (4x - 5yi) - (7x - 2yi)$
12. $5 - yi = x + 4i$
13. $(3 - 2i) - (x + yi) = (2x + 5i) - (4 + i)$
14. $(5 + 2i) - (3i) - (x - 3i) = (x + yi) - (3 + 2yi)$

7.14 Multiplying Complex Numbers

Multiplication is distributive over addition for the set of complex numbers, as it is for the set of real numbers. For this reason and because $i^2 = -1$, the definition of the multiplication of two complex numbers is

$$(a + bi)(c + di) = (ac - bd) + (ad + bc)i$$

However, instead of memorizing this definition, you can apply the following technique.

Multiplying two complex numbers

1. Multiply the complex numbers as you would multiply two binomials.
2. Replace i^2 with -1.
3. Collect and combine like terms, and write the result in the form $a + bi$.

EXAMPLE 1 Multiply $(4 + 3i)(-5 + 2i)$.

SOLUTION

$$\begin{aligned}(4 + 3i)(-5 + 2i) &= -20 + 8i - 15i + 6i^2 \\ &= -20 - 7i + 6(-1) && \text{Replacing } i^2 \text{ with } -1 \\ &= -20 - 7i - 6 \\ &= -26 - 7i && \text{The product is in the form } a + bi\end{aligned}$$

EXAMPLE 2 Find and simplify $(5 + 3i)^2$.

SOLUTION 1 We square $(5 + 3i)$ as we would square a binomial:

$$\begin{aligned}(5 + 3i)^2 &= 5^2 + 2(5)(3i) + (3i)^2\\ &= 25 + 30i + 9i^2\\ &= 25 + 30i + 9(-1)\\ &= 25 + 30i - 9\\ &= 16 + 30i\end{aligned}$$

SOLUTION 2 Using the FOIL method, we have

$$\begin{aligned}(5 + 3i)^2 &= (5 + 3i)(5 + 3i)\\ &= 5 \cdot 5 + 15i + 15i + (3i)(3i)\\ &= 25 + 30i + 9i^2\\ &= 25 + 30i + 9(-1)\\ &= 25 + 30i - 9\\ &= 16 + 30i\end{aligned}$$

As we mentioned earlier, if a and b are both negative, then $\sqrt{a}\sqrt{b} \neq \sqrt{ab}$. Both factors must be expressed in the form bi before we can multiply them. Before we can multiply $\sqrt{-25}\sqrt{-9}$, for example, we must write $\sqrt{-25}$ as $5i$ and $\sqrt{-9}$ as $3i$; this gives

$$\sqrt{-25}\sqrt{-9} = (5i)(3i) = 15i^2 = 15(-1) = -15$$

Notice that $\sqrt{(-25)(-9)} = \sqrt{225} = 15$, *not* -15. Therefore, $\sqrt{-25}\sqrt{-9} \neq \sqrt{(-25)(-9)}$.

EXAMPLE 3 Multiply $(2 - \sqrt{-25})(-3 + \sqrt{-9})$.

SOLUTION We first convert the numbers to the form $a + bi$.

$$\begin{aligned}(2 - \sqrt{-25})(-3 + \sqrt{-9}) &= (2 - 5i)(-3 + 3i)\\ &= -6 + 21i - 15i^2\\ &= -6 + 21i - 15(-1)\\ &= -6 + 21i + 15\\ &= 9 + 21i\end{aligned}$$

Simplifying Powers of i

Let's consider a few powers of i:

$i = i$ — Any number is equal to itself (the reflexive property of equality)

$i^2 = -1$ — By definition

$i^3 = i^2 \cdot i = (-1)i = -i$

$i^4 = i^2 \cdot i^2 = (-1)(-1) = 1$

Also, $i^0 = 1$ — By definition

In fact, any power of i can and should be rewritten as one of the following: 1, -1, i, or $-i$.

> An algebraic expression is not considered to be in *simplest form* if it contains any powers of i greater than 1.

We can find powers of i as follows:

Finding powers of i

1. Divide the exponent of i by 4, identifying the *quotient* and the *remainder.*
2. Write the power of i as $(i^4)^{(\text{quotient})} \cdot i^{(\text{remainder})}$.
3. The factor that is $(i^4)^{(\text{quotient})}$ can be written as $(1)^{(\text{quotient})}$ (because $i^4 = 1$), and then $(1)^{\text{quotient})} = 1$ (because any power of 1 is 1). The factor that is $i^{(\text{remainder})}$ must be simplified.

For example, consider i^{75}. If we divide 75 by 4, the *quotient* is 18 and the *remainder* is 3. Therefore, we can write

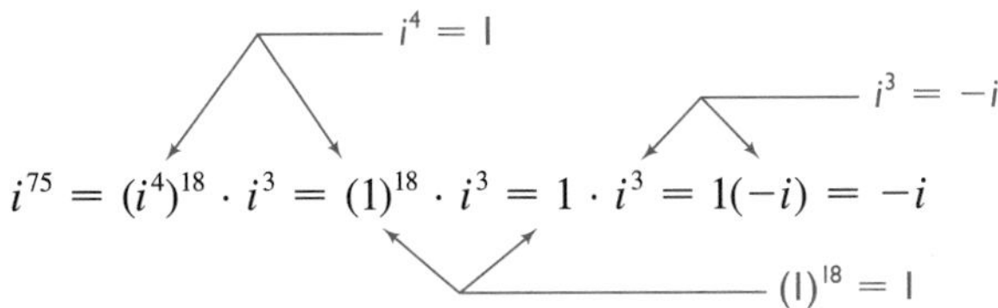

$$i^{75} = (i^4)^{18} \cdot i^3 = (1)^{18} \cdot i^3 = 1 \cdot i^3 = 1(-i) = -i$$

EXAMPLE 4 Examples of simplifying powers of i:

a. $i^{13} = (i^4)^3 \cdot i^1 = (1)^3 \cdot i = 1 \cdot i = i$ — If we divide 13 by 4, the quotient is 3 and the remainder is 1

b. $i^{50} = (i^4)^{12} \cdot i^2 = (1)^{12} \cdot i^2 = 1(-1) = -1$ — If we divide 50 by 4, the quotient is 12 and the remainder is 2

c. $i^{100} = (i^4)^{25} \cdot i^0 = (1)^{25} \cdot 1 = 1 \cdot 1 = 1$ — If we divide 100 by 4, the quotient is 25 and the remainder is 0

Note It is not necessary to show the step in which we write $(i^4)^n$ as $(1)^n$, as we have done in Example 4.

In Example 5, we raise the product $2i$ to a power. In such a problem, we use the rule for raising a product to a power, raising the coefficient of i and also i itself to the power.

EXAMPLE 5 Simplify $(2i)^7$.

SOLUTION $(2i)^7 = 2^7 i^7 = 128(i^4)^1(i^3) = 128(i^3) = 128(-i) = -128i$

★ **EXAMPLE 6** Simplify $(1 + i)^6$.

SOLUTION We will use the binomial theorem. The coefficients are 1, 6, 15, 20, 15, 6, and 1. Therefore,

$$\begin{aligned}(1 + i)^6 &= 1(1)^6 + 6(1)^5 i + 15(1)^4 i^2 + 20(1)^3 i^3 + 15(1)^2 i^4 + 6(1)i^5 + 1i^6 \\ &= 1 + 6i + 15(-1) + 20(-i) + 15(1) + 6i + (-1) \\ &= 1 + 6i - 15 - 20i + 15 + 6i - 1 \\ &= 0 - 8i, \quad \text{or} \quad -8i\end{aligned}$$

Exercises 7.14
Set I

In Exercises 1–36, perform the indicated operations. Express each answer in the form $a + bi$.

1. $(1 + i)(1 - i)$
2. $(3 + 2i)(3 - 2i)$
3. $(4 - i)(3 + 2i)$
4. $(5 + 2i)(2 - 3i)$
5. $(6 - 2i)(2 - 3i)$
6. $(4 + 7i)(3 + 2i)$
7. $(\sqrt{5} + 2i)(\sqrt{5} - 2i)$
8. $(\sqrt{7} - 3i)(\sqrt{7} + 3i)$
9. $5i(i - 2)$
10. $6i(2i - 1)$
11. $(2 + 5i)^2$
12. $(3 - 4i)^2$
13. i^{10}
14. i^{23}
15. i^{87}
16. i^{73}
17. $(3i)^3$
18. $(2i)^3$
19. $(2i)^4$
20. $(3i)^4$
21. $(3 - \sqrt{-4})(4 + \sqrt{-25})$
22. $(5 + \sqrt{-64})(2 - \sqrt{-36})$
23. $(2 - \sqrt{-1})^2$
24. $(3 + \sqrt{-1})^2$
25. $[3 + i^6]^2$
26. $[4 - i^{10}]^2$
27. $i^{10}(i^{23})$
28. $i^{34}(i^{16})$
29. $i^{15} + i^7$
30. $i^{27} + i^{14}$
31. $[2 + (-i)^{11}]^2$
32. $[3 - (-i)^5]^2$
33. ★ $(1 - i)^5$
34. ★ $(1 + i)^4$
35. ★ $(2 - i)^4$
36. ★ $(4 - i)^3$
37. Show that $2i$ is a solution of the equation $x^2 = -4$.
38. Show that $-3i$ is a solution of the equation $x^2 = -9$.
39. Show that $2 + i\sqrt{3}$ is a solution of the equation $x^2 - 4x + 7 = 0$.
40. Show that $3 - i\sqrt{5}$ is a solution of the equation $x^2 - 6x + 14 = 0$.

Writing Problems

1. Make up an example different from the one in the text, using numbers, to show that $\sqrt{a}\sqrt{b} \neq \sqrt{ab}$ if a and b are both negative.

Exercises 7.14
Set II

In Exercises 1–36, perform the indicated operations. Express each answer in the form $a + bi$.

1. $(8 + 5i)(3 - 4i)$
2. $(7 - i)(3 + 2i)$
3. $(6 - 3i)(4 + 5i)$
4. $(8 - i)(7 + 2i)$
5. $(7 + 2i)(1 - i)$
6. $(1 + 3i)(1 - 3i)$
7. $(4 - 2i)(4 + 2i)$
8. $(1 + i)(1 + i)$
9. $8i(2 - 3i)$
10. $-4i(3 - 2i)$
11. $(-4 + 3i)^2$
12. $(-7 - 2i)^2$
13. i^{13}
14. i^{83}
15. i^{91}
16. i^{74}
17. $(2i)^5$
18. $(-3i)^3$
19. $(-5i)^2$
20. $(-3i)^4$
21. $(5 + \sqrt{-36})(3 - \sqrt{-100})$
22. $(2 - \sqrt{-49})(4 + \sqrt{-81})$
23. $(4 + \sqrt{-1})^2$
24. $(6 - \sqrt{-25})^2$
25. $(3 - i^{13})^2$
26. $(3 + [-i]^7)^2$
27. $i^{24}(i^{18})$
28. $i^{37}(i^{23})$
29. $i^{14} + i^{25}$
30. $i^{25} - i^{17}$
31. $[8 - (-i)^{13}]^2$
32. $[6 + 5(-i)^7]^2$
33. ★ $(1 + i)^5$
34. ★ $(1 + 2i)^4$
35. ★ $(3 - i)^3$
36. ★ $(1 - i)^6$
37. Show that $-4i$ is a solution of the equation $x^2 = -16$.
38. Show that $5i$ is a solution of the equation $x^2 + 25 = 0$.
39. Show that $4 - i\sqrt{5}$ is a solution of the equation $x^2 - 8x + 21 = 0$.
40. Show that $-3 + 2i\sqrt{5}$ is a solution of the equation $x^2 + 6x + 29 = 0$.

7.15 Dividing Complex Numbers

Recall that the conjugate of $\sqrt{a} + \sqrt{b}$ is $\sqrt{a} - \sqrt{b}$. Similarly, the conjugate of a complex number is found by changing the sign of its imaginary part. That is,

the conjugate of $a + bi$ is $a - bi$

and the conjugate of $a - bi$ is $a + bi$

EXAMPLE 1 Examples of the conjugates of complex numbers:

a. The conjugate of $3 - 2i$ is $3 + 2i$.
b. The conjugate of $-5 + 4i$ is $-5 - 4i$.
c. The conjugate of $7i$ is $-7i$, because $7i = 0 + 7i$ with conjugate $0 - 7i = -7i$.
d. The conjugate of 5 is 5, because $5 = 5 + 0i$ with conjugate $5 - 0i = 5$.

> The product of a complex number and its conjugate is always a positive real number.

PROOF $(a + bi)(a - bi) = a^2 - b^2i^2 = a^2 + b^2$ A positive real number ■

In order to divide one complex number by another, we must convert the expression to the form $\frac{a}{c} + \frac{b}{c}i$; in other words, we must make the divisor (or denominator) a *real* number. Because the product of a complex number and its conjugate is a real number, we use the following procedure for dividing complex numbers:

Dividing one complex number by another

1. Write the division in fraction form if it's not already in that form.
2. Multiply both numerator and denominator by the conjugate of the denominator:

$$\frac{(a + bi)\,(c - di)}{(c + di)\,(c - di)}$$

3. Simplify, and write the result in the form $a + bi$ or $\frac{a}{c} + \frac{b}{c}i$.

This procedure is similar to the procedure for dividing radicals when the divisor (or denominator) contains radicals and has two terms. It is natural that the two procedures should be similar, since $i = \sqrt{-1}$.

EXAMPLE 2 Examples of dividing complex numbers:

a. $$\frac{10i}{1 - 3i} = \frac{(10i)\,(1 + 3i)}{(1 - 3i)\,(1 + 3i)} = \frac{10i + 30i^2}{1 - 9i^2} = \frac{10i - 30}{1 + 9} = \frac{-30 + 10i}{10}$$

$$= \frac{10(-3 + i)}{10} = -3 + i$$

The denominator is now a real number

b. $(2+i) \div (3-2i) = \dfrac{2+i}{3-2i} = \dfrac{(2+i)(3+2i)}{(3-2i)(3+2i)}$

$= \dfrac{6+7i+2i^2}{9-4i^2} = \dfrac{6+7i-2}{9+4} = \dfrac{4+7i}{13}$

$= \frac{4}{13} + \frac{7}{13}i$ The quotient is in the form $\frac{a}{c} + \frac{b}{c}i$

c. $(3+i) \div i = \dfrac{(3+i)(-i)}{i\ (-i)} = \dfrac{-3i-i^2}{-i^2} = \dfrac{-3i+1}{1} = 1-3i$

The conjugate of i is $-i$, because $i = 0 + i$ and its conjugate is $0 - i = -i$

d. $5 \div (-2i) = \dfrac{5}{-2i} = \dfrac{5 \cdot i}{-2i \cdot i} = \dfrac{5i}{-2i^2} = \dfrac{5i}{2} = \dfrac{5}{2}i = 0 + \dfrac{5}{2}i$

Even though the conjugate of $-2i$ is $2i$, we were able to make the denominator a real number by multiplying by i instead of $2i$.

Exercises 7.15 Set I

In Exercises 1–6, write the conjugate of each complex number.

1. $3 - 2i$
2. $5 + 4i$
3. $5i$
4. $-7i$
5. 10
6. -8

In Exercises 7–18, perform the indicated operation, and write each answer in the form $a + bi$.

7. $\dfrac{10}{1+3i}$
8. $\dfrac{5}{1+2i}$
9. $\dfrac{1+i}{1-i}$
10. $\dfrac{1-i}{1+i}$
11. $\dfrac{8+i}{i}$
12. $\dfrac{4-i}{i}$
13. $\dfrac{3}{2i}$
14. $\dfrac{4}{5i}$
15. $\dfrac{15i}{1-2i}$
16. $\dfrac{20i}{1-3i}$
17. $\dfrac{4+3i}{2-i}$
18. $\dfrac{3+2i}{4+2i}$

Writing Problems

Express the answers in your own words and in complete sentences.

1. Describe what your first step would be in performing the division $\dfrac{3-4i}{6+i}$, and explain why you would do that first.
2. Describe what your first step would be in performing the division $\dfrac{2+3i}{i}$, and explain why you would do that first.

Exercises 7.15 Set II

In Exercises 1–6, write the conjugate of each complex number.

1. $5 - 3i$
2. $-2 + 3i$
3. $-3i$
4. 0
5. 7
6. $3i$

In Exercises 7–18, perform the indicated operation, and write each answer in the form $a + bi$.

7. $\dfrac{8}{2-3i}$
8. $\dfrac{2+6i}{2+i}$
9. $\dfrac{3+i}{3-i}$
10. $\dfrac{8+i}{6i}$
11. $\dfrac{4+2i}{3i}$
12. $\dfrac{5+3i}{5-3i}$
13. $\dfrac{9}{4i}$
14. $\dfrac{8-3i}{8+3i}$
15. $\dfrac{29i}{2+5i}$
16. $\dfrac{3}{7i}$
17. $\dfrac{2i+5}{3+2i}$
18. $\dfrac{6i-1}{6i+1}$

Sections 7.10–7.15 REVIEW

Solving a Radical Equation 7.10

1. Find the domain of the variable.
2. Arrange the terms so that one term with a radical is by itself on one side of the equal sign.
3. Raise each side of the equation to a power equal to the index of the radical.
4. Simplify each side of the equation.

   If a radical still remains, repeat steps 2, 3, and 4.
5. Solve the resulting equation.
6. Check apparent solutions in the original equation.

The Pythagorean Theorem 7.11

The square of the hypotenuse of a right triangle is equal to the sum of the squares of the two legs. If c represents the length of the hypotenuse and a and b represent the lengths of the legs, then

$$c^2 = a^2 + b^2$$

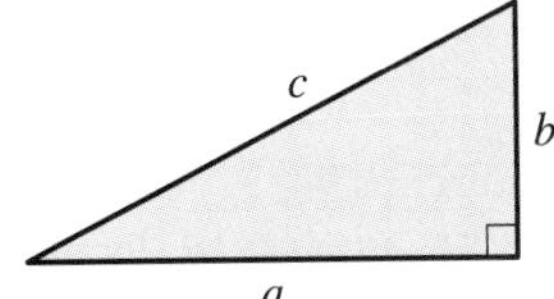

Solving Equations of the Form $x^2 = p$ 7.11

The solutions of an equation that can be written in the form $x^2 = p$ are $\sqrt{p}$ and $-\sqrt{p}$.

Complex Numbers 7.12

A **complex number** is a number that can be expressed in the form $a + bi$, where a and b are real numbers and, by definition, $i^2 = -1$, so $i = \sqrt{-1}$.

The set of complex numbers is $C = \{a + bi \mid a, b \in R, i = \sqrt{-1}\}$.

The real part → a + bi ← The imaginary part

where $i^2 = -1$, or $i = \sqrt{-1}$.

When the real part is zero, the complex number is a *pure imaginary number.*

When the imaginary part is zero, the complex number is a *real number.*

Equality of Complex Numbers 7.12

If $a + bi = c + di$, then $a = c$ (the real parts are equal) and $b = d$ (the coefficients of i are equal).

Adding and Subtracting Complex Numbers 7.13

$$(a + bi) + (c + di) = (a + c) + (b + d)i$$

$$(a + bi) - (c + di) = (a - c) + (b - d)i$$

Multiplying Two Complex Numbers 7.14

1. Multiply the complex numbers as you would multiply two binomials.
2. Replace i^2 with -1.
3. Collect and combine like terms, and write the result in the form $a + bi$.

Simplifying Powers of i 7.14

Any power of i should be rewritten as one of the following: 1, -1, i, or $-i$.

Dividing Complex Numbers 7.15

1. Write the division in fraction form if it's not already in that form.
2. Multiply both numerator and denominator by the conjugate of the denominator:

$$\frac{(a + bi)\,(c - di)}{(c + di)\,(c - di)}$$

3. Simplify, and write the result in the form $a + bi$ or $\dfrac{a}{c} + \dfrac{b}{c}i$.

Sections 7.10–7.15 REVIEW EXERCISES Set I

In Exercises 1–6, find the solution set of each equation.

1. $\sqrt{x-5} = \sqrt{3x+8}$
2. $\sqrt{3x-5} + 3 = x$
3. $\sqrt{5x-4} - \sqrt{2x+1} = 1$
4. $\sqrt[5]{5x+4} = -2$
5. $x^{5/6} = 32$
6. $x^{3/4} = 27$

In Exercises 7 and 8, solve for x and y.

7. $5 - yi = x + 6i$
8. $(2 - 5i) - (x + yi) = (x - 7i) + (4 - 2yi)$

In Exercises 9–15, perform the indicated operations, and write the answers in the form $a + bi$.

9. $(3i + 2)(4 - 2i)$
★10. $(3 - i)^4$
11. $\dfrac{2+i}{1+3i}$
12. $\dfrac{3+i}{1-2i}$
13. i^{82}
14. $(3 + \sqrt{-1})(5 - \sqrt{-121})$
15. $(4 + \sqrt{-27}) + (2 - \sqrt{-12}) - (1 - \sqrt{-3})$

In Exercises 16–20, set up each problem algebraically, solve, and check.

16. The length of a rectangle is 2 ft more than its width. Its diagonal is $\sqrt{34}$ ft. Find the dimensions of the rectangle.
17. The width of a rectangle is 3 m less than its length. Its diagonal is $\sqrt{45}$ m. Find the dimensions of the rectangle.
18. Using the formula $\sigma = \sqrt{npq}$, find p, the probability of success, if σ, the standard deviation, is $2\sqrt{6}$, q, the probability of failure, is $\frac{2}{5}$, and n, the number of trials, is 100.
19. Two cars left a desert intersection at the same time, one traveling east and the other traveling south. Twenty minutes later, the cars were 25 mi apart. The car traveling south was going 15 mph faster than the other car. What was the speed of the faster car?
20. The formula for the length of a diagonal of a rectangular box is $d = \sqrt{l^2 + w^2 + h^2}$, where d represents the length of the diagonal and where l is the length, w is the width, and h is the height of the box. Find the width of the box if the length of the box is 12 cm, the height is 5 cm, and the length of a diagonal is $\sqrt{233}$ cm.

REVIEW EXERCISES Set II

Name ____________________

ANSWERS

1. ____________
2. ____________
3. ____________
4. ____________
5. ____________
6. ____________
7. ____________
8. ____________
9. ____________
10. ____________
11. ____________

In Exercises 1–6, find the solution set of each equation.

1. $x = \sqrt{3x + 10}$

2. $\sqrt[3]{4x - 1} = -2$

3. $\sqrt{3x + 7} - \sqrt{x - 2} = 3$

4. $x^{3/5} = -8$

5. $x = 2\sqrt{x + 6} - 3$

6. $x^{1/3} = -2$

In Exercises 7 and 8, solve for x and y.

7. $3x - 3i = 8 - yi$

8. $(4 - 3i) - (x + 2yi) = (x + yi) - (4 - 5i)$

In Exercises 9–15, perform the indicated operations, and write the answers in the form $a + bi$.

9. $(8 + 7i) - (3 - 2i) + (\sqrt{2} - 3i)$

10. $4i(6 - 2i)$

★11. $(2 + i)^5$

12. $(3 - \sqrt{-50}) - (1 + \sqrt{-18}) + \sqrt{-32}$

12. ______________

13. ______________

14. ______________

13. $\sqrt{-16}\sqrt{-4}$

14. i^{53}

15. $\dfrac{8 + 5i}{8 - 5i}$

15. ______________

16. ______________

In Exercises 16–20, set up each problem algebraically, solve, and check.

17. ______________

16. The length of a rectangle is 7 ft more than its width. If the length of its diagonal is 13 ft, find the dimensions of the rectangle.

18. ______________

19. ______________

20. ______________

17. The longer leg of a right triangle is 3 in. longer than the shorter leg, and the hypotenuse is 6 in. longer than the shorter leg. Find the lengths of the three sides.

18. Using the formula $I = \sqrt{\dfrac{P}{R}}$, find P, the amount of power consumed, if an appliance has a resistance, R, of 36 ohms and draws a current, I, of 3 amps.

19. Eduardo and Miguel left a certain point at the same time, bicycling along paths that were at right angles to each other. Miguel, who was going downhill, was bicycling 6 mph faster than Eduardo. Two hours later, they were 60 mi apart. How fast was Eduardo bicycling?

20. Market research indicates that the *demand equation* for a certain product is $p = 20 - \sqrt{0.01x + 3}$, $x \geq 0$, where x is the number of units demanded per day and p is the price per unit. Use the given demand equation and a calculator to find the demand, x, if $p = \$16.50$.

Chapter 7 DIAGNOSTIC TEST

The purpose of this test is to see how well you understand operations involving radicals, rational exponents, and complex numbers. We recommend that you work this diagnostic test *before* your instructor tests you on this chapter. Allow yourself about 50 minutes.

Complete solutions for all the problems on this test, together with section references, are given in the answer section at the end of the book. For the problems you do incorrectly, we suggest that you study the sections cited.

In Problems 1–5, perform the indicated operations. Express the answers in exponential form with positive exponents. Assume that the variables represent positive numbers.

1. $x^{1/2}x^{-1/4}$
2. $(R^{-4/3})^3$
3. $\dfrac{a^{5/6}}{a^{1/3}}$
4. $\left(\dfrac{x^{-2/3}y^{3/5}}{x^{1/3}y}\right)^{-5/2}$
5. $\dfrac{b^{2/3}}{b^{-1/5}}$

In Problems 6–10, write each expression in simplest radical form.

6. $\sqrt[3]{54x^6y^7}$
7. $\dfrac{4xy}{\sqrt{2x}}$, $x > 0$
8. $\sqrt[6]{a^3}$
9. $\sqrt{40} + \sqrt{9}$
10. $\sqrt{x}\sqrt[3]{x}$, $x \geq 0$
11. Evaluate the expression $(-27)^{2/3}$.

In Problems 12–18, perform the indicated operations. Express the answers in simplest form. Assume that $x > 0$ and $y \geq 0$.

12. $4\sqrt{8y} + 3\sqrt{32y}$
13. $3\sqrt{\dfrac{5x^2}{2}} - 5\sqrt{\dfrac{x^2}{10}}$
14. $\sqrt{2x^4}\sqrt{8x^3}$
15. $\sqrt{2x}(\sqrt{8x} - 5\sqrt{2})$
16. $\dfrac{\sqrt{10x} + \sqrt{5x}}{\sqrt{5x}}$
17. $\dfrac{5}{\sqrt{7} + \sqrt{2}}$

★ 18. $(1 - \sqrt[3]{x})^3$

In Problems 19–22, perform the indicated operations; express the answers in the form $a + bi$.

19. $(5 - \sqrt{-8}) - (3 - \sqrt{-18})$
20. $(3 + i)(2 - 5i)$
21. $\dfrac{10}{1 - 3i}$

★ 22. $(2 - i)^3$

23. Find the solution set of $x^{3/2} = 8$.
24. Find the solution set of $\sqrt{x - 3} + 5 = x$.
25. Set up the following problem algebraically, solve, and check:

    The longer leg of a right triangle is $\sqrt{12}$ cm long, and the hypotenuse is 2 cm longer than the shorter leg. Find the length of the shorter leg.

Chapters 1–7 CUMULATIVE REVIEW EXERCISES

In Exercises 1–6, perform the indicated operations and simplify the results. Assume that any values of the variables that make a denominator zero are excluded.

1. $\dfrac{6}{a^2 - 9} - \dfrac{2}{a^2 - 4a + 3}$
2. $\dfrac{x^3 - 8}{2x^2 + x - 10} \div \dfrac{3x^3 - x^2}{6x^2 + 13x - 5}$
3. $\dfrac{\dfrac{4}{x} - \dfrac{8}{x^2}}{\dfrac{1}{x} - \dfrac{2}{x^2}}$
4. $\sqrt{5x}(\sqrt{5x} + 2)$, $x \geq 0$
5. $(\sqrt{26} - \sqrt{10})^2$
6. $\dfrac{6}{2 - \sqrt{5}}$

In Exercises 7–11, find the solution set of each equation or inequality.

7. $3(4x - 1) - (3 - 2x) = 4(2x - 3)$
8. $\dfrac{3x}{4} - \dfrac{5}{6} \leq \dfrac{7x}{12}$
9. $\sqrt{2x + 2} = 1 + \sqrt{3x - 12}$
10. $\dfrac{3}{x + 2} - \dfrac{11}{2x + 4} = \dfrac{5}{2}$
11. $6x^2 + 11x = 10$
12. Solve for a in terms of the other variables: $\dfrac{1}{a} + \dfrac{1}{b} = \dfrac{1}{c}$
13. Find the solution set of $|2x - 5| > 3$.

In Exercises 14–18, set up each problem algebraically, solve, and check.

14. Virginia can do a job in 5 hr. Adolph can do the same job in 6 hr. How long will the job take if Virginia and Adolph work together?
15. The area of a rectangle is 54 sq. cm. Find its dimensions if its sides are in the ratio of 2 to 3.
16. Min invested $26,500, placing some of the money in an account that earned 7.6% interest annually and the rest of the money in an account that earned 7.3% annually. If she earned $1,977.10 in the first year, how much did she invest at each rate?
17. A boat's engine is running so that the boat would go 27 mph in still water. It can go 120 mi upstream (against the current) in the same amount of time that it can go 150 mi downstream. What is the speed of the current?
18. Amir's living room is square. If the room were 2 yd longer and 1 yd narrower, the cost of carpeting it with carpeting that costs $35 per square yard would be $3,080. What are the dimensions of his living room?

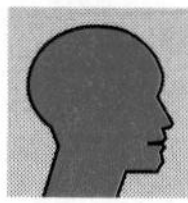

Critical Thinking and Problem-Solving Exercises

1. Predict what the next three symbols should be for the following list.

   A, 1, D, 3, G, 5, J, __, __, __,

2. Loreen is thinking of a number. Guess what that number is by using the following clues.

   The number is a prime number.

   The number is between 20 and 70.

   The sum of the digits of the number is 10.

3. Roger, David, and John are traveling sales associates who share an office. John is in the office every sixth day, David is in the office every eighth day, and Roger is in the office every tenth day.

   a. How often are Roger and David in the office on the same day?

   b. How often are Roger and John in the office on the same day?

   c. How often are David and John in the office on the same day?

   d. How often are all three men in the office on the same day?

4. Jan, Merv, Bert, and Kuen met for lunch. Each was in a different type of business, and each ordered a different lunch. The items ordered were quiche, a Cobb salad, a taco salad, and French onion soup. Using the figure and the following clues, determine who sat in which chair, what business each person was in, and what each person ordered for lunch.

   The accountant was not sitting in seat 3 or 4.

   The pharmacist sat opposite the lawyer, who ordered a taco salad.

   Jan sat in seat 4.

   Kuen sat in an odd-numbered seat, opposite the person who ordered quiche.

   The pharmacist sat in seat 2 and ate French onion soup.

   Merv is an accountant.

   One person is a physician.

Figure for Exercise 4

5. As you know, we rationalize the denominator of an expression such as $\dfrac{d}{\sqrt{a}+\sqrt{b}}$ by multiplying both numerator and denominator by $\sqrt{a}-\sqrt{b}$, the conjugate of the denominator. This procedure forces the denominator to become a rational number because $(\sqrt{a}+\sqrt{b})(\sqrt{a}-\sqrt{b}) = (\sqrt{a})^2 - (\sqrt{b})^2$, or $a - b$, a rational number.

   Now consider an expression such as $\dfrac{d}{\sqrt[3]{a}+\sqrt[3]{b}}$. What is the result of multiplying both numerator and denominator by $\sqrt[3]{a}-\sqrt[3]{b}$? Do you see that this procedure will *not* rationalize the denominator? Can you find the correct expression to multiply both numerator and denominator by so that the new denominator will be a rational number? (*Hint*: The expression $(\sqrt[3]{a})^3 + (\sqrt[3]{b})^3$ is a rational number, since it equals $a + b$.)

Graphing; Linear Equations and Inequalities

CHAPTER

8

Graphs are mathematical pictures that help us see relationships between variables. In this chapter, we discuss graphing ordered pairs and graphing the solution sets of equations and inequalities in the rectangular coordinate system. We also discuss mathematical relations, finding the slope of a straight line, and writing the equation of a straight line.

8.1 The Rectangular Coordinate System; Mathematical Relations

The Rectangular Coordinate System

The **rectangular coordinate system** in the plane* usually consists of a horizontal number line, called the **horizontal axis** or ***x*-axis,** and a vertical number line, called the **vertical axis** or ***y*-axis.** These lines intersect at a point we call the **origin.** The units of measure on these two lines are usually equal, although they don't *have* to be equal.

The two axes divide the plane into four **quadrants.** Some of the terms commonly used with a rectangular coordinate system are shown in Figure 1.

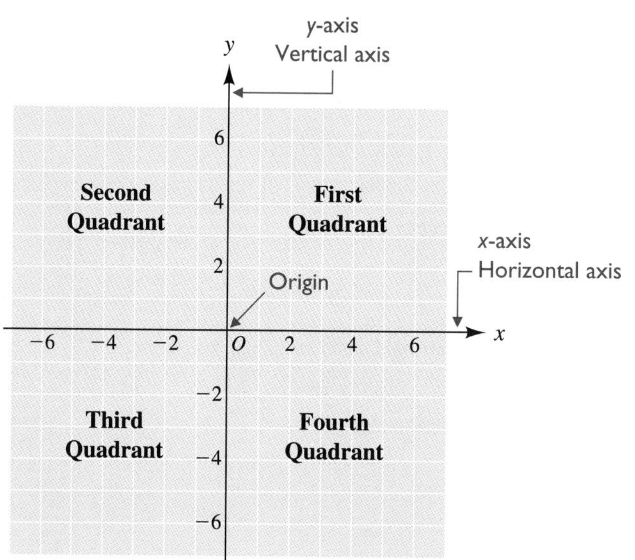

Figure 1 The rectangular coordinate system

Ordered Pairs

An **ordered pair** is a pair of numbers in which the *order* the two numbers are listed in makes a difference; to indicate that a pair of numbers is an *ordered pair,* we put a comma between the two numbers, and we enclose the numbers within *parentheses,* never braces or brackets. Thus, $(3, 2)$ represents an ordered pair; *because* the order makes a difference, $(2, 3)$ represents a *different* ordered pair—that is, $(3, 2) \neq (2, 3)$. In general, $(a, b) \neq (b, a)$. $\{5, 7\}$ and $[4, -1]$ do *not* represent ordered pairs. The *first* number of an ordered pair is called the ***x*-coordinate**, and the *second* number is called the ***y*-coordinate**. (Other names for the coordinates are given in Figure 2.)

*We assume that you recall from geometry that a *plane* is a flat surface.

The Graph of an Ordered Pair Whereas points on the real number line correspond to a *single* real number, points in the plane correspond to an ordered pair of real numbers. In fact, there is a one-to-one correspondence between the set of ordered pairs of real numbers and the set of points in the plane (that is, there is exactly one point in the plane that corresponds to each ordered pair of real numbers, and there is exactly one ordered pair of real numbers that corresponds to each point in the plane). The origin is the point that corresponds to the ordered pair $(0, 0)$.

The x-coordinate tells us how far the point is from the y-axis. A *positive* x-coordinate indicates that the point is to the *right* of the y-axis; a *negative* x-coordinate indicates that the point is to the *left* of the y-axis.

The y-coordinate tells us how far the point is from the x-axis. A *positive* y-coordinate indicates that the point is *above* the x-axis; a *negative* y-coordinate indicates that the point is *below* the x-axis.

The ordered pair $(3, 2)$ is graphed (or plotted) in Figure 2.

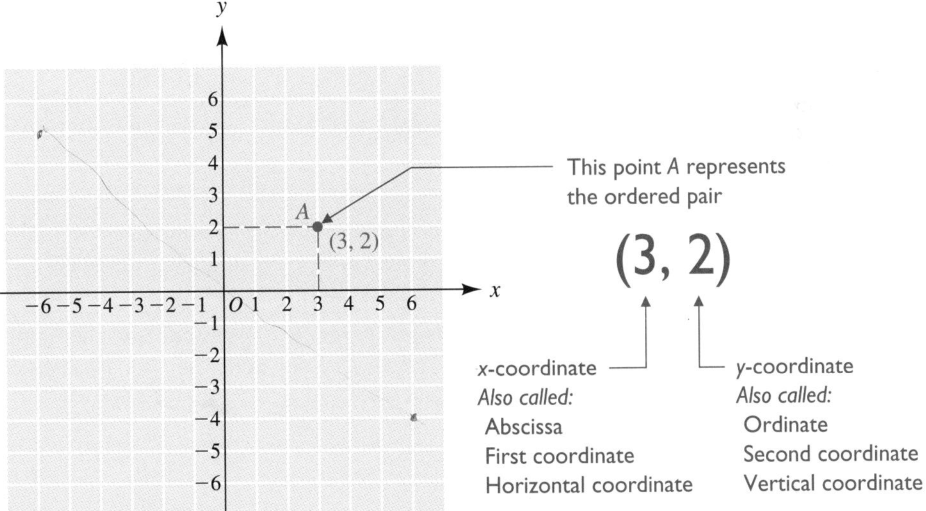

Figure 2 The graph of an ordered pair

EXAMPLE 1 Graph (plot) the points $A(1, 4)$, $B(4, 1)$, $C(-1, 3)$, $D(3, -1)$, and $E(0, 2)$.

SOLUTION

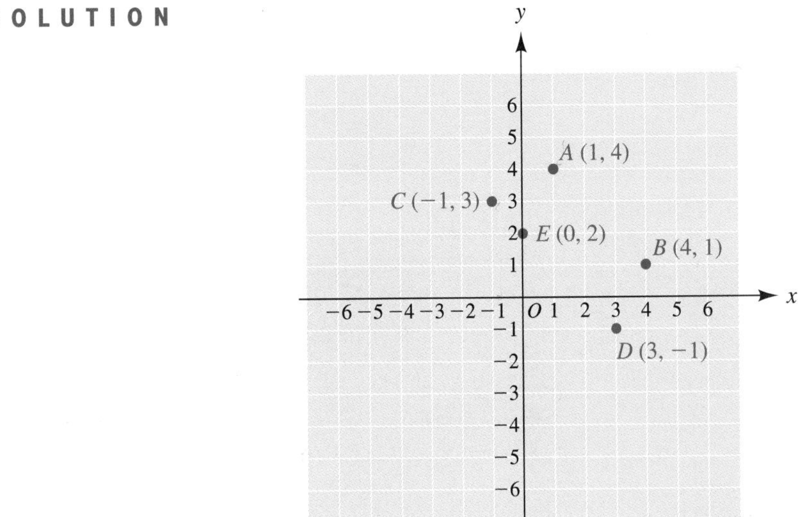

Note When the numbers in an ordered pair are interchanged, we get a different point. We see in the figure for Example 1 that points A and B are different from each other, as are points C and D.

Recall that we use *subscripts* to distinguish between different constants; recall, too, that x_1 is read "x sub one." We use subscripts when we're discussing two or more ordered pairs that could be anywhere in the plane—for example, $P_1(x_1, y_1)$ represents a fixed point (the *name* of the point is P_1, and the coordinates of that point are x_1 and y_1) that is different from the fixed point $P_2(x_2, y_2)$. When subscripts are used with points, y_1 is usually the y-value that corresponds to x_1, y_2 is usually the y-value that corresponds to x_2, y_3 is usually the y-value that corresponds to x_3, and so forth.

If two points have equal y-coordinates, they must lie on the same *horizontal* line (and if two points lie on the same horizontal line, they must have equal y-coordinates); therefore, $P_1(x_1, y_1)$ and $P_2(x_2, y_1)$ lie on the same horizontal line because both y-coordinates are y_1. Similarly, if two points have equal x-coordinates, they must lie on the same *vertical* line (and vice versa); therefore, $P_1(x_1, y_1)$ and $P_2(x_1, y_2)$ lie on the same vertical line because both x-coordinates are x_1.

The Distance Between Two Points

We use the notation $|P_1P_2|$ to represent the **distance** between points P_1 and P_2.

Finding the distance between points P_1 and P_2

1. If P_1 is the point (x_1, y_1) and P_2 is the point (x_2, y_1), the points lie on the same *horizontal* line and the distance between them is defined to be $|P_1P_2| = |x_2 - x_1|$.
2. If P_1 is the point (x_1, y_1) and P_2 is the point (x_1, y_2), the points lie on the same *vertical* line and the distance between them is defined to be $|P_1P_2| = |y_2 - y_1|$.
3. If P_1 is the point (x_1, y_1) and P_2 is the point (x_2, y_2), the points lie anywhere in the plane and the distance, d, between them is

$$d = |P_1P_2| = \sqrt{(x_2 - x_1)^2 + (y_2 - y_1)^2}$$

PROOF OF (3) Suppose $P_1(x_1, y_1)$ and $P_2(x_2, y_2)$ are two points that do not lie on the same horizontal or vertical line. Let's draw a horizontal line through $P_1(x_1, y_1)$ and a vertical line through $P_2(x_2, y_2)$. These two lines meet at a point we will call C (see Figure 3); C *must* be the point (x_2, y_1). The triangle formed by joining P_1, C, and P_2 is a right triangle with P_1P_2 as its hypotenuse.

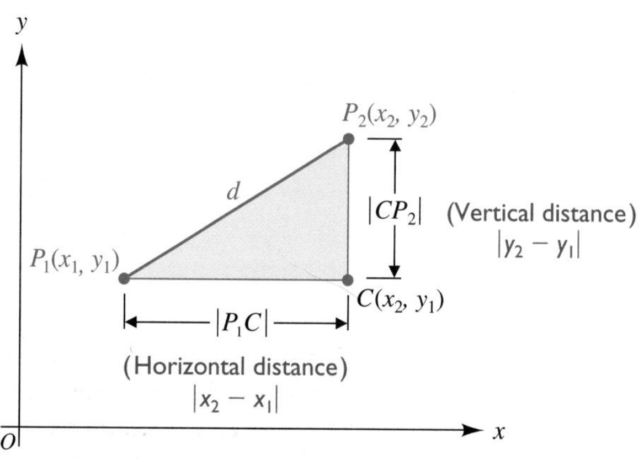

Figure 3

The horizontal distance $|P_1C| = |x_2 - x_1|$, and the vertical distance $|CP_2| = |y_2 - y_1|$. Because $|x_2 - x_1|^2 = (x_2 - x_1)^2$ and $|y_2 - y_1|^2 = (y_2 - y_1)^2$ (can you verify this?), using the Pythagorean theorem to find d gives

$$(d)^2 = (|x_2 - x_1|)^2 + (|y_2 - y_1|)^2$$
$$= (x_2 - x_1)^2 + (y_2 - y_1)^2$$

Solving for d, we have

$$d = \pm\sqrt{(x_2 - x_1)^2 + (y_2 - y_1)^2}$$

However, since d represents the length of a line segment, we reject the negative answer. Therefore,

$$d = \sqrt{(x_2 - x_1)^2 + (y_2 - y_1)^2}$$

■

Note This formula *can* be used even if the points lie on the same vertical line or on the same horizontal line.

EXAMPLE 2 Find the distance between the points $(-6, 5)$ and $(6, -4)$.

SOLUTION Let $P_1 = (-6, 5)$ and $P_2 = (6, -4)$. Then

$$d = \sqrt{(x_2 - x_1)^2 + (y_2 - y_1)^2}$$
$$= \sqrt{[6 - (-6)]^2 + [-4 - 5]^2}$$
$$= \sqrt{(12)^2 + (-9)^2} = \sqrt{144 + 81}$$
$$= \sqrt{225} = 15$$

The distance is not changed if the points P_1 and P_2 are interchanged.

Let $P_1 = (6, -4)$ and $P_2 = (-6, 5)$. Then

$$d = \sqrt{[-6 - 6]^2 + [5 - (-4)]^2}$$
$$= \sqrt{(-12)^2 + (9)^2} = \sqrt{144 + 81}$$
$$= \sqrt{225} = 15$$

Mathematical Relations

A **mathematical relation** is a set of ordered pairs.

EXAMPLE 3 Examples of relations:

$\mathcal{R}_1$ represents the relation
↓
a. $\mathcal{R}_1 = \{(-1, 2), (3, -4), (0, 5), (4, 3)\}$
b. $\mathcal{R}_2 = \{(3, 5), (-5, 2), (-5, -4), (0, -3), (4, -6)\}$

The **domain** *of a relation* is the set of all the first coordinates of the ordered pairs of that relation. We represent the domain of the relation $\mathcal{R}$ by the symbol $D_{\mathcal{R}}$.

The **range** *of a relation* is the set of all the second coordinates of the ordered pairs of that relation. We represent the range of the relation $\mathcal{R}$ by the symbol $R_{\mathcal{R}}$.

The **graph** *of a relation* is the graph of all the ordered pairs of that relation.

EXAMPLE 4 Find the domain, range, and graph of the relation

$$\mathcal{R} = \{(-1, 2), (3, -4), (0, 5), (4, 3)\}$$

SOLUTION

The domain of $\{(\,-1\,, 2), (\,3\,, -4), (\,0\,, 5), (\,4\,, 3)\}$ is $\{-1, 3, 0, 4\} = D_{\mathcal{R}}$.

The range of $\{(-1,\,2\,), (3,\,-4\,), (0,\,5\,), (4,\,3\,)\}$ is $\{2, -4, 5, 3\} = R_{\mathcal{R}}$.

The graph of $\{(-1, 2), (3, -4), (0, 5), (4, 3)\}$ is shown below.

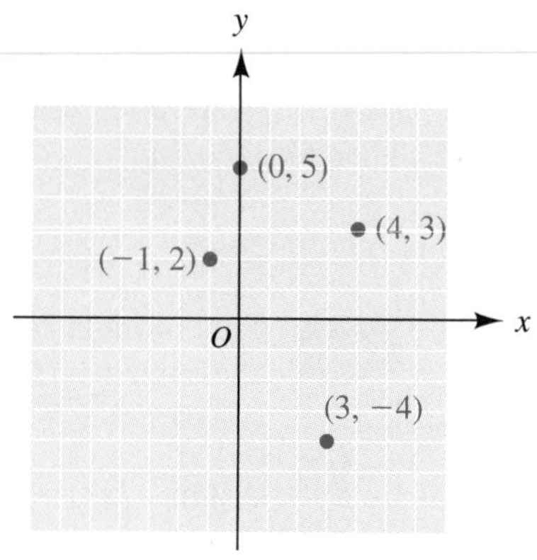

Exercises 8.1
Set I

For Exercises 1 and 2, use the following graph.

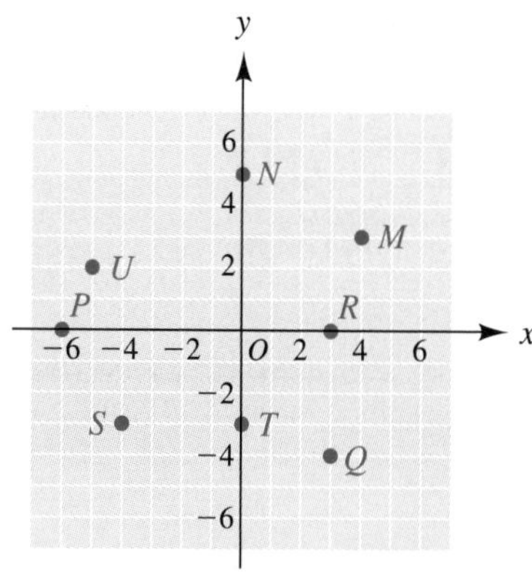

1. Find the ordered pair that corresponds to each of the following points as shown on the graph.

   a. R b. N c. U d. S

2. Find the ordered pair that corresponds to each of the following points as shown on the graph.

   a. M b. P c. Q d. T

3. Draw the triangle that has vertices* at the following points:

$$A(0, 0),\ B(3, 2),\ C(-4, 5)$$

4. Draw the triangle that has vertices at the following points:

$$A(-2, -3),\ B(-2, 4),\ C(3, 5)$$

5. Find the distance between each pair of points.

   a. $(-2, -2)$ and $(2, 1)$ b. $(-3, 3)$ and $(3, -1)$
   c. $(5, 3)$ and $(-2, 3)$ d. $(2, -2)$ and $(2, -5)$
   e. $(4, 6)$ and $(0, 0)$

6. Find the distance between each pair of points.

   a. $(-4, -3)$ and $(8, 2)$ b. $(-3, 2)$ and $(4, -3)$
   c. $(-3, -4)$ and $(-3, 2)$ d. $(-1, -2)$ and $(5, -2)$
   e. $(-6, 9)$ and $(0, 0)$

7. Find the perimeter of the triangle that has vertices at $A(-2, 2)$, $B(4, 2)$, and $C(6, 8)$.

8. Find the perimeter of the triangle that has vertices at $A(0, 2)$, $B(11, 2)$, and $C(8, 6)$.

9. Use the distance formula to discover whether the triangle with vertices $A(-3, -2)$, $B(5, -1)$, and $C(3, 2)$ is or is not a right triangle.

10. Use the distance formula to discover whether the triangle with vertices $A(3, 2)$, $B(-5, 1)$, and $C(-3, -2)$ is or is not a right triangle.

*Recall from geometry that the points where the sides of a triangle meet are called the *vertices* of the triangle. (*One* such point is called a *vertex*.)

11. Find the domain and range of the relation $\{(2, -1), (3, 4), (0, 2), (-3, -2)\}$, and graph the relation.

12. Find the domain and range of the relation $\{(-4, 0), (0, 0), (3, -2), (1, 5), (-3, -3)\}$, and graph the relation.

Writing Problems

Express the answers in your own words and in complete sentences.

1. Explain the difference in meaning between $(5, 1)$ and $\{5, 1\}$.
2. Explain what a relation is.
3. Explain what the domain of a relation is.
4. Explain what the range of a relation is.

Exercises 8.1 Set II

For Exercises 1 and 2, use the following graph.

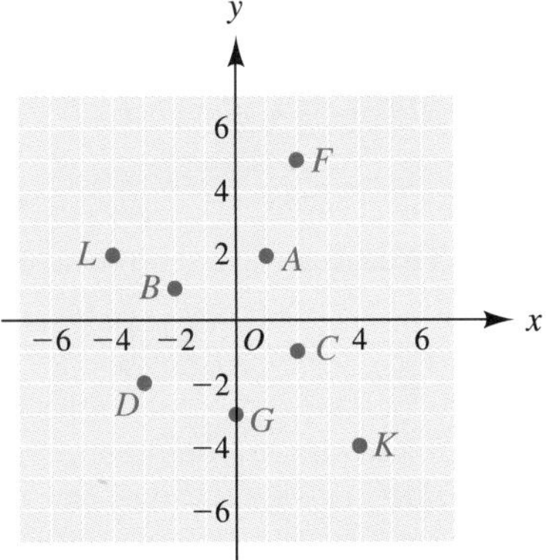

1. Find the ordered pair that corresponds to each of the following points as shown on the graph.

   a. F b. G c. K d. L

2. Find the ordered pair that corresponds to each of the following points as shown on the graph.

   a. A b. B c. C d. D

3. Draw the triangle that has vertices at the following points:

$$A(4, -4),\ B(2, 3),\ C(-4, 1)$$

4. Draw the parallelogram that has vertices at the following points:

$$A(-2, -5),\ B(3, -5),\ C(0, 4),\ D(5, 4)$$

5. Find the distance between each pair of points.

   a. $(7, -3)$ and $(15, 3)$ b. $(-14, 6)$ and $(-14, -13)$

   c. $(9, -8)$ and $(-6, -5)$

   d. $(-11, -18)$ and $(-17, -18)$

   e. $(2, 12)$ and $(-6, 16)$

6. Find the distance between each pair of points.

   a. $(3, 0)$ and $(-1, 4)$ b. $(3, -1)$ and $(-3, 1)$

   c. $(8, 3)$ and $(-2, 3)$ d. $(-4, 2)$ and $(3, -1)$

   e. $(7, 2)$ and $(7, -5)$

7. Find the perimeter of the triangle that has vertices at $R(4, -7)$, $S(9, 5)$, and $T(-8, -2)$.

8. Use the distance formula to discover whether the triangle with vertices $A(-2, -5)$, $B(-4, -2)$, and $C(5, 0)$ is or is not a right triangle.

9. Use the distance formula to discover whether the triangle with vertices $A(-2, -3)$, $B(2, -2)$, and $C(0, 3)$ is or is not a right triangle.

10. Do the points $P(10, -6)$, $Q(-5, 3)$, and $R(-15, 9)$ lie in a straight line? Why or why not?

11. Find the domain and range of the relation $\{(-4, 3), (-2, 0), (-3, -5), (4, -2), (4, 4)\}$, and graph the relation.

12. Find the domain and range of the relation $\{(-6, 3), (2, -4), (0, 5), (-1, -1)\}$, and graph the relation.

8.2 Graphing Solution Sets of Linear Equations in Two Variables

A Solution of an Equation in x and y An ordered pair (a, b) is a **solution** of an equation in x and y if we get a *true statement* when we substitute a for x and b for y in that equation; if (a, b) is a solution of an equation, we often say that the point (a, b) *satisfies* the equation. The solution *set* of an equation in two variables is the set of all the ordered pairs that satisfy the equation, and the solution set of such an equation is a *relation*. There are always many ordered pairs that are solutions of an equation in two variables and many others that are not solutions.

The Graph of the Solution Set of an Equation in Two Variables A point (a, b) lies on the **graph** of the solution set of an equation in two variables if its coordinates satisfy that equation (and if a point lies on the graph of the solution set of an equation, then its coordinates satisfy that equation).

The graph of the solution set of an equation must fulfill the following conditions:

1. It must contain *only* those points whose coordinates satisfy the equation.
2. It must include, within the limits of the size of the graph, *all* the points whose coordinates satisfy the equation. (Because there are infinitely many points whose coordinates satisfy an equation in one or two variables, we usually cannot graph *all* of these points—our graph would have to be infinitely large! However, we use arrowheads to indicate that the graph extends beyond the portion of the line that we have drawn.)

Note We commonly say that we are *graphing the equation* when we really mean that we are graphing the solution set of that equation.

Linear Equations, Their Solutions, and the Graphs of Their Solution Sets

A *first-degree equation* is also called a **linear equation**. (Notice the word *line* in the word *line*ar.) A linear equation in two variables can be expressed in the form

$$Ax + By + C = 0 \qquad \text{or} \qquad 0 = Ax + By + C$$

where A, B, and C are any real numbers but A and B are not *both* zero.

To find a solution of $Ax + By + C = 0$, we first choose a value for x or for y; then we use the equation to find the corresponding value of the other variable. Students often worry about what values to choose for x or for y. Usually *any* values can be chosen. However, in the following two cases, we *cannot* choose any value we wish for either variable:

1. If $A = 0$ (that is, if the equation contains no x-term), we *cannot* let y have just any value whatever. We can, however, choose any value we wish for x.
2. If $B = 0$ (that is, if the equation contains no y-term), we *cannot* let x have just any value whatever. We can, however, choose any value we wish for y.

In Example 4, we graph the solution set of an equation in which $A = 0$, and in Example 3, we graph the solution set of an equation in which $B = 0$. First, however, let's consider some cases in which neither A nor B is 0.

Let's find some solutions of $4x + 3y = 12$, or $4x + 3y - 12 = 0$; in this case, $A = 4$ and $B = 3$. To find one solution, we can let x have *any* value whatever, because $B \neq 0$; for example, we can let $x = -3$. Then we find the corresponding y-value by substituting -3 for x and solving the resulting equation for y. That is, we solve $4(-3) + 3y = 12$ for y, getting $y = 8$. Therefore, one solution of $4x + 3y = 12$ is $(-3, 8)$.

We can also find a solution of $4x + 3y = 12$ by letting y have *any* value whatever, because $A \neq 0$; for example, we can let $y = 4$. Then we find the corresponding x-value by substituting 4 for y in the equation and solving the resulting equation for x. That is, we solve $4x + 3(4) = 12$ for x, getting $x = 0$. Therefore, another solution of $4x + 3y = 12$ is $(0, 4)$. (See Example 1 for another solution and the complete graph of $4x + 3y = 12$.)

The following statement must be memorized; it will not be proved.

The graph of a linear (first-degree) equation

> The graph of the solution set of any linear (first-degree) equation in one or two variables is the *straight line* that contains, within the limits of the size of the graph, all the ordered pairs—and only those ordered pairs—that satisfy the given equation.

Therefore, when we graph a first-degree equation in one or two variables, after we have graphed a few points that lie on the line (see the Note below), *we must connect those points with a straight line*, and *we must draw an arrowhead at each end of the line*. Only in this manner can we indicate that we're including *all* the ordered pairs that satisfy the equation.

Note A straight line can be drawn if we know any *two* points on that line. Although only two points are *necessary*, it is advisable to graph a third point as a *checkpoint*. If the equation is a first-degree equation and if the three points do *not* lie in a straight line, some mistake has been made either in the calculation of the coordinates of one or more of the points or in graphing the points.

The Intercepts of a Line The **x-intercept** of a line is the point where the line crosses the x-axis. We find it by letting $y = 0$; it will be the point $(a, 0)$, where a is the x-coordinate of the x-intercept. The **y-intercept** of a line is the point where the line crosses the y-axis. We find it by letting $x = 0$; it will be the point $(0, b)$, where b is the y-coordinate of the y-intercept (see Figure 4).

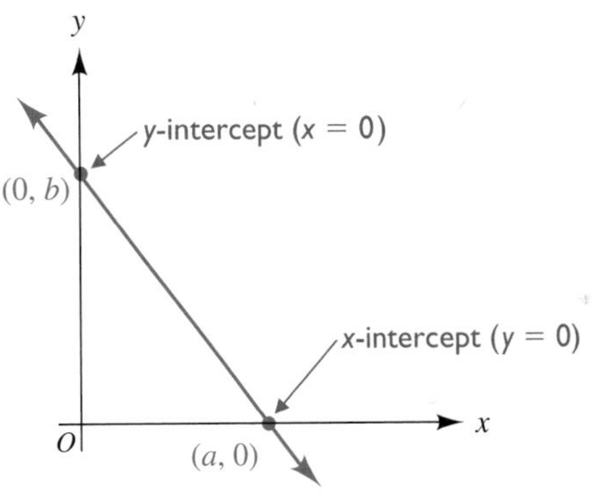

Figure 4

The Intercept Method of Graphing a Straight Line To use the **intercept method of graphing the line** $Ax + By + C = 0$ (assuming $A \neq 0$ and $B \neq 0$), we find the x- and y-intercepts; we usually also find one other point on the line, the *checkpoint*. We graph the three points, draw a straight line through them, and draw an arrowhead at each end of the line (see Example 1). We often enter the coordinates of the points we find into a *table of values* instead of writing them as ordered pairs of numbers.

EXAMPLE 1 Graph the relation $4x + 3y = 12$.

SOLUTION

***x*-intercept** Let $y = 0$.

Then $4x + 3y = 12$

becomes $4x + 3(0) = 12$

$4x = 12$

$x = 3$

The x-intercept is $(3, 0)$

Table of values

x	y
3	0

We sometimes say that "the x-intercept is 3" (naming only the x*-coordinate* of the point where the line crosses the x-axis, instead of the point itself).

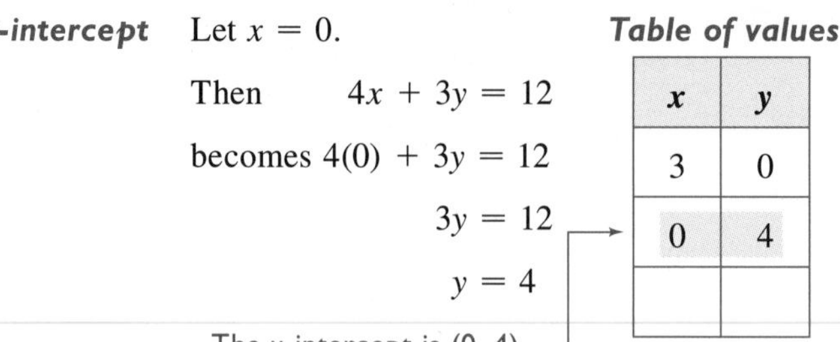

***y*-intercept** Let $x = 0$.

Then $4x + 3y = 12$

becomes $4(0) + 3y = 12$

$3y = 12$

$y = 4$

The y-intercept is (0, 4)

Table of values

x	y
3	0
0	4

We sometimes say that "the y-intercept is 4" (naming only the y*-coordinate* of the point where the line crosses the y-axis, instead of the point itself).

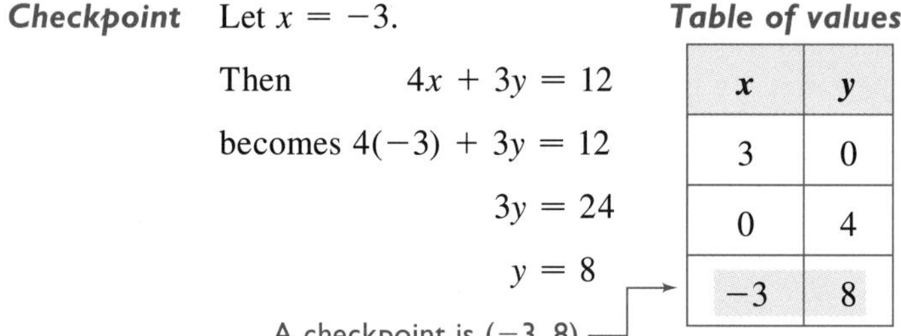

Checkpoint Let $x = -3$.

Then $4x + 3y = 12$

becomes $4(-3) + 3y = 12$

$3y = 24$

$y = 8$

A checkpoint is $(-3, 8)$

Table of values

x	y
3	0
0	4
−3	8

In Figure 5, we graph the x- and y-intercepts and the third point and note that these three points *do* appear to lie on the same straight line. These three points *alone* do not constitute the graph of $4x + 3y = 12$, however. There are *infinitely* many points—points such as $\left(1, \frac{8}{3}\right)$, $\left(\frac{9}{4}, 1\right)$, and $(-6, 12)$—whose coordinates satisfy that equation. Therefore, we must now connect these points with a *straight line*. The line actually extends infinitely far in both directions; to indicate this, we draw an arrowhead at each end of the line. The graph shown in Figure 6 *is* the graph of $4x + 3y = 12$.

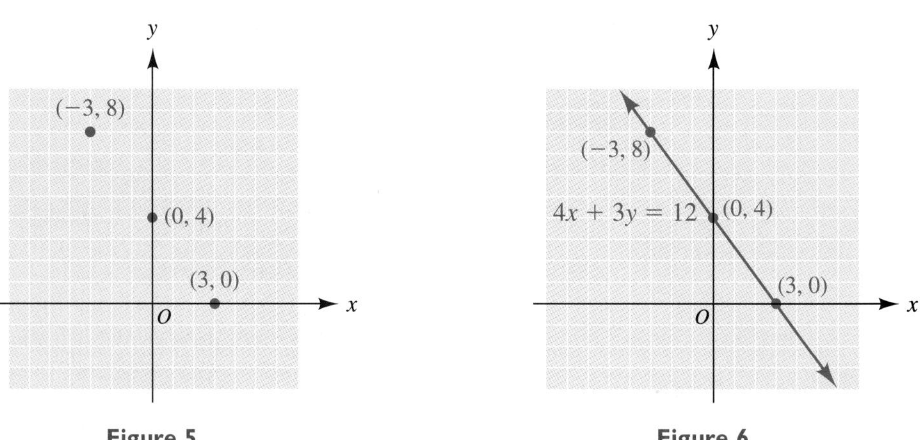

Figure 5 **Figure 6**

An Alternative Method of Graphing a Straight Line Some people find it helpful to first solve the equation for y (using the method discussed earlier for solving literal equations) and then let x have three different values. (The method of solving the equation for y *must* be used when we graph straight lines with a graphics calculator, as we do in Examples 7, 8, and 9.)

If we solve $4x + 3y = 12$ (the equation from Example 1) for y, the equation becomes $y = -\frac{4}{3}x + 4$ (you might verify this), and if we let x be 3, 0, and -3, the table of values remains unchanged. With the equation written as $y = -\frac{4}{3}x + 4$, it's easy to see that if we let the values of x be multiples of 3, we will avoid fractions.

EXAMPLE 2 Graph the relation $3x - 4y = 0$.

SOLUTION

x-intercept Let $y = 0$; if $y = 0$, $x = 0$. Therefore, the x-intercept is $(0, 0)$—the origin. Since the line goes through the origin, the *y-intercept* is also $(0, 0)$.

We have found only one point on the line: $(0, 0)$. Now we will find two other points.

Second point Let $y = 3$; if $y = 3$, $x = 4$. This gives the point $(4, 3)$.

Third point Let $x = -4$; if $x = -4$, $y = -3$. This gives the point $(-4, -3)$.

We now graph the points $(0, 0)$, $(4, 3)$, and $(-4, -3)$ and connect them with a straight line.

Table of values

Both intercepts are the same point

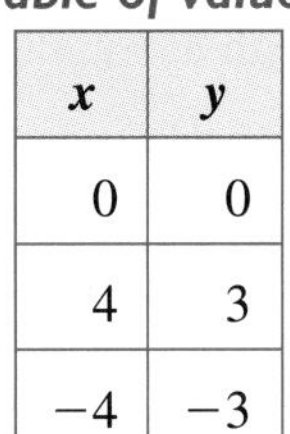

x	y
0	0
4	3
−4	−3

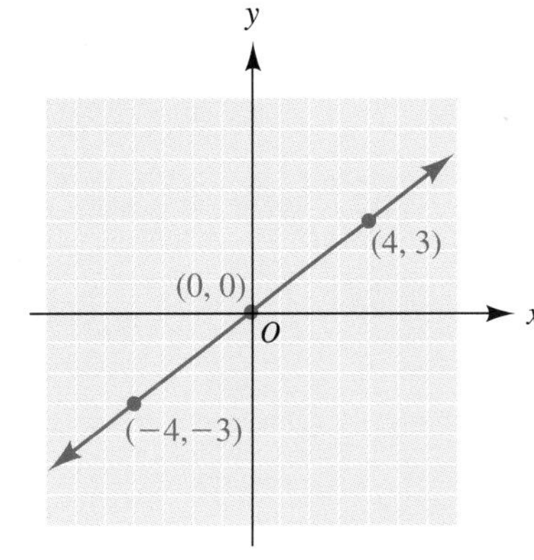

Note You might verify that if we solve $3x - 4y = 0$ (the equation in Example 2) for y, we get $y = \frac{3}{4}x$; if we then let $x = 0, 4$, and -4, we get the same table of values as in Example 2.

Note If the x- and y-intercepts are very close together, the checkpoint should be far enough away from the intercepts that an accurate line can be drawn.

Graphing Vertical and Horizontal Lines As we have mentioned, sometimes the equation of a line contains only one variable. If so, the graph is always either a vertical line or a horizontal line. See Examples 3 and 4.

EXAMPLE 3 Graph the relation $x = 3$. (The equation $x = 3$ is equivalent to $x + 0 \cdot y = 3$ and to $x - 3 = 0$.)

SOLUTION The domain of the relation is $\{3\}$, since 3 is the only value that x can have. There is no y-term in the equation $x = 3$; that is, $B = 0$. Although y is not mentioned in the equation, it is understood that $x = 3$ for *all y-values.* We needn't make a table of values, since x must *always* equal 3. We arbitrarily select y-values of 0, -2, and 4. Therefore, we graph the points $(3, 0)$, $(3, -2)$, and $(3, 4)$ and connect them with a straight line. The graph of the relation $x = 3$ is shown below.

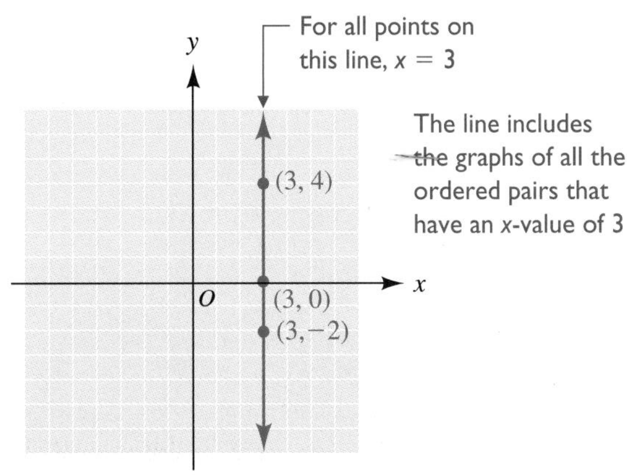

> The graph of $x = k$ or of $x - k = 0$, where k is any real number, is always a vertical line that passes through $(k, 0)$.

EXAMPLE 4 Graph the relation $y + 4 = 0$. (The equation $y + 4 = 0$ is equivalent to $0 \cdot x + y + 4 = 0$ and to $y = -4$.)

SOLUTION The domain of the relation is the set of all real numbers. There is no x-term in the equation $y + 4 = 0$; that is, $A = 0$. Although x is not mentioned in the equation, it is understood that $y = -4$ for *all x-values.* We arbitrarily select x-values of 0, 3, and -2. Therefore, we graph the points $(0, -4)$, $(3, -4)$, and $(-2, -4)$ and connect them with a straight line. The graph of the relation $y + 4 = 0$ is shown below.

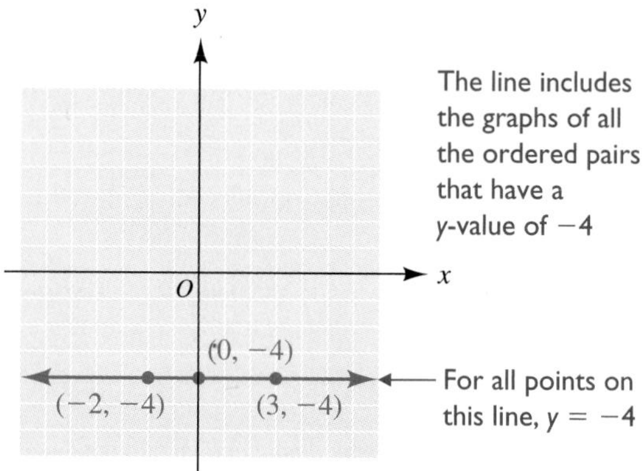
y
The line includes the graphs of all the ordered pairs that have a y-value of −4
x
O
(0, −4)
(−2, −4)
(3, −4)
For all points on this line, y = −4

> The graph of $y = b$ or of $y - b = 0$, where b is any real number, is always a horizontal line that passes through $(0, b)$.

The following example illustrates the use of different scales on the two axes.

EXAMPLE 5 Graph the relation $\{(x, y) \mid 50y - x = 100\}$.

SOLUTION The statement in set-builder notation means that the relation consists of all of the ordered pairs (x, y) that satisfy the equation $50y - x = 100$. Since the domain of the relation is not specified, it is the set of all real numbers, and the graph is a straight line.

Using the intercept method

x-intercept If $y = 0$,

$$50(0) - x = 100$$
$$-x = 100$$
$$x = -100$$

y-intercept If $x = 0$,

$$50y - 0 = 100$$
$$50y = 100$$
$$y = 2$$

Checkpoint If $y = 3$,

$$50(3) - x = 100$$
$$150 - x = 100$$
$$-x = -50$$
$$x = 50$$

Solving the equation for y

$$50y - x = 100$$
$$50y = x + 100$$
$$y = \frac{x + 100}{50}$$
$$y = \tfrac{1}{50}x + 2$$

If $x = 0$, $y = \frac{1}{50}(0) + 2$
$y = 2$

If $x = 50$, $y = \frac{1}{50}(50) + 2$
$y = 3$

If $x = -100$, $y = \frac{1}{50}(-100) + 2$
$y = 0$

It would be impractical to have the units on the two axes be of equal lengths. The graph of the relation is shown below.

Table of values

x	y
0	2
−100	0
50	3

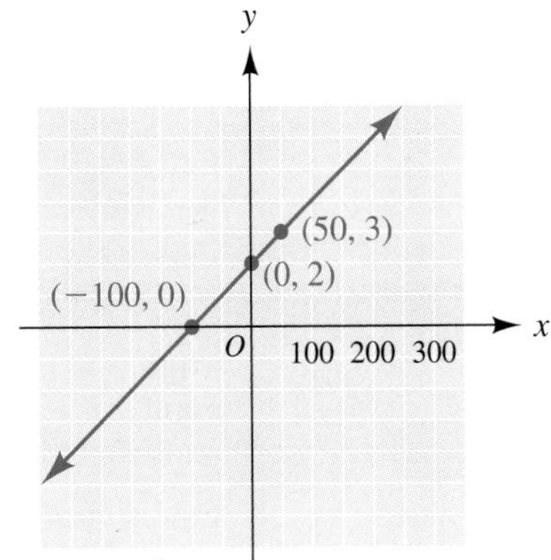

EXAMPLE 6 Find the domain and range of the relation $\{(x, y) \mid 2x + 3y = 6, x = -3, 0, 3, 6\}$, and graph the relation.

SOLUTION The domain is *not* the set of real numbers; it is $\{-3, 0, 3, 6\}$.

Table of values

	x	y
If $x = -3$, $y = 4$.	−3	4
If $x = 0$, $y = 2$.	0	2
If $x = 3$, $y = 0$.	3	0
If $x = 6$, $y = -2$.	6	−2

The domain (the x column) — The range (the y column)

Therefore, the range $R_{\mathcal{R}} = \{4, 2, 0, -2\}$. The graph of the relation is shown below.

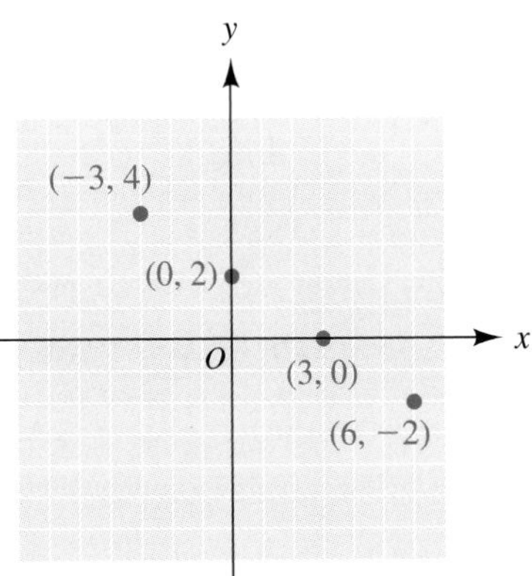

This relation can also be written $\{(-3, 4), (0, 2), (3, 0), (6, -2)\}$.

In Example 6, it would be *incorrect* to connect the four points with a straight line. However, if the domain of the relation were changed from $\{-3, 0, 3, 6\}$ to the set of *all* real numbers, then the graph would be the straight line through the four points graphed in the example.

The methods we've demonstrated in the examples are summarized in the box below:

Graphing a straight line

The intercept method

1. Let $y = 0$, and solve for x to find the x-intercept.* Graph this point.
2. Let $x = 0$, and solve for y to find the y-intercept.† [If both intercepts are $(0, 0)$, find a second point by letting x or y have any nonzero value.] Graph this point.
3. Find a checkpoint by letting x or y have any value not already used. Graph this point.

Solving the equation for y

1. Solve the given equation for y.
2. Let x have *three* different values, and solve for the corresponding y-values.
3. Graph the points found in step 2. Then proceed with step 4, below.

4. Draw a straight line through the points that have been graphed in the preceding steps, and draw an arrowhead at each end of the line.

Horizontal and vertical lines: The graph of $y = b$ (or $y - b = 0$), where b is any real number, is always a horizontal line that passes through $(0, b)$. The graph of $x = k$ (or $x - k = 0$), where k is any real number, is always a vertical line that passes through $(k, 0)$.

*If the equation can be expressed in the form $y = b$, it will be a horizontal line, and if $b \neq 0$, it will not have an x-intercept.

†If the equation can be expressed in the form $x = k$, it will be a vertical line, and if $k \neq 0$, it will not have a y-intercept.

It is also possible to graph a straight line by letting x (or y) have *any* values.

Graphing Straight Lines with a Graphics Calculator (Optional)

We show the procedure for graphing straight lines with the TI-81 graphics calculator. In the examples, we assume that you have turned the calculator on and have set the *mode* for *function*, *connected*, and *rect*, with *grid off*. (If you select *grid on*, you will see grid marks in the background of the graph.)

EXAMPLE 7 Graph $y = 3x + 5$ with a TI-81 graphics calculator.

SOLUTION We press the following keys, in order:

[Y=] [3] [X|T] [+] [5] [GRAPH]

The display is

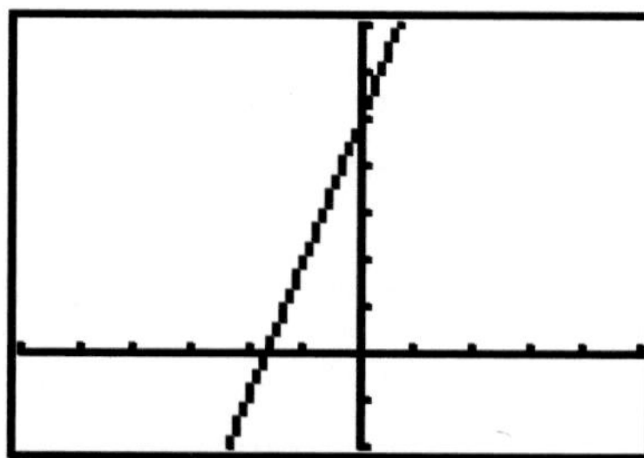

Note The key Y= stands for "y =," and the key X|T is the key to use for the variable x.

To "trace" the coordinates of points on the line, press TRACE. A cursor (a flashing dot) moves as you press the ▶ and ◀ keys, and the coordinates of the points are shown on the screen. The display might be

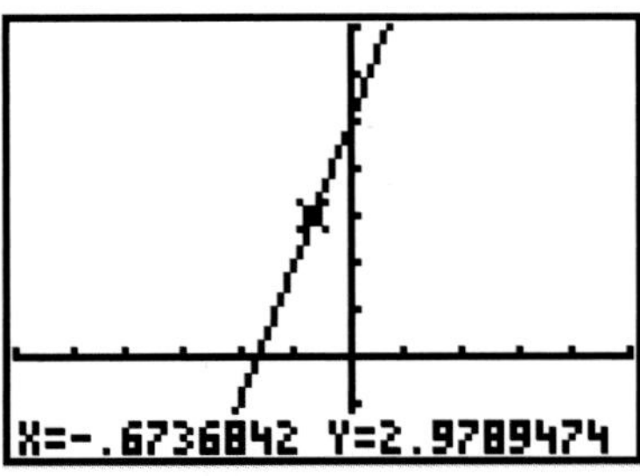

Note If there's no graph in the display, you will need to reset the *range.* To graph $y = 3x + 5$, we set the range as follows: Press RANGE; set $x$min to -6 by pressing (−) 6; press ▼ to move to the next line of the menu; and set $x$max to 5. In a similar way, set $y$min to -2 and $y$max to 7. Press GRAPH again to see the graph.

To "erase" the graph, press CLEAR. To "erase" the equation $y = 3x + 5$ from the calculator's memory, press Y= CLEAR.

EXAMPLE 8 Graph $3x - 5y = 15$ with a TI-81 graphics calculator.

SOLUTION We must first solve the equation for y. You can verify that, when we do this, we get $y = \frac{3}{5}x - 3$. Because $\frac{3}{5} = 0.6$, we can then press these keys, in order:

Y= . 6 X|T − 3 GRAPH

The display is

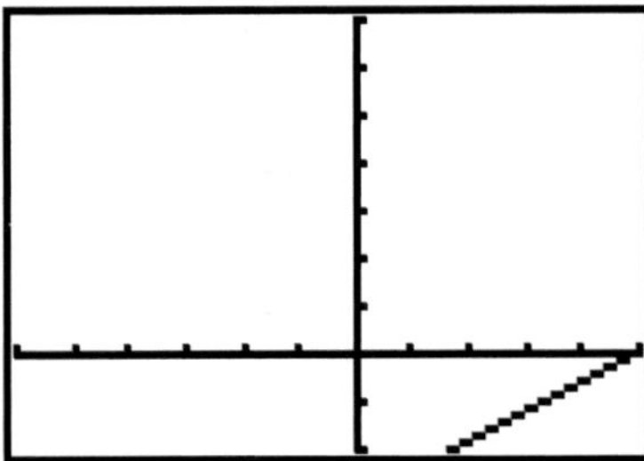

If we change the range so that $x$min is -5, $x$max is 10, $y$min is -6, and $y$max is 5, the axes move and the display changes to

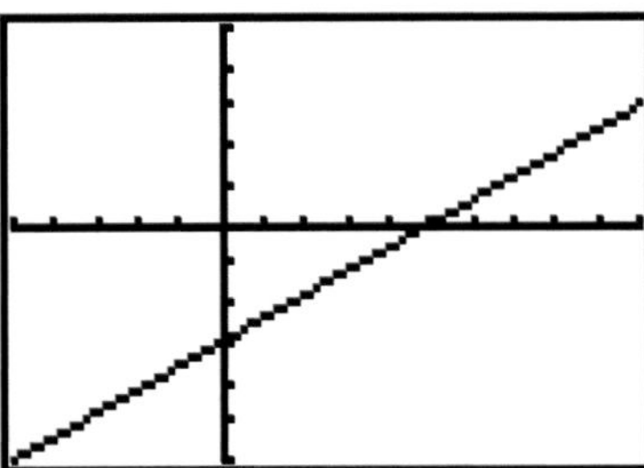

EXAMPLE 9 Graph $3x + 7y - 350 = 0$ with a TI-81 graphics calculator.

SOLUTION Solving the given equation for y, we get $y = -\frac{3}{7}x + 50$. If the range is set so that $x$min is -10, $x$max is 10, $y$min is -10, and $y$max is 10 and if we press

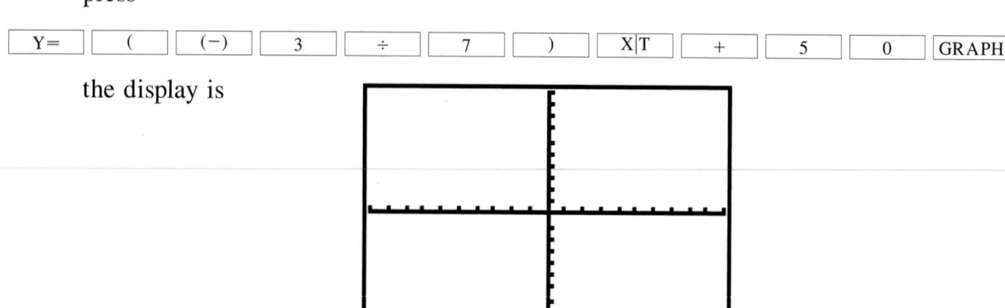

the display is

We don't see the graph of the line at all, because the graph is outside the range. To correct the problem, let's find the *intercepts* for $3x + 7y - 350 = 0$. You can verify that if $x = 0$, $y = 50$, and if $y = 0$, $x \approx 116.7$. Therefore, let's change the range so that $x$min is -10, $x$max is 130, $y$min is -10, $y$max is 60, and $x$scl and $y$scl are 10; the display changes to

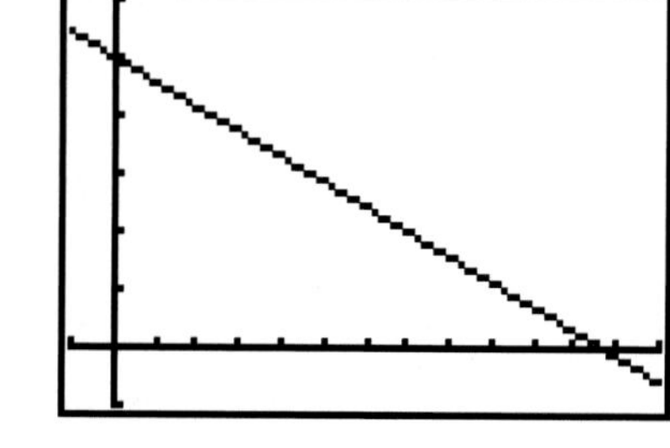

(Changing $x$scl and $y$scl to 10 changes the number of marks on the axes; on each axis there is now just one mark corresponding to every *ten* units.) We now see the graph, including both intercepts.

Your instructor may or may not allow you to use a graphics calculator to graph the lines in the following exercises.

Exercises 8.2
Set I

In Exercises 1–20, sketch the graph of each relation.

1. $3x + 2y = 6$
2. $4x - 3y = 12$
3. $5x - 3y = 15$
4. $2x + 5y = 10$
5. $9x + 5y = 18$
6. $6x - 11y = 22$
7. $10x = 21 + 7y$
8. $13y = 40 - 8x$
9. $9y = 25 - 7x$
10. $17x = 31 + 6y$
11. $8x - 41 = 14y$
12. $5y - 33 = -15x$
13. $6x + 11y = 0$
14. $3x + 2y = 0$
15. $4y = -8x$
16. $9x = 3y$
17. $x + 5 = 0$
18. $x = 4$
19. $y = -3$
20. $y - 2 = 0$

In Exercises 21–34, graph each given relation.

21. $\{(x, y) \mid 7x + 5y = 2\}$
22. $\{(x, y) \mid 3x + 8y = 4\}$
23. $\left\{(x, y) \mid y = \frac{1}{2}x - 1\right\}$
24. $\left\{(x, y) \mid y = \frac{1}{3}x + 2\right\}$
25. $\{(x, y) \mid 3(x - 5) = 7y\}$
26. $\{(x, y) \mid 4x = 3(y - 6)\}$
27. $\{(x, y) \mid 50x + y = -100\}$
28. $\{(x, y) \mid x - 30y = 90\}$
29. $\{(x, y) \mid x = 50y\}$
30. $\{(x, y) \mid y = 70x\}$
31. $\{(x, y) \mid y = -2x, x = 1, 3, 5\}$
32. $\{(x, y) \mid y = 3x, x = -1, 3, 4\}$
33. $\{(x, y) \mid x - y = 5, x = 1, 4\}$
34. $\{(x, y) \mid 2x - y = 3, x = 0, 2\}$

In Exercises 35 and 36, (a) graph the two relations for each exercise on the same set of axes, and (b) find the ordered pair corresponding to the point where the two lines intersect.

35. $\begin{cases} 5x - 7y = 18 \\ 2x + 3y = -16 \end{cases}$
36. $\begin{cases} 4x + 9y = 3 \\ 2x - 5y = 11 \end{cases}$

Writing Problems

Express the answers in your own words and in complete sentences.

1. Explain what a solution of a linear equation in two variables *is*.
2. Explain why $(3, -2)$ is a solution of the equation $x + y = 1$.
3. Explain how to find an ordered pair that is a solution of $3x - y = 2$. (Don't *find* the ordered pair.)
4. Describe the method you would use in graphing $3x - y = 2$, and explain why you would use that method.
5. Describe the method you would use in graphing $y = -1$, and explain why you would use that method.
6. Describe the graph of $y = 0$.
7. Describe the graph of $x = -2$.

Exercises 8.2 Set II

In Exercises 1–20, sketch the graph of each relation.

1. $3x + 8y = 24$
2. $4x - 12y = 24$
3. $4x - 5y = 20$
4. $x = 5$
5. $15x + 8y = 32$
6. $y = -6$
7. $7x - 28 = 9y$
8. $-2x + 5y = 10$
9. $19y = 35 - 8x$
10. $y = 3x + 4$
11. $7y = 16x - 47$
12. $y = \frac{2}{3}x + 4$
13. $5x - 13y = 0$
14. $3x = 5y$
15. $2x = 8y$
16. $y = \frac{2}{3}x - 1$
17. $x + 4 = 0$
18. $y - 6 = 0$
19. $y = 4$
20. $x + y = 0$

In Exercises 21–34, graph each given relation.

21. $\{(x, y) \mid 11x - 4y = 2\}$
22. $\{(x, y) \mid x = 0\}$
23. $\left\{(x, y) \mid y = \frac{2}{5}x - 3\right\}$
24. $\left\{(x, y) \mid y = 0\right\}$
25. $\{(x, y) \mid 4(y + 7) = 9x\}$
26. $\{(x, y) \mid x = -4\}$
27. $\{(x, y) \mid 20y - x = 80\}$
28. $\left\{(x, y) \mid y = \frac{2}{3}x\right\}$
29. $\{(x, y) \mid y = 100x\}$
30. $\left\{(x, y) \mid x = \frac{2}{3}y\right\}$
31. $\{(x, y) \mid y = 4x, x = -2, 2\}$
32. $\{(x, y) \mid y = 5x - 2, x = 2, 4, 6\}$
33. $\{(x, y) \mid x - 2y = 4, x = 2, 4, 6\}$
34. $\{(x, y) \mid x = 5y, x = -5, 0, 5\}$

In Exercises 35 and 36, (a) graph the two relations for each exercise on the same set of axes, and (b) find the ordered pair corresponding to the point where the two lines intersect.

35. $\begin{cases} 7x + 10y = -1 \\ 3x - 8y = -25 \end{cases}$
36. $\begin{cases} 8x - y = 4 \\ x + 2y = 9 \end{cases}$

8.3 The Slope of a Line

The **slope of a line** is the measure of the *steepness* of the line. To find the slope of a nonvertical line, we select two points on the line and we divide the vertical change (the difference between the two y-values) by the horizontal change (the difference between the two corresponding x-values) (see Figure 7).

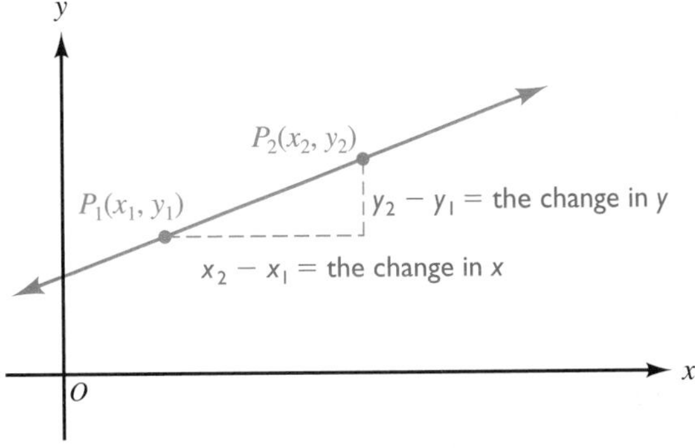

Figure 7

The slope of a line is defined as follows:

The slope of a line

If $P_1(x_1, y_1)$ and $P_2(x_2, y_2)$ are two distinct points on a nonvertical line (see Figure 7) and if m represents the slope of the line, then

$$m = \frac{y_2 - y_1}{x_2 - x_1} \quad \text{or} \quad m = \frac{y_1 - y_2}{x_1 - x_2}$$

The slope of a vertical line *does not exist.*

A Word of Caution It is *incorrect* to write $\frac{x_2 - x_1}{y_2 - y_1}$, $\frac{x_1 - x_2}{y_1 - y_2}$, $\frac{y_2 - y_1}{x_1 - x_2}$, or $\frac{y_1 - y_2}{x_2 - x_1}$ as the slope of a line.

Note The slope is often thought of as $\frac{\text{“the rise”}}{\text{“the run”}}$. 

EXAMPLE 1 Find the slope of the line through the points $(-3, 5)$ and $(6, -1)$.

SOLUTION Let $P_1 = (-3, 5)$ and $P_2 = (6, -1)$. Then

$$m = \frac{y_2 - y_1}{x_2 - x_1} = \frac{-1 - 5}{6 - (-3)} = \frac{-6}{9} = -\frac{2}{3}$$

The slope is not changed if the points P_1 and P_2 are interchanged. Let $P_1 = (6, -1)$ and $P_2 = (-3, 5)$. Then

$$m = \frac{y_2 - y_1}{x_2 - x_1} = \frac{5 - (-1)}{-3 - 6} = \frac{6}{-9} = -\frac{2}{3}$$

The line is graphed in Figure 8.

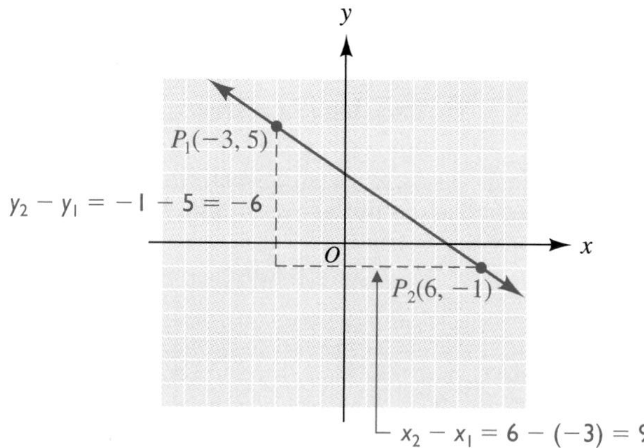

Figure 8

Notice in Example 1 that the slope of the line is *negative* and that a point moving along the line toward the right *falls.* Observe, in fact, that for every three units we move toward the right, we move two units downward.

EXAMPLE 2 Find the slope of the line through the points $A(-2, -4)$ and $B(5, 1)$.

SOLUTION

$$m = \frac{y_2 - y_1}{x_2 - x_1} = \frac{1 - (-4)}{5 - (-2)} = \frac{5}{7}$$

The line is graphed in Figure 9.

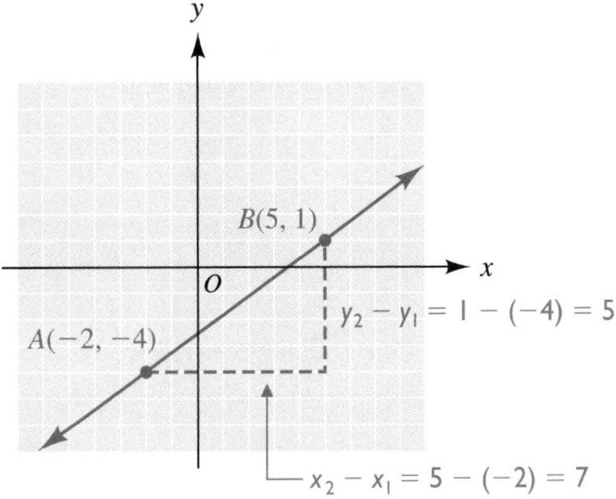

Figure 9

Notice in Example 2 that the slope of the line is *positive* and that a point moving along the line toward the right *rises.* Observe, in fact, that for every seven units we move toward the right, we move five units upward.

EXAMPLE 3 Find the slope of the line through the points $E(-4, -3)$ and $F(2, -3)$.

SOLUTION

$$m = \frac{y_2 - y_1}{x_2 - x_1} = \frac{-3 - (-3)}{2 - (-4)} = \frac{0}{6} = 0$$

The line is graphed in Figure 10.

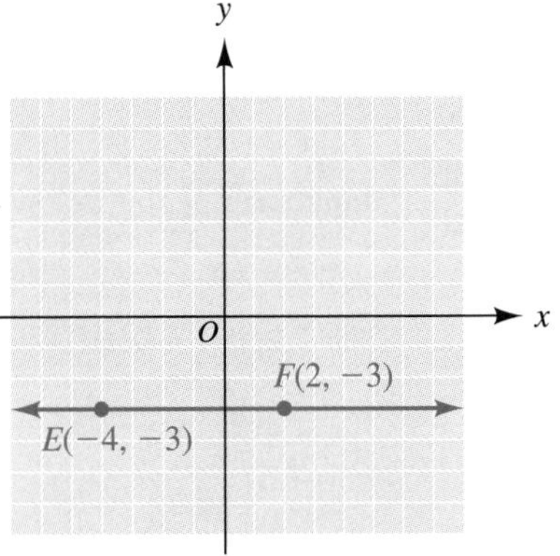

Figure 10

Notice in Example 3 that the line is *horizontal* and that the slope of the line is *zero.*

Whenever a line is horizontal, its slope is zero; and whenever the slope of a line is zero, the line is horizontal.

EXAMPLE 4 Find the slope of the line that passes through the points $R(4, 5)$ and $S(4, -2)$ (see Figure 11).

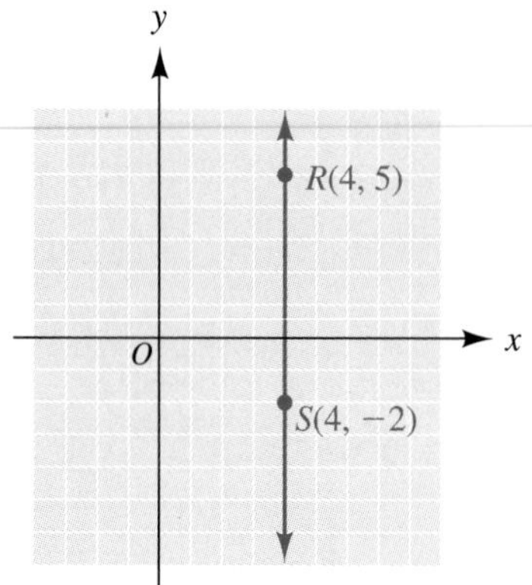

Figure 11

SOLUTION Because the x-coordinates of both points are 4, the line must be a vertical line. Therefore, *the slope does not exist.*

If we had not realized this and had substituted 4 for x_1, 5 for y_1, 4 for x_2, and -2 for y_2, we would have had

$$m = \frac{y_2 - y_1}{x_2 - x_1}$$

$$= \frac{-2 - 5}{4 - 4} = \frac{-7}{0} \longleftarrow \text{Not defined}$$

Our conclusion *still* would have been that the slope does not exist.

Whenever a line is vertical, its slope does not exist; and whenever the slope of a line does not exist, the line is vertical.

The facts about the slope of a line are summarized as follows:

The slope of a line is positive if a point moving along the line toward the right *rises* (Figure 9).

The slope of a line is negative if a point moving along the line toward the right *falls* (Figure 8).

The slope of a line is zero if the line is horizontal (Figure 10).

The slope of a line does not exist if the line is vertical (Figure 11).

If we're given the equation of a line, we can find the slope of the line by finding the coordinates of *any* two points on the line (see Example 5).

EXAMPLE 5 Graph each of the following lines, and find the slope of each line.

a. $y = -\frac{2}{3}x$

SOLUTION

x	y	
-3	2	A
0	0	B
3	-2	C

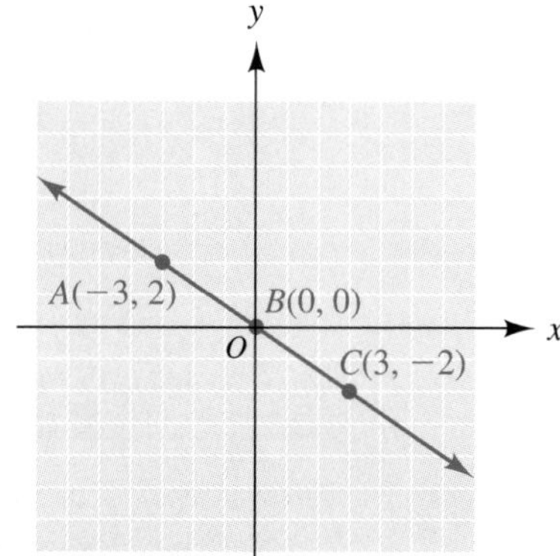

For the slope, if we use the points $A(-3, 2)$ and $B(0, 0)$, we have

$$m = \frac{0 - 2}{0 - (-3)} = \frac{-2}{3} = -\frac{2}{3}$$

If we use the points $B(0, 0)$ and $C(3, -2)$, we have

$$m = \frac{-2 - 0}{3 - 0} = \frac{-2}{3} = -\frac{2}{3}$$

b. $y = -\frac{2}{3}x - 2$

SOLUTION

x	y	
-6	2	D
-3	0	E
0	-2	F

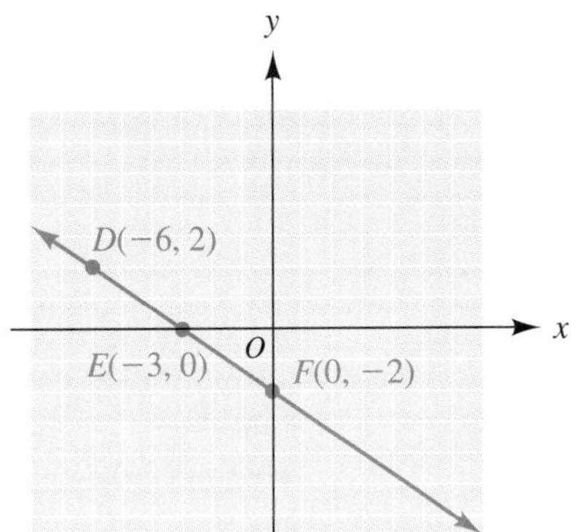

For the slope, if we use the points $D(-6, 2)$ and $E(-3, 0)$, we have

$$m = \frac{0 - 2}{-3 - (-6)} = \frac{-2}{3} = -\frac{2}{3}$$

If we use the points $E(-3, 0)$ and $F(0, -2)$, we have

$$m = \frac{-2 - 0}{0 - (-3)} = \frac{-2}{3} = -\frac{2}{3}$$

c. $y = \frac{3}{2}x$

SOLUTION

x	y	
-2	-3	G
0	0	H
2	3	I

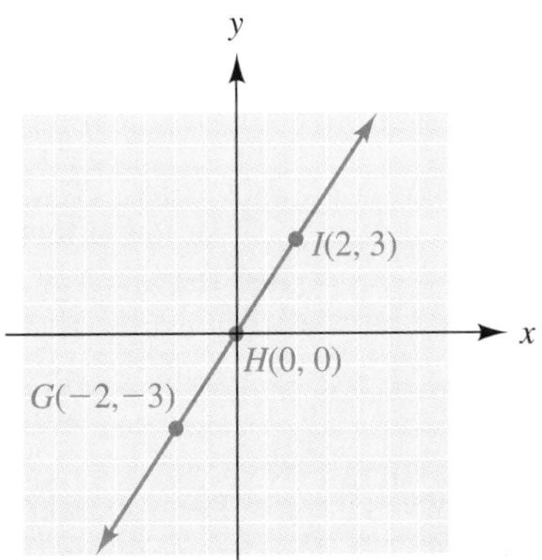

For the slope, if we use the points $G(-2, -3)$ and $H(0, 0)$, we have

$$m = \frac{0 - (-3)}{0 - (-2)} = \frac{3}{2}$$

If we use the points $H(0, 0)$ and $I(2, 3)$, we have

$$m = \frac{3 - 0}{2 - 0} = \frac{3}{2}$$

Note You can verify for each of the lines that if any other points on the line had been used, the slope would have remained unchanged.

EXAMPLE 6 Graph the lines from Example 5a and Example 5b on the same axes.

SOLUTION

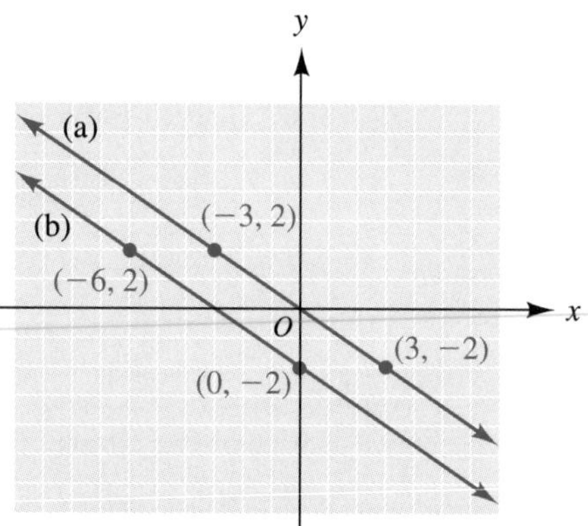

We found in Examples 5a and 5b that the slopes of both lines are $-\frac{2}{3}$ (that is, both lines have the same slope), and we see in the graph above that the lines appear to be parallel.

EXAMPLE 7 Graph the lines from Example 5a and Example 5c on the same axes.

SOLUTION

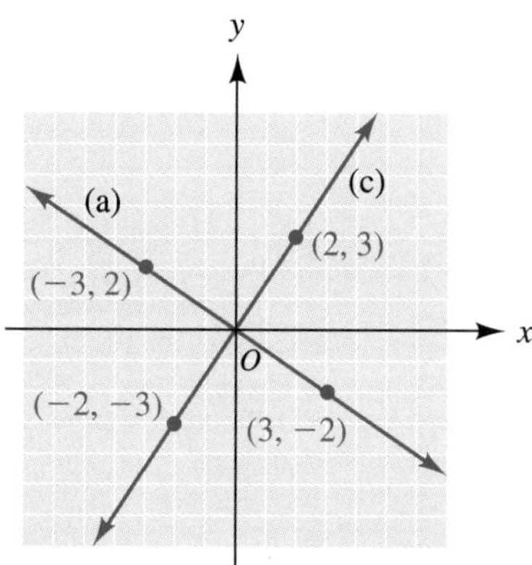

We found that the slope of the line given in Example 5a is $-\frac{2}{3}$ and that the slope of the line given in Example 5c is $\frac{3}{2}$. Notice that $\frac{3}{2}$ is the negative reciprocal of $-\frac{2}{3}$, or, in other words, the product of the two slopes is -1. Notice, also, that the lines in the graph above appear to be perpendicular.

The facts about parallel and perpendicular lines (which are stated without proof) are summarized in the following box:

> Parallel lines have the same slope; conversely, any lines that have the same slope are parallel.
>
> Perpendicular lines (if neither is vertical) have slopes whose product is -1; equivalently, if one line with a slope of m_1 is perpendicular to another line with a slope of m_2, then
>
> $$m_1 = -\frac{1}{m_2}$$
>
> A vertical line is parallel to any other vertical line and is perpendicular to any horizontal line.

Notice that our observations about the lines discussed in Examples 5, 6, and 7 are consistent with the facts stated in the box. That is, the lines shown in Example 6 appear to be parallel, and the slopes that we found in Examples 5a and 5b are the same. Furthermore, the lines shown in Example 7 appear to be perpendicular, and the slopes that we found in Examples 5a and 5c are the negative reciprocals of each other.

Graphing a Line When Its Slope and One Point on the Line Are Known

By making use of the fact that the slope of a line is often thought of as $\frac{\text{the rise}}{\text{the run}}$, we can graph a line if we're given the slope of the line and a point on the line. The numerator of the slope tells us how far up or down from the given point to move; we'll move *upward* if the numerator is *positive* and *downward* if the numerator is *negative.* The denominator of the slope tells us how far right or left from the given point to move; we'll move *right* if the denominator is *positive* and *left* if the denominator is *negative* (see Example 8).

EXAMPLE 8 Graph the line that passes through the point $(3, 4)$ and has a slope of $-\frac{2}{3}$.

SOLUTION If we interpret the slope as $\frac{-2}{3}$, the line should fall 2 units for each 3 units it moves toward the right. We know that one point the line passes through is $(3, 4)$. To find a second point, we move 2 units *downward* from $(3, 4)$ (downward because the numerator is negative) and 3 units toward the *right* (toward the right because the denominator is positive). Therefore, the line is as shown in the graph below.

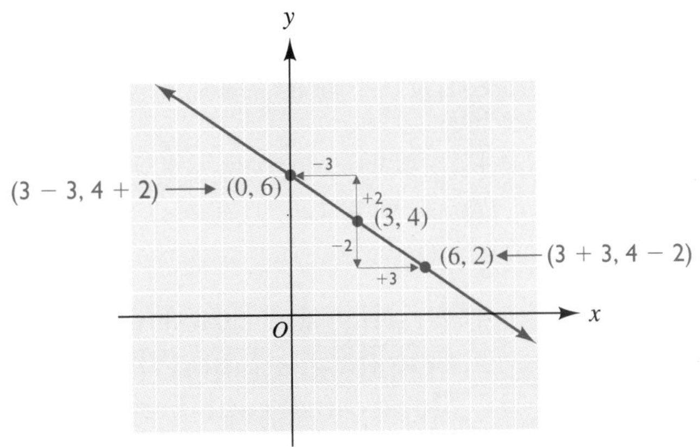

If we interpret the slope as $\frac{2}{-3}$, the line should rise 2 units for each 3 units it moves toward the left. To find another point, we move 2 units *upward* from $(3, 4)$ (upward because the numerator is positive) and 3 units toward the *left* (toward the left because the denominator is negative). We see that when we interpret the slope as $\frac{2}{-3}$, we simply get a different point on the same line.

Exercises 8.3
Set I

In Exercises 1–8, find the slope of the line that passes through each given pair of points.

1. $(1, 4)$ and $(10, 6)$
2. $(-3, -5)$ and $(3, 0)$
3. $(-5, -5)$ and $(1, -7)$
4. $(-1, 0)$ and $(7, -4)$
5. $(-7, -5)$ and $(2, -5)$
6. $(0, -2)$ and $(5, -2)$
7. $(-4, 3)$ and $(-4, -2)$
8. $(-5, 2)$ and $(-5, 7)$

In Exercises 9 and 10, graph each set of lines on the same axes, and find the slope of each line.

9. a. $y = \frac{3}{4}x$ b. $y = \frac{3}{4}x + 5$ c. $y = \frac{3}{4}x - 2$

10. a. $y = x$ b. $y = x - 3$ c. $y = x + 4$

In Exercises 11–14, graph the line described.

11. Passes through $(-1, 3)$ and has a slope of 3
12. Passes through $(2, -4)$ and has a slope of 5
13. Passes through $(0, 4)$ and has a slope of $-\frac{2}{5}$
14. Passes through $(1, -3)$ and has a slope of $-\frac{5}{3}$

Writing Problems

Express the answers in your own words and in complete sentences.

1. Explain what the slope of a line is.
2. Explain why the slope of a vertical line does not exist.
3. Explain how to recognize the equation of a vertical line.

Exercises 8.3 Set II

In Exercises 1–8, find the slope of the line that passes through each given pair of points.

1. $(-5, -8)$ and $(-9, 4)$
2. $(9, 15)$ and $(-6, 3)$
3. $(-2, -11)$ and $(-7, -11)$
4. $(14, -9)$ and $(14, -21)$
5. $(-3, 2)$ and $(4, -3)$
6. $(8, 1)$ and $(8, 6)$
7. $(0, 5)$ and $(-3, 5)$
8. $(4, -2)$ and $(-2, 4)$

In Exercises 9 and 10, graph each set of lines on the same axes, and find the slope of each line.

9. a. $y = -\frac{1}{2}x$ b. $y = -\frac{1}{2}x + 2$ c. $y = -\frac{1}{2}x - 3$

10. a. $y = -4x$ b. $y = -4x + 3$ c. $y = -4x - 2$

In Exercises 11–14, graph the line described.

11. Passes through $(5, 0)$ and has a slope of -3
12. Passes through $(-2, 4)$ and has a slope of $\frac{3}{5}$
13. Passes through $(0, 0)$ and has a slope of 0
14. Passes through $(3, -1)$ and has a slope of $-\frac{1}{4}$

8.4 Writing the Equation of a Straight Line; Graphing a Line Whose Equation Is in the Slope-Intercept Form

In this section, we show how to convert the equation of a straight line into one of several different forms, and we show how to write the *equation* of a line when certain facts about the line are known.

Three forms of the equation of a straight line are especially useful: the *general form*, the *point-slope form*, and the *slope-intercept form.*

The General Form of the Equation of a Line

The **general form of the equation of a line**, *as used in this text*, is defined as follows:

The general form of the equation of a line

$$Ax + By + C = 0^* \quad \text{or} \quad 0 = Ax + By + C$$

where A, B, and C are integers, $A \geq 0$, and A and B are not both 0.

*Some authors give $Ax + By = C$ as the general form. We use the form found in most higher-level mathematics books.

A is the coefficient of x, and the x-term is written to the left of the y-term; B is the coefficient of y; C is the constant, and it is written to the right of the y-term. Zero must be the only term on one side of the equation.

To express the equation of a straight line in the *general form*, we must remove any grouping symbols, remove any denominators by multiplying both sides of the equation by the LCD, move all terms to one side of the equation, and combine all like terms. If A is then negative, we change *all* the signs of the equation (or multiply both sides of the equation by -1).

EXAMPLE 1 Write $-\frac{2}{3}x + \frac{1}{2}y = 1$ in the general form.

SOLUTION The LCD is 6.

$$-\tfrac{2}{3}x + \tfrac{1}{2}y = 1$$

$$6\left(-\tfrac{2}{3}x + \tfrac{1}{2}y\right) = (6)\,(1) \quad \text{Multiplying both sides by 6}$$

$$\tfrac{6}{1}\cdot\left(-\tfrac{2}{3}x\right) + \tfrac{6}{1}\cdot\left(\tfrac{1}{2}y\right) = 6 \quad \text{Using the distributive property on the left}$$

$$-4x + 3y = 6$$

$$-4x + 3y - 6 = 0 \quad \text{Next, we will change all the signs of the equation}$$

$$4x - 3y + 6 = 0 \quad \text{The general form}$$

The Point-Slope Form of the Equation of a Line

To derive the **point-slope form of the equation of a line**, we let $P_1(x_1, y_1)$ be a known, fixed point on a line whose slope is m, and we let $P(x, y)$ be *any* other point on that line (see Figure 12). Using the definition of the slope, we have

$$\frac{y - y_1}{x - x_1} = m$$

$$\frac{(x - x_1)}{1}\cdot\frac{y - y_1}{x - x_1} = \frac{(x - x_1)}{1}\cdot m \quad \text{Multiplying both sides by the LCD}$$

$$y - y_1 = m(x - x_1)$$

The point-slope form

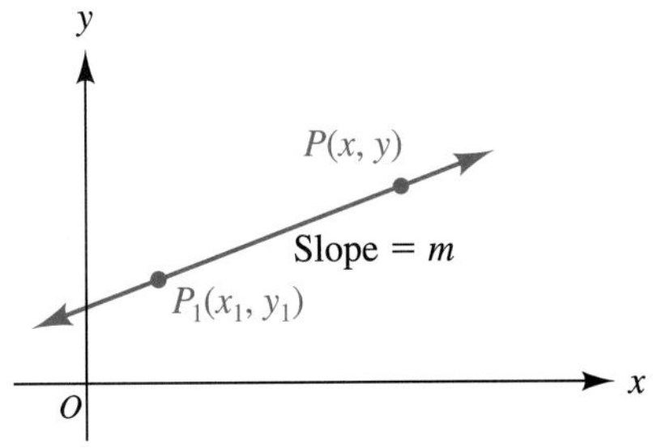

Figure 12

The point-slope form of the equation of a line

$$y - y_1 = m(x - x_1)$$

where m is the slope of the line and $P_1(x_1, y_1)$ is a known, fixed point on the line.

We might think of the *point-slope form* as a *formula.* We *use* this formula—the point-slope form of the equation of a line—to write the equation of a line when we are given the *slope* of the line and a *point* through which the line passes. We substitute (or "plug in") numerical values for y_1, m, and x_1. The x that has no subscript *remains* x, and the y that has no subscript *remains* y (see Example 2).

EXAMPLE 2 Write the equation of the line that passes through $(-1, 4)$ and has a slope of $-\frac{2}{3}$. Express the final answer in the general form.

SOLUTION Because we're given a *point* on the line and the *slope* of the line, we first use the *point-slope* form of the equation of a line. We substitute -1 for x_1, 4 for y_1, and $-\frac{2}{3}$ for m.

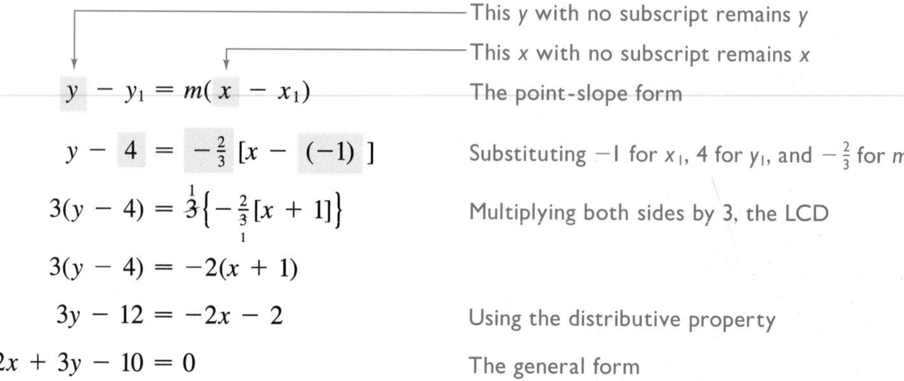

This y with no subscript remains y
This x with no subscript remains x
$y - y_1 = m(x - x_1)$ The point-slope form
$y - 4 = -\frac{2}{3}[x - (-1)]$ Substituting −1 for x_1, 4 for y_1, and $-\frac{2}{3}$ for m
$3(y - 4) = \overset{1}{3}\{-\frac{2}{\underset{1}{3}}[x + 1]\}$ Multiplying both sides by 3, the LCD
$3(y - 4) = -2(x + 1)$
$3y - 12 = -2x - 2$ Using the distributive property
$2x + 3y - 10 = 0$ The general form

The Slope-Intercept Form of the Equation of a Line

Suppose $(0, b)$ is given as the *y-intercept* of a line whose *slope* is m (see Figure 13). If we let $P_1(x_1, y_1)$ be $(0, b)$, then $x_1 = 0$ and $y_1 = b$. Substituting into the point-slope form of the equation of a line, we have

$y - y_1 = m(x - x_1)$ The point-slope form

$y - b = m(x - 0)$ Substituting b for y_1 and 0 for x_1

$y - b = mx$

$y = mx + b$ ← The slope-intercept form

The slope (m) — The y-intercept* (b)

Figure 13

The slope-intercept form of the equation of a line

$$y = mx + b$$

where m is the slope of the line and b is the y-intercept of the line.

We *use* the *slope-intercept form* of the equation of a line when we are given the *slope* of a line and its *y-intercept* and are asked to write the equation of the line (see Example 3). The point-slope form could be used instead, if we remember that the ordered pair corresponding to the y-intercept is $(0, b)$.

EXAMPLE 3 Write the equation of the line that has a slope of $-\frac{3}{4}$ and a y-intercept of -2. Express the final answer in the general form.

SOLUTION Because we're given the *slope* and the *y-intercept*, we use the *slope-intercept* form of the equation of a line. We substitute $-\frac{3}{4}$ for m and -2 for b.

---

*When we say that the y-intercept is b, we really mean that the y-intercept is the point $(0, b)$.

This y remains y — This x remains x

$y = mx + b$	The slope-intercept form
$y = -\frac{3}{4}x + (-2)$	Substituting
$4y = 4\left(-\frac{3}{4}x - 2\right)$	Multiplying both sides by 4, the LCD
$4y = \overset{1}{4}\left(-\frac{3}{\underset{1}{4}}x\right) + 4(-2)$	Using the distributive property on the right
$4y = -3x - 8$	Simplifying
$3x + 4y + 8 = 0$	The general form

Note You might verify that we obtain the same answer for Example 3 if we use the point-slope form of the equation, remembering that the ordered pair corresponding to a y-intercept of -2 is $(0, -2)$.

Writing the Equation of a Horizontal Line

Because a horizontal line *has* a slope, its equation can be written by using the point-slope form or the slope-intercept form, *or* the following statement can be memorized (see Figure 14):

> The equation of any horizontal line can be expressed as $y = b$ or as $y - b = 0$, where b is the y-coordinate of every point on the line.

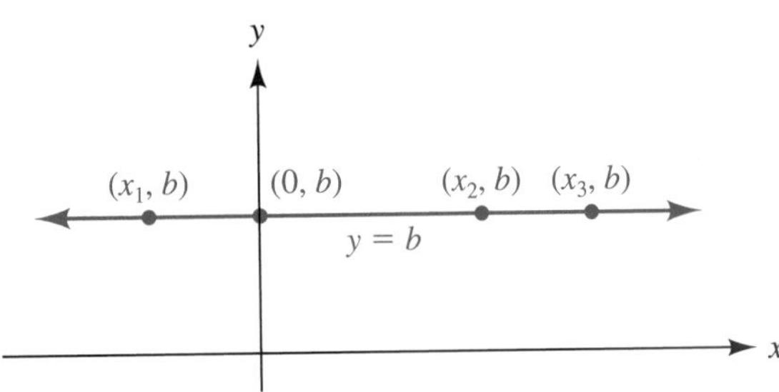

Figure 14

EXAMPLE 4 Write the equation of the horizontal line that passes through $(-2, 4)$ (see Figure 15).

SOLUTION 1 To use the rule in the box, we note that the y-coordinate of the given point is 4. Therefore, the equation is

$$y = 4$$

or $\quad y - 4 = 0$ The general form

SOLUTION 2 To use the point-slope form, we note that the slope of every horizontal line is 0. Therefore, we have

$$y - y_1 = m(x - x_1)$$

$$y - 4 = 0[x - (-2)]$$

$$y - 4 = 0 \quad \text{The general form}$$

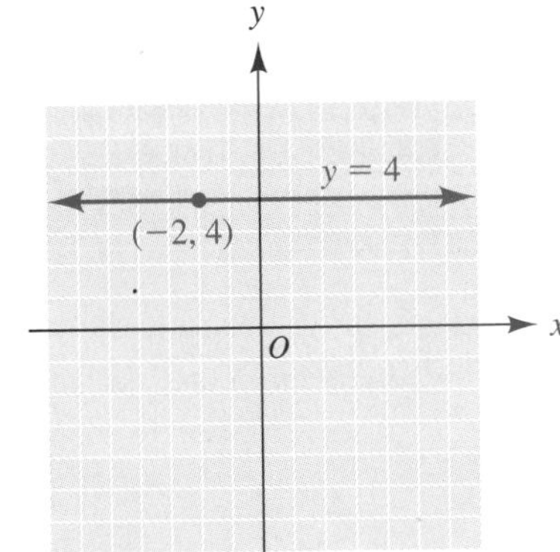

Figure 15

Writing the Equation of a Vertical Line

Because a vertical line *has no slope*, we cannot use the point-slope form or the slope-intercept form to write its equation. The only way to write the equation of a vertical line is to memorize the following statement (see Figure 16):

> The equation of any vertical line can be expressed as $x = k$ or as $x - k = 0$, where k is the x-coordinate of every point on the line.

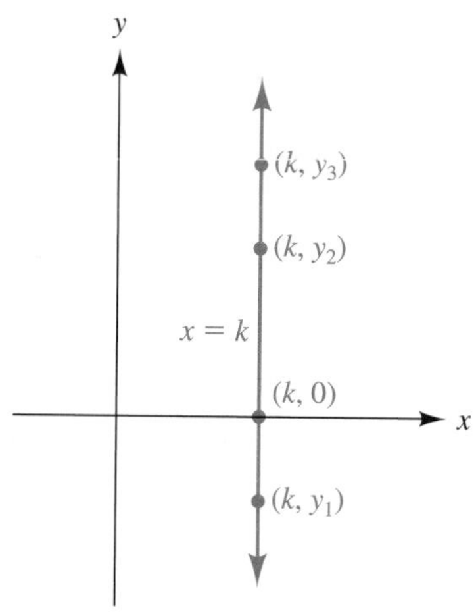

Figure 16

EXAMPLE 5 Write the equation of the vertical line that passes through $(-2, 3)$ (see Figure 17).

SOLUTION Because the x-coordinate of the given point is -2, the equation of the line must be

$$x = -2$$

or

$$x + 2 = 0 \quad \text{The general form}$$

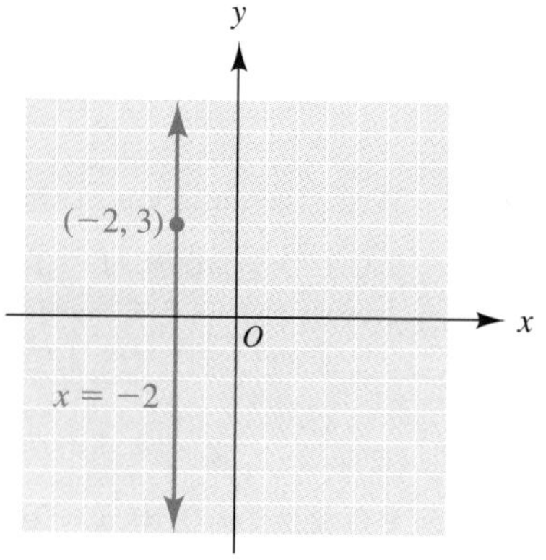

Figure 17

Writing the Equation of a Line When Two Points on the Line Are Known

Example 6 illustrates the technique for writing the equation of a line when we are given two points on the line.

EXAMPLE 6 Find the equation of the line that passes through the points $(-15, -9)$ and $(-5, 3)$.

SOLUTION

Step 1. We first find the slope from the two given points:

$$m = \frac{3 - (-9)}{-5 - (-15)} = \frac{12}{10} = \frac{6}{5}$$

Step 2. We then use this slope with *either* given point to find the equation of the line.

Using the point $(-5, 3)$	*Using the point* $(-15, -9)$
$y - y_1 = m(x - x_1)$	$y - y_1 = m(x - x_1)$
$y - 3 = \frac{6}{5}[x - (-5)]$	$y - (-9) = \frac{6}{5}[x - (-15)]$
$5y - 15 = 6(x + 5)$	$5(y + 9) = 6(x + 15)$
$5y - 15 = 6x + 30$	$5y + 45 = 6x + 90$
$0 = 6x - 5y + 45$	$0 = 6x - 5y + 45$

Notice that the same equation is obtained no matter which of the two given points is used as the known, fixed point.

We are sometimes asked to express the equation of a line in the slope-intercept form. To do this, we solve the equation for y by using the method given earlier for solving literal equations (see Example 7). When an equation is expressed in the slope-intercept form, we can easily identify the slope (it is the coefficient of x) and the y-intercept (it is the constant) of the line.

EXAMPLE 7 Express $2x + 3y + 6 = 0$ in the slope-intercept form, find the slope of the line, and find the y-intercept of the line.

SOLUTION We must solve the given equation for y.

$$2x + 3y + 6 = 0 \quad \text{The given equation}$$

$$3y = -2x - 6$$

$$y = -\frac{2}{3}x + (-2) \quad \text{This is in the form } y = mx + b, \text{ the slope-intercept form}$$

The slope: $-\frac{2}{3}$; The y-intercept: (-2)

The slope-intercept form of the equation is $y = -\frac{2}{3}x - 2$, the slope is $-\frac{2}{3}$, and the y-intercept is -2.

Recall that if two lines are parallel, they have the same slope; and if two lines are perpendicular, the *product* of their slopes must be -1 (that is, the slopes must be the *negative reciprocals* of each other). We use these facts in Examples 8 and 9.

EXAMPLE 8

Write the equation of the line that passes through the point $(-4,7)$ and is parallel to the line $2x + 3y + 6 = 0$.

SOLUTION We found in Example 7 that the slope of $2x + 3y + 6 = 0$ is $-\frac{2}{3}$; therefore, we must write the equation of a line whose slope *equals* $-\frac{2}{3}$ (we want the slopes to be equal, since the lines are to be parallel), and the line must pass through the point $(-4,7)$.

$y - y_1 = m(x - x_1)$	We use the point-slope form, since we know a point and the slope
$y - 7 = -\frac{2}{3}[x - (-4)]$	Substituting -4 for x_1, 7 for y_1, and $-\frac{2}{3}$ for m
$3(y - 7) = \overset{1}{\cancel{3}}\left\{-\frac{2}{\underset{1}{\cancel{3}}}[x + 4]\right\}$	Multiplying both sides by 3, the LCD
$3y - 21 = -2x - 8$	Using the distributive property
$2x + 3y - 13 = 0$	The general form

EXAMPLE 9

Write the equation of the line that passes through the point $(-4,7)$ and is perpendicular to the line $2x + 3y + 6 = 0$.

SOLUTION Because the slope of $2x + 3y + 6 = 0$ is $-\frac{2}{3}$, we must write the equation of the line whose slope is $\frac{3}{2}$ (we use the *negative reciprocal* of $-\frac{2}{3}$, since the lines are to be perpendicular), and the line must pass through the point $(-4,7)$.

$y - y_1 = m(x - x_1)$	We use the point-slope form, since we know a point and the slope
$y - 7 = \frac{3}{2}[x - (-4)]$	Substituting -4 for x_1, 7 for y_1, and $\frac{3}{2}$ for m
$2(y - 7) = \overset{1}{\cancel{2}}\left\{\frac{3}{\underset{1}{\cancel{2}}}[x + 4]\right\}$	Multiplying both sides by 2, the LCD
$2y - 14 = 3x + 12$	Using the distributive property
$-3x + 2y - 26 = 0$	Next, we must change all the signs
$3x - 2y + 26 = 0$	The general form

We summarize here the important forms of the equation of a line.

The forms of the equation of a line

The general form	$Ax + By + C = 0$	where A, B, and C are integers, $A \geq 0$, and A and B are not both 0
The point-slope form	$y - y_1 = m(x - x_1)$	where m is the slope and $P_1(x_1, y_1)$ is a known, fixed point on the line
The slope-intercept form	$y = mx + b$	where m is the slope and b is the y-intercept
The equation of a horizontal line	$y = b$	where b is the y-coordinate of every point on the line
The equation of a vertical line	$x = k$	where k is the x-coordinate of every point on the line

We can graph a line whose equation is in the slope-intercept form by using the techniques for graphing a line when its slope and one point on the line are given, as was done in Example 8 of the previous section. In the slope-intercept form, the known point is $(0, b)$, and the slope is m (see Example 10).

EXAMPLE 10 Graph $y = -3x + 2$.

SOLUTION The equation is in the slope-intercept form. The y-intercept is 2, or $(0, 2)$, and the slope is -3, which we can interpret as $\frac{-3}{+1}$. We graph the point $(0, 2)$. To find another point on the line, we move 3 units *downward* from $(0, 2)$, and then, from that point, we move 1 unit toward the *right*. This gives the point $(1, -1)$. If we interpret the slope as $\frac{+3}{-1}$, we move 3 units *upward* and 1 unit toward the *left* from $(0, 2)$. This gives the point $(-1, 5)$, another point on the same line. We draw a straight line through these points and draw an arrowhead at each end of the line.

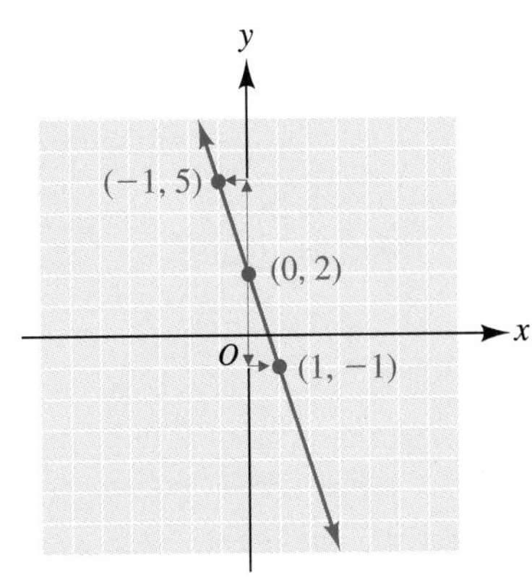

Exercises 8.4
Set I

In Exercises 1–8, write each equation in the general form.

1. $5x = 3y - 7$
2. $4x = -9y + 3$
3. $\frac{x}{2} - \frac{y}{5} = 1$
4. $\frac{y}{6} - \frac{x}{7} = 1$
5. $y = -\frac{5}{3}x + 4$
6. $y = -\frac{3}{8}x - 5$
7. $4(x - y) = 11 - 2(x + 3y)$
8. $15 - 6(3x + y) = 7(x - 2y)$

In Exercises 9–14, write an equation of the line that passes through the given point and has the indicated slope. Express the answer in the general form.

9. $(4, -3)$, $m = \frac{1}{5}$
10. $(-2, -1)$, $m = -\frac{5}{6}$
11. $(-6, 5)$, $m = \frac{1}{4}$
12. $(-3, -2)$, $m = -\frac{4}{5}$
13. $(-1, 3)$, $m = -4$
14. $(3, 0)$, $m = 2$

In Exercises 15–20, write an equation of the line that has the indicated slope and y-intercept. Express the answer in the general form.

15. $m = \frac{5}{7}$, y-intercept $= -3$
16. $m = -\frac{1}{4}$, y-intercept $= -2$
17. $m = -\frac{4}{3}$, y-intercept $= \frac{1}{2}$
18. $m = -\frac{3}{5}$, y-intercept $= \frac{3}{4}$
19. $m = 0$, y-intercept $= 5$
20. $m = 0$, y-intercept $= 7$

21. Write an equation of the horizontal line that passes through the point $(-4, 3)$.
22. Write an equation of the horizontal line that passes through the point $(2, -5)$.
23. Write an equation of the vertical line that passes through the point $(7, -2)$.
24. Write an equation of the vertical line that passes through the point $(-6, 4)$.

In Exercises 25–28, (a) write the given equation in the slope-intercept form, (b) find the slope of the line, and (c) find the y-intercept of the line.

25. $4x - 5y + 20 = 0$
26. $8x + 3y - 24 = 0$
27. $\frac{2}{3}x + 3y + 5 = 0$
28. $3x - \frac{5}{6}y - 2 = 0$

In Exercises 29–34, write an equation of the line that passes through the given points. Express the answer in the general form.

29. $(8, -1)$ and $(6, 4)$
30. $(7, -2)$ and $(5, 1)$
31. $(10, 0)$ and $(7, 4)$
32. $(-4, 0)$ and $(-6, -5)$
33. $(-9, 3)$ and $(-3, -1)$
34. $(-11, 4)$ and $(-3, -2)$

In Exercises 35–42, express the final answer in the general form.

35. Write an equation of the line that passes through $(-4, 7)$ and is parallel to the line $3x - 5y = 6$.

36. Write an equation of the line that passes through $(8, -5)$ and is parallel to the line $7x + 4y + 3 = 0$.

37. Write an equation of the line that passes through $(6, 2)$ and is perpendicular to the line $2x + 4y = 3$.

38. Write an equation of the line that passes through $(5, -1)$ and is perpendicular to the line $3x - 6y = 2$.

39. Write an equation of the line that has an x-intercept of 4 and is parallel to the line $3x + 5y - 12 = 0$.

40. Write an equation of the line that has an x-intercept of -3 and is parallel to the line $9x - 14y + 6 = 0$.

41. Write an equation of the line that has an x-intercept of -6 and a y-intercept of 4.

42. Write an equation of the line that has an x-intercept of 15 and a y-intercept of -12.

In Exercises 43–48, graph each line by using the slope and the y-intercept of the line.

43. $y = -4x - 2$
44. $y = -3x + 1$
45. $y = 2x + 3$
46. $y = 3x + 1$
47. $y = \frac{2}{3}x + 1$
48. $y = \frac{3}{2}x + 2$

Writing Problems

Express the answers in your own words and in complete sentences.

1. Describe what your first step would be in writing the equation of a line if you were given two points on the line, and explain why you would do that first.
2. Explain how you would know that the lines $y = \frac{7}{9}x + 8$ and $y = \frac{7}{9}x - 5$ are parallel.
3. Explain how you would know that the lines $y = \frac{7}{9}x + 8$ and $y = -\frac{9}{7}x - 5$ are perpendicular.
4. Explain why you would *not* write the equation of a line by using the slope-intercept form of the equation if it was given that the slope of the line was $-\frac{1}{5}$ and that the line passed through $(1, 4)$.
5. Describe how you would write the equation of a line parallel to the line $4x - 7y = 3$.
6. Describe how you would write the equation of a line perpendicular to the line $4x - 7y = 3$.

Exercises 8.4 Set II

In Exercises 1–8, write each equation in the general form.

1. $6y = 8 - 13x$
2. $8y = 2x - 4$
3. $\dfrac{x}{-3} - \dfrac{y}{6} = 1$
4. $5x = 2y$
5. $y = \frac{2}{9}x - 13$
6. $x = \frac{1}{5}y + 3$
7. $17 - 5(2x - y) = 8(x + 4y)$
8. $3x - 5(2y - 5x) = 4y + 7x - 1$

In Exercises 9–14, write an equation of the line that passes through the given point and has the indicated slope. Express the answer in the general form.

9. $(7, -4)$, $m = \frac{1}{6}$
10. $(8, 3)$, $m = -4$
11. $(-5, -8)$, $m = -\frac{3}{4}$
12. $(0, 0)$, $m = \frac{2}{5}$
13. $(2, -5)$, $m = -1$
14. $(-3, 0)$, $m = 8$

In Exercises 15–20, write an equation of the line that has the indicated slope and y-intercept. Express the answer in the general form.

15. $m = \frac{5}{6}$, y-intercept $= -4$
16. $m = -\frac{2}{9}$, y-intercept $= \frac{1}{3}$
17. $m = 0$, y-intercept $= -3$
18. $m = 3$, y-intercept $= 0$
19. $m = -\frac{1}{4}$, y-intercept $= 1$
20. $m = -5$, y-intercept $= -3$
21. Write an equation of the horizontal line that passes through the point $(-3, 7)$.
22. Write an equation of the horizontal line that passes through the point $(-9, -7)$.
23. Write an equation of the vertical line that passes through the point $(-5, -1)$.
24. Write an equation of the vertical line that passes through the point $(0, -5)$.

In Exercises 25–28, (a) write the given equation in the slope-intercept form, (b) find the slope of the line, and (c) find the y-intercept of the line.

25. $3x - 5y + 30 = 0$
26. $8x + 3y = 5$
27. $\frac{4}{5}x - 6y + 3 = 0$
28. $3x + \frac{4}{7}y - 5 = 0$

In Exercises 29–34, write an equation of the line that passes through the given points. Express the answer in the general form.

29. $(12, -7)$ and $(8, -9)$
30. $(-17, 3)$ and $(-5, -12)$

31. (14, −6) and (−1, 4)
32. (−3, 5) and (−3, 2)
33. (4, 7) and (7, 4)
34. (0, 4) and (−3, 5)

In Exercises 35–42, express the final answer in the general form.

35. Write an equation of the line that passes through (−7, −13) and is parallel to the line $6x - 8y = 15$.

36. Write an equation of the line that passes through (−7, −13) and is parallel to the line $6y - 4x = 5$.

37. Write an equation of the line that passes through (0, 0) and is perpendicular to the line $2x + 5y = 10$.

38. Write an equation of the line that passes through (0, 3) and is perpendicular to the line $2y + 5x = 1$.

39. Write an equation of the line that has an x-intercept of −8 and is parallel to the line $12x - 9y - 7 = 0$.

40. Write an equation of the line that has a y-intercept of −5 and is parallel to the line $8x + 3y = 5$.

41. Write an equation of the line that has an x-intercept of −14 and a y-intercept of −6.

42. Write an equation of the line that has an x-intercept of 3 and a y-intercept of 5.

In Exercises 43–48, graph each line by using the slope and the *y*-intercept of the line.

43. $y = -2x + 4$
44. $y = -4x - 1$
45. $y = 3x + 2$
46. $y = \frac{1}{4}x + 1$
47. $y = \frac{3}{4}x + 2$
48. $y = -\frac{4}{3}x - 3$

8.5 Graphing Solution Sets of Linear Inequalities in Two Variables

Earlier, we solved linear inequalities in one variable, and the graph of the solution set was generally some portion of the number line, *not* just a point. In this section, we graph the solution sets of linear inequalities in two variables *in the plane* of the rectangular coordinate system. The graph must contain, within the limits of the size of the graph, the set of *all* ordered pairs—and only those ordered pairs—that satisfy the given inequality.

Half-Planes Any line in a plane divides that plane into two **half-planes.** For example, in Figure 18, the line $\overleftrightarrow{AB}$ separates the plane into the two half-planes shown.

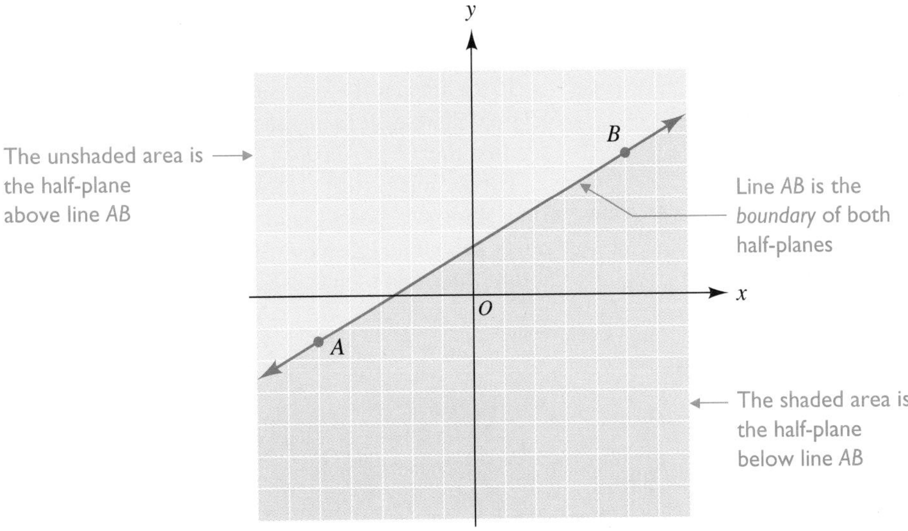
y
B
The unshaded area is the half-plane above line AB
Line AB is the boundary of both half-planes
x
O
A
The shaded area is the half-plane below line AB

Figure 18

The following statement must be memorized; it will not be proved.

> The graph of the solution set of a linear inequality in one or two variables is a half-plane.

Graphing the Boundary Line of a Half-Plane To graph a half-plane, we first find the equation of the boundary line. This equation is obtained by replacing the inequality symbol with an equal sign. For example, if the inequality is $-3x + 4y > 12$, the equation of the boundary line is $-3x + 4y = 12$. Next, we graph the boundary line. We discuss below whether to graph the boundary line as a solid line or as a dashed line.

How to Determine Whether the Boundary Is a Solid Line or a Dashed Line If the inequality symbol is $\geq$ or $\leq$, then points *on* the boundary line make the original statement *true*, and they must be part of the solution set; therefore, the boundary line must be a *solid* line.

For example, if the inequality is $3x - 4y \leq 12$, all the points on the boundary line $3x - 4y = 12$ [points such as $(4, 0)$ and $(0, -3)$] make $3x - 4y \leq 12$ true (you can verify this), and these points *are* part of the solution set. They should be graphed as *solid dots*, and then those dots should be connected with a solid line (see Figure 19).

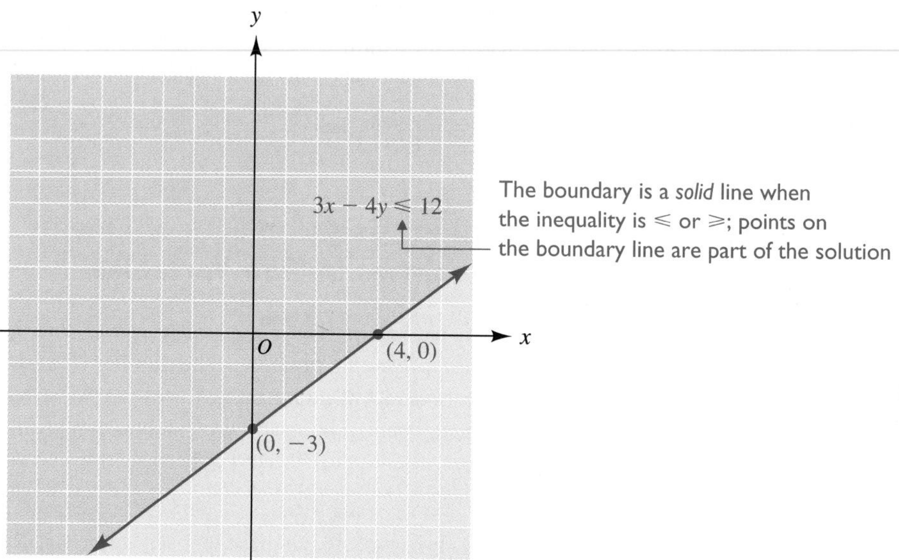
y
x
3x − 4y ≤ 12
O
(4, 0)
(0, −3)
The boundary is a solid line when the inequality is ≤ or ≥; points on the boundary line are part of the solution

Figure 19

If the inequality symbol is $>$ or $<$, then points on the boundary line make the original statement *false*, and they are *not* part of the solution set; therefore, the boundary line must be a *dashed* line.

For example, if the inequality is $3x - 4y < 12$, all the points on the boundary line $3x - 4y = 12$ [points such as $(4, 0)$ and $(0, -3)$] make $3x - 4y < 12$ *false* (you might verify this), and these points are *not* part of the solution set. They should be graphed as *hollow dots*, and then those dots should be connected with a dashed line (see Figure 20).

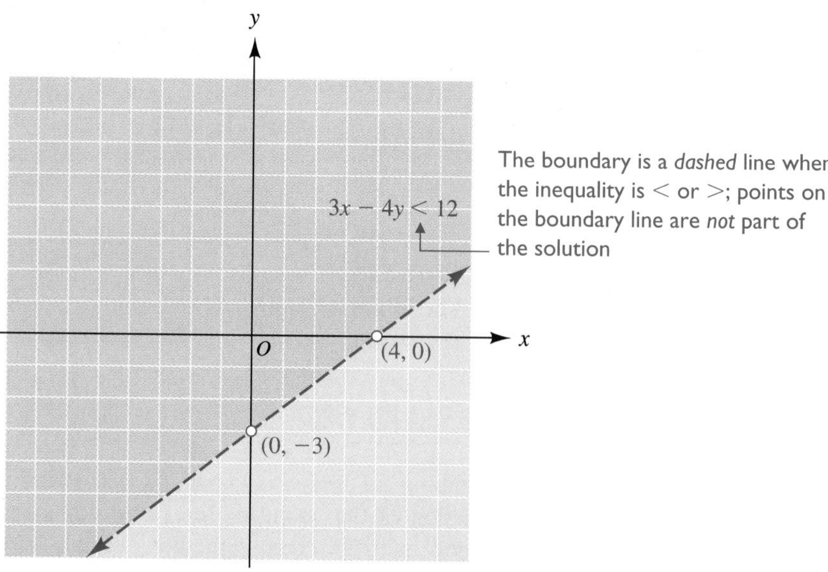
y
x
3x − 4y < 12
O
(4, 0)
(0, −3)
The boundary is a dashed line when the inequality is < or >; points on the boundary line are not part of the solution

Figure 20

Finally, we shade one of the two half-planes. We next discuss how to determine which half-plane to shade.

How to Determine the Correct Half-Plane by Using a Test-Point Select any point that does *not* lie on the boundary line, and substitute the coordinates of that point into the inequality. (If the boundary line does not go through the origin, then the origin is the best point to select.)

If the resulting statement is *true*, the half-plane that contains the point selected should be shaded (see Example 1). The graph of the solution set is the shaded half-plane. That is, an ordered pair that corresponds to *any* point that is in the shaded region is in the solution set.

If the resulting statement is *false*, the half-plane that does *not* contain the point selected should be shaded (see Example 2). The graph of the solution set is the shaded half-plane.

The procedure can be summarized as follows:

Graphing the solution set of a linear inequality in one or two variables in the plane

1. Find the equation of the boundary line by replacing the inequality symbol with an equal sign.
2. Graph the boundary line. It is a *solid* line if the inequality symbol is $\geq$ or $\leq$. It is a *dashed* line if the inequality symbol is $>$ or $<$.
3. Select any point that does *not* lie on the boundary line, and substitute the coordinates of that point into the inequality. If the resulting statement is *true*, shade the half-plane that contains the point selected. If the resulting statement is *false*, shade the half-plane that does *not* contain the point selected.

EXAMPLE 1 Graph the solution set of $2x - 3y < 6$.

SOLUTION The graph of the solution set of a linear inequality in two variables is a half-plane.

Step 1. We replace the inequality symbol with an equal sign to get the equation of the boundary line. The boundary line is $2x - 3y = 6$.

Step 2. We graph the boundary line $2x - 3y = 6$ by first finding three points on it.

		x	y
x-intercept	Let $y = 0$; if $y = 0$, $x = 3$.	3	0
y-intercept	Let $x = 0$; if $x = 0$, $y = -2$.	0	-2
Checkpoint	Let $x = -3$; if $x = -3$, $y = -4$.	-3	-4

Because the inequality symbol is $<$ (equality is *not* included), we graph the points $(3, 0)$, $(0, -2)$, and $(-3, -4)$ as *hollow dots*, and we connect them with a *dashed* line. (The points on this line are *not* in the solution set of $2x - 3y < 6$.)

Step 3. To select the correct half-plane, we'll choose the origin as the test-point. (The boundary line does not pass through the origin.) Substituting the coordinates of the origin, (0, 0), into the inequality, we have

$$2x - 3y < 6$$
$$2(0) - 3(0) < 6$$
$$0 < 6 \quad \text{True}$$

Therefore, we shade the half-plane that contains the origin. Any ordered pair that corresponds to a point in the shaded region of the graph is in the solution set.

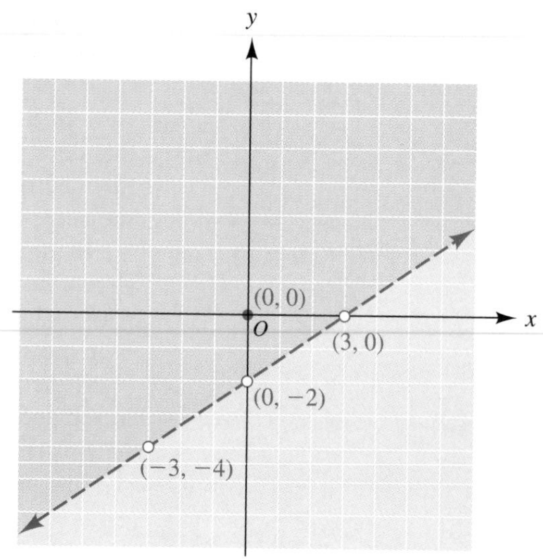

EXAMPLE 2 Graph the solution set of $3x + 4y \le -12$.

SOLUTION The graph is a half-plane because we're graphing a linear inequality in two variables.

Step 1. We replace the inequality symbol with an equal sign to get the equation of the boundary line, so the boundary line is $3x + 4y = -12$.

Step 2. We graph the boundary line by first finding three points on it.

x-intercept Let $y = 0$; if $y = 0$, $x = -4$.

y-intercept Let $x = 0$; if $x = 0$, $y = -3$.

Checkpoint Let $x = 4$; if $x = 4$, $y = -6$.

x	y
−4	0
0	−3
4	−6

Because the inequality symbol is $\le$ (equality *is* included), we graph the points $(-4, 0)$, $(0, -3)$, and $(4, -6)$ as *solid dots*, and we connect them with a *solid* line. (The points on this line *are* in the solution set of $3x + 4y \le -12$.)

Step 3. To select the correct half-plane, we'll choose the origin as the test-point. (The boundary line does not pass through the origin.) Substituting the coordinates of the origin, (0, 0), into the inequality, we have

$$3x + 4y \le -12$$
$$3(0) + 4(0) \le -12$$
$$0 \le -12 \quad \text{False}$$

Therefore, we shade the half-plane that does *not* contain the origin. Any ordered pair that corresponds to a point in the shaded region of the graph is in the solution set.

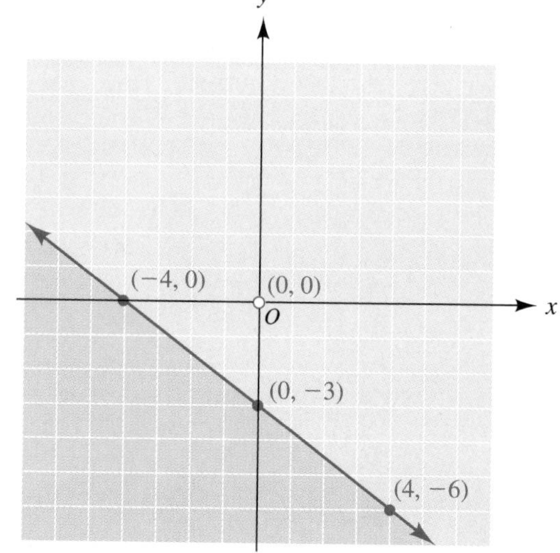

Some inequalities contain only one variable. Such inequalities have graphs whose boundaries are either *vertical* lines or *horizontal* lines.

EXAMPLE 3

Graph the solution set of $x + 4 < 0$.

SOLUTION The graph will be a half-plane.

Step 1. We replace the inequality symbol with an equal sign to get the equation of the boundary line, so the boundary line is $x + 4 = 0$, or $x = -4$, which is a vertical line.

Step 2. Because the inequality symbol is $<$ (equality is *not* included), we graph $(-4, 0)$ as a *hollow dot*, and we graph the vertical line that passes through $(-4, 0)$ as a *dashed* line. (The points on this line are *not* in the solution set of $x + 4 < 0$.)

Step 3. To select the correct half-plane, we choose the origin as the test-point. (The boundary line does not pass through the origin.) Substituting the coordinates of the origin, $(0, 0)$, into the inequality, we have

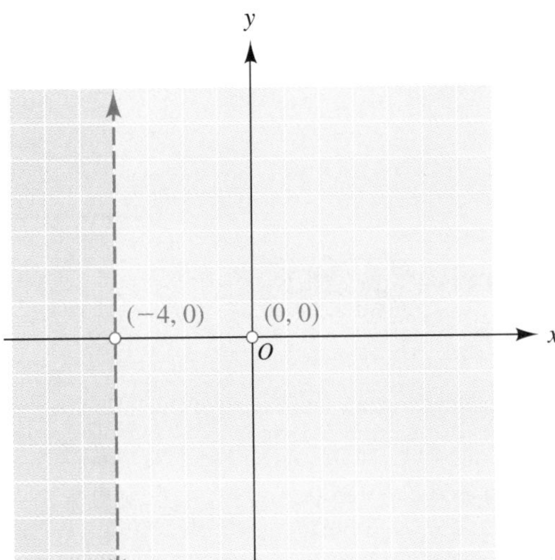

$x + 4 < 0$

$0 + 4 < 0$

$4 < 0$ False

Therefore, we shade the half-plane that does *not* contain the origin. Any ordered pair that corresponds to a point in the shaded region of the graph is in the solution set.

Note Early in this book, we were working in one-dimensional space, so when we graphed $x < -4$, the graph was a portion of a straight line. In this section, however, we are working in two-dimensional space, and Example 3 of this section shows that the graph of the solution set of the inequality $x + 4 < 0$ is an entire half-plane when it is graphed in the rectangular coordinate system. (The inequality is viewed as $x + 0y + 4 < 0$.)

EXAMPLE 4

Graph the solution set of $2y - 5x \geq 0$.

SOLUTION The graph will be a half-plane.

Step 1. We replace the inequality symbol with an equal sign to get the equation of the boundary line, so the boundary line is $2y - 5x = 0$.

Step 2. We graph the boundary line by first finding three points on it.

x-intercept Let $y = 0$; if $y = 0$, $x = 0$.

Second point Let $x = 2$; if $x = 2$, $y = 5$.

Checkpoint Let $x = -2$; if $x = -2$, $y = -5$.

x	y
0	0
2	5
−2	−5

[Both intercepts were $(0, 0)$, so we had to find two other points on the line.] Because the inequality symbol is $\geq$ (equality *is* included), we graph the points $(0, 0)$, $(2, 5)$, and $(-2, -5)$ as *solid dots*, and we connect them with a *solid* line. (The points on this line *are* in the solution set of $2y - 5x \geq 0$.)

Step 3. We *cannot* choose the origin as the test-point, since the boundary line passes through the origin. Therefore, to select the correct half-plane, we'll try (1, 0). Substituting 1 for x and 0 for y, we have

$$2y - 5x \geq 0$$
$$2(0) - 5(1) \geq 0$$
$$-5 \geq 0 \quad \text{False}$$

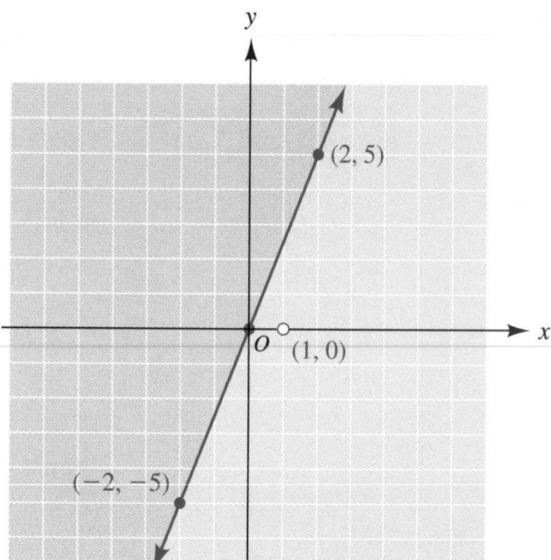

[Any other point *not* on the boundary could have been used instead of (1, 0).] Therefore, we shade the half-plane that does *not* contain the point (1, 0). The graph of the solution set is the shaded region.

Exercises 8.5
Set I

Graph the solution set of each of the following inequalities.

1. $4x + 5y < 20$
2. $5x - 3y > 15$
3. $3x - 8y > -16$
4. $6x + 5y < -18$
5. $9x + 7y \leq -27$
6. $5x - 14y \geq -28$
7. $x \geq -1$
8. $y \leq -4$
9. $6x - 13y > 0$
10. $4x + 9y < 0$
11. $14x + 3y \leq 17$
12. $10x - 4y \geq 23$
13. $y \geq -4x - 2$
14. $y \leq -3x + 1$
15. $y > 2x + 3$
16. $y < 3x + 1$
17. $\frac{x}{4} - \frac{y}{2} > 1$
18. $\frac{x}{3} + \frac{y}{5} < 1$
19. $4(x + 2) + 7 \leq 3(5 - 2x)$
20. $3(y + 4) - 8 \geq 4(1 - 2y)$
21. $\frac{2x + y}{3} - \frac{x - y}{2} \geq \frac{5}{6}$
22. $\frac{4x - 3y}{5} - \frac{2x - y}{2} \geq \frac{2}{5}$

Writing Problems

Express the answers in your own words and in complete sentences.

1. Explain how you would decide whether the boundary line for $5x - 3y \geq 15$ should be a solid line or a dashed line.
2. Describe what your first step would be in graphing $4x - 3y \geq 12$, and explain why you would do that first.

Exercises 8.5
Set II

Graph the solution set of each of the following inequalities.

1. $3x - 4y > 12$
2. $7x - 2y < 14$
3. $8x + 11y < -24$
4. $x - 5y \geq 0$
5. $12x - 17y \geq -36$
6. $x > 3$
7. $y \leq 3$
8. $x + y < 0$
9. $5x + 8y < 0$
10. $x \leq -2$
11. $12x - 7y \geq 19$
12. $4y + 2x > 4$
13. $y \leq -2x + 4$
14. $y < -4x - 1$
15. $y > 3x + 2$
16. $y \geq \frac{1}{4}x + 1$
17. $\frac{x}{2} + \frac{y}{6} < -1$
18. $\frac{y}{3} - \frac{x}{2} \geq 2$
19. $6(4 - y) + 11 \geq 5(6y + 7)$
20. $8(3 + x) + 2 < 3(x - 4)$
21. $\frac{3x - 5y}{3} - \frac{4x + 7y}{12} < \frac{3}{4}$
22. $\frac{2}{3} - \frac{4x - y}{2} > y$

8.6 Miscellaneous Applications

In real-life situations, many applications lead to equations whose graphs are straight lines. In Example 1, you are asked to find such an equation.

EXAMPLE 1

Suppose it is known that the graph of the equation for predicting the cost, y, of producing x widgets is a straight line. Suppose, also, that the cost of producing 100 widgets is \$450 and the cost of producing 200 widgets is \$600. Find the equation that shows the relationship between the cost, y, and the number of widgets produced, x. Write the equation in the *slope-intercept form.* Then use the equation to find the cost of producing 800 widgets.

SOLUTION We know that the graph is a straight line, and we know that when x is 100, $y = 450$, and when $x = 200$, $y = 600$. Therefore, we know two points on the line: (100, 450) and (200, 600).

The slope of the line is $\frac{600 - 450}{200 - 100}$, or $\frac{150}{100}$, or $\frac{3}{2}$. If we use the point (100, 450) as the known, fixed point, we have

$$y - 450 = \tfrac{3}{2}(x - 100)$$

$$y - 450 = \tfrac{3}{2}x - 150 \qquad \text{Using the distributive property on the right}$$

$$y = \tfrac{3}{2}x + 300 \qquad \text{Adding 450 to both sides}$$

Therefore, the equation that shows the relationship between the cost, y, and the number of widgets produced, x, is $y = \frac{3}{2}x + 300$.

To find the cost of producing 800 widgets, we simply substitute 800 for x in the equation:

$$y = \tfrac{3}{2}x + 300$$

$$y = \tfrac{3}{2}(800) + 300 = 1{,}200 + 300 = 1{,}500$$

It will cost \$1,500 to produce 800 widgets.

The only way to learn how to solve problems that are expressed in words is to attempt to solve lots of them! In this section, we want you to try your skills at using algebra in problem solving, so we provide you with a number of different kinds of applied problems to solve. Some are money problems, mixture problems, digit problems (see Example 2), and so on, but some are different from any problems you have seen so far. You can (and should) solve *all* these problems by using the *algebraic methods* that have been covered in this book.

EXAMPLE 2

The sum of the digits of a two-digit number is 8. The number formed by reversing the digits is 36 more than the original number. Find the original number.

SOLUTION Recall that a two-digit number equals the tens digit times ten, *plus* the units digit. If we *reverse* the digits of the number, the new number will be ten times the *old* units digit, plus the *old* tens digit.

Step 1. Let $x =$ the units digit of the original number
Then $8 - x =$ the tens digit of the original number

Think Notice that $x + (8 - x)$, the sum of the digits, is 8.

The value of the original number: $10(8 - x) + x$

The value of the number with the digits reversed: $10x + (8 - x)$

Reread Value of the original number $+\ 36 =$ Value of the number with digits reversed

Step 2. $[10(8 - x) + x] + 36 = 10x + (8 - x)$

Step 3. $80 - 10x + x + 36 = 10x + 8 - x$

$-9x + 116 = 9x + 8$

$108 = 18x$ Adding $9x - 8$ to both sides

Step 4. $x = 6$ The units digit

$8 - x = 8 - 6 = 2$ The tens digit

The number appears to be 26.

Step 5. *Check* The sum of the digits of 26 is 2 + 6, or 8. The number formed by reversing the digits is 62, and 62 is 36 more than 26.

Step 6. Therefore, the original number is 26.

The odd- and even-numbered problems in the following exercise set are not necessarily "matched."

Exercises 8.6

Set I

In each exercise, set up the problem algebraically, solve, and check. Be sure to state what your variables represent.

1. The graph of the equation that describes the relationship between degrees Celsius (C) and degrees Fahrenheit (F) is known to be a straight line. On the Fahrenheit scale, water freezes at 32°F and boils at 212°F. On the Celsius scale, water freezes at 0°C and boils at 100°C. Find an equation that shows the relationship between C and F, and solve the equation for F (that is, express F in terms of C). Then use that equation to find the Fahrenheit temperature that corresponds to a Celsius reading of 37°.

2. Suppose it is known that the graph of the equation for predicting the cost, y, of producing x tie-dyed T-shirts is a straight line. Suppose, also, that the cost of producing 150 tie-dyed T-shirts is \$2,250 and the cost of producing 225 tie-dyed T-shirts is \$2,925. Find the equation that shows the relationship between the cost, y, and the number of tie-dyed T-shirts produced, x. Write the equation in the *slope-intercept form*. Then use the equation to find the cost of producing 300 tie-dyed T-shirts.

3. The sum of the digits of a two-digit number is 13. The number formed by reversing the digits is 27 more than the original number. Find the original number.

4. Karla can reach Rachelle's house from her house in 3 hr if she averages 45 mph. If Susan lives 35 mi closer to Karla than Rachelle does, what is the distance between Karla's house and Susan's house?

5. Heather has an octagonally shaped aquarium that contains some water. Right now it is one-third full. If she adds 3,000 cubic inches of water, it will be three-fourths full. What is the volume of the tank (in cubic inches)?

6. The sum of the digits of a two-digit number is 7. The number formed by reversing the digits is 27 less than the original number. Find the original number.

7. The units digit of a three-digit number is twice the hundreds digit. The sum of the digits is 6. If the digits are reversed, the new number is 198 more than the original number. Find the original number.

8. Rebecca is supposed to practice the piano for a certain number of minutes each day. So far today, she has practiced two-fifths of that time. If she spends 12 min more at the piano, she will have practiced two-thirds of her alloted time. How many minutes is she supposed to practice each day?

For Exercises 9 and 10, use the following: Nickel coinage is composed of copper and nickel in the ratio of 3 to 1.

9. How much pure copper should be melted with 20 lb of a copper-nickel alloy containing 60% copper to obtain the alloy for nickel coinage?

10. How much pure copper should be melted with 30 lb of a copper-nickel alloy containing 40% copper to obtain the alloy for nickel coinage?

11. Farah invested \$13,000. Part of the money was placed in an account that paid 5.25% interest per year, and the rest was put into an account that paid 5.75% interest per year. If the interest Farah earned the first year was \$702.50, how much did she place in each account?

12. The units digit of a three-digit number is twice the hundreds digit. The sum of the digits is 11. If the digits are reversed, the new number is 297 more than the original number. Find the original number.

13. A merchant purchased 18 cameras. If she had purchased 10 cameras of a higher quality, she would have paid \$48 more per camera for the same total expenditure. Find the price of each type of camera.

14. Fifty-six percent of the children enrolled in kindergarten at Jill's school are boys. Last Monday, 12.5% of the boys and 25% of the girls went on a field trip. If 246 children did *not* go on the field trip, how many children are enrolled in kindergarten at Jill's school?

15. A certain force, y, is required to stretch a spring x inches beyond its natural length, and the graph of the equation that describes the relationship between x and y is known to be a straight line. Suppose that a force of 28 lb is required to stretch a spring 8 in. and that a force of 17.5 lb is required to stretch the same spring 5 in. Find an equation that shows the relationship between x and y, writing the equation in the slope-intercept form. Then find the force necessary to stretch the spring 10 in.

Exercises 8.6
Set II

In each exercise, set up the problem algebraically, solve, and check. Be sure to state what your variables represent.

1. The sum of the digits of a two-digit number is 12. The number formed by reversing the digits is 36 more than the original number. Find the original number.

2. The sum of the digits of a two-digit number is 10. The ratio of the tens digit to the units digit is 3 to 2. What is the number?

3. The units digit of a three-digit number is 3 times the hundreds digit. The sum of the digits is 12. If the digits are reversed, the new number is 594 more than the original number. Find the original number.

4. The graph of the equation that describes the relationship between degrees Celsius (C) and degrees Fahrenheit (F) is known to be a straight line. Water boils at 100°C and at 212°F, and water freezes at 0°C and at 32°F. Find an equation that shows the relationship between C and F, and solve the equation for C (that is, express C in terms of F). Then find the Celsius temperature that corresponds to a Fahrenheit reading of 86°.

5. Items that are purchased for businesses are generally *depreciated* over a period of years for accounting purposes. Suppose it is known that the graph of the equation for predicting the value, y, of an item x years after it was purchased is a straight line. If the value of a machine that Frank bought for his machine shop was \$5,500 two years after he purchased it and \$3,700 four years after he purchased it, find the equation that shows the relationship between the value, y, and the number of years after the original purchase, x. Write the equation in the slope-intercept form. Then use the equation to find the value of the machine seven years after the original purchase.

6. The tens digit of a two-digit number is 4 times the units digit. If the digits are reversed, the new number is 54 less than the original number. Find the original number.

7. How much pure copper should be melted with 50 lb of a copper-nickel alloy containing 40% copper to obtain an alloy that is 50% copper?

8. George has received exam scores of 83, 75, 62, and 87. What score must he receive on the fifth exam in order to have an average score of 80?

9. Jim invested \$20,000. Some of the money was invested at 7.4% interest per year, and the rest was invested at 6.8%. The amount of interest he earned for the year was \$1,469.20. How much did he invest at each rate?

10. Tickets for some seats at a concert sold for \$9 each, and tickets for better seats sold for \$12 each. If 1,000 tickets were sold in all and if the amount of money brought in was \$10,320, how many of the \$9 tickets were sold?

11. A camera dealer bought 22 cameras. If she had bought 15 cameras of a higher quality, she would have paid \$35 more per camera for the same total expenditure. Find the price of each type of camera.

12. Scott is 6 times as old as Sharon. In 20 years, Scott will be only twice as old as Sharon. How old is Scott now?

13. Jason wants to enclose a rectangular area with 336 ft of fencing. If the rectangle is to be three times as long as it is wide, what must the length and the width of the rectangle be?

14. Kevin had a basket that contained only red balls and green balls, 40% of which were green. Then 25% of the red balls and 75% of the green balls were removed. If 33 balls remained in the basket, how many balls did Kevin have in the basket originally?

15. Suppose that a large company had a profit of \$96,500 when it had 31,000 customers and that it had a profit of \$127,500 when it had 35,000 customers. Suppose, also, that the graph of the equation that shows the relationship between the profit, y, and the number of customers, x, is a straight line. Find the equation that shows the relationship between the profit and the number of customers. Write the equation in the *slope-intercept form*. Then use the equation to find the profit if the number of customers increases to 40,000.

Sections 8.1–8.6 REVIEW

Ordered Pairs 8.1

An **ordered pair** is a pair of numbers in which the *order* the two numbers are listed in makes a difference; to indicate that a pair of numbers is an *ordered pair*, we put a comma between the two numbers, and we enclose the numbers within *parentheses*:

$$(a, b)$$

x-coordinate (for a), y-coordinate (for b)

Distance Between Two Points 8.1

The distance between $P_1(x_1, y_1)$ and $P_2(x_2, y_2)$ is

$$d = \sqrt{(x_2 - x_1)^2 + (y_2 - y_1)^2}$$

Relations 8.1

A **mathematical relation** is a set of ordered pairs (x, y).

The **domain** of a relation is the set of all the first coordinates of the ordered pairs of that relation.

The **range** of a relation is the set of all the second coordinates of the ordered pairs of that relation.

The **graph** of a relation is the graph of all the ordered pairs of that relation.

Graphing Straight Lines 8.2

The intercept method

1. Let $y = 0$, and solve for x to find the x-intercept, if it exists. Graph this point.
2. Let $x = 0$, and solve for y to find the y-intercept, if it exists. [If both intercepts are $(0, 0)$, find an additional point by letting x or y have any nonzero value.] Graph this point.
3. Find a checkpoint by letting x or y have any value not already used. Graph this point.

Solving the equation for y

1. Solve the given equation for y.
2. Let x have *three* different values, and solve for the corresponding y-values.
3. Graph the points found in step 2. Then proceed with step 4, below.

4. Draw a straight line through the points that have been graphed in the preceding steps, and draw an arrowhead at each end of the line.

Horizontal and vertical lines: The graph of $y = b$ (or $y - b = 0$), where b is any real number, is always a horizontal line that passes through $(0, b)$. The graph of $x = k$ (or $x - k = 0$), where k is any real number, is always a vertical line that passes through $(k, 0)$.

8.2 Straight lines can also be graphed on a graphics calculator.

Slope of a Line 8.3

If $x_2 \neq x_1$, the slope of the line through $P_1(x_1, y_1)$ and $P_2(x_2, y_2)$ is

$$m = \frac{y_2 - y_1}{x_2 - x_1}$$

The slope of a vertical line does not exist. Parallel lines have the same slope. If two lines are perpendicular (and neither is a vertical line), the product of their slopes is -1.

Equations of Lines
8.4

1. *General form*: $Ax + By + C = 0$ or $0 = Ax + By + C$, where A, B, and C are integers, $A \geq 0$, and A and B are not both zero
2. *Point-slope form*: $y - y_1 = m(x - x_1)$, where (x_1, y_1) is a known, fixed point on the line and m is the slope of the line
3. *Slope-intercept form*: $y = mx + b$, where m is the slope and b is the y-intercept of the line
4. The equation of any *horizontal line* is $y = b$, or $y - b = 0$, where b is the y-coordinate of every point on the line.
5. The equation of any *vertical line* is $x = k$, or $x - k = 0$, where k is the x-coordinate of every point on the line.

Graphing Linear Inequalities in the Plane
8.5

The graph of the solution set of a linear inequality in one or two variables is a half-plane. To graph a linear inequality:

1. Find the equation of the boundary line by replacing the inequality symbol with an equal sign.
2. Graph the boundary line. It is a *solid* line if the inequality symbol is $\geq$ or $\leq$. It is a *dashed* line if the inequality symbol is $>$ or $<$.
3. Select any point that does *not* lie on the boundary line, and substitute the coordinates of that point into the inequality. If the resulting statement is *true*, shade the half-plane that contains the point selected. If the resulting statement is *false*, shade the half-plane that does *not* contain the point selected.

Applications
8.6

Many applications in real-life situations lead to equations whose graphs are straight lines.

Sections 8.1–8.6 REVIEW EXERCISES Set I

1. Find the distance between each given pair of points.
   a. $(-3, -4)$ and $(2, -4)$
   b. $(-2, -3)$ and $(4, 1)$
2. Find the domain and range of the relation $\{(0, 5), (-2, 3), (3, -4), (0, 0)\}$.
3. Find the domain and range of $\{(x, y)|3x - 2y = 6, x = -2, 0, 6, 8\}$, and graph the relation.

In Exercises 4–7, graph each relation.

4. $3y + 2x = -12$
5. $x + 2y = 0$
6. $y = 2x$
7. $\dfrac{2x + 3y}{5} - \dfrac{x - 3y}{4} = \dfrac{9}{10}$

In Exercises 8 and 9, find the slope of the line that passes through each given pair of points.

8. $(3, -5)$ and $(-2, 4)$
9. $(-6, 2)$ and $(3, 2)$
10. Write an equation of the line that passes through $(-8, 4)$ and has a slope of $-\frac{3}{4}$.
11. Write an equation of the line that has a slope of $-\frac{1}{2}$ and a y-intercept of 6.
12. Write an equation of the line that passes through the points $(-2, -4)$ and $(1, 3)$.
13. Find the slope and y-intercept of the line $\dfrac{2x}{5} - \dfrac{3y}{2} = 3$.

In Exercises 14–16, graph the solution set of each inequality in the plane.

14. $y > 3$
15. $2x - 5y > 10$
16. $3x - 2y \geq 0$
17. A certain force, y, is required to stretch a spring x inches beyond its natural length, and the graph of the equation that describes the relationship between x and y is known to be a straight line. Suppose that a force of 18 lb is required to stretch a spring 6 in. and that a force of 24 lb is required to stretch the same spring 8 in. Find an equation that shows the relationship between x and y, writing the equation in the slope-intercept form. Then find the force necessary to stretch the spring 12 in.

In Exercises 18–20, set up each problem algebraically, solve, and check. Be sure to state what your variables represent.

18. The sum of the digits of a two-digit number is 11. If the digits are reversed, the new number is 45 less than the original number. What is the original number?

19. Mr. Curtis invested part of $27,000 at 12% interest and the remainder at 8%. His income the first year from these two investments was $2,780. How much was invested at each rate?

20. It takes Ms. Sontag 30 min to drive to work in the morning, but 45 min to return home over the same route during the evening rush hour. If her average morning speed is 10 mph faster than her average evening speed, how far is it from her home to her work?

Sections 8.1–8.6 REVIEW EXERCISES Set II

Name ______________________

ANSWERS

1a. ______________

b. ______________

2. ______________

____

3. ______________

____

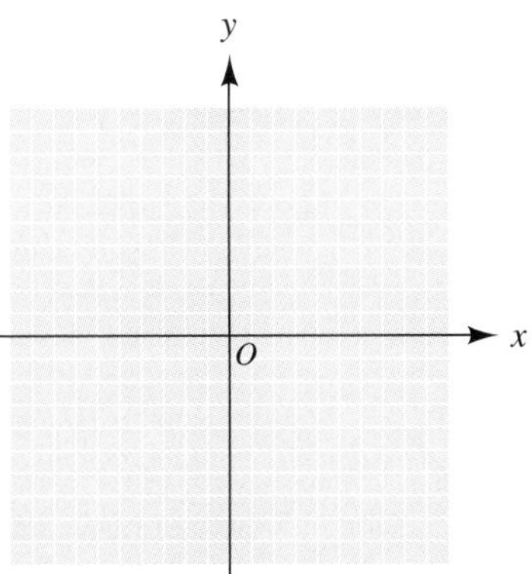

4.

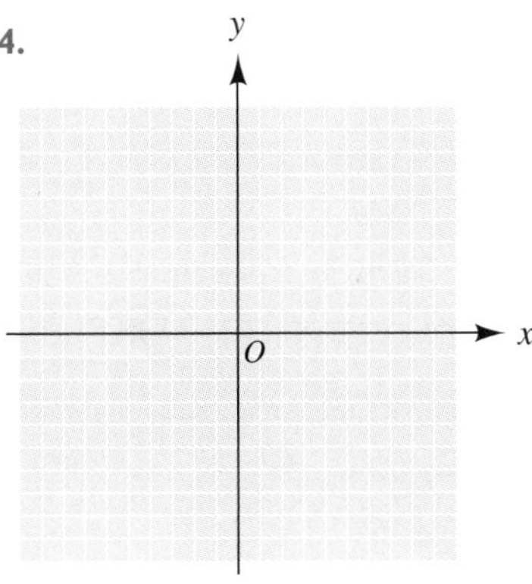

1. Find the distance between each given pair of points.

   a. (8, −6) and (−12, −2) b. (−7, 5) and (7, −13)

2. Find the domain and range of the relation $\{(10, 6), (-5, -2), (0, 7), (-9, -6)\}$.

3. Find the domain and range of $\{(x, y)|4x + 3y = 12, x = -3, 0, 3\}$, and graph the relation.

4. Graph the relation $5x - 3y = 0$.

5. Graph the relation $\frac{x - 4y}{7} - \frac{3x - 7y}{6} = \frac{20}{21}$.

6. Graph the relation $x = 5$.

5.

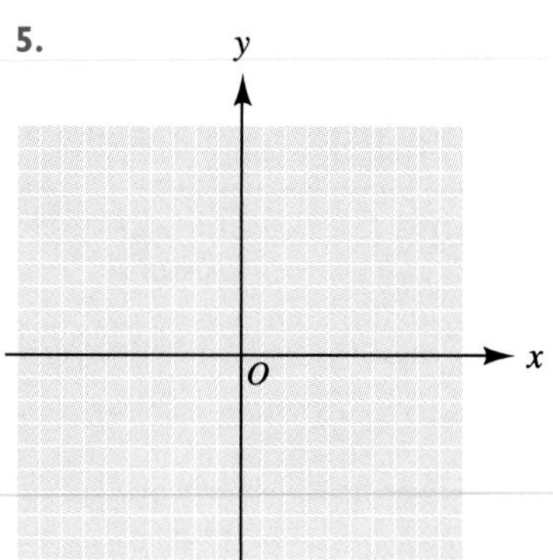

6.

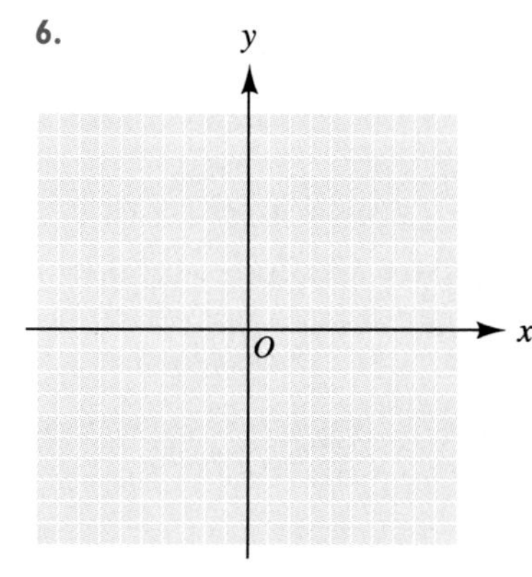

In Exercises 7 and 8, find the slope of the line that passes through each given pair of points.

7. $(-12, 11)$ and $(-25, -28)$

8. $(-17, -14)$ and $(35, -22)$

9. Write an equation of the line that passes through $(-7, 12)$ and has a slope of $-\frac{4}{5}$.

10. Write an equation of the line that has a slope of $-\frac{7}{4}$ and a y-intercept of -6.

11. Write an equation of the line that passes through the points $(-16, 15)$ and $(-38, -25)$.

7. ______________________

8. ______________________

9. ______________________

10. ______________________

11. ______________________

12. Find the slope and y-intercept of the line $\frac{4x}{7} - \frac{6y}{5} = 2$.

12. ______________________

____

13.

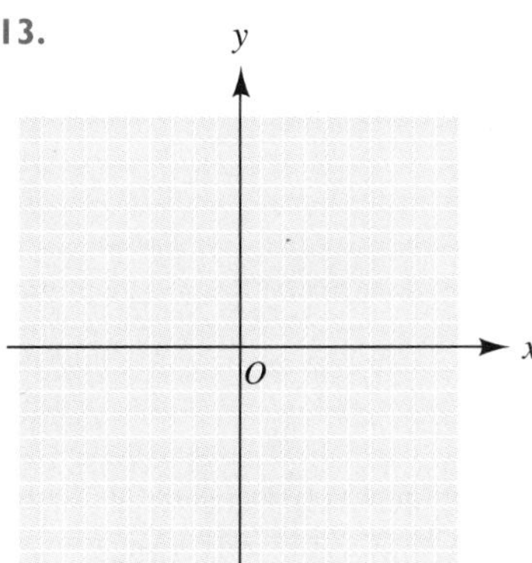

In Exercises 13–15, graph the solution set of each inequality in the plane.

13. $x + 4y \geq 0$

14.

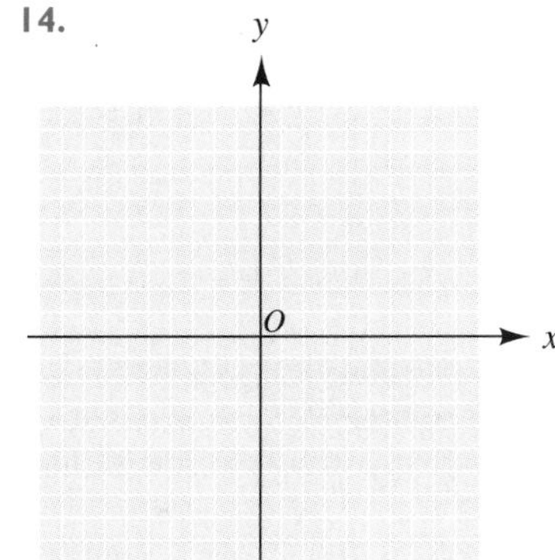

14. $y \leq -1$

15.

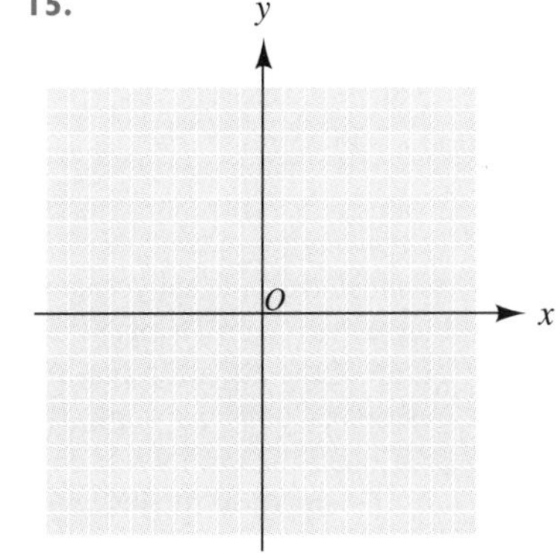

15. $3x - 9y > 9$

In Exercises 16–18, set up each problem algebraically, solve, and check. Be sure to state what your variables represent.

16. ______________________

17. ______________________

18. ______________________

____

19. ______________________

____

16. The sum of the digits of a two-digit number is 10. If the digits are reversed, the new number is 72 less than the original number. Find the original number.

17. Lupe is twice as old as Juana. In 9 yr, Juana will be five-sevenths as old as Lupe is then. How old is Juana now?

18. Mr. Reid had \$15,000 to invest. Part of the money was invested at 7.8% interest per year, and the rest was invested at 7.2% interest per year. If the amount of money earned the first year was \$1,138.80, how much was invested at each rate?

19. Suppose that a large company had a profit of \$132,600 when it had 43,000 customers and that it had a profit of \$153,200 when it had 47,000 customers. Suppose, also, that the graph of the equation that shows the relationship between the profit, y, and the number of customers, x, is a straight line. Find the equation that shows the relationship between the profit and the number of customers. Write the equation in the slope-intercept form. Then use the equation to find the profit if the number of customers increases to 52,000.

Chapter 8 DIAGNOSTIC TEST

The purpose of this test is to see how well you understand graphing solution sets of linear equations and inequalities and how well you can write equations of lines. We recommend that you work this diagnostic test *before* your instructor tests you on this chapter. Allow yourself about 50 minutes.

Complete solutions for all the problems on this test, together with section references, are given in the answer section at the end of the book. For the problems you do incorrectly, we suggest that you study the sections cited.

For Problems 1–3, use $A(-4, 2)$, $B(1, -3)$, and $C(5, 3)$.

1. Draw the triangle with vertices at A, B, and C.
2. Find the slope of the line through A and B, and write an equation of that line. (Express the equation in the general form.)
3. Find the x- and y-intercepts of the line through A and B.
4. Find an equation of the line that has a slope of $\frac{6}{5}$ and a y-intercept of -4. (Express the equation in the general form.)
5. Write the equation $3x - 5y = 15$ in the slope-intercept form. Find the slope and the y-intercept.
6. Write an equation of a line that passes through the point $(-1, 6)$ and is parallel to the line from Problem 5.
7. Graph the relation $x - 2y = 6$.
8. Graph the relation $4x - 3y \le -12$.
9. Given the relation $\{(-4, -5), (2, 4), (4, -2), (-2, 3), (2, -1)\}$:
   a. Find its domain.
   b. Find its range.
   c. Graph the relation.
10. A certain force, y, is required to stretch a spring x inches beyond its natural length, and the graph of the equation that describes the relationship between x and y is known to be a straight line. Suppose that a force of 20 lb is required to stretch a spring 8 in. and that a force of 25 lb is required to stretch the same spring 10 in. Find an equation that shows the relationship between x and y, writing the equation in the slope-intercept form. Then find the force necessary to stretch the spring 11 in.

Chapters 1–8 CUMULATIVE REVIEW EXERCISES

In Exercises 1 and 2, factor the expression completely.

1. $2x^4 - 24x^2 + 54$
2. $8a^4b - 27ab^4$

In Exercises 3 and 4, perform the indicated operations, and express the answers in simplest radical form.

3. $\sqrt[3]{-4a^4} \cdot \sqrt[3]{2a}$
4. $4\sqrt{18x^4} + \sqrt{32x^5} - \sqrt{50x^4}$

In Exercises 5–7, perform the indicated operations, and write the answers in the form $a + bi$.

5. $\dfrac{3 + 2i}{2 - i}$
6. $(7 - \sqrt{-16}) - (5 - \sqrt{-36})$

* 7. $(3 + i)^4$

8. Solve for r: $S = \dfrac{a}{1 - r}$.
9. Find the solution set of $x^{-1/3} = 3$.
10. Subtract $(2x^2 - 3x + 4)$ from the sum of $(9x^2 + x - 7)$ and $(8 - 6x - 4x^2)$.
11. Write the equation $7x + 2y = 14$ in the slope-intercept form; then find the slope and the y-intercept.
12. Graph $2x - y = 4$.
13. Write an equation of the straight line that passes through the points $(-3, 4)$ and $(1, -2)$.

In Exercises 14–18, set up each problem algebraically, solve, and check. Be sure to state what your variables represent.

14. The manager of a discount store bought 25 stereos. If she had purchased 14 stereos of a higher quality, she would have paid \$55 more per stereo for the same total expenditure. Find the price of each type of stereo.
15. A grocer is going to make up a 60-lb mixture of cashews and peanuts. If the cashews are worth \$7.40 per lb and the peanuts are worth \$2.80 per lb, how many pounds of each kind of nut must be used in order for the mixture to be worth \$4.18 per lb?
16. Dorothy can drive from A to B at 55 mph, but in driving from B to C, she can travel only 50 mph. B is 8 mi farther from A than it is from C. If it takes her the same length of time to get from A to B as it takes her to get from B to C, what is the distance from A to B?
17. One leg of a right triangle is 7 cm longer than the other leg, and the hypotenuse is 9 cm longer than the shorter leg. What are the lengths of the three sides of the triangle?
18. Gwen has \$2.20 in nickels, dimes, and quarters. If she has 3 more dimes than quarters and 2 fewer nickels than quarters, how many of each kind of coin does she have?

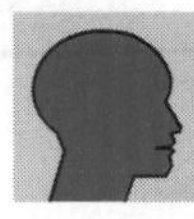

Critical Thinking and Problem-Solving Exercises

1. Ben went to a used car lot and noticed six cars parked in a row: a Buick, a Ford, a Chevrolet, a Dodge, a Toyota, and a Mercury. He faced the row of cars and made the following observations:

   The Mercury was between the Chevrolet and the Ford.

   The Toyota and the Chevrolet were in the two center positions.

   The Buick was to the left of, but not next to, the Chevrolet.

   The Ford and the Dodge were on the two ends.

   List (in order from left to right) the order in which the cars were parked.

2. Teri learned that she could buy bleach and window cleaner on sale if she bought them by the case. Each case of bleach contains 4 bottles, and each case of window cleaner holds 6 bottles. Teri wants to buy a total of 60 bottles of these two items. If she buys at least one case of each, what are the possible combinations of *cases* of bleach and window cleaner that will give her a total of 60 bottles?

3. Suppose that you needed to memorize the following area code and phone number: (909) 936-1827. What, if any, patterns do you see that would help you to remember the number?

4. Suppose that a large company had a profit of \$132,600 when it had 43,000 customers and a profit of \$153,200 when it had 47,000 customers. Suppose, also, that the graph of the equation that shows the relationship between the profit, y, and the number of customers, x, is a straight line. Graph the line that shows the relationship between the profit and the number of customers, and find the minimum number of customers needed to have a profit. (Round off the answer to the nearest unit.)

Functions

CHAPTER

9

We live in a world of functions. Our salary is often a function of the number of hours we work per week or per month; the demand for a product is often a function of its price; the cost per day of renting a car is generally a function of the number of miles the car is driven that day; the amount of interest earned when a fixed amount of money is invested at a fixed rate of interest is a function of time. Functions can be described graphically, numerically, algebraically, and in words. In this chapter, we define and discuss functions and their graphs, functional notation, the algebra of functions, and inverse functions. We also discuss proportions and variation.

9.1 Basic Definitions

The Definition of a Function

A **function** is a rule, or a correspondence, that assigns to each element of some set X (called the **domain**) *exactly* one element of some set Y (called the **range**).

Thus, if $x \in X$ and $y \in Y$, then the set of ordered pairs (x, y) is a *function* if no two ordered pairs with the same first coordinate have second coordinates that are different from each other. For example, the relation $\{(3,7), (1,2), (3,9)\}$ is *not* a function, because two *different* y-values are assigned to an x-value of 3; in other words, the ordered pairs $(3,7)$ and $(3,9)$ have the same first coordinate but different second coordinates. The relation $\{(3,7), (1,7), (6,7)\}$ *is* a function, because each x-value is assigned exactly one corresponding y-value.

A function can be described in a number of different ways: by a graph, by a set of ordered pairs, by an equation, by a written statement, or by a table of values.

The following property allows us to look at the graph of a relation and determine whether the relation is a function:

No vertical line can intersect the graph of a function in more than one point.

EXAMPLE 1 Determine from each graph whether the relation is a function.

a.

y
x
O

SOLUTION

This *is* the graph of a function, because no vertical line can intersect the graph in more than one point.

b.

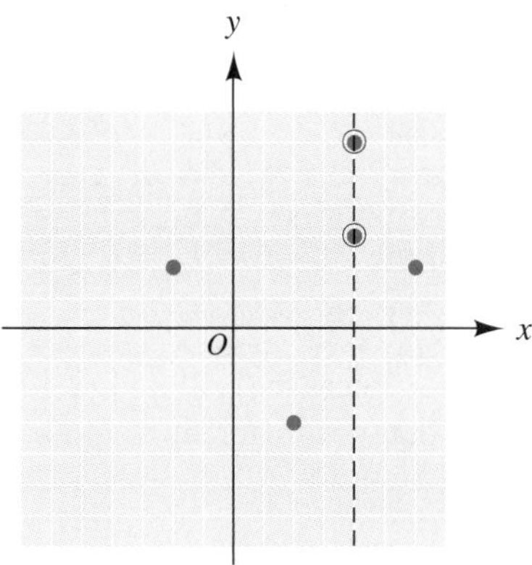

This is *not* the graph of a function, because one vertical line intersects the graph in two points.

c.

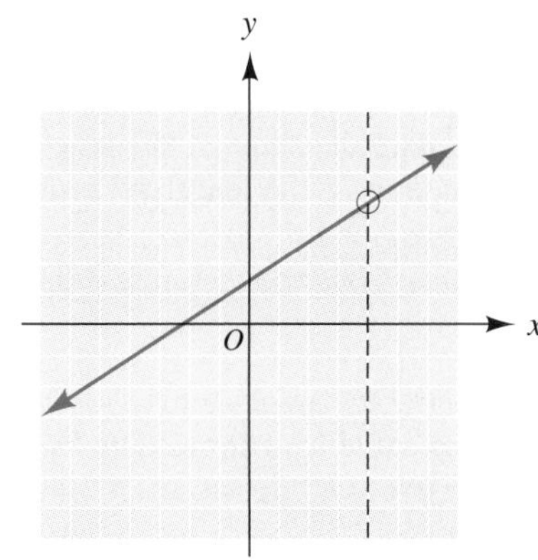

This *is* the graph of a function, because no vertical line can intersect the graph in more than one point.

d.

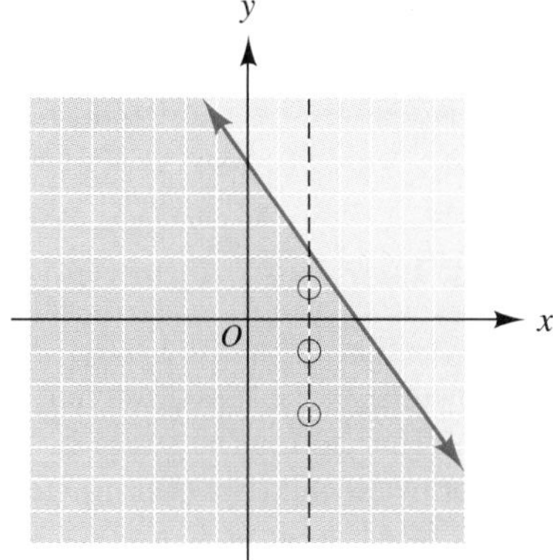

This is *not* the graph of a function, because at least one vertical line intersects the graph in more than one point.

e.

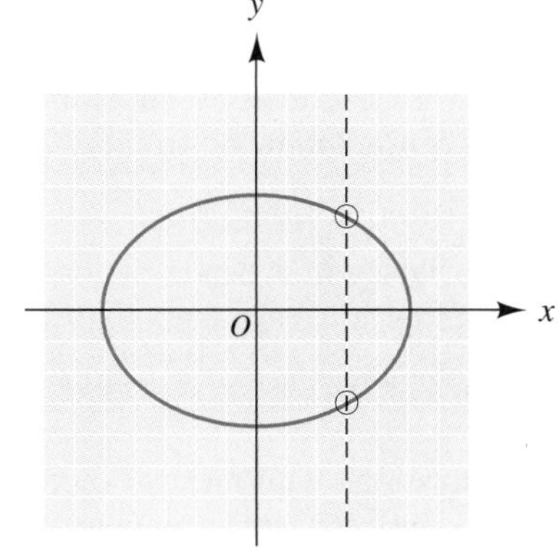

This is *not* the graph of a function, because at least one vertical line intersects the graph in two points.

EXAMPLE 2 Determine whether the relation is a function.

a. $\{(1,7), (3,6), (4,7)\}$

SOLUTION Since no two ordered pairs have the same first coordinate, the relation *is* a function.

b. $\{(2,8), (2,9), (3,7)\}$

SOLUTION Because $(2,8)$ and $(2,9)$ have the same first coordinate but second coordinates that are different from each other, the relation is *not* a function.

EXAMPLE 3 Examples of functions:

a. $\{(x, y) \mid y = 3x + 1\}$
For each value of x, there is just one "answer" for y; therefore, this is a function.

b. "The area of a square depends on the length of a side of the square."
The area is a function of the length of a side of a square, because there is one and only one area corresponding to each different length of a side.

c. The following table gives the distance d through which a freely falling object falls in t seconds:

t	1	2	3	4	5	6	7	8
d	16	64	144	256	400	576	784	1,024

This is a function, because for each value of t, there is one and only one corresponding value of d.

The Domain and Range of a Function

Because a function is a set of ordered pairs, it is a relation, and the meanings of the domain and range are the same for a function as for a relation.

The **domain** of a function is the set of all the permitted first coordinates (that is, the domain is the set of all the first coordinates for which the second coordinates will be defined).

When a function is described by a *graph*, we find the domain by looking "right and left" to see how much of the plane is "used up" (see Example 4).

When a function is described by a set of ordered pairs or by a table, the domain is the set of all the first coordinates (see Example 5).

When a function is described by an *equation*, the domain will be the set of real numbers *unless* one of the following occurs:

1. The domain is restricted by some statement accompanying the equation.
2. There are variables in a denominator (or, equivalently, there are variable expressions with negative exponents).
3. Variables occur under a radical sign when the index of the radical is an even number.
4. Variable expressions with rational (noninteger) exponents occur where the denominator of the exponent is an even number.

(In Example 6, we will see functions that are described by equations.)

The **range** of a function is the set of all the second coordinates.

If the function is graphed, then the range can be found by looking "up and down" to see how much of the plane is "used up" (see Example 4).

If the function is described by a set of ordered pairs or by a table, the range is the set of all the second coordinates (see Example 5).

When a function is described by an equation, it is often more difficult to find its range than to find its domain (see Example 6). In fact, in some cases, the range cannot be found without using techniques that are beyond the scope of this book.

EXAMPLE 4 Find the domain and the range of each of the following functions.

SOLUTION

a.

y
(0, 1)
(−2, 0) O (2, 0)
x

Because no part of the graph is to the *left* of $x = -2$ and no part is to the *right* of $x = 2$, the domain is $\{x \mid -2 \le x \le 2\}$. Because no part of the graph is *above* $y = 1$ and no part is *below* $y = 0$, the range is $\{y \mid 0 \le y \le 1\}$.

b.

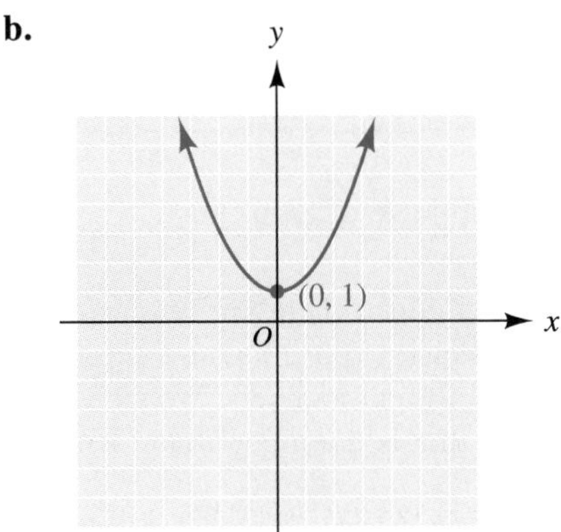

Because the graph extends infinitely far left and right, the domain is $\{x \mid x \in R\}$. Because the graph extends infinitely far upward and because no part of the graph is *below* $y = 1$, the range is $\{y \mid y \ge 1\}$.

c.

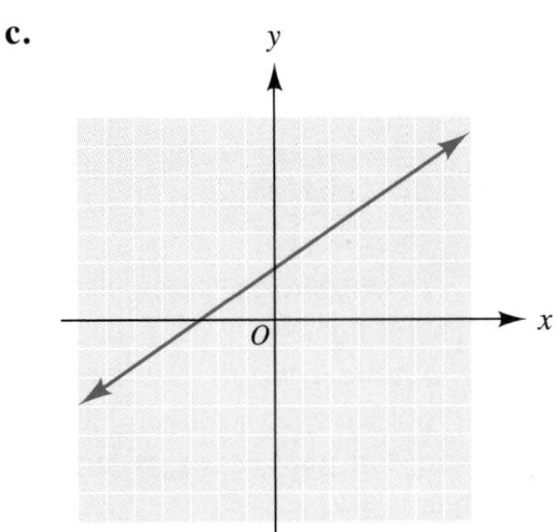

Because the graph extends infinitely far left and right, the domain is $\{x \mid x \in R\}$. Because the graph extends infinitely far up and down, the range is $\{y \mid y \in R\}$.

d.

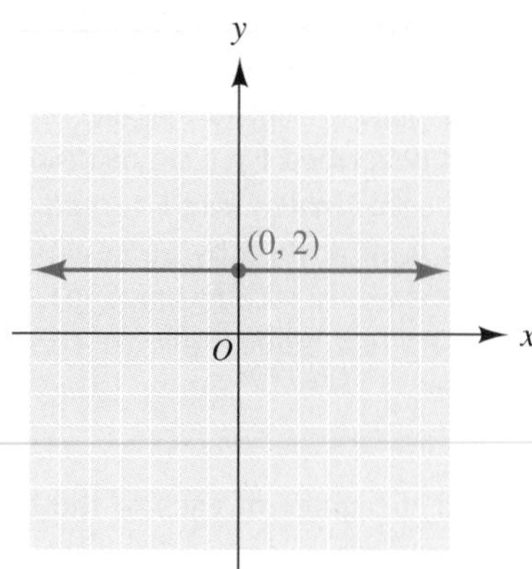

Because the graph extends infinitely far left and right, the domain is $\{x \mid x \in R\}$. Because the only y-value that is "used up" is $y = 2$, the range is $\{2\}$.

e.

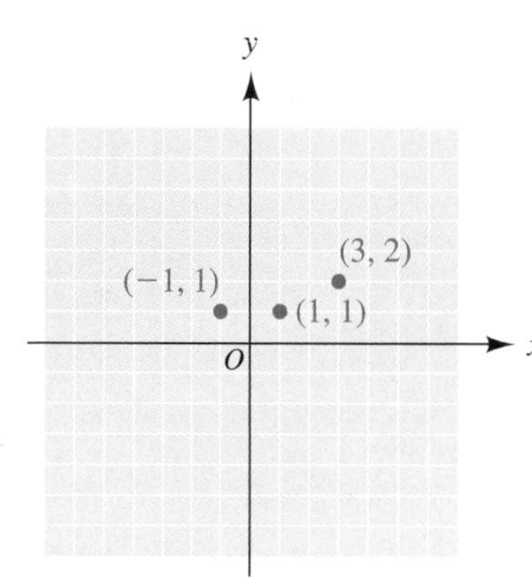

Because the only x-values that are "used up" are -1, 1, and 3, the domain is $\{-1, 1, 3\}$. Because the only y-values that are "used up" are 1 and 2, the range is $\{1, 2\}$.

EXAMPLE 5 Find the domain and range of the function $\{(-2, 7), (3, 5), (4, 6), (8, 5)\}$.

SOLUTION The domain is $\{-2, 3, 4, 8\}$, the set of all the first coordinates. The range is $\{7, 5, 6\}$, or $\{5, 6, 7\}$, the set of all the second coordinates.

EXAMPLE 6 Find the domain and range of each function.

a. $y = 3x + 2$

SOLUTION Since there are no denominators, no radical signs, no negative exponents, no rational exponents, and no other restrictions on the domain, the domain is the set of all real numbers. We recognize the equation as being the equation of a straight line that is *not* a horizontal line; therefore, the range is also the set of all real numbers.

b. $\{(x, y) \mid y = 2x, x = -1, 4\}$

SOLUTION The domain is restricted to $\{-1, 4\}$. From the equation, we see that when $x = -1$, $y = -2$, and when $x = 4$, $y = 8$. Therefore, the range is $\{-2, 8\}$.

c. $y = 7$

SOLUTION The equation is that of a horizontal line. Therefore, the domain is $\{x \mid x \in R\}$, and the range is $\{7\}$.

d. $y = \sqrt{2 + x}$

SOLUTION Because the equation contains a square root, the radicand must be nonnegative. Therefore, $2 + x \geq 0$, or $x \geq -2$. The domain is $\{x \mid x \geq -2\}$. Finding the range is more difficult. We know that y can never be negative, because a principal square root is never negative. If $x = -2$, $y = \sqrt{0} = 0$; therefore, y *can* equal 0. We can see, too, that as x gets larger and larger, y gets larger and larger. Therefore, the range is $\{y \mid y \geq 0\}$.

Because the second coordinate of an ordered pair of a function *depends on* the first coordinate, we call the first coordinate (often the x-coordinate) the **independent variable** and the second coordinate (often the y-coordinate) the **dependent variable**.

We graph a function the same way we graph a relation: We simply graph the set of ordered pairs of the function.

Exercises 9.1
Set I

1. Which of the following are graphs of functions?

a.

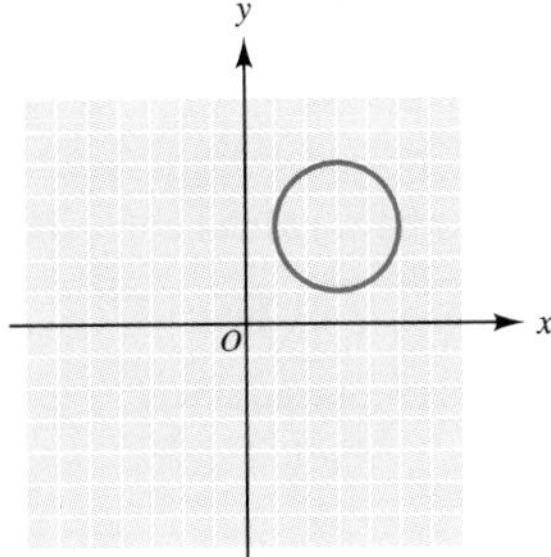

b.

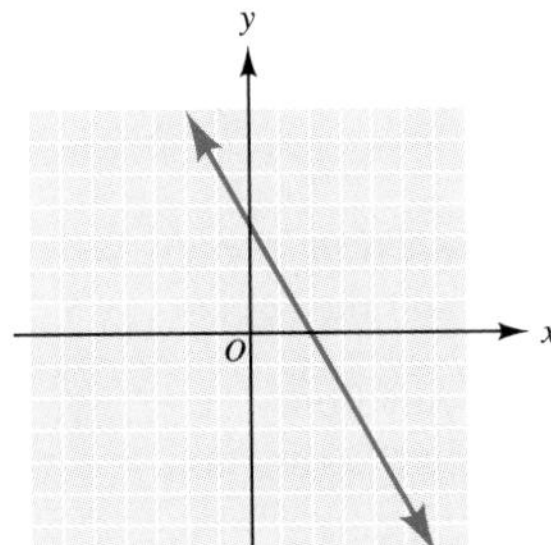

c.

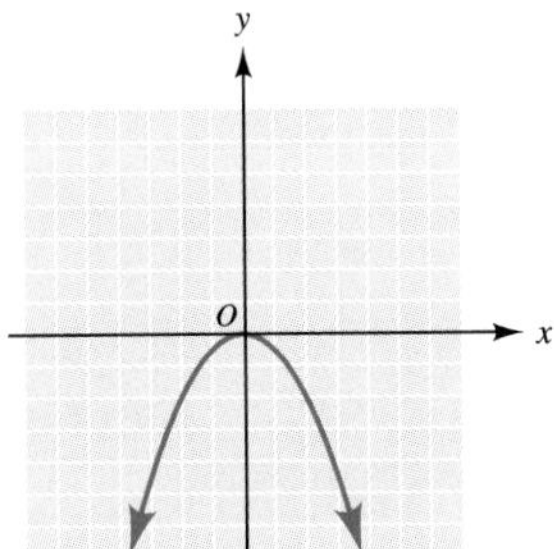

d.

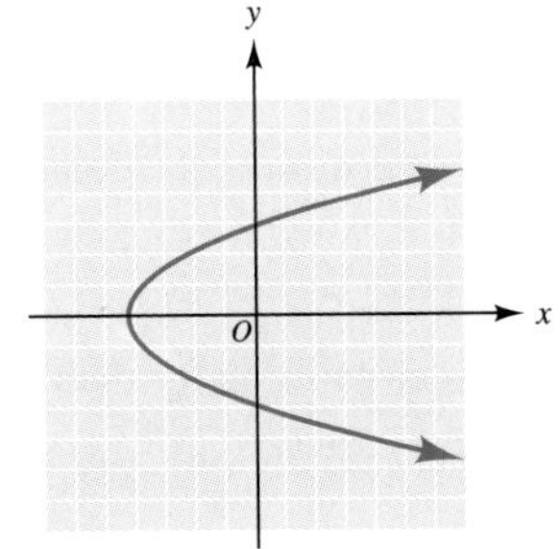

2. Which of the following are graphs of functions?

a.

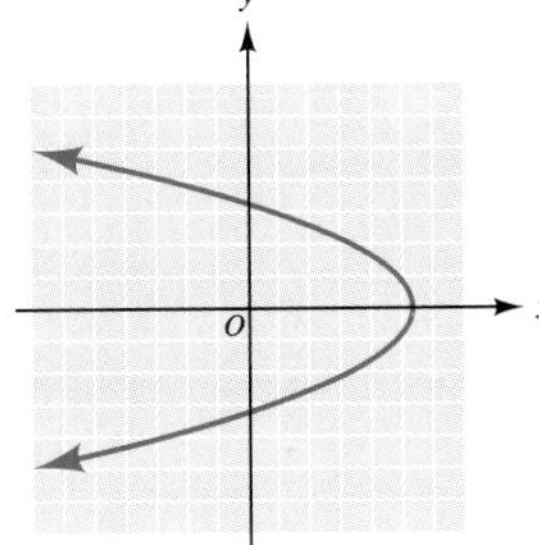

b.

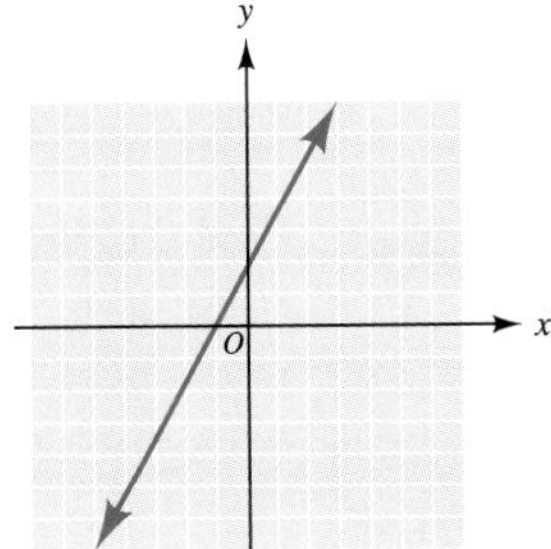

c.

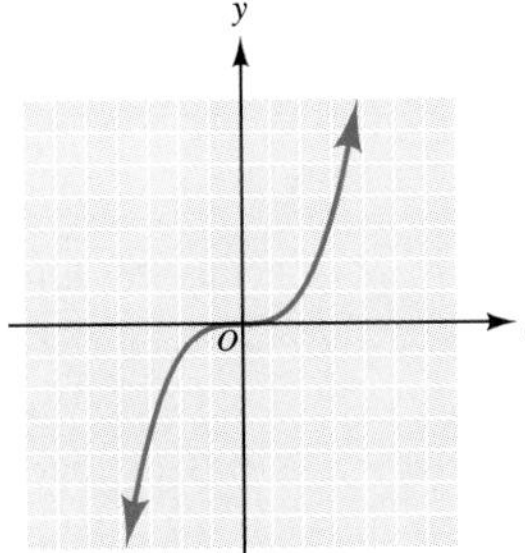

d.

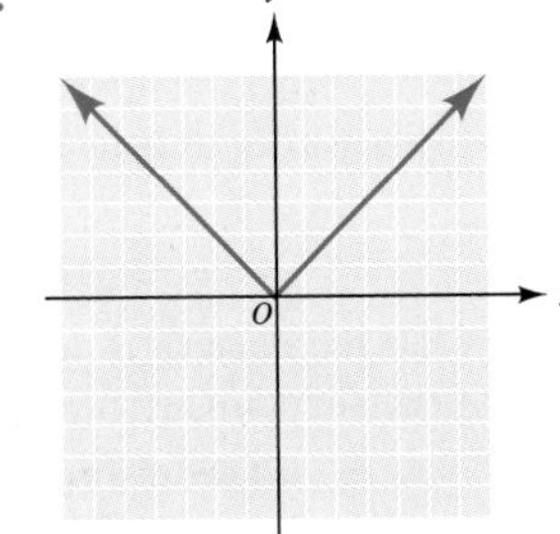

In Exercises 3 and 4, determine whether each relation is a function.

3. a. $\{(2, 3), (5, 3), (1, 6)\}$ b. $\{(-1, 6), (-1, 8)\}$
   c. $\{(x, y) \mid x = 7\}$ d. $\{(x, y) \mid y = 3\}$
4. a. $\{(-3, 6), (-3, -8), (4, 2)\}$ b. $\{(2, -8), (5, -8)\}$
   c. $\{(x, y) \mid y = 3x + 4\}$ d. $\{(x, y) \mid x = 0\}$

In Exercises 5 and 6, find the domain and range of each function.

5. a. The graph is

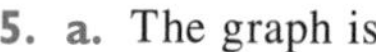

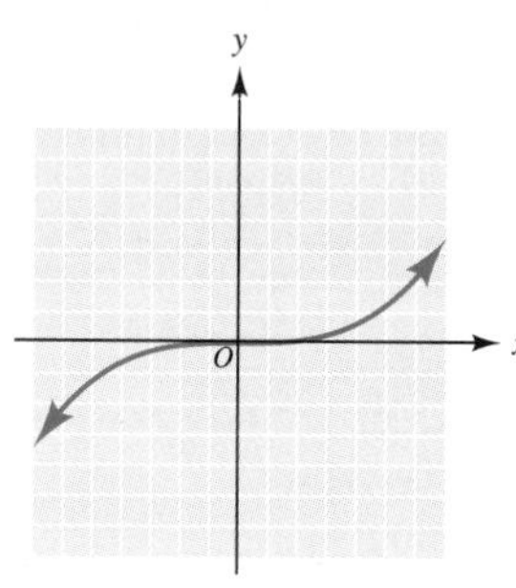

   b. $y = -2x + 1$
   c. $\{(x, y) \mid y = \sqrt{x - 5}\}$
   d. $\{(x, y) \mid y = -x + 1, x = 1, 4, 7\}$
   e. $\{(3, -6), (4, 2), (2, 4)\}$
6. a. The graph is

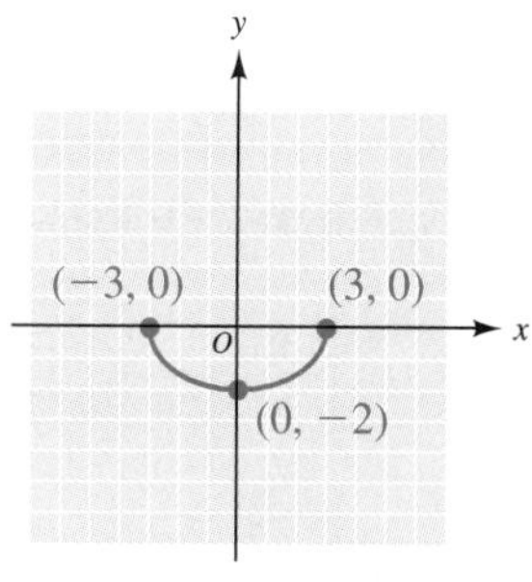

   b. $\{(x, y) \mid y = \sqrt{3 + x}\}$
   c. $y = \frac{2}{3}x - 4$
   d. $\{(x, y) \mid y = 3x - 4, x = 0, 1\}$
   e. $\{(6, 1), (8, 1), (3, 1), (5, 1)\}$
7. The following table gives corresponding weights, w, for various numbers of widgets, n:

n	1	2	3	4	5	6	7	8
w	4	8	12	16	20	24	28	32

   Explain why w is a function of n. Find the domain and the range of the function.
8. The following table gives the y-values corresponding to various x-values:

x	1	2	3	4	5	6	7	8
y	1	4	8	16	32	64	128	256

   Explain why y is a function of x. Find the domain and the range of the function.
9. Find the range of the function $\{(x, y) \mid 2x + 5y = 10, x = -5, -1, 0, 2\}$, and graph the function.
10. Find the range of the function $\{(x, y) \mid 3x - 6 = 2y, x = -2, 0, 2, 4\}$, and graph the function.
11. Graph the function $\{(x, y) \mid y = 2x - 3\}$, with the domain being the set of real numbers.
12. Graph the function $\left\{(x, y) \,\middle|\, \frac{x}{2} - \frac{y}{3} = 1\right\}$, with the domain being the set of real numbers.

Writing Problems

Express the answers in your own words and in complete sentences.

1. Explain what a function is.
2. Explain how to determine what the domain of a function is if you're given a graph of the function.
3. Explain how to determine what the range of a function is if you're given a graph of the function.
4. Describe your first step in determining the domain of the function $f(x) = \sqrt{5 - x}$, and explain why you would do that first.
5. Describe your first step in determining the domain of the function $g(x) = \frac{x + 8}{3 - x}$, and explain why you would do that first.

Exercises 9.1
Set II

1. Which of the following are graphs of functions?

a.

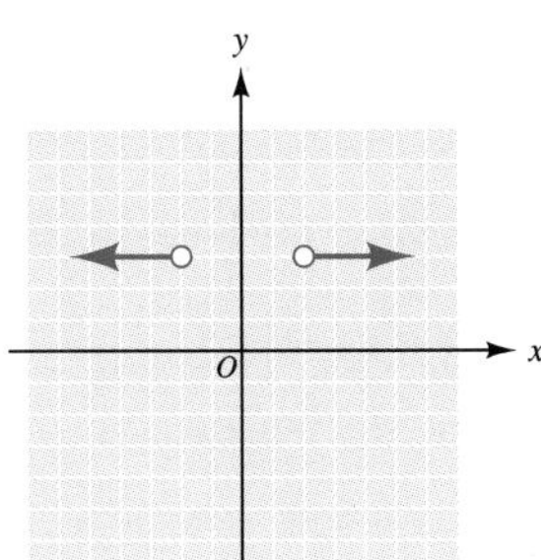

b.

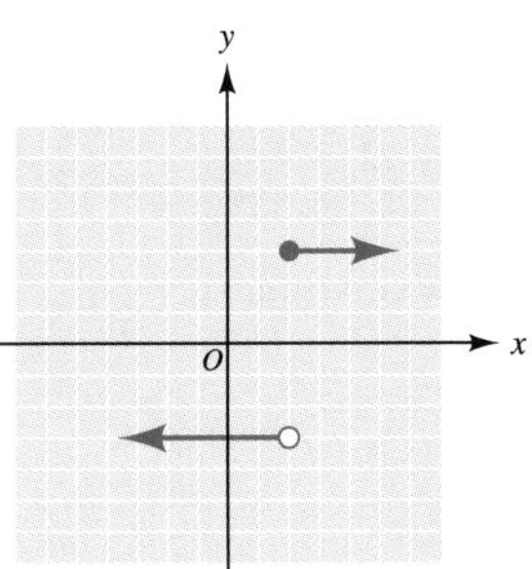

c.

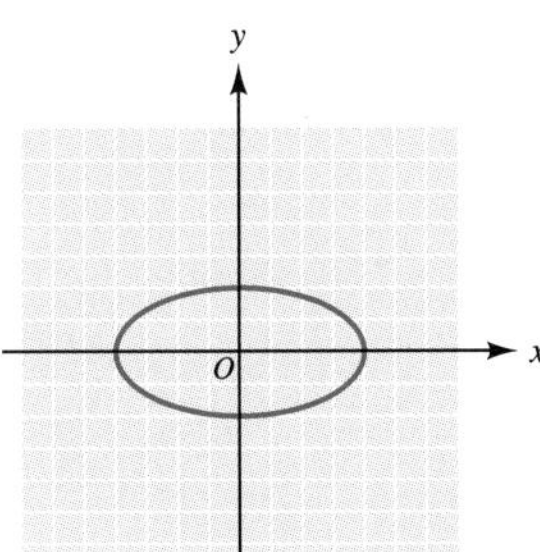

d.

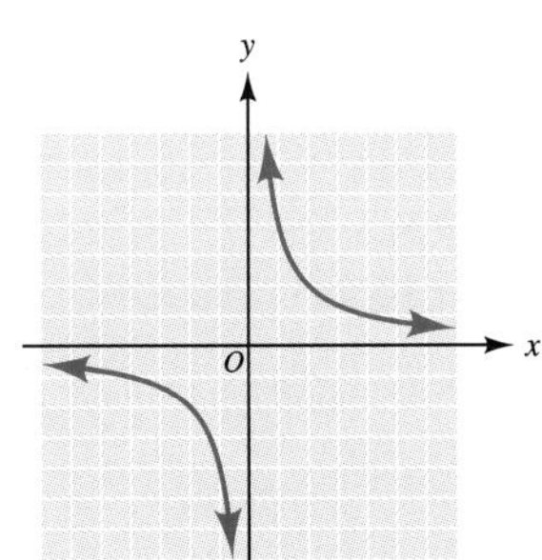

2. Which of the following are graphs of functions?

a.

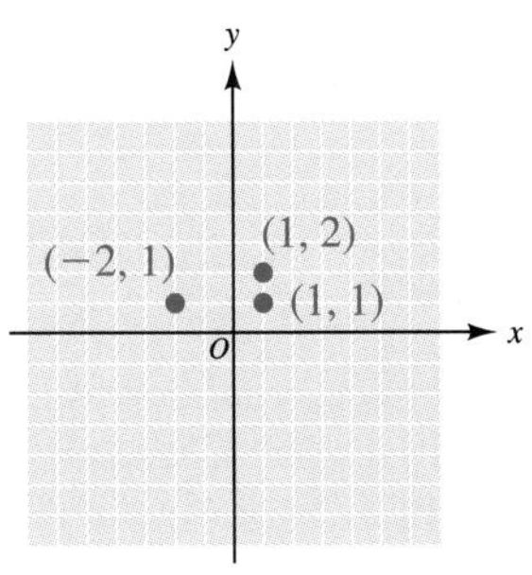

b.

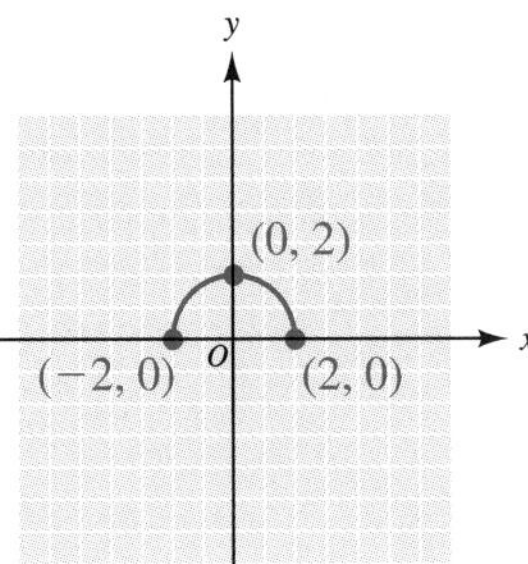

c.

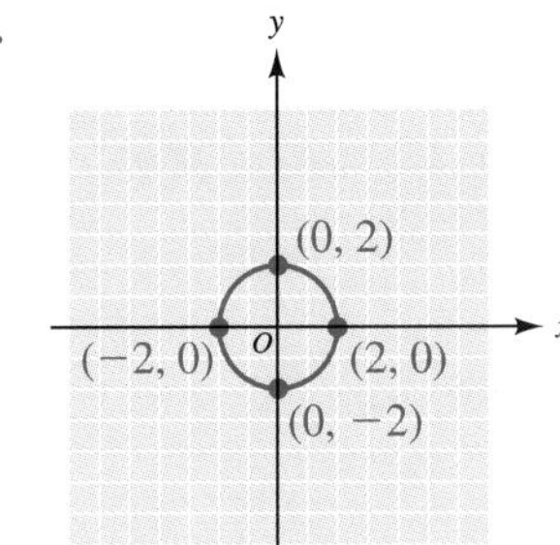

d.

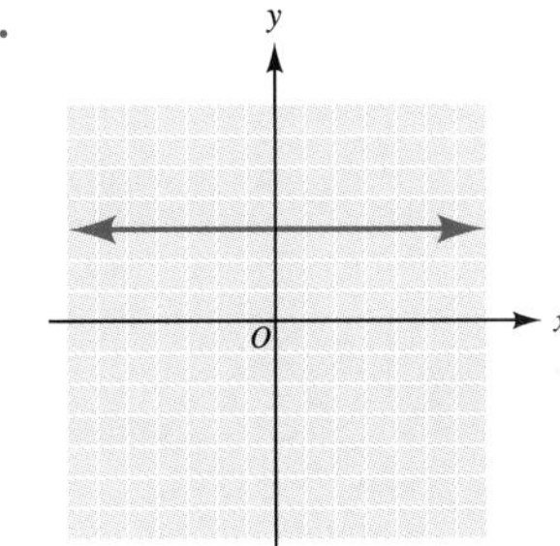

In Exercises 3 and 4, determine whether each relation is a function.

3. a. $\{(0, 3), (0, 7), (1, 2)\}$ b. $\{(1, 6), (2, 6)\}$
   c. $\{(x, y) \mid y = x\}$ d. $\{(x, y) \mid x = 6\}$
4. a. $\{(3, 5), (4, 5), (7, 5)\}$ b. $\{(1, 0), (1, 1)\}$
   c. $\{(x, y) \mid x = 3\}$ d. $\{(x, y) \mid y = -1\}$

In Exercises 5 and 6, find the domain and range of each function.

5. a. The graph is

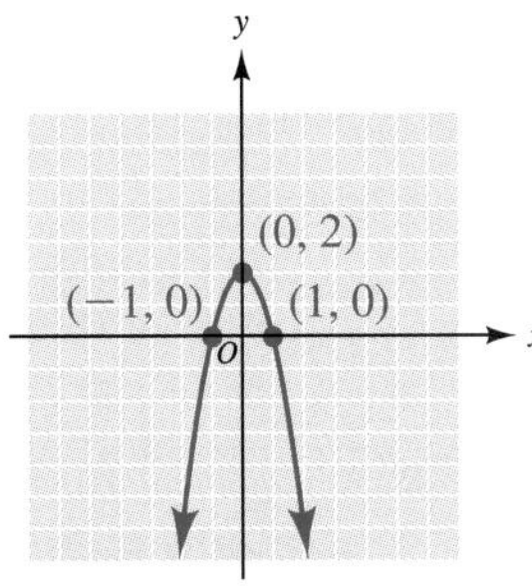

   b. $\{(x, y) \mid y = \sqrt{1 - x}\}$
   c. $y = 4x$
   d. $\{(x, y) \mid y = -x, x = -1, 2, 5\}$
   e. $\{(1, 4), (6, 5), (5, 6), (4, 1)\}$

6. a. The graph is

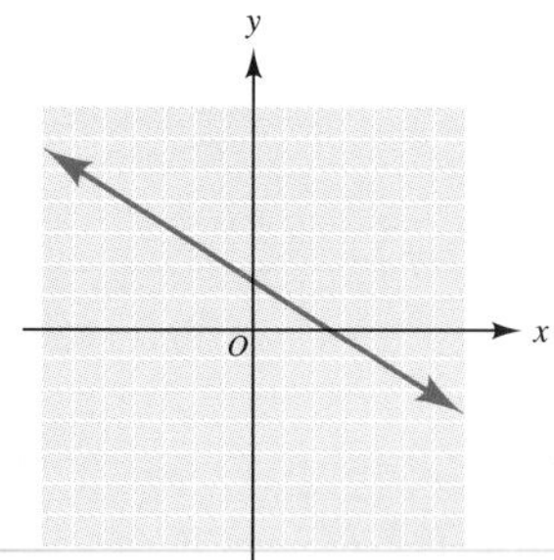

b. $y = 3x - 2$

c. $\{(x, y) \mid y = \sqrt{6 - 3x}\}$

d. $\{(x, y) \mid y = \frac{2}{3}x - 1, x = -3, 0, 3\}$

e. $\{(4, 4), (5, 5), (9, 9), (7, 7)\}$

7. The following table gives the y-values corresponding to various x-values:

x	-3	-2	-1	0	1	2	3	4
y	9	4	1	0	1	4	9	16

Explain why y is a function of x. Find the domain and the range of the function.

8. The following table gives the s-values corresponding to various t-values:

t	1	2	3	4	5	6	7	8
s	24	72	152	264	408	584	792	1,032

Explain why s is a function of t. Find the domain and the range of the function.

9. Find the range of the function $\{(x, y) \mid 10x - 7y = 25, x = -1, 2, 4, 6\}$, and graph the function.

10. Find the range of the function $\{(x, y) \mid 3x - 9y = 6, x = -1, 0, 1, 2\}$, and graph the function.

11. Graph the function $\{(x, y) \mid 4x - 12y = 12\}$, with the domain being the set of real numbers.

12. Graph the function $\left\{(x, y) \middle| \frac{x}{2} + \frac{y}{5} = -1\right\}$, with the domain being the set of real numbers.

9.2 Functional Notation

It is often desirable to *name* a function, and when y is a function of x, we name the function by using the notation $y = f(x)$, read y equals f of x. This is called **functional notation**. The *name* of the function is f, and the symbol $f(x)$ stands for the *value* of the function at x. The symbol $f(3)$ is read "f of 3," and it stands for the value of y when x is 3. For example, if $f(x) = 6x + 2$, then $f(3) = 6(3) + 2 = 20$; we say "f of 3 is 20," which means that y is 20 when x is 3 or that $f(x)$ is 20 when x is 3.

A Word of Caution The notation $f(x)$ does *not* mean f times x. It is simply an alternative way of writing y when y is a function of x.

The equations $y = 5x - 7$ and $f(x) = 5x - 7$ mean the same thing, and we often combine the equations as follows:

$$y = f(x) = 5x - 7$$

Evaluating a Function

To **evaluate a function** means to find the value of y [or of $f(x)$] that corresponds to a particular value of x. If a function $f(x)$ is to be evaluated at $x = a$, we write this as $f(a)$, where $f(a)$ is read as "f of a," and we find $f(a)$ by substituting a for x (see Example 1).

EXAMPLE 1 If $y = f(x) = 5x - 7$, find $f(2)$.

SOLUTION We will substitute 2 for x:

$$f(x) = 5x - 7$$

$$f(2) = 5(2) - 7 \quad \text{Substituting 2 for } x$$

$$f(2) = 10 - 7 = 3 \quad \text{Simplifying}$$

The statement $f(2) = 3$ means that $f(x) = 3$ when $x = 2$, or that $y = 3$ when $x = 2$.

EXAMPLE 2 Given $f(x) = \dfrac{2x^2 - 7x}{5x - 3}$, find $f(-2)$, $f(0)$, $f(h)$, and $f(a + b)$.

SOLUTION

$$f(x) = \frac{2(x)^2 - 7(x)}{5(x) - 3}$$

$$f(-2) = \frac{2(-2)^2 - 7(-2)}{5(-2) - 3} = \frac{8 + 14}{-10 - 3} = -\frac{22}{13} \quad \text{Substituting } -2 \text{ for every } x \text{ in } f(x)$$

$$f(0) = \frac{2(0)^2 - 7(0)}{5(0) - 3} = \frac{0 - 0}{0 - 3} = 0 \quad \text{Substituting 0 for every } x \text{ in } f(x)$$

$$f(h) = \frac{2(h)^2 - 7(h)}{5(h) - 3} = \frac{2h^2 - 7h}{5h - 3} \quad \text{Substituting } h \text{ for every } x \text{ in } f(x)$$

$$f(a + b) = \frac{2(a + b)^2 - 7(a + b)}{5(a + b) - 3} \quad \text{Substituting } (a + b) \text{ for every } x \text{ in } f(x)$$

$$= \frac{2a^2 + 4ab + 2b^2 - 7a - 7b}{5a + 5b - 3}$$

EXAMPLE 3 Given $f(x) = 2x + 1$, evaluate $\dfrac{f(-5) - f(2)}{4}$.

SOLUTION We will separately find $f(-5)$ and $f(2)$ and then substitute these values into the expression $\dfrac{f(-5) - f(2)}{4}$.

$$f(x) = 2(x) + 1$$

$$f(-5) = 2(-5) + 1 = -10 + 1 = -9$$

$$f(2) = 2(2) + 1 = 4 + 1 = 5$$

Therefore,

$$\frac{f(-5) - f(2)}{4} = \frac{-9 - 5}{4} = \frac{-14}{4} = -\frac{7}{2}$$

EXAMPLE 4 Given $f(x) = 3x^2 + 2x - 3$, find and simplify $\dfrac{f(x + h) - f(x)}{h}$.

SOLUTION To find $f(x + h)$, we substitute $x + h$ for every x in the given equation:

$$f(x + h) = 3(x + h)^2 + 2(x + h) - 3$$

Then

The parentheses around $3x^2 + 2x - 3$ are essential

$$\frac{f(x+h)-f(x)}{h} = \frac{\overbrace{[3(x+h)^2 + 2(x+h) - 3]}^{f(x+h)} - \overbrace{(3x^2+2x-3)}^{f(x)}}{h}$$

$$= \frac{[3(x^2+2xh+h^2) + 2x + 2h - 3] - 3x^2 - 2x + 3}{h}$$

$$= \frac{\cancel{3x^2} + 6xh + 3h^2 \cancel{+2x} + 2h \cancel{-3} \cancel{-3x^2} \cancel{-2x} \cancel{+3}}{h}$$

$$= \frac{6xh + 3h^2 + 2h}{h}$$

$$= \frac{\overset{1}{\cancel{h}}(6x + 3h + 2)}{\underset{1}{\cancel{h}}}$$

$$= 6x + 3h + 2$$

A Word of Caution It is important to realize that $f(x+h) \neq f(x) + h$. We saw as part of Example 4 that if $f(x) = 3x^2 + 2x - 3$, $f(x+h) = 3x^2 + 6xh + 3h^2 + 2x + 2h - 3$. However, $f(x) + h = 3x^2 + 2x - 3 + h$; it is *completely* different from $f(x+h)$.

★ **EXAMPLE 5** Given $f(x) = x^5$, find and simplify $\dfrac{f(x+h)-f(x)}{h}$.

SOLUTION If $f(x) = x^5$, then $f(x+h) = (x+h)^5$. Using the binomial theorem, we have

$$f(x+h) = (x+h)^5 = x^5 + 5x^4h + 10x^3h^2 + 10x^2h^3 + 5xh^4 + h^5$$

Therefore,

$$f(x+h) - f(x) = \overbrace{x^5 + 5x^4h + 10x^3h^2 + 10x^2h^3 + 5xh^4 + h^5}^{f(x+h)} - \overbrace{(x^5)}^{f(x)}$$

and

$$\frac{f(x+h)-f(x)}{h} = \frac{\cancel{x^5} + 5x^4h + 10x^3h^2 + 10x^2h^3 + 5xh^4 + h^5 - \cancel{x^5}}{h}$$

$$= \frac{5x^4h + 10x^3h^2 + 10x^2h^3 + 5xh^4 + h^5}{h}$$

$$= \frac{\overset{1}{\cancel{h}}(5x^4 + 10x^3h + 10x^2h^2 + 5xh^3 + h^4)}{\underset{1}{\cancel{h}}}$$

$$= 5x^4 + 10x^3h + 10x^2h^2 + 5xh^3 + h^4$$

When we're discussing two or more different functions in the same problem, we use different letters to name the functions; we might call one function *f*, one *g*, and one *h*, or one *f* and another *F*. We can use letters other than *x* for the independent variable and letters other than *y* for the dependent variable (see Example 6).

EXAMPLE 6 Examples of using various letters in functional notation:

a. $s = g(t) = 16t^2$ s is a function of t; t is the independent variable, and s is the dependent variable; the name of the function is g

b. $C = h(r) = 2\pi r$ C is a function of r; r is the independent variable, and C is the dependent variable; the name of the function is h

c. $V = F(e) = e^3$ V is a function of e; e is the independent variable, and V is the dependent variable; the name of the function is F

If y is a function of x, then the symbols $f(x)$ and y are interchangeable; therefore, we can call the vertical axis the $f(x)$-axis rather than the y-axis (see Figure 1a). If $v = g(t)$, we can call the horizontal axis the t-axis and the vertical axis the $g(t)$-axis (see Figure 1b); and if $u = h(s)$, we can call the horizontal axis the s-axis and the vertical axis the $h(s)$-axis (see Figure 1c).

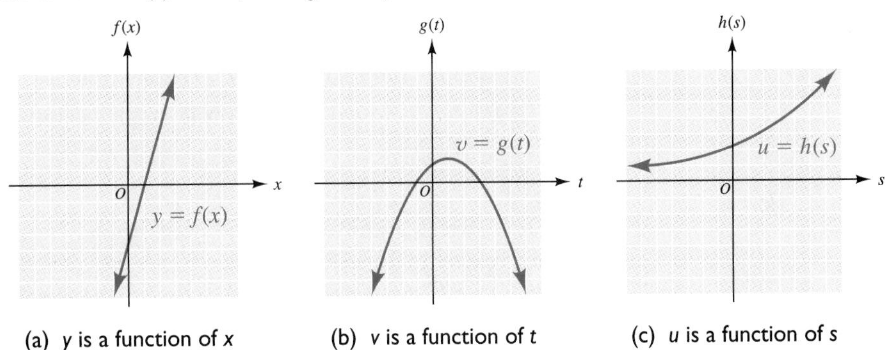

(a) y is a function of x (b) v is a function of t (c) u is a function of s

Figure 1

EXAMPLE 7 Given $h(r) = 5r^2$ and $g(s) = 4s - 1$, find $h(a + 1) - g(3a)$.

SOLUTION We first find $h(a + 1)$ and $g(3a)$; then we subtract $g(3a)$ from $h(a + 1)$.

$$h(r) = 5(r)^2$$
$$h(a + 1) = 5(a + 1)^2 = 5(a^2 + 2a + 1) = 5a^2 + 10a + 5$$
$$g(s) = 4(s) - 1$$
$$g(3a) = 4(3a) - 1 = 12a - 1$$

Therefore,

$$\begin{aligned} h(a + 1) - g(3a) &= (5a^2 + 10a + 5) - (12a - 1) \\ &= 5a^2 + 10a + 5 - 12a + 1 \\ &= 5a^2 - 2a + 6 \end{aligned}$$

Applications Involving Functions

There are many applications of functions and functional notation in business, statistics, the sciences, and so on.

EXAMPLE 8 If h is measured in feet and t in seconds and if the formula for finding the height of an arrow that is shot upward from ground level is $h(t) = 64t - 16t^2$, $0 \le t \le 4$, find the height of the arrow 1 sec after it is shot into the air.

SOLUTION We need to find h when $t = 1$. Therefore, we must find $h(1)$:

$$h(1) = 64(1) - 16(1^2) = 64 - 16 = 48$$

The arrow is 48 ft above the ground 1 sec after it is shot upward.

Functions of More Than One Variable

All the functions discussed so far in this section have been functions of *one* variable. Many functions are functions of more than one variable. For example, the volume of a rectangular box is a function of the length, the width, and the height of the box.

EXAMPLE 9 An example of a function of more than one variable:
If we put \$100 in principal (P) into a bank account that pays $\frac{1}{4}\%$ interest per month (i), we find the amount of money in the account (A) after 24 months (n) by using the formula

$$A = f(P, i, n) = P(1 + i)^n$$

The notation $A = f(P, i, n)$ is read A equals f of P, i, and n or A is a function of P, i, and n. (It is understood that the interest rate must be converted to decimal form—or to common fraction form—when we enter it into the formula.) The *dependent* variable, A, is a function of *three* independent variables, P, i, and n. For the given facts in this problem,

$$A = f(100, 0.0025, 24) = 100(1 + 0.0025)^{24} = 100(1.0025)^{24} \approx 106.18$$

Therefore, there would be \$106.18 in the account after 24 months.

If we're asked to find the domain and range of a function when the function is expressed in functional notation, we use exactly the same techniques we use when there's a y in the equation rather than an $f(x)$ (see Exercises 25 and 26 in the following exercise sets).

Exercises 9.2
Set I

1. Given $f(x) = 3x - 1$, find the following:
   a. $f(2)$ b. $f(0)$
   c. $f(a - 2)$ d. $f(x + 2)$
2. Given $f(x) = 4x + 1$, find the following:
   a. $f(3)$ b. $f(-5)$
   c. $f(0)$ d. $f(x - 2)$
3. Given $f(x) = 2x^2 - 3$, evaluate $\dfrac{f(5) - f(2)}{6}$.
4. Given $f(x) = 5x^2 - 2$, evaluate $\dfrac{f(3) - f(1)}{5}$.
5. Given $f(x) = 3x^2 - 2x + 4$, find $2f(3) + 4f(1) - 3f(0)$.
6. Given $f(x) = (x + 1)(x^2 - x + 1)$, find $3f(0) + 5f(2) - f(1)$.
7. If $f(x) = x^3$ and $g(x) = \dfrac{1}{x}$, find $f(-3) - 6g(2)$.
8. If $f(x) = x^2$ and $g(x) = \dfrac{1}{x}$, find $2f(-4) - 9g(3)$.
9. If $H(x) = 3x^2 - 2x + 4$ and $K(x) = x - x^2$, find $2H(2) - 3K(3)$.
10. If $P(a) = a^2 - 4$ and $Q(a) = 2 - a$, find $3P(2) + 2Q(4)$.

In Exercises 11–14, find and simplify $\dfrac{f(x + h) - f(x)}{h}$ for the given functions.

11. $f(x) = x^2 - x$
12. $f(x) = 3x^2$
*13. $f(x) = x^4$
*14. $f(x) = x^3$
15. If $A(r) = \pi r^2$ and $C(r) = 2\pi r$, find $\dfrac{3A(r) - 2C(r)}{\pi r}$.
16. If $A(r) = \pi r^2$ and $C(r) = 2\pi r$, find $\dfrac{5A(r) + 3C(r)}{\pi r}$.
17. If the formula for finding the height of an arrow that was shot upward from ground level is $h(t) = 64t - 16t^2$, $0 \le t \le 4$, find the height of the arrow 2 sec after it is shot into the air.
18. If the formula for finding the height of an arrow that was shot upward from ground level is $h(t) = 128t - 16t^2$, $0 \le t \le 8$, find the height of the arrow 1 sec after it is shot into the air.
19. If the formula for finding the cost of manufacturing x bushings per day is $C(x) = 500 + 20x - 0.1x^2$ (where C is in dollars), find the cost of manufacturing 100 bushings per day.
20. If the formula for finding the cost of manufacturing x brackets per day is $C(x) = 400 + 10x - 0.1x^2$ (where C is in dollars), find the cost of manufacturing 100 brackets per day.

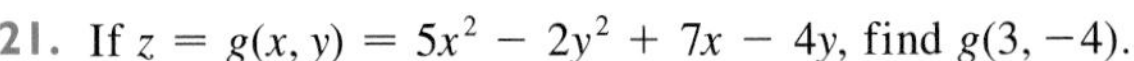

21. If $z = g(x, y) = 5x^2 - 2y^2 + 7x - 4y$, find $g(3, -4)$.

22. If $z = h(x, y) = 2x^2 - 6xy + 5y^2$, find $h(-1, -2)$.

23. Given that $A = f(P, i, n) = P(1 + i)^n$, use a calculator to approximate $f(100, 0.08, 12)$. Round off the answer to two decimal places.

24. Given that $F = f(A, v) = 58.6Av^2$, find $f(126.5, 634)$.

25. Find the domain and range of the function $f(x) = \sqrt{x - 4}$.

26. Find the domain and range of the function $f(x) = -\sqrt{9x - 16}$.

Writing Problems

Express the answers in your own words and in complete sentences.

1. Explain how to find $f(-4)$ if it's given that $f(x) = (x + 6)^5$.

2. Explain how to find $f(2, -3)$ if it's given that $f(x, y) = x^3 - 2y^2$.

3. Explain the difference between $f(x + h)$ and $f(x) + h$.

4. Find and describe the error in the following: If $f(x) = 3x^2$, then

$$\frac{f(x + h) - f(x)}{h} = \frac{3x^2 + h - 3x^2}{h} = \frac{h}{h} = 1$$

Explain how to evaluate $\dfrac{f(x + h) - f(x)}{h}$ correctly.

Exercises 9.2 Set II

1. Given $f(x) = 13 - 5x$, find the following:
   a. $f(9)$
   b. $f(-7)$
   c. $f(0)$
   d. $f(x - 6)$

2. Given $f(x) = 8 + 2x$, find the following:
   a. $f(0)$
   b. $f(-1)$
   c. $f(x + 2)$
   d. $f(x) + 2$

3. Given $f(x) = 8 - 3x^2$, evaluate $\dfrac{f(4) - f(-6)}{12}$.

4. Given $f(x) = x^2 - 2$, evaluate $\dfrac{f(0) + f(2)}{3}$.

5. Given $f(x) = (3x - 5)(x^2 - 9)$, find $6f(8) - 13f(0) - f(-5)$.

6. Given $f(x) = (x - 1)(2x + 1)(x + 1)$, find $3f(0) - 7f(-1) + 2f(4)$.

7. If $f(x) = 2x^2$ and $g(x) = \dfrac{1}{x^2}$, find $\dfrac{8}{f(-6)} - 12g(4)$.

8. If $f(x) = x + 3$ and $g(x) = \sqrt{x - 1}$, find $\dfrac{1}{f(0)} + g(5)$.

9. If $h(x) = \dfrac{x}{x - 2}$ and $k(x) = x^2 + 4$, find $h(0) + 3h(3) - k(1)$.

10. If $F(e) = \dfrac{3e - 1}{e^2}$ and $G(w) = \dfrac{1}{5 - w}$, find $\dfrac{4F(-3)}{5G(4)}$.

In Exercises 11–14, find and simplify $\dfrac{f(x + h) - f(x)}{h}$ for the given functions.

11. $f(x) = 5x^2 - 3$

12. $f(x) = 1 - x^2$

*13. $f(x) = -2x^3$

*14. $f(x) = -x^3$

15. If $h(x) = 3x^2$ and $k(x) = x^2 - 2x$, find $\dfrac{h(x) - k(x)}{x}$.

16. If $A(r) = \pi r^2$ and $C(r) = 2\pi r$, find $\dfrac{6A(r) + 3C(r)}{12\pi r}$.

17. If the formula for finding the height of an arrow that was shot upward from ground level is $h(t) = 128t - 16t^2$, $0 \le t \le 8$, find the height of the arrow 3 sec after it is shot into the air.

18. If the formula for finding the height of an arrow that was shot upward from ground level is $h(t) = 64t - 16t^2$, $0 \le t \le 4$, find the height of the arrow 3 sec after it is shot into the air.

19. If the formula for finding the cost of manufacturing x handles per day is $C(x) = 600 + 30x - 0.1x^2$ (where C is in dollars), find the cost of manufacturing 100 handles per day.

20. If the formula for finding the cost of manufacturing x faucets per day is $C(x) = 1{,}000 + 20x - 0.1x^2$ (where C is in dollars), find the cost of manufacturing 100 faucets per day.

21. If $w = f(x, y, z) = \dfrac{5xyz - x^2}{2z}$, find $f(-3, 4, -15)$.

22. If $w = f(x, y, z) = \dfrac{3x + y^2}{z}$, find $f(-2, -3, 5)$.

23. Given that $A = f(P, i, n) = P(1 + i)^n$, use a calculator to approximate $f(500, 0.11, 20)$. Round off the answer to two decimal places.

24. Given that $w = f(x, y, z) = \dfrac{x - z}{2y}$, find $f(3, -2, -4)$.

25. Find the domain and range of the function $f(x) = \sqrt{15 - 4x}$.

26. Find the domain and range of the function $f(x) = \sqrt{5x - 2}$.

9.3 Graphing $y = |x|$; Graphing $y = f(x) + h$ and $y = f(x + h)$ from the Graph of $y = f(x)$

In this section, we consider graphing the absolute value function, $y = |x|$, and we also discuss how the graphs of $y = f(x) + h$ and $y = f(x + h)$ are related to the graph of $y = f(x)$.

Graphing $y = f(x) = |x|$

EXAMPLE 1 Find the domain of the function $y = |x|$; then graph the function and find its range.

SOLUTION There are no restrictions on x; therefore, the domain is $\{x \mid x \in R\}$. Because $f(0) = 0$, one point on the graph will be the point $(0, 0)$.

By the definition of absolute value, if $x > 0$, $|x| = x$. Therefore, to the *right* of the y-axis (that is, in the region where $x > 0$), we will graph the line $y = x$ (see Figure 2).

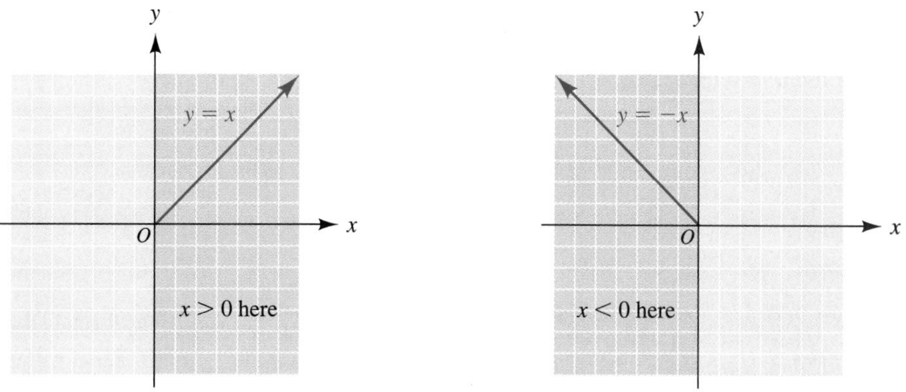

Figure 2 Figure 3

Also by the definition of absolute value, if $x < 0$, $|x| = -x$. Therefore, to the *left* of the y-axis (that is, in the region where $x < 0$), we will graph the line $y = -x$ (see Figure 3).

Putting these two graphs together with the point $(0, 0)$, we have the graph of $y = f(x) = |x|$ (see Figure 4). Notice that the graph is shaped like a V.

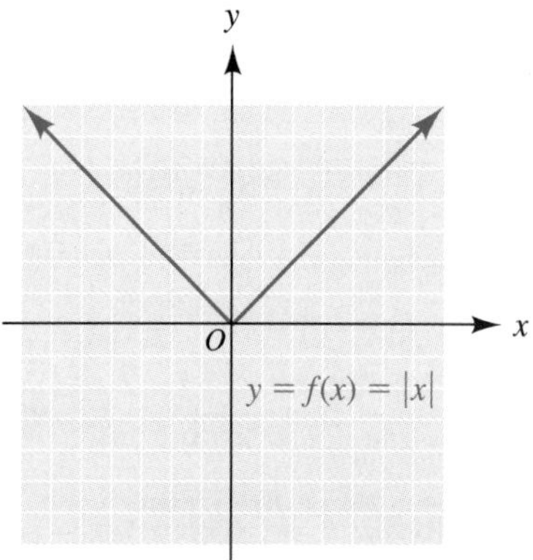

Figure 4

Note To graph $y = f(x) = |x|$ with a TI-81 calculator, set xmin to -10, xmax to 10, ymin to -2, and ymax to 10; then press these keys, in order:

ABS D

[Y=] [2nd] [x^{-1}] [X|T] [GRAPH]

Graphing $y = f(x) + h$ from the Graph of $y = f(x)$

Because $f(x) + h$ indicates that h is to be added to $f(x)$ [and because $f(x) = y$], if we know what the graph of $y = f(x)$ looks like, we can graph $y = f(x) + h$ by simply adding h units to each y-value; that is, we can simply shift the graph of $y = f(x)$ *upward* h units if h is positive and *downward* $|h|$ units if h is negative (see Example 2).

EXAMPLE 2 Graph $y = |x| + 3$.

SOLUTION We can graph $y = |x| + 3$ by shifting the graph of $y = |x|$ *upward* three units. You can verify that the graph is correct by examining the table of values.

Table of values

x	y
-2	5
-1	4
0	3
1	4
2	5

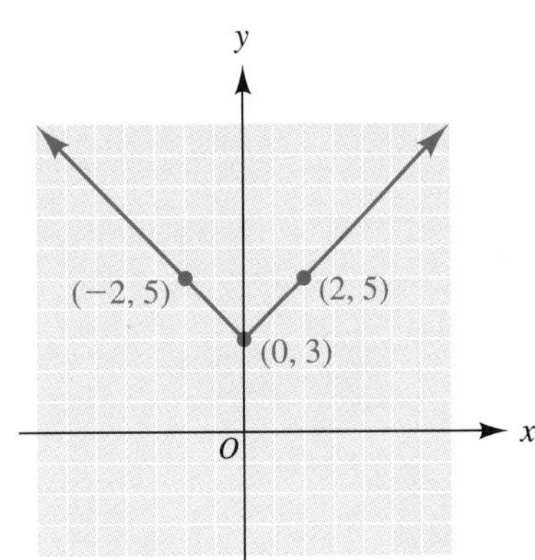

Note To graph $y = f(x) = |x| + 3$ with a TI-81 calculator, press these keys, in order:

ABS D

[Y=] [2nd] [x^{-1}] [X|T] [+] [3] [GRAPH]

To see the graphs of $y = |x|$ and $y = |x| + 3$ *together*, press the [Y=] key. You'll see ":y_1 ="; let $y_1 = |x|$, as described in the note following Example 1. Then press the [▼] key to go to the next line of the menu. You'll see ":y_2 ="; let $y_2 = |x| + 3$, as described above. Press [GRAPH], and you'll see both graphs on the same axes.

Graphing $y = f(x + h)$ from the Graph of $y = f(x)$

Because $x + h$ means that we're to add h units to x *before* we find the corresponding y-value, if we know what the graph of $y = f(h)$ looks like, we can graph $y = f(x + h)$ by shifting the graph of $y = f(x)$ *to the left* h units if h is positive and *to the right* $|h|$ units if h is negative (see Example 3).

EXAMPLE 3 Graph $y = |x + 3|$.

SOLUTION We can graph $y = |x + 3|$ by shifting the graph of $y = |x|$ three units to the *left*. (We shift to the left because *h is positive*.) You can verify that the graph is correct by examining the table of values.

Table of values

x	y
−7	4
−6	3
−5	2
−4	1
−3	0
−2	1
−1	2
0	3
1	4

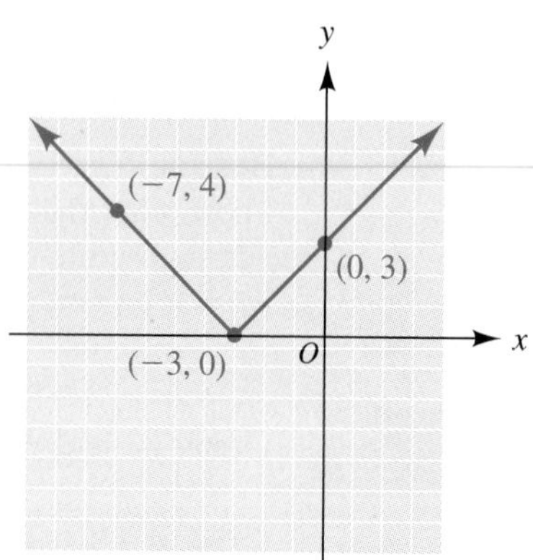

Note To graph $y = f(x) = |x + 3|$ with a TI-81 calculator, press these keys, in order:

ABS D

[Y=] [2nd] [x^{-1}] [(] [X|T] [+] [3] [)] [GRAPH]

Note The graph of the function $f(x) = |ax + b|$ is always shaped like a V, and the graph of $f(x) = -|ax + b|$ is always shaped like an upside-down V.

Exercises 9.3
Set I

Sketch the graph of each function. In each exercise, compare the graphs of the a, b, and c parts.

1. a. $f(x) = |x| - 2$ b. $f(x) = |x - 2|$ c. $f(x) = -|x - 2|$
2. a. $f(x) = |x| + 5$ b. $f(x) = |x + 5|$ c. $f(x) = -|x + 5|$
3. a. $f(x) = -3x$ b. $f(x) = -3x + 2$ c. $f(x) = -3(x + 2)$
4. a. $f(x) = 2x$ b. $f(x) = 2x - 3$ c. $f(x) = 2(x - 3)$

Writing Problems

Express the answers in your own words and in complete sentences.

1. Explain how the graph of $y = f(x) + h$ is related to the graph of $y = f(x)$.
2. Explain how the graph of $y = f(x + h)$ is related to the graph of $y = f(x)$.

Exercises 9.3
Set II

Sketch the graph of each function. In each exercise, compare the graphs of the a, b, and c parts.

1. a. $f(x) = |x| + 4$ b. $f(x) = |x + 4|$ c. $f(x) = -|x + 4|$
2. a. $f(x) = |x| - 3$ b. $f(x) = |x - 3|$ c. $f(x) = -|x - 3|$
3. a. $f(x) = 4x$ b. $f(x) = 4x + 2$ c. $f(x) = 4(x + 2)$
4. a. $f(x) = -3x$ b. $f(x) = -3x + 2$ c. $f(x) = -3(x + 2)$

9.4 The Algebra of Functions; Composition of Functions

The Algebra of Functions

Functions can be added, subtracted, multiplied, and divided—we call this the **algebra of functions**. The *sum* $(f + g)$, the *difference* $(f - g)$, the *product* $(f \cdot g)$, and the *quotient* $\left(\frac{f}{g}\right)$ are defined as follows:

> Let $f(x)$ and $g(x)$ be two functions, and let D be the *intersection* of the domains of $f(x)$ and $g(x)$. Then,
>
> $(f + g)(x) = f(x) + g(x)$, and the domain of $f + g$ is all of D.
>
> $(f - g)(x) = f(x) - g(x)$, and the domain of $f - g$ is all of D.
>
> $(f \cdot g)(x) = (f(x))(g(x))$, or $f(x)g(x)$, and the domain of $f \cdot g$ is all of D.
>
> $\left(\frac{f}{g}\right)(x) = \frac{f(x)}{g(x)}$, and the domain of $\frac{f}{g}$ is the set of those values of D for which $g(x)$ is *not* 0.

EXAMPLE 1 Given $f(x) = 3x + 2$ and $g(x) = -2x + 6$, find the domain of f, the domain of g, and D, the intersection of those domains. Then find $f + g, f - g, f \cdot g$, and $\frac{f}{g}$, and find the domain of each of those functions.

SOLUTION For both f and g, there are no restrictions on the domain, no denominators, no radical signs, and so on. Therefore, the domain of f is the set of all real numbers, which is also the domain of g. Consequently, D is the set of all real numbers, R.

$(f + g)(x) = f(x) + g(x) = (3x + 2) + (-2x + 6) = x + 8$, and the domain is R.

$(f - g)(x) = f(x) - g(x) = (3x + 2) - (-2x + 6) = 5x - 4$, and the domain is R.

$(f \cdot g)(x) = f(x)g(x) = (3x + 2)(-2x + 6) = -6x^2 + 14x + 12$, and the domain is R.

$\left(\frac{f}{g}\right)(x) = \frac{f(x)}{g(x)} = \frac{3x + 2}{-2x + 6}$, and the domain is the set of all R *except* those numbers that would make $-2x + 6$ (the denominator) 0. Therefore, the domain is the set of all real numbers except 3.

EXAMPLE 2 Given $f(x) = \sqrt{x+2}$ and $g(x) = \sqrt{1-x}$, find the domain of f, the domain of g, and D, the intersection of those domains. Then find $f + g$, $f - g$, $f \cdot g$, and $\frac{f}{g}$, and find the domain of each of those functions.

SOLUTION For f, $x + 2 \geq 0$, or $x \geq -2$. Therefore, the domain of f is $\{x \mid x \geq -2\}$, or, in interval notation, $[-2, +\infty)$. For g, $1 - x \geq 0$, or $1 \geq x$, or $x \leq 1$. Therefore, the domain of g is $\{x \mid x \leq 1\}$, or, in interval notation, $(-\infty, 1]$. D is the intersection of these two sets, and you can verify that the intersection is $\{x \mid -2 \leq x \leq 1\}$, or, in interval notation, $[-2, 1]$.

$(f + g)(x) = f(x) + g(x) = \sqrt{x+2} + \sqrt{1-x}$, and the domain is $[-2, 1]$.

$(f - g)(x) = f(x) - g(x) = \sqrt{x+2} - \sqrt{1-x}$, and the domain is $[-2, 1]$.

$(f \cdot g)(x) = f(x)g(x) = \sqrt{x+2}\sqrt{1-x} = \sqrt{2 - x - x^2}$, and the domain is $[-2, 1]$.

$\left(\frac{f}{g}\right)(x) = \frac{f(x)}{g(x)} = \frac{\sqrt{x+2}}{\sqrt{1-x}}$, and the domain is the set of all the numbers in $[-2, 1]$ *except* those that would make $\sqrt{1-x}$ (the denominator) 0. Therefore, we must exclude 1 from $[-2, 1]$, so the domain of $\frac{f}{g}$ is $[-2, 1)$.

Composition of Functions

Another operation that can be performed on two functions is called **composition**. The symbol for the composition of f with g is $f \circ g$, which is read f composed with g or the composition of f with g. The operation is defined as follows:

> Let $f(x)$ and $g(x)$ be two functions.
>
> If $g(x)$ is in the domain of f, then the composition of f with g is
>
> $$(f \circ g)(x) = f(g(x))$$
>
> The domain of $f \circ g$ is the set of all x in the domain of g for which $g(x)$ is in the domain of f.
>
> Also, if $f(x)$ is in the domain of g, then the composition of g with f is
>
> $$(g \circ f)(x) = g(f(x))$$
>
> The domain of $g \circ f$ is the set of all x in the domain of f for which $f(x)$ is in the domain of g.

A Word of Caution The domain of $f \circ g$ can never contain any numbers that were not in the domain of g (it sometimes contains *fewer numbers* than were in the domain of g). Similarly, the domain of $g \circ f$ can never contain any numbers that were not in the domain of f (it sometimes contains *fewer numbers* than were in the domain of f).

EXAMPLE 3 Given $f(x) = x^2$ and $g(x) = \sqrt{x-3}$, find the domains of f and g, find $f \circ g$ and its domain, and find $g \circ f$. Do not try to find the domain of $g \circ f$.*

SOLUTION The domain of f is R, and the domain of g is $\{x \mid x \geq 3\}$, or, in interval notation, $[3, +\infty)$.

*We discuss finding the domains of such functions in a later chapter.

To find $f \circ g$, we will substitute $\sqrt{x-3}$ for every x that appears in $f(x)$. Therefore, since $f(x) = x^2$, $(f \circ g)(x) = f(g(x)) = (\sqrt{x-3})^2 = x - 3$; however, the domain of $f \circ g$ cannot contain any number *not* in the set $\{x \mid x \geq 3\}$, and that statement must be *part of* the answer for $f \circ g$. (That is, it is *incorrect* to write $(f \circ g)(x) = x - 3$, since the domain of $x - 3$ is the set of all real numbers.) Therefore, the correct answer is $(f \circ g)(x) = x - 3, x \geq 3$.

(Notice that g *is* in the domain of f, or $\sqrt{x-3}$ is in the domain of $f(x) = x^2$, since $\sqrt{x-3}$ is a real number for all $x \geq 3$.)

To find $g \circ f$, we will substitute x^2 for every x that appears in $g(x)$. Therefore, since $g(x) = \sqrt{x-3}$, $(g \circ f)(x) = g(f(x)) = \sqrt{x^2 - 3}$.

Note Notice that in Example 3, $f \circ g$ is *not* equal to $g \circ f$.

EXAMPLE 4 Given $f(x) = 3x + 2$ and $g(x) = \dfrac{x-2}{3}$, find the domains of f and g, find $f \circ g$ and its domain, and find $g \circ f$ and its domain.

SOLUTION The domain of f is R, and the domain of g is R. Also, $g(x)$ is in the domain of f for all x, and $f(x)$ is in the domain of g for all x.

To find $f(g(x))$, we will substitute $\dfrac{x-2}{3}$ for every x that's in $f(x)$:

$$(f \circ g)(x) = f(g(x)) = 3\left(\frac{x-2}{3}\right) + 2 = x - 2 + 2 = x$$

Therefore, $(f \circ g)(x) = x$, and the domain of $f \circ g$ is R.

To find $g(f(x))$, we will substitute $3x + 2$ for every x that's in $g(x)$:

$$(g \circ f)(x) = g(f(x)) = \frac{(3x+2) - 2}{3} = \frac{3x}{3} = x$$

Therefore, $(g \circ f)(x) = x$, and the domain of $g \circ f$ is R.

Exercises 9.4
Set I

In Exercises 1–6, for each pair of functions, verify that the domains of f and g are R, so the intersection of those domains, D, is also R. Then find $f + g$, $f - g$, $f \cdot g$, and $\dfrac{f}{g}$, and find the domain of $\dfrac{f}{g}$.

1. $f(x) = 7x - 1$; $g(x) = 2 - 5x$
2. $f(x) = -3x - 4$; $g(x) = 2 - 3x$
3. $f(x) = 2 - x$; $g(x) = x^2 + 3$
4. $f(x) = 6 + x$; $g(x) = 6 - x^2$
5. $f(x) = 7x - 1$; $g(x) = 7x - 1$
6. $f(x) = 12x$; $g(x) = \dfrac{x}{12}$

In Exercises 7–12, for each pair of functions, find the domain of f and the domain of g, and find $f \circ g$ and $g \circ f$; also, find the domains of $f \circ g$ and $g \circ f$, unless you're told *not* to find a domain.

7. $f(x) = \sqrt{7 - x}$; $g(x) = x^2 + 3$; do not find the domain of $f \circ g$.

★ 8. $f(x) = x^3 + 1$; $g(x) = 1 - \sqrt[3]{x}$

9. $f(x) = 5x - 1$; $g(x) = \dfrac{x+1}{5}$
10. $f(x) = \dfrac{x+8}{2}$; $g(x) = 2x - 8$
11. $f(x) = 3x + 2$; $g(x) = \dfrac{1}{3x+2}$
12. $f(x) = 7 - 2x$; $g(x) = \dfrac{1}{7-2x}$

Writing Problems

Express the answer in your own words and in complete sentences.

1. Discuss whether finding the composition of functions f and g is a commutative operation.

Exercises 9.4 Set II

In Exercises 1–6, for each pair of functions, verify that the domains of f and g are R, so the intersection of those domains, D, is also R. Then find $f + g, f - g, f \cdot g$, and $\frac{f}{g}$, and find the domain of $\frac{f}{g}$.

1. $f(x) = 6x - 2$; $g(x) = 3 - 9x$
2. $f(x) = -4x - 3$; $g(x) = 3 - 2x$
3. $f(x) = 6 - x$; $g(x) = x^2 + 5$
4. $f(x) = 7 + x$; $g(x) = 7 - x^2$
5. $f(x) = 8x - 3$; $g(x) = 8x - 3$
6. $f(x) = 15x$; $g(x) = \frac{x}{15}$

In Exercises 7–12, for each pair of functions, find the domain of f and the domain of g, and find $f \circ g$ and $g \circ f$; also, find the domains of $f \circ g$ and $g \circ f$, unless you're told *not* to find a domain.

7. $f(x) = \sqrt{9 - x}$; $g(x) = x^2 + 5$; do not find the domain of $f \circ g$.

★ 8. $f(x) = x^5 + 3$; $g(x) = 3 - \sqrt[5]{x}$

9. $f(x) = 4x - 3$; $g(x) = \frac{x + 3}{4}$
10. $f(x) = \frac{x + 6}{3}$; $g(x) = 3x - 6$
11. $f(x) = 8x + 3$; $g(x) = \frac{1}{8x + 3}$
12. $f(x) = 1 - 4x$; $g(x) = \frac{1}{1 - 4x}$

9.5 Inverse Functions

One-to-One Functions

As mentioned earlier, a function is a relation in which no two ordered pairs with the same first coordinate have second coordinates that are different from each other. If, within a function, there *also* are no two ordered pairs with the same *second* coordinate but *first* coordinates that are different from each other, then we say that the function is a **one-to-one function**.

For example, because no ordered pairs of the relation $\{(3, 1), (4, 2), (5, 3)\}$ have the same first coordinate but different second coordinates and no ordered pairs have the same second coordinate but different first coordinates, the relation is a *one-to-one function.*

On the other hand, the relation $\{(7, 5), (6, 1), (3, 5)\}$ is a *function* (no two ordered pairs have the same first coordinate but different second coordinates), but it is *not* a one-to-one function, because the ordered pairs $(7, 5)$ and $(3, 5)$ have the same second coordinate but different first coordinates.

Graphically, a relation is a one-to-one function if no *vertical* line can intersect the graph in more than one point *and* if no *horizontal* line can intersect the graph in more than one point.

Inverse Functions

A function will have an **inverse** if and only if it is a one-to-one function, and we then call the function **invertible**. The notation for the inverse of a function f is f^{-1}, read f inverse or the inverse of f.

A Word of Caution The superscript -1 in f^{-1} is *not* an exponent, and f^{-1} does *not* mean $\frac{1}{f}$. For example, if $f(x) = 3x + 2$, $f^{-1}(x) \neq \frac{1}{3x + 2}$.

Inverse functions are defined as follows:

> If f is a one-to-one function with domain X and range Y and if g is a one-to-one function with domain Y and range X, then g is the inverse function of f if and only if
>
> $$(f \circ g)(x) = f(g(x)) = x \quad \text{for all } x \text{ in } Y$$
>
> and $$(g \circ f)(x) = g(f(x)) = x \quad \text{for all } x \text{ in } X$$

If g is the inverse of f, then it is also true that f is the inverse of g; we also can say $g = f^{-1}$.

The functions $f(x) = 3x + 2$ and $g(x) = \frac{x - 2}{3}$ from Example 4 of the last section are inverses of each other, because, as we found in that example, $(f \circ g)(x) = x$ and $(g \circ f)(x) = x$, and the domains of f and g are both R.

EXAMPLE 1 Verify that $f(x) = x - 2$ and $g(x) = x + 2$ are inverses of each other.

SOLUTION The domain of f is R, and the domain of g is R.

$$(f \circ g)(x) = f(g(x)) = (x + 2) - 2 = x$$

and $$(g \circ f)(x) = g(f(x)) = (x - 2) + 2 = x$$

Therefore, f and g are inverses of each other.

We can also think of inverse functions another way: If f is the set of ordered pairs (x, y), then f^{-1}, the inverse of f, is the set of ordered pairs (y, x). That is, we obtain the inverse function by *interchanging* the coordinates of each of the ordered pairs. For example, suppose f is the function $f(x) = x - 2$, with domain $\{3, 4, 5\}$; then

$$f = \{(3, 1), (4, 2), (5, 3)\}$$

and the inverse is $$f^{-1} = \{(1, 3), (2, 4), (3, 5)\}$$

We could describe the inverse function as $f^{-1}(x) = x + 2$, with domain $\{1, 2, 3\}$. Notice that the *domain* of the inverse function equals the *range* of the original function and that the *range* of the inverse function equals the *domain* of the original function. Notice also that the two functions "undo" each other. That is, f subtracts 2 from each x-value to get the corresponding y-value, and f^{-1} adds 2 to each x-value to get the corresponding y-value.

We can see from the definition of the inverse of a function that the *domain* of the inverse function must always equal the *range* of the original function and that the *range* of the inverse function must always equal the *domain* of the original function.

Graphs of Inverse Functions

If we graph a function and its inverse on the same set of axes, corresponding points will always be symmetric with respect to the line $y = x$. (We can think of corresponding points as being the "mirror images" of each other, with the line $y = x$ being the mirror.)

EXAMPLE 2 Given $f = \{(2, -3), (-3, -1), (-4, 5), (4, 0)\}$, find f^{-1}, find the domain and range of f, find the domain and range of f^{-1}, and graph f and f^{-1} on the same set of axes.

SOLUTION

$$f = \{(2, -3), (-3, -1), (-4, 5), (4, 0)\}$$

$$f^{-1} = \{(-3, 2), (-1, -3), (5, -4), (0, 4)\}$$

The *domain* of f is $\{2, -3, -4, 4\}$, and this is also the *range* of f^{-1}. The *range* of f is $\{-3, -1, 5, 0\}$, and this is also the *domain* of f^{-1}. The graphs of f and its inverse, f^{-1}, are shown in Figure 5.

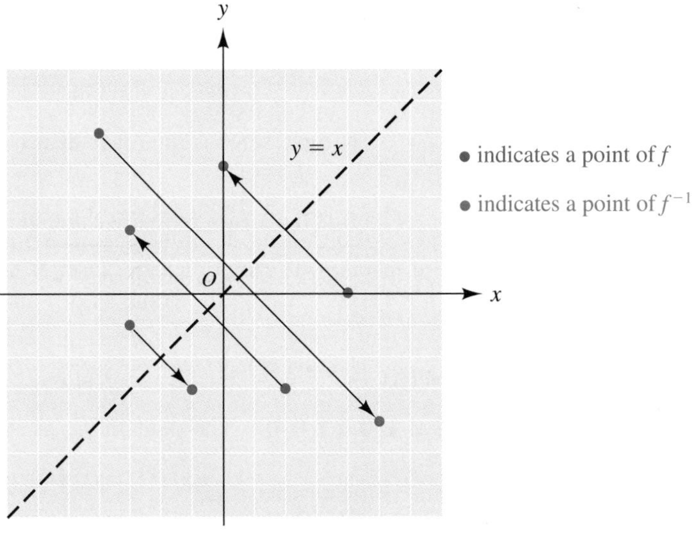

Figure 5

Finding the Inverse of a Function When Its Equation Is Given

The following procedure can be used to find the inverse of a function when its equation is given.

1. Substitute y for $f(x)$, if necessary.
2. Interchange x and y in the equation.
3. Solve the new equation for y.
4. Replace y with $f^{-1}(x)$.

The *domain* of f equals the *range* of f^{-1}, and the *range* of f equals the *domain* of f^{-1}.

EXAMPLE 3 Find the inverse function of $y = f(x) = 2x - 3$, and graph $y = f(x)$ and its inverse on the same set of axes.

SOLUTION Step 1 does not apply.

$y = 2x - 3$

Step 2. $x = 2y - 3$ Interchanging x and y

Step 3. $x + 3 = 2y$ Solving for y

$$y = \frac{x+3}{2}$$

Replacing y with $f^{-1}(x)$

Step 4. $f^{-1}(x) = \dfrac{x+3}{2}$ is the inverse function of $f(x) = 2x - 3$.

Graphing the given function

$y = f(x) = 2x - 3$

	x	y
If $y = 0$, $x = \frac{3}{2}$.	$\frac{3}{2}$	0
If $x = 0$, $y = -3$.	0	-3
If $x = 3$, $y = 3$.	3	3

Graphing the inverse function

$y = f^{-1}(x) = \dfrac{x+3}{2}$

	x	y
If $y = 0$, $x = -3$.	-3	0
If $x = 0$, $y = \frac{3}{2}$.	0	$\frac{3}{2}$
If $x = 3$, $y = 3$.	3	3

It was not really necessary to calculate the entries in the inverse table. We could have used the entries from the function table, entering the x-values in the y-column and the y-values in the x-column. The graphs are shown in Figure 6.

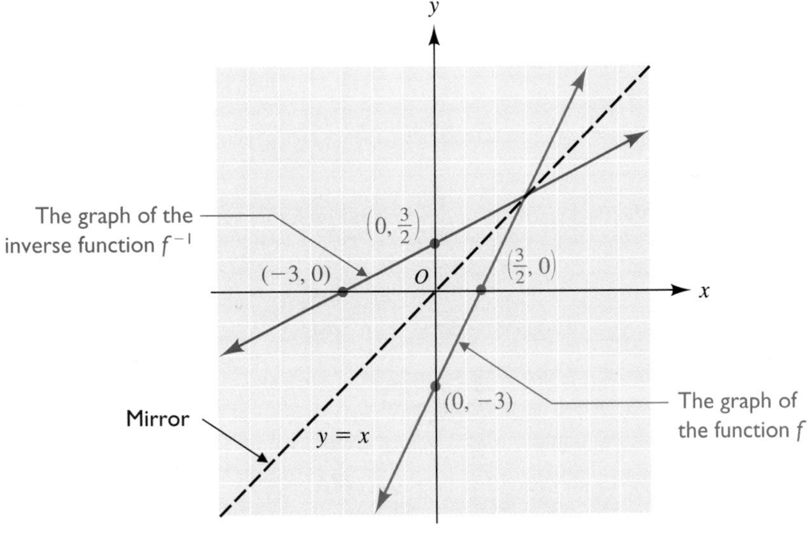

Figure 6

In Example 4, we cannot find the range of f simply by examining the equation. However, we *can* find the range of f by first finding the *domain* of f^{-1}, since the range of f *equals* the domain of f^{-1}.

EXAMPLE 4 Find the inverse function of $f(x) = \dfrac{5x - 1}{x}$, and find the domain and range of both functions.

SOLUTION

Step 1. $y = \dfrac{5x - 1}{x}$ Substituting y for $f(x)$

Step 2. $x = \dfrac{5y - 1}{y}$ Interchanging x and y

Step 3. $xy = 5y - 1$ Solving for y

$1 = 5y - xy$

$1 = y(5 - x)$

$y = \dfrac{1}{5 - x}$

Replacing y with $f^{-1}(x)$

Step 4. $f^{-1}(x) = \dfrac{1}{5 - x}$ is the inverse function of $f(x) = \dfrac{5x - 1}{x}$.

The domain of f is the set of all real numbers except 0, since the denominator cannot equal 0. Then, because the domain of f equals the range of f^{-1}, the range of f^{-1} must be the set of all real numbers except 0. To find the range of f, we must first find the domain of f^{-1}. The domain of $f^{-1}(x) = \dfrac{1}{5 - x}$ is the set of all real numbers except 5, since the denominator cannot equal zero. Because the range of f equals the domain of f^{-1}, we *now* know that the range of f must be the set of all real numbers except 5.

In summary,

$$\text{The domain of } f = \{x \mid x \in R, x \neq 0\}$$
$$\text{The range of } f = \{y \mid y \in R, y \neq 5\}$$
$$\text{The domain of } f^{-1} = \{x \mid x \in R, x \neq 5\}$$
$$\text{The range of } f^{-1} = \{y \mid y \in R, y \neq 0\}$$

Exercises 9.5
Set I

1. Verify that $f(x) = 5x - 1$ and $g(x) = \dfrac{x + 1}{5}$ are inverses of each other.

2. Verify that $f(x) = \dfrac{x + 8}{2}$ and $g(x) = 2x - 8$ are inverses of each other.

3. Show that $f(x) = 3x + 2$ and $g(x) = \dfrac{1}{3x + 2}$ are *not* inverses of each other by showing that $(f \circ g)(x) \neq x$.

4. Show that $f(x) = 7 - 2x$ and $g(x) = \dfrac{1}{7 - 2x}$ are *not* inverses of each other by showing that $(f \circ g)(x) \neq x$.

5. Given $f = \{(-2, 5), (3, -3), (-5, -4), (4, -2)\}$, determine whether f is one-to-one. If it is, find f^{-1}, find the domain and range of f, find the domain and range of f^{-1}, and graph f and f^{-1} on the same set of axes.

6. Given $f = \{(4, -4), (0, 3), (3, 4), (-2, 1), (5, -2)\}$, determine whether f is one-to-one. If it is, find f^{-1}, find the domain and range of f, find the domain and range of f^{-1}, and graph f and f^{-1} on the same set of axes.

7. Find the inverse function of $y = f(x) = 5 - 2x$. Graph $y = f(x)$ and its inverse on the same set of axes.

8. Find the inverse function of $y = f(x) = 3x - 10$. Graph $y = f(x)$ and its inverse on the same set of axes.

9. Find the inverse function of $y = f(x) = \dfrac{4x - 3}{5}$.

10. Find the inverse function of $y = f(x) = \dfrac{2x - 7}{3}$.

11. Find the inverse function of $y = f(x) = \dfrac{5}{x + 2}$.

12. Find the inverse function of $y = f(x) = \dfrac{10}{2x - 1}$.

Writing Problems

Express the answers in your own words and in complete sentences.

1. Explain how to find the inverse of a one-to-one function when the function is given as a set of ordered pairs.

2. Explain why $y = f(x) = \dfrac{x}{x+2}$ and $y = g(x) = \dfrac{x+2}{x}$ are not inverses of each other.

Exercises 9.5 Set II

1. Verify that $f(x) = 4x - 3$ and $g(x) = \dfrac{x+3}{4}$ are inverses of each other.

2. Verify that $f(x) = \dfrac{x+6}{3}$ and $g(x) = 3x - 6$ are inverses of each other.

3. Show that $f(x) = 8x + 3$ and $g(x) = \dfrac{1}{8x+3}$ are *not* inverses of each other by showing that $(f \circ g)(x) \neq x$.

4. Show that $f(x) = 1 - 4x$ and $g(x) = \dfrac{1}{1-4x}$ are *not* inverses of each other by showing that $(f \circ g)(x) \neq x$.

5. Given $f = \{(-5, 2), (4, 5), (-6, -4), (1, -3), (-2, 3)\}$, determine whether f is one-to-one. If it is, find f^{-1}, find the domain and range of f, find the domain and range of f^{-1}, and graph f and f^{-1} on the same set of axes.

6. Given $f = \{(-3, 6), (5, 6), (2, 3)\}$, determine whether f is one-to-one. If it is, find f^{-1}, find the domain and range of f, find the domain and range of f^{-1}, and graph f and f^{-1} on the same set of axes.

7. Find the inverse function of $y = f(x) = \frac{1}{5}x + 2$. Graph $y = f(x)$ and its inverse on the same set of axes.

8. Find the inverse function of $y = f(x) = \frac{2}{3}x - 4$. Graph $y = f(x)$ and its inverse on the same set of axes.

9. Find the inverse function of $y = f(x) = \dfrac{3x-8}{4}$.

10. Find the inverse function of $y = f(x) = \dfrac{3}{x-2}$.

11. Find the inverse function of $y = f(x) = \dfrac{6}{4-3x}$.

12. Find the inverse function of $y = f(x) = \dfrac{8}{4-x}$.

9.6 Polynomial Functions and Their Graphs

It is shown in calculus that the graph of the solution set of an equation that can be expressed in the form

$$y = a_n x^n + a_{n-1}x^{n-1} + \cdots + a_2x^2 + a_1x + a_0$$

is always a *continuous, smooth curve*; that is, it has *no holes, no jumps,* and *no sharp corners.* Because the right side of the above equation is a polynomal, we call the function a **polynomial function**.

EXAMPLE 1 Examples of polynomial functions:

a. $f(x) = 3x - 5$

b. $f(x) = x^2 - x - 2$

c. $f(x) = x^3 - 4x$

d. $f(x) = 7x^5 - 2x^3 + 6$

Special Polynomial Functions Some polynomial functions have special names.

The function $f(x) = k$, where k is some real number, is called the **constant function**; its graph is always a horizontal line. It is *not* a one-to-one function, and so it does not have an inverse.

The function $f(x) = x$ is called the **identity function**. When we find the composition of a function with its inverse, the result is always the identity function.

The function $f(x) = ax + b$, where $a, b \in R$ and $a \neq 0$, is a **linear function**. The function given in Example 1a is a linear function. (The *identity function* is also a linear function.) All linear functions are one-to-one functions; therefore, they have inverses.

The function $f(x) = ax^2 + bx + c$, where $a, b, c \in R$ and $a \neq 0$, is a **quadratic function**. The function given in Example 1b is a quadratic function. We *graph* a few quadratic functions in this section, and then we study quadratic functions and their graphs in much more detail in a later chapter.

The function $f(x) = ax^3 + bx^2 + cx + d$, where $a, b, c, d \in R$ and $a \neq 0$, is a **cubic function**, but often it's just called a polynomial function. The function given in Example 1c is a cubic function.

If the polynomial is of a higher degree than third, it's usually just called a **polynomial function**. The function given in Example 1d is a fifth-degree polynomial function.

Functions for which $f(-x) = f(x)$ are called **even functions**; for example, $f(x) = x^2$ is an even function because $f(-x) = (-x)^2 = x^2 = f(x)$. Functions for which $f(-x) = -f(x)$ are called **odd functions**; for example, $f(x) = x^3$ is an odd function because $f(-x) = (-x)^3 = -x^3 = -f(x)$. Most functions are neither even nor odd; $f(x) = x^2 + x$, for example, is neither even nor odd, because $f(-x) = (-x)^2 + (-x) = x^2 - x$, and $x^2 - x \neq f(x)$ and $x^2 - x \neq -f(x)$.

Graphing Nonlinear Polynomial Functions

In this section, we show how to graph polynomial functions that are not linear. Their graphs are curves, not straight lines. Two points are all that we really need to graph a straight line; to draw a curve, however, we first find a number of points on the curve, and then we connect those points in order from left to right with a smooth curve (see Examples 2–6).

EXAMPLE 2 Graph the function $y = f(x) = x^2$.

SOLUTION

x	y
-3	9
-2	4
-1	1
$-\frac{1}{2}$	$\frac{1}{4}$
$-\frac{1}{4}$	$\frac{1}{16}$
0	0
$\frac{1}{4}$	$\frac{1}{16}$
$\frac{1}{2}$	$\frac{1}{4}$
1	1
2	4
3	9

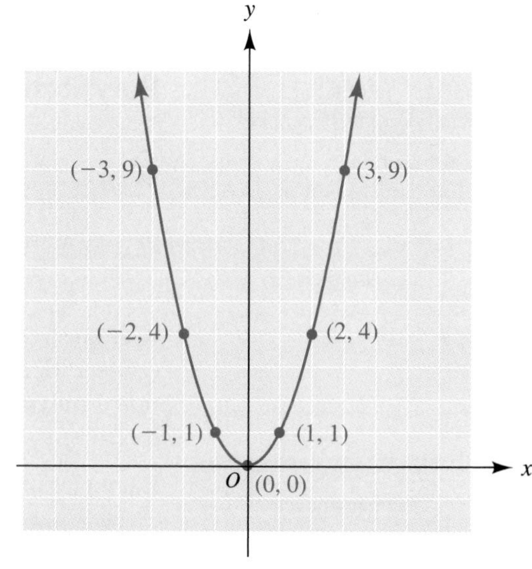

In drawing the smooth curve, we start with the point in the table of values having the smallest x-value, and we draw to the point having the next larger value of x. We continue in this way, drawing a smooth curve through all the points. The graph of the equation $y = x^2$ is called a **parabola**.

A parabola is one of a family of curves of quadratic polynomials called *conic sections*; they are discussed further in a later chapter.

Note The graph of an *even function* is always symmetric with respect to the y-axis. Notice that the graph in Example 2 is symmetric with respect to the y-axis and that the function $f(x) = x^2$ is an *even function.*

A Word of Caution The graph in Example 2 has *no sharp corners* and thus does *not* look like this:

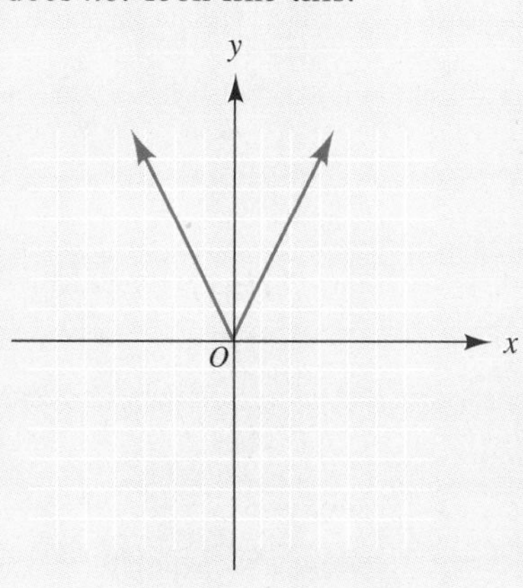

Note To graph $y = f(x) = x^2$ with a TI-81 graphics calculator, press the following keys, in order:

[Y=] [X|T] [x^2] [GRAPH]

To get a better look at what the graph looks like near the origin, you can "zoom in" on that region. To do so, press [ZOOM], select "1:Box," and press [ENTER]. By pressing the left, right, up, and/or down arrow keys, move the cursor to one corner of the "box" (or rectangle) you'd like to view more closely; when you're satisfied with one corner, press [ENTER]. Move the cursor to the *opposite corner* of the rectangle you want to magnify, and press [ENTER] again. You will see an enlarged view of the curve. You can then "trace" the curve to see what the x- and y-coordinates of points on the curve are, if you desire.

A Word of Caution If you do zoom in, the *range* will be changed. To get back to a full-screen view, you must manually change the range back to what it was before you zoomed in (or at least to something close to that range).

A Word of Caution If you *do* trace the curve, you will probably *not* see $x = 0$, $y = 0$ as a point on the curve, because the calculator does not give the coordinates of *all* points on the curve. If you see $x = .00110803$, $y = 1.2277E - 6$, you should be aware of the fact that the y-coordinate is in scientific notation; $1.2277E - 6$ means 1.2277×10^{-6}, or 0.000 001 227 7, which is *quite* close to zero, and, of course, 0.001 108 03 is also close to zero.

EXAMPLE 3 Graph the function $y = g(x) = x^2 - 2$.

SOLUTION We will use the method discussed earlier in the chapter, where we graphed $y = f(x) + h$ by shifting the graph of $y = f(x)$ upward or downward $|h|$ units. To graph $y = x^2 - 2$, we shift the graph of $y = x^2$ two units *downward*; we move downward because h is negative (h is -2). You might also make a table of values to verify that the points shown on the graph in Figure 7 are, indeed, points on the graph of $y = x^2 - 2$.

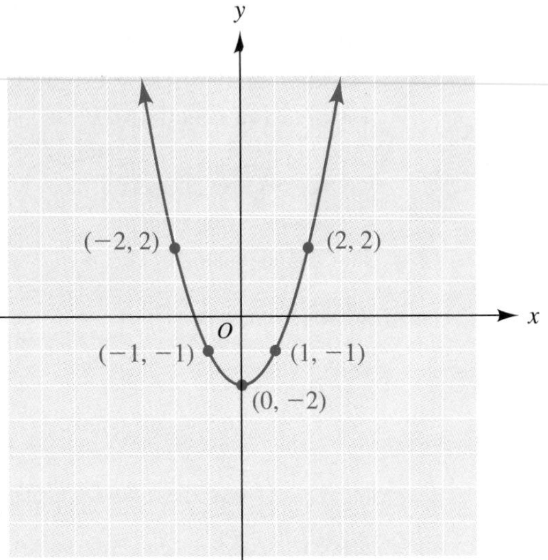

Figure 7

Note To graph $y = x^2 - 2$ with a TI-81 graphics calculator, press the following keys, in order: [Y=] [X|T] [x^2] [−] [2] [GRAPH].

EXAMPLE 4 Graph the function $y = h(x) = (x - 2)^2$.

SOLUTION We will use the method discussed earlier in the chapter, where we graphed $y = f(x + h)$ by shifting the graph of $y = f(x)$ right or left $|h|$ units. To graph $y = (x - 2)^2$, we shift the graph of $y = x^2$ two units *to the right*; we move toward the right because h is negative (h is -2). You might also make a table of values to verify that the points shown on the graph in Figure 8 are, indeed, points on the graph of $y = (x - 2)^2$.

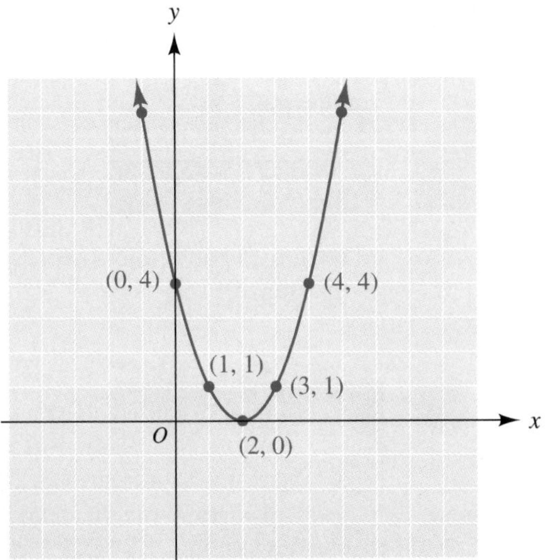

Figure 8

Note To graph $y = (x - 2)^2$ with a TI-81 graphics calculator, press the following keys, in order:

[Y=] [(] [X|T] [−] [2] [)] [x^2] [GRAPH]

EXAMPLE 5 Graph the function $y = x^2 - x - 2$.

SOLUTION We make a table of values by substituting several values for x in the equation and finding the corresponding values of y. Because the domain of the variable is the set of all real numbers, we must be sure to include some negative values for x, some positive values, and zero. (We'll use integer values, although we could include some noninteger values for x, also.)

If $x = -2$, then $y = (-2)^2 - (-2) - 2 = 4 + 2 - 2 = 4$.

If $x = -1$, then $y = (-1)^2 - (-1) - 2 = 1 + 1 - 2 = 0$.

If $x = 0$, then $y = (0)^2 - (0) - 2 = -2$.

If $x = 1$, then $y = (1)^2 - (1) - 2 = 1 - 1 - 2 = -2$.

If $x = 2$, then $y = (2)^2 - (2) - 2 = 4 - 2 - 2 = 0$.

If $x = 3$, then $y = (3)^2 - (3) - 2 = 9 - 3 - 2 = 4$.

x	y
−2	4
−1	0
0	−2
1	−2
2	0
3	4

We graph these points and connect them in order from left to right with a smooth curve.

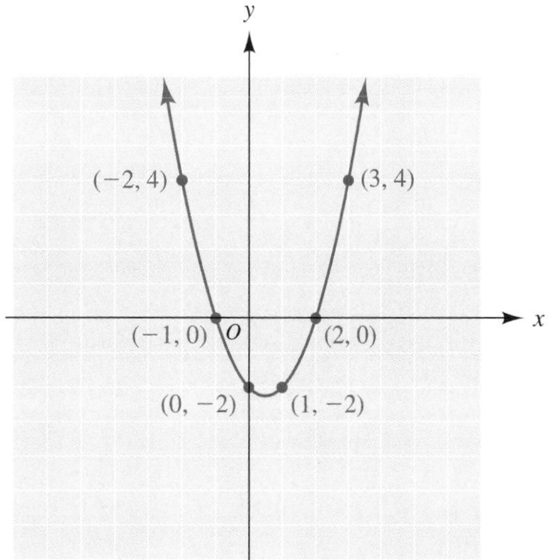

Note To graph $y = x^2 - x - 2$ with a TI-81 graphics calculator, press the following keys, in order:

[Y=] [X|T] [x^2] [−] [X|T] [−] [2] [GRAPH]

EXAMPLE 6 Graph the function $y = x^3 - 4x$.

SOLUTION As we mentioned at the beginning of this section, it is shown in calculus that the graph of any polynomial function *is a smooth curve.* Since $y = x^3 - 4x$ is a polynomial function, its graph must be a smooth curve. We'll find seven points on the curve by making a table of values, letting x have integral values from -3 to 3. (We could also, of course, let x have noninteger values.)

If $x = -3$, then $y = (-3)^3 - 4(-3) = -27 + 12 = -15$.

If $x = -2$, then $y = (-2)^3 - 4(-2) = -8 + 8 = 0$.

If $x = -1$, then $y = (-1)^3 - 4(-1) = -1 + 4 = 3$.

If $x = 0$, then $y = (0)^3 - 4(0) = 0$.

If $x = 1$, then $y = (1)^3 - 4(1) = 1 - 4 = -3$.

If $x = 2$, then $y = (2)^3 - 4(2) = 8 - 8 = 0$.

If $x = 3$, then $y = (3)^3 - 4(3) = 27 - 12 = 15$.

x	y
−3	−15
−2	0
−1	3
0	0
1	−3
2	0
3	15

We graph these points and then connect them with a smooth curve, taking the points in order from left to right.

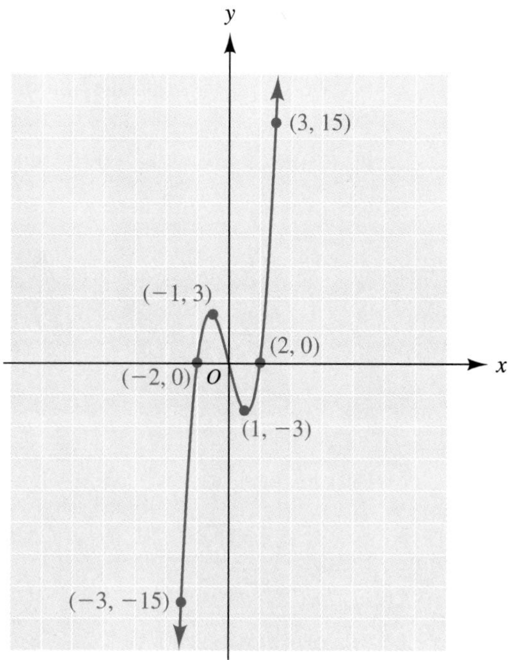

Each line of the grid represents 2 units

Note To graph $y = x^3 - 4x$ with a TI-81 graphics calculator, press the following keys, in order:

Finding the Table of Values by Using Synthetic Division (Optional)

Graphing polynomial functions by using synthetic division to find the y-values is often simpler than finding the y-values by direct substitution. To find y-values by synthetic division, we use the remainder theorem, which is stated below and proved in Appendix A.

The remainder theorem

> If we divide a polynomial in x by $x - a$, the remainder is equal to the value of the polynomial when $x = a$. In symbols, if the remainder is R and the function is $y = f(x)$, then $f(a) = R$.

EXAMPLE 7 Given $f(x) = 2x^4 - 11x^3 + 7x^2 - 12x + 14$, find $f(5)$.

SOLUTION 1 Using substitution, we have

$$f(x) = 2x^4 - 11x^3 + 7x^2 - 12x + 14$$
$$f(5) = 2(5)^4 - 11(5)^3 + 7(5)^2 - 12(5) + 14$$
$$= 1250 - 1375 + 175 - 60 + 14 = 4$$

SOLUTION 2 To use synthetic division and the remainder theorem in finding $f(5)$, we want to divide the polynomial by $x - 5$. Therefore, when we set up the problem for synthetic division, we write 5 (that is, *positive* 5) to the left of the coefficients of the terms of the dividend.

5	2	−11	7	−12	14	
		10	−5	10	−10	← These products need not be written
	2	−1	2	−2	4	← The remainder is 4

Because the remainder is 4, the remainder theorem tells us that $f(5) = 4$. (Notice that this result is the same as the result we obtained in Solution 1.)

To use synthetic division to find a *table* of values for graphing a function, we usually write each number that we're dividing by immediately to the left of the *sums*, and we do *not* write the *products*—that is, we do the multiplying and adding mentally (see Example 8).

EXAMPLE 8 Use synthetic division to find the table of values for the function given in Example 6.

SOLUTION $y = f(x) = x^3 - 4x$

$$= 1x^3 + 0x^2 - 4x + 0$$

x	1	0	−4	0 (y)
−3	1	−3	5	−15
−2	1	−2	0	0
−1	1	−1	−3	3
0	1	0	−4	0
1	1	1	−3	−3
2	1	2	0	0
3	1	3	5	15

The products are not written

The numbers we're dividing *by* (the x column)

The remainders (the y column)

The table above and the remainder theorem tell us that $f(-3) = -15$ [or, in other words, one point on the curve is $(-3, -15)$], $f(-2) = 0$ [or another point on the curve is $(-2, 0)$], and so on.

Exercises 9.6 Set I

1. Graph $y = x^2 + 3$ by using integer values of x from -3 to $+3$ to make a table of values.
2. Graph $y = x^2 + 1$ by using integer values of x from -3 to $+3$ to make a table of values.
3. Graph $y = (x + 1)^2$ by using integer values of x from -4 to $+2$ to make a table of values.
4. Graph $y = (x + 3)^2$ by using integer values of x from -6 to 0 to make a table of values.
5. Graph $y = x^2 - 2x$ by using integer values of x from -2 to $+4$ to make a table of values.
6. Graph $y = 3x - x^2$ by using integer values of x from -2 to $+4$ to make a table of values.
7. Graph $y = 2x - x^2$ by using integer values of x from -2 to $+4$ to make a table of values.
8. Graph $y = 2x + x^2$ by using integer values of x from -3 to $+1$ to make a table of values.
9. Graph $y = x^3$ by using integer values of x from -2 to $+2$ to make a table of values.
10. Graph $y = x^3 + 2$ by using integer values of x from -2 to $+2$ to make a table of values.
11. Graph $y = (x + 1)^3$ by using integer values of x from -3 to $+1$ to make a table of values.
12. Graph $y = (x - 1)^3$ by using integer values of x from -1 to $+3$ to make a table of values.
13. Find the range of the function $f(x) = x^3 - 2x^2 - 13x + 20$ if the domain is $\{-4, -3, -2, 0, 2, 3, 4, 5\}$.
14. Find the range of the function $f(x) = x^3 - 3x^2 - 20x + 12$ if the domain is $\{-3, -2, 0, 2, 3, 4, 5, 6\}$.

Exercises 9.6 Set II

1. Graph $y = x^2 - 1$ by using integer values of x from -3 to $+3$ to make a table of values.
2. Graph $y = (x - 1)^2$ by using integer values of x from -2 to $+4$ to make a table of values.
3. Graph $y = x - \frac{1}{2}x^2$ by using integer values of x from -2 to $+4$ to make a table of values.
4. Graph $y = 2x^2$ by using integer values of x from -3 to $+3$ to make a table of values.
5. Graph $y = 2x^2 + 3$ by using integer values of x from -3 to $+3$ to make a table of values.
6. Graph $y = 2x^2 - 4x$ by using integer values of x from -2 to $+4$ to make a table of values.
7. Graph $y = 3x - x^2$ by using integer values of x from -1 to $+4$ to make a table of values.
8. Graph $y = 3x + x^2$ by using integer values of x from -4 to $+1$ to make a table of values.
9. Graph $y = x^3 - 1$ by using integer values of x from -2 to $+2$ to make a table of values.
10. Graph $y = x^4$ by using integer values of x from -2 to $+2$ to make a table of values.
11. Graph $y = x^4 - 1$ by using integer values of x from -2 to $+2$ to make a table of values.
12. Graph $y = x^3 - 3x + 4$ by using integer values of x from -2 to $+2$ to make a table of values.
13. Find the range of the function $f(x) = x^3 + 3x^2 - x - 3$ if the domain is $\{-3, -2, -1, 0, 1, 2\}$.
14. Find the range of the function $f(x) = x^3 + x^2 - 4x - 4$ if the domain is $\{-2, -1, 0, 1, 2, 3\}$.

9.7 Direct Variation; Proportion

A **variation** is an equation that relates one variable to one or more other variables by means of multiplication or division or both.

Direct Variation

In a **direct variation**, one variable is some constant multiple of the other variable. Thus, if a function can be represented by an equation of the form $y = kx^n$, where k is some constant and n is some positive number, then that function is called a *direct variation*. We call k the **constant of proportionality**.

In a direct variation, as one variable increases, the other increases also, and as one variable decreases, the other decreases also.

Many examples of direct variation exist in daily life:

While a train is traveling at a constant speed, the number of miles it covers is directly proportional to the number of hours it travels.

If someone is working at a certain wage per hour (k), then as the number of hours (h) changes, the salary (s) changes according to the formula $s = kh$. Therefore, s is directly proportional to h, and s is a function of h.

In circles, as the diameter (d) changes, the circumference (C) changes according to the formula $C = \pi d$, where $\pi \approx 3.14$. In this case, π is the constant of proportionality. C is a function of d, and we can say that C varies as d or that C is directly proportional to d.

Direct variation

If $n > 0$ and if k is the constant of proportionality, then the following sentences *all* translate to the equation $y = kx^n$:

y varies directly as (or *with*) x^n.

y varies as (or *with*) x^n.

y is directly proportional to x^n.

y is proportional to x^n.

Notice that the relationship between x and y in a direct variation can be linear, or x can be raised to any positive power. Thus, $y = kx$, $y = kx^2$, and $y = kx^3$ are all examples of direct variation.

EXAMPLE 1 Given: y varies directly with x. Write the variation equation. If $y = 10$ when $x = 2$, find k (the constant of proportionality), and find y when $x = 3$.

SOLUTION Since y varies directly with x, the variation equation is $y = kx$. Substituting 10 for y and 2 for x, we have $10 = k(2)$. Therefore, $k = 5$.

The equation is now $y = 5x$. Substituting 3 for x, we have $y = 5(3) = 15$.

You might try some other values of x in Example 1 to see that as x increases, y increases also, and as x decreases, y decreases also.

We will give more examples involving direct variation after we have discussed proportions. Before we discuss proportions, however, we need to discuss using subscripts to indicate that two numbers are a *pair of corresponding values*. In Example 1, we consider $y = 10$ and $x = 2$ to be a pair of corresponding values, because we are told that $y = 10$ *when* $x = 2$. To indicate by subscripted variables that these are corresponding values, we use the *same* subscript. That is,

$$\text{Condition 1:} \quad x_1 = 2 \text{ and } y_1 = 10$$

In Example 1, we also consider $y = 15$ and $x = 3$ to be a pair of corresponding values. Therefore,

$$\text{Condition 2:} \quad x_2 = 3 \text{ and } y_2 = 15$$

In general, for the direct variation whose equation is $y = kx$,

Condition 1: If $y = y_1$ when $x = x_1$, then $y_1 = kx_1$.

Condition 2: If $y = y_2$ when $x = x_2$, then $y_2 = kx_2$.

Proportions

When an equation can be expressed in the form $\dfrac{a}{b} = \dfrac{c}{d}$, it is called a **proportion**. (Stated differently, a proportion is a statement that two ratios or two rates are equal.)

Related proportions for the direct variation whose equation is $y = kx^n$

$$\frac{y_1}{y_2} = \frac{x_1^n}{x_2^n} \quad \text{and} \quad \frac{y_1}{x_1^n} = \frac{y_2}{x_2^n}$$

The proportions in the box above are derived as follows: If the equation $y_1 = kx_1^n$ is divided by the equation $y_2 = kx_2^n$, we have

$$\frac{y_1}{y_2} = \frac{kx_1^n}{kx_2^n} = \frac{x_1^n}{x_2^n}$$

Alternatively, if the equation $y_1 = kx_1^n$ is solved for k, we have $k = \dfrac{y_1}{x_1^n}$; and if the equation $y_2 = kx_2^n$ is solved for k, we have $k = \dfrac{y_2}{x_2^n}$. Then, since $\dfrac{y_1}{x_1^n}$ and $\dfrac{y_2}{x_2^n}$ both equal k, they must equal each other:

$$\frac{y_1}{x_1^n} = \frac{y_2}{x_2^n}$$

Note There are several proportions associated with each direct variation. Two other proportions equivalent to those in the box above are

$$\frac{y_2}{y_1} = \frac{x_2^n}{x_1^n} \quad \text{and} \quad \frac{x_1^n}{y_1} = \frac{x_2^n}{y_2}$$

Because a proportion is simply a rational equation, it can be solved by using the techniques discussed earlier. An alternative method of solution is based on the following fact: When we multiply both sides of the equation $\dfrac{a}{b} = \dfrac{c}{d}$ by the LCD, we obtain the new equation $ad = bc$. Therefore, the equation $ad = bc$ is equivalent to the equation $\dfrac{a}{b} = \dfrac{c}{d}$ *except* that in the equation $\dfrac{a}{b} = \dfrac{c}{d}$, neither b nor d can equal zero. When we rewrite $\dfrac{a}{b} = \dfrac{c}{d}$ as $ad = bc$, we say that we are **cross-multiplying**.

We can, then, cross-multiply as the first step in solving a proportion. However, since we may introduce extraneous roots when we do this, we must be sure to reject any apparent solutions that aren't in the domain of the variable.

In solving variation problems, we can, if we wish, solve the related proportion without first finding k. This method will be designated as the alternative method in most of the examples in this section.

Solving applied problems that lead to proportions

1. Represent the unknown quantity by a variable, and use the given conditions to form two ratios or two rates.
2. Form a proportion by setting the two ratios or two rates equal to each other, being sure to write the *units* next to the numbers when you write the proportion and being sure that the units occupy corresponding positions in the two ratios (or rates) of the proportion. For example, if the problem involves miles and hours:

Correct arrangements	*Incorrect arrangements*
$\frac{\text{miles}}{\text{hours}} = \frac{\text{miles}}{\text{hours}}$	$\frac{\text{miles}}{\text{hours}} = \frac{\text{hours}}{\text{miles}}$
$\frac{\text{hours}}{\text{miles}} = \frac{\text{hours}}{\text{miles}}$	$\frac{\text{hours}}{\text{miles}} = \frac{\text{miles}}{\text{hours}}$
$\frac{\text{miles}}{\text{miles}} = \frac{\text{hours}}{\text{hours}}$	

   Both numerators must correspond to condition 1

   Both denominators must correspond to condition 2

3. Once the numbers have been correctly entered into the proportion by using the units as a guide, drop the units and solve for the variable.

EXAMPLE 2

The number of miles driven is proportional to the number of gallons of gasoline used. Sherma knows that she can drive 2,220 mi on 60 gal of gasoline. Find k, the constant of proportionality, and find how many miles she can expect to drive on 17 gal of gasoline.

SOLUTION Because the number of miles (m) driven is proportional to the number of gallons of gasoline (g), the variation equation is $m = kg$.

Let m_2 = the number of mi on 17 gal of gasoline.

$$\text{Condition 1: } \begin{cases} m_1 = 2{,}220 \text{ mi} \\ g_1 = 60 \text{ gal} \end{cases} \qquad \text{Condition 2: } \begin{cases} m_2 = ? \\ g_2 = 17 \text{ gal} \end{cases}$$

From condition 1,

$$m_1 = kg_1$$

$$2{,}220 = k(60)$$

$$k = \frac{2{,}220}{60} = 37 \quad \text{The constant of proportionality}$$

From condition 2,

$$m_2 = kg_2 = 37(17) = 629$$

ALTERNATIVE METHOD Using a proportion, we have

$$\frac{m_1}{g_1} = \frac{m_2}{g_2}$$

$$\frac{2{,}220 \text{ mi}}{60 \text{ gal}} = \frac{m_2 \text{ mi}}{17 \text{ gal}}$$

$$\frac{2{,}220}{60} = \frac{m_2}{17}$$

$$2{,}220(17) = 60m_2$$

$$m_2 = \frac{2{,}220(17)}{60} = 629$$

Sherma can expect to drive 629 mi on 17 gal of gasoline.

The following theorem from geometry leads to equations involving proportions: If two triangles are *similar*, their sides are directly proportional. (Recall that two triangles are similar if their corresponding angles are equal.) Example 3 uses this theorem and shows the alternative method only.

EXAMPLE 3

If a 6-ft man casts a $4\frac{1}{2}$-ft shadow, how tall is a nearby tree that casts a $13\frac{1}{2}$-ft shadow?

SOLUTION We use the fact that the length of a shadow is determined by the angle of a ray of the sun. (We assume that the tree and the man are both vertical; you should verify that the triangles are similar.)

Note The ground can be, but does not have to be, horizontal.

Let x = the height of the tree (in ft).

A 6-ft man casts a $4\frac{1}{2}$-ft shadow	How tall is a tree that casts a $13\frac{1}{2}$-ft shadow?
Man	*Tree*

Height ⟶ 6 ft / Shadow ⟶ $4\frac{1}{2}$ ft = x ft ⟵ Height / $13\frac{1}{2}$ ft ⟵ Shadow

$$\frac{6 \text{ ft}}{4\frac{1}{2} \text{ ft}} = \frac{x \text{ ft}}{13\frac{1}{2} \text{ ft}}$$

$$\frac{6}{4\frac{1}{2}} = \frac{x}{13\frac{1}{2}}$$

$$\left(4\tfrac{1}{2}\right)x = 6\left(13\tfrac{1}{2}\right)$$

$$\tfrac{9}{2}x = 81$$

$$\tfrac{2}{9} \cdot \tfrac{9}{2}x = \tfrac{2}{9} \cdot \tfrac{81}{1}$$

$$x = 18$$

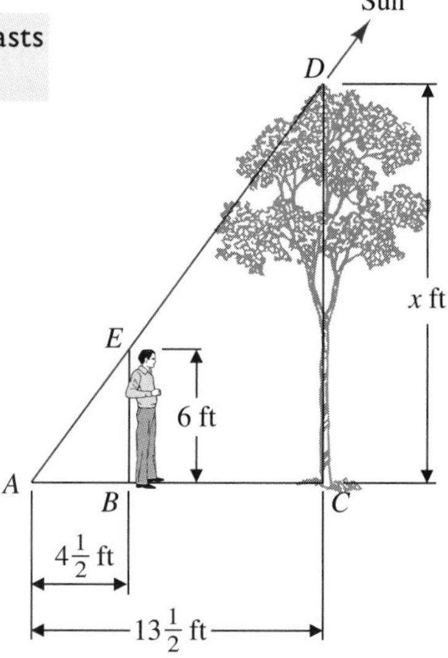

Checking will verify that the tree is 18 ft tall.

EXAMPLE 4

Given: The area (A) of a circle varies directly with the square of the radius (r). Write the variation equation. If $A = 200.96$ when $r = 8$, find k, the constant of proportionality, and find A when $r = 5$.

SOLUTION Because A varies directly as r^2, the variation equation is $A = kr^2$.

Condition 1: $\begin{cases} A_1 = 200.96 \\ r_1 = 8 \end{cases}$ Condition 2: $\begin{cases} A_2 = ? \\ r_2 = 5 \end{cases}$

From condition 1,

$$A_1 = kr_1^{\,2}$$

$$200.96 = k(8^2)$$

$$k = \tfrac{200.96}{64} = 3.14 \quad \text{The constant of proportionality}$$

From condition 2, $A_2 = kr_2^{\,2} = 3.14(5^2) = 78.5$

ALTERNATIVE METHOD Using a proportion, we have

$$\frac{A_1}{r_1^{\,2}} = \frac{A_2}{r_2^{\,2}}$$

$$\frac{200.96}{8^2} = \frac{A_2}{5^2}$$

$$A_2 = \frac{200.96}{8^2}(5^2) = 78.5$$

The area is 78.5 when the radius is 5.

Exercises 9.7
Set I

1. Given: y varies directly with x. Write the variation equation. If $y = 12$ when $x = 3$, find k, the constant of proportionality, and find y when $x = 5$.

2. Given: y varies directly with x. Write the variation equation. If $y = 10$ when $x = 4$, find k, the constant of proportionality, and find y when $x = -6$.

3. Given: A varies directly as the square of s. Write the variation equation. If $A = -27$ when $s = -3$, find k, the constant of proportionality, and find A when $s = 5$.

4. Given: R varies directly as the square of x. Write the variation equation. If $R = 32$ when $x = -4$, find k, the constant of proportionality, and find R when $x = 10$.

5. The *change* in the length (l) of a spring is directly proportional to the force (F) applied to the spring. Write the variation equation. If a 5-lb force stretches a spring 4 in., find the change in the length of the spring when a 2-lb force is applied.

6. If the number of miles per gallon of gasoline obtained by a certain car is a constant, then the number of miles (m) driven in that car is directly proportional to the number of gallons of gasoline (g) used. Write the variation equation. If a car can go 238 mi on 14 gal of gasoline, find how many miles that car could go on 23 gal of gasoline.

7. The distance (s) through which an object falls (disregarding air resistance, etc.) varies directly with the square of the time (t). Write the variation equation. If an object falls 64 ft in 2 sec, how far will it fall in 5 sec?

8. The air resistance (R) on a car varies directly with the square of the car's velocity (v). Write the variation equation. If the air resistance is 400 lb at 60 mph, find the air resistance at 80 mph.

In Exercises 9 and 10, assume that the number of miles driven is proportional to the number of gallons of gasoline used.

9. Leon knows that he can drive his motorhome 161 mi on 23 gal of gasoline. Find k, the constant of proportionality. If his tanks are full and together hold 50 gal, how far can he expect to drive before stopping for more gasoline?

10. Nick knows that he can drive his motorhome 85 mi on 17 gal of gasoline. Find k, the constant of proportionality. If his tanks are full and together hold 75 gal, how far can he expect to drive before stopping for more gasoline?

11. If a 5-ft woman casts a 3-ft shadow, how tall is a nearby tree that casts a 27-ft shadow?

12. If a 6-ft man casts a 4-ft shadow, how tall is a nearby tree that casts a 22-ft shadow?

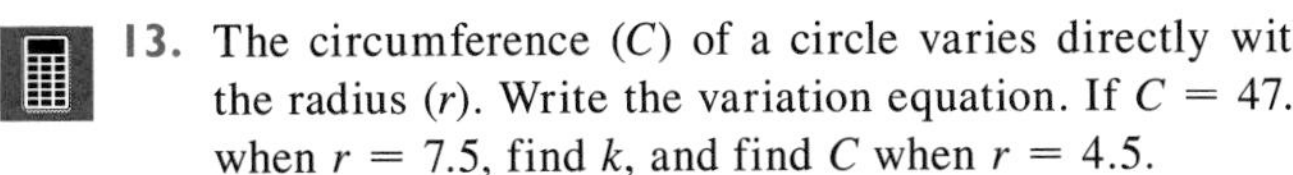

13. The circumference (C) of a circle varies directly with the radius (r). Write the variation equation. If $C = 47.1$ when $r = 7.5$, find k, and find C when $r = 4.5$.

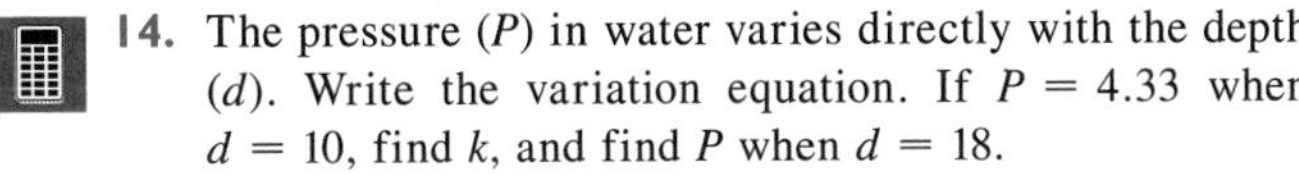

14. The pressure (P) in water varies directly with the depth (d). Write the variation equation. If $P = 4.33$ when $d = 10$, find k, and find P when $d = 18$.

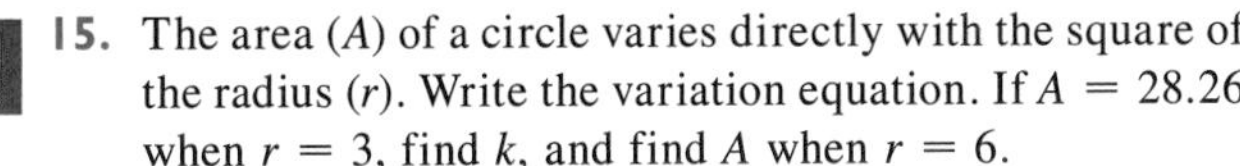

15. The area (A) of a circle varies directly with the square of the radius (r). Write the variation equation. If $A = 28.26$ when $r = 3$, find k, and find A when $r = 6$.

16. The surface area (S) of a sphere varies directly with the square of the radius (r). Write the variation equation. If $S = 50.24$ when $r = 2$, find k, and find S when $r = 4$.

17. The amount of sediment a stream will carry is directly proportional to the sixth power of the speed of the current. If a stream carries 1 unit of sediment when the speed of the current is 2 mph, how many units of sediment will it carry when the speed of the current is 4 mph?

18. The salary John earns is directly proportional to the number of hours he works. One week he worked 18 hours, and his salary was \$153.54. The next week he worked 25 hours. How much did he earn the second week?

Writing Problems

Express the answers in your own words and in complete sentences.

1. Explain what a direct variation is.

2. Make up an applied problem that involves direct variation.

Exercises 9.7 Set II

1. Given: y varies directly with x. Write the variation equation. If $y = 3$ when $x = -6$, find k, the constant of proportionality, and find y when $x = 8$.

2. Given: y varies directly as x. Write the variation equation. If $y = 4$ when $x = -2$, find k, the constant of proportionality, and find y when $x = 7$.

3. Given: d varies directly as the square of s. Write the variation equation. If $d = 64$ when $s = 2$, find k, the constant of proportionality, and find d when $s = 5$.

4. Given: y varies directly as the cube of x. Write the variation equation. If $y = 192$ when $x = 4$, find k, the constant of proportionality, and find y when $x = 3$.

5. In a business, if the price of an item is kept fixed, the revenue (R) varies directly as the number of items sold (n). Write the variation equation. If the revenue is $15,000 when 400 items are sold, what is the revenue when 600 items are sold?

6. Elaine's weekly salary (S) is directly proportional to the number of hours (h) she works during the week. Write the variation equation. If she made $268.80 for working 24 hr, how many hours would she have to work in order to earn $369.60?

7. The resistance (R) of a boat moving through water varies directly with the square of its speed (s). Write the variation equation. If the resistance of a boat is 50 lb at a speed of 10 knots, what is the resistance at 15 knots?

8. In a business, if the price of an item is kept fixed, the revenue (R) varies directly with the number of items sold (n). Write the variation equation. If the revenue is $12,000 when 900 items are sold, how many items must be sold for the revenue to be $16,000?

In Exercises 9 and 10, assume that the number of miles driven is proportional to the number of gallons of gasoline used.

9. Ted knows that he can drive his motorhome 117 mi on 13 gal of gasoline. Find k, the constant of proportionality. If his tanks are full and together hold 50 gal, how far can he expect to drive before stopping for more gasoline?

10. Lee knows that she can drive her car 494 mi on 13 gal of gasoline. Find k, the constant of proportionality. If her tank is full and holds 15 gal, how far can she expect to drive before stopping for more gasoline?

11. If a 6-ft man casts a 4-ft shadow, how tall is a nearby building that casts a 24-ft shadow?

12. Ruth drove 1,008 mi in $3\frac{1}{2}$ days. At this rate, how far should she be able to drive in 5 days? Assume that the number of miles driven varies directly with the number of days.

13. The circumference (C) of a circle varies directly with the diameter (d). Write the variation equation. If $C = 9.42$ when $d = 3$, find k, and find C when $d = 15$.

14. The pressure (P) in water varies directly with the depth (d). Write the variation equation. If $P = 8.66$ when $d = 20$, find k, and find P when $d = 50$.

15. Given: y varies directly as the square of x. Write the variation equation. If $y = 20$ when $x = 10$, find k, and find y when $x = 15$.

16. The area (A) of a circle varies directly with the square of the radius (r). Write the variation equation. If $A = 13.8474$ when $r = 2.1$, find k, and find A when $r = 1.2$.

17. The distance (d) in miles that a person can see to the horizon from a point h ft above the surface of the earth varies approximately as the square root of the height (h). If, for a height of 600 ft, the horizon is 30 mi distant, how far is the horizon from a point that is 1,174 ft high? (Round off to the nearest mile.)

18. Given: y is directly proportional to the cube of x. If $y = 72$ when $x = 6$, find k, and find y when $x = -3$.

9.8 Inverse Variation

In an **inverse variation**, as one variable *increases*, the other *decreases*. If a function can be represented by an equation of the form $y = \dfrac{k}{x^n}$, where k is some constant, $x \neq 0$, and n is some positive number, then that function is called an *inverse variation*. We call k the constant of proportionality.

Two examples of inverse variation follow:

If the area of a rectangle is held constant, then the length of the rectangle varies inversely as its width. For example, suppose that the area of a rectangle has to be held constant at 72 sq. cm. Then the width of the rectangle varies inversely as the length. You can verify that if the length of the rectangle increases from 12 cm to 18 cm, the width must decrease from 6 cm to 4 cm.

If the amount of work to be done is held constant, then the amount of time required to do that work is inversely proportional to the rate, or speed, of the worker or machine (see Example 2).

Inverse variation

If $n > 0$, if $x \neq 0$, and if k is the constant of proportionality, then the following sentences translate to the equation $y = \frac{k}{x^n}$:

y varies inversely as (or *with*) x^n.

y is inversely proportional to x^n.

(Notice that x^n is in the *denominator* of the rational expression $\frac{k}{x^n}$.)

EXAMPLE 1

Given: y varies inversely with the square of x. If $y = -3$ when $x = 4$, find k, and find y when $x = -6$.

SOLUTION Condition 1: $\begin{cases} x_1 = 4 \\ y_1 = -3 \end{cases}$ Condition 2: $\begin{cases} x_2 = -6 \\ y_2 = ? \end{cases}$

Since y varies inversely with x^2, the variation equation is $y = \frac{k}{x^2}$. From condition 1, we have

$$y_1 = \frac{k}{x_1^{\,2}}$$

$$-3 = \frac{k}{4^2}$$

$$k = -3(4^2) = -48 \quad \text{The constant of proportionality}$$

From condition 2,

$$y_2 = \frac{k}{x_2^{\,2}} = \frac{-48}{(-6)^2} = \frac{-48}{36} = -\frac{4}{3}$$

EXAMPLE 2

The amount of time necessary to complete a job is inversely proportional to the rate, or speed, of the worker. Ruth types at the rate of $5\frac{1}{2}$ pages per hr. If she spends 8 hr on a certain job, find k, the constant of proportionality. How long would it take Mike, who types at the rate of $3\frac{2}{3}$ pages per hr, to do the same job?

SOLUTION Because *time* is inversely proportional to *rate*, the variation equation is $t = \frac{k}{r}$. Let $t_2 =$ the number of hr Mike works.

Condition 1: $\begin{cases} r_1 = 5\frac{1}{2} \\ t_1 = 8 \end{cases}$ Condition 2: $\begin{cases} r_2 = 3\frac{2}{3} \\ t_2 = ? \end{cases}$

From condition 1,

$$t_1 = \frac{k}{r_1}$$

$$8 = \frac{k}{5.5}$$

$$k = 8(5.5) = 44 \quad \text{The constant of proportionality}$$

From condition 2,

$$t_2 = \frac{k}{r_2} = \frac{44}{3\frac{2}{3}} = 12$$

It would take Mike 12 hr to do the job.

An alternative way to solve inverse variation problems is to use the equation $x_1{}^n y_1 = x_2{}^n y_2$. This method will be shown in the examples in this section as the alternative method. The equation is derived as follows:

Condition 1: If $y = y_1$ when $x = x_1$, then $y_1 = \dfrac{k}{x_1{}^n}$.

If we solve this equation for k, we have $k = x_1{}^n y_1$.

Condition 2: If $y = y_2$ when $x = x_2$, then $y_2 = \dfrac{k}{x_2{}^n}$.

If we solve this equation for k, we have $k = x_2{}^n y_2$. Then, since $x_1{}^n y_1$ and $x_2{}^n y_2$ both equal k, they must equal each other:

$$x_1{}^n y_1 = x_2{}^n y_2$$

EXAMPLE 3

For the facts given in Example 1 (y is inversely proportional to the square of x, and $y = -3$ when $x = 4$), find y when $x = -6$ by using the alternative method.

SOLUTION We use the equation $x_1{}^2 y_1 = x_2{}^2 y_2$. Therefore,

$$x_1{}^2 y_1 = x_2{}^2 y_2$$

$$4^2(-3) = (-6)^2 y_2$$

$$y_2 = \frac{4^2(-3)}{(-6)^2} = \frac{-48}{36} = -\frac{4}{3}$$

EXAMPLE 4

Under certain conditions, the pressure (P) of a gas varies inversely with the volume (V). If $P = 30$ when $V = 500$, find k, the constant of proportionality, and find P when $V = 200$.

SOLUTION Condition 1: $\begin{cases} P_1 = 30 \\ V_1 = 500 \end{cases}$ Condition 2: $\begin{cases} P_2 = ? \\ V_2 = 200 \end{cases}$

Since P varies *inversely* with V, the variation equation is $P = \dfrac{k}{V}$. From condition 1, we have

$$P_1 = \frac{k}{V_1}$$

$$30 = \frac{k}{500}$$

$$k = 30(500) = 15{,}000 \quad \text{The constant of proportionality}$$

We can then find P when $V = 200$:

$$P_2 = \frac{k}{V_2} = \frac{15{,}000}{200} = 75$$

ALTERNATIVE METHOD The equation $P_1 V_1 = P_2 V_2$ can be used:

$$P_1 V_1 = P_2 V_2$$

$$30(500) = P_2(200)$$

$$P_2 = \frac{30(500)}{200} = 75$$

The pressure is 75 when the volume is 200.

Exercises 9.8
Set I

1. Given: y varies inversely with x. Write the variation equation. If $y = 7$ when $x = 2$, find k, the constant of proportionality, and find y when $x = -7$.

2. Given: z varies inversely with w. Write the variation equation. If $z = 4$ when $w = -12$, find k, and find z when $w = 8$.

3. Given: Pressure (P) varies inversely with volume (V). Write the variation equation. If $P = 18$ when $V = 15$, find k, and find P when $V = 10$.

4. Given: s varies inversely with t. Write the variation equation. If $s = 8$ when $t = 5$, find k, and find s when $t = 4$.

In Exercises 5 and 6, assume that the *time* is inversely proportional to the *rate*.

5. Machine A works for 3 hr to complete one order. It makes a certain part at the rate of 375 parts per hr. Find k, and find how long it would take machine B, which makes the parts at the rate of 225 parts per hr, to complete the same order.

6. Machine C works for 9 hr to complete one order. It makes a certain part at the rate of 275 parts per hr. Find k, and find how long it would take machine D, which makes the parts at the rate of 330 parts per hr, to complete the same order.

7. Given: y varies inversely with the square of x. Write the variation equation. If $y = 9$ when $x = 4$, find k, and find y when $x = 3$.

8. Given: C varies inversely with the square of v. Write the variation equation. If $C = 8$ when $v = 3$, find k, and find C when $v = 6$.

9. Given: F varies inversely with the square of d. Write the variation equation. If $F = 3$ when $d = 4$, find k, and find F when $d = 2$.

10. The light intensity (I) received from a light source varies inversely with the square of the distance (d) from the source. If the light intensity is 15 foot-candles at a distance of 10 ft from the light source, what is the light intensity at a distance of 15 ft?

11. a. Write the variation equation for the following statement: When the area of a rectangle is held constant, the length of the rectangle varies inversely as its width.

    b. If the area of a rectangle is held constant and if the length is 15 cm when the width is 3 cm, find k, the constant of proportionality, and find the length when the width is 5 cm.

12. a. Write the variation equation for the following statement: When the area of a parallelogram is held constant, the altitude of the parallelogram varies inversely as the length of its base.

    b. If the area of a parallelogram is held constant and if the altitude is 12 cm when the base is 5 cm, find k, the constant of proportionality, and find the altitude when the base is 10 cm.

13. Sound intensity (I), or loudness, is inversely proportional to the square of the distance (d) from the source of the sound. Write the variation equation. If an electric sander has a sound intensity of 75 decibels at 50 ft, find its sound intensity at 125 ft.

14. The time (t) it takes to get a certain job done varies inversely with the rate (r) at which the work is done. Write the variation equation. If it takes 15 hr to get a certain job done when the rate is $7 \frac{\text{jobs}}{\text{hr}}$, how long will it take to get that job done when the rate is $9 \frac{\text{jobs}}{\text{hr}}$?

Writing Problems

Express the answers in your own words and in complete sentences.

1. Explain what an inverse variation is.
2. Make up an applied problem that involves inverse variation.

Exercises 9.8 Set II

1. Given: y varies inversely with x. Write the variation equation. If $y = 5$ when $x = 4$, find k, the constant of proportionality, and find y when $x = 2$.
2. Given: u varies inversely with v. Write the variation equation. If $u = 15$ when $v = 7$, find k, and find u when $v = 3$.
3. Given: V varies inversely with P. Write the variation equation. If $V = 340$ when $P = 30$, find k, and find V when $P = 85$.
4. Given: w varies inversely with x. Write the variation equation. If $w = 38$ when $x = 8$, find k, and find w when $x = 19$.

In Exercises 5 and 6, assume that the *time* is inversely proportional to the *rate*.

5. Machine A works for 4 hr to complete one order. It makes a certain part at the rate of 130 parts per hr. Find k, and find how long it would take machine B, which makes the parts at the rate of 120 parts per hr, to complete the same order.
6. Machine A, which duplicates negative at the rate of 290 ft per min, works for 3.5 hr to complete one job. Find k, the constant of proportionality. How long would it take machine B, which duplicates negative at the rate of 350 ft per min, to do the same job?
7. Given: L varies inversely with the square of r. Write the variation equation. If $L = 9$ when $r = 5$, find k, and find L when $r = 3$.
8. Given: y varies inversely with the cube of x. Write the variation equation. If $y = 960$ when $x = 4$, find k, and find y when $x = 2$.
9. Given: x varies inversely with the cube of y. Write the variation equation. If $x = \frac{1}{4}$ when $y = 2$, find k, and find x when $y = 4$.

10. The gravitational attraction (F) between two bodies varies inversely with the square of the distance (d) separating them. If the attraction measures 36 when the distance is 4 cm, find the attraction when the distance is 80 cm.
11. a. Write the variation equation for the following statement: When the area of a rectangle is held constant, the width of the rectangle varies inversely as its length.
    b. If the area of a rectangle is held constant and if the width is 15 m when the length is 16 m, find k, the constant of proportionality, and find the width when the length is 12 m.
12. a. Write the variation equation for the following statement: When the area of a triangle is held constant, the altitude of the triangle varies inversely as the length of its base.
    b. If the area of a triangle is held constant and if the altitude is 18 in. when the base is 12 in., find k, the constant of proportionality, and find the altitude when the base is 36 in.

13. The electrical resistance (R) of a wire varies inversely with the square of the diameter (d) of the wire. Write the variation equation. If the resistance of a wire that has a diameter of 0.0201 in. is 3.85 ohms, find the resistance of a wire of the same length and material that has a diameter of 0.0315 in. (Round off the answer to two decimal places.)
14. The rate (r) at which work is done varies inversely as the time (t) in which the work is done. Write the variation equation. If it takes 16 hr to get the job done when the rate is $6 \dfrac{\text{jobs}}{\text{hr}}$, what will the rate be when it takes 9 hr to get the job done?

9.9 Joint Variation; Combined Variation

Direct and inverse variation are functions of *one* variable; variations also exist that are functions of *several* variables. We will now discuss variations of this kind.

Joint Variation

Joint variation is a type of variation that relates one variable to the *product* of two (or more) other variables.

Joint variation

> If $n > 0$ and $m > 0$ and if k is the constant of proportionality, then the following sentences translate to the equation $z = kx^n y^m$:
>
> z varies jointly as (or *with*) x^n and y^m.
>
> z is jointly proportional to x^n and y^m.

EXAMPLE 1

Examples of joint variation:

Given: z varies jointly with x and y according to the formula $z = 2xy$.

a. If $x = 3$ and $y = 5$, then $z = 2(3)(5) = 30$.
b. If $x = 4$ and $y = 5$, then $z = 2(4)(5) = 40$.
c. If $x = 3$ and $y = 4$, then $z = 2(3)(4) = 24$.

Notice that as x *increases* from 3 to 4 while y is held constant at 5, z *increases* from 30 to 40. Also, as y *decreases* from 5 to 4 while x is held constant at 3, z *decreases* from 30 to 24. The value of z depends on the *product* xy. That is, when the product xy *increases* from $(3)(5) = 15$ to $(4)(5) = 20$, z *increases* from 30 to 40; and when the product *decreases* from $(4)(5) = 20$ to $(3)(4) = 12$, z *decreases* from 40 to 24.

EXAMPLE 2

If the principal is held constant, then the amount of (simple) interest (I) earned in time (t) varies jointly with t and the interest rate (r). If a certain amount of money earns \$440 in 2 yr when it is invested at 5.5% interest (per yr), find k, the constant of proportionality, and find how much interest would be earned if the money were invested at 5% interest for 3 yr.

SOLUTION Condition 1: $\begin{cases} I_1 = 440 \\ r_1 = 0.055 \\ t_1 = 2 \end{cases}$ Condition 2: $\begin{cases} I_2 = ? \\ r_2 = 0.05 \\ t_2 = 3 \end{cases}$

Because I varies jointly with t and r, the variation equation is $I = krt$. From condition 1,

$$I_1 = kr_1t_1$$

$$440 = k(0.055)(2)$$

$$k = \frac{440}{0.055(2)} = 4{,}000$$

From condition 2, $I_2 = kr_2t_2 = 4{,}000(0.05)(3) = 600$

Therefore, the amount of interest earned in 3 yr at 5% would be \$600. In this example, k is the principal, \$4,000.

EXAMPLE 3

The heat (H) generated by an electric heater is jointly proportional to R and the square of I. If $H = 1{,}200$ when $I = 8$ and $R = 15$, find k, and find H when $I = 5.5$ and $R = 20$.

SOLUTION Condition 1: $\begin{cases} I_1 = 8 \\ R_1 = 15 \\ H_1 = 1{,}200 \end{cases}$ Condition 2: $\begin{cases} I_2 = 5.5 \\ R_2 = 20 \\ H_2 = ? \end{cases}$

Since H is jointly proportional to R and I^2, the variation equation is $H = kRI^2$. From condition 1,

$$H_1 = kR_1I_1^2$$

$$1{,}200 = k(15)(8^2)$$

$$k = \frac{1{,}200}{(15)(8^2)} = 1.25$$

From condition 2, $\quad H_2 = kR_2I_2^2 = 1.25(20)(5.5)^2 = 756.25$

Combined Variation

A variable can be *directly* proportional to one variable (or *jointly* proportional to two or more variables) at the same time that it is *inversely* proportional to other variables. The equations resulting from such circumstances are called **combined variations.**

EXAMPLE 4 Examples of combined variation, where k is the constant of proportionality:

a. $w = \dfrac{kx}{y}$ $\quad$ w varies directly with x and inversely with y

b. $R = \dfrac{kL}{d^2}$ $\quad$ R varies directly as L and inversely as d^2

c. $F = \dfrac{kMm}{d^2}$ $\quad$ F varies jointly with M and m and inversely with d^2

EXAMPLE 5

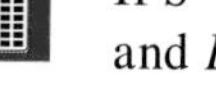

The strength (S) of a rectangular beam varies jointly with b and the square of d and inversely with L, where b is the breadth, d is the depth, and L is the length of the beam. If $S = 2{,}000$ when $b = 2$, $d = 10$, and $L = 15$, find k, and find S when $b = 4$, $d = 8$, and $L = 12$.

SOLUTION

Condition 1: $\begin{cases} b_1 = 2 \\ d_1 = 10 \\ L_1 = 15 \\ S_1 = 2{,}000 \end{cases}$ $\quad$ Condition 2: $\begin{cases} b_2 = 4 \\ d_2 = 8 \\ L_2 = 12 \\ S_2 = ? \end{cases}$

Since the problem involves a combined variation, the variation equation is $S = \dfrac{kbd^2}{L}$.

From condition 1,

$$S_1 = \frac{kb_1d_1^2}{L_1}$$

$$2{,}000 = \frac{k(2)(10^2)}{15}$$

$$k = \frac{2{,}000(15)}{(2)(10^2)} = 150$$

When $b = 4$, $d = 8$, and $L = 12$,

$$S_2 = \frac{kb_2d_2^2}{L_2} = \frac{150(4)(8^2)}{12} = 3{,}200$$

Exercises 9.9
Set I

1. Given: z varies jointly with x and y. Write the variation equation. If $z = -36$ when $x = -3$ and $y = 2$, find k, the constant of proportionality, and find z when $x = 4$ and $y = 3$.

2. Given: A varies jointly with L and W. Write the variation equation. If $A = 120$ when $L = 6$ and $W = 5$, find k, and find A when $L = 7$ and $W = 3$.

3. If the time is held constant, the (simple) interest (I) varies jointly as the principal (P) and the interest rate (r). If \$880 earns \$115.50 when invested at 8.75% interest (so use $r_1 = 0.0875$), find k, and find what the interest would be if \$760 were invested at 9.25% interest (using $r_2 = 0.0925$) for the same length of time.

4. If the time is held constant, the simple interest (I) varies jointly as the principal (P) and the interest rate (r). If \$860 earns \$199.95 when invested at 7.75% interest (so use $r_1 = 0.0775$), find k, and find what the interest would be if \$1,250 were invested at 9.5% interest (using $r_2 = 0.095$) for the same length of time.

5. The wind force (F) on a vertical surface varies jointly as the area (A) of the surface and the square of the wind velocity (V). When the wind is blowing at 20 mph, the force on 1 sq. ft of surface area is 1.8 lb. Find the force exerted on a surface area of 2 sq. ft when the wind velocity is 60 mph.

6. The pressure (P) in a liquid varies jointly with the depth (h) and density (D) of the liquid. If $P = 204$ when $h = 163.2$ and $D = 1.25$, find P when $h = 182.5$ and $D = 13.56$.

7. Given: z varies directly as x and inversely as y. Write the variation equation. If $z = 12$ when $x = 6$ and $y = 2$, find k, and find z when $x = -8$ and $y = -4$.

8. The electrical resistance (R) of a wire varies directly as L and inversely as the square of d. If $R = 2$ when $L = 8$ and $d = 4$, find k, and find R when $L = 10$ and $d = 5$.

9. The elongation (e) of a wire varies jointly as P and L and inversely as A. If $e = 3$ when $L = 45$, $P = 2.4$, and $A = 0.9$, find k, and find e when $L = 40$, $P = 1.5$, and $A = 0.75$.

10. When a horizontal beam with a rectangular cross-section is supported at both ends, its strength (S) varies jointly as the breadth (b) and the square of the depth (d) and inversely as the length (L). A 2- by 4-in. beam 8 ft long resting on the 2-in. side will safely support 600 lb. What is the safe load when the beam is resting on the 4-in. side?

Exercises 9.9
Set II

1. Given: V varies jointly with L and H. Write the variation equation. If $V = 144$ when $L = 3$ and $H = 8$, find k, the constant of proportionality, and find V when $L = 2$ and $H = 5$.

2. Given: C varies jointly with L and W. Write the variation equation. If $C = 7{,}500$ when $L = 25$ and $W = 20$, find k, and find C when $L = 18$ and $W = 23$.

3. If the time is held constant, the (simple) interest (I) varies jointly as the principal (P) and the interest rate (r). If \$1,250 earns \$487.50 when invested at 9.75% interest (so use $r_1 = 0.0975$), find k, and find what the interest would be if \$1,500 were invested at 8.25% interest (using $r_2 = 0.0825$) for the same length of time.

4. On a certain truck line, it costs \$56.80 to send 5 tons of goods 8 mi. How much will it cost to send 14 tons a distance of 15 mi if the cost varies jointly with the weight and the distance?

5. Given: H varies jointly with R and the square of I. Write the variation equation. If $H = 1{,}458$ when $R = 24$ and $I = 4.5$, find k, and find H when $R = 22$ and $I = 5.5$.

6. Given: y varies jointly with x and the square of z. Write the variation equation. If $y = 72$ when $x = 2$ and $z = 3$, find k, and find y when $x = 5$ and $z = 2$.

7. Given: z varies directly as x and inversely as y. Write the variation equation. If $z = 2$ when $x = 4$ and $y = 10$, find k, and find z when $x = 3$ and $y = 6$.

8. Given: P varies directly as T and inversely as V. If $P = 10$ when $T = 250$ and $V = 400$, find k, and find P when $T = 280$ and $V = 350$.

9. Given: W varies jointly with x and y and inversely with the square of z. If $W = 1{,}200$ when $x = 8$, $y = 6$, and $z = 2$, find k, and find W when $x = 5.6$, $y = 3.8$, and $z = 1.5$.

10. The gravitational attraction (F) between two masses varies jointly with M and m and inversely with the square of d. If $F = 1{,}000$ when $d = 100$, $m = 50$, and $M = 2{,}000$, find k, and find F when $d = 66$, $m = 125$, and $M = 1{,}450$. (Round off to the nearest unit.)

Sections 9.1–9.9 REVIEW

Functions 9.1

A **function** is a rule, or a correspondence, that assigns to each element of some set X (called the domain) *exactly* one element of some set Y (called the range). No vertical line can intersect the graph of a function in more than one point.

The **domain** of a function is the set of all permitted first coordinates of the ordered pairs, and the **range** of a function is the set of all the second coordinates of the ordered pairs.

Functional Notation 9.2

The notation $f(x)$ is read "f of x" and is used to name a function. $f(a)$ stands for the value of y [or of $f(x)$] when $x = a$.

The Graphs of $y = f(x) + h$ and $y = f(x + h)$ 9.3

The graph of $y = f(x) + h$ can be found by shifting the graph of $y = f(x)$ upward $|h|$ units if h is positive and downward $|h|$ units if h is negative.

The graph of $y = f(x + h)$ can be found by shifting the graph of $y = f(x)$ to the left $|h|$ units if h is positive and to the right $|h|$ units if h is negative.

The Algebra of Functions 9.4

Let $f(x)$ and $g(x)$ be two functions, and let D be the *intersection* of the domains of $f(x)$ and $g(x)$. Then

$(f + g)(x) = f(x) + g(x)$, and the domain of $f + g$ is all of D.

$(f - g)(x) = f(x) - g(x)$, and the domain of $f - g$ is all of D.

$(f \cdot g)(x) = (f(x))(g(x))$, and the domain of $f \cdot g$ is all of D.

$\left(\frac{f}{g}\right)(x) = \frac{f(x)}{g(x)}$, and the domain of $\frac{f}{g}$ is the set of those values of D for which $g(x)$ is *not* 0.

Composition of Functions 9.4

Let $f(x)$ and $g(x)$ be two functions. If $g(x)$ is in the domain of f, then the **composition** of f with g is $(f \circ g)(x) = f(g(x))$. The domain of $f \circ g$ is the set of all x in the domain of g for which $g(x)$ is in the domain of f.

The Inverse of a Function 9.5

If f is a one-to-one function with domain X and range Y and if g is a one-to-one function with domain Y and range X, then g is the **inverse function** of f if and only if $(f \circ g)(x) = f(g(x)) = x$ for all x in Y and $(g \circ f)(x) = g(f(x)) = x$ for all x in X.

A function is **invertible** if for each y-value there is only one x-value. If f is given as a set of ordered pairs, we find its inverse, f^{-1}, by interchanging the coordinates of the ordered pairs. If f is described by an equation, we can find its inverse as follows:

1. Substitute y for $f(x)$, if necessary.
2. Interchange x and y in the equation.
3. Solve the new equation for y.
4. Replace y by $f^{-1}(x)$.

The *domain* of f equals the *range* of f^{-1}, and the *range* of f equals the *domain* of f^{-1}.

Polynomial Functions 9.6

$f(x) = k$, where k is some real number, is the **constant function**.

$f(x) = x$ is the **identity function**; it is also a linear function.

$f(x) = ax + b$, where $a, b \in R$ and $a \neq 0$, is a **linear function**.

$f(x) = ax^2 + bx + c$, where $a, b, c \in R$ and $a \neq 0$, is a **quadratic function**.

$f(x) = ax^3 + bx^2 + cx + d$, where $a, b, c, d \in R$ and $a \neq 0$, is a **cubic function**.

To graph a polynomial function other than a linear function, use the equation to make a table of values, graph the points, and connect them in order from left to right with a *smooth* curve.

Direct Variation 9.7 In **direct variation**, one variable is some constant multiple of the other variable, and as one variable *increases*, the other increases also. The statements "y varies directly as (or *with*) x" and "y is directly proportional to x" are statements of direct variation, and they translate to the equation $y = kx$, where k is the constant of proportionality.

Proportions 9.7 A **proportion** is a statement that two ratios or two rates are equal; that is, it is an equation that can be expressed in the form $\frac{a}{b} = \frac{c}{d}$.

Inverse Variation 9.8 In **inverse variation**, as one variable *increases*, the other *decreases*. The statements "y varies inversely as (or *with*) x" and "y is inversely proportional to x" are statements of inverse variation, and they translate into the equation $y = \frac{k}{x}$, where k is the constant of proportionality.

Joint Variation 9.9 In **joint variation**, one variable is directly proportional to the product of two or more other variables.

Combined Variation 9.9 In **combined variation**, one variable varies directly with another variable (or jointly with the product of two or more variables) and inversely with still other variables.

Sections 9.1–9.9 REVIEW EXERCISES Set I

1. Which of the following are graphs of functions?

a.

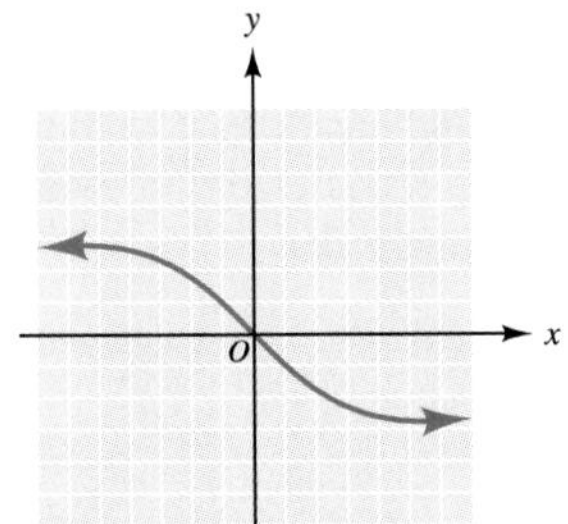

b.

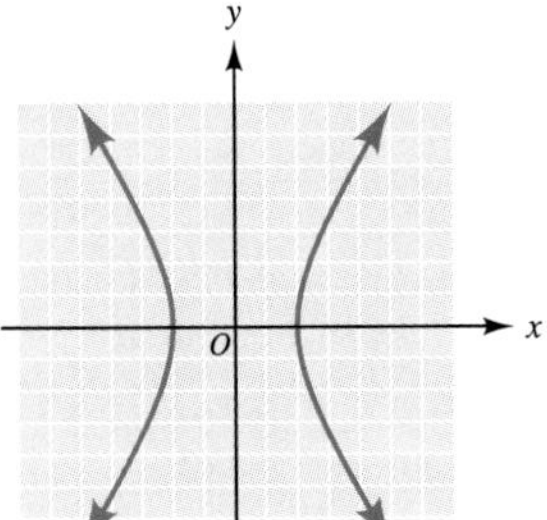

c.

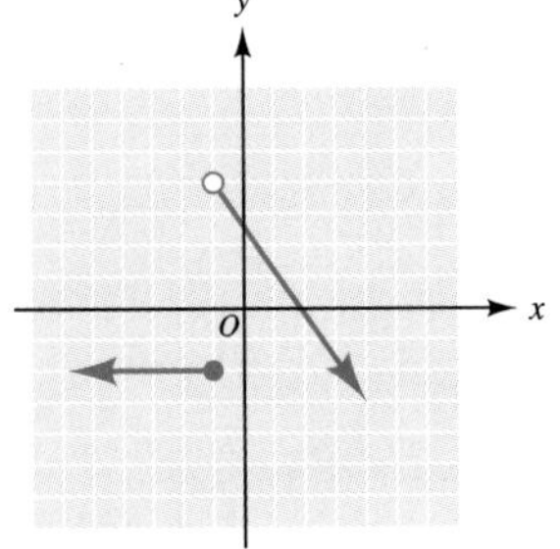

2. Which of the following are graphs of functions?

a.

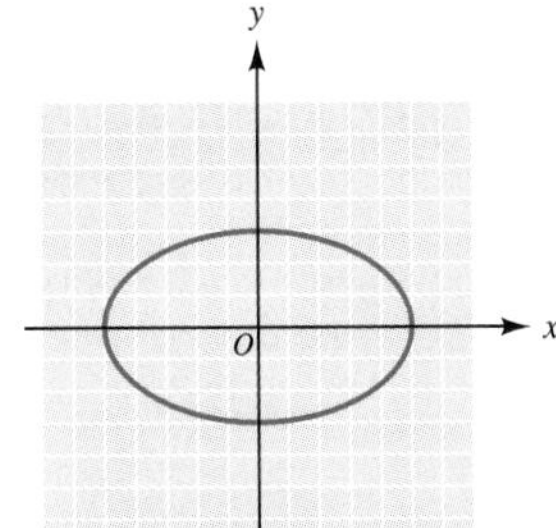

b.

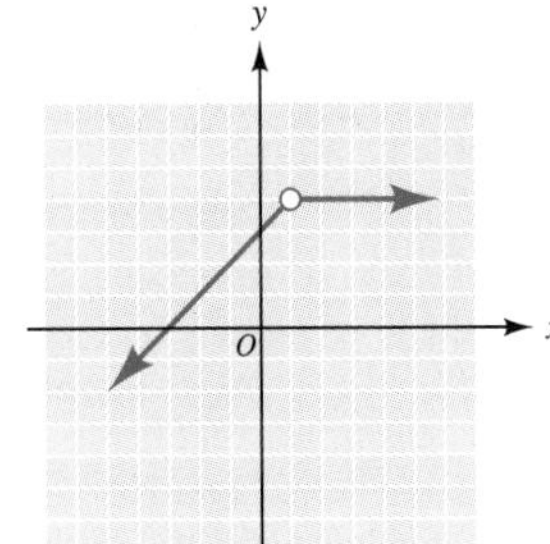

c.

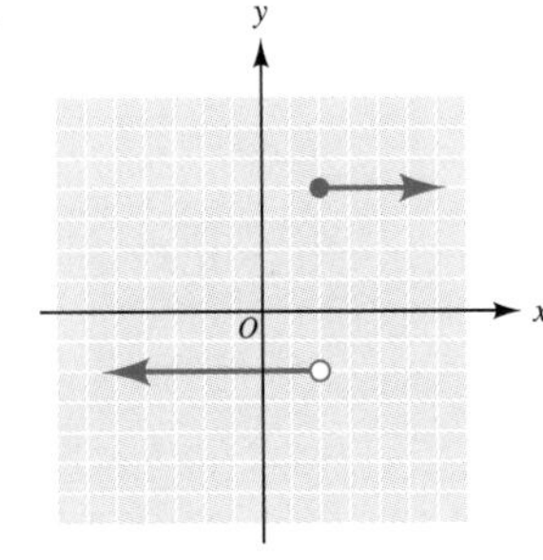

3. Given: $\mathcal{R} = \{(-1, -2), (-4, 4), (5, 0), (4, -2), (-5, -3)\}$.

   a. Is $\mathcal{R}$ a function?

   b. Is $\mathcal{R}$ a one-to-one function?

   c. If $\mathcal{R}$ *is* a one-to-one function, find its inverse.

4. Determine whether $\{(x, y) \mid y = 5x + 3\}$ is a function.

5. Given $f(x) = 3x^2 - 5x + 4$, find the following:

   a. $f(2)$

   b. $f(0)$

   c. $f(x - 1)$

6. If $F(x) = x^3 - 2x^2 + 6$ and $g(x) = 2x^2 - 7x$, find $\frac{1}{3}F(2) - 6g(4)$.

In Exercises 7–9, graph each function.

7. $y = f(x) = |x| + 1$

8. $y = f(x) = |x| - 1$

9. $y = f(x) = |x + 1|$

10. If $f(x) = 4x^2 + 6$ and $g(x) = -3x + 5$, find $f + g$, $f - g, f \cdot g$, and $\dfrac{f}{g}$.

11. If $f(x) = 4x^2 + 6$ and $g(x) = -3x + 5$, find $f \circ g$ and $g \circ f$.

12. Show that $f(x) = 7x + 9$ and $g(x) = \dfrac{x - 9}{7}$ are inverses of each other.

In Exercises 13 and 14, find the inverse function of each function. Graph each function and its inverse on the same set of axes. Find the domain and range of f and of f^{-1}.

13. $y = f(x) = \dfrac{x + 6}{3}$

14. $y = f(x) = -\frac{5}{2}x + 1$

In Exercises 15 and 16, find the inverse function of each function.

15. $y = f(x) = \dfrac{9}{2 - 7x}$

16. $y = f(x) = \dfrac{11}{2(4x - 5)}$

17. Graph $y = \dfrac{x^2}{2}$ by using x-values of $-4, -2, -1, 0, 1, 2$, and 4.

18. Graph $y = \dfrac{x^2}{2} + 1$ by using x-values of $-4, -2, -1, 0$, 1, 2, and 4.

19. Graph $y = x^3 - 3x$ by using integer values of x from -3 to $+3$.

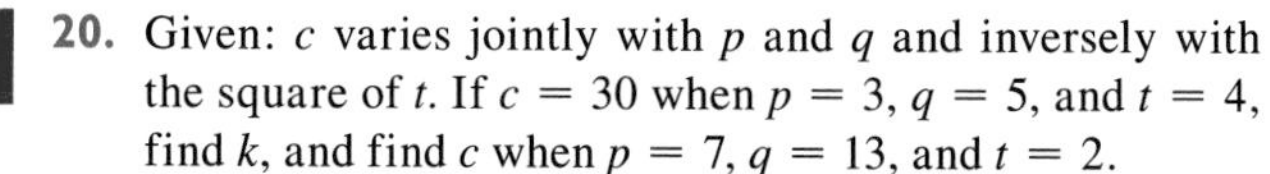

20. Given: c varies jointly with p and q and inversely with the square of t. If $c = 30$ when $p = 3$, $q = 5$, and $t = 4$, find k, and find c when $p = 7$, $q = 13$, and $t = 2$.

21. The surface area (S) of a sphere varies directly with the square of the radius (r). If $S = 615.44$ when $r = 7$, find k, and find S when $r = 10$.

22. (For this exercise, assume that the *time* is inversely proportional to the *rate*.) Machine A works for 4 hr to complete one order, making a certain part at the rate of 125 parts per hr. Find k, and find how long it would take machine B, which makes the parts at the rate of 75 parts per hr, to complete the same order.

Sections 9.1–9.9 REVIEW EXERCISES Set II

Name

1. Which of the following are graphs of functions?

a.

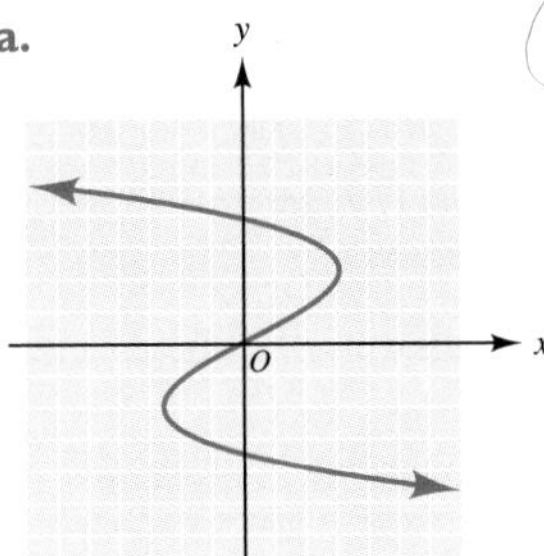

b.

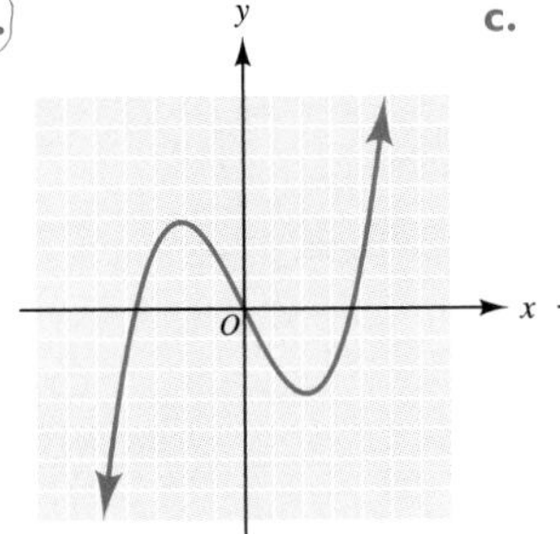

c.

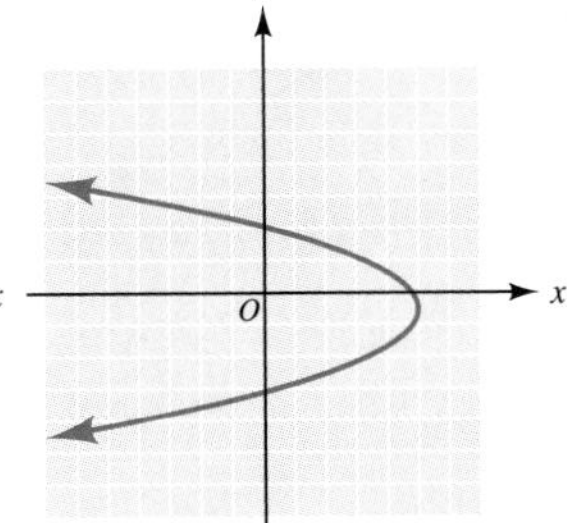

2. Given $f(x) = 5x^2 - 12x + 16$, find the following.

a. $f(-1)$ b. $f(4)$ c. $f(2a - 1)$

3. If $H(x) = 5x^3 + 7$ and $g(y) = 4y - 2y^2$, find $\frac{5}{11}H(-2) - \frac{1}{6}g(5)$.

4. If $F(x) = 3x - x^2$ and $G(w) = 4 - w^2$, find $\dfrac{F(1 - c)}{G(c)}$.

In Exercises 5 and 6, graph each function.

5. $y = f(x) = |x| - 4$

6. $y = f(x) = |x - 4|$

ANSWERS

1. ____________

2a. ____________

b. ____________

c. ____________

3. ____________

4. ____________

5.

y

x

O

6.

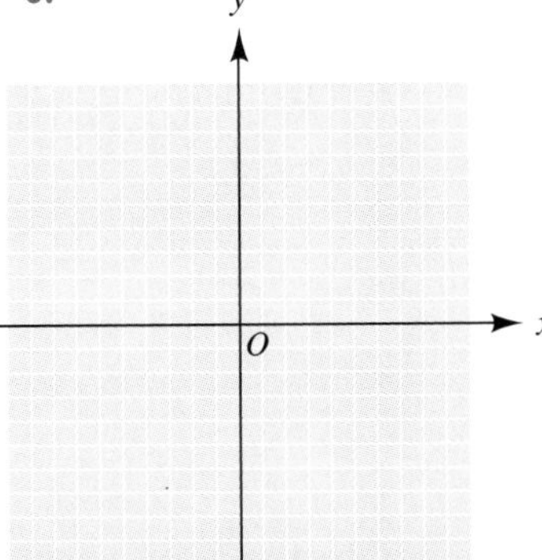

7. Given: $\mathcal{R} = \{(-2, 4), (-1, 2), (3, -1), (0, 5)\}$.

a. Is $\mathcal{R}$ a function?

b. Is $\mathcal{R}$ a one-to-one function?

c. If $\mathcal{R}$ *is* a one-to-one function, find its inverse.

7a. ______

b. ______

c. ______

8. ______

9a. ______

b. ______

c. ______

d. ______

10a. ______

b. ______

c. ______

11.

8. Determine whether $\{(x, y) \mid y = -4x + 9\}$ is a function. (Answer yes or no.)

9. Given: $f(x) = -6x + 4$ and $g(x) = 12x^2 + 1$.

a. Find $f + g$.

b. Find $f - g$.

c. Find $f \cdot g$.

d. Find $\frac{f}{g}$.

10. Given: $f(x) = -\frac{1}{3}x + 4$ and $g(x) = -3x + 12$.

a. Find $f \circ g$.

b. Find $g \circ f$.

c. If f and g are inverses of each other, explain *why* they are inverses.

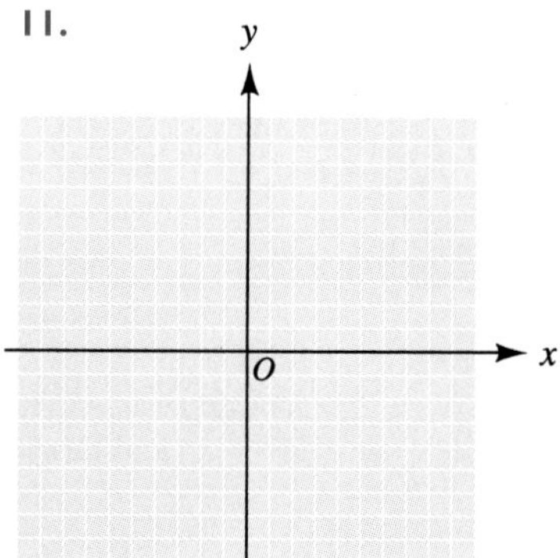

11. For $y = f(x) = \frac{4}{7}x - 2$, find the inverse function, $f^{-1}(x)$. Graph the function and its inverse on the same set of axes. Find the domain and range of f and of f^{-1}.

For Exercises 12 and 13, find the inverse function, $y = f^{-1}(x)$.

12. $y = f(x) = \dfrac{1}{x}$

13. $y = f(x) = \dfrac{x - 2}{5}$

14. Given: y varies directly as the cube of x, and $y = 54$ when $x = 3$.

a. Write the variation equation.

b. Find k, the constant of proportionality.

c. Find y when $x = -4$.

15. Given: y varies inversely as the square of x, and $y = 5$ when $x = 3$.

a. Write the variation equation.

b. Find k, the constant of proportionality.

c. Find y when $x = 5$.

16. Graph $y = \dfrac{(x + 1)^2}{2}$ by using integer values of x from -4 to $+2$.

17. Graph $y = 5x - x^3$ by using integer values of x from -3 to $+3$.

ANSWERS

12. ______

13. ______

14a. ______

b. ______

c. ______

15a. ______

b. ______

c. ______

16.

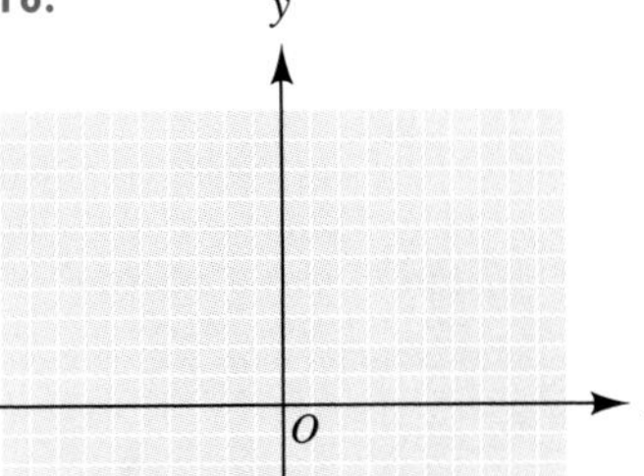

17.

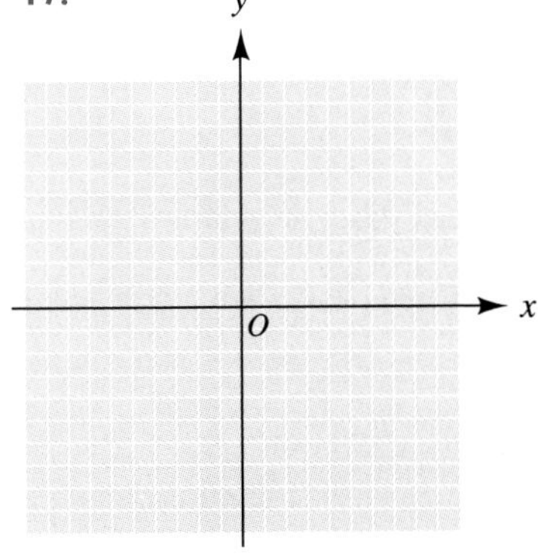

18. Given: $f(x) = x^2 - 3x + 5$, find and simplify $\dfrac{f(x + h) - f(x)}{h}$.

18. ______________________

19. ______________________

20. ______________________

____

19. Given: y varies directly with the square of x and inversely with the cube of z. If $y = \frac{4}{9}$ when $x = 2$ and $z = 3$, find k.

21a. ______________________

b. ______________________

c. ______________________

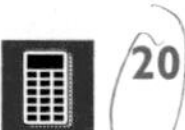

20. The pressure (P) in water varies directly with the depth (d). If $P = 4.33$ when $d = 10$, find k, and find P when $d = 25$.

21. a. Write the variation equation for the following statement: When the area of a rectangle is held constant, the width of the rectangle varies inversely as its length. Use this equation in b and c below.

b. If the width is 8 m when the length is 12 m, find k.

c. Find the width when the length is 16 m.

Chapter 9 DIAGNOSTIC TEST

The purpose of this test is to see how well you understand functions and their graphs, functional notation, the algebra of functions, and variation. We recommend that you work this diagnostic test *before* your instructor tests you on this chapter. Allow yourself about 50 minutes.

Complete solutions for all the problems on this test, together with section references, are given in the answer section at the end of the book. For the problems you do incorrectly, we suggest that you study the sections cited.

1. Which of the following graphs are graphs of functions?

a.

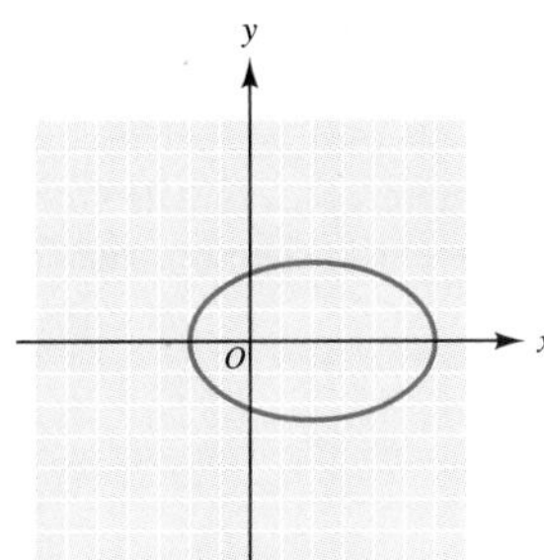

b.

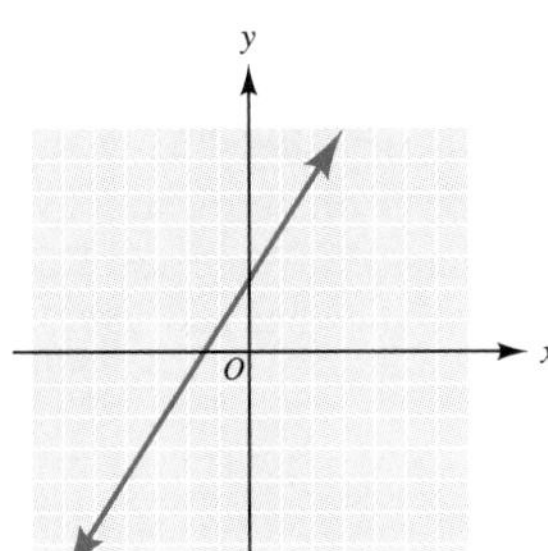

c.

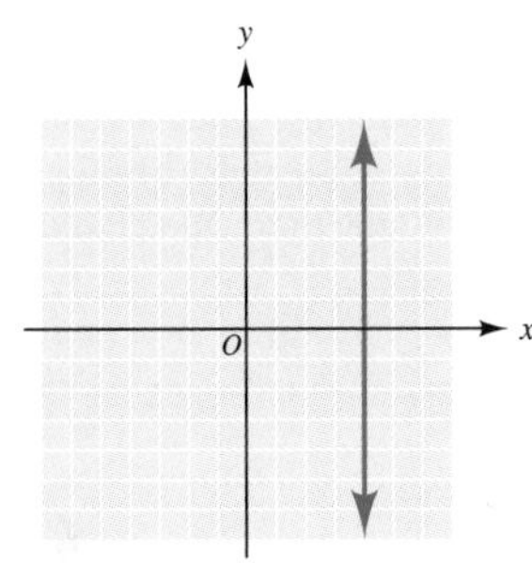

2. Given: $\mathcal{R} = \{(1, 5), (3, 7), (5, -2), (0, 0)\}$.

   a. Is $\mathcal{R}$ a function?

   b. Is $\mathcal{R}$ a one-to-one function?

   c. If $\mathcal{R}$ *is* a one-to-one function, find its inverse.

3. Is $\{(x, y) \mid y = -4x + 9\}$ a function? If so, what is its domain?

4. Given $f(x) = 3x^2 - 5$, find the following.

   a. $f(-2)$ b. $f(4)$ c. $\dfrac{f(4) - f(-2)}{6}$

5. a. Graph $y = |x| - 4$. b. Graph $y = |x - 4|$.

6. Given: $f(x) = 7x - 3$ and $g(x) = 5x^2 + 2$.

   a. Find $f + g$. b. Find $f - g$.

   c. Find $f \cdot g$. d. Find $\dfrac{f}{g}$.

7. Given: $f(x) = 8x - 1$ and $g(x) = \dfrac{x + 1}{8}$.

   a. Find $f \circ g$. b. Find $g \circ f$.

   c. Are f and g inverses of each other? Why or why not?

8. Use integer values of x from -2 to 3 to graph $y = 1 + x - x^2$.

9. a. Find the inverse function for $y = f(x) = -\frac{3}{2}x + 1$. Graph $y = f(x)$ and $y = f^{-1}(x)$ on the same axes.

   b. Find the inverse function for $y = f(x) = \dfrac{15}{4(2 - 5x)}$. Find the domain and range for $y = f(x)$ and $y = f^{-1}(x)$.

10. Given: w varies jointly with x and y and inversely with the square of z. If $w = 20$ when $x = 8$, $y = 6$, and $z = 12$, find k, the constant of proportionality, and find w when $x = 6$, $y = 10$, and $z = 5$.

Chapters 1–9 CUMULATIVE REVIEW EXERCISES

1. Write "true" if the statement is always true; otherwise, write "false."

   a. $\sqrt{5}$ is a real number.

   b. $8x^{-1} = \dfrac{1}{8x}$

   c. $3x$ is a term of the expression $3xy$.

   d. Subtraction is commutative.

   e. $6(5 \cdot 3) = (6 \cdot 5) \cdot (6 \cdot 3)$

   f. $(x + y)^2 = x^2 + y^2$

   g. $8 < 9 > 7$ is a valid continued inequality.

   h. If $x = -2$, then $|x| = x$.

   i. If $f(x) = 3x + 5$, then $f^{-1}(x) = \dfrac{1}{3x + 5}$.

   j. The least common multiple of 12 and 18 is 6.

2. Write the equation of the straight line that passes through the points $(-2, 5)$ and $(6, -3)$.

3. Given the formula $s = \frac{1}{2}gt^2$, find s if $g = 32$ and $t = 2$.

In Exercises 4 and 5, remove grouping symbols and combine like terms.

4. $2xy(4xy - 5x + 2) - 3y(2x^2y - x)$

5. $6 - \{4 - [3x - 2(5 - 3x)]\}$

In Exercises 6–10, find the solution set of each equation or inequality.

6. $\frac{x}{2} + \frac{x+3}{3} = \frac{1}{6}$

7. $12 - 3(4x - 5) \geq 9(2 - x) - 6$

8. $-3 < 2x + 1 < 5$

9. $\left|\frac{2x-6}{4}\right| = 1$

10. $|4x - 2| < 10$

In Exercises 11–14, set up each problem algebraically, solve, and check. Be sure to state what your variables represent.

11. The ratio of the length of a rectangle to its width is 7:4. The perimeter is 66 in. Find the length and the width.

12. The tens digit of a three-digit number is twice the hundreds digit, and the sum of the digits is 17. If the digits are reversed, the new number is 99 more than the original number. Find the original number.

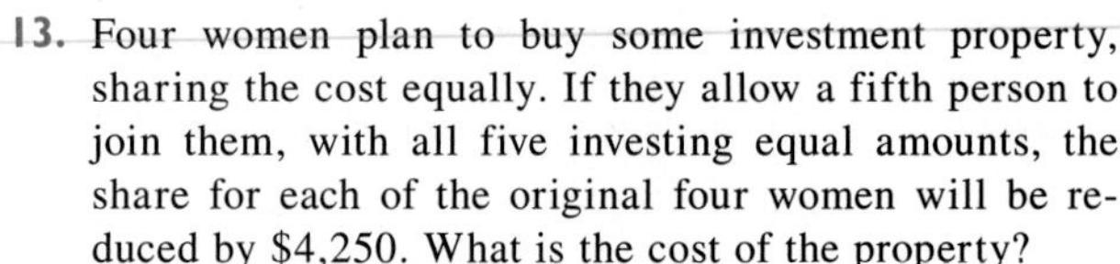

13. Four women plan to buy some investment property, sharing the cost equally. If they allow a fifth person to join them, with all five investing equal amounts, the share for each of the original four women will be reduced by $4,250. What is the cost of the property?

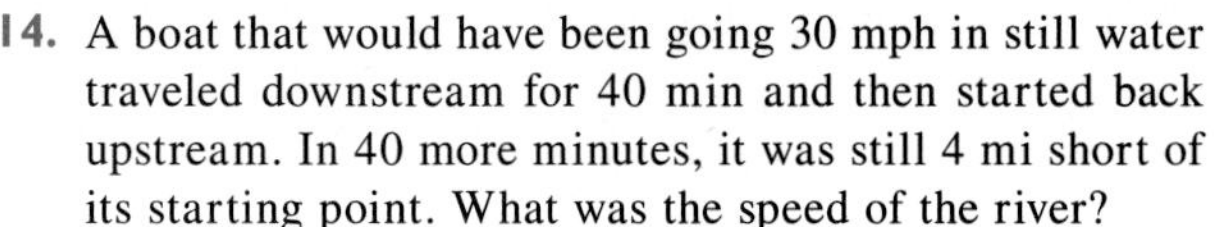

14. A boat that would have been going 30 mph in still water traveled downstream for 40 min and then started back upstream. In 40 more minutes, it was still 4 mi short of its starting point. What was the speed of the river?

Critical Thinking and Problem-Solving Exercises

1. The sum of the measures of the angles of a three-sided polygon is 180°, the sum of the measures of the angles of a four-sided polygon is 360°, and the sum of the measures of the angles of a five-sided polygon is 540°. If this same pattern continues, what is the sum of the measures of the angles of a seven-sided polygon? Of an n-sided polygon?

2. In one of David's computer games, the character Etta lives in a cave and can move horizontally or vertically (on the computer screen), but not diagonally. On a dark and stormy night, Etta left her cave and traveled 4 mi west, then 3 mi north, then 2 mi east, then 6 mi north, then 8 mi east, and finally 12 mi south.
   a. How far did Etta travel on that dark and stormy night?
   b. Moving only horizontally or vertically, what would be the shortest distance Etta would have to travel to return to her cave?
   c. If Etta *could* move diagonally, what would be the shortest distance Etta would have to travel to return to her cave?

3. Larry, Jiao, and Mike are friends. One of them is a college administrator, one is a professor of languages, and one is a statistician. Their last names are Hayes, Mayes, and Tayes. By using the following clues, can you match the first name, last name, and profession of each of these three men?

   Mr. Tayes is a college administrator.

   Mr. Hayes played tennis with the language professor.

   Mike is a statistician.

   Larry is not a college administrator.

4. Roger, Wing, and Bob wanted to attend a football game, and, in checking to see whether they had enough money for tickets, hot dogs, and so on, they discovered that Roger and Wing together had $58, Wing and Bob together had $61, and Roger and Bob together had $59. How much did each man have individually?

5. In the field of economics, the cost (C) of producing x units of some item is often assumed to be a linear function of x, so $C(x) = ax + b$. Suppose that, because of overhead costs and so forth, it costs a company $450 when *no* items are produced, and suppose that it costs the company $1,000 to produce 110 items. Write the cost function [that is, determine what a and b are for the equation $C(x) = ax + b$]. Then determine the cost of producing 200 items.

Exponential and Logarithmic Functions and Equations

CHAPTER

10

In this chapter, we discuss exponential and logarithmic functions and some of their applications. Exponential functions are very important in business and in the sciences, as we shall see. Logarithms have many applications in mathematics and science—for example, they are used to measure the magnitude of earthquakes and the pH factor of solutions.

10.1 Exponential and Logarithmic Functions and Their Graphs

The Exponential Function and Its Graph

An **exponential function** is a function that can be expressed in the form $y = f(x) = b^x$, where b is any real number greater than zero, except $b \neq 1$, and where x is any real number.

The *domain* of the exponential function is the set of all real numbers. It is proved in higher mathematics courses that the graph of the exponential function is a smooth curve, that the *range* of the exponential function $y = b^x$ is the set of all real numbers greater than zero, and that there is only one x-value corresponding to each y-value. The technique of graphing an exponential function is demonstrated in Example 1.

EXAMPLE 1 Graph the exponential function $y = 2^x$, and discuss its range.

SOLUTION We calculate the values of the function for integral values of x from -2 to $+3$.

$y = 2^x$	x	y
$y = 2^{-2} = \frac{1}{4}$	-2	$\frac{1}{4}$
$y = 2^{-1} = \frac{1}{2}$	-1	$\frac{1}{2}$
$y = 2^0 = 1$	0	1
$y = 2^1 = 2$	1	2
$y = 2^2 = 4$	2	4
$y = 2^3 = 8$	3	8

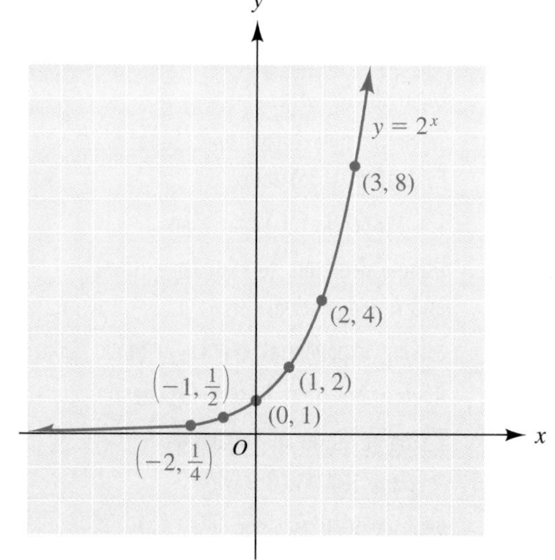

Next, we graph the points listed in the table of values and draw a *smooth curve* through them, taking the points in order from left to right. It may appear from the graph that the curve touches the x-axis, but if we could draw graphs more accurately, we would see that it does not. To verify this, let's consider some *very* small values of x: If $x = -10$, $y = 2^{-10} \approx 0.00098$, a small *but positive* number; if $x = -100$, $y = 2^{-100} \approx 7.9 \times 10^{-31}$, a still smaller *but still positive* number. Hence, in theory, the graph never touches the x-axis and is never below it. Thus, the range of the function $y = 2^x$ is, indeed, the set of all real numbers greater than 0. We can see that no horizontal line can touch the graph in more than one point.

Note To graph $y = 2^x$ with a TI-81 graphics calculator, press these keys, in order:

[Y=] [2] [∧] [X|T] [GRAPH]

We recommend setting the range so that xmin is -5, xmax is 5, ymin is -1, and ymax is 5.

EXAMPLE 2 Graph $y = \left(\frac{1}{3}\right)^x$.

SOLUTION

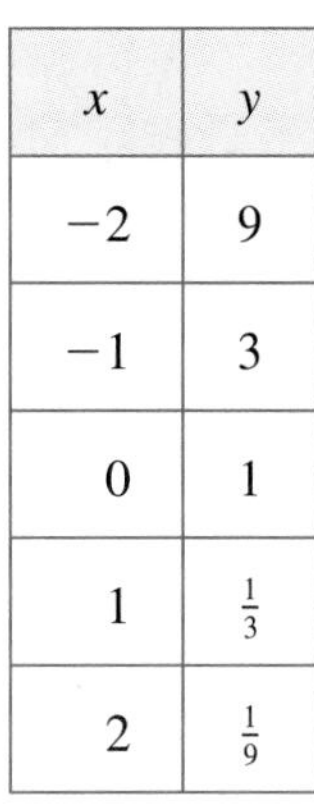

x	y
-2	9
-1	3
0	1
1	$\frac{1}{3}$
2	$\frac{1}{9}$

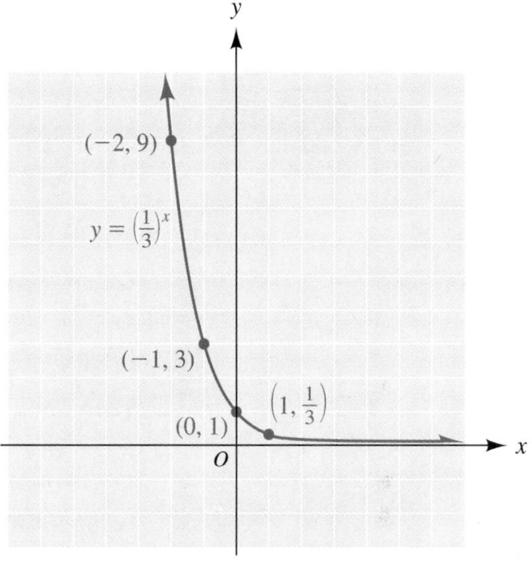

We can verify that the graph (in theory) can *never* touch the x-axis and can never be below it, so the *range* is the set of all real numbers greater than zero.

Note To graph $y = \left(\frac{1}{3}\right)^x$ with a TI-81 graphics calculator, press these keys, in order:

[Y=] [3] [x^{-1}] [∧] [X|T] [GRAPH]

or [Y=] [(] [1] [÷] [3] [)] [∧] [X|T] [GRAPH]

We recommend the same range settings as for graphing $y = 2^x$.

If $b > 1$, the graph of $y = f(x) = b^x$ will always look like the graph in Figure 1. The curve *rises* as we move toward the right and lies entirely above the x-axis. It passes through the point (0, 1).

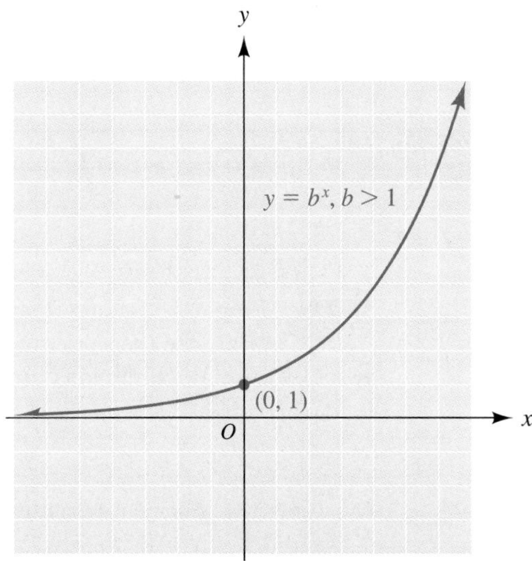

Figure 1

The Number e

An *irrational number* designated by the letter e is particularly important as a base for exponential functions.* It is so important, in fact, that special tables are made up for the function $y = f(x) = e^x$ (see Appendix C). On most scientific calculators, we find the values for exponential functions with the base e by using the inverse of the key labeled [ln] or [LN]. Usually e^x is written above the [ln] key, and we press [INV] (or [2nd] or [SHIFT]) before pressing [ln].

EXAMPLE 3 Using a calculator, find e^1. Round off the answer to two decimal places.

SOLUTION

	Keys	Display
Scientific (algebraic)	[1] [INV] [ln] (e^x)†	2.718281828
Scientific (RPN)	[1] [ENTER] [e^x]	2.71828182846
Graphics	[2nd] [LN] (e^x S) [1] [ENTER]	2.718281828

Therefore, $e^1 \approx 2.72$, which means that $e \approx 2.72$.

EXAMPLE 4 Using a calculator, find e^{-5}. Round off the answer to four decimal places.

SOLUTION

	Keys	Display
Scientific (algebraic)	[5] [+/−] [INV] [ln] (e^x)†	6.737946999 E − 03
Scientific (RPN)	[5] [+/−] [ENTER] [e^x]	6.737947 E ⁻3
Graphics	[2nd] [LN] (e^x S) [(−)] [5] [ENTER]	.006737947

Therefore, $e^{-5} \approx 0.0067$.

We state here (without proof) two other properties of exponential functions; see Examples 1, 3, 4, 5, and 6 in Section 10.3 for examples of the use of these properties.

PROPERTY 1

For $b > 0$, but $b \neq 1$,

$$\text{if} \quad b^x = b^y, \quad \text{then} \quad x = y$$

PROPERTY 2

For $a > 0$ and $b > 0$,

$$\text{if} \quad a^x = b^x, \quad \text{then} \quad a = b$$

Note Property 1 states that when two exponential expressions are equal, if the *bases* are the same, the exponents must be equal. Property 2 states that when two exponential expressions are equal, if the *exponents* are the same, the bases must be equal.

*An explanation of how e is calculated is beyond the scope of this text.

†On some scientific calculators, we must press [2nd] [ln] (e^x) before the number and [=] after it.

The Logarithmic Function and Its Graph

Because there is only one x-value corresponding to each y-value for the function $y = f(x) = b^x$, that function must have an *inverse*. Recall that the graph of the inverse of a function is always the mirror image, with respect to the line $y = x$, of the original function. Therefore, we even know what the graph of the inverse function looks like—its graph is the mirror image, with respect to the line $y = x$, of the graph of $y = b^x$ (see Figure 2). The domain of the inverse function must be the set of all real numbers greater than zero, because the domain of the inverse of a function equals the range of the original function. The range of the inverse function must be the set of all real numbers, because the range of an inverse function equals the domain of the original function. In theory, the graph of the *inverse* function never touches the y-axis and is never to the left of the y-axis.

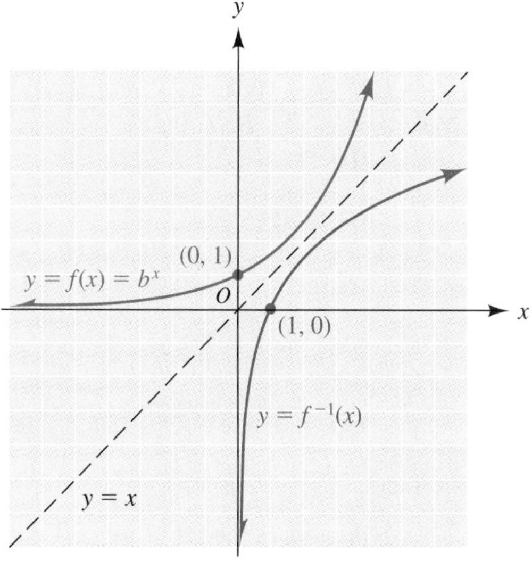

Figure 2

To find the inverse of $y = f(x) = b^x$, we use the method discussed earlier for finding the inverse of a function; that is, we first interchange x and y and write $x = b^y$. Then we try to solve this new equation for y. However, we need new notation and a definition before we can solve that equation for y.

Logarithms

The definition of a logarithm is as follows:

The **logarithm** of a positive number* x is the **exponent** y to which the base b ($b > 0$, $b \neq 1$) must be raised to give x.

In symbols: $y = \log_b x$ if and only if $b^y = x$

where x is any real number greater than zero.

*The logarithm of a negative number exists, but it is a complex number. We will not discuss the logarithm of a negative number in this book.

  $y = \log_b x$ is read y is the logarithm of x to the base b or, informally, y is the log of x to the base b.

The Argument of a Logarithm In the expression $\log_b x$, x is called the **argument** of the logarithmic function. The argument can be any expression, such as $(x + 1)$, $\dfrac{2x - 1}{5 - x}$, or $(x^2 - 3x + 2)$. Because the argument must *always* be positive, however, the domain is generally some subset of the set of real numbers. We discuss the argument further in a later section of this chapter.

EXAMPLE 5 Graph the logarithmic function $y = \log_2 x$.

SOLUTION We rewrite $y = \log_2 x$ in the form $x = 2^y$ and let y be the independent variable; we'll choose some negative values for y as well as some nonnegative ones.

$x = 2^y$	x	y
If $y = -3$, $x = 2^{-3} = \frac{1}{8}$	$\frac{1}{8}$	-3
If $y = -2$, $x = 2^{-2} = \frac{1}{4}$	$\frac{1}{4}$	-2
If $y = -1$, $x = 2^{-1} = \frac{1}{2}$	$\frac{1}{2}$	-1
If $y = 0$, $x = 2^0 = 1$	1	0
If $y = 1$, $x = 2^1 = 2$	2	1
If $y = 2$, $x = 2^2 = 4$	4	2
If $y = 3$, $x = 2^3 = 8$	8	3

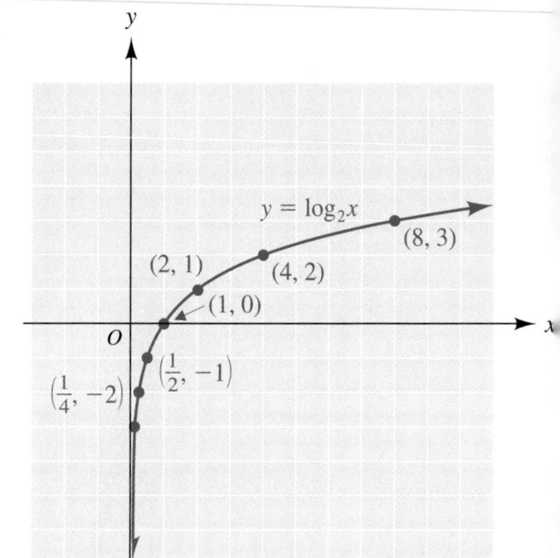

You might let y have some values less than -3 in order to convince yourself that the *domain* is the set of all real numbers greater than zero and that the *range* is the set of all real numbers.

Note To graph $y = \log_2 x$ with a TI-81 graphics calculator, press the following keys, in order:

Y= | LOG | X|T | ÷ | LOG | 2 | GRAPH

(The reason for this sequence of keystrokes will be seen at the end of this chapter.) We recommend setting the range so that *x*min is -5, *x*max is 10, *y*min is -5, and *y*max is 5.

EXAMPLE 6 Graph $y = 2^x$ and $y = \log_2 x$ on the same axes.

SOLUTION Combining Examples 1 and 5, we get the following graph:

y
The exponential function $y = 2^x$
The logarithmic function $y = \log_2 x$
$y = x$
x

 **Note** To do the problem in Example 6 with a TI-81 graphics calculator, let $y_1 = 2^x$ and let $y_2 = \log_2 x$ (in the Note following Example 2 of Section 9.3, we explained how to let y_1 be one function and y_2 be a different function), and then press GRAPH.

In general, the graph of any logarithmic function with a base greater than 1 has the appearance and characteristics of the curve shown and described in Figure 3.

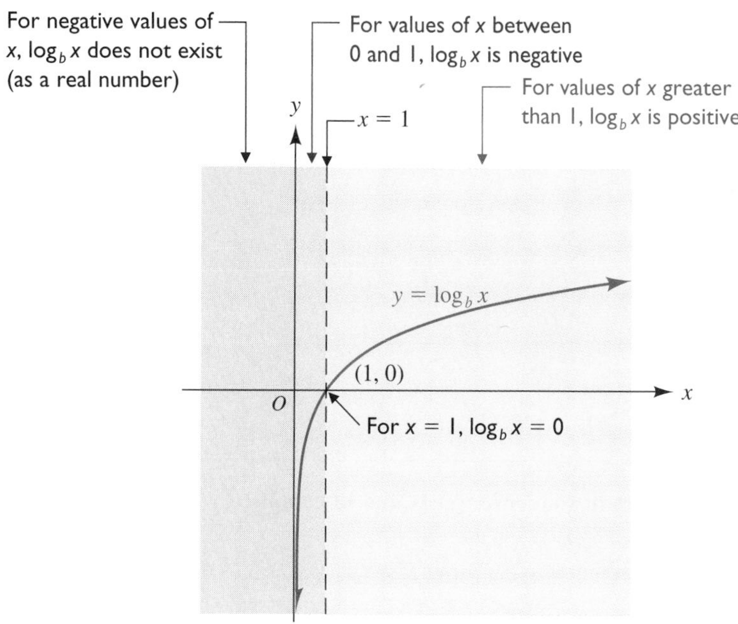
For negative values of x, $\log_b x$ does not exist (as a real number)
For values of x between 0 and 1, $\log_b x$ is negative
For values of x greater than 1, $\log_b x$ is positive
y
$x = 1$
$y = \log_b x$
(1, 0)
O
x
For $x = 1$, $\log_b x = 0$

Figure 3

The following three properties of logarithms are stated without proof.

1. The logarithmic function is an *increasing function*. This means that as x gets larger, $\log_b x$ also gets larger.
2. If two numbers are equal, then their logarithms (taken to the same base) must be equal, because $\log_b x$ is a *function*. This fact is used in a later section in solving exponential equations and is stated in symbols in Property 3.

PROPERTY 3

$$\text{If} \quad M = N, \quad \text{then} \quad \log_b M = \log_b N$$

3. If the logarithms of two numbers are equal (when the bases are the same), then the numbers must be equal. This fact is used later in solving logarithmic equations and is stated in symbols in Property 4.

PROPERTY 4

$$\text{If} \quad \log_b M = \log_b N, \quad \text{then} \quad M = N$$

 ***A Word of Caution*** We can neither *multiply* nor *divide* both sides of an equation by "$\log_b$"; "$\log_b$" by itself has absolutely no meaning. When we rewrite the equation $M = N$ as $\log_b M = \log_b N$, we are using Property 3; *we are not multiplying both sides by* "$\log_b$."

Similarly, when we rewrite $\log_b M = \log_b N$ as $M = N$, we are using Property 4; *we are not dividing both sides by* "$\log_b$."

Remember: "$\log_b$" by itself is meaningless.

In the following exercises, the even-numbered problems are not "matched" to the odd-numbered problems.

Exercises 10.1 Set I

In Exercises 1–4, find the decimal approximation of each expression. Use a calculator and round off the answers to four decimal places, or use Table I of Appendix C.

1. e^3 2. $e^{0.5}$ 3. e^{-3} 4. $e^{-0.5}$

In Exercises 5–10, graph each function.

5. $y = 4^x$ 6. $y = 5^x$

7. $y = e^x$ (Use a calculator or Table I, Appendix C, for the y-values.)

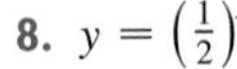

8. $y = \left(\frac{1}{2}\right)^x$ 9. $y = \left(\frac{3}{2}\right)^x$ 10. $y = \left(\frac{2}{3}\right)^x$

11. Graph $y = \log_2 x$ and $y = \log_{10} x$ on the same set of axes.

12. Graph the exponential function $y = 3^x$ and its inverse function, $y = \log_3 x$, on the same set of axes.

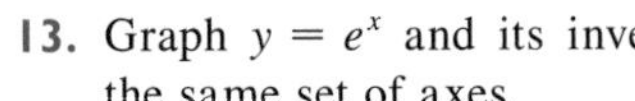

13. Graph $y = e^x$ and its inverse function, $y = \log_e x$, on the same set of axes.

Writing Problems

Express the answers in your own words and in complete sentences.

1. Describe the graph of $y = b^x$, where $b > 1$.
2. Describe the graph of $y = b^x$, where $0 < b < 1$.
3. Describe the graph of $y = \log_b x$.

Exercises 10.1 Set II

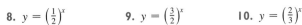

In Exercises 1–4, find the decimal approximation of each expression. Use a calculator and round off the answers to four decimal places, or use Table I of Appendix C.

1. $e^{2.2}$ 2. $e^{-4.6}$ 3. $e^{-0.03}$ 4. $e^{0.24}$

In Exercises 5–10, graph each function.

5. $y = 6^x$ 6. $y = -3^x$

7. $y = -e^x$ (Use a calculator or Table I, Appendix C, for the y-values.)

8. $y = \left(\frac{5}{2}\right)^x$ 9. $y = \left(\frac{1}{4}\right)^x$ 10. $y = \left(\frac{3}{4}\right)^x$

11. Graph $y = \log_2 x$ and $y = \log_3 x$ on the same set of axes.

12. Graph $y = 4^x$ and its inverse function, $y = \log_4 x$, on the same set of axes.

13. Graph $y = 5^x$ and its inverse function, $y = \log_5 x$, on the same set of axes.

10.2 Converting Between Exponential Form and Logarithmic Form

Because of the definition of logarithms, *logarithms are exponents*. Every exponential statement can be written in logarithmic form, and every logarithmic statement can be written in exponential form.

The Symbol ⇔ The symbol ⇔, read is equivalent to, is used between two statements to indicate that those statements have exactly the same meaning. We often use this symbol between a statement expressed in exponential form and the equivalent statement expressed in logarithmic form.

Exponential form		*Logarithmic form*
$b^y = x$	$\Leftrightarrow$	$y = \log_b x$

Remembering that *logarithms are exponents* is helpful in converting an equation from exponential form to logarithmic form and vice versa (see Examples 1, 2, and 3).

EXAMPLE 1 Convert $10^2 = 100$ to logarithmic form.

SOLUTION

The logarithm (the exponent 2)

$$10^2 = 100$$

The base (10) — The number (100)

The *logarithm* is the *exponent* (2) to which the base (10) must be raised to give the number (100). This relationship can be written $\log_{10} 100 = 2$ and is read "the logarithm of 100 to the base 10 equals 2."

Therefore, when we convert $10^2 = 100$ to logarithmic form, the result is

$$\log_{10} 100 = 2$$

EXAMPLE 2 Examples of converting exponential statements to logarithmic form:

	Exponential form		*Logarithmic form*
a.	$5^2 = 25$	$\Leftrightarrow$	$\log_5 25 = 2$
b.	$2^3 = 8$	$\Leftrightarrow$	$\log_2 8 = 3$

EXAMPLE 3 Examples of converting logarithmic statements to exponential form:

	Logarithmic form		*Exponential form*
a.	$\log_4 16 = 2$	$\Leftrightarrow$	$4^2 = 16$
b.	$\log_{25} 5 = \frac{1}{2}$	$\Leftrightarrow$	$25^{1/2} = 5$
c.	$\log_{10} 10{,}000 = 4$	$\Leftrightarrow$	$10^4 = 10{,}000$
d.	$\log_{10} 0.01 = -2$	$\Leftrightarrow$	$10^{-2} = 0.01$

Exercises 10.2
Set I

In Exercises 1–12, write each statement in logarithmic form.

1. $3^2 = 9$
2. $4^3 = 64$
3. $10^3 = 1{,}000$
4. $10^5 = 100{,}000$
5. $2^4 = 16$
6. $4^2 = 16$
7. $3^{-2} = \frac{1}{9}$
8. $2^{-3} = \frac{1}{8}$
9. $12^0 = 1$
10. $8^0 = 1$
11. $16^{1/2} = 4$
12. $8^{1/3} = 2$

In Exercises 13–24, write each statement in exponential form.

13. $\log_8 64 = 2$
14. $\log_2 32 = 5$
15. $\log_7 49 = 2$
16. $\log_4 64 = 3$
17. $\log_5 1 = 0$
18. $\log_6 1 = 0$
19. $\log_9 3 = \frac{1}{2}$
20. $\log_8 2 = \frac{1}{3}$
21. $\log_{10} 100 = 2$
22. $\log_{10} 1{,}000 = 3$
23. $\log_{10} 0.001 = -3$
24. $\log_{10} 0.01 = -2$

Writing Problems

Express the answers in your own words and in complete sentences.

1. Explain how to convert an exponential statement to logarithmic form.
2. Explain how to convert a logarithmic statement to exponential form.

Exercises 10.2 Set II

In Exercises 1–12, write each statement in logarithmic form.

1. $2^3 = 8$
2. $10^4 = 10{,}000$
3. $4^{-2} = \frac{1}{16}$
4. $5^0 = 1$
5. $4^{1/2} = 2$
6. $3^4 = 81$
7. $5^2 = 25$
8. $3^{-3} = \frac{1}{27}$
9. $6^{-3} = \frac{1}{216}$
10. $17^0 = 1$
11. $64^{1/3} = 4$
12. $2^5 = 32$

In Exercises 13–24, write each statement in exponential form.

13. $\log_4 64 = 3$
14. $\log_{16} 4 = \frac{1}{2}$
15. $\log_3 1 = 0$
16. $\log_7 7 = 1$
17. $\log_{10} 100{,}000 = 5$
18. $\log_b a = c$
19. $\log_2 64 = 6$
20. $\log_7 1 = 0$
21. $\log_6 36 = 2$
22. $\log_{13} 13 = 1$
23. $\log_{10} 0.0001 = -4$
24. $\log_{16} 2 = \frac{1}{4}$

10.3 Solving Simple Logarithmic Equations; Evaluating Logarithms

Solving Simple Logarithmic Equations

A **logarithmic equation** is an equation in which the variable appears in the argument of a logarithm. We can *sometimes* solve a logarithmic equation by rewriting the equation in exponential form and then solving the resulting exponential equation (see Examples 1, 2, and 3). We solve other logarithmic equations in a later section.

EXAMPLE 1 Solve $\log_b 64 = 3$ for b.

SOLUTION

Logarithmic form		*Exponential form*
$\log_b 64 = 3$	$\Leftrightarrow$	$b^3 = 64$

Because $64 = 4^3$, we can write $b^3 = 64 = 4^3$. Then, by Property 2, if $b^3 = 4^3$, $b = 4$.

EXAMPLE 2 Solve $\log_2 N = 4$ for N.

SOLUTION

Logarithmic form		*Exponential form*
$\log_2 N = 4$	$\Leftrightarrow$	$2^4 = N$

If $N = 2^4$, then $N = 16$.

EXAMPLE 3 Solve $\log_2 32 = x$ for x.

SOLUTION

Logarithmic form		*Exponential form*
$\log_2 32 = x$	$\Leftrightarrow$	$2^x = 32$

We know that $32 = 2^5$. Then, by Property 1, if $2^x = 2^5$, $x = 5$.

Evaluating Logarithms

It is possible to *evaluate* certain logarithms as follows:

1. Let the logarithmic expression equal x.
2. Rewrite the resulting logarithmic equation in exponential form.
3. Solve the exponential equation from step 2 for x.

See Examples 4–7.

EXAMPLE 4 Find $\log_{10} 1{,}000$.

SOLUTION Let $x = \log_{10} 1{,}000$.

	Logarithmic form		*Exponential form*
Then	$x = \log_{10} 1{,}000$	$\Leftrightarrow$	$10^x = 1{,}000 = 10^3$

Then, by Property 1, if $10^x = 10^3$, $x = 3$. Therefore, $\log_{10} 1{,}000 = 3$.

EXAMPLE 5 Find $\log_8 4$.

SOLUTION Let $x = \log_8 4$.

	Logarithmic form		*Exponential form*	
Then	$x = \log_8 4$	$\Leftrightarrow$	$8^x = 4$	
			$(2^3)^x = 2^2$	Writing both 8 and 4 as powers of 2
			$2^{3x} = 2^2$	
			$3x = 2$	Using Property 1
			$x = \frac{2}{3}$	

Therefore, $\log_8 4 = \frac{2}{3}$.

EXAMPLE 6 Find $\log_7 \frac{1}{49}$.

SOLUTION Let $x = \log_7 \frac{1}{49}$.

	Logarithmic form		*Exponential form*	
Then	$x = \log_7 \dfrac{1}{49}$	$\Leftrightarrow$	$7^x = \dfrac{1}{49} = \dfrac{1}{7^2} = 7^{-2}$	
			$x = -2$	Using Property 1

Therefore, $\log_7 \frac{1}{49} = -2$.

Note We see from Example 6 that the logarithm of a positive number can be negative. However, as was mentioned previously, the logarithm of a *negative* number is *complex* and is not discussed in this book.

EXAMPLE 7 Find $\log_{10} 0$.

SOLUTION Let $x = \log_{10} 0$.

	Logarithmic form		*Exponential form*
Then	$x = \log_{10} 0$	$\Leftrightarrow$	$10^x = 0$

There is no solution for the equation $10^x = 0$, because *no* value of x can make $10^x = 0$. Therefore, $\log_{10} 0$ *does not exist.* (Remember, too, that the domain of the logarithmic function is $\{x \mid x > 0\}$, so 0 is not in the domain.)

The result stated in Example 7 is generalized as follows:

> The logarithm of 0 to any base b ($b > 0$, $b \neq 1$) does not exist.

Exercises 10.3 Set I

In Exercises 1–18, find the value of the unknown, b, N, or x.

1. $\log_5 N = 2$
2. $\log_2 N = 5$
3. $\log_3 9 = x$
4. $\log_2 16 = x$
5. $\log_b 27 = 3$
6. $\log_b 81 = 4$
7. $\log_5 125 = x$
8. $\log_3 81 = x$
9. $\log_{10} 10^{-4} = x$
10. $\log_{10} 10^{-3} = x$
11. $\log_{3/2} N = 2$
12. $\log_{5/3} N = 3$
13. $\log_9 \frac{1}{3} = x$
14. $\log_8 \frac{1}{64} = x$
15. $\log_b 8 = 1.5$
16. $\log_b 125 = 1.5$
17. $\log_2 N = -2$
18. $\log_{10} N = -2$

In Exercises 19–30, find the value of each logarithm.

19. $\log_5 25$
20. $\log_2 16$
21. $\log_{10} 10{,}000$
22. $\log_{10} 100{,}000$
23. $\log_4 8$
24. $\log_{27} 9$
25. $\log_3 3^4$
26. $\log_2 2^5$
27. $\log_{16} 16$
28. $\log_{20} 20$
29. $\log_8 1$
30. $\log_7 1$

Writing Problems

Express the answers in your own words and in complete sentences.

1. Describe what your first step would be in solving $\log_{16} 32 = N$, and explain why you would do that first.
2. Describe what your first step would be in evaluating $\log_{16} 32$, and explain why you would do that first.

Exercises 10.3 Set II

In Exercises 1–18, find the value of the unknown, b, N, or x.

1. $\log_{25} N = 1.5$
2. $\log_b 9 = \frac{2}{3}$
3. $\log_{16} 32 = x$
4. $\log_{16} N = \frac{5}{4}$
5. $\log_4 8 = x$
6. $\log_{625} N = 0.5$
7. $\log_b \frac{1}{2} = -\frac{1}{3}$
8. $\log_b \frac{27}{8} = 3$
9. $\log_b 32 = \frac{5}{3}$
10. $\log_8 4 = x$
11. $\log_{1/2} N = 2$
12. $\log_b 32 = 5$
13. $\log_{25} \frac{1}{5} = x$
14. $\log_{3/4} N = -1$
15. $\log_b \frac{1}{6} = -1$
16. $\log_{27} 3 = x$
17. $\log_4 N = 0$
18. $\log_{25} 5 = x$

In Exercises 19–30, find the value of each logarithm.

19. $\log_4 64$
20. $\log_{13} 13$
21. $\log_4 1$
22. $\log_{64} 8$
23. $\log_9 27$
24. $\log_5 1$
25. $\log_{81} 3$
26. $\log_7 7^8$
27. $\log_{16} 4$
28. $\log_{16} 8$
29. $\log_3 1$
30. $\log_{49} 7$

10.4 The Properties of Logarithms

Since logarithms are exponents, we can derive the properties of logarithms from the properties of exponents.

PROPERTY 5

If $b > 0$, but $b \neq 1$, then

$$\log_b b = 1$$

In words: The logarithm of b to the base b equals 1.

PROOF By definition, $b = b^1$. If we rewrite this exponential statement in logarithmic form, we have $\log_b b = 1$. ■

EXAMPLE 1 Find $\log_6 6$.

SOLUTION By Property 5, $\log_6 6 = 1$.

PROPERTY 6

If $b > 0$, but $b \neq 1$, then

$$\log_b 1 = 0$$

In words: The logarithm of 1 to any base b equals 0.

PROOF By definition, $b^0 = 1$. If we rewrite this exponential statement in logarithmic form, we have $\log_b 1 = 0$. ■

EXAMPLE 2 Find $\log_5 1$.

SOLUTION By Property 6, $\log_5 1 = 0$.

Note In the statements of Properties 7–10, it is understood that $b > 0$, but $b \neq 1$, and that $M > 0$ and $N > 0$.

PROPERTY 7

$$\log_b MN = \log_b M + \log_b N$$

In words: The logarithm of a *product* equals the *sum* of the logarithms of the factors.

PROOF Let $x = \log_b M$ and $y = \log_b N$.

Then $b^x = M$ and $b^y = N$	Writing $x = \log_b M$ and $y = \log_b N$ in exponential form
and $b^x b^y = MN$	Substituting M for b^x and N for b^y in $b^x b^y$
But $b^x b^y = b^{x+y}$	Using the product rule
so $MN = b^{x+y}$	Substituting MN for $b^x b^y$
Also, $\log_b MN = x + y$	Writing $MN = b^{x+y}$ in logarithmic form
or $\log_b MN = \log_b M + \log_b N$	Substituting $\log_b M$ for x and $\log_b N$ for y

■

Note The notation $\log_b MN$ means $\log_b(MN)$, not $(\log_b M)N$.

A Word of Caution A common error is to think that

$$\log_b(M + N) = \log_b M + \log_b N \quad \text{Incorrect}$$

In general, $M + N \neq MN$, so $\log_b(M + N) \neq \log_b MN$. Then, substituting $\log_b M + \log_b N$ for $\log_b MN$, we have

$$\log_b(M + N) \neq \log_b M + \log_b N$$

Warning $\log_b(M + N)$ cannot be rewritten as a sum of terms.

PROPERTY 8

$$\log_b \frac{M}{N} = \log_b M - \log_b N$$

In words: The logarithm of a *quotient* equals the logarithm of the dividend (or numerator) *minus* the logarithm of the divisor (or denominator).

PROOF Let $x = \log_b M$ and $y = \log_b N$.

Then	$b^x = M$ and $b^y = N$	Writing $x = \log_b M$ and $y = \log_b N$ in exponential form
and	$\dfrac{b^x}{b^y} = \dfrac{M}{N}$	Substituting M for b^x and N for b^y in $\dfrac{b^x}{b^y}$
But	$\dfrac{b^x}{b^y} = b^{x-y}$	Using the quotient rule
so	$\dfrac{M}{N} = b^{x-y}$	Substituting $\dfrac{M}{N}$ for $\dfrac{b^x}{b^y}$
Also,	$\log_b \dfrac{M}{N} = x - y$	Writing $\dfrac{M}{N} = b^{x-y}$ in logarithmic form
or	$\log_b \dfrac{M}{N} = \log_b M - \log_b N$	Substituting $\log_b M$ for x and $\log_b N$ for y

■

A Word of Caution $\log_b(M - N) \neq \log_b M - \log_b N$.

Warning $\log_b(M - N)$ *cannot* be rewritten as a difference of two terms.

A Word of Caution $\log_b \dfrac{M}{N} \neq \dfrac{\log_b M}{\log_b N}$. The rule is $\log_b \dfrac{M}{N} = \log_b M - \log_b N$.

PROPERTY 9

$$\log_b N^p = p \log_b N$$

In words: The logarithm of a number to a *power* equals the *product* of the *exponent* and the *logarithm of the number.*

PROOF Let $y = \log_b N$

Then $py = p \log_b N$ Multiplying both sides by p

Furthermore, $b^y = N$ Writing $y = \log_b N$ in exponential form

so $(b^y)^p = N^p$ Using Property 2 (raising both sides of $b^y = N$ to the pth power)

or $b^{py} = N^p$ $(b^y)^p = b^{yp} = b^{py}$

Also, $\log_b N^p = py$ Writing $b^{py} = N^p$ in logarithmic form

or $\log_b N^p = p \log_b N$ Substituting $\log_b N$ for y ■

The properties of logarithms can be summarized as follows:

Properties of logarithms

If $M = N$, then $\log_b M = \log_b N$.	Property 3
If $\log_b M = \log_b N$, then $M = N$.	Property 4
$\log_b b = 1$	Property 5
$\log_b 1 = 0$	Property 6
$\log_b MN = \log_b M + \log_b N$	Property 7
$\log_b \dfrac{M}{N} = \log_b M - \log_b N$	Property 8
$\log_b N^p = p \log_b N$	Property 9

where M and N are positive real numbers, $b > 0$, $b \neq 1$, and p is any real number.

A Word of Caution We must be careful about the use of parentheses, as usual; for example, $\log_b(M + N) \neq \log_b M + N$.

EXAMPLE 3 Examples of transforming logarithmic expressions by using the properties of logarithms:

a. $\log_{10}(56)(107) = \log_{10} 56 + \log_{10} 107$ Using Property 7

b. $\log_{10} \frac{275}{89} = \log_{10} 275 - \log_{10} 89$ Using Property 8

c. $\log_{10}(37)^2 = 2 \log_{10} 37$ Using Property 9

d. $\log_{10} \sqrt{5} = \log_{10} 5^{1/2} = \frac{1}{2} \log_{10} 5$ Using Property 9

e. $\log_{10} \dfrac{(57)(23)}{101} = \log_{10}(57)(23) - \log_{10} 101$ Using Property 8

$= \log_{10} 57 + \log_{10} 23 - \log_{10} 101$ Using Property 7

f. $\log_{10} \dfrac{(47)(19)^3}{(1.04)^7} = \log_{10}(47)(19)^3 - \log_{10}(1.04)^7$ Using Property 8

$= \log_{10} 47 + \log_{10}(19)^3 - \log_{10}(1.04)^7$ Using Property 7

$= \log_{10} 47 + 3 \log_{10} 19 - 7 \log_{10} 1.04$ Using Property 9

g. $\log_{10} \dfrac{\sqrt[5]{21.4}}{(3.5)^4} = \log_{10}(21.4)^{1/5} - \log_{10}(3.5)^4$ Using Property 8

$= \frac{1}{5} \log_{10} 21.4 - 4 \log_{10} 3.5$ Using Property 9

In Example 4, we use the properties of logarithms (including Property 5, which tells us that $\log_{10} 10 = 1$) to find approximations for several logarithms when decimal approximations for just *two* logarithms are given. (The methods for finding that $\log_{10} 2 \approx 0.301$ and that $\log_{10} 3 \approx 0.477$ are shown in the next section and in Appendix B; in *this* section, we just accept these values as given.) To solve the problems in Example 4, the numbers whose logarithms we want to find must be expressed as *products, quotients,* and/or *powers* of 2, 3, and 10.

EXAMPLE 4 Given that $\log_{10} 2 \approx 0.301$ and $\log_{10} 3 \approx 0.477$, approximate the following logarithms.

SOLUTION

a. $\log_{10} 6$ $\quad \log_{10} 6 = \log_{10}(2)(3) = \log_{10} 2 + \log_{10} 3$ $\quad$ Using Property 7

$\approx 0.301 + 0.477 = 0.778$

b. $\log_{10} 1.5$ $\quad \log_{10} 1.5 = \log_{10} \frac{3}{2} = \log_{10} 3 - \log_{10} 2$ $\quad$ Using Property 8

$\approx 0.477 - 0.301 = 0.176$

c. $\log_{10} 8$ $\quad \log_{10} 8 = \log_{10} 2^3 = 3 \log_{10} 2$ $\quad$ Using Property 9

$\approx 3(0.301) = 0.903$

d. $\log_{10} \sqrt{3}$ $\quad \log_{10} \sqrt{3} = \log_{10} 3^{1/2} = \frac{1}{2} \log_{10} 3$ $\quad$ Using Property 9

$\approx \frac{1}{2}(0.477) = 0.2385$

e. $\log_{10} 5$ $\quad \log_{10} 5 = \log_{10} \frac{10}{2} = \log_{10} 10 - \log_{10} 2$ $\quad$ Using Property 8

$\approx 1 - 0.301 = 0.699$

The properties of logarithms can be used to simplify some logarithmic expressions, as in Example 5.

EXAMPLE 5 Examples of writing logarithmic expressions that contain two terms as single logarithms:

a. $\log_b 5x + 2\log_b x = \log_b 5x + \log_b x^2$ $\quad$ Using Property 9

$= \log_b[(5x)(x^2)]$ $\quad$ Using Property 7

$= \log_b 5x^3$ $\quad$ Simplifying

b. $\frac{1}{2}\log_b x - 4\log_b y = \log_b x^{1/2} - \log_b y^4$ $\quad$ Using Property 9

$= \log_b \dfrac{x^{1/2}}{y^4} = \log_b \dfrac{\sqrt{x}}{y^4}$ $\quad$ Using Property 8

Exercises 10.4
Set I

In Exercises 1–10, transform each expression by using the properties of logarithms, as was done in Example 3.

1. $\log_{10}(31)(7)$
2. $\log_{10}(17)(29)$
3. $\log_{10} \frac{41}{13}$
4. $\log_{10} \frac{19}{23}$
5. $\log_{10}(19)^3$
6. $\log_{10}(7)^4$
7. $\log_{10} \sqrt[5]{75}$
8. $\log_{10} \sqrt[4]{38}$
9. $\log_{10} \dfrac{35\sqrt{73}}{(1.06)^8}$
10. $\log_{10} \dfrac{27\sqrt{31}}{(1.03)^{10}}$

In Exercises 11–20, approximate the value of each logarithm, given that $\log_{10} 2 \approx 0.301$, $\log_{10} 3 \approx 0.477$, and $\log_{10} 7 \approx 0.845$. (*Remember*: $\log_{10} 10 = 1$.)

11. $\log_{10} 14$
12. $\log_{10} 21$
13. $\log_{10} \frac{9}{7}$
14. $\log_{10} \frac{7}{4}$
15. $\log_{10} \sqrt{27}$
16. $\log_{10} \sqrt{8}$
17. $\log_{10}(36)^2$
18. $\log_{10}(98)^2$
19. $\log_{10} 6{,}000$
20. $\log_{10} 1{,}400$

In Exercises 21–30, write each expression as a single logarithm and simplify.

21. $\log_b x + \log_b y$
22. $4\log_b x + 2\log_b y$
23. $2\log_b x - 3\log_b y$
24. $\frac{1}{2}\log_b x^4$
25. $3(\log_b x - 2\log_b y)$
26. $\frac{1}{3}\log_b y^3$
27. $\log_b(x^2 - y^2) - 3\log_b(x + y)$
28. $\log_b(x^2 - z^2) - \log_b(x - z)$
29. $2\log_b 2xy - \log_b 3xy^2 + \log_b 3x$
30. $2\log_b 3xy - \log_b 6x^2y^2 + \log_b 2y^2$

Writing Problems

Express the answers in your own words and in complete sentences.

1. Find and describe the error in the following:

$$\log_b(x + y) = \log_b x + \log_b y$$

2. Find and describe the error in the following:

$$\frac{\log_b x}{\log_b y} = \log_b \frac{x}{y}$$

3. Find and describe the error in the following:

$$\log_b(x - y) = \log_b x - \log_b y$$

4. Find and describe the error in the following:

$$\log_b MN = (\log_b M)(\log_b N)$$

Exercises 10.4 Set II

In Exercises 1–10, transform each expression by using the properties of logarithms, as was done in Example 3.

1. $\log_{10}(27)(11)$
2. $\log_{10}(8)(12)(4)$
3. $\log_{10} \frac{5}{14}$
4. $\log_{10} \frac{83}{7}$
5. $\log_{10}(24)^5$
6. $\log_{10}(18)^{-4}$
7. $\log_{10} \frac{(17)(31)}{29}$
8. $\log_{10} \frac{(7)(11)}{13}$
9. $\log_{10} \frac{53}{(11)(19)^2}$
10. $\log_{10} \frac{29}{(31)^3(47)}$

In Exercises 11–20, approximate the value of each logarithm, given that $\log_{10} 2 \approx 0.301$, $\log_{10} 3 \approx 0.477$, and $\log_{10} 7 \approx 0.845$. (*Remember*: $\log_{10} 10 = 1$.)

11. $\log_{10} 40$
12. $\log_{10} 90$
13. $\log_{10} \sqrt[4]{3}$
14. $\log_{10} \sqrt[5]{7}$
15. $\log_{10} 7^3$
16. $\log_{10} 32$
17. $\log_{10} \sqrt{14}$
18. $\log_{10} \frac{3}{7}$
19. $\log_{10} \frac{20}{3}$
20. $\log_{10} 8^5$

In Exercises 21–30, write each expression as a single logarithm and simplify.

21. $\log_b x - \log_b y$
22. $2(3\log_b x - \log_b y)$
23. $\frac{1}{2}\log_b(x - a) - \frac{1}{2}\log_b(x + a)$
24. $3\log_b v + 2\log_b v^2 - \log_b v$
25. $\log_b \frac{6}{7} - \log_b \frac{27}{4} + \log_b \frac{21}{16}$
26. $\log_b x - \log_b y - \log_b z$
27. $5\log_b xy^3 - \log_b x^2y$
28. $2\log_b x - 3\log_b y + \log_b z$
29. $2\log_b x + 2\log_b y$
30. $\log_b(x^3 + y^3) - \log_b(x + y)$

10.5 Using Calculators to Find Logarithms and Antilogarithms

Finding Logarithms with a Calculator

Two systems of logarithms are in such widespread use that there are special tables and special calculator keys just for these two systems:

Common logarithms (base 10)

Natural logarithms (base e, where $e \approx 2.71828$)

When the logarithm of a number x is written as $\log x$, *without* a base being specified, the base of the logarithm is understood to be 10. For example, log 8 is understood to mean $\log_{10} 8$. The special abbreviation $\ln x$ is reserved for *natural* logarithms. For example, ln 6 means $\log_e 6$.

Scientific calculators use symbols that reflect these abbreviations. Keys marked [log] or [LOG] are used for common logarithms, and keys marked [ln] or [LN] are used for natural logarithms. (On some RPN calculators, the same key is used for both. The key itself is labeled [LN], and it is used for natural logarithms; then we see LOG *above* that key, so we press [↰] [LN] (LOG L.R.) to find common logarithms.)

EXAMPLE 1

Using a calculator, find log 5. Round off the answer to four decimal places.

SOLUTION

		Display
Scientific (algebraic)	[5] [log]*	0.698970004
Scientific (RPN)	[5] [ENTER] [↰] [LN] (LOG L.R.)	6.98970004E⁻1
Graphics	[LOG] [5] [ENTER]	0.6989700043

Therefore, $\log 5 \approx 0.6990$.

Nearly all the logarithms found by using a calculator or by using tables are approximations. Nevertheless, we commonly say "Find the logarithm" rather than "Approximate the logarithm."

In Example 2, we show only the keystrokes to use on most scientific calculators.

EXAMPLE 2

Using a calculator, find the logarithm of each number, rounding off the answer to four decimal places.

a. log 2

SOLUTION We press these keys, in this order: [2] [log]. When we round off the answer, we see that $\log 2 \approx 0.3010$.

b. log 20

SOLUTION We press these keys, in this order: [2] [0] [log]. When we round off the answer, we see that $\log 20 \approx 1.3010$.

*On some scientific calculators, we must press [log] *before* the argument and [=] after the argument.

c. log 200

SOLUTION We press these keys, in this order: [2] [0] [0] [log]; log 200 ≈ 2.3010.

d. log 0.002

SOLUTION We press these keys, in this order: [.] [0] [0] [2] [log]; log 0.002 ≈ −2.6990.

e. 3 + log 0.002

SOLUTION 3 + log 0.002 ≈ 3 + (−2.6990) = 0.3010

f. log 0

SOLUTION When we press [0] [log], we get an error message. This is to be expected, since $\log_{10} 0$ does not exist.

g. log(−3)

SOLUTION When we press [3] [+/−] [log], we get an error message, because $\log_{10}(-3)$ is not a real number.

Note Observe that the answers to Examples 2b, 2c, and 2d are consistent with what we know about the properties of logarithms. That is, because 20 = 10 × 2,

$$\log 20 = \log(10 \times 2) = \log 10 + \log 2 = 1 + \log 2 \approx 1 + 0.3010 = 1.3010$$

Similarly,

$$\log 200 = \log(10^2 \times 2) = \log 10^2 + \log 2 = 2 \log 10 + \log 2 = 2(1) + \log 2$$
$$= 2 + \log 2 \approx 2 + 0.3010 = 2.3010$$

and

$$\log 0.002 = \log(10^{-3} \times 2) = \log 10^{-3} + \log 2 = -3 \log 10 + \log 2$$
$$= -3(1) + \log 2 \approx -3 + 0.3010 = -2.6990$$

For further discussion, see Appendix B. 

EXAMPLE 3 Using a calculator, find ln 5. Round off the answer to four decimal places.

SOLUTION

		Display
Scientific (algebraic)	[5] [ln]*	1.609437912
Scientific (RPN)	[5] [ENTER] [LN]	1.60943791243
Graphics	[LN] [5] [ENTER]	1.609437912

Therefore, ln 5 ≈ 1.6094.

In Example 4, we show only the keystrokes to use on most scientific calculators.

*On some scientific calculators, we must press [ln] *before* the argument and [=] after the argument.

EXAMPLE 4 Using a calculator, find the logarithm of each number, rounding off the answer to four decimal places.

a. ln 2

SOLUTION We press these keys, in this order: [2] [ln]. When we round off, we see that $\ln 2 \approx 0.6931$.

b. ln 20

SOLUTION We press these keys, in this order: [2] [0] [ln]; $\ln 20 \approx 2.9957$.

c. ln 0.002

SOLUTION When we press [.] [0] [0] [2] [ln], we see that $\ln 0.002 \approx -6.2146$.

d. ln 0

SOLUTION When we press [0] [ln], we get an error message, because ln 0 does not exist.

e. $\ln(-5)$

SOLUTION When we press [5] [+/−] [ln], we get an error message, because $\ln(-5)$ is not a real number.

We see from Examples 4a, 4b, and 4c that *natural* logarithms do not exhibit the consistent decimal part we saw for *common* logarithms in Examples 2a, 2b, 2c, and 2e.

Significant Digits

When we use a calculator to find a logarithm or to find a number whose logarithm is known, we almost never get an exact answer. For that reason, we will give here a brief explanation of how to find the number of **significant digits** of a number.

To determine the number of significant digits of a number, read from left to right and start counting with the *first nonzero* digit. If the number does not have a decimal point or if it has a decimal point but no digits to the right of the decimal point, *stop counting* when all the remaining digits are zeros. If the number has a decimal point and one or more digits to the right of the decimal point, start counting with the first nonzero digit and continue counting to the end of the number.

EXAMPLE 5 Find the number of significant digits of each number.

a. 6,080,000

SOLUTION

We start counting here

6,080,000

We end here—this is the last nonzero digit

6,080,000 has three significant digits.

b. 300

SOLUTION

We start counting here

300

We end here—this is the last nonzero digit

300 has one significant digit.

c. 300.00

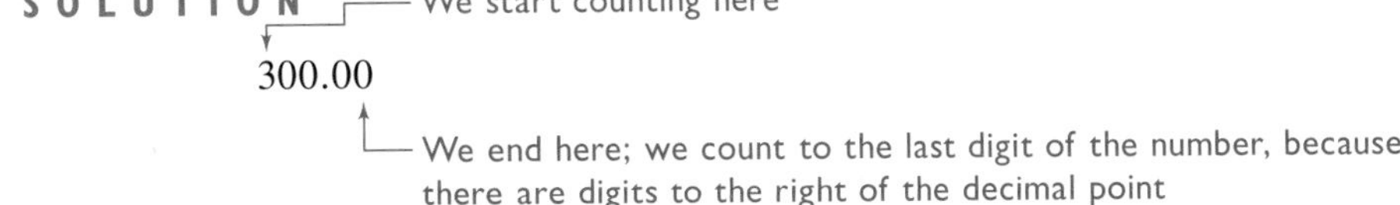

SOLUTION
We start counting here
300.00
We end here; we count to the last digit of the number, because there are digits to the right of the decimal point

300.00 has five significant digits.

d. 0.00030

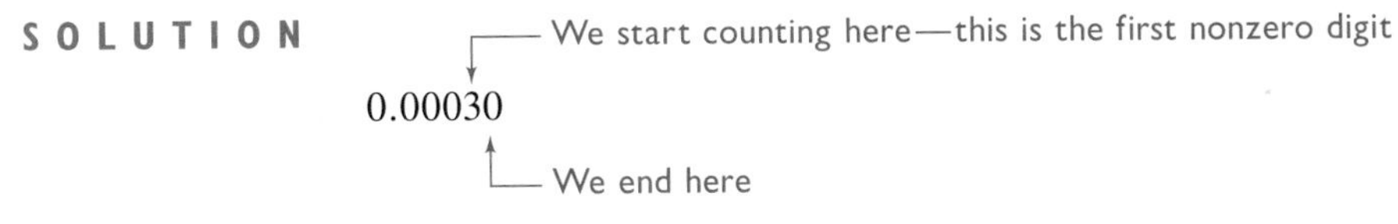

SOLUTION
We start counting here—this is the first nonzero digit
0.00030
We end here

0.00030 has two significant digits.

Another way of thinking of significant digits is as follows: A nonzero digit is always significant. Zeros are significant if they are between nonzero digits *or* if they are to the right of a nonzero digit when the number includes digits to the right of a decimal point.

Antilogarithms

We now consider the inverse operation of finding the logarithm of a number; that is, we now consider finding a *number* when its logarithm is known.

Because logarithmic functions and exponential functions are the inverses of each other, on *most* calculators we use the same key for finding the inverse of a logarithm as for finding the logarithm itself—we simply press the [INV] or [2nd] or [SHIFT] key first. (This is *not* true for some RPN calculators.)

EXAMPLE 6 Find N if $\log N = 0.301$. Round off the answer to three significant digits.

SOLUTION

	Keystrokes	*Display*
Scientific (algebraic)	[.] [3] [0] [1] [INV] [log] (10^x)*	1.99986187
Scientific (RPN)	[.] [3] [0] [1] [ENTER] [↰] [e^x] (10^x PROB)	1.99986186963
Graphics	[2nd] [LOG] (10^x N) [.] [3] [0] [1] [ENTER]	1.99986187

Therefore, $N \approx 2.00$, rounded off to three significant digits.

When we're given the logarithm of a number and are asked to find what the *number* is, we *can* rewrite the logarithmic statement in exponential form and solve the resulting exponential equation. For example, for the problem in Example 6, the equation $\log N = 0.301$ is equivalent to the equation $\log_{10} N = 0.301$, which, in turn, is equivalent to the exponential equation $10^{0.301} = N$. Then to solve *that* equation, we use *exactly* the same keystrokes as shown in Example 6, since those are the keystrokes we would use to find $10^{0.301}$; the result is 2.00.

*On some scientific calculators, we press [2nd] [log] (10^x) .301 [=].

When the base is 10 (that is, for common logarithms), the term **antilogarithm** (abbreviated **antilog**) is often used to indicate the inverse of the logarithm. In symbols:

If	$\log N = L$
then	$N = \text{antilog } L$

If a base is not mentioned, it is understood to be 10. Problems in which we need to find an antilogarithm can be worded in either of two ways:

Find antilog 3.6263.
Find N if $\log N = 3.6263$.
} Both statements mean the same thing

In Example 7, we show only the keystrokes to use on most scientific calculators.

EXAMPLE 7 Find the (common) antilogs, and round off each answer to three significant digits.

a. antilog 3.6263

SOLUTION Press these keys, in this order:

								Display
3	.	6	2	6	3	INV	log (10^x)	4229.606843

When we round off the calculator display to three significant digits, we find that antilog $3.6263 \approx 4{,}230$; that statement is equivalent to the statements $\log 4{,}230 \approx 3.6263$ and $10^{3.6263} \approx 4{,}230$. The *answer* is approximately 4,230.

b. Find N if $\log N = -1.186$.

SOLUTION Press these keys, in this order:

								Display
1	.	1	8	6	+/−	INV	log (10^x)	0.065162839

When we round off the calculator display to three significant digits, we find that $N \approx 0.0652$. This means that $10^{-1.186} \approx 0.0652$ and that antilog$(-1.186) \approx 0.0652$. The *answer* is that $N \approx 0.0652$.

In Example 8, we find a number when its *natural* logarithm is known.

EXAMPLE 8 Find N if $\ln N = 0.489$. Round off the answer to three significant digits.

SOLUTION

								Display
Scientific (algebraic)	.	4	8	9	INV	ln (e^x)*		1.63068472
Scientific (RPN)	.	4	8	9	ENTER	e^x		1.63068471962
Graphics	2nd	LN (e^x S)	.	4	8	9	ENTER	1.63068472

Therefore, $N \approx 1.63$, rounded off to three significant digits.

*On some scientific calculators, we press 2nd ln (e^x) .489 = .

For the problem in Example 8, we could rewrite the equation in exponential form; $\ln N = 0.489$ is equivalent to $\log_e N = 0.489$, which is equivalent to $e^{0.489} = N$. Then, to evaluate $e^{0.489}$, we would use *exactly* the same keystrokes we used in Example 8, and we would still find that $N \approx 1.63$.

Exercises 10.5 Set I

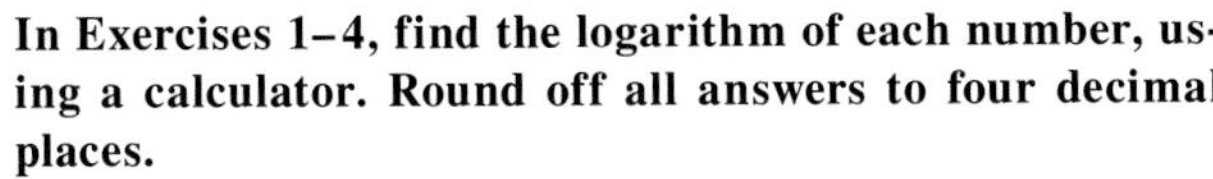

In Exercises 1–4, find the logarithm of each number, using a calculator. Round off all answers to four decimal places.

1. a. log 3 b. log 3,000 c. log 30
   d. ln 3 e. ln 3,000 f. ln 30
2. a. log 10 b. log 100 c. log 100,000
   d. ln 10 e. ln 100 f. ln 100,000
3. a. log 0.1 b. log 0.01 c. log 0.001
   d. ln 0.1 e. ln 0.01 f. ln 0.001
4. a. log 0.3 b. log 0.0003 c. log 0.000 003
   d. ln 0.3 e. ln 0.0003 f. ln 0.000 003

In Exercises 5–8, find the common antilogarithms. Round off the answers to three significant digits.

5. antilog 0.3711
6. antilog 0.6085
7. antilog(−0.0701)
8. antilog(−0.1221)

In Exercises 9–16, find *N*. Round off the answers to three significant digits.

9. $\log N = 3.4082$
10. $\log N = 1.8136$
11. $\log N = -2.5143$
12. $\log N = -3.3883$
13. $\ln N = 2.7183$
14. $\ln N = 5.4366$
15. $\ln N = -2.5849$
16. $\ln N = -5.1534$

Writing Problems

Express the answers in your own words and in complete sentences.

1. If you try to find log 0 with a calculator, you get an error message. Explain why.
2. If you try to find log(−3.631) with a calculator, you get an error message. Explain why.
3. If you try to find antilog(−3.631), you do *not* get an error message; you get the (positive) answer 0.000233884. Explain why this is the correct answer.

Exercises 10.5 Set II

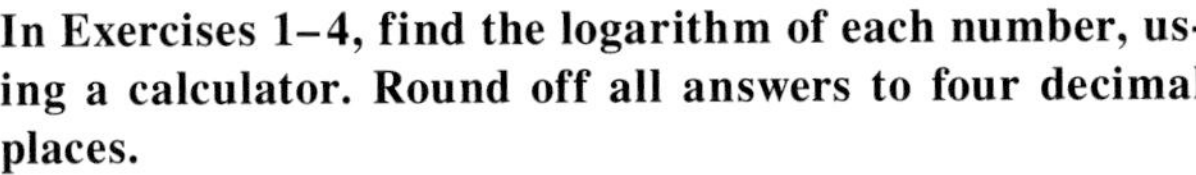

In Exercises 1–4, find the logarithm of each number, using a calculator. Round off all answers to four decimal places.

1. a. log 5 b. log 5,000,000 c. log 500
   d. ln 5 e. ln 5,000,000 f. ln 500
2. a. log 0.5 b. log 0.000 005 c. log 0.005
   d. ln 0.5 e. ln 0.000 005 f. ln 0.005
3. a. log 12 b. log 120,000 c. log 0.00012
   d. ln 12 e. ln 120,000 f. ln 0.00012
4. a. log 0.16 b. log 0.0016 c. log 0.000 016
   d. ln 0.16 e. ln 0.0016 f. ln 0.000 016

In Exercises 5–8, find the common antilogarithms. Round off the answers to three significant digits.

5. antilog 1.6990
6. antilog(−1.6990)
7. antilog 4.7275
8. antilog(−2.2381)

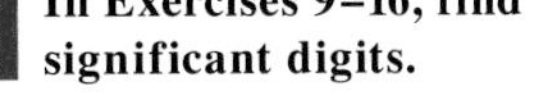

In Exercises 9–16, find *N*. Round off the answers to three significant digits.

9. $\log N = 2.5119$
10. $\log N = 4.5465$
11. $\log N = -2.6271$
12. $\log N = -3.2652$
13. $\ln N = 3.0204$
14. $\ln N = 5.8435$
15. $\ln N = -2.7790$
16. $\ln N = -7.0436$

10.6 Solving Exponential and Logarithmic Equations

In this section, we discuss solving exponential and logarithmic equations. Although the use of logarithms in performing arithmetic calculations has become obsolete because of the widespread use of handheld calculators and computers, the use of exponential and logarithmic functions will continue to be important. Exponential functions and exponential equations are seen often in business and the sciences, and natural logarithms and logarithmic equations are used extensively in higher-level mathematics, especially in calculus. Therefore, it is important that you understand the properties of logarithms and how to solve exponential and logarithmic equations.

Exponential Equations

An **exponential equation** is an equation in which the variable appears in one or more exponents.

EXAMPLE 1 Examples of exponential equations:

a. $3^x = 17$ **b.** $(5.26)^{x+1} = 75.4$

It is possible to solve *some* exponential equations by expressing both sides as powers of the same base (see Example 2).

EXAMPLE 2 Solve $16^x = \frac{1}{4}$.

SOLUTION

$$16^x = \frac{1}{4}$$

$$\left.\begin{aligned}(2^4)^x &= \frac{1}{2^2}\\ 2^{4x} &= 2^{-2}\end{aligned}\right\}\quad \text{Expressing both sides as powers of the same base, 2}$$

$$4x = -2 \quad \text{Using Property 1}$$

$$x = -\frac{2}{4} = -\frac{1}{2}$$

When both sides of an exponential equation *cannot* be expressed as powers of the same base, we solve the equation by using Property 3 and *taking the logarithm of both sides of the equation*. In fact, *most* exponential equations are solved this way. We can take either the *common* logarithm or the *natural* logarithm of both sides (see Example 3). However, if the base is 10, it's easiest to take the *common* logarithm of both sides, and if the base is e, it's easiest to take the *natural* logarithm of both sides (see Example 4).

EXAMPLE 3 Solve $3^x = 17$. Round off the answer to four significant digits.

SOLUTION 1 We first solve the equation by taking the common logarithm of both sides.

$$3^x = 17$$

$$\log 3^x = \log 17 \leftarrow \text{Using Property 3—taking the common logarithm of both sides}$$

$$x \log 3 = \log 17 \quad \text{Using Property 9}$$

$$x = \frac{\log 17}{\log 3} \quad \text{Dividing both sides of the equation by log 3}$$

$$x \approx \frac{1.2304}{0.4771} \approx 2.579$$

A Word of Caution A common error is to think that

$$\frac{\log 17}{\log 3} = \log \frac{17}{3} = \log 17 - \log 3 \quad \text{Incorrect}$$

However, $\dfrac{\log 17}{\log 3} \approx \dfrac{1.2304}{0.4771} \approx 2.579$ whereas $\log \dfrac{17}{3} = \log 17 - \log 3 \approx 1.2304 - 0.4771 = 0.7533$

Unequal

SOLUTION 2 Now we solve the equation by taking the natural logarithm of both sides.

$$3^x = 17$$

$$\ln 3^x = \ln 17 \leftarrow \text{Using Property 3—taking the natural logarithm of both sides}$$

$$x \ln 3 = \ln 17 \quad \text{Using Property 9}$$

$$x = \frac{\ln 17}{\ln 3} \quad \text{Dividing both sides of the equation by ln 3}$$

$$x \approx \frac{2.8332}{1.0986} \approx 2.579$$

Notice that we obtain the same answer when we take the *natural* logarithm of both sides as when we take the *common* logarithm of both sides.

EXAMPLE 4 Solve $e^{4x} = 7$. Round off the answer to four significant digits.

SOLUTION We could take the common logarithm of both sides of the equation, but because the base of the exponential function is e, we will, instead, take the natural logarithm of both sides:

$$e^{4x} = 7$$

$$\ln e^{4x} = \ln 7 \leftarrow \text{Using Property 3—taking the natural logarithm of both sides}$$

$$4x \ln e = \ln 7 \quad \text{Using Property 9}$$

$$4x(1) = \ln 7 \quad \text{Using Property 5 } (\ln e = \log_e e = 1)$$

$$4x = \ln 7$$

$$x = \frac{\ln 7}{4} \approx \frac{1.9459101}{4} \approx 0.4865$$

A completely different method can also be used. We can rewrite the given exponential equation in logarithmic form:

$$e^{4x} = 7 \Leftrightarrow \log_e 7 = 4x$$

$$\ln 7 = 4x \quad \text{Writing } \log_e 7 \text{ as } \ln 7$$

$$x = \frac{\ln 7}{4} \approx 0.4865$$

Logarithmic Equations

Recall that in the expression $\log_b x$, x is called the *argument* of the logarithmic function. A **logarithmic equation** is an equation in which the variable appears in the argument of a logarithm.

EXAMPLE 5

Examples of logarithmic equations:

a. $\ln(2x + 1) + \ln 5 = \ln(x + 6)$ **b.** $\log(x - 1) + \log(x + 2) = 1$

If every term of a logarithmic equation contains a logarithm, as in Example 5a, we can solve the equation by simplifying each side of the equation and then using Property 4: If $\log_b M = \log_b N$, then $M = N$ (see Example 6).

If any term of a logarithmic equation contains *no* logarithms, as in Example 5b, we can solve the equation by using either of two methods: The equation can be simplified to the form $\log_b x = k$, rewritten in exponential form, and then solved (see Example 7, Solution 1), *or* the term that does *not* contain a logarithm can be rewritten as a logarithm (see Example 7, Solution 2, and Example 8).

Apparent solutions of logarithmic equations *must* be checked, because we often obtain extraneous roots when we solve logarithmic equations (see Example 7). Any solution that would result in zero or a negative number as the *argument* of a logarithm must be rejected.

EXAMPLE 6

Solve $\ln(2x + 1) + \ln 5 = \ln(x + 6)$.

SOLUTION

$$\ln(2x + 1) + \ln 5 = \ln(x + 6)$$

$$\ln(2x + 1)(5) = \ln(x + 6) \quad \text{Using Property 7}$$

$$(2x + 1)(5) = x + 6 \quad \text{Using Property 4 (if } \log_b M = \log_b N, \text{ then } M = N)$$

$$10x + 5 = x + 6$$

$$9x = 1$$

$$x = \tfrac{1}{9}$$

√ ***Check***

$$\ln(2x + 1) + \ln 5 = \ln(x + 6)$$

$$\ln[2(\tfrac{1}{9}) + 1] + \ln 5 \stackrel{?}{=} \ln(\tfrac{1}{9} + 6)$$

$$\ln(\tfrac{11}{9}) + \ln 5 \stackrel{?}{=} \ln(\tfrac{55}{9})$$

$$\ln(\tfrac{11}{9})(5) \stackrel{?}{=} \ln(\tfrac{55}{9})$$

$$\ln(\tfrac{55}{9}) = \ln(\tfrac{55}{9}) \quad \text{True}$$

Therefore, $\frac{1}{9}$ is the solution.

EXAMPLE 7

Solve $\log(x - 1) + \log(x + 2) = 1$.

SOLUTION 1

$$\log(x - 1) + \log(x + 2) = 1$$

$$\log_{10}(x - 1)(x + 2) = 1 \quad \leftarrow \text{Property 7}$$

$$(x - 1)(x + 2) = 10^1 \quad \leftarrow \text{Rewriting the equation in exponential form}$$

$$x^2 + x - 2 = 10$$

$$x^2 + x - 12 = 0$$

$$(x - 3)(x + 4) = 0$$

$$x - 3 = 0 \quad \text{or} \quad x + 4 = 0$$

$$x = 3 \quad \text{or} \quad x = -4$$

SOLUTION 2

$$\log(x - 1) + \log(x + 2) = 1 \quad \leftarrow \text{Rewriting 1 as log 10}$$

$$\text{Property 7} \rightarrow \log(x - 1)(x + 2) = \log 10$$

$$(x - 1)(x + 2) = 10 \quad \text{Property 4}$$

(See Solution 1 for the remainder of the solution.)

√ ***Check for*** $x = 3$

$$\log(x - 1) + \log(x + 2) = 1$$
$$\log(3 - 1) + \log(3 + 2) \stackrel{?}{=} 1$$
$$\log 2 + \log 5 \stackrel{?}{=} 1$$
$$\log(2)(5) \stackrel{?}{=} 1$$
$$\log 10 = 1 \quad \text{True}$$

Check for $x = -4$

$$\log(x - 1) + \log(x + 2) = 1$$
$$\log(-4 - 1) + \log(-4 + 2) \stackrel{?}{=} 1$$
$$\log(-5) + \log(-2) \stackrel{?}{=} 1$$

Logarithms of negative numbers are not real numbers

Therefore, 3 is a solution, whereas -4 is *not* a solution; -4 is an extraneous root. The only solution is 3.

EXAMPLE 8 Solve $\log(2x + 3) - \log(x - 1) = 0.73$. Round off the answer to four significant digits.

SOLUTION We can first write 0.73 as a logarithm; that is, we can find antilog 0.73. Antilog $0.73 \approx 5.370$. Therefore, $0.73 \approx \log 5.370$. We then have

$$\log(2x + 3) - \log(x - 1) \approx \log 5.370$$
$$\log \frac{2x + 3}{x - 1} \approx \log 5.370$$
$$\frac{2x + 3}{x - 1} \approx 5.370 \quad \text{Using Property 4}$$
$$2x + 3 \approx 5.370(x - 1)$$
$$2x + 3 \approx 5.370x - 5.370$$
$$8.370 \approx 3.370x$$
$$\frac{8.370}{3.370} \approx x$$
$$x \approx 2.484$$

(We can also solve the equation by writing the equation $\log\left(\frac{2x + 3}{x - 1}\right) = 0.73$ in exponential form—that is, in the form $10^{0.73} = \frac{2x + 3}{x - 1}$—and then solving for x.)

√ ***Check for*** $x \approx 2.484$

$$\log(2x + 3) - \log(x - 1) \approx 0.73$$
$$\log[2(2.484) + 3] - \log[2.484 - 1] \stackrel{?}{\approx} 0.73$$
$$\log 7.968 - \log 1.484 \stackrel{?}{\approx} 0.73$$
$$0.9013 - 0.1714 \stackrel{?}{\approx} 0.73$$
$$0.7299 \approx 0.73 \quad \text{The numbers are close}$$

Therefore, $x \approx 2.484$.

Exercises 10.6 Set I

In Exercises 1–6, solve each equation by expressing both sides as powers of the same base.

1. $27^x = \frac{1}{9}$
2. $8^x = \frac{1}{16}$
3. $4^x = \frac{1}{8}$
4. $125^x = \frac{1}{25}$
5. $25^{2x+3} = 5^{x-1}$
6. $27^{3x-1} = 9^{x+2}$

In Exercises 7–16, solve each equation by taking the logarithm of both sides of the equation. Round off the answers to three significant digits.

7. $2^x = 3$
8. $5^x = 4$
9. $e^x = 8$
10. $e^x = 20$
11. $(7.43)^{x+1} = 9.55$
12. $(5.14)^{x-1} = 7.08$
13. $(8.71)^{2x+1} = 8.57$
14. $(9.55)^{3x-1} = 3.09$
15. $e^{3x+4} = 5$
16. $3^{2x-1} = 23$

In Exercises 17–30, solve and check each logarithmic equation. (Round off Exercises 21 and 22 to three significant digits.)

17. $\log(3x - 1) + \log 4 = \log(9x + 2)$
18. $\log(2x - 1) + \log 3 = \log(4x + 1)$
19. $\ln(x + 4) - \ln 3 = \ln(x - 2)$
20. $\ln(2x + 1) - \ln 5 = \ln(x - 1)$
21. $\log(5x + 2) - \log(x - 1) = 0.7782$
22. $\log(8x + 11) - \log(x + 1) = 0.9542$
23. $\log x + \log(7 - x) = \log 10$
24. $\log x + \log(11 - x) = \log 10$
25. $\ln x + \ln(x - 3) = \ln 4$
26. $\ln x + \ln(x + 2) = \ln 8$
27. $\log(x + 1) + \log(x - 2) = 1$
28. $\log(x + 6) + \log(x - 3) = 1$
29. $\log 10x - \log(x - 450) = 2$
30. $\log(x + 48) - \log(x - 6) = 1$

Writing Problems

Express the answers in your own words and in complete sentences.

1. Describe what your first step would be in solving $e^x = 7$, and explain why you would do that first.
2. Describe what your first step would be in solving $3^x = 5$, and explain why you would do that first.
3. Describe what your first step would be in solving $\log(x - 2) = \log 5 + \log x$, and explain why you would do that first.
4. Find and describe the error in the following:

$$x = \frac{\ln 5}{\ln 3} = \ln 5 - \ln 3$$

Exercises 10.6 Set II

In Exercises 1–6, solve each equation by expressing both sides as powers of the same base.

1. $2^{-3x} = \frac{1}{8}$
2. $9^x = \frac{1}{3^{-2}}$
3. $5^{3x-2} = 25^x$
4. $4^{3x} = 8$
5. $27^{5x} = 9^2$
6. $25^x = \frac{1}{125}$

In Exercises 7–16, solve each equation by taking the logarithm of both sides of the equation. Round off the answers to three significant digits.

7. $3^x = 50$
8. $8^x = 17$
9. $e^x = 29$
10. $e^x = 14$
11. $(4.6)^{x+1} = 100$
12. $(34.7)^{2x} = (12.5)^{3x-2}$
13. $(13.5)^{4x-2} = 7.12$
14. $(2.03)^{2x-1} = 142$
15. $e^{2x-3} = 60$
16. $3^{3x+5} = 25$

In Exercises 17–30, solve and check each logarithmic equation.

17. $\log(x + 4) - \log 10 = \log 6 - \log x$
18. $\log(x - 2) - \log 5 = \log 3 - \log x$
19. $\ln(5x - 7) = \ln(2x - 3) + \ln 3$
20. $\ln x + \ln(x + 4) = \ln 21$
21. $\log(2x + 1) = \log 1 + \log(x + 2)$
22. $2 \log(x + 3) = \log(7x + 1) + \log 2$
23. $\log x = \log(7x + 12) - \log(x + 3)$
24. $\log 2x + \log(x + 2) = \log(12 - x)$
25. $\ln(2x + 3) - \ln(x - 2) = \ln 5$
26. $\ln x + \ln(3x + 8) = \ln 3$
27. $\log(x + 4) + \log(x + 1) = 1$
28. $\log(x + 13) + \log(x - 8) = 2$
29. $\log 25x - \log(x - 60) = 2$
30. $\log(x + 28) - \log(x - 44) = 1$

10.7 Applications That Involve Exponential Equations

Many applied problems involve using exponential and logarithmic functions. The formulas for calculating bacterial growth and radioactive decay contain exponential functions, as do the formulas for calculating compound interest in business. Applying many formulas from the sciences, including the social sciences, earth science, and astronomy, requires the ability to solve exponential and logarithmic equations.

We have selected just a few applications for this section. Listed below are the formulas we will use.

Formulas 1–3 allow us to determine the amount of money in a savings account that earns compound interest. In these formulas, P is the amount originally invested, r is the annual interest rate (expressed in decimal or common fraction form), t is the number of years, and A is the amount in the account at the end of t years.

(1) $A = P(1 + r)^t$ If interest is compounded annually

(2) $A = P\left(1 + \dfrac{r}{k}\right)^{kt}$ If interest is compounded k times per year

(3) $A = Pe^{rt}$ If interest is compounded continuously ($e \approx 2.71828$)

Under certain conditions, a population (including a "population" of bacteria in a culture) increases exponentially. Formula 4 allows us to predict the size of a population at various times. In this formula, C is the initial size of the population (the number of people, animals, bacteria, or whatever), $e \approx 2.71828$, k is some positive constant (which varies from situation to situation), t is the time, and y is the size of the population after time t.

(4) $y = Ce^{kt}$ The formula for exponential growth

Radioactive substances decompose over time; Formula 5 enables us to determine the amount of a radioactive substance present at various times. In this formula, C is the amount of the substance initially present, $e \approx 2.71828$, k is some positive constant (it varies from substance to substance), t is the time, and y is the amount of the substance present after time t.

(5) $y = Ce^{-kt}$ The formula for radioactive decay

In Example 1, we use Formulas 1–3 to determine the amount that would be in a savings account that earns compound interest. You should be aware that calculator answers sometimes differ slightly from each other when very large exponents are involved, as in Example 1c.

EXAMPLE 1 If \$1,000 is invested at $5\frac{1}{4}\%$ annual interest, find the amount that will be in the account after 10 yr if the interest is compounded (a) annually, (b) monthly, (c) daily, and (d) continuously. (Round off answers to the nearest cent.)

SOLUTION For part a only, we show the calculator keystrokes to use for raising a quantity to a power when the base is neither 10 nor e. (See the Note that follows part a.) For all parts of Example 1, we convert $5\frac{1}{4}\%$ to 0.0525, and we let A = the amount in the account at the end of 10 yr.

a. Because the interest is being compounded annually, we use Formula 1. Letting $P = \$1{,}000$, $r = 0.0525$, and $t = 10$, we have

$A = P(1 + r)^t$ Formula 1

$A = \$1{,}000(1 + 0.0525)^{10}$

$A = \$1{,}000(1.0525)^{10}$ See the Note following for calculator keystrokes

$A \approx \$1{,}000(1.668096016)$

$A \approx \$1{,}668.10$ Rounded off to the nearest cent

  Here's how to find $(1.0525)^{10}$ with a calculator. (We do not show the individual keystrokes to use for entering the numbers themselves.)

		Display
Scientific (algebraic)	1.0525 [y^x]* 10 [=]	1.668096016
Scientific (RPN)	1.0525 [ENTER] 10 [y^x]	1.66809601586
Graphics	1.0525 [∧] 10 [ENTER]	1.668096016

b. Because the interest is being compounded 12 times per year, we use Formula 2. Letting $P = \$1{,}000$, $r = 0.0525$, $k = 12$, and $t = 10$, we have

$$A = P\left(1 + \frac{r}{k}\right)^{kt} \quad \text{Formula 2}$$

$$A = \$1{,}000\left(1 + \frac{0.0525}{12}\right)^{(12)(10)}$$

$$A = \$1{,}000(1 + 0.004375)^{120}$$

$$A = \$1{,}000(1.004375)^{120}$$

$$A \approx \$1{,}000(1.688524214)$$

$$A \approx \$1{,}688.52$$

c. Because the interest is being compounded 365 times per year, we use Formula 2. Letting $P = \$1{,}000$, $r = 0.0525$, $k = 365$, and $t = 10$, we have

$$A = P\left(1 + \frac{r}{k}\right)^{kt} \quad \text{Formula 2}$$

$$A = \$1{,}000\left(1 + \frac{0.0525}{365}\right)^{(365)(10)}$$

$$A = \$1{,}000(1 + 0.000143836)^{3650}$$

$$A = \$1{,}000(1.000143836)^{3650}$$

$$A \approx \$1{,}000(1.6903949)$$

$$A \approx \$1{,}690.39^{\dagger}$$

d. Because the interest is being compounded continuously, we use Formula 3. Letting $P = \$1{,}000$, $r = 0.0525$, and $t = 10$, we have

$$A = Pe^{rt} \quad \text{Formula 3}$$

$$A = \$1{,}000e^{0.0525(10)}$$

$$A = \$1{,}000e^{0.525}$$

$$A \approx \$1{,}000(1.6904588) \quad \text{See Section 10.1 for calculator keystrokes for finding powers of e}$$

$$A \approx \$1{,}690.46$$

Therefore, the amount in the account at the end of 10 yr will be \$1,668.10 if the interest is compounded annually, \$1,688.52 if it is compounded monthly, \$1,690.39 if it is compounded daily, and \$1,690.46 if it is compounded continuously.

---

*You may have to press [INV] or [2nd] or [SHIFT] before pressing [y^x].
†On some calculators, the answer rounds off to \$1,690.40.

EXAMPLE 2 If \$1,000 is invested at $5\frac{1}{4}\%$ annual interest, compounded continuously, how long will it take for the money to double? Round off the answer to three significant digits.

SOLUTION Let t = the number of years for the amount of money to double.

Because the interest is being compounded continuously, we use Formula 3. Letting P = \$1,000, r = 0.0525, and A = \$1,000, we must solve for t.

$$A = Pe^{rt} \qquad \text{Formula 3}$$

$$\$2{,}000 = \$1{,}000e^{0.0525t}$$

$$2 = e^{0.0525t} \qquad \text{Dividing both sides by \$1,000}$$

We show two ways to finish the problem:

Method 1

We can take the natural logarithm of both sides of $2 = e^{0.0525t}$:

$$\ln 2 = \ln e^{0.0525t}$$

$$0.693147 \approx 0.0525t(\ln e)$$

$$0.693147 \approx 0.0525t$$

Method 2

We can rewrite the equation $2 = e^{0.0525t}$ in logarithmic form:

$$\log_e 2 = 0.0525t$$

$$\ln 2 = 0.0525t$$

$$0.693147 \approx 0.0525t$$

Then $\qquad \frac{0.693147}{0.0525} \approx t$

or $\qquad t \approx 13.2$

It will take about 13.2 yr for the money to double.

EXAMPLE 3 Suppose that a certain culture of bacteria increases according to the formula $y = Ce^{0.04t}$, where t is in hours. How many hours will it take for the bacteria count to grow from 1,000 to 4,000? Round off the answer to four significant digits.

SOLUTION Let t = the number of hours before the bacteria increases to 4,000.

Using Formula 4, with C = 1,000 and y = 4,000, we must solve for t.

$$y = Ce^{0.04t} \qquad \text{Formula 4}$$

$$4{,}000 = 1{,}000e^{0.04t}$$

$$4 = e^{0.04t} \qquad \text{Dividing both sides by 1,000}$$

Method 1

Taking the natural logarithm of both sides, we have

$$\ln 4 = \ln e^{0.04t}$$

$$1.3862944 \approx 0.04t(\ln e)$$

$$1.3862944 \approx 0.04t$$

Method 2

Rewriting the equation in logarithmic form, we have

$$\log_e 4 = 0.04t$$

$$1.3862944 \approx 0.04t$$

Then $\qquad t \approx \frac{1.3862944}{0.04}$

or $\qquad t \approx 34.66$

It will take about 34.66 hr for the number of bacteria to increase to 4,000.

Exercises 10.7 Set I

In all the exercises, round off each answer to the nearest cent if it is an amount of money. Otherwise, round it off to four significant digits.

1. $1,250 is invested at $5\frac{1}{2}\%$ annual interest. Find the amount that will be in the account after 20 yr if the interest is compounded:

   a. annually b. monthly c. daily d. continuously

2. $2,500 is invested at $5\frac{3}{4}\%$ annual interest. Find the amount that will be in the account after 10 yr if the interest is compounded:

   a. annually b. monthly c. daily d. continuously

3. $1,500 is invested at $5\frac{3}{4}\%$ annual interest. How long will it take for the money to increase to $2,000 if the interest is compounded continuously?

4. $2,500 is invested at $5\frac{1}{2}\%$ annual interest. How long will it take for the money to increase to $4,000 if the interest is compounded continuously?

5. It is known that a certain type of bacteria increases according to the formula $y = Ce^{0.035t}$, where t is in days.

   a. How many bacteria will there be after 3 days if 500 bacteria were present initially?

   b. How many days will it take for the bacteria count to grow from 500 to 800?

6. It is known that a certain type of bacteria increases according to the formula $y = Ce^{0.025t}$, where t is in days.

   a. How many bacteria will there be after 2 days if 900 bacteria were present initially?

   b. How many days will it take for the bacteria count to grow from 900 to 1,600?

7. A certain radioactive substance decomposes according to the formula $y = Ce^{-0.3t}$, where t is in years. How many years will it take for 100 g of that substance to decompose to 80 g?

8. A certain radioactive substance decomposes according to the formula $y = Ce^{-0.4t}$, where t is in years. How many years will it take for 150 g of that substance to decompose to 110 g?

Exercises 10.7 Set II

In all the exercises, round off each answer to the nearest cent if it is an amount of money. Otherwise, round it off to four significant digits.

1. $2,000 is invested at $5\frac{1}{4}\%$ annual interest. Find the amount that will be in the account after 8 yr if the interest is compounded:

   a. annually b. monthly c. daily d. continuously

2. The formula from physics for measuring sound intensity, N, in decibels is $N = 10 \log \dfrac{I}{I_0}$, where I_0 is a constant and I is the power of the sound being measured. Find the number of decibels in a sound whose power is 3×10^{-10}, if $I_0 = 10^{-16}$.

3. $2,000 is invested at $5\frac{1}{2}\%$ annual interest. How long will it take for the money to increase to $2,800 if the interest is compounded continuously?

4. In chemistry, the symbol pH is used for the measure of the acidity or alkalinity of a solution. If (H+) is the hydronium ion concentration measured in moles per liter, then $\text{pH} = -\log(\text{H}+)$. Find the pH of a solution with a hydronium ion concentration of 4.0×10^{-3} moles per liter.

5. It is known that a certain type of bacteria increases according to the formula $y = Ce^{0.04t}$, where t is in hours.

   a. How many bacteria will there be after 4 hr if 800 bacteria were present initially?

   b. How many hours will it take for the bacteria count to grow from 800 to 1,100?

6. The magnitude, M, of an earthquake, as measured on the Richter scale, is calculated as follows: $M = \log \dfrac{a}{a_0}$, where a_0 is a constant and a is the amplitude of the seismic wave. If $a_0 = 10^{-3}$, find

   a. the amplitude of the seismic wave if the magnitude of the earthquake is 6

   b. the amplitude of the seismic wave if the magnitude of the earthquake is 5?

7. A certain radioactive substance decomposes according to the formula $y = Ce^{-0.3t}$, where t is in years. How many years will it take for 120 g of that substance to decompose to 100 g?

8. The formula for finding the monthly payment on a homeowner's mortgage is
$$R = \frac{Ai(1+i)^n}{(1+i)^n - 1}$$
where R is the monthly payment, i is the interest rate *per month* expressed as a decimal, n is the number of months, and A is the original amount of the mortgage. Find the monthly payment on a 25-yr, $40,000 loan at 12% *annual* interest.

10.8 Changing Bases of Logarithms

In some applications, logarithms to a base different from 10 or e are given or obtained. To evaluate such logarithms, we must be able to *change the base* of a logarithm. The formula for changing bases is derived as follows:

Let $x = \log_b N$.

$$b^x = N$$ Writing the equation $x = \log_b N$ in exponential form

$$\log_a b^x = \log_a N$$ Property 3—taking logarithms of both sides *to the base a*

$$x \log_a b = \log_a N$$ Property 9

$$x = \frac{\log_a N}{\log_a b}$$ Dividing both sides of the equation by $\log_a b$

$$\log_b N = \frac{\log_a N}{\log_a b}$$ Substituting $\log_b N$ for x

PROPERTY 10
Changing the base of a logarithm

If $b > 0$ but $b \neq 1$ and if $a > 0$ but $a \neq 1$, then

$$\log_b N = \frac{\log_a N}{\log_a b}$$

In words: To find the logarithm of N to the base b when we have calculator (or table) values to the base a, we divide the logarithm of N to the base a by the logarithm of b to the base a.

Property 10 makes it possible to use a calculator (or a standard table of logarithms) to find the logarithm of a number to any base.

EXAMPLE 1 Find $\log_5 51.7$. (Round off the answer to four significant digits.)

SOLUTION

$$\log_b N = \frac{\log_a N}{\log_a b}$$ Using Property 10

$$\log_5 51.7 = \frac{\log_{10} 51.7}{\log_{10} 5}$$ Letting $b = 5$, $a = 10$, and $N = 51.7$

$$\log_5 51.7 \approx \frac{1.71349}{0.698970} \approx 2.451$$

EXAMPLE 2 Find $\log_e 51.7$. (Use $e \approx 2.71828$, and round off the answer to four significant digits.)

SOLUTION

$$\log_b N = \frac{\log_a N}{\log_a b}$$ Using Property 10

$$\log_e 51.7 \approx \frac{\log_{10} 51.7}{\log_{10} 2.71828} \approx \frac{1.71349}{0.434294} \approx 3.945$$

Note We can now explain why we graphed $y = \log_2 x$ with a TI-81 graphics calculator by using this sequence of keystrokes:

[Y=] [LOG] [X|T] [÷] [LOG] [2] [GRAPH]

Calculators are programmed with only two bases for logarithms—base 10 and base e. Therefore, to graph $y = \log_2 x$, it was necessary to change the base. According to Property 10, $\log_2 x = \dfrac{\log_{10} x}{\log_{10} 2}$; the keystrokes shown above indicate that we divided the logarithm of x to the base 10 by the logarithm of 2 to the base 10.

Exercises 10.8 Set I

Find each logarithm, rounding off the answer to four significant digits.

1. $\log_2 156$
2. $\log_3 231$
3. $\log_{12} 7.54$
4. $\log_{20} 9.75$
5. $\log_e 3.04$
6. $\log_e 4.08$
7. $\log_{6.8} 0.507$
8. $\log_{8.3} 0.0304$

Writing Problems

Express the answer in your own words and in complete sentences.

1. Describe how you would graph $y = \log_5 x$ with a graphics calculator. Explain why you would use that method.

Exercises 10.8 Set II

Find each logarithm, rounding off the answer to four significant digits.

1. $\log_5 29.8$
2. $\log_e 53.7$
3. $\log_{14} 0.842$
4. $\log_{5.2} 0.926$
5. $\log_e 16.1$
6. $\log_e 0.076$
7. $\log_8 12$
8. $\log_3 0.333$

Sections 10.1–10.8 REVIEW

Exponential Functions 10.1

An **exponential function** is a function of the form $y = f(x) = b^x$, where $b > 0$, $b \neq 1$, and $x =$ any real number.

Properties of exponents:

For $b > 0$ but $b \neq 1$, if $b^x = b^y$, then $x = y$. Property 1

For $a > 0$ and $b > 0$, if $a^x = b^x$, then $a = b$. Property 2

Logarithmic Functions 10.1

A **logarithmic function** is a function of the form $y = f(x) = \log_b x$, where $b > 0$, $b \neq 1$, and $x > 0$.

The logarithm of a (positive) number x is the exponent y to which the base b ($b > 0$, $b \neq 1$) must be raised to give x. The logarithm of a number can be found either by using a calculator or by using tables (Appendix B).

Exponential and Logarithmic Forms 10.2	*Logarithmic form*	*Exponential form*
	$y = \log_b x$	$x = b^y$

Properties of Logarithms 10.4

If $M = N$, then $\log_b M = \log_b N$.	Property 3
If $\log_b M = \log_b N$, then $M = N$.	Property 4
$\log_b b = 1$	Property 5
$\log_b 1 = 0$	Property 6
$\log_b MN = \log_b M + \log_b N$	Property 7
$\log_b \dfrac{M}{N} = \log_b M - \log_b N$	Property 8
$\log_b N^p = p \log_b N$	Property 9
10.8 $\log_b N = \dfrac{\log_a N}{\log_a b}$	Property 10 (change of base)

where M and N are positive real numbers, $b > 0$, $b \neq 1$, p is any real number, and $a > 0$, $a \neq 1$.

Logarithms and the Calculator 10.5

Use the key marked [log] or [LOG] to find common logarithms and the key marked [ln] or [LN] to find natural logarithms.

Exponential Equations 10.6

An **exponential equation** is an equation in which the variable appears in one or more exponents.

To solve an exponential equation:
When both sides of the equation can be expressed as powers of the same base:

1. Express both sides as powers of the same base.
2. Use Property 1 (if $b^x = b^y$, then $x = y$).
3. Solve the resulting equation.

When both sides of the equation cannot be expressed as powers of the same base:

1. Take the logarithm (to the same base) of both sides.
2. Use Property 9 to rewrite the equation with no exponents.
3. Solve the resulting equation.

Logarithmic Equations 10.3, 10.6

A **logarithmic equation** is an equation in which the variable appears in the argument of a logarithm.

Some logarithmic equations can be solved by writing the equation in exponential form and solving the resulting equation.

To solve a logarithmic equation when every term contains a logarithm:

1. Use the properties of logarithms to write each side as a single logarithm.
2. Use Property 4 (if $\log_b M = \log_b N$, then $M = N$).
3. Solve the resulting equation.
4. Check apparent solutions in the given logarithmic equation, rejecting any extraneous roots.

To solve a logarithmic equation when one term doesn't contain a logarithm:

Method 1. The equation can be simplified to the form $\log_b x = k$, rewritten in exponential form, and then solved.

Method 2. The term that doesn't contain a logarithm can be rewritten as a logarithm. Then we can follow the procedure for solving a logarithmic equation when every term contains a logarithm.

Sections 10.1–10.8 REVIEW EXERCISES Set I

1. Write $3^4 = 81$ in logarithmic form.

2. Write $\log_4 0.0625 = -2$ in exponential form.

In Exercises 3–7, find the value of the unknown, *b*, *N*, or *x*.

3. $\log_{10} 1{,}000 = x$

4. $\log_{10} 0.01 = x$

5. $\log_9 N = \frac{3}{2}$

6. $\log_b \frac{1}{8} = -3$

7. $\log_{10} 145.6 = x$ (Round off the answer to four decimal places.)

In Exercises 8–10, write each expression as a single logarithm and simplify.

8. $\frac{1}{5} \log x^5 + 3 \log x^4$

9. $\log \frac{3}{5} + \log \frac{5}{3}$

10. $\log x^4y^4 - \log 6x + \log 3 - 4 \log xy$

In Exercises 11–14, find each logarithm, rounding off the answer to four decimal places.

11. $\log 25.48$

12. $\log 0.000\,800\,5$

13. $\ln 0.0342$

14. $\ln 5{,}300$

In Exercises 15 and 16, find each antilogarithm, rounding off the answer to four significant digits.

15. antilog 3.4072

16. Find N if $\ln N = 2.4849$.

In Exercises 17 and 18, solve each equation.

17. $81^{x-1} = \frac{1}{9}$

18. $\log 2 + \log(3x - 1) = \log(4x + 1)$

19. Find $\log_4 75$. Round off the answer to three decimal places.

20. Graph $y = 6^x$ and its inverse logarithmic function on the same set of axes.

21. If \$2,000 is invested at 5.8% interest, how long will it take for the money to increase to \$2,500 if the interest is compounded continuously? Round off the answer to two significant digits.

Sections 10.1–10.8 REVIEW EXERCISES Set II

Name ____________________

ANSWERS

1. ______
2. ______
3. ______
4. ______
5. ______
6. ______
7. ______
8. ______
9. ______
10. ______

1. Write $4^2 = 16$ in logarithmic form.

2. Write $\log_{1.3} 1.69 = 2$ in exponential form.

In Exercises 3–7, find the value of the unknown, *b*, *N*, or *x*.

3. $\log_{10} 0.001 = x$

4. $\log_{27} N = \frac{2}{3}$

5. $\log_b \frac{1}{16} = -4$

6. $\log_7 N = 0$

7. $\log_b \frac{1}{16} = -2$

In Exercises 8–10, write each expression as a single logarithm and simplify.

8. $\frac{1}{4} \log x^4 + 2 \log x^2$

9. $\log \frac{14}{3} - \log \frac{7}{3}$

10. $\log(x^2 - x - 12) - \log(x - 4)$

In Exercises 11–14, find each logarithm correct to four decimal places.

11. log 28.25

12. log 0.000 368 4

13. ln 0.00235

14. ln 12

In Exercises 15 and 16, find each antilogarithm correct to two decimal places.

15. antilog 2.6551

16. Find N if $\ln N = 2.70805$.

In Exercises 17 and 18, solve each equation.

17. $(4.55)^{x+1} = 8.45$
(Round off to three decimal places.)

18. $\log(x + 3) + \log(x - 2) = \log 6$

19. Find $\log_6 148$ correct to three decimal places.

20. If $1,600 is invested at 5.7% interest, how long will it take for the money to increase to $2,000 if the interest is compounded continuously? Round off the answer to two significant digits.

21. Graph $y = 3^x$ and its inverse logarithmic function on the same set of axes.

11. ______

12. ______

13. ______

14. ______

15. ______

16. ______

17. ______

18. ______

19. ______

20. ______

21.

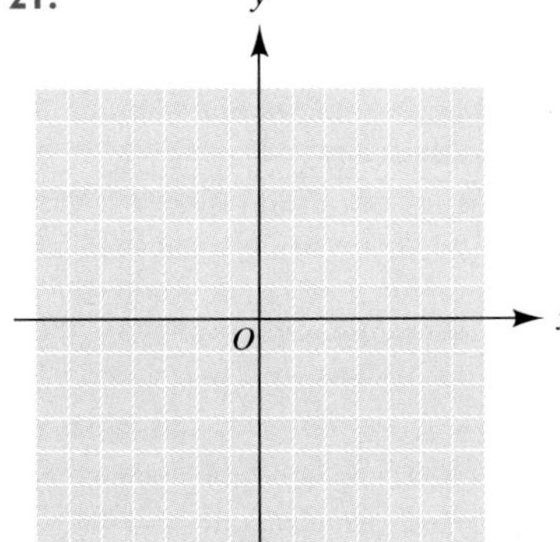

Chapter 10 DIAGNOSTIC TEST

The purpose of this test is to see how well you understand exponential and logarithmic functions and equations. We recommend that you work this diagnostic test *before* your instructor tests you on this chapter. Allow yourself about 50 minutes.

Complete solutions for all the problems on this test, together with section references, are given in the answer section at the end of the book. For the problems you do incorrectly, we suggest that you study the sections cited.

1. a. Write $2^4 = 16$ in logarithmic form.

   b. Write $\log_{2.5} 6.25 = 2$ in exponential form.

2. Find the value of the unknown, b, N, or x.

   a. $\log_4 N = 3$

   b. $\log_{10} 10^{-2} = x$

   c. $\log_b 6 = 1$

   d. $\log_5 1 = x$

   e. $\log_{0.5} N = -2$

In Problems 3 and 4, write each expression as a single logarithm, and simplify the result.

3. a. $\log x + \log y - \log z$

   b. $\frac{1}{2} \log x^4 + 2 \log x$

4. $\log(x^2 - 9) - \log(x - 3)$

5. Solve $\log(3x + 5) - \log 7 = \log(x - 1)$.

6. Solve $\log(x + 8) + \log(x - 2) = \log 11$.

In Problems 7 and 8, express each answer correct to three significant digits.

7. Solve $e^{3x-4} = 8$.

8. Find $\log_2 718$.

9. Graph $y = 7^x$ and its inverse function, $y = \log_7 x$, on the same set of axes.

10. Suppose a certain culture of bacteria increases according to the formula $y = Ce^{0.05t}$, where t is in hours. If there were initially 1,500 bacteria, find

    a. the number of bacteria present after 2 hr (round off the answer to four significant digits).

    b. the number of hours it will take for the bacteria count to triple (round off the answer to two significant digits).

Chapters 1–10 CUMULATIVE REVIEW EXERCISES

In Exercises 1 and 2, perform the indicated operations and simplify.

1. $\dfrac{7}{x^2 - x - 12} - \dfrac{3}{x^2 - 5x + 4}$

2. $\dfrac{\dfrac{2}{x} - \dfrac{2}{y}}{\dfrac{y^2 - x^2}{xy}}$

In Exercises 3 and 4, graph the solution set of each equation.

3. $2x - 7y = 7$

4. $y = x^2 - 2x - 3$

In Exercises 5–8, find the solution set of each equation or inequality.

5. $\dfrac{2x}{3} - 1 \le \dfrac{x + 2}{5}$

6. $\sqrt{1 - 4x} = x + 5$

7. $\log(5x + 2) + \log 3 = \log(12x + 15)$

8. $e^{3x} = 5$ (Round off the answer to three significant digits.)

9. List any prime numbers greater than 7 and less than 31 that yield a remainder of 1 when divided by 5.

10. Find the length of the diagonal of a rectangle that is 2 m long and 1 m wide.

11. Find the distance between $(3, -4)$ and $(1, 2)$.

12. Write an equation of the line that has an x-intercept of -14 and a y-intercept of -6. Express the answer in the general form.

13. Write an equation of the line that passes through $(9, -13)$ and is parallel to the line $10x - 6y + 15 = 0$. Express the answer in the general form.

14. What is the multiplicative identity element?

15. What is the multiplicative inverse of -3?

16. Write $2^7 = 128$ in logarithmic form.

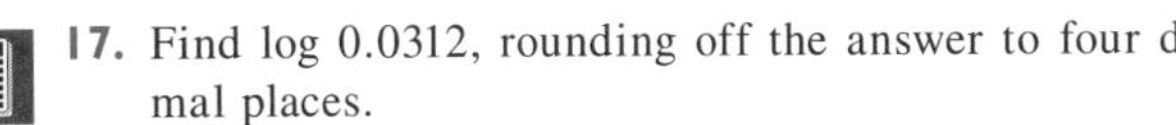

17. Find $\log 0.0312$, rounding off the answer to four decimal places.

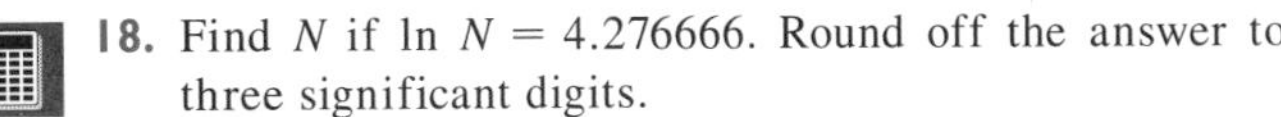

18. Find N if $\ln N = 4.276666$. Round off the answer to three significant digits.

In Exercises 19–21, set up each problem algebraically and solve.

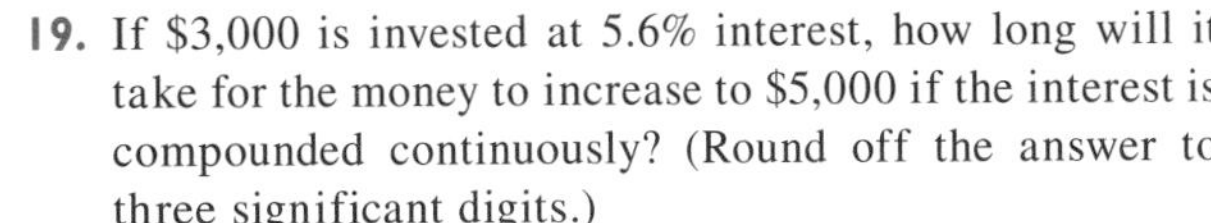

19. If \$3,000 is invested at 5.6% interest, how long will it take for the money to increase to \$5,000 if the interest is compounded continuously? (Round off the answer to three significant digits.)

20. How many liters of water must be added to 5 L of a 60% solution of alcohol to make a 40% solution?

21. It takes Claudio 2 hr longer to do a certain job than it does Maria. After Maria has worked on the job for 2 hr, Claudio joins her, and together they finish the job in $3\frac{1}{3}$ hr. How long would it take each of them, working alone, to do the entire job?

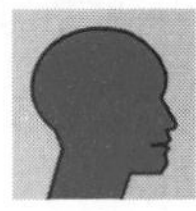

Critical Thinking and Problem-Solving Exercises

1. Brett has decided to start exercising. He plans to walk 5 blocks the first day of his exercise program, 9 blocks the second day, 13 blocks the third day, 17 blocks the fourth day, and so on. If he continues in this manner, how many blocks will he walk on the fifth day? How many blocks on the sixth day? How long will it take him to reach his goal of walking 45 blocks in one day?

2. At 7 A.M., Travis the Frog started trying to climb out of a well that was $18\frac{1}{2}$ ft deep. He managed to climb up 3 ft every hour, but unfortunately, he always immediately slid back 1 ft. Travis finally got out of the well between __ and __ in the $\underline{\hspace{4cm}}$ (morning or afternoon).

3. Todd asked his son, Nathan, how many baseball cards he had. Nathan, being in a playful mood, replied, "The number of cards is more than 225 but fewer than 315, and the number of cards is divisible by 5 and also divisible by 9." How many cards does Nathan have?

4. Suppose that we want to find the area of a circle that has a radius of 6.25 in., and suppose that we use a calculator and the formula $A = \pi r^2$. The calculator display is 122.718463. Would it be reasonable to say that we are sure that the area is 122.718463 sq. in.? Why or why not?

5. In Example 2 of Section 10.5, we found that $\log 2 \approx 0.3010$, $\log 20 \approx 1.3010$, $\log 200 \approx 2.3010$, and $3 + \log 0.002 \approx 0.3010$. Why should the fractional parts of these answers all have been the same?

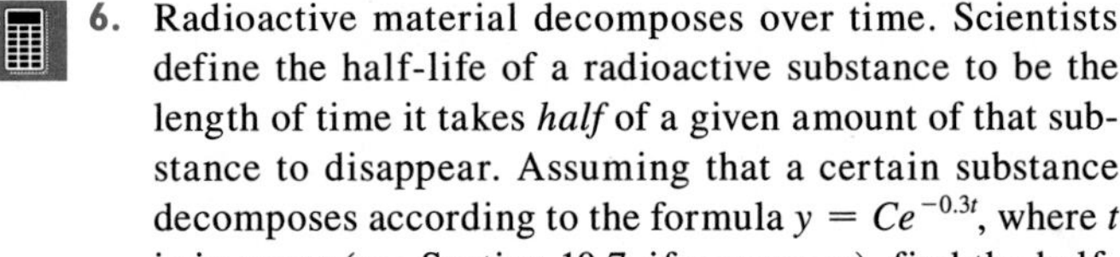

6. Radioactive material decomposes over time. Scientists define the half-life of a radioactive substance to be the length of time it takes *half* of a given amount of that substance to disappear. Assuming that a certain substance decomposes according to the formula $y = Ce^{-0.3t}$, where t is in years (see Section 10.7, if necessary), find the half-life of the substance. (Round off the answer to one decimal place.)

Nonlinear Equations and Inequalities

CHAPTER

11

In this chapter, we discuss several methods of solving nonlinear equations and inequalities, with emphasis on solving quadratic equations and inequalities in one and two variables. (We have already solved quadratic equations in one variable by *factoring*.) We also discuss graphing quadratic functions (parabolas) in more detail, and we graph conic sections.

11.1 Quadratic Equations in One Variable; Solving Incomplete Quadratic Equations

Quadratic Equations in One Variable

As you know, a **quadratic equation in one variable** is a polynomial equation (in one variable) in which the highest-degree term is of the second degree. It is proved in higher mathematics courses that the number of solutions (or *zeros*, or *roots*) of a polynomial equation equals the degree of the equation, although the solutions are not necessarily *real* numbers (the solutions of a quadratic equation could be $3i$ and $-3i$, for example) and they are not necessarily distinct (the two solutions of a quadratic equation might both be 4, for example).

EXAMPLE 1 Examples of quadratic equations in one variable:

a. $3x^2 + 7x + 2 = 0$ **b.** $\frac{1}{2}x^2 = \frac{2}{3}x - 4$

c. $x^2 + 4 = 0$ **d.** $5x^2 - 15x = 0$

The General Form of a Quadratic Equation The **general form of a quadratic equation**, *as used in this text*, is as follows:

The general form of a quadratic equation

$$ax^2 + bx + c = 0 \quad \text{or} \quad 0 = ax^2 + bx + c$$

where a, b, and c are integers and $a > 0$.

Earlier, we converted quadratic equations into the general form just before we factored one side of the equation; we use the same procedure now. That is, we remove any grouping symbols, move *all* terms to one side of the equation (getting zero on the other side), combine any like terms, and, finally, arrange the terms in descending powers.

In this section, we will also then identify a, b, and c: In the equation $ax^2 + bx + c = 0$, a is always the coefficient of x^2, b is always the coefficient of x, and c is always the constant.

EXAMPLE 2 Convert each of the quadratic equations into the general form, and then identify a, b, and c.

a. $7x = 5 - 2x^2$

SOLUTION $2x^2 + 7x - 5 = 0$ The general form

$a = 2, b = 7, c = -5$

b. $5x^2 = 3$

SOLUTION $5x^2 + 0x - 3 = 0$ The general form

$a = 5, b = 0, c = -3$

c. $6x = 11x^2$

SOLUTION $0 = 11x^2 - 6x + 0$ The general form

$a = 11, b = -6, c = 0$

d. $\frac{2}{3}x^2 - 5x = \frac{1}{2}$

SOLUTION The LCD = 6.

$6\left(\frac{2}{3}x^2 - 5x\right) = 6\left(\frac{1}{2}\right)$ Multiplying both sides by 6

$6\left(\frac{2}{3}x^2\right) + 6(-5x) = 6\left(\frac{1}{2}\right)$ Using the distributive property on the left

$4x^2 - 30x = 3$

$4x^2 - 30x - 3 = 0$ The general form

$a = 4, b = -30, c = -3$

e. $(x + 2)(2x - 3) = 3x - 7$

SOLUTION $2x^2 + x - 6 = 3x - 7$

$2x^2 - 2x + 1 = 0$ The general form

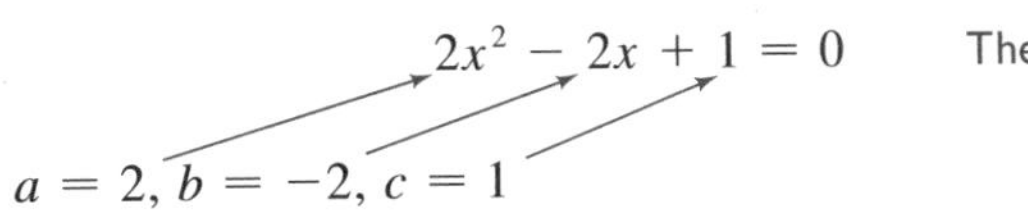

$a = 2, b = -2, c = 1$

Solving Incomplete Quadratic Equations

An incomplete quadratic equation is a quadratic equation in which $b = 0$ or $c = 0$ (or $b = 0$ *and* $c = 0$).

EXAMPLE 3 Examples of incomplete quadratic equations:

a. $12x^2 + 5 = 0$ $(b = 0)$ **b.** $7x^2 - 2x = 0$ $(c = 0)$
c. $3x^2 = 0$ $(b = 0 \text{ and } c = 0)$

An incomplete quadratic equation in which $c = 0$ is solved by factoring (see Example 4).

EXAMPLE 4 Find the solution set of $12x^2 = 3x$.

SOLUTION

$12x^2 - 3x = 0$ The general form

$3x(4x - 1) = 0$ The GCF = $3x$

$3x = 0$ | $4x - 1 = 0$ Using the zero-factor property (setting each factor equal to 0)

$x = 0$ | $4x = 1$

$x = \frac{1}{4}$

Checking will verify that the solution set is $\left\{0, \frac{1}{4}\right\}$.

A Word of Caution In Example 4, a common error is to divide both sides of the equation by x.

$$12x^2 = 3x$$

$$12x = 3 \quad \text{Dividing both sides by } x$$

$$x = \tfrac{1}{4}$$

Using this method, we find only the solution $\frac{1}{4}$. The equations $12x^2 = 3x$ and $12x = 3$ are *not* equivalent equations. By dividing both sides of the equation by x, we lose the solution 0.

We must not divide both sides of an equation by an expression containing the variable because we may lose solutions.

We have already solved incomplete quadratic equations in which $b = 0$ *and* $p > 0$. To refresh your memory, we repeat here the property we sometimes used for solving such equations.

PROPERTY 1

If $x^2 = p$, then $x = \sqrt{p}$ or $x = -\sqrt{p}$

It happens to be true that this property holds even when p is negative; therefore, we can solve an equation that can be expressed in the form $ax^2 + c = 0$ as follows:

Solving the quadratic equation $ax^2 + c = 0$

1. Add $-c$ to both sides of the equation.
2. Divide both sides of the equation by a, the coefficient of x^2.
3. Use Property 1: If $x^2 = p$, then $x = \pm\sqrt{p}$. (When we use this property, we often say that we are "taking the square root of both sides of the equation.")
4. Express the square roots in simplest form.

When the radicand is *positive* or *zero*, the roots are real numbers.
When the radicand is *negative*, the roots are complex numbers.

Note When the solutions are irrational, they can be expressed in simplest radical form or we can find decimal approximations for them or both.

EXAMPLE 5 Find the solution set of $3x^2 - 5 = 0$.

SOLUTION

$$3x^2 = 5 \quad \text{Adding 5 to both sides}$$

$$x^2 = \tfrac{5}{3} \quad \text{Dividing both sides by 3}$$

$$x = \pm\sqrt{\tfrac{5}{3}} \quad \text{Using Property 1}$$

$$x = \pm\sqrt{\frac{5 \cdot 3}{3 \cdot 3}} = \pm\frac{\sqrt{15}}{3} = \pm\frac{1}{3}\sqrt{15} \quad \text{Simplifying } \sqrt{\tfrac{5}{3}}$$

Checking will confirm that the solution set is $\{\frac{1}{3}\sqrt{15}, -\frac{1}{3}\sqrt{15}\}$.

EXAMPLE 6 Find the solution set of $x^2 + 25 = 0$.

SOLUTION

$$x^2 + 25 = 0$$
$$x^2 = -25$$
$$x = \pm\sqrt{-25} \quad \text{Using Property 1}$$
$$x = \pm 5i \quad \text{Simplifying } \sqrt{-25}$$

✓ **Check**

$$x^2 + 25 = 0$$
$$(\pm 5i)^2 + 25 \stackrel{?}{=} 0$$
$$25i^2 + 25 \stackrel{?}{=} 0$$
$$-25 + 25 = 0$$

The solution set is $\{5i, -5i\}$.

Earlier, we solved equations with rational exponents by raising both sides of the equation to the same power. However, the problems we did then were carefully selected to ensure that the power was never a fraction with a denominator that was an even number. Raising both sides of an equation to a fractional power in which the *denominator* of the exponent is an *even* number is equivalent to taking an *even root* of both sides of the equation. When we do take an even root of both sides of an equation, there should be both a positive and a negative solution, so we must put a $\pm$ sign in front of the constant. (This is equivalent to using Property 1.) Therefore, in Example 7, we must insert a $\pm$ sign in front of the constant when we raise both sides to the $-\frac{3}{2}$ power.

EXAMPLE 7 Find the solution set of $x^{-2/3} = 4$.

SOLUTION

$$x^{-2/3} = 2^2$$
$$(x^{-2/3})^{-3/2} = \pm(2^2)^{-3/2}$$

We raise both sides to the $-\frac{3}{2}$ power in order to make the exponent of x equal to 1: $\left(-\frac{2}{3}\right)\left(-\frac{3}{2}\right) = 1$

$$x^{(-2/3)(-3/2)} = \pm 2^{2(-3/2)}$$
$$x = \pm 2^{-3} = \pm\frac{1}{8}$$

✓ ***Check for*** $x = \frac{1}{8}$

$$x^{-2/3} = 4$$
$$\left(\tfrac{1}{8}\right)^{-2/3} \stackrel{?}{=} 4$$
$$(8)^{2/3} \stackrel{?}{=} 4$$
$$(\sqrt[3]{8})^2 \stackrel{?}{=} 4$$
$$2^2 \stackrel{?}{=} 4$$
$$4 = 4$$

Check for $x = -\frac{1}{8}$

$$x^{-2/3} = 4$$
$$\left(-\tfrac{1}{8}\right)^{-2/3} \stackrel{?}{=} 4$$
$$(-8)^{2/3} \stackrel{?}{=} 4$$
$$\sqrt[3]{(-8)^2} \stackrel{?}{=} 4$$
$$\sqrt[3]{64} \stackrel{?}{=} 4$$
$$4 = 4$$

The solution set is $\left\{-\frac{1}{8}, \frac{1}{8}\right\}$.

Later in this chapter, we will need to solve equations of the form $(x - h)^2 = p$, and we can also use Property 1 in solving such equations. When p is not the square of an integer, we can express the solutions in simplest radical form or we can find decimal approximations for them or both (see Example 8).

EXAMPLE 8 Examples of solving equations of the form $(x - h)^2 = p$:

a. $(x - 4)^2 = 13$

$x - 4 = \pm\sqrt{13}$ Using Property I

$x = 4 \pm \sqrt{13}$ Adding 4 to both sides

$x \approx 4 \pm 3.606$ $\sqrt{13} \approx 3.606$; $4 + 3.606 = 7.606$ and $4 - 3.606 = 0.394$

The solution set is $\{4 + \sqrt{13}, 4 - \sqrt{13}\}$, or approximately $\{7.606, 0.394\}$.

b. $(x + 3)^2 = -7$

$x + 3 = \pm\sqrt{-7}$ Using Property I

$x = -3 \pm i\sqrt{7}$ Adding -3 to both sides and simplifying $\sqrt{-7}$

$x \approx -3 \pm 2.646i$ $\sqrt{7} \approx 2.646$

The solution set is $\{-3 + i\sqrt{7}, -3 - i\sqrt{7}\}$, or approximately $\{-3 \pm 2.646i\}$.

Exercises 11.1
Set I

In Exercises 1–10, write each quadratic equation in the general form; then identify *a*, *b*, and *c*.

1. $3x^2 + 5x = 2$
2. $3x + 5 = 2x^2$
3. $3x^2 = 4$
4. $16 = x^2$
5. $\frac{4x}{3} = 4 + x^2$
6. $\frac{3x}{2} - 5 = x^2$
7. $x(x + 2) = 4$
8. $4(x - 5) = x^2$
9. $3x(x - 2) = (x + 1)(x - 5)$
10. $7x(2x + 3) = (x - 3)(x + 4)$

In Exercises 11–30, find the solution set of each equation. Express any irrational solutions in simplest radical form.

11. $x^2 - 27 = 0$
12. $x^2 - 8 = 0$
13. $x^2 + 16 = 0$
14. $x^2 + 81 = 0$
15. $x^2 = 7$
16. $x^2 = 13$
17. $x^2 = -12$
18. $x^2 = -75$
19. $12x = 8x^3$
20. $9x = 12x^3$
21. $5x^2 + 4 = 0$
22. $3x^2 + 25 = 0$
23. $\frac{2x^2}{3} = 4x$
24. $\frac{3x}{5} = 6x^2$
25. $x(x - 2) = (2x + 3)x$
26. $x(x - 3) = x(2x - 8)$
27. $\frac{x + 2}{3x} = \frac{x + 1}{x}$
28. $\frac{3x - 2}{4x} = \frac{x + 1}{3x}$
29. $3x^{-2/3} = 48$
30. $x^{2/5} = 4$

In Exercises 31–38, follow the method shown in Example 8 to solve the equations. Express the solutions in simplest radical form and also give decimal approximations to those solutions, rounding off the approximations to two decimal places.

31. $(x - 3)^2 = 35$
32. $(x - 7)^2 = 3$
33. $(x + 2)^2 = 5$
34. $(x + 11)^2 = 13$
35. $(x + 6)^2 = 18$
36. $(x + 1)^2 = 50$
37. $(x - 5)^2 = -15$
38. $(x - 12)^2 = -17$

In Exercises 39–42, set up each problem algebraically and solve. In Exercises 40–42, express each answer in simplest radical form.

39. The length of the diagonal of a square is $\sqrt{32}$ cm. What is the length of a side of the square?
40. The length of the diagonal of a square is 18 in. What is the length of a side of the square?
41. A rectangle is 7 cm wide and 10 cm long. Find the length of its diagonal.
42. A rectangle is 12 cm long and 8 cm wide. Find the length of its diagonal.

Writing Problems

Express the answers in your own words and in complete sentences.

1. Describe what your first two steps would be in solving the equation $7x^2 = 5x$, and explain why you would do those things first.
2. Describe what your first two steps would be in solving the equation $6x^2 - 5 = 0$, and explain why you would do those things first.
3. Describe what your first step would be in solving the equation $x^{4/3} = 81$, and explain why you would do that first.

Exercises 11.1
Set II

In Exercises 1–10, write each quadratic equation in the general form; then identify *a*, *b*, and *c*.

1. $2x - 5 = 6x^2$
2. $6 - 3x = 5x^2$
3. $8 = 2x^2$
4. $x^2 - 1 = 3x$
5. $\frac{x^2}{2} - 3 = 7x$
6. $\frac{3 - x^2}{4} = x$
7. $2x(3 - x) = 5$
8. $x(x - 4) = 3$
9. $5x(x + 3) = (4x - 5)(6x + 3)$
10. $(2x - 1)(3x + 5) = 2x - 1$

In Exercises 11–30, find the solution set of each equation. Express any irrational solutions in simplest radical form.

11. $x^2 - 50 = 0$
12. $x^2 - 48 = 0$
13. $x^2 + 144 = 0$
14. $x^2 = -5$
15. $x^2 = 3$
16. $x^2 - 12 = 0$
17. $x^2 = -20$
18. $x^2 = 3x$
19. $8x = 10x^3$
20. $5x = 3x^2$
21. $3x^2 + 7 = 0$
22. $2x^2 = 11$
23. $\frac{5x}{2} = 15x^2$
24. $3x = \frac{2x^2}{5}$
25. $2x(x - 1) = 3x(2x + 1)$
26. $4x(3x + 2) = x(x - 5)$
27. $\frac{2x - 1}{3x} = \frac{x - 3}{x}$
28. $\frac{4x - 3}{5x} = \frac{2x - 1}{x}$
29. $x^{4/5} = 16$
30. $x^{4/3} = 81$

In Exercises 31–38, follow the method shown in Example 8 to solve the equations. Express the solutions in simplest radical form and also give decimal approximations to those solutions, rounding off the approximations to two decimal places.

31. $(x - 12)^2 = 14$
32. $(x + 6)^2 = 11$
33. $(x + 3)^2 = 5$
34. $(x - 4)^2 = 19$
35. $(x + 7)^2 = 75$
36. $(x + 10)^2 = 23$
37. $(x - 8)^2 = -27$
38. $(x + 5)^2 = -32$

In Exercises 39–42, set up each problem algebraically and solve. In Exercises 39–41, express each answer in simplest radical form.

39. The length of the diagonal of a square is 100 in. What is the length of a side of the square?
40. The area of a square is 75 sq. ft. What is the length of a side of the square?
41. A rectangle is 12 m long and 4 m wide. Find the length of its diagonal.
42. The area of a certain square is numerically equal to its perimeter. Find the length of a side of the square.

11.2 Solving Equations That Are Quadratic in Form

Sometimes equations that are not quadratic equations can be solved like quadratic equations after an appropriate substitution is made. If an equation can be written in the form $aX^2 + bX^1 + c = 0$, where X represents any algebraic expression, we say that the equation is **quadratic in form** (see Example 1).

EXAMPLE 1 Examples of equations that are quadratic in form:

a. $9 + 5y^{-8} - 19y^{-4} = 0$ can be written as $5(y^{-4})^{2} - 19(y^{-4})^{1} + 9 = 0$; therefore, it is quadratic in form.

b. $h^{-2/3} - h^{-1/3} = 0$ can be written as $(h^{-1/3})^{2} - (h^{-1/3})^{1} = 0$; therefore, it is quadratic in form.

c. $12 + (x^2 - 2x)^2 - 7(x^2 - 2x) = 0$ can be written as $(x^2 - 2x)^{2} - 7(x^2 - 2x)^{1} + 12 = 0$; therefore, it is quadratic in form.

Using substitution to solve an equation that is written as $a(X)^2 + b(X) + c = 0$

1. Let some variable (*other than* the variable in the expression X) equal X.
2. Substitute that variable for X in the given equation.
3. Solve the resulting quadratic equation.
4. Substitute back; that is, set each solution of the quadratic equation equal to X.
5. Solve each resulting equation.
6. Check all answers in the *original* equation, because there may be extraneous roots.

EXAMPLE 2 Find the solution set of $x^4 - 29x^2 + 100 = 0$.

SOLUTION (Because this is a fourth-degree equation, we expect to find four roots.) $x^4 - 29x^2 + 100 = 0$ can be written as $(x^2)^2 - 29(x^2)^1 + 100 = 0$. Let $z = x^2$; if $z = x^2$, then $z^2 = x^4$. (The coefficients and the constants remain unchanged.)

Therefore, $x^4 - 29x^2 + 100 = 0$

becomes $z^2 - 29z + 100 = 0$ This equation is quadratic in z

$(z - 4)(z - 25) = 0$

$z - 4 = 0$	$z - 25 = 0$	Setting each factor equal to 0
$z = 4$	$z = 25$	
$x^2 = 4$	$x^2 = 25$	Substituting back (replacing z with x^2)
$x = \pm 2$	$x = \pm 5$	Using Property 1

√

Check for $x = \pm 2$	***Check for*** $x = \pm 5$
$(\pm 2)^4 - 29(\pm 2)^2 + 100 \stackrel{?}{=} 0$	$(\pm 5)^4 - 29(\pm 5)^2 + 100 \stackrel{?}{=} 0$
$16 - 29(4) + 100 \stackrel{?}{=} 0$	$625 - 29(25) + 100 \stackrel{?}{=} 0$
$16 - 116 + 100 \stackrel{?}{=} 0$	$625 - 725 + 100 \stackrel{?}{=} 0$
$0 = 0$	$0 = 0$

The solution set is $\{\pm 2, \pm 5\}$.

A Word of Caution In making a substitution, we should not use the variable that was in the original equation as the variable of substitution. For example, in Example 2, we should *not* let $x = x^2$; some letter *other than* x must be used. (We used z, but *any* letter other than x could have been used.)

A Word of Caution A common error in using substitution is to forget to *substitute back* (step 4). In Example 2, we let $z = x^2$, and the solutions for z were 4 and 25. It is incorrect to state that the solution set of the *original* equation is $\{4, 25\}$.

EXAMPLE 3 Find the solution set of $h^{-2/3} - h^{-1/3} = 0$.

SOLUTION $h^{-2/3} - h^{-1/3} = 0$ can be written as $(h^{-1/3})^2 - (h^{-1/3})^1 = 0$. Let $z = h^{-1/3}$; if $z = h^{-1/3}$, then $z^2 = h^{-2/3}$.

Therefore, $$h^{-2/3} - h^{-1/3} = 0$$

becomes $$z^2 - z = 0$$

$$z(z - 1) = 0$$

$z = 0$	$z - 1 = 0$
	$z = 1$
$h^{-1/3} = 0$	$h^{-1/3} = 1$ Substituting back (replacing z with $h^{-1/3}$)
$(h^{-1/3})^{-3} = (0)^{-3}$ ← Not defined	$(h^{-1/3})^{-3} = (1)^{-3}$
	$h = 1$

Checking will confirm that the solution set is $\{1\}$.

EXAMPLE 4 Find the solution set of $(x^2 - 2x)^2 - 11(x^2 - 2x) + 24 = 0$.

SOLUTION Let $z = x^2 - 2x$; then $z^2 = (x^2 - 2x)^2$.

Therefore, $$(x^2 - 2x)^2 - 11(x^2 - 2x) + 24 = 0$$

becomes $$z^2 - 11z + 24 = 0$$

$$(z - 3)(z - 8) = 0$$

$z - 3 = 0$	$z - 8 = 0$
$z = 3$	$z = 8$
$x^2 - 2x = 3$	$x^2 - 2x = 8$ Substituting back (replacing z with $x^2 - 2x$)
$x^2 - 2x - 3 = 0$	$x^2 - 2x - 8 = 0$
$(x + 1)(x - 3) = 0$	$(x + 2)(x - 4) = 0$

$x + 1 = 0$	$x - 3 = 0$	$x + 2 = 0$	$x - 4 = 0$
$x = -1$	$x = 3$	$x = -2$	$x = 4$

Checking will verify that the solution set is $\{-1, 3, -2, 4\}$.

Note It is not *necessary* to use a substitution in solving equations that are quadratic in form.

Exercises II.2 Set I

In Exercises 1–12, find the solution set of each equation by factoring, after making appropriate substitutions.

1. $x^4 - 37x^2 + 36 = 0$
2. $y^4 - 13y^2 + 36 = 0$
3. $z^{-4} - 10z^{-2} + 9 = 0$
4. $x^{-4} - 5x^{-2} + 4 = 0$
5. $y^{2/3} - 5y^{1/3} = -4$
6. $x^{2/3} - 10x^{1/3} = -9$
7. $z^{-4} - 4z^{-2} = 0$
8. $R^{-4} - 9R^{-2} = 0$
9. $K^{-2/3} + 2K^{-1/3} + 1 = 0$
10. $M^{-1} - 2M^{-1/2} + 1 = 0$
11. $(x^2 - 4x)^2 - (x^2 - 4x) - 20 = 0$
12. $(x^2 - 2x)^2 - 2(x^2 - 2x) - 3 = 0$

In Exercises 13–16, set up each problem algebraically, solve, and check.

13. The length of a rectangle is twice its width. If the numerical sum of its area and perimeter is 80, find the length and width of the rectangle.
14. The length of a rectangle is three times its width. If the numerical sum of its area and perimeter is 80, find the dimensions of the rectangle.
15. Bruce drove from Los Angeles to the Mexican border and back to Los Angeles, a total distance of 240 mi. His average speed returning to Los Angeles was 20 mph faster than his average speed going to Mexico. If his total driving time was 5 hr, what was his average speed driving from Los Angeles to Mexico?
16. Ruth drove from Creston to Des Moines, a distance of 90 mi. Then she continued on from Des Moines to Omaha, a distance of 120 mi. Her average speed was 10 mph faster on the second part of the journey than on the first part. If the total driving time was 6 hr, what was her average speed on the first leg of the journey?

Writing Problems

Express the answer in your own words and in complete sentences.

1. Describe what your first step would be in solving the equation $x^{-2/3} - 6x^{-1/3} + 9 = 0$, and explain why you would do that first.

Exercises II.2 Set II

In Exercises 1–12, find the solution set of each equation by factoring, after making appropriate substitutions.

1. $y^4 - 26y^2 + 25 = 0$
2. $x^4 + 49 = 50x^2$
3. $x^{-4} - 17x^{-2} = -16$
4. $x^{-4} - 29x^{-2} + 100 = 0$
5. $x^{2/3} - 4x^{1/3} + 4 = 0$
6. $y^{2/3} - 2y^{1/3} = 3$
7. $x^{-4} - 16x^{-2} = 0$
8. $x^4 = 16x^2$
9. $x^{-2/3} - 6x^{-1/3} + 9 = 0$
10. $x^{-2/3} + x^{-1/3} = 2$
11. $(x^2 - 6x)^2 + 17(x^2 - 6x) + 72 = 0$
12. $(x^2 + 2x)^2 - 7(x^2 + 2x) = 8$

In Exercises 13–16, set up each problem algebraically, solve, and check.

13. The length of a rectangle is 2 cm more than twice its width. If its diagonal is 3 cm more than twice its width, what are the dimensions of the rectangle?
14. The tens digit of a two-digit number is 4 more than its units digit. If the product of the units digit and tens digit is 21, find the number.
15. Jeff jogged from his home to a park 15 mi away, and then he walked back home. He jogged 2 mph faster than he walked. If his total traveling time was 8 hr, how fast did he jog?
16. If the product of two consecutive even integers is increased by 4, the result is 84. Find the integers.

11.3 Completing the Square of a Quadratic Equation; Using the Quadratic Formula

The methods shown in previous sections for solving quadratic equations can be used to solve only *some* quadratic equations. The methods we show in this section—completing the square and using the quadratic formula—can be used to solve *all* quadratic equations.

Completing the Square

Before we try to solve quadratic equations by **completing the square**, let's look at some factoring problems in order to decide what *numbers* to use for completing the square.*

$$x^2 + 6x + 9 = (x + 3)^2$$
$$x^2 - 10x + 25 = (x - 5)^2$$
$$x^2 - 14x + 49 = (x - 7)^2$$

What if the constants had been missing? That is, suppose we had

$$x^2 + 6x + ? = (x + ?)^2$$
$$x^2 - 10x + ? = (x - ?)^2 \quad \text{See Example 1a}$$
$$x^2 - 14x + ? = (x - ?)^2 \quad \text{See Example 1b}$$

and suppose we needed to determine what numbers should replace the question marks. What number needs to be added to $x^2 + 6x + \square$ in order to make the expression the square of some binomial? A little experimenting and some thought will show that if we find $\frac{1}{2}$ of 6 (that is, find half of the coefficient of x) and square that number, we'll have

$$\tfrac{1}{2}(6) = 3 \text{ and } 3^2 = 9, \qquad \text{which gives} \qquad x^2 + 6x + \boxed{9} = (x + \boxed{3})^2$$

When we find the constant to add to $x^2 + 6x$ (for example) to make the expression a *perfect square trinomial* (that is, the *square* of a binomial), we say that we are *completing the square* of the polynomial.

EXAMPLE 1 Examples of finding the numbers for completing the square:

a. $x^2 - 10x + ? = (x - ?)^2$

One-half of -10 is -5, and $(-5)^2 = 25$

$x^2 - 10x + 25 = (x - 5)^2$

b. $x^2 - 14x + ? = (x - ?)^2$

One-half of -14 is -7, and $(-7)^2 = 49$

$x^2 - 14x + 49 = (x - 7)^2$

---

*For the benefit of those who may have skipped Section 5.6, we repeat here some of the discussion given in that section.

Solving Quadratic Equations in One Variable by Completing the Square

In completing the square to solve quadratic equations, we want to get the equation into the form $(x - h)^2 = p$. Then we will use Property 1 and the method shown in Example 8 of Section 11.1 to solve the equation for x.

Therefore, to solve an *equation* by completing the square, we decide what number must be added to the x^2- and x-terms of the polynomial to complete the square and then add that number to *both sides of the equation*, using the addition property of equality.

The entire procedure for solving an equation by completing the square is summarized in the following box.

Solving a quadratic equation (in x) by completing the square

1. If a, the coefficient of x^2, is not 1, divide both sides of the equation by a. (It is understood that $a \neq 0$.)
2. Move the x^2-term and the x-term to the left side of the equal sign and the constant to the right side.
3. Determine what number should be added to the left side to make the left side the square of a binomial (it is found by squaring one-half of the coefficient of x), and add this number *to both sides of the equation.*
4. Factor the left side of the equation. [The equation is now in the form $(x - h)^2 = p$.]
5. Use Property 1 to solve the equation for $x - h$.
6. Add h to both sides of the equation to solve for x.
7. Simplify the right side of the equation, and, if the two solutions are *rational*, separate and simplify them.
8. Check the apparent solutions in the original equation.

EXAMPLE 2 Find the solution set of $x^2 - 4x + 1 = 0$ by completing the square.

SOLUTION Because the coefficient of x^2 is already 1, we start with step 2.

$$x^2 - 4x + 1 = 0$$

Step 2. $x^2 - 4x + 1 - 1 = 0 - 1$ Adding -1 to both sides

$x^2 - 4x = -1$ One-half of -4 is -2, and $(-2)^2 = 4$

Step 3. $x^2 - 4x + 4 = -1 + 4$ Adding 4 to both sides to complete the square

Step 4. $(x - 2)^2 = 3$ Factoring the left side

Step 5. $x - 2 = \pm\sqrt{3}$ Using Property 1

Step 6. $x = 2 \pm \sqrt{3}$ Adding 2 to both sides of the equation

Step 7. The right side is in simplest form

Step 8.

Check for $x = 2 + \sqrt{3}$	*Check for $x = 2 - \sqrt{3}$*
$x^2 - 4x + 1 = 0$	$x^2 - 4x + 1 = 0$
$(2 + \sqrt{3})^2 - 4(2 + \sqrt{3}) + 1 \stackrel{?}{=} 0$	$(2 - \sqrt{3})^2 - 4(2 - \sqrt{3}) + 1 \stackrel{?}{=} 0$
$4 + 4\sqrt{3} + 3 - 8 - 4\sqrt{3} + 1 \stackrel{?}{=} 0$	$4 - 4\sqrt{3} + 3 - 8 + 4\sqrt{3} + 1 \stackrel{?}{=} 0$
True $0 = 0$	True $0 = 0$

Therefore, the solution set is $\{2 + \sqrt{3}, 2 - \sqrt{3}\}$.

EXAMPLE 3 Find the solution set of $25x^2 - 30x + 11 = 0$ by completing the square.

SOLUTION

$$25x^2 - 30x + 11 = 0$$

Step 1. $\dfrac{25x^2 - 30x + 11}{25} = \dfrac{0}{25}$ — Dividing both sides by 25, the coefficient of x^2

$x^2 - \frac{30}{25}x + \frac{11}{25} = 0$ — $\frac{30}{25}$ reduces to $\frac{6}{5}$

Step 2. $x^2 - \frac{6}{5}x + \frac{11}{25} - \frac{11}{25} = 0 - \frac{11}{25}$ — Adding $-\frac{11}{25}$ to both sides

$x^2 - \frac{6}{5}x = -\frac{11}{25}$ — One-half of $-\frac{6}{5}$ is $\frac{1}{2}\left(-\frac{6}{5}\right) = -\frac{3}{5}$, and $\left(-\frac{3}{5}\right)^2 = \frac{9}{25}$

Step 3. $x^2 - \frac{6}{5}x + \frac{9}{25} = -\frac{11}{25} + \frac{9}{25}$ — Adding $\frac{9}{25}$ to both sides to complete the square

Step 4. $\left(x - \frac{3}{5}\right)^2 = -\frac{2}{25}$ — Factoring the left side

Step 5. $x - \frac{3}{5} = \pm\sqrt{-\frac{2}{25}}$ — Using Property I

Step 6. $x = \frac{3}{5} \pm \sqrt{-\frac{2}{25}}$ — Adding $\frac{3}{5}$ to both sides

Step 7. $x = \frac{3}{5} \pm \frac{\sqrt{2}}{5}i$ — Simplifying the radical

√ **Step 8.** ***Check for*** $x = \dfrac{3}{5} - \dfrac{\sqrt{2}}{5}i = \dfrac{3 - i\sqrt{2}}{5}$

$$25\left(\frac{3 - i\sqrt{2}}{5}\right)^2 - \overset{6}{\cancel{30}}\left(\frac{3 - i\sqrt{2}}{\underset{1}{\cancel{5}}}\right) + 11 \stackrel{?}{=} 0$$

$$\overset{1}{\cancel{25}}\left(\frac{9 - 6i\sqrt{2} + 2i^2}{\underset{1}{\cancel{25}}}\right) - 6(3 - i\sqrt{2}) + 11 \stackrel{?}{=} 0$$

$$9 - \cancel{6i\sqrt{2}} - 2 - 18 + \cancel{6i\sqrt{2}} + 11 \stackrel{?}{=} 0$$

$$0 = 0 \quad \text{True}$$

The check for $x = \frac{3}{5} + \frac{\sqrt{2}}{5}i$ is left to the student. The solution set is $\left\{\dfrac{3}{5} + \dfrac{\sqrt{2}}{5}i, \dfrac{3}{5} - \dfrac{\sqrt{2}}{5}i\right\}$. (The solution set can also be expressed as $\left\{\dfrac{3 \pm i\sqrt{2}}{5}\right\}$ or as $\left\{\frac{1}{5}(3 \pm i\sqrt{2})\right\}$.)

In Example 4, we use the method of completing the square to solve the quadratic equation that arises from an applied problem. The equation *can* be solved by factoring, but many people find using a calculator and completing the square easier in problems with numbers such as the ones in this example.

EXAMPLE 4 The length of a rectangle is 20 cm more than its width. If the area of the rectangle is 4,125 sq. cm, find the width of the rectangle.

SOLUTION Let x = the width of the rectangle (in cm)
Then $x + 20$ = the length of the rectangle (in cm)
and $x(x + 20)$ = the area of the rectangle (in sq. cm)

The area of the rectangle | is | 4,125

$$x(x + 20) = 4{,}125$$

Step 2. $x^2 + 20x = 4{,}125$ — $\frac{1}{2}$ of 20 is 10, and $10^2 = 100$

Step 3. $x^2 + 20x + 100 = 4{,}125 + 100$ — Completing the square

Step 4. $(x + 10)^2 = 4{,}225$ — Factoring the left side

Step 5. $x + 10 = \pm\sqrt{4{,}225}$ — Using Property I

Step 6. $x = -10 \pm \sqrt{4{,}225}$ — Adding -10 to both sides

Step 7. $x = -10 \pm 65$ — Using a calculator: $\sqrt{4{,}225} = 65$

Separating the two answers

$x = -10 + 65 = 55$ | $x = -10 - 65 = -75$ We reject this solution. The width of a rectangle cannot be negative

$x + 20 = 55 + 20 = 75$

√ **Step 8.** ***Check*** If the width is 55 cm and the length is 75 cm, the area is

$$(55 \text{ cm})(75 \text{ cm}) = 4{,}125 \text{ cm}^2$$

Therefore, the width of the rectangle is 55 cm.

Solving a Quadratic Equation by Using the Quadratic Formula

The method of completing the square can be used to solve *any* quadratic equation. We now use it to solve the *general form* of the quadratic equation, and in this way we derive the **quadratic formula**.

$$ax^2 + bx + c = 0$$ The general form; $a \neq 0$

Step 1. $\dfrac{ax^2 + bx + c}{a} = \dfrac{0}{a}$ — Dividing both sides by a

$$x^2 + \frac{b}{a}x + \frac{c}{a} = 0$$

Step 2. $x^2 + \dfrac{b}{a}x = -\dfrac{c}{a}$ — Adding $-\dfrac{c}{a}$ to both sides

One-half of $\dfrac{b}{a}$ is $\dfrac{1}{2}\left(\dfrac{b}{a}\right) = \dfrac{b}{2a}$, and $\left(\dfrac{b}{2a}\right)^2 = \dfrac{b^2}{4a^2}$

Step 3. $x^2 + \dfrac{b}{a}x + \dfrac{b^2}{4a^2} = \dfrac{b^2}{4a^2} - \dfrac{c}{a}$ — Adding $\dfrac{b^2}{4a^2}$ to both sides to complete the square

Step 4. $\left(x + \dfrac{b}{2a}\right)^2 = \dfrac{b^2 - 4ac}{4a^2}$ — Factoring the left side and adding the rational expressions on the right side

Step 5. $x + \dfrac{b}{2a} = \pm\sqrt{\dfrac{b^2 - 4ac}{4a^2}}$ — Using Property I

$$x + \frac{b}{2a} = \pm\frac{\sqrt{b^2 - 4ac}}{\sqrt{4a^2}} = \pm\frac{\sqrt{b^2 - 4ac}}{2a}$$ Simplifying the radical

Step 6. $x = -\frac{b}{2a} \pm \frac{\sqrt{b^2 - 4ac}}{2a}$ Adding $-\frac{b}{2a}$ to both sides

Step 7. $x = \frac{-b \pm \sqrt{b^2 - 4ac}}{2a}$ The quadratic formula

The quadratic formula

If $ax^2 + bx + c = 0$ and if $a \neq 0$, then

$$x = \frac{-b \pm \sqrt{b^2 - 4ac}}{2a}$$

The procedure for using the quadratic formula can be summarized as follows:

Solving a quadratic equation by using the quadratic formula

1. Express the equation in the general form:
   $$ax^2 + bx + c = 0 \qquad \text{or} \qquad 0 = ax^2 + bx + c$$
2. Identify a, b, and c, and substitute these values into the *quadratic formula:*
   $$x = \frac{-b \pm \sqrt{b^2 - 4ac}}{2a} \quad (a \neq 0)$$
3. Simplify the apparent solutions.
4. If the solutions are *rational* (that is, if the radicand, $b^2 - 4ac$, is a perfect square), write the two solutions separately.
5. Check the apparent solutions in the original equation.

A Word of Caution A common error is to make the fraction bar too short. The formula is *not* $x = -b \pm \frac{\sqrt{b^2 - 4ac}}{2a}$, nor is it $x = -b \pm \frac{\sqrt{b^2 - 4ac}}{2a}$. Both of these are *incorrect.*

Another common error is to make the bar of the radical sign too short. The formula is *not* $x = \frac{-b \pm \sqrt{b^2} - 4ac}{2a}$.

EXAMPLE 5 Find the solution set of $x^2 - 5x + 6 = 0$ by using the quadratic formula.

SOLUTION

Step 1. The equation is in the general form.

Step 2. $a = 1$, $b = -5$, and $c = 6$. We substitute these values into the quadratic formula:

$$x = \frac{-b \pm \sqrt{b^2 - 4ac}}{2a}$$

$$x = \frac{-(-5) \pm \sqrt{(-5)^2 - 4(1)(6)}}{2(1)}$$

Step 3. $x = \frac{5 \pm \sqrt{25 - 24}}{2} = \frac{5 \pm \sqrt{1}}{2} = \frac{5 \pm 1}{2}$ Next we must write the two solutions separately

Step 4. $x = \frac{5 + 1}{2} = \frac{6}{2} = 3$ | $x = \frac{5 - 1}{2} = \frac{4}{2} = 2$

Step 5. Checking will confirm that the solution set is $\{3, 2\}$.

You might verify that the equation in Example 5 *could* have been solved by factoring. In fact, we can often solve a quadratic equation more quickly by factoring than by using the quadratic formula or by completing the square. Therefore, after you have mastered the technique of using the quadratic formula (and/or the technique of completing the square), we recommend that you always try to solve a quadratic equation first by factoring; if you can't easily factor the polynomial, then use one of the other methods.

Note If a and c are unusually large or if they have several pairs of possible factors, it may save time to use the quadratic formula (or the method of completing the square) immediately.

EXAMPLE 6 Find the solution set of $\frac{1}{4}x^2 = 1 - x$.

SOLUTION

$$\frac{1}{4}x^2 = 1 - x$$

$$4\left(\frac{1}{4}x^2\right) = 4(1 - x) \quad \text{Multiplying both sides by 4 to clear fractions}$$

$$x^2 = 4 - 4x$$

Step 1. $x^2 + 4x - 4 = 0$ The general form

Step 2. $a = 1$, $b = 4$, and $c = -4$. We substitute these values into the quadratic formula:

$$x = \frac{-b \pm \sqrt{b^2 - 4ac}}{2a}$$

$$x = \frac{-(4) \pm \sqrt{(4)^2 - 4(1)(-4)}}{2(1)}$$

Step 3. $x = \dfrac{-4 \pm \sqrt{16 + 16}}{2} = \dfrac{-4 \pm \sqrt{32}}{2} = \dfrac{-4 \pm \sqrt{16 \cdot 2}}{2} = \dfrac{-4 \pm 4\sqrt{2}}{2}$

$$= \frac{\overset{2}{\cancel{4}}(-1 \pm \sqrt{2})}{\underset{1}{\cancel{2}}} \quad \text{Factoring the numerator and reducing the expression}$$

$$= 2(-1 \pm \sqrt{2}), \text{ or } -2 \pm 2\sqrt{2}$$

Step 4. The solutions are not rational.

Step 5.

Check for $x = -2 + 2\sqrt{2}$	***Check for*** $x = -2 - 2\sqrt{2}$
$\frac{1}{4}x^2 = 1 - x$	$\frac{1}{4}x^2 = 1 - x$
$\frac{1}{4}(-2 + 2\sqrt{2})^2 \stackrel{?}{=} 1 - (-2 + 2\sqrt{2})$	$\frac{1}{4}(-2 - 2\sqrt{2})^2 \stackrel{?}{=} 1 - (-2 - 2\sqrt{2})$
$\frac{1}{4}(4 - 8\sqrt{2} + 8) \stackrel{?}{=} 1 + 2 - 2\sqrt{2}$	$\frac{1}{4}(4 + 8\sqrt{2} + 8) \stackrel{?}{=} 1 + 2 + 2\sqrt{2}$
$\frac{1}{4}(12 - 8\sqrt{2}) \stackrel{?}{=} 3 - 2\sqrt{2}$	$\frac{1}{4}(12 + 8\sqrt{2}) \stackrel{?}{=} 3 + 2\sqrt{2}$
$3 - 2\sqrt{2} = 3 - 2\sqrt{2}$ True	$3 + 2\sqrt{2} = 3 + 2\sqrt{2}$ True

The solution set is $\{-2 + 2\sqrt{2}, -2 - 2\sqrt{2}\}$.

We will not show Step 1, Step 2, and so on, in the remaining examples.

EXAMPLE 7 Find the solution set of $x^2 - 6x + 13 = 0$.

SOLUTION The equation is in the general form, and $a = 1$, $b = -6$, and $c = 13$. We substitute these values into the quadratic formula:

$$x = \frac{-b \pm \sqrt{b^2 - 4ac}}{2a}$$

$$x = \frac{-(-6) \pm \sqrt{(-6)^2 - 4(1)(13)}}{2(1)}$$

$$x = \frac{6 \pm \sqrt{36 - 52}}{2} = \frac{6 \pm \sqrt{-16}}{2} = \frac{6 \pm 4i}{2} = \frac{6}{2} \pm \frac{4i}{2} = 3 \pm 2i$$

√ **Check for** $x = 3 + 2i$

$$x^2 - 6x + 13 = 0$$

$$(3 + 2i)^2 - 6(3 + 2i) + 13 \stackrel{?}{=} 0$$

$$9 + 12i + 4i^2 - 18 - 12i + 13 \stackrel{?}{=} 0$$

$$9 + 12i - 4 - 18 - 12i + 13 \stackrel{?}{=} 0$$

$$0 = 0 \quad \text{True}$$

We leave the check for $3 - 2i$ to the student. The solution set is $\{3 \pm 2i\}$.

EXAMPLE 8 Find the solution set of $8x^2 - 20x - 3 = 0$. Express the solutions in simplest radical form *and* approximate the solutions correct to two decimal places.

SOLUTION The equation is in the general form, and $a = 8$, $b = -20$, and $c = -3$. Substituting these values into the quadratic formula, we have

$$x = \frac{-b \pm \sqrt{b^2 - 4ac}}{2a}$$

$$x = \frac{-(-20) \pm \sqrt{(-20)^2 - 4(8)(-3)}}{2(8)}$$

$$x = \frac{20 \pm \sqrt{400 + 96}}{16} = \frac{20 \pm \sqrt{496}}{16} \qquad \sqrt{496} = \sqrt{2^4 \cdot 31} = 4\sqrt{31}$$

$$x = \frac{20 \pm 4\sqrt{31}}{16} = \frac{\overset{1}{\cancel{4}}(5 \pm \sqrt{31})}{\underset{4}{\cancel{16}}} = \frac{5 \pm \sqrt{31}}{4} \qquad \sqrt{31} \approx 5.568$$

$$x \approx \frac{5 + 5.568}{4} = \frac{10.568}{4} \approx 2.64 \qquad \Bigg| \qquad x \approx \frac{5 - 5.568}{4} = \frac{-0.568}{4} \approx -0.14$$

Checking will verify that the solution set is exactly $\left\{\frac{5 \pm \sqrt{31}}{4}\right\}$ or about $\{2.64, -0.14\}$.

It is important to realize that although lengths of lines, objects, and so forth, cannot be negative and cannot be complex, *they can be irrational.* When lengths are irrational, we often want decimal approximations for the solutions, and we can use a calculator (or Table I) for finding the approximations. For example, the length of a line might be $\frac{5 + \sqrt{31}}{4}$, with a decimal approximation of 2.64. However, the length of a line *cannot* be $\frac{5 - \sqrt{31}}{4}$, since $\frac{5 - \sqrt{31}}{4}$ is negative (the decimal approximation is -0.14).

EXAMPLE 9 The width of a rectangle is 5 m less than its length. If the area of the rectangle is 7 sq. m, find the width of the rectangle. Give an exact answer and also a decimal approximation, rounded off to two decimal places.

SOLUTION Let $x =$ the width of the rectangle (in m)

Then $x + 5 =$ the length of the rectangle (in m)

and $x(x + 5) =$ the area of the rectangle (in sq. m)

The area of the rectangle	is	7
$x(x+5)$	$=$	7

$$x^2 + 5x - 7 = 0$$

$a = 1$, $b = 5$, and $c = -7$. We substitute these values into the quadratic formula:

$$x = \frac{-b \pm \sqrt{b^2 - 4ac}}{2a}$$

$$x = \frac{-(5) \pm \sqrt{(5)^2 - 4(1)(-7)}}{2(1)}$$

$$x = \frac{-5 \pm \sqrt{25 + 28}}{2} = \frac{-5 \pm \sqrt{53}}{2}$$

Separating the two answers

$$x = \frac{-5 + \sqrt{53}}{2} \approx \frac{-5 + 7.280}{2} = \frac{2.280}{2} = 1.14$$

$$x + 5 = \frac{-5 + \sqrt{53}}{2} + \frac{10}{2} = \frac{5 + \sqrt{53}}{2} \approx \frac{5 + 7.280}{2}$$

$$= \frac{12.280}{2} = 6.14$$

$$x = \frac{-5 - \sqrt{53}}{2}$$

We reject this solution

The width of a rectangle cannot be negative

Check If the width is $\frac{-5 + \sqrt{53}}{2}$ m and the length is $\frac{5 + \sqrt{53}}{2}$ m, then the area is

$$\left(\frac{-5 + \sqrt{53}}{2} \text{ m}\right)\left(\frac{5 + \sqrt{53}}{2} \text{ m}\right) = \frac{(-5 + \sqrt{53})(5 + \sqrt{53})}{4} \text{ sq. m}$$

$$= \frac{-25 + 53}{4} \text{ sq. m} = \frac{28}{4} \text{ sq. m} = 7 \text{ sq. m}$$

Also, if the width is about 1.14 m and the length is close to 6.14 m, then the area is approximately (1.14 m)(6.14 m) = 6.9996 sq. m, which is close to 7 sq. m.

Therefore, the width of the rectangle is $\frac{-5 + \sqrt{53}}{2}$ m, or about 1.14 m.

Exercises 11.3

Set I

In Exercises 1–10, find each solution set by completing the square. Express any irrational solutions in simplest radical form.

1. $x^2 = 6x + 11$
2. $x^2 = 10x - 13$
3. $x^2 - 13 = 4x$
4. $x^2 + 20 = 8x$
5. $x^2 - 2x - 2 = 0$
6. $4x^2 - 8x + 1 = 0$
7. $x^2 + 24x - 2{,}881 = 0$

8. $x^2 + 28x - 2{,}613 = 0$

9. $x^2 + 6x - 8{,}091 = 0$

10. $x^2 - 8x - 4{,}608 = 0$

In Exercises 11–30, find each solution set by using the quadratic formula. Express any irrational solutions in simplest radical form. In Exercises 29 and 30, show the check.

11. $3x^2 - x - 2 = 0$

12. $2x^2 + 3x - 2 = 0$

13. $x^2 - 4x + 1 = 0$

14. $x^2 - 4x - 1 = 0$

15. $x^2 - 4x + 2 = 0$

16. $x^2 - 2x - 2 = 0$

17. $x^2 + x + 5 = 0$

18. $x^2 + x + 7 = 0$

19. $3x^2 + 2x + 1 = 0$

20. $4x^2 + 3x + 2 = 0$

21. $2x^2 = 8x - 9$

22. $3x^2 = 6x - 4$

23. $x + \frac{1}{3} = \frac{-1}{3x}$

24. $x + \frac{1}{4} = \frac{-1}{4x}$

25. $2x^2 - 5x = -7$

26. $3x^2 - 5x = -6$

27. $5x^2 - 102x = 891$

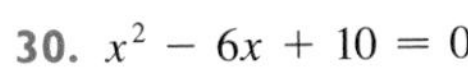

28. $4x^2 - 97x = 992$

29. $x^2 - 4x + 5 = 0$

30. $x^2 - 6x + 10 = 0$

In Exercises 31–38, set up each problem algebraically and solve. Be sure to state what your variables represent. Express any irrational solutions in simplest radical form, and check them. *Also* use a calculator to approximate each irrational solution, rounding off the answer to two decimal places.

31. The length of a rectangle is 30 cm more than its width. If its area is 6,664 sq. cm, what is the width of the rectangle?

32. The width of a rectangle is 32 in. less than its length. If its area is 1,769 sq. in., what is the length of the rectangle?

33. Typing at a constant rate, Bill typed 825 words in a certain length of time. If he had typed 20 words per min faster, he would have cut the time for typing the 825 words by 4 min. At what rate was he typing?

34. A solution was pumped into a 667-gal tank at a constant rate, and it filled the tank in a certain length of time. If the rate of flow had been 6 gal per min slower, it would have taken 6 min longer to fill the tank. At what rate was the pump working?

35. The length of a rectangle is 2 yd more than its width. If its area is 2 sq. yd, find the dimensions of the rectangle.

36. The length of a rectangle is 4 cm more than its width. If its area is 6 sq. cm, find the dimensions of the rectangle.

37. The perimeter of a square is numerically 4 more than its area. Find the length of a side of the square.

38. The area of a square is numerically 2 more than its perimeter. Find the length of a side of the square.

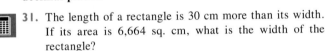

Writing Problems

Express the answers in your own words and in complete sentences.

1. Determine whether you would use the quadratic formula or the method of completing the square in solving the equation $x^2 + 6x + 800 = 0$, and explain why you would use that method.

2. Determine whether you would use the quadratic formula or the method of completing the square in solving the equation $5x^2 + 8x - 3 = 0$, and explain why you would use that method.

Exercises 11.3 Set II

In Exercises 1–10, find each solution set by completing the square. Express any irrational solutions in simplest radical form.

1. $x^2 = 4x + 10$

2. $x^2 - 6x - 5 = 0$

3. $x^2 - 3 = 8x$

4. $x^2 + x = 5$

5. $x^2 - 3x - 3 = 0$

6. $3x^2 - 6x + 1 = 0$

7. $x^2 + 38x - 3{,}608 = 0$

8. $x^2 - 70x + 741 = 0$

9. $x^2 + 12x - 2{,}365 = 0$

10. $x^2 - 54x + 713 = 0$

In Exercises 11–30, find each solution set by using the quadratic formula. Express any irrational solutions in simplest radical form. In Exercises 29 and 30, show the check.

11. $4x^2 = 12x - 7$
12. $3x^2 + 2x = -5$
13. $3x^2 = 4x + 1$
14. $5x^2 = 1 - 2x$
15. $2x^2 = 3 - 5x$
16. $3x = 1 - x^2$
17. $x^2 + x + 4 = 0$
18. $x^2 - x + 4 = 0$
19. $4x^2 + 3x + 1 = 0$
20. $6x^2 = 2 - x$
21. $x^2 + 6 = 2x$
22. $x^2 + 3x = -4$
23. $\frac{x}{2} + \frac{6}{x} = \frac{5}{2}$
24. $\frac{x}{3} + \frac{3}{x} = 2$
25. $4x^2 - 5x = -2$
26. $5x^2 - x = -3$
27. $6x^2 - 181x = 782$
28. $5x^2 + 133x = 864$
29. $x^2 - x + 1 = 0$
30. $x^2 + x + 1 = 0$

In Exercises 31–38, set up each problem algebraically and solve. Be sure to state what your variables represent. Express any irrational solutions in simplest radical form, and check them. *Also* use a calculator to approximate each irrational solution, rounding off the answer to two decimal places.

31. The length of a rectangle is 22 in. more than its width. If its area is 1,995 sq. in., what is the width of the rectangle?

32. The length of a rectangle is 26 more than its width. If its area is (numerically) 503 more than its perimeter, what is the width of the rectangle?

33. Painting at a constant rate, Esther painted 900 sq. ft of wall in a certain length of time. If she had painted 10 sq. ft per min faster, she could have cut the time for painting 900 sq. ft by 1 min. At what rate was she painting?

34. Typing at a constant rate, Lyndsey typed 1,680 words in a certain length of time. If she had typed 14 words per min faster, she could have cut the time for typing the 1,680 words by 10 min. At what rate was she typing?

35. The length of a rectangle is 4 m more than its width. If its area is 1 sq. m, find the dimensions of the rectangle.

36. The length of each side of a square is 4 cm. Find the length of the diagonal.

37. The area of a square is numerically 12 more than its perimeter. Find the length of a side of the square.

38. The length of a rectangle is 2 ft, and its diagonal is 4 ft long. Find the width of the rectangle.

11.4 The Nature of Quadratic Roots

Of the quadratic equations solved so far, some have had real roots, some complex roots, some rational roots, and some irrational roots. Some have had equal roots, and some have had unequal roots. In this section, we will show how to determine what kinds of roots a quadratic equation has *without actually solving the equation.*

The Discriminant

We know that the roots of the quadratic equation $ax^2 + bx + c = 0$ are $\frac{-b \pm \sqrt{b^2 - 4ac}}{2a}$. If, for some equation, the roots are $\frac{-3 \pm \sqrt{-24}}{8}$ (that is, if $b^2 - 4ac = -24$), then, because $\sqrt{-24}$ is not real, the two roots of the equation are *complex conjugates.* If, for another equation, the roots are $\frac{5 \pm \sqrt{13}}{6}$ (that is, if $b^2 - 4ac = 13$), then, because $\sqrt{13}$ is irrational, the roots of that equation are *real, irrational conjugates.* If, for still another equation, the roots are $\frac{-1 \pm \sqrt{25}}{2}$ (that is, if $b^2 - 4ac = 25$), then, because $\sqrt{25}$ is a *rational* number ($\sqrt{25} = 5$), the roots of that equation are *real* and *rational.* The expression $b^2 - 4ac$ is so important that it has a special name; it is called the **discriminant**.

We can see that the value of the discriminant determines the *nature* of the roots (whether they are *real*, *complex*, and so on).

The relationship between the quadratic discriminant and the roots of an equation is summarized as follows:

The relation between the discriminant and the roots of a quadratic equation

For the equation $ax^2 + bx + c = 0$,

If $b^2 - 4ac$ *is*	*There will be*
Positive and a perfect square	Two distinct, real roots; both are rational
Positive but not a perfect square	Two distinct, real roots; both are irrational
Zero	One real (rational) root *of multiplicity two**
Negative	Two distinct complex roots

*When the discriminant is zero, we sometimes consider the one real root of multiplicity two as being two *equal* roots; thus, we can say that all quadratic equations have *two* roots.

It is shown in higher-level mathematics courses that when a, b, and c are *integers*, then any *real*, *irrational* roots of $ax^2 + bx + c = 0$ occur in *conjugate pairs*. This means that if $2 + \sqrt{3}$ is a root, then another root must be $2 - \sqrt{3}$. Similarly, when a, b, and c are integers, any *complex* roots occur in conjugate pairs; thus, if $2 - i$ is a root, $2 + i$ must also be a root.

EXAMPLE 1 Determine the nature of the roots of each of the following equations without solving the equation.

a. $2x^2 + 5x - 12 = 0$

SOLUTION $a = 2$, $b = 5$, and $c = -12$. Therefore,

$$b^2 - 4ac = (5)^2 - 4(2)(-12) = 25 + 96 = 121$$

Because the discriminant, 121, is a positive, perfect square, there are two distinct, real roots; both are rational.

b. $x^2 - 2x - 2 = 0$

SOLUTION $a = 1$, $b = -2$, and $c = -2$. Therefore,

$$b^2 - 4ac = (-2)^2 - 4(1)(-2) = 4 + 8 = 12$$

Because the discriminant, 12, is positive but not a perfect square, there are two distinct real roots; they are irrational, and they are the conjugates of each other.

c. $9x^2 - 6x + 1 = 0$

SOLUTION $a = 9$, $b = -6$, and $c = 1$. Therefore,

$$b^2 - 4ac = (-6)^2 - 4(9)(1) = 36 - 36 = 0$$

Because the discriminant is zero, there is one real (rational) root of multiplicity two (or we can say that there are two *equal* real, rational roots).

d. $x^2 - 6x + 11 = 0$

SOLUTION $a = 1$, $b = -6$, and $c = 11$. Therefore,

$$b^2 - 4ac = (-6)^2 - 4(1)(11) = 36 - 44 = -8$$

Because the discriminant, -8, is negative, there are two distinct complex roots (and they are the conjugates of each other).

Writing an Equation When Its Roots Are Known

Sometimes it is necessary to find an equation when we are given the roots of the equation. We do this by *reversing* the procedure used to solve an equation by factoring; we first set each root *equal to x* (see Examples 2–5).

EXAMPLE 2 Find a quadratic equation that has the roots -3 and 5.

SOLUTION If the roots are -3 and 5, then

$x = -3$ | $x = 5$ — Setting each root equal to x

Then $x + 3 = 0$ | $x - 5 = 0$ — Rewriting each equation with zero on one side of the equal sign

so $(x + 3)(x - 5) = 0$ — If two numbers each equal 0, their product must be 0

or $x^2 - 2x - 15 = 0$

Therefore, $x^2 - 2x - 15 = 0$ is a quadratic equation that has -3 and 5 as its roots.

Note The steps we used in Example 2 are exactly the reverse of the steps we would have used in solving the equation $x^2 - 2x - 15 = 0$.

Note The answers for problems such as these are not unique. We could have multiplied both sides of $x^2 - 2x - 15 = 0$ by *any* nonzero number and obtained an equivalent equation, or we could have added 15 (for example) to both sides, obtaining the equation $x^2 - 2x = 15$, and so on.

EXAMPLE 3 Find an equation of lowest degree *that has integral coefficients* and that has $1 + \sqrt{2}$ as a root.

SOLUTION Because the coefficients are to be integers, any irrational roots must occur in conjugate pairs. Therefore, since $1 + \sqrt{2}$ is a root, $1 - \sqrt{2}$ must also be a root.

$x = 1 + \sqrt{2}$ | $x = 1 - \sqrt{2}$ ← Setting each root equal to x

Then $x - 1 - \sqrt{2} = 0$ | $x - 1 + \sqrt{2} = 0$ ← Rewriting each equation with zero on one side of the equal sign

so $(x - 1 - \sqrt{2}) \cdot (x - 1 + \sqrt{2}) = 0$ — If two numbers each equal 0, their product must be 0

$[(x - 1) - \sqrt{2}] \cdot [(x - 1) + \sqrt{2}] = 0$* — Enclosing $(x - 1)$ in parentheses so that the left side will be in the form $[a - b][a + b]$

$$(x - 1)^2 - (\sqrt{2})^2 = 0$$
$$x^2 - 2x + 1 - 2 = 0$$
$$x^2 - 2x - 1 = 0$$

Therefore, $x^2 - 2x - 1 = 0$ is an equation that has integral coefficients and that has $1 + \sqrt{2}$ as a root.

*The multiplication can be done in other ways, but we think this is the easiest way.

EXAMPLE 4 Find an equation of lowest degree *that has integral coefficients* and that has $2 - 3i$ as a root.

SOLUTION Because the coefficients are to be integers, any complex roots must occur in conjugate pairs. Therefore, since $2 - 3i$ is a root, $2 + 3i$ must also be a root.

$x = 2 - 3i$ | $x = 2 + 3i$ ← Setting each root equal to x

Then $x - 2 + 3i = 0$ | $x - 2 - 3i = 0$ ← Rewriting each equation with zero on one side of the equal sign

so $(x - 2 + 3i) \cdot (x - 2 - 3i) = 0$ — If two numbers each equal 0, their product must be 0

$[(x - 2) + 3i] \cdot [(x - 2) - 3i] = 0$ — Enclosing $(x - 2)$ in parentheses so that the left side will be in the form $[a + b][a - b]$

$$(x - 2)^2 - (3i)^2 = 0$$

$$x^2 - 4x + 4 - 9i^2 = 0$$

$$x^2 - 4x + 4 - 9(-1) = 0$$ Replacing i^2 with -1

$$x^2 - 4x + 4 + 9 = 0$$

$$x^2 - 4x + 13 = 0$$

Therefore, $x^2 - 4x + 13 = 0$ is an equation that has integral coefficients and that has $2 - 3i$ as a root.

EXAMPLE 5 Find a cubic equation that has roots $\frac{1}{2}$, -3, and $\frac{2}{3}$.

SOLUTION

$x = \frac{1}{2}$	$x = -3$	$x = \frac{2}{3}$	Setting each root equal to x
$2x = 1$	$x = -3$	$3x = 2$	Clearing fractions
$2x - 1 = 0$	$x + 3 = 0$	$3x - 2 = 0$	

$$[(2x - 1) \cdot (x + 3)] \cdot (3x - 2) = 0$$

$$(2x^2 + 5x - 3)(3x - 2) = 0$$

$$6x^3 + 11x^2 - 19x + 6 = 0$$

Therefore, $6x^3 + 11x^2 - 19x + 6 = 0$ is an equation that has $\frac{1}{2}$, -3, and $\frac{2}{3}$ as its roots.

Exercises 11.4

Set I

In Exercises 1–8, use the quadratic discriminant to determine the nature of the roots without solving the equation.

1. $x^2 - x - 12 = 0$
2. $x^2 + 3x - 10 = 0$
3. $6x^2 - 7x = 2$
4. $10x^2 - 11x = 5$
5. $x^2 - 4x = -4$
6. $x^2 + 8x = -16$
7. $9x^2 + 2 = 6x$
8. $2x^2 + 6x + 5 = 0$

In Exercises 9–18, find a quadratic equation that has the given roots.

9. 4 and -2
10. -3 and 2
11. 0 and 5
12. 6 and 0
13. $2 + \sqrt{3}$ and $2 - \sqrt{3}$
14. $3 + \sqrt{5}$ and $3 - \sqrt{5}$
15. $\frac{1}{2}$ and $\frac{2}{3}$
16. $\frac{3}{5}$ and $\frac{2}{3}$
17. $\dfrac{1 + i\sqrt{3}}{2}$ and $\dfrac{1 - i\sqrt{3}}{2}$
18. $\dfrac{1 - i\sqrt{2}}{3}$ and $\dfrac{1 + i\sqrt{2}}{3}$

In Exercises 19–22, find a cubic equation that has the given roots.

19. 1, 3, and 4
20. 2, 1, and 5
21. 3, $-2i$, and $+2i$
22. 5, $-3i$, and $+3i$

In Exercises 23–28, find and simplify an equation of lowest degree that has integral coefficients and that has the given roots.

23. $5 - i$
24. $2 + 3i$
25. $1 - 2\sqrt{5}$
26. $3 + 5\sqrt{2}$
27. 2 and $-3i$
28. -3 and $6i$

Writing Problems

Express the answers in your own words and in complete sentences.

1. Explain how you can use the discriminant to determine the nature of the roots of a quadratic equation.
2. Describe how to write a quadratic equation if you're given the two roots of the equation.

Exercises 11.4 Set II

In Exercises 1–8, use the quadratic discriminant to determine the nature of the roots without solving the equation.

1. $x^2 + 25 = 10x$
2. $2x^2 + 3x = 2$
3. $x^2 - 2x = 2$
4. $x^2 - x + 1 = 0$
5. $x^2 + 4x = 0$
6. $x^2 - 8x = -16$
7. $3x^2 + 5x = 2$
8. $x^2 = 8$

In Exercises 9–18, find a quadratic equation that has the given roots.

9. 3 and -4
10. $3\sqrt{2}$ and $-3\sqrt{2}$
11. -8 and 0
12. $3 - \sqrt{5}$ and $3 + \sqrt{5}$
13. $4 - \sqrt{7}$ and $4 + \sqrt{7}$
14. $2i$ and $-2i$
15. $\frac{3}{4}$ and $\frac{1}{8}$
16. 3 and $\frac{2}{5}$
17. $\dfrac{3 + 5i}{3}$ and $\dfrac{3 - 5i}{3}$
18. $3 - i\sqrt{5}$ and $3 + i\sqrt{5}$

In Exercises 19–22, find a cubic equation that has the given roots.

19. 0, 2, and 5
20. 2, $1 + \sqrt{2}$, and $1 - \sqrt{2}$
21. 4, $5i$, and $-5i$
22. 0, $2 - 3i$, and $2 + 3i$

In Exercises 23–28, find and simplify an equation of lowest degree that has integral coefficients and that has the given roots.

23. $6 + i$
24. $2 - 4i$
25. $5 + 3\sqrt{7}$
26. 3 and $\sqrt{5}$
27. 3 and $7i$
28. 0 and i

Sections 11.1–11.4 REVIEW

Quadratic Equations 11.1 A **quadratic equation** is a polynomial equation that has a second-degree term as its highest-degree term.

The General Form 11.1 The general form of a quadratic equation is $ax^2 + bx + c = 0$, where a, b, and c are real numbers ($a \neq 0$). In this book, we will not consider the quadratic equation to be in the general form unless a, b, and c are *integers* and a is positive.

Solving Incomplete Quadratics
11.1

When $c = 0$, find the greatest common factor (GCF); then solve by factoring.

When $b = 0$, we can use the following method to solve the equation $ax^2 + c = 0$:

1. Add $-c$ to both sides of the equation.
2. Divide both sides of the equation by a, the coefficient of x^2.
3. Use Property 1: If $x^2 = p$, then $x = \pm\sqrt{p}$.
4. Express the square roots in simplest form.

Using Substitution in Solving an Equation That Is Written as $a(X)^2 + b(X)^1 + c = 0$
11.2

1. Let some variable (*other than* the variable in the expression X) equal X.
2. Substitute that variable for X in the given equation.
3. Solve the resulting quadratic equation.
4. Substitute back; that is, set each solution of the quadratic equation equal to X.
5. Solve each resulting equation.
6. Check all answers in the original equation, because there may be extraneous roots.

Solving Quadratic Equations by Completing the Square
11.3

1. If a, the coefficient of x^2, is not 1, divide both sides of the equation by a. (It is understood that $a \neq 0$.)
2. Move the x^2-term and the x-term to the left side of the equal sign and the constant to the right side.
3. Determine what number should be added to the left side to make the left side the square of a binomial, and add this number *to both sides of the equation.*
4. Factor the left side of the equation.
5. Use Property 1 to solve the equation for $x - h$.
6. Add h to both sides of the equation to solve for x.
7. Simplify the right side of the equation, and, if the two solutions are *rational*, separate and simplify them.
8. Check the apparent solutions in the original equation.

Solving Quadratic Equations by Using the Quadratic Formula
11.3

1. Express the equation in the general form:

$$ax^2 + bx + c = 0 \qquad \text{or} \qquad 0 = ax^2 + bx + c$$

2. Identify a, b, and c, and substitute these values into the *quadratic formula*:

$$x = \frac{-b \pm \sqrt{b^2 - 4ac}}{2a} \qquad (a \neq 0)$$

3. Simplify the apparent solutions.
4. If the solutions are *rational* (that is, if the radicand, $b^2 - 4ac$, is a perfect square), write the two solutions separately.
5. Check the apparent solutions in the original equation.

The Relation Between the Quadratic Discriminant and Roots
11.4

If $b^2 - 4ac$ is	*There will be*
Positive and a perfect square	Two distinct, real, rational roots
Positive but not a perfect square	Two distinct, real, irrational roots
Zero	One real, rational root of multiplicity two
Negative	Two distinct complex roots

Using Roots to Find the Equation
11.4

Reverse the procedure used for solving an equation by factoring.

Sections 11.1–11.4 REVIEW EXERCISES Set I

In Exercises 1–16, find the solution set of each equation by any convenient method.

1. $x^2 + x = 6$
2. $x^2 = 3x + 10$
3. $x^2 - 2x - 4 = 0$
4. $x^2 - 4x + 2 = 0$
5. $x^2 - 2x + 5 = 0$
6. $x^2 + 8x - 425 = 0$
7. $\dfrac{2x}{3} = \dfrac{3}{8x}$
8. $\dfrac{3x}{5} = \dfrac{5}{12x}$
9. $\dfrac{x+2}{3} = \dfrac{1}{x-2} + \dfrac{2}{3}$
10. $\dfrac{x+2}{4} = \dfrac{2}{x+2} + \dfrac{1}{2}$
11. $(x + 5)(x - 2) = x(3 - 2x) + 2$
12. $(2x - 1)(3x + 5) = x(x + 7) + 4$
13. $(x^2 - 4x)^2 + 5(x^2 - 4x) + 4 = 0$
14. $(x^2 + 6x)^2 + 13(x^2 + 6x) + 36 = 0$
15. $x^4 - 65x^2 + 64 = 0$
16. $x^4 - 82x^2 + 81 = 0$

In Exercises 17 and 18, use the quadratic discriminant to determine the nature of the roots without solving the equation.

17. $x^2 - 6x + 7 = 0$
18. $x^2 - 8x + 13 = 0$

In Exercises 19–22, set up each problem algebraically and solve. If answers are not rational, express them in simplest radical form *and* give decimal approximations rounded off to two decimal places.

19. If the product of two consecutive odd integers is decreased by 14, the result is 85. Find the integers.
20. If the product of two consecutive integers is increased by 4, the result is 60. Find the integers.
21. The length of a rectangle is 3 in. more than its width. If its area is 8 sq. in., find the dimensions of the rectangle.
22. The length of one leg of a right triangle is 2 cm more than the other leg. If its hypotenuse is 3 cm, find the length of each leg.
23. Find a quadratic equation that has the roots $1 - \sqrt{7}$ and $1 + \sqrt{7}$.
24. Find an equation of lowest degree that has integral coefficients and that has the roots 2 and $-i\sqrt{6}$.

Sections 11.1–11.4 REVIEW EXERCISES Set II

Name ______________________

ANSWERS

1. ______
2. ______
3. ______
4. ______
5. ______
6. ______
7. ______
8. ______
9. ______
10. ______
11. ______
12. ______
13. ______
14. ______
15. ______
16. ______

In Exercises 1–16, find the solution set of each equation by any convenient method.

1. $x^2 + x = 20$

2. $x^2 - 13 = 0$

3. $x^2 - 2x - 15 = 0$

4. $4x^2 + 3x = 0$

5. $x^2 - x + 7 = 0$

6. $x^2 + 46x + 465 = 0$

7. $\dfrac{3x}{5} = \dfrac{5}{3x}$

8. $\dfrac{5x}{2} = \dfrac{1}{x - 2}$

9. $\dfrac{3}{x} - \dfrac{x}{x + 2} = 2$

10. $\dfrac{2}{x} + \dfrac{x}{x + 1} = 5$

11. $(x + 3)(x - 4) = x(3 - x) + 3$

12. $(3x - 4)(x - 1) = x(2 - 4x) - 2$

13. $(x^2 - 6x)^2 + 2(x^2 - 6x) - 63 = 0$

14. $(x^2 - 3x)^2 - 14(x^2 - 3x) + 40 = 0$

15. $y^4 - 17y^2 + 16 = 0$

16. $x^{2/3} - 3x^{1/3} + 2 = 0$

In Exercises 17 and 18, use the quadratic discriminant to determine the nature of the roots without solving the equation.

17. $x^2 + x + 5 = 0$

18. $3x^2 - 5 = 0$

17. ______

18. ______

19. ______

______

In Exercises 19–22, set up each problem algebraically and solve. If answers are not rational, express them in simplest radical form *and* give decimal approximations rounded off to two decimal places.

20. ______

21. ______

22. ______

______

23. ______

24. ______

19. If the product of two consecutive even integers is decreased by 8, the result is 40. Find the integers.

20. The area of a square is 5 sq. m. Find the length of a side of the square.

21. The length of a rectangle is 4 cm more than its width. If the area is 3 sq. cm, find the width of the rectangle.

22. One leg of a right triangle is 3 units shorter than the other. The hypotenuse is $\sqrt{11}$ units. Find the length of each leg.

23. Find a quadratic equation that has the roots $-2 - 3i$ and $-2 + 3i$.

24. Find an equation of lowest degree that has integral coefficients and that has the roots $2i$ and $-\sqrt{5}$.

11.5 Graphing Quadratic Functions of One Variable

Earlier, we graphed a few quadratic functions of one variable. Recall that a quadratic function of one variable is a function that can be written in the form $f(x) = ax^2 + bx + c$ and that the graph of $y = f(x)$ is a curve called a *parabola*.

In this section, we include additional information about the graphs of quadratic functions of one variable—information that will simplify the drawing of the graphs—and we show several methods of graphing parabolas.

From studying Section 9.6, you probably have a good mental picture of what a parabola looks like. (Two parabolas are shown in Figure 1.) If the equation of a parabola is written in the form $y = f(x) = ax^2 + bx + c$, then the parabola opens *upward* if $a > 0$ and the parabola opens *downward* if $a < 0$ (see Figure 1).

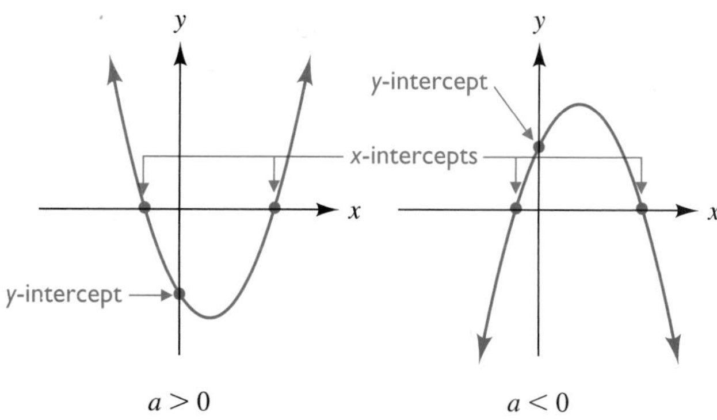

Figure 1

The y-intercept

The **y-intercept** is the point where the curve intersects the y-axis; it is found by letting $x = 0$. For the parabola $y = f(x) = ax^2 + bx + c$, there is always exactly one y-intercept—the point $(0, c)$:

$$y = f(x) = ax^2 + bx + c$$

$$y = f(0) = a(0)^2 + b(0) + c$$

$$y = c$$

The y-coordinate of the y-intercept of a quadratic function of one variable is *always equal to the constant term.*

The x-intercepts

The **x-intercepts** are the points where the curve intersects the x-axis; they are found by letting $y = 0$ and solving the resulting equation for x:

$$y = f(x) = ax^2 + bx + c = 0$$

If there are two real solutions for that equation, then there are two x-intercepts; if there is just one real solution, then there is just one x-intercept; if the solutions are complex, then there are no x-intercepts.

EXAMPLE 1 Determine whether each parabola opens upward or downward, and find all the intercepts.

a. $y = f(x) = 6 - x - x^2$

SOLUTION $a = -1 < 0$. Therefore, the parabola opens downward.

Because $c = 6$, the y-intercept is $(0, 6)$. Since the x-coordinate of the y-intercept is *always* zero, we can simply say that the y-intercept is 6.

We let $y = 0$ to find the x-intercepts:

$$y = 6 - x - x^2 = 0$$

$$(2 - x)(3 + x) = 0$$

$$2 - x = 0 \quad\Big|\quad 3 + x = 0$$

$$x = 2 \quad\Big|\quad x = -3$$

The x-intercepts are $(2, 0)$ and $(-3, 0)$. Since the y-coordinates of the x-intercepts are *always* zero, we can simply say that the x-intercepts are 2 and -3.

b. $y = f(x) = x^2 - 2x - 1$

SOLUTION $a = 1 > 0$. Therefore, the parabola opens upward.

Because $c = -1$, the y-intercept is $(0, -1)$.

We let $y = 0$ to find the x-intercepts:

$$y = f(x) = x^2 - 2x - 1 = 0$$

Because $x^2 - 2x - 1$ won't factor, we use the quadratic formula; $a = 1$, $b = -2$, and $c = -1$. Substituting into the quadratic formula, we have

$$x = \frac{-b \pm \sqrt{b^2 - 4ac}}{2a}$$

$$x = \frac{-(-2) \pm \sqrt{(-2)^2 - 4(1)(-1)}}{2(1)}$$

$$= \frac{2 \pm \sqrt{4 + 4}}{2} = \frac{2 \pm \sqrt{8}}{2} = \frac{2 \pm 2\sqrt{2}}{2} = 1 \pm \sqrt{2}$$

The x-coordinates of the x-intercepts are $1 + \sqrt{2}$ and $1 - \sqrt{2}$. These can be approximated. Using a calculator, we find that $\sqrt{2} \approx 1.4$ so

$$1 + \sqrt{2} \approx 1 + 1.4 = 2.4 \quad\Big|\quad 1 - \sqrt{2} \approx 1 - 1.4 = -0.4$$

Therefore, the x-intercepts are exactly $(1 + \sqrt{2}, 0)$ and $(1 - \sqrt{2}, 0)$ or approximately $(2.4, 0)$ and $(-0.4, 0)$.

c. $y = f(x) = x^2 - 4x + 5$

SOLUTION $a = 1 > 0$; the parabola opens upward.

Because $c = 5$, the y-intercept is $(0, 5)$.

We let $y = 0$ to find the x-intercepts:

$$y = f(x) = x^2 - 4x + 5 = 0$$

Because $x^2 - 4x + 5$ won't factor, we use the quadratic formula; $a = 1$, $b = -4$, and $c = 5$. Substituting into the quadratic formula, we have

$$x = \frac{-b \pm \sqrt{b^2 - 4ac}}{2a}$$

$$x = \frac{-(-4) \pm \sqrt{(-4)^2 - 4(1)(5)}}{2(1)}$$

$$= \frac{4 \pm \sqrt{16 - 20}}{2} = \frac{4 \pm \sqrt{-4}}{2} = \frac{4 \pm 2i}{2} = 2 \pm i$$

Because the solutions are complex numbers, not real numbers, *the parabola does not cross the x-axis.* There are no x-intercepts in this case.

Note The *graphs* of the functions discussed in Example 1 are shown in Example 4.

Symmetric Points, the Axis of Symmetry, and the Vertex

If two points are **symmetric** with respect to a line, then that line must be the perpendicular bisector of the line segment that joins the two points, and the line is called the **axis of symmetry**. The axis of symmetry of a parabola is also called **the axis of the parabola** (see Figure 2).

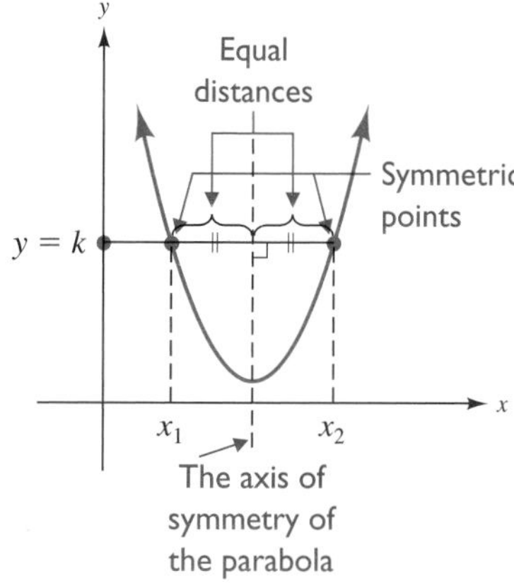

Figure 2

The axis of symmetry of the parabola $y = f(x) = ax^2 + bx + c$ is a *vertical* line midway between any pair of symmetric points on that parabola. It is shown in Appendix A that the *equation* of the axis of symmetry for $y = f(x) = ax^2 + bx + c$ can be found by substituting into the following formula.

> The equation of the axis of symmetry for the parabola
> $$y = f(x) = ax^2 + bx + c$$
> is
> $$x = -\frac{b}{2a}$$

EXAMPLE 2 Find the equation of the axis of symmetry for the graph of the quadratic function $y = f(x) = 2x^2 + 7x - 4$.

SOLUTION Substituting 2 for a and 7 for b in the formula, we have

$$x = -\frac{b}{2a} = -\frac{7}{2(2)} = -\frac{7}{4}$$

Therefore, $x = -\frac{7}{4}$ is the equation of the axis of symmetry.

A Word of Caution A common error in solving a problem like the one in Example 2 is to say that the equation of the axis of symmetry is $-\frac{7}{4}$. *This is incorrect*; $-\frac{7}{4}$ is not an equation at all! An equation must contain an equal sign *and* must have some expression on each side of the equal sign. The *equation* of the axis of symmetry is $x = -\frac{7}{4}$.

The Vertex of a Parabola The point at which a parabola intersects its axis of symmetry is called the **vertex of the parabola**. The vertex is the point

$$\left(-\frac{b}{2a}, f\left(-\frac{b}{2a}\right)\right)$$

Although this notation may look difficult and confusing, it is simply telling you how to find the coordinates of the vertex: First find the x-coordinate of the vertex by using the formula $x = -\frac{b}{2a}$, and then use that value of x in the original equation to find its corresponding y-value (see Example 3).

If the parabola opens upward, the vertex will be the *lowest* point on the parabola, and y will have its *minimum* value there. If the parabola opens downward, the vertex will be the *highest* point on the parabola, and y will have its *maximum* value there (see Figure 3).

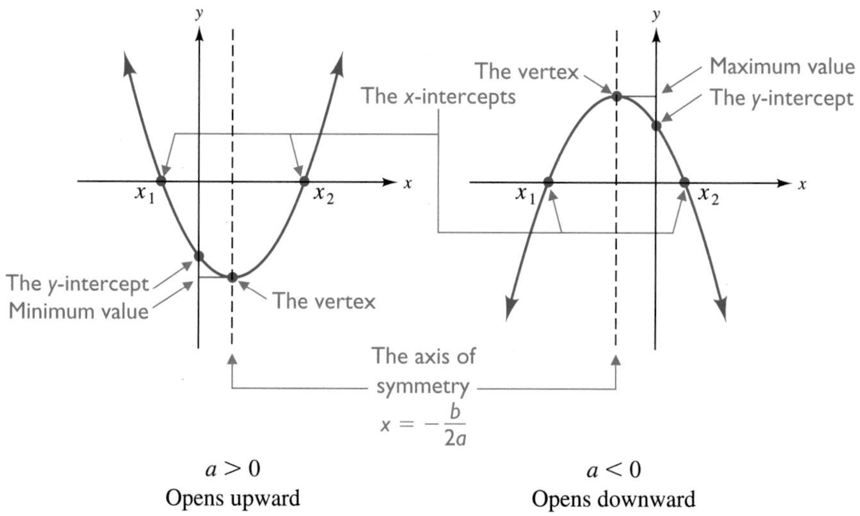

Figure 3

EXAMPLE 3 For each parabola from Example 1, find the equation of the axis of symmetry, and find the ordered pair that represents the vertex.

a. $y = f(x) = 6 - x - x^2$

SOLUTION $a = -1$ and $b = -1$. The equation of the axis of symmetry is $x = -\frac{-1}{2(-1)}$, or $x = -\frac{1}{2}$.

This means that the x-coordinate of the vertex is $-\frac{1}{2}$. The y-coordinate of the vertex is $f\left(-\frac{1}{2}\right)$.

$$f\left(-\tfrac{1}{2}\right) = 6 - \left(-\tfrac{1}{2}\right) - \left(-\tfrac{1}{2}\right)^2 = 6 + \tfrac{1}{2} - \tfrac{1}{4} = 6\tfrac{1}{4}$$

Therefore, the ordered pair that represents the vertex is $\left(-\frac{1}{2}, 6\frac{1}{4}\right)$.

A Word of Caution A common error in solving problems like the one in part a is to think that the equations $y = 6 - x - x^2$ and $y = x^2 + x - 6$ are equivalent (probably because we know that the equations $0 = 6 - x - x^2$ and $0 = x^2 + x - 6$ *are* equivalent). However, if we multiply both sides of $y = 6 - x - x^2$ by -1, we get $-y = x^2 + x - 6$, *not* $y = x^2 + x - 6$. The equations $y = 6 - x - x^2$ and $y = x^2 + x - 6$ *are not equivalent*, and they do not have the same vertex or the same graph.

b. $y = f(x) = x^2 - 2x - 1$

SOLUTION $a = 1$ and $b = -2$. The equation of the axis of symmetry is $x = -\frac{-2}{2(1)}$, or $x = 1$.

Then $$f(1) = 1^2 - 2(1) - 1 = -2$$

Therefore, the ordered pair that represents the vertex is $(1, -2)$.

c. $y = f(x) = x^2 - 4x + 5$

SOLUTION $a = 1$ and $b = -4$. The equation of the axis of symmetry is $x = -\frac{-4}{2(1)}$, or $x = 2$.

Then $$f(2) = (2)^2 - 4(2) + 5 = 4 - 8 + 5 = 1$$

Therefore, the ordered pair that represents the vertex is $(2, 1)$.

Finding Symmetric Points In Example 1a, we found that the y-intercept of the curve $y = 6 - x - x^2$ is $(0, 6)$, and in Example 3a, we found that the axis of symmetry for the same curve is $x = -\frac{1}{2}$. The point $(0, 6)$ lies one-half of a unit to the *right* of the axis of symmetry, and so if we move one-half of a unit to the *left* of the axis of symmetry *along the same horizontal line* (that is, along the line $y = 6$), we'll find another point on the curve. This point is $(-1, 6)$ (see Example 4a).

Similarly, in Example 1b, we found that the y-intercept of the curve $y = x^2 - 2x - 1$ is $(0, -1)$, and in Example 3b, we found that the axis of symmetry for the same curve is $x = 1$. The point $(0, -1)$ lies 1 unit to the *left* of the axis of symmetry, and so if we move 1 unit to the *right* of the axis of symmetry along the same horizontal line (that is, along the line $y = -1$), we'll find another point on the curve. This point is $(2, -1)$ (see Example 4b).

Examining Examples 1c and 3c, we see that the point $(0, 5)$ lies 2 units to the *left* of the axis of symmetry, and so if we move 2 units to the *right* of the axis of symmetry along the same horizontal line (that is, along the line $y = 5$), we find the point $(4, 5)$, another point on the curve (see Example 4c).

Graphing Quadratic Functions of One Variable

In graphing a parabola that has two x-intercepts, we can usually draw a fairly accurate graph by plotting the x- and y-intercepts and the vertex and then connecting these points with a smooth curve. If the parabola has fewer than two x-intercepts, we need to find additional points on the curve. The procedure is summarized below:

Graphing a quadratic function of one variable

To graph $$y = f(x) = ax^2 + bx + c,$$

1. Determine whether the parabola opens upward or downward.
2. Find the y-intercept: $(0, c)$.
3. Find the x-intercepts: Let $y = 0$, and solve the resulting equation for x.
4. Find the equation of the axis of symmetry by using the formula $x = -\frac{b}{2a}$.
5. Find the vertex: $\left(-\frac{b}{2a}, f\left(-\frac{b}{2a}\right)\right)$.
6. Graph the points found in steps 2, 3, and 5. Then graph the points symmetrical to those points with respect to the axis of symmetry. If necessary, graph additional points.
7. Draw the parabola as a smooth curve through all the points, taking the points in order from left to right.

EXAMPLE 4 Graph the parabolas from Example 1; label the intercepts, the vertex, and any known symmetric points, and sketch the axis of symmetry.

a. $y = 6 - x - x^2$

SOLUTION The parabola opens downward.

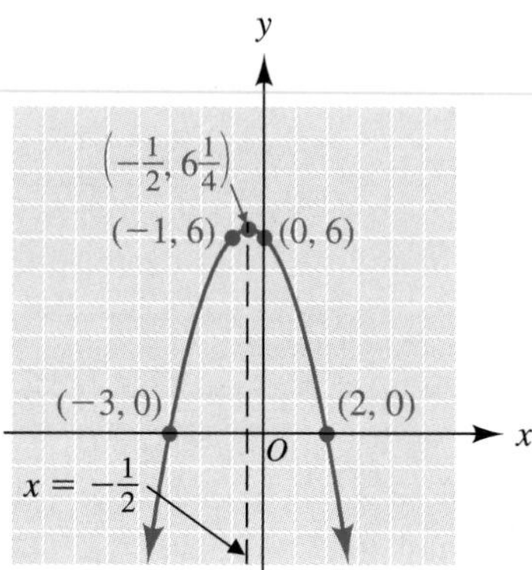

b. $y = x^2 - 2x - 1$

SOLUTION The parabola opens upward.

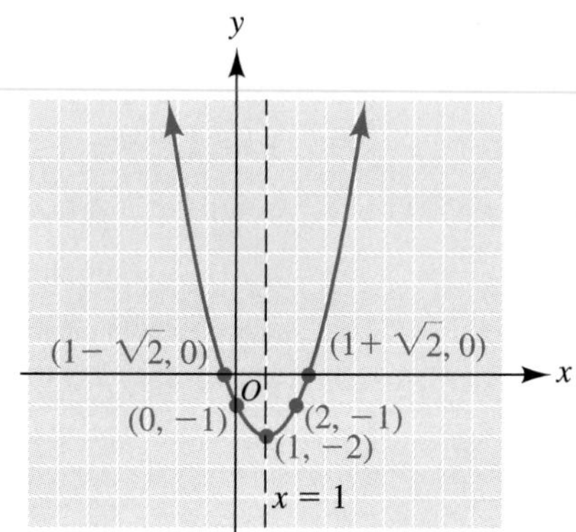

c. $y = x^2 - 4x + 5$

SOLUTION The parabola opens upward.

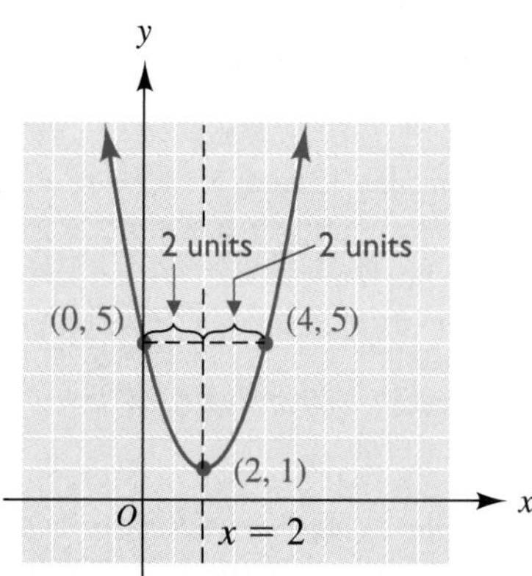

Note To graph the curves from Example 4 with a TI-81 graphics calculator, proceed as follows:

a. To graph $y = 6 - x - x^2$, press the following keys, in order:

Y= | 6 | – | X|T | – | X|T | x^2 | GRAPH

Set the range so that *x*min is −5, *x*max is 5, *y*min is −2, and *y*max is 7.

b. To graph $y = x^2 - 2x - 1$, press the following keys, in order:

Y= | X|T | x^2 | – | 2 | X|T | – | 1 | GRAPH

Set the range so that *x*min is −4, *x*max is 6, *y*min is −3, and *y*max is 5.

c. To graph $y = x^2 - 4x + 5$, press the following keys, in order:

Y= | X|T | x^2 | – | 4 | X|T | + | 5 | GRAPH

Set the range so that *x*min is −3, *x*max is 7, *y*min is −1, and *y*max is 7. 

EXAMPLE 5 Discuss and graph $y = f(x) = 9x^2 - 6x + 1$.

SOLUTION

Step 1. The parabola opens upward, since $a = 9 > 0$.

Step 2. The y-intercept is $(0, 1)$, since $c = 1$.

Step 3. $y = f(x) = 9x^2 - 6x + 1 = 0$ Letting $y = 0$

$$(3x - 1)(3x - 1) = 0$$

$3x - 1 = 0$	$3x - 1 = 0$
$x = \frac{1}{3}$	$x = \frac{1}{3}$ $\frac{1}{3}$ is a root of multiplicity two

The x-intercepts are both the same, which means that the parabola touches the x-axis at only one point, $x = \frac{1}{3}$ (see Figure 4); that is, the parabola is *tangent* to the x-axis.

Step 4. The equation of the axis of symmetry is $x = -\dfrac{-6}{2(9)}$, or $x = \dfrac{1}{3}$.

Step 5. The ordered pair that represents the vertex is $\left(\frac{1}{3}, 0\right)$.

Step 6. We graph the points found in steps 2, 3, and 5.

The y-intercept, which is $(0, 1)$, lies one-third of a unit to the *left* of the axis of symmetry; to find the point symmetric to $(0, 1)$, we need to move one-third of a unit (along the line $y = 1$) to the *right* of $x = \frac{1}{3}$. This gives the point $\left(\frac{2}{3}, 1\right)$.

Because there is only one x-intercept and because the three points found so far are quite close together, we'll find another point on the curve and its symmetric point. Let $x = 1$; then

$$f(1) = 9(1)^2 - 6(1) + 1 = 4$$

Therefore, another point on the curve is $(1, 4)$. This point lies two-thirds of a unit to the *right* of the axis of symmetry; to find its symmetric point, we must move two-thirds of a unit (along the line $y = 4$) to the *left* of $x = \frac{1}{3}$. This gives the point $\left(-\frac{1}{3}, 4\right)$. We graph these additional points.

Step 7. From left to right, we connect all the points with a smooth curve. See Figure 4.

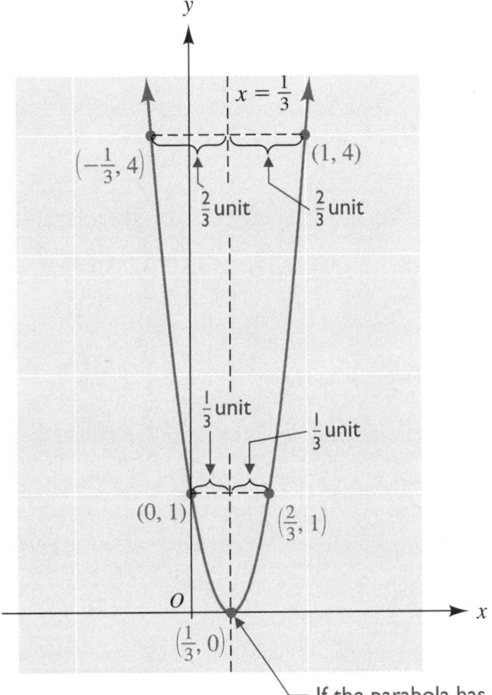

Figure 4

Note We will no longer show the keystrokes or the range to use for graphing quadratic functions of one variable with a graphics calculator. We assume that, by now, you can graph such functions without help.

EXAMPLE 6 Graph the function $f(x) = x^2 - x - 2$.

SOLUTION

Step 1. The parabola opens upward, since $a = 1 > 0$.
Step 2. The y-intercept is $(0, -2)$, since $c = -2$.
Step 3. The x-intercepts: $y = f(x) = x^2 - x - 2 = 0$

$$(x + 1)(x - 2) = 0$$

$$x + 1 = 0 \quad\Big|\quad x - 2 = 0$$

$$x = -1 \quad\Big|\quad x = 2$$

Step 4. The axis of symmetry: $x = -\dfrac{b}{2a} = -\dfrac{-1}{2(1)} = \dfrac{1}{2}$

Step 5. Because the axis of symmetry is $x = \frac{1}{2}$, the x-coordinate of the vertex is $\frac{1}{2}$, and the y-coordinate is $f\left(\frac{1}{2}\right)$.

$$f\left(\tfrac{1}{2}\right) = \left(\tfrac{1}{2}\right)^2 - \left(\tfrac{1}{2}\right) - 2 = -2\tfrac{1}{4}$$

The vertex is $\left(\frac{1}{2}, -2\frac{1}{4}\right)$.

Step 6. We graph the points found in steps 2, 3, and 5. The point symmetric to $(0, -2)$ is $(1, -2)$. To find an additional point to help draw the graph, we let $x = 3$; then

$$f(3) = (3)^2 - (3) - 2 = 9 - 3 - 2 = 4$$

Therefore, an additional point is $(3, 4)$; its symmetric point is $(-2, 4)$. We graph these points.

Step 7. From left to right, we draw a smooth curve through all the points (Figure 5).

Optionally, a table of values can be found by using synthetic division (Section 9.6), although we may not find the vertex this way. The table is shown below.

$f(x) = x^2 \quad -x \quad -2$

x	1	−1	−2 $\;y = f(x)$
−2	1	−3	4
−1	1	−2	0
0	1	−1	−2
1	1	0	−2
2	1	1	0
3	1	2	4

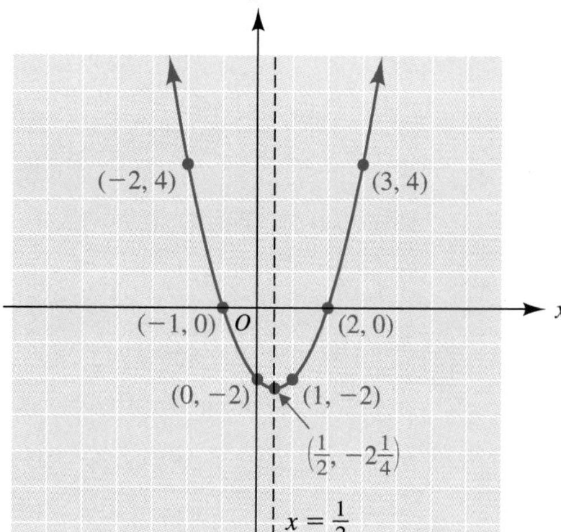

Figure 5

Alternative Methods for Graphing Parabolas

Recall that the graph of $y = f(x) + h$ can be obtained from the graph of $y = f(x)$ by shifting the graph of $y = f(x)$ upward or downward and that the graph of $y = f(x + h)$ can be obtained from the graph of $y = f(x)$ by shifting the graph of $y = f(x)$ toward the right or the left. Therefore, the graph of $y - k = (x - h)^2$ has the exact same *size* and *shape* as the graph of $y = x^2$; however, the *vertex* of $y - k = (x - h)^2$ is at the point (h, k) instead of at $(0, 0)$. The equation of the axis of symmetry is $x = h$.

EXAMPLE 7 Graph $y - 2 = (x - 3)^2$.

SOLUTION Since $y - 2 = (x - 3)^2$ fits the pattern $y - k = (x - h)^2$ if $k = 2$ and $h = 3$, the vertex of the parabola is at $(3, 2)$ and the parabola opens upward. The equation of the axis of symmetry is $x = 3$. By letting x equal zero, we find that the y-intercept is $(0, 11)$; $(0, 11)$ lies 3 units to the *left* of the axis of symmetry. To find its symmetric point, we must move 3 units (along the line $y = 11$) to the *right* of $x = 3$; this gives the point $(6, 11)$. The curve is shown in Figure 6.

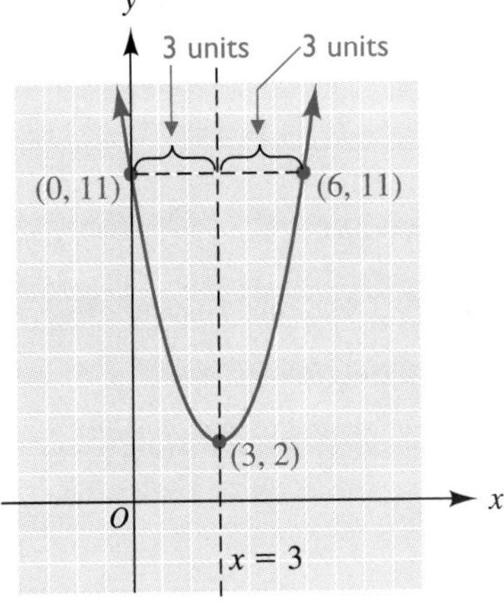

Figure 6

Note The equation for the parabola from Example 7 *could* have been expressed as $y = x^2 - 6x + 11$. In a later section, we discuss the form $y - k = (x - h)^2$ in more detail, and we discuss how to complete the square of $y = x^2 - 6x + 11$ to get it into the form $y - k = (x - h)^2$.

EXAMPLE 8 Graph $y = (x - 4)^2 - 1$.

SOLUTION To convert the given equation into the form $y - k = (x - h)^2$, we add 1 to both sides of the equation, obtaining $y + 1 = (x - 4)^2$, or $y - (-1) = (x - 4)^2$. Therefore, $k = -1$ and $h = 4$, so the vertex is at $(4, -1)$ and the equation of the axis of symmetry is $x = 4$. [If $x = 0$, $y = (0 - 4)^2 - 1 = 15$, so the y-intercept is $(0, 15)$; we will not show this point on the graph.] You can verify that the x-intercepts are $(3, 0)$ and $(5, 0)$. If we let $x = 2$, $y = (2 - 4)^2 - 1 = 3$, so another point on the curve is $(2, 3)$; its symmetric point is $(6, 3)$. The curve is shown in Figure 7.

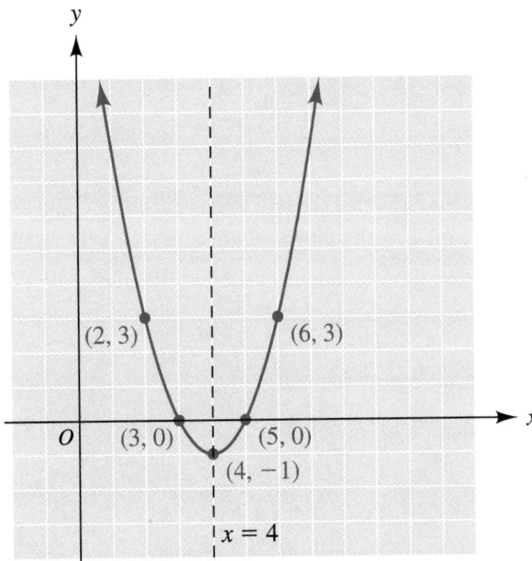

Figure 7

Minimum and Maximum Values

If a parabola opens upward, then the vertex will be the lowest point on the curve. In this case, the y-coordinate of the vertex is known as the **minimum value** of the function.

If a parabola opens downward, then the vertex will be the highest point on the curve, in which case the y-coordinate of the vertex is known as the **maximum value** of the function.

In other words, the quadratic function $y = f(x) = ax^2 + bx + c$ has a *minimum* if a is positive and a *maximum* if a is negative. The minimum or maximum is $f\left(-\dfrac{b}{2a}\right)$.

Note The *maximum* or *minimum* refers to the value of the y-coordinate of the vertex *only*.

EXAMPLE 9 Find the maximum or minimum value of $f(x) = 3x^2 - 6x + 2$.

SOLUTION Because a is *positive*, $f(x)$ has a *minimum* value.

$$\text{Minimum value} = f\left(-\frac{b}{2a}\right) = f\left(-\frac{-6}{2(3)}\right) = f(1)$$

$$= 3(1)^2 - 6(1) + 2 = 3 - 6 + 2 = -1$$

EXAMPLE 10 Find the maximum or minimum value of $f(x) = 5 - 2x - 4x^2$.

SOLUTION In the general form, the equation is $f(x) = -4x^2 - 2x + 5$. Because a is *negative*, $f(x)$ has a *maximum* value.

$$\text{Maximum value} = f\left(-\frac{b}{2a}\right) = f\left(-\frac{-2}{2(-4)}\right) = f\left(-\frac{1}{4}\right)$$

$$= -4\left(-\tfrac{1}{4}\right)^2 - 2\left(-\tfrac{1}{4}\right) + 5 = -\tfrac{1}{4} + \tfrac{2}{4} + 5 = 5\tfrac{1}{4} = \tfrac{21}{4}$$

Exercises 11.5
Set I

In Exercises 1–26, analyze and graph the functions.

1. $y = f(x) = x^2 - 2x - 3$
2. $y = f(x) = x^2 - 2x - 15$
3. $y = f(x) = x^2 - 2x - 13$
4. $y = f(x) = x^2 - 4x - 8$
5. $y = f(x) = 3 + x^2 - 4x$
6. $y = f(x) = 2x - 8 + x^2$
7. $y = f(x) = x^2 + 3x - 10$
8. $y = f(x) = 2x^2 - 7x - 4$
9. $f(x) = x^2 - 8x + 12$
10. $f(x) = x^2 - 4x - 6$
11. $f(x) = 2x - x^2 + 3$
12. $f(x) = 5 - 4x - x^2$
13. $y = f(x) = x^2 - 6x + 10$
14. $y = f(x) = x^2 - 6x + 11$
15. $y = f(x) = 4x^2 - 9$
16. $y = f(x) = 9x^2 - 4$
17. $f(x) = 6 + x^2 - 4x$
18. $f(x) = 2x - x^2 - 2$
19. $y = x^2 - 6$
20. $y = x^2 - 3$
21. $y - 3 = (x + 2)^2$
22. $y - 1 = (x + 3)^2$
23. $y + 4 = (x - 1)^2$
24. $y + 2 = (x - 1)^2$
25. $y = (x - 1)^2 + 3$
26. $y = (x - 3)^2 + 1$

In Exercises 27–32, find the maximum or minimum value of and the vertex of each quadratic function.

27. $f(x) = x^2 - 6x + 7$
28. $f(x) = x^2 - 4x + 5$
29. $f(x) = 8x - 2x^2 - 3$
30. $f(x) = 4 + 6x - 3x^2$
31. $f(x) = -\frac{1}{2}x^2 + x + \frac{3}{2}$
32. $f(x) = -\frac{2}{3}x^2 - \frac{8}{3}x + \frac{1}{3}$

Writing Problems

Express the answers in your own words and in complete sentences.

1. Explain how to find the vertex of a parabola.
2. Explain how to find the x-intercepts of a parabola.
3. Explain how it might be possible for a parabola to have no x-intercepts.
4. Explain how the graph of $y - k = (x - h)^2$ is related to the graph of $y = x^2$.

Exercises 11.5 Set II

In Exercises 1–26, analyze and graph the functions.

1. $y = f(x) = x^2 + x - 2$
2. $y = f(x) = 2 + x - x^2$
3. $y = f(x) = 2x^2 - x - 6$
4. $y = f(x) = 3x^2 - 3x - 6$
5. $y = f(x) = 3x^2 - 10x + 6$
6. $y = f(x) = 3 + 5x - 2x^2$
7. $y = f(x) = x^2 - 2x - 8$
8. $y = f(x) = x^2 + 4x + 4$
9. $y = f(x) = 6x^2 - 7x - 3$
10. $y = f(x) = x^2 + 2x + 4$
11. $y = f(x) = 2x^2 + 7x - 4$
12. $y = f(x) = 2x - x^2 - 1$
13. $y = f(x) = 30x - 9x^2 - 25$
14. $y = f(x) = 3 + 2x - x^2$
15. $y = f(x) = x^2 - 1$
16. $y = f(x) = 4 - x^2$
17. $f(x) = 2x^2 - 4x + 1$
18. $f(x) = 3x^2 - 6x + 1$
19. $y = x^2 - 2$
20. $y = (x - 2)^2$
21. $y - 2 = (x + 1)^2$
22. $y = (x + 2)^2 - 2$
23. $y + 3 = (x - 2)^2$
24. $y = (x + 1)^2 - 4$
25. $y = (x - 2)^2 + 5$
26. $y = (x + 5)^2 - 1$

In Exercises 27–32, find the maximum or minimum value of and the vertex of each quadratic function.

27. $y = f(x) = 4x^2 - 8x + 7$
28. $y = f(x) = x^2$
29. $y = f(x) = 5x^2 + 10x - 3$
30. $y = f(x) = 3x^2 - x$
31. $y = f(x) = x^2 + 2x - 8$
32. $y = f(x) = 5x^2 - 1$

11.6 Graphing Conic Sections

Conic sections are curves formed when a plane cuts through a cone or a pair of cones. (See Figures 8, 9, 10, and 11.) Parabolas, circles, ellipses, and hyperbolas are all conic sections. Circles, ellipses, and hyperbolas are never functions; a parabola is a function only if its axis of symmetry is a vertical line.

In this book, we will consider only those conic sections that have a vertical or horizontal axis of symmetry and only those ellipses and hyperbolas that have their centers at the origin.

The Equations of the Conic Sections

General equations of parabolas

Vertical axis of symmetry: $y = f(x) = ax^2 + bx + c$ ← The axis is $x = -\dfrac{b}{2a}$

or $y - k = a(x - h)^2$ ← The vertex is at (h, k) and the axis is $x = h$

The parabola opens upward if $a > 0$ and downward if $a < 0$.

Horizontal axis of symmetry: $x = ay^2 + by + c$ ← The axis is $y = -\dfrac{b}{2a}$

or $x - h = a(y - k)^2$ ← The vertex is at (h, k) and the axis is $y = k$

The parabola opens to the right if $a > 0$ and to the left if $a < 0$.

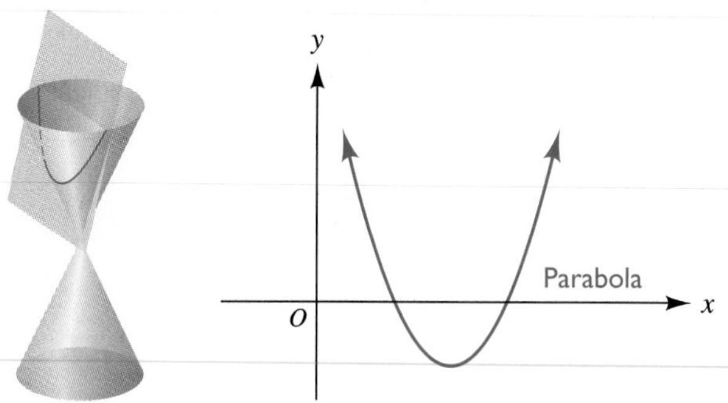

Figure 8

General equations of circles

Center at the origin:	$x^2 + y^2 = r^2$
Center at (h, k):	$(x - h)^2 + (y - k)^2 = r^2$
The radius is r in both cases.	

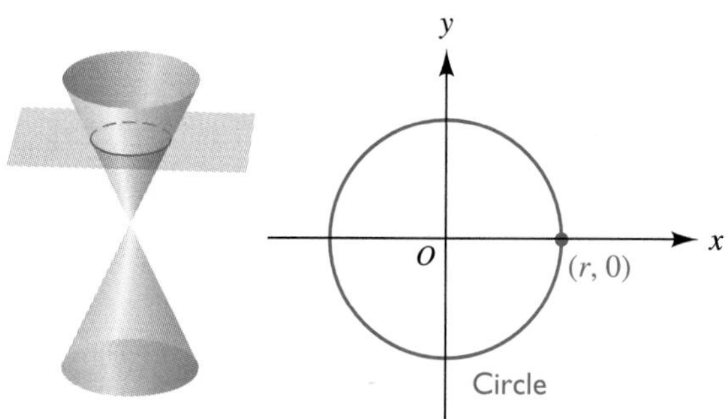

Figure 9

The general equation of an ellipse

Center at the origin:	$\dfrac{x^2}{a^2} + \dfrac{y^2}{b^2} = 1$	The x-intercepts are $(\pm a, 0)$ The y-intercepts are $(0, \pm b)$
where $a \neq 0$ and $b \neq 0$.		

Note If c, d, and e all have the same sign (and if $c \neq 0$, $d \neq 0$, and $e \neq 0$), then the graph of an equation that can be expressed in the form $cx^2 + dy^2 = e$ is always an ellipse; we can convert the equation to the general form by dividing both sides of the equation by e.

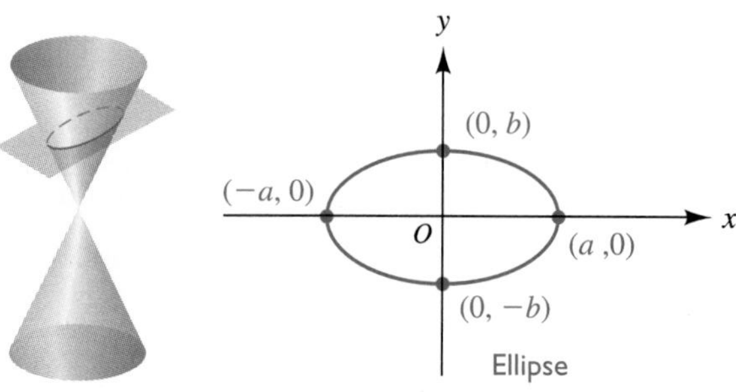

Figure 10

General equations of hyperbolas

Center at the origin:	$\dfrac{x^2}{a^2} - \dfrac{y^2}{b^2} = 1$	The x-intercepts are $(\pm a, 0)$ There are no y-intercepts
where $a \neq 0$ and $b \neq 0$.		
Center at the origin:	$\dfrac{y^2}{c^2} - \dfrac{x^2}{d^2} = 1$	The y-intercepts are $(0, \pm c)$ There are no x-intercepts
where $c \neq 0$ and $d \neq 0$.		

Note If $f > 0$, $g > 0$, and $h > 0$, then the graph of an equation that can be expressed in the form $fx^2 - gy^2 = h$ is always a hyperbola, and so is the graph of an equation that can be expressed in the form $gy^2 - fx^2 = h$; we can convert these equations to the general form by dividing both sides of each equation by h.

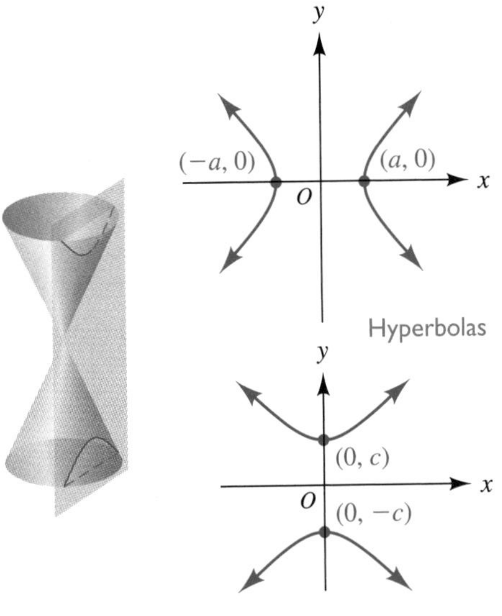

Figure 11

How to Determine Whether an Equation Is That of a Parabola, a Circle, an Ellipse, or a Hyperbola Notice that in the equation of a parabola, either x is squared or y is squared, *but not both.* In the equations of the other conic sections, x and y are *both* squared. Both equations of the hyperbola have a *negative* sign between the x^2-term and the y^2-term, whereas the circle and the ellipse both have a *plus* sign there. How can you tell the equation of a circle from that of an ellipse? In the equation for a circle, the coefficients of x^2 and y^2 are equal; in the equation for an ellipse, the coefficients are *not* equal.

Graphing Conic Sections

EXAMPLE 1 Name and graph $9x^2 + 4y^2 = 36$.

SOLUTION Because both x and y are squared, because there is a plus sign between the x^2-term and the y^2-term, and because the coefficients of x^2 and y^2 are *not* numerically equal, the curve is an *ellipse*. If we divide both sides of the equation by 36, we have

$$\frac{9x^2}{36} + \frac{4y^2}{36} = \frac{36}{36}$$

or $$\frac{x^2}{4} + \frac{y^2}{9} = 1 \qquad \text{This equation is in the form } \frac{x^2}{a^2} + \frac{y^2}{b^2} = 1$$

Then $a^2 = 4$ and $b^2 = 9$, so the x-intercepts are $(2, 0)$ and $(-2, 0)$ and the y-intercepts are $(0, 3)$ and $(0, -3)$. (The intercepts could also be found by letting $y = 0$ and solving for x and then letting $x = 0$ and solving for y.) The graph is shown in Figure 12.

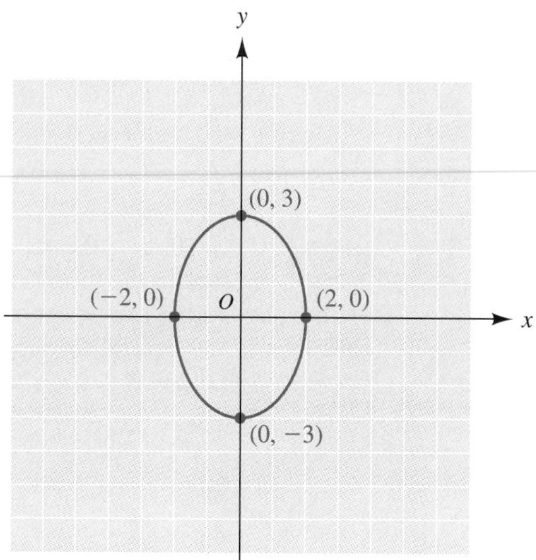

Figure 12

Note $9x^2 + 4y^2 = 36$ is *not* a function, and with the graphics calculator, we can graph only functions (we have to tell the calculator what y equals). We can, however, solve the equation for y and get two separate equations that *are* functions, as follows:

$$9x^2 + 4y^2 = 36$$

$$4y^2 = 36 - 9x^2$$

$$y^2 = \frac{36 - 9x^2}{4} = \frac{9(4 - x^2)}{4}$$

Using Property 1

$$y = \pm\sqrt{\frac{9(4 - x^2)}{4}} = \pm\frac{3}{2}\sqrt{4 - x^2} = \pm 1.5\sqrt{4 - x^2}$$

We can now let $y_1 = 1.5\sqrt{4 - x^2}$ and $y_2 = -1.5\sqrt{4 - x^2}$ and graph these functions together, as follows:

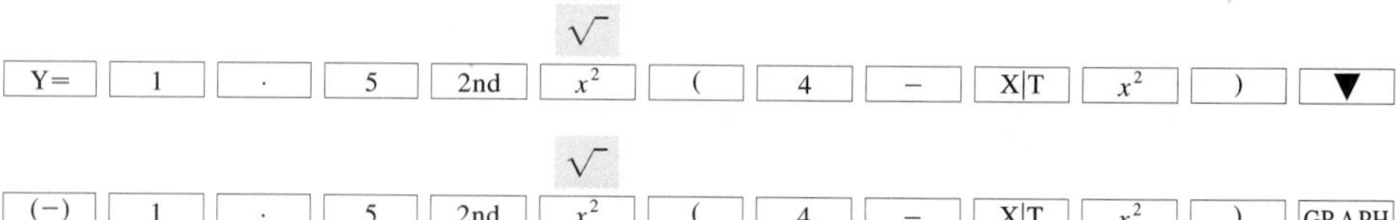

(The parentheses are very important. Without them, the calculator would interpret the equations as $y_1 = 1.5\sqrt{4} - x^2 = 3 - x^2$ and $y_2 = -1.5\sqrt{4} - x^2 = -3 - x^2$ and would graph the wrong curves!) Set the range so that $x$min is -6, $x$max is 6, $y$min is -4, and $y$max is 4.

If we want to *trace* the curve, we must press ▼ and/or ▲ to get from one branch of the curve to the other (that is, to get from the graph of $y_1 = 1.5\sqrt{4 - x^2}$ to the graph of $y_2 = -1.5\sqrt{4 - x^2}$).

EXAMPLE 2 Name and graph $2y^2 - 4y + x = 0$.

SOLUTION Because y is squared and x is not, we know that the curve is a *parabola* with a horizontal axis of symmetry. We want to get the equation into the form

$$x - h = a(y - k)^2$$

Therefore, we will *complete the square* for the y-terms. We start by getting a *positive x* on one side of the equation and all other terms on the other side.

$2y^2 - 4y + x = 0$

$x = -2y^2 + 4y$ — Adding $-2y^2 + 4y$ to both sides

$x = -2(y^2 - 2y + \quad)$ — Factoring out the coefficient of y^2

We now want to complete the square for $y^2 - 2y$; therefore, we must add $\left(\frac{1}{2}[-2]\right)^2$, or $+1$, *inside* the parentheses

$x - 2 = -2(y^2 - 2y + 1)$ — (Because of the -2 *outside* the parentheses, we're really adding -2 to the right side and must add -2 to the left side also)

$x - 2 = -2(y - 1)^2$

This equation fits the pattern $x - h = a(y - k)^2$, where $h = 2$, $k = 1$, and $a = -2$. Therefore, the curve is a parabola that has its vertex at (2, 1); since a is negative, the parabola opens to the left. The axis of symmetry is the line $y = k$, or $y = 1$. The parabola passes through the point (0, 0), since $x = 0$ if $y = 0$. By symmetry, one other point must be (0, 2). The graph is shown in Figure 13.

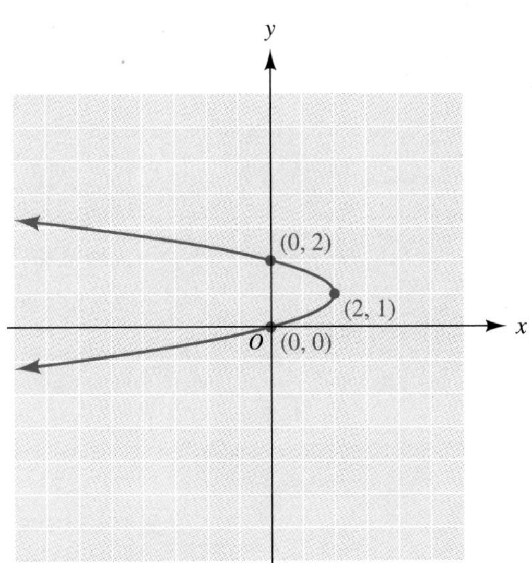

Figure 13

Note $2y^2 - 4y + x = 0$ is not a function, so to graph it with a calculator, we must solve for y, obtaining two equations. There are two ways to do this.

Method 1 We can complete the square to find y; when we completed the square in Example 2, our last equation was

$$x - 2 = -2(y - 1)^2$$

Then $(y - 1)^2 = \dfrac{2 - x}{2}$ — Dividing both sides by -2

$y - 1 = \pm\sqrt{\dfrac{2 - x}{2}}$ — Using Property 1

$y - 1 = \pm\sqrt{\dfrac{(2 - x)(2)}{2(2)}}$ — Rationalizing the denominator

Therefore, $y = 1 \pm \dfrac{\sqrt{4 - 2x}}{2}$ — Adding 1 to both sides to solve for y

Method 2 The equation $2y^2 - 4y + x = 0$ is quadratic in y; therefore, we can solve the equation for y by using the quadratic formula. (We have not previously used the quadratic formula when there were *two* variables in the problem; in this case, we treat x as if it were the constant.) $a = 2$, $b = -4$, and $c = x$. *Remember*: The *variable* is y.

$$y = \frac{-b \pm \sqrt{b^2 - 4ac}}{2a}$$

$$y = \frac{-(-4) \pm \sqrt{(-4)^2 - 4(2)(x)}}{2(2)}$$

$$y = \frac{4 \pm \sqrt{16 - 8x}}{4} = \frac{4 \pm \sqrt{8(2 - x)}}{4} = \frac{4 \pm 2\sqrt{2(2 - x)}}{4} = \frac{4}{4} \pm \frac{2\sqrt{2(2 - x)}}{4}$$

or $y = 1 \pm \dfrac{\sqrt{4 - 2x}}{2}$

Now that we have solved for y, we can separate the two functions. We can let $y_1 = 1 + \dfrac{\sqrt{4 - 2x}}{2}$ and let $y_2 = 1 - \dfrac{\sqrt{4 - 2x}}{2}$ and graph these functions together, as follows:

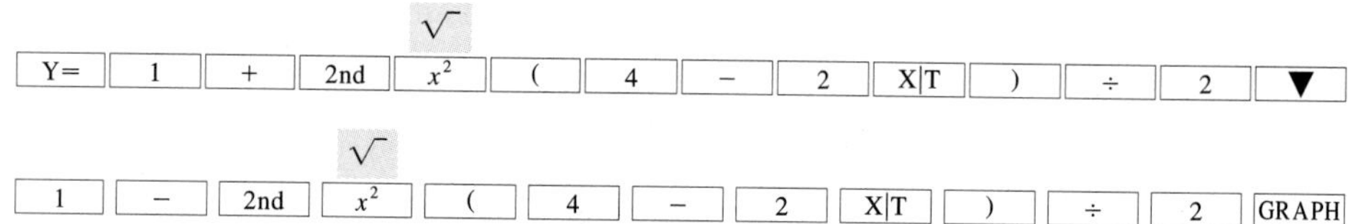

Y= | 1 | + | 2nd | x² (√) | (| 4 | − | 2 | X|T |) | ÷ | 2 | ▼
1 | − | 2nd | x² (√) | (| 4 | − | 2 | X|T |) | ÷ | 2 | GRAPH

(Again, the parentheses are very important. Without them, the calculator would interpret the equations as $y_1 = 1 + \sqrt{4} - \dfrac{2x}{2}$ and $y_2 = 1 - \sqrt{4} - \dfrac{2x}{2}$ and would graph two straight lines!) Set the range so that $x$min is -7, $x$max is 5, $y$min is -3, and $y$max is 5.

EXAMPLE 3 Name and graph $x^2 + y^2 - 6x + 2y - 6 = 0$.

SOLUTION Because x and y are both squared and because the coefficients of the squared terms are equal, this conic section must be a *circle*. We want to get the equation into the form

$$(x - h)^2 + (y - k)^2 = r^2$$

Therefore, we will complete the squares for both x and y. We start by getting the x-term immediately to the right of the x^2-term, the y-term immediately to the right of the y^2-term, and the constant on the right side of the equation.

$$x^2 + y^2 - 6x + 2y - 6 = 0$$

$$(x^2 - 6x + \quad) + (y^2 + 2y + \quad) = 6 \quad \leftarrow \text{Rearranging terms and adding 6 to both sides}$$

$$\left(\tfrac{1}{2}[-6]\right)^2 = 9$$

$$(x^2 - 6x + 9) + (y^2 + 2y + 1) = 6 + 9 + 1 \quad \text{Adding 9 and 1 to both sides to complete the squares}$$

$$\left(\tfrac{1}{2}[2]\right)^2 = 1$$

$$(x - 3)^2 + (y + 1)^2 = 16 = 4^2$$

This equation fits the pattern $(x - h)^2 + (y - k)^2 = r^2$, where $h = 3$, $k = -1$, and $r = 4$. Therefore, the graph will be a circle of radius 4 with its center (indicated on the graph by the hollow dot) at the point $(3, -1)$ (see Figure 14).

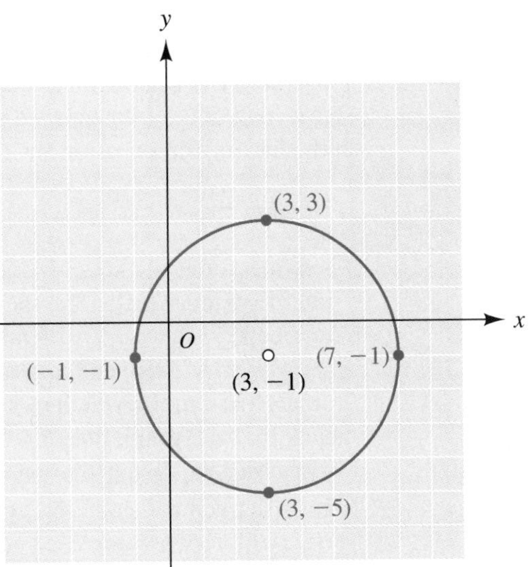

Figure 14

Note To graph $x^2 + y^2 - 6x + 2y - 6 = 0$ with a graphics calculator, we must solve the equation for y. You can verify that when we do this (by using Property 1 after we have completed the square or by using the quadratic formula), we get $y = -1 \pm \sqrt{7 + 6x - x^2}$. We can then graph the functions $y_1 = -1 + \sqrt{7 + 6x - x^2}$ and $y_2 = -1 - \sqrt{7 + 6x - x^2}$ together.

If the range is set so that $x$min is -4, $x$max is 10, $y$min is -6, and $y$max is 4, the curve will look like a circle. Otherwise, it may look like an ellipse. (Be sure to put parentheses around the radicands as you enter the equations into the calculator.)

The Asymptotes of a Hyperbola Associated with each hyperbola are two straight lines called the **asymptotes** of the hyperbola. The hyperbola gets closer to these straight lines as the points of the graph get farther from the center of the hyperbola, but the curve never touches the asymptotes. The asymptotes are helpful in sketching the graph of the hyperbola, but they are *not* part of the graph itself.

If the equation of a hyperbola can be expressed in the form $\frac{x^2}{a^2} - \frac{y^2}{b^2} = 1$, one asymptote passes through the origin and through the point (a, b) while the other passes through the origin and through the point $(-a, b)$. The rectangle with vertices at (a, b), $(-a, b)$, $(-a, -b)$, and $(a, -b)$ is called the **rectangle of reference** for the hyperbola $\frac{x^2}{a^2} - \frac{y^2}{b^2} = 1$; the asymptotes of the hyperbola are the diagonals of this rectangle.

If the equation of a hyperbola can be expressed in the form $\frac{y^2}{c^2} - \frac{x^2}{d^2} = 1$, one asymptote passes through the origin and through the point (d, c) while the other passes through the origin and through the point $(-d, c)$. The rectangle with vertices at (d, c), $(-d, c)$, $(-d, -c)$, and $(d, -c)$ is called the rectangle of reference for the hyperbola $\frac{y^2}{c^2} - \frac{x^2}{d^2} = 1$.

EXAMPLE 4 Name and graph $3y^2 - 4x^2 = 12$.

SOLUTION In this equation, x and y are both squared, but there is a *negative* sign between the two squared terms; therefore, the curve is a *hyperbola*. If we divide both sides of the equation by 12, we have

$$\frac{3y^2}{12} - \frac{4x^2}{12} = \frac{12}{12} \quad \text{or} \quad \frac{y^2}{4} - \frac{x^2}{3} = 1 \longleftarrow \text{This equation is in the form } \frac{y^2}{c^2} - \frac{x^2}{d^2} = 1$$

Therefore, $c^2 = 4$, so $c = \pm 2$, and the y-intercepts are (0, 2) and (0, −2). (The y-intercepts could also be found by letting $x = 0$ and solving for y.) There are *no* x-intercepts. (If we try to find the x-intercepts by letting $y = 0$, we obtain the solutions $x = \pm i\sqrt{3}$; since these are imaginary numbers, they cannot be graphed in a coordinate system in which points on both axes represent only real numbers.) This means that the graph does not intersect the x-axis.

In order to sketch the asymptotes, we need to identify d. Comparing our equation to the general equation again, we see that $d^2 = 3$, or $d = \pm\sqrt{3}$. Therefore, the vertices of the rectangle of reference are $(\sqrt{3}, 2)$, $(-\sqrt{3}, 2)$, $(-\sqrt{3}, -2)$, and $(\sqrt{3}, -2)$. We can now sketch the asymptotes of the hyperbola, as they are the diagonals of this rectangle (see Figure 15).

A Word of Caution The hyperbola does *not* pass through the points $(\sqrt{3}, 2)$, $(-\sqrt{3}, 2)$, $(-\sqrt{3}, -2)$, and $(\sqrt{3}, -2)$.

Let's find some other points on the graph. If $x = 3$, $y = \pm 4$; therefore, the graph passes through the points (3, 4) and (3, −4). If $x = -3$, $y = \pm 4$, so two more points on the curve are (−3, 4) and (−3, −4). We know what a hyperbola should look like (see Figure 11), so we can draw the graph. See Figure 16.

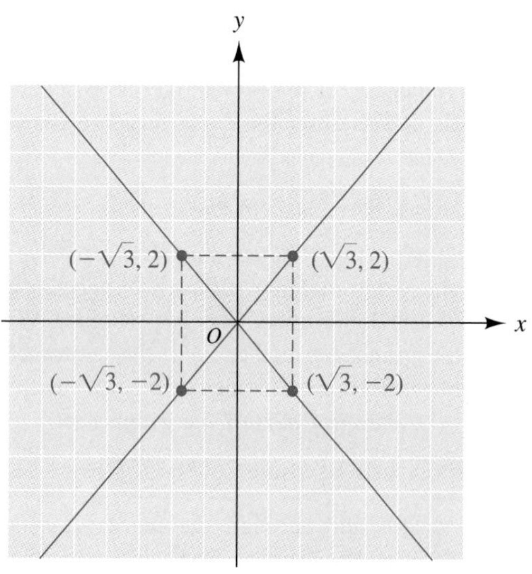

Figure 15 The asymptotes of the hyperbola

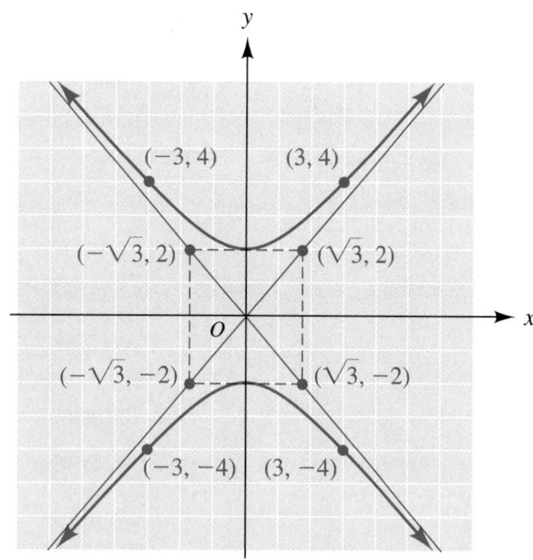

Figure 16 The hyperbola and its asymptotes

Note As usual, in order to graph $3y^2 - 4x^2 = 12$ with a graphics calculator, we must solve for y; you can verify that when we do this, we get $y = \pm 2\sqrt{\dfrac{x^2 + 3}{3}}$. Then we can graph the functions $y_1 = 2\sqrt{\dfrac{x^2 + 3}{3}}$ and $y_2 = -2\sqrt{\dfrac{x^2 + 3}{3}}$ together. (Be sure to use parentheses around the numerator of the fraction *and* around the radicand.)

Exercises 11.6
Set I

Identify (name) and sketch each conic section.

1. $x^2 + y^2 = 9$
2. $x^2 + y^2 = 4$
3. $y = x^2 + 2x - 3$
4. $y = 6 - x - x^2$
5. $3x^2 + 3y^2 = 21$
6. $3x^2 + 4y^2 = 12$
7. $4x^2 + 9y^2 = 36$
8. $16x^2 + y^2 = 16$
9. $9x^2 - 4y^2 = 36$
10. $x^2 - 4y^2 = 4$
11. $y^2 = 8x$
12. $y^2 = -4x$
13. $x^2 + y^2 + 2x - 4y + 1 = 0$
14. $x^2 + y^2 - 4x + 6y + 12 = 0$
15. $y^2 + 2y + x = 0$
16. $2y^2 - 4y - x = 0$

Writing Problems

Express the answers in your own words and in complete sentences.

1. Explain how to tell *from the equation* which of the conic sections $4x^2 - y^2 = 4$ is.
2. Explain how to tell *from the equation* which of the conic sections $4x^2 - y = 4$ is.
3. Explain how to tell *from the equation* which of the conic sections $4x^2 + y^2 = 4$ is.
4. Explain how to tell *from the equation* which of the conic sections $x^2 + y^2 = 4$ is.

Exercises 11.6
Set II

Identify (name) and sketch each conic section.

1. $x^2 + y^2 = 20$
2. $\dfrac{x^2}{9} + \dfrac{y^2}{4} = 1$
3. $y = 5 + 4x - x^2$
4. $x^2 - \dfrac{y^2}{4} = 1$
5. $2x^2 + 2y^2 = 8$
6. $4x^2 + y^2 = 4$
7. $9x^2 + 16y^2 = 144$
8. $x^2 = 6y$
9. $4y^2 - 16x^2 = 64$
10. $x^2 + y^2 = 9$
11. $y^2 = 4x$
12. $x^2 - y^2 = 9$
13. $x^2 + y^2 + 10x - 4y + 20 = 0$
14. $x^2 + y^2 - 4x + 2y - 11 = 0$
15. $y^2 - 4y - x = 0$
16. $y^2 + 4y - 2x = 0$

11.7 Solving Nonlinear Inequalities in One Variable

Earlier, we discussed solving linear inequalities in one variable and graphing their solution sets on the real number line. In this section, we show two different methods of solving *nonlinear* inequalities in one variable. The solution sets can be written in set-builder notation or in interval notation, and they can also be graphed on the real number line.

Solving a nonlinear inequality in one variable

1. For *either* method, temporarily substitute an equal sign for the inequality symbol.
2. Solve the resulting equation. (For method 1, the solutions will be the x-intercepts of the curve that will be sketched. For method 2, the solutions will be the *critical points* that separate the real number line into several intervals.) If there are variables in a denominator, *set the denominator equal to zero* to find the additional critical points, and use method 2 (see Examples 5 and 6).
3. Proceed with method 1 or method 2. (Both methods are demonstrated and explained in the following examples.)

EXAMPLE 1

Find the solution set of $x^2 + x - 2 > 0$, and graph the solution set on the number line.

SOLUTION

Step 1. $x^2 + x - 2 = 0$ Substituting $=$ for $>$

Step 2. $(x + 2)(x - 1) = 0$

$x + 2 = 0 \quad | \quad x - 1 = 0$

$x = -2 \quad | \quad x = 1$

Step 3. We will show both methods of solution.

Method 1 We first substitute "$= y$" for "> 0." Thus, we obtain the equation of the related quadratic function $x^2 + x - 2 = y$. This is the equation of a parabola that opens upward. Because $c = -2$, the y-intercept is -2. Moreover, we found in step 2 that the x-intercepts are -2 and 1. With this information, we can make a rough sketch of the parabola (see Figure 17). Because the inequality is $>$, not $\geq$, we graph the x-intercepts as hollow dots; these points will *not* be in the final solution set of the given inequality.

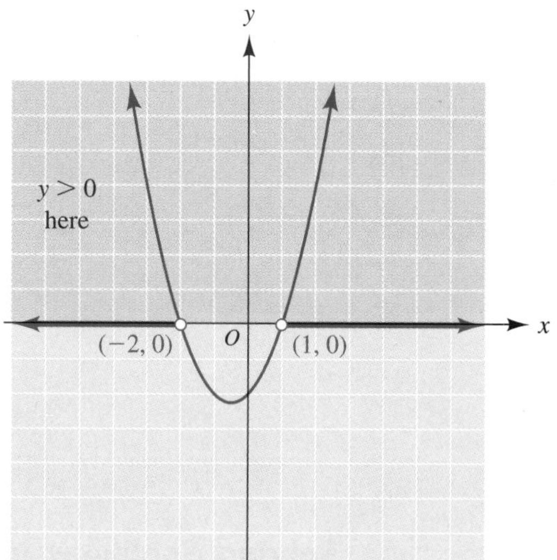

Figure 17

Because we have let $y = x^2 + x - 2$ and because we are solving $x^2 + x - 2 > 0$, we now need to decide which x-values make $y > 0$. We see from Figure 17 that the parabola lies *above* the x-axis for $x < -2$ and for $x > 1$. These values are shown by the heavy horizontal arrows in Figure 17.

Method 2 We found the critical points (-2 and 1) in step 2; these are the only points where the sign of the function can change. We graph these points on the real number line; we use *hollow* dots for now because the inequality is $>$. We see below that these two points separate the number line into three intervals: $x < -2$, $-2 < x < 1$, and $x > 1$, or, in interval notation, $(-\infty, -2)$, $(-2, 1)$, and $(1, +\infty)$.

$x < -2$ $\quad -2 < x < 1 \quad$ $x > 1$

$-4 \ -3 \ -2 \ -1 \ 0 \ 1 \ 2 \ 3 \ 4$

There are two ways to decide whether or not numbers in each of these intervals are in the solution set:

1. We can indicate above the number line what the sign of each factor is. The factor $x + 2$ is negative for all $x < -2$, so we write a string of negative signs to the left of -2; $x + 2$ is positive for all $x > -2$, so we write a string of plus signs to the right of -2. The factor $x - 1$ is negative for all $x < 1$, so we write a string of negative signs to the left of 1; $x - 1$ is positive for all $x > 1$, so we write a string of plus signs to the right of 1. This method is shown below.

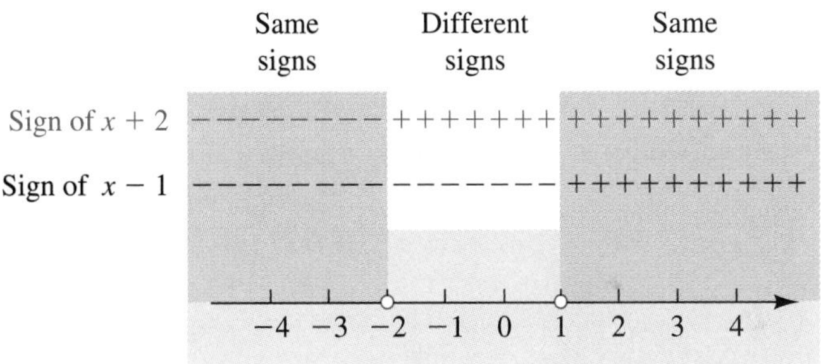

Because we're interested in the interval(s) on which the *product* $(x + 2)(x - 1)$ is *positive*, we select the interval(s) where both signs are the same.

2. We can, instead, *test* one number in each interval. For the interval $(-\infty, -2)$, we can let $x = -3$; if $x = -3$, $(x + 2)$ is negative, as is $(x - 1)$, so their *product is positive*. We indicate this on the sketch by writing $(-)(-) = +$ above the lefthand portion of the line. [You might test several other numbers less than -2 to see that the product $(x + 2)(x - 1)$ is positive for *all* $x < -2$.] For the interval $(-2, 1)$, we can let $x = 0$; if $x = 0$, $(x + 2)$ is positive and $(x - 1)$ is negative, so their *product* is negative. For the interval $(1, +\infty)$, we can let $x = 2$; if $x = 2$, $(x + 2)$ is positive, as is $(x - 1)$, so their *product is positive*. Because we're solving $(x + 2)(x - 1) > 0$, the solution set must include only the intervals on which the *product is positive*—the intervals $(-\infty, -2)$ and $(1, +\infty)$.

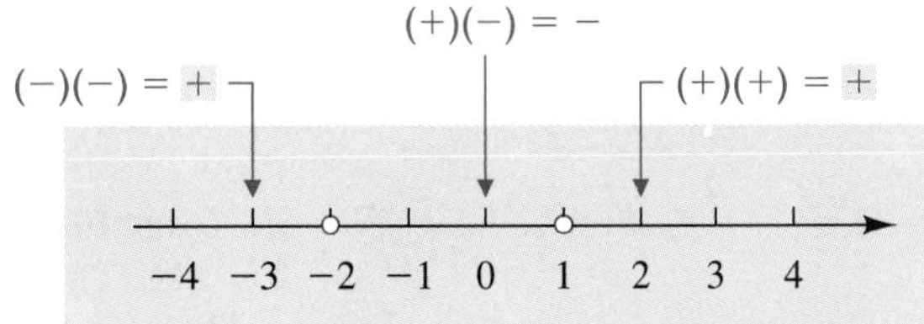

By *either* method, the final solution set is $\{x \mid x < -2\} \cup \{x \mid x > 1\}$, or, in interval notation, $(-\infty, -2) \cup (1, +\infty)$. This means that the solution set consists of all $x < -2$, *as well as* all $x > 1$. The graph of the solution set is as follows:

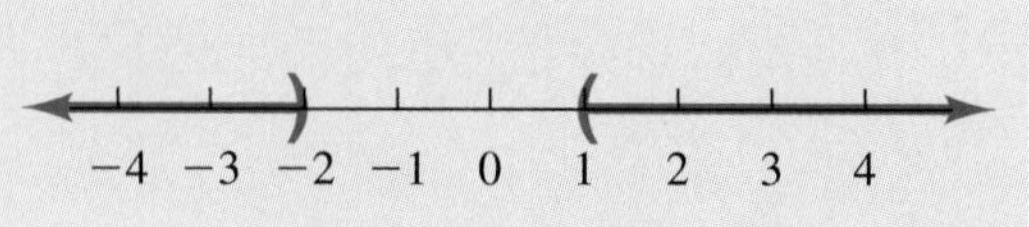

EXAMPLE 2 Find the domain of the function $y = f(x) = \sqrt{8 - 10x - 3x^2}$.

SOLUTION We know that the radicand must be greater than or equal to 0. Therefore, we must solve the inequality $8 - 10x - 3x^2 \geq 0$.

Step 1. $8 - 10x - 3x^2 = 0$ Substituting = for ≥

Step 2. $(2 - 3x)(4 + x) = 0$

$$2 - 3x = 0 \qquad 4 + x = 0$$
$$x = \tfrac{2}{3} \qquad x = -4$$

Step 3. ***Method I*** To solve $8 - 10x - 3x^2 \geq 0$, we first substitute "$= y$" for "≥ 0." Thus, we obtain the equation of the related quadratic function $y = 8 - 10x - 3x^2$. This is the equation of a parabola that opens downward and that has a y-intercept of 8 and x-intercepts of $\frac{2}{3}$ and -4. With this information, we can make a rough sketch of the parabola (see Figure 18). Because the inequality is $\geq$, we graph the x-intercepts as solid dots; these points *will* be in the final solution set of the given inequality.

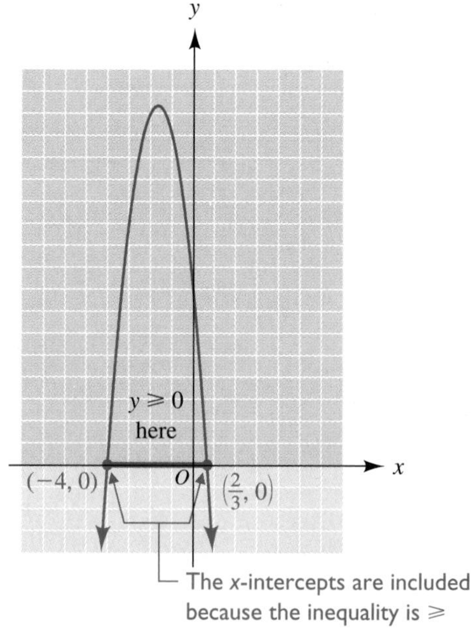

Figure 18

Because we have let $y = 8 - 10x - 3x^2$ and because we are solving $8 - 10x - 3x^2 \geq 0$, we now need to decide which x-values make $y \geq 0$. We see from Figure 18 that the parabola lies *above* or *on* the x-axis for $-4 \leq x \leq \frac{2}{3}$. These values are shown by the heavy horizontal line segment in Figure 18.

Method 2 We found the *critical points* $\left(-4 \text{ and } \frac{2}{3}\right)$ in step 2. Because the inequality is $\geq$, we graph these points as *solid dots* for now, indicating that they *will* be in the solution set. The sketch below shows strings of signs above the number line.

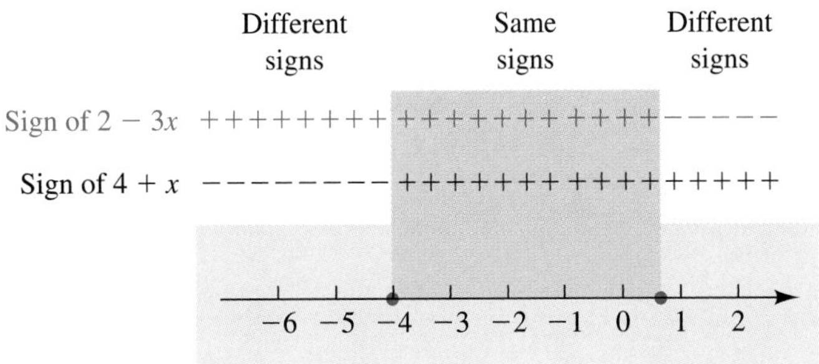

Because we want the product $(2 - 3x)(4 + x)$ to be *positive*, we select the interval on which the signs of both factors are the *same*. We must also include -4 and $\frac{2}{3}$ in the interval.

The following sketch illustrates finding the solution set by using a test-point in each interval. For the interval $(-\infty, -4)$, we let $x = -5$; if $x = -5$, $(4 + x)$ is negative and $(2 - 3x)$ is positive, so their *product* is negative. For the interval $\left(-4, \frac{2}{3}\right)$, we let $x = 0$; if $x = 0$, $(4 + x)$ is positive and $(2 - 3x)$ is positive, so their *product is positive*. For the interval $\left(\frac{2}{3}, +\infty\right)$, we let $x = 1$; if $x = 1$, $(4 + x)$ is positive and $(2 - 3x)$ is negative, so their *product* is negative.

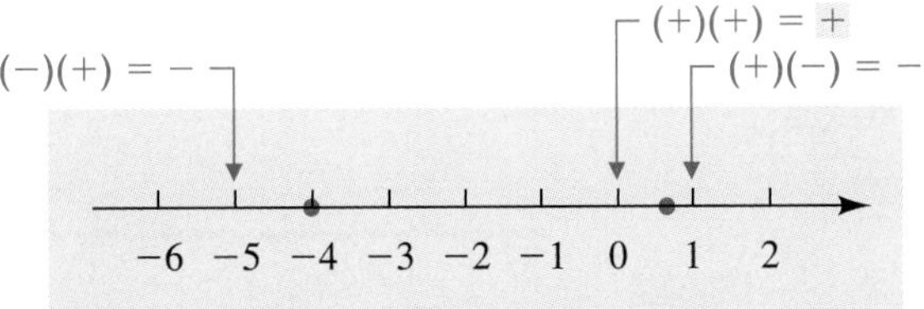

Because we want the product to be *positive* and because the endpoints are to be included in the solution set, we select the interval $\left[-4, \frac{2}{3}\right]$.

By either method, the final solution set of the inequality is $\left\{x \mid -4 \leq x \leq \frac{2}{3}\right\}$, or $\left[-4, \frac{2}{3}\right]$. The graph is as follows:

Therefore, the domain of the function $y = \sqrt{8 - 10x - 3x^2}$ is $\left\{x \mid -4 \leq x \leq \frac{2}{3}\right\}$.

Note Now that we can solve quadratic inequalities, we can find domains that we could not find in Section 9.4. In Example 3 of that section, $g \circ f$ was found to be $\sqrt{x^2 - 3}$; you can now verify that the domain is $(-\infty, -\sqrt{3}\,] \cup [\sqrt{3}, +\infty)$.

When a quadratic function cannot be factored, we can use the *quadratic formula* to find the x-intercepts for method 1 or the critical points for method 2 (see Example 3).

EXAMPLE 3 Find the solution set of $x^2 - 2x - 4 < 0$, and graph the solution set on the real number line.

SOLUTION

Step 1. $x^2 - 2x - 4 = 0$ Substituting = for <

Step 2. For the quadratic formula, $a = 1$, $b = -2$, and $c = -4$.

$$x = \frac{-b \pm \sqrt{b^2 - 4ac}}{2a} = \frac{-(-2) \pm \sqrt{(-2)^2 - 4(1)(-4)}}{2(1)}$$

$$= \frac{2 \pm \sqrt{4 + 16}}{2} = \frac{2 \pm \sqrt{20}}{2} = \frac{2 \pm 2\sqrt{5}}{2} = 1 \pm \sqrt{5} \qquad \sqrt{5} \approx 2.2$$

$$x = 1 - \sqrt{5} \approx 1 - 2.2 = -1.2 \qquad x = 1 + \sqrt{5} \approx 1 + 2.2 = 3.2$$

Step 3. We will show only method 1 for this example. We graph $y = x^2 - 2x - 4$, which is a parabola that opens upward. The y-intercept is -4, and the x-intercepts are approximately -1.2 and 3.2. (See Figure 19.)

Because we have let $y = x^2 - 2x - 4$ and because we are solving $x^2 - 2x - 4 < 0$, we now need to decide which x-values make $y < 0$. We see from Figure 19 that the parabola lies *below* the x-axis for $-1.2 < x < 3.2$. These values are shown by the heavy line segment in Figure 19. Therefore, the solution set is $\{x \mid 1 - \sqrt{5} < x < 1 + \sqrt{5}\}$, or $(1 - \sqrt{5}, 1 + \sqrt{5})$. The graph of the solution set is as follows:

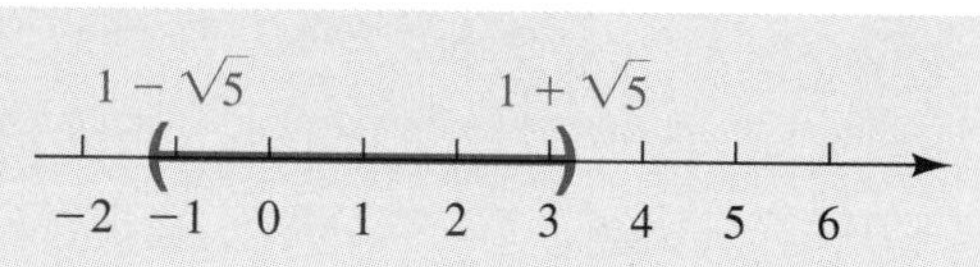

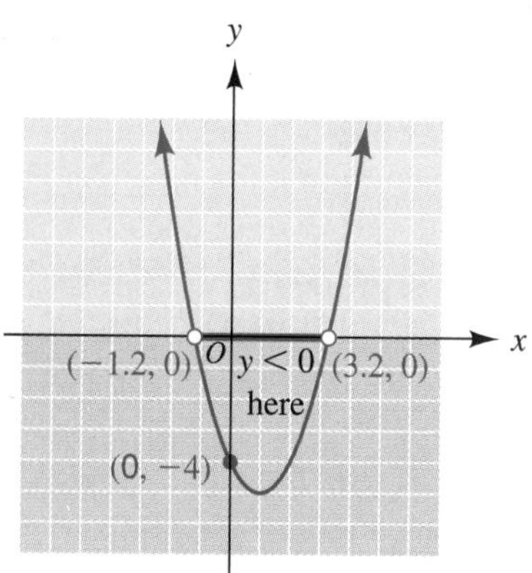

Figure 19

EXAMPLE 4 Find the solution set of $(x - 2)^2(x + 3) \geq 0$, and graph the solution set on the real number line.

SOLUTION

Step 1. $(x - 2)^2(x + 3) = 0$ Substituting = for ≥

Step 2.

$$x - 2 = 0 \qquad x + 3 = 0$$

$$x = 2 \qquad x = -3$$

Step 3. We will show only method 2 for this example. The critical points are -3 and 2. We graph them as *solid* dots, since the inequality (≥) includes the equal sign. Note that $(x - 2)^2$ is never negative.

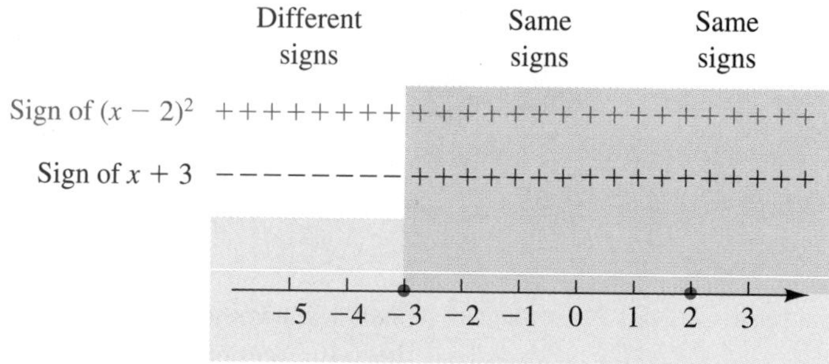

Because the inequality symbol is $\geq$, we select the interval where both signs are *the same*. The solution set, then, is $\{x \mid x \geq -3\}$, or $[-3, +\infty)$. The graph of the solution set is as follows:

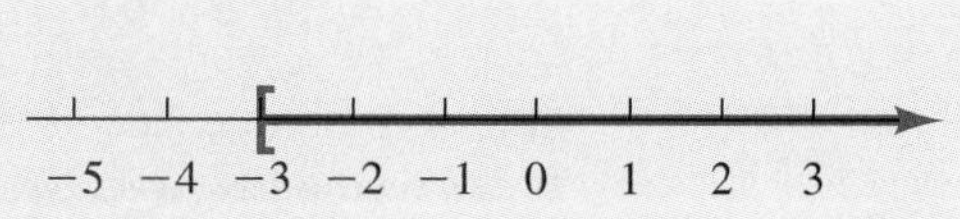

In Examples 5 and 6, the inequality contains a denominator with variables. When the denominator contains variables, *the factors of the denominator*, as well as those of the numerator, *must be set equal to zero* in order to find critical points.

EXAMPLE 5 Find the solution set of $\dfrac{x+5}{x-3} \leq 0$, and graph the solution set on the real number line.

SOLUTION Note that 3 is not in the domain.

Step 1. $\dfrac{x+5}{x-3} = 0$ Substituting = for ≤

Step 2. $x + 5 = 0$ Setting the *numerator* equal to zero

$x = -5$

The equation has only one solution. However, we must find other critical points by setting the denominator equal to zero; we then find that 3 is also a *critical point.*

Step 3. We will not use method 1 because we cannot easily graph $y = \dfrac{x+5}{x-3}$. Using method 2, we graph the critical point -5 with a *solid* dot, because the inequality symbol includes the equal sign. However, because 3 is *not in the domain* of the variable, we must graph the critical point 3 with a hollow dot.

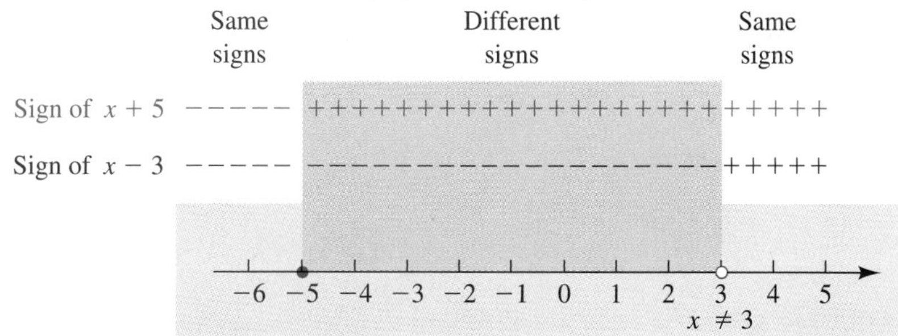

Because we're solving $\dfrac{x+5}{x-3} \leq 0$ and because a fraction is negative if the numerator and denominator have different signs, we select the interval on which the signs are *different*. The solution set is $\{x \mid -5 \leq x < 3\}$, or $[-5, 3)$, and the graph of the solution set is as follows:

A Word of Caution For the problem in Example 5, the solution set is *not* $\{x \mid -5 \leq x \leq 3\}$. The 3 *cannot* be included, since $x - 3$ is in the denominator.

EXAMPLE 6 Solve $x \leq \frac{3}{x+2}$, and graph the solution set on the real number line.

SOLUTION We can solve this inequality by moving $\frac{3}{x+2}$ to the left so that we have zero on one side of the inequality symbol. We then proceed in a manner similar to the method used in Example 5.

$$x - \frac{3}{x+2} \leq 0$$

$$\frac{x(x+2) - 3}{x+2} \leq 0 \quad \text{Rewriting the left side as a single fraction}$$

$$\frac{x^2 + 2x - 3}{x+2} \leq 0$$

$$\frac{(x+3)(x-1)}{x+2} \leq 0 \quad \text{Note that } -2 \text{ is not in the domain}$$

We set each factor of the numerator *and* of the denominator equal to zero to find the critical points:

$x + 3 = 0$	$x - 1 = 0$	$x + 2 = 0$
$x = -3$	$x = 1$	$x = -2$

The critical points are -3, 1, and -2, and these three points separate the number line into four intervals. We must graph -2 as a hollow dot because -2 is *not in the domain* of the variable.

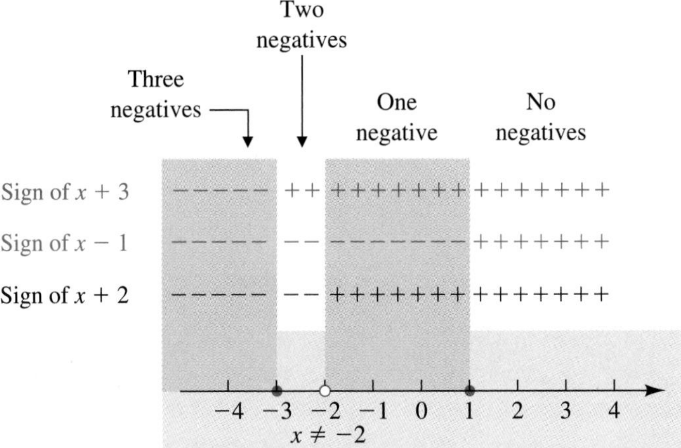

Because the fraction is to be negative or 0, there must be an *odd* number of negative signs in the expression. We have an odd number of negative signs to the left of -3 and also between -2 and 1.

The solution set is $\{x \mid x \leq -3\} \cup \{x \mid -2 < x \leq 1\}$, or $(-\infty, -3] \cup (-2, 1]$. This means that the solution set consists of all $x \leq -3$, *as well as* all x between -2 and 1, including 1. The graph of the solution set is as follows:

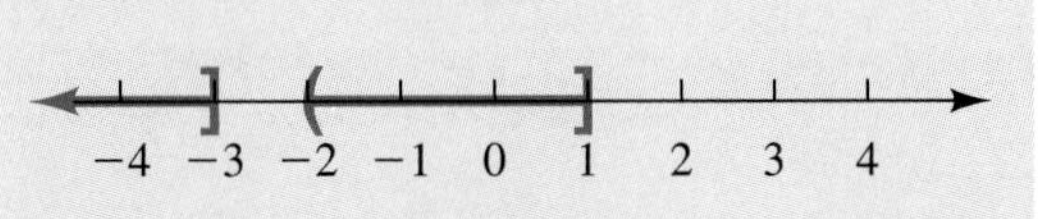

An alternative method of solution is shown below. To remove fractions in *equations*, we usually multiply both sides by the LCD. Since we're solving an *inequality*, however, we must multiply by $(x+2)^2$ to make sure that both sides are being multiplied by a *positive* number so that the *sense of the inequality does not change*.

$$x \le \frac{3}{x+2}$$

$$\frac{(x+2)^2}{1} \cdot \frac{x}{1} \le \frac{(x+2)^2}{1} \cdot \frac{3}{(x+2)} \qquad (x+2)^2 \text{ is positive for } \textit{all } x$$

$$x(x+2)^2 \le 3(x+2)$$

$$x(x+2)^2 - 3(x+2) \le 0 \longleftarrow (x+2) \text{ is a common factor}$$

$$(x+2)[x(x+2) - 3] \le 0$$

$$(x+2)[x^2 + 2x - 3] \le 0$$

$$(x+2)(x+3)(x-1) \le 0$$

-2, -3, and 1 are the critical points. Even though -2 is a critical point, we must *not* include it in the final solution set, since it is not in the domain of the variable.

The problem can then be finished by using method 2; the results will be the same.

A Word of Caution Let's see what happens if we multiply both sides of the original inequality from Example 6 by $(x + 2)$:

$$\frac{x+2}{1} \cdot \frac{x}{1} \le \frac{x+2}{1} \cdot \frac{3}{x+2}$$

$$x^2 + 2x \le 3$$

$$x^2 + 2x - 3 \le 0$$

$$(x+3)(x-1) \le 0$$

The critical points are -3 and 1, and we graph these as solid dots.

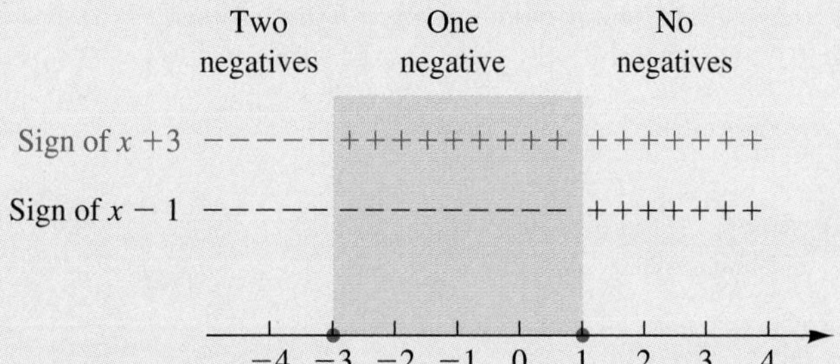

The solution set appears to be $\{x \mid -3 \le x \le 1\}$. We can easily verify, however, that this is not correct. We saw in Example 6 that any numbers *less than* -3 should satisfy the original inequality. Let's try -4, which is *not* in $\{x \mid -3 \le x \le 1\}$:

$$x \le \frac{3}{x+2} \qquad -4 \overset{?}{\le} \frac{3}{-4+2} \qquad -4 \le \frac{3}{-2} \quad \text{True}$$

We also saw in Example 6 that numbers between -3 and -2 should *not* satisfy the original inequality. Let's try $-2\frac{1}{2}$, which *is* in $\{x \mid -3 \le x \le 1\}$:

$$x \le \frac{3}{x+2} \qquad -2\tfrac{1}{2} \overset{?}{\le} \frac{3}{-2\frac{1}{2}+2} \qquad -2\tfrac{1}{2} \overset{?}{\le} \frac{3}{-\frac{1}{2}} \qquad -2\tfrac{1}{2} \le -6 \quad \text{False}$$

We see, then, that the solution we obtained by multiplying both sides of the inequality by $x + 2$ is incorrect. We should not multiply both sides of an *inequality* by any expression that is not *always* positive, and $x + 2$ is positive for some values of x but negative for others.

Exercises II.7
Set I

In Exercises 1–24, solve each inequality, writing the solution set in set-builder notation *and* in interval notation, and graph the solution set on the real number line.

1. $(x + 1)(x - 2) < 0$
2. $(x + 1)(x - 2) > 0$
3. $3 - 2x - x^2 \geq 0$
4. $2 - x - x^2 \leq 0$
5. $x^2 - 3x - 4 < 0$
6. $x^2 + 4x - 5 > 0$
7. $x^2 + 7 > 6x$
8. $x^2 + 13 < 8x$
9. $x^2 \leq 5x$
10. $x^2 \geq 3x$
11. $3x - x^2 > 0$
12. $7x - x^2 < 0$
13. $x^3 - 3x^2 - x + 3 > 0$
14. $x^3 + 2x^2 - x - 2 > 0$
15. $(x + 1)(x - 2)^2 < 0$
16. $(x - 1)(x + 2)^2 < 0$
17. $\dfrac{x - 2}{x + 3} \geq 0$
18. $\dfrac{x + 4}{x - 2} \leq 0$
19. $\dfrac{2x + 5}{3x - 1} \leq 0$
20. $\dfrac{3x - 4}{2x + 5} \geq 0$
21. $x < \dfrac{2}{x + 1}$
22. $x > \dfrac{4}{x + 3}$
23. $x \geq \dfrac{8}{x + 2}$
24. $x \leq \dfrac{4}{x - 3}$

In Exercises 25–28, find the domain of each function.

25. $y = f(x) = \sqrt{x^2 - 4x - 12}$
26. $y = f(x) = \sqrt{x^2 + 3x - 10}$
27. $y = f(x) = \sqrt{4 + 3x - x^2}$
28. $y = f(x) = \sqrt{5 - 4x - x^2}$

Writing Problems

Express the answers in your own words and in complete sentences.

1. Describe the method you would use in solving $x^2 - 3x + 6 \geq 0$, and explain why you would use that method.
2. Describe the method you would use in solving $\dfrac{x + 1}{x - 4} \leq 0$, and explain why you would use that method.
3. Explain why 4 is not in the solution set of $\dfrac{x + 1}{x - 4} \leq 0$.

Exercises II.7
Set II

In Exercises 1–24, solve each inequality, writing the solution set in set-builder notation *and* in interval notation, and graph the solution set on the real number line.

1. $(x + 4)(x - 1) < 0$
2. $(2x - 5)(x + 2) > 0$
3. $8 - 2x - x^2 \geq 0$
4. $x^2 - 2x \leq 3$
5. $x^2 - 5x - 6 < 0$
6. $(x - 1)^2(x - 5) < 0$
7. $x^2 + 8 > 7x$
8. $x^2 < 9$
9. $x^2 \geq x$
10. $(x + 1)^2(x - 4) < 0$
11. $3x - x^2 < 0$
12. $1 + 4x + 3x^2 \geq 0$
13. $x^3 + x^2 - 4x - 4 > 0$
14. $(x - 1)^2(x - 3) < 0$
15. $(x - 2)(x + 1)(x + 3) < 0$
16. $x^2 + 9 < 6x$
17. $\dfrac{x + 2}{x - 4} \geq 0$
18. $\dfrac{3x - 1}{2 - x} \leq 0$
19. $\dfrac{4x - 7}{2x + 3} \leq 0$
20. $\dfrac{6x - 3}{2x + 4} \geq 0$
21. $x < \dfrac{6}{x + 5}$
22. $x \geq \dfrac{5}{6 - x}$
23. $x \leq \dfrac{12}{x + 1}$
24. $x > \dfrac{15}{x - 2}$

In Exercises 25–28, find the domain of each function.

25. $y = f(x) = \sqrt{x^2 + 6x - 7}$
26. $y = f(x) = \sqrt{x^2 + 4x + 3}$
27. $y = f(x) = \sqrt{8 + 2x - x^2}$
28. $y = f(x) = \sqrt{10 - x^2 + 3x}$

11.8 Graphing Quadratic Inequalities in Two Variables

A parabola separates the plane into two regions: the region "inside" the parabola and the region "outside" the parabola. Similarly, circles and ellipses separate the plane into the regions "inside" them and the regions "outside" them. In Figures 20 and 21, the region inside a parabola and the region inside an ellipse are shaded. A hyperbola separates the plane into *three* regions (see Figure 22).

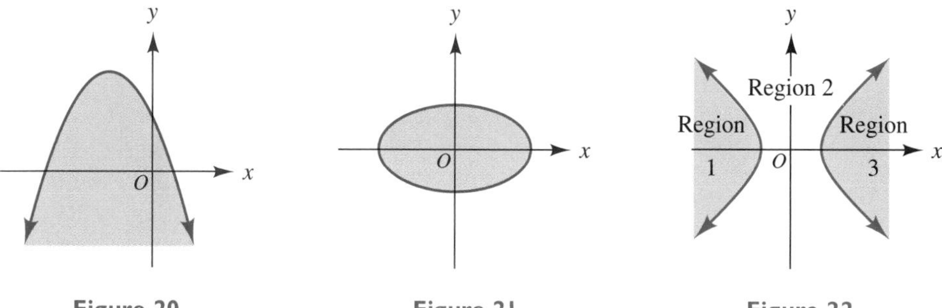

Figure 20 Figure 21 Figure 22

Graphing a quadratic inequality in two variables

1. Substitute an equal sign for the inequality symbol, and graph the resulting conic section. Graph it with a *solid curve* if the inequality is $\leq$ or $\geq$ and with a *dotted curve* if the inequality is $<$ or $>$.
2. Select any point *not* on the conic section and substitute its coordinates into the original inequality. If the resulting statement is
   a. *true*, shade the region of the plane that contains the point selected;
   b. *false*, shade the region of the plane that does *not* contain the point selected.

Any ordered pair that corresponds to a point that lies in the shaded region(s) or on a *solid* curve is in the solution set; the coordinates of any such point will satisfy the inequality.

EXAMPLE 1 Graph $y \geq x^2 - 5x - 6$.

SOLUTION We first sketch the graph of $y = x^2 - 5x - 6$. It is a parabola that opens upward; the x-intercepts are 6 and -1, and the y-intercept is -6. We draw the graph as a *solid curve* because the inequality symbol is $\geq$. (See Figure 23.)

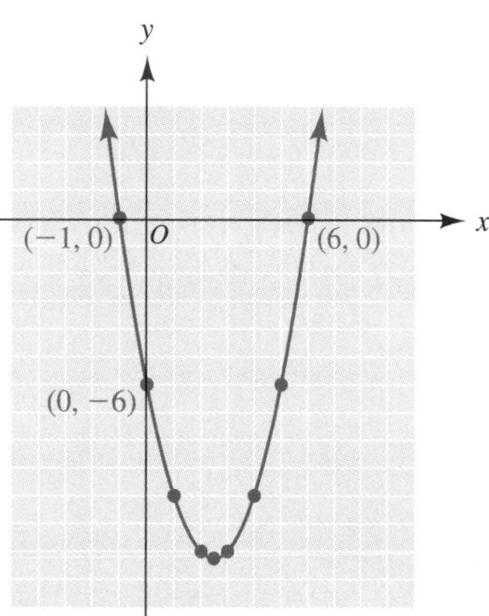

Figure 23

Now we must select a point *not* on the parabola. Let's use (0, 0). Substituting 0 for x and 0 for y into the original inequality, we have

$$0 \geq 0^2 - 5(0) - 6$$

$$0 \geq -6 \quad \text{A } \textit{true} \text{ statement}$$

Therefore, we shade the region that contains the point (0, 0). (See Figure 24.)

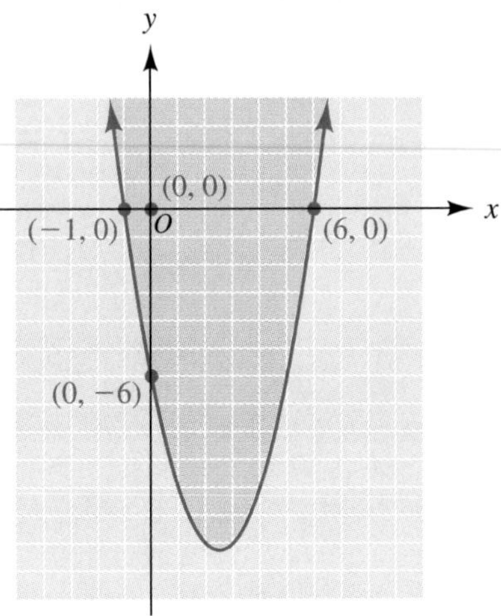

Figure 24

Any ordered pair that corresponds to a point that lies in the shaded region is in the solution set.

EXAMPLE 2 Graph $(x + 1)^2 + (y - 3)^2 > 4$.

SOLUTION We temporarily replace $>$ with $=$, getting $(x + 1)^2 + (y - 3^2) = 4$. This is the equation of a circle with its center at the point $(-1, 3)$ and with radius 2. Because the inequality is $>$, we sketch the circle with a *dotted line*. (See Figure 25.)

We now select any point not *on* the circle. Again, let's choose (0, 0). Substituting 0 for x and 0 for y into the inequality, we have

$$(0 + 1)^2 + (0 - 3)^2 > 4$$

$$1 + 9 > 4$$

$$10 > 4 \quad \text{A } \textit{true} \text{ statement}$$

Therefore, we shade the region that contains the point (0, 0). (See Figure 26.)

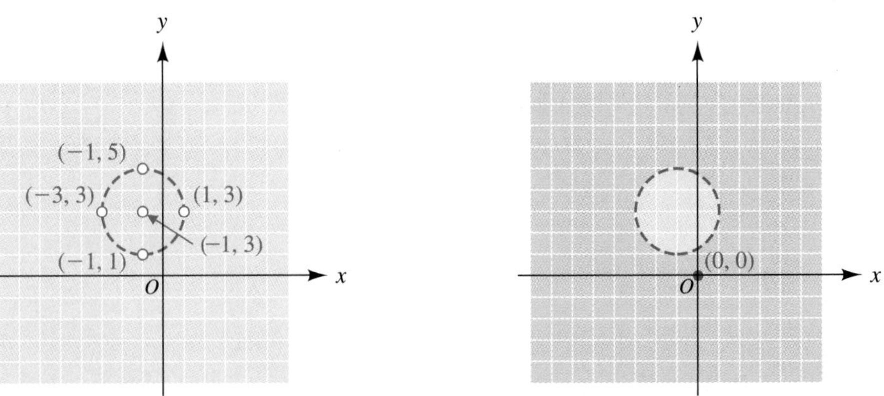

Figure 25

Figure 26

Any ordered pair that corresponds to a point that lies in the shaded region is in the solution set.

EXAMPLE 3 Graph $x^2 - y^2 > 1$.

SOLUTION We temporarily replace $>$ with $=$, getting $x^2 - y^2 = 1$. This is a *hyperbola*. If $y = 0$, $x^2 = 1$. Therefore, the x-intercepts are $(1, 0)$ and $(-1, 0)$. Because the inequality is $>$, we sketch the hyperbola with a dotted curve. (See Figure 27.) We can see that the hyperbola separates the plane into three regions.

Again, let's choose $(0, 0)$ as the point to try in the inequality.

$$0^2 - 0^2 > 1$$

$$0 > 1 \quad \text{A } \textit{false} \text{ statement}$$

Therefore, we shade the portions of the plane that do *not* contain $(0, 0)$. (See Figure 28.)

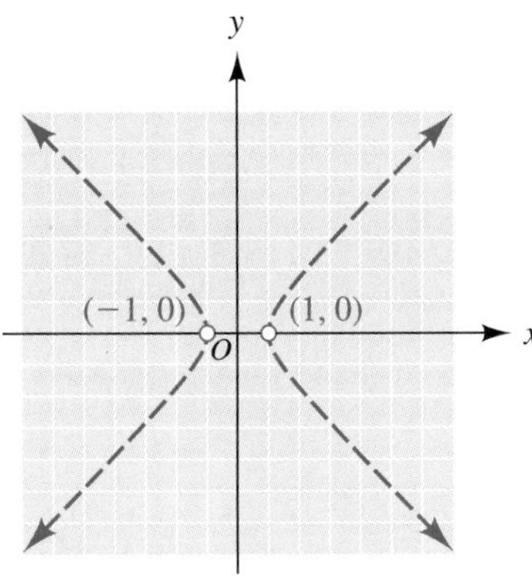

Figure 27

Figure 28

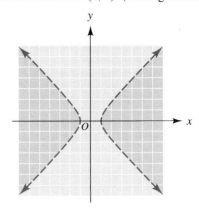

Any ordered pair that corresponds to a point that lies in either shaded region is in the solution set. (You might verify that several ordered pairs in both shaded regions do satisfy the inequality.)

Because the graphs of the solution sets of quadratic inequalities in two variables occupy portions of the xy-plane (rather than just portions of the real number line), it is not possible to write such solution sets in interval notation.

Exercises 11.8
Set I

Graph each of the following inequalities.

1. $y \le x^2 + 5x + 6$
2. $y \le x^2 + 4x + 3$
3. $x^2 + 4y^2 > 4$
4. $9x^2 + y^2 > 9$
5. $(x + 2)^2 + (y - 3)^2 \le 9$
6. $(x - 1)^2 + (y + 2)^2 \le 4$
7. $4x^2 - y^2 < 4$
8. $9x^2 - y^2 < 9$

Writing Problems

Express the answer in your own words and in complete sentences.

1. Explain how you would decide which region of the plane to shade if you were asked to graph $x^2 + 9y^2 > 9$.

Exercises 11.8 Set II

Graph each of the following inequalities.

1. $y \leq x^2 + 3x + 2$
2. $(x + 1)^2 + (y + 1)^2 \leq 1$
3. $16x^2 + y^2 > 16$
4. $y < x^2 - 3x + 2$
5. $(x - 3)^2 + (y - 2)^2 \leq 16$
6. $x^2 - 9y^2 > 9$
7. $16x^2 - y^2 < 16$
8. $x^2 + 9y^2 \geq 9$

Sections 11.5–11.8 REVIEW

Quadratic Functions 11.5

A **quadratic function** is a second-degree function:

$$y = f(x) = ax^2 + bx + c \quad \text{The general form of a quadratic function}$$

Its graph is called a *parabola*.

Its *x-intercepts* are found by setting y equal to zero and then solving the resulting quadratic equation for x.

Its *y-intercept* is found by setting x equal to 0 and then solving the resulting equation for y. The y-coordinate of the y-intercept of a quadratic function of one variable is always equal to the constant term.

The parabola opens *downward* and a *maximum value* occurs if $a < 0$.

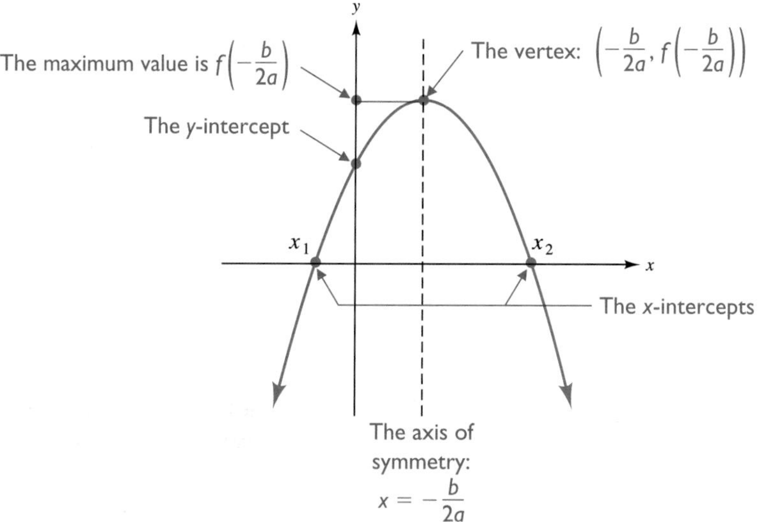

The parabola opens *upward* and a *minimum value* occurs if $a > 0$.

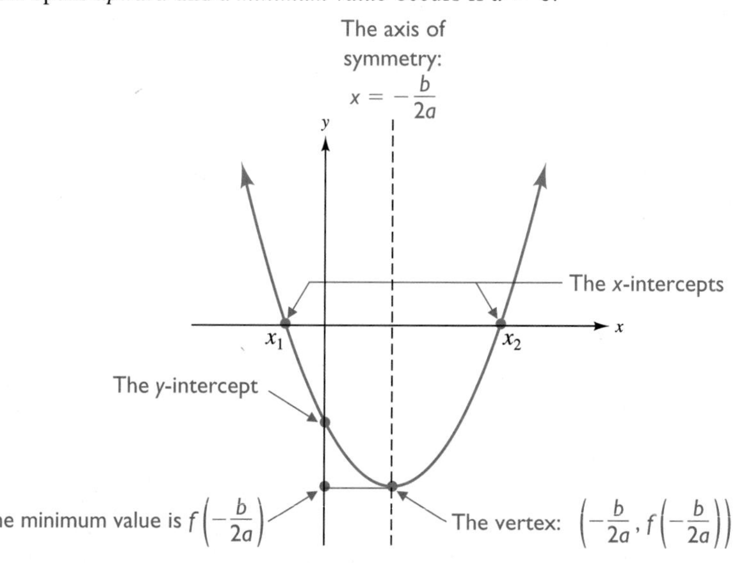

Conic Sections
11.6

Conic sections are curves formed when a plane cuts through a cone or a pair of cones. Parabolas, circles, ellipses, and hyperbolas are conic sections. A quick sketch of the conics covered in this book can usually be made by using the *intercept method.*

General equations of parabolas with vertex (h, k) are

$y - k = a(x - h)^2$ The parabola opens upward if $a > 0$ and downward if $a < 0$

and

$x - h = a(y - k)^2$ The parabola opens to the right if $a > 0$ and to the left if $a < 0$

General equations of circles are

$$x^2 + y^2 = r^2$$

where the center is at the origin and the radius is r, and

$$(x - h)^2 + (y - k)^2 = r^2$$

where the center is at (h, k) and the radius is r.

The general equation of an ellipse with its center at the origin is

$$\frac{x^2}{a^2} + \frac{y^2}{b^2} = 1$$

where the x-intercepts are $(\pm a, 0)$, and the y-intercepts are $(0, \pm b)$.

General equations of hyperbolas with centers at the origin are

$$\frac{x^2}{a^2} - \frac{y^2}{b^2} = 1$$

where the x-intercepts are $(\pm a, 0)$ and there are no y-intercepts, and

$$\frac{y^2}{c^2} - \frac{x^2}{d^2} = 1$$

where the y-intercepts are $(0, \pm c)$ and there are no x-intercepts.

Solving a Quadratic Inequality in One Variable
11.7

1. Substitute an equal sign for the inequality symbol.
2. Solve the resulting equation. The solutions will be the x-intercepts if method 1 is used, and they will be the critical points if method 2 is used. Set any denominators equal to zero to find more critical points.
3. *Method 1:* Sketch the graph of the related function that is obtained by substituting y for 0 and an equal sign for the inequality symbol in the inequality. Read the solution set from the graph.

   Method 2: Separate the real number line into intervals by graphing the critical points. Determine the sign of the function in each of the intervals.

Graphing a Quadratic Inequality in Two Variables
11.8

1. Substitute an equal sign for the inequality symbol, and graph the resulting conic section.
2. Select any point *not on* the conic section, and substitute its coordinates into the *original inequality*. If the statement is
   a. *true*, shade the region of the plane that contains the point selected;
   b. *false*, shade the region of the plane that does *not* contain the point selected.

Sections 11.5–11.8 REVIEW EXERCISES Set I

In Exercises 1 and 2, (a) find the x-intercepts, if they exist; (b) find the y-intercept; (c) find the coordinates of the vertex; and (d) graph the function.

1. $f(x) = x^2 - 2x - 8$
2. $f(x) = 6x - 9 - x^2$

In Exercises 3–5, solve each inequality, writing the solution set in set-builder notation *and* in interval notation.

3. $(4 - x)(2 + x) > 0$
4. $x + \dfrac{2}{x - 3} < 0$
5. $x > \dfrac{3}{x + 2}$
6. Find the domain of the function

$$y = f(x) = \sqrt{6 + 5x - x^2}$$

In Exercises 7–9, identify and sketch each conic section.

7. $3x^2 + 3y^2 = 27$
8. $4y^2 - x^2 = 16$
9. $9x^2 + 4y^2 = 36$
10. Graph $y \geq x^2 + 2x - 3$.

Sections 11.5–11.8 REVIEW EXERCISES Set II

Name

ANSWERS

In Exercises 1 and 2, graph the function, using the grid.

1. $f(x) = 2x + 3 - x^2$

2. $f(x) = x^2 + 4x + 4$

1.

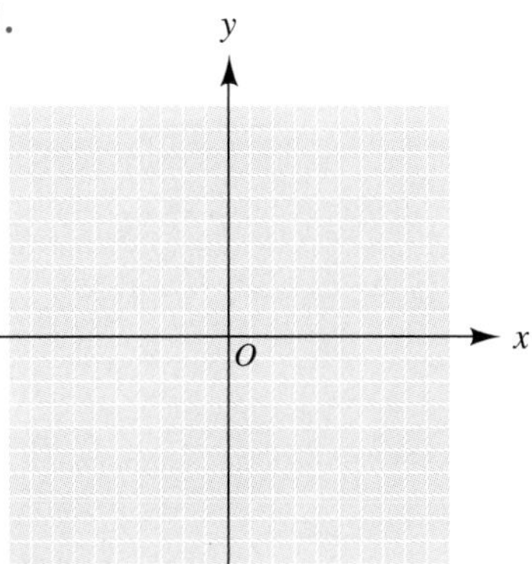

In Exercises 3–5, solve each inequality, writing the solution set in set-builder notation *and* in interval notation.

3. $x^2 \leq 3x + 10$

4. $x^2 \geq 1 - 2x$

2.

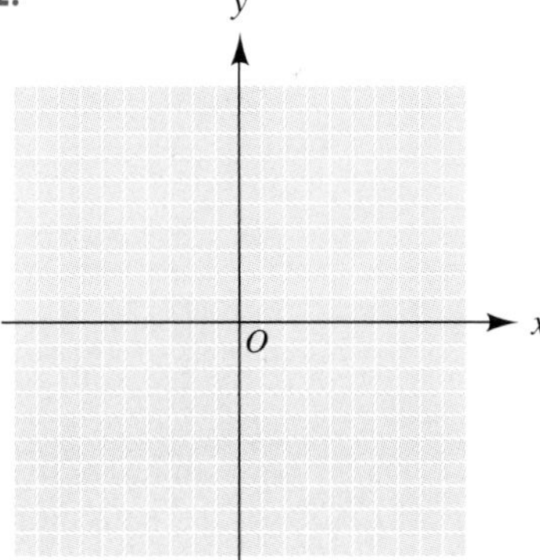

5. $x < \dfrac{4}{x + 3}$

6. Find the domain of the function $y = f(x) = \sqrt{x^2 + 7x - 8}$.

3. ______________________

____

4. ______________________

____

5. ______________________

____

6. ______________________

In Exercises 7 and 8, identify and sketch each conic section.

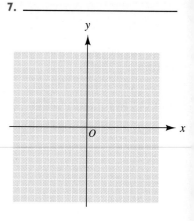

7. $x^2 + 16y^2 = 16$

7. ______________________

8. ______________________

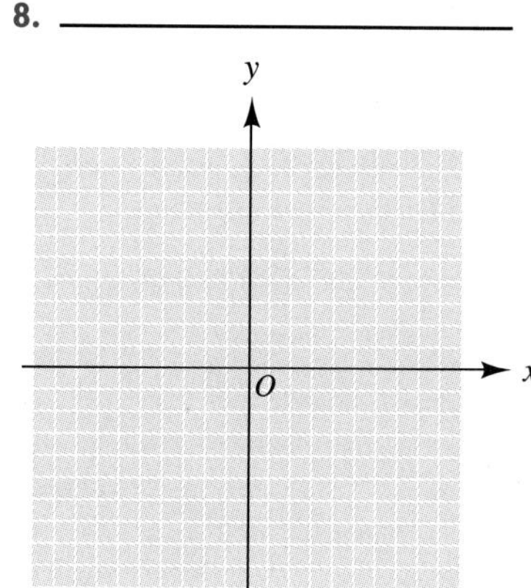

8. $x^2 - 4y^2 = 4$

9.

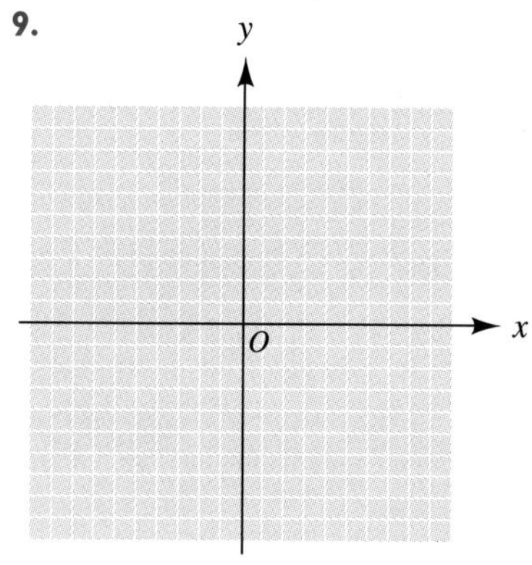

9. Graph $y \geq x^2 + 5x + 4$.

Chapter 11 DIAGNOSTIC TEST

The purpose of this test is to see how well you understand nonlinear equations and inequalities. We recommend that you work this diagnostic test *before* your instructor tests you on this chapter. Allow yourself about 50 minutes.

Complete solutions for all the problems on this test, together with section references, are given in the answer section at the end of the book. For the problems you do incorrectly, we suggest that you study the sections cited.

In Problems 1–5, find the solution set of each equation by any convenient method.

1. a. $2x^2 = 6x$ b. $2x^2 = 18$

2. $\dfrac{x-1}{2} + \dfrac{4}{x+1} = 2$

3. $2x^{2/3} + 3x^{1/3} = 2$

4. a. $x^2 = 6x - 7$ b. $3x^2 + 7x = 1$

5. $x^2 + 6x + 10 = 0$

6. a. Use the quadratic discriminant to determine the nature of the roots of the equation $25x^2 - 20x + 7 = 0$. Do *not* solve for the roots of the equation.

 b. Write an equation of least degree with integral coefficients that has roots 3 and $2 + i\sqrt{3}$.

7. Identify and sketch the graph of the conic section whose equation is $25x^2 + 16y^2 = 400$.

8. Solve each inequality, expressing the solution set in set-builder notation *and* in interval notation.

 a. $x^2 + 5 < 6x$

 b. $x^2 + 2x \geq 8$

9. Given the quadratic function $f(x) = x^2 - 4x$, find:

 a. the equation of the axis of symmetry

 b. the coordinates of the vertex

 c. the x-intercepts (if they exist)

 Draw the graph.

Set up the following problem algebraically and solve.

10. The length of a rectangle is 18 cm more than its width. If its area is 8,944 sq. cm, what is the width of the rectangle?

Chapters 1–11 CUMULATIVE REVIEW EXERCISES

1. Find the slope of the line through $(4, -1)$ and $(2, 5)$.

2. Find an equation of the line with slope $-\frac{1}{2}$ that passes through $(-2, 3)$. Write the answer in the general form.

3. Solve $|4 - 3x| \geq 10$.

4. Find the solution set of $\sqrt{2-x} - 4 = x$.

5. Graph $4x - 2y < 8$.

6. Find the inverse function $y = f(x) = 2x - 3$. Graph $y = f(x)$ and its inverse.

In Exercises 7–9, perform the indicated operations, and express the answers in exponential form with positive exponents or no exponents.

7. $\left(\dfrac{x^{-3}y^4}{x^2y}\right)^{-2}$ 8. $(8x^{-3/5})^{2/3}$ 9. $a^{1/3} \cdot a^{-1/2}$

10. Find the solution set of $x^2 - x + 7 = 0$.

11. Find the solution set of $3x^2 + x = 3$.

12. Solve $7^x = 45$. (Round off the answer to three decimal places.)

13. Solve $\log(x + 11) - \log(x + 1) = \log 6$.

In Exercises 14 and 15, find the value of the variable.

14. $\log_b 2 = \frac{1}{4}$ 15. $\log_7 \frac{1}{7} = x$

16. Graph $9x^2 + 16y^2 = 144$.

17. Find the domain of $y = f(x) = \sqrt{4 - 3x - x^2}$.

In Exercises 18–22, set up each problem algebraically and solve.

18. The length of a rectangle is 2 yd more than its width. If the length of its diagonal is 10 yd, find the dimensions of the rectangle.

19. It takes Mina 3 hr longer to do a job than it does Merwin. After Mina has worked on the job for 5 hr, Merwin joins her. Together they finish the job in 3 hr. How long would it take each of them working alone to do the entire job?

20. Eric and Colin left an intersection at the same time, Eric walking east and Colin walking south. Eric was walking 3 mph faster than Colin, and at the end of an hour the two men were 5 mi apart. How fast was Colin walking? (Express the answer in simplest radical form and also give a decimal approximation, rounded off to one decimal place.)

21. If \$1,000 is invested at $6\frac{1}{4}\%$ annual interest, how long will it take for the money to double if the interest is compounded continuously? Round off the answer to the nearest year.

22. The three sides of a triangle are in the ratio 2:4:5. The perimeter is 88 cm. Find the lengths of the three sides.

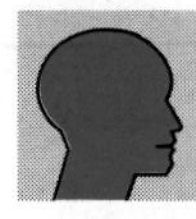

Critical Thinking and Problem-Solving Exercises

1. Gloria's son is 11 years old and her daughter is 16 years old, so the *sum of the digits* of their ages is 9. How old will Gloria's son be the next time that the *sum of the digits* of their ages is 9?

2. John has three pairs of slacks that he is particularly fond of (one pair is kelly green, one pair is brown, and one pair is gray) and four shirts that he likes really well (one is kelly green, one is tan, one is black, and one is white). If he can wear any of the four shirts with any of the three pairs of slacks, how many different combinations are possible?

3. Victoria has one compact disc of each of her six favorite composers, and she has put the six CDs into two stacks. She put the Mozart CD directly under the Beethoven CD. She put the Debussy CD to the right of the Mozart CD, but not on top. She put the Liszt CD on top of the Beethoven CD. Finally, she put the Chopin CD between the Grieg CD and the Debussy CD. Figure out which CD is in which position.

4. One day recently, Kevin, Jason, and Nathan were eating hamburgers. Kevin and Nathan each ate the same number of hamburgers, and they each ate at least one hamburger; Jason ate the most hamburgers, and he ate fewer than 10. The *product* of the numbers of hamburgers eaten by the three young men was 12. How many hamburgers did each man eat?

5. Dick and Helen have 80 ft of fencing with which to fence their vegetable garden. They want their garden to have a rectangular shape, and it must be fenced on all four sides. Verify that an equation that expresses the area, A, of the garden as a function of x, where x is the length of one side of the garden, is $A(x) = 40x - x^2$. *By graphing this function*, determine the value of x that gives the maximum area, *and explain how you know* that that value of x gives the maximum area.

Systems of Equations and Inequalities

CHAPTER 12

In previous chapters, we showed how to solve a single equation for a single variable. In this chapter, we show how to solve systems of two or more equations in two or more variables. We also include brief discussions of matrices, determinants, and systems of inequalities.

12.1 Basic Definitions; Solutions of Systems of Equations

A System of Two Equations in Two Variables and Its Solution

An example of a **system of two equations in two variables** is $\left\{\begin{matrix} x + y = 6 \\ x - y = 2 \end{matrix}\right\}$. A **solution** of a system of two equations in two variables is an **ordered pair** that satisfies *both* equations.

EXAMPLE 1 Verify that (4, 2) is a solution of the system $\left\{\begin{matrix} x + y = 6 \\ x - y = 2 \end{matrix}\right\}$.

SOLUTION Substituting (4, 2) into each equation, we have

First equation		*Second equation*	
$x + y = 6$		$x - y = 2$	
$4 + 2 = 6$	True	$4 - 2 = 2$	True

Therefore, (4, 2) satisfies both equations and is a solution of the given system.

A System of Three Equations in Three Variables and Its Solution

To be able to solve a system of three equations in three variables, we need the following definition.

An Ordered Triple An **ordered triple** is a set of three numbers in which the *order* the numbers are listed in makes a difference. To indicate that a set of three numbers is an *ordered triple,* we put commas between the numbers and we enclose the numbers within *parentheses*, never braces or brackets. Thus, (4, 1, 7) represents an ordered triple. Then, because the order makes a difference, (7, 4, 1) represents an ordered triple that is *different* from (4, 1, 7); that is, $(4, 1, 7) \neq (7, 4, 1)$. In general, $(a, b, c) \neq (b, c, a)$. (Because the numbers are not enclosed within parentheses, {4, 1, 7} and [4, 1, 7] do *not* represent ordered triples.) The *first* number of an ordered triple is called the ***x*-coordinate**, the *second* number is called the ***y*-coordinate**, and the *third* number is called the ***z*-coordinate**.

Note It is possible to graph ordered triples in a *three-dimensional* coordinate system. However, we do not graph ordered triples in this book.

An example of a **system of three equations in three variables** is

$$\left\{\begin{matrix} 2x - 3y + z = 1 \\ x + 2y + z = -1 \\ 3x - y + 3z = 4 \end{matrix}\right\}$$

A **solution** of a system of three equations in three variables is an **ordered triple** that satisfies *all three* equations.

EXAMPLE 2 Verify that the ordered triple $(-3, -1, 4)$ is a solution of the system

$$\begin{Bmatrix} 2x - 3y + z = 1 \\ x + 2y + z = -1 \\ 3x - y + 3z = 4 \end{Bmatrix}$$

SOLUTION Substituting $(-3, -1, 4)$ into each equation, we have

First equation	*Second equation*	*Third equation*
$2x - 3y + z = 1$	$x + 2y + z = -1$	$3x - y + 3z = 4$
$2(-3) - 3(-1) + 4 \stackrel{?}{=} 1$	$-3 + 2(-1) + 4 \stackrel{?}{=} -1$	$3(-3) - (-1) + 3(4) \stackrel{?}{=} 4$
$-6 + 3 + 4 = 1$	$-3 - 2 + 4 = -1$	$-9 + 1 + 12 = 4$
True	True	True

Therefore, the ordered triple $(-3, -1, 4)$ satisfies all three equations and is a solution of the given system of equations.

Types of Systems of Equations and Inequalities

If each equation of a system is a first-degree equation, the system is called a **linear system.**

If the highest-degree equation of a system is second degree, then the system is called a **quadratic system.**

If a system contains inequalities rather than equations, it is called a **system of inequalities.**

EXAMPLE 3 Examples of systems of equations and inequalities:

a. $\begin{Bmatrix} 2x - 3y + z = 1 \\ x + 2y + z = -1 \\ 3x - y + 3z = 4 \end{Bmatrix}$ is a linear system (a system of linear equations).

b. $\begin{Bmatrix} x^2 + y^2 = 25 \\ x - y = 4 \end{Bmatrix}$ is a quadratic system.

c. $\begin{Bmatrix} 3x - 2y > 6 \\ x - 2y < 4 \end{Bmatrix}$ is a system of linear *inequalities.*

Solving a System of Equations To *solve* a system of two equations in two variables means to find an ordered pair (if one exists) that satisfies *both* equations. To solve a system of three equations in three variables means to find an ordered triple (if one exists) that satisfies *all three* equations. In this text, we will consider solving only those systems with the same number of equations as variables.

Exercises 12.1
Set I

In Exercises 1 and 2, determine whether each ordered pair is a solution of the given system of equations.

1. $\begin{Bmatrix} x + y = 8 \\ x - y = 2 \end{Bmatrix}$ a. (1, 7) b. (0, 8) c. (5, 3)

2. $\begin{Bmatrix} x + 2y = 5 \\ x - 3y = 10 \end{Bmatrix}$ a. (3, 1) b. (7, −1) c. (5, 0)

In Exercises 3 and 4, determine whether each ordered triple is a solution of the given system of equations.

3. $\begin{Bmatrix} 2x - 3y + 7z = 14 \\ 4x + 3y - 5z = 0 \\ 3x - 6y - 2z = 10 \end{Bmatrix}$
a. (5, 1, 1) b. (2, −1, 1) c. (0, 0, 2)

4. $\begin{Bmatrix} 3x + 2y + z = 2 \\ 5x + 3y - 2z = -7 \\ 2x - 4y + 3z = 19 \end{Bmatrix}$
a. (0, 1, 0) b. (1, 0, −1) c. (1, −2, 3)

Writing Problems

Express the answers in your own words and in complete sentences.

1. Explain what the solution of a system of two equations in two variables is.

2. Explain why $(2, -1)$ is a solution of the system $\begin{Bmatrix} x - 3y = 5 \\ 4x - 2y = 10 \end{Bmatrix}$.

3. Explain why $(4, 4)$ is not a solution of the system $\begin{Bmatrix} x - 3y = -8 \\ 4x - 2y = 10 \end{Bmatrix}$.

Exercises 12.1 Set II

In Exercises 1 and 2, determine whether each ordered pair is a solution of the given system of equations.

1. $\begin{Bmatrix} x - y = 7 \\ x + y = 3 \end{Bmatrix}$ a. $(8, 1)$ b. $(8, 0)$ c. $(5, -2)$

2. $\begin{Bmatrix} 2x + 3y = 6 \\ 4x + 6y = 12 \end{Bmatrix}$ a. $(0, 2)$ b. $(3, 0)$ c. $(-3, 4)$

In Exercises 3 and 4, determine whether each ordered triple is a solution of the given system of equations.

3. $\begin{Bmatrix} x - 2y + 5z = 4 \\ 5x + 2y - 6z = 1 \\ 4x - 3y - z = 0 \end{Bmatrix}$
   a. $(13, 2, -1)$ b. $(0, -2, 0)$ c. $(1, 1, 1)$

4. $\begin{Bmatrix} 5x + 3y + z = -7 \\ 2x - 3y - 2z = -1 \\ 2x - 4y + 3z = -1 \end{Bmatrix}$
   a. $(-1, -1, 1)$ b. $(1, 0, -1)$ c. $(1, -2, 3)$

12.2 Solving a Linear System of Two Equations in Two Variables Graphically

The **graphical method of solving a linear system of two equations in two variables** is not an exact method of solution, but it can sometimes be used successfully in solving such systems.

Solving a linear system of two equations in two variables graphically

1. Graph the solution set of each equation on the same set of axes.
2. Three outcomes are possible:
   a. *The lines intersect in exactly one point* (see Figure 1 of Example 1). The *solution* is the ordered pair that represents the point of intersection of the two lines; the *set* that contains that ordered pair is the *solution set* of the system.
   b. *The lines are parallel*; they will never intersect, even if they are extended indefinitely (see Figure 2 of Example 2). There is *no solution* for the system of equations; the *solution set* is { }.
   c. The lines coincide; that is, the equations are both equations of the same line (see Figure 3 of Example 3). There are *infinitely many solutions*. For the line $ax + by = c$, the *solution set* is the set $\left\{\left(t, \dfrac{c - at}{b}\right)^{*}, t \in R\right\}$ or the set $\left\{\left(\dfrac{c - bs}{a}, s\right), s \in R\right\}$.
3. If the lines graphed in step 1 intersect in exactly one point, check the coordinates of that point in *both* equations.

*If, in $ax + by = c$, we let $x = t$, then $at + by = c$, so $by = c - at$, or $y = \dfrac{c - at}{b}$; see also Example 3.

When the lines intersect in exactly one point, we say that

1. There is a *unique* solution for the system.
2. The system of equations is *consistent* and *independent.*

When the lines are parallel, we say that

1. There is *no* solution for the system.
2. The system of equations is *inconsistent* and *independent.*

When the lines coincide (that is, when both equations have the same line for their graph), we say that

1. There are *infinitely* many solutions for the system.
2. The system of equations is *consistent* and *dependent.*

EXAMPLE 1 Find the solution set of the system $\left\{\begin{matrix} x + y = 6 \\ x - y = 2 \end{matrix}\right\}$ graphically.

SOLUTION We graph the solution set of each equation on the same set of axes (see Figure 1).

Line 1 $x + y = 6$ has intercepts (6, 0) and (0, 6).

Line 2 $x - y = 2$ has intercepts (2, 0) and (0, −2).

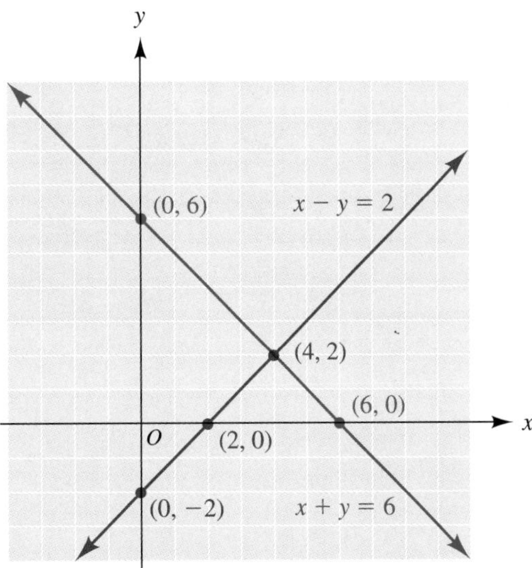

Figure 1

The lines appear to intersect at the point (4, 2). We can check to see whether they do, in fact, intersect there by substituting 4 for x and 2 for y in *both* equations:

$x + y = 6$		$x - y = 2$	
$4 + 2 = 6$	True	$4 - 2 = 2$	True

Therefore, the solution (4, 2) is correct, and we can say that the lines intersect in exactly one point, that the system of equations is *consistent* and *independent*, and that there is a *unique solution* for the system. The solution set is {(4, 2)}.

Note To solve a system of two equations in two variables with a TI-81 graphics calculator, we first solve each equation for y. For the system in Example 1, we can let $y_1 = 6 - x$ and $y_2 = x - 2$. We graph these two lines together and then *zoom in* (see the second note following Example 2 in Section 9.6, if necessary). We can trace one of the lines until we are close to the point of intersection, and then we can zoom in again and trace again. We can keep doing this until we are very close to the point of intersection; we will probably *not* see $x = 4$, $y = 2$ in the display, but rather something like $x = 3.9999816$, $y = 2.0000184$. As we have mentioned previously, numerical answers on graphics calculators are not always exact.

EXAMPLE 2 Find the solution set of the system $\begin{Bmatrix} 2x - 3y = 6 \\ 6x - 9y = 36 \end{Bmatrix}$ graphically.

SOLUTION We graph the solution set of each equation on the same set of axes (Figure 2).

Line 1 $2x - 3y = 6$ has intercepts $(3, 0)$ and $(0, -2)$.

Line 2 $6x - 9y = 36$ has intercepts $(6, 0)$ and $(0, -4)$.

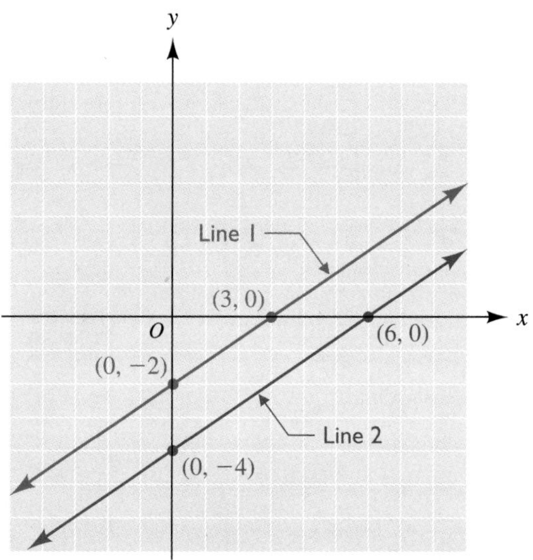

Figure 2

The lines *appear* to be parallel; to be *sure* that they are parallel, we check the slopes and the y-intercepts:

Line 1 $2x - 3y = 6$

$$-3y = -2x + 6$$

$$y = \frac{2}{3}x - 2$$

Line 2 $6x - 9y = 36$

$$-9y = -6x + 36$$

$$y = \frac{2}{3}x - 4$$

The slopes of both lines are $\frac{2}{3}$, but the y-intercepts are different; therefore, the lines are parallel, the system of equations is *inconsistent* and *independent*, and there is *no solution* for the system of equations. The solution set is { }.

EXAMPLE 3 Find the solution set of $\begin{Bmatrix} 3x + 5y = 15 \\ 6x + 10y = 30 \end{Bmatrix}$ graphically, if possible.

SOLUTION We graph the solution set of each equation on the same set of axes (Figure 3).

Line 1 $3x + 5y = 15$ has intercepts (5, 0) and (0, 3).

Line 2 $6x + 10y = 30$ has intercepts (5, 0) and (0, 3).

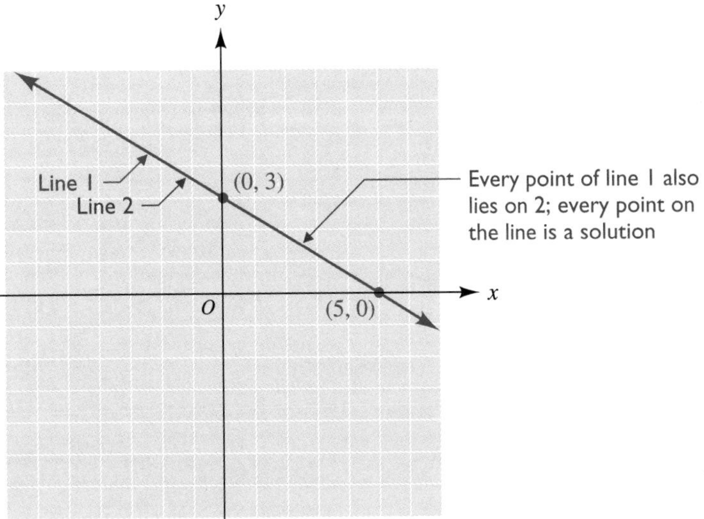

Figure 3

Since both lines go through the same two points, they must be the same line. We can say that the lines *coincide*, that the system of equations is *consistent* and *dependent*, and that there are *infinitely many solutions* for the system of equations.

We can find the solution set algebraically by substituting some variable, such as t, for x in one of the equations and then solving for y:

$$3t + 5y = 15 \qquad \text{Substituting } t \text{ for } x \text{ in the first equation}$$

$$5y = 15 - 3t$$

$$y = \frac{15 - 3t}{5}$$

The solution set is a set of ordered pairs that can be expressed as $\left\{\left(t, \dfrac{15 - 3t}{5}\right), t \in R\right\}$. You might verify that if we let $t = 0$, we get the point (0, 3); if we let $t = 5$, we get (5, 0); and if we let $t = -5$, we get $(-5, 6)$, *another* point on the line. In fact, when we let t be *any* real number, we always get a point on the line.

If, instead of substituting t for x in the first equation, we substitute s for y and then solve for x, we find that the solution set can also be expressed as $\left\{\left(\dfrac{15 - 5s}{3}, s\right), s \in R\right\}$. You can verify that if we let $s = 0$, we get the point (5, 0); if we let $s = 3$, we get (0, 3); and so on.

Example 4 illustrates the drawbacks of the graphical method and shows the need for the algebraic methods of solving systems of equations, given in the next two sections of this chapter. Nevertheless, the graphical method helps you understand what it means to *solve* a system of equations—especially what it means when a system has no solution or has many solutions.

EXAMPLE 4 Try to solve the system $\left\{\begin{matrix} -x + y = 3 \\ x + 2y = -2 \end{matrix}\right\}$ graphically.

SOLUTION We graph the solution set of each equation on the same set of axes (Figure 4).

Line 1 $-x + y = 3$ has intercepts $(0, 3)$ and $(-3, 0)$.

Line 2 $x + 2y = -2$ has intercepts $(0, -1)$ and $(-2, 0)$.

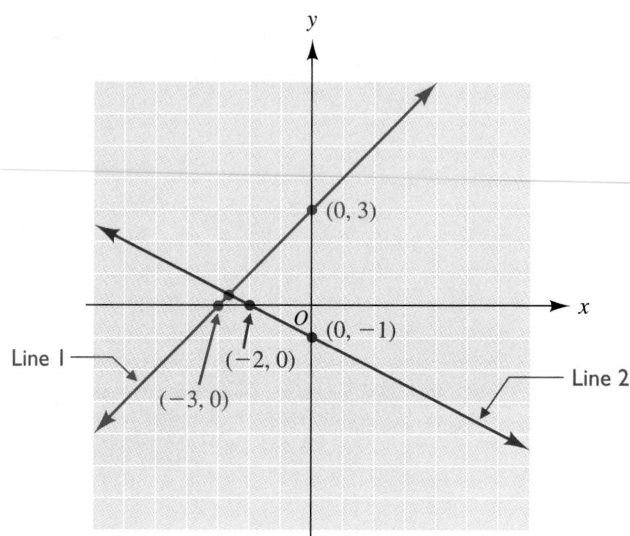

Figure 4

It's difficult to name the coordinates of the point of intersection of the two lines. One guess might be $\left(-2\frac{1}{2}, \frac{1}{2}\right)$; however, this ordered pair does *not* check in the second equation (you can verify this). We'll find in Example 1 of the next section that the correct solution is $\left(-2\frac{2}{3}, \frac{1}{3}\right)$. (You might verify that this ordered pair does satisfy *both* equations.)

Exercises 12.2
Set I

Find the solution set of each system graphically. If the system is dependent, find the solution set by using the method shown in Example 3.

1. $\begin{Bmatrix} 2x + y = 6 \\ x - y = 0 \end{Bmatrix}$

2. $\begin{Bmatrix} 2x - y = -4 \\ x + y = -2 \end{Bmatrix}$

3. $\begin{Bmatrix} x + 2y = 3 \\ 3x - y = -5 \end{Bmatrix}$

4. $\begin{Bmatrix} 5x + 4y = 12 \\ x + 3y = -2 \end{Bmatrix}$

5. $\begin{Bmatrix} x + 2y = 0 \\ 2x - y = 0 \end{Bmatrix}$

6. $\begin{Bmatrix} 2x + y = 0 \\ x - 3y = 0 \end{Bmatrix}$

7. $\begin{Bmatrix} 8x - 5y = 15 \\ 10y - 16x = 16 \end{Bmatrix}$

8. $\begin{Bmatrix} 3x + 9y = 18 \\ 2x + 6y = -24 \end{Bmatrix}$

9. $\begin{Bmatrix} 8x - 10y = 16 \\ 15y - 12x = -24 \end{Bmatrix}$

10. $\begin{Bmatrix} 14x + 30y = -70 \\ 15y + 7x = -35 \end{Bmatrix}$

Writing Problems

Express the answers in your own words and in complete sentences.

1. Explain how to solve a system of two equations in two variables graphically.

2. If there is *no* solution for a system of equations, describe the *graph* of the system.

3. If there are *many* solutions for a system of equations, describe the *graph* of the system.

Exercises 12.2
Set II

Find the solution set of each system graphically. If the system is dependent, find the solution set by using the method shown in Example 3.

1. $\begin{cases} 8x - 5y = 30 \\ 2x - 7y = -4 \end{cases}$

2. $\begin{cases} x + 2y = -5 \\ x - y = 4 \end{cases}$

3. $\begin{cases} 3x + 2y = -8 \\ 5x - 2y = -8 \end{cases}$

4. $\begin{cases} 3x - 2y = 0 \\ 2x + 5y = 0 \end{cases}$

5. $\begin{cases} 2x + 3y = -12 \\ x - y = -1 \end{cases}$

6. $\begin{cases} 3x + 8y = 9 \\ 9x + 5y = -30 \end{cases}$

7. $\begin{cases} 4x + 8y = -20 \\ 10y + 5x = 10 \end{cases}$

8. $\begin{cases} 5x + 8y = -22 \\ 4x - 3y = 20 \end{cases}$

9. $\begin{cases} 15x + 7y = -35 \\ 14y + 30x = -70 \end{cases}$

10. $\begin{cases} 5x - 8y = 14 \\ 9x - 2y = -12 \end{cases}$

12.3 Solving a Linear System of Two Equations in Two Variables by Elimination

The graphical method of solving a system of equations has several disadvantages. It is slow, and it is not an exact method of solution. Furthermore, the lines might intersect at some point that is far off the grid (or even off the paper). The **elimination method** has none of these disadvantages. We need one definition, however, before we proceed.

Equivalent systems of equations

Two systems of equations are said to be **equivalent systems** if they both have the same solution set.

The operations in the following box always lead to a system of equations equivalent to the system we started with; we usually use the second and third operations when we use the elimination method of solving a system of equations.

Operations that yield equivalent systems of equations

1. The equations can be interchanged.
2. Any of the equations can be rewritten by using the multiplication property of equality.
3. Any two of the equations can be added together; that is, the following property of equality can be used:

$$\text{If} \quad a = b \quad \text{and} \quad c = d, \quad \text{then} \quad a + c = b + d$$

In solving a system of two equations in two variables algebraically, our objective is to find an *equivalent* system in which one of the equations is in the form $x = a$ and the other is in the form $y = b$. That is, we hope to find an equivalent system in which each equation contains only one variable. Then the *solution* is the ordered pair (a, b), and the *solution set* is $\{(a, b)\}$.

We will be able to find such an equivalent system if the system of equations is consistent and independent. In the other two cases, *both* variables will drop out. We will be left with a *false* statement (such as $0 = 5$) if the system of equations is inconsistent and with a *true* statement (such as $0 = 0$) if the system is dependent.

We must begin by *eliminating one of the variables* from the equations (hence the method's name). The complete procedure for the elimination method (sometimes called the *addition method*) is as follows.

Solving a linear system of two equations in two variables by elimination

1. Express both equations in the form $ax + by = c$.
2. If necessary, multiply one or both equations by numbers that make the coefficients of one of the variables equal numerically but opposite in sign. Write the equations one under the other, with the equal signs aligned and *with like terms aligned*.
3. Add the equations from step 1 together. (This will eliminate at least one of the variables.)
4. Three outcomes are possible:
   a. *One variable remains.* Solve *the resulting equation for that variable.* Then substitute this value into either equation to find the value of the other variable. There is a unique solution for the system.
   b. *Both variables are eliminated, and a false statement results.* There is no solution; the solution set is { }.
   c. *Both variables are eliminated, and a true statement results.* There are infinitely many solutions. For the line $ax + by = c$, the solution set is $\left\{\left(t, \frac{c - at}{b}\right), t \in R\right\}$.
5. If a unique solution was found in step 4, check it in *both* equations.

We will give examples of each type of system. In Example 1, we solve algebraically the system that we tried to solve graphically in Example 4 of the last section.

Systems That Have Only One Solution (Consistent, Independent Systems)

EXAMPLE 1 Find the solution set of the system $\begin{Bmatrix} -x + y = 3 \\ x + 2y = -2 \end{Bmatrix}$.

SOLUTION We want to eliminate one of the variables. In this case, if we simply *add* the two equations together, the x's will be eliminated:

$$\begin{aligned} -x + y &= 3 \\ x + 2y &= -2 \\ \hline 3y &= 1 \\ y &= \tfrac{1}{3} \end{aligned}$$

We must add *both* sides of both equations

We must not drop the equal sign!

Dividing both sides by 3

By eliminating the x's, we were able to solve for y. Now that we know what the y-coordinate of the point of intersection is, we can substitute that value $\left(\frac{1}{3}\right)$ for y into *either* of the first two equations to find the corresponding x-coordinate; we will substitute into the second equation, obtaining

$$x + 2\left(\tfrac{1}{3}\right) = -2$$
$$x + \tfrac{2}{3} = -2$$
$$x = -2\tfrac{2}{3}$$

√ **Check**

$-x + y = 3$	$x + 2y = -2$
$-\left(-2\frac{2}{3}\right) + \frac{1}{3} \stackrel{?}{=} 3$	$-2\frac{2}{3} + 2\left(\frac{1}{3}\right) \stackrel{?}{=} -2$
$2\frac{2}{3} + \frac{1}{3} \stackrel{?}{=} 3$	$-2\frac{2}{3} + \frac{2}{3} \stackrel{?}{=} -2$
$3 = 3$ True	$-2 = -2$ True

Therefore, the solution set of the system is $\left\{\left(-2\frac{2}{3}, \frac{1}{3}\right)\right\}$.

When we have found the value of one variable, that value may be substituted into *either* of the original equations to find the value of the second variable. For Example 1, you might verify that if we substituted $\frac{1}{3}$ for y in the *first* equation, we would still get $-2\frac{2}{3}$ for x.

EXAMPLE 2 Find the solution set of the system $\begin{Bmatrix} 3x + 2y = 13 \\ 3x - 4y = 1 \end{Bmatrix}$.

SOLUTION If we multiply the second equation by -1, the coefficients of x will be equal numerically and opposite in sign:

$$\begin{aligned} 3x + 2y &= 13 \\ -3x + 4y &= -1 \\ \hline 6y &= 12 \\ y &= 2 \end{aligned}$$

We add *both* sides of both equations

We must not drop the equal sign

Substituting 2 for y in the second equation, we have

$$\begin{aligned} 3x - 4y &= 1 \\ 3x - 4(2) &= 1 \\ 3x &= 9 \\ x &= 3 \end{aligned}$$

✓ **Check**

$3x + 2y = 13$	$3x - 4y = 1$
$3(3) + 2(2) \stackrel{?}{=} 13$	$3(3) - 4(2) \stackrel{?}{=} 1$
$9 + 4 \stackrel{?}{=} 13$	$9 - 8 \stackrel{?}{=} 1$
$13 = 13$ True	$1 = 1$ True

Therefore, the solution set is $\{(3, 2)\}$.

The Symbol ⇒ The symbol ⇒, read **implies**, means that the second statement is true if the first statement is true. For example,

$$3x + 4y = 6 \Rightarrow -6x - 8y = -12$$

is read "$3x + 4y = 6$ implies $-6x - 8y = -12$" and means that $-6x - 8y = -12$ is true if $3x + 4y = 6$ is true. We will use this symbol between equations in the next several examples.

In some cases, it is easier to eliminate one variable than the other, but in all cases, *either* variable can be eliminated (see Example 3).

EXAMPLE 3 Find the solution set of $\begin{Bmatrix} 3x + 4y = 6 \\ 2x + 3y = 5 \end{Bmatrix}$ by eliminating (a) the x's and (b) the y's.

SOLUTION

a. If the first equation is multiplied by -2 and the second equation is multiplied by 3, the coefficients of x will be equal numerically but opposite in sign.

This symbol means that the equation is to be multiplied by -2

$$\begin{aligned} -2\,]\quad 3x + 4y = 6 &\Rightarrow -6x - 8y = -12 \\ 3\,]\quad 2x + 3y = 5 &\Rightarrow \underline{\;6x + 9y = 15\;} \\ & \qquad\qquad y = 3 \end{aligned}$$

We add *both* sides of both equations

This pair of numbers is found by interchanging the coefficients of x and making one of them negative

Substituting 3 for y in the second equation, we have

$$\begin{aligned} 2x + 3y &= 5 \\ 2x + 3(3) &= 5 \\ 2x + 9 &= 5 \\ 2x &= -4 \\ x &= -2 \end{aligned}$$

Therefore, the solution of the system is apparently $(-2, 3)$.

b. This pair is found by interchanging the coefficients of y and making one of them negative

$$\begin{aligned} 3\,]\quad 3x + 4y = 6 &\Rightarrow 9x + 12y = 18 \\ -4\,]\quad 2x + 3y = 5 &\Rightarrow -8x - 12y = -20 \\ & \quad\; x = -2 \end{aligned}$$

We add *both* sides of both equations

Substituting -2 for x in the first equation, we have

$$\begin{aligned} 3x + 4y &= 6 \\ 3(-2) + 4y &= 6 \\ -6 + 4y &= 6 \\ 4y &= 12 \\ y &= 3 \end{aligned}$$

Therefore, we get the same solution as before: $(-2, 3)$. Checking will confirm that the solution set is $\{(-2, 3)\}$.

For each of the systems in Examples 1, 2, and 3, we can say that there is a unique solution for the system, that the system of equations is consistent and independent, and that there is exactly one point of intersection for each pair of lines.

We can always use the method shown in Example 3 and find the pair of numbers to use as multipliers by interchanging the coefficients of one of the variables. Sometimes, however, smaller numbers can be used; the *smallest* possible absolute value for the new coefficients is the *least common multiple* (LCM) of both of the original coefficients (see Example 4).

EXAMPLE 4 For the system $\begin{Bmatrix} 10x - 9y = 5 \\ 15x + 6y = 4 \end{Bmatrix}$, find the smallest numbers that can be used to eliminate (a) the x's and (b) the y's. (Do not solve the system.)

SOLUTION

a. Because the coefficients of the x's are 10 and 15 and because the LCM of 10 and 15 is 30, we want the new coefficients to be -30 and 30. We can get these numbers by multiplying the first equation by -3 and the second by 2. However, instead of determining the LCM beforehand, we *can* do the work as follows:

This pair of numbers is found by interchanging the coefficients of x and making one of them negative

$$\begin{aligned} \cancel{-15}\,]\quad -3\,]\quad 10x - 9y = 5 &\Rightarrow -30x + 27y = -15 \\ \cancel{10}\,]\quad 2\,]\quad 15x + 6y = 4 &\Rightarrow 30x + 12y = 8 \end{aligned}$$

This pair is found by dividing both -15 and 10 by 5, their GCF

Notice in the last step that the coefficients of x are -30 and 30. If we added the last two equations, the x's would be eliminated.

b. Because the coefficients of the y's are -9 and 6 and because the LCM of 9 and 6 is 18, we want the new coefficients to be -18 and 18. We can get these numbers by multiplying the first equation by 2 and the second by 3. (Because the coefficients of y already have different signs, we do *not* need to make one of the multipliers negative.) Instead of finding the LCM first, we *can* do the work as follows:

This pair of numbers is found by interchanging the absolute values of the coefficients of y

$$\begin{array}{ll|l} 6\,] & 2\,] & 10x - 9y = 5 \Rightarrow 20x - 18y = 10 \\ 9\,] & 3\,] & 15x + 6y = 4 \Rightarrow 45x + 18y = 12 \end{array}$$

This pair is found by dividing both 6 and 9 by 3, their GCF

Notice in the last step that the coefficients of y are -18 and 18. If we added the last two equations, the y's would be eliminated.

Systems That Have No Solution (Inconsistent, Independent Systems)

If, when we attempt to solve a system of equations by elimination, *both* variables drop out and a *false statement* results, there is no solution for the system. (The graphs of the equations of the system would be parallel lines.) The solution set is { }.

EXAMPLE 5 Find the solution set of the system $\begin{Bmatrix} 2x - 3y = 6 \\ 6x - 9y = 36 \end{Bmatrix}$.

SOLUTION

$$\begin{array}{ll|l} 9\,] & -3\,] & 2x - 3y = 6 \Rightarrow -6x + 9y = -18 \\ 3\,] & 1\,] & 6x - 9y = 36 \Rightarrow \underline{\;\;6x - 9y = \;\;36} \\ & & \qquad\qquad\qquad\qquad 0 = \;\;18 \end{array}$$

← A false statement

There is no solution for this system of equations. (We saw in Example 2 of the last section that the slopes of the lines are the same and that the lines are parallel.) The system is inconsistent and independent. The solution set is { }.

Note It is possible to predict that a system will have no solution by *mentally* examining the coefficients. In Example 5, we could *mentally* multiply both sides of the first equation by 3, obtaining the *new* equation $6x - 9y = 18$. Comparing this equation with the second equation, we could see that the coefficient of x in both equations would be 6 and that the coefficient of y in both equations would be -9, but the *constants* in the two equations would be *different* (one 18, the other 36). This tells us that there is no solution for the system.

Systems That Have Many Solutions (Consistent, Dependent Systems)

If, when we attempt to solve a system of equations algebraically, *both* variables drop out and a *true* statement results, there are infinitely many solutions for the system. (The graphs of the equations of the system would be lines that coincide.)

EXAMPLE 6 Find the solution set of the system $\begin{Bmatrix} 4x + 6y = 4 \\ 6x + 9y = 6 \end{Bmatrix}$.

SOLUTION

$$\begin{array}{ll|l} -9\,] & -3\,] & 4x + 6y = 4 \Rightarrow -12x - 18y = -12 \\ 6\,] & 2\,] & 6x + 9y = 6 \Rightarrow \underline{\;\;12x + 18y = \;\;12} \\ & & \qquad\qquad\qquad\qquad\quad 0 = \;\;0 \end{array}$$

← A true statement

Therefore, there are infinitely many solutions for this system of equations. (The system is consistent and dependent, and the graphs of the lines would coincide.) We can find the solution set of the system by letting $x = t$ in either equation and solving for y:

$$4t + 6y = 4 \quad \text{Substituting } t \text{ for } x \text{ in the first equation}$$

$$6y = 4 - 4t$$

$$y = \frac{4 - 4t}{6} = \frac{\overset{1}{\cancel{2}}(2 - 2t)}{\underset{3}{\cancel{6}}} = \frac{2 - 2t}{3}$$

Therefore, the solution set can be expressed as $\left\{\left(t, \dfrac{2 - 2t}{3}\right), t \in R\right\}$. If we had, instead, let $y = s$, the solution set would have been expressed as $\left\{\left(\dfrac{2 - 3s}{2}, s\right), s \in R\right\}$.

Note It is possible to predict that a system will have many solutions by *mentally* examining the equations. Whenever the two original equations are equivalent to the *same* equation, the system of equations is dependent. In Example 6, for instance, we could mentally divide both sides of $4x + 6y = 4$ by 2, getting $2x + 3y = 2$, and mentally divide both sides of $6x + 9y = 6$ by 3, getting $2x + 3y = 2$; because both given equations can be rewritten as the *same* equation, the system is dependent.

Exercises 12.3
Set I

Find the solution set of each system by the elimination method.

1. $\begin{cases} 3x - y = 11 \\ 3x + 2y = -4 \end{cases}$

2. $\begin{cases} 6x + 5y = 2 \\ 2x - 5y = -26 \end{cases}$

3. $\begin{cases} 8x + 15y = 11 \\ 4x - y = 31 \end{cases}$

4. $\begin{cases} x + 6y = 24 \\ 5x - 3y = 21 \end{cases}$

5. $\begin{cases} 7x - 3y = 3 \\ 20x - 9y = 12 \end{cases}$

6. $\begin{cases} 10x + 7y = -1 \\ 2x + y = 5 \end{cases}$

7. $\begin{cases} 6x + 5y = 0 \\ 4x - 3y = 38 \end{cases}$

8. $\begin{cases} 7x + 4y = 12 \\ 2x - 3y = -38 \end{cases}$

9. $\begin{cases} 4x + 6y = 5 \\ 8x + 12y = 7 \end{cases}$

10. $\begin{cases} 5x - 2y = 3 \\ 15x - 6y = 4 \end{cases}$

11. $\begin{cases} 7x - 3y = 5 \\ 14x - 6y = 10 \end{cases}$

12. $\begin{cases} 8x - 12y = 16 \\ 2x - 3y = 4 \end{cases}$

13. $\begin{cases} 9x + 4y = -4 \\ 15x - 6y = 25 \end{cases}$

14. $\begin{cases} 16x - 25y = -38 \\ 8x + 5y = -12 \end{cases}$

15. $\begin{cases} 9x + 10y = -3 \\ 14y = 7 - 15x \end{cases}$

16. $\begin{cases} 35x + 18y = 30 \\ 7x = 24y - 17 \end{cases}$

Writing Problems

Express the answers in your own words and in complete sentences.

1. Explain why it would not be helpful to add (just as they are) the two equations for the system $\begin{cases} 4x + 3y = 12 \\ 8x + 3y = -4 \end{cases}$.

2. In Problem 1, if we *do* add the two equations, we obtain the equation $12x + 6y = 8$. Explain how the graph of $12x + 6y = 8$ is related to the graphs of the lines of the original system of equations.

Exercises 12.3
Set II

Find the solution set of each system by the elimination method.

1. $\left\{\begin{array}{l} 2x - 7y = 6 \\ 4x + 7y = -30 \end{array}\right\}$

2. $\left\{\begin{array}{l} x - 2y = -32 \\ 7x + 8y = -4 \end{array}\right\}$

3. $\left\{\begin{array}{l} 12x + 17y = 30 \\ 3x - y = 39 \end{array}\right\}$

4. $\left\{\begin{array}{l} 3x - 2y = -51 \\ 2x + 3y = -21 \end{array}\right\}$

5. $\left\{\begin{array}{l} 3x - 5y = -2 \\ 2x + y = 16 \end{array}\right\}$

6. $\left\{\begin{array}{l} 3x - y = -14 \\ x + 2y = 7 \end{array}\right\}$

7. $\left\{\begin{array}{l} 2x - y = 0 \\ 3x + 4y = -22 \end{array}\right\}$

8. $\left\{\begin{array}{l} 2x - 5y = 15 \\ 3x + 2y = 13 \end{array}\right\}$

9. $\left\{\begin{array}{l} 12x - 9y = 28 \\ 20x - 15y = 35 \end{array}\right\}$

10. $\left\{\begin{array}{l} 7x - 2y = 4 \\ 14x - 4y = 8 \end{array}\right\}$

11. $\left\{\begin{array}{l} 21x + 35y = 28 \\ 12x + 20y = 16 \end{array}\right\}$

12. $\left\{\begin{array}{l} 8x - 5y = 3 \\ 24x - 15y = 6 \end{array}\right\}$

13. $\left\{\begin{array}{l} 3x + 10y = 4 \\ 9x - 20y = -18 \end{array}\right\}$

14. $\left\{\begin{array}{l} 7x + 2y = 4 \\ 2x + 3y = -11 \end{array}\right\}$

15. $\left\{\begin{array}{l} 3x + 4y = 2 \\ 4y - 3x = 0 \end{array}\right\}$

16. $\left\{\begin{array}{l} 22x - 15y = -29 \\ 33x = 10y - 11 \end{array}\right\}$

12.4 Solving a Linear System of Two Equations in Two Variables by Substitution

All linear systems of two equations in two variables can be solved by the elimination (or addition) method shown in the last section. However, if one equation has already been solved for one variable, the **substitution method** is easier to use than the elimination method. Also, you should learn the substitution method of solution at this time so that you can apply it later, when we solve quadratic systems.

Solving a linear system of two equations in two variables by substitution

1. Solve one equation for one of the variables in terms of the other by using the method of solving literal equations.
2. Substitute the value of the variable from step 1 into the *other* equation, and simplify both sides of the resulting equation.
3. Three outcomes are possible:
   a. *One variable remains.* Solve the equation resulting from step 2 for its variable, and substitute that value into either equation to find the value of the other variable. There is a unique solution for the system.
   b. *Both variables are eliminated, and a false statement results.* There is no solution for the system; the solution set is { }.
   c. *Both variables are eliminated, and a true statement results.* There are many solutions for the system; the solution set can be found by using the method demonstrated in Example 6 of the last section.
4. If a unique solution was found in step 4, check it in *both* equations.

Examples 1 through 3 show how to decide which equation to use and which variable to solve for in step 1 of the substitution method.

EXAMPLE 1 An example in which one of the equations is already solved for a variable:

$$\begin{Bmatrix} 2x - 5y = 4 \\ y = 3x + 7 \end{Bmatrix}$$

Already solved for y

Use $y = 3x + 7$.

EXAMPLE 2 An example in which one of the variables has a coefficient of 1:

$$\begin{Bmatrix} 2x + 6y = 3 \\ x - 4y = 2 \end{Bmatrix} \Rightarrow x = 4y + 2$$

x has a coefficient of 1

Use $x = 4y + 2$.

EXAMPLE 3 An example of choosing the variable with the smallest coefficient:

$$\begin{Bmatrix} 11x - 7y = 10 \\ 14x + 2y = 9 \end{Bmatrix} \Rightarrow y = \frac{9 - 14x}{2}$$

The smallest of the four coefficients

The smallest possible denominator in this case

Use $y = \dfrac{9 - 14x}{2}$.

Systems That Have a Unique Solution (Consistent, Independent Systems)

We now give two examples of using the substitution method with systems that have a unique solution.

EXAMPLE 4 Find the solution set of the system $\begin{Bmatrix} 2x - 5y = 4 \\ y = 3x + 7 \end{Bmatrix}$.

SOLUTION

Step 1. The second equation has already been solved for y.

Step 2. We'll substitute $3x + 7$ for y in the first equation.

$$\begin{aligned} 2x - 5y &= 4 \\ 2x - 5(3x + 7) &= 4 \\ 2x - 15x - 35 &= 4 \\ -13x &= 39 \\ x &= -3 \end{aligned}$$

Step 3. Next, we substitute -3 for x in $y = 3x + 7$:

$$y = 3(-3) + 7 = -9 + 7 = -2$$

Therefore, $(-3, -2)$ is the apparent solution of the system.

√ **Step 4.** ***Check***

$2x - 5y = 4$	$y = 3x + 7$
$2(-3) - 5(-2) \stackrel{?}{=} 4$	$-2 \stackrel{?}{=} 3(-3) + 7$
$-6 + 10 \stackrel{?}{=} 4$	$-2 \stackrel{?}{=} -9 + 7$
$4 = 4$ True	$-2 = -2$ True

The solution set is $\{(-3, -2)\}$.

EXAMPLE 5 Find the solution set of the system $\begin{Bmatrix} x - 2y = 11 \\ 3x + 5y = -11 \end{Bmatrix}$.

SOLUTION

Step 1. We solve the first equation for x:

$$x - 2y = 11$$
$$x = 11 + 2y$$

Step 2. Next, we substitute $11 + 2y$ for x in the second equation:

$$3x + 5y = -11$$
$$3(11 + 2y) + 5y = -11$$
$$33 + 6y + 5y = -11$$
$$11y = -44$$
$$y = -4$$

Step 3. Then we substitute -4 for y in $x = 11 + 2y$:

$$x = 11 + 2(-4)$$
$$x = 11 - 8$$
$$x = 3$$

√ **Step 4.** ***Check***

$x - 2y = 11$	$3x + 5y = -11$
$3 - 2(-4) \stackrel{?}{=} 11$	$3(3) + 5(-4) \stackrel{?}{=} -11$
$3 + 8 \stackrel{?}{=} 11$	$9 - 20 \stackrel{?}{=} -11$
$11 = 11$ True	$-11 = -11$ True

Therefore, $\{(3, -4)\}$ is the solution set.

Systems That Have No Solution (Inconsistent, Independent Systems)

If, when we attempt to solve a system of equations by substitution, *both* variables drop out and a *false* statement results, there is no solution for the system.

EXAMPLE 6 Find the solution set of the system $\begin{Bmatrix} 2x - 3y = 6 \\ 6x - 9y = 36 \end{Bmatrix}$.

(The smallest coefficient: 2)

SOLUTION

Step 1. We solve the first equation for x:

$$2x - 3y = 6 \Rightarrow 2x = 3y + 6$$
$$x = \frac{3y + 6}{2}$$

Step 2. Next, we substitute $\frac{3y + 6}{2}$ for x in the second equation:

$$6x \qquad - 9y = 36$$

$$\overset{3}{\cancel{6}}\left(\frac{3y + 6}{\underset{1}{\cancel{2}}}\right) - 9y = 36$$

$$3(3y + 6) - 9y = 36$$

$$9y + 18 - 9y = 36$$

$$18 = 36 \leftarrow \text{A false statement}$$

Step 3. *No* values for x and y can make $18 = 36$. Therefore, there is *no solution* for this system of equations; the system is inconsistent and independent. The solution set is { }.

Systems That Have Many Solutions (Consistent, Dependent Systems)

If, when we attempt to solve a system of equations by substitution, *both* variables drop out and a *true* statement results, there are infinitely many solutions for the system.

EXAMPLE 7 Find the solution set of the system $\begin{Bmatrix} 9x + 6y = 6 \\ 6x + 4y = 4 \end{Bmatrix}$.

The smallest coefficient (4)

SOLUTION

Step 1. We solve the second equation for y:

$$6x + 4y = 4 \Rightarrow 4y = 4 - 6x$$

$$y = \frac{4 - 6x}{4} = \frac{\overset{1}{\cancel{2}}(2 - 3x)}{\underset{2}{\cancel{4}}} = \frac{2 - 3x}{2}$$

Step 2. Next, we substitute $\frac{2 - 3x}{2}$ for y in the first equation:

$$9x + 6y = 6$$

$$9x + \overset{3}{\cancel{6}}\left(\frac{2 - 3x}{\underset{1}{\cancel{2}}}\right) = 6$$

$$9x + 3(2 - 3x) = 6$$

$$9x + 6 - 9x = 6$$

$$6 = 6 \leftarrow \text{A true statement}$$

Step 3. There are infinitely many solutions; the system is consistent and dependent. To find the solution set, we let $x = t$. Then $y = \frac{2 - 3t}{2}$. Therefore, the solution set can be expressed as $\left\{\left(t, \frac{2 - 3t}{2}\right), t \in R\right\}$.

Exercises 12.4 Set I

Find the solution set of each system.

In Exercises 1–10, use the substitution method.

1. $\begin{Bmatrix} 7x + 4y = 4 \\ y = 6 - 3x \end{Bmatrix}$

2. $\begin{Bmatrix} 2x + 3y = -5 \\ x = y - 10 \end{Bmatrix}$

3. $\begin{Bmatrix} 5x - 4y = -1 \\ 3x + y = -38 \end{Bmatrix}$

4. $\begin{Bmatrix} 3x - 5y = 5 \\ x - 6y = 19 \end{Bmatrix}$

5. $\begin{Bmatrix} 8x - 5y = 4 \\ x - 2y = -16 \end{Bmatrix}$

6. $\begin{Bmatrix} 12x - 6y = 24 \\ 3x - 2y = -2 \end{Bmatrix}$

7. $\begin{Bmatrix} 15x + 5y = 8 \\ 6x + 2y = -10 \end{Bmatrix}$

8. $\begin{Bmatrix} 15x - 5y = 30 \\ 12x - 4y = 11 \end{Bmatrix}$

9. $\begin{Bmatrix} 20x - 10y = 70 \\ 6x - 3y = 21 \end{Bmatrix}$

10. $\begin{Bmatrix} 2x - 10y = 18 \\ 5x - 25y = 45 \end{Bmatrix}$

In Exercises 11–14, solve by any convenient method.

11. $\begin{Bmatrix} 8x + 4y = 7 \\ 3x + 6y = 6 \end{Bmatrix}$

12. $\begin{Bmatrix} 5x - 4y = 2 \\ 15x + 12y = 12 \end{Bmatrix}$

13. $\begin{Bmatrix} 4x + 4y = 3 \\ 6x + 12y = -6 \end{Bmatrix}$

14. $\begin{Bmatrix} 4x + 9y = -11 \\ 10x + 6y = 11 \end{Bmatrix}$

Writing Problems

Express the answers in your own words and in complete sentences.

1. Determine whether you would use the elimination method or the substitution method in solving the system $\begin{Bmatrix} 3x + 7y = 10 \\ 4x - 5y = -1 \end{Bmatrix}$, and explain why you would use that method.

2. Determine whether you would use the elimination method or the substitution method in solving the system $\begin{Bmatrix} x = 10 - 6y \\ 4x - 5y = 2 \end{Bmatrix}$, and explain why you would use that method.

Exercises 12.4 Set II

Find the solution set of each system.

In Exercises 1–10, use the substitution method.

1. $\begin{Bmatrix} 4x + 3y = -7 \\ y = 2x - 9 \end{Bmatrix}$

2. $\begin{Bmatrix} 13y - 7x = 17 \\ 2x - y = -13 \end{Bmatrix}$

3. $\begin{Bmatrix} 12x - 16y = -3 \\ 8x - 4y = 8 \end{Bmatrix}$

4. $\begin{Bmatrix} y = x + 8 \\ 2y - 3x = 18 \end{Bmatrix}$

5. $\begin{Bmatrix} x + 4y = 1 \\ 2x + 9y = 1 \end{Bmatrix}$

6. $\begin{Bmatrix} 3x + y = 5 \\ -6x - 2y = -10 \end{Bmatrix}$

7. $\begin{Bmatrix} 3x - 9y = 15 \\ 4x - 12y = 7 \end{Bmatrix}$

8. $\begin{Bmatrix} x + 4y = 8 \\ 3y - 2x = -5 \end{Bmatrix}$

9. $\begin{Bmatrix} 8x - 2y = 26 \\ 16x - 4y = 52 \end{Bmatrix}$

10. $\begin{Bmatrix} x - 4y = 3 \\ 4y - x = 3 \end{Bmatrix}$

In Exercises 11–14, solve by any convenient method.

11. $\begin{Bmatrix} -19x + 10y = 26 \\ 12x - 5y = 2 \end{Bmatrix}$

12. $\begin{Bmatrix} 7x - 4y = 5 \\ 4y - 7x = 5 \end{Bmatrix}$

13. $\begin{Bmatrix} 3x + 2y = -1 \\ 15x + 14y = -23 \end{Bmatrix}$

14. $\begin{Bmatrix} 3x - 2y = 3 \\ 8x + 5y = 8 \end{Bmatrix}$

12.5 Solving Third- and Fourth-Order Linear Systems

A system that contains three linear equations in three variables is called a **third-order linear system,** a system that contains four linear equations in four variables is called a **fourth-order linear system,** and so on. In this section, we consider solving third- and fourth-order linear systems.

Solving a Third-Order Linear System

To solve a third-order linear system, we want to find, if it exists, an *equivalent* system of equations in which one equation is in the form $x = a$, another is in the form $y = b$, and the third is in the form $z = c$. The *solution set* of the system, then, will be $\{(a, b, c)\}$—the set that contains the ordered triple (a, b, c).

The following method can be used for solving a third-order system:

Solving a third-order linear system

1. Eliminate one variable from two of the equations by using either elimination or substitution.
2. Eliminate the *same* variable from a *different* pair of equations, using either elimination or substitution.
3. The equation in two variables obtained in step 1 and the equation in the *same* two variables obtained in step 2 form a system of two equations in two variables; solve that system for *its* two variables by using either elimination or substitution.
4. Substitute the values found in step 3 into (usually) any of the original three equations to find the value of the third variable.
5. Check the apparent solution in *all three* equations.

In Example 1, we use elimination in steps 1, 2, and 3; substitution could be used instead.

EXAMPLE 1

Find the solution set of the system
$$\begin{array}{l} (1) \\ (2) \\ (3) \end{array} \left\{ \begin{array}{rcrcrcr} 2x &-& 3y &+& z &=& 1 \\ x &+& 2y &+& z &=& -1 \\ 3x &-& y &+& 3z &=& 4 \end{array} \right\}.$$

SOLUTION

Step 1. We first eliminate z from Equations 1 and 2. (Although any one of the variables could be eliminated, we choose to eliminate z in this case.)

$$\begin{array}{lrl} (1) & 1\,] & 2x - 3y + z = 1 \Rightarrow 2x - 3y + z = 1 \\ (2) & -1\,] & x + 2y + z = -1 \Rightarrow \underline{-x - 2y - z = 1} \\ (4) & & \qquad\qquad\qquad\qquad x - 5y \quad = 2 \end{array}$$

We will add the equations

Step 2. Next, we eliminate z from Equations 1 and 3:

$$\begin{array}{lrl} (1) & 3\,] & 2x - 3y + z = 1 \Rightarrow 6x - 9y + 3z = 3 \\ (3) & -1\,] & 3x - y + 3z = 4 \Rightarrow \underline{-3x + y - 3z = -4} \\ (5) & & \qquad\qquad\qquad\qquad 3x - 8y \quad = -1 \end{array}$$

We will add the equations

(We could have eliminated z from Equations 2 and 3 instead.)

Step 3. Equations 4 and 5 form a second-order system that we solve by elimination. (The system *could* be solved by substitution.)

$$\begin{array}{lrl} (4) & 3\,] & x - 5y = 2 \Rightarrow 3x - 15y = 6 \\ (5) & -1\,] & 3x - 8y = -1 \Rightarrow \underline{-3x + 8y = 1} \\ & & \qquad\qquad\qquad\qquad -7y = 7 \\ & & \qquad\qquad\qquad\qquad\quad y \quad -1 \end{array}$$

(4) $x - 5y = 2$

$x - 5(-1) = 2$ Substituting -1 for y in Equation 4

$x + 5 = 2$

$x = -3$

Step 4. The values found for x and y can be substituted into any one of the original equations to find the value of the third variable. We'll substitute into Equation 2.

(2) $x + 2y + z = -1$

$(-3) + 2(-1) + z = -1$ Substituting -3 for x and -1 for y in Equation 2

$-3 - 2 + z = -1$

$z = 4$

√ **Step 5.** ***Check*** The check for the solution $(-3, -1, 4)$ was shown in Example 2 of the first section of this chapter.

Therefore, the solution set of the system is $\{(-3, -1, 4)\}$.

Note If one of the original equations had contained only two variables (or only one), then in step 4 we might *not* have been able to substitute into that equation to find the value of the third variable.

Solving a Fourth-Order Linear System

The method of solving a third-order system can be extended to a fourth-order system, as follows:

Solving a fourth-order linear system

1. Eliminate one variable from a pair of equations. This gives one equation in three variables.
2. Eliminate the *same* variable from a *different* pair of equations. This gives a second equation in three variables.
3. Eliminate the *same* variable from a third different pair of equations. This gives a third equation in three variables. (Note that each equation of the system must be used in at least one pair of equations.)
4. The three equations obtained in steps 1, 2, and 3 form a third-order system that is then solved by the method given earlier in this section.

The *solution* of the system will be an ordered 4-tuple, written as (a, b, c, d), where the fourth coordinate, d, is the w-coordinate. The *solution set* is $\{(a, b, c, d)\}$.

This same method can be extended to solve systems of *any* order. However, because of the amount of work involved in solving higher-order systems, solutions of such systems are usually carried out by computer.

Higher-order systems can be inconsistent, and they can be dependent, as was true for second-order systems. In this book, however, we consider only higher-order systems that have a single solution (consistent, independent systems).

The graph of a linear equation in three variables is a *plane* in three-dimensional space. Therefore, we don't attempt to solve third-order linear systems graphically. Graphical solutions for *higher* than third-order linear systems are not possible, as we cannot visualize (or represent on paper) four-dimensional space, five-dimensional space, and so on.

Exercises 12.5 Set I

Find the solution set of each system.

1. $\begin{cases} 2x + y + z = 4 \\ x - y + 3z = -2 \\ x + y + 2z = 1 \end{cases}$

2. $\begin{cases} x + y + z = 1 \\ 2x + y - 2z = -4 \\ x + y + 2z = 3 \end{cases}$

3. $\begin{cases} x + 2y + 2z = 0 \\ 2x - y + z = -3 \\ 4x + 2y + 3z = 2 \end{cases}$

4. $\begin{cases} 2x + y - z = 0 \\ 3x + 2y + z = 3 \\ x - 3y - 5z = 5 \end{cases}$

5. $\begin{cases} x + 2z = 7 \\ 2x - y = 5 \\ 2y + z = 4 \end{cases}$

6. $\begin{cases} x - 2y = 4 \\ y + 3z = 8 \\ 2x - z = 1 \end{cases}$

7. $\begin{cases} 2x + 3y + z = 7 \\ 4x - 2z = -6 \\ 6y - z = 0 \end{cases}$

8. $\begin{cases} 4x + 5y + z = 4 \\ 10y - 2z = 6 \\ 8x + 3z = 3 \end{cases}$

9. $\begin{cases} x + y + z + w = 5 \\ 2x - y + 2z - w = -2 \\ x + 2y - z - 2w = -1 \\ -x + 3y + 3z + w = 1 \end{cases}$

10. $\begin{cases} x + y + z + w = 4 \\ x - 2y - z - 2w = -2 \\ 3x + 2y + z + 3w = 4 \\ 2x + y - 2z - w = 0 \end{cases}$

11. $\begin{cases} 6x + 4y + 9z + 5w = -3 \\ 2x + 8y - 6z + 15w = 8 \\ 4x - 4y + 3z - 10w = -3 \\ 2x - 4y + 3z - 5w = -1 \end{cases}$

12. $\begin{cases} 12x + 9y + 4z - 8w = 1 \\ 6x + 15y + 2z + 4w = 2 \\ 3x + 6y + 4z + 2w = 5 \\ 4x + 4y + 4z + 4w = 3 \end{cases}$

Writing Problems

Express the answer in your own words and in complete sentences.

1. Describe what your first two steps would be in solving the following system of equations by using the elimination method, and explain why you would do those things first.

$$\begin{cases} 3x - y + 2z = 4 \\ 5x + y + 3z = 9 \\ 2x + y - 4z = -1 \end{cases}$$

Exercises 12.5 Set II

Find the solution set of each system.

1. $\begin{cases} x + y - z = 0 \\ 2x - y + 3z = 1 \\ 3x + y + z = 2 \end{cases}$

2. $\begin{cases} 2x + y + z = -1 \\ 3x + 5y + z = 0 \\ 7x - y + 2z = 1 \end{cases}$

3. $\begin{cases} x + 2y - 3z = 5 \\ x + y + z = 0 \\ 3x + 4y + 2z = -1 \end{cases}$

4. $\begin{cases} x + 2y + z = 0 \\ 2x + 3y - 5z = 1 \\ -3x + y + 4z = -7 \end{cases}$

5. $\begin{cases} 3x + 2z = 0 \\ 5y - z = 6 \\ 2x - 3y = 8 \end{cases}$

6. $\begin{cases} x + 2y = 0 \\ y - 2z = 0 \\ x - 4z = 0 \end{cases}$

7. $\begin{cases} 5x + y + 6z = -2 \\ 2y - 3z = 3 \\ 5x + 6z = -4 \end{cases}$

8. $\begin{cases} x + y + 2z = 7 \\ 5x - y + z = 2 \\ 3x + 3y - z = 0 \end{cases}$

9. $\begin{cases} x + y + z + w = 5 \\ 3x - y + 2z - w = 0 \\ 2x + y - z + 2w = 4 \\ 2x + y + z + 2w = 2 \end{cases}$

10. $\begin{cases} 2x + 5y + 3z - 4w = 0 \\ -x + y + 6w = -1 \\ 3x - z - 2w = 1 \\ 4x + 2y - z = 3 \end{cases}$

11. $\begin{cases} x + 2y + 2z + w = 0 \\ -x + y + 3z + w = 2 \\ 2x + 3y + 2z - w = -5 \\ 3x - y - 7z + w = 2 \end{cases}$

12. $\begin{cases} 3x + 2y + 4z + 5w = 5 \\ 9x - 8z + 10w = -4 \\ 6y + 12z + 5w = 5 \\ -6x - 4y + 15w = -1 \end{cases}$

Sections 12.1–12.5 REVIEW

Systems of Equations 12.1

In a **linear system of equations**, each equation is a first-degree equation.

In a **quadratic system of equations**, the highest-degree equation is a second-degree equation.

A **solution of a system of two equations in two variables** is an *ordered pair* that satisfies both equations.

A **solution of a system of three equations in three variables** is an *ordered triple* that satisfies all three equations.

Methods of Solving a System of Linear Equations

Second-order systems can be solved by

12.2 **1.** The **graphical method**

12.3 **2.** The **elimination** (or addition) **method**

12.4 **3.** The **substitution method**

When we solve a system of two linear equations in two variables, three outcomes are possible:

1. *There is a unique solution.* The solution set consists of a single ordered pair.

- **a.** Graphical method: The lines intersect in exactly one point.
- **b.** Algebraic methods (elimination or substitution): An equivalent system can be found in which one equation is of the form $x = a$ and the other is of the form $y = b$.

2. *There is no solution.* The solution set is { }.

- **a.** Graphical method: The lines are parallel.
- **b.** Algebraic methods (elimination or substitution): Both variables are eliminated, and a *false* statement results.

3. *There are many solutions.* The solution set is a set of infinitely many ordered pairs.

- **a.** Graphical method: The lines coincide.
- **b.** Algebraic methods (elimination or substitution): Both variables are eliminated, and a *true* statement results.

12.5 *Third- and fourth-order systems* can be solved by the elimination method or the substitution method.

Sections 12.1–12.5 REVIEW EXERCISES Set I

In Exercises 1–3, find the solution set of each system graphically. If the system is dependent, find the solution set by using algebraic methods.

1. $\begin{Bmatrix} 4x + 5y = 22 \\ 3x - 2y = 5 \end{Bmatrix}$

2. $\begin{Bmatrix} 9x - 12y = 3 \\ 12x - 16y = 4 \end{Bmatrix}$

3. $\begin{Bmatrix} 2x - 3y = 3 \\ 3y - 2x = 6 \end{Bmatrix}$

In Exercises 4–10, find the solution set of each system by any convenient method.

4. $\begin{Bmatrix} 6x + 4y = 13 \\ 8x + 10y = 1 \end{Bmatrix}$

5. $\begin{Bmatrix} 4x - 8y = 4 \\ 3x - 6y = 3 \end{Bmatrix}$

6. $\begin{Bmatrix} 4x - 7y = 28 \\ 7y - 4x = 20 \end{Bmatrix}$

7. $\begin{Bmatrix} x = y + 2 \\ 4x - 5y = 3 \end{Bmatrix}$

8. $\begin{Bmatrix} 7x - 3y = 1 \\ y = x + 5 \end{Bmatrix}$

9. $\begin{Bmatrix} 2x + y - z = 1 \\ 3x - y + 2z = 3 \\ x + 2y + 3z = -6 \end{Bmatrix}$

10. $\begin{Bmatrix} x - 2y + z = -3 \\ 3x + 4y + 2z = 4 \\ 2x - 4y - z = 3 \end{Bmatrix}$

Name

ANSWERS

In Exercises 1–3, find the solution set of each system graphically.

1. $\begin{Bmatrix} 2x + 3y = -5 \\ 4x - 5y = 23 \end{Bmatrix}$

2. $\begin{Bmatrix} 5y - 3x = 10 \\ 3x - 5y = 15 \end{Bmatrix}$

3. $\begin{Bmatrix} 7x - 8y = -9 \\ 5x + 6y = 17 \end{Bmatrix}$

1. __________

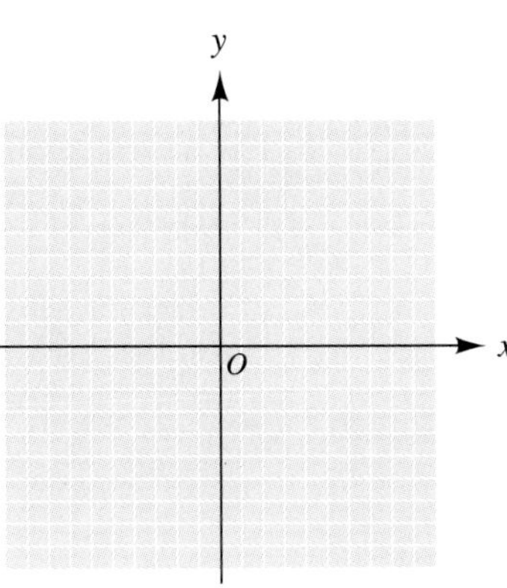

2. __________

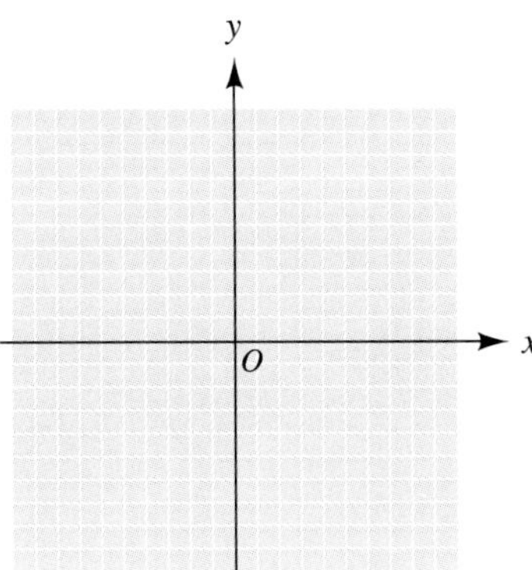

3. __________

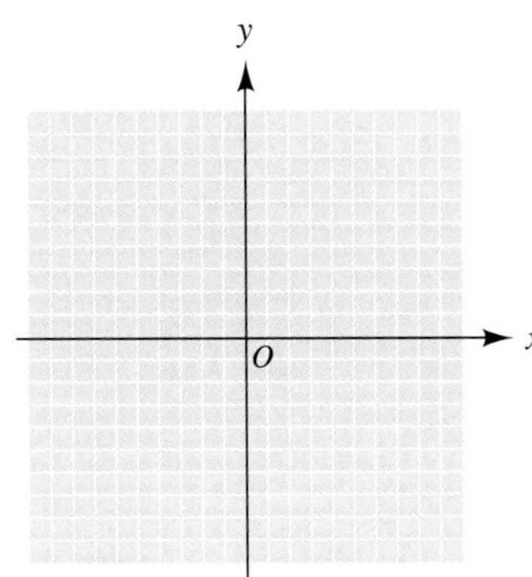

In Exercises 4–9, find the solution set of each system by any convenient method.

4. ______

5. ______

6. ______

7. ______

8. ______

9. ______

4. $\begin{Bmatrix} 6x + 4y = -1 \\ 4x + 6y = -9 \end{Bmatrix}$

5. $\begin{Bmatrix} 8x - 4y = 12 \\ 6x - 3y = 9 \end{Bmatrix}$

6. $\begin{Bmatrix} 3x + 2y = 1 \\ 7x + 3y = 9 \end{Bmatrix}$

7. $\begin{Bmatrix} 4x + 6y = -9 \\ 2x - 8y = 23 \end{Bmatrix}$

8. $\begin{Bmatrix} 3x + 2y + z = 5 \\ 2x - 3y - 3z = 1 \\ 4x + 3y + 2z = 8 \end{Bmatrix}$

9. $\begin{Bmatrix} 4x + 3y + 2z = 6 \\ 8x + 3y - 3z = -1 \\ -4x + 6y + 5z = 5 \end{Bmatrix}$

12.6 A Matrix Method of Solving Linear Systems (Optional)

In this section, we discuss a **matrix method** of solving a system of linear equations; it simplifies the solution of linear systems in the same way that the use of synthetic division simplifies long division of polynomials. The matrix method can be used for solving any system of linear equations, and it is based on those operations that can be performed on a system of equations to yield an *equivalent system*—the operations listed in Section 12.3.

A Matrix

A **matrix** (plural: **matrices**) is a rectangular array of numbers. We will use brackets around a matrix; some authors use parentheses. A **row** of a matrix is a *horizontal* line of its elements. A **column** of a matrix is a *vertical* line of its elements. An $\boldsymbol{m \times n}$ (read "m by n") **matrix** has m rows and n columns. The following are examples of matrices:

A 3 × 4 matrix

$$\begin{bmatrix} 2 & -3 & 1 & 1 \\ 1 & 2 & 1 & -1 \\ 3 & -1 & 3 & 4 \end{bmatrix}$$

(3 rows, 4 columns)

A 4 × 5 matrix

$$\begin{bmatrix} 2 & -1 & 3 & 1 & -4 \\ 1 & 2 & -1 & 2 & -4 \\ 3 & 1 & -1 & -2 & 4 \\ -2 & 3 & 1 & 3 & 0 \end{bmatrix}$$

(4 rows, 5 columns)

A Word of Caution We *cannot* use vertical bars around the array of numbers that forms a matrix; vertical bars are reserved for use with *determinants*, which are discussed in the next section.

The Main Diagonal The **main diagonal** of a matrix is the diagonal of numbers that goes from the upper left-hand corner downward toward the right.

Triangular Form We say that a matrix is in **triangular form** if the numbers *below* the main diagonal are all zeros; that is, a matrix is in triangular form if it is in the form

$$\begin{bmatrix} a & b & c & d \\ 0 & e & f & g \\ 0 & 0 & h & k \end{bmatrix}$$

The main diagonal

where a, b, c, and so forth, represent any constants.

Row-Echelon Form If a matrix is in triangular form *and* has only 1's down the main diagonal, we say that it is in **row-echelon form**.

Note The actual definition of the row-echelon form of a matrix is slightly more complicated than we've indicated here; however, our definition of row-echelon form will suffice for the problems in this text.

The Augmented Matrix Associated with a System of Equations

An **augmented matrix** is associated with each system of linear equations. In writing the augmented matrix of a system of equations, it's helpful to first write the equations as follows:

1. Write the equations one under the other, with all the terms containing variables on the left side of each equation and with *like terms aligned*; write the constant on the right side of each equation.
2. Write 0 for any missing terms, and *write* 1 as the coefficient of any term that has an understood coefficient of 1.

The augmented matrix is then formed from that system of equations as follows:

1. Drop (delete) all the variables, all the plus signs, *and* all the equal signs. (Do not drop any negative signs.)
2. Put brackets around the resulting array of numbers.

Thus, if we're solving a system of four equations in x, y, z, and w, for example, and if the variables are listed in that order, then the first *column* of the augmented matrix will be the coefficients of x, the second column will be the coefficients of y, and so forth. The rightmost column will be the column of constants.

The augmented matrix of a system of n equations in n unknowns will always be an $n \times (n + 1)$ matrix.

Note When writing an augmented matrix, some authors insert a solid vertical line or a dashed vertical line before the column of constants; we do not follow that practice.

EXAMPLE 1 Write the augmented matrix of each system of equations.

a. $\begin{cases} 2x - 3y + 1z = 1 \\ 1x + 2y + 1z = -1 \\ 3x - 1y + 3z = 4 \end{cases}$ Three equations in three unknowns

SOLUTION There are no 0's or 1's to write, all like terms are aligned, and the constants are on the right. The augmented matrix is

The coefficients of x
The coefficients of y
The coefficients of z

$$\begin{bmatrix} 2 & -3 & 1 & 1 \\ 1 & 2 & 1 & -1 \\ 3 & -1 & 3 & 4 \end{bmatrix}$$ A 3×4 matrix

The constants

b. $\begin{cases} x + y + 2z = 5 \\ 2x = z - 1 \\ 3y = -2z \end{cases}$

SOLUTION Writing all the terms with variables on the left, aligning like terms, and inserting the 0's and 1's, we have

$$\begin{cases} 1x + 1y + 2z = 5 \\ 2x + 0y - 1z = -1 \\ 0x + 3y + 2z = 0 \end{cases}$$ Three equations in three unknowns

The augmented matrix is $\begin{bmatrix} 1 & 1 & 2 & 5 \\ 2 & 0 & -1 & -1 \\ 0 & 3 & 2 & 0 \end{bmatrix}$ A 3×4 matrix

EXAMPLE 2 Assume that each of the following matrices is the augmented matrix of a system of linear equations. Find the system of linear equations associated with each matrix.

a. $\begin{bmatrix} 1 & 3 & 2 & 4 \\ 2 & -5 & -4 & 0 \\ 3 & 3 & 2 & 3 \end{bmatrix}$

SOLUTION A 3×4 matrix is associated with a system of equations that contains *three* variables. Therefore, the system of equations is

$$\begin{cases} x + 3y + 2z = 4 \\ 2x - 5y - 4z = 0 \\ 3x + 3y + 2z = 3 \end{cases}$$

b. $\begin{bmatrix} 5 & 1 & 0 & 2 & 7 \\ -3 & 4 & 2 & 6 & 4 \\ 1 & 0 & 1 & 0 & -3 \\ 5 & 2 & -7 & 1 & 0 \end{bmatrix}$

SOLUTION A 4×5 matrix is associated with a system of equations that contains *four* variables. Therefore, the system of equations is

$$\begin{cases} 5x + y + 2w = 7 \\ -3x + 4y + 2z + 6w = 4 \\ x + z = -3 \\ 5x + 2y - 7z + w = 0 \end{cases}$$

Suppose that a matrix representing a system of three equations in three variables is in row-echelon form; that is, suppose it is in the following form:

$$\begin{bmatrix} 1 & a & b & c \\ 0 & 1 & d & e \\ 0 & 0 & 1 & k \end{bmatrix}$$

where a, b, c, d, e, and k represent any constants. Then the last row represents the equation $0x + 0y + 1z = k$, or $z = k$, and the system has been solved for z. The second row represents an equation in y and z, so we can find y. After finding y, we can find x by substitution, and the system is solved.

We see, then, that it is easy to solve a system of equations if the augmented matrix of the system is in row-echelon form. Before we can discuss converting a matrix to row-echelon form, we need the following information about row-equivalent matrices.

Row-Equivalent Matrices and the Elementary Row Operations

If we perform any of the following three operations on a matrix, the new matrix is defined to be **row-equivalent** to the given matrix.

The elementary row operations

The following operations yield a matrix that is row-equivalent to a given matrix:

1. Any two rows of the matrix can be interchanged.
2. Each element in any row can be multiplied by some nonzero constant.
3. A (nonzero) multiple of the elements of one row can be added to the corresponding elements of another row.

When we use any of these operations, we use the symbol $\sim$ between the two matrices to indicate that we have performed one or more of the elementary row operations. We do *not* use an equal sign! The two matrices *are not equal*—they are row-equivalent.

Performing any of these operations on the augmented matrix of a system of equations is equivalent to performing the corresponding operation (listed in Section 12.3) on a system of equations. For instance, interchanging two rows of the matrix is equivalent to interchanging two equations of the system of equations.

Converting a Matrix to Row-Echelon Form

There are several different ways to convert a matrix to row-echelon form. We discuss only one of the methods in this text.

Converting a matrix to row-echelon form

In this order, use the elementary row operations to

1. Move any row that consists of all zeros to the bottom of the matrix.
2. Get a 1 in the upper left-hand corner, usually by interchanging two rows or by multiplying the elements of row 1 by the multiplicative inverse of the element in row 1, column 1—assuming that that element is not zero.
3. Get 0's beneath that 1, usually by using the *third* elementary row operation.

Without performing any operation that will change the first column or the first row, use the elementary row operations to

4. Get a 1 in row 2, column 2, if possible.* (See step 2.)
5. Get 0's beneath that 1. (See step 3.)

Without performing any operation that will change the first two columns or the first two rows, use the elementary row operations to

6. Get a 1 in row 3, column 3, if possible. (See step 2.)
7. Get 0's beneath that 1. (See step 3.)

Continue in this manner, trying to get 1's down the main diagonal and 0's beneath that diagonal, to the last row of the matrix.

*If you can't get a 1 in row 2, column 2, try to get a 1 in row 2, column 3; if not there, then in row 2, column 4; and so forth.

Although this procedure may appear to be very complicated, you will soon see that it is as easy as arithmetic and that it is a *very* useful tool. In Example 3, we show how to use this procedure.

Note When we're using the third elementary row operation and multiplying one row by some constant, *that row must remain unchanged* in the new matrix; the row we're adding *to* changes.

EXAMPLE 3 Convert the matrix from Example 1b to row-echelon form.

SOLUTION No row consists of all zeros, so we begin with step 2. There is already a 1 in the upper left-hand corner; we want 0's beneath that 1.

This notation means that we'll multiply row 1 by -2 and add these products to row 2

There is already a 1 in the upper left-hand corner; if we multiply row 1 by -2 and add those products to row 2, *leaving row 1 unchanged*, we'll have 0's beneath that 1

Step 2. $\begin{matrix} -2] \\ \rightarrow \\ \end{matrix} \begin{bmatrix} 1 & 1 & 2 & 5 \\ 2 & 0 & -1 & -1 \\ 0 & 3 & 2 & 0 \end{bmatrix}$

Notice that row 1 *was unchanged*

Now we have 0's beneath the 1 in the upper left-hand corner

Step 3. $\sim -\frac{1}{2}] \begin{bmatrix} 1 & 1 & 2 & 5 \\ 0 & -2 & -5 & -11 \\ 0 & 3 & 2 & 0 \end{bmatrix}$

Next, we want a 1 in row 2, column 2, so we'll multiply row 2 by $-\frac{1}{2}$

This is what we want in column 1

Now we have a 1 in row 2, column 2

Step 4. $\sim \begin{matrix} \\ -3] \\ \rightarrow \end{matrix} \begin{bmatrix} 1 & 1 & 2 & 5 \\ 0 & 1 & \frac{5}{2} & \frac{11}{2} \\ 0 & 3 & 2 & 0 \end{bmatrix}$

Next, we want a 0 beneath that 1; we'll multiply row 2 by -3 and add those products to row 3, *leaving row 2 unchanged*

Notice that rows 1 and 2 *were unchanged*

Now we have a 0 beneath the 1 in row 2, column 2

Step 5. $\sim \begin{matrix} \\ \\ -\frac{2}{11}] \end{matrix} \begin{bmatrix} 1 & 1 & 2 & 5 \\ 0 & 1 & \frac{5}{2} & \frac{11}{2} \\ 0 & 0 & -\frac{11}{2} & -\frac{33}{2} \end{bmatrix}$

Next, we want a 1 in row 3, column 3, so we'll multiply row 3 by $-\frac{2}{11}$

This is what we want in column 2

Step 6. $\sim \begin{bmatrix} 1 & 1 & 2 & 5 \\ 0 & 1 & \frac{5}{2} & \frac{11}{2} \\ 0 & 0 & 1 & 3 \end{bmatrix}$

The matrix is in row-echelon form

The Gaussian Elimination Method of Solving a System of Equations

We can solve a system of equations by forming its augmented matrix, reducing that matrix to row-echelon form, and then writing the system of equations that corresponds to the final matrix. If no errors have been made, that system will be equivalent to the original system. This method is called the **Gaussian elimination method**.

If the number of equations is the same as the number of variables (as it will be in this text), then three possibilities exist:

1. If the next-to-last number of the last row of the augmented matrix is 1, the system is consistent and independent.
2. If any row of the augmented matrix is of the form $0\ 0 \cdots 0\ k$, where k is some nonzero number, the system is inconsistent and independent.
3. If no row of the augmented matrix is of the form $0\ 0 \cdots 0\ k$, where k is some nonzero number, and if the last row of the matrix consists of all zeros, the system is consistent and dependent.

To see that statement 2 is true, consider this case: Suppose we're solving a system of four equations in four variables and suppose that some row of the augmented matrix is

$$0 \quad 0 \quad 0 \quad 0 \quad 5$$

This is equivalent to the equation $0x + 0y + 0z + 0w = 5$, or $0 = 5$, which can't be true for *any* values of x, y, z, and w. Therefore, the system is inconsistent.

To see that statement 3 is true, consider this case: Suppose we're solving a system of four equations in four variables and suppose that *no* row is of the form

$$0 \quad 0 \quad 0 \quad 0 \quad k$$

where k is some nonzero number, and suppose that the last row of the augmented matrix is

$$0 \quad 0 \quad 0 \quad 0 \quad 0$$

This is equivalent to the equation $0x + 0y + 0z + 0w = 0$, or $0 = 0$. Since this statement is true for all x, y, z, and w, there are infinitely many solutions for the system of equations.

If the system is consistent and independent, the last equation will already be solved for one of the variables.

EXAMPLE 4 Find the solution set of the following system of equations by using the Gaussian elimination method:

$$\begin{Bmatrix} 2x - y + 3z + w = -4 \\ x + 2y - z + 2w = -4 \\ 3x + y - z - 2w = 4 \\ -2x + 3y + z + 3w = 0 \end{Bmatrix}$$

SOLUTION The augmented matrix of this system of equations is

$$\begin{bmatrix} 2 & -1 & 3 & 1 & -4 \\ 1 & 2 & -1 & 2 & -4 \\ 3 & 1 & -1 & -2 & 4 \\ -2 & 3 & 1 & 3 & 0 \end{bmatrix}$$

We reduce this matrix to row-echelon form as follows:

This notation means that we will interchange rows 1 and 2; we will then have a 1 in the upper left-hand corner

We'll multiply row 1 by −2 and add those products to row 2 to get a 0 in row 2, column 1

$$\begin{bmatrix} 2 & -1 & 3 & 1 & -4 \\ 1 & 2 & -1 & 2 & -4 \\ 3 & 1 & -1 & -2 & 4 \\ -2 & 3 & 1 & 3 & 0 \end{bmatrix} \sim -2] \begin{bmatrix} 1 & 2 & -1 & 2 & -4 \\ \textcircled{2} & -1 & 3 & 1 & -4 \\ 3 & 1 & -1 & -2 & 4 \\ -2 & 3 & 1 & 3 & 0 \end{bmatrix}$$

We'll multiply row 1 by −3 and add those products to row 3 to get a 0 in row 3, column 1

We'll multiply row 1 by 2 and add those products to row 4 to get a 0 in row 4, column 1

$$\sim -3] \begin{bmatrix} 1 & 2 & -1 & 2 & -4 \\ 0 & -5 & 5 & -3 & 4 \\ \textcircled{3} & 1 & -1 & -2 & 4 \\ -2 & 3 & 1 & 3 & 0 \end{bmatrix} \sim 2] \begin{bmatrix} 1 & 2 & -1 & 2 & -4 \\ 0 & -5 & 5 & -3 & 4 \\ 0 & -5 & 2 & -8 & 16 \\ \textcircled{-2} & 3 & 1 & 3 & 0 \end{bmatrix}$$

We'll multiply row 2 by $-\frac{1}{5}$ to get a 1 in row 2, column 2

We'll multiply row 2 by 5 and add those products to row 3 to get a 0 in row 3, column 2

$$\sim -\tfrac{1}{5}] \begin{bmatrix} 1 & 2 & -1 & 2 & -4 \\ 0 & \textcircled{-5} & 5 & -3 & 4 \\ 0 & -5 & 2 & -8 & 16 \\ 0 & 7 & -1 & 7 & -8 \end{bmatrix} \sim 5] \begin{bmatrix} 1 & 2 & -1 & 2 & -4 \\ 0 & 1 & -1 & \frac{3}{5} & -\frac{4}{5} \\ 0 & \textcircled{-5} & 2 & -8 & 16 \\ 0 & 7 & -1 & 7 & -8 \end{bmatrix}$$

This is what we want in column 1

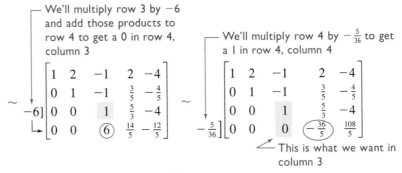

We'll multiply row 2 by −7 and add those products to row 4 to get a 0 in row 4, column 2

$$\sim \begin{bmatrix} 1 & 2 & -1 & 2 & -4 \\ 0 & 1 & -1 & \frac{3}{5} & -\frac{4}{5} \\ 0 & 0 & -3 & -5 & 12 \\ 0 & 7 & -1 & 7 & -8 \end{bmatrix}$$

We'll multiply row 3 by $-\frac{1}{3}$ to get a 1 in row 3, column 3

$$\sim \begin{bmatrix} 1 & 2 & -1 & 2 & -4 \\ 0 & 1 & -1 & \frac{3}{5} & -\frac{4}{5} \\ 0 & 0 & -3 & -5 & 12 \\ 0 & 0 & 6 & \frac{14}{5} & -\frac{12}{5} \end{bmatrix}$$

This is what we want in column 2

We'll multiply row 3 by −6 and add those products to row 4 to get a 0 in row 4, column 3

$$\sim \begin{bmatrix} 1 & 2 & -1 & 2 & -4 \\ 0 & 1 & -1 & \frac{3}{5} & -\frac{4}{5} \\ 0 & 0 & 1 & \frac{5}{3} & -4 \\ 0 & 0 & 6 & \frac{14}{5} & -\frac{12}{5} \end{bmatrix}$$

We'll multiply row 4 by $-\frac{5}{36}$ to get a 1 in row 4, column 4

$$\sim \begin{bmatrix} 1 & 2 & -1 & 2 & -4 \\ 0 & 1 & -1 & \frac{3}{5} & -\frac{4}{5} \\ 0 & 0 & 1 & \frac{5}{3} & -4 \\ 0 & 0 & 0 & -\frac{36}{5} & \frac{108}{5} \end{bmatrix}$$

This is what we want in column 3

$$\sim \begin{bmatrix} 1 & 2 & -1 & 2 & -4 \\ 0 & 1 & -1 & \frac{3}{5} & -\frac{4}{5} \\ 0 & 0 & 1 & \frac{5}{3} & -4 \\ 0 & 0 & 0 & 1 & -3 \end{bmatrix}$$

The matrix is in row-echelon form. The last row is equivalent to the equation

$$0x + 0y + 0z + 1w = -3$$

Therefore, $w = -3$

The third row is equivalent to the equation

$$0x + 0y + 1z + \tfrac{5}{3}w = -4$$

Substituting −3 for w, we have

$$z + \tfrac{5}{3}(-3) = -4$$

or $z = 1$

The second row is equivalent to the equation

$$0x + 1y - 1z + \tfrac{3}{5}w = -\tfrac{4}{5}$$

Substituting −3 for w and 1 for z, we have

$$y - 1(1) + \tfrac{3}{5}(-3) = -\tfrac{4}{5}$$

or $y = 2$

The first row is equivalent to the equation

$$1x + 2y - 1z + 2w = -4$$

Substituting −3 for w, 1 for z, and 2 for y, we have

$$1x + 2(2) - 1(1) + 2(-3) = -4$$

or $x = -1$

Therefore, the solution *appears* to be (−1, 2, 1, −3). We must check it in all *four* equations.

√ **Check** (We will not write the original equations in this check.)

$$2(-1) - 2 + 3(1) + (-3) \stackrel{?}{=} -4 \qquad -4 = -4 \quad \text{True}$$
$$-1 + 2(2) - 1 + 2(-3) \stackrel{?}{=} -4 \qquad -4 = -4 \quad \text{True}$$
$$3(-1) + 2 - 1 - 2(-3) \stackrel{?}{=} 4 \qquad 4 = 4 \quad \text{True}$$
$$-2(-1) + 3(2) + 1 + 3(-3) \stackrel{?}{=} 0 \qquad 0 = 0 \quad \text{True}$$

Therefore, the solution set *is* $\{(-1, 2, 1, -3)\}$.

Note This matrix method can also be used for solving a system of two linear equations in two variables. However, the methods discussed in previous sections are probably easier to use in solving such systems.

Note It is possible to work with matrices (from 2 × 2 to 6 × 6 matrices) and to perform the elementary row operations on them with a TI-81 graphics calculator; consult the calculator manual for details.

Exercises 12.6 Set I

Find the solution set of each of the following systems of equations by using the Gaussian elimination method.

1. $\begin{Bmatrix} 2x + y + z = 4 \\ x - y + 3z = -2 \\ x + y + 2z = 1 \end{Bmatrix}$

2. $\begin{Bmatrix} x + y + z = 1 \\ 2x + y - 2z = -4 \\ x + y + 2z = 3 \end{Bmatrix}$

3. $\begin{Bmatrix} 2x + 3y + z = 7 \\ 4x - 2z = -6 \\ 6y - z = 0 \end{Bmatrix}$

4. $\begin{Bmatrix} x - 2y = 4 \\ y + 3z = 8 \\ 2x - z = 1 \end{Bmatrix}$

5. $\begin{Bmatrix} x + y + z + w = 5 \\ 2x - y + 2z - w = -2 \\ x + 2y - z - 2w = -1 \\ -x + 3y + 3z + w = 1 \end{Bmatrix}$

6. $\begin{Bmatrix} x + y + z + w = 4 \\ x - 2y - z - 2w = -2 \\ 3x + 2y + z + 3w = 4 \\ 2x + y - 2z - w = 0 \end{Bmatrix}$

Writing Problems

Express the answer in your own words and in complete sentences.

1. Describe what your first step would be in solving the following system of equations by using the Gaussian elimination method and explain why you would do that first.

$$\begin{Bmatrix} 3x - y + 2z = 4 \\ 5x + y + 3z = 9 \\ 2x + y - 4z = -1 \end{Bmatrix}$$

Exercises 12.6 Set II

Find the solution set of each of the following systems of equations by using the Gaussian elimination method.

1. $\begin{Bmatrix} x + 2z = 7 \\ 2x - y = 5 \\ 2y + z = 4 \end{Bmatrix}$

2. $\begin{Bmatrix} 4x + 5y + z = 4 \\ 10y - 2z = 6 \\ 8x + 3z = 3 \end{Bmatrix}$

3. $\begin{Bmatrix} 2x + y - 3z = 3 \\ x + 2y = 0 \\ -2x - 2y + 5z = -2 \end{Bmatrix}$

4. $\begin{Bmatrix} 3x - y + 2z = 4 \\ x - 2y - 3z = -4 \\ x + y = 2 \end{Bmatrix}$

5. $\begin{Bmatrix} x + y + z + w = 5 \\ 3x - y + 2z - w = 0 \\ 2x + y - z + 2w = 4 \\ 2x + y + z + 2w = 2 \end{Bmatrix}$

6. $\begin{Bmatrix} 2x + 5y + 3z - 4w = 0 \\ -x + y + 6w = -1 \\ 3x - z - 2w = 1 \\ 4x + 2y - z = 3 \end{Bmatrix}$

12.7 Determinants (Optional)

In the next section, we will discuss using Cramer's rule for solving systems of linear equations. In order to do this, we must first discuss *determinants*. We will begin with some definitions that apply to determinants of any size.

A **determinant** is a *real number* that is represented by a *square* array of numbers enclosed between *two vertical bars*. (Students who have studied the previous section should note the difference in notation between a *matrix* and a *determinant*: A *matrix* is enclosed within *brackets*, whereas a *determinant* is enclosed between straight vertical bars.)

The **elements** of a determinant are the numbers in its array. The **principal diagonal** of a determinant is the line of its elements from the upper left-hand corner to the lower right-hand corner. The **secondary diagonal** of a determinant is the line of elements from the lower left-hand corner to the upper right-hand corner.

Second-Order Determinants

A **second-order determinant** is a determinant that has two rows and two columns of elements.

Evaluating a Second-Order Determinant The **value** of a second-order determinant is the product of the elements in its principal diagonal minus the product of the elements in its secondary diagonal.

$$\begin{vmatrix} a & b \\ c & d \end{vmatrix} = ad - bc$$

The secondary diagonal
The value of the second-order determinant
The principal diagonal

EXAMPLE 1 Examples of finding the value of second-order determinants:

a. $\begin{vmatrix} -5 & -6 \\ 2 & 3 \end{vmatrix} = (-5)(3) - (2)(-6) = -15 + 12 = -3$

b. $\begin{vmatrix} 6 & -7 \\ 4 & 0 \end{vmatrix} = (6)(0) - (4)(-7) = 0 + 28 = 28$

Third-Order Determinants

A **third-order determinant** is a determinant that has three rows and three columns.

	Column 1	Column 2	Column 3
Row 1	a_1	b_1	c_1
Row 2	a_2	b_2	c_2
Row 3	a_3	b_3	c_3

Minor The **minor** of an element of a determinant is the determinant that remains after we strike out the row and column in which that element appears. For example, for the third-order determinant above, the minor of element a_2 is the determinant $\begin{vmatrix} b_1 & c_1 \\ b_3 & c_3 \end{vmatrix}$.

Cofactor The **cofactor** of an element of a determinant is the *signed* minor of that element; in particular, the cofactor of an element in the ith row and the jth column of a determinant is $(-1)^{i+j}$ times the minor of that element. For the third-order determinant on the previous page, the cofactor of element a_2 is

$$(-1)^{2+1}\begin{vmatrix} b_1 & c_1 \\ b_3 & c_3 \end{vmatrix} = (-1)^3\begin{vmatrix} b_1 & c_1 \\ b_3 & c_3 \end{vmatrix} = -1\begin{vmatrix} b_1 & c_1 \\ b_3 & c_3 \end{vmatrix}$$

EXAMPLE 2 Find the required minors and cofactors, given the determinant $\begin{vmatrix} 2 & 0 & -1 \\ 5 & -4 & 6 \\ -3 & 1 & 7 \end{vmatrix}$.

a. Find the minor of 1.

SOLUTION Striking out the row and column that contain 1, we have

$$\begin{vmatrix} 2 & 0 & -1 \\ 5 & -4 & 6 \\ -3 & \textcircled{1} & 7 \end{vmatrix}$$

Therefore, the minor of 1 is $\begin{vmatrix} 2 & -1 \\ 5 & 6 \end{vmatrix}$.

b. Find the cofactor of 1.

SOLUTION Because 1 is in the *third* row and the *second* column, the sign of its cofactor is $(-1)^{3+2}$, or $(-1)^5$, or (-1). Therefore, the cofactor of 1 is $(-1)\begin{vmatrix} 2 & -1 \\ 5 & 6 \end{vmatrix}$.

c. Find the minor of 7.

SOLUTION Striking out the row and column that contain 7, we have

$$\begin{vmatrix} 2 & 0 & -1 \\ 5 & -4 & 6 \\ -3 & 1 & \textcircled{7} \end{vmatrix}$$

Therefore, the minor of 7 is $\begin{vmatrix} 2 & 0 \\ 5 & -4 \end{vmatrix}$.

d. Find the cofactor of 7.

SOLUTION Because 7 is in the *third* row and the *third* column, the sign of its cofactor is $(-1)^{3+3}$, or $(-1)^6$, or $+1$. Therefore, the cofactor of 7 is $\begin{vmatrix} 2 & 0 \\ 5 & -4 \end{vmatrix}$.

An easier way to find the sign of the cofactor of any element is to memorize the pattern of signs shown below. (The pattern can be extended to a determinant of any size.)

$$\begin{vmatrix} + & - & + \\ - & + & - \\ + & - & + \end{vmatrix}$$

Evaluating a Third-Order Determinant The **value** of the third-order determinant

$$\begin{vmatrix} a_1 & b_1 & c_1 \\ a_2 & b_2 & c_2 \\ a_3 & b_3 & c_3 \end{vmatrix}$$

is, by definition,

$$a_1b_2c_3 - a_1b_3c_2 - a_2b_1c_3 + a_2b_3c_1 + a_3b_1c_2 - a_3b_2c_1$$

You can verify that this value could be rewritten as

$$a_1(b_2c_3 - b_3c_2) - a_2(b_1c_3 - b_3c_1) + a_3(b_1c_2 - b_2c_1)$$

or as

$$a_1\begin{vmatrix} b_2 & c_2 \\ b_3 & c_3 \end{vmatrix} - a_2\begin{vmatrix} b_1 & c_1 \\ b_3 & c_3 \end{vmatrix} + a_3\begin{vmatrix} b_1 & c_1 \\ b_2 & c_2 \end{vmatrix}$$

Notice the negative sign, and notice that the cofactor of a_2 is negative

When we rewrite the value of a third-order determinant in this last way, we say that we are *expanding by cofactors of the elements of the first column.*

We can expand a determinant by cofactors of *any* row or *any* column. To expand by cofactors of a *row,* we multiply each element in that row by its cofactor and then add the products. To expand by cofactors of a *column,* we multiply each element in that column by its cofactor and then add the products. *The same value is obtained no matter what row or column is used.*

EXAMPLE 3 Find the value of $\begin{vmatrix} 1 & 3 & 2 \\ 2 & -1 & 1 \\ -4 & 1 & -3 \end{vmatrix}$ by expansion by (a) column 1 and (b) row 2.

SOLUTION

a.

The elements of column 1

$$+(1)\begin{vmatrix} -1 & 1 \\ 1 & -3 \end{vmatrix} - (2)\begin{vmatrix} 3 & 2 \\ 1 & -3 \end{vmatrix} + (-4)\begin{vmatrix} 3 & 2 \\ -1 & 1 \end{vmatrix}$$

The shaded signs are the signs of the cofactors of the elements in the first column

$$\begin{vmatrix} + & - & + \\ - & + & - \\ + & - & + \end{vmatrix}$$

$$= 1(3 - 1) - 2(-9 - 2) - 4(3 + 2) = 1(2) - 2(-11) - 4(5)$$

$$= 2 + 22 - 20 = 4$$

b.

The elements of row 2

$$-(2)\begin{vmatrix} 3 & 2 \\ 1 & -3 \end{vmatrix} + (-1)\begin{vmatrix} 1 & 2 \\ -4 & -3 \end{vmatrix} - (1)\begin{vmatrix} 1 & 3 \\ -4 & 1 \end{vmatrix}$$

The shaded signs are the signs of the cofactors of the elements in the second row

$$\begin{vmatrix} + & - & + \\ - & + & - \\ + & - & + \end{vmatrix}$$

$$= -2(-9 - 2) - 1(-3 + 8) - 1(1 + 12) = -2(-11) - 1(5) - 1(13)$$

$$= 22 - 5 - 13 = 4$$

Notice that the same value, 4, was obtained from *both* expansions.

If zeros appear anywhere in a determinant, expanding the determinant by the row or column that contains the most zeros minimizes the work involved in evaluating the determinant.

EXAMPLE 4 Examples of selecting the best row or column to expand the determinant by:

a. $\begin{vmatrix} -2 & 1 & 0 \\ 4 & -2 & 1 \\ 3 & -1 & 5 \end{vmatrix}$ We will expand by row 1 because it contains a zero

Expanding by row 1, we have

$$\begin{vmatrix} -2 & 1 & 0 \\ 4 & -2 & 1 \\ 3 & -1 & 5 \end{vmatrix} = (-2)\begin{vmatrix} -2 & 1 \\ -1 & 5 \end{vmatrix} - (1)\begin{vmatrix} 4 & 1 \\ 3 & 5 \end{vmatrix} + (0)\begin{vmatrix} 4 & -2 \\ 3 & -1 \end{vmatrix}$$

$$= (-2)(-10 + 1) - 1(20 - 3) + 0$$

$$= (-2)(-9) - 1(17) = 18 - 17 = 1$$

b. $\begin{vmatrix} -4 & 0 & -1 \\ 1 & 2 & -3 \\ -2 & 0 & 5 \end{vmatrix}$ We will expand by column 2 because it contains two zeros

Expanding by column 2, we have

It is not necessary to write the minor when it is to be multiplied by zero

$$\begin{vmatrix} -4 & 0 & -1 \\ 1 & 2 & -3 \\ -2 & 0 & 5 \end{vmatrix} = -(0)\begin{vmatrix} \end{vmatrix} + (2)\begin{vmatrix} -4 & -1 \\ -2 & 5 \end{vmatrix} - (0)\begin{vmatrix} \end{vmatrix}$$

$$= \quad 0 \quad + 2(-20 - 2) - \quad 0$$

$$= 2(-22) = -44$$

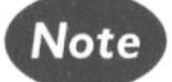

Note It is possible to evaluate determinants with a TI-81 graphics calculator; consult the calculator manual for details.

Exercises 12.7

Set I

In Exercises 1–6, find the value of each second-order determinant.

1. $\begin{vmatrix} 3 & 4 \\ 2 & 5 \end{vmatrix}$
2. $\begin{vmatrix} 4 & 3 \\ 2 & 7 \end{vmatrix}$
3. $\begin{vmatrix} 2 & -4 \\ 5 & -3 \end{vmatrix}$
4. $\begin{vmatrix} 5 & 0 \\ -9 & 8 \end{vmatrix}$
5. $\begin{vmatrix} -7 & -3 \\ 5 & 8 \end{vmatrix}$
6. $\begin{vmatrix} 2 & -4 \\ -3 & 6 \end{vmatrix}$

In Exercises 7 and 8, solve for x.

7. $\begin{vmatrix} 2 & -4 \\ 3 & x \end{vmatrix} = 20$
8. $\begin{vmatrix} 3 & x \\ -4 & 5 \end{vmatrix} = 27$

9. For the determinant $\begin{vmatrix} 1 & 2 & 3 \\ 4 & 5 & -1 \\ -3 & -5 & 0 \end{vmatrix}$, find the following.

   a. the minor of 2
   b. the cofactor of 2
   c. the minor of -5
   d. the cofactor of -5
   e. the minor of -3
   f. the cofactor of -3

10. For the determinant $\begin{vmatrix} 2 & 0 & 1 \\ 4 & 1 & 5 \\ -3 & -2 & 4 \end{vmatrix}$, find the following.

   a. the minor of -3
   b. the cofactor of -3
   c. the minor of 5
   d. the cofactor of 5
   e. the minor of -2
   f. the cofactor of -2

In Exercises 11–14, find the value of each determinant by expanding by cofactors of the indicated row or column.

11. $\begin{vmatrix} 1 & 2 & 1 \\ 3 & 1 & 2 \\ 4 & 2 & 0 \end{vmatrix}$; column 3
12. $\begin{vmatrix} 2 & 1 & 1 \\ 1 & 3 & 1 \\ 0 & 2 & 4 \end{vmatrix}$; column 1
13. $\begin{vmatrix} 1 & 3 & -2 \\ -1 & 2 & -3 \\ 0 & 4 & 1 \end{vmatrix}$; row 3
14. $\begin{vmatrix} 2 & -1 & 1 \\ 0 & 5 & -3 \\ 1 & 2 & -2 \end{vmatrix}$; row 2

In Exercises 15–18, expand by any row or column.

15. $\begin{vmatrix} 1 & -2 & 3 \\ -3 & 4 & 0 \\ 2 & 6 & 5 \end{vmatrix}$

16. $\begin{vmatrix} 2 & -4 & -1 \\ 5 & 0 & -6 \\ -3 & 4 & -2 \end{vmatrix}$

17. $\begin{vmatrix} 6 & 7 & 8 \\ -6 & 7 & -9 \\ 0 & 0 & -2 \end{vmatrix}$

18. $\begin{vmatrix} 0 & 0 & -3 \\ 5 & 6 & 9 \\ -5 & 6 & -8 \end{vmatrix}$

In Exercises 19 and 20, solve for x.

19. $\begin{vmatrix} x & 0 & 1 \\ 0 & 2 & 3 \\ 4 & -1 & -2 \end{vmatrix} = 6$

20. $\begin{vmatrix} 0 & x & 1 \\ -3 & 2 & 4 \\ -2 & 1 & 0 \end{vmatrix} = 17$

Writing Problems

Express the answer in your own words and in complete sentences.

1. Explain the difference between the minor of an element of a determinant and the cofactor of that element.

Exercises 12.7 Set II

In Exercises 1–6, find the value of each second-order determinant.

1. $\begin{vmatrix} 6 & 7 \\ -5 & -2 \end{vmatrix}$

2. $\begin{vmatrix} -9 & 8 \\ -3 & 2 \end{vmatrix}$

3. $\begin{vmatrix} 1 & -3 \\ 2 & -6 \end{vmatrix}$

4. $\begin{vmatrix} 3 & 4 \\ 1 & 5 \end{vmatrix}$

5. $\begin{vmatrix} -1 & 4 \\ 3 & 2 \end{vmatrix}$

6. $\begin{vmatrix} 5 & -1 \\ 6 & -2 \end{vmatrix}$

In Exercises 7 and 8, solve for x.

7. $\begin{vmatrix} -2 & 3 \\ x & 4 \end{vmatrix} = 10$

8. $\begin{vmatrix} 2x & 5 \\ 2 & 3x \end{vmatrix} = -11x$

9. For the determinant $\begin{vmatrix} -3 & 1 & 2 \\ 0 & 4 & -1 \\ 5 & -2 & 6 \end{vmatrix}$, find the following.

a. the minor of 4

b. the cofactor of 4

c. the minor of 2

d. the cofactor of 2

e. the minor of 1

f. the cofactor of 1

10. For the determinant $\begin{vmatrix} 2 & 5 & 6 \\ 1 & 8 & 4 \\ 3 & 7 & 9 \end{vmatrix}$, find the following.

a. the minor of 2

b. the cofactor of 2

c. the minor of 1

d. the cofactor of 1

e. the minor of 7

f. the cofactor of 7

In Exercises 11–14, find the value of each determinant by expanding by cofactors of the indicated row or column.

11. $\begin{vmatrix} 2 & -1 & -3 \\ 4 & 0 & -2 \\ -5 & 2 & 3 \end{vmatrix}$; column 2

12. $\begin{vmatrix} 0 & 5 & -2 \\ -3 & 4 & -1 \\ 2 & -4 & 6 \end{vmatrix}$; row 1

13. $\begin{vmatrix} 3 & -2 & 4 \\ 5 & 1 & 0 \\ 0 & -1 & 2 \end{vmatrix}$; row 3

14. $\begin{vmatrix} 1 & 5 & -1 \\ 2 & 0 & -2 \\ 3 & 1 & -1 \end{vmatrix}$; column 2

In Exercises 15–18, expand by any row or column.

15. $\begin{vmatrix} -1 & 3 & -2 \\ -4 & 2 & 5 \\ 0 & 1 & -3 \end{vmatrix}$

16. $\begin{vmatrix} 4 & -5 & 2 \\ -2 & 0 & -3 \\ 6 & 0 & -1 \end{vmatrix}$

17. $\begin{vmatrix} 3 & -2 & 0 \\ -2 & 1 & -2 \\ 2 & -1 & 2 \end{vmatrix}$

18. $\begin{vmatrix} 3 & 0 & 5 \\ -1 & -1 & 3 \\ 2 & -3 & 1 \end{vmatrix}$

In Exercises 19 and 20, solve for x.

19. $\begin{vmatrix} 0 & -4 & 3 \\ x & 2 & 0 \\ -1 & 5 & x \end{vmatrix} = 31$

20. $\begin{vmatrix} 2 & x & 1 \\ -1 & x & 2 \\ x & 0 & 1 \end{vmatrix} = 18$

12.8 Using Cramer's Rule for Solving Second- and Third-Order Linear Systems (Optional)

Terminology Associated with Cramer's Rule

We define the determinant of coefficients, the column of constants, and the columns of x- and y-coefficients for the system

$$\begin{cases} a_1x + b_1y = c_1 \\ a_2x + b_2y = c_2 \end{cases}$$

as follows:

The determinant of coefficients → $\begin{vmatrix} a_1 & b_1 \\ a_2 & b_2 \end{vmatrix}$ $\begin{matrix} c_1 \\ c_2 \end{matrix}$ ← The column of constants

The column of x-coefficients → $\begin{matrix} a_1 \\ a_2 \end{matrix}$ $\begin{matrix} b_1 \\ b_2 \end{matrix}$ ← The column of y-coefficients

Cramer's Rule

Consider the linear system $\begin{cases} a_1x + b_1y = c_1 \\ a_2x + b_2y = c_2 \end{cases}$. Solving for x by elimination, we have

$$\begin{array}{lrcl} b_2\,]\ \ a_1x + b_1\,y = c_1 \Rightarrow & a_1b_2x + b_1b_2y &=& c_1b_2 \\ -b_1\,]\ \ a_2x + b_2\,y = c_2 \Rightarrow & -a_2b_1x - b_1b_2y &=& -c_2b_1 \\ \hline & a_1b_2x - a_2b_1x &=& c_1b_2 - c_2b_1 \end{array}$$

We will add the equations

or

$$(a_1b_2 - a_2b_1)x = c_1b_2 - c_2b_1$$

Therefore,

$$x = \frac{c_1b_2 - c_2b_1}{a_1b_2 - a_2b_1}$$

which can be written

$$x = \frac{c_1b_2 - c_2b_1}{a_1b_2 - a_2b_1} = \frac{\begin{vmatrix} c_1 & b_1 \\ c_2 & b_2 \end{vmatrix}}{\begin{vmatrix} a_1 & b_1 \\ a_2 & b_2 \end{vmatrix}}$$

You can verify that if we solve for y by elimination, we find that

$$y = \frac{a_1c_2 - a_2c_1}{a_1b_2 - a_2b_1}$$

which can be written

$$y = \frac{a_1c_2 - a_2c_1}{a_1b_2 - a_2b_1} = \frac{\begin{vmatrix} a_1 & c_1 \\ a_2 & c_2 \end{vmatrix}}{\begin{vmatrix} a_1 & b_1 \\ a_2 & b_2 \end{vmatrix}}$$

The formulas derived above make up Cramer's rule.

Cramer's rule

In the solution of a linear system of equations, each variable is equal to the ratio of two determinants. The denominator for *every* variable is the determinant of coefficients. The numerator for each variable is the determinant of coefficients *with the column of coefficients for that variable replaced by the column of constants.*

Solving a second-order system by using Cramer's rule

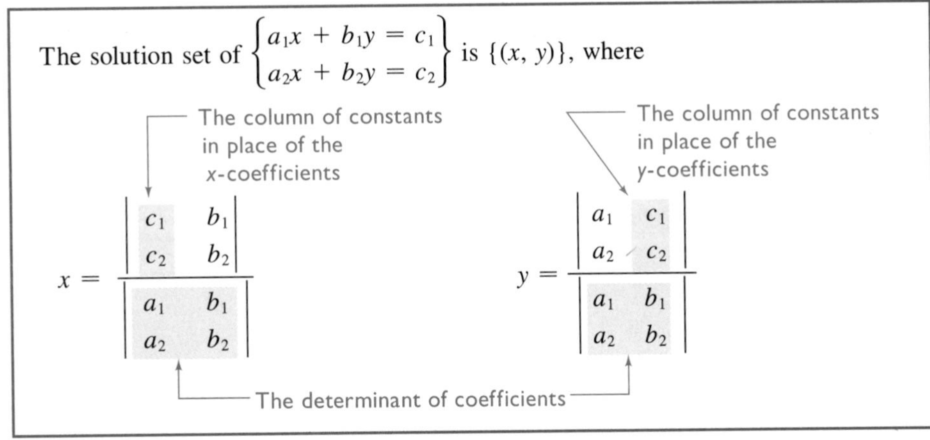
The solution set of $\begin{cases} a_1x + b_1y = c_1 \\ a_2x + b_2y = c_2 \end{cases}$ is $\{(x, y)\}$, where
The column of constants in place of the x-coefficients
The column of constants in place of the y-coefficients
$x = \dfrac{\begin{vmatrix} c_1 & b_1 \\ c_2 & b_2 \end{vmatrix}}{\begin{vmatrix} a_1 & b_1 \\ a_2 & b_2 \end{vmatrix}}$
$y = \dfrac{\begin{vmatrix} a_1 & c_1 \\ a_2 & c_2 \end{vmatrix}}{\begin{vmatrix} a_1 & b_1 \\ a_2 & b_2 \end{vmatrix}}$
The determinant of coefficients

When a system of equations (any order) is solved by using Cramer's rule, three outcomes are possible:

1. *There is a unique solution* (the system is *consistent* and *independent*): The determinant of coefficients is not equal to zero.
2. *There is no solution* (the system is *inconsistent* and *independent*): The determinant of coefficients is equal to zero, *and* the determinants serving as numerators for the variables are *not* equal to zero.
3. *There are many solutions* (the system is *consistent* and *dependent*): The determinant of coefficients is equal to zero, *and* the determinants serving as numerators for the variables are *also* equal to zero.

Notice that systems that do not have unique solutions *always* contain zero denominators. For the system $\begin{cases} a_1x + b_1y = c_1 \\ a_2x + b_2y = c_2 \end{cases}$, the determinant of coefficients, $\begin{vmatrix} a_1 & b_1 \\ a_2 & b_2 \end{vmatrix}$, will equal zero *if and only if* $a_1b_2 - a_2b_1 = 0$, or, in other words, if $a_1b_2 = a_2b_1$. Therefore, we can mentally scan the coefficients of the variables to see whether the denominator will be zero.

EXAMPLE 1 Find the solution set of $\begin{cases} x + 4y = 7 \\ 3x + 5y = 0 \end{cases}$ by using Cramer's rule.

SOLUTION

The column of constants in place of the x-coefficients

$$x = \frac{\begin{vmatrix} 7 & 4 \\ 0 & 5 \end{vmatrix}}{\begin{vmatrix} 1 & 4 \\ 3 & 5 \end{vmatrix}} = \frac{(7)(5) - (0)(4)}{(1)(5) - (3)(4)} = \frac{35}{-7} = -5$$

The determinant of coefficients

The column of constants in place of the y-coefficients

$$y = \frac{\begin{vmatrix} 1 & 7 \\ 3 & 0 \end{vmatrix}}{\begin{vmatrix} 1 & 4 \\ 3 & 5 \end{vmatrix}} = \frac{(1)(0) - (3)(7)}{(1)(5) - (3)(4)} = \frac{-21}{-7} = 3$$

The determinant of coefficients

Checking will confirm that the solution set is $\{(-5, 3)\}$. This is a consistent, independent system.

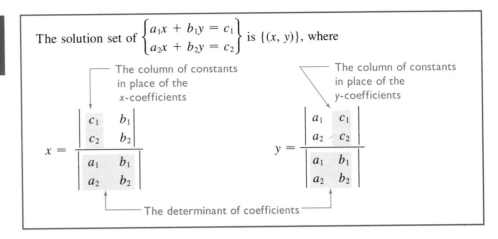

EXAMPLE 2 Find the solution set of $\left\{\begin{array}{l} 2x - 3y = 6 \\ 6x - 9y = 36 \end{array}\right\}$ by using Cramer's rule.

SOLUTION

$$x = \frac{\begin{vmatrix} 6 & -3 \\ 36 & -9 \end{vmatrix}}{\begin{vmatrix} 2 & -3 \\ 6 & -9 \end{vmatrix}} = \frac{(6)(-9) - (36)(-3)}{(2)(-9) - (6)(-3)} = \frac{-54 + 108}{-18 + 18} = \frac{54}{0} \quad \text{Not defined}$$

The denominator is zero and the numerator nonzero; we need not try to solve for y. This system has *no solution*. It is an inconsistent, independent system. The solution set is { }.

EXAMPLE 3 Find the solution set of $\left\{\begin{array}{l} 3x + 5y = 15 \\ 6x + 10y = 30 \end{array}\right\}$ by using Cramer's rule, if possible.

SOLUTION

$$x = \frac{\begin{vmatrix} 15 & 5 \\ 30 & 10 \end{vmatrix}}{\begin{vmatrix} 3 & 5 \\ 6 & 10 \end{vmatrix}} = \frac{(15)(10) - (30)(5)}{(3)(10) - (6)(5)} = \frac{150 - 150}{30 - 30} = \frac{0}{0} \quad \text{Not defined}$$

The denominator and numerator are both zero. (We need not use Cramer's rule to try to solve for y.) This is a consistent, dependent system with many solutions. The solution set cannot be found by using Cramer's rule. To find the solution set, we would need to use the algebraic method discussed early in this chapter.

The method of solving a third-order linear system by using Cramer's rule can be summarized as follows (the rule is stated without proof):

Solving a third-order system by using Cramer's rule

The solution set of $\left\{\begin{array}{l} a_1x + b_1y + c_1z = d_1 \\ a_2x + b_2y + c_2z = d_2 \\ a_3x + b_3y + c_3z = d_3 \end{array}\right\}$ is $\left\{\left(\frac{D_x}{D}, \frac{D_y}{D}, \frac{D_z}{D}\right)\right\}$, where

$$D = \begin{vmatrix} a_1 & b_1 & c_1 \\ a_2 & b_2 & c_2 \\ a_3 & b_3 & c_3 \end{vmatrix} \quad \text{The determinant of coefficients}$$

The column of constants in place of the x-coefficients; The column of constants in place of the y-coefficients; The column of constants in place of the z-coefficients

$$D_x = \begin{vmatrix} d_1 & b_1 & c_1 \\ d_2 & b_2 & c_2 \\ d_3 & b_3 & c_3 \end{vmatrix}, \quad D_y = \begin{vmatrix} a_1 & d_1 & c_1 \\ a_2 & d_2 & c_2 \\ a_3 & d_3 & c_3 \end{vmatrix}, \quad D_z = \begin{vmatrix} a_1 & b_1 & d_1 \\ a_2 & b_2 & d_2 \\ a_3 & b_3 & d_3 \end{vmatrix}$$

EXAMPLE 4 Find the solution set of $\begin{Bmatrix} x - 3y + 2z = 3 \\ 3x - 4y + 2z = -2 \\ x + 5y - z = -1 \end{Bmatrix}$ by using Cramer's rule.

SOLUTION

$$\begin{Bmatrix} x - 3y + 2z = & 3 \\ 3x - 4y + 2z = & -2 \\ x + 5y - z = & -1 \end{Bmatrix}$$

The column of constants in place of the coefficients of the variable being solved for

$$x = \frac{D_x}{D} = \frac{\begin{vmatrix} 3 & -3 & 2 \\ -2 & -4 & 2 \\ -1 & 5 & -1 \end{vmatrix}}{\begin{vmatrix} 1 & -3 & 2 \\ 3 & -4 & 2 \\ 1 & 5 & -1 \end{vmatrix}}, \quad y = \frac{D_y}{D} = \frac{\begin{vmatrix} 1 & 3 & 2 \\ 3 & -2 & 2 \\ 1 & -1 & -1 \end{vmatrix}}{\begin{vmatrix} 1 & -3 & 2 \\ 3 & -4 & 2 \\ 1 & 5 & -1 \end{vmatrix}}, \quad z = \frac{D_z}{D} = \frac{\begin{vmatrix} 1 & -3 & 3 \\ 3 & -4 & -2 \\ 1 & 5 & -1 \end{vmatrix}}{\begin{vmatrix} 1 & -3 & 2 \\ 3 & -4 & 2 \\ 1 & 5 & -1 \end{vmatrix}}$$

(In evaluating D, D_x, D_y, and D_z, we will shade the row or column we're expanding by.)

$$D = \begin{vmatrix} 1 & -3 & 2 \\ 3 & -4 & 2 \\ 1 & 5 & -1 \end{vmatrix} = +(1)\begin{vmatrix} -4 & 2 \\ 5 & -1 \end{vmatrix} - (-3)\begin{vmatrix} 3 & 2 \\ 1 & -1 \end{vmatrix} + (2)\begin{vmatrix} 3 & -4 \\ 1 & 5 \end{vmatrix}$$

$$= 1(4 - 10) + 3(-3 - 2) + 2(15 + 4) = 17$$

$$D_x = \begin{vmatrix} 3 & -3 & 2 \\ -2 & -4 & 2 \\ -1 & 5 & -1 \end{vmatrix} = +(3)\begin{vmatrix} -4 & 2 \\ 5 & -1 \end{vmatrix} - (-2)\begin{vmatrix} -3 & 2 \\ 5 & -1 \end{vmatrix} + (-1)\begin{vmatrix} -3 & 2 \\ -4 & 2 \end{vmatrix}$$

$$= 3(4 - 10) + 2(3 - 10) - 1(-6 + 8) = -34$$

$$D_y = \begin{vmatrix} 1 & 3 & 2 \\ 3 & -2 & 2 \\ 1 & -1 & -1 \end{vmatrix} = -(3)\begin{vmatrix} 3 & 2 \\ 1 & -1 \end{vmatrix} + (-2)\begin{vmatrix} 1 & 2 \\ 1 & -1 \end{vmatrix} - (-1)\begin{vmatrix} 1 & 2 \\ 3 & 2 \end{vmatrix}$$

$$= -3(-3 - 2) - 2(-1 - 2) + 1(2 - 6) = 17$$

$$D_z = \begin{vmatrix} 1 & -3 & 3 \\ 3 & -4 & -2 \\ 1 & 5 & -1 \end{vmatrix} = +(3)\begin{vmatrix} 3 & -4 \\ 1 & 5 \end{vmatrix} - (-2)\begin{vmatrix} 1 & -3 \\ 1 & 5 \end{vmatrix} + (-1)\begin{vmatrix} 1 & -3 \\ 3 & -4 \end{vmatrix}$$

$$= 3(15 + 4) + 2(5 + 3) - 1(-4 + 9) = 68$$

Therefore, $x = \frac{D_x}{D} = \frac{-34}{17} = -2$, $y = \frac{D_y}{D} = \frac{17}{17} = 1$, and $z = \frac{D_z}{D} = \frac{68}{17} = 4$. Checking will confirm that the solution set is $\{(-2, 1, 4)\}$.

Cramer's rule can be used to solve linear systems whenever we have the same number of equations as variables.

Exercises 12.8
Set I

For Exercises 1–12, use Cramer's rule to work Exercises 1–12 from Set I, Exercises 12.3.

(If the system is dependent, write "dependent"; do not find the solution set.)

For Exercises 13–16, use Cramer's rule to work Exercises 1–4 from Set I, Exercises 12.6.

Writing Problems

Express the answer in your own words and in complete sentences.

1. Describe what your first step would be in solving the following system of equations by using Cramer's rule, and explain why you would do that first.

$$\begin{Bmatrix} 3x - y + 2z = 4 \\ 5x + y + 3z = 9 \\ 2x + y - 4z = -1 \end{Bmatrix}$$

Exercises 12.8
Set II

For Exercises 1–12, use Cramer's rule to work Exercises 1–12 from Set II, Exercises 12.3.

(If the system is dependent, write "dependent"; do not find the solution set.)

For Exercises 13–16, use Cramer's rule to work Exercises 1–4 from Set II, Exercises 12.6.

12.9 Solving Applied Problems by Using Systems of Equations

In solving applied problems that involve more than one unknown, it is sometimes difficult to represent each unknown in terms of a single variable. In this section, we eliminate that difficulty by using a different variable for each unknown.

Solving an applied problem by using a system of equations

Read First read the problem very carefully.

Think Determine what *type* of problem it is, if possible. Determine what is unknown.

Sketch Draw a sketch *with labels*, if it would be helpful.

Step 1. Represent each unknown number by a *different* variable.

Reread Reread the problem, breaking it up into small pieces.

Step 2. Translate each English phrase into an algebraic expression; fit these expressions together into *two or more different equations. There must be as many equations as variables.*

Step 3. Solve the *system* of equations for *all* the variables, using any of the methods discussed in this chapter.

Step 4. Check the solution(s) *in the word statements.*

Step 5. State the results clearly, being sure to answer *all* the questions asked.

EXAMPLE 1

Doris has 17 coins in her purse with a total value of \$1.15. If she has only nickels and dimes, how many of each are there?

SOLUTION We'll express *all* the values in cents, and \$1.15 = 115¢.

Step 1. Let d = the number of dimes
Let n = the number of nickels

Reread Doris has 17 coins She has only nickels and dimes

(1) $17 = n + d$

The value of the dimes + the value of the nickels = the total value of the money

(2) $10d + 5n = 115$

Step 2. $\begin{cases} (1) & n + d = 17 \\ (2) & 5n + 10d = 115 \end{cases}$ We solve this system by elimination

Step 3.

$$\begin{aligned} (1)\quad -5]\quad n + d &= 17 \Rightarrow -5n - 5d = -85 \\ (2)\quad 1]\quad 5n + 10d &= 115 \Rightarrow \underline{5n + 10d = 115} \\ 5d &= 30 \\ d &= 6 \end{aligned}$$

The number of dimes

$n + 6 = 17$ Substituting into Equation 1
$n = 11$ The number of nickels

√ **Step 4.** ***Check*** If there are 6 dimes and 11 nickels, there are 17 coins altogether. The value of 6 dimes is \$0.60, and the value of 11 nickels is \$0.55; the total value is \$1.15.

Step 5. Therefore, Doris has 6 dimes and 11 nickels in her purse.

EXAMPLE 2

Jeff left Riverside at 5 A.M., traveling toward Stockton. His friend George left Riverside at 5:40 A.M., also traveling toward Stockton. George drove 10 mph faster than Jeff and overtook Jeff at 9 A.M. Find the average speed of each car, and find the total distance traveled before they met.

SOLUTION Note that Jeff traveled 4 hr and George traveled 3 hr 20 min before they met, and remember that 3 hr 20 min can be expressed as $3\frac{1}{3}$ hr. or as $\frac{10}{3}$ hr.

Step 1. Let x = Jeff's average speed (in mph)
Let y = George's average speed (in mph)

Then $4x$ = Jeff's distance $\left(x\frac{\text{mi}}{\cancel{\text{hr}}}\right)(4\ \cancel{\text{hr}}) = 4x$ mi

and $\frac{10}{3}y$ = George's distance $\left(y\frac{\text{mi}}{\cancel{\text{hr}}}\right)\left(\frac{10}{3}\ \cancel{\text{hr}}\right) = \frac{10}{3}y$ mi

Step 2. $\begin{cases} (1) & 4x = \frac{10}{3}y \\ (2) & y = x + 10 \end{cases}$ The distances were equal; George's speed was 10 mph faster than Jeff's

Step 3. Because Equation 2 has already been solved for y, we'll use substitution. Clearing fractions in Equation 1, we have

$12x = 10y$
$12x = 10(x + 10)$ Substituting $x + 10$ for y in $12x = 10y$
$12x = 10x + 100$
$2x = 100$
$x = 50$ Jeff's average speed in $\frac{\text{mi}}{\text{hr}}$
$y = x + 10 = 60$ George's average speed in $\frac{\text{mi}}{\text{hr}}$
$4x = 4(50) = 200$ The number of miles Jeff traveled before they met

√ **Step 4.** ***Check*** 60 mph = 50 mph + 10 mph — George's speed was 10 mph more than Jeff's

$$\left(50\frac{\text{mi}}{\cancel{\text{hr}}}\right)(4\ \cancel{\text{hr}}) = 200\text{ mi}$$ Jeff's distance

$$\left(60\frac{\text{mi}}{\cancel{\text{hr}}}\right)\left(\frac{10}{3}\ \cancel{\text{hr}}\right) = 200\text{ mi}$$ George's distance

George's distance and Jeff's distance were equal.

Step 5. Therefore, Jeff's average speed was 50 mph, George's average speed was 60 mph, and they traveled 200 mi before they met.

EXAMPLE 3

Seven cars can be completed if crew A works 9 hr, crew B works 4 hr, and crew C works 10 hr. If crew A works 3 hr and crew B works 8 hr, then 5 cars can be completed. If crew B works 6 hr and crew C works 5 hr, then 4 cars can be completed. How long would it take each crew, working alone, to complete 1 car?

SOLUTION We think it's easiest to let the variables equal the *rates*. This means, however, that we will not be finished when we have found the values of the variables.

Step 1. Let crew A's rate $= a\,\frac{\text{cars}}{\text{hr}}$

Let crew B's rate $= b\,\frac{\text{cars}}{\text{hr}}$

Let crew C's rate $= c\,\frac{\text{cars}}{\text{hr}}$

Recall that the basic relationships used in solving work problems are as follows:

> Rate × time = amount of work done
>
> (Amount of work done by A) + (amount of work done by B) + (amount of work done by C) = (total amount of work done)

Crew A works 9 hr and crew B works 4 hr and crew C works 10 hr — 7 cars are completed

$$(1)\quad 9\ \cancel{\text{hr}}\cdot a\frac{\text{cars}}{\cancel{\text{hr}}} + 4\ \cancel{\text{hr}}\cdot b\frac{\text{cars}}{\cancel{\text{hr}}} + 10\ \cancel{\text{hr}}\cdot c\frac{\text{cars}}{\cancel{\text{hr}}} = 7\text{ cars}$$

Crew A works 3 hr and crew B works 8 hr — 5 cars are completed

$$(2)\quad 3\ \cancel{\text{hr}}\cdot a\frac{\text{cars}}{\cancel{\text{hr}}} + 8\ \cancel{\text{hr}}\cdot b\frac{\text{cars}}{\cancel{\text{hr}}} = 5\text{ cars}$$

Crew B works 6 hr and crew C works 5 hr — 4 cars are completed

$$(3)\quad 6\ \cancel{\text{hr}}\cdot b\frac{\text{cars}}{\cancel{\text{hr}}} + 5\ \cancel{\text{hr}}\cdot c\frac{\text{cars}}{\cancel{\text{hr}}} = 4\text{ cars}$$

Step 2.

$$\begin{array}{l} (1)\\ (2)\\ (3) \end{array}\left\{\begin{array}{l} 9a + 4b + 10c = 7\\ 3a + 8b \qquad\quad\;\; = 5\\ \qquad\;\; 6b + 5c = 4 \end{array}\right\}$$ We solve this system by elimination

Step 3. We first eliminate the variable c from Equations 1 and 3, as follows:

$$\begin{array}{llrcl}
(1) & 1] & 9a + 4b + 10c = 7 & \Rightarrow & 9a + 4b + 10c = 7 \\
(3) & -2] & 6b + 5c = 4 & \Rightarrow & -12b - 10c = -8 \\ \hline
(4) & & & & 9a - 8b = -1
\end{array}$$

Next, we will add Equations 2 and 4.

$$\begin{array}{lrcl}
(2) & 3a + 8b & = & 5 \\
(4) & 9a - 8b & = & -1 \\ \hline
 & 12a & = & 4
\end{array}$$

$$a = \tfrac{4}{12} = \tfrac{1}{3}$$

$$(2) \quad 3a + 8b = 5$$

$$3\left(\tfrac{1}{3}\right) + 8b = 5 \quad \text{Substituting } \tfrac{1}{3} \text{ for } a \text{ in Equation 2}$$

$$1 + 8b = 5$$

$$8b = 4$$

$$b = \tfrac{4}{8} = \tfrac{1}{2}$$

$$(3) \quad 6b + 5c = 4$$

$$6\left(\tfrac{1}{2}\right) + 5c = 4 \quad \text{Substituting } \tfrac{1}{2} \text{ for } b \text{ in Equation 3}$$

$$3 + 5c = 4$$

$$5c = 1$$

$$c = \tfrac{1}{5}$$

Step 4. The check will not be shown.

Step 5. If crew A's rate is $\frac{1}{3}\frac{\text{car}}{\text{hr}}$, or $\frac{1 \text{ car}}{3 \text{ hr}}$, then crew A completes 1 car in 3 hr. If crew B's rate is $\frac{1}{2}\frac{\text{car}}{\text{hr}}$, then crew B completes 1 car in 2 hr. If crew C's rate is $\frac{1}{5}\frac{\text{car}}{\text{hr}}$, then crew C completes 1 car in 5 hr.

Exercises 12.9
Set I

Set up each problem algebraically, *using at least two variables*, solve, and check. Be sure to state what your variables represent.

1. The sum of two numbers is 30. Their difference is 12. What are the numbers?

2. The sum of two angles is 180°. Their difference is 70°. Find the angles.

3. Beatrice has 15 coins with a total value of \$1.75. If the coins are nickels and quarters, how many of each kind are there?

4. Raul has 22 coins with a total value of \$5.00. If the coins are dimes and half-dollars, how many of each kind are there?

5. A fraction has a value of two-thirds. If 10 is added to its numerator and 5 is subtracted from its denominator, the value of the fraction becomes 1. What was the original fraction?

6. A fraction has a value of one-half. If 8 is added to its numerator and 6 is added to its denominator, the value of the fraction becomes two-thirds. What was the original fraction?

7. The sum of the digits of a three-digit number is 20. The tens digit is 3 more than the units digit. The sum of the hundreds digit and the tens digit is 15. Find the number.

8. The sum of the digits of a three-digit number is 21. The units digit is 1 less than the tens digit. Twice the hundreds digit plus the tens digit is 17. Find the number.

9. Albert, Bill, and Carlos, working together, can do a job in 2 hr. Bill and Carlos together can do the same job in 3 hr. Albert and Bill together can do the same job in 4 hr. How long would it take each man, working alone, to do the entire job?

10. Crews A and C, working together, can assemble 1 machine in 3 hr. It takes crews A and B, working together, 8 hr to assemble 3 machines. It takes crews B and C, working together, 24 hr to assemble 5 machines. How long would it take each crew, working alone, to assemble 1 machine?

11. It took a pilot $5\frac{1}{2}$ hr to fly 2,750 mi *against* the wind but only 5 hr to return to her starting point, flying *with* the same wind. Find the average speed of the plane in still air, and find the average speed of the wind.

12. It took a pilot $2\frac{1}{2}$ hr to fly 1,200 mi *against* the wind but only 2 hr to return to his starting point, flying *with* the same wind. Find the average speed of the plane in still air, and find the average speed of the wind.

13. A 90-lb mixture of two different grades of coffee is worth $338.90. If grade A is worth $3.85 per pound and grade B is worth $3.65 per pound, how many pounds of each grade were used?

14. An 80-lb mixture of two different grades of coffee is worth $305.25. If grade A is worth $3.95 per pound and grade B is worth $3.70 per pound, how many pounds of each grade were used?

15. The sum of the digits of a two-digit number is 12. If the digits are reversed, the new number is 54 less than the original number. Find the original number.

16. The sum of the digits of a two-digit number is 14. If the digits are reversed, the new number is 18 less than the original number. Find the original number.

17. Tom spent $7.21 on 29 stamps, buying 18¢, 22¢, and 45¢ stamps. He bought twice as many 22¢ stamps as 18¢ stamps. How many of each kind did he buy?

18. Sherma spent $5.07 on 21 stamps, buying 18¢, 22¢, and 45¢ stamps. She bought twice as many 18¢ stamps as 45¢ stamps. How many of each kind did she buy?

19. A tie and a pin cost $1.10. The tie costs $1.00 more than the pin. What is the cost of each?

20. A number of birds were resting on two limbs of a tree. One limb was above the other. A bird on the lower limb said to the birds on the upper limb, "If one of you will come down here, we will have an equal number on each limb." A bird from above replied, "If one of you will come up here, we will have twice as many up here as you have down there." How many birds were sitting on each limb?

Exercises 12.9 Set II

Set up each problem algebraically, *using at least two variables*, solve, and check. Be sure to state what your variables represent.

1. The sum of two numbers is 50. Their difference is 22. What are the numbers?

2. The sum of two angles is 90°. Their difference is 40°. Find the angles.

3. Carol has 18 coins with a total value of $3.45. If the coins are dimes and quarters, how many of each kind are there?

4. Don spent $3.70 for 22 stamps. If he bought only 15¢ stamps and 20¢ stamps, how many of each kind did he buy?

5. A fraction has a value of three-fourths. If 4 is subtracted from the numerator and 2 is added to the denominator, the value of the new fraction is one-half. What was the original fraction?

6. Several families went to a movie together. They spent $24.75 for 8 tickets. If adult tickets cost $4.50 and children's tickets cost $2.25, how many of each kind of ticket were purchased?

7. The sum of the digits of a three-digit number is 15. The tens digit is 5 more than the units digit. The sum of the hundreds digit and the units digit is 9. Find the number.

8. One-third the sum of two numbers is 12. Twice their difference is 12. Find the numbers.

9. A refinery tank has one fill pipe and two drain pipes. Pipe 1 can fill the tank in 3 hr. Pipe 2 takes 6 hr longer to drain the tank than pipe 3 does. If it takes 12 hr to fill the tank when the valves of all three pipes are open, how long does it take pipe 3, alone, to drain the tank?

10. The sum of the digits of a three-digit number is 11. The tens digit is twice the units digit. If the digits are reversed, the new number is 297 less than the original number. Find the original number.

11. It took Jerry 6 hr to ride his bicycle 30 mi *against* the wind, but it took him only 2 hr to return to his starting point *with* the same wind. Find Jerry's average riding speed in still air, and find the average speed of the wind.

12. A fraction has a value of two-thirds. If 5 is added to the numerator and 4 is added to the denominator, the value of the new fraction is three-fourths. What was the original fraction?

13. A 60-lb mixture of two different grades of coffee is worth \$218.50. If grade A is worth \$3.80 per pound and grade B is worth \$3.55 per pound, how many pounds of each grade were used?

14. Find two numbers such that 5 times the larger plus 3 times the smaller is 47, and 4 times the larger minus twice the smaller is 20.

15. The sum of the digits of a two-digit number is 11. If the digits are reversed, the new number is 27 less than the original number. Find the original number.

16. The sum of the digits of a three-digit number is 13. The tens digit is 4 times the units digit. If the digits are reversed, the new number is 99 less than the original number. Find the original number.

17. Gloria spent \$6.31 on 26 stamps, buying 18¢, 22¢, and 45¢ stamps. She bought twice as many 18¢ stamps as 22¢ stamps. How many of each kind did she buy?

18. \$26,000 was divided and placed in three different accounts. Twice as much was put into an account earning 7% per year as was put into an account earning 6% per year. The rest was put into an account earning 5% per year. The interest earned in one year by all three accounts together was \$1,650. How much was invested at each rate?

19. Al, Chet, and Muriel are two brothers and a sister. Ten years ago, Al was twice as old as Chet was then. Three years ago, Muriel was three-fourths Chet's age at that time. In 15 years, Al will be 8 years older than Muriel is then. Find their ages now.

20. Tickets for a certain concert sold for \$6, \$8, and \$12 each. There were 280 more \$6 tickets sold than the number of \$8 and \$12 tickets together, and the total revenue from 840 tickets was \$6,080. How many tickets were sold at each price?

12.10 Solving Quadratic Systems of Equations

All the systems studied in this chapter so far have been *linear systems* of equations. In this section, we discuss the solution of *quadratic systems*. A quadratic system is one that has a second-degree equation as its highest-degree equation.

Most of the quadratic equations that appear in the quadratic systems of this section are the conic sections discussed and graphed earlier in this book.

Systems That Contain One Quadratic Equation and One Linear Equation

When the system has one quadratic equation and one linear equation, the graphs may intersect in two points, one point, or no points (see Figure 5). Thus, the system may have two real solutions, one real solution, or no real solutions.

1. Two points of intersection (two real solutions)

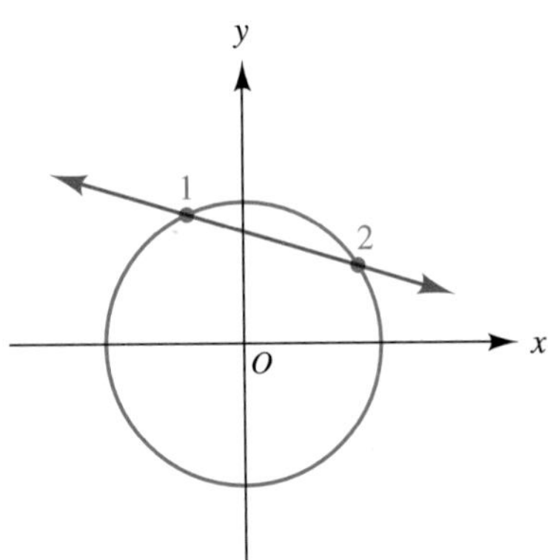

2. One point of intersection (one real solution)

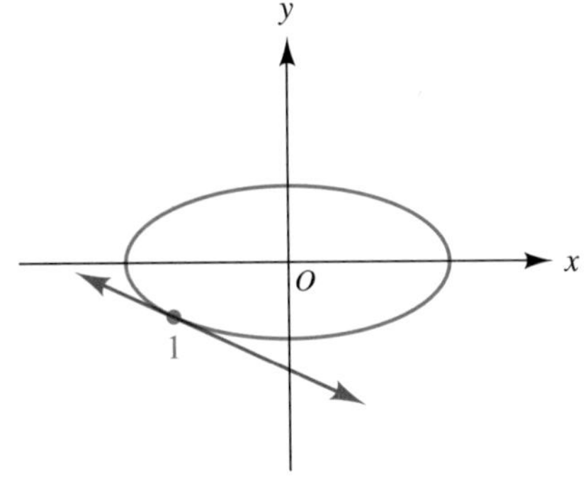

3. No points of intersection (no real solutions; there are two *complex* solutions)

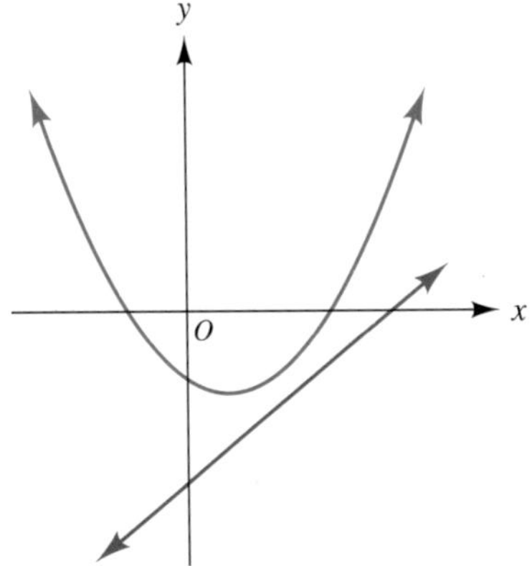

Figure 5

Solving a system that has one quadratic and one linear equation

1. Solve the linear equation for one variable.
2. Substitute the value of the variable from step 1 into the quadratic equation; then solve the resulting equation.
3. Substitute the solutions obtained in step 2 into the equation obtained in step 1 to solve for the remaining variable.*

---

*If you substitute into the *quadratic* equation, you may get extraneous roots.

EXAMPLE 1 Find the solution set of the quadratic system $\left\{\begin{matrix} x - 2y = -4 \\ x^2 = 4y \end{matrix}\right\}$.

SOLUTION

Step 1. We solve the linear equation for x: $x - 2y = -4$

$$x = 2y - 4$$

Step 2. We substitute $2y - 4$ for x in the quadratic equation:

$$x^2 = 4y$$
$$(2y - 4)^2 = 4y$$
$$4y^2 - 16y + 16 = 4y$$
$$4y^2 - 20y + 16 = 0$$
$$y^2 - 5y + 4 = 0 \quad \text{Dividing both sides by 4}$$
$$(y - 1)(y - 4) = 0$$
$$y - 1 = 0 \quad \Big| \quad y - 4 = 0$$
$$y = 1 \quad \Big| \quad y = 4$$

Step 3. We substitute the values found in step 2 into $x = 2y - 4$.

If $y = 1$, then $x = 2y - 4 = 2(1) - 4 = -2$. Therefore, one solution is $(-2, 1)$.

If $y = 4$, then $x = 2y - 4 = 2(4) - 4 = 8 - 4 = 4$. Therefore, the second solution is $(4, 4)$.

When the graphs of both of the original equations are drawn on the same set of axes, we see that the two solutions, $(-2, 1)$ and $(4, 4)$, are the points where the graphs *intersect* (see Figure 6).

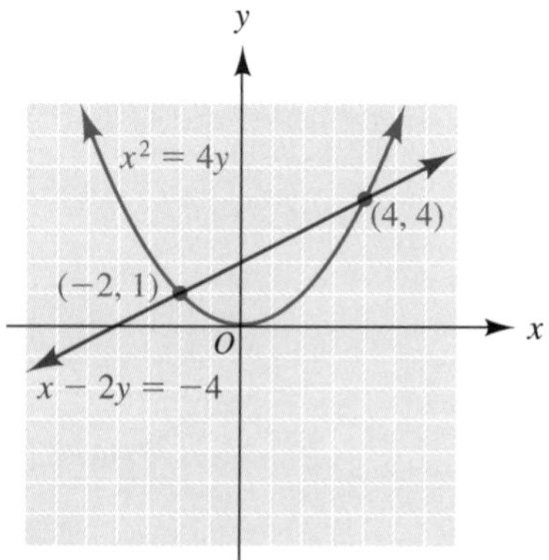

Figure 6

Checking will confirm that the solution set is $\{(-2, 1), (4, 4)\}$.

Systems That Contain Two Quadratic Equations

When the system has two quadratic equations, the curves can intersect in from zero to four points (see Figure 7).

1. No points of intersection (no real solution; there are *complex* solutions)

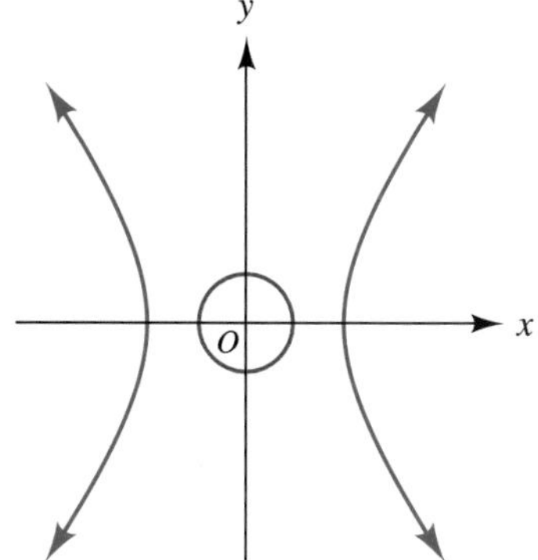

2. One point of intersection (one real solution)

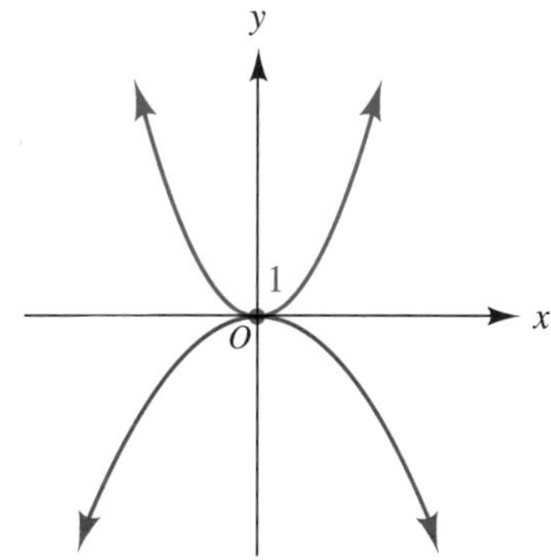

3. Two points of intersection (two real solutions)

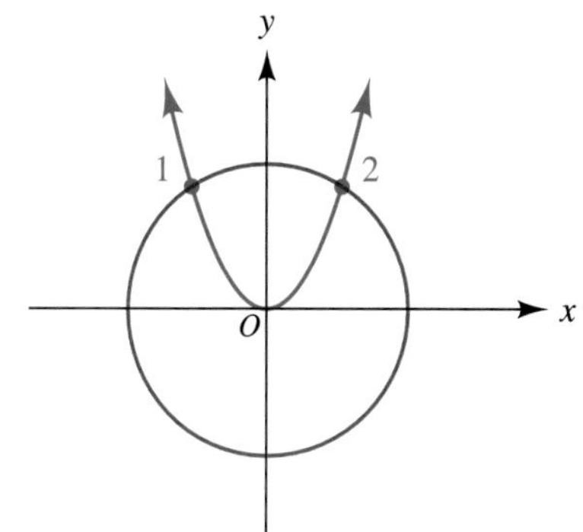

4. Three points of intersection (three real solutions)

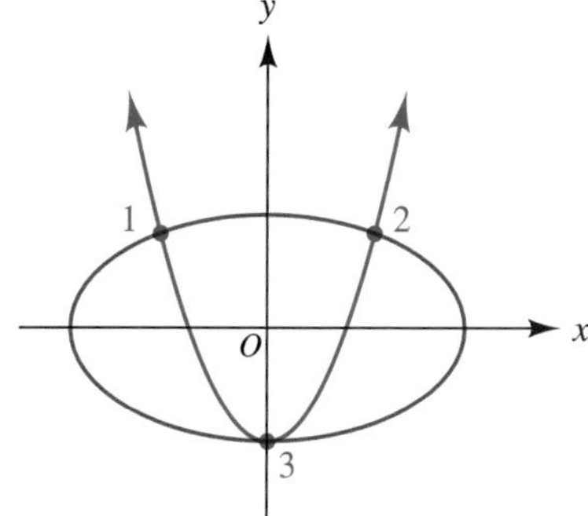

5. Four points of intersection (four real solutions)

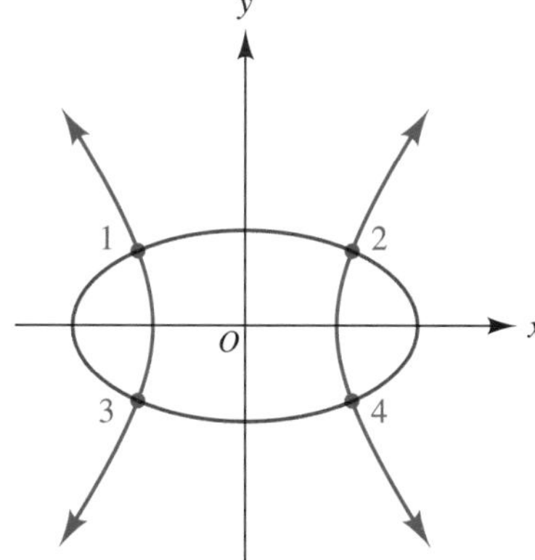

Figure 7

If both equations of a system are quadratic equations that have like terms, we can find the solution set of the system by using elimination (see Example 2). Otherwise, we use substitution (see Example 3).

EXAMPLE 2 Find the solution set of the quadratic system $\left\{\begin{aligned} x^2 + y^2 &= 13 \\ 2x^2 + 3y^2 &= 30 \end{aligned}\right\}$.

SOLUTION Because the two equations have like terms, we can use the elimination method.

$$\begin{aligned} -2]\quad x^2 + y^2 = 13 &\Rightarrow -2x^2 - 2y^2 = -26 \\ 1]\quad 2x^2 + 3y^2 = 30 &\Rightarrow \underline{2x^2 + 3y^2 = 30} \\ &\qquad\qquad\quad y^2 = 4 \\ &\qquad\qquad\quad y = \pm 2 \end{aligned}$$

We will add the equations

We find the corresponding x-values by substitution, as follows:

Substituting 2 for y in the first equation	*Substituting −2 for y in the first equation*
$x^2 + y^2 = 13$	$x^2 + y^2 = 13$
$x^2 + (2)^2 = 13$	$x^2 + (-2)^2 = 13$
$x^2 + 4 = 13$	$x^2 + 4 = 13$
$x^2 = 9$	$x^2 = 9$
$x = \pm 3$	$x = \pm 3$
Two solutions are (3, 2) and (−3, 2).	Two solutions are (3, −2) and (−3, −2).

When the graphs of both equations are drawn on the same set of axes, we see that the four solutions—(3, 2), (3, −2), (−3, 2), and (−3, −2)—are the points where the two graphs *intersect* (Figure 8). The solutions should *not* be written as (±3, ±2), as this notation does not make it clear that there are four distinct solutions.

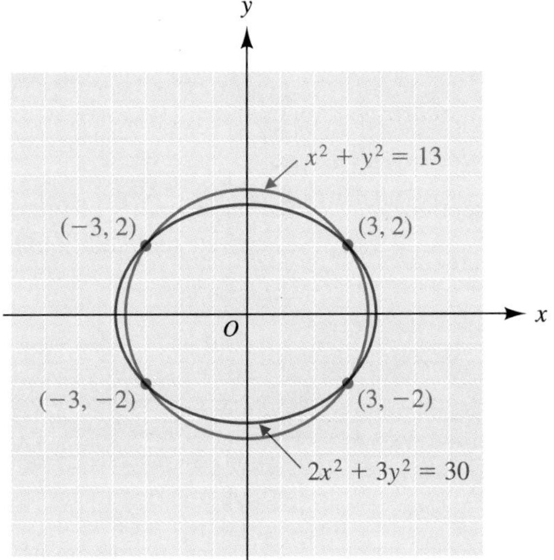

Figure 8

Checking will confirm that the solution set is {(3, 2), (3, −2), (−3, 2), (−3, −2)}.

EXAMPLE 3 Find the solution set of the system $\left\{\begin{aligned} xy &= 1 \\ x^2 + y^2 &= 2 \end{aligned}\right\}$.

SOLUTION We have not previously discussed an equation such as $xy = 1$. It *is* a quadratic equation, because it is of degree 2. (Remember that to find the degree of the term xy, we add the exponents on the variables: $xy = x^1y^1$, and $1 + 1 = 2$.) The graph of the solution set happens to be a *hyperbola* with the x- and y-axes as its asymptotes. [You can easily verify that the coordinates of the points $\left(\frac{1}{4}, 4\right)$, $\left(\frac{1}{3}, 3\right)$, $\left(\frac{1}{2}, 2\right)$, $\left(-\frac{1}{2}, -2\right)$, $\left(-\frac{1}{3}, -3\right)$, and $\left(-\frac{1}{4}, -4\right)$, which are on the graph in Figure 9, satisfy $xy = 1$.]

Because the equations do not contain like terms, we must use substitution in solving this system. If we solved the second equation for y (*or* for x), the resulting equation would involve radicals. Solving the *first* equation for y avoids this difficulty:

$$xy = 1$$

$$y = \frac{1}{x}$$

Next, we substitute $\frac{1}{x}$ for y in the second equation:

$$x^2 + y^2 = 2$$

$$x^2 + \left(\frac{1}{x}\right)^2 = 2$$

$$x^2 + \frac{1}{x^2} = 2$$

$$x^4 + 1 = 2x^2 \quad \text{Multiplying both sides by } x^2$$

$$x^4 - 2x^2 + 1 = 0$$

$$(x^2 - 1)^2 = 0$$

$$x^2 - 1 = 0$$

$$x^2 = 1$$

$$x = \pm 1$$

If $x = 1$, then $y = \frac{1}{x} = \frac{1}{1} = 1$. Therefore, one solution is (1, 1).

If $x = -1$, then $y = \frac{1}{x} = \frac{1}{-1} = -1$. Therefore, the second solution is $(-1, -1)$.

The graph is shown in Figure 9.

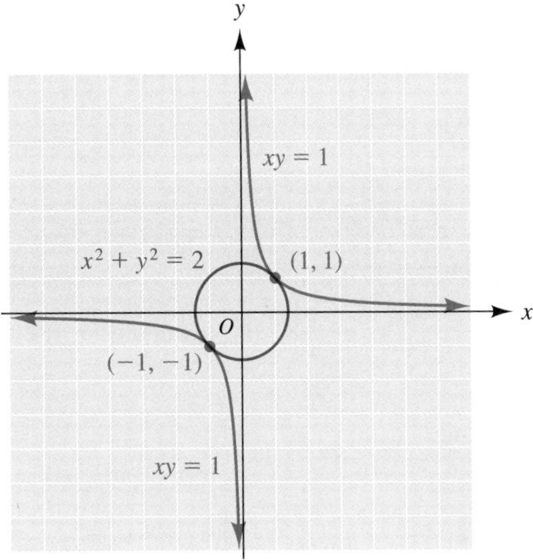

Figure 9

You can verify that if we substitute 1 for x in the second equation, we obtain the solutions $y = \pm 1$, and if we substitute -1 for x in the second equation, we obtain the solutions $y = \pm 1$. Therefore, we might think that (1, 1), (1, −1), (−1, 1), and (−1, −1) were all points of intersection. However, checking these solutions in the first equation (*or* examining the graph) shows that (−1, 1) and (1, −1) are extraneous solutions. The solutions should *not* be written as (±1, ±1). The solution set is {(1, 1), (−1, −1)}.

Note When a quadratic equation contains an xy-term, we cannot determine which of the conic sections it is by simply looking at the equation.

Note It is possible to solve quadratic systems with a TI-81 graphics calculator; to do so, we combine the techniques for graphing the conic sections (shown in the notes following the examples in Section 11.6) with the technique for solving a system of equations graphically (shown in the note following Example 1 in Section 12.2).

Exercises 12.10
Set I

Find the solution set of each system of equations. (When the curves do not intersect, the solution set will contain complex numbers.)

1. $\begin{Bmatrix} x^2 = 2y \\ x - y = -4 \end{Bmatrix}$

2. $\begin{Bmatrix} x^2 = 4y \\ x + 2y = 4 \end{Bmatrix}$

3. $\begin{Bmatrix} x^2 = 4y \\ x - y = 1 \end{Bmatrix}$

4. $\begin{Bmatrix} x^2 + y^2 = 25 \\ x - 3y = -5 \end{Bmatrix}$

5. $\begin{Bmatrix} xy = 4 \\ x - 2y = 2 \end{Bmatrix}$

6. $\begin{Bmatrix} xy = 3 \\ x + y = 0 \end{Bmatrix}$

7. $\begin{Bmatrix} x^2 + y^2 = 61 \\ x^2 - y^2 = 11 \end{Bmatrix}$

8. $\begin{Bmatrix} x^2 + 4y^2 = 4 \\ x^2 - 4y = 4 \end{Bmatrix}$

9. $\begin{Bmatrix} 2x^2 + 3y^2 = 21 \\ x^2 + 2y^2 = 12 \end{Bmatrix}$

10. $\begin{Bmatrix} 4x^2 - 5y^2 = 62 \\ 5x^2 + 8y^2 = 106 \end{Bmatrix}$

Writing Problems

Express the answers in your own words and in complete sentences.

1. Determine which method you would use in solving the system $\begin{Bmatrix} y = 5 - 3x \\ 4x^2 - y^2 = 4 \end{Bmatrix}$, and explain why you would use that method.

2. Determine which method you would use in solving the system $\begin{Bmatrix} 9x^2 - y^2 = 9 \\ x^2 + y^2 = 1 \end{Bmatrix}$, and explain why you would use that method.

Exercises 12.10
Set II

Find the solution set of each system of equations.

1. $\begin{Bmatrix} x^2 = 2y \\ 2x + y = -2 \end{Bmatrix}$

2. $\begin{Bmatrix} xy = 6 \\ x - y = 1 \end{Bmatrix}$

3. $\begin{Bmatrix} x^2 + y^2 = 25 \\ 3x + y = -5 \end{Bmatrix}$

4. $\begin{Bmatrix} 4x^2 + 9y^2 = 36 \\ 2x^2 - 9y = 18 \end{Bmatrix}$

5. $\begin{Bmatrix} 3x^2 + 4y^2 = 35 \\ 2x^2 + 5y^2 = 42 \end{Bmatrix}$

6. $\begin{Bmatrix} y = 3 - 2x - x^2 \\ y = x^2 + 2x + 3 \end{Bmatrix}$

7. $\begin{Bmatrix} 9x^2 + 16y^2 = 144 \\ 3x + 4y = 12 \end{Bmatrix}$

8. $\begin{Bmatrix} x^2 - y^2 = 4 \\ x^2 + y^2 = 4 \end{Bmatrix}$

9. $\begin{Bmatrix} x + y = 4 \\ y = x^2 - 4x + 4 \end{Bmatrix}$

10. $\begin{Bmatrix} 9x^2 + y^2 = 9 \\ 3x - y = -3 \end{Bmatrix}$

12.11 Graphing Solution Sets of Systems of Inequalities

In this section, we discuss graphing solution sets of systems of *inequalities*.

Graphing the solution set of a system of two inequalities

1. Graph the first inequality, shading the region that represents its solution set.
2. Graph the second inequality on the same set of axes. Use a different type of shading for the region that represents its solution set.
3. Heavily shade the region that contains *both* types of shading, if such a region exists.
4. Any ordered pair that corresponds to a point that lies in the heavily shaded region is in the solution set of the system; the coordinates of any such point will satisfy *both* inequalities.

In Example 1, we graph the solution set of a system of *linear* inequalities. In this case, the region that represents the solution set of each separate inequality is a *half-plane*.

EXAMPLE 1 Graph the solution set of the system $\begin{Bmatrix} 2x - 3y < 6 \\ 3x + 4y \le -12 \end{Bmatrix}$.

SOLUTION

Step 1. To graph $2x - 3y < 6$, we first graph the boundary line $2x - 3y = 6$. This is a straight line that has intercepts (3, 0) and (0, −2); because the inequality is $<$, we graph these points as *hollow dots* and graph the boundary line as a dashed line. Because the line does not pass through the origin, we can determine which half-plane to shade by testing (0, 0) in $2x - 3y < 6$.

$$2x - 3y < 6$$
$$2(0) - 3(0) < 6$$
$$0 < 6 \quad \text{True}$$

Therefore, the solution set includes (0, 0), and we shade the half-plane that contains that point, using *diagonal* lines (see Figure 10).

Step 2. To graph $3x + 4y \le -12$, we first graph the boundary line $3x + 4y = -12$. This is a straight line that has intercepts (−4, 0) and (0, −3); because the inequality is $\le$, we graph these points as *solid dots* and graph the boundary line as a solid line. Again, the line does not pass through the origin, so we can determine which half-plane to shade by testing (0, 0) in $3x + 4y \le -12$.

$$3x + 4y \le -12$$
$$3(0) + 4(0) \le -12$$
$$0 \le -12 \quad \text{False}$$

Therefore, the solution set does *not* include (0, 0), and we shade the half-plane that does *not* contain that point, using *vertical* lines (see Figure 10).

Step 3. Finally, we heavily shade the region that contains both types of shading (see Figure 10).

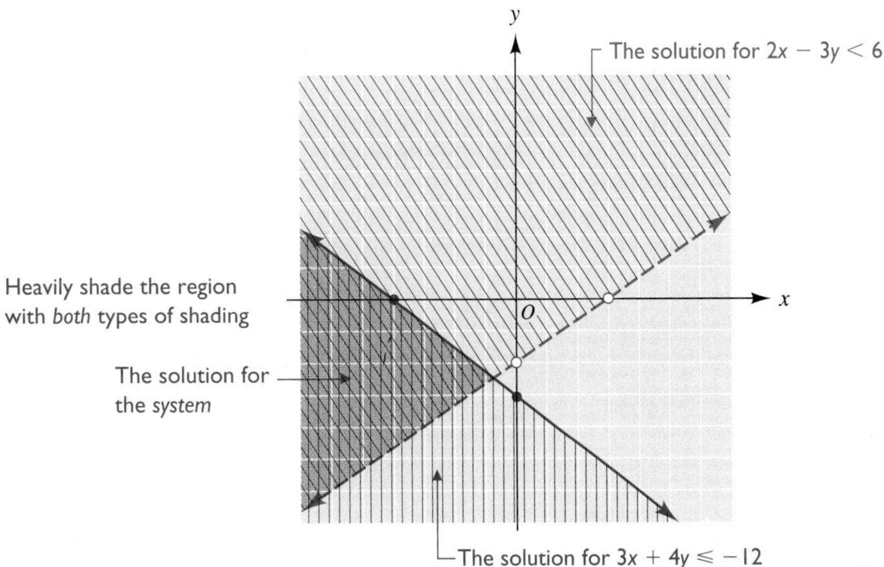

Figure 10

Step 4. Any ordered pair that corresponds to a point that lies in the heavily shaded region is in the solution set of the system; its coordinates will satisfy both inequalities.

Next, we graph the solution set of a system in which both inequalities are quadratic.

EXAMPLE 2 Graph the solution set of the system $\begin{Bmatrix} y \geq x^2 - 2x - 3 \\ x^2 + 4y^2 \leq 4 \end{Bmatrix}$.

SOLUTION To graph $y \geq x^2 - 2x - 3$, we first graph $y = x^2 - 2x - 3$, which is a parabola that opens upward. Its intercepts are (3, 0), (−1, 0), and (0, −3), and its vertex is at (1, −4). We graph the parabola with a solid curve, since the inequality is $\geq$. The parabola does not pass through the origin, so we test (0, 0) in $y \geq x^2 - 2x - 3$; when we do this, we obtain the *true* statement $0 \geq -3$. Thus, the solution set contains the origin, so we shade the region *inside* the parabola, using *diagonal* lines (see Figure 11).

To graph $x^2 + 4y^2 \leq 4$, we first graph $x^2 + 4y^2 = 4$, or $\frac{x^2}{4} + y^2 = 1$, which is an ellipse. Its intercepts are (2, 0), (−2, 0), (0, 1), and (0, −1). Because the inequality is $\leq$, we graph the ellipse with a solid curve. We can use (0, 0) as a test point in $x^2 + 4y^2 \leq 4$, and when we do so, we obtain the statement $0 \leq 4$, a *true* statement. Therefore, the solution set contains the origin, so we shade the region *inside* the ellipse, using *vertical* lines (see Figure 11).

Finally, we heavily shade the region that contains *both* types of shading (see Figure 11).

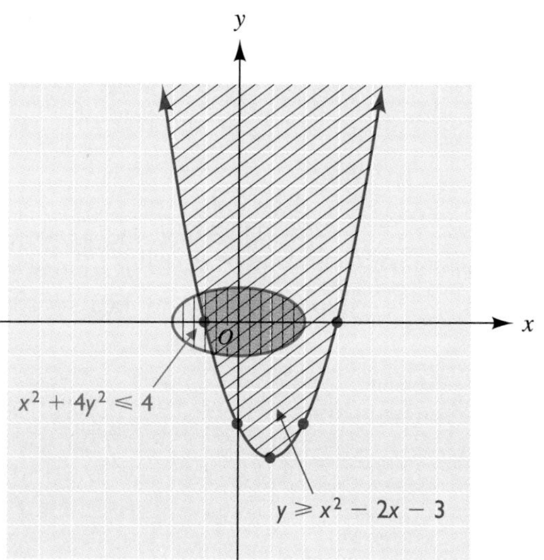

Figure 11

Any ordered pair that corresponds to a point that lies in the heavily shaded region is in the solution set of the system of inequalities.

In Example 3, we graph the solution set of a system in which one inequality is linear and the other is quadratic.

EXAMPLE 3 Graph the solution set of the system $\begin{Bmatrix} y < 4 - x^2 \\ y \leq x + 3 \end{Bmatrix}$.

SOLUTION To graph $y < 4 - x^2$, we first graph $y = 4 - x^2$, which is a parabola that opens downward. Its intercepts are (2, 0), (−2, 0), and (0, 4), and its vertex is at (0, 4). We graph the parabola with a dashed curve, since the inequality is $<$. We can use (0, 0) as a test point in $y < 4 - x^2$, and when we do so, we obtain the *true* statement $0 < 4$. Therefore, we shade the region that contains the origin—the region *inside* the parabola—using *diagonal* lines (see Figure 12).

To graph $y \leq x + 3$, we first graph $y = x + 3$, which is a straight line that has intercepts (0, 3) and (−3, 0). Because the inequality is $\leq$, we graph these points as *solid dots* and graph the boundary line as a solid line. We again use the origin as a test point and obtain the statement $0 \leq 3$, which is *true*. Therefore, we shade the half-plane that contains the origin, using *vertical* lines (see Figure 12).

Finally, we shade that portion of the plane that contains both kinds of shading (see Figure 12).

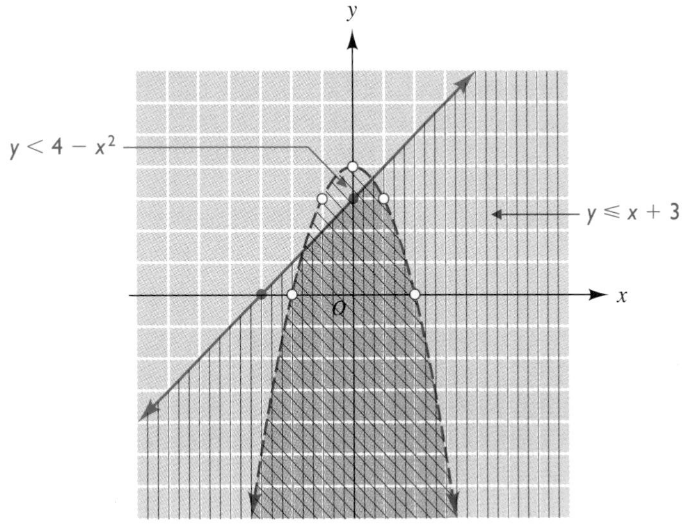

Figure 12

Any ordered pair that corresponds to a point that lies in the heavily shaded region is in the solution set of the system of inequalities.

To graph the solution set of a system of *more* than two inequalities, we simply use a different type (or color) of shading for the graph of the solution set of each separate inequality and then heavily shade the region that contains *all* the types (or colors) of shading that we have used. The coordinates of any point that lies in the heavily shaded region will be in the solution set of the system of inequalities.

Exercises 12.11
Set I

Graph the solution set of each system of inequalities.

1. $\begin{Bmatrix} 4x - 3y > -12 \\ y > 2 \end{Bmatrix}$

2. $\begin{Bmatrix} x - y > -2 \\ 2x + y > 2 \end{Bmatrix}$

3. $\begin{Bmatrix} 2x - y \leq 2 \\ x + y \geq 5 \end{Bmatrix}$

4. $\begin{Bmatrix} x - y \leq 4 \\ x + y \geq -2 \end{Bmatrix}$

5. $\begin{Bmatrix} 2x + y < 0 \\ x - y \geq -3 \end{Bmatrix}$

6. $\begin{Bmatrix} 3x > 6 - y \\ y + 3x \leq 0 \end{Bmatrix}$

7. $\begin{Bmatrix} 3x - 2y < 6 \\ x + 2y \leq 4 \\ 6x + y > -6 \end{Bmatrix}$

8. $\begin{Bmatrix} 3x + 4y \leq 15 \\ 2x + y \leq 2 \\ -4 < x < 1 \\ -2 < y < 6 \end{Bmatrix}$

9. $\begin{Bmatrix} \dfrac{x^2}{9} + \dfrac{y^2}{4} < 1 \\ \dfrac{x^2}{4} + \dfrac{y^2}{9} < 1 \end{Bmatrix}$

10. $\begin{Bmatrix} \dfrac{x^2}{4} + y^2 \leq 1 \\ x + y < 1 \end{Bmatrix}$

11. $\begin{Bmatrix} y > 1 - x^2 \\ x^2 + y^2 < 4 \end{Bmatrix}$

12. $\begin{Bmatrix} x^2 + y^2 \geq 1 \\ x^2 + y^2 \leq 9 \end{Bmatrix}$

13. $\begin{Bmatrix} x^2 + y^2 \leq 9 \\ y \geq x - 2 \end{Bmatrix}$

14. $\begin{Bmatrix} y > 2x + 2 \\ 4x^2 + y^2 > 4 \end{Bmatrix}$

Writing Problems

Express the answer in your own words and in complete sentences.

1. Describe what your first two steps would be in solving the system

$$\begin{Bmatrix} y > 3 - 5x \\ 4x^2 - y^2 \le 4 \end{Bmatrix}$$

and explain why you would do those things first.

Exercises 12.11 Set II

Graph the solution set of each system of inequalities.

1. $\begin{Bmatrix} 3x - 5y < 15 \\ x < 3 \end{Bmatrix}$

2. $\begin{Bmatrix} 2x + 3y < 6 \\ x - 3y < 3 \end{Bmatrix}$

3. $\begin{Bmatrix} 3x - y \ge 0 \\ 3x + 2y < 6 \end{Bmatrix}$

4. $\begin{Bmatrix} 2x \ge y - 4 \\ y - 2x > 0 \end{Bmatrix}$

5. $\begin{Bmatrix} 3x - 2y < 6 \\ x + 2y \ge 2 \end{Bmatrix}$

6. $\begin{Bmatrix} 4x + y \le 4 \\ 2x - y > 2 \end{Bmatrix}$

7. $\begin{Bmatrix} 4x + 3y \ge 12 \\ y - x > 1 \\ 9x - 28y \ge 42 \end{Bmatrix}$

8. $\begin{Bmatrix} 4x - 3y \ge 12 \\ x < 3 \end{Bmatrix}$

9. $\begin{Bmatrix} x^2 - \dfrac{y^2}{9} \le 1 \\ \dfrac{x^2}{9} + \dfrac{y^2}{4} \le 1 \end{Bmatrix}$

10. $\begin{Bmatrix} y \le x^2 - 1 \\ 4x^2 + y^2 \le 4 \end{Bmatrix}$

11. $\begin{Bmatrix} x^2 + \dfrac{y^2}{9} \le 1 \\ y \le 1 - x^2 \end{Bmatrix}$

12. $\begin{Bmatrix} \dfrac{x^2}{9} + y^2 \le 1 \\ x^2 + y^2 \ge 4 \end{Bmatrix}$

13. $\begin{Bmatrix} x^2 + y^2 \le 16 \\ x + y > 2 \end{Bmatrix}$

14. $\begin{Bmatrix} y > -\frac{1}{2}x + 1 \\ x^2 + 4y^2 > 4 \end{Bmatrix}$

Sections 12.6–12.11 REVIEW

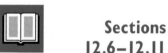

A Matrix 12.6 A **matrix** (plural: **matrices**) is a rectangular array of numbers; we use brackets around a matrix. A *row* of a matrix is a *horizontal* line of its elements. A *column* of a matrix is a *vertical* line of its elements. An $m \times n$ (read "m by n") matrix has m rows and n columns.

Augmented Matrix for a System of Equations 12.6 An **augmented matrix** is associated with each system of linear equations. The augmented matrix for a system of n equations in n unknowns is always an $n \times (n + 1)$ matrix.

Row-Echelon Form 12.6 If a matrix is in the form $\begin{bmatrix} 1 & a & b & c \\ 0 & 1 & d & e \\ 0 & 0 & 1 & k \end{bmatrix}$, where a, b, c, d, e, and k are any constants, we say that it is in **row-echelon form**.

Elementary Row Operations 12.6 If any of the following three operations are performed on a matrix, we say that the new matrix is **row-equivalent** to the given matrix.

1. Any two rows of the matrix can be interchanged.
2. Each element in any row can be multiplied by some nonzero constant.
3. A (nonzero) multiple of the elements of one row can be added to the corresponding elements of another row.

Gaussian Elimination 12.6 To use **Gaussian elimination** to solve a system of linear equations, we form the augmented matrix, reduce it to row-echelon form, and then write the system of equations that corresponds to the final matrix.

Determinants 12.7 A **determinant** is a *real number* that is represented as a square array of numbers enclosed between two vertical bars.

The value of the second-order determinant $\begin{vmatrix} a & b \\ c & d \end{vmatrix}$ is $ad - cb$.

To find the value of a third-order determinant, see Section 12.7.

Cramer's Rule 12.8 **Cramer's rule** can be used to solve linear systems of equations when the system contains the same number of equations as variables; however, if the system is dependent, Cramer's rule does not give the solutions.

Solving an Applied Problem by Using a System of Equations 12.9

Read First read the problem very carefully.

Think Determine what *type* of problem it is, if possible. Determine what is unknown.

Sketch Draw a sketch *with labels*, if it might be helpful.

Step 1. Represent each unknown number by a *different* variable.

Reread Reread the problem, breaking it up into small pieces.

Step 2. Translate each English phrase into an algebraic expression; fit these expressions together into *two or more different equations. There must be as many equations as variables.*

Step 3. Solve the *system* of equations for *all* the variables, using any of the methods discussed in this chapter.

Step 4. Check the solution(s) *in the word statements.*

Step 5. State the results clearly, being sure to answer *all* the questions asked.

Methods of Solving a Quadratic System of Equations 12.10

1. *One quadratic and one linear equation*: Solve the linear equation for one variable; then substitute the expression so obtained into the quadratic equation.
2. *Two quadratic equations*: Elimination can be used if the equations have like terms. Otherwise, use substitution.

Graphing a System of Inequalities 12.11

1. Graph the first inequality, shading the region that represents its solution.
2. Graph the second inequality, using a different type of shading.
3. Heavily shade the region that contains *both* types of shading.
4. Any ordered pair that corresponds to a point that lies in the heavily shaded region is in the solution set of the system.

Sections 12.6–12.11 REVIEW EXERCISES Set I

In Exercises 1–3, use the matrix method or Cramer's rule to find the solution set of each system.

1. $\begin{Bmatrix} 3x - 2y = 8 \\ 2x + 5y = -1 \end{Bmatrix}$

2. $\begin{Bmatrix} 3x + y + 2z = 4 \\ 2x - 3y + 3z = -5 \\ 5x + 2y + 3z = 7 \end{Bmatrix}$

3. $\begin{Bmatrix} 5x + 2y = 1 \\ 7x - 6y = 8 \end{Bmatrix}$

In Exercises 4–6, find the solution set of each system by any convenient method.

4. $\begin{Bmatrix} 2x + 3y - 4z = 4 \\ 3x + 2y - 5z = 0 \\ 4x + 5y + 2z = -4 \end{Bmatrix}$

5. $\begin{Bmatrix} x - 2y = -1 \\ 2x^2 - 3y^2 = 6 \end{Bmatrix}$

6. $\begin{Bmatrix} x + 3y = -2 \\ x^2 - 2y^2 = 8 \end{Bmatrix}$

In Exercises 7–9, graph the solution set of each system of inequalities.

7. $\begin{Bmatrix} x + 4y \le 4 \\ 3x + 2y > 2 \end{Bmatrix}$

8. $\begin{Bmatrix} 3x + 6 \ge 2y \\ 3x + y < 3 \end{Bmatrix}$

9. $\begin{Bmatrix} y \le 12 + 4x - x^2 \\ x^2 + y^2 \le 16 \end{Bmatrix}$

In Exercises 10–12, set up algebraically, solve, and check, using at least two variables.

10. The difference of two numbers is −7. The difference of their squares is 7. Find the numbers.

11. The purchasing agent of a mail-order office paid \$730 for a total of 80 rolls of stamps in two denominations. If one kind of stamp cost \$10 per roll and the other kind cost \$8 per roll, how many rolls of each kind were purchased?

12. Jennifer motored 64 mi upstream in 4 hr and motored back in half that time. Find the average speed of her boat in still water, and find the average speed of the river.

Name

ANSWERS

1. ________

2. ________

3. ________

4. ________

5. ________

6. ________

In Exercises 1–3, use the matrix method or Cramer's rule to find the solution set of each system.

1. $\begin{Bmatrix} 3x - 6y = -4 \\ -5x + 8y = 6 \end{Bmatrix}$

2. $\begin{Bmatrix} 3x - 2y = 10 \\ 5x + 4y = 24 \end{Bmatrix}$

3. $\begin{Bmatrix} x + 2y + z = 3 \\ 3x + y - 2z = -1 \\ 2x - 3y - z = 4 \end{Bmatrix}$

In Exercises 4–6, find the solution set of each system by any convenient method.

4. $\begin{Bmatrix} 25x^2 + y^2 = 25 \\ 5x + y = -5 \end{Bmatrix}$

5. $\begin{Bmatrix} 4x + 3y + 2z = 6 \\ 8x + 3y - 3z = -1 \\ -4x + 6y + 5z = 5 \end{Bmatrix}$

6. $\begin{Bmatrix} x^2 + y^2 = 5 \\ x^2 - 3y^2 = 1 \end{Bmatrix}$

In Exercises 7 and 8, graph the solution set of each system of inequalities.

7. $\begin{Bmatrix} 5x + 2y \geq 0 \\ 5y - 6x > 15 \end{Bmatrix}$

7.

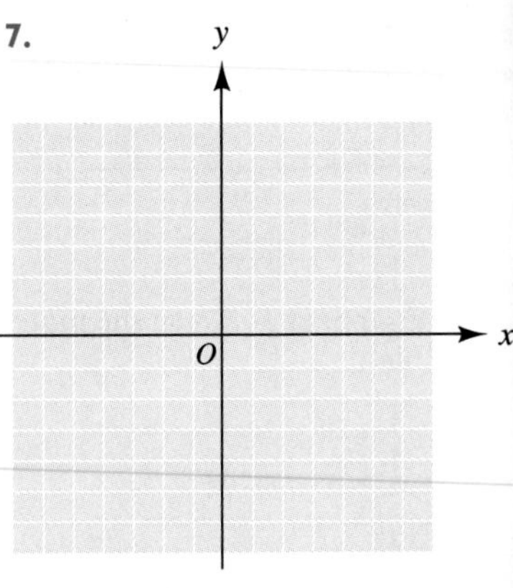

8. $\begin{Bmatrix} x^2 + y^2 \geq 4 \\ x^2 + 9y^2 \leq 9 \end{Bmatrix}$

8.

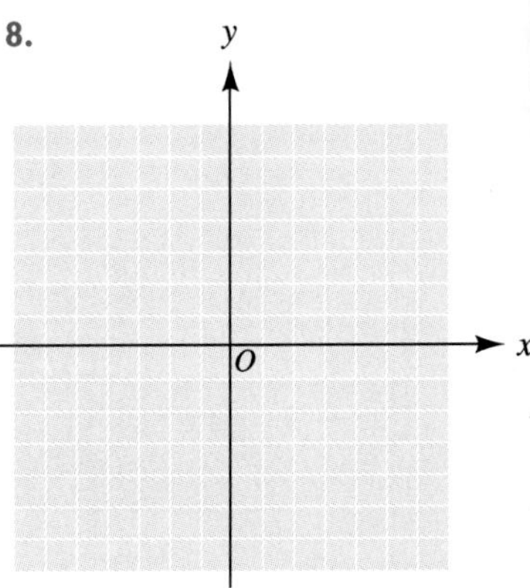

In Exercises 9 and 10, set up algebraically, solve, and check, using at least two variables.

9. Leticia worked at two jobs during the week for a total of 26 hr. For this she received a total of $118.25. If she was paid $4.75 per hour as a lab assistant and $4.40 per hour as a clerk-typist, how many hours did she work at each job?

9. ____________________

____

10. ____________________

____

10. Fausto motored 21 mi upstream in 3 hr and motored back to his starting point in one-third of that time. Find the average speed of the boat in still water, and find the average speed of the river.

Chapter 12 DIAGNOSTIC TEST

The purpose of this test is to see how well you understand systems of equations and systems of inequalities. We recommend that you work this diagnostic test *before* your instructor tests you on this chapter. Allow yourself about 50 minutes.

Complete solutions for all the problems on this test, together with section references, are given in the answer section at the end of the book. For the problems you do incorrectly, we suggest that you study the sections cited.

1. Find the solution set of the system $\begin{Bmatrix} 3x + 2y = 4 \\ x - y = 3 \end{Bmatrix}$ graphically.

In Problems 2–7, find the solution set of each system by any convenient method.

2. $\begin{Bmatrix} 4x - 3y = 13 \\ 5x - 2y = 4 \end{Bmatrix}$

3. $\begin{Bmatrix} 5x + 4y = 23 \\ 3x + 2y = 9 \end{Bmatrix}$

4. $\begin{Bmatrix} 15x + 8y = -18 \\ 9x + 16y = -8 \end{Bmatrix}$

5. $\begin{Bmatrix} 35y - 10x = -18 \\ 4x - 14y = 8 \end{Bmatrix}$

6. $\begin{Bmatrix} x + y + z = 0 \\ 2x \quad -3z = 5 \\ 3y + 4z = 3 \end{Bmatrix}$

7. $\begin{Bmatrix} y^2 = 8x \\ 3x + y = 2 \end{Bmatrix}$

8. Graph the solution set of the system $\begin{Bmatrix} 2x + 3y \le 6 \\ y - 2x < 2 \end{Bmatrix}$.

9. Find the solution set of the system $\begin{Bmatrix} x - y - z = 0 \\ x + 3y + z = 4 \\ 7x - 2y - 5z = 2 \end{Bmatrix}$.

   Use Gaussian elimination or Cramer's rule if you studied either of those optional sections.

Set up the following problem algebraically, *using two variables*, solve, and check.

10. Barney motored 30 mi upstream in 2 hr and motored back to his starting point in half that time. Find the average speed of his boat in still water, and find the average speed of the river.

Chapters 1–12 CUMULATIVE REVIEW EXERCISES

1. Add $\dfrac{x}{xy - y^2} + \dfrac{y}{xy - x^2}$.

2. Divide $(x^3 - 5x^2 + 9x + 20)$ by $(x - 3)$.

In Exercises 3 and 4, simplify each radical.

3. $\sqrt[6]{4x^2}$

4. $\sqrt{\dfrac{3a^3b^7}{8a^5b^4}}$

5. Write an equation of the line through $(-1, 2)$ and $(2, 4)$.

In Exercises 6–9, find each solution set.

6. $x^2 - 8x + 12 < 0$

7. $4x^2 - 25 = 0$

8. $\dfrac{x+2}{3} + \dfrac{5}{x-2} = 4$

9. $x^2 - 4x + 5 = 0$

10. Is division associative?

11. What is the additive inverse of $-\frac{3}{5}$?

12. What is the multiplicative inverse of $-\frac{3}{5}$?

In Exercises 13–15, find the solution set of each system by any convenient method.

13. $\begin{Bmatrix} 4x + 3y = 8 \\ 8x + 7y = 12 \end{Bmatrix}$

14. $\begin{Bmatrix} 2x - 3y = 3 \\ 3y - 2x = 6 \end{Bmatrix}$

15. $\begin{Bmatrix} 2x + 3y + z = 4 \\ x + 4y - z = 0 \\ 3x + y - z = -5 \end{Bmatrix}$

16. Graph the solution set of the system $\begin{Bmatrix} 9x^2 + 4y^2 \ge 36 \\ 3x + 2y \ge 6 \end{Bmatrix}$.

17. Find the domain of the function $y = f(x) = \sqrt{12 - 4x - x^2}$.

In Exercises 18–21, set up each problem algebraically, solve, and check.

18. The length of a rectangle is 7 cm more than its width. If the area of the rectangle is 60 sq. cm, find its dimensions.

19. It takes Darryl 2 hr longer to do a job than it does Jeannie. After Darryl has worked on the job for 1 hr, Jeannie joins him. Together they finish the job in 3 hr. How long would it take each of them working alone to do the entire job?

20. On a bicycle tour, it took Barbara 1 hr longer to ride 45 mi than it took Pat to ride 48 mi. If Pat averaged 3 mph more than Barbara, find the average speed of each.

21. How much pure alcohol should be mixed with 3 L of a 50% solution of alcohol to make a 70% solution?

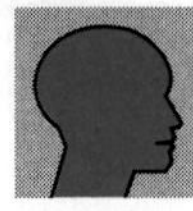

Critical Thinking and Problem-Solving Exercises

1. Can you predict what the next two numbers should be for the following list?

   5, 10, 7, 14, 9, 18, ___, ___

2. Kelly has four favorite English teachers, three favorite math teachers, and one physics teacher whom she likes very well. If she takes one English class, one math class, and one physics class next semester, how many different combinations of her favorite teachers are possible?

3. Four friends, Bud, Al, Ray, and Mike, went to a fast-food restaurant. They each ordered one item, and all the items were different. One ordered a taco, one a hamburger, one a slice of pizza, and one a hot dog. Neither Al nor Mike ordered pizza or a hot dog; Bud didn't order a hamburger or a hot dog, and Al ordered a taco. What did each order?

4. Michelle is a Head Start teacher, and she has 24 students in her class. The favorite activity of one-sixth of her students is painting; the favorite activity of one-half of the *remaining* students is playing ball; the favorite activity of one-half of the *remaining* students is working puzzles; the favorite activity of one-fifth of the *remaining* students is telling stories; the favorite activity of one student is working with clay. How many students do not have *any* favorite activity?

5. Chris has a total of 17 pets; some are fish, some are birds, and some are cats. Chris's pets have a total of 26 legs; 12 of his pets cannot fly. First solve this problem by guessing-and-checking (that is, *without* setting up a system of equations). Then solve the problem by setting up and solving a system of equations.

Sequences and Series (Optional)

CHAPTER

13

In this chapter, we introduce *sequences* and *series*. There are many applications of sequences and series in the sciences and in the mathematics of finance. For example, formulas for the calculation of interest, annuities, and mortgage loans are derived using series. The numbers in the logarithm tables in this book were calculated using series. This chapter is just a brief introduction to a very extensive and important part of mathematics.

13.1 Basic Definitions

Sequences

When we examine the following lists of numbers, it's easy to determine what the next number in each list should be:

$$30, 40, 50, 60, 70, \ldots$$

Recall that the three dots (called an ellipsis) indicate that the list never ends

$$7, 10, 13, 16, \ldots$$

These lists of numbers are examples of sequences of numbers. A **sequence** of numbers is a list of numbers in which the numbers follow a definite pattern and are arranged in a definite order. The numbers that make up the sequence are called the **terms** of the sequence.

A general sequence can be written as

$$a_1, a_2, a_3, \ldots, a_n, \ldots$$

where a_1 represents the *first* term of the sequence, a_2 represents the *second* term of the sequence, a_3 represents the *third* term of the sequence, and a_n represents the *n*th term, or the general term, of the sequence.

In general, the *subscript* of each term represents the *term number*; thus, a_7 represents the *seventh* term of the sequence.

Finite Sequences and Infinite Sequences

If, in counting the terms of a sequence, the counting comes to an end, the sequence is called a *finite* sequence. The last term of a finite sequence with n terms is represented by the symbol a_n. (Some authors represent the last term by ℓ.)

If, in counting the terms of a sequence, the counting never comes to an end, the sequence is called an *infinite* sequence.

EXAMPLE 1 Examples of finite and infinite sequences:

a. $0, 1, 2, 3, 4, 5, 6, 7, 8, 9$ is a *finite* sequence. (We know that 9 is the last number in the sequence because there is no ellipsis after the 9.) Each succeeding term is found by adding 1 to the preceding term. This sequence could also be written $0, 1, 2, \ldots, 9$.

b. $0, 1, 2, 3, \ldots$ is an *infinite* sequence. Each term is found by adding 1 to the preceding term.

c. $15, 10, 5, 0, -5, -10$ is a *finite* sequence. Each term is found by adding -5 to the preceding term.

Each term of a sequence is actually a function of n, where n, the term number, is any *natural number.* That is,

$$a_n = f(n)$$

This means that we can think of a sequence as being a *function* in which the *domain* is the set of natural numbers.

EXAMPLE 2 Examples of sequences written in functional form:

a. If $f(n) = \dfrac{n}{2}$

$$\begin{aligned} \text{then } a_1 &= f(1) = \tfrac{1}{2} \\ a_2 &= f(2) = \tfrac{2}{2} = 1 \\ a_3 &= f(3) = \tfrac{3}{2} \\ &\vdots \\ a_{11} &= f(11) = \tfrac{11}{2} \\ &\vdots \end{aligned}$$

b. If $f(n) = \dfrac{n+1}{4}$

$$\begin{aligned} \text{then } a_1 &= f(1) = \frac{1+1}{4} = \frac{2}{4} = \frac{1}{2} \\ a_2 &= f(2) = \frac{2+1}{4} = \frac{3}{4} \\ a_3 &= f(3) = \frac{3+1}{4} = \frac{4}{4} = 1 \\ &\vdots \\ a_{11} &= f(11) = \frac{11+1}{4} = \frac{12}{4} = 3 \\ &\vdots \end{aligned}$$

Series

A **series** is the indicated *sum* of a finite or infinite sequence of terms. It is a *finite* or an *infinite series* according to whether the number of terms is finite or infinite. An infinite series is usually written

$$a_1 + a_2 + a_3 + \cdots + a_n + \cdots$$

A **partial sum of a series** is the sum of a *finite* number of consecutive terms of the series. The symbol we use for the nth partial sum is S_n. (We use the symbol S_1 for the *first* "partial sum," even though it consists of one term only and, thus, is not really a sum.)

$S_1 = a_1$	The *first* partial sum
$S_2 = a_1 + a_2$	The *second* partial sum
$S_3 = a_1 + a_2 + a_3$	The *third* partial sum
$\vdots$	
$S_n = a_1 + a_2 + \cdots + a_n$	The nth partial sum
$\vdots$	

EXAMPLE 3 Examples of the sequence, series, and partial sums associated with the infinite *sequence* in which the nth term is $f(n) = \dfrac{1}{n}$:

The *sequence* is

$\frac{1}{1}, \frac{1}{2}, \frac{1}{3}, \ldots$ A sequence is a *list* of numbers

The *series* associated with this sequence is

$1 + \frac{1}{2} + \frac{1}{3} + \frac{1}{4} + \cdots$ A series is a *sum*

The first three *partial sums* are

$$\begin{aligned} S_1 &= 1 &&= 1 \\ S_2 &= 1 + \tfrac{1}{2} &&= \tfrac{3}{2} \\ S_3 &= 1 + \tfrac{1}{2} + \tfrac{1}{3} &&= \tfrac{11}{6} \end{aligned}$$

Summation Notation (or Sigma Notation) We often use the symbol Σ, the Greek (capital) letter sigma (standing for *sum*), to indicate a series, and we call the notation **summation notation**, or **sigma notation**. For example, summation notation for the general finite series $f(1) + f(2) + f(3) + \cdots + f(n)$ is as follows:

This number, called the upper limit, tells us what natural number to use *last* for n

$\sum_{i=1}^{n} f(i)$ Read "the sum of the f of i's as i goes from 1 to n"

This number, called the lower limit, tells us what integer to use *first* for n*

This variable is called the index of summation; any other letter (except n) could be used instead of i

By definition,

$$\sum_{i=1}^{n} f(i) = f(1) + f(2) + f(3) + \cdots + f(n)$$

Summation notation for the infinite series $g(1) + g(2) + g(3) + \cdots$ is as follows:

$$\sum_{k=1}^{\infty} g(k) = g(1) + g(2) + g(3) + \cdots$$

$\sum_{k=1}^{\infty} g(k)$ is read "the sum of the g of k's as k goes from 1 to infinity."

EXAMPLE 4 Write $\sum_{i=2}^{6} i^2$ in expanded form.

SOLUTION The function is i^2. We first substitute 2 for i, getting 2^2. Next we write a plus sign (remember, Σ stands for *sum*), substitute 3 for i, then write a plus sign, and so forth.

The *last* number we substitute for i is 6

$$\sum_{i=2}^{6} i^2 = 2^2 + 3^2 + 4^2 + 5^2 + 6^2$$ The right side is in expanded form

The *first* number we substitute for i is 2

EXAMPLE 5 Write $\sum_{k=3}^{\infty} \frac{1}{k}$ in expanded form.

SOLUTION Because there is no last term, we will write an ellipsis $(\cdots)$ after we've written a few terms of the series.

$$\sum_{k=3}^{\infty} \frac{1}{k} = \frac{1}{3} + \frac{1}{4} + \frac{1}{5} + \frac{1}{6} + \frac{1}{7} + \cdots$$

3 is the *first* number we substitute for k

*The lower limit of the sum is sometimes an integer rather than a natural number.

Exercises 13.1 Set I

In Exercises 1–4, determine, by inspection, the next three terms for each given sequence.

1. 10, 15, 20, __, __, __, ...
2. 8, 11, 14, __, __, __, ...
3. 15, 13, 11, __, __, __, ...
4. $1, \frac{4}{3}, \frac{5}{3}$, __, __, __, ...

In Exercises 5–10, use the given general term to write the terms specified.

5. $a_n = f(n) = n + 4;$ first three terms
6. $a_n = f(n) = 2n + 1;$ first four terms
7. $a_n = \dfrac{1 - n}{n};$ first four terms
8. $a_n = \dfrac{n(n - 1)}{2};$ first three terms
9. $a_n = n^2 - 1;$ first three terms
10. $a_n = n^3 + 1;$ first four terms

In Exercises 11–14, find the indicated partial sum by using the given general term.

11. Given $a_n = 2n - 3$, find S_4.
12. Given $a_n = \dfrac{2n - 1}{5 - n}$, find S_4.
13. Given $a_n = \dfrac{n - 1}{n + 1}$, find S_3.
14. Given $a_n = 3^n + 2$, find S_3.

In Exercises 15–18, write each series in expanded form.

15. $\sum_{i=1}^{5} \dfrac{i}{7}$
16. $\sum_{k=2}^{6} \dfrac{k - 1}{11}$
17. $\sum_{i=1}^{\infty} (2i + 1)^2$
18. $\sum_{k=3}^{\infty} k^3$

Writing Problems

Express the answers in your own words and in complete sentences.

1. Explain what a sequence is.
2. Explain the difference between a sequence and a series.

Exercises 13.1 Set II

In Exercises 1–4, determine, by inspection, the next three terms for each given sequence.

1. 20, 16, 12, __, __, __, ...
2. $\frac{1}{2}, \frac{3}{4}, 1$, __, __, __, ...
3. 17, 19, 21, __, __, __, ...
4. $4\frac{2}{3}, 4, 3\frac{1}{3}$, __, __, __, ...

In Exercises 5–10, use the given general term to write the terms specified.

5. $a_n = f(n) = 1 - 2n;$ first four terms
6. $a_n = \dfrac{n}{1 - 2n};$ first three terms
7. $a_n = f(n) = 3n - 1;$ first four terms
8. $a_n = f(n) = 1 - n;$ first four terms
9. $a_n = f(n) = n^2 + 1;$ first four terms
10. $a_n = \dfrac{2 - n}{1 + n^2};$ first three terms

In Exercises 11–14, find the indicated partial sum by using the given general term.

11. Given $a_n = 4 - n$, find S_5.
12. Given $a_n = \dfrac{1 - 2^n}{3}$, find S_4.
13. Given $a_n = \dfrac{2n - 1}{n}$, find S_3.
14. Given $a_n = \dfrac{n^2 - 1}{2^n}$, find S_3.

In Exercises 15–18, write each series in expanded form.

15. $\sum_{k=1}^{7} 3k$
16. $\sum_{i=1}^{5} \dfrac{2i - 1}{13}$
17. $\sum_{i=2}^{\infty} (5i + 1)^3$
18. $\sum_{k=0}^{\infty} \dfrac{1}{k + 1}$

13.2 Arithmetic Sequences and Series

Arithmetic Sequences

When each term (except the first term) of a sequence can be obtained from the preceding term by *adding* some fixed number, we call the sequence an **arithmetic sequence**, or an **arithmetic progression (AP)**. In other words, the *difference* between any two consecutive terms of an arithmetic sequence is some fixed number; that difference is called the **common difference**, d.

To determine whether a sequence is an arithmetic sequence, we *subtract* each term from the term that follows it. If all these differences are the same, the sequence is an arithmetic sequence.

EXAMPLE 1 Examples of determining whether or not a sequence is an arithmetic sequence:

a. $-8, -3, 2, 7, 12$

$$\left.\begin{aligned} 12 - 7 &= 5 \\ 7 - 2 &= 5 \\ 2 - (-3) &= 5 \\ -3 - (-8) &= 5 \end{aligned}\right\}$$ All these differences are the same (5)

The sequence *is* an arithmetic sequence, and the common difference is 5.

b. $1, 5, 9, 12, 16$

$$\left.\begin{aligned} 16 - 12 &= 4 \\ 12 - 9 &= 3 \end{aligned}\right\}$$ These differences are *not* the same; we need not find the remaining differences

The sequence is *not* an arithmetic sequence.

c. $(5x - 3), (7x - 4), (9x - 5), \ldots$

$$\left.\begin{aligned} (9x - 5) - (7x - 4) &= 2x - 1 \\ (7x - 4) - (5x - 3) &= 2x - 1 \end{aligned}\right\}$$ The differences are the same $(2x - 1)$

The sequence *is* an arithmetic sequence, and the common difference is $(2x - 1)$.

EXAMPLE 2 Write the first five terms of the infinite arithmetic sequence in which a_1 (the first term) is 3 and d (the common difference) is 2.

SOLUTION

$$a_1 = 3$$

$$a_2 = a_1 + d = 3 + 2 = 5$$

$$a_3 = a_2 + d = 5 + 2 = 7$$

$$a_4 = a_3 + d = 7 + 2 = 9$$

$$a_5 = a_4 + d = 9 + 2 = 11$$

Therefore, the first five terms are 3, 5, 7, 9, and 11. (The *entire* sequence is 3, 5, 7, 9, 11, ...)

EXAMPLE 3

Write a six-term (finite) arithmetic sequence in which the first term is 15 and the common difference is -7.

SOLUTION

$$a_1 = 15$$
$$a_2 = a_1 + d = 15 + (-7) = 8$$
$$a_3 = a_2 + d = 8 + (-7) = 1$$
$$a_4 = a_3 + d = 1 + (-7) = -6$$
$$a_5 = a_4 + d = -6 + (-7) = -13$$
$$a_6 = a_5 + d = -13 + (-7) = -20$$

Therefore, the sequence is 15, 8, 1, -6, -13, -20.

In general, an arithmetic sequence that has a_1 as its first term and d as its common difference has the following terms:

$$a_1 = a_1$$
$$a_2 = a_1 + d$$
$$a_3 = a_2 + d = (a_1 + d) + d = a_1 + 2d$$
$$a_4 = a_3 + d = (a_1 + 2d) + d = a_1 + 3d$$
$$a_5 = a_4 + d = (a_1 + 3d) + d = a_1 + 4d$$
$$\vdots$$
$$a_n = ?$$

We can discover a formula for determining a_n (for an arithmetic sequence) by comparing the terms listed above with the term *numbers*, as follows:

Term number	1	2	3	4	...	n
		↘	↘	↘		↘
Term	a_1,	$a_1 + 1d$,	$a_1 + 2d$,	$a_1 + 3d, \ldots,$		$a_1 + (n - 1)d$

We see that the coefficient of d is always 1 less than the term number. Therefore, we have the following formula to use for finding a_n:

The *n*th term of an arithmetic sequence

$$a_n = a_1 + (n - 1)d$$

where n represents a natural number, a_n represents the nth term of an arithmetic sequence, a_1 represents the first term, and d represents the common difference.

EXAMPLE 4

Find the twenty-first term of the arithmetic sequence in which $a_1 = 23$ and $d = -2$.

SOLUTION

$$a_n = a_1 + (n - 1)d$$
$$a_{21} = 23 + (21 - 1)(-2) \quad \text{Substituting 23 for } a_1\text{, 21 for } n\text{, and } -2 \text{ for } d$$
$$a_{21} = 23 + (-40) = -17$$

Therefore, the twenty-first term is -17.

EXAMPLE 5 Write the first five terms of the arithmetic sequence in which $a_7 = -10$ and $a_{12} = 5$.

SOLUTION We must first find a_1 and d. According to the formula, for $n = 7$, $a_7 = a_1 + (7 - 1)d = a_1 + 6d$. But it's given that $a_7 = -10$; therefore, one equation we can work with is $a_1 + 6d = -10$. Also, according to the formula, for $n = 12$, $a_{12} = a_1 + (12 - 1)d = a_1 + 11d$. But it's given that $a_{12} = 5$; therefore, another equation is $a_1 + 11d = 5$. We now solve this system of equations:

$$-1]\ \begin{cases} a_1 + 6d = -10 \\ a_1 + 11d = 5 \end{cases} \Rightarrow \begin{aligned} -a_1 - 6d &= 10 \\ a_1 + 11d &= 5 \end{aligned} \quad \text{We will add}$$

$$5d = 15$$

$$d = 3$$

To find a_1, we substitute 3 for d in the first equation:

$$a_1 + 6d = -10$$

$$a_1 + 6(3) = -10$$

$$a_1 = -28$$

Therefore, the first five terms of the sequence are $-28, -25, -22, -19$, and -16. (You might verify that a_7 is -10 and that a_{12} is 5.)

Arithmetic Series

An **arithmetic series** is the sum of the terms of an arithmetic sequence. An infinite arithmetic series can be written

$$a_1 + (a_1 + d) + (a_1 + 2d) + (a_1 + 3d) + \cdots + a_n + \cdots$$

and a finite arithmetic series can be written

$$a_1 + (a_1 + d) + (a_1 + 2d) + (a_1 + 3d) + \cdots + a_n$$

EXAMPLE 6 Find the sum of the first 100 integers.

SOLUTION The story is told that the famous German mathematician Carl Friedrich Gauss at the age of ten very quickly solved this problem when it was first presented in his arithmetic class.

We are to find

$$\sum_{i=1}^{100} i = S_{100} = \underset{\text{1st term}}{1} + \underset{\text{2nd term}}{2} + \underset{\text{3rd term}}{3} + \cdots + 98 + 99 + \underset{\text{100th term}}{100}$$

If we choose to, we can also write the equation as

$$S_{100} = 100 + 99 + 98 + \cdots + 3 + 2 + 1$$

Now let us add these two equations together, adding the right sides term by term. We then have

$$S_{100} + S_{100} = \underbrace{\underset{\text{1st term}}{101} + \underset{\text{2nd term}}{101} + \underset{\text{3rd term}}{101} + \cdots + 101 + 101 + \underset{\text{100th term}}{101}}_{\text{100 terms}}$$

Therefore, $2S_{100} = 101(100)$, and

$$\sum_{i+1}^{100} i = S_{100} = \frac{101(100)}{2} = 5{,}050$$

If the last term of a finite arithmetic series with n terms is a_n, then we *subtract d* from this term to get the next-to-last term, obtaining $(a_n - d)$; the term before *that* is $(a_n - 2d)$; and so on. Therefore, the *sum*, S_n, of the n terms of an arithmetic series can be written as

$$S_n = a_1 + (a_1 + d) + (a_1 + 2d) + \cdots + (a_n - 2d) + (a_n - d) + a_n$$

Now, by using the same reasoning as in Example 6, let's find a general formula for the sum of a finite arithmetic series. In the general case, we have

	1st term		2nd term		3rd term			(n − 2)nd term		(n − 1)st term		nth term
$S_n =$	a_1	+	$(a_1 + d)$	+	$(a_1 + 2d)$	+	$\cdots$ +	$(a_n - 2d)$	+	$(a_n - d)$	+	a_n
$S_n =$	a_n	+	$(a_n - d)$	+	$(a_n - 2d)$	+	$\cdots$ +	$(a_1 + 2d)$	+	$(a_1 + d)$	+	a_1

(Observe that the terms on the right side of the second equation are identical to the terms on the right side of the first equation; they are just in the reverse order.) Adding the two equations (and adding the right sides term by term), we have

$$2S_n = \underbrace{(a_1 + a_n) + (a_1 + a_n) + (a_1 + a_n) + \cdots + (a_1 + a_n) + (a_1 + a_n) + (a_1 + a_n)}_{n \text{ terms}}$$

The right side of the last equation has n terms of $(a_1 + a_n)$; therefore,

$$2S_n = n(a_1 + a_n)$$

or

$$S_n = \frac{n(a_1 + a_n)}{2}$$

If we substitute $a_1 + (n - 1)d$ for a_n, we also have the alternative form

$$S_n = \frac{n[2a_1 + (n - 1)d]}{2}$$

The sum of the first n terms of an arithmetic series

$$S_n = \frac{n(a_1 + a_n)}{2} \quad \text{or} \quad S_n = \frac{n[2a_1 + (n - 1)d]}{2}$$

where n represents a natural number, S_n represents the sum of the first n terms of an arithmetic series, a_1 represents the first term, a_n represents the last term, and d represents the common difference.

We recommend that you find $\sum_{i=1}^{100} i$ by using these formulas and compare your answers with the answer for Example 6.

Note that for $\sum_{i=1}^{100} i$, if we try to find n by subtracting 1 (the lower limit of the sum) from 100 (the upper limit of the sum), we get 99. Yet if we expand the sum to $1 + 2 + 3 + \cdots + 100$, we see that there are 100 terms, not 99 terms; therefore, n is not 99—it is 100. To use the lower and upper limits of the sum to find n, we have to *add 1* to the difference between those two limits (see Example 7).

EXAMPLE 7 Find $\sum_{i=7}^{92} 4i$ by using the formula for S_n.

SOLUTION The first term, a_1, is 4(7), or 28, and the last term, a_n, is 4(92), or 368; n is $(92 - 7 + 1)$, or 86. Substituting into the formula $S_n = \frac{n(a_1 + a_n)}{2}$, we have

$$\sum_{i=7}^{92} 4i = S_{86} = \frac{86(28 + 368)}{2} = 43(396) = 17{,}028$$

EXAMPLE 8 Given an arithmetic series with $a_1 = -8$, $a_n = 20$, and $S_n = 30$, find d and n.

SOLUTION When we substitute the given values for a_1, a_n, and S_n into the formula for S_n, we'll be able to solve for n:

$$S_n = \frac{n(a_1 + a_n)}{2} \Rightarrow 30 = \frac{n(-8 + 20)}{2} \Rightarrow 30 = \frac{n(12)}{2} \Rightarrow 30 = 6n \Rightarrow n = 5$$

Then we can substitute into the formula for a_n to find d:

$$a_n = a_1 + (n - 1)d \Rightarrow 20 = -8 + (5 - 1)d$$

$$20 = -8 + 4d$$

$$28 = 4d$$

$$7 = d$$

Therefore, $n = 5$ and $d = 7$.

Exercises 13.2
Set I

In Exercises 1–8, determine whether or not each sequence is an arithmetic sequence. If it is, find d, the common difference.

1. 3, 8, 13, 18
2. 7, 11, 15, 19
3. 7, 4, 1, −2, . . .
4. 9, 4, −1, −6, . . .
5. 4, $5\frac{1}{2}$, 7, 9
6. 3, $4\frac{1}{4}$, $5\frac{1}{2}$, $6\frac{1}{2}$
7. $2x - 1, x, 1, -x + 2$
8. $3 - 2x, 2 - x, 1, x$
9. Write the first four terms of the arithmetic sequence in which $a_1 = 5$ and $d = -7$.
10. Write the first five terms of the arithmetic sequence in which $a_1 = 4$ and $d = -5$.
11. Write the thirty-first term of this arithmetic sequence: −8, −2, 4, ...
12. Write the forty-first term of this arithmetic sequence: −5, −1, 3, ...
13. Write the eleventh term of this arithmetic sequence: $x, 2x + 1, 3x + 2, \ldots$.
14. Write the ninth term of this arithmetic sequence: $2z + 1, 3z, 4z - 1, \ldots$.
15. Write the first five terms of the arithmetic sequence in which $a_1 = 7$ and $a_5 = 31$.
16. Write the first six terms of the arithmetic sequence in which $a_1 = 6$ and $a_6 = 51$.
17. Find the sum of the even integers from 2 to 100, inclusive.
18. Find the sum of the odd integers from 1 to 99, inclusive.

In Exercises 19–24, verify that each series is an arithmetic series, and find the indicated sum by using the formula for S_n.

19. $\sum_{i=1}^{100} 2i$
20. $\sum_{i=1}^{50} (2i - 1)$
21. $\sum_{k=1}^{10} (3k - 4)$
22. $\sum_{k=1}^{20} (5 - 3k)$
23. $\sum_{i=3}^{8} 5i$
24. $\sum_{j=5}^{10} 6j$
25. Given an arithmetic sequence in which $a_6 = 15$ and $a_{12} = 39$, find a_1 and d.
26. Given an arithmetic sequence in which $a_5 = 12$ and $a_{14} = 57$, find a_1 and d.
27. Given an arithmetic series in which $a_1 = -5$, $d = 3$, and $a_n = 16$, find n and S_n.
28. Given an arithmetic series in which $a_1 = -7$, $d = 4$, and $a_n = 25$, find n and S_n.
29. Given an arithmetic series in which $a_1 = 5$, $a_n = 17$, and $S_n = 44$, find d and n.
30. Given an arithmetic series in which $a_1 = 3$, $a_n = 42$, and $S_n = 180$, find d and n.
31. Given an arithmetic series in which $d = \frac{3}{2}$, $n = 9$, and $S_n = -\frac{9}{4}$, find a_1 and a_n.
32. Given an arithmetic series in which $d = \frac{3}{4}$, $n = 7$, and $S_n = \frac{21}{4}$, find a_1 and a_n.

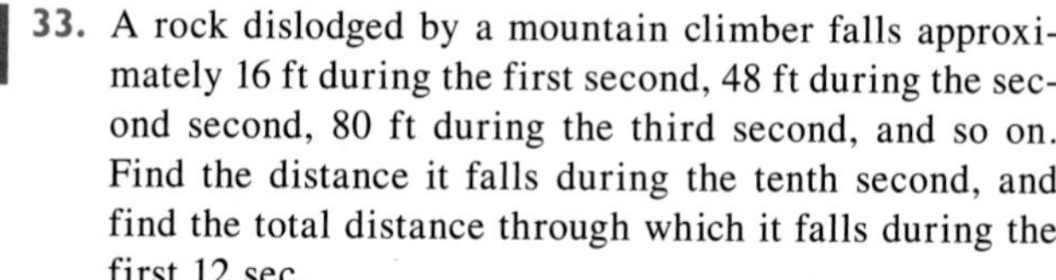

33. A rock dislodged by a mountain climber falls approximately 16 ft during the first second, 48 ft during the second second, 80 ft during the third second, and so on. Find the distance it falls during the tenth second, and find the total distance through which it falls during the first 12 sec.

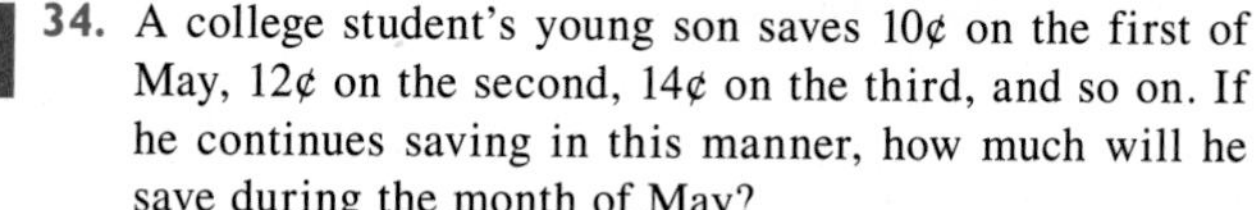

34. A college student's young son saves 10¢ on the first of May, 12¢ on the second, 14¢ on the third, and so on. If he continues saving in this manner, how much will he save during the month of May?

Writing Problems

Express the answer in your own words and in complete sentences.

1. Explain what an arithmetic sequence is.

Exercises 13.2 Set II

In Exercises 1–8, determine whether or not each sequence is an arithmetic sequence. If it is, find d, the common difference.

1. $-41, -24, -7, 10$
2. $13, 5, -3, -11, \ldots$
3. $-1, \frac{1}{2}, 1, 2\frac{1}{2}$
4. $-2 + x, -1 - x, -3x, 1 - 5x, \ldots$
5. $8, 4, 2, 1, \ldots$
6. $17, 14, 11, 8, \ldots$
7. $-9, -5, -1, 3, \ldots$
8. $\frac{13}{16}, \frac{1}{2}, \frac{3}{16}, -\frac{1}{8}, \ldots$
9. Write the first four terms of the arithmetic sequence in which $a_1 = -17$ and $d = -3$.
10. Write the first five terms of the arithmetic sequence in which $a_1 = 19$ and $d = -24$.
11. Write the twenty-ninth term of this arithmetic sequence: $-26, -22, -18, \ldots$.
12. Write the eighty-third term of this arithmetic sequence: $28, 34, 40, \ldots$.
13. Write the seventh term of this arithmetic sequence: $-x, x + 3, 3x + 6, \ldots$.
14. Write the eighth term of this arithmetic sequence: $-5x + 1, -3x - 1, -x - 3, \ldots$.
15. Write the first four terms of the arithmetic sequence in which $a_1 = 83$ and $a_6 = 48$.
16. Write the first five terms of the arithmetic sequence in which $a_1 = 7$ and $a_7 = -11$.
17. Find the sum of the even integers from 100 to 200, inclusive.
18. Find the sum of the odd integers from 101 to 201, inclusive.

In Exercises 19–24, verify that each series is an arithmetic series, and find the indicated sum by using the formula for S_n.

19. $\sum_{i=1}^{40} 3i$
20. $\sum_{j=1}^{35} (2j + 1)$
21. $\sum_{k=1}^{23} (2k - 7)$
22. $\sum_{i=3}^{31} (3 - 2i)$
23. $\sum_{k=7}^{22} 2k$
24. $\sum_{j=2}^{13} 8j$
25. Given an arithmetic sequence in which $a_5 = 46$ and $a_9 = 74$, find a_1 and d.
26. Given an arithmetic sequence in which $a_7 = 11$ and $a_{13} = 29$, find a_1 and d.
27. Given an arithmetic series in which $a_1 = 18$, $d = -3$, and $a_n = -6$, find n and S_n.
28. Given an arithmetic series in which $a_1 = -37$, $d = 4$, and $a_n = 7$, find n and S_n.
29. Given an arithmetic series in which $a_1 = 13$, $a_n = 49$, and $S_n = 217$, find d and n.
30. Given an arithmetic series in which $a_1 = 17$, $d = -12$, and $a_n = -103$, find n and S_n.
31. Given an arithmetic series in which $d = \frac{2}{3}$, $n = 15$, and $S_n = 35$, find a_1 and a_n.
32. Given an arithmetic series in which $a_1 = 7$, $a_n = -83$, and $S_n = -722$, find d and n.
33.  If we put one penny on the first square of a chessboard, three pennies on the second square, and five pennies on the third square, continuing in this way until all the squares are covered, how much money will there be on the board? (A chessboard has 64 squares.)
34. Jason saves \$1 the first week of the year, \$2 the second week of the year, \$3 the third week of the year, and so on. If he continues saving in this manner, how much will he save during one year?

13.3 Geometric Sequences and Series

Geometric Sequences

A **geometric sequence**, or **geometric progression (GP)**, is a sequence in which each term after the first is found by *multiplying* the preceding term by some fixed number, called the **common ratio**, r.

To determine whether a sequence is a geometric sequence, we *divide* each term by the preceding term. If all the resulting ratios are the same, the sequence is a geometric sequence, and that ratio is the common ratio.

EXAMPLE 1 Examples of determining whether or not a given sequence is a geometric sequence:

a. $24, -12, 6, -3$

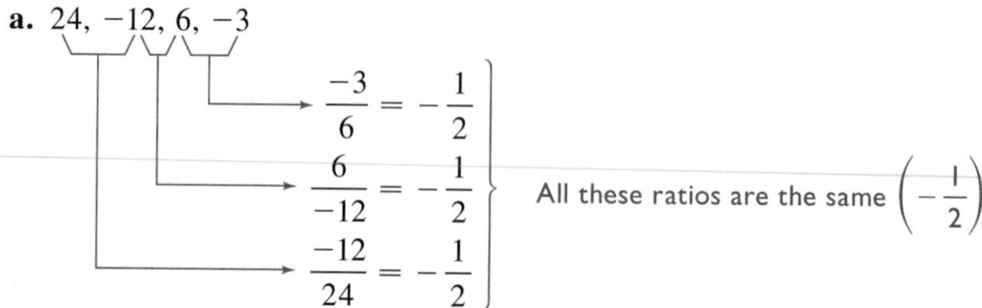

-3/6 = -1/2
6/-12 = -1/2
-12/24 = -1/2
All these ratios are the same (-1/2)

The sequence *is* a geometric sequence, and the common ratio is $-\frac{1}{2}$.

b. $36, 9, 3, 1$

$$\frac{1}{3} = \frac{1}{3},\quad \frac{3}{9} = \frac{1}{3},\quad \frac{9}{36} = \frac{1}{4}$$ These ratios are *not* all the same

The sequence is *not* a geometric sequence.

c. $-\frac{3x}{y^2}, \frac{9x^2}{y}, -27x^3, \ldots$

$$-27x^3 \div \frac{9x^2}{y} = \frac{-27x^3}{1} \cdot \frac{y}{9x^2} = -3xy$$

$$\frac{9x^2}{y} \div \left(-\frac{3x}{y^2}\right) = \frac{9x^2}{y} \cdot \left(-\frac{y^2}{3x}\right) = -3xy$$

These ratios are the same $(-3xy)$

The sequence *is* a geometric sequence, and the common ratio is $-3xy$.

EXAMPLE 2 Write the first four terms of the geometric sequence in which a_1 (the first term) is 5 and r (the common ratio) is 2.

SOLUTION

$$a_1 = 5$$

$$a_2 = a_1 r = 5 \cdot 2 = 10$$

$$a_3 = a_2 r = 10 \cdot 2 = 20$$

$$a_4 = a_3 r = 20 \cdot 2 = 40$$

Therefore, the first four terms of the geometric sequence are 5, 10, 20, and 40.

In general, a geometric sequence that has a_1 as the first term and r as the common ratio has the following terms:

$$a_1 = a_1$$

$$a_2 = a_1 r$$

$$a_3 = a_2\, r = (a_1 r)\, r = a_1 r^2$$

$$a_4 = a_3\, r = (a_1 r^2)\, r = a_1 r^3$$

$$a_5 = a_4\, r = (a_1 r^3)\, r = a_1 r^4$$

$$\vdots$$

$$a_n = ?$$

We can discover a formula for determining a_n (for a geometric sequence) by comparing the terms listed above with the term *numbers,* as follows:

Term number	1	2	3	4 ...	n
Term	a_1,	a_1r^1,	a_1r^2,	$a_1r^3,\ldots,$	a_1r^{n-1}

We see that the exponent of r is always 1 less than the term number. Therefore, we have the following formula to use for finding a_n:

The *n*th term of a geometric sequence

$$a_n = a_1r^{n-1}$$

where n represents a natural number, a_n represents the nth term of a geometric sequence, a_1 represents the first term, and r represents the common ratio.

EXAMPLE 3 Find the fifth term of the geometric sequence in which $a_1 = 18$ and $r = -\frac{1}{3}$.

SOLUTION

$$a_n = a_1r^{n-1}$$

$$a_5 = 18\left(-\tfrac{1}{3}\right)^4 \quad \text{Substituting 5 for } n, \text{ 18 for } a_1, \text{ and } -\tfrac{1}{3} \text{ for } r$$

$$a_5 = \frac{18}{1}\cdot\frac{1}{81} = \frac{2}{9}$$

Therefore, the fifth term is $\frac{2}{9}$.

EXAMPLE 4 Write the first five terms of the geometric sequence in which $a_2 = 12$ and $a_5 = 96$.

SOLUTION We must first find a_1 and r. According to the formula, for $n = 2$, $a_2 = a_1r^{2-1} = a_1r^1$. But it's given that $a_2 = 12$; therefore, one equation we can work with is

(1) $$a_1r = 12$$

Also, according to the formula, for $n = 5$, $a_5 = a_1r^{5-1} = a_1r^4$. But it's given that $a_5 = 96$; therefore, another equation is

(2) $$a_1r^4 = 96$$

If we now *divide* Equation 2 by Equation 1, we have

$$\frac{a_1r^4}{a_1r} = \frac{96}{12} \quad \text{Dividing Equation 2 by Equation 1}$$

$$r^3 = 8 \quad \text{Simplifying both sides of the equation}$$

$$r = 2 \quad \text{Taking the cube root of both sides of the equation}$$

To find a_1, we substitute 2 for r in Equation 1:

$$a_1r = 12$$

$$a_1(2) = 12$$

$$a_1 = 6$$

Therefore, the first five terms of the geometric sequence are 6, 12, 24, 48, and 96.

Geometric Series

A **geometric series** is the sum of the terms of a geometric sequence. An infinite geometric series can be written

$$\sum_{k=1}^{\infty} a_1 r^{k-1} = a_1 + a_1 r + a_1 r^2 + a_1 r^3 + \cdots + a_n + \cdots$$

The sum of a finite geometric series can be written

$$(1) \qquad S_n = \sum_{k=1}^{n} a_1 r^{k-1} = a_1 + a_1 r + a_1 r^2 + a_1 r^3 + \cdots + a_1 r^{n-1}$$

To find a *formula* for finding the sum of a finite geometric series, we can first multiply both sides of Equation 1 by r, then subtract the resulting equation *from* Equation 1, and finally solve for S_n:

$$\begin{aligned}
(1) \qquad S_n &= a_1 + a_1 r^1 + a_1 r^2 + \cdots + a_1 r^{n-1} \\
(2) \qquad rS_n &= \phantom{a_1 +{}} a_1 r^1 + a_1 r^2 + \cdots + a_1 r^{n-1} + a_1 r^n \\
\hline
S_n - rS_n &= a_1 \qquad\qquad\qquad\qquad\qquad\qquad - a_1 r^n
\end{aligned}$$

← Multiplying both sides of Equation 1 by r

← Subtracting Equation 2 from Equation 1

$$(1 - r)S_n = a_1(1 - r^n) \qquad \text{Factoring both sides}$$

$$S_n = \frac{a_1(1 - r^n)}{1 - r} \qquad \text{Dividing both sides by } 1 - r$$

Therefore, we have the following formula for finding S_n:

The sum of the first n terms of a geometric series

If $r \neq 1$, $\qquad S_n = \dfrac{a_1(1 - r^n)}{1 - r}$

where n represents a natural number, S_n represents the sum of the first n terms of a geometric series, a_1 represents the first term, and r represents the common ratio.

EXAMPLE 5 Write $\sum_{i=1}^{6} 24\left(\frac{1}{2}\right)^{i-1}$ in expanded form, and then find the sum of the series by using the formula.

SOLUTION

$$\begin{aligned}
\sum_{i=1}^{6} 24\left(\tfrac{1}{2}\right)^{i-1} &= 24\left(\tfrac{1}{2}\right)^{1-1} + 24\left(\tfrac{1}{2}\right)^{2-1} + 24\left(\tfrac{1}{2}\right)^{3-1} + 24\left(\tfrac{1}{2}\right)^{4-1} + 24\left(\tfrac{1}{2}\right)^{5-1} + 24\left(\tfrac{1}{2}\right)^{6-1} \\
&= 24\left(\tfrac{1}{2}\right)^{0} + 24\left(\tfrac{1}{2}\right)^{1} + 24\left(\tfrac{1}{2}\right)^{2} + 24\left(\tfrac{1}{2}\right)^{3} + 24\left(\tfrac{1}{2}\right)^{4} + 24\left(\tfrac{1}{2}\right)^{5} \\
&= 24 + \tfrac{24}{2} + \tfrac{24}{4} + \tfrac{24}{8} + \tfrac{24}{16} + \tfrac{24}{32} \\
&= 24 + 12 + 6 + 3 + \tfrac{3}{2} + \tfrac{3}{4}
\end{aligned}$$

To find the *sum*, we note that a_1 is 24 and that the common ratio is $\frac{1}{2}$.

$$S_n = \frac{a_1(1 - r^n)}{1 - r}$$

and $$S_6 = \frac{24\left[1 - \left(\frac{1}{2}\right)^6\right]}{1 - \frac{1}{2}} = \frac{24\left[1 - \left(\frac{1}{64}\right)\right]}{\frac{1}{2}} = \frac{24\left[\frac{63}{64}\right]}{\frac{1}{2}} = \frac{189}{4}$$

You might find the sum of the expanded series arithmetically and verify that the answer for Example 5 is correct.

EXAMPLE 6 Given the geometric series in which $r = -2$, $a_n = 80$, and $S_n = 55$, find a_1 and n.

SOLUTION We will need to use the formula for a_n *and* the formula for S_n. Let's start with the formula for S_n, substituting 55 for S_n and -2 for r:

$$S_n = \frac{a_1(1 - r^n)}{1 - r}$$

$$55 = \frac{a_1(1 - [-2]^n)}{1 - [-2]}$$

(1) $$55 = \frac{a_1 - a_1[-2]^n}{3}$$ Using the distributive property in the numerator

Next, we need to know what $a_1[-2]^n$ equals. Let's turn to the formula for a_n, substituting 80 for a_n and -2 for r:

$$a_n = a_1 r^{n-1}$$

(2) $$80 = a_1(-2)^{n-1}$$

If we multiply $(-2)^{n-1}$ by -2, we have $(-2)^{n-1}(-2)^1 = (-2)^{n-1+1} = (-2)^n$, which is what we need to use in Equation 1. Therefore, we will multiply both sides of Equation 2 by -2:

$$(-2)(80) = a_1(-2)^{n-1}(-2)$$

(3) $$-160 = a_1(-2)^n$$

We will now substitute -160 for $a_1(-2)^n$ in Equation 1:

$$55 = \frac{a_1 - [-160]}{3}$$

$$165 = a_1 + 160$$

$$a_1 = 5$$

To find n, we substitute 5 for a_1 in Equation 3:

$$-160 = 5(-2)^n$$

$$-32 = (-2)^n$$

$$(-2)^5 = (-2)^n$$

$$5 = n$$

Therefore, $a_1 = 5$ and $n = 5$.

EXAMPLE 7 Determine whether 1, 1, 2, 3, 5, 8, 13,... is an arithmetic sequence, a geometric sequence, or neither.

SOLUTION To see whether or not the sequence is arithmetic, we *subtract*:

1, 1, 2, 3, 5, 8, 13, ...

0 1 We need not go further

The differences are not all the same, so the sequence is not arithmetic.

To see whether or not the sequence is geometric, we *divide*:

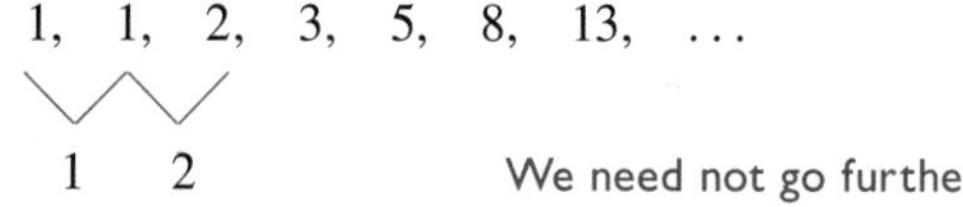

1, 1, 2, 3, 5, 8, 13, ...

1 2 We need not go further

The ratios are not all the same, so the sequence is not geometric.

The given sequence is neither arithmetic nor geometric. As a matter of fact, this is an important sequence known as a Fibonacci sequence.

Exercises 13.3
Set I

In Exercises 1–8, determine whether or not each sequence is a geometric sequence. If it is, find r, the common ratio.

1. 4, 12, 36, 108
2. 7, 14, 28, 56
3. $-5, 15, -45, 135, \ldots$
4. $-6, 24, -96, 384, \ldots$
5. $2, \frac{1}{2}, \frac{1}{8}, \frac{1}{16}$
6. $3, \frac{1}{2}, \frac{1}{8}, \frac{1}{48}$
7. $5x, 10xy, 20xy^2, 40xy^3, \ldots$
8. $4xz, 12xz^2, 36xz^3, 108xz^4, \ldots$

In Exercises 9–12, determine whether each sequence is an arithmetic sequence, a geometric sequence, or neither.

9. $36, 12, 4, \frac{4}{3}, \ldots$
10. $-18, -16, -13, -9, -4, \ldots$
11. $5, 7, 10, 14, 19, \ldots$
12. $-1, 2, -4, 8, -16, \ldots$
13. Write the first five terms of the geometric sequence in which $a_1 = 12$ and $r = \frac{1}{3}$.
14. Write the first five terms of the geometric sequence in which $a_1 = 8$ and $r = \frac{3}{2}$.
15. Write the seventh term of this geometric sequence: -9, $-6, -4, \ldots$.
16. Write the eighth term of this geometric sequence: -12, $-18, -27, \ldots$.
17. Write the eighth term of this geometric sequence: $16x$, $8xy, 4xy^2, \ldots$.
18. Write the seventh term of this geometric sequence: $27y$, $9x^2y, 3x^4y, \ldots$.
19. Write the five-term geometric sequences in which $a_1 = \frac{2}{3}$ and $a_5 = 54$. (There are two answers.)
20. Write the five-term geometric sequences in which $a_1 = \frac{3}{25}$ and $a_5 = 75$. (There are two answers.)

In Exercises 21–24, verify that each series is a geometric series, and find the indicated sum by using the formula for S_n.

21. $\sum_{i=1}^{5} 3\left(\frac{1}{3}\right)^{i-1}$
22. $\sum_{k=1}^{7} 5\left(\frac{1}{2}\right)^{k-1}$
23. $\sum_{j=1}^{6} \left(\frac{3}{4}\right)^{j}$

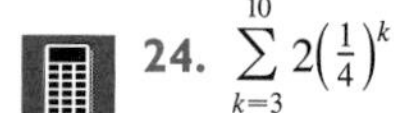

24. $\sum_{k=3}^{10} 2\left(\frac{1}{4}\right)^{k}$

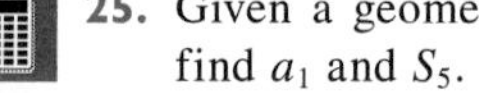

25. Given a geometric series in which $a_5 = 80$ and $r = \frac{2}{3}$, find a_1 and S_5.

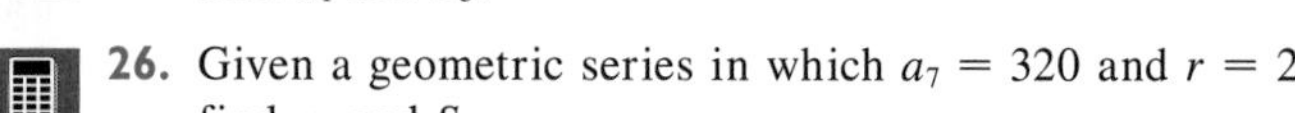

26. Given a geometric series in which $a_7 = 320$ and $r = 2$, find a_1 and S_5.

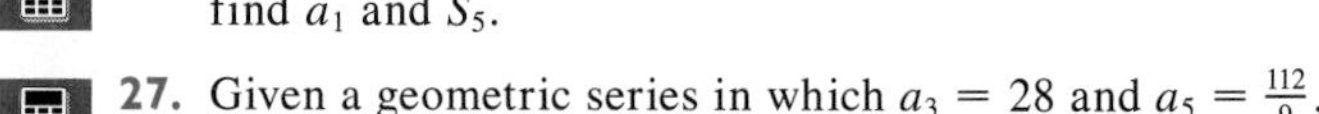

27. Given a geometric series in which $a_3 = 28$ and $a_5 = \frac{112}{9}$, find a_1, r, and S_5. (There are two answers.)
28. Given a geometric series in which $a_2 = 384$ and $a_4 = 24$, find a_1, r, and S_4. (There are two answers.)
29. Given a geometric series in which $r = \frac{1}{2}$, $a_n = 3$, and $S_n = 189$, find a_1 and n.
30. Given a geometric series in which $r = \frac{1}{3}$, $a_n = 4$, and $S_n = 160$, find a_1 and n.
31. A woman invested a certain amount of money that earned $1\frac{1}{5}$ times as much in the second year as in the first year, $1\frac{1}{5}$ times as much in the third year as in the second year, and so on. If the investment earned $22,750 in the first 3 years, how much would it earn in the fifth year?
32. A man invested a certain amount of money that earned $1\frac{1}{4}$ times as much in the second year as in the first year, $1\frac{1}{4}$ times as much in the third year as in the second year, and so on. If the investment earned $9,760 in the first 3 years, how much would it earn in the fifth year?

33. Suppose you were to take a job that paid 1¢ the first day, 2¢ the second day, and 4¢ the third day, with the pay continuing to increase in this manner for a month of 31 days.

a. How much would you make on the tenth day?

b. How much would you make on the thirty-first day?

c. What would be your total earnings for the month?

Writing Problems

Express the answers in your own words and in complete sentences.

1. Explain what a geometric sequence is.

2. Do some research on Fibonacci sequences, and write a short report about them.

Exercises 13.3 Set II

In Exercises 1–8, determine whether or not each sequence is a geometric sequence. If it is, find r, the common ratio.

1. $-11, -44, -176, -704$

2. $-6, 30, -150, 750, \ldots$

3. $1, -\frac{1}{2}, -\frac{1}{4}, -\frac{1}{8}$

4. $20ab, -5a^3b, \frac{5}{4}a^5b, -\frac{5}{16}a^7b, \ldots$

5. $36, 18, 9, \frac{9}{2}, \ldots$

6. $-16, -8, 0, 8, \ldots$

7. $12x^5, 4x^3, \dfrac{4}{3}x, \dfrac{4}{9x}, \ldots$

8. $2x^4, -8x^3, 16x^2, -32x, \ldots$

In Exercises 9–12, determine whether each sequence is an arithmetic sequence, a geometric sequence, or neither.

9. $15, 12, 9, 6, \ldots$

10. $-5, -1, -\frac{1}{5}, -\frac{1}{25}, \ldots$

11. $15, 13, 10, 6, 1, \ldots$

12. $16, -8, 4, -2, 1, \ldots$

13. Write the first five terms of the geometric sequence in which $a_1 = 4$ and $r = 3$.

14. Write the first four terms of the geometric sequence in which $a_1 = -9$ and $r = -\frac{1}{3}$.

15. Write the seventh term of this geometric sequence: $-\frac{25}{54}, \frac{5}{18}, -\frac{1}{6}, \ldots$.

16. Write the ninth term of this geometric sequence: $-16, -8, -4, \ldots$.

17. Write the eighth term of this geometric sequence: $\dfrac{24}{hk}, -\dfrac{12k}{h}, \dfrac{6k^3}{h}, \ldots$.

18. Write the seventh term of this geometric sequence: $-32, -8x, -2x^2, \ldots$.

19. Write the five-term geometric sequences in which $a_1 = -\frac{3}{4}$ and $a_5 = -12$. (There are two answers.)

20. Write the five-term geometric sequences in which $a_1 = \frac{1}{4}$ and $a_5 = 4$. (There are two answers.)

In Exercises 21–24, verify that each series is a geometric series, and find the indicated sum by using the formula for S_n.

21. $\sum_{i=1}^{4} 8\left(\frac{1}{5}\right)^{i-1}$

22. $\sum_{k=1}^{8} 2\left(\frac{1}{10}\right)^{k-1}$

23. $\sum_{j=1}^{7} \left(\frac{1}{2}\right)^{j}$

24. $\sum_{k=4}^{10} 7\left(\frac{1}{3}\right)^{k}$

25. Given a geometric series in which $a_5 = 40$ and $r = -\frac{2}{3}$, find a_1 and S_5.

26. Given a geometric series in which $a_4 = -\frac{10}{3}$ and $r = -\frac{1}{3}$, find a_1 and S_5.

27. Given a geometric series in which $a_3 = 16$ and $a_5 = 9$, find a_1, r, and S_5. (There are two answers.)

28. Given a geometric series in which $a_7 = 192$ and $a_4 = 24$, find a_1, r, and S_5.

29. Given a geometric series in which $r = -\frac{1}{2}$, $a_n = 5$, and $S_n = 55$, find a_1 and n.

30. Given a geometric series in which $a_n = 972$, $r = 3$, and $S_n = 1{,}456$, find a_1 and n.

31. If we put one penny on the first square of a chessboard, two pennies on the second square, and four pennies on the third square, continuing in this way until all squares are covered, how much money will there be on the board? (A chessboard has 64 squares.)

32. If it takes 1 sec for a certain type of microbe to split into two microbes, how long will it take a colony of 1,500 such microbes to exceed 6 million?

33. Suppose you were to take a job that paid 1¢ the first day, 3¢ the second day, and 9¢ the third day, with the pay continuing to increase in this manner for a month of 31 days.

a. How much would you make on the eighth day?

b. How much would you make on the thirty-first day?

c. What would be your total earnings for the month?

13.4 Infinite Geometric Series

Let's consider again the formula for finding the sum of n terms of a geometric series: If $r \neq 1$,

$$S_n = \frac{a_1(1 - r^n)}{1 - r}$$

In particular, let's consider r^n. If $|r| < 1$, then r^n gets smaller and smaller as n gets larger and larger. For example, suppose $r = \frac{1}{2}$. Then

$$\left(\frac{1}{2}\right)^1 = \frac{1}{2} = 0.5$$

$$\left(\frac{1}{2}\right)^2 = \frac{1}{4} = 0.25$$

$$\left(\frac{1}{2}\right)^3 = \frac{1}{8} = 0.125$$

$$\vdots$$

$$\left(\frac{1}{2}\right)^{10} = \frac{1}{1{,}024} \approx 0.001$$

$$\vdots$$

$$\left(\frac{1}{2}\right)^{20} \approx 0.000\,001$$

$$\vdots$$

$$\left(\frac{1}{2}\right)^{100} \approx 8 \times 10^{-31} \approx 0$$

We see that $\left(\frac{1}{2}\right)^n$ gets closer and closer to zero as n gets larger and larger. How does this affect S_{100}?

$$S_{100} \approx \frac{1}{2} + \frac{1}{4} + \frac{1}{8} + \cdots + \frac{1}{1{,}024} + \cdots + \underbrace{0}_{\left(\frac{1}{2}\right)^{98}} + \underbrace{0}_{\left(\frac{1}{2}\right)^{99}} + \underbrace{0}_{\left(\frac{1}{2}\right)^{100}}$$

In general, when $|r| < 1$, r^n is so close to zero that it contributes essentially nothing to the sum S_n when n is large.

Therefore, for very large values of n, the formula for S_n changes as follows:

This term contributes essentially nothing when n becomes infinitely large if $|r| < 1$

$$S_n = \frac{a_1(1 - r^n)}{1 - r} \approx \frac{a_1}{1 - r} \quad \text{if } |r| < 1 \text{ and } n \text{ is large}$$

The symbol S_∞ (read "S sub infinity") represents S_n when n becomes infinitely large, and we have the following formula:

The sum of an infinite geometric series

If $|r| < 1$, then

$$S_\infty = \sum_{i=1}^{\infty} a_1 r^{i-1} = \frac{a_1}{1 - r}$$

where S_∞ represents the sum of an infinite geometric series, a_1 represents the first term, and r represents the common ratio.

In higher-level mathematics courses, it is proved that S_∞ is *exactly* equal to, not approximately equal to, $\dfrac{a_1}{1 - r}$.

A Word of Caution The formula for S_∞ *cannot* be used if $|r| \geq 1$. The reason that $|r|$ must be less than 1 in $S_\infty = \dfrac{a_1}{1 - r}$ is that for $r = 1$, the fraction is undefined (the denominator equals zero), and for $|r| > 1$, the absolute values of the succeeding terms of the geometric series become larger and larger and S_∞ becomes unpredictable.

EXAMPLE 1 Evaluate the infinite geometric series $1 + \frac{1}{2} + \frac{1}{4} + \cdots$.

SOLUTION To find r, we divide the second term by the first, and $\frac{1}{2} \div 1 = \frac{1}{2}$, so $r = \frac{1}{2}$. Because $\left|\frac{1}{2}\right| < 1$, we can use the formula for S_∞.

$$S_\infty = \frac{a_1}{1 - r} = \frac{1}{1 - \frac{1}{2}} = \frac{1}{\frac{1}{2}} = 2$$

Therefore, $1 + \frac{1}{2} + \frac{1}{4} + \cdots = 2$.

Note You might evaluate the first few *partial sums* for this series and see that as you use more and more terms, the partial sums get closer and closer to 2. (For example, S_5 is closer to 2 than S_4 is, and so on.)

EXAMPLE 2 Evaluate $\sum_{i=1}^{\infty} 6\left(-\frac{2}{3}\right)^{i-1}$.

SOLUTION This series fits the pattern $\sum_{i=1}^{\infty} a_1 r^{i-1}$. Therefore, it is a geometric series, and $r = -\frac{2}{3}$, which is between -1 and 1. To find a_1, we let $i = 1$, finding that the first term is $6\left(-\frac{2}{3}\right)^{1-1} = 6\left(-\frac{2}{3}\right)^0$, or 6.

$$S_\infty = \frac{a_1}{1 - r} = \frac{6}{1 - \left(-\frac{2}{3}\right)} = \frac{6}{1 + \left(+\frac{2}{3}\right)} = \frac{6}{\frac{5}{3}} = \frac{18}{5}$$

Therefore, the sum is $\frac{18}{5}$.

EXAMPLE 3 Write the repeating decimal $0.\overline{25}$ as a common fraction reduced to lowest terms.

SOLUTION $0.\overline{25} = 0.252525\ldots$, and we can rewrite this expression as the *sum of an infinite number of terms*, as follows:

$$0.252525\ldots = 0.25 + 0.0025 + 0.000025 + \cdots$$

We can find the common ratio *and* see that this is a geometric series if we divide the second term by the first, divide the third term by the second, and so on: $\frac{0.0025}{0.25} = 0.01$; $\frac{0.000025}{0.0025} = 0.01; \ldots$.

Then $r = 0.01$, which is between -1 and 1, and $a_1 = 0.25$.

$$S_\infty = \frac{a_1}{1 - r} = \frac{0.25}{1 - 0.01} = \frac{0.25}{0.99} = \frac{25}{99}$$

Therefore, $0.\overline{25} = \frac{25}{99}$.

You can check the result of Example 3 by converting $\frac{25}{99}$ to decimal form; you will get the repeating decimal $0.252525\ldots$, or $0.\overline{25}$.

Exercises 13.4
Set I

In Exercises 1–6, find the sum of each geometric series.

1. $3 + 1 + \frac{1}{3} + \cdots$
2. $9 - 1 + \frac{1}{9} - \cdots$
3. $\frac{4}{3} + 1 + \frac{3}{4} + \cdots$
4. $10^{-1} + 10^{-2} + 10^{-3} + \cdots$
5. $-6 - 4 - \frac{8}{3} - \cdots$
6. $-49 - 35 - 25 - \cdots$

In Exercises 7–12, write each repeating decimal as a common fraction reduced to lowest terms.

7. $0.\overline{2}$
8. $0.\overline{21}$
9. $0.0\overline{54}$
10. $0.0\overline{39}$
11. $8.6\overline{4}$
12. $5.2\overline{6}$
13. Evaluate $\sum_{i=1}^{\infty} 5\left(-\frac{3}{4}\right)^{i-1}$.
14. Evaluate $\sum_{i=1}^{\infty} 2\left(-\frac{1}{5}\right)^{i-1}$.
15. A rubber ball is dropped from a height of 9 ft. Each time it strikes the floor, it rebounds to a height that is two-thirds the height from which it last fell. Find the total distance the ball travels *vertically* before coming to rest.
16. A ball bearing is dropped from a height of 10 ft. Each time it strikes the metal floor, it rebounds to a height that is three-fifths the height from which it last fell. Find the total distance the bearing travels *vertically* before coming to rest.
17. The first swing of a pendulum is 12 in. Each succeeding swing is nine-tenths as long as the preceding one. Find the total distance traveled by the pendulum before it comes to rest.

Writing Problems

Express the answer in your own words and in complete sentences.

1. Explain what it means to say that the *sum* of the infinite geometric sequence $1 + \frac{1}{2} + \frac{1}{4} + \cdots$ is 2.

Exercises 13.4
Set II

In Exercises 1–6, find the sum of each geometric series.

1. $5 - 1 + \frac{1}{5} - \cdots$
2. $10^{-2} + 10^{-4} + 10^{-6} + \cdots$
3. $\frac{6}{5} + 1 + \frac{5}{6} + \cdots$
4. $\frac{1}{4} + \frac{1}{8} + \frac{1}{16} + \cdots$
5. $2 - \frac{2}{3} + \frac{2}{9} - \cdots$
6. $4 + 2 + 1 + \cdots$

In Exercises 7–12, write each repeating decimal as a common fraction reduced to lowest terms.

7. $0.\overline{26}$
8. $0.0\overline{143}$
9. $0.\overline{70}$
10. $0.\overline{12}$
11. $0.00\overline{136}$
12. $3.7\overline{65}$
13. Evaluate $\sum_{i=1}^{\infty} 3\left(-\frac{1}{8}\right)^{i-1}$.
14. Evaluate $\sum_{i=1}^{\infty} 3\left(\frac{1}{2}\right)^{i-1}$.
15. When Bob stops his car suddenly, the 3-ft radio antenna oscillates back and forth. Each swing is three-fourths as great as the previous one. If the initial travel of the antenna tip is 4 in., how far will the tip travel before it comes to rest?
16. If a rabbit moves 10 yd in the first second and 5 yd in the second second, continuing forever to move one-half as far in each succeeding second as it did in the preceding second, how many yards does the rabbit travel?
17. The first swing of a pendulum is 10 in. Each succeeding swing is eight-ninths as long as the preceding one. Find the total distance traveled by the pendulum before it comes to rest.

Sections 13.1–13.4 REVIEW

Sequences 13.1 A **sequence** of numbers is a list of numbers in which the numbers follow a definite pattern and are arranged in a definite order. The numbers that make up the sequence are called the *terms* of the sequence. If, in counting the terms of a sequence, the counting comes to an end, the sequence is a *finite sequence*; if the counting never ends, the sequence is an *infinite sequence*. The usual notation for the *general term* of a sequence is a_n.

Series 13.1 A **series** is the indicated sum of a finite or infinite sequence of terms. It is a finite or an infinite series according to whether the number of terms is finite or infinite. A *partial sum of a series* is the sum of a finite number of consecutive terms of the series, beginning with the first term.

Sigma Notation 13.1 The symbol Σ is used to indicate the sum of a series. **Sigma notation** for the series $f(1) + f(2) + f(3) + \cdots + f(n)$ is $\sum_{i=1}^{n} f(i)$, and for the infinite series $g(1) + g(2) + g(3) + \cdots$ it is $\sum_{k=1}^{\infty} g(k)$.

Arithmetic Sequences and Series 13.2 An **arithmetic sequence** (or **arithmetic progression, AP**) is a sequence in which each term (after the first term) can be obtained by *adding* some fixed number, d, to the preceding term; d can be found by subtracting any term from the term that follows it. If the first term is a_1, then

The nth term is $a_n = a_1 + (n - 1)d$.

The sum of n terms is $S_n = \dfrac{n(a_1 + a_n)}{2} = \dfrac{n[2a_1 + (n - 1)d]}{2}$.

Geometric Sequences and Series 13.3 A **geometric sequence** (or **geometric progression, GP**) is a sequence in which each term (after the first term) can be obtained by *multiplying* the preceding term by some fixed number, r; r can be found by dividing any term by the term that precedes it. If the first term is a_1, then

The nth term is $a_n = a_1 r^{n-1}$.

The sum of n terms is $S_n = \dfrac{a_1(1 - r^n)}{1 - r}$ if $r \neq 1$.

The Sum of an Infinite Geometric Series 13.4

$$S_\infty = \frac{a_1}{1 - r} \quad \text{if} \quad |r| < 1$$

Sections 13.1–13.4 REVIEW EXERCISES Set I

In Exercises 1–5, determine whether each sequence is an arithmetic sequence, a geometric sequence, or neither. If it is an arithmetic sequence, name d, and if it is a geometric sequence, name r.

1. $7, 5, 3, \ldots$
2. $-2, -6, -18, \ldots$
3. $\frac{1}{2}, \frac{1}{4}, \frac{1}{6}, \ldots$
4. $\frac{3}{5}, -\frac{1}{5}, \frac{1}{15}, \ldots$
5. $3x - 2, 2x - 1, x, \ldots$
6. Write the first four terms of the sequence associated with the arithmetic series in which $a_n = f(n) = 2n - 6$. Then find d, a_{30}, and S_{30}.
7. Write the first three terms of the sequence associated with the geometric series in which $a_n = f(n) = \left(\frac{1}{2}\right)^n$. Then find S_5 and S_∞.
8. Given an arithmetic series in which $a_1 = -5$, $a_n = 7$, and $S_n = 16$, find d and n.
9. Given an arithmetic series in which $d = \frac{3}{2}$, $n = 7$, and $S_n = \frac{7}{2}$, find a_1 and a_n.

10. Given a geometric series in which $a_3 = \frac{9}{4}$ and $r = -\frac{2}{3}$, find a_1 and S_6.

11. Given a geometric series in which $a_3 = 8$ and $a_5 = \frac{32}{9}$, find a_1, r, and S_5. (There are two answers.)

12. Write $3.2\overline{845}$ as a common fraction reduced to lowest terms.

13. A rubber ball is dropped from a height of 8 ft. Each time it strikes the floor, it rebounds to a height that is three-fourths the height from which it last fell. Find the total distance the ball travels *vertically* before coming to rest.

14. Write the series $\sum_{i=2}^{6} 3\left(\frac{i+1}{11}\right)$ in expanded form.

15. Evaluate $\sum_{i=1}^{6} 2\left(\frac{1}{4}\right)^{i-1}$.

Sections 13.1–13.4 REVIEW EXERCISES Set II

Name ______________________

ANSWERS

1. ______

2. ______

3. ______

4. ______

5. ______

6. ______

7. ______

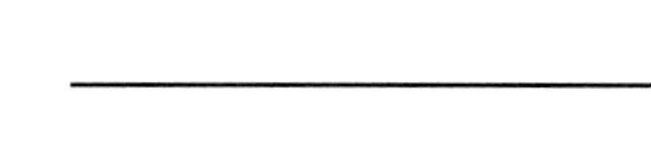

8. ______

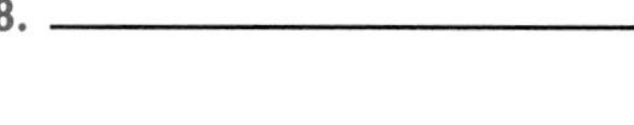

In Exercises 1–5, determine whether each sequence is an arithmetic sequence, a geometric sequence, or neither. If it is an arithmetic sequence, name d, and if it is a geometric sequence, name r.

1. $23, 4, -15, \ldots$

2. $-5, -35, -245, \ldots$

3. $\frac{1}{15}, \frac{1}{20}, \frac{1}{25}, \ldots$

4. $\frac{10}{3}, -5, \frac{15}{2}, \ldots$

5. $6 - x, 3, x, \ldots$

6. Write the first four terms of the sequence associated with the arithmetic series in which $a_n = f(n) = 5n - 2$. Then find d, a_{25}, and S_{25}.

7. Write the first three terms of the sequence associated with the geometric series in which $a_n = f(n) = \left(\frac{2}{5}\right)^n$. Then find S_5 and S_∞.

8. Given an arithmetic series in which $a_1 = 2$, $a_n = -1$, and $S_n = 2\frac{1}{2}$, find d and n.

9. Given an arithmetic series in which $d = -\frac{4}{3}$, $n = 7$, and $S_n = 7$, find a_1 and a_n.

9. ____________

____

10. ____________

10. Given a geometric series in which $a_4 = 27$ and $r = -\frac{3}{2}$, find a_1 and S_6.

____

11. ____________

____

11. Given a geometric series in which $a_3 = 18$ and $a_5 = 32$, find a_1, r, and S_5. (There are two answers.)

____

____

____

12. Write $2.9\overline{048}$ as a common fraction reduced to lowest terms.

____

12. ____________

13. ____________

13. A mine's output in the first month was \$10,000 in gold. If each succeeding monthly output is $\frac{9}{10}$ the previous month's output, what is the most that this mine's total output can be?

14. ____________

15. ____________

14. Write the series $\sum_{i=1}^{4} 3\left(\frac{i^2}{2}\right)$ in expanded form.

15. Evaluate $\sum_{i=1}^{8} 5\left(\frac{1}{3}\right)^{i-1}$.

Chapter 13 DIAGNOSTIC TEST

The purpose of this test is to see how well you understand sequences and series. We recommend that you work this diagnostic test *before* your instructor tests you on this chapter. Allow yourself about 50 minutes.

Complete solutions for all the problems on this test, together with section references, are given in the answer section at the end of the book. For the problems you do incorrectly, we suggest that you study the sections referred to.

1. Given $a_n = \dfrac{2n-1}{n}$, find S_4.

2. Determine whether each sequence is an arithmetic sequence, a geometric sequence, or neither. If it is an arithmetic sequence, name d, and if it is a geometric sequence, name r.

   a. $8, -20, 50, \ldots$

   b. $\frac{1}{2}, \frac{3}{4}, 1, \frac{5}{4}, \frac{3}{2}, \ldots$

   c. $2x - 1, 3x, 4x + 2, \ldots$

   d. $\dfrac{c^4}{16}, -\dfrac{c^3}{8}, \dfrac{c^2}{4}, \ldots$

3. a. Write the first five terms of the arithmetic sequence in which $a_1 = x + 1$ and $d = x - 1$.

   b. Write the five-term arithmetic sequence in which $a_1 = 2$ and $a_5 = -2$.

   c. Write the fifteenth term of the following arithmetic sequence: $1 - 6h, 2 - 4h, 3 - 2h$.

4. a. Given an arithmetic sequence in which $a_3 = 1$ and $a_7 = 2$, find a_1 and d.

   b. Given an arithmetic sequence in which $a_1 = 10$, $a_n = -8$, and $S_n = 7$, find d and n.

5. Write the first five terms of the geometric sequence in which $a_1 = \dfrac{c^4}{16}$ and $r = -\dfrac{2}{c}$.

6. Given a geometric series in which $a_2 = -18$ and $a_4 = -8$, find a_1, r, and S_4. (There are two answers.)

7. Given a geometric series in which $r = -\frac{2}{3}$, $a_n = -16$, and $S_n = 26$, find a_1 and n.

8. Find the sum of $8, -4, 2, -1, \ldots$.

9. Write $3.0\overline{3}$ as a common fraction reduced to lowest terms.

10. A ball bearing is dropped from a height of 6 ft. Each time it strikes the floor, it rebounds to a height that is two-thirds the height from which it last fell. Find the total distance the ball bearing travels *vertically* before it comes to rest.

Chapters 1–13 CUMULATIVE REVIEW EXERCISES

In Exercises 1 and 2, perform the indicated operations, and write the answers using only positive exponents or no exponents.

1. $\left(\dfrac{x^3y^{-4}z^{-1}}{x^{-2}z^{-3}}\right)^{-2}$

2. $(32a^{-1/2} \cdot a^{1/3})^{-6/5}$

In Exercises 3 and 4, perform the indicated operations, and write the answers in simplest radical form. (Assume $x > 0$.)

3. $\sqrt{27x^3y^4} + 3xy^2\sqrt{12x} - x^2y^2\sqrt{\dfrac{3}{x}}$

4. $\dfrac{15}{\sqrt{7} - \sqrt{2}}$

5. Perform the indicated operation, and write the answer in the form $a + bi$: $\dfrac{6 - 5i}{6 + 5i}$.

6. Reduce $\dfrac{x^3 + 2x^2 - x - 2}{x^3 + 8}$ to lowest terms.

In Exercises 7–10, perform the indicated operations, and write the answers in simplest form.

7. $\dfrac{x}{x-3} - \dfrac{9}{x+3} - \dfrac{1}{2x^2 - 18}$

8. $(x^4 + 3x^3 - 6x^2 - 2x + 1) \div (x^2 - x + 1)$

9. $\dfrac{x+1}{3x^2 + 14x - 5} \div \dfrac{2x^2 - x - 3}{6x^2 - 11x + 3}$

★10. $(x + 2)^5$

In Exercises 11–19, find the solution set of each equation or inequality or system of equations.

11. $\dfrac{6-x}{x^2 - 4} - \dfrac{x}{x+2} = 2$

12. $\log(x + 3) + \log(x - 2) = \log 6$

13. $x = \sqrt{x + 9} + 3$

14. $x^2 + 2x - 15 \ge 0$

15. a. $\log_5 \frac{1}{25} = x$ b. $\log_b 4 = \frac{1}{2}$

16. $x^2 - 4x = 8$

17. $\begin{Bmatrix} 2x + 4y = 0 \\ 5x - 3y = 13 \end{Bmatrix}$

18. $\begin{Bmatrix} x + y + z = 1 \\ 2x + y - 2z = -4 \\ x + y + 2z = 3 \end{Bmatrix}$

19. $\left|\dfrac{x}{2} + 3\right| < 4$

20. Graph $y = x^2 - 2x - 3$.

21. Graph the solution set of the system $\begin{Bmatrix} 2x - 5y \geq 10 \\ 5x + y < 5 \end{Bmatrix}$.

For Exercises 22 and 23, use the function $y = f(x) = \dfrac{2}{3x - 5}$.

22. a. Find the domain. b. Find $f(x + h)$.

23. Find the inverse function, $f^{-1}(x)$.

For Exercises 24 and 25, use the points (3, −4) and (1, 6).

24. a. Find the distance between the points.

b. Find the slope of the line through the points.

25. Write the equation of the line through the points.

26. Given an arithmetic series in which $a_1 = -7$, $a_n = 11$, and $S_n = 10$, find d and n.

27. Given a geometric series in which $a_4 = -1$ and $r = -\frac{2}{7}$, find a_1 and S_5.

In Exercises 28–30, set up each problem algebraically, solve, and check.

28. Find three consecutive integers such that the product of the first two is 10 more than 4 times the third.

29. Coffee that is worth \$3.50 per lb is to be mixed with coffee that is worth \$4.25 per lb to obtain 30 lb of a blend worth \$4.00 per lb. How much of each type of coffee should be used?

30. Roberta and John left an intersection at the same time. Roberta, jogging west, was jogging 3 mph faster than John, who was jogging north. Twenty minutes later, they were 5 mi apart. What were their rates?

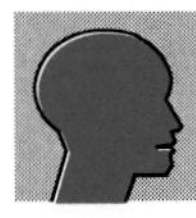

Critical Thinking and Problem-Solving Exercises

1. If you are to receive change of 31¢ in some combination of quarters, dimes, nickels, and pennies, how many different combinations are possible? (For example, 1 dime and 21 pennies counts as one combination.)

2. At noon, Nick finished a batch of cookies for his grandchildren. By 3 P.M., half of the cookies had been eaten; by 4 P.M., one-third of the remaining cookies had been eaten; by 5 P.M., one-fourth of the remaining cookies were gone; by 6 P.M., one-fifth of the remaining cookies had disappeared; by 7 P.M., one-sixth of the remaining cookies were gone, and there were only ten cookies left. How many cookies did Nick bake?

3. Find the next three numbers for this sequence:

3, 4, 7, 11, 18, ____, ____, ____

4. Carol, Charlie, Marilyn, and Richard, four college students, studied at the same table in the library. Each had a different major, and each was in the library to study for a different subject. The subjects being studied were English, music, French, and biology. Using the following clues, determine who sat in which chair, what the major of each student was, and what subject each student was studying in the library.

One student is a business major.

Carol is a geography major.

Charlie sat in seat 3.

The geography major was not sitting in seat 3 or 4.

The physics major sat opposite the math major, who was studying French.

Richard sat in an even-numbered seat, opposite the student who was studying music.

The physics major sat in seat 1 and studied English.

Figure for Exercise 4

Proofs

APPENDIX A

The negative exponent rule: Form 2

$$\frac{1}{x^{-n}} = x^n \qquad \text{(Section 2.2)}$$

PROOF $\dfrac{1}{x^{-n}} = \dfrac{1}{\dfrac{1}{x^n}} = 1 \div \dfrac{1}{x^n} = 1 \cdot \dfrac{x^n}{1} = x^n$ ■

The negative exponent rule: Form 3

$$\left(\frac{x}{y}\right)^{-n} = \left(\frac{y}{x}\right)^n \qquad \text{(Section 2.2)}$$

PROOF $\left(\dfrac{x}{y}\right)^{-n} = \dfrac{1}{\left(\dfrac{x}{y}\right)^n} = 1 \div \left(\dfrac{x}{y}\right)^n = 1 \div \dfrac{x^n}{y^n} = 1 \cdot \dfrac{y^n}{x^n} = \dfrac{y^n}{x^n} = \left(\dfrac{y}{x}\right)^n$ ■

PROPERTY 1 for absolute values

If $k \ge 0$ and if $|X| = k$, then $X = k$ or $X = -k$, where X represents any algebraic expression (Section 3.4).

PROOF

Case I, $X \ge 0$:

If $X \ge 0$, $|X| = X$. Then $|X| = k \Rightarrow X = k$.

Case II, $X < 0$:

If $X < 0$, $|X| = -X$. Then $|X| = k \Rightarrow -X = k$. But if $-X = k$, then $X = -k$.

The final answer is the *union* of the solution sets of Cases I and II. Therefore, if $|X| = k$, then $X = k$ or $X = -k$. ■

PROPERTY 2 for absolute values

If k is a positive real number,

1. if $|X| < k$, then $-k < X < k$, and

2. if $|X| \le k$, then $-k \le X \le k$,

where X represents any algebraic expression (Section 3.4).

PROOF OF 1

Case I, $X \ge 0$:

If $X \ge 0$, $|X| = X$. Therefore, $|X| < k \Rightarrow X < k$. We must find the intersection between $X \ge 0$ and $X < k$. Because $k > 0$, the intersection is $0 \le X < k$:

Case II, $X < 0$:

If $X < 0$, $|X| = -X$. Therefore, $|X| < k \Rightarrow -X < k$. If $-X < k$, then $X > -k$. We must find the intersection between $X < 0$ and $X > -k$. Because $k > 0$, the intersection is $-k < X < 0$:

The final answer is the *union* of the solution sets of Cases I and II, which is $-k < X < k$:

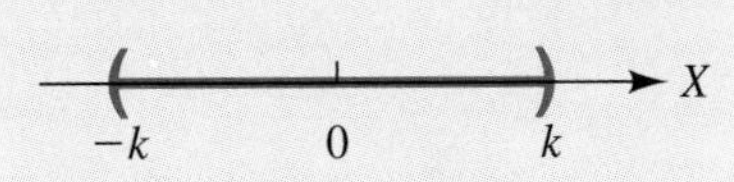

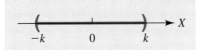

Therefore, if $|X| < k$, then $-k < X < k$. ■

The proof of 2 is similar to the proof of 1 and will not be shown.

PROPERTY 3 for absolute values

> If k is a positive real number,
>
> **1.** if $|X| > k$, then $X > k$ or $X < -k$, and
>
> **2.** if $|X| \geq k$, then $X \geq k$ or $X \leq -k$,
>
> where X represents any algebraic expression (Section 3.4).

PROOF OF 1

Case I, $X \geq 0$:

If $X \geq 0$, $|X| = X$. Therefore, $|X| > k \Rightarrow X > k$. We must find the intersection between $X \geq 0$ and $X > k$. Because $k > 0$, the intersection is $X > k$:

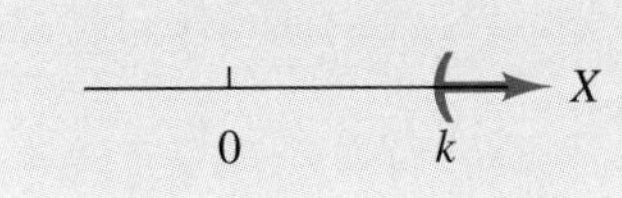

Case II, $X < 0$:

If $X < 0$, $|X| = -X$. Therefore, $|X| > k \Rightarrow -X > k$. If $-X > k$, then $X < -k$. We must find the intersection between $X < 0$ and $X < -k$. Because $k > 0$, the intersection is $X < -k$:

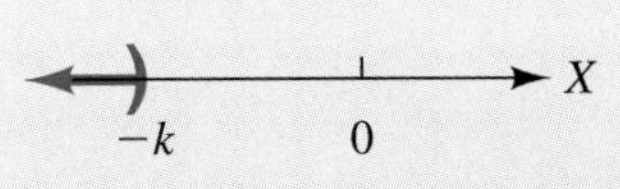

The final answer is the *union* of the solution sets of Cases I and II, which is $\{X \mid X > k\} \cup \{X \mid X < -k\}$:

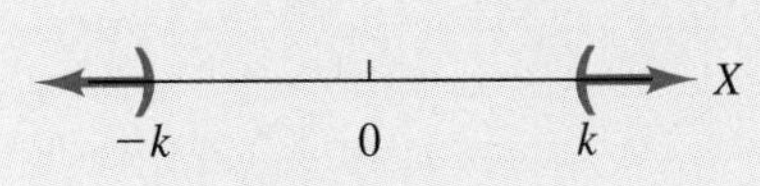

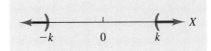

Therefore, if $|X| > k$, then $X > k$ or $X < -k$. ■

The proof of 2 is similar to the proof of 1 and will not be shown.

The remainder theorem

> If $f(x)$ is a polynomial function and if $f(x)$ is divided by $(x - a)$, then the remainder equals $f(a)$ (Section 9.6).

PROOF Let us call the quotient $q(x)$ and the remainder R. Then if $f(x)$ is a polynomial of degree n, $q(x)$ is a polynomial of degree $n - 1$, and $f(x) = (x - a)\,q(x) + R$ must be a true statement for all x. Let $x = a$. We then have

$$f(a) = (a - a)q(a) + R$$

But since $a - a = 0$, $f(a) = R$. Therefore, if $f(x)$ is divided by $x - a$, $R = f(a)$. ■

The axis of symmetry of a parabola

> The equation of the axis of symmetry of the parabola $y = ax^2 + bx + c$ is $x = -\dfrac{b}{2a}$ (Section 11.5).

PROOF For simplicity, let us assume that the parabola crosses the x-axis. The x-intercepts are $\dfrac{-b + \sqrt{b^2 - 4ac}}{2a}$ and $\dfrac{-b - \sqrt{b^2 - 4ac}}{2a}$. The axis of symmetry passes through a point whose x-coordinate is the average of these two values—that is, the x-coordinate of a point on the axis of symmetry is

$$\frac{\dfrac{-b + \sqrt{b^2 - 4ac}}{2a} + \dfrac{-b - \sqrt{b^2 - 4ac}}{2a}}{2} = \frac{\dfrac{-2b}{2a}}{2} = -\frac{b}{2a}$$

The equation of the vertical line that passes through the point $\left(-\dfrac{b}{2a}, 0\right)$ is $x = -\dfrac{b}{2a}$. ■

The Use of Tables for Logarithms; Computations with Logarithms

APPENDIX B

B.1 Common Logarithms: The Characteristic

We first consider logarithms of powers of 10. *The logarithm of a number is the exponent to which the base must be raised to give that number.*

$\log_{10} 10^{3} = 3$ The base 10 must be raised to the exponent 3 to give the number 10^3

$\log_{10} 10^{2} = 2$ The base 10 must be raised to the exponent 2 to give the number 10^2

$\log_{10} 10^{1} = 1$ And so on

$\log_{10} 10^{0} = 0$

$\log_{10} 10^{-1} = -1$

$\log_{10} 10^{-2} = -2$

$\vdots$

In general,

$$\log_{10} 10^k = k$$

The logarithm of a number that is *not* an integral power of 10 is made up of two parts:

1. *An integer part called the* **characteristic**. The characteristic of the logarithm of a number equals the exponent on the 10 when that number is written in scientific notation.
2. *A decimal part called the* **mantissa** (found in Table II, Appendix C). *The mantissas in the table are never negative.* We discuss finding the mantissa in the next section.

Finding the Characteristic

Although the characteristic of the logarithm of a number is the exponent of 10 when the number is written in scientific notation, it is better for computational purposes to write *negative* characteristics as the *difference* of two numbers, as shown in Examples 1b, 1c, 1d, and 1f.

EXAMPLE 1 Examples of writing the characteristic of the logarithm of a number:

	Number	*Number in scientific notation*	*Characteristic*	*Also written as*
a.	76.3	7.63×10^1	1	1. ▭
b.	0.506	5.06×10^{-1}	−1	9. ▭ − 10
c.	0.0932	9.32×10^{-2}	−2	8. ▭ − 10
d.	0.000 004 79	4.79×10^{-6}	−6	4. ▭ − 10
e.	1.83×10^4	1.83×10^4	4	4. ▭
f.	236×10^{-9}	2.36×10^{-7}	−7	3. ▭ − 10

The mantissa will go here

The characteristic of the logarithm of a number that is written in standard decimal notation can also be found as follows:

If the number is greater than or equal to 1, the characteristic of its logarithm is *one less than the number of digits to the left of the decimal point.*

If the number is less than 1, the characteristic of its logarithm is negative, and the absolute value of the characteristic equals the *number of decimal places between the actual decimal point and standard position for the decimal point.*

Exercises B.1
Set I

Write the characteristic of the logarithm of each number.

1. 386
2. 27
3. 5.67
4. 30.4
5. 0.516
6. 0.089
7. 93,000,000
8. 186,000
9. 0.0000806
10. 0.000777
11. 78,000
12. 1,400
13. 2.06×10^5
14. 3.55×10^4
15. 7.14×10^{-3}
16. 8.96×10^{-5}

Exercises B.1
Set II

Write the characteristic of the logarithm of each number.

1. 784
2. 8.99
3. 0.314
4. 0.000578
5. 2.56×10^4
6. 3.14×10^{-3}
7. 7.0005
8. 0.000109
9. 3,480
10. 0.00437
11. 2,300,000
12. 8.32
13. 2.78×10^3
14. 8.62×10^6
15. 3.58×10^{-4}
16. 9.51×10^{-2}

B.2 Common Logarithms: The Mantissa

In this section, we discuss common logarithms and the use of the Table of Common Logarithms (Table II, Appendix C). Table II gives the *mantissas* (the decimal parts) of the logarithms of the numbers from 1.00 to 9.99. Except for the first one (0.0000), the mantissas are all approximations. *The mantissas in the table are never negative.*

Tables of common logarithms come in different accuracies. Some tables have mantissas rounded off to four decimal places, others to five places, and so on. Table II is a four-place table. If you use other than a four-place table to solve the problems in this appendix, you may get slightly different answers.

Finding the Mantissa

Hereafter, *when the notation log N is used, the base is understood to be 10*; that is, $\log N = \log_{10} N$.

EXAMPLE 1 Find the mantissa for log 5.74.

SOLUTION Shown below is a portion of Table II, Appendix C. The number 5.74 is already in scientific notation, so we look up $5_\wedge 74$ in the table. Notice that the caret shows us where the decimal point belongs. We look down the left-hand column for the first two digits ($5_\wedge 7$) and across the top row for the third digit (4).

The *first two digits* of 5.7 4 — The *third digit* of 5.7 4

N	0	1	2	3	4	5	6	7	8	9
$5_\wedge 5$	0.7404	0.7412	0.7419	0.7427	0.7435	0.7443	0.7451	0.7459	0.7466	0.7474
$5_\wedge 6$	0.7482	0.7490	0.7497	0.7505	0.7513	0.7520	0.7528	0.7536	0.7543	0.7551
$5_\wedge 7$	0.7559	0.7566	0.7574	0.7582	0.7589	0.7597	0.7604	0.7612	0.7619	0.7627
$5_\wedge 8$	0.7634	0.7642	0.7649	0.7657	0.7664	0.7672	0.7679	0.7686	0.7694	0.7701
$5_\wedge 9$	0.7709	0.7716	0.7723	0.7731	0.7738	0.7745	0.7752	0.7760	0.7767	0.7774
$6_\wedge 0$	0.7782	0.7789	0.7796	0.7803	0.7810	0.7818	0.7825	0.7832	0.7839	0.7846
$6_\wedge 1$	0.7853	0.7860	0.7868	0.7875	0.7882	0.7889	0.7896	0.7903	0.7910	0.7917

Therefore, 0.7589 is the *mantissa* of log 5.74.

EXAMPLE 2 Find log 8,360.

SOLUTION We first write 8,360 in scientific notation.

$$\log 8{,}360 = \log(8.36 \times 10^3) = \log 8.36 + \log 10^3 = \log 8.36 + 3$$

We find log 8.36 in Table II, a portion of which is shown below.

N	0	1	2	3	4	5	6	7	8	9
$7_\wedge 8$	0.8921	0.8927	0.8932	0.8938	0.8943	0.8949	0.8954	0.8960	0.8965	0.8971
$7_\wedge 9$	0.8976	0.8982	0.8987	0.8993	0.8998	0.9004	0.9009	0.9015	0.9020	0.9025
$8_\wedge 0$	0.9031	0.9036	0.9042	0.9047	0.9053	0.9058	0.9063	0.9069	0.9074	0.9079
$8_\wedge 1$	0.9085	0.9090	0.9096	0.9101	0.9106	0.9112	0.9117	0.9122	0.9128	0.9133
$8_\wedge 2$	0.9138	0.9143	0.9149	0.9154	0.9159	0.9165	0.9170	0.9175	0.9180	0.9186
$8_\wedge 3$	0.9191	0.9196	0.9201	0.9206	0.9212	0.9217	0.9222	0.9227	0.9232	0.9238
$8_\wedge 4$	0.9243	0.9248	0.9253	0.9258	0.9263	0.9269	0.9274	0.9279	0.9284	0.9289

Therefore,

$$\log 8{,}360 \approx 3\,.\,9222$$

The mantissa: 9222; The characteristic: 3

(The characteristic is 3, because $8{,}360 = 8.36 \times 10^3$.) In practice, we usually determine and write down the characteristic and a decimal point and then look up the mantissa and write the decimal part to the right of the decimal point. That is, we usually do *not* write all the steps found at the beginning of this example; we simply write

$$\log 8{,}360 \approx 3.9222$$

EXAMPLE 3 Find log 727.

SOLUTION The characteristic is 2, because $727 = 7.27 \times 10^2$. The mantissa (from Table II) is 0.8615. Therefore, $\log 727 \approx 2.8615$.

Note The characteristic and the mantissa may be positive, negative, or zero; however, the mantissas *in the table* are never negative. Therefore, if we were going to find a logarithm, perform further calculations using that logarithm, *and then use Table II* again, we would want the mantissa to be *positive*.

EXAMPLE 4 Find log 0.0438.

SOLUTION

$$\log 0.0438 = \log(4.38 \times 10^{-2}) = \log 4.38 + \log 10^{-2} = \log 4.38 + (-2)$$

From Table II, we find that $\log 4.38 \approx 0.6415$. Therefore,

$$\log 0.0438 \approx 0.6415 - 2$$

but we usually write -2 as $8. \quad -10$ and write

$$\log 0.0438 \approx 8.6415 - 10$$

If we performed either of the subtractions, we would have $\log 0.0438 \approx -1.3585$, which would equal the rounded-off calculator display for the logarithm. As previously mentioned, if we were going to perform any further calculations using this logarithm *and then use Table II* again, we would want the mantissa to be *positive* and we would leave the answer as $0.6415 - 2$, or, more commonly, as $8.6415 - 10$.

EXAMPLE 5 Find log 0.0000429.

SOLUTION The characteristic is -5, which we will write as $5. \quad -10$. The mantissa is 0.6325. Therefore, $\log 0.0000429 \approx 5.6325 - 10$.

Exercises B.2 Set I

Find each logarithm by using Table II.

1. log 754
2. log 186
3. log 17
4. log 29
5. log 3,350
6. log 4,610
7. log 7,000
8. log 200
9. log 0.0604
10. log 0.0186
11. $\log(5.64 \times 10^3)$
12. $\log(2.14 \times 10^{-4})$

Exercises B.2 Set II

Find each logarithm by using Table II.

1. log 0.905
2. log 0.306
3. log 58.9
4. log 36.7
5. $\log(5.77 \times 10^{-4})$
6. $\log(3.96 \times 10^3)$
7. log 15,000
8. log 0.0123
9. log 0.0013
10. log 5
11. $\log(3.16 \times 10^5)$
12. $\log(4.1 \times 10^{-3})$

B.3 Natural Logarithms

A special abbreviation, ln, is used for logarithms to the base e. That is, $\ln x = \log_e x$.

A brief table of natural logarithms is given in Table III, Appendix C. Natural logarithms do not have a characteristic or a mantissa. To find the natural logarithm of a number in Table III, we look for that number under the column headed n. If we find the desired number in that column, we read the value of its logarithm directly from the table, under the column headed $\log_e n$. If we do *not* find the desired number in the column headed n, we must use the properties of logarithms before we proceed (see Example 2). (Except for ln 1.0, all the logarithms in Table III are approximations.)

EXAMPLE 1 Find ln 4.7.

SOLUTION Shown below is a portion of Table III, Appendix C. We *do* find 4.7 in the table under the column headed n. Therefore, we read $\ln 4.7 \approx 1.5476$.

n	$\log_e n$
4.5	1.5041
4.6	1.5261
→ 4.7	1.5476 ←
4.8	1.5686
4.9	1.5892

EXAMPLE 2 Find ln 2,000.

SOLUTION We do *not* find 2,000 in the column headed n. We can, however, rewrite 2,000 as 20(100). Then

$$\begin{aligned}\ln 2{,}000 &= \ln 20(100)\\ &= \ln 20 + \ln 100\\ &\approx 2.9957 + 4.6052\\ &= 7.6009\end{aligned}$$

We could, instead, write 2,000 as $2(10^3)$. Then $\ln 2{,}000 = \ln 2 + 3 \ln 10$, and the final result is the same.

Exercises B.3 Set I

Find the logarithms by using Table III.

1. ln 3.6
2. ln 5.2
3. ln 8.1
4. ln 7.5
5. ln 83
6. ln 62
7. ln 0.002
8. ln 0.006

Exercises B.3 Set II

Find the logarithms by using Table III.

1. ln 2.8
2. ln 7.6
3. ln 1.6
4. ln 140
5. ln 78
6. ln 14,000
7. ln 0.008
8. ln 0.035

B.4 Interpolation

Sometimes we need to find the logarithm of a number that lies *between* two consecutive numbers in Table II. The process of finding such a number is called **interpolation**.

When we interpolate to find a logarithm, the answer should not contain more decimal places than the mantissas in the table have.

EXAMPLE 1 Find log 29.38.

SOLUTION 2‸9.38 is *between* 2‸9.3 and 2‸9.4, and the characteristic of log 29.38 is 1.

N	0	1	2	3	4	5	6	7	8	9
2‸8	0.4472	0.4487	0.4502	0.4518	0.4533	0.4548	0.4564	0.4579	0.4594	0.4609
2‸9	0.4624	0.4639	0.4654	0.4669	0.4683	0.4698	0.4713	0.4728	0.4742	0.4757
3‸0	0.4771	0.4786	0.4800	0.4814	0.4829	0.4843	0.4857	0.4871	0.4886	0.4900
3‸1	0.4914	0.4928	0.4942	0.4955	0.4969	0.4983	0.4997	0.5011	0.5024	0.5038
3‸2	0.5051	0.5065	0.5079	0.5092	0.5105	0.5119	0.5132	0.5145	0.5159	0.5172
3‸3	0.5185	0.5198	0.5211	0.5224	0.5237	0.5250	0.5263	0.5276	0.5289	0.5302
3‸4	0.5315	0.5328	0.5340	0.5353	0.5366	0.5378	0.5391	0.5403	0.5416	0.5428

The mantissa for 2‸9.38 is a number *between* 0.4669 and 0.4683

We generally arrange the work and indicate the differences as follows:

We ignore the decimal points and find the differences 2938 − 2930 and 2940 − 2930

$$\log 29.30 \approx 1.4669$$
$$\log 29.38 \approx$$
$$\log 29.40 \approx 1.4683$$

8, 10

The *difference* is 0.0014

Since 29.38 is eight-tenths of the way from 29.30 to 29.40, we assume that log 29.38 is eight-tenths of the way from 1.4669 to 1.4683. We first subtract 1.4669 from 1.4683, getting 0.0014. Next we find $\frac{8}{10}$ of 0.0014, which is 0.00112. We round that answer off to four decimal places (to 0.0011) and then add 0.0011 to 1.4669 (the smaller number). It is customary to omit the decimal point and the zeros in the numbers 0.0014, 0.00112, and 0.0011 and to show the work as follows:

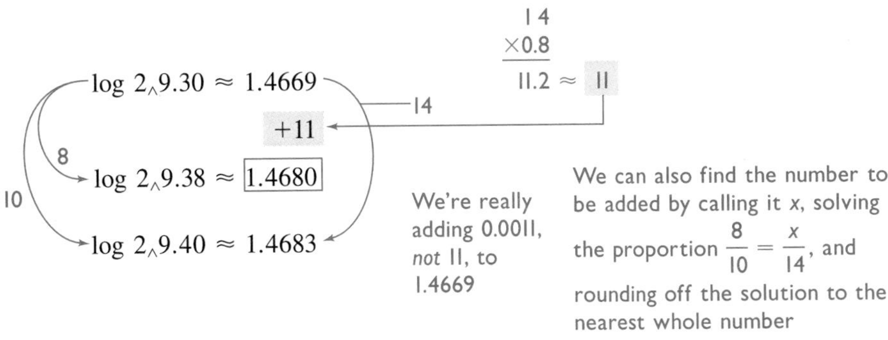

14
×0.8
11.2 ≈ 11
log 2‸9.30 ≈ 1.4669
+11
14
8
log 2‸9.38 ≈ 1.4680
10
log 2‸9.40 ≈ 1.4683
We're really adding 0.0011, not 11, to 1.4669
We can also find the number to be added by calling it x, solving the proportion 8/10 = x/14, and rounding off the solution to the nearest whole number

Therefore, log 29.38 ≈ 1.4680.

EXAMPLE 2 Find log 0.002749.

SOLUTION $0.002749 = 2.749 \times 10^{-3}$, and 2‸749 is *between* 2‸740 and 2‸750. The characteristic of log 0.002749 is −3, or 7 − 10.

N	0	1	2	3	4	5	6	7	8	9
2‸0	0.3010	0.3032	0.3054	0.3075	0.3096	0.3118	0.3139	0.3160	0.3181	0.3201
2‸1	0.3222	0.3243	0.3263	0.3284	0.3304	0.3324	0.3345	0.3365	0.3385	0.3404
2‸2	0.3424	0.3444	0.3464	0.3483	0.3502	0.3522	0.3541	0.3560	0.3579	0.3598
2‸3	0.3617	0.3636	0.3655	0.3674	0.3692	0.3711	0.3729	0.3747	0.3766	0.3784
2‸4	0.3802	0.3820	0.3838	0.3856	0.3874	0.3892	0.3909	0.3927	0.3945	0.3962
2‸5	0.3979	0.3997	0.4014	0.4031	0.4048	0.4065	0.4082	0.4099	0.4116	0.4133
2‸6	0.4150	0.4166	0.4183	0.4200	0.4216	0.4232	0.4249	0.4265	0.4281	0.4298
2‸7	0.4314	0.4330	0.4346	0.4362	0.4378	0.4393	0.4409	0.4425	0.4440	0.4456

The mantissa for 2‸749 is a number *between* 0.4378 and 0.4393

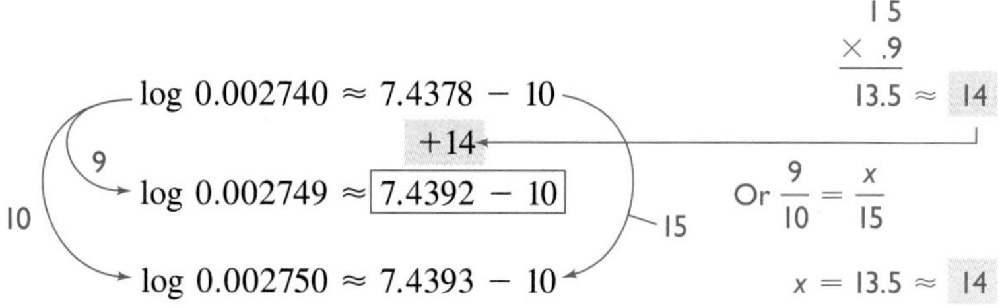

Therefore, log 0.002749 ≈ 7.4392 − 10.

Logarithms of numbers can be found by using a calculator with a [log] key as well as by using tables. If you have a calculator with a [log] key, we suggest that you use your calculator to check logarithms found by using the tables. One difference between logarithms obtained by calculator and those obtained from Table II occurs when the characteristic of the logarithm is negative. In Example 2,

$$\log 0.002749 \approx 7.4392 - 10$$

was found by using Table II. If this same logarithm were found by pressing the [log] key of a calculator, we would get

$$\log 0.002749 \approx -2.5608$$

But $$7.4392 - 10 = -2.5608$$

Exercises B.4
Set I

Find each logarithm by using Table II, and check each answer by using a calculator.

1. log 23.35
2. log 27.85
3. log 3.062
4. log 4.098
5. log 0.06644
6. log 0.5839
7. log 150.7
8. log 20.88
9. $\log(8.375 \times 10^6)$
10. $\log(3.875 \times 10^{-5})$
11. log 324.38
12. log 75.062

Exercises B.4
Set II

Find each logarithm by using Table II, and check each answer by using a calculator.

1. log 186,300
2. log 92,840,000
3. log 0.8006
4. log 0.07093
5. log 0.004003
6. $\log(1.756 \times 10^{-6})$
7. log 62.34
8. log 0.005326
9. log 2,168
10. log 0.0003152
11. $\log(1.425 \times 10^{4})$
12. $\log(4.257 \times 10^{-3})$

B.5 Finding Common Antilogarithms

Note The term *antilog* is discussed in Section 10.5 as is finding N when $\ln N$ (its natural logarithm) is known.

In the preceding sections, we discussed finding the logarithm of a given number. In this section, we discuss finding a number when its logarithm is known. The two statements that follow are equivalent.

$$\log N = L$$
$$N = \text{antilog } L$$

EXAMPLE 1 If $\log N = 3.6263$, find N.

SOLUTION First, we locate the mantissa (0.6263) in the body of Table II and find the number $4_{\wedge}23$ as shown in the table below.

The *first two digits* of N — The *third digit* of N

N	0	1	2	3	4	5	6	7	8	9
$4_{\wedge}0$	0.6021	0.6031	0.6042	0.6053	0.6064	0.6075	0.6085	0.6096	0.6107	0.6117
$4_{\wedge}1$	0.6128	0.6138	0.6149	0.6160	0.6170	0.6180	0.6191	0.6201	0.6212	0.6222
$4_{\wedge}2$	0.6232	0.6243	0.6253	0.6263	0.6274	0.6284	0.6294	0.6304	0.6314	0.6325
$4_{\wedge}3$	0.6335	0.6345	0.6355	0.6365	0.6375	0.6385	0.6395	0.6405	0.6415	0.6425
$4_{\wedge}4$	0.6435	0.6444	0.6454	0.6464	0.6474	0.6484	0.6493	0.6503	0.6513	0.6522

The mantissa

Then we use 3, the characteristic in 3 .6263, to locate the actual decimal point (the decimal point must go three places to the *right* of standard position). Therefore,

$$N \approx 4.230 \times 10^{3} = 4{,}230$$

The characteristic

EXAMPLE 2 If $N = \text{antilog}(7.8675 - 10)$, find N.

SOLUTION $\log N = 7.8675 - 10$

First, we locate the mantissa (0.8675) in the body of Table II and find the number $7_\wedge37$ as shown in the table that follows.

The *first two digits* of N — The *third digit* of N

N	0	1	2	3	4	5	6	7	8	9
$7_\wedge0$	0.8451	0.8457	0.8463	0.8470	0.8476	0.8482	0.8488	0.8494	0.8500	0.8506
$7_\wedge1$	0.8513	0.8519	0.8525	0.8531	0.8537	0.8543	0.8549	0.8555	0.8561	0.8567
$7_\wedge2$	0.8573	0.8579	0.8585	0.8591	0.8597	0.8603	0.8609	0.8615	0.8621	0.8627
$7_\wedge3$	0.8633	0.8639	0.8645	0.8651	0.8657	0.8663	0.8669	0.8675	0.8681	0.8686
$7_\wedge4$	0.8692	0.8698	0.8704	0.8710	0.8716	0.8722	0.8727	0.8733	0.8739	0.8745

The mantissa

Then we use −3, the characteristic in 7 .8675 − 10 , to locate the actual decimal point (the decimal point must go three places to the *left* of standard position). Therefore,

The characteristic is −3

$$N \approx 7.37 \times 10^{-3} = 0.007\,37$$

When the mantissa of log N falls between two consecutive numbers in Table II, we must *interpolate* to find a fourth significant digit for N. (See Examples 3 and 4.)

EXAMPLE 3 Find antilog 4.7129.

SOLUTION Let $N = \text{antilog } 4.7129$. Then $\log N = 4.7129$.

The *first two digits* of N — The *third digit* of N

N	0	1	2	3	4	5	6	7	8	9
$5_\wedge0$	0.6990	0.6998	0.7007	0.7016	0.7024	0.7033	0.7042	0.7050	0.7059	0.7067
$5_\wedge1$	0.7076	0.7084	0.7093	0.7101	0.7110	0.7118	0.7126	0.7135	0.7143	0.7152
$5_\wedge2$	0.7160	0.7168	0.7177	0.7185	0.7193	0.7202	0.7210	0.7218	0.7226	0.7235
$5_\wedge3$	0.7243	0.7251	0.7259	0.7267	0.7275	0.7284	0.7292	0.7300	0.7308	0.7316
$5_\wedge4$	0.7324	0.7332	0.7340	0.7348	0.7356	0.7364	0.7372	0.7380	0.7388	0.7396

The mantissa 0.7129 is *between* 0.7126 and 0.7135

The mantissa 0.7129 is *between* 0.7126 and 0.7135. The mantissa 0.7126 has $5_\wedge16$ as its antilog, and the mantissa 0.7135 has $5_\wedge17$ as its antilog; we append a fourth digit, 0, to each of these numbers as shown below.

Again omitting the decimal points, we find that the difference 7,129 − 7,126 is 3 and that the difference 7,135 − 7,126 is 9.

$$\log 5_\wedge160 \approx 0.7126$$

$$\tfrac{3}{9} \times 10 \approx 3 \qquad + 3 \qquad 3$$

$$\log \boxed{5_\wedge163} \approx 0.7129$$

$$10 \qquad 9$$

$$\log 5_\wedge170 \approx 0.7135$$

Since the mantissa 0.7129 is three-ninths of the way from 0.7126 to 0.7135, we assume that N is three-ninths of the way from $5_\wedge 160$ to $5_\wedge 170$. Therefore, we add $\frac{3}{9} \times 10 \approx 3$ to the *last* place of $5_\wedge 160$ and get $5_\wedge 163$. Then, since we want to find antilog 4 .7129, we locate the actual decimal point by using the characteristic, 4; that is, the decimal point must be four places to the right of standard position:

$$N \approx 5_\wedge 1630. = 51{,}630$$

4 ← The characteristic

Therefore, antilog 4.7129 ≈ 51,630.

EXAMPLE 4 Find antilog(8.7385 − 10).

SOLUTION Let N = antilog(8.7385 − 10).

$$\log N = 8.7385 - 10$$

The mantissa 0.7385 is *between* 0.7380 and 0.7388. The mantissa 0.7380 has $5_\wedge 47$ as its antilog, and the mantissa 0.7388 has $5_\wedge 48$ as its antilog; we append a fourth digit, 0, to each of these numbers. The difference 7,385 − 7,380 is 5, and the difference 7,388 − 7,380 is 8.

$\frac{5}{8} \times 10 = 6.25$ $\log 5_\wedge 470 \approx 0.7380$

≈ 6 → $+\,6$ 5

$\log \boxed{5_\wedge 476} \approx 0.7385$ 10   8

$\log 5_\wedge 480 \approx 0.7388$

Then, since the characteristic of log N is 8 − 10, or −2, the decimal point must go two places to the left of standard position:

$$N \approx 0.05_\wedge 476 = 0.054\ 76$$

−2 ← The characteristic = 8 − 10 = −2

Therefore, antilog(8.7385 − 10) ≈ 0.054 76.

The antilogarithms of numbers can be found with a calculator by using the [INV] [log] keys or the [10^x] key.

Exercises B.5
Set I

In Exercises 1–12, find each antilogarithm by using Table II.

1. antilog 3.5478
2. antilog 2.4409
3. antilog 0.9605
4. antilog 0.8848
5. antilog(9.2529 − 10)
6. antilog(8.1271 − 10)
7. antilog 3.5051
8. antilog 2.6335
9. antilog 4.0588
10. antilog 3.0846
11. antilog(6.9900 − 10)
12. antilog(7.9596 − 10)

In Exercises 13–16, find N by using Table II.

13. $\log N = 7.7168 - 10$
14. $\log N = 4.9410 - 10$
15. $\log N = 1.7120$
16. $\log N = 4.9873$

Exercises B.5 Set II

In Exercises 1–12, find each antilogarithm by using Table II.

1. antilog(7.6117 − 10)
2. antilog(6.6010 − 10)
3. antilog 1.1685
4. antilog 2.5470
5. antilog(8.6908 − 10)
6. antilog(9.7995 − 10)
7. antilog 4.7388
8. antilog(7.6096 − 10)
9. antilog 2.0486
10. antilog(8.8136 − 10)
11. antilog(6.2963 − 10)
12. antilog 6.0925

In Exercises 13–16, find *N* by using Table II.

13. $\log N = 3.4084$
14. $\log N = 0.5011$
15. $\log N = 0.6860$
16. $\log N = 8.0367 - 10$

B.6 Calculating with Logarithms

Logarithms can be used to convert multiplication problems to addition problems (see Example 1), to convert division problems to subtraction problems (see Example 2), and to convert problems involving raising to powers and extracting roots to multiplication problems (see Examples 3 and 4). In this section, we will express all answers correct to three significant digits.

The procedure for calculating with logarithms can be summarized as follows:

Calculating with logarithms

1. Let N equal the given expression.
2. Take the logarithm of both sides of the resulting equation.
3. Analyze the problem.
4. Make a blank outline.
5. Write all the characteristics in the blank outline. Use Table II to find all the mantissas; insert them in the blank outline. Carry out the calculations indicated in the outline.
6. Solve for N. That is, find the antilog of the final result from step 5.

Because all answers are to be expressed correct to three significant digits, no interpolation will be necessary in step 6. That is, we simply look in the table for the mantissa *closest* to the mantissa of $\log N$, and we find N by reading the digits corresponding to that closest mantissa.

EXAMPLE 1 Find (37.5) (0.008 42).

SOLUTION

Step 1. Let $N = (37.5)(0.008\ 42)$

Step 2. $\log N = \log(37.5)(0.008\ 42)$ Using Property 3 of Section 10.1

Step 3. According to Property 7 of Section 10.4, the logarithm of a product equals the sum of the logarithms of the factors. Therefore, we have

$$\log N = \log 37.5 + \log 0.008\ 42$$ Using Property 7 of Section 10.4

We will be adding the logarithms. Therefore, we want both mantissas to be positive, so in step 5, we will write the characteristic of log 0.008 42 (which is −3) as $7 - 10$.

Step 4. *Blank outline of procedure*		**Step 5.** *Actual calculations*	
$\log 37.5 \approx$		$\log 37.5 \approx$	$1\ .5740$
$+ \log 0.008\ 42 \approx +$		$+ \log 0.008\ 42 \approx +$	$7\ .9253 - 10$
$\log N \approx$		$\log N \approx$	$9\ .4993 - 10$
		or $\log N \approx$	$0\ .4993 + (-1)$
$N \approx$		**Step 6.** $N \approx$	$0.3_{\wedge}16$

Because the characteristic of log *N* is −1, the decimal point in *N* goes one place to the left of standard position

Therefore, $(37.5)(0.008\ 42) \approx 0.316$.

We will not write "Step 1," "Step 2," and so forth, for the remaining examples.

EXAMPLE 2 Divide $\frac{6.74}{0.0391}$.

SOLUTION Let $N = \frac{6.74}{0.0391}$

$\log N = \log \frac{6.74}{0.0391}$ Using Property 3 of Section 10.1

$\log N = \log 6.74 - \log 0.0391$ Using Property 8 of Section 10.4

The characteristic of the logarithm of 6.74, 0, is written as 10 − 10 to help with the subtraction

Blank outline of procedure		*Actual calculations*	
$\log 6.74 \approx$		$\log 6.74 \approx$	$10\ .8287 - 10$
$- \log 0.0391 \approx -$		$- \log 0.0391 \approx$	$-(8\ .5922 - 10)$
$\log N \approx$		$\log N \approx$	$2\ .2365$
$N \approx$		$N \approx$	$1_{\wedge}72. = 172$

Therefore, $\frac{6.74}{0.0391} \approx 172$.

EXAMPLE 3 Find $(1.05)^{10}$.

SOLUTION Let $N = (1.05)^{10}$

$\log N = \log(1.05)^{10}$ Using Property 3 of Section 10.1

$\log N = 10 \log 1.05$ Using Property 9 of Section 10.4

Blank outline		*Actual calculations*
$\log 1.05 \approx$		$\log 1.05 \approx 0.0212$
$\log N = 10 \log 1.05 \approx$		$\log N = 10 \log 1.05 \approx 0.2120$
$N \approx$		$N \approx 1.63$
		$N \approx 1\hat{.}63$

Therefore, $(1.05)^{10} \approx 1.63$.

EXAMPLE 4 Find $\sqrt[3]{0.506}$.

SOLUTION Let

$N = \sqrt[3]{0.506} = (0.506)^{1/3}$ Rewriting the cube root in exponential form

$\log N = \log(0.506)^{1/3}$ Using Property 3 of Section 10.1

$\log N = \frac{1}{3} \log 0.506$ Using Property 9 of Section 10.4

The characteristic 9 − 10 is written as 29 − 30 so that the second term, −30, is exactly divisible by 3

Blank outline	*Actual calculations*
$\log 0.506 \approx$	$\log 0.506 \approx 9.7042 - 10$
	$\approx 29.7042 - 30$
$\log N = \frac{1}{3}\log 0.506 \approx$	$\log N = \frac{1}{3}\log 0.506 \approx 9.9014 - 10$
$N \approx$	$N \approx 0.7_\wedge 97$
	$N \approx 0.797$

Therefore, $\sqrt[3]{0.506} \approx 0.797$.

EXAMPLE 5 Find N if $N = \dfrac{(1.16)^5(31.7)}{\sqrt{481}\,(0.629)}$, or $\dfrac{(1.16)^5(31.7)}{(481)^{1/2}(0.629)}$.

SOLUTION In the blank outline, we will represent the logarithm of the numerator by "log(num)" and the logarithm of the denominator by "log(den)."

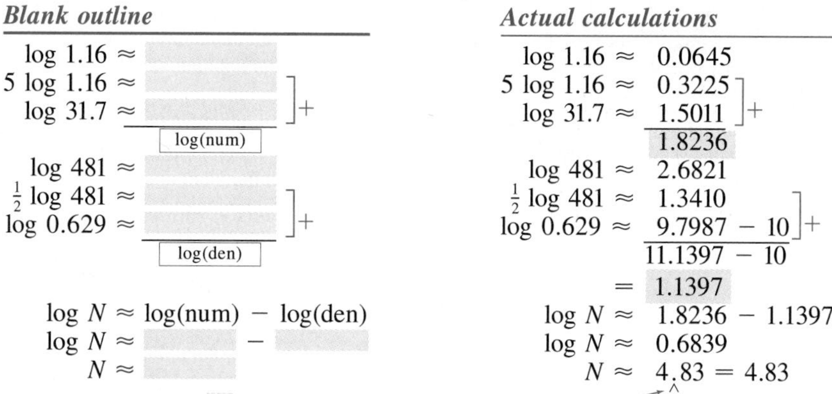

Blank outline	*Actual calculations*
$\log 1.16 \approx$	$\log 1.16 \approx 0.0645$
$5\log 1.16 \approx$	$5\log 1.16 \approx 0.3225$
$\log 31.7 \approx$ (+)	$\log 31.7 \approx 1.5011$ (+)
log(num)	1.8236
$\log 481 \approx$	$\log 481 \approx 2.6821$
$\frac{1}{2}\log 481 \approx$	$\frac{1}{2}\log 481 \approx 1.3410$
$\log 0.629 \approx$ (+)	$\log 0.629 \approx 9.7987 - 10$ (+)
log(den)	$11.1397 - 10$
	$= 1.1397$
$\log N \approx \log(\text{num}) - \log(\text{den})$	$\log N \approx 1.8236 - 1.1397$
$\log N \approx \quad - \quad$	$\log N \approx 0.6839$
$N \approx$	$N \approx 4_\wedge 83 = 4.83$

The characteristic in 0 .6839 is 0; therefore, the decimal point goes in standard position

Therefore, $N \approx 4.83$.

In the following exercises, we suggest that you use a calculator to verify the results obtained by the logarithmic calculations. There may be slight differences in the answers obtained by using a calculator and those found by using the table because of their different levels of accuracy.

Exercises B.6 Set I

Use logarithms to perform the calculations. Express all answers correct to three significant digits.

1. 74.3×0.618
2. 0.314×14.9
3. $\dfrac{562}{21.4}$
4. $\dfrac{651}{30.6}$
5. $(1.09)^5$
6. $(3.4)^4$
7. $\sqrt[3]{0.444}$
8. $\sqrt[4]{0.897}$
9. $2.863 + \log 38.46$
10. $\dfrac{\log 7.86}{\log 38.4}$
11. $\sqrt[5]{\dfrac{(5.86)(17.4)}{\sqrt{450}}}$
12. $\dfrac{(5.65)\sqrt[6]{175}}{(2.4)^4}$

Exercises B.6 Set II

Use logarithms to perform the calculations. Express all answers correct to three significant digits.

1. $\dfrac{(4.92)(25.7)}{388}$
2. $\dfrac{(2.04)^5}{(5.9)(0.66)}$
3. $\log 786.4 + 3.154$
4. $\sqrt[3]{0.564}$
5. $\dfrac{\log 58.4}{\log 2.50}$
6. $\sqrt[4]{\dfrac{(39.4)(7.86)}{\sqrt[3]{704}}}$
7. $\sqrt{0.00378}$
8. $\dfrac{1.26}{(0.0245)(0.00143)}$
9. $(2.87)^{12}$
10. $\sqrt{0.792}$
11. $\dfrac{\sqrt{4.63}}{\sqrt{0.0732}}$
12. $\dfrac{\log 1.25}{\log 1.05}$

Tables

APPENDIX

C

Table I Exponential Functions

x	e^x	e^{-x}	x	e^x	e^{-x}
0.00	1.0000	1.0000	1.5	4.4817	0.2231
0.01	1.0101	0.9900	1.6	4.9530	0.2019
0.02	1.0202	0.9802	1.7	5.4739	0.1827
0.03	1.0305	0.9704	1.8	6.0496	0.1653
0.04	1.0408	0.9608	1.9	6.6859	0.1496
0.05	1.0513	0.9512	2.0	7.3891	0.1353
0.06	1.0618	0.9418	2.1	8.1662	0.1225
0.07	1.0725	0.9324	2.2	9.0250	0.1108
0.08	1.0833	0.9231	2.3	9.9742	0.1003
0.09	1.0942	0.9139	2.4	11.023	0.0907
0.10	1.1052	0.9048	2.5	12.182	0.0821
0.11	1.1163	0.8958	2.6	13.464	0.0743
0.12	1.1275	0.8869	2.7	14.880	0.0672
0.13	1.1388	0.8781	2.8	16.445	0.0608
0.14	1.1503	0.8694	2.9	18.174	0.0550
0.15	1.1618	0.8607	3.0	20.086	0.0498
0.16	1.1735	0.8521	3.1	22.198	0.0450
0.17	1.1853	0.8437	3.2	24.533	0.0408
0.18	1.1972	0.8353	3.3	27.113	0.0369
0.19	1.2092	0.8270	3.4	29.964	0.0334
0.20	1.2214	0.8187	3.5	33.115	0.0302
0.21	1.2337	0.8106	3.6	36.598	0.0273
0.22	1.2461	0.8025	3.7	40.447	0.0247
0.23	1.2586	0.7945	3.8	44.701	0.0224
0.24	1.2712	0.7866	3.9	49.402	0.0202
0.25	1.2840	0.7788	4.0	54.598	0.0183
0.30	1.3499	0.7408	4.1	60.340	0.0166
0.35	1.4191	0.7047	4.2	66.686	0.0150
0.40	1.4918	0.6703	4.3	73.700	0.0136
0.45	1.5683	0.6376	4.4	81.451	0.0123
0.50	1.6487	0.6065	4.5	90.017	0.0111
0.55	1.7333	0.5769	4.6	99.484	0.0101
0.60	1.8221	0.5488	4.7	109.95	0.0091
0.65	1.9155	0.5220	4.8	121.51	0.0082
0.70	2.0138	0.4966	4.9	134.29	0.0074
0.75	2.1170	0.4724	5.0	148.41	0.0067
0.80	2.2255	0.4493	5.5	244.69	0.0041
0.85	2.3396	0.4274	6.0	403.43	0.0025
0.90	2.4596	0.4066	6.5	665.14	0.0015
0.95	2.5857	0.3867	7.0	1096.6	0.0009
1.0	2.7183	0.3679	7.5	1808.0	0.0006
1.1	3.0042	0.3329	8.0	2981.0	0.0003
1.2	3.3201	0.3012	8.5	4914.8	0.0002
1.3	3.6693	0.2725	9.0	8103.1	0.0001
1.4	4.0552	0.2466	10.0	22026.5	0.00005

Table II Common Logarithms

N	0	1	2	3	4	5	6	7	8	9
1‸0	0.0000	0.0043	0.0086	0.0128	0.0170	0.0212	0.0253	0.0294	0.0334	0.0374
1‸1	0.0414	0.0453	0.0492	0.0531	0.0569	0.0607	0.0645	0.0682	0.0719	0.0755
1‸2	0.0792	0.0828	0.0864	0.0899	0.0934	0.0969	0.1004	0.1038	0.1072	0.1106
1‸3	0.1139	0.1173	0.1206	0.1239	0.1271	0.1303	0.1335	0.1367	0.1399	0.1430
1‸4	0.1461	0.1492	0.1523	0.1553	0.1584	0.1614	0.1644	0.1673	0.1703	0.1732
1‸5	0.1761	0.1790	0.1818	0.1847	0.1875	0.1903	0.1931	0.1959	0.1987	0.2014
1‸6	0.2041	0.2068	0.2095	0.2122	0.2148	0.2175	0.2201	0.2227	0.2253	0.2279
1‸7	0.2304	0.2330	0.2355	0.2380	0.2405	0.2430	0.2455	0.2480	0.2504	0.2529
1‸8	0.2553	0.2577	0.2601	0.2625	0.2648	0.2672	0.2695	0.2718	0.2742	0.2765
1‸9	0.2788	0.2810	0.2833	0.2856	0.2878	0.2900	0.2923	0.2945	0.2967	0.2989
2‸0	0.3010	0.3032	0.3054	0.3075	0.3096	0.3118	0.3139	0.3160	0.3181	0.3201
2‸1	0.3222	0.3243	0.3263	0.3284	0.3304	0.3324	0.3345	0.3365	0.3385	0.3404
2‸2	0.3424	0.3444	0.3464	0.3483	0.3502	0.3522	0.3541	0.3560	0.3579	0.3598
2‸3	0.3617	0.3636	0.3655	0.3674	0.3692	0.3711	0.3729	0.3747	0.3766	0.3784
2‸4	0.3802	0.3820	0.3838	0.3856	0.3874	0.3892	0.3909	0.3927	0.3945	0.3962
2‸5	0.3979	0.3997	0.4014	0.4031	0.4048	0.4065	0.4082	0.4099	0.4116	0.4133
2‸6	0.4150	0.4166	0.4183	0.4200	0.4216	0.4232	0.4249	0.4265	0.4281	0.4298
2‸7	0.4314	0.4330	0.4346	0.4362	0.4378	0.4393	0.4409	0.4425	0.4440	0.4456
2‸8	0.4472	0.4487	0.4502	0.4518	0.4533	0.4548	0.4564	0.4579	0.4594	0.4609
2‸9	0.4624	0.4639	0.4654	0.4669	0.4683	0.4698	0.4713	0.4728	0.4742	0.4757
3‸0	0.4771	0.4786	0.4800	0.4814	0.4829	0.4843	0.4857	0.4871	0.4886	0.4900
3‸1	0.4914	0.4928	0.4942	0.4955	0.4969	0.4983	0.4997	0.5011	0.5024	0.5038
3‸2	0.5051	0.5065	0.5079	0.5092	0.5105	0.5119	0.5132	0.5145	0.5159	0.5172
3‸3	0.5185	0.5198	0.5211	0.5224	0.5237	0.5250	0.5263	0.5276	0.5289	0.5302
3‸4	0.5315	0.5328	0.5340	0.5353	0.5366	0.5378	0.5391	0.5403	0.5416	0.5428
3‸5	0.5441	0.5453	0.5465	0.5478	0.5490	0.5502	0.5514	0.5527	0.5539	0.5551
3‸6	0.5563	0.5575	0.5587	0.5599	0.5611	0.5623	0.5635	0.5647	0.5658	0.5670
3‸7	0.5682	0.5694	0.5705	0.5717	0.5729	0.5740	0.5752	0.5763	0.5775	0.5786
3‸8	0.5798	0.5809	0.5821	0.5832	0.5843	0.5855	0.5866	0.5877	0.5888	0.5899
3‸9	0.5911	0.5922	0.5933	0.5944	0.5955	0.5966	0.5977	0.5988	0.5999	0.6010
4‸0	0.6021	0.6031	0.6042	0.6053	0.6064	0.6075	0.6085	0.6096	0.6107	0.6117
4‸1	0.6128	0.6138	0.6149	0.6160	0.6170	0.6180	0.6191	0.6201	0.6212	0.6222
4‸2	0.6232	0.6243	0.6253	0.6263	0.6274	0.6284	0.6294	0.6304	0.6314	0.6325
4‸3	0.6335	0.6345	0.6355	0.6365	0.6375	0.6385	0.6395	0.6405	0.6415	0.6425
4‸4	0.6435	0.6444	0.6454	0.6464	0.6474	0.6484	0.6493	0.6503	0.6513	0.6522
4‸5	0.6532	0.6542	0.6551	0.6561	0.6571	0.6580	0.6590	0.6599	0.6609	0.6618
4‸6	0.6628	0.6637	0.6646	0.6656	0.6665	0.6675	0.6684	0.6693	0.6702	0.6712
4‸7	0.6721	0.6730	0.6739	0.6749	0.6758	0.6767	0.6776	0.6785	0.6794	0.6803
4‸8	0.6812	0.6821	0.6830	0.6839	0.6848	0.6857	0.6866	0.6875	0.6884	0.6893
4‸9	0.6902	0.6911	0.6920	0.6928	0.6937	0.6946	0.6955	0.6964	0.6972	0.6981
5‸0	0.6990	0.6998	0.7007	0.7016	0.7024	0.7033	0.7042	0.7050	0.7059	0.7067
5‸1	0.7076	0.7084	0.7093	0.7101	0.7110	0.7118	0.7126	0.7135	0.7143	0.7152
5‸2	0.7160	0.7168	0.7177	0.7185	0.7193	0.7202	0.7210	0.7218	0.7226	0.7235
5‸3	0.7243	0.7251	0.7259	0.7267	0.7275	0.7284	0.7292	0.7300	0.7308	0.7316
5‸4	0.7324	0.7332	0.7340	0.7348	0.7356	0.7364	0.7372	0.7380	0.7388	0.7396

(continued)

Table II (continued)

N	0	1	2	3	4	5	6	7	8	9
5‸5	0.7404	0.7412	0.7419	0.7427	0.7435	0.7443	0.7451	0.7459	0.7466	0.7474
5‸6	0.7482	0.7490	0.7497	0.7505	0.7513	0.7520	0.7528	0.7536	0.7543	0.7551
5‸7	0.7559	0.7566	0.7574	0.7582	0.7589	0.7597	0.7604	0.7612	0.7619	0.7627
5‸8	0.7634	0.7642	0.7649	0.7657	0.7664	0.7672	0.7679	0.7686	0.7694	0.7701
5‸9	0.7709	0.7716	0.7723	0.7731	0.7738	0.7745	0.7752	0.7760	0.7767	0.7774
6‸0	0.7782	0.7789	0.7796	0.7803	0.7810	0.7818	0.7825	0.7832	0.7839	0.7846
6‸1	0.7853	0.7860	0.7868	0.7875	0.7882	0.7889	0.7896	0.7903	0.7910	0.7917
6‸2	0.7924	0.7931	0.7938	0.7945	0.7952	0.7959	0.7966	0.7973	0.7980	0.7987
6‸3	0.7993	0.8000	0.8007	0.8014	0.8021	0.8028	0.8035	0.8041	0.8048	0.8055
6‸4	0.8062	0.8069	0.8075	0.8082	0.8089	0.8096	0.8102	0.8109	0.8116	0.8122
6‸5	0.8129	0.8136	0.8142	0.8149	0.8156	0.8162	0.8169	0.8176	0.8182	0.8189
6‸6	0.8195	0.8202	0.8209	0.8215	0.8222	0.8228	0.8235	0.8241	0.8248	0.8254
6‸7	0.8261	0.8267	0.8274	0.8280	0.8287	0.8293	0.8299	0.8306	0.8312	0.8319
6‸8	0.8325	0.8331	0.8338	0.8344	0.8351	0.8357	0.8363	0.8370	0.8376	0.8382
6‸9	0.8388	0.8395	0.8401	0.8407	0.8414	0.8420	0.8426	0.8432	0.8439	0.8445
7‸0	0.8451	0.8457	0.8463	0.8470	0.8476	0.8482	0.8488	0.8494	0.8500	0.8506
7‸1	0.8513	0.8519	0.8525	0.8531	0.8537	0.8543	0.8549	0.8555	0.8561	0.8567
7‸2	0.8573	0.8579	0.8585	0.8591	0.8597	0.8603	0.8609	0.8615	0.8621	0.8627
7‸3	0.8633	0.8639	0.8645	0.8651	0.8657	0.8663	0.8669	0.8675	0.8681	0.8686
7‸4	0.8692	0.8698	0.8704	0.8710	0.8716	0.8722	0.8727	0.8733	0.8739	0.8745
7‸5	0.8751	0.8756	0.8762	0.8768	0.8774	0.8779	0.8785	0.8791	0.8797	0.8802
7‸6	0.8808	0.8814	0.8820	0.8825	0.8831	0.8837	0.8842	0.8848	0.8854	0.8859
7‸7	0.8865	0.8871	0.8876	0.8882	0.8887	0.8893	0.8899	0.8904	0.8910	0.8915
7‸8	0.8921	0.8927	0.8932	0.8938	0.8943	0.8949	0.8954	0.8960	0.8965	0.8971
7‸9	0.8976	0.8982	0.8987	0.8993	0.8998	0.9004	0.9009	0.9015	0.9020	0.9025
8‸0	0.9031	0.9036	0.9042	0.9047	0.9053	0.9058	0.9063	0.9069	0.9074	0.9079
8‸1	0.9085	0.9090	0.9096	0.9101	0.9106	0.9112	0.9117	0.9122	0.9128	0.9133
8‸2	0.9138	0.9143	0.9149	0.9154	0.9159	0.9165	0.9170	0.9175	0.9180	0.9186
8‸3	0.9191	0.9196	0.9201	0.9206	0.9212	0.9217	0.9222	0.9227	0.9232	0.9238
8‸4	0.9243	0.9248	0.9253	0.9258	0.9263	0.9269	0.9274	0.9279	0.9284	0.9289
8‸5	0.9294	0.9299	0.9304	0.9309	0.9315	0.9320	0.9325	0.9330	0.9335	0.9340
8‸6	0.9345	0.9350	0.9355	0.9360	0.9365	0.9370	0.9375	0.9380	0.9385	0.9390
8‸7	0.9395	0.9400	0.9405	0.9410	0.9415	0.9420	0.9425	0.9430	0.9435	0.9440
8‸8	0.9445	0.9450	0.9455	0.9460	0.9465	0.9469	0.9474	0.9479	0.9484	0.9489
8‸9	0.9494	0.9499	0.9504	0.9509	0.9513	0.9518	0.9523	0.9528	0.9533	0.9538
9‸0	0.9542	0.9547	0.9552	0.9557	0.9562	0.9566	0.9571	0.9576	0.9581	0.9586
9‸1	0.9590	0.9595	0.9600	0.9605	0.9609	0.9614	0.9619	0.9624	0.9628	0.9633
9‸2	0.9638	0.9643	0.9647	0.9652	0.9657	0.9661	0.9666	0.9671	0.9675	0.9680
9‸3	0.9685	0.9689	0.9694	0.9699	0.9703	0.9708	0.9713	0.9717	0.9722	0.9727
9‸4	0.9731	0.9736	0.9741	0.9745	0.9750	0.9754	0.9759	0.9763	0.9768	0.9773
9‸5	0.9777	0.9782	0.9786	0.9791	0.9795	0.9800	0.9805	0.9809	0.9814	0.9818
9‸6	0.9823	0.9827	0.9832	0.9836	0.9841	0.9845	0.9850	0.9854	0.9859	0.9863
9‸7	0.9868	0.9872	0.9877	0.9881	0.9886	0.9890	0.9894	0.9899	0.9903	0.9908
9‸8	0.9912	0.9917	0.9921	0.9926	0.9930	0.9934	0.9939	0.9943	0.9948	0.9952
9‸9	0.9956	0.9961	0.9965	0.9969	0.9974	0.9978	0.9983	0.9987	0.9991	0.9996

Table III Natural Logarithms

n	$\log_e n$	n	$\log_e n$	n	$\log_e n$
		4.5	1.5041	9.0	2.1972
0.1	7.6974*	4.6	1.5261	9.1	2.2083
0.2	8.3906*	4.7	1.5476	9.2	2.2192
0.3	8.7960*	4.8	1.5686	9.3	2.2300
0.4	9.0837*	4.9	1.5892	9.4	2.2407
0.5	9.3069*	5.0	1.6094	9.5	2.2513
0.6	9.4892*	5.1	1.6292	9.6	2.2618
0.7	9.6433*	5.2	1.6487	9.7	2.2721
0.8	9.7769*	5.3	1.6677	9.8	2.2824
0.9	9.8946*	5.4	1.6864	9.9	2.2925
1.0	0.0000	5.5	1.7047	10	2.3026
1.1	0.0953	5.6	1.7228	11	2.3979
1.2	0.1823	5.7	1.7405	12	2.4849
1.3	0.2624	5.8	1.7579	13	2.5649
1.4	0.3365	5.9	1.7750	14	2.6391
1.5	0.4055	6.0	1.7918	15	2.7081
1.6	0.4700	6.1	1.8083	16	2.7726
1.7	0.5306	6.2	1.8245	17	2.8332
1.8	0.5878	6.3	1.8405	18	2.8904
1.9	0.6419	6.4	1.8563	19	2.9444
2.0	0.6931	6.5	1.8718	20	2.9957
2.1	0.7419	6.6	1.8871	25	3.2189
2.2	0.7885	6.7	1.9021	30	3.4012
2.3	0.8329	6.8	1.9169	35	3.5553
2.4	0.8755	6.9	1.9315	40	3.6889
2.5	0.9163	7.0	1.9459	45	3.8067
2.6	0.9555	7.1	1.9601	50	3.9120
2.7	0.9933	7.2	1.9741	55	4.0073
2.8	1.0296	7.3	1.9879	60	4.0943
2.9	1.0647	7.4	2.0015	65	4.1744
3.0	1.0986	7.5	2.0149	70	4.2485
3.1	1.1314	7.6	2.0281	75	4.3175
3.2	1.1632	7.7	2.0412	80	4.3820
3.3	1.1939	7.8	2.0541	85	4.4427
3.4	1.2238	7.9	2.0669	90	4.4998
3.5	1.2528	8.0	2.0794	100	4.6052
3.6	1.2809	8.1	2.0919	110	4.7005
3.7	1.3083	8.2	2.1041	120	4.7875
3.8	1.3350	8.3	2.1163	130	4.8676
3.9	1.3610	8.4	2.1282	140	4.9416
4.0	1.3863	8.5	2.1401	150	5.0106
4.1	1.4110	8.6	2.1518	160	5.0752
4.2	1.4351	8.7	2.1633	170	5.1358
4.3	1.4586	8.8	2.1748	180	5.1930
4.4	1.4816	8.9	2.1861	190	5.2470

*Subtract 10 for $n < 1$. Thus, $\log_e 0.1 = 7.6974 - 10 = -2.3026$.

ANSWERS

Answers to Set I Exercises (including Solutions to Odd-Numbered Exercises), Diagnostic Tests, and Cumulative Review Exercises

Exercises 1.1 (page 7)

1. a. True; a set is a collection of things.
b. False; $\{1, 2, 3\} \neq \{1, 2, 3, \ldots\}$. **c.** True
d. False; "23" does not denote a set.
e. False; 0 is *not* a natural number.
f. False; $\{2\}$ denotes a *set*, not an element. **g.** True
h. True; $\{\ \}$ is a subset of every set.
i. False; 10 is not a digit.
j. False; the empty set has *no* elements. **k.** True
l. True; it is the empty set.
m. False; the two symbols in the "middle" are not identical.

2. a. True **b.** False **c.** True **d.** True **e.** False
f. False **g.** True **h.** True **i.** True **j.** False
k. True **l.** True **m.** False

3. a. Infinite **b.** Finite **c.** Finite

4. a. Infinite **b.** Finite **c.** Finite

5. $\{\ \}, \{a\}, \{b\}, \{c\}, \{a,b\}, \{a,c\}, \{b,c\}, \{a,b,c\}$

6. $\{\ \}, \{1\}, \{2\}, \{1,2\}$

7. a. False; $12 \in B$ but $12 \notin A$.
b. True; all the elements of C are also elements of B.
c. False; all the elements of C are also elements of A.

8. a. True **b.** False **c.** True

9. a. $\{0, 2, 4, 6, 8\}$ **b.** $\{5, 10, 15, \ldots\}$

10. a. $\{1, 3, 5, \ldots\}$ **b.** $\{11, 22, 33, \ldots\}$

11. a. $\{x|x$ is one of the first 5 letters of the alphabet$\}$
b. $\{x|x$ is a multiple of 4 and $x \in N$ and x is less than 13$\}$
c. $\{x|x$ is a natural number that is divisible by 10$\}$

12. a. $\{x|x$ is a digit that is divisible by 2$\}$
b. $\{x|x$ is one of the last three letters of the alphabet$\}$
c. $\{x|x$ is a natural number that is divisible by 5$\}$

Exercises 1.2 (page 9)

1. a. $\{1, 2, 3, 4, 5\}$ **b.** $\{2, 4\}$ **c.** $\{1, 2, 3, 4, 5\}$
d. $\{2, 4\}$

2. a. $\{\ \}$, or $\emptyset$ **b.** $\{2, 5, 6, 7, 8, 9, 12\}$ **c.** $\{\ \}$, or $\emptyset$
d. $\{2, 5, 6, 7, 8, 9, 12\}$

3. a. $\{5, 11\}$ **b.** $\{0, 3, 4, 5, 6, 7, 11, 13\}$ **c.** $\{6\}$
d. $\{\ \}$, or $\emptyset$ **e.** $\{2, 5, 6, 7, 11, 13\}$
f. $\{0, 3, 4, 5, 6, 7, 11, 13\}$ **g.** $\{5, 11\} \cap \{0, 3, 4, 6\} = \{\ \}$
h. $\{2, 5, 6, 11\} \cap \{\ \} = \{\ \}$
i. $\{2, 5, 6, 7, 11, 13\} \cup \{0, 3, 4, 6\} = \{0, 2, 3, 4, 5, 6, 7, 11, 13\}$
j. $\{2, 5, 6, 11\} \cup \{0, 3, 4, 5, 6, 7, 11, 13\}$
$= \{0, 2, 3, 4, 5, 6, 7, 11, 13\}$

4. a. $\{4, b\}$ **b.** $\{a, b, m, 4, 6, 7\}$ **c.** $\{a, b, n, t, 3, 4, 5, 7\}$
d. $\{\ \}$, or $\emptyset$ **e.** $\{\ \}$, or $\emptyset$ **f.** $\{b, m, n, t, 3, 4, 5, 6\}$
g. $\{\ \}$, or $\emptyset$ **h.** $\{\ \}$, or $\emptyset$
i. $\{a, b, m, n, t, 3, 4, 5, 6, 7\}$ **j.** $\{a, b, m, n, t, 3, 4, 5, 6, 7\}$

Exercises 1.3 (page 14)

1.

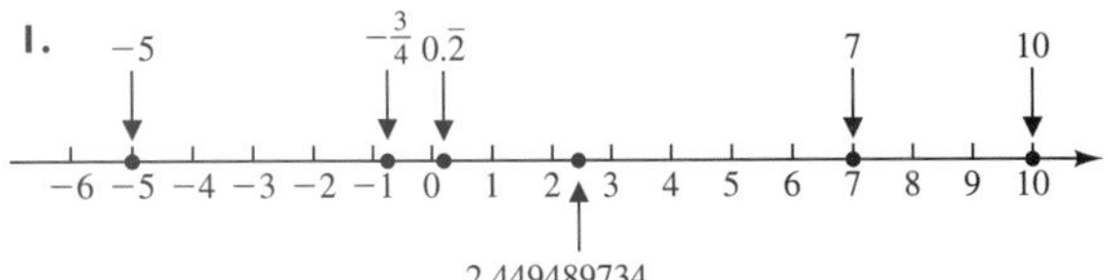

a. All are real. **b.** $7, -5, 10$ **c.** $7, 10$
d. $2.449\,489\,734\ldots$ **e.** $7, -5, 10, -\frac{3}{4}, 0.\bar{2}$

2.

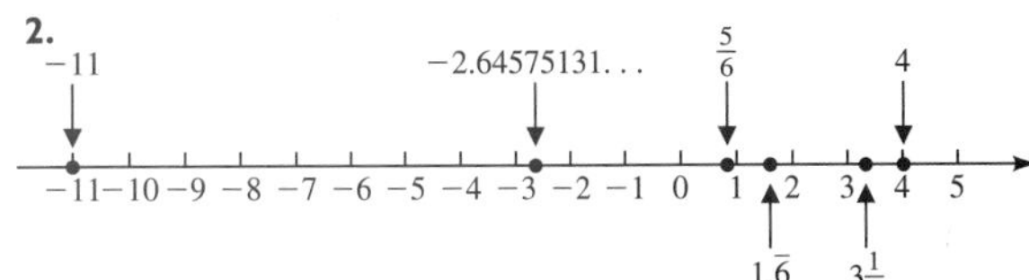

a. All **b.** $-11, 4$ **c.** 4 **d.** $-2.645\,751\,31\ldots$
e. $-11, \frac{5}{6}, 3\frac{1}{3}, 1.\bar{6}, 4$

3. a. $\subseteq$ **b.** $\in$ **c.** $\subseteq$ **d.** $\in$ **e.** $\in$ **f.** $\subseteq$

4. a. $\subseteq$ **b.** $\subseteq$ **c.** $\in$ **d.** $\in$ **e.** $\subseteq$ **f.** $\subseteq$

5. a. False; irrationals are real. **b.** True; $J \subseteq R$.
c. False; fractions are real numbers but not whole numbers.
d. True; 3 is an element of the set of integers.
e. False; 3 is not an irrational number.
f. True; the irrational numbers form a subset of the set of real numbers.
g. True; a nonrepeating, nonterminating decimal is irrational.
h. True; a nonrepeating, nonterminating decimal is real.

6. a. True **b.** False **c.** False **d.** True **e.** True
f. True **g.** False **h.** True

Exercises 1.4 (page 17)

1. a. 3 **b.** x, y **2. a.** 4, −2 **b.** y, z

3. a. Three terms **b.** $5F$ **c.** E, 1

4. a. Three terms **b.** $2T$ **c.** R, 1

5. a. Two terms **b.** $-2(x + y)$ **c.** $(R + S)$, 1

6. a. Two terms **b.** $-5(W + V)$ **c.** $(A + 2B)$, 1

7. a. Two terms **b.** 4 **c.** 3, X, Y, Z

8. a. Two terms **b.** $3x$ **c.** 4, a, b

9. a. Two terms **b.** $\dfrac{3B - C}{DE}$ **c.** 2, A

10. a. Two terms **b.** $\dfrac{w + z}{xyz}$ **c.** 3, s, t **11. a.** 2 **b.** RT

12. a. 4 **b.** xy **13. a.** −1 **b.** y **14. a.** −1 **b.** b

Exercises 1.5 (page 21)

1. a. $<$ **b.** $<$ **c.** $>$ **d.** $<$

2. a. $>$ **b.** $<$ **c.** $>$ **d.** $<$

3. a. Valid; $3 < 5$, $5 < 10$, and $3 < 10$ are all true.

b. Invalid; $-2 > 1$, $1 > 7$, and $-2 > 7$ are all false.

c. Invalid; one is "<" and one is ">".

d. Valid; $8 > 3$, $3 > -1$, and $8 > -1$ are all true.

e. Invalid; one is "<" and one is ">".

f. Invalid; $0 > 3$, $3 > 8$, and $0 > 8$ are all false.

4. a. Valid **b.** Invalid; one is "<" and one is ">".

c. Invalid; one is "<" and one is ">". **d.** Valid

e. Invalid; $4 > 7$, $7 > 11$, and $4 > 11$ are false.

f. Invalid; $5 < 0$, $0 < -2$, and $5 < -2$ are false.

5. False **6.** True

7. a. 7 **b.** $|-5| = -(-5) = 5$

c. $-|-12| = -(12) = -12$

8. a. 34 **b.** −5 **c.** −3

Exercises 1.6 (page 29)

1. $-(|-9| - |5|) = -(9 - 5) = -4$ **2.** −3

3. $-(|-12| + |-7|) = -(12 + 7) = -19$ **4.** $-\frac{5}{4}$

5. $-\left(\left|-\frac{7}{12}\right| + \left|-\frac{1}{6}\right|\right) = -\left(\frac{7}{12} + \frac{1}{6}\right) = -\frac{9}{12} = -\frac{3}{4}$

6. $-\frac{9}{8}$, or $-1\frac{1}{8}$ **7.** $\dfrac{21}{4} + \left(\dfrac{-5}{2}\right) = \dfrac{21}{4} + \left(\dfrac{-10}{4}\right) = \dfrac{11}{4}$ **8.** $1\frac{7}{10}$

9. $-13.5 + 8.06 = -5.44$ **10.** −0.37

11. $2.4 + 13 = 15.4$ **12.** 96.2 **13.** $\frac{1}{3} + \frac{1}{2} = \frac{5}{6}$ **14.** $26\frac{1}{6}$

15. $\left(-5\frac{3}{4}\right) + \left(-2\frac{2}{4}\right) = -7\frac{5}{4} = -8\frac{1}{4}$ **16.** −11.26

17. $\begin{array}{r} 16.71 \\ +\ 18.90 \\ \hline 35.61 \end{array}$ **18.** 12.61 **19.** −3 **20.** 0

21. 260 (When the signs are the same, the product is positive.)

22. 7.7

23. $\dfrac{\overset{1}{\cancel{3}}}{\underset{2}{\cancel{8}}} \cdot \dfrac{\overset{-1}{\cancel{-4}}}{\underset{3}{\cancel{9}}} = -\dfrac{1}{6}$ (Unlike signs give a negative product.)

24. −15 **25.** $-\frac{1}{3}$ **26.** $\frac{5}{2}$, or $2\frac{1}{2}$

27. $\left(\dfrac{15}{2}\right) \cdot \left(\dfrac{-2}{1}\right) = -15$

28. Undefined

29. 0 (0 divided by a nonzero number *is* possible, and the answer is 0.)

30. Undefined **31.** $\left(-\frac{5}{4}\right)\left(-\frac{8}{3}\right) = \frac{10}{3}$, or $3\frac{1}{3}$ **32.** 250

33. $-8 - (-3) = -8 + 3 = -5$ **34.** −1.56

35. $5 + \frac{2}{3} = 5\frac{2}{3}$ **36.** −7 **37.** $10 + 25 = 35$ **38.** −7

39. $-3(4) = -12$ **40.** 11

Exercises 1.7 (page 33)

1. True; commutative property of addition

2. True; commutative property of addition

3. True; commutative properties of addition and multiplication

4. True; commutative properties of addition and multiplication

5. True; distributive property **6.** True; distributive property

7. False **8.** False **9.** False **10.** False **11.** False

12. False **13.** True; commutative property of multiplication

14. True; commutative properties of addition and multiplication

15. True; additive identity property

16. True; additive identity property

17. True; associative property of addition **18.** False

19. False **20.** False **21.** True; distributive property

22. True; distributive property

23. True; multiplicative identity property

24. True; multiplicative identity property **25.** False

26. True; commutative property of addition

27. True; additive inverse property

28. True; additive inverse property

29. True; commutative property of addition (twice)

30. True; commutative property of addition (twice)

31. True; multiplicative inverse property

32. True; additive inverse property

33. $-5(-4)$ **34.** $-3(8)$ **35.** $[7 + (-2)] + 8$

36. $(-9 + 12) + (-3)$ **37.** $[-3(2)] + [(-3)(7)]$

38. $[5(-3)] + [5(6)]$ **39.** $8 + (-7)$ **40.** $-4 + (-5)$

41. $[9(-4)](7)$ **42.** $[9(-3)](8)$ **43.** $-8(6 + 3)$

44. $9(6 + 8)$

Exercises 1.8 (page 38)

1. $4 \cdot 4 \cdot 4 = 64$ **2.** 49 **3.** $(-3)(-3)(-3)(-3) = 81$

4. 16 **5.** $-2^4 = -(2 \cdot 2 \cdot 2 \cdot 2) = -16$ **6.** −81 **7.** 0

8. 0 **9.** −1 (An odd power of a negative number is negative.)

10. 1 **11.** $\left(\frac{1}{2}\right)\left(\frac{1}{2}\right)\left(\frac{1}{2}\right)\left(\frac{1}{2}\right) = \frac{1}{16}$ **12.** $\frac{49}{64}$

13. $(-0.1)(-0.1)(-0.1)(-0.1)(-0.1) = -0.000\,01$

14. 0.000 001 **15.** $\left(\frac{1}{10}\right)\left(\frac{1}{10}\right)\left(\frac{1}{10}\right) = \frac{1}{1{,}000}$ **16.** −5

17. 5,728.9761 **18.** 103.823 **19.** 18,106.3936 **20.** 299.29

21. 778.688 **22.** 5.0625 **23.** 39.0625 **24.** −148.877

Exercises 1.9 (page 42)

1. −6 **2.** 7 **3.** −5 **4.** Not real **5.** −10

6. −12 **7.** 9 **8.** 11 **9.** −16 **10.** −4

11. 458 **12.** 624 **13.** 3.464 **14.** 4.123 **15.** 13.565

16. 13.820 **17.** 1.673 **18.** 3.063 **19.** 2 **20.** 1

21. The index is even, and the radicand is negative; not real

22. Not real **23.** −4 **24.** −3 **25.** −2 **26.** −4

27. −10 **28.** −6 **29.** $-(\sqrt[5]{-32}) = -(-2) = 2$

30. 5

Exercises 1.10 (page 47)

1. $(16 - 9) - 4 = 7 - 4 = 3$ **2.** 9
3. $(12 \div 6) \div 2 = 2 \div 2 = 1$ **4.** 2
5. $(10 \div 2)(-5) = 5(-5) = -25$ **6.** -9
7. $3 \times 16 = 48$ **8.** 45 **9.** $8 + 30 = 38$ **10.** 11
11. $7 + \frac{5}{3} = 7 + 1\frac{2}{3} = 8\frac{2}{3}$ **12.** $6\frac{1}{2}$ **13.** $10 - 6 = 4$
14. 4 **15.** $10(225) - 64 = 2{,}250 - 64 = 2{,}186$ **16.** 32
17. $\frac{1}{2} - 20 = -19\frac{1}{2}$ **18.** $-3\frac{2}{3}$, or $-\frac{11}{3}$
19. $(100)(5)(4) = 2{,}000$ **20.** 1,500
21. $2 + 300 \div 25 = 2 + 12 = 14$ **22.** 53
23. $28 + 2 = 30$ **24.** 42
25. $2(8 - 5)(3) = 2(3)(3) = 6 \times 3 = 18$ **26.** 240
27. $6(-6) = -36$ **28.** -8
29. $-1{,}000 - 5(100)(-3) = -1{,}000 + 1{,}500 = 500$
30. 10,100 **31.** $20 - [5 - (-3)] = 20 - [8] = 12$ **32.** 3
33. $\dfrac{-5}{5} = -1$ **34.** -4
35. $8 - [5(-8) - 4] = 8 - [-40 - 4] = 8 - [-44] = 52$
36. -12
37. $(3 \cdot 25 - 5) \div (-7) = (75 - 5) \div (-7) = 70 \div (-7) = -10$
38. -20
39.
$$\begin{aligned} 15 - \{4 - [2 - 3(2)]\} &= 15 - \{4 - [2 - 6]\} \\ &= 15 - \{4 - [-4]\} \\ &= 15 - \{4 + 4\} = 15 - 8 = 7 \end{aligned}$$
40. 10 **41.** $\dfrac{49.1}{0.64} = 76.718\ 75$ **42.** 185.9375
43. $\dfrac{6.79}{30.697\ 88} \approx 0.221\ 187\ 912$ **44.** 0.044 563 049
45. $\dfrac{2.728}{11.1965} \approx 0.243\ 647\ 568$ **46.** 0.934 920 634
47. $39.4384 + 2.25 = 41.6884$ **48.** 53.6356
49. $(7.78)^2 = 60.5284$ **50.** 72.9316
51. $39.4384 + 18.84 + 2.25 = 60.5284$ **52.** 72.9316
53. $\sqrt{21.16 + 13.69} = \sqrt{34.85} \approx 5.903\ 388\ 857$
54. 9.492 101 98

Exercises 1.11 (page 52)

1.
```
2 | 28
2 | 14
     7
```
$28 = 2^2 \cdot 7$

2. $2 \cdot 3 \cdot 5$

3.
```
2 | 32
2 | 16
2 |  8
2 |  4
     2
```
$32 = 2^5$

4. $3 \cdot 11$

Work will not be shown in finding the prime factorizations for Exercises 5–12.

5. 43 is prime; 43 **6.** $5 \cdot 7$ **7.** $2^2 \cdot 3 \cdot 7$ **8.** $3 \cdot 5^2$
9. $2^4 \cdot 3^2$ **10.** $2^2 \cdot 3^2 \cdot 5$ **11.** $2^2 \cdot 3 \cdot 13$ **12.** $13 \cdot 17$
13. P; ±1, ±5 **14.** C; ±1, ±2, ±4, ±8 **15.** P; ±1, ±13
16. C; ±1, ±3, ±5, ±15 **17.** C; ±1, ±2, ±3, ±4, ±6, ±12
18. P; ±1, ±11 **19.** C; 1, 3, 17, 51
20. C; 1, 2, 3, 6, 7, 14, 21, 42
21. C; 1, 3, 37, 111 **22.** P; 1, 101
23.
$$\begin{aligned} 144 &= 2^4 \cdot 3^2 \\ 360 &= 2^3 \cdot 3^2 \cdot 5 \\ \text{LCM} &= 2^4 \cdot 3^2 \cdot 5 = 16 \cdot 9 \cdot 5 = 720 \end{aligned}$$
24. 420
25.
$$\begin{aligned} 150 &= 2 \cdot 3 \cdot 5^2 \\ 525 &= 3 \cdot 5^2 \cdot 7 \\ \text{LCM} &= 2 \cdot 3 \cdot 5^2 \cdot 7 = 1{,}050 \end{aligned}$$
26. 4,200
27.
$$\begin{aligned} 78 &= 2 \cdot 3 \cdot 13 \\ 130 &= 2 \cdot 5 \cdot 13 \\ 156 &= 2^2 \cdot 3 \cdot 13 \\ \text{LCM} &= 2^2 \cdot 3 \cdot 5 \cdot 13 = 780 \end{aligned}$$
28. 660
29.
$$\begin{aligned} 270 &= 2 \cdot 3^3 \cdot 5 \\ 900 &= 2^2 \cdot 3^2 \cdot 5^2 \\ 75 &= 3 \cdot 5^2 \\ \text{LCM} &= 2^2 \cdot 3^3 \cdot 5^2 = 2{,}700 \end{aligned}$$
30. 2,940
31. The prime numbers between 17 and 37 are 19, 23, 29, and 31; the numbers between 17 and 37 that leave a remainder of 5 when divided by 7 are 14 + 5, or 19; 21 + 5, or 26; and 28 + 5, or 33. The required number is 19.
32. 19, 31, 37

Sections 1.1–1.11 Review Exercises (page 56)

1. Infinite **2.** Finite **3.** $\{0, 1, 2, 3, 4\}$ **4.** No
5. a. $\{2, 5, 7, 8\}$ **b.** $\{5, 7\}$ **c.** $\{\ \}$, or $\emptyset$ **d.** $\{\ \}$, or $\emptyset$
6. a. None **b.** -2 and 0 **c.** -2, 4.53, $0.\overline{16}$, $\frac{2}{3}$, 0
d. 2.645 751 3 . . . **e.** All
7. 4 **8.** 0 **9.** 10 **10.** 36 **11.** $-(5)(5) = -25$
12. 0 **13.** Undefined **14.** 0
15. $6 + 2 \cdot 4 - 8 = 6 + 8 - 8 = 6$ **16.** 34
17. $6 + (18 \div 6) \div 3 = 6 + 3 \div 3 = 6 + 1 = 7$
18. a. ±1, ±2, ±4, ±7, ±14, ±28 **b.** $2^2 \cdot 7$ **c.** $2^3 \cdot 3 \cdot 7$
d. $2^3 \cdot 3 \cdot 7 = 168$
19. Yes **20.** No; $1 > 5$, $5 > 7$, and $1 > 7$ are false.
21. No; the inequalities are not both $>$ or both $<$.
22. False **23.** True; multiplicative identity property
24. True; associative property of addition
25. True; additive inverse property **26.** False
27. True; additive identity property
28. True; commutative property of addition **29.** False
30. True; distributive property

Chapter 1 Diagnostic Test (page 59)

Following each problem number is the number (in parentheses) of the textbook section where that kind of problem is discussed.

1. (1.1) True; repeated elements of a set can be disregarded.
2. (1.3) True **3.** (1.3) True
4. (1.6) False; division *by* zero is undefined.
5. (1.7) False; $(7 \cdot 5) \cdot 2 = 7 \cdot (5 \cdot 2)$ illustrates the associative property of multiplication.
6. (1.7) False **7.** (1.1) False **8.** (1.5) True
9. (1.3) False **10.** (1.4) True **11.** (1.1) True
12. (1.6) False; the multiplicative inverse of $-\frac{2}{5}$ is $-\frac{5}{2}$.
13. (1.4) False; x is a *factor* of $7xy$.
14. (1.5) False; the inequality symbols have opposite senses.
15. (1.7) False **16.** (1.9) False; $\sqrt{-9}$ is not a real number.
17. (1.9) True **18.** (1.1) False; $y \in B$, but $y \notin A$.
19. (1.1) True **20.** (1.1) False; k is not a set at all.
21. (1.3) 5 **22.** (1.3) -3, 0, 5
23. (1.3) 2.4, -3, 0, 5, $\frac{1}{2}$, $0.\overline{18}$ **24.** (1.3) 2.865 291 6 . . .
25. (1.3) All **26.** (1.2) $\{r, s, w, x, y, z\}$ **27.** (1.2) $\{\ \}$

28. (1.2) $\{w, x\}$

29. (1.2) $B \cap C = \{y\}$; therefore, $A \cup (B \cap C) = A \cup \{y\} = \{w, x, y, z\}$

30. (1.5) 17 **31.** (1.8) $(-5)^2 = (-5)(-5) = 25$

32. (1.6) $\frac{5}{-30} = -\frac{1}{6}$ **33.** (1.6) -37 **34.** (1.6) 5

35. (1.6) 72 **36.** (1.6) 0 **37.** (1.6) $-27 + (+17) = -10$

38. (1.6) -22 **39.** (1.8) $-(8^2) = -(8)(8) = -64$

40. (1.5) $-(+3) = -3$ **41.** (1.9) -3

42. (1.10) $(16 \div 4)(2) = (4)(2) = 8$

43. (1.10) $2(3) - 5 = 6 - 5 = 1$ **44.** (1.9) 9

45. (1.9) Not real

46. (1.10) $(15.3 \div 3)(7) = (5.1)(7) = 35.7$. On algebraic calculators, keystrokes are 15.3 [÷] 3 [×] 7 [=].

47. (1.10) $\dfrac{5.6088}{3.1592} \approx 1.775\ 386\ 174$. On algebraic calculators, keystrokes are 4.56 [×] 1.23 [÷] 2.872 [÷] 1.1 [=].

48. (1.11) $2 \cdot 3 \cdot 13$ **49.** (1.11) $5 \cdot 13$

50. (1.11) $2 \cdot 3 \cdot 5 \cdot 13 = 390$

Exercises 2.1 (page 64)

1. $10^{2+4} = 10^6$ **2.** 2^5 **3.** $x^{2+5} = x^7$ **4.** y^9

5. Cannot be simplified **6.** Cannot be simplified

7. Cannot be simplified **8.** Cannot be simplified

9. $a^{8-3} = a^5$ **10.** x^3 **11.** Cannot be simplified

12. Cannot be simplified **13.** $10^{3 \cdot 2} = 10^6$ **14.** 5^8

15. $3^{a \cdot b} = 3^{ab}$ **16.** 2^{mn} **17.** x^5y^5 **18.** u^4v^4

19. 2^6x^6, or $64x^6$ **20.** 3^4x^4 **21.** Cannot be simplified

22. Cannot be simplified **23.** $3^{42+73} = 3^{115}$ **24.** 7^{113}

25. $g^{9-4} = g^5$ **26.** x^3 **27.** Cannot be simplified

28. Cannot be simplified **29.** $\dfrac{x^3}{y^3}$ **30.** $\dfrac{u^7}{v^7}$ **31.** $\dfrac{3^4}{x^4}$, or $\dfrac{81}{x^4}$

32. $\dfrac{x^2}{5^2}$, or $\dfrac{x^2}{25}$

Exercises 2.2 (page 69)

1. $\dfrac{1}{a^3}$ **2.** $\dfrac{1}{x^2}$ **3.** z^7 **4.** b^9 **5.** $\dfrac{5}{b^7}$ **6.** $\dfrac{3}{y^2}$

7. $\dfrac{1}{(5b)^2} = \dfrac{1}{5^2b^2} = \dfrac{1}{25b^2}$ **8.** $\dfrac{1}{9y^2}$ **9.** $\dfrac{1}{x^3} \cdot \dfrac{y^2}{1} \cdot 1 = \dfrac{y^2}{x^3}$

10. $\dfrac{r^3}{s^4}$ **11.** $\dfrac{x}{1} \cdot \dfrac{1}{y^2} \cdot \dfrac{1}{z^3} \cdot 1 = \dfrac{x}{y^2z^3}$ **12.** $\dfrac{b}{c^5z^4}$

13. $a^3 \cdot \dfrac{b^4}{1} = a^3b^4$ **14.** c^4d^5 **15.** $\dfrac{1}{x^3} \cdot \dfrac{y^2}{1} = \dfrac{y^2}{x^3}$ **16.** $\dfrac{Q^4}{P^2}$

17. $x^{-3+(-4)} = x^{-7} = \dfrac{1}{x^7}$ **18.** $\dfrac{1}{y^5}$ **19.** $a^{3(-2)} = a^{-6} = \dfrac{1}{a^6}$

20. $\dfrac{1}{b^8}$ **21.** $x^{8+(-2)} = x^6$ **22.** a^2 **23.** $x^{2a(-3)} = x^{-6a} = \dfrac{1}{x^{6a}}$

24. $\dfrac{1}{y^{6c}}$ **25.** $y^{3-(-2)} = y^5$ **26.** x^{12} **27.** $x^{3a-(-a)} = x^{4a}$

28. a^{6x} **29.** 1 **30.** 1 **31.** $5 \cdot 1 = 5$ **32.** 2

33. $\dfrac{1}{x^3} + \dfrac{1}{x^5}$ **34.** $\dfrac{1}{y^2} + \dfrac{1}{y^6}$ **35.** $x^7 - \dfrac{1}{x^5}$ **36.** $y^{10} - \dfrac{1}{y^3}$

37. $1 + 1 = 2$ **38.** 3 **39.** 1 **40.** 1

41. $10^{5+(-2)} = 10^3 = 1{,}000$ **42.** 4 **43.** $3^{(-2)(-2)} = 3^4 = 81$

44. 1,000 **45.** $(1)^5 = 1$ **46.** 1

47. $10^{2+(-1)-(-3)} = 10^{2-1+3} = 10^4 = 10{,}000$ **48.** 8 **49.** $x^{-3}y$

50. xy^{-2} **51.** a^4x **52.** m^2n^3 **53.** $x^4y^{-3}z^2$ **54.** $a^{-1}b^3c^4$

Exercises 2.3 (page 73)

1. $\dfrac{1}{(5x)^3} = \dfrac{1}{5^3x^3} = \dfrac{1}{125x^3}$ **2.** $\dfrac{1}{16y^2}$ **3.** $\dfrac{7}{1} \cdot \dfrac{1}{x^2} = \dfrac{7}{x^2}$ **4.** $\dfrac{3}{y^4}$

5. $\dfrac{27}{x^3}$ **6.** $\dfrac{7^2}{y^2}$, or $\dfrac{49}{y^2}$ **7.** $\dfrac{8^2}{z}$, or $\dfrac{64}{z}$ **8.** $\dfrac{8}{x}$

9. $a^{2 \cdot 2}b^{3 \cdot 2} = a^4b^6$ **10.** $x^{12}y^{15}$ **11.** $m^{(-2)4}n^{1 \cdot 4} = m^{-8}n^4 = \dfrac{n^4}{m^8}$

12. $\dfrac{r^5}{p^{15}}$ **13.** $x^{(-2)(-4)}y^{3(-4)} = x^8y^{-12} = \dfrac{x^8}{y^{12}}$ **14.** $\dfrac{w^6}{z^8}$

15. $(1k^{-4})^{-2} = k^{(-4)(-2)} = k^8$ **16.** z^{10}

17. $2^{1 \cdot 3}x^{2 \cdot 3}y^{(-4)(3)} = 8x^6y^{-12} = \dfrac{8x^6}{y^{12}}$ **18.** $\dfrac{9b^{10}}{a^2}$

19. $5^{1(-2)}m^{(-3)(-2)}n^{5(-2)} = 5^{-2}m^6n^{-10} = \dfrac{m^6}{25n^{10}}$ **20.** $\dfrac{y^2}{8x^8}$

21. $\dfrac{x^{1 \cdot 2}y^{4 \cdot 2}}{z^{2 \cdot 2}} = \dfrac{x^2y^8}{z^4}$ **22.** $\dfrac{a^9b^3}{c^6}$

23. $\dfrac{M^{(-2)(4)}}{N^{3 \cdot 4}} = \dfrac{M^{-8}}{N^{12}} = \dfrac{1}{M^8N^{12}}$ **24.** $R^{15}S^{12}$

25. $\dfrac{x^{(-5)(-2)}}{y^{4(-2)}z^{(-3)(-2)}} = \dfrac{x^{10}}{y^{-8}z^6} = \dfrac{x^{10}y^8}{z^6}$ **26.** $\dfrac{a^{12}b^6}{c^{15}}$ **27.** 1

28. 1 **29.** $\dfrac{3^{1 \cdot 2}x^{2 \cdot 2}}{y^{3 \cdot 2}} = \dfrac{9x^4}{y^6}$ **30.** $\dfrac{16a^{16}}{b^8}$

31. $\dfrac{4^{1(-1)}a^{(-2)(-1)}}{b^{3(-1)}} = \dfrac{4^{-1}a^2}{b^{-3}} = \dfrac{a^2b^3}{4}$ **32.** $25m^8n^6$

33. $(x^{-1-4}y^2)^{-2} = (x^{-5}y^2)^{-2} = x^{(-5)(-2)}y^{2(-2)} = x^{10}y^{-4} = \dfrac{x^{10}}{y^4}$

34. $u^{12}v^3$

35. $(s^{-4-3}t^{-5-(-4)})^{-3} = (s^{-7}t^{-1})^{-3} = s^{(-7)(-3)}t^{(-1)(-3)} = s^{21}t^3$

36. $\dfrac{1}{u^{50}v^5}$ **37.** $(2)(-3)(x^5 \cdot x^4)(y^2) = -6x^9y^2$ **38.** $-28a^2b^5$

39. $(-4)(-2)(x^{-2}xy^3y^{-1}) = 8x^{-1}y^2 = \dfrac{8y^2}{x}$ **40.** $\dfrac{10b}{a}$

41. $(2)(3)(-1)(s^3 \cdot s \cdot s)(1)(u^{-4} \cdot u^3) = -6s^5u^{-1} = -\dfrac{6s^5}{u}$

42. $-\dfrac{8x^4}{z}$

Exercises 2.4 (page 78)

1. 2.856×10^1 **2.** 3.754×10^2 **3.** 6.184×10^{-2}

4. 3.056×10^{-3} **5.** 7.8×10^4 **6.** 1.4×10^3

7. 2.006×10^{-1} **8.** 9.5×10^{-5}

9. $(3.62 \times 10^{-1}) \times 10^{-2} = 3.62 \times 10^{-3}$ **10.** 6.314×10^{-4}

11. $(2.452 \times 10^2) \times 10^{-5} = 2.452 \times 10^{-3}$ **12.** 3.17×10^{-3}

13.
$$\frac{(6 \times 10^{-5}) \times (8 \times 10^8)}{(5 \times 10^7) \times (3 \times 10^{-4})} = \frac{6 \times 8}{5 \times 3} \times \frac{10^{-5} \times 10^8}{10^7 \times 10^{-4}} = \frac{16}{5} \times \frac{10^3}{10^3} = 3.2 \times 10^0, \text{ or } 3.2$$

14. 2.5×10^4, or 25,000

15.
$$\frac{(6.3 \times 10^{-6}) \times (5.5 \times 10^6)}{(3.5 \times 10^5) \times (3.3 \times 10^{-5})} = \frac{(6.3 \times 5.5) \times (10^{-6} \times 10^6)}{(3.5 \times 3.3) \times (10^5 \times 10^{-5})} = \frac{\overset{\overset{3}{\cancel{9}}}{\cancel{63}} \times \overset{\overset{1}{\cancel{5}}}{\cancel{55}}}{\underset{\underset{1}{\cancel{5}}}{\cancel{35}} \times \underset{\underset{1}{\cancel{3}}}{\cancel{33}}} \times \frac{10^0}{10^0} = 3 \times 1 = 3, \text{ or } 3 \times 10^0$$

16. 1.43×10^6 **17.** 1.288×10^{10} **18.** 3×10^{-9}

19. 1.6×10^{-3} **20.** 9×10^{-4}

21. $\left(6.02 \times 10^{23}\ \frac{\text{molecules}}{\text{mole}}\right)(600 \text{ moles}) = 3{,}612 \times 10^{23}$ molecules
$= (3.612 \times 10^3) \times 10^{23}$ molecules $= 3.612 \times 10^{26}$ molecules

22. \$9,360,000,000

23. $\frac{4{,}400{,}000{,}000 \text{ mi}}{12 \text{ years}}\left(\frac{1 \text{ year}}{365 \text{ days}}\right)\left(\frac{1 \text{ day}}{24 \text{ hr}}\right) \approx 41{,}857\ \frac{\text{mi}}{\text{hr}}$

24. $1.607\,04 \times 10^{10}$ mi $= 16{,}070{,}400{,}000$ mi

Exercises 2.5 (page 81)

1. $2(-6)^2 + 3(5) = 2(36) + 15 = 72 + 15 = 87$ **2.** 5

3. $(-5) - 12\left(\frac{1}{3}\right)^2 = -5 - 12\left(\frac{1}{9}\right) = -5 - \frac{4}{3} = -\frac{19}{3} = -6\frac{1}{3}$

4. $-\frac{26}{3}$, or $-8\frac{2}{3}$ **5.** $(-5)^2 - 4(5)(-6) = 25 - (-120) = 145$

6. 61 **7.** $(5 + [-6])^2 = (-1)^2 = 1$ **8.** 36

9. $(5^2) + 2(5)(-6) + (-6)^2 = 25 - 60 + 36 = 1$ **10.** 36

11. $5^2 + (-6)^2 = 25 + 36 = 61$ **12.** 26

13. $\frac{3(0)}{5 + (-15)} = \frac{0}{-10} = 0$ **14.** 0

15. $2(-1) - [5 - (0 - 5(-15))] = -2 - [5 - 75]$
$= -2 - [-70] = 68$

16. -32

17. $-(-1) - \sqrt{(-1)^2 - 4(-4)(5)} = 1 - \sqrt{1 + 80}$
$= 1 - \sqrt{81} = 1 - 9 = -8$

18. 12 **19.** $5b^2 - 4b + 8$ **20.** $4c + 3$

21. $2(x^2 - 4x)^2 - 3(x^2 - 4x) + 7$ **22.** $(y^4 + 2)^2 - 2(y^4 + 2)$

Exercises 2.6 (page 83)

1. $q = \frac{DQ}{H} = \frac{5(420)}{30} = 70$ **2.** 125

3. $A = P(1 + rt) = 500[1 + 0.09(2.5)]$
$= 500(1 + 0.225) = 500(1.225) = 612.50$

4. 498

5. $A = P(1 + i)^n = 600(1.085)^2 = 600(1.177\,225) \approx 706.34$

6. $808.9375 \approx 808.94$

7. $C = \frac{5}{9}(F - 32) = \frac{5}{9}(-10 - 32) = \frac{5}{9}(-42) = -23\frac{1}{3}$,
or ≈ -23.33

8. $-21\frac{2}{3}$, or ≈ -21.67

9. $S = \frac{1}{2}gt^2 = \frac{1}{2}(32)\left(8\frac{1}{2}\right)^2 = (16)\left(\frac{17}{2}\right)^2 = 1{,}156$ **10.** 361

11. $Z = \frac{Rr}{R + r} = \frac{22(8)}{22 + 8} = \frac{176}{30} = 5\frac{13}{15} \approx 5.87$

12. $17\frac{3}{16} \approx 17.19$

13. $S = 2\pi r^2 + 2\pi rh \approx 2(3.14)(3)^2 + 2(3.14)(3)(20)$
$= 2(3.14)(9) + 376.8 = 56.52 + 376.8 = 433.32$

14. 602.88

Exercises 2.7 (page 84)

1. $2x^{2/2} = 2x^1 = 2x$ **2.** $3y$ **3.** $m^{4/2}n^{2/2} = m^2n^1 = m^2n$

4. u^5v^3 **5.** $5a^{4/2}b^{2/2} = 5a^2b^1 = 5a^2b$ **6.** $10b^2c$

7. $x^{10/2}y^{4/2} = x^5y^2$ **8.** x^6y^4 **9.** $10a^{10/2}y^{2/2} = 10a^5y^1 = 10a^5y$

10. $11a^{12}b^2$ **11.** $9m^{8/2}n^{16/2} = 9m^4n^8$ **12.** $7c^9d^5$

Exercises 2.8 (page 88)

1. $10 + 4x - y$ **2.** $8 + 3a - b$ **3.** $2x + 7y + 3z - 6$

4. $5z - 3w + 4x + 2y$ **5.** $(3a)(6) + (3a)(x) = 18a + 3ax$

6. $35b + 5by$ **7.** $(x)(-4) + (-5)(-4) = -4x + 20$

8. $-5y + 10$

9. $(-3)(x) - (-3)(2y) + (-3)(2) = -3x - (-6y) + (-6)$
$= -3x + 6y - 6$

10. $-2x + 6y - 8$ **11.** $(x)(xy) - (x)(3) = x^2y - 3x$

12. $a^2b - 4a$ **13.** $(3a)(ab) + (3a)(-2a^2) = 3a^2b - 6a^3$

14. $12x^2 - 8xy^2$

15. $(3x^3)(-2xy) + (-2x^2y)(-2xy) + (y^3)(-2xy)$
$= -6x^4y + 4x^3y^2 - 2xy^4$

16. $-8yz^4 + 2y^2z^3 + 2y^4z$

17. $(-2)(3)(6)(a \cdot a^2 \cdot a)(b \cdot b \cdot b)(c^3) = -36a^4b^3c^3$

18. $-30x^4y^5z^2$

19. $(4xy^2)(3x^3y^2) - (4xy^2)(2x^2y^3) + (4xy^2)(5xy^4)$
$= 12x^4y^4 - 8x^3y^5 + 20x^2y^6$

20. $-10x^6y + 4x^5y^2 + 6x^3y^3$

21. $(3mn^2)(-2m^2n)(5m^2) + (3mn^2)(-2m^2n)(-n^2)$
$= (3)(-2)(5)m^{1+2+2}n^{2+1} + (3)(-2)(-1)m^{1+2}n^{2+1+2}$
$= -30m^5n^3 + 6m^3n^5$

22. $-36a^5b^3 + 18a^3b^5$

23. $-8x^3y^2z(3xy - 2xz + 5yz)$
$= (-8x^3y^2z)(3xy) - (-8x^3y^2z)(2xz) + (-8x^3y^2z)(5yz)$
$= -24x^4y^3z - (-16x^4y^2z^2) + (-40x^3y^3z^2)$
$= -24x^4y^3z + 16x^4y^2z^2 - 40x^3y^3z^2$

24. $21a^4b^3 - 14a^3b^4 - 7a^3b^3c$ **25.** $7x + 3y - 2x^2$

26. $3a - b - 4c$

27. $-1(3x - 2y) = (-1)(3x) + (-1)(-2y) = -3x + 2y$

28. $-7z - 2w$ **29.** $7 - 1(-4R - S) = 7 + 4R + S$

30. $9 + 3m + n$ **31.** $6 - 2a + 6b$ **32.** $12 - 6R + 3S$

33. $3 - 2x^2 + 8xy$ **34.** $2 - 10x^2 + 15xy$

35. $-x + y + 2 - a$ **36.** $-a + b + x - 3$

37. $2a - 2b - 6$ **38.** $3x - 3y - 5$

39. $x - [a + y - b] = x - a - y + b$ **40.** $y - m - x + n$

41. $5 - 3[a - 8x + 4y] = 5 - 3a + 24x - 12y$

42. $7 - 5x + 30a - 15b$

43. $2 - [a - 1(b - c)] = 2 - 1[a - b + c] = 2 - a + b - c$

44. $5 - x + y + z$

45. $9 - 2[-3a - 8x + 4y] = 9 + 6a + 16x - 8y$

46. $P - x + y - 4 + z$

Exercises 2.9 (page 92)

1. $-2x$ **2.** $-a$ **3.** $6x^2y$ **4.** $7ab^2$ **5.** $6xy^2 + 8x^2y$

6. $5a^2b - 4ab^2$ **7.** $-2xy$ **8.** $4mn$ **9.** $5xyz^2 - 2x^2y^2z^2$

10. $3a^3b^3c^3 - 4abc^3$ **11.** $0xyz^2 = 0$ **12.** $1a^2bc$, or a^2bc

13. $3x^2y - 2xy^2$ **14.** $2xy^2 - 5x^2y$ **15.** $2ab - a + b$

16. $6xy - x - y$ **17.** $2x^3 - 2x^2 - 2x$

18. $-3y^3 + 5y^2 - 2y$ **19.** $2x - 9y + 11$

20. $-7a - 2b - 5$ **21.** $-2a^2b - ab + 7ab^2$

22. $4xy^2 + y - 4x^2y$

23. $6h^3 - 2hk - hk + 3k^4 = 6h^3 - 3hk + 3k^4$

24. $11xy^2 - 14x^2$ **25.** $3x - 4 - 5x = -2x - 4$

26. $-3x - 7$ **27.** $(-5x)(3x) - (-5x)(4) = -15x^2 + 20x$

28. $-40x^2 + 56x$ **29.** $2 + 3x$ **30.** $5 + 8y$

31. $3x - [5y - 2x + 4y] = 3x - [9y - 2x] = 3x - 9y + 2x$
$= 5x - 9y$

32. $5x - 9y$

33. $-10[-6x + 10 + 17] - 4x = -10[-6x + 27] - 4x$
$= 60x - 270 - 4x = 56x - 270$

2

34. $115x - 640$

35. $8 - 2(x - y + 3x) = 8 - 2(4x - y) = 8 - 8x + 2y$

36. $9 - 12u + 4t$ **37.** $8x + 10x^2 - 4x = 4x + 10x^2$

38. $4y + 25y^2$

39. $3u - v - \{2u - 10 + v - 20\} - 8v$
$= 3u - v - \{2u + v - 30\} - 8v$
$= 3u - v - 2u - v + 30 - 8v = u - 10v + 30$

40. $-5x - 2y + 23$

41. $50 - \{-2t - [5t - 6 + 2t]\} + 7^0$
$= 50 - \{-2t - [7t - 6]\} + 1 = 50 - \{-2t - 7t + 6\} + 1$
$= 50 - \{-9t + 6\} + 1 = 50 + 9t - 6 + 1 = 45 + 9t$

42. $22 + 11x$

43. $100v - 3\{-4[8 + 2v - 5v]\} = 100v - 3\{-4[8 - 3v]\}$
$= 100v - 3\{-32 + 12v\}$
$= 100v + 96 - 36v$
$= 64v + 96$

44. $72z + 96$ **45.** $w^4 - 4w^2 + 4w^2 - 16 = w^4 - 16$

46. $x^4 - 81$ **47.** $(3)(5)(4)(2)(x \cdot x^2 \cdot x^3) = 120x^6$ **48.** $60y^8$

49. $5X^{-4+6} + 3(1) = 5X^2 + 3$ **50.** $3Y^2 + 2$

51. $13^{3x+6x} = 13^{9x}$ **52.** 7^{13m} **53.** $5^{x+3x} = 5^{4x}$ **54.** 6^{13x}

55. $3^{9x-6x} = 3^{3x}$ **56.** 5^{4t}

57. $(x^{3n-4n-5n})^3 = (x^{-6n})^3 = x^{-18n} = \dfrac{1}{x^{18n}}$ **58.** $\dfrac{1}{w^{6m}}$

59. $3(x + 2y) - 5(3x - y) = 3x + 6y - 15x + 5y = -12x + 11y$

60. $-2x + 10y$

61. $2[(3)(x + 2y) - (3x - y)] = 2[3x + 6y - 3x + y] = 2(7y)$
$= 14y$

62. $42x - 21y$

Sections 2.1–2.9 Review Exercises (page 94)

1. $x^{3+5} = x^8$ **2.** Cannot be simplified **3.** $N^{2 \cdot 3} = N^6$

4. Cannot be simplified **5.** $a^{5-2} = a^3$

6. Cannot be simplified **7.** $\dfrac{2^3a^3}{b^{2 \cdot 3}} = \dfrac{8a^3}{b^6}$ **8.** $\dfrac{x^4}{y^2}$

9. $\left(\dfrac{x^2y}{x^4}\right)^{-1} = \left(\dfrac{y}{x^2}\right)^{-1} = \left(\dfrac{x^2}{y}\right)^1 = \dfrac{x^2}{y}$ **10.** 2 **11.** 1

12. Cannot be simplified **13.** $3c^4d^2 - 12c^3d^3$

14. $8 - 6x + 2y$

15. $(-10)(-8)(-1)(x^2x^3x)(y^3y^2)(z^4) = -80x^6y^5z^4$

16. $-8x^4 - 4x^2 + 2xy$

17. $5 - 2[3 - 5x + 5y + 4x - 6] = 5 - 2[-3 - x + 5y]$
$= 5 + 6 + 2x - 10y$
$= 11 + 2x - 10y$

18. $4x^3 - y$ **19.** 1.486×10^2 **20.** 0.003 17

21. $A = P(1 + rt) = 550[1 + (0.09)(2.5)] = 550(1 + 0.225)$
$= 550(1.225) = 673.75$

22. $C = 40$

23. $S = R\left[\dfrac{(1 + i)^n - 1}{i}\right] = 750\left[\dfrac{(1.09)^3 - 1}{0.09}\right] = 2{,}458.575$

24. 472.5

Chapter 2 Diagnostic Test (page 97)

Following each problem number is the number (in parentheses) of the textbook section where that kind of problem is discussed.

1. (2.1) $2^5 \cdot 2^7 = 2^{5+7} = 2^{12}$

2. (2.2, 2.1) $x^2 \cdot x^{-5} = x^{2+(-5)} = x^{-3} = \dfrac{1}{x^3}$

3. (2.1) $(N^2)^4 = N^{2 \cdot 4} = N^8$

4. (2.3) $\left(\dfrac{2X^3}{Y}\right)^2 = \dfrac{2^2X^{3 \cdot 2}}{Y^{1 \cdot 2}} = \dfrac{4X^6}{Y^2}$

5. (2.3) $\left(\dfrac{xy^{-2}}{y^{-3}}\right)^2 = (xy^{-2-(-3)})^2 = (xy^{-2+(+3)})^2 = (xy^1)^2 = (xy)^2$
$= x^2y^2$

6. (2.2) $\dfrac{1}{a^{-3}} = a^3$ **7.** (2.2) $(15x)^0 = 1$

8. (2.2) $15(x^0) = 15(1) = 15$ **9.** (2.2) $5(z^{-2}) = 5 \cdot \dfrac{1}{z^2} = \dfrac{5}{z^2}$

10. (2.2) $(5z)^{-2} = \dfrac{1}{(5z)^2} = \dfrac{1}{5^2z^2} = \dfrac{1}{25z^2}$ **11.** (2.2) $(-2)^0 = 1$

12. (2.2) $(3^{-2})^{-1} = 3^{(-2)(-1)} = 3^2 = 9$

13. (2.2) $10^{-3} \cdot 10^5 = 10^{-3+5} = 10^2 = 100$

14. (2.2) $\dfrac{2^{-4}}{2^{-7}} = 2^{-4-(-7)} = 2^{-4+(+7)} = 2^3 = 8$

15. (2.7, 2.9) $7x - 2(5 - x) + \sqrt{81x^2}$
$= 7x + (-2)(5) + (-2)(-x) + 9x$
$= 7x - 10 + 2x + 9x = 18x - 10$

16. (2.8, 2.9) $(3x^2y - 2x) - 3x = 3x^2y - 2x - 3x = 3x^2y - 5x$

17. (2.8) $(3x^2y - 2x)(-3x) = (3x^2y)(-3x) + (-2x)(-3x)$
$= -9x^3y + 6x^2$

18. (2.3) $(3x^2y)(-2x)(-3x) = 3(-2)(-3)(x^2xx)(y) = 18x^4y$

19. (2.9) $6x(2xy^2 - 3x^3) - 3x^2(2y^2 - 6x^2)$
$= 6x(2xy^2) + 6x(-3x^3) + (-3x^2)(2y^2) + (-3x^2)(-6x^2)$
$= 12x^2y^2 - 18x^4 - 6x^2y^2 + 18x^4 = 6x^2y^2$

20. (2.9) $7x - 2\{6 - 3[8 - 2(x - 3) - 2(6 - x)]\}$
$= 7x - 2\{6 - 3[8 - 2x + 6 - 12 + 2x]\}$
$= 7x - 2\{6 - 3[2]\} = 7x - 2\{6 - 6\}$
$= 7x - 2\{0\} = 7x - 0 = 7x$

21. a. (2.4) $81{,}300{,}000 = 8.13 \times 10^7$
b. (2.4) $0.000\,000\,000\,38 = 3.8 \times 10^{-10}$

22. (2.4) $(8.13 \times 10^7)(3.8 \times 10^{-10}) = 0.030\,894$

23. (2.5) $\dfrac{-b - \sqrt{b^2 - 4ac}}{2a} = \dfrac{-(-1) - \sqrt{(-1)^2 - 4(3)(-2)}}{2(3)}$
$= \dfrac{1 - \sqrt{1 + 24}}{6} = \dfrac{1 - \sqrt{25}}{6} = \dfrac{1 - 5}{6} = \dfrac{-4}{6} = -\dfrac{2}{3}$

24. (2.6) $S = \dfrac{a(1 - r^n)}{1 - r} = \dfrac{-8[1 - (3)^2]}{1 - (3)} = \dfrac{-8[1 - 9]}{1 - 3} = \dfrac{-8[-8]}{-2}$
$= \dfrac{64}{-2} = -32$

25. (2.6) $P = 3{,}400{,}000{,}000 = 3.4 \times 10^9$, $r = 0.075$, $t = 3$
$I = Prt = (3.4 \times 10^9)(0.075)(3) = 765{,}000{,}000$
The interest is $765,000,000.

Chapters 1–2 Cumulative Review Exercises (page 97)

1. True **2.** True **3.** True **4.** False **5.** False
6. False **7.** False **8.** True **9.** False **10.** False
11. False **12.** False **13.** True **14.** False **15.** True

16. 24 **17.** $-(3^4) = -81$ **18.** 81 **19.** $3^{-4} = \dfrac{1}{3^4} = \dfrac{1}{81}$

20. $-\dfrac{1}{81}$ **21.** $(-3)^{-4} = \dfrac{1}{(-3)^4} = \dfrac{1}{81}$ **22.** $-\dfrac{1}{32}$

23. Not defined **24.** 2^7 **25.** $3^{2+(-6)} = 3^{-4} = \dfrac{1}{3^4} = \dfrac{1}{81}$

26. 7 **27.** -2 **28.** 5^3, or 125 **29.** Not defined

30. 0 **31.** $\frac{1}{4^{-2}} = 4^2 = 16$ **32.** 1 **33.** $a^{-5+3} = a^{-2} = \frac{1}{a^2}$

34. $2 + 3x - 9y + 3z$ **35.** $(8)(-4)(-2)(xy^2xx) = 64x^3y^2$

36. $-16x^2y^2 + 8x^2$ **37.** $8xy^2 - 4x - 2x = 8xy^2 - 6x$

38. $\frac{1}{d^5} + \frac{1}{d^3}$ **39.** $\frac{1}{b^4} \cdot \frac{1}{b^6} = \frac{1}{b^4b^6} = \frac{1}{b^{10}}$ **40.** 1

Exercises 3.1 (page 105)

Checks will not be shown.

1. $4x + 12 + 9x = -1$
$13x + 12 = -1$
$13x = -13$
$x = -1$
Solution set: $\{-1\}$; graph: −3 −2 −1 0 1

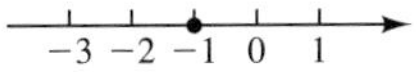

2. $\{-2\}$; graph: −3 −2 −1 0 1

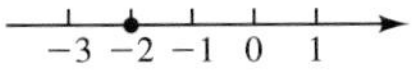

3. $7y - 10 - 8y = 8$
$-y = 18$
$y = -18$
Solution set: $\{-18\}$; graph: −20 −19 −18 −17 −16 −15

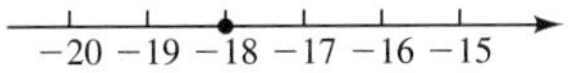

4. $\{-15\}$; graph: −17 −16 −15 −14 −13 −12

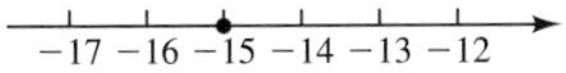

5. $4x + 12 = 12 + 4x$
$12 = 12$ True
$\{x \mid x \in R\}$; identity

6. $\{x \mid x \in R\}$; identity

7. $3[5 - 10 + 2z] = 6z + 14$
$3[-5 + 2z] = 6x + 14$
$-15 + 6z = 6z + 14$
$-15 = 14$ False
{ }; no solution

8. { }; no solution

9. LCM = 6
$(6)\left(\frac{x}{3} - \frac{x}{6}\right) = (6)(18)$
$(\overset{2}{\cancel{6}})\left(\frac{x}{\underset{1}{\cancel{3}}}\right) - (\overset{1}{\cancel{6}})\left(\frac{x}{\underset{1}{\cancel{6}}}\right) = 108$
$2x - x = 108$
$x = 108$
$\{108\}$; graph: 105 106 107 108 109 110

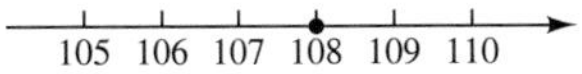

10. $\{128\}$; graph: 125 126 127 128 129 130 131

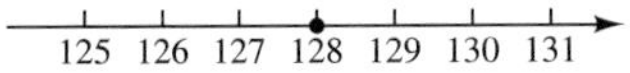

11. LCD = 40
$\frac{\overset{5}{\cancel{40}}}{1}\left(\frac{y+3}{\underset{1}{\cancel{8}}}\right) - \frac{\overset{10}{\cancel{40}}}{1}\left(\frac{3}{\underset{1}{\cancel{4}}}\right) = \frac{\overset{4}{\cancel{40}}}{1}\left(\frac{y+6}{\underset{1}{\cancel{10}}}\right)$
$5y + 15 - 30 = 4y + 24$
$5y - 15 = 4y + 24$
$y = 39$
$\{39\}$

12. $\{-25\}$

13. $5z - 6 - 9z = 6$
$-6 - 4z = 6$
$-4z = 12$
$z = -3$
$-3 \notin N$; { }; no solution

14. { }; no solution

15. $7x - 10 - 8x = 8$
$-x = 18$
$x = -18$
$x \in J$; $\{-18\}$

16. $\{0\}$

17. $6x - 12 - 15x - 12 = 35x - 40$
$-9x - 24 = 35x - 40$
$16 = 44x$
$x = \frac{16}{44} = \frac{4}{11}$
$\frac{4}{11} \notin J$; { }; no solution

18. { }; no solution

19. LCD = 10
$\frac{\overset{2}{\cancel{10}}}{1} \cdot \frac{2(y-3)}{\underset{1}{\cancel{5}}} - \frac{\overset{5}{\cancel{10}}}{1} \cdot \frac{3(y+2)}{\underset{1}{\cancel{2}}} = \frac{\overset{1}{\cancel{10}}}{1} \cdot \frac{7}{\underset{1}{\cancel{10}}}$
$4y - 12 - 15y - 30 = 7$
$-11y = 49$
$y = \frac{-49}{11}$, or $-4\frac{5}{11}$
$\left\{-4\frac{5}{11}\right\}$

20. $\left\{\frac{81}{11}\right\}$, or $\left\{7\frac{4}{11}\right\}$

21. $6.23x + 2.5(3.08 - 8.2x) = -14.7$
$6.23x + 7.7 - 20.5x = -14.7$
$-14.27x + 7.7 = -14.7$
$-14.27x = -22.4$
$x \approx 1.57$
$\{1.57\}$

22. $\{0.97\}$

Exercises 3.2 (page 114)

1. $[5, +\infty)$; 4 5 6 7 8

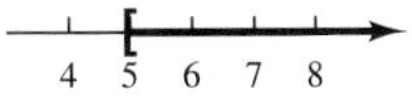

2. $(-\infty, 2]$; −1 0 1 2 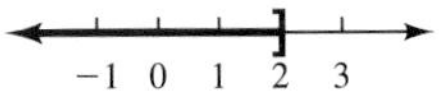 3

3. $(-\infty, -3)$; −6 −5 −4 −3 −2

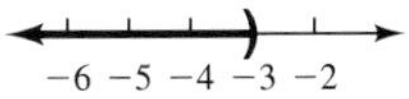

4. $(-1, +\infty)$; −2 −1 0 1 2

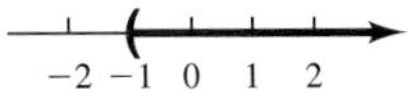

5. $\{x \mid x \ge 1\}$; $[1, +\infty)$ **6.** $\{x \mid x \ge -2\}$; $[-2, +\infty)$

7. $\{x \mid x < 4\}$; $(-\infty, 4)$ **8.** $\{x \mid x < 6\}$; $(-\infty, 6)$

9. $3x - 1 < 11$
$3x < 12$
$x < 4$
$\{x \mid x < 4\}$; $(-\infty, 4)$;
1 2 3 4 5

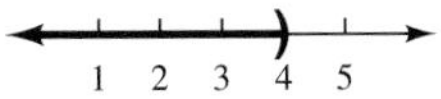

10. $\{x \mid x < 6\}$; $(-\infty, 6)$;
3 4 5 6 7

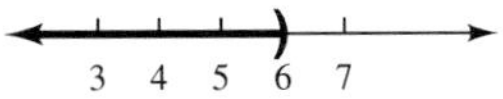

11. $17 \ge 2x - 9$
$26 \ge 2x$
$13 \ge x$
$x \le 13$
$\{x \mid x \le 13\}$; $(-\infty, 13]$;
11 12 13 14 15

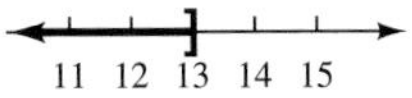

12. $\{x \mid x \ge -7\}$; $[-7, +\infty)$;
−9 −8 −7 −6 −5

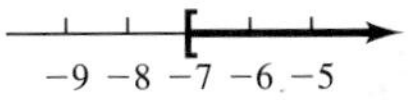

13. $2y - 16 > 17 + 5y$
$-16 > 17 + 3y$
$-33 > 3y$
$-11 > y$
$\{y \mid y < -11\}$; $(-\infty, -11)$;
−13 −12 −11 −10

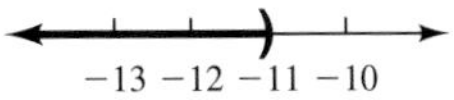

14. $\{y \mid y > -5\}$; $(-5, +\infty)$;
−6 −5 −4 −3 −2

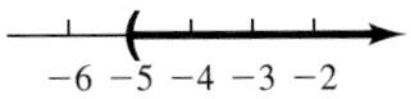

15. $4z - 22 < 6z - 42$
$-22 < 2z - 42$
$20 < 2z$
$10 < z$
$\{z \mid z > 10\}$; $(10, +\infty)$;
9 10 11 12 13

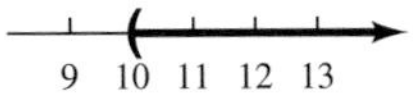

16. $\{a \mid a < -2\}$; $(-\infty, -2)$;
−4 −3 −2 −1 0

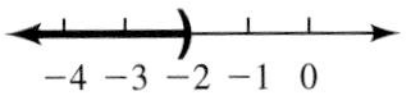

17. $18 - 45m - 4 \geq 13m + 24 - 56m$
$14 - 45m \geq -43m + 24$
$14 \geq 2m + 24$
$-10 \geq 2m$
$-5 \geq m$
$\{m \mid m \leq -5\}$; $(-\infty, -5]$;
−7 −6 −5 −4 −3

18. $\{k \mid k \leq 1\}$; $(-\infty, 1]$;
−1 0 1 2 3

19. $10 - 5x > 2[3 - 5x + 20]$
$10 - 5x > 2[23 - 5x]$
$10 - 5x > 46 - 10x$
$10 + 5x > 46$
$5x > 36$
$x > \frac{36}{5}$, or $x > 7\frac{1}{5}$
$\{x \mid x > 7\frac{1}{5}\}$; $(7\frac{1}{5}, +\infty)$;
5 6 7 8 9

20. $\{y \mid y < -6\}$; $(-\infty, -6)$;
−8 −7 −6 −5 −4 −3

21. LCD = 12
$$\frac{\overset{4}{\cancel{12}}}{1} \cdot \frac{z}{\underset{1}{\cancel{3}}} > \frac{12}{1} \cdot \frac{7}{1} - \frac{\overset{3}{\cancel{12}}}{1} \cdot \frac{z}{\underset{1}{\cancel{4}}}$$
$4z > 84 - 3z$
$7z > 84$
$z > 12$
$\{z \mid z > 12\}$; $(12, +\infty)$;
10 11 12 13 14

22. $\{t \mid t > 15\}$; $(15, +\infty)$;
13 14 15 16 17

23. LCD = 15
$$\frac{\overset{5}{\cancel{15}}}{1} \cdot \frac{1}{\underset{1}{\cancel{3}}} + \frac{\overset{3}{\cancel{15}}}{1} \cdot \frac{(w + 2)}{\underset{1}{\cancel{5}}} \geq \frac{\overset{5}{\cancel{15}}}{1} \frac{(w - 5)}{\underset{1}{\cancel{3}}}$$
$5 + 3w + 6 \geq 5w - 25$
$11 \geq 2w - 25$
$36 \geq 2w$
$18 \geq w$, or $w \leq 18$
$\{w \mid w \leq 18\}$; $(-\infty, 18]$;
16 17 18 19 20

24. $\{u \mid u \geq 6\}$; $[6, +\infty)$;
4 5 6 7 8

25. $14.73(2.65x - 11.08) - 22.51x \geq 13.94x(40.27)$
$39.0345x - 163.2084 - 22.51x \geq 561.3638x$
$-544.8393x \geq 163.2084$
$x \leq -0.300$ (approx.)
$\{x \mid x \leq -0.300\}$ (approx.)

26. $\{x \mid x \geq 0.723\}$ (approx.)

27. $x + 3 < 10$
$x < 7$
$x \in N$
The natural numbers < 7 are 1, 2, 3, 4, 5, and 6.
$\{1, 2, 3, 4, 5, 6\}$;
0 1 2 3 4 5 6 7 8

28. $\{1, 2\}$;
0 1 2 3 4

29. $2(x + 3) \leq 11$
$2x + 6 \leq 11$
$2x \leq 5$
$x \leq \frac{5}{2}$
The natural numbers $\leq \frac{5}{2}$ are 1 and 2.
$\{1, 2\}$;
0 1 2 3 4 5

30. $\{1, 2, 3, 4\}$;
0 1 2 3 4

Exercises 3.3 (page 121)

1. $\{x \mid 0 \leq x < 3\}$; $[0, 3)$
2. $\{x \mid 1 < x \leq 4\}$; $(1, 4]$
3. $\{x \mid 3 < x \leq 6\}$; $(3, 6]$
4. $\{x \mid 6 \leq x < 8)\}$; $[6, 8)$
5. $\{x \mid -1 \leq x \leq 1\}$; $[-1, 1]$
6. $\{x \mid 1 \leq x \leq 2\}$; $[1, 2]$
7. $\{x \mid -2 < x < 1\}$; $(-2, 1)$
8. $\{x \mid -3 < x < -1\}$; $(-3, -1)$

9. $5 > x - 2 \geq 3$
$5 + 2 > x - 2 + 2 \geq 3 + 2$
$7 > x \geq 5$
or $5 \leq x < 7$
$[5, 7)$;
4 5 6 7 8 9

10. $[7, 10)$;
6 7 8 9 10 11

11. "$-5 \geq 2$" is false; therefore, the solution set is { }; there is no graph.

12. The solution set is { }; there is no graph.

13. $-4 < 3x - 1 \leq 7$
$-4 + 1 < 3x - 1 + 1 \leq 7 + 1$
$-3 < 3x \leq 8$
$\frac{-3}{3} < \frac{3x}{3} \leq \frac{8}{3}$
$-1 < x \leq \frac{8}{3}$
$(-1, \frac{8}{3}]$;
−2 −1 0 1 2 3 4

14. $(-1, \frac{7}{4}]$;
−2 −1 0 1 2

15. $x - 1 > 3$ or $x - 1 < -3$
$x - 1 + 1 > 3 + 1$ or $x - 1 + 1 < -3 + 1$
$x > 4$ or $x < -2$
$(-\infty, -2) \cup (4, +\infty)$;
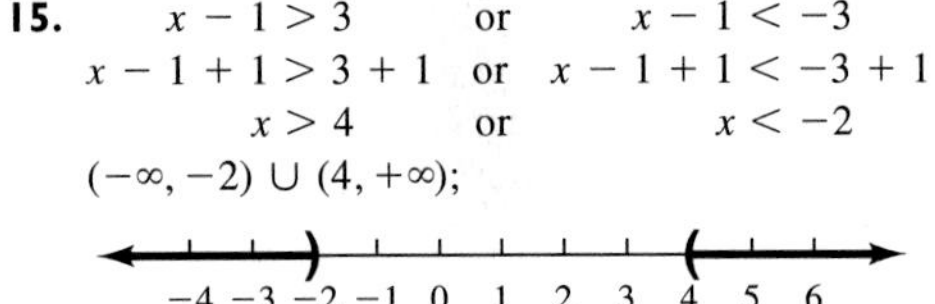

16. $(-\infty, -3) \cup (7, +\infty)$;
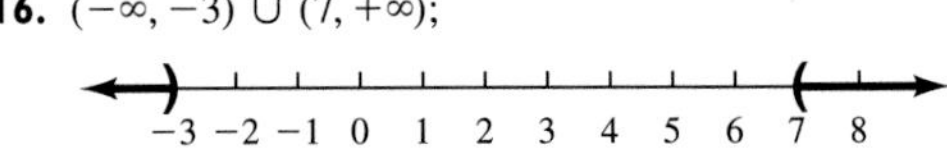

17. $2x + 1 \geq 3$ or $2x + 1 \leq -3$
$2x + 1 - 1 \geq 3 - 1$ or $2x + 1 - 1 \leq -3 - 1$
$2x \geq 2$ or $2x \leq -4$
$x \geq 1$ or $x \leq -2$
$(-\infty, -2] \cup [1, +\infty)$;
−4 −3 −2 −1 0 1 2 3 4

18. $(-\infty, -1] \cup [\frac{7}{3}, +\infty)$;
−4 −3 −2 −1 0 1 2 3 4 5 6

19.
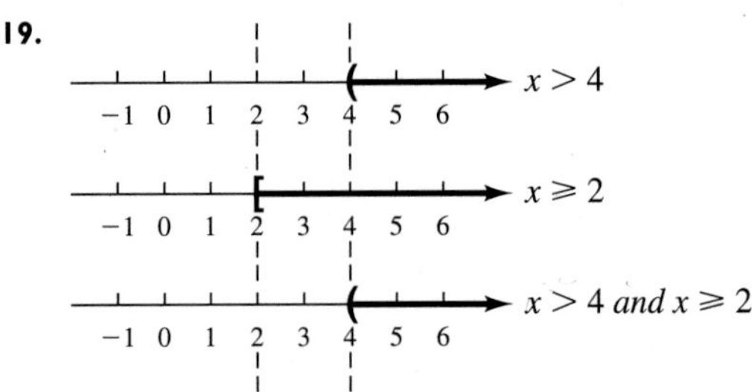

Final inequality: $x > 4$; $(4, +\infty)$

20. $(-\infty, -1)$;
−4 −3 −2 −1 0 1

21. $x > 4$ or $x \geq 2$;

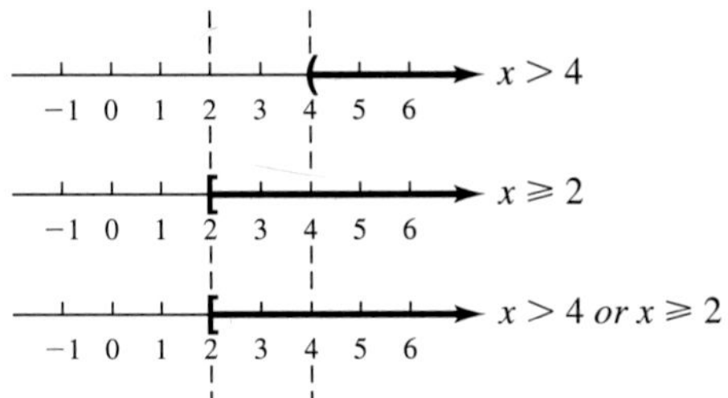

Final inequality: $x \geq 2$; $[2, +\infty)$

22. $(-\infty, 3)$;

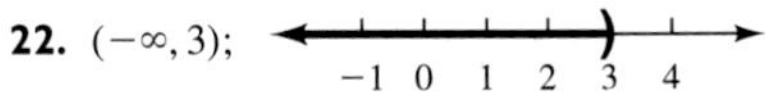

23.
$$\begin{aligned} -5 &\leq x - 3 \leq 2 \\ -5 + 3 &\leq x - 3 + 3 \leq 2 + 3 \\ -2 &\leq x \leq 5 \end{aligned}$$
but $x \in N$; the solution set is $\{1, 2, 3, 4, 5\}$.

24. $\{1, 2, 3, 4, 5, 6\}$;

25.
$$\begin{aligned} 4 &\geq x - 3 > -5 \\ 4 + 3 &\geq x - 3 + 3 > -5 + 3 \\ 7 &\geq x > -2 \end{aligned}$$
but $x \in J$; the solution set is $\{-1, 0, 1, 2, 3, 4, 5, 6, 7\}$.

26. $\{-1, 0, 1, 2, 3, 4, 5, 6, 7, 8\}$;

27.
$$\begin{aligned} -3 &\leq 2x + 1 \leq 7 \\ -3 - 1 &\leq 2x + 1 - 1 \leq 7 - 1 \\ -4 &\leq 2x \leq 6 \\ -2 &\leq x \leq 3 \end{aligned}$$
but $x \in N$; the solution set is $\{1, 2, 3\}$.

28. $\{1\}$;

Exercises 3.4 (page 129)

1. If $|x| = 3$, $x = 3$ or $x = -3$.
$\{3, -3\}$;

2. $\{5, -5\}$;

3. If $|3x| = 12$,
$3x = 12$ or $3x = -12$
$x = 4$ or $x = -4$
$\{4, -4\}$;

4. $\{5, -5\}$;

5. If $|x| < 2$, $-2 < x < 2$.
$\{x \mid -2 < x < 2\}$;

6. $\{x \mid -7 < x < 7\}$;

7. If $|4x| < 12$,
$-12 < 4x < 12$
$-3 < x < 3$
$\{x \mid -3 < x < 3\}$;

8. $\{x \mid -3 < x < 3\}$;

9. If $|5x| \leq 25$,
$-25 \leq 5x \leq 25$
$-5 \leq x \leq 5$
$\{x \mid -5 \leq x \leq 5\}$;

10. $\{x \mid -1 \leq x \leq 1\}$;

11. If $|x| > 2$, $x > 2$ or $x < -2$.
$\{x \mid x > 2 \text{ or } x < -2\}$;

12. $\{x \mid x > 3 \text{ or } x < -3\}$;

13. Since $|3x|$ is always ≥ 0, $|3x|$ will always be > -3. Therefore, the solution set is $\{x \mid x \in R\}$.

14. $\{x \mid x \in R\}$;

15. If $|x + 2| = 5$,
$x + 2 = 5$ or $x + 2 = -5$
$x = 3$ or $x = -7$
$\{3, -7\}$;

16. $\{-6, 0\}$;

17. If $|x - 3| < 2$, then
$$\begin{aligned} -2 &< x - 3 < 2 \\ -2 + 3 &< x - 3 + 3 < 2 + 3 \\ 1 &< x < 5 \end{aligned}$$
$\{x \mid 1 < x < 5\}$;

18. $\{x \mid 3 < x < 5\}$;

19. Since $|x + 4|$ is always ≥ 0, $|x + 4|$ can *never* be ≤ -3. Therefore, there is no solution; the solution set is $\{\ \}$; (no graph).

20. Solution set is $\{\ \}$; no graph

21. If $|x + 1| > 3$, then
$x + 1 > 3$ or $x + 1 < -3$
$x > 2$ or $x < -4$
$\{x \mid x > 2 \text{ or } x < -4\}$;

22. $\{x \mid x > 6 \text{ or } x < -2\}$;

23. If $|x + 5| \geq 2$,
$x + 5 \geq 2$ or $x + 5 \leq -2$
$x \geq -3$ or $x \leq -7$
$\{x \mid x \leq -7 \text{ or } x \geq -3\}$;

24. $\{x \mid x \geq -3 \text{ or } x \leq -5\}$;

25. If $|3x + 4| = 3$,
$3x + 4 = 3$ or $3x + 4 = -3$
$3x = -1$ or $3x = -7$
$x = -\frac{1}{3}$ or $x = -\frac{7}{3}$
$\left\{-\frac{1}{3}, -\frac{7}{3}\right\}$;

26. $\left\{-2, \frac{1}{2}\right\}$;

27. If $|2x - 3| < 4$, then
$$\begin{aligned} -4 &< 2x - 3 < 4 \\ -4 + 3 &< 2x - 3 + 3 < 4 + 3 \\ -1 &< 2x < 7 \\ -\tfrac{1}{2} &< \frac{2x}{2} < \tfrac{7}{2} \end{aligned}$$
$\left\{x \mid -\frac{1}{2} < x < \frac{7}{2}\right\}$;

3

28. $\{x \mid -\frac{4}{3} < x < 2\}$;

−2 −1 0 1 2 3 4

29. If $|3x - 5| \geq 6$,

$3x - 5 \geq 6$ or $3x - 5 \leq -6$

$3x \geq 11$ or $3x \leq -1$

$x \geq \frac{11}{3}$ or $x \leq -\frac{1}{3}$

$\{x \mid x \geq \frac{11}{3}$ or $x \leq -\frac{1}{3}\}$;

−2 −1 0 1 2 3 4 5

30. $\{x \mid x \geq 1$ or $x \leq -\frac{1}{2}\}$;

−2 −1 0 1 2

31. If $|1 - 2x| \leq 5$, then

$-5 \leq 1 - 2x \leq 5$

$-1 - 5 \leq -1 + 1 - 2x \leq -1 + 5$

$-6 \leq -2x \leq 4$

$\frac{-6}{-2} \geq \frac{-2x}{-2} \geq \frac{4}{-2}$

$3 \geq x \geq -2$

or $-2 \leq x \leq 3$

$\{x \mid -2 \leq x \leq 3\}$;

−3 −2 −1 0 1 2 3 4

32. $\{x \mid -\frac{2}{3} < x < 2\}$;

−2 −1 0 1 2 3

33. If $|2 - 3x| > 4$,

$2 - 3x > 4$ or $2 - 3x < -4$

$-3x > 2$ or $-3x < -6$

$x < -\frac{2}{3}$ or $x > 2$

$\{x \mid x < -\frac{2}{3}$ or $x > 2\}$;

−2 −1 0 1 2 3 4

34. $\{x \mid x \leq -\frac{1}{2}$ or $x \geq \frac{11}{2}\}$;

−2 −1 0 1 2 3 4 5 6 7

35. If $\left|\frac{5x + 2}{3}\right| \geq 2$,

$\frac{5x + 2}{3} \geq 2$ or $\frac{5x + 2}{3} \leq -2$

$5x + 2 \geq 6$ or $5x + 2 \leq -6$

$5x \geq 4$ or $5x \leq -8$

$x \geq \frac{4}{5}$ or $x \leq -\frac{8}{5}$

$\{x \mid x \geq \frac{4}{5}$ or $x \leq -\frac{8}{5}\}$;

−3 −2 −1 0 1 2

36. $\{x \mid x \geq 3$ or $x \leq -\frac{9}{2}\}$;

−6 −5 −4 −3 −2 −1 0 1 2 3 4

37. If $\left|\frac{3x - 4}{5}\right| < 1$,

$-1 < \frac{3x - 4}{5} < 1$

$-5 < 3x - 4 < 5$

$-1 < 3x < 9$

$-\frac{1}{3} < x < 3$

$\{x \mid -\frac{1}{3} < x < 3\}$;

−2 −1 0 1 2 3 4 5

38. $\{x \mid -\frac{3}{5} < x < 1\}$;

−1 0 1 2

39. If $\left|\frac{1 - x}{2}\right| = 3$, then

$\frac{1 - x}{2} = 3$ or $\frac{1 - x}{2} = -3$

$1 - x = 6$ or $1 - x = -6$

$-x = 5$ or $-x = -7$

$x = -5$ or $x = 7$

$\{-5, 7\}$;

−5 −4 −3 −2 −1 0 1 2 3 4 5 6 7 8

40. $\{-1, 7\}$;

−3 −2 −1 0 1 2 3 4 5 6 7 8

41. If $\left|\frac{5 - x}{2}\right| \leq 2$, then

$-2 \leq \frac{5 - x}{2} \leq 2$

$-4 \leq 5 - x \leq 4$

$-9 \leq -x \leq -1$

$\frac{-9}{-1} \geq \frac{-x}{-1} \geq \frac{-1}{-1}$

$9 \geq x \geq 1$

or $1 \leq x \leq 9$

$\{x \mid 1 \leq x \leq 9\}$;

−1 0 1 2 3 4 5 6 7 8 9 10

42. $\{x \mid 1 \leq x \leq 7\}$;

−1 0 1 2 3 4 5 6 7 8 9

43. If $\left|3 - \frac{x}{2}\right| > 2$, then

$3 - \frac{x}{2} > 2$ or $3 - \frac{x}{2} < -2$

$6 - x > 4$ or $6 - x < -4$

$-x > -2$ or $-x < -10$

$x < 2$ or $x > 10$

$\{x \mid x < 2\} \cup \{x \mid x > 10\}$;

−1 0 1 2 3 4 5 6 7 8 9 10 11 12 13

44. $\{x \mid x < 9$ or $x > 15\}$;

7 8 9 10 11 12 13 14 15 16 17

45. If $\left|4 - \frac{x}{2}\right| < 2$, then

$-2 < 4 - \frac{x}{2} < 2$

$-4 < 8 - x < 4$

$-12 < -x < -4$

$\frac{-12}{-1} > \frac{-x}{-1} > \frac{-4}{-1}$

$12 > x > 4$

or $4 < x < 12$

$\{x \mid 4 < x < 12\}$;

0 1 2 3 4 5 6 7 8 9 10 11 12 13

46. $\{x \mid 9 < x < 21\}$;

8 9 10 11 12 13 14 15 16 17 18 19 20 21 22

Exercises 3.5 (page 134)

The checks for these exercises will not be shown.

1. Let x = unknown number

$2x + 7 = 23$

$2x = 16$

$x = 8$ The number is 8.

2. 9

3. Let x = unknown number

$4x - 7 = 25$

$4x = 32$

$x = 8$ The number is 8.

4. 11

5. Let x = unknown number

$x - 7 = \frac{1}{2}x$

$2(x - 7) = 2\left(\frac{1}{2}x\right)$

$2x - 14 = x$

$x = 14$ The number is 14.

6. 24

7. Let x = number of cm in second piece of string

Then $3x$ = number of cm in first piece of string

$x + 3x = 12$

$4x = 12$

$x = 3$

$3x = 9$

The first piece is 9 cm long, and the second piece is 3 cm long.

8. 8 m, 4m

9. Let x = first integer
Then $x + 1$ = second integer
and $x + 2$ = third integer

$$x + (x + 1) + (x + 2) = 19$$
$$3x + 3 = 19$$
$$3x = 16$$
$$x = \tfrac{16}{3}$$

$\tfrac{16}{3}$ is not an integer; no solution

10. No solution

11. Let x = length of second piece
Then $x + 8$ = length of first piece

$$x + (x + 8) = 42$$
$$2x + 8 = 42$$
$$2x = 34$$
$$x = 17$$
$$x + 8 = 25$$

The first piece is 25 cm long, and the second piece is 17 cm long.

12. 28 m, 22 m

13. Let x = first integer
Then $x + 1$ = second integer
and $x + 2$ = third integer

$$x + (x + 1) - (x + 2) = 10$$
$$x + x + 1 - x - 2 = 10$$
$$x - 1 = 10$$
$$x = 11$$
$$x + 1 = 12$$
$$x + 2 = 13$$

The integers are 11, 12, and 13.

14. 18, 19, and 20

15. Let x = first odd integer
Then $x + 2$ = second odd integer
and $x + 4$ = third odd integer

$$3[(x + 2) + (x + 4)] = 40 + 5(x)$$
$$3(2x + 6) = 40 + 5x$$
$$6x + 18 = 40 + 5x$$
$$x = 22$$

22 is not an odd integer; no solution

16. No solution

17. Let x = number of cans of peas
Then $x + 4$ = number of cans of corn
and $3x$ = number of cans of green beans

$$x + (x + 4) + (3x) = 24$$
$$5x + 4 = 24$$
$$5x = 20$$
$$x = 4$$
$$x + 4 = 8$$
$$3x = 12$$

David bought 4 cans of peas, 8 cans of corn, and 12 cans of green beans.

18. 3 cans of pears, 9 cans of peaches, 9 cans of cherries

19. Let x = unknown number

$$3(8 + x) = 2(x + 7)$$
$$24 + 3x = 2x + 14$$
$$x = -10$$

The number is -10.

20. 7

21. Let x = unknown number

$$8x - 2(5 + x) = 4(8 + 2x)$$
$$8x - 10 - 2x = 32 + 8x$$
$$-42 = 2x$$
$$x = -21$$

The number is -21.

22. 2

23. Let x = first even integer
Then $x + 2$ = second even integer
and $x + 4$ = third even integer

$$21 < x + (x + 2) + (x + 4) < 45$$
$$21 < 3x + 6 < 45$$
$$15 < 3x < 39$$
$$5 < x < 13$$

The even integers between 5 and 13 are 6, 8, 10, and 12. Therefore, the answers are 6, 8, and 10; 8, 10, and 12; 10, 12, and 14; and 12, 14, and 16.

24. 26, 28, and 30; and 28, 30, and 32

25. Let x = score needed

$$x \le 200$$
$$560 \le 396 + x \le 640$$
$$164 \le x \le 244$$

But $x \le 200$, so $164 \le x \le 200$. To receive a C, Clarke must have a score between 164 and 200 on the final exam.

26. 136 to 200

Sections 3.1–3.5 Review Exercises (page 137)

1.

$$7 - 2M + 8 = 5$$
$$-2M + 15 = 5$$
$$\frac{-2M}{-2} = \frac{-10}{-2}$$
$$M = 5$$

$\{5\}$

2. No solution; { }

3. LCD = 12

$$\frac{\overset{3}{\cancel{12}}}{1} \cdot \frac{3(x + 3)}{\underset{1}{\cancel{4}}} - \frac{\overset{4}{\cancel{12}}}{1} \cdot \frac{2(x - 3)}{\underset{1}{\cancel{3}}} = \frac{12}{1} \cdot 1$$
$$9(x + 3) - 8(x - 3) = 12$$
$$9x + 27 - 8x + 24 = 12$$
$$x = -39$$

$\{-39\}$

4. $\{x \mid x \in R\}$; identity

5.

$$2[-7y - 15 + 12y + 10] = 10y - 12$$
$$2[5y - 5] = 10y - 12$$
$$10y - 10 = 10y - 12$$
$$0 = -2 \quad \text{False}$$

{ }; no solution

6. $\left\{x \mid x \le \tfrac{1}{3}\right\}$

7. LCD = 30

$$\frac{\overset{6}{\cancel{30}}}{1} \cdot \frac{3z}{\underset{1}{\cancel{5}}} + \frac{\overset{10}{\cancel{30}}}{1} \cdot \frac{-2z}{\underset{1}{\cancel{3}}} < \frac{\overset{15}{\cancel{30}}}{1} \cdot \frac{1}{\underset{1}{\cancel{2}}}$$
$$18z - 20z < 15$$
$$-2z < 15$$
$$z > -\tfrac{15}{2}$$

$\left\{z \mid z > -\tfrac{15}{2}\right\}$

8. $\left\{w \mid w > -3\tfrac{1}{2}\right\}$

9. LCD = 20

$$\frac{\overset{2}{\cancel{20}}}{1} \cdot \frac{2(x + 6)}{\underset{1}{\cancel{10}}} + \frac{\overset{1}{\cancel{20}}}{1} \cdot \frac{3x}{\underset{1}{\cancel{20}}} < \frac{20}{1} \cdot \frac{3}{1}$$
$$4(x + 6) + 3x < 60$$
$$4x + 24 + 3x < 60$$
$$7x < 36$$
$$x < 5\tfrac{1}{7}$$

$\left\{x \mid x < 5\tfrac{1}{7}\right\}$

10. $\left\{x \mid x > -9\tfrac{1}{11}\right\}$

11.

$$3x + 2 = 11$$
$$3x = 9$$
$$x = 3$$

$\{3\}$; 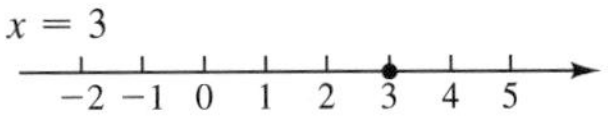

12. $\{3, -3\}$; 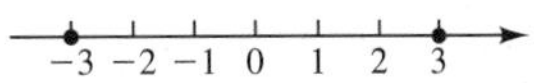

13. Since $|x|$ is always ≥ 0, $|x|$ will *never* be ≤ -3. Therefore, there is no solution; the solution set is { }, and there are no points to graph.

14. $\{x|x \geq 2\} \cup \{x|x \leq -2\}$; $(-\infty, -2] \cup [2, +\infty)$;

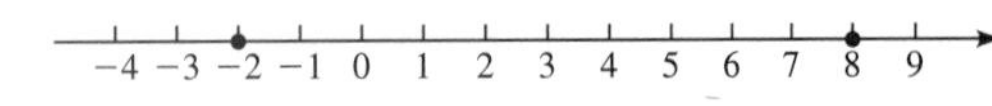

15. If $|6 - 2x| = 10$,

$$\begin{aligned} 6 - 2x = 10 &\text{ or } 6 - 2x = -10 \\ -2x = 4 &\text{ or } -2x = -16 \\ x = -2 &\text{ or } x = 8 \end{aligned}$$

$\{-2, 8\}$;

16. $\{x|0 \leq x \leq 4\}$; $[0, 4]$;

17.

$$\begin{aligned} 2|x - 3| &\leq 4 \\ |x - 3| &\leq 2 \\ -2 \leq x - 3 &\leq 2 \\ 1 \leq x &\leq 5 \end{aligned}$$

$\{x|1 \leq x \leq 5\}$; $[1, 5]$;

18. $\left\{x\middle|-1 < x < \frac{9}{2}\right\}$; $\left(-1, \frac{9}{2}\right)$;

19. $-2 < x + 1 < 0$

$-3 < x < -1$

$\{x|-3 < x < -1, x \in J\} = \{-2\}$;

20. $\{3\}$;

21.

$$\frac{\overset{2}{\cancel{8}}}{1} \cdot \frac{3(x-1)}{\underset{1}{\cancel{4}}} + \frac{\overset{1}{\cancel{8}}}{1} \cdot \frac{x}{\underset{1}{\cancel{8}}} = \frac{8}{1} \cdot \frac{1}{1}$$

$$\begin{aligned} 6(x - 1) + x &= 8 \\ 6x - 6 + x &= 8 \\ 7x - 6 &= 8 \\ 7x &= 14 \\ x &= 2 \end{aligned}$$

$\{2\}$;

22. $\{x|3 \leq x \leq 9\}$; $[3, 9]$;

23. If $\left|\dfrac{2x - 4}{3}\right| \geq 2$,

$$\frac{2x-4}{3} \geq 2 \quad \text{or} \quad \frac{2x-4}{3} \leq -2$$

$$\frac{\overset{1}{\cancel{3}}}{1} \cdot \frac{2x-4}{\underset{1}{\cancel{3}}} \geq \frac{3}{1} \cdot \frac{2}{1} \quad \text{or} \quad \frac{\overset{1}{\cancel{3}}}{1} \cdot \frac{2x-4}{\underset{1}{\cancel{3}}} \leq \frac{3}{1}\left(\frac{-2}{1}\right)$$

$$\begin{aligned} 2x - 4 \geq 6 &\quad \text{or} \quad 2x - 4 \leq -6 \\ 2x \geq 10 &\quad \text{or} \quad 2x \leq -2 \\ x \geq 5 &\quad \text{or} \quad x \leq -1 \end{aligned}$$

$\{x|x \geq 5 \text{ or } x \leq -1\}$; $(-\infty, -1] \cup [5, +\infty)$;

24. $\{x|x < 1\}$; $(-\infty, 1)$ **25.** $\{x|-3 < x \leq 1\}$; $(-3, 1]$

26. $\{x|x \geq 4\}$; $[4, +\infty)$

The check will not be shown for 27.

27. Let x = unknown number **28.** No solution

$$\frac{28 - 5x}{3} = 3x$$

$$\frac{\overset{1}{\cancel{3}}}{1} \cdot \frac{28 - 5x}{\underset{1}{\cancel{3}}} = \frac{3}{1}(3x)$$

$$\begin{aligned} 28 - 5x &= 9x \\ 28 &= 14x \\ x &= 2 \end{aligned}$$

The number is 2.

Chapter 3 Diagnostic Test (page 141)

Following each problem number is the number (in parentheses) of the textbook section where that kind of problem is discussed. (No checks will be shown.)

1. (3.1)

$$\begin{aligned} 12(3x - 5) &= 8[5 - 2(x + 4)] \\ 36x - 60 &= 8[5 - 2x - 8] \\ 36x - 60 &= 8[-3 - 2x] \\ 36x - 60 &= -24 - 16x \\ 52x &= 36 \\ x &= \tfrac{36}{52} = \tfrac{9}{13} \end{aligned}$$

$\left\{\frac{9}{13}\right\}$

2. (3.1)

$$\begin{aligned} 2[7x - 4(1 + 3x)] &= 5(3 - 2x) - 23 \\ 2[7x - 4 - 12x] &= 15 - 10x - 23 \\ 2[-5x - 4] &= -8 - 10x \\ -10x - 8 &= -8 - 10x \\ 0 &= 0 \quad \text{True} \end{aligned}$$

$\{x|x \in R\}$; identity

3. (3.1) LCD = 12

$$12\left(\frac{x}{6} - \frac{x+2}{4}\right) = 12\left(\frac{1}{3}\right)$$

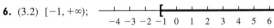

$$12\left(\frac{x}{6}\right) - 12\left(\frac{x+2}{4}\right) = 4$$

$$\begin{aligned} 2x - 3(x + 2) &= 4 \\ 2x - 3x - 6 &= 4 \\ -x - 6 &= 4 \\ -x &= 10 \\ x &= -10 \end{aligned}$$

$\{-10\}$

4. (3.1)

$$\begin{aligned} 3(x - 6) &= 5(1 + 2x) - 7(x - 4) \\ 3x - 18 &= 5 + 10x - 7x + 28 \\ 3x - 18 &= 3x + 33 \\ -18 &= 33 \quad \text{False} \end{aligned}$$

{ }; no solution

5. (3.2) $(-\infty, -3)$;

6. (3.2) $[-1, +\infty)$;

7. (3.3) $\{x|2 < x \leq 6\}$; $(2, 6]$ **8.** (3.2) $\{x|x \geq 5\}$; $[5, +\infty)$

9. (3.2)

$$\begin{aligned} 5w + 2 &\leq 10 - w \\ 6w &\leq 8 \\ w &\leq \tfrac{8}{6} \\ w &\leq \tfrac{4}{3} \end{aligned}$$

$\left\{w\middle|w \leq \frac{4}{3}\right\}$; $\left(-\infty, \frac{4}{3}\right]$;

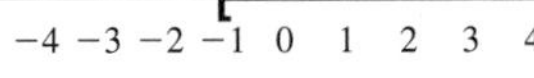

10. (3.2)

$$\begin{aligned} 13h - 4(2 + 3h) &\geq 0 \\ 13h - 8 - 12h &\geq 0 \\ h - 8 &\geq 0 \\ h &\geq 8 \end{aligned}$$

$\{h|h \geq 8\}$; $[8, +\infty)$;

11. (3.3) $-3 < x + 1 < 5$

$$-3 - 1 < x + 1 - 1 < 5 - 1$$
$$-4 < x < 4$$

$\{x \mid -4 < x < 4, x \in J\} = \{-3, -2, -1, 0, 1, 2, 3\}$;

12. (3.3) $4 \geq 3x + 7 > -2$

$$4 - 7 \geq 3x + 7 - 7 > -2 - 7$$
$$-3 \geq 3x > -9$$
$$-1 \geq x > -3$$

$\{x \mid -3 < x \leq -1, x \in R\}$; $(-3, -1]$;

13. (3.2) LCD = 12

$$\frac{\overset{4}{\cancel{12}}}{1} \cdot \frac{5(x-2)}{\underset{1}{\cancel{3}}} + \frac{\overset{3}{\cancel{12}}}{1} \cdot \frac{x}{\underset{1}{\cancel{4}}} \leq \frac{12}{1} \cdot \frac{12}{1}$$
$$20(x - 2) + 3x \leq 144$$
$$20x - 40 + 3x \leq 144$$
$$23x \leq 184$$
$$x \leq 8$$

$\{x \mid x \leq 8, x \in R\}$; $(-\infty, 8]$;

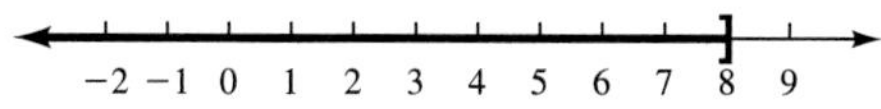

14. (3.4) If $\left|\frac{2x+3}{5}\right| = 1$,

$$\frac{2x+3}{5} = 1 \quad \text{or} \quad \frac{2x+3}{5} = -1$$
$$2x + 3 = 5 \quad \text{or} \quad 2x + 3 = -5$$
$$2x = 2 \quad \text{or} \quad 2x = -8$$
$$x = 1 \quad \text{or} \quad x = -4$$

$\{-4, 1\}$;

15. (3.4) If $|3x - 1| > 2$,

$$3x - 1 > 2 \quad \text{or} \quad 3x - 1 < -2$$
$$3x > 3 \quad \text{or} \quad 3x < -1$$
$$x > 1 \quad \text{or} \quad x < -\tfrac{1}{3}$$

$\{x \mid x > 1\} \cup \left\{x \mid x < -\tfrac{1}{3}\right\}$; $\left(-\infty, -\tfrac{1}{3}\right) \cup (1, +\infty)$;

16. (3.4) If $|7 - 3x| \geq 6$,

$$7 - 3x \geq 6 \quad \text{or} \quad 7 - 3x \leq -6$$
$$7 \geq 3x + 6 \quad \text{or} \quad 7 \leq 3x - 6$$
$$1 \geq 3x \quad \text{or} \quad 13 \leq 3x$$
$$\tfrac{1}{3} \geq x \quad \text{or} \quad \tfrac{13}{3} \leq x$$

$\left\{x \mid x \geq 4\tfrac{1}{3}\right\} \cup \left\{x \mid x \leq \tfrac{1}{3}\right\}$; $\left(-\infty, \tfrac{1}{3}\right] \cup \left[4\tfrac{1}{3}, +\infty\right)$;

17. (3.4) If $\left|\frac{5x+1}{2}\right| \leq 7$,

$$-7 \leq \frac{5x+1}{2} \leq 7$$
$$-14 \leq 5x + 1 \leq 14$$
$$-14 - 1 \leq 5x + 1 - 1 \leq 14 - 1$$
$$-15 \leq 5x \leq 13$$
$$-3 \leq x \leq \tfrac{13}{5}$$

$\left\{x \mid -3 \leq x \leq \tfrac{13}{5}\right\}$; $\left[-3, \tfrac{13}{5}\right]$;

18. (3.4) If $|2x - 5| < 11$,

$$-11 < 2x - 5 < 11$$
$$-11 + 5 < 2x - 5 + 5 < 11 + 5$$
$$-6 < 2x < 16$$
$$-3 < x < 8$$

$\{x \mid -3 < x < 8\}$; $(-3, 8)$;

19. (3.5) Let x = unknown number

If 23	is added to	four times an unknown number	the sum is	31 .
23	+	$4 \cdot x$	=	31

$$23 + 4x = 31$$
$$4x = 8$$
$$x = 2$$

The unknown number is 2.

20. (3.5) Let x = first integer
Then $x + 1$ = second integer
Their sum is 55 .

$$x + (x + 1) = 55$$
$$x + x + 1 = 55$$
$$2x = 54$$
$$x = 27 \quad \text{First integer}$$
$$x + 1 = 27 + 1 = 28 \quad \text{Second integer}$$

The integers are 27 and 28.

Chapters 1–3 Cumulative Review Exercises (page 141)

1. $(-14) + (+22) = 8$ **2.** 48 **3.** -1 **4.** -14
5. $(-2)(-2)(-2)(-2) = 16$ **6.** 8
7. Undefined **8.** 0 **9.** $\frac{20}{-5} = -4$ **10.** -25
11. -4 **12.** -36 **13.** -5 **14.** 0 **15.** -1
16. 1,000 **17.** $1^3 = 1$ **18.** 8
19. $-3 - 4 \cdot 6 = -3 - 24 = -27$ **20.** 11 **21.** 4
22. -28 **23.** 0 **24.** 9 **25.** 3 **26.** 13
27. Undefined **28.** 1 **29.** 0 **30.** 4 **31.** $2 \cdot 3 \cdot 13$
32. 19, 23 **33.** 0 **34.** $(3 + 2) + 7$
35. $y - 2x + 2y - 3 + 3y - 2y = 4y - 2x - 3$
36. $x^3 - 3x^2 + 4x$ **37.** $5^2x^2(3^3x^6) = 25 \cdot 27x^{2+6} = 675x^8$
38. $-3x^3 - 4y^2$ **39.** $8x - 8 - 12x = 12$

$$-4x - 8 = 12$$
$$-4x = 20$$
$$x = -5$$
$$\{-5\}$$

40. $\{y \mid y > 3\}$ **41.** $18x - 30 + 7 = 27 + 18x - 1$

$$18x - 23 = 18x + 26$$
$$-23 = 26 \quad \text{False}$$

{ }; no solution

42. $\left\{x \mid x \geq -\tfrac{11}{3}\right\}$ **43.** $-5 < x + 4 \leq 3$

$$-9 < x \leq -1$$
$$\{x \mid -9 < x \leq -1\}$$

44. $\{x \mid x \in R\}$; identity
45. If $|2x + 3| > 1$,

$$2x + 3 > 1 \quad \text{or} \quad 2x + 3 < -1$$
$$2x > -2 \quad \text{or} \quad 2x < -4$$
$$x > -1 \quad \text{or} \quad x < -2$$

$\{x \mid x > -1 \text{ or } x < -2\}$

46. $\{x \mid 1 \leq x \leq 7\}$

4

47. Let x = Juana's age now
Then $2x$ = Lupe's age now
Also, $x + 9$ = Juana's age in 9 yr
and $2x + 9$ = Lupe's age in 9 yr

$$x + 9 = \tfrac{5}{7}(2x + 9)$$
$$7(x + 9) = 5(2x + 9)$$
$$7x + 63 = 10x + 45$$
$$18 = 3x$$
$$x = 6 \quad \text{Juana's age now}$$
$$2x = 12 \quad \text{Lupe's age now}$$

48. \$13,500

Exercises 4.1 (page 146)

1. 2nd degree **2.** 2nd degree
3. Not a polynomial because the exponents are not positive integers
4. Not a polynomial **5.** 0 degree **6.** 0 degree
7. Not a polynomial because a variable is in a denominator
8. Not a polynomial **9.** 1st degree **10.** 1st degree
11. Not a polynomial because a variable is under a radical sign
12. Not a polynomial
13. Not a polynomial because a variable is in a denominator
14. Not a polynomial **15.** 6th degree **16.** 6th degree
17. $8x^5 + 7x^3 - 4x - 5$; 8 **18.** $-3y^5 - 2y^3 + 4y^2 + 10$; -3
19. $-y^5 + y^3 + 3x^2y + 8x^3$; -1 **20.** $6y^3 - 4y^2 + y + 7x^2$; 6

Exercises 4.2 (page 149)

1. $-3x^4 - 2x^3 + 5 + 2x^4 + x^3 - 7x - 12 = -x^4 - x^3 - 7x - 7$
2. $-2y^3 + 3y^2 - 4y + 4$
3. $7 - 8v^3 + 9v^2 + 4v - 9v^3 - 6 + 8v^2 - 4v = -17v^3 + 17v^2 + 1$
4. $13x^3 - 4x^2 + 8x - 7$ **5.** $6x^3 + 2x^2 - 2$
6. $-2y^4 + 8y + 4$
7. $4x^3 + 6 + x - 6 - 3x^5 + 4x^2 = -3x^5 + 4x^3 + 4x^2 + x$
8. $4x^4 - 2x^3 - 3x$
9. $2y^4 + 4y^3 + 8 - 3y^4 - 2y^3 + 5y = -y^4 + 2y^3 + 5y + 8$
10. $5y^3 + 8y^2 - 5y - 4$ **11.** $-5x^4 - 7x^3 + x^2 + 4x - 2$
12. $-2x^4 + x^3 + 5x^2 - 2x + 6$ **13.** $7x^3 + 4x^2 + 7x - 11$
14. $7x^3 - 2x^2 - x - 12$
15. $6m^2n^2 - 8mn + 9 - 10m^2n^2 + 18mn - 11 + 3m^2n^2 - 2mn + 7$
$= -m^2n^2 + 8mn + 5$
16. $27u^2v + 4uv^2 + 17$
17. $[5 + xy^2 + x^3y - 6 - 3xy^2 + 4x^3y]$
$- [x^3y + 3xy^2 - 4 + 2x^3y - xy^2 + 5]$
$= [-1 - 2xy^2 + 5x^3y] - [3x^3y + 2xy^2 + 1]$
$= -1 - 2xy^2 + 5x^3y - 3x^3y - 2xy^2 - 1$
$= 2x^3y - 4xy^2 - 2$
18. $6m^2n + 6$
19. $8.586x^2 - 9.030x + 6.976 - [1.946x^2 - 41.45x - 7.468 + 3.914x^2]$
$= 8.586x^2 - 9.030x + 6.976 - 1.946x^2 + 41.45x + 7.468$
$- 3.914x^2$
$= 2.726x^2 + 32.42x + 14.444$
20. $-40.06x^2 - 117.32x + 88.55$

Exercises 4.3 (page 153)

1. $x^2 + 9x + 20$ **2.** $y^2 + 10y + 21$ **3.** $y^2 - y - 72$
4. $z^2 + 7z - 30$ **5.** $6x^2 - 7x - 20$ **6.** $8y^2 + 14y - 15$
7. $16x^2 - 25y^2$ **8.** $s^2 - 4t^2$ **9.** $10x^2 - 17xy + 3y^2$
10. $18u^2 - 27uv + 4v^2$ **11.** $56wz + 21z - 16w - 6$
12. $36xz + 45z - 4x - 5$ **13.** $64x^2 - 81y^2$
14. $49w^2 - 4x^2$ **15.** $2s^3 + 2s^2 + 5s + 5$
16. $3u^3 + 3u^2 + 2u + 2$ **17.** $4x^2 - 4xy + y^2$
18. $25z^2 - 10wz + w^2$ **19.** $6xy - 8x + 9y - 12$
20. $12uv - 20u - 6v + 10$ **21.** $49x^4 - 42x^2 + 9$
22. $16x^4 - 40x^2 + 25$ **23.** $6x^3 - 8x^2 + 15x - 20$
24. $15y^3 - 6y^2 + 5y - 2$
25. $3x[2x^2 - 6x + 4] = 6x^3 - 18x^2 + 12x$
26. $14x^3 + 7x^2 - 105x$
27.
$$\begin{array}{r} 4h^2 - 5h + 7 \\ 2h - 3 \\ \hline -12h^2 + 15h - 21 \\ 8h^3 - 10h^2 + 14h \qquad \\ \hline 8h^3 - 22h^2 + 29h - 21 \end{array}$$
28. $10k^3 + 23k^2 - 57k + 18$
29.
$$\begin{array}{r} a^4 + 3a^2 - 2a + 4 \\ a + 3 \\ \hline 3a^4 \qquad + 9a^2 - 6a + 12 \\ a^5 \qquad + 3a^3 - 2a^2 + 4a \qquad \\ \hline a^5 + 3a^4 + 3a^3 + 7a^2 - 2a + 12 \end{array}$$
30. $b^5 - 7b^4 + 10b^3 + 3b^2 - 20b + 25$
31.
$$\begin{array}{r} -3z^3 + z^2 - 5z + 4 \\ -z + 4 \\ \hline -12z^3 + 4z^2 - 20z + 16 \\ 3z^4 - z^3 + 5z^2 - 4z \qquad \\ \hline 3z^4 - 13z^3 + 9z^2 - 24z + 16 \end{array}$$
32. $v^4 - 4v^3 + 5v + 6$
33.
$$\begin{array}{r} 3u^2 - u + 5 \\ 2u^2 + 4u - 1 \\ \hline -3u^2 + u - 5 \\ 12u^3 - 4u^2 + 20u \qquad \\ 6u^4 - 2u^3 + 10u^2 \qquad\qquad \\ \hline 6u^4 + 10u^3 + 3u^2 + 21u - 5 \end{array}$$
34. $10w^4 - w^3 - 40w^2 + 20w + 7$ **35.** $6x^3y^2 - 2x^2y^3 + 8xy^4$
36. $12x^2y^3 + 3x^3y^2 - 9x^4y$
37. $-5a^5b + 15a^4b^2 - 15a^3b^3 + 5a^2b^4$
38. $-4m^4p^2 + 12m^3p^3 - 12m^2p^4 + 4mp^5$
39.
$$\begin{array}{r} x^2 + 2x + 3 \\ x^2 + 2x + 3 \\ \hline 3x^2 + 6x + 9 \\ 2x^3 + 4x^2 + 6x \qquad \\ x^4 + 2x^3 + 3x^2 \qquad\qquad \\ \hline x^4 + 4x^3 + 10x^2 + 12x + 9 \end{array}$$
40. $z^4 - 6z^3 + z^2 + 24z + 16$
41.
$$\begin{array}{r} x^2 - xy + y^2 \\ x + y \\ \hline x^2y - xy^2 + y^3 \\ x^3 - x^2y + xy^2 \qquad \\ \hline x^3 \qquad\qquad + y^3 \end{array} \qquad \begin{array}{r} x^2 + xy + y^2 \\ x - y \\ \hline -x^2y - xy^2 - y^3 \\ x^3 + x^2y + xy^2 \qquad \\ \hline x^3 \qquad\qquad - y^3 \end{array}$$
Then
$[(x + y)(x^2 - xy + y^2)][(x - y)(x^2 + xy + y^2)]$
$= [x^3 + y^3][x^3 - y^3] = x^6 - y^6$
42. $a^6 - 1$

Exercises 4.4 (page 157)

1. $4u^2 - 25v^2$ **2.** $9m^2 - 49n^2$ **3.** $4x^4 - 81$
4. $100y^4 - 9$ **5.** $x^{10} - y^{12}$ **6.** $a^{14} - b^8$
7. $49m^2n^2 - 4r^2s^2$ **8.** $64h^2k^2 - 25e^2f^2$ **9.** $144x^8y^6 - u^{14}v^2$

10. $121a^{10}b^4 - 81c^6d^{12}$

11. $([a + b] + 2)([a + b] - 2) = (a + b)^2 - (2)^2$
$= (a + b)(a + b) - 4$
$= a^2 + 2ab + b^2 - 4$

12. $x^2 + 2xy + y^2 - 25$

13. $([x^2 + y] + 5)([x^2 + y] - 5) = [x^2 + y]^2 - 5^2$
$= [x^2 + y][x^2 + y] - 25 = x^4 + 2x^2y + y^2 - 25$

14. $u^4 - 2u^2v + v^2 - 49$ **15.** $7x(x^2 - 1) = 7x^3 - 7x$

16. $12y^3 - 75y$

17. $(2x)^2 + 2(2x)(3) + 3^2 = 4x^2 + 12x + 9$

18. $36x^2 + 60x + 25$

19. $(5x)^2 - 2(5x)(3) + 3^2 = 25x^2 - 30x + 9$

20. $81x^2 - 108x + 36$

21. $(7x)^2 - 2(7x)(10y) + (10y)^2 = 49x^2 - 140xy + 100y^2$

22. $16u^2 - 72uv + 81v^2$

23. $(7x^2)^2 + 2(7x^2)(9y) + (9y)^2 = 49x^4 + 126x^2y + 81y^2$

24. $64s^2 + 176st^3 + 121t^6$

25. $(15u)^2 + 2(15u)(-v^2) + (-v^2)^2 = 225u^2 - 30uv^2 + v^4$

26. $256w^2 - 32wt^2 + t^4$

27. $s^2 + 2s(t + u) + (t + u)^2 = s^2 + 2st + 2su + t^2 + 2tu + u^2$

28. $v^2 + 2vx + 2sv + x^2 + 2sx + s^2$

29. $(u + v)^2 + 2(u + v)(7) + (7)^2 = u^2 + 2uv + v^2 + 14u + 14v + 49$

30. $s^2 + 2st + t^2 + 8s + 8t + 16$

31. $[2x - y]^2 - 2[2x - y][3] + 3^2$
$= (2x)^2 - 2(2x)(y) + y^2 - 12x + 6y + 9$
$= 4x^2 - 4xy + y^2 - 12x + 6y + 9$

32. $9z^2 - 6wz + w^2 - 42z + 14w + 49$

33. $x^2 + 2x(u + v) + (u + v)^2 = x^2 + 2ux + 2vx + u^2 + 2uv + v^2$

34. $s^2 + 2sy + 4s + y^2 + 4y + 4$

35. $y^2 - 2y[x - 2] + [x - 2]^2 = y^2 - 2xy + 4y + x^2 - 4x + 4$

36. $x^2 - 2xz + 10x + z^2 - 10z + 25$

37. $5[x^2 - 2(x)(3) + 3^2] = 5(x^2 - 6x + 9) = 5x^2 - 30x + 45$

38. $7x^2 + 14x + 7$

Exercises 4.5 (page 168)

1. $4! = 4 \cdot 3 \cdot 2 \cdot 1 = 24$ **2.** 720

3. $\binom{3}{2} = \frac{3!}{2!(3-2)!} = \frac{3(2!)}{(2!)(1)} = 3$ **4.** 35

5. $\binom{6}{4} = \frac{6!}{4!(6-4)!} = \frac{6 \cdot 5 \cdot 4!}{4!(2!)} = \frac{6 \cdot 5}{2} = 15$ **6.** 1

7. $x^5 + 5x^4y + \frac{5!}{(2!)(3!)}x^3y^2 + \frac{5!}{(3!)(2!)}x^2y^3 + 5xy^4 + y^5$
$= x^5 + 5x^4y + 10x^3y^2 + 10x^2y^3 + 5xy^4 + y^5$

8. $r^4 + 4r^3s + 6r^2s^2 + 4rs^3 + s^4$

9. $x^5 - 5x^4(2) + \frac{5!}{(2!)(3!)}x^3(2)^2 - \frac{5!}{(3!)(2!)}x^2(2)^3 + 5x(2)^4 - (2)^5$
$= x^5 - 10x^4 + 10(4)x^3 - 10(8)x^2 + 5(16)x - 32$
$= x^5 - 10x^4 + 40x^3 - 80x^2 + 80x - 32$

10. $y^4 - 12y^3 + 54y^2 - 108y + 81$

11. $(3r + s)^6 = (3r)^6 + 6(3r)^5s + \frac{6!}{(2!)(4!)}(3r)^4s^2 + \frac{6!}{(3!)(3!)}(3r)^3s^3$
$+ 15(3r)^2s^4 + 6(3r)s^5 + s^6$
$= 729r^6 + 6(243)r^5s + 15(81)r^4s^2 + 20(27)r^3s^3 + 15(9)r^2s^4$
$+ 18rs^5 + s^6$
$= 729r^6 + 1{,}458r^5s + 1{,}215r^4s^2 + 540r^3s^3 + 135r^2s^4 + 18rs^5 + s^6$

12. $64x^6 + 192x^5y + 240x^4y^2 + 160x^3y^3 + 60x^2y^4 + 12xy^5 + y^6$

13. $(x + y^2)^4 = x^4 + 4x^3(y^2) + 6x^2(y^2)^2 + 4x(y^2)^3 + (y^2)^4$
$= x^4 + 4x^3y^2 + 6x^2y^4 + 4xy^6 + y^8$

14. $u^5 + 5u^4v^2 + 10u^3v^4 + 10u^2v^6 + 5uv^8 + v^{10}$

15. $(2x)^5 + 5(2x)^4\left(-\frac{1}{2}\right) + \frac{5!}{(2!)(3!)}(2x)^3\left(-\frac{1}{2}\right)^2 + 10(2x)^2\left(-\frac{1}{2}\right)^3$
$+ 5(2x)\left(-\frac{1}{2}\right)^4 + \left(-\frac{1}{2}\right)^5$
$= 32x^5 + 5(16)\left(-\frac{1}{2}\right)x^4 + 10(8)\left(\frac{1}{4}\right)x^3 + 10(4)\left(-\frac{1}{8}\right)x^2$
$+ 5(2)\left(\frac{1}{16}\right)x + \left(-\frac{1}{32}\right)$
$= 32x^5 - 40x^4 + 20x^3 - 5x^2 + \frac{5}{8}x - \frac{1}{32}$

16. $81x^4 - 36x^3 + 6x^2 - \frac{4}{9}x + \frac{1}{81}$

17. $\left(\frac{x}{3}\right)^4 + 4\left(\frac{x}{3}\right)^3\left(\frac{3}{2}\right) + \frac{4!}{(2!)(2!)}\left(\frac{x}{3}\right)^2\left(\frac{3}{2}\right)^2 + 4\left(\frac{x}{3}\right)\left(\frac{3}{2}\right)^3 + \left(\frac{3}{2}\right)^4$
$= \frac{1}{81}x^4 + 4\left(\frac{1}{27}\right)\left(\frac{3}{2}\right)x^3 + 6\left(\frac{1}{9}\right)\left(\frac{9}{4}\right)x^2 + 4\left(\frac{1}{3}\right)\left(\frac{27}{8}\right)x + \frac{81}{16}$
$= \frac{1}{81}x^4 + \frac{2}{9}x^3 + \frac{3}{2}x^2 + \frac{9}{2}x + \frac{81}{16}$

18. $\frac{1}{625}x^4 + \frac{2}{25}x^3 + \frac{3}{2}x^2 + \frac{25}{2}x + \frac{625}{16}$

19. $(4x^2 - 3y^2)^5 = (4x^2)^5 + 5(4x^2)^4(-3y^2) + 10(4x^2)^3(-3y^2)^2$
$+ 10(4x^2)^2(-3y^2)^3 + 5(4x^2)(-3y^2)^4 + (-3y^2)^5$
$= 1{,}024x^{10} - 3{,}840x^8y^2 + 5{,}760x^6y^4$
$- 4{,}320x^4y^6 + 1{,}620x^2y^8 - 243y^{10}$

20. $243x^{10} - 810x^8y^3 + 1{,}080x^6y^6 - 720x^4y^9 + 240x^2y^{12} - 32y^{15}$

21. $(x + x^{-1})^4 = (x)^4 + 4(x)^3(x^{-1}) + 6(x)^2(x^{-1})^2 + 4(x)(x^{-1})^3 + (x^{-1})^4$
$= x^4 + 4x^2 + 6 + 4x^{-2} + x^{-4}$
$= x^4 + 4x^2 + 6 + \frac{4}{x^2} + \frac{1}{x^4}$

22. $\frac{1}{x^4} - \frac{4}{x^2} + 6 - 4x^2 + x^4$

23. The coefficient of the 6th term is
$\binom{9}{5} = \frac{9!}{5!(4!)} = \frac{9 \cdot 8 \cdot 7 \cdot 6 \cdot 5!}{5!(4 \cdot 3 \cdot 2 \cdot 1)} = 126$
The 6th term is $126a^4b^5$.

24. $35a^4b^3$

25. The coefficient of the 8th term is
$\binom{10}{7} = \frac{10!}{7!(3!)} = \frac{10 \cdot 9 \cdot 8 \cdot 7!}{7!(3 \cdot 2 \cdot 1)} = 120$
The 8th term is $120(3x^2)^3(-y)^7$, or $-3{,}240x^6y^7$.

26. $326{,}592x^{10}w^4$

27. $(x)^{10} + 10(x)^9(2y^2) + \frac{10!}{(2!)(8!)}(x)^8(2y^2)^2 + \frac{10!}{(3!)(7!)}(x)^7(2y^2)^3 + \cdots$
$= x^{10} + 20x^9y^2 + 180x^8y^4 + 960x^7y^6 + \cdots$

28. $x^8 + 24x^7y^2 + 252x^6y^4 + 1{,}512x^5y^6 + \cdots$

29. $(x - 3y^2)^{10} = (x)^{10} + 10(x)^9(-3y^2) + \frac{10!}{(2!)(8!)}(x)^8(-3y^2)^2$
$+ \frac{10!}{(3!)(7!)}(x)^7(-3y^2)^3 + \cdots$
$= x^{10} - 30x^9y^2 + 405x^8y^4 - 3{,}240x^7y^6 + \cdots$

30. $x^{11} - 22x^{10}y^3 + 220x^9y^6 - 1{,}320x^8y^9 + \cdots$

Exercises 4.6 (page 173)

1. $\frac{18x^5}{6x^2} + \frac{-24x^4}{6x^2} + \frac{-12x^3}{6x^2} = 3x^3 - 4x^2 - 2x$

2. $-4y^2 + 9y - 5$ **3.** $\frac{55a^4b^3}{-11ab} + \frac{-33ab^2}{-11ab} = -5a^3b^2 + 3b$

4. $-2mn^3 + 4m^2$ **5.** $\frac{-15x^2y^2z^2}{-5xyz} + \frac{-30xyz}{-5xyz} = 3xyz + 6$

6. $3abc + 2$

7. $\frac{13x^3y^2}{13x^2y^2} + \frac{-26x^5y^3}{13x^2y^2} + \frac{39x^4y^6}{13x^2y^2} = x - 2x^3y + 3x^2y^4$

8. $3n - 5m - 2mn^3$

4

9.
$$\begin{array}{r} x+3 \\ x+7\overline{)x^2+10x-5} \\ \underline{x^2+7x} \\ 3x-5 \\ \underline{3x+21} \\ -26 \end{array}$$
Answer: $x + 3 - \dfrac{26}{x + 7}$

10. $x + 5 - \dfrac{25}{x + 4}$

11.
$$\begin{array}{r} x-4 \\ x+7\overline{)x^2+3x-10} \\ \underline{x^2+7x} \\ -4x-10 \\ \underline{-4x-28} \\ +18 \end{array}$$
Answer: $x - 4 + \dfrac{18}{x + 7}$

12. $x + 8$

13.
$$\begin{array}{r} x+1 \\ x-3\overline{)x^2-2x-5} \\ \underline{x^2-3x} \\ x-5 \\ \underline{x-3} \\ -2 \end{array}$$
Answer: $x + 1 - \dfrac{2}{x - 3}$

14. $x + 1 + \dfrac{4}{x - 6}$

15.
$$\begin{array}{r} 2z^2-z-3 \\ 3z-5\overline{)6z^3-13z^2-4z+15} \\ \underline{6z^3-10z^2} \\ -3z^2-4z \\ \underline{-3z^2+5z} \\ -9z+15 \\ \underline{-9z+15} \\ 0 \end{array}$$
Answer: $2z^2 - z - 3$

16. $3y^2 - y - 4$

17.
$$\begin{array}{r} 4x^2+8x+8 \\ -x+2\overline{)-4x^3+8x+10} \\ \underline{-4x^3+8x^2} \\ -8x^2+8x \\ \underline{-8x^2+16x} \\ -8x+10 \\ \underline{-8x+16} \\ -6 \end{array}$$
Answer: $4x^2 + 8x + 8 - \dfrac{6}{2 - x}$

18. $x^2 + 3x - 3 + \dfrac{-6}{3 - x}$

19.
$$\begin{array}{r} x^2+5x-3 \\ -x+5\overline{)-x^3+0x^2+28x-10} \\ \underline{-x^3+5x^2} \\ -5x^2+28x \\ \underline{-5x^2+25x} \\ 3x-10 \\ \underline{3x-15} \\ 5 \end{array}$$
Answer: $x^2 + 5x - 3 + \dfrac{5}{5 - x}$

20. $y^2 + 4y - 3 + \dfrac{16}{4 - y}$

21.
$$\begin{array}{r} x^3+2x^2+4x+1 \\ x-2\overline{)x^4+0x^3+0x^2-7x+6} \\ \underline{x^4-2x^3} \\ 2x^3+0x^2 \\ \underline{2x^3-4x^2} \\ 4x^2-7x \\ \underline{4x^2-8x} \\ x+6 \\ \underline{x-2} \\ 8 \end{array}$$
Answer: $x^3 + 2x^2 + 4x + 1 + \dfrac{8}{x - 2}$

22. $x^3 + x^2 + x + 7 + \dfrac{10}{x - 1}$

23.
$$\begin{array}{r} 2x^2+0x+3 \\ x^2+2\overline{)2x^4+0x^3+7x^2+0x+5} \\ \underline{2x^4+4x^2} \\ 3x^2+5 \\ \underline{3x^2+6} \\ -1 \end{array}$$
Answer: $2x^2 + 3 - \dfrac{1}{x^2 + 2}$

24. $3x^2 + 2 + \dfrac{-5}{x^2 + 3}$

25.
$$\begin{array}{r} v^3-4v^2-4v-5 \\ v+1\overline{)v^4-3v^3-8v^2-9v-5} \\ \underline{v^4+v^3} \\ -4v^3-8v^2 \\ \underline{-4v^3-4v^2} \\ -4v^2-9v \\ \underline{-4v^2-4v} \\ -5v-5 \\ \underline{-5v-5} \\ 0 \end{array}$$
Answer: $v^3 - 4v^2 - 4v - 5$

26. $w^3 - 4w^2 - 4w - 5$

27.
$$\begin{array}{r} 4x^2-xy+5y^2 \\ 3x-2y\overline{)12x^3-11x^2y+17xy^2-10y^3} \\ \underline{12x^3-8x^2y} \\ -3x^2y+17xy^2 \\ \underline{-3x^2y+2xy^2} \\ 15xy^2-10y^3 \\ \underline{15xy^2-10y^3} \\ 0 \end{array}$$
Answer: $4x^2 - xy + 5y^2$

28. $5x^2 - xy - 4y^2$

29.
$$\begin{array}{r} x^2-x+2 \\ 4x^2+x+1\overline{)4x^4-3x^3+8x^2+x+2} \\ \underline{4x^4+x^3+x^2} \\ -4x^3+7x^2+x \\ \underline{-4x^3-x^2-x} \\ 8x^2+2x+2 \\ \underline{8x^2+2x+2} \\ 0 \end{array}$$
Answer: $x^2 - x + 2$

30. $x^2 - x + 5$

31.
$$\begin{array}{r} x^2-5 \\ 2x^3+3\overline{)2x^5+0x^4-10x^3+3x^2+0x-15} \\ \underline{2x^5+3x^2} \\ -10x^3-15 \\ \underline{-10x^3-15} \\ 0 \end{array}$$
Answer: $x^2 - 5$

32. $x^2 - 2$

33.

$$\begin{array}{rl} & x^2 + 5x + 2 \\ 3x^2 - x + 1 & \overline{)3x^4 + 14x^3 + 2x^2 + 3x + 2} \\ & \underline{3x^4 - x^3 + x^2} \\ & 15x^3 + x^2 + 3x \\ & \underline{15x^3 - 5x^2 + 5x} \\ & 6x^2 - 2x + 2 \\ & \underline{6x^2 - 2x + 2} \\ & 0 \end{array}$$

Answer: $x^2 + 5x + 2$

34. $x^2 + 7x - 2$

Exercises 4.7 (page 178)

The checks for these exercises will not be shown.

1.

$$\begin{array}{r|rrr} 3 & 1 & 2 & -18 \\ & & 3 & 15 \\ \hline & 1 & 5 & \boxed{-3} \end{array}$$

Answer: $x + 5 - \dfrac{3}{x - 3}$

2. $x + 6 + \dfrac{2}{x - 2}$

3.

$$\begin{array}{r|rrrr} -4 & 1 & 3 & -5 & 6 \\ & & -4 & 4 & 4 \\ \hline & 1 & -1 & -1 & \boxed{10} \end{array}$$

Answer: $x^2 - x - 1 + \dfrac{10}{x + 4}$

4. $x^2 + x - 1 + \dfrac{-2}{x + 5}$

5.

$$\begin{array}{r|rrrrr} -6 & 1 & 6 & 0 & -1 & -4 \\ & & -6 & 0 & 0 & 6 \\ \hline & 1 & 0 & 0 & -1 & \boxed{2} \end{array}$$

Answer: $x^3 - 1 + \dfrac{2}{x + 6}$

6. $2x^3 - x^2 + 3x + 1 + \dfrac{-5}{x + 3}$

7.

$$\begin{array}{r|rrrrr} 2 & 1 & 0 & 0 & 0 & -16 \\ & & 2 & 4 & 8 & 16 \\ \hline & 1 & 2 & 4 & 8 & \boxed{0} \end{array}$$

Answer: $x^3 + 2x^2 + 4x + 8$

8. $x^6 + x^5 + x^4 + x^3 + x^2 + x + 1$

9.

$$\begin{array}{r|rrrrrrr} 3 & 1 & -3 & 0 & 0 & -2 & 3 & 5 \\ & & 3 & 0 & 0 & 0 & -6 & -9 \\ \hline & 1 & 0 & 0 & 0 & -2 & -3 & \boxed{-4} \end{array}$$

Answer: $x^5 - 2x - 3 + \dfrac{-4}{x - 3}$

10. $x^5 + 2x^4 + x^3 + 2x^2 + 4x + 1$

11.

$$\begin{array}{r|rrrrr} \frac{1}{3} & 3 & -1 & 9 & 0 & -1 \\ & & 1 & 0 & 3 & 1 \\ \hline & 3 & 0 & 9 & 3 & \boxed{0} \end{array}$$

Answer: $3x^3 + 9x + 3$

12. $3x^3 - 6x^2 + 3x - 3$

13.

$$\begin{array}{r|rrrrrr} -3 & 1 & 1 & -45 & -45 & 324 & 324 \\ & & -3 & 6 & 117 & -216 & -324 \\ \hline & 1 & -2 & -39 & 72 & 108 & \boxed{0} \end{array}$$

Answer: $x^4 - 2x^3 - 39x^2 + 72x + 108$

14. $x^4 - 5x^3 - 15x^2 + 45x + 54$

15.

$$\begin{array}{r|rrrrr} 2 & 4 & 0 & -45 & 3 & 100 \\ & & 8 & 16 & -58 & -110 \\ \hline & 4 & 8 & -29 & -55 & \boxed{-10} \end{array}$$

Answer: $4x^3 + 8x^2 - 29x - 55 + \dfrac{-10}{x - 2}$

16. $9x^3 + 9x^2 - 4x - 2 + \dfrac{4}{x - 1}$

17.

$$\begin{array}{r|rrrr} 1.5 & 2.6 & 0 & 1.8 & -6.4 \\ & & 3.9 & 5.85 & 11.475 \\ \hline & 2.6 & 3.9 & 7.65 & \boxed{5.075} \end{array}$$

Answer: $2.6x^2 + 3.9x + 7.65 + \dfrac{5.075}{x - 1.5}$

18. $3.8x^2 + 8.1x + 20.25 + \dfrac{26.725}{x - 2.5}$

19.

$$\begin{array}{r|rrrr} 1.2 & 2.7 & 0 & -1.6 & 3.289 \\ & & 3.24 & 3.888 & 2.7456 \\ \hline & 2.7 & 3.24 & 2.288 & \boxed{6.0346} \end{array}$$

Answer: $2.7x^2 + 3.24x + 2.288 + \dfrac{6.0346}{x - 1.2}$

20. $3x^2 + 6x + 9.6 + \dfrac{13.86}{x - 1.6}$

Exercises 4.8 (page 182)

The checks for these exercises will not be shown.

1. Let x = number of lb of Colombian coffee
Then $100 - x$ = number of lb of Brazilian coffee

$$\begin{aligned} 3.90x + 3.60(100 - x) &= 3.72(100) \\ 39x + 36(100 - x) &= 37.2(100) \\ 39x + 3{,}600 - 36x &= 3{,}720 \\ 3x &= 120 \\ x &= 40 \\ 100 - x &= 60 \end{aligned}$$

40 lb of Colombian coffee and 60 lb of Brazilian coffee should be used.

2. 16 lb of cashews, 34 lb of peanuts

3. Let x = number of lb of almonds
Then $10 - x$ = number of lb of walnuts

$$\begin{aligned} 3.00x + 4.50(10 - x) &= 39.72 \\ 300x + 450(10 - x) &= 3{,}972 \\ 300x + 4{,}500 - 450x &= 3{,}972 \\ -150x &= -528 \\ x &= 3.52 \\ 10 - x &= 6.48 \end{aligned}$$

3.52 lb of almonds and 6.48 lb of walnuts should be used.

4. 4.5 lb of granola, 1.5 lb of apple chunks

5. Let x = width of the room (in ft)
Then $x + 3$ = length of the room (in ft)
The distance *around* the room is $[2(x + 3) + 2x]$ ft, or $[2x + 6 + 2x]$ ft, or $[4x + 6]$ ft. The cost of the molding is $\left(\dfrac{\$1.20}{1\text{ ft}}\right)[(4x + 6)\text{ ft}]$, or $\$(1.2)(4x + 6)$.

$$\begin{aligned} (1.2)(4x + 6) &= 79.2 \\ (12)(4x + 6) &= 792 \\ 48x + 72 &= 792 \\ 48x &= 720 \\ x &= 15 \end{aligned}$$

Margaretha's living room is 15 ft wide.

6. The dog-run is 6.5 m wide.

7. Let D = number of dimes
Then $17 - D$ = number of nickels
Value of dimes + value of nickels = 115¢

$$\begin{aligned} 10D \quad + \quad 5(17 - D) &= 115 \\ 10D + 85 - 5D &= 115 \\ 5D &= 30 \\ D &= 6 \\ 17 - D = 17 - 6 &= 11 \end{aligned}$$

Doris has 6 dimes and 11 nickels.

8. 11 quarters, 6 dimes

9. Let x = number of quarters
Then $x + 7$ = number of dimes
and $3x$ = number of nickels

$$\begin{aligned} 0.25x + 0.10(x + 7) + 0.05(3x) &= 3.20 \\ 25x + 10(x + 7) + 5(3x) &= 320 \\ 25x + 10x + 70 + 15x &= 320 \\ 50x &= 250 \\ x &= 5 \\ x + 7 &= 12 \\ 3x &= 15 \end{aligned}$$

Dianne has 5 quarters, 12 dimes, and 15 nickels.

10. 6 dimes, 11 nickels, 12 quarters

11. Let x = number of lb of peanut brittle

$$\begin{aligned} 6.60(15) + 4.20x &= 5.10(15 + x) \\ 66(15) + 42x &= 51(15 + x) \\ 990 + 42x &= 765 + 51x \\ 225 &= 9x \\ x &= 25 \end{aligned}$$

25 lb of peanut brittle should be used.

12. 16 lb

13. Let x = amount invested at 5% interest
Then $7{,}000 - x$ = amount invested at 4% interest
The amount of interest earned at 5%: $I = x(0.05)(1)$, or $0.05x$
The amount of interest earned at 4%: $I = (7{,}000 - x)(0.04)(1)$, or $0.04(7{,}000 - x)$

$$\begin{aligned} 0.05x + 0.04(7{,}000 - x) &= 306.50 \\ 5x + 4(7{,}000 - x) &= 30{,}650 \\ 5x + 28{,}000 - 4x &= 30{,}650 \\ x + 28{,}000 &= 30{,}650 \\ x &= 2{,}650 \\ 7{,}000 - x &= 4{,}350 \end{aligned}$$

George invested \$2,650 at 5% and \$4,350 at 4%.

14. Henrietta invested \$1,640 at 4% and \$6,360 at 5%.

15. Let x = number of children's tickets
Then $23 - x$ = number of adults' tickets
Each child's ticket is worth \$3.50, so total value = $3.50x$.
Each adult's ticket is worth \$5.25, so total value = $5.25(23 - x)$.

$$\begin{aligned} 3.50x + 5.25(23 - x) &= 108.50 \\ 350x + 525(23 - x) &= 10{,}850 \\ 350x + 12{,}075 - 525x &= 10{,}850 \\ -175x &= -1{,}225 \\ x &= 7 \\ 23 - x &= 16 \end{aligned}$$

She bought 7 children's tickets and 16 adults' tickets.

16. Twenty-seven 22¢ stamps, eighteen 18¢ stamps

Exercises 4.9 (page 187)

The checks for these exercises will not be shown.

1. Let x = number of mL of water

$$\begin{aligned} 0.40(500) + 0x &= 0.25(500 + x) \\ 40(500) &= 25(500 + x) \\ 20{,}000 &= 12{,}500 + 25x \\ 7{,}500 &= 25x \\ x &= 300 \end{aligned}$$

300 mL of water must be added.

2. 5 L

3. Let x = number of L of pure alcohol

$$\begin{aligned} 0.20(10) + x &= 0.50(10 + x) \\ 2(10) + 10x &= 5(10 + x) \\ 20 + 10x &= 50 + 5x \\ 5x &= 30 \\ x &= 6 \end{aligned}$$

6 L of pure alcohol should be added.

4. 500 mL

5. Let x = number of cc of 20% solution

$$\begin{aligned} 0.20x + 0.50(100) &= 0.25(100 + x) \\ 20x + 50(100) &= 25(100 + x) \\ 20x + 5{,}000 &= 2{,}500 + 25x \\ 2{,}500 &= 5x \\ x &= 500 \end{aligned}$$

500 cc of 20% solution should be added.

6. 20 pt

7. Let x = number of gal of 30% solution
Then $100 - x$ = number of gal of 90% solution

$$\begin{aligned} 0.30(x) + 0.90(100 - x) &= 100(0.75) \\ 30x + 90(100 - x) &= 100(75) \\ 30x + 9{,}000 - 90x &= 7{,}500 \\ -60x &= -1{,}500 \\ x &= 25 \\ 100 - x &= 75 \end{aligned}$$

25 gal of the 30% solution and 75 gal of the 90% solution should be used.

8. 400 g of 65% alloy, 800 g of 20% alloy

9. Let x = number of L of 40% solution
Then $10 - x$ = number of L of 90% solution

$$\begin{aligned} 0.40x + 0.90(10 - x) &= 0.50(10) \\ 40x + 90(10 - x) &= 50(10) \\ 40x + 900 - 90x &= 500 \\ -50x &= -400 \\ x &= 8 \\ 10 - x &= 2 \end{aligned}$$

8 L of the 40% solution and 2 L of the 90% solution should be used.

10. 1,200 mL

Sections 4.1–4.9 Review Exercises (page 189)

1. 3rd degree, since the sum of the exponents on the variables in the second term is 3

2. Not a polynomial **3.** $6! = 6 \cdot 5 \cdot 4 \cdot 3 \cdot 2 \cdot 1 = 720$

4. 35 **5.** $-2x^3 - 38x^2 + 24x - 3$

6. $9x^2y + xy^2 + 2y^2 - 12$ **7.** $-2x^3 - 2x^2 - 5x + 7$

8. $x^4 - x^3 + x^2 - x + 1$

9. $\underline{2k^3} - 7k + 11 + \underline{4k^3} + k^2 - 9k - 3k^2 + 5k + 6$
$= 6k^3 - 2k^2 - 11k + 17$

10. $-15a^3b^3 - 9a^2b^2 + 6ab^2c$ **11.** $35x^2 - 38xy + 8y^2$

12. $x^8 + 12x^6 + 54x^4 + 108x^2 + 81$

13.

$$\begin{array}{r} 4x^2 - 5x + 1 \\ 2x^2 + x - 3 \\ \hline -12x^2 + 15x - 3 \\ 4x^3 - 5x^2 + x \\ 8x^4 - 10x^3 + 2x^2 \\ \hline 8x^4 - 6x^3 - 15x^2 + 16x - 3 \end{array}$$

14. $4x^5 - 4xy^4$

15. $(2x^2)^5 + 5(2x^2)^4\left(\frac{1}{2}\right)^1 + \dfrac{5!}{(2!)(3!)}(2x^2)^3\left(\frac{1}{2}\right)^2$
$+ 10(2x^2)^2\left(\frac{1}{2}\right)^3 + 5(2x^2)^1\left(\frac{1}{2}\right)^4 + \left(\frac{1}{2}\right)^5$
$= 32x^{10} + 5(16)\left(\frac{1}{2}\right)x^8 + 10(8)\left(\frac{1}{4}\right)x^6 + 10(4)\left(\frac{1}{8}\right)x^4$
$+ 5(2)\left(\frac{1}{16}\right)x^2 + \frac{1}{32}$
$= 32x^{10} + 40x^8 + 20x^6 + 5x^4 + \frac{5}{8}x^2 + \frac{1}{32}$

16. $x^2 + 2xy + y^2 + 6x + 6y + 9$

17. $a^2 - 2a(5) + 5^2 = a^2 - 10a + 25$

18. $z^3 + 27$

19. $\dfrac{-15a^2b^3}{-5ab} + \dfrac{20a^4b^2}{-5ab} + \dfrac{-10ab}{-5ab} = 3ab^2 - 4a^3b + 2$

20. $3x + \dfrac{10}{2x - 3}$

21.
$$\begin{array}{r}
2a + 5b \\
5a - b\,\overline{)\,10a^2 + 23ab - 5b^2} \\
10a^2 - 2ab \\
\hline
25ab - 5b^2 \\
25ab - 5b^2 \\
\hline
0
\end{array}$$

Answer: $2a + 5b$

22. $x^2 - 1 + \dfrac{-10}{3x^2 - 2x + 5}$

23.
$$\begin{array}{r}
x^3 - 3x^2 + 9x - 27 \\
x + 3\,\overline{)\,x^4 + 0x^3 + 0x^2 + 0x - 81} \\
x^4 + 3x^3 \\
\hline
-3x^3 + 0x^2 \\
-3x^3 - 9x^2 \\
\hline
9x^2 + 0x \\
9x^2 + 27x \\
\hline
-27x - 81 \\
-27x - 81 \\
\hline
0
\end{array}$$

Answer: $x^3 - 3x^2 + 9x - 27$

24. $3x^4 + x^3 - 2x^2 + \dfrac{4}{x + 2}$

25.
$$\begin{array}{r|rrrrrr}
\frac{2}{5} & 5 & -2 & 10 & -4 & 0 & 2 \\
 & & 2 & 0 & 4 & 0 & 0 \\
\hline
 & 5 & 0 & 10 & 0 & 0 & \boxed{2}
\end{array}$$

Answer: $5x^4 + 10x^2 + \dfrac{2}{x - \frac{2}{5}}$

26. 840 cc

27. Let x = number of lb at \$2.20
Then $15 - x$ = number of lb at \$2.60
$$\begin{aligned}
2.20x + 2.60(15 - x) &= 2.36(15) \\
220x + 260(15 - x) &= 236(15) \\
220x + 3{,}900 - 260x &= 3{,}540 \\
-40x &= -360 \\
x &= 9 \text{ lb at \$2.20} \\
15 - x &= 6 \text{ lb at \$2.60}
\end{aligned}$$
9 lb of the cheaper candy and 6 lb of the more expensive candy should be used.

Chapter 4 Diagnostic Test (page 193)

Following each problem number is the number (in parentheses) of the textbook section where that kind of problem is discussed. (No checks will be shown.)

1. (4.1) **a.** Degree of $-\frac{1}{3}x^1y^1$ is $1 + 1 = 2$.
b. The degree of a polynomial is the degree of the highest-degree term; the degree of $2x^2y^3$ is $2 + 3 = 5$. Therefore, the polynomial is of 5th degree.

2. (4.2) $-x^3 - x^2 + 4x + 3$

3. (4.2) $-2x^4 - 2x^3 - x^2 + 4x + 3$

4. (4.2) $(-3z^2 - 6z + 8) - (8 - z + 4z^2)$
$= -3z^2 - 6z + 8 - 8 + z - 4z^2 = -7z^2 - 5z$

5. (4.2) $3ab^2 - 5ab - a^3 - 2ab + 4ab - 7ab^2$
$= -4ab^2 - 3ab - a^3$

6. (4.3) $(-3ab)(6a^2) + (-3ab)(-2ab^2) + (-3ab)(5b)$
$= -18a^3b + 6a^2b^3 - 15ab^2$

7. (4.6) $\dfrac{9z^3w}{3zw} + \dfrac{6z^2w^2}{3zw} - \dfrac{12zw^3}{3zw} = 3z^2 + 2zw - 4w^2$

8. (4.3)
$$\begin{array}{r}
x^2 + 2x + 4 \\
x - 2 \\
\hline
-2x^2 - 4x - 8 \\
x^3 + 2x^2 + 4x \phantom{{}- 8} \\
\hline
x^3 \phantom{{}+ 2x^2 + 4x} - 8
\end{array}$$

9. (4.4) $4x^8 - 9$

10. (4.3) $15m^2 + 14m - 8$

11. (4.4) $(3R^2)^2 - 2(3R^2)(5) + 5^2 = 9R^4 - 30R^2 + 25$

12. (4.6)
$$\begin{array}{r}
4y - 4 \\
3y + 2\,\overline{)\,12y^2 - 4y + 1} \\
12y^2 + 8y \phantom{{}+ 1} \\
\hline
-12y + 1 \\
-12y - 8 \\
\hline
9
\end{array}$$

Answer: $4y - 4 + \dfrac{9}{3y + 2}$

13. (4.7) $2x^4 + 3x^3 - 7x^2 + 0x - 5$
$$\begin{array}{r|rrrrr}
-3 & 2 & 3 & -7 & 0 & -5 \\
 & & -6 & 9 & -6 & 18 \\
\hline
 & 2 & -3 & 2 & -6 & \boxed{13}
\end{array}$$

Answer: $2x^3 - 3x^2 + 2x - 6 + \dfrac{13}{x + 3}$

14. (4.3)
$$\begin{array}{r}
m^2 - 2m + 5 \\
m^2 - 2m + 5 \\
\hline
5m^2 - 10m + 25 \\
-2m^3 + 4m^2 - 10m \phantom{{}+ 25} \\
m^4 - 2m^3 + 5m^2 \phantom{{}- 10m + 25} \\
\hline
m^4 - 4m^3 + 14m^2 - 20m + 25
\end{array}$$

15. (4.6)
$$\begin{array}{r}
3x^2 + 0x - 5 \\
2x^2 - x + 4\,\overline{)\,6x^4 - 3x^3 + 2x^2 + 5x - 7} \\
6x^4 - 3x^3 + 12x^2 \phantom{{}+ 5x - 7} \\
\hline
-10x^2 + 5x - 7 \\
-10x^2 + 5x - 20 \\
\hline
13
\end{array}$$

Answer: $3x^2 - 5 + \dfrac{13}{2x^2 - x + 4}$

16. (4.5) **a.** $7! = 7 \cdot 6 \cdot 5 \cdot 4 \cdot 3 \cdot 2 \cdot 1$ **b.** $\dbinom{7}{5} = \dfrac{7!}{5!(2!)} = \dfrac{7 \cdot 6 \cdot 5!}{5! \cdot 2 \cdot 1} = 21$

17. (4.5) $(2x + 1)^5 = (2x)^5 + 5(2x)^4(1) + \dfrac{5!}{(2!)(3!)}(2x)^3(1)^2$
$+ \dfrac{5!}{(3!)(2!)}(2x)^2(1)^3 + 5(2x)(1)^4 + 1^5$
$= 32x^5 + 5(16)x^4 + 10(8)x^3 + 10(4)x^2 + 10x + 1$
$= 32x^5 + 80x^4 + 80x^3 + 40x^2 + 10x + 1$

18. (4.8) Let x = number of dimes
Then $14 - x$ = number of quarters
Value of dimes + value of quarters = total value
$$\begin{aligned}
10x + 25(14 - x) &= 215 \\
10x + 350 - 25x &= 215 \\
-15x &= -135 \\
x &= 9 \\
14 - x &= 5
\end{aligned}$$
Linda has 9 dimes and 5 quarters.

19. (4.8) Let x = number of lb of cashews
Then $60 - x$ = number of lb of peanuts
$$\begin{aligned}
7.40x + 2.80(60 - x) &= 4.18(60) \\
740x + 280(60 - x) &= 418(60) \\
740x + 16{,}800 - 280x &= 25{,}080 \\
460x &= 8{,}280 \\
x &= 18 \\
60 - x &= 42
\end{aligned}$$
The grocer needs to mix 18 lb of cashews and 42 lb of peanuts.

20. (4.9) Let x = number of cc of water
Then $600 + x$ = number of cc in mixture

$$\begin{pmatrix}\text{Amount of potassium}\\ \text{chloride in 20\% solution}\end{pmatrix} = \begin{pmatrix}\text{amount of potassium}\\ \text{chloride in 15\% solution}\end{pmatrix}$$

$$\begin{aligned} 0.20(600) + 0x &= 0.15(600 + x)\\ 20(600) &= 15(600 + x)\\ 12{,}000 &= 9{,}000 + 15x\\ 3{,}000 &= 15x\\ x &= 200 \end{aligned}$$

200 cc of water should be added.

Chapters 1–4 Cumulative Review Exercises (page 193)

5

1. a. 10 b. All c. $-2, \frac{1}{2}, 4.5, 10, 0, 0.\overline{234}$
d. $-2, 10, 0$ e. $1.414\,213\,6\ldots$

2. $\dfrac{1}{x^2y^3}$

3. $108 = 2^2 \cdot 3^3$
$360 = 2^3 \cdot 3^2 \cdot 5$
$\text{LCM} = 2^3 \cdot 3^3 \cdot 5 = 1{,}080$

4. $6a - 3$

5. $$\begin{aligned} 4x - 12 &= 4 - x - 6\\ 5x &= 10\\ x &= 2 \end{aligned}$$
$\{2\}$

6. $\{x \mid x \geq 3\}$

7. If $|5 - x| = 6$,
$5 - x = 6$ or $5 - x = -6$
$-x = 1$ $\quad -x = -11$
$x = -1$ $\quad x = 11$
$\{-1, 11\}$

8. $\{x \mid x > 4\} \cup \left\{x \mid x < -\frac{2}{3}\right\}$

9. If $|2x - 1| \leq 4$,
$$\begin{aligned} -4 \leq 2x - 1 &\leq 4\\ -3 \leq 2x &\leq 5\\ -\tfrac{3}{2} \leq x &\leq \tfrac{5}{2} \end{aligned}$$
$\left\{x \mid -\frac{3}{2} \leq x \leq \frac{5}{2}\right\}$

10. $3x^5 + 6x^4 + 7x^3 + 12x^2 - 4x$

11. $x^6 - 6x^5 + \dfrac{6!}{(2!)(4!)}x^4 - \dfrac{6!}{(3!)(3!)}x^3 + 15x^2 - 6x + 1$
$= x^6 - 6x^5 + 15x^4 - 20x^3 + 15x^2 - 6x + 1$

12. $x^2 - 2x - 5 + \dfrac{4}{x + 4}$

13. $$\begin{aligned} -3 < 2x + 1 &< 5\\ -3 - 1 < 2x + 1 - 1 &< 5 - 1\\ -4 < 2x &< 4\\ -2 < x &< 2 \end{aligned}$$

14. 4.5 lb

15. Let x = amount invested at 7.2%
Then $13{,}500 - x$ = amount invested at 7.4%
$$\begin{aligned} 0.072x + 0.074(13{,}500 - x) &= 986.60\\ 72x + 74(13{,}500 - x) &= 986{,}600\\ 72x + 999{,}000 - 74x &= 986{,}600\\ 999{,}000 - 2x &= 986{,}600\\ 2x &= 12{,}400\\ x &= 6{,}200\\ 13{,}500 - x &= 7{,}300 \end{aligned}$$
Consuelo invested \$6,200 at 7.2% and \$7,300 at 7.4%.

16. 8 bags of potting soil, 3 bags of perlite

17. Let x = amount invested at 10%
Then $23{,}000 - x$ = amount invested at 8%
$$\begin{aligned} 0.10x + 0.08(23{,}000 - x) &= 2{,}170\\ 10x + 8(23{,}000 - x) &= 217{,}000\\ 10x + 184{,}000 - 8x &= 217{,}000\\ 2x &= 33{,}000\\ x &= 16{,}500\\ 23{,}000 - x &= 6{,}500 \end{aligned}$$
Mrs. Rice invested \$16,500 at 10% and \$6,500 at 8%.

Exercises 5.1 (page 199)

1. $18xy(3x^2z^4 - 4y^2)$ 2. $15ab^2c(15b^3 - 7a^2c^5)$
3. $4x(4x^2 - 2x + 1)$ 4. $3a(9a^3 - 3a + 1)$
5. Not factorable 6. Not factorable 7. $3(2my + 5mz - 3n)$
8. $4(nx + 2ny + 3z)$ 9. $-5r^7s^5(7t^4 + 11rs^4u^4 - 8p^8r^2s^3)$, or $5r^7s^5(-7t^4 - 11rs^4u^4 + 8p^8r^2s^3)$
10. $-40a^4c^3(3a^4b^7c^2 - d^9 + 2ac^2)$, or $40a^4c^3(-3a^4b^7c^2 + d^9 - 2ac^2)$
11. Not factorable 12. Not factorable
13. $-12x^4y^3(2x^4 + x^3y - 4xy^2 - 5y^3)$
14. $16y^4z^5(4y^5 + 3y^4z - y^3z^2 - 5z^3)$
15. $\text{GCF} = (a + b)$
$m(a + b) + n(a + b) = (a + b)(m + n)$
16. $(a - 2b)(3a + 2)$
17. $\text{GCF} = (y + 1)$
$x(y + 1) - (y + 1) = (y + 1)(x - 1)$
18. $(3e - f)(2e - 3)$
19. $\text{GCF} = (x - y)$
$$\begin{aligned} 5(x - y) - (x - y)^2 &= (x - y)[5 - (x - y)]\\ &= (x - y)(5 - x + y) \end{aligned}$$
20. $(a + b)(4 - a - b)$
21. $$\begin{aligned} 8x(y^2 + 3z)^2 - 6x^4(y^2 + 3z) &= 2x(y^2 + 3z)[4(y^2 + 3z) - 3x^3]\\ &= 2x(y^2 + 3z)(4y^2 + 12z - 3x^3) \end{aligned}$$
22. $3a^2(b - 2c^5)^2(4ab - 8ac^5 - 5)$
23. $5(x + y)^2(a + b)^5([x + y] + 3[a + b])$
$= 5(x + y)^2(a + b)^5(x + y + 3a + 3b)$
24. $7(s + t)^4(u + v)^6(2u + 2v + s + t)$

Exercises 5.2 (page 203)

1. $x(m - n) - y(m - n) = (m - n)(x - y)$
2. $(h - k)(a - b)$ 3. $x(y + 1) - 1(y + 1) = (y + 1)(x - 1)$
4. $(a - 1)(d + 1)$
5. $3a(a - 2b) + 2(a - 2b) = (a - 2b)(3a + 2)$
6. $(h - 3k)(2h + 5)$
7. $2e(3e - f) - 3(3e - f) = (3e - f)(2e - 3)$
8. $(2m - n)(4m - 3)$
9. $x^2(x + 3) - 2(x + 3) = (x + 3)(x^2 - 2)$
10. $(a - 1)(a^2 - 2)$ 11. $b^2(b + 4) + 5(b - 4)$ Not factorable
12. Not factorable
13. $2a^2(a + 4) - 3(a + 4) = (a + 4)(2a^2 - 3)$
14. $(y - 2)(5y^2 + 2)$
15. $$\begin{aligned} c[am + bm + an + bn] &= c[m(a + b) + n(a + b)]\\ &= c(a + b)(m + n) \end{aligned}$$
16. $k(u + v)(c + d)$
17. $x(a^2 + 2a + 5) + y(a^2 + 2a + 5) = (a^2 + 2a + 5)(x + y)$
18. $(x^2 + 3x + 7)(a + b)$
19. $x(s^2 - s + 4) + y(s^2 - s + 4) = (s^2 - s + 4)(x + y)$
20. $(t^2 - t + 3)(a + b)$
21. $a(x^2 + x + 1) - (x^2 + x + 1) = (x^2 + x + 1)(a - 1)$
22. $(y^2 + y + 2)(x - 1)$

Exercises 5.3 (page 205)

1. $2(x^2 - 4y^2) = 2(x - 2y)(x + 2y)$ 2. $3(x + 3y)(x - 3y)$
3. $2(49u^4 - 36v^4) = 2(7u^2 + 6v^2)(7u^2 - 6v^2)$
4. $3(9m^3 + 10n^2)(9m^3 - 10n^2)$
5. $(x^2 + y^2)(x^2 - y^2) = (x^2 + y^2)(x + y)(x - y)$
6. $(a^2 + 4)(a + 2)(a - 2)$ 7. Not factorable

8. Not factorable 9. $(2h^2k^2 + 1)(2h^2k^2 - 1)$
10. $(3x^2 + 1)(3x^2 - 1)$
11. $a^2b^2(25a^2 - b^2) = a^2b^2(5a + b)(5a - b)$
12. $x^2y^2(y + 10x)(y - 10x)$ 13. $8x(2x + 1)$ 14. $5y(5y + 1)$
15. Not factorable 16. Not factorable
17. $[(x + y) + 2][(x + y) - 2] = (x + y + 2)(x + y - 2)$
18. $(a + b + 3)(a + b - 3)$
19. $(a + b)[x^2 - y^2] = (a + b)(x + y)(x - y)$
20. $(x + y)(a + b)(a - b)$ 21. Not factorable
22. Not factorable
23. $(x + 3y)(x - 3y) + (x - 3y) = (x - 3y)([x + 3y] + 1)$
$= (x - 3y)(x + 3y + 1)$
24. $(x - y)(x + y + 1)$
25. $(x + 2y) + (x + 2y)(x - 2y) = (x + 2y)(1 + x - 2y)$
26. $(a + b)(1 + a - b)$

Exercises 5.4A (page 209)

1. $(t - 3)(t + 10)$ 2. $(m + 2)(m - 15)$
3. $(m + 1)(m + 12)$ 4. $(x + 14)(x + 1)$ 5. Not factorable
6. Not factorable 7. Not factorable 8. Not factorable
9. $x^2 + 10x = x(x + 10)$ 10. $y(y + 10)$
11. $(u^2 - 14)(u^2 - 1) = (u^2 - 14)(u - 1)(u + 1)$
12. $(y - 1)(y + 1)(y^2 - 15)$ 13. $(u - 4)(u + 16)$
14. $(v + 2)(v - 32)$ 15. $x^2 + 3x + 2 = (x + 2)(x + 1)$
16. $(a + 5)(a + 2)$ 17. $x^2(x^2 - 6x + 2)$
18. $y^2(y^4 - 2y^2 + 2)$ 19. $(x + 2y)(x - y)$ 20. $(x + 3y)^2$
21. $[(a + b) + 2][(a + b) + 4] = (a + b + 2)(a + b + 4)$
22. $(m + n + 1)(m + n + 8)$
23. $[(x + y) + 2][(x + y) - 15] = (x + y + 2)(x + y - 15)$
24. $(x + y + 2)(x + y - 12)$

Exercises 5.4B (page 216)

1. $(x + 1)(5x + 4)$ 2. $(5x + 2)(x + 2)$
3. $(7 - b)(1 - 3b)$ or $(3b - 1)(b - 7)$
4. $(1 - u)(7 - 3u)$ or $(3u - 7)(u - 1)$ 5. Not factorable
6. Not factorable 7. $(3n - 1)(n + 5)$ 8. $(3n + 5)(n - 1)$
9. $(3t + z)(t - 6z)$ 10. $(3x + 2y)(x - 3y)$
11. Not factorable 12. Not factorable
13. $4(2 + 3z - 2z^2) = 4(1 + 2z)(2 - z)$
14. $3(1 + 3z)(3 - 2z)$ 15. $(2a + 1)^2$ 16. $(3b + 1)^2$
17. $2(x^2 - 9) = 2(x - 3)(x + 3)$ 18. $5(y + 4)(y - 4)$
19. $7h^2 - 11h + 4 = (7h - 4)(h - 1)$ 20. $(h - 2)(7h - 2)$
21. $2(36x^2 + 12xy + y^2) = 2(6x + y)^2$ 22. $2(4x + y)^2$
23. Not factorable 24. Not factorable
25. $2y(x^2 + 4xy + 4y^2) = 2y(x + 2y)^2$ 26. $3x(x + y)^2$
27. $(3e^2 + 4)(2e^2 - 5)$ 28. $(5f^2 + 3)(2f^2 - 7)$
29. $3x^2(4x^2 - 25y^2) = 3x^2(2x + 5y)(2x - 5y)$
30. $4x^2(3x + 2y)(3x - 2y)$ 31. $(a^2 + b^2)^2$ 32. $(x^2 + 3y^2)^2$
33. Let $(a + b) = x$
Then $2(a + b)^2 + 7(a + b) + 3 = 2x^2 + 7x + 3$
$= (2x + 1)(x + 3)$
But since $x = (a + b)$,
$2(a + b)^2 + 7(a + b) + 3 = [2(a + b) + 1][(a + b) + 3]$
$= (2a + 2b + 1)(a + b + 3)$
34. $(3a - 3b + 1)(a - b + 2)$
35. Let $(x - y) = a$
Then $4(x - y)^2 - 8(x - y) - 5 = 4a^2 - 8a - 5$
$= (2a + 1)(2a - 5)$
But since $a = (x - y)$,
$4(x - y)^2 - 8(x - y) - 5 = (2[x - y] + 1)(2[x - y] - 5)$
$= (2x - 2y + 1)(2x - 2y - 5)$
36. $(2x + 2y + 1)(2x + 2y - 3)$
37. $5x^2 + 10xy + 5y^2 - 21x - 21y + 4$
$= 5(x^2 + 2xy + y^2) - 21(x + y) + 4$
$= 5(x + y)^2 - 21(x + y) + 4$
Let $a = (x + y)$
Then $5(x + y)^2 - 21(x + y) + 4 = 5a^2 - 21a + 4$
$= (5a - 1)(a - 4)$
But since $a = (x + y)$,
$5(x + y)^2 - 21(x + y) + 4 = (5[x + y] - 1)([x + y] - 4)$
$= (5x + 5y - 1)(x + y - 4)$
38. $(5x - 5y - 2)(x - y - 2)$
39. $(2x - y)^2 - (3a + b)^2$ Difference of two squares
$= [(2x - y) + (3a + b)][(2x - y) - (3a + b)]$
$= (2x - y + 3a + b)(2x - y - 3a - b)$
40. $(2x + 3y + a - b)(2x + 3y - a + b)$
41. $\underbrace{x^2 + 10xy + 25y^2} - 9$
$= (x + 5y)^2 \quad - 9$
$= (x + 5y - 3)(x + 5y + 3)$
42. $(x + 4y + 5)(x + 4y - 5)$
43. $a^2 - 4x^2 - 4xy - y^2$
$= a^2 - (4x^2 + 4xy + y^2)$
$= a^2 - (2x + y)^2$ Difference of two squares
$= (a + [2x + y])(a - [2x + y])$
$= (a + 2x + y)(a - 2x - y)$
44. $(x + 3a + b)(x - 3a - b)$
45. $(4x^2 - 9)(x^2 - 1) = (2x - 3)(2x + 3)(x - 1)(x + 1)$
46. $(3x - 2)(3x + 2)(x + 1)(x - 1)$
47. $\underbrace{3x^2 - 7xy - 6y^2} \quad \underbrace{- x + 3y}$
$= (3x + 2y)(x - 3y) - (x - 3y)$
$= (x - 3y)([3x + 2y] - 1) = (x - 3y)(3x + 2y - 1)$
48. $(3t + z)(t - 6z - 1)$
49. $\underbrace{3n^2 + 2mn - 5m^2} \quad \underbrace{+ 3n + 5m}$
$= (3n + 5m)(n - m) + (3n + 5m)$
$= (3n + 5m)(n - m + 1)$
50. $(3n - m)(n + 5m + 1)$

Sections 5.1–5.4 Review Exercises (page 218)

1. $13xy(5xy^2 - 3y^3 - 1)$ 2. Not factorable
3. $3x(x^2 + 3x - 4) = 3x(x + 4)(x - 1)$ 4. Not factorable
5. $(x + 5)(x + 8)$ 6. $(x - 4)(x + 5)$ 7. Not factorable
8. $(x - 9)(x - 2)$ 9. $(x - 16)(x + 16)$
10. $(x - 13)(x - 1)$ 11. $(3x - 7)(x + 2)$
12. $(2x + 5)(x - 8)$
13. $3x(y + 4) + 2(y + 4) = (y + 4)(3x + 2)$
14. $(x + 5)(x + 1)(x - 1)$ 15. Not factorable
16. Not factorable 17. $2x(4x^2 - 1) = 2x(2x - 1)(2x + 1)$
18. Not factorable 19. $(4x - 1)(x + 3)$
20. $(8a - 5)(a + 2)$

5

Exercises 5.5 (page 224)

1. $(x)^3 - (2)^3 = (x-2)[(x)^2 + (2)(x) + (2)^2]$
$= (x-2)(x^2+2x+4)$

2. $(x-3)(x^2+3x+9)$

3. $(4)^3 + (a)^3 = (4+a)[(4)^2 - (4)(a) + (a)^2]$
$= (4+a)(16-4a+a^2)$

4. $(2+b)(4-2b+b^2)$

5. $(5)^3 - x^3 = (5-x)[(5)^2 + (5)(x) + (x)^2]$
$= (5-x)(25+5x+x^2)$

6. $(1-a)(1+a+a^2)$

7. $2x(4x^2-1) = 2x(2x-1)(2x+1)$

8. $3x(3x+1)(3x-1)$

9. $c^3 - (3ab)^3 = (c-3ab)[(c)^2 + (c)(3ab) + (3ab)^2]$
$= (c-3ab)(c^2+3abc+9a^2b^2)$

10. $(c-4ab)(c^2+4abc+16a^2b^2)$

11. $(2xy^2)^3 + (3)^3 = (2xy^2+3)[(2xy^2)^2 - (2xy^2)(3) + (3)^2]$
$= (2xy^2+3)(4x^2y^4 - 6xy^2 + 9)$

12. $(4a^2b+5)(16a^4b^2 - 20a^2b + 25)$

13. Not factorable **14.** Not factorable

15. $a[a^3+b^3] = a(a+b)(a^2-ab+b^2)$

16. $y(x+y)(x^2-xy+y^2)$

17. $3(27-x^3) = 3(3-x)[(3)^2 + (3)(x) + (x)^2]$
$= 3(3-x)(9+3x+x^2)$

18. $5(2-b)(4+2b+b^2)$

19. $(x+y)^3 + (1)^3 = [(x+y)+(1)][(x+y)^2 - (x+y)(1) + (1)^2]$
$= (x+y+1)(x^2+2xy+y^2-x-y+1)$

20. $(1+x-y)(1-x+y+x^2-2xy+y^2)$

21. $(4x)^3 - (y^2)^3 = (4x-y^2)[(4x)^2 + (4x)(y^2) + (y^2)^2]$
$= (4x-y^2)(16x^2+4xy^2+y^4)$

22. $(5w-v^2)(25w^2+5wv^2+v^4)$

23. $4(a^3b^3+27c^6) = 4[(ab)^3 + (3c^2)^3]$
$= 4[ab+3c^2][(ab)^2 - (ab)(3c^2) + (3c^2)^2]$
$= 4(ab+3c^2)(a^2b^2 - 3abc^2 + 9c^4)$

24. $5(xy^2+2z^3)(x^2y^4 - 2xy^2z^3 + 4z^6)$

25. A binomial that is both the difference of two squares and the difference of two cubes should be treated *first* as the difference of two squares.
$x^6 - 729 = (x^3-27)(x^3+27)$
$= (x-3)(x^2+3x+9)(x+3)(x^2-3x+9)$

26. $(y-2)(y^2+2y+4)(y+2)(y^2-2y+4)$

27. $(x+1)^3 - (y-z)^3$
$= [(x+1)-(y-z)][(x+1)^2 + (x+1)(y-z) + (y-z)^2]$
$= (x+1-y+z)$
$\cdot (x^2+2x+1+xy-xz+y-z+y^2-2yz+z^2)$

28. $(x-y-a-b)$
$\cdot (x^2-2xy+y^2+ax+bx-ay-by+a^2+2ab+b^2)$

Exercises 5.6 (page 230)

1. $x^2+12x+36 = (x+6)^2$ **2.** $x^2+18x+81 = (x+9)^2$

3. $x^2+8x+16 = (x+4)^2$ **4.** $x^2+22x+121 = (x+11)^2$

5. $x^2 - 14x + 49 - 49 - 480 = (x-7)^2 - 529$
$= [(x-7)+23][(x-7)-23)] = (x+16)(x-30)$

6. $(x+20)(x-32)$

7. $x^2+6x+9-9-1{,}015 = (x+3)^2 - 1{,}024$
$= [(x+3)+32][(x+3)-32] = (x+35)(x-29)$

8. $(x-29)(x+37)$

9. $x^2-2x+1-1-532 = (x-1)^2 - 533$; not factorable, since 533 is not the square of an integer

10. Not factorable

11. $x^2+16x+64-64+793 = (x+8)^2 + 729$; not factorable, since we do not have a *difference* of two squares

12. Not factorable

Steps 1–3 will not be shown.

13. Try $(x^2+2)^2$.
Add and subtract x^2.
$x^4+3x^2+x^2+4-x^2 = x^4+4x^2+4-x^2$
$= (x^2+2)^2 - x^2$
$= (x^2+2-x)(x^2+2+x)$
$= (x^2-x+2)(x^2+x+2)$

14. $(x^2+x+3)(x^2-x+3)$

15. Try $(2m^2+1)^2$.
Add and subtract m^2.
$4m^4+3m^2+m^2+1-m^2 = (4m^4+4m^2+1) - m^2$
$= (2m^2+1)^2 - m^2$
$= (2m^2+1-m)(2m^2+1+m)$
$= (2m^2-m+1)(2m^2+m+1)$

16. $(3u^2+u+1)(3u^2-u+1)$

17. Add and subtract $16a^2b^2$.
$64a^4+16a^2b^2+b^4-16a^2b^2$
$= (8a^2+b^2)^2 - 16a^2b^2$
$= (8a^2+b^2-4ab)(8a^2+b^2+4ab)$
$= (8a^2-4ab+b^2)(8a^2+4ab+b^2)$

18. $(a^2+2ab+2b^2)(a^2-2ab+2b^2)$

19. Add and subtract $9x^2$.
$x^4-3x^2+9x^2+9-9x^2 = x^4+6x^2+9-9x^2$
$= (x^2+3)^2 - 9x^2$
$= (x^2+3-3x)(x^2+3+3x)$
$= (x^2-3x+3)(x^2+3x+3)$

20. $(x^2+3x+4)(x^2-3x+4)$

21. Add and subtract $9a^2b^2$.
$a^4 - 17a^2b^2 + 9a^2b^2 + 16b^4 - 9a^2b^2$
$= (a^4 - 8a^2b^2 + 16b^4) - 9a^2b^2$
$= (a^2-4b^2)^2 - 9a^2b^2$
$= (a^2-4b^2-3ab)(a^2-4b^2+3ab)$
$= (a^2-3ab-4b^2)(a^2+3ab-4b^2)$
$= (a-4b)(a+b)(a+4b)(a-b)$

22. $(a+1)(a-1)(a+6)(a-6)$

23. Not factorable **24.** Not factorable

25. Add and subtract $9a^2b^2$.
$a^4 - 3a^2b^2 + 9a^2b^2 + 9b^4 - 9a^2b^2$
$= (a^4 + 6a^2b^2 + 9b^4) - 9a^2b^2$
$= (a^2+3b^2)^2 - 9a^2b^2$
$= (a^2+3b^2-3ab)(a^2+3b^2+3ab)$
$= (a^2-3ab+3b^2)(a^2+3ab+3b^2)$

26. $(a^2+5ab+5b^2)(a^2-5ab+5b^2)$

27. Not factorable **28.** Not factorable

29. $2(25x^4 - 6x^2y^2 + y^4)$
Add and subtract $16x^2y^2$.
$2[25x^4 - 6x^2y^2 + 16x^2y^2 + y^4 - 16x^2y^2]$
$= 2[25x^4 + 10x^2y^2 + y^4 - 16x^2y^2]$
$= 2[(5x^2+y^2)^2 - 16x^2y^2]$
$= 2[(5x^2+y^2-4xy)(5x^2+y^2+4xy)]$
$= 2(5x^2-4xy+y^2)(5x^2+4xy+y^2)$

30. $2(4x^2+3xy+y^2)(4x^2-3xy+y^2)$

31. $2n(4m^4+n^4)$
Add and subtract $4m^2n^2$.
$2n[(4m^4+4m^2n^2+n^4) - 4m^2n^2]$
$= 2n[(2m^2+n^2)^2 - 4m^2n^2]$
$= 2n(2m^2+2mn+n^2)(2m^2-2mn+n^2)$

32. $3m(m^2+2mn+2n^2)(m^2-2mn+2n^2)$

33. $2y(25x^4+16x^2y^2+4y^4)$
Add and subtract $4x^2y^2$.

$$\begin{aligned} &2y(25x^4 + 16x^2y^2 + 4x^2y^2 + 4y^4 - 4x^2y^2) \\ &= 2y(25x^4 + 20x^2y^2 + 4y^4 - 4x^2y^2) \\ &= 2y[(5x^2 + 2y^2)^2 - 4x^2y^2] \\ &= 2y(5x^2 + 2xy + 2y^2)(5x^2 - 2xy + 2y^2) \end{aligned}$$

34. $3x(4x^2 + xy + y^2)(4x^2 - xy + y^2)$

Exercises 5.7 (page 235)

1. Factors of the constant term are ± 1, ± 3.

$$\begin{array}{r|rrrr} 1 & 1 & 1 & 1 & -3 \\ & & 1 & 2 & 3 \\ \hline & 1 & 2 & 3 & \boxed{0} \end{array}$$

Remainder is zero; therefore, $(x - 1)$ is a factor

$x^2 + 2x + 3$ Will not factor

Therefore, $x^3 + x^2 + x - 3 = (x - 1)(x^2 + 2x + 3)$.

2. $(x - 2)(x^2 + 3x + 1)$

3. Factors of the constant term are ± 1, ± 2, ± 3, ± 4, ± 6, ± 12.

$$\begin{array}{r|rrrr} 2 & 1 & -3 & -4 & 12 \\ & & 2 & -2 & -12 \\ \hline & 1 & -1 & -6 & \boxed{0} \end{array}$$

Remainder is zero; therefore, $(x - 2)$ is a factor

$x^2 - x - 6$ Quotient is another factor

$(x - 3)(x + 2)$ Factors of quotient

Therefore, $x^3 - 3x^2 - 4x + 12 = (x - 2)(x - 3)(x + 2)$.

4. $(x - 1)(x + 2)(x - 3)$

5. $2(x^3 - 4x^2 + x + 6)$

We now factor $x^3 - 4x^2 + x + 6$.

Factors of the constant term are ± 1, ± 2, ± 3, ± 6.

$$\begin{array}{r|rrrr} 1 & 1 & -4 & 1 & 6 \\ & & 1 & -3 & -2 \\ \hline & 1 & -3 & -2 & \boxed{4} \end{array}$$

Remainder is not zero; therefore, $(x - 1)$ is not a factor

$$\begin{array}{r|rrrr} -1 & 1 & -4 & 1 & 6 \\ & & -1 & 5 & -6 \\ \hline & 1 & -5 & 6 & \boxed{0} \end{array}$$

Remainder is zero; therefore, $(x + 1)$ is a factor

$x^2 - 5x + 6$ Quotient is another factor

$(x - 2)(x - 3)$ Factors of quotient

Therefore, $x^3 - 4x^2 + x + 6 = (x + 1)(x - 2)(x - 3)$ and $2x^3 - 8x^2 + 2x + 12 = 2(x + 1)(x - 2)(x - 3)$.

6. $2(x + 1)(x + 2)(x + 3)$

7. Factors of the constant term are ± 1, ± 2, ± 4.

$$\begin{array}{r|rrrr} 2 & 6 & -13 & 0 & +4 \\ & & 12 & -2 & -4 \\ \hline & 6 & -1 & -2 & \boxed{0} \end{array}$$

Remainder is zero; therefore $(x - 2)$ is a factor

$$\begin{aligned} \text{Therefore, } 6x^3 - 13x^2 + 4 &= (x - 2)(6x^2 - x - 2) \\ &= (x - 2)(3x - 2)(2x + 1) \end{aligned}$$

8. $(x + 1)(x + 2)(x - 3)$

9. Factors of the constant term are ± 1, ± 2, ± 4.

$$\begin{array}{r|rrrrr} -2 & 1 & 0 & -3 & 4 & 4 \\ & & -2 & 4 & -2 & -4 \\ \hline & 1 & -2 & 1 & 2 & \boxed{0} \end{array}$$

Remainder is zero; therefore, $(x + 2)$ is a factor

$x^3 - 2x^2 + x + 2$ Quotient is another factor; it does not factor

Therefore, $x^4 - 3x^2 + 4x + 4 = (x + 2)(x^3 - 2x^2 + x + 2)$.

10. $(x + 2)(x^3 - 3x^2 + x + 3)$

11. Factors of the constant term are ± 1, ± 2, ± 4.

$$\begin{array}{r|rrrrr} 1 & 1 & 2 & -3 & -8 & -4 \\ & & 1 & 3 & 0 & -8 \\ \hline & 1 & 3 & 0 & -8 & \boxed{-12} \end{array}$$

Remainder is not zero; therefore, $(x - 1)$ is not a factor

$$\begin{array}{r|rrrrr} 2 & 1 & 2 & -3 & -8 & -4 \\ & & 2 & 8 & 10 & 4 \\ \hline & 1 & 4 & 5 & 2 & \boxed{0} \end{array}$$

Remainder is zero; therefore, $(x - 2)$ is a factor

We *now* work with the coefficients 1, 4, 5, 2.

Factors of the constant term are ± 1 and ± 2. However, since $+1$ did not work for the original coefficients, it will not work for the new ones. We must try $+2$ a second time:

$$\begin{array}{r|rrrr} 2 & 1 & 4 & 5 & 2 \\ & & 2 & 12 & 34 \\ \hline & 1 & 6 & 17 & \boxed{36} \end{array}$$

Remainder is not zero; therefore, $(x - 2)$ is not a factor of $x^3 + 4x^2 + 5x + 2$

$$\begin{array}{r|rrrr} -1 & 1 & 4 & 5 & 2 \\ & & -1 & -3 & -2 \\ \hline & 1 & 3 & 2 & \boxed{0} \end{array}$$

Remainder is zero; therefore, $(x + 1)$ is a factor

$$\begin{aligned} \text{Therefore, } x^4 + 2x^3 - 3x^2 - 8x - 4 &= (x - 2)(x + 1)(x^2 + 3x + 2) \\ &= (x - 2)(x + 1)(x + 1)(x + 2) \text{ or} \\ &= (x - 2)(x + 1)^2(x + 2) \end{aligned}$$

12. $(x - 1)(x + 1)(x + 2)^2$

13. Factors of the constant term are ± 1. We may also have to try $\pm\frac{1}{3}$.

$$\begin{array}{r|rrrrr} 1 & 3 & -4 & 0 & 0 & -1 \\ & & 3 & -1 & -1 & -1 \\ \hline & 3 & -1 & -1 & -1 & \boxed{-2} \end{array}$$

Remainder is not zero; therefore, $(x - 1)$ is not a factor

$$\begin{array}{r|rrrrr} -1 & 3 & -4 & 0 & 0 & -1 \\ & & -3 & 7 & -7 & 7 \\ \hline & 3 & -7 & 7 & -7 & \boxed{6} \end{array}$$

Remainder is not zero; therefore, $(x + 1)$ is not a factor

$$\begin{array}{r|rrrrr} \frac{1}{3} & 3 & -4 & 0 & 0 & -1 \\ & & 1 & -1 & -\frac{1}{3} & -\frac{1}{9} \\ \hline & 3 & -3 & -1 & -\frac{1}{3} & \boxed{-\frac{10}{9}} \end{array}$$

Remainder is not zero; therefore, $\left(x - \frac{1}{3}\right)$ is not a factor

$$\begin{array}{r|rrrrr} -\frac{1}{3} & 3 & -4 & 0 & 0 & -1 \\ & & -1 & \frac{5}{3} & -\frac{5}{9} & \frac{5}{27} \\ \hline & 3 & -5 & \frac{5}{3} & -\frac{5}{9} & \boxed{-\frac{22}{27}} \end{array}$$

Remainder is not zero; therefore, $\left(x + \frac{1}{3}\right)$ is not a factor

Since none of the possible factors work, the expression is not factorable.

14. Not factorable

15. Factors of the constant term are ± 1, ± 2, ± 3, ± 4, ± 6, ± 8, ± 12, ± 24.

$$\begin{array}{r|rrrrr} 1 & 1 & -4 & -7 & 34 & -24 \\ & & 1 & -3 & -10 & 24 \\ \hline 2 & 1 & -3 & -10 & 24 & \boxed{0} \\ & & 2 & -2 & -24 & \\ \hline & 1 & -1 & -12 & \boxed{0} & \end{array}$$

Remainder is zero; therefore, $(x - 1)$ is a factor

Remainder is zero; therefore, $(x - 2)$ is a factor

$$\begin{aligned} \text{Therefore, } x^4 - 4x^3 - 7x^2 + 34x - 24 &= (x - 1)(x - 2)(x^2 - x - 12) \\ &= (x - 1)(x - 2)(x - 4)(x + 3) \end{aligned}$$

16. $(x - 2)(x + 1)(x + 3)(x + 4)$

17. Factors of the constant term are ± 1, ± 2, ± 3, ± 6. (We may also have to try fractions whose denominators are factors of 6.)

$$\begin{array}{r|rrrr} 1 & 6 & 1 & -11 & -6 \\ & & 6 & 7 & -4 \\ \hline & 6 & 7 & -4 & \boxed{-10} \end{array}$$

Remainder is not zero; therefore, $(x - 1)$ is not a factor

$$\begin{array}{r|rrrr} 2 & 6 & 1 & -11 & -6 \\ & & 12 & 26 & 30 \\ \hline & 6 & 13 & 15 & \boxed{24} \end{array}$$

Remainder is not zero; therefore, $(x - 2)$ is not a factor

$$\begin{array}{r|rrrr} -1 & 6 & 1 & -11 & -6 \\ & & -6 & 5 & 6 \\ \hline & 6 & -5 & -6 & \boxed{0} \end{array}$$

Remainder is zero; $(x + 1)$ is a factor

$$\begin{aligned} \text{Therefore, } 6x^3 + x^2 - 11x - 6 &= (x + 1)(6x^2 - 5x - 6) \\ &= (x + 1)(2x - 3)(3x + 2) \end{aligned}$$

18. $(x - 3)(3x - 2)(2x + 1)$

5

Exercises 5.8 (page 237)

1. $(4e - 5)(3e + 7)$ **2.** $(6f + 7)(5f - 3)$

3. $6(ac - bd + bc - ad) = 6[a(c - d) + b(c - d)]$
$= 6(a + b)(c - d)$

4. $(2c + d)(5y - 3z)$

5. $2xy(y^2 - 2y - 15) = 2xy(y + 3)(y - 5)$

6. $3yz(z - 4)(z + 2)$

7. $3(x^3 + 8h^3) = 3(x + 2h)(x^2 - 2xh + 4h^2)$

8. $2(3f - g)(9f^2 + 3fg + g^2)$ **9.** $(3e - 5f)^2$

10. $(4m + 7p)^2$

11. $x^2(x + 3) - 4(x + 3) = (x + 3)(x^2 - 4)$
$= (x + 3)(x + 2)(x - 2)$

12. $(a - 2)(a + 3)(a - 3)$

13. $(a + b)(a - b) - 1(a - b) = (a - b)(a + b - 1)$

14. $(x + y)(x - y - 1)$ **15.** Not factorable

16. Not factorable

17. $\underbrace{x^3 - 8y^3} + \underbrace{x^2 - 4y^2}$
$= (x - 2y)(x^2 + 2xy + 4y^2) + (x - 2y)(x + 2y)$
$= (x - 2y)([x^2 + 2xy + 4y^2] + [x + 2y])$
$= (x - 2y)(x^2 + 2xy + 4y^2 + x + 2y)$

18. $(a - b)(a^2 + ab + b^2 + a + b)$

19. Not factorable **20.** Not factorable

21. $x^2 - 4xy + 4y^2 - 5x + 10y + 6 = (x - 2y)^2 - 5(x - 2y) + 6$
Let $a = (x - 2y)$
Then $(x - 2y)^2 - 5(x - 2y) + 6 = a^2 - 5a + 6 = (a - 2)(a - 3)$
But since $a = (x - 2y)$,
$(a - 2)(a - 3) = ([x - 2y] - 2)([x - 2y] - 3)$
Therefore,
$x^2 - 4xy + 4y^2 - 5x + 10y + 6 = (x - 2y - 2)(x - 2y - 3)$

22. $(x - 3y - 3)(x - 3y - 5)$

23. $(x^2 - 6xy + 9y^2) - 25 = (x - 3y)^2 - 25$
$= (x - 3y - 5)(x - 3y + 5)$

24. $(a - 4b + 1)(a - 4b - 1)$

Exercises 5.9 (page 241)

1. $3x = 0$ | $x - 4 = 0$
$x = 0$ | $x = 4$
$\{0, 4\}$

2. $\{0, -6\}$

3. $4x^2 - 12x = 0$
$4x(x - 3) = 0$
$4x = 0$ | $x - 3 = 0$
$x = 0$ | $x = 3$
$\{0, 3\}$

4. $\left\{0, \frac{3}{2}\right\}$

5. $x^2 - 4x = 12$
$x^2 - 4x - 12 = 0$
$(x - 6)(x + 2) = 0$
$x - 6 = 0$ | $x + 2 = 0$
$x = 6$ | $x = -2$
$\{6, -2\}$

6. $\{-3, 5\}$

7. $2x^3 + x^2 - 3x = 0$
$x(2x^2 + x - 3) = 0$
$x(x - 1)(2x + 3) = 0$
$x = 0$ | $x - 1 = 0$ | $2x + 3 = 0$
$x = 1$ | $2x = -3$
$x = -\frac{3}{2}$
$\left\{0, 1, -\frac{3}{2}\right\}$

8. $\left\{0, -5, \frac{1}{2}\right\}$

9. $2x^2 + 7x - 15 = 0$
$(2x - 3)(x + 5) = 0$
$2x - 3 = 0$ | $x + 5 = 0$
$2x = 3$ | $x = -5$
$x = \frac{3}{2}$
$\left\{\frac{3}{2}, -5\right\}$

10. $\left\{\frac{2}{3}, -5\right\}$

11. $4x^2 - 12x + 9 = 0$
$(2x - 3)(2x - 3) = 0$
$2x - 3 = 0$
$2x = 3$
$x = \frac{3}{2}$
$\left\{\frac{3}{2}\right\}$

12. $\left\{\frac{2}{5}\right\}$

13. $18x^3 - 21x^2 - 60x = 0$
$3x(6x^2 - 7x - 20) = 0$
$3x(2x - 5)(3x + 4) = 0$
$3x = 0$ | $2x - 5 = 0$ | $3x + 4 = 0$
$x = 0$ | $2x = 5$ | $3x = -4$
$x = \frac{5}{2}$ | $x = -\frac{4}{3}$
$\left\{0, \frac{5}{2}, -\frac{4}{3}\right\}$

14. $\left\{0, \frac{3}{5}, \frac{5}{3}\right\}$

15. $4x = 0$ | $2x - 1 = 0$ | $3x + 7 = 0$
$x = 0$ | $2x = 1$ | $3x = -7$
$x = \frac{1}{2}$ | $x = -\frac{7}{3}$
$\left\{0, \frac{1}{2}, -\frac{7}{3}\right\}$

16. $\left\{0, \frac{3}{4}, \frac{6}{7}\right\}$

17. $x^3 + 3x^2 - 4x - 12 = 0$
$x^2(x + 3) - 4(x + 3) = 0$
$(x + 3)(x^2 - 4) = 0$
$(x + 3)(x + 2)(x - 2) = 0$
$x + 3 = 0$ | $x + 2 = 0$ | $x - 2 = 0$
$x = -3$ | $x = -2$ | $x = 2$
$\{-3, -2, 2\}$

18. $\{-3, 3, -1\}$

19. $(x^2 - 9)(x^2 - 1) = 0$
$(x - 3)(x + 3)(x - 1)(x + 1) = 0$
$x - 3 = 0$ | $x + 3 = 0$ | $x - 1 = 0$ | $x + 1 = 0$
$x = 3$ | $x = -3$ | $x = 1$ | $x = -1$
$\{3, -3, 1, -1\}$

20. $\{2, -2, 3, -3\}$

21. $x^3 + 3x^2(3) + 3x(3^2) + 3^3 = x^3 + 63$
$x^3 + 9x^2 + 27x + 27 = x^3 + 63$
$9x^2 + 27x - 36 = 0$
$9(x^2 + 3x - 4) = 0$
$9(x + 4)(x - 1) = 0$
$9 \neq 0$ | $x + 4 = 0$ | $x - 1 = 0$
$x = -4$ | $x = 1$
$\{-4, 1\}$

22. $\{3, -4\}$

23. $x^4 + 4x^3(3) + 6x^2(3^2) + 4x(3^3) + 3^4 = x^4 + 108x + 81$
$x^4 + 12x^3 + 54x^2 + 108x + 81 = x^4 + 108x + 81$
$12x^3 + 54x^2 = 0$
$6x^2(2x + 9) = 0$
$6x^2 = 0$ | $2x + 9 = 0$
$x = 0$ | $x = -\frac{9}{2}$
$\left\{0, -\frac{9}{2}\right\}$

24. $\{0, -3\}$

Exercises 5.10 (page 244)

The checks for these exercises will not be shown.

1. Let x = first even integer
Then $x + 2$ = second even integer
and $x + 4$ = third even integer
$x(x + 2) = 38 + (x + 4)$
$x^2 + 2x = 38 + x + 4$
$x^2 + x - 42 = 0$
$(x - 6)(x + 7) = 0$

$x - 6 = 0$ | $x + 7 = 0$
$x = 6$ | $x = -7$ Not an even integer
$x + 2 = 8$
$x + 4 = 10$
The integers are 6, 8, and 10.

2. 7, 9, and 11

3. Let x = length of a side of the cube
Then x = height of box (in in.)
and $x + 3$ = width of box (in in.)
and $4x$ = length of box (in in.)
x^3 = volume of cube
$x(x + 3)(4x)$ = volume of box
$x(x + 3)(4x) = 8x^3$
$4x^3 + 12x^2 = 8x^3$
$0 = 4x^3 - 12x^2$
$0 = 4x^2(x - 3)$
$4 \neq 0$ | $x^2 = 0$ | $x - 3 = 0$
 | $x = 0$ | $x = 3$
 | Not in the domain ↑ |
The height of the box is 3 in., the width is 6 in., and the length is 12 in. The volume of the cube is $(3 \text{ in.})^3$, or 27 cu. in.

4. Height of box is 5 cm; width is 10 cm; length is 15 cm. Volume of cube is 125 cc.

5. Let x = length (in m)
Then $x - 5$ = width (in m)
Area = perimeter + 46
$x(x - 5) = 4x - 10 + 46$
$x^2 - 5x = 4x + 36$
$x^2 - 9x - 36 = 0$
$(x - 12)(x + 3) = 0$
$x - 12 = 0$ | $x + 3 = 0$
$x = 12$ | $x = -3$ Not in the domain
$x - 5 = 7$ |
The length is 12 m, and the width is 7 m.

6. Length is 10 m; width is 3 m.

7. Let h = altitude (in cm)
Then $7 + h$ = base (in cm)
$\frac{1}{2}h(7 + h)$ = area
$\frac{1}{2}h(7 + h) = 39$
$h(7 + h) = 78$
$h^2 + 7h - 78 = 0$
$(h + 13)(h - 6) = 0$
$h + 13 = 0$ | $h - 6 = 0$
$h = -13$ | $h = 6$
Not in the domain ↑ | $7 + h = 13$
The altitude is 6 cm, and the base is 13 cm.

8. Altitude is 8 m; base is 12 m.

9. Let x = length of side of smaller square (in cm)
Then $x + 6$ = length of side of larger square (in cm)
Area of smaller square = x^2
Area of larger square = $(x + 6)^2$
$9x^2 = (x + 6)^2$
$9x^2 = x^2 + 12x + 36$
$8x^2 - 12x - 36 = 0$
$4(2x^2 - 3x - 9) = 0$
$2x^2 - 3x - 9 = 0$
$(2x + 3)(x - 3) = 0$
$2x + 3 = 0$ | $x - 3 = 0$
$x = -\frac{3}{2}$ | $x = 3$
↑
Not in the domain
The length of a side of the smaller square is 3 cm, and the length of a side of the larger square is 9 cm.

10. Side of larger square is 8 cm; side of smaller square is 4 cm.

11. Let x = length of side of smaller cube (in cm)
Then $x + 3$ = length of side of larger cube (in cm)
x^3 = volume of smaller cube
$(x + 3)^3$ = volume of larger cube
$(x + 3)^3 = 63 + x^3$
$x^3 + 9x^2 + 27x + 27 = 63 + x^3$
$9x^2 + 27x - 36 = 0$
$9(x^2 + 3x - 4) = 0$
$9(x + 4)(x - 1) = 0$
$9 \neq 0$ | $x + 4 = 0$ | $x - 1 = 0$
 | $x = -4$ | $x = 1$
 | Not in the domain ↗ | $x + 3 = 4$
The length of a side of the smaller cube is 1 cm, and the length of a side of the larger cube is 4 cm.

12. Length of side of smaller cube is 3 cm; length of side of larger cube is 4 cm.

13. Let x = width (in in.)
Then $x + 3$ = length (in in.)
(length)(width)(depth) = volume
$(x + 3)(x)(2) = 80$
$(x + 3)(x) = 40$
$x^2 + 3x - 40 = 0$
$(x + 8)(x - 5) = 0$
$x + 8 = 0$ | $x - 5 = 0$
$x = -8$ | $x = 5$
↑ | $x + 3 = 8$
Not in the domain

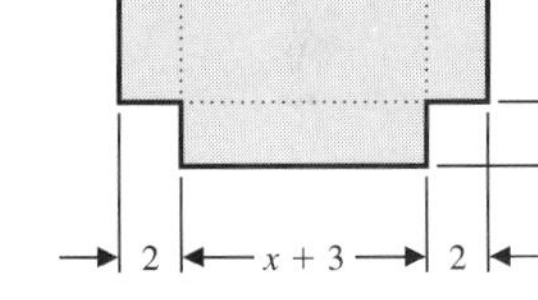

a. The dimensions of the metal sheet are 9 in. by 12 in.
b. The dimensions of the box are depth = 2 in., width = 5 in., length = 8 in.

14. **a.** 8 in. by 11 in. **b.** 2 in. by 5 in. by 3 in.

15. Let x = width of rectangular room (in yd)
Then $2x - 3$ = length of rectangular room (in yd)
$x(2x - 3) = 35$
$2x^2 - 3x = 35$
$2x^2 - 3x - 35 = 0$
$(2x + 7)(x - 5) = 0$
$2x + 7 = 0$ | $x - 5 = 0$
$x = -\frac{7}{2}$ | $x = 5$
Not in the domain ↑ | $2x - 3 = 7$
The width is 5 yd, and the length is 7 yd.

16. 6 yd by 9 yd

17. Let x = number of in. in a side of the square. If the picture were 3 in. longer and 2 in. narrower, its area would be $(x + 3)(x - 2)$ sq. in.; the cost would be $\$0.30(x + 3)(x - 2)$.
$0.30(x + 3)(x - 2) = 45$
$(x + 3)(x - 2) = 150$ Dividing both sides by 0.30
$x^2 + x - 6 = 150$
$x^2 + x - 156 = 0$
$(x - 12)(x + 13) = 0$
$x - 12 = 0$ | $x + 13 = 0$
$x = 12$ | $x = -13$ Reject
The picture is 12 in. by 12 in.

18. 9 yd by 9 yd

Sections 5.5–5.10 Review Exercises (page 246)

1. $3uv(5u - 1)$ **2.** $(3n + 1)(n + 5)$ **3.** $4xy(x - 2y + 1)$
4. $(5x + 1)(x + 2)$ **5.** $5u^2 + 17u - 12 = (5u - 3)(u + 4)$
6. $(1 + 5x)(4 + x)$ **7.** $3uv(2u^2v - 3v^2 - 4)$

5

8. $15(a + 2b)(a - b)$ **9.** $9(9 - m^2) = 9(3 - m)(3 + m)$

10. Not factorable **11.** $(2x + 3y)(5x - 8y)$

12. $(4x - 3y)(7x + 2y)$

13. $(2a)^3 - (3b)^3 = (2a - 3b)[(2a)^2 + (2a)(3b) + (3b)^2]$
$= (2a - 3b)(4a^2 + 6ab + 9b^2)$

14. $(4h - 5k)(16h^2 + 20hk + 25k^2)$ **15.** $(1 + 2y)^2$

16. $(2 + 3x)^2$

17. $(x)^3 + (2y)^3 = (x + 2y)[(x)^2 - (x)(2y) + (2y)^2]$
$= (x + 2y)(x^2 - 2xy + 4y^2)$

18. $(3x + y)(9x^2 - 3xy + y^2)$

19. $x^2 - y^2 + x - y = (x + y)(x - y) + 1(x - y)$
$= (x - y)(x + y + 1)$

20. $(x - y)(x + y - 1)$ **21.** $2(4a^2 - 4ab + b^2) = 2(2a - b)^2$

22. $2(3h - k)^2$

23. $x^2(x - 4) - 4(x - 4) = (x - 4)(x^2 - 4)$
$= (x - 4)(x + 2)(x - 2)$

24. Not factorable

25. $x^2(x - 2) - 9(x - 2) = (x - 2)(x^2 - 9)$
$= (x - 2)(x + 3)(x - 3)$

26. $(x + 4 + 5y)(x + 4 - 5y)$

27. Let $a = (2x + 3y)$. Then
$(2x + 3y)^2 + (2x + 3y) - 6 = a^2 + a - 6 = (a + 3)(a - 2)$
But since $a = (2x + 3y)$,
$(a + 3)(a - 2) = ([2x + 3y] + 3)([2x + 3y] - 2)$
Therefore,
$(2x + 3y)^2 + (2x + 3y) - 6 = (2x + 3y + 3)(2x + 3y - 2)$

28. $(a + 3b - 3)(a + 3b - 4)$

29. $x^2 + 4xy + 4y^2 - 5x - 10y + 6 = (x + 2y)^2 - 5(x + 2y) + 6$
Let $a = (x + 2y)$. Then
$(x + 2y)^2 - 5(x + 2y) + 6 = a^2 - 5a + 6 = (a - 2)(a - 3)$
But since $a = (x + 2y)$,
$(a - 2)(a - 3) = ([x + 2y] - 2)([x + 2y] - 3)$
Therefore,
$x^2 + 4xy + 4y^2 - 5x - 10y + 6 = (x + 2y - 2)(x + 2y - 3)$

30. $\{1, \frac{7}{5}\}$ **31.** $x^2 = 18 + 3x$ **32.** $\{0, 36\}$
$x^2 - 3x - 18 = 0$
$(x - 6)(x + 3) = 0$
$x - 6 = 0 \quad | \quad x + 3 = 0$
$x = 6 \quad | \quad x = -3$
$\{6, -3\}$

33. $6x^2 - 13x - 5 = 0$ **34.** $\{2, 3, -3\}$
$(3x + 1)(2x - 5) = 0$
$3x + 1 = 0 \quad | \quad 2x - 5 = 0$
$x = -\frac{1}{3} \quad | \quad x = \frac{5}{2}$
$\{-\frac{1}{3}, \frac{5}{2}\}$

35. $x^3 + 3x^2(3) + 3x(3^2) + 3^3 = x^3 + 27$ **36.** $\{-2, -\frac{2}{5}\}$
$x^3 + 9x^2 + 27x + 27 = x^3 + 27$
$9x^2 + 27x = 0$
$9x(x + 3) = 0$
$9 \neq 0 \quad | \quad x = 0 \quad | \quad x + 3 = 0$
$x = -3$
$\{0, -3\}$

37. $x^4 + 12x^3 + 54x^2 + 108x + 81 = x^4 + 12x^3 + 81$
$54x^2 + 108x = 0$
$54x(x + 2) = 0$
$54 \neq 0 \quad | \quad x = 0 \quad | \quad x + 2 = 0$
$x = -2$
$\{0, -2\}$

38. $\{0, -\frac{4}{3}\}$

39. Let $x =$ first odd integer
Then $x + 2 =$ second odd integer
and $x + 4 =$ third odd integer
$x(x + 4) = 5 + 8(x + 2)$
$x^2 + 4x = 5 + 8x + 16$
$x^2 - 4x - 21 = 0$
$(x - 7)(x + 3) = 0$
$x - 7 = 0 \quad | \quad x + 3 = 0$
$x = 7 \quad | \quad x = -3$
$x + 2 = 9 \quad | \quad x + 2 = -1$
$x + 4 = 11 \quad | \quad x + 4 = 1$
Therefore, one set of such numbers is 7, 9, and 11; another is -3, -1, and 1.

40. Width is 15 ft; length is 20 ft.

41. Let $x =$ width (in yd)
Then $x + 1 =$ length (in yd)
and $x(x + 1) =$ area (in sq. yd)
$26x(x + 1) = 520$
$x(x + 1) = 20$ Dividing both sides by 26
$x^2 + x = 20$
$x^2 + x - 20 = 0$
$(x - 4)(x + 5) = 0$
$x - 4 = 0 \quad | \quad x + 5 = 0$
$x = 4 \quad | \quad x = -5$ Not in the domain
$x + 1 = 5$
The width is 4 yd, and the length is 5 yd.

42. 6 by 13

43. Let $x =$ length of side of smaller cube (in cm)
Then $x + 4 =$ length of side of larger cube (in cm)
and $x^3 =$ volume of smaller cube
and $(x + 4)^3 =$ volume of larger cube
$(x + 4)^3 = 316 + x^3$
$x^3 + 12x^2 + 48x + 64 = 316 + x^3$
$12x^2 + 48x - 252 = 0$
$12(x^2 + 4x - 21) = 0$
$12(x + 7)(x - 3) = 0$
$12 \neq 0 \quad | \quad x + 7 = 0 \quad | \quad x - 3 = 0$
$x = -7 \quad | \quad x = 3$
Not in the domain $\quad | \quad x + 4 = 7$
The length of a side of the smaller cube is 3 cm, and the length of a side of the larger cube is 7 cm.

44. Length of side of smaller square is 1 cm; length of side of larger square is 4 cm.

Chapter 5 Diagnostic Test (page 251)

Following each problem number is the number (in parentheses) of the textbook section where that kind of problem is discussed.

1. (5.1, 5.3) $4x - 16x^3 = 4x(1 - 4x^2) = 4x(1 - 2x)(1 + 2x)$

2. (5.1) $43 + 7x^2 + 6 = 7x^2 + 49 = 7(x^2 + 7)$

3. (5.4B) $7x^2 + 23x + 6 = (7x + 2)(x + 3)$

4. (5.3) $x^2 + 81$ is a sum of two squares and cannot be factored.

5. (5.1) $2x^3 + 4x^2 + 16x = 2x(x^2 + 2x + 8)$

6. (5.4B) $6x^2 - 5x - 6 = (3x + 2)(2x - 3)$

7. (5.5) $y^3 - 1 = (y)^3 - (1)^3 = (y - 1)(y^2 + y + 1)$

8. (5.2) $3ac + 6bc - 5ad - 10bd = 3c(a + 2b) - 5d(a + 2b)$
$= (a + 2b)(3c - 5d)$

9. (5.3) $(4x^2 + 4x + 1) - y^2 = (2x + 1)^2 - y^2$
$= ([2x + 1] + y)([2x + 1] - y)$
$= (2x + 1 + y)(2x + 1 - y)$

10. (5.3) $y^2 - 4 = (y + 2)(y - 2)$

11. (5.4) $4z^2 + z + 1$ is not factorable.

12. (5.5) $8x^3 + y^3 = (2x + y)[(2x)^2 - (2x)(y) + (y)^2]$
$= (2x + y)(4x^2 - 2xy + y^2)$

13. (5.9)
$$x^2 - 16 = 0$$
$$(x+4)(x-4) = 0$$
$x + 4 = 0 \quad | \quad x - 4 = 0$
$x = -4 \quad | \quad x = 4$
$\{-4, 4\}$

14. (5.9)
$$z^2 + 7z = 0$$
$$z(z+7) = 0$$
$z = 0 \quad | \quad z + 7 = 0$
$\quad | \quad z = -7$
$\{0, -7\}$

15. (5.9) $(x+3)(2x-5) = 0$
$x + 3 = 0 \quad | \quad 2x - 5 = 0$
$x = -3 \quad | \quad x = \frac{5}{2}$
$\left\{-3, \frac{5}{2}\right\}$

16. (5.9)
$$8y^2 - 4y = 0$$
$$4y(2y-1) = 0$$
$4 \neq 0 \quad | \quad y = 0 \quad | \quad 2y - 1 = 0$
$\quad | \quad \quad | \quad y = \frac{1}{2}$
$\left\{0, \frac{1}{2}\right\}$

17. (5.9)
$$2x^3 - 7x^2 + 3x = 0$$
$$x(2x^2 - 7x + 3) = 0$$
$$x(2x-1)(x-3) = 0$$
$x = 0 \quad | \quad 2x - 1 = 0 \quad | \quad x - 3 = 0$
$\quad | \quad 2x = 1 \quad | \quad x = 3$
$\quad | \quad x = \frac{1}{2} \quad |$
$\left\{0, \frac{1}{2}, 3\right\}$

18. (5.9)
$$(x+1)^3 = x^3 + 1$$
$$x^3 + 3x^2 + 3x + 1 = x^3 + 1$$
$$3x^2 + 3x = 0$$
$$3x(x+1) = 0$$
$3 \neq 0 \quad | \quad x = 0 \quad | \quad x + 1 = 0$
$\quad | \quad \quad | \quad x = -1$
$\{0, -1\}$

19. (5.10) Let x = first even integer
Then $x + 2$ = second even integer
and $x + 4$ = third even integer
$$x(x+2) = 68 + (x+4)$$
$$x^2 + 2x = 68 + x + 4$$
$$x^2 + x - 72 = 0$$
$$(x+9)(x-8) = 0$$
$x + 9 = 0 \quad | \quad x - 8 = 0$
$x = -9 \quad | \quad x = 8$
Not an even integer ↑ $\quad | \quad x + 2 = 10$
$\quad | \quad x + 4 = 12$
The integers are 8, 10, and 12.

20. (5.10) Let x = altitude (in cm)
Then $x + 8$ = base (in cm)
$$\tfrac{1}{2}x(x+8) = \text{area}$$
$$\tfrac{1}{2}x(x+8) = 64$$
$$x(x+8) = 128$$
$$x^2 + 8x - 128 = 0$$
$$(x+16)(x-8) = 0$$
$x + 16 = 0 \quad | \quad x - 8 = 0$
$x = -16 \quad | \quad x = 8$
Not in the domain ↗ $\quad | \quad x + 8 = 16$
The altitude is 8 cm, and the base is 16 cm.

Chapters 1–5 Cumulative Review Exercises (page 251)

1. $18 \div 2 \cdot 3 - 16 \cdot 3 = 9 \cdot 3 - 16 \cdot 3 = 27 - 48 = -21$

2. 32

3.
$$2x - 6 - 5 = 6 - 3x - 12$$
$$2x - 11 = -3x - 6$$
$$5x = 5$$
$$x = 1$$
$\{1\}$

4. $\{x \mid -1 < x < 3\}$

5. If $|2x - 3| \geq 7$,
$2x - 3 \geq 7$ or $2x - 3 \leq -7$
$2x \geq 10$ or $2x \leq -4$
$x \geq 5$ or $x \leq -2$
$\{x \mid x \geq 5 \text{ or } x \leq -2\}$

6. All are real numbers.

7.
$$2x^2 - 9x - 5 = 0$$
$$(2x+1)(x-5) = 0$$
$2x + 1 = 0$ or $x - 5 = 0$
$x = -\frac{1}{2}$ or $x = 5$
$\left\{-\frac{1}{2}, 5\right\}$

8. $3(x+3)(x^2 - 3x + 9)$

9.
$$x^3 + 5x^2 - x - 5 = x^2(x+5) - (x+5)$$
$$= (x+5)(x^2-1)$$
$$= (x+5)(x-1)(x+1)$$

10. $6x^2 + x + 13$

11. $(a-4)(a^2 - 2a + 5)$
$$a^3 - 2a^2 + 5a - 4a^2 + 8a - 20$$
$$a^3 - 6a^2 + 13a - 20$$

12. $9x^2 + 30x + 25$

13.
$$(2x-1)^5 = (2x)^5 - 5(2x)^4(1) + \frac{5!}{(2!)(3!)}(2x)^3(1^2)$$
$$- 10(2x)^2(1^3) + 5(2x)(1^4) - (1)^5$$
$$= 32x^5 - 80x^4 + 80x^3 - 40x^2 + 10x - 1$$

14. $x^2 - 4x + 2 + \dfrac{4}{x+2}$

15. factors **16.** dividend **17.** subset

18. additive inverses **19.** rational **20.** No solution

21. Let x = number of quarts of antifreeze to be added
$$0.20(10) + 1.00(x) = 0.50(10 + x)$$
$$20(10) + 100(x) = 50(10 + x)$$
$$200 + 100x = 500 + 50x$$
$$50x = 300$$
$$x = 6$$
6 qt of antifreeze should be added.

22. 3 yr old

23. Let x = number of quarters
Then $4x$ = number of nickels
and $27 - (x + 4x)$ = number of dimes
$$x(2[27 - 5x]) = 4x$$
$$54x - 10x^2 = 4x$$
$$0 = 10x^2 - 50x$$
$$0 = 10x(x-5)$$
$x - 5 = 0 \quad | \quad 10x = 0$
$x = 5 \quad | \quad x = 0$
$4x = 20 \quad | \quad 4x = 0$
$27 - 5x = 2 \quad | \quad 27 - 5x = 27$
(The checks will not be shown.) There are two answers: Yang has 5 quarters, 20 nickels, and 2 dimes, *or* he has 27 dimes, no nickels, and no quarters.

24. 5 yd wide

Exercises 6.1 (page 255)

1. Yes; $\dfrac{3 \div 3}{6y \div 3} = \dfrac{1}{2y}$ **2.** Yes **3.** No **4.** No

5. Yes; $\dfrac{8(2+x) \div 8}{8(3-y) \div 8} = \dfrac{2+x}{3-y}$ **6.** Yes

7. Yes; $\dfrac{-1(x-y)}{-1(y-x)} = \dfrac{y-x}{x-y}$ **8.** Yes

6

Exercises 6.2 (page 258)

1. $3m^2$ 2. $-\frac{k^3}{4}$ 3. $-\frac{3a^3c}{7b^2}$ 4. $\frac{5f}{2e^2g^2}$

5. $\frac{8x(5-x)}{5x(x+2)} = \frac{8(5-x)}{5(x+2)}$ 6. $\frac{2y^2}{3}$

7. Cannot be reduced 8. Cannot be reduced

9. $\frac{\cancel{8}wx^3\cancel{(3w-2x)}}{\cancel{6}w^2x\cancel{(3w-2x)}} = \frac{4x^2}{3w}$ 10. $\frac{3c^3}{2d^2}$

11. $\frac{(x+4)\cancel{(x-4)}}{\cancel{(x-4)}(x-5)} = \frac{x+4}{x-5}$ 12. $\frac{x-5}{x-3}$

13. $\frac{(2x+y)(x-y)}{(y-3x)(y-x)} = \frac{(2x+y)(-1)\cancel{(y-x)}}{(y-3x)\cancel{(y-x)}} = -\frac{2x+y}{y-3x}$, or $\frac{2x+y}{3x-y}$

14. $\frac{2k+5h}{4k-3h}$

15. $\frac{(x-3)(2x+3)}{(4-x)(3-x)} = \frac{(-1)\cancel{(3-x)}(2x+3)}{(4-x)\cancel{(3-x)}} = -\frac{2x+3}{4-x}$, or $\frac{2x+3}{x-4}$

16. $\frac{3+2y}{1-4y}$

17. $\frac{(2y-3x)(y+2x)}{(3x-2y)(x+y)} = \frac{-\cancel{(3x-2y)}(y+2x)}{\cancel{(3x-2y)}(x+y)} = -\frac{2x+y}{x+y}$

18. $-\frac{3x+2y}{2x+3y}$ 19. $\frac{(2x+1)(x-5)}{(2x+3)(x+1)}$ Cannot be reduced

20. Cannot be reduced

21. $\frac{(a-1)(a^2+a+1)}{(1-a)(1+a)} = \frac{\cancel{(a-1)}(a^2+a+1)}{(-1)\cancel{(a-1)}(a+1)} = -\frac{a^2+a+1}{a+1}$

22. $\frac{x^2-xy+y^2}{y-x}$ 23. $\frac{x^2+4}{(x+2)^2}$ Cannot be reduced

24. Cannot be reduced

25. $\frac{13x^3y^2}{13x^2y^2}+\frac{-26xy^3}{13x^2y^2}+\frac{39xy}{13x^2y^2} = x-\frac{2y}{x}+\frac{3}{xy}$, or $\frac{x^2y-2y^2+3}{xy}$

26. $3n-5m-\frac{2}{mn}$, or $\frac{3mn^2-5m^2n-2}{mn}$

27. $\frac{6a^2bc^2}{6abc}+\frac{-4ab^2c^2}{6abc}+\frac{12bc}{6abc} = ac-\frac{2bc}{3}+\frac{2}{a}$, or $\frac{3a^2c-2abc+6}{3a}$

28. $2a^2b-a-\frac{5}{2b}$, or $\frac{4a^2b^2-2ab-5}{2b}$

Exercises 6.3 (page 265)

1. $\frac{\cancel{27}x^4y^3}{\cancel{22}x^5yz}\cdot\frac{\cancel{55}x^2z^2}{\cancel{9}y^3z} = \frac{15x}{2y}$ 2. $\frac{5c}{18a^2}$ 3. $\frac{mn^3}{\cancel{18}n^2}\cdot\frac{\cancel{24}m^3n}{5m^4} = \frac{4n^2}{15}$

4. $-\frac{36}{5h^2}$ 5. $\frac{3u\cancel{(5-2u)}}{\cancel{10}u^2}\cdot\frac{\cancel{15}u^3}{7\cancel{(5-2u)}} = \frac{9u^2}{14}$ 6. $-\frac{4v}{15}$

7. $\frac{-\cancel{15}c^4}{8c^2\cancel{(5c-3)}}\cdot\frac{\cancel{7}c\cancel{(5c-3)}}{\cancel{35}c} = -\frac{3c^2}{8}$ 8. 20

9. $\frac{d^2e\cancel{(e-d)}}{\cancel{12}e^2d}\cdot\frac{\cancel{3}e^2(d+e)}{de^2\cancel{(d-e)}} = -\frac{d+e}{4e}$ 10. $-\frac{15(3m+n)}{8m}$

11. $\frac{\cancel{(w-4)}\cancel{(w+2)}}{6\cancel{(w-4)}}\cdot\frac{5w^2}{\cancel{(w+2)}(w-5)} = \frac{5w^2}{6(w-5)}$ 12. $\frac{3k}{8}$

13. $\frac{\cancel{4}\cancel{(a+b)}\cancel{(a+b)}}{\cancel{(a+b)}\cancel{(a-b)}}\cdot\frac{\cancel{(b-a)}}{\cancel{6}b\cancel{(a+b)}} = -\frac{2}{3b}$ 14. $-\frac{1}{u}$

15. $\frac{\cancel{2}\cancel{(2-a)}}{\cancel{2}\cancel{(a+1)}}\cdot\frac{\cancel{(a+1)}(a+1)}{2\cancel{(a-2)}(a^2+2a+4)} = -\frac{a+1}{2(a^2+2a+4)}$

16. $-\frac{6}{a^2-3a+9}$

17. $\frac{(x+y)\cancel{(x^2-xy+y^2)}}{2\cancel{(x-y)}}\cdot\frac{(x+y)\cancel{(x-y)}}{\cancel{(x^2-xy+y^2)}} = \frac{(x+y)^2}{2}$

18. $\frac{(x-y)^2}{3}$ 19. $\frac{\cancel{(e+5f)}\cancel{(e+5f)}}{\cancel{(e+5f)}\cancel{(e-5f)}}\cdot\frac{3\cancel{(e-f)}}{\cancel{(f-e)}}\cdot\frac{\cancel{(5f-e)}}{\cancel{(e+5f)}} = 3$

20. 1

21. $\frac{\cancel{(x+y)}(x+y+1)}{\cancel{(x-y)}(x-y-1)}\cdot\frac{\cancel{(x-y)}\cancel{(x-y)}}{\cancel{(x+y)}\cancel{(x+y)}}\cdot\frac{\cancel{x+y}}{\cancel{x-y}} = \frac{x+y+1}{x-y-1}$

22. $\frac{c-d}{a+b}$ 23. $\left(\frac{\cancel{7}x^2}{\cancel{15}y^3}\cdot\frac{\cancel{5}x}{\cancel{14}y}\right)\div\frac{8x^2}{3y} = \frac{\cancel{x^3}}{\cancel{6y^4}}\cdot\frac{\cancel{3y}}{8\cancel{x^2}} = \frac{x}{16y^3}$

24. $\frac{s}{15t}$ 25. $\frac{\cancel{11}x^3}{\cancel{7xy}}\cdot\frac{\cancel{3}x}{\cancel{22}y}\cdot\frac{\cancel{14xy^2}}{\cancel{9y^2}} = \frac{x^4}{3y}$ 26. $\frac{3a^4}{5b^4}$

27. $\frac{\cancel{x-5}}{\cancel{x+3y}}\cdot\frac{\cancel{(x+3y)}\cancel{(x-3y)}}{\cancel{x-3y}}\cdot\frac{x}{y\cancel{(x-5)}} = \frac{x}{y}$ 28. $\frac{y}{x}$

29. $-\frac{+5}{+8} = +\frac{+5}{-8}$ The missing term is −8. 30. 6

31. $+\frac{-x}{+5} = +\frac{+x}{-5}$ The missing term is −5. 32. −6

33. $+\frac{+(8-y)}{+(4y-7)} = +\frac{-(8-y)}{-(4y-7)} = \frac{y-8}{7-4y}$ 34. $2-w$
The missing term is $7-4y$.

35. $+\frac{+(u-v)}{+(a-b)} = +\frac{-(u-v)}{-(a-b)} = \frac{v-u}{b-a}$ The missing term is $b-a$.

36. $x-2$

37. $+\frac{+(a-b)}{[+(3a+2b)][+(a-5b)]} = +\frac{-(a-b)}{[+(3a+2b)][-(a-5b)]} = \frac{b-a}{(3a+2b)(5b-a)}$
The missing term is $b-a$.

38. $f-2e$

Exercises 6.4 (page 268)

1. (1) $5^2\cdot a^3$; $3\cdot 5\cdot a$ Denominators in factored form
(2) 3, 5, a All the different bases
(3) $3^1, 5^2, a^3$ Highest power of each base
(4) LCD $= 3^1\cdot 5^2\cdot a^3 = 75a^3$

2. $36b^4$

3. (1) $2^2\cdot 3\cdot 5\cdot h\cdot k^3$; $2\cdot 3^2\cdot 5\cdot h^2\cdot k^4$
(2) 2, 3, 5, h, k
(3) $2^2, 3^2, 5, h^2, k^4$
(4) LCD $= 2^2\cdot 3^2\cdot 5\cdot h^2\cdot k^4 = 180h^2k^4$

4. $294x^3y^2$ (or $147x^3y^2$, if fractions were reduced first)

5. (1) $2(w-5)$; 2^2w
(2) $2, w, (w-5)$
(3) $2^2, w^1, (w-5)^1$
(4) LCD $= 2^2 \cdot w^1 \cdot (w-5)^1 = 4w(w-5)$

6. $8m^2(m-6)$

7. (1) $(3b+c)(3b-c)$; $(3b-c)^2$
(2) $(3b+c), (3b-c)$
(3) $(3b+c)^1, (3b-c)^2$
(4) LCD $= (3b+c)(3b-c)^2$

8. $(2e+5f)(2e-5f)^2$

9. (1) $2 \cdot g^3$; $(g-3)^2$; $2^2 \cdot g \cdot (g-3)$
(2) $2, g, (g-3)$
(3) $2^2, g^3, (g-3)^2$
(4) LCD $= 2^2 \cdot g^3 \cdot (g-3)^2 = 4g^3(g-3)^2$
[The LCD is $2g^3(g-3)$ if fractions were reduced first.]

10. $9y^2(y-6)^2$ [or $9y^2(y-6)$, if fractions were reduced first.]

11. (1) $2 \cdot (x-4)^2$; $(x-4) \cdot (x+5)$
(2) $2, (x-4), (x+5)$
(3) $2^1, (x-4)^2, (x+5)^1$
(4) LCD $= 2(x-4)^2(x+5)$

12. $5(k+7)(k-3)^2$

13. (1) $3 \cdot e^2$; $(e+3)(e-3)$; $2^2(e-3)$
(2) $2, 3, e, (e+3)(e-3)$
(3) $2^2, 3^1, e^2, (e+3)^1, (e-3)^1$
(4) LCD $= 2^2 \cdot 3^1 \cdot e^2 \cdot (e+3)(e-3) = 12e^2(e+3)(e-3)$

14. $24u^3(u+3)(u-8)$

15. (1) $2^2 \cdot 3 \cdot x^2 \cdot (x+2)$; $(x-2)^2$; $(x+2)(x-2)$
(2) $2, 3, x, (x+2), (x-2)$
(3) $2^2, 3^1, x^2, (x+2)^1, (x-2)^2$
(4) LCD $= 2^2 \cdot 3 \cdot x^2 \cdot (x+2) \cdot (x-2)^2 = 12x^2(x+2)(x-2)^2$

16. $8y(y+3)^2(y-3)$

Exercises 6.5 (page 275)

1. $\dfrac{5a+10}{a+2} = \dfrac{5\cancel{(a+2)}^{1}}{\cancel{(a+2)}_{1}} = 5$ **2.** 6

3. $\dfrac{8m-12n}{2m-3n} = \dfrac{4\cancel{(2m-3n)}^{1}}{\cancel{(2m-3n)}_{1}} = 4$ **4.** 7

5. $\dfrac{x-3-(6x+3)}{5x+7} = \dfrac{x-3-6x-3}{5x+7} = \dfrac{-5x-6}{5x+7}$

6. $\dfrac{-4y+2}{8y+3}$

7. $\dfrac{6z-5-(2z-4)}{3z-5} = \dfrac{6z-5-2z+4}{3z-5} = \dfrac{4z-1}{3z-5}$

8. $\dfrac{-4t-2}{7t-3}$

9. $\dfrac{15w}{5w-1} - \dfrac{3}{5w-1} = \dfrac{15w-3}{5w-1} = \dfrac{3\cancel{(5w-1)}^{1}}{\cancel{(5w-1)}_{1}} = 3$ **10.** 5

11. $\dfrac{7z}{8z-4} + \dfrac{5z-6}{8z-4} = \dfrac{7z+5z-6}{8z-4} = \dfrac{12z-6}{8z-4} = \dfrac{\cancel{6}^{3}\cancel{(2z-1)}^{1}}{\cancel{4}_{2}\cancel{(2z-1)}_{1}} = \dfrac{3}{2}$

12. 2

13. $\dfrac{12x-31}{12x-28} + \dfrac{18x-39}{12x-28} = \dfrac{30x-70}{12x-28} = \dfrac{\cancel{10}^{5}\cancel{(3x-7)}^{1}}{\cancel{4}_{2}\cancel{(3x-7)}_{1}} = \dfrac{5}{2}$

14. 2

15. LCD $= 75a^3$
$\dfrac{9}{25a^3} \cdot \dfrac{3}{3} + \dfrac{7}{15a} \cdot \dfrac{5a^2}{5a^2} = \dfrac{27}{75a^3} + \dfrac{35a^2}{75a^3} = \dfrac{27+35a^2}{75a^3}$

16. $\dfrac{26b^2+33}{36b^4}$

17. LCD $= 180h^2k^4$
$\dfrac{49}{60h^2k^2} \cdot \dfrac{3k^2}{3k^2} - \dfrac{71}{90hk^4} \cdot \dfrac{2h}{2h} = \dfrac{147k^2}{180h^2k^4} - \dfrac{142h}{180h^2k^4} = \dfrac{147k^2-142h}{180h^2k^4}$

18. $\dfrac{154x-135y}{147x^3y^2}$

19. LCD $= t(t-4)$
$\dfrac{5}{t} \cdot \dfrac{t-4}{t-4} + \dfrac{2t}{t-4} \cdot \dfrac{t}{t} = \dfrac{5t-20}{t(t-4)} + \dfrac{2t^2}{t(t-4)} = \dfrac{2t^2+5t-20}{t(t-4)}$

20. $\dfrac{6r^2-11r+88}{r(r-8)}$

21. LCD $= 12k(2k-1)$
$\dfrac{3k}{4(2k-1)} - \dfrac{7}{6k} = \dfrac{9k^2}{12k(2k-1)} - \dfrac{7 \cdot 2(2k-1)}{12k(2k-1)}$
$= \dfrac{9k^2}{12k(2k-1)} - \dfrac{28k-14}{12k(2k-1)}$
$= \dfrac{9k^2-28k+14}{12k(2k-1)}$

22. $\dfrac{2(3j^2+3j+2)}{9j(3j+2)}$

23. LCD $= x(x-3)$
$\dfrac{x^2(x)(x-3)}{1(x)(x-3)} + \dfrac{-3(x-3)}{x(x-3)} + \dfrac{5(x)}{(x-3)(x)}$
$= \dfrac{x^4-3x^3-3x+9+5x}{x(x-3)}$
$= \dfrac{x^4-3x^3+2x+9}{x(x-3)}$

24. $\dfrac{y^4-5y^3+y+10}{y(y-5)}$

25. LCD $= b(2a-3b)$
$\dfrac{(2a+3b)(2a-3b)}{b(2a-3b)} + \dfrac{b^2}{b(2a-3b)}$
$= \dfrac{4a^2-9b^2}{b(2a-3b)} + \dfrac{b^2}{b(2a-3b)}$
$= \dfrac{4a^2-8b^2}{b(2a-3b)} = \dfrac{4(a^2-2b^2)}{b(2a-3b)}$

26. $\dfrac{9x^2-24y^2}{y(3x+5y)}$

27. LCD $= (a+3)(a-1)$
$\dfrac{2(a-1)}{(a+3)(a-1)} + \dfrac{-4(a+3)}{(a-1)(a+3)}$
$= \dfrac{2a-2-4a-12}{(a+3)(a-1)} = \dfrac{-2a-14}{(a+3)(a-1)}$, or $\dfrac{14+2a}{(3+a)(1-a)}$

28. $\dfrac{2b+26}{(b-2)(b+4)}$

29. LCD $= (x-3)(x-2)$
$\dfrac{(x+2)(x-2)}{(x-3)(x-2)} + \dfrac{-(x+3)(x-3)}{(x-2)(x-3)}$
$= \dfrac{x^2-4}{(x-3)(x-2)} + \dfrac{-(x^2-9)}{(x-3)(x-2)} = \dfrac{x^2-4-x^2+9}{(x-3)(x-2)}$
$= \dfrac{5}{(x-3)(x-2)}$

30. $\dfrac{20}{(x+6)(x+4)}$

31. $\dfrac{\cancel{(x+2)}^{1}}{(x-1)\cancel{(x+2)}_{1}} + \dfrac{3}{(x+1)(x-1)}$ LCD $= (x+1)(x-1)$
$\dfrac{1(x+1)}{(x-1)(x+1)} + \dfrac{3}{(x+1)(x-1)} = \dfrac{x+1+3}{(x+1)(x-1)} = \dfrac{x+4}{x^2-1}$

32. $\dfrac{x+7}{(x+2)(x-2)}$

33. LCD $= (x-3)(x+3)$

$$\frac{2x(x+3)}{(x-3)(x+3)} + \frac{-2x(x-3)}{(x+3)(x-3)} + \frac{36}{(x+3)(x-3)}$$
$$= \frac{2x^2+6x-2x^2+6x+36}{(x-3)(x+3)} = \frac{12x+36}{(x-3)(x+3)}$$
$$= \frac{12\cancel{(x+3)}}{(x-3)\cancel{(x+3)}} = \frac{12}{x-3}$$

34. $\dfrac{12}{6-m}$

35. LCD $= (x+2)^2(x-2)$

$$\frac{x-2}{(x+2)(x+2)}\cdot\frac{x-2}{x-2} - \frac{x+1}{(x+2)(x-2)}\cdot\frac{x+2}{x+2}$$
$$= \frac{x^2-4x+4}{(x+2)^2(x-2)} - \frac{x^2+3x+2}{(x+2)^2(x-2)}$$
$$= \frac{x^2-4x+4-x^2-3x-2}{(x+2)^2(x-2)} = \frac{2-7x}{(x+2)^2(x-2)}$$

36. $\dfrac{1-5x}{(x+1)(x-1)^2}$

37. LCD $= (x^2+2x+4)(x+2)$

$$\frac{4(x+2)}{(x^2+2x+4)(x+2)} + \frac{(x-2)(x^2+2x+4)}{(x+2)(x^2+2x+4)}$$
$$= \frac{4x+8}{(x^2+2x+4)(x+2)} + \frac{x^3-8}{(x^2+2x+4)(x+2)}$$
$$= \frac{4x+x^3}{(x^2+2x+4)(x+2)}$$

38. $\dfrac{x^3+3x}{(x-9)(x^2-3x+9)}$

39.
$$\frac{5}{2g^3} - \frac{3\cancel{(g-3)}}{(g-3)\cancel{(g-3)}} + \frac{\cancel{12g}^{\,3}}{\cancel{4g}(g-3)}$$
$$= \frac{5}{2g^3} - \cancel{\frac{3}{g-3}} + \cancel{\frac{3}{g-3}} = \frac{5}{2g^3}$$

40. $\dfrac{7}{9y^2}$

41. LCD $= 2(x-4)^2(x+5)$

$$\frac{(2x-5)(x+5)}{2(x-4)^2(x+5)} + \frac{(4x+7)(2)(x-4)}{(x+5)(x-4)(2)(x-4)}$$
$$= \frac{2x^2+5x-25+8x^2-18x-56}{2(x-4)^2(x+5)}$$
$$= \frac{10x^2-13x-81}{2(x-4)^2(x+5)}$$

42. $\dfrac{23k^2-10k+53}{5(k-3)^2(k+7)}$

43. LCD $= 12e^2(e+3)(e-3)$

$$\frac{35(4)(e+3)(e-3)}{3e^2(4)(e+3)(e-3)} - \frac{2e(12e^2)}{(e+3)(e-3)(12e^2)} - \frac{3(3e^2)(e+3)}{4(e-3)(3e^2)(e+3)}$$
$$= \frac{140(e^2-9)-2e(12e^2)-9e^2(e+3)}{12e^2(e+3)(e-3)}$$
$$= \frac{140e^2-1260-24e^3-9e^3-27e^2}{12e^2(e+3)(e-3)} = \frac{-33e^3+113e^2-1260}{12e^2(e+3)(e-3)}$$
$$= -\frac{33e^3-113e^2+1260}{12e^2(e+3)(e-3)}$$

44. $\dfrac{-164u^4-4u^3-23u^2-45u-216}{24u^3(u+3)(u-8)}$, or $\dfrac{164u^4+4u^3+23u^2+45u+216}{24u^3(u+3)(8-u)}$

45. LCD $= 12x^2(x-2)^2(x+2)$

$$\frac{x^2+1}{12x^2(x+2)} - \frac{4x+3}{(x-2)^2} - \frac{1}{(x-2)(x+2)}$$
$$= \frac{(x^2+1)(x-2)^2-(4x+3)(12x^2)(x+2)-12x^2(x-2)}{12x^2(x-2)^2(x+2)}$$
$$= \frac{x^4-4x^3+5x^2-4x+4-48x^4-132x^3-72x^2-12x^3+24x^2}{12x^2(x-2)^2(x+2)}$$
$$= \frac{-47x^4-148x^3-43x^2-4x+4}{12x^2(x-2)^2(x+2)},$$
or $-\dfrac{47x^4+148x^3+43x^2+4x-4}{12x^2(x-2)^2(x+2)}$

46. $\dfrac{72y^3+149y^2-186y-99}{8y(y+3)^2(y-3)}$

47. LCD $= 3y(y+4)^2(y^2-4y+16)$

$$\frac{7}{3y(y^2-4y+16)} + \frac{y^2+4}{(y+4)(y^2-4y+16)} - \frac{y}{(y+4)^2}$$
$$= \frac{7(y+4)^2}{3y(y^2-4y+16)(y+4)^2} + \frac{(y^2+4)(3y)(y+4)}{3y(y^2-4y+16)(y+4)(y+4)} - \frac{y(3y)(y^2-4y+16)}{3y(y^2-4y+16)(y+4)^2}$$
$$= \frac{7(y^2+8y+16)+3y(y^3+4y^2+4y+16)-3y^2(y^2-4y+16)}{3y(y+4)^2(y^2-4y+16)}$$
$$= \frac{24y^3-29y^2+104y+112}{3y(y+4)^2(y^2-4y+16)}$$

48. $\dfrac{12x^3+5x^2+84x+45}{2x(x+3)^2(x^2-3x+9)}$

49. LCD $= (x+1)(x-1)(x+3)(x-3)$

$$\frac{x-1}{x^2(x+1)-9(x+1)} - \frac{x+3}{x^2(x-3)-(x-3)}$$
$$= \frac{x-1}{(x+1)(x^2-9)} - \frac{x+3}{(x-3)(x^2-1)}$$
$$= \frac{x-1}{(x+1)(x+3)(x-3)} - \frac{x+3}{(x-3)(x+1)(x-1)}$$
$$= \frac{(x-1)(x-1)}{(x+1)(x+3)(x-3)(x-1)} - \frac{(x+3)(x+3)}{(x-3)(x+1)(x-1)(x+3)}$$
$$= \frac{(x^2-2x+1)-(x^2+6x+9)}{(x+1)(x+3)(x-3)(x-1)}$$
$$= \frac{\cancel{x^2}-2x+1-\cancel{x^2}-6x-9}{(x+1)(x+3)(x-3)(x-1)}$$
$$= \frac{-8x-8}{(x+1)(x+3)(x-3)(x-1)}$$
$$= \frac{-8\cancel{(x+1)}}{\cancel{(x+1)}(x+3)(x-3)(x-1)}$$
$$= \frac{-8}{(x+3)(x-3)(x-1)}$$

50. $\dfrac{6x+3}{(x+1)(x-1)(x+2)(x-2)}$

51. Division must be done before addition.

$$\frac{x+6}{x-5} + \frac{1}{x+4}\cdot\frac{x^2-x-20}{x+6} = \frac{x+6}{x-5} + \frac{1}{\cancel{x+4}}\cdot\frac{(x-5)\cancel{(x+4)}}{x+6}$$
$$= \frac{x+6}{x-5} + \frac{x-5}{x+6} \quad \text{LCD} = (x-5)(x+6)$$
$$= \frac{(x+6)(x+6)}{(x-5)(x+6)} + \frac{(x-5)(x-5)}{(x+6)(x-5)}$$
$$= \frac{x^2+12x+36+x^2-10x+25}{(x-5)(x+6)}$$
$$= \frac{2x^2+2x+61}{(x-5)(x+6)}$$

52. $\dfrac{5x^2-4x+25}{(x+4)(2x-3)}$

53. $\dfrac{3}{8x^2}-\left(\dfrac{\cancel{x+3}}{4\cancel{x}(\cancel{x-2})}\cdot\dfrac{\cancel{x}(\cancel{x-2})}{(\cancel{x+3})(x-1)}\right)=\dfrac{3}{8x^2}-\dfrac{1}{4(x-1)}$

LCD $=8x^2(x-1)$

$=\dfrac{3(x-1)}{8x^2(x-1)}-\dfrac{1(2x^2)}{4(x-1)(2x^2)}=\dfrac{3x-3-2x^2}{8x^2(x-1)}$,

or $\dfrac{-2x^2+3x-3}{8x^2(x-1)}$

54. $\dfrac{-3x^2+8x+24}{9x^2(x+3)}$

55. $\dfrac{x+3}{x-5}-\left(\dfrac{x-2}{(\cancel{x-4})(x+3)}\cdot\dfrac{(\cancel{x+7})(\cancel{x-4})}{\cancel{x+7}}\right)=\dfrac{x+3}{x-5}-\dfrac{x-2}{x+3}$

LCD $=(x-5)(x+3)$

$=\dfrac{(x+3)(x+3)}{(x-5)(x+3)}-\dfrac{(x-2)(x-5)}{(x-5)(x+3)}$

$=\dfrac{x^2+6x+9-(x^2-7x+10)}{(x-5)(x+3)}$

$=\dfrac{x^2+6x+9-x^2+7x-10}{(x-5)(x+3)}=\dfrac{13x-1}{(x-5)(x+3)}$

56. $\dfrac{16x+4}{(x-2)(x+4)}$

Exercises 6.6 (page 281)

1. $\dfrac{\overset{3}{\cancel{21}}m^3n}{\underset{2}{\cancel{14}}mn^2}\cdot\dfrac{\overset{2}{\cancel{8}}mn^3}{\underset{5}{\cancel{20}}m^2n^2}=\dfrac{3m}{5}$ **2.** $\dfrac{8}{3ab}$

3. $\dfrac{\overset{1}{\cancel{3}}\overset{1}{(\cancel{5h-2})}}{\underset{3}{\cancel{18}}h}\cdot\dfrac{\overset{2}{\cancel{8}}h}{\underset{3}{\cancel{6}}h\underset{1}{(\cancel{5h-2})}}=\dfrac{2}{9h}$ **4.** $\dfrac{3k}{5}$

5. $\dfrac{d^2}{d^2}\cdot\dfrac{\left(\dfrac{c}{d}+2\right)}{\left(\dfrac{c^2}{d^2}-4\right)}=\dfrac{d^2\left(\dfrac{c}{d}\right)+d^2(2)}{d^2\left(\dfrac{c^2}{d^2}\right)-d^2(4)}$ **6.** $\dfrac{x+y}{y}$

$=\dfrac{cd+2d^2}{c^2-4d^2}=\dfrac{d(\cancel{c+2d})}{(c-2d)(\cancel{c+2d})}=\dfrac{d}{c-2d}$

7. $\dfrac{\dfrac{(a+2)}{1}\cdot\dfrac{(a+2)}{1}+\dfrac{(a+2)}{1}\left(\dfrac{-9}{a+2}\right)}{\dfrac{(a+2)}{1}\cdot\dfrac{(a+1)}{1}+\dfrac{(a+2)}{1}\left(\dfrac{a-7}{a+2}\right)}$ **8.** $\dfrac{x+3}{x+5}$

$=\dfrac{a^2+4a+4-9}{a^2+3a+2+a-7}=\dfrac{a^2+4a-5}{a^2+4a-5}=1$

9. $\dfrac{\dfrac{y(x-y)}{1}\cdot\dfrac{(x+y)}{y}+\dfrac{y(x-y)}{1}\cdot\dfrac{y}{x-y}}{\dfrac{y(x-y)}{1}\cdot\dfrac{y}{x-y}}=\dfrac{x^2-y^2+y^2}{y^2}=\dfrac{x^2}{y^2}$

10. $-\dfrac{1}{a}$

11. $\dfrac{\dfrac{x(x+1)}{1}\cdot\dfrac{x}{x+1}+\dfrac{x(x+1)}{1}\cdot\dfrac{4}{x}}{\dfrac{x(x+1)}{1}\cdot\dfrac{x}{x+1}+\dfrac{x(x+1)}{1}\cdot\dfrac{(-2)}{1}}$ **12.** $\dfrac{2x+1}{4x}$

$=\dfrac{x^2+4x+4}{x^2-2x^2-2x}=\dfrac{(x+2)(\cancel{x+2})}{-x(\cancel{x+2})}=-\dfrac{x+2}{x}$

13. $\dfrac{\dfrac{x(x-1)}{1}\cdot\dfrac{(x+4)}{x}+\dfrac{x(x-1)}{1}\cdot\left(-\dfrac{3}{x-1}\right)}{\dfrac{x(x-1)}{1}\cdot\dfrac{(x+1)}{1}+\dfrac{x(x-1)}{1}\cdot\left(\dfrac{2x+1}{x-1}\right)}$

$=\dfrac{x^2+3x-4-3x}{x^3-x+2x^2+x}=\dfrac{x^2-4}{x^3+2x^2}=\dfrac{(\cancel{x+2})(x-2)}{x^2(\cancel{x+2})}=\dfrac{x-2}{x^2}$

14. $\dfrac{x-2}{(x-6)(x-4)}$

15. $\dfrac{\dfrac{4}{x^2}-\dfrac{1}{y^2}}{\dfrac{2}{x}+\dfrac{1}{y}}=\dfrac{\dfrac{x^2y^2}{1}\cdot\dfrac{4}{x^2}+\dfrac{x^2y^2}{1}\left(-\dfrac{1}{y^2}\right)}{\dfrac{x^2y^2}{1}\cdot\dfrac{2}{x}+\dfrac{x^2y^2}{1}\cdot\dfrac{1}{y}}$

$=\dfrac{4y^2-x^2}{2xy^2+x^2y}=\dfrac{(\cancel{2y+x})(2y-x)}{xy(\cancel{2y+x})}=\dfrac{2y-x}{xy}$

16. $\dfrac{3x+y}{xy}$

17. $\dfrac{\dfrac{(x+2)(x-2)}{1}\left(\dfrac{x-2}{x+2}\right)-\dfrac{(x+2)(x-2)}{1}\left(\dfrac{x+2}{x-2}\right)}{\dfrac{(x+2)(x-2)}{1}\left(\dfrac{x-2}{x+2}\right)+\dfrac{(x+2)(x-2)}{1}\left(\dfrac{x+2}{x-2}\right)}$

$=\dfrac{(x-2)(x-2)-(x+2)(x+2)}{(x-2)(x-2)+(x+2)(x+2)}$

$=\dfrac{x^2-4x+4-(x^2+4x+4)}{x^2-4x+4+x^2+4x+4}=\dfrac{-8x}{2x^2+8}=-\dfrac{4x}{x^2+4}$

18. $\dfrac{m^2+9}{6m}$

19. $\dfrac{\dfrac{xy(x-y)}{1}\cdot\dfrac{2x+y}{x}-\dfrac{xy(x-y)}{1}\cdot\dfrac{3x+y}{(x-y)}}{\dfrac{xy(x-y)}{1}\cdot\dfrac{x+y}{y}+\dfrac{xy(x-y)}{1}\cdot\dfrac{2(x+y)}{(x-y)}}$

$=\dfrac{y(x-y)(2x+y)-xy(3x+y)}{x(x+y)(x-y)+2xy(x+y)}$

$=\dfrac{y[(x-y)(2x+y)-x(3x+y)]}{x(x+y)[(x-y)+2y]}$

$=\dfrac{y[2x^2-xy-y^2-3x^2-xy]}{x(x+y)(x+y)}$

$=\dfrac{-y[x^2+2xy+y^2]}{x(x+y)^2}=\dfrac{-y(x+y)^2}{x(x+y)^2}=-\dfrac{y}{x}$

20. $\dfrac{a}{b}$

21. $\dfrac{1}{x+\dfrac{1}{x+\dfrac{1}{2x}}}=\dfrac{1}{x+\dfrac{1(2x)}{\left(x+\dfrac{1}{2x}\right)(2x)}}=\dfrac{1}{x+\dfrac{2x}{2x^2+1}}$

$=\dfrac{1(2x^2+1)}{\left(x+\dfrac{2x}{2x^2+1}\right)(2x^2+1)}=\dfrac{2x^2+1}{x(2x^2+1)+2x}$

$=\dfrac{2x^2+1}{2x^3+x+2x}=\dfrac{2x^2+1}{2x^3+3x}$

22. $\dfrac{2y^2+2}{y^3+3y}$

23. $\dfrac{x+\dfrac{1(3)}{\left(2+\dfrac{x}{3}\right)(3)}}{x-\dfrac{3(2)}{\left(4+\dfrac{x}{2}\right)(2)}}=\dfrac{x+\dfrac{3}{6+x}}{x-\dfrac{6}{8+x}}=\dfrac{\left(x+\dfrac{3}{6+x}\right)(6+x)(8+x)}{\left(x-\dfrac{6}{8+x}\right)(6+x)(8+x)}$

$=\dfrac{x(48+14x+x^2)+3(8+x)}{x(48+14x+x^2)-6(6+x)}=\dfrac{48x+14x^2+x^3+24+3x}{48x+14x^2+x^3-36-6x}$

$=\dfrac{x^3+14x^2+51x+24}{x^3+14x^2+42x-36}$

24. $\dfrac{x^3+8x^2+22x+60}{x^3+8x^2-24}$

Sections 6.1–6.6 Review Exercises (page 283)

1. Yes, since $\frac{2(x+3)}{2(5)} = \frac{2x+6}{10}$ **2.** No

3. $\frac{9(-1)}{(x-3)(-1)} = \frac{-9}{3-x}$ The missing term is -9.

4. The missing term is $x - 2$.

5. $\frac{\overset{2}{\cancel{4}}z(z^2+z-6)}{\underset{1}{\cancel{2}}(z^2+2z-3)} = \frac{2z\cancel{(z+3)}(z-2)}{\cancel{(z+3)}(z-1)} = \frac{2z(z-2)}{z-1}$

6. $\frac{2k(k+1)}{k+4}$

7. $\frac{\cancel{(a-3b)}(a^2+3ab+9b^2)}{\cancel{(a-3b)}(a+2)} = \frac{a^2+3ab+9b^2}{a+2}$

8. $\frac{1}{4x^2-2x+1}$

9. $\frac{-\overset{5}{\cancel{35}}mn^2p^2}{\underset{2}{\cancel{14}}m^3p^3}\cdot\frac{\overset{1}{\cancel{13}}m^4n}{\underset{4}{\cancel{52}}n^3p} = -\frac{5m^2}{8p^2}$ **10.** $-\frac{4c^2}{3ab^3}$

11. $\frac{(z+2)\cancel{(z+1)}}{\cancel{(z+1)}\cancel{(z+1)}}\cdot\frac{\cancel{(z+1)}\cancel{(z-1)}}{(z+3)\cancel{(z-1)}} = \frac{z+2}{z+3}$ **12.** 1

13. $\frac{\cancel{(x+y)}\cancel{(x^2-xy+y^2)}}{\underset{1}{\cancel{3}}\cancel{(x^2-xy+y^2)}}\cdot\frac{\cancel{(x+2y)}\cancel{(x-y)}}{\cancel{(x+y)}\cancel{(x-y)}}\cdot\frac{\overset{3}{\cancel{15}}x^2y}{\underset{1}{\cancel{5}}xy\cancel{(x+2y)}} = x$

14. 2

15. $\frac{20y-7}{6y-8} - \frac{17+2y}{6y-8} = \frac{20y-7-17-2y}{6y-8}$

$= \frac{18y-24}{6y-8} = \frac{\overset{3}{\cancel{6}}\cancel{(3y-4)}}{\underset{1}{\cancel{2}}\cancel{(3y-4)}} = 3$

16. -2 **17.** $\frac{11}{30e^3f}\cdot\frac{3f}{3f} - \frac{7}{45e^2f^2}\cdot\frac{2e}{2e} = \frac{33f-14e}{90e^3f^2}$

18. $\frac{30u^2-49v}{280u^4v^2}$

19. $\frac{a+1}{(a+1)(a-2)} - \frac{a-2}{(a+3)(a-2)} = \frac{1}{a-2} - \frac{1}{a+3}$

LCD $= (a-2)(a+3)$

$\frac{(a+3)}{(a+3)}\cdot\frac{1}{a-2} + \frac{(a-2)}{(a-2)}\cdot\frac{-1}{a+3} = \frac{a+3-a+2}{(a-2)(a+3)}$

$= \frac{5}{(a-2)(a+3)}$

20. $\frac{3}{(x-1)(x+2)}$

21. $\frac{\overset{3}{\cancel{15}}x}{\underset{1}{\cancel{5}}x(x+4)} - \frac{7}{3x^2} - \frac{3\cancel{(x+4)}}{(x+4)^{\cancel{2}}} = \cancel{\frac{3}{x+4}} - \frac{7}{3x^2} - \cancel{\frac{3}{x+4}} = -\frac{7}{3x^2}$

22. $-\frac{5}{11y^3}$

23. $\frac{\frac{y+1}{1}}{\frac{y+1}{1}}\cdot\frac{\frac{x}{y+1}+2}{\frac{x}{y+1}-2} = \frac{\frac{y+1}{1}\cdot\frac{x}{y+1}+\frac{y+1}{1}\cdot\frac{2}{1}}{\frac{y+1}{1}\cdot\frac{x}{y+1}+\frac{y+1}{1}\cdot\frac{-2}{1}}$

$= \frac{x+(y+1)(2)}{x+(y+1)(-2)} = \frac{x+2y+2}{x-2y-2}$

24. $\frac{3b-a+6}{2b+a+4}$

25. $\frac{\frac{R^3T^3}{1}}{\frac{R^3T^3}{1}}\cdot\frac{\frac{8}{R^3}+\frac{1}{T^3}}{\frac{4}{R^2}-\frac{1}{T^2}} = \frac{\frac{R^3T^3}{1}\cdot\frac{8}{R^3}+\frac{R^3T^3}{1}\cdot\frac{1}{T^3}}{\frac{R^3T^3}{1}\cdot\frac{4}{R^2}-\frac{R^3T^3}{1}\cdot\frac{1}{T^2}}$

$= \frac{8T^3+R^3}{4RT^3-R^3T} = \frac{\cancel{(2T+R)}(4T^2-2RT+R^2)}{RT\cancel{(2T+R)}(2T-R)}$

$= \frac{4T^2-2RT+R^2}{RT(2T-R)}$

26. $\frac{mn(n+4m)}{n^2+4mn+16m^2}$

Exercises 6.7 (page 293)

1. If $y+4=0$, $y=-4$; the set of all real numbers except -4

2. All real numbers except 5

3. No number can make the denominator zero; the set of all real numbers

4. All real numbers except 0

5. If $a^2-25=0$, then $a^2=25$, so $a=\pm 5$; the set of all real numbers except 5 and -5

6. All real numbers except 3 and -2

7. If $c^4-13c^2+36=0$, then $(c^2-4)(c^2-9)=0$. Therefore, $c^2-4=0$ or $c^2-9=0$, so $c=\pm 2$ or $c=\pm 3$; the set of all real numbers except 2, -2, 3, and -3

8. All real numbers except 5, 3, and -3

The checks for Exercises 9–32 will usually not be shown.

9. LCD $= 4k(k-5)$

$\frac{4k\cancel{(k-5)}}{1}\cdot\frac{2}{\cancel{k-5}} - \frac{4\cancel{k}(k-5)}{1}\cdot\frac{5}{\cancel{k}} = \frac{\cancel{4k}(k-5)}{1}\cdot\frac{3}{\cancel{4k}}$

$4k(2) - 20(k-5) = (k-5)(3)$

$8k - 20k + 100 = 3k - 15$

$-15k = -115$

$k = \frac{115}{15} = \frac{23}{3}$

$\left\{\frac{23}{3}\right\}$

10. $\left\{-\frac{35}{4}\right\}$

11. LCD $= x-2$ **12.** $\{\ \}$

$\frac{\cancel{x-2}}{1}\cdot\frac{x}{\cancel{x-2}} = \frac{\cancel{x-2}}{1}\cdot\frac{2}{\cancel{x-2}} + \frac{x-2}{1}(5)$

$x = 2 + 5x - 10$

$8 = 4x$

$x = 2$

Not in the domain; $\{\ \}$

13. LCD $= 2m-3$

$\frac{\cancel{2m-3}}{1}\cdot\frac{12m}{\cancel{2m-3}} = \frac{2m-3}{1}\cdot\frac{6}{1} + \frac{\cancel{2m-3}}{1}\cdot\frac{18}{\cancel{2m-3}}$

$12m = (2m-3)(6) + 18$

$12m = 12m - 18 + 18$

$12m = 12m$

$0 = 0$ Identity

Solution set: all real numbers except $\frac{3}{2}$

14. Identity; solution set: all real numbers except $-\frac{6}{5}$

15. $2y(3y) = 7y + 5$ **16.** $\left\{\frac{1}{2}, -\frac{3}{2}\right\}$

$6y^2 = 7y + 5$

$6y^2 - 7y - 5 = 0$

$(2y+1)(3y-5) = 0$

$2y + 1 = 0$ | $3y - 5 = 0$

$y = -\frac{1}{2}$ | $y = \frac{5}{3}$

$\left\{-\frac{1}{2}, \frac{5}{3}\right\}$

17.
$$\begin{aligned} (3e-5)(2e+3) &= 4e(e) \\ 6e^2 - e - 15 &= 4e^2 \\ 2e^2 - e - 15 &= 0 \\ (e-3)(2e+5) &= 0 \end{aligned}$$
$$e - 3 = 0 \quad\Big|\quad 2e + 5 = 0$$
$$e = 3 \quad\Big|\quad e = -\tfrac{5}{2}$$
$\left\{3, -\tfrac{5}{2}\right\}$

18. $\left\{1, -\tfrac{7}{4}\right\}$

19. LCD $= 2x^2$
$$\overset{1}{\cancel{2}}x^2 \cdot \frac{1}{\underset{1}{\cancel{2}}} - 2x^{\cancel{2}} \cdot \frac{1}{\underset{1}{\cancel{x}}} = \overset{1}{2}\cancel{x^2} \cdot \frac{4}{\underset{1}{\cancel{x^2}}}$$
$$\begin{aligned} x^2 - 2x &= 8 \\ x^2 - 2x - 8 &= 0 \\ (x-4)(x+2) &= 0 \end{aligned}$$
$$x - 4 = 0 \quad\Big|\quad x + 2 = 0$$
$$x = 4 \quad\Big|\quad x = -2$$
$\{4, -2\}$

20. $\left\{3, -\tfrac{4}{3}\right\}$

21. LCD $= 15x(x+1)$
$$\frac{15x(\overset{1}{\cancel{x+1}})}{1} \cdot \frac{4}{\underset{1}{\cancel{x+1}}} = \frac{15x(x+1)}{1} \cdot \frac{3}{x} + \frac{\overset{1}{\cancel{15}}x(x+1)}{1} \cdot \frac{1}{\underset{1}{\cancel{15}}}$$
$$\begin{aligned} 60x &= 45x + 45 + x^2 + x \\ 0 &= x^2 - 14x + 45 \\ 0 &= (x-5)(x-9) \end{aligned}$$
$$x - 5 = 0 \quad\Big|\quad x - 9 = 0$$
$$x = 5 \quad\Big|\quad x = 9$$
$\{5, 9\}$

22. $\{-2, -3\}$

23. LCD $= x(x+3)(x+4)$
$$\frac{x(x+3)(\overset{1}{\cancel{x+4}})}{1} \cdot \frac{6}{\underset{1}{\cancel{x+4}}}$$
$$= \frac{x(\overset{1}{\cancel{x+3}})(x+4)}{1} \cdot \frac{5}{\underset{1}{\cancel{x+3}}} + \frac{\overset{1}{\cancel{x}}(x+3)(x+4)}{1} \cdot \frac{4}{\underset{1}{\cancel{x}}}$$
$$\begin{aligned} x(x+3)(6) &= x(x+4)(5) + (x+3)(x+4)(4) \\ 6x^2 + 18x &= 5x^2 + 20x + 4x^2 + 28x + 48 \\ 0 &= 3x^2 + 30x + 48 \\ 0 &= 3(x^2 + 10x + 16) \\ 0 &= 3(x+2)(x+8) \end{aligned}$$
$$x + 2 = 0 \quad\Big|\quad x + 8 = 0$$
$$x = -2 \quad\Big|\quad x = -8$$
$\{-2, -8\}$

24. $\left\{2, -\tfrac{5}{4}\right\}$

25. LCD $= 5(x+3)(x-3)$

Domain: all real numbers except -3 and 3
$$\frac{5(\overset{1}{\cancel{x+3)(x-3}})}{1} \cdot \frac{6}{\underset{1}{\cancel{x^2-9}}} + \frac{\overset{1}{\cancel{5}}(x+3)(x-3)}{1} \cdot \frac{1}{\underset{1}{\cancel{5}}}$$
$$= \frac{5(x+3)(\overset{1}{\cancel{x-3}})}{1} \cdot \frac{1}{\underset{1}{\cancel{x-3}}}$$
$$\begin{aligned} 30 + x^2 - 9 &= 5x + 15 \\ x^2 - 5x + 6 &= 0 \\ (x-2)(x-3) &= 0 \end{aligned}$$
$$x - 2 = 0 \quad\Big|\quad x - 3 = 0$$
$$x = 2 \quad\Big|\quad x = 3 \text{ Not in the domain}$$

Check for $x = 2$
$$\begin{aligned} \frac{6}{x^2-9} + \frac{1}{5} &= \frac{1}{x-3} \\ \frac{6}{2^2-9} + \frac{1}{5} &\overset{?}{=} \frac{1}{2-3} \\ -\frac{6}{5} + \frac{1}{5} &\overset{?}{=} -1 \\ -1 &= -1 \end{aligned}$$
$\{2\}$

26. $\left\{\tfrac{7}{3}\right\}$

27. LCD $= (x+2)(x-2)$

Domain: all real numbers except 2 and -2
$$(x+2)(x-2)\left(\frac{x+2}{x-2} - \frac{x-2}{x+2}\right) = \overset{1}{\cancel{(x+2)(x-2)}}\left(\frac{16}{\underset{1}{\cancel{x^2-4}}}\right)$$
$$(x+2)(\overset{1}{\cancel{x-2}})\left(\frac{x+2}{\underset{1}{\cancel{x-2}}}\right) - (\overset{1}{\cancel{x+2}})(x-2)\left(\frac{x-2}{\underset{1}{\cancel{x+2}}}\right) = 16$$
$$\begin{aligned} (x+2)(x+2) - (x-2)(x-2) &= 16 \\ x^2 + 4x + 4 - (x^2 - 4x + 4) &= 16 \\ x^2 + 4x + 4 - x^2 + 4x - 4 &= 16 \\ 8x &= 16 \\ x &= 2 \end{aligned}$$
But 2 is not in the domain of the variable. Therefore, the solution set is { }.

28. { }

29. LCD $= (2x-5)(x-3)(x+3)$
$$(2x-5)(x-3)(x+3)\left(\frac{1}{(2x-5)(x-3)} + \frac{x-1}{(2x-5)(x+3)}\right)$$
$$= (2x-5)(x-3)(x+3)\left(\frac{-4}{(x+3)(x-3)}\right)$$
$$\begin{aligned} (x+3) + (x-3)(x-1) &= (2x-5)(-4) \\ x + 3 + x^2 - 4x + 3 &= -8x + 20 \\ x^2 + 5x - 14 &= 0 \\ (x+7)(x-2) &= 0 \end{aligned}$$
$$x + 7 = 0 \quad\Big|\quad x - 2 = 0$$
$$x = -7 \quad\Big|\quad x = 2$$
$\{2, -7\}$

30. $\{1, 10\}$

31. LCD $= (x+4)(x^2 - 4x + 16)(x-4)$
$$(x+4)(x^2-4x+16)(x-4)\left(\frac{8}{(x+4)(x^2-4x+16)} + \frac{3}{x^2-16}\right)$$
$$= (x+4)(x^2-4x+16)(x-4)\left(\frac{-1}{x^2-4x+16}\right)$$
$$\begin{aligned} 8(x-4) + 3(x^2 - 4x + 16) &= -1(x+4)(x-4) \\ 8x - 32 + 3x^2 - 12x + 48 &= -(x^2 - 16) \\ 3x^2 - 4x + 16 &= -x^2 + 16 \\ 4x^2 - 4x &= 0 \\ 4x(x-1) &= 0 \end{aligned}$$
$$4x = 0 \quad\Big|\quad x - 1 = 0$$
$$x = 0 \quad\Big|\quad x = 1$$
$\{0, 1\}$

32. $\{2\}$

Exercises 6.8 (page 296)

1.
$$\begin{aligned} 6x - 2y &= xy - 12 \\ 6x + 12 &= xy + 2y \\ 6(x+2) &= (x+2)y \\ \frac{6(\overset{1}{\cancel{x+2}})}{(\underset{1}{\cancel{x+2}})} &= y \\ y &= 6 \end{aligned}$$

2. $x = -2$

3.
$$\begin{aligned} zs &= x - m \\ m &= x - zs \end{aligned}$$

4. $N = \dfrac{n - s^2}{1 - s^2}$

5.
$$\begin{aligned} (5yz)\left(\frac{2x}{5yz}\right) &= (5yz)(z + x) \\ 2x &= 5yz^2 + 5xyz \\ 2x - 5xyz &= 5yz^2 \\ x(2 - 5yz) &= 5yz^2 \\ x &= \frac{5yz^2}{2 - 5yz} \end{aligned}$$

6. $y = \dfrac{xz}{x - z}$

7. $9C = 5(F - 32)$
$9C = 5F - 160$
$9C + 160 = 5F$
$F = \dfrac{9C + 160}{5}$

8. $B = \dfrac{2A - hb}{h}$

9. $s = c + \dfrac{\overset{1}{\cancel{c}}a}{\underset{1}{\cancel{c}}}$
$s = c + a$
$c = s - a$

10. $R = \dfrac{rZ}{r - Z}$

11. $A = P + Prt$
$A - P = Prt$
$r = \dfrac{A - P}{Pt}$

12. $t = \dfrac{2S + g}{2g}$

13. $\dfrac{ra}{1} \cdot \dfrac{v^2}{1} = \dfrac{\overset{1}{\cancel{r}}a}{1} \cdot \dfrac{2}{\underset{1}{\cancel{r}}} - \dfrac{r\overset{1}{\cancel{a}}}{1} \cdot \dfrac{1}{\underset{1}{\cancel{a}}}$
$rav^2 = 2a - r$
$r = 2a - rav^2$
$r = a(2 - rv^2)$
$a = \dfrac{r}{2 - rv^2}$

14. $s = \dfrac{p}{1 - p}$

15. $S(1 - r) = a$
$S - Sr = a$
$S - a = Sr$
$r = \dfrac{S - a}{S}$

16. $R = \dfrac{E - Ir}{I}$

17. $Fuv\left(\dfrac{1}{F}\right) = Fuv\left(\dfrac{1}{u} + \dfrac{1}{v}\right)$
$uv = Fuv\left(\dfrac{1}{u}\right) + Fuv\left(\dfrac{1}{v}\right)$
$uv = Fv + Fu$
$uv = F(v + u)$
$F = \dfrac{uv}{u + v}$

18. $b = \dfrac{ac}{a - c}$

19. $L = a + nd - d$
$L - a + d = nd$
$n = \dfrac{L - a + d}{d}$

20. $h = \dfrac{A - 2\pi r^2}{2\pi r}$

21. $C = \dfrac{\pi A \cdot a}{\pi A \cdot 1 + \dfrac{\overset{1}{\cancel{\pi A}}}{1} \cdot \dfrac{a}{\underset{1}{\cancel{\pi A}}}}$
$\dfrac{C}{1} = \dfrac{\pi Aa}{\pi A + a}$
$C(\pi A + a) = \pi Aa$
$C\pi A + Ca = \pi Aa$
$C\pi A = \pi Aa - Ca$
$C\pi A = a(\pi A - C)$
$a = \dfrac{C\pi A}{\pi A - C}$

22. $v = \dfrac{c^2(R - V)}{c^2 - RV}$

Exercises 6.9 (page 299)

The checks for these exercises will not be shown.

1. Let $4x$ = smaller number
Let $5x$ = larger number
$4x + 5x = 81$
$9x = 81$
$x = 9$
$4x = 4(9) = 36$
$5x = 5(9) = 45$
The numbers are 36 and 45.

2. The numbers are 56 and 21.

3. Let $3x$ = length of first side (in m)
Let $4x$ = length of second side (in m)
Let $5x$ = length of third side (in m)
The perimeter is 108 m.
$3x + 4x + 5x = 108$
$12x = 108$
$x = 9$
$3x = 27$
$4x = 36$
$5x = 45$
The lengths of the sides are 27 m, 36 m, and 45 m.

4. 32 in., 40 in., 48 in.

5. Let $4x$ = hr spent studying
Let $2x$ = hr spent in class
Let $3x$ = hr spent at work
$54 = 4x + 2x + 3x$
$54 = 9x$
$6 = x$
$4x = 4(6) = 24$
$2x = 2(6) = 12$
$3x = 3(6) = 18$
The student spends 24 hr studying, 12 hr in class, and 18 hr working.

6. Study, 16 hr; class, 8 hr; work, 24 hr

7. Let $7x$ = length of rectangle (in ft)
Let $6x$ = width of rectangle (in ft)
$2(7x) + 2(6x) = 78$
$14x + 12x = 78$
$26x = 78$
$x = 3$
$7x = 21$
$6x = 18$
The length is 21 ft, and the width is 18 ft.

8. Length = 63 cm, width = 35 cm

9. Let $3x$ = length (in m)
Let $2x$ = width (in m)
$A = (\text{length})(\text{width})$
$150 = (3x)(2x)$
$150 = 6x^2$
$25 = x^2$
$25 - x^2 = 0$
$(5 - x)(5 + x) = 0$
$5 - x = 0 \mid 5 + x = 0$
$5 = x \mid x = -5$ Not in the domain
$3x = 15$
$2x = 10$
The length is 15 m, and the width is 10 m.

10. Width = 12 in.; length = 16 in.

11. Let $7x$ = one number
Let $6x$ = other number
$143 = 7x + 6x$
$143 = 13x$
$x = 11$
$7x = 77$
$6x = 66$
One number is 77, and the other is 66.

12. 136, 85

13. Let $4x$ = shortest side
Let $5x$ = second side
Let $6x$ = third side
$0 < 4x + 5x + 6x < 60$
$0 < 15x < 60$
$0 < x < 4$
$0 < 4x < 16$
The shortest side must be between 0 and 16.

14. The shortest side must be between 0 and 24.

Exercises 6.10 (page 305)

The checks for these exercises will not be shown.

1. Let x = speed of Malone car (in mph)
Then $x + 9$ = speed of King car (in mph)

$6x =$ distance for Malone car
$5(x + 9) =$ distance for King car
$6x = 5(x + 9)$ Distances are equal.
$6x = 5x + 45$

a. $x = 45$
$x + 9 = 45 + 9 = 54$
The speed of the Malone car was 45 mph, and the speed of the King car was 54 mph.

b. $45 \times 6 = 54 \times 5 = 270$ The distance traveled was 270 mi.

2. **a.** Duran car, 40 mph; Silva car, 50 mph **b.** 400 mi

3. Let $x =$ hours required to return from lake
Then $x + 3 =$ hours required to hike to lake
$2(x + 3) =$ distance going to lake
$5x =$ distance returning from lake
$5x = 2(x + 3)$ Distances are equal.
$5x = 2x + 6$
$3x = 6$
$x = 2$

a. $x + 3 = 5$
It took Eric 5 hr to hike to the lake.

b. The distance is $\left(2\frac{\text{mi}}{\cancel{\text{hr}}}\right)(5\ \cancel{\text{hr}}) = 10$ mi.

4. **a.** 3 hr **b.** 6 mi

5. Let $x =$ Fran's speed (in mph)
Then $\frac{4}{5}x =$ Ron's speed (in mph)
$3x =$ distance traveled by Fran
$3\left(\frac{4}{5}x\right) =$ distance traveled by Ron
$3x + 3\left(\frac{4}{5}x\right) = 54$ Sum of the distances is 54; LCD = 5
$\frac{5}{1}\left(\frac{3}{1}x\right) + \frac{5}{1}\left(\frac{3}{1}\right)\left(\frac{4}{5}x\right) = \frac{5}{1}\left(\frac{54}{1}\right)$
$15x + 12x = 270$
$27x = 270$
$x = 10$
$\frac{4}{5}x = \frac{4}{5}(10)$
$= 8$
Fran's speed is 10 mph, and Ron's speed is 8 mph.

6. Tran's rate is 9 mph; Atour's rate is 6 mph.

7. Let $x =$ speed of boat in still water (in mph)
Then $x + 2 =$ speed of boat downstream (in mph)
and $x - 2 =$ speed of boat upstream (in mph)
$5(x - 2) =$ distance traveled upstream
$3(x + 2) =$ distance boat traveled downstream
$$\begin{pmatrix}\text{Distance traveled}\\ \text{downstream}\end{pmatrix} = (6\text{ mi}) + \begin{pmatrix}\text{Distance traveled}\\ \text{upstream}\end{pmatrix}$$
$$3(x + 2) = 6 + 5(x - 2)$$
$$3x + 6 = 6 + 5x - 10$$
$$10 = 2x$$
$$x = 5$$
$3(x + 2) = 3(5 + 2) = 3(7) = 21$
The speed of the boat is 5 mph in still water, and Colin traveled 21 mi downstream.

8. 3 mph in still water; 16 mi downstream

9. Rate with a head wind is $(140 - 28)$ mph, or 112 mph.
Rate with a tail wind is $(140 + 28)$ mph, or 168 mph.
Let $t =$ time with a head wind (in hr)
Then $2.5 - t =$ time with a tail wind (in hr)
$112t =$ distance with a head wind
$168(2.5 - t) =$ distance with a tail wind
$112t = 168(2.5 - t)$ The distances are equal.
$112t = 420 - 168t$
$280t = 420$
$t = 1.5$ Time with a head wind
The storage facility is $\left(112\frac{\text{mi}}{\text{hr}}\right)(1.5\text{ hr})$, or 168 mi, from the office.

10. 6 mi

11. Let $d =$ distance in mi
Slow plane's time $= \frac{d}{400}$ $\left(t = \frac{d}{r}\right)$
Fast plane's time $= \frac{d}{500}$
$$\begin{pmatrix}\text{Slow plane's}\\ \text{time}\end{pmatrix} - \begin{pmatrix}\text{Fast plane's}\\ \text{time}\end{pmatrix} = \frac{1}{2}\text{ hr}$$
$$\frac{d}{400} - \frac{d}{500} = \frac{1}{2};\quad \text{LCD} = 2{,}000$$
$$\frac{2{,}000}{1}\cdot\frac{d}{400} - \frac{2{,}000}{1}\cdot\frac{d}{500} = \frac{2{,}000}{1}\cdot\frac{1}{2}$$
$$5d - 4d = 1{,}000$$
$$d = 1{,}000$$
The distance is 1,000 mi.

12. 6 mi

13. Let $x =$ actual speed of cyclist (in mph)
Then $x + 3 =$ faster speed (in mph)
Also, $\frac{90}{x} =$ actual time (in hr)
and $\frac{90}{x + 3} =$ time (in hr) if he went faster
$$\frac{90}{x} = \frac{90}{x + 3} + 1$$
$$x(x + 3)\left(\frac{90}{x}\right) = x(x + 3)\left(\frac{90}{x + 3} + 1\right)$$
$$90x + 270 = 90x + x^2 + 3x$$
$$0 = x^2 + 3x - 270$$
$$0 = (x - 15)(x + 18)$$
$x - 15 = 0$ | $x + 18 = 0$
$x = 15$ | $x = -18$ Reject
The actual speed of the cyclist was 15 mph.

14. The actual speed of the boat was 30 mph.

15. Let $x =$ number of hr until planes will be 1,400 mi apart
Then $210x =$ distance first plane flies
and $190x =$ distance second plane flies
$210x + 190x = 1{,}400$
$400x = 1{,}400$
$x = 3.5$
$\leftarrow 210x$ mi $\rightarrow\leftarrow 190x$ mi $\rightarrow$
The planes will be 1,400 mi apart in 3.5 hr.

16. The boats will be 117 mi apart in 2.25 hr.

Exercises 6.11 (page 311)

The checks will not be shown.

1. Henry's rate $= \frac{1\text{ house}}{5\text{ days}} = \frac{1}{5}$ house per day
Teri's rate $= \frac{1\text{ house}}{4\text{ days}} = \frac{1}{4}$ house per day
Let $x =$ number of days to paint the house
$$\begin{pmatrix}\text{Amount}\\ \text{Henry paints}\end{pmatrix} + \begin{pmatrix}\text{Amount}\\ \text{Teri paints}\end{pmatrix} = \begin{pmatrix}\text{Amount}\\ \text{painted together}\end{pmatrix}$$
$$\frac{x}{5} + \frac{x}{4} = 1$$ One house painted; LCD = 20
$$\frac{20}{1}\cdot\frac{x}{5} + \frac{20}{1}\cdot\frac{x}{4} = \frac{20}{1}\cdot 1$$
$$4x + 5x = 20$$
$$9x = 20$$
$$x = \frac{20}{9} = 2\tfrac{2}{9}$$
It will take $2\frac{2}{9}$ days to paint the house.

2. $3\frac{3}{5}$ days

3. Let x = number of hr for David to type 80 pages

Trisha's rate $= \dfrac{100 \text{ pages}}{3 \text{ hr}} = \dfrac{100}{3}$ pages per hr

David's rate $= \dfrac{80 \text{ pages}}{x \text{ hr}} = \dfrac{80}{x}$ pages per hr

Since they both type for 10 hr to produce 500 pages,

$$\begin{pmatrix}\text{Amount}\\ \text{Trisha types}\end{pmatrix} + \begin{pmatrix}\text{Amount}\\ \text{David types}\end{pmatrix} = 500 \text{ pages}$$

$$\frac{100}{3}(10) + \frac{80}{x}(10) = 500; \quad \text{LCD} = 3x$$

$$\frac{3x}{1}\cdot\frac{1{,}000}{3} + \frac{3x}{1}\cdot\frac{800}{x} = \frac{3x}{1}\cdot\frac{500}{1}$$

$$1{,}000x + 2{,}400 = 1{,}500x$$

$$2{,}400 = 500x$$

$$x = \frac{2{,}400}{500} = \frac{24}{5} = 4\frac{4}{5}$$

It takes David $4\frac{4}{5}$ hr to type 80 pages.

4. $49\frac{1}{2}$ min

5. Machine A's rate $= \dfrac{1 \text{ job}}{36 \text{ hr}} = \dfrac{1}{36}$ job per hr

Machine B's rate $= \dfrac{1 \text{ job}}{24 \text{ hr}} = \dfrac{1}{24}$ job per hr

Let x = number of hr for both machines to run together
Since machine A runs for 12 hr before machine B is turned on, machine A runs for $x + 12$ hr.

$$\begin{pmatrix}\text{Amount of job}\\ \text{done by machine A}\end{pmatrix} + \begin{pmatrix}\text{Amount of job}\\ \text{done by machine B}\end{pmatrix} = 1 \text{ job}$$

$$\tfrac{1}{36}(x + 12) + \tfrac{1}{24}(x) = 1; \quad \text{LCD} = 72$$

$$\frac{72}{1}\cdot\frac{x + 12}{36} + \frac{72}{1}\cdot\frac{x}{24} = 72 \cdot 1$$

$$2x + 24 + 3x = 72$$

$$5x = 48$$

$$x = \tfrac{48}{5} = 9\tfrac{3}{5}$$

It will take $9\frac{3}{5}$ hr to finish the job.

6. $6\frac{9}{11}$ hr

7. Let x = smaller number
Then $x + 8$ = larger number

One-fourth the larger number	is	one more than	one-third the smaller number

$$\tfrac{1}{4}(x + 8) = 1 + \tfrac{1}{3}(x); \quad \text{LCD} = 12$$

$$\frac{12}{1}\cdot\frac{x + 8}{4} = \frac{12}{1}\cdot 1 + \frac{12}{1}\cdot\frac{x}{3}$$

$$3x + 24 = 12 + 4x$$

$$12 = x$$

$$x = 12$$

$$x + 8 = 12 + 8 = 20$$

The numbers are 12 and 20.

8. Smaller number is -20; larger number is -14.

9. Let x = number of qt of antifreeze to be drained and replaced

$$0.45(14) - 0.45(x) + 1.00x = 0.50(14)$$

$$6.3 + 0.55x = 7$$

$$630 + 55x = 700$$

$$55x = 70$$

$$x = \tfrac{70}{55} = \tfrac{14}{11} = 1\tfrac{3}{11}$$

$1\frac{3}{11}$ qt must be drained and replaced.

10. $3\frac{3}{7}$ qt

11. Let u = units digit
Then $u + 1$ = tens digit

$$\frac{(u + 1)u}{u + 1 + u} = \frac{6}{5}$$

$$\frac{u^2 + u}{2u + 1} = \frac{6}{5}$$

$$5(u^2 + u) = 6(2u + 1)$$

$$5u^2 + 5u = 12u + 6$$

$$5u^2 - 7u - 6 = 0$$

$$(u - 2)(5u + 3) = 0$$

$u - 2 = 0$ | $5u + 3 = 0$
$u = 2$ | $u = -\frac{3}{5}$ Not in the domain

$u + 1 = 3$

The number is 32.

12. 34

13. Let x = number of hr for pipe 1 to fill tank
Then $x + 1$ = number of hr for pipe 2 to fill tank

Rate of pipe 1 $= \dfrac{1 \text{ tank}}{x \text{ hr}} = \dfrac{1}{x}$ tank per hr

Rate of pipe 2 $= \dfrac{1 \text{ tank}}{(x + 1) \text{ hr}} = \dfrac{1}{x + 1}$ tank per hr

Rate of pipe 3 $= \dfrac{1 \text{ tank}}{2 \text{ hr}} = \dfrac{1}{2}$ tank per hr

$$\begin{pmatrix}\text{Amount 1}\\ \text{does in 3 hr}\end{pmatrix} + \begin{pmatrix}\text{Amount 2}\\ \text{does in 3 hr}\end{pmatrix} - \begin{pmatrix}\text{Amount 3}\\ \text{does in 3 hr}\end{pmatrix} = 1 \text{ full tank}$$

$$\frac{1}{x}(3) + \frac{1}{x + 1}(3) - \frac{1}{2}(3) = 1$$

$$\frac{2x(x + 1)}{1}\cdot\frac{3}{x} + \frac{2x(x + 1)}{1}\cdot\frac{3}{x + 1} - \frac{2x(x + 1)}{1}\cdot\frac{3}{2} = \frac{2x(x + 1)}{1}\cdot\frac{1}{1}$$

$$6(x + 1) + 6x - 3x(x + 1) = 2x(x + 1)$$

$$6x + 6 + 6x - 3x^2 - 3x = 2x^2 + 2x$$

$$5x^2 - 7x - 6 = 0$$

$$(x - 2)(5x + 3) = 0$$

$x - 2 = 0$ | $5x + 3 = 0$
$x = 2$ | $x = -\frac{3}{5}$ Not in the domain

It takes 2 hr for pipe 1 to fill the tank.

14. 2 hr

15. Let x = number of hr for Sandra to proofread 60 pages

Ruth's rate $= \dfrac{230 \text{ pages}}{4 \text{ hr}}$

Sandra's rate $= \dfrac{60 \text{ pages}}{x \text{ hr}}$

$$\begin{pmatrix}\text{Number of pages}\\ \text{Ruth reads in 6 hr}\end{pmatrix} + \begin{pmatrix}\text{Number of pages}\\ \text{Sandra reads in 6 hr}\end{pmatrix} = 525 \text{ pages}$$

$$\left(\frac{230}{4}\right)(6) + \left(\frac{60}{x}\right)(6) = 525; \quad \text{LCD} = x$$

$$x\left(345 + \frac{360}{x}\right) = 525x$$

$$345x + 360 = 525x$$

$$360 = 180x$$

$$x = 2$$

It takes 2 hr for Sandra to proofread 60 pages.

16. 28 hr

17. Let x = number of min for machine B to process 4,300 ft of film

Rate of machine A $= \dfrac{5{,}700 \text{ ft}}{60 \text{ min}}$

Rate of machine B $= \dfrac{4{,}300 \text{ ft}}{x \text{ min}}$

$$\begin{pmatrix}\text{Amount of}\\ \text{film done}\\ \text{by machine A}\\ \text{in 50 min}\end{pmatrix} + \begin{pmatrix}\text{Amount of}\\ \text{film done}\\ \text{by machine B}\\ \text{in 50 min}\end{pmatrix} = \begin{pmatrix}\text{Amount of}\\ \text{film done}\\ \text{together}\\ \text{in 50 min}\end{pmatrix}$$

$$\frac{5{,}700}{60}\cdot(50) + \frac{4{,}300}{x}\cdot(50) = 15{,}500$$

$$\begin{aligned} 4{,}750 + \frac{215{,}000}{x} &= 15{,}500 \\ 4{,}750x + 215{,}000 &= 15{,}500x \\ 215{,}000 &= 10{,}750x \\ x &= 20 \end{aligned}$$

It takes 20 min for machine B to process 4,300 ft of film.

18. 30 min

Sections 6.7–6.11 Review Exercises (page 313)

1.
$$\begin{aligned} 20 - 45a^2 &= 0 \\ 5(4 - 9a^2) &= 0 \\ 5(2 + 3a)(2 - 3a) &= 0 \end{aligned}$$
$2 + 3a = 0$, $a = -\frac{2}{3}$ | $2 - 3a = 0$, $a = \frac{2}{3}$

The domain is the set of all real numbers except $-\frac{2}{3}$ and $\frac{2}{3}$.

2. All real numbers except 6 and $-\frac{5}{2}$

3.
$$\begin{aligned} (5a - 4)(3a + 10) &= (6a)(-2) \\ 15a^2 + 38a - 40 &= -12a \\ 15a^2 + 50a - 40 &= 0 \\ 3a^2 + 10a - 8 &= 0 \\ (a + 4)(3a - 2) &= 0 \end{aligned}$$
$a + 4 = 0$, $a = -4$ | $3a - 2 = 0$, $a = \frac{2}{3}$

$\left\{-4, \frac{2}{3}\right\}$

4. $\left\{2, \frac{7}{5}\right\}$

5. LCD $= 4x^2$
$$\frac{4x^2}{1} \cdot \frac{3}{x} + \frac{4x^2}{1} \cdot \frac{-8}{x^2} = \frac{4x^2}{1} \cdot \frac{1}{4}$$
$$\begin{aligned} 12x - 32 &= x^2 \\ 0 &= x^2 - 12x + 32 \\ 0 &= (x - 4)(x - 8) \end{aligned}$$
$x - 4 = 0$, $x = 4$ | $x - 8 = 0$, $x = 8$

$\{4, 8\}$

6. $\left\{-2, \frac{4}{5}\right\}$

7. LCD $= x(x + 1)(x - 1)$
$$\frac{x(x + 1)(x - 1)}{1} \cdot \frac{9}{(x + 1)} = \frac{x(x + 1)(x - 1)}{1} \cdot \frac{4}{x} + \frac{x(x + 1)(x - 1)}{1} \cdot \frac{1}{(x - 1)}$$
$$\begin{aligned} 9x(x - 1) &= 4(x + 1)(x - 1) + x(x + 1) \\ 9x^2 - 9x &= 4x^2 - 4 + x^2 + x \\ 4x^2 - 10x + 4 &= 0 \\ 2(2x^2 - 5x + 2) &= 0 \\ 2(x - 2)(2x - 1) &= 0 \end{aligned}$$
$x - 2 = 0$, $x = 2$ | $2x - 1 = 0$, $x = \frac{1}{2}$

$\left\{2, \frac{1}{2}\right\}$

8. $\left\{1, -\frac{1}{5}\right\}$

9. LCD $= x + 5$; the domain is the set of all real numbers except -5.
$$\frac{x + 5}{1} \cdot \frac{x}{x + 5} = \frac{x + 5}{1} \cdot \frac{8}{1} - \frac{x + 5}{1} \cdot \frac{5}{x + 5}$$
$$\begin{aligned} x &= 8(x + 5) - 5 \\ x &= 8x + 40 - 5 \\ x &= 8x + 35 \\ -7x &= 35 \\ x &= -5 \quad \text{Not in the domain} \end{aligned}$$
No solution; { }

10. $\{x \mid x \text{ is any real number except 0 or 3}\}$

11.
$$\begin{aligned} 5x - 10y &= 14 + 6x - 3y \\ -14 - x &= 7y \\ y &= \frac{-x - 14}{7} = -\frac{x + 14}{7} \end{aligned}$$

12. $x = \dfrac{y - 6}{7}$

13.
$$\begin{aligned} R(R_1 + R_2) &= R_1R_2 \\ RR_1 + RR_2 &= R_1R_2 \\ RR_2 &= R_1R_2 - RR_1 \\ RR_2 &= R_1(R_2 - R) \\ R_1 &= \frac{RR_2}{R_2 - R} \end{aligned}$$

14. $v = \dfrac{uF}{u - F}$

15. Let $3x$ = length of shortest side
Let $7x$ = length of next side
Let $8x$ = length of longest side
$$\begin{aligned} 0 < 3x + 7x + 8x &< 36 \\ 0 < 18x &< 36 \\ 0 < x &< 2 \\ 0 < 3x &< 6 \end{aligned}$$
The length of the shortest side is $3x$. The shortest side can be between 0 and 6 units long.

16. Karla walks at the rate of 6 ft per sec.

17. A's rate $= \dfrac{1 \text{ job}}{20 \text{ hr}} = \dfrac{1}{20}$ job per hr

B's rate $= \dfrac{1 \text{ job}}{15 \text{ hr}} = \dfrac{1}{15}$ job per hr

Let x = number of hr both tractors worked together

Since tractor A starts working 5 hr before tractor B starts, the time tractor A works is $(x + 5)$ hr.

$$\left(\begin{matrix}\text{Amount} \\ \text{A does}\end{matrix}\right) + \left(\begin{matrix}\text{Amount} \\ \text{B does}\end{matrix}\right) = 1 \text{ job done}$$
$$\tfrac{1}{20}(x + 5) + \tfrac{1}{15}(x) = 1; \quad \text{LCD} = 60$$
$$\frac{60}{1} \cdot \frac{x + 5}{20} + \frac{60}{1} \cdot \frac{x}{15} = 60 \cdot 1$$
$$\begin{aligned} 3x + 15 + 4x &= 60 \\ 7x &= 45 \\ x &= \tfrac{45}{7} = 6\tfrac{3}{7} \end{aligned}$$
It will take $6\frac{3}{7}$ hr to finish the job.

18. $5\frac{1}{7}$ min

19. Let x = numerator; Then $x + 10$ = denominator $\Big\}$ Original fraction $= \dfrac{x}{x + 10}$

If 1 is added to both the numerator and denominator	the value of the new fraction is	$\frac{2}{3}$

$$\frac{(x) + 1}{(x + 10) + 1} = \frac{2}{3}$$
$$\frac{x + 1}{x + 11} = \frac{2}{3}; \quad \text{LCD} = 3(x + 11)$$
$$\frac{3(x + 11)}{1} \cdot \frac{x + 1}{(x + 11)} = \frac{3(x + 11)}{1} \cdot \frac{2}{3}$$
$$\begin{aligned} 3(x + 1) &= (x + 11)(2) \\ 3x + 3 &= 2x + 22 \\ x &= 19 \\ x + 10 &= 29 \end{aligned}$$
Therefore, the original fraction is $\frac{19}{29}$.

20. $4\frac{8}{13}$ hr

Chapter 6 Diagnostic Test (page 317)

Following each problem number is the number (in parentheses) of the textbook section where that kind of problem is discussed. (No checks will be shown.)

1. (6.1) No; we can't get the second fraction from the first by multiplying or dividing both numerator and denominator by the same number.

2. (6.2) $\dfrac{f^2+5f+6}{f^2-9} = \dfrac{(f+2)\overset{1}{\cancel{(f+3)}}}{(f-3)\underset{1}{\cancel{(f+3)}}} = \dfrac{f+2}{f-3}$

3. (6.2) $\dfrac{x^4-2x^3+5x^2-10x}{x^3-8}$ ← Factor by grouping / ← Difference of two cubes

$= \dfrac{x^3(x-2)+5x(x-2)}{(x-2)(x^2+2x+4)}$

$= \dfrac{(x-2)(x^3+5x)}{(x-2)(x^2+2x+4)} = \dfrac{x(x^2+5)}{x^2+2x+4}$

4. (6.3) $\dfrac{\cancel{(x+4)}\cancel{(x-6)}}{\cancel{(x-6)}\cancel{(x+6)}} \cdot \dfrac{\cancel{(x+1)}\cancel{(x+6)}}{(x-3)\cancel{(x+4)}} \cdot \dfrac{1}{\cancel{(x+1)}(x^2-x+1)}$

$= \dfrac{1}{(x-3)(x^2-x+1)}$

5. (6.3) $\dfrac{3\cancel{(m+n)}}{(m-n)\cancel{(m^2+mn+n^2)}} \cdot \dfrac{\cancel{(m^2+mn+n^2)}}{\cancel{(m+n)}(m-n)} = \dfrac{3}{(m-n)^2}$

6. (6.5) $\dfrac{20a+27b}{12a-20b} + \dfrac{13b-44a}{12a-20b} = \dfrac{20a+27b+13b-44a}{12a-20b}$

$= \dfrac{-24a+40b}{12a-20b} = \dfrac{-\overset{2}{\cancel{8}}\overset{1}{\cancel{(3a-5b)}}}{\underset{1}{\cancel{4}}\underset{1}{\cancel{(3a-5b)}}} = -2$

7. (6.5) $\dfrac{x(x-4)}{(x+4)(x-4)} - \dfrac{x(x+4)}{(x-4)(x+4)} - \dfrac{32}{(x-4)(x+4)}$

$= \dfrac{x(x-4)-x(x+4)-32}{(x+4)(x-4)} = \dfrac{x^2-4x-x^2-4x-32}{(x+4)(x-4)}$

$= \dfrac{-8x-32}{(x+4)(x-4)} = \dfrac{-8\cancel{(x+4)}}{\cancel{(x+4)}(x-4)}$

$= \dfrac{-8}{x-4}$, or $-\dfrac{8}{x-4}$, or $\dfrac{8}{4-x}$

8. (6.5) $\dfrac{3(x+2)}{(x-2)(x+3)(x+2)} - \dfrac{2(x+3)}{(x-2)(x+2)(x+3)} - \dfrac{3(x-2)}{(x+2)(x+3)(x-2)}$

$= \dfrac{3(x+2)-2(x+3)-(3)(x-2)}{(x-2)(x+3)(x+2)}$

$= \dfrac{3x+6-2x-6-3x+6}{(x-2)(x+3)(x+2)} = \dfrac{-2x+6}{(x-2)(x+3)(x+2)}$

9. (6.3) Multiply both numerator and denominator of the first rational expression by -1:

$\dfrac{4(-1)}{-h(-1)} = \dfrac{-4}{h}$ The missing term is h.

10. (6.3) Change the signs of the denominator and the rational expression:

$-\dfrac{3}{k-2} = +\dfrac{3}{-(k-2)} = \dfrac{3}{2-k}$

The missing term is $2-k$.

11. (6.6) $\dfrac{6-\dfrac{4}{w}}{\dfrac{3w}{w-2}+\dfrac{1}{w}} \cdot \dfrac{w(w-2)}{w(w-2)}$

$= \dfrac{\dfrac{6}{1}\cdot\dfrac{w(w-2)}{1} - \dfrac{4}{\cancel{w}}\cdot\dfrac{\cancel{w}(w-2)}{1}}{\dfrac{3w}{\cancel{w-2}}\cdot\dfrac{w\cancel{(w-2)}}{1} + \dfrac{1}{\cancel{w}}\cdot\dfrac{\cancel{w}(w-2)}{1}}$

$= \dfrac{6w(w-2)-4(w-2)}{3w^2+w-2} = \dfrac{2(w-2)(3w-2)}{(w+1)(3w-2)}$

$= \dfrac{2(w-2)}{w+1}$

12. (6.7) The values that make the denominator zero must be excluded:

$x^2-4x=0$

$x(x-4)=0$

$x=0 \quad | \quad x-4=0$

$x=4$ Exclude 0 and 4

The domain is the set of all real numbers except 0 and 4.

13. (6.7) $3y^2-y-10=0$

$(3y+5)(y-2)=0$

$3y+5=0 \quad | \quad y-2=0$

$y=-\frac{5}{3} \quad | \quad y=2$ Exclude $-\frac{5}{3}$ and 2

The domain is the set of all real numbers except $-\frac{5}{3}$ and 2.

14. (6.7) The domain is the set of all real numbers except $-\frac{5}{3}$ and 2.

$\dfrac{\cancel{(3a+5)}(a-2)}{1}\cdot\dfrac{2}{\cancel{3a+5}} - \dfrac{(3a+5)\cancel{(a-2)}}{1}\cdot\dfrac{6}{\cancel{a-2}}$

$= \dfrac{(3a+5)(a-2)}{1}\cdot\dfrac{3}{1}$

$(a-2)(2)-(3a+5)(6) = (3a^2-a-10)(3)$

$2a-4-18a-30 = 9a^2-3a-30$

$0 = 9a^2+13a+4$

$0 = (9a+4)(a+1)$

$9a+4=0 \quad | \quad a+1=0$

$a=-\frac{4}{9} \quad | \quad a=-1$

$\{-\frac{4}{9}, -1\}$

15. (6.7) The domain is the set of all real numbers except -7.

$(x+7)\left(\dfrac{x}{x+7}\right) = (x+7)\left(3-\dfrac{7}{x+7}\right)$

$x = (x+7)(3)-(x+7)\left(\dfrac{7}{x+7}\right)$

$x = 3x+21-7$

$x = 3x+14$

$-14 = 2x$

$x = -7$

But -7 is not in the domain of the variable. Therefore, there is no solution. The solution set is { }.

16. (6.7) The domain is the set of all real numbers except 0.

$\dfrac{\overset{3}{\cancel{6}}z^2}{1}\cdot\dfrac{3}{\underset{1}{\cancel{2}}z} + \dfrac{\overset{1}{\cancel{6z^2}}}{1}\cdot\dfrac{3}{\underset{1}{\cancel{z^2}}} = \dfrac{\overset{1}{\cancel{6}}z^2}{1}\left(-\dfrac{1}{\underset{1}{\cancel{6}}}\right)$

$9z+18 = -z^2$

$z^2+9z+18=0$

$(z+3)(z+6)=0$

$z+3=0 \quad | \quad z+6=0$

$z=-3 \quad | \quad z=-6$

$\{-3, -6\}$

17. (6.8) $I = \dfrac{E}{R+r}$

$I(R+r) = E$

$IR+Ir = E$

$Ir = E-IR$

$r = \dfrac{E-IR}{I}$

18. (6.9) Let $5x$ = length (in ft)

Let $4x$ = width (in ft)

Perimeter is 90.

$2(5x)+2(4x) = 90$

$10x+8x = 90$

$18x = 90$

$x = 5$

$5x = 25$ Length

$4x = 20$ Width

The length is 25 ft, and the width is 20 ft.

19. (6.10) Let t = time to hike from lake (in hr)

$$\begin{aligned} \text{Then } t+2 &= \text{time to hike to lake (in hr)}\\ 3(t+2) &= \text{distance to lake}\\ 5t &= \text{distance from lake}\\ 5t &= 3(t+2) \quad \text{Distances are equal.}\\ 5t &= 3t+6\\ 2t &= 6\\ t &= 3 \qquad \text{Time to hike } \textit{from} \text{ the lake}\\ t+2 &= 5 \qquad \text{Time to hike } \textit{to} \text{ the lake} \end{aligned}$$

$$\text{Distance} = (\text{rate})(\text{time}) = \left(3\frac{\text{mi}}{\cancel{\text{hr}}}\right)(5\ \cancel{\text{hr}}) = 15\text{ mi}$$

a. It takes Roy 5 hr to hike to the lake.
b. The lake is 15 mi from Roy's camp.

20. (6.11) Let x = number of hr for Melissa to make 8 bushings

$$\text{Rachelle's rate} = \frac{24 \text{ bushings}}{8 \text{ hr}} = \frac{3}{1} \text{ bushings per hr}$$

$$\text{Melissa's rate} = \frac{8 \text{ bushings}}{x \text{ hr}} = \frac{8}{x} \text{ bushings per hr}$$

$$\begin{aligned} \frac{3}{1}(4) + \frac{8}{x}(4) &= 14\\ \frac{x}{1}\cdot\frac{12}{1} + \frac{x}{1}\cdot\frac{32}{x} &= \frac{x}{1}\cdot\frac{14}{1}\\ 12x + 32 &= 14x\\ 32 &= 2x\\ 16 &= x \end{aligned}$$

It takes Melissa 16 hr to make 8 bushings.

Chapters 1–6 Cumulative Review Exercises (page 317)

1. $(x^{4-(-2)}y^{-3})^{-2} = (x^6y^{-3})^{-2} = x^{6(-2)}y^{-3(-2)} = x^{-12}y^6 = \dfrac{y^6}{x^{12}}$

2. $x = -9$ **3.** $9 - 2m + 12 \le 12 - 3m + 5$
$21 - 2m \le 17 - 3m$
$m \le -4$
4. Yes

5. Yes **6.** $2x^3 - 13x^2 + 26x - 15$

7.
$$\begin{array}{r} 4a - 3 \quad \text{R } 10\\ 2a+3\overline{)8a^2 + 6a + 1}\\ \underline{8a^2 + 12a\phantom{{}+1}}\\ -6a + 1\\ \underline{-6a - 9}\\ +10 \end{array}$$

Answer: $4a - 3 + \dfrac{10}{2a+3}$

8. $x^6 - 12x^5 + 60x^4 - 160x^3 + 240x^2 - 192x + 64$

9. $$\frac{\cancel{x+3}}{(\cancel{2x-1})(\cancel{x+2})}\cdot\frac{(\cancel{2x-1})(\cancel{x-4})}{(3x+1)(\cancel{x+3})}\cdot\frac{\cancel{x+2}}{\cancel{x-4}} = \frac{1}{3x+1}$$

10. $\dfrac{3x^2 - 5x + 28}{(2x+1)(x+7)(x-3)}$

11. LCD $= (a+3)(a^2 - 3a + 9)(a - 3)$

$$\begin{aligned} &\frac{a-3}{(a+3)(a^2-3a+9)} - \frac{1}{(a+3)(a-3)}\\ &= \frac{(a-3)(a-3)}{(a+3)(a^2-3a+9)(a-3)} + \frac{-1(a^2-3a+9)}{(a+3)(a^2-3a+9)(a-3)}\\ &= \frac{a^2-6a+9-a^2+3a-9}{(a+3)(a^2-3a+9)(a-3)} = \frac{-3a}{(a+3)(a^2-3a+9)(a-3)} \end{aligned}$$

12. $(2x+3y)(5x-6y)$
13. $(x^2+3)(x^2-4) = (x^2+3)(x-2)(x+2)$
14. $(x-2)(x^2+2x+4)$
15. $3x(2a+b) - y(2a+b) = (2a+b)(3x-y)$

16. Length = 12 in.; width = 5 in.

17. Let x = number of 20¢ stamps

$$\begin{aligned} \text{Then } 50 - x &= \text{number of 18¢ stamps}\\ 20x + 18(50 - x) &= 970\\ 20x + 900 - 18x &= 970\\ 2x &= 70\\ x &= 35\\ 50 - x &= 15 \end{aligned}$$

(The check will not be shown.)
Mrs. Kishinami bought thirty-five 20¢ stamps and fifteen 18¢ stamps.

18. \$600 at 6.25%; \$800 at 5.75%

19. The boat will go (28 − 4) mph, or 24 mph, when going upstream and (28 + 4) mph, or 32 mph, when going downstream.

Let x = *time* it will take to go upstream (in hr)
Then $3\frac{1}{2} - x$ = *time* it will take to go downstream (in hr)

$$\left(24\frac{\text{mi}}{\cancel{\text{hr}}}\right)(x\ \cancel{\text{hr}}) = \text{distance the boater can go upstream}$$

$$\left(32\frac{\text{mi}}{\cancel{\text{hr}}}\right)\left\{\left(\frac{7}{2} - x\right)\cancel{\text{hr}}\right\} = \text{distance the boater will go downstream}$$

$$\begin{aligned} 24x &= 32\left(\tfrac{7}{2} - x\right) \quad \text{The distances are equal.}\\ 24x &= 112 - 32x\\ 56x &= 112\\ x &= 2 \qquad \text{Number of hr for boat to go upstream} \end{aligned}$$

$$\left(24\frac{\text{mi}}{\cancel{\text{hr}}}\right)(2\ \cancel{\text{hr}}) = 48\text{ mi} \qquad \text{Distance boater can go upstream}$$

(The check will not be shown.) The boater can go 48 mi upstream.

20. 7.5 L of 2% solution; 4.5 L of 10% solution

21. Let x = number of hr Carlos travels (before overtaking Alberto)
Then $x + 2$ = number of hr Alberto travels

$$\left(55\frac{\text{mi}}{\cancel{\text{hr}}}\right)(x\ \cancel{\text{hr}}) = \text{distance Carlos travels (before overtaking Alberto)}$$

$$\left(45\frac{\text{mi}}{\cancel{\text{hr}}}\right)[(x+2)\cancel{\text{hr}}] = \text{distance Alberto travels (before being overtaken by Carlos)}$$

$$\begin{aligned} 55x &= 45(x+2) \quad \text{The distances are equal.}\\ 55x &= 45x + 90\\ 10x &= 90\\ x &= 9 \end{aligned}$$

(The check will not be shown.) Carlos will overtake Alberto in 9 hr.

22. The room is 5 yd by 9 yd.

Exercises 7.1 (page 324)

1. $5^{1/2}$ **2.** $7^{1/2}$ **3.** $z^{1/3}$ **4.** $x^{1/4}$ **5.** $x^{3/4}$ **6.** $y^{2/5}$
7. $(x^{2/3})^2 = x^{4/3}$ **8.** $x^{3/2}$ **9.** $x^{2n/n} = x^2$ **10.** y^5
11. $\sqrt{7}$ **12.** $\sqrt[3]{5}$ **13.** $\sqrt[5]{a^3}$ **14.** $\sqrt[3]{b^2}$ **15.** $\sqrt[n]{x^m}$
16. $\sqrt[a]{x^{a+b}}$, or $x\sqrt[a]{x^b}$ **17.** $(2^3)^{1/3} = 2^1 = 2$ **18.** 3
19. $[(-3)^3]^{2/3} = (-3)^2 = 9$ **20.** 4
21. $4^{3/2} = (\sqrt{4})^3 = 2^3 = 8$ **22.** 27
23. $(\sqrt[4]{-16})^3$ The index is even and the radicand is negative; the number is not real.
24. Not real **25.** $-(4^{1/2}) = -\sqrt{4} = -2$ **26.** −11
27. $-(64^{5/6}) = -(\sqrt[6]{64})^5 = -2^5 = -32$ **28.** −27

Exercises 7.2 (page 326)

1. $x^{1/2+3/2} = x^{4/2} = x^2$ **2.** y^2 **3.** $a^{3/4+(-1/2)} = a^{3/4-2/4} = a^{1/4}$
4. $b^{1/2}$ **5.** $z^{-3/6}z^{4/6} = z^{1/6}$ **6.** $N^{5/12}$ **7.** $H^{(3/4)(2)} = H^{3/2}$
8. $s^{5/2}$ **9.** $x^{(-3/4)(1/3)} = x^{-1/4} = \dfrac{1}{x^{1/4}}$ **10.** $\dfrac{1}{y^{1/3}}$
11. $a^{3/4-1/2} = a^{3/4-2/4} = a^{1/4}$ **12.** $b^{1/2}$
13. $x^{1/2-(-1/3)} = x^{3/6+2/6} = x^{5/6}$ **14.** $z^{7/12}$
15. $x^{4/6}x^{6/6}x^{-3/6} = x^{4/6+6/6-3/6} = x^{7/6}$ **16.** $x^{17/12}$
17. $x^{-3/2}x^{4/3} = x^{-9/6+8/6} = x^{-1/6} = \dfrac{1}{x^{1/6}}$ **18.** $\dfrac{1}{x^{1/2}}$
19. $u^{1/2-(-1/4)}v^{-2/3-(-1)} = u^{2/4+1/4}v^{-2/3+1} = u^{3/4}v^{1/3}$
20. $uv^{2/5}$ **21.** $(2^4)^{3/2}x^{(-2/5)(3/2)}y^{(4/9)(3/2)} = 2^6x^{-3/5}y^{2/3} = \dfrac{64y^{2/3}}{x^{3/5}}$
22. $\dfrac{4x^{3/4}}{y^{1/2}}$ **23.** $\dfrac{a^{6(7/3)}d^{0(7/3)}}{b^{-9(7/3)}c^{3(7/3)}} = \dfrac{a^{14}d^0}{b^{-21}c^7} = \dfrac{a^{14}b^{21}}{c^7}$ **24.** $\dfrac{y^{14}w^{21}}{z^7}$
25. $\left(\dfrac{x^4}{2^5z^3}\right)^{2/5} = \dfrac{x^{8/5}}{2^2z^{6/5}} = \dfrac{x^{8/5}}{4z^{6/5}}$ **26.** $\dfrac{a^{10/3}}{4b^{10/3}}$
27. $\dfrac{x^{(-1)(-3/5)}y^{(2/3)(-3/5)}}{z^{(-5)(-3/5)}} = \dfrac{x^{3/5}y^{-2/5}}{z^3} = \dfrac{x^{3/5}}{y^{2/5}z^3}$ **28.** $\dfrac{b^{2/3}}{a^{1/2}c^{4/3}}$
29. $(3^2x^{-2/3-(-2)}y^{2/9})^{-3/2} = 3^{2(-3/2)}x^{(4/3)(-3/2)}y^{(2/9)(-3/2)}$
$= 3^{-3}x^{-2}y^{-1/3} = \dfrac{1}{27x^2y^{1/3}}$
30. $\dfrac{1}{9R^{2/5}S}$ **31.** $(2^3)^{-2/3} = 2^{3(-2/3)} = 2^{-2} = \dfrac{1}{2^2} = \dfrac{1}{4}$ **32.** $\frac{1}{9}$
33. $(2^2)^{1/2}(3^2)^{-3/2} = 2(3^{-3}) = \frac{2}{27}$ **34.** $\frac{1}{32}$
35. $(10^2)^{-1/2}(-2^{3(1/3)}) = 10^{-1}(-2) = -\frac{2}{10} = -\frac{1}{5}$ **36.** $-\frac{3}{100}$

Exercises 7.3 (page 332)

1. $2|x|$ **2.** $3|y|$ **3.** $2x$ **4.** $3y$
5. $\sqrt[4]{2^4x^4y^8} = 2\cdot|x|\cdot y^2 = 2|x|y^2$ **6.** $3u^2|v|$
7. $\sqrt{2^4\cdot 2} = 2^2\sqrt{2} = 4\sqrt{2}$ **8.** $3\sqrt{3}$
9. $\sqrt{(-2)^2} = |-2| = 2$ **10.** 3
11. $-\sqrt[3]{3^3\cdot 3^2} = -3\sqrt[3]{3^2} = -3\sqrt[3]{9}$ **12.** $-4\sqrt[3]{4}$
13. $\sqrt[4]{2^5} = \sqrt[4]{2^4\cdot 2} = 2\sqrt[4]{2}$ **14.** $2\sqrt[4]{3}$
15. $-\sqrt[5]{x^5x^2} = -x\sqrt[5]{x^2}$ **16.** $-z\sqrt[5]{z^3}$
17. $\sqrt{2^3a^4b^2} = \sqrt{2^2\cdot 2a^4b^2} = 2a^2|b|\sqrt{2}$ **18.** $2m^4|u|\sqrt{5}$
19. $3m\sqrt{2\cdot 3^2m^{12}n^6} = 3m(3m^6|n^3|\sqrt{2}) = 9m^7|n^3|\sqrt{2}$
Notice that both expressions are positive for $m > 0$ and negative for $m < 0$. ($9|m^7n^3|\sqrt{2}$ is *incorrect*.)
20. $10h^5|k^3|\sqrt{2}$
21. $5\sqrt[3]{-2^3\cdot 3a^5b^2} = -5\sqrt[3]{2^3\cdot 3a^3a^2b^2} = -5(2a)\sqrt[3]{3a^2b^2}$
$= -10a\sqrt[3]{3a^2b^2}$
22. $-18c\sqrt[3]{2cd}$ **23.** $\sqrt[5]{2^5\cdot 2m^{10}mp^{15}u} = 2m^2p^3\sqrt[5]{2mu}$
24. $2v^2w^3\sqrt[5]{4u^4w}$ **25.** $\sqrt[3]{2^3(a+b)^3} = 2(a+b)$
26. $3(x-y)^2$ **27.** $\sqrt{2^2(x-y)^2} = 2|x-y|$ **28.** $3|3z-w|$

Exercises 7.4 (page 337)

1. $\dfrac{\sqrt{16}}{\sqrt{25}} = \dfrac{4}{5}$ **2.** $\frac{4}{5}$ **3.** $\dfrac{\sqrt[3]{-27}}{\sqrt[3]{64}} = -\dfrac{3}{4}$ **4.** $-\frac{2}{5}$
5. $\dfrac{\sqrt[4]{a^4b^8}}{\sqrt[4]{16}} = \dfrac{|a|b^2}{2}$ **6.** $\dfrac{c^2|d^3|}{3}$ **7.** $\sqrt{\dfrac{4x^2}{y^2}} = \dfrac{\sqrt{4x^2}}{\sqrt{y^2}} = \left|\dfrac{2x}{y}\right|$
8. $\dfrac{x^2}{3|y|}$ **9.** $\dfrac{10}{\sqrt{5}}\cdot\dfrac{\sqrt{5}}{\sqrt{5}} = \dfrac{10\sqrt{5}}{\sqrt{5}\sqrt{5}} = \dfrac{10\sqrt{5}}{5} = 2\sqrt{5}$
10. $7\sqrt{2}$ **11.** $\dfrac{5}{\sqrt{3}}\cdot\dfrac{\sqrt{3}}{\sqrt{3}} = \dfrac{5\sqrt{3}}{\sqrt{3}\sqrt{3}} = \dfrac{5\sqrt{3}}{3}$ **12.** $\dfrac{8\sqrt{7}}{7}$
13. $\dfrac{9}{\sqrt[3]{3}}\cdot\dfrac{\sqrt[3]{3^2}}{\sqrt[3]{3^2}} = \dfrac{9\sqrt[3]{3^2}}{\sqrt[3]{3^3}} = \dfrac{9\sqrt[3]{9}}{3} = 3\sqrt[3]{9}$ **14.** $2\sqrt[3]{25}$
15. $\dfrac{8}{\sqrt[5]{2^2}}\cdot\dfrac{\sqrt[5]{2^3}}{\sqrt[5]{2^3}} = \dfrac{8\sqrt[5]{2^3}}{\sqrt[5]{2^5}} = \dfrac{\overset{4}{\cancel{8}}\sqrt[5]{8}}{\underset{1}{\cancel{2}}} = 4\sqrt[5]{8}$ **16.** $\dfrac{5\sqrt[3]{27}}{3}$
17. $-\sqrt[3]{\dfrac{m^5}{3}\cdot\dfrac{3^2}{3^2}} = -\sqrt[3]{\dfrac{m^3(9m^2)}{3^3}} = -\dfrac{m\sqrt[3]{9m^2}}{\sqrt[3]{3^3}} = -\dfrac{m\sqrt[3]{9m^2}}{3}$
18. $-\dfrac{k\sqrt[3]{25k}}{5}$ **19.** $\dfrac{n}{2m}\sqrt[3]{\dfrac{4m^2n}{n^3}} = \dfrac{n}{2m}\cdot\dfrac{1}{n}\sqrt[3]{4m^2n} = \dfrac{1}{2m}\sqrt[3]{4m^2n}$
20. $\dfrac{1}{3y}\sqrt[3]{9xy^2}$ **21.** $\sqrt[4]{\dfrac{3m^4}{2^2p^2}\cdot\dfrac{2^2p^2}{2^2p^2}} = \dfrac{\sqrt[4]{12m^4p^2}}{\sqrt[4]{2^4p^4}} = \dfrac{|m|\sqrt[4]{12p^2}}{|2p|}$
22. $\dfrac{|a|\sqrt[4]{40b^2}}{2|b|}$ **23.** $\sqrt[5]{\dfrac{5y^5}{2^3x^2}\cdot\dfrac{2^2x^3}{2^2x^3}} = \dfrac{\sqrt[5]{20x^3y^5}}{\sqrt[5]{2^5x^5}} = \dfrac{y\sqrt[5]{20x^3}}{2x}$
24. $\dfrac{p\sqrt[5]{20m^2}}{2m}$ **25.** $\dfrac{4x^2}{5y^3}\sqrt[3]{\dfrac{3y^2}{8x^3}} = \dfrac{4x^2}{5y^3}\cdot\dfrac{1}{2x}\sqrt[3]{3y^2} = \dfrac{2x}{5y^3}\sqrt[3]{3y^2}$
26. $\dfrac{1}{4y}\sqrt[3]{18xy}$
27. $\dfrac{2x^3}{y}\cdot\sqrt[3]{\dfrac{5}{2^4x^6xz^6z}} = \dfrac{\overset{1\ x}{\cancel{2}\cancel{x^3}}}{y}\cdot\dfrac{\sqrt[3]{5}}{\underset{1\ 1}{\cancel{2}\cancel{x^2}}z^2\sqrt[3]{2xz}}$
$= \dfrac{x\sqrt[3]{5}}{yz^2\sqrt[3]{2xz}}\cdot\dfrac{\sqrt[3]{2^2x^2z^2}}{\sqrt[3]{2^2x^2z^2}} = \dfrac{\overset{1}{\cancel{x}}\sqrt[3]{20x^2z^2}}{yz^2\cdot 2\underset{1}{\cancel{x}}z} = \dfrac{\sqrt[3]{20x^2z^2}}{2yz^3}$
28. $\dfrac{\sqrt[3]{5bc^2d^2}}{c}$

Exercises 7.5 (page 339)

1. $x^{3/6} = x^{1/2} = \sqrt{x}$ **2.** $\sqrt[3]{x}$ **3.** $a^{6/8} = a^{3/4} = \sqrt[4]{a^3}$
4. $\sqrt[3]{a}$ **5.** $(3^3b^3)^{1/6} = 3^{3/6}b^{3/6} = 3^{1/2}b^{1/2} = (3b)^{1/2} = \sqrt{3b}$
6. $\sqrt[3]{2b^2}$ **7.** $(7^2a^2)^{1/6} = 7^{2/6}a^{2/6} = 7^{1/3}a^{1/3} = (7a)^{1/3} = \sqrt[3]{7a}$
8. $2\sqrt{3x}$ **9.** $(3^4x^4z^{12})^{1/8} = 3^{1/2}x^{1/2}z^{3/2} = \sqrt{3xz^3} = z\sqrt{3xz}$
10. $x\sqrt{3xy}$ **11.** $(2^8x^8y^4z^{10})^{1/6} = 2^{8/6}x^{8/6}y^{4/6}z^{10/6} = 2^{4/3}x^{4/3}y^{2/3}z^{5/3}$
$= 2^{3/3}\cdot 2^{1/3}\cdot x^{3/3}\cdot x^{1/3}\cdot y^{2/3}\cdot z^{3/3}z^{2/3}$
$= 2xz(2xy^2z^2)^{1/3} = 2xz\sqrt[3]{2xy^2z^2}$
12. $2x^2z\sqrt{2yz}$ **13.** $\left(\dfrac{x^3}{3^3}\right)^{1/6} = \dfrac{x^{1/2}}{3^{1/2}} = \sqrt{\dfrac{x}{3}} = \sqrt{\dfrac{x(3)}{3(3)}} = \dfrac{\sqrt{3x}}{3}$
14. $\dfrac{\sqrt[3]{4x}}{2}$ **15.** $\dfrac{1}{a^{3/6}} = \dfrac{1}{a^{1/2}} = \dfrac{1}{\sqrt{a}} = \dfrac{1\sqrt{a}}{\sqrt{a}\sqrt{a}} = \dfrac{\sqrt{a}}{a}$ **16.** $\dfrac{\sqrt{x}}{x}$

Exercises 7.6 (page 342)

1. $11\sqrt{2}$ **2.** $18\sqrt{5}$ **3.** $4\sqrt{6}$ **4.** $6\sqrt{7}$
5. $\sqrt{15}+\sqrt{10}$ **6.** $\sqrt{2}+\sqrt{14}$ **7.** $2\sqrt{5}$ **8.** $3\sqrt{3}$
9. $3+2\sqrt{3}$ **10.** $5+3\sqrt{5}$ **11.** $2\sqrt{3}-2\sqrt{2}$
12. $2\sqrt{7}-\sqrt{3}$ **13.** $3\sqrt{2}+\sqrt{2} = 4\sqrt{2}$ **14.** $2\sqrt{3}$
15. $7\sqrt[3]{xy}$ **16.** $10\sqrt[4]{ab}$
17. $2\sqrt{25\cdot 2} - \sqrt{16\cdot 2} = 2\cdot 5\sqrt{2} - 4\sqrt{2}$
$= 10\sqrt{2} - 4\sqrt{2} = 6\sqrt{2}$
18. $3\sqrt{6}$
19. $3\sqrt{16\cdot 2x} - \sqrt{4\cdot 2x} = 12\sqrt{2x} - 2\sqrt{2x} = 10\sqrt{2x}$
20. $6\sqrt{3y}$
21. $\sqrt{25\cdot 5M} + \sqrt{4\cdot 5M} - \sqrt{9\cdot 5M}$
$= 5\sqrt{5M} + 2\sqrt{5M} - 3\sqrt{5M} = 4\sqrt{5M}$ **22.** $4\sqrt{3P}$
23. $3\sqrt[3]{x} + \frac{2}{2}\sqrt[3]{x} = 3\sqrt[3]{x} + \sqrt[3]{x} = 4\sqrt[3]{x}$ **24.** $6\sqrt[3]{a}$

25. $\sqrt[3]{a^3 \cdot a} + 2a\sqrt[3]{8 \cdot a} = a\sqrt[3]{a} + 4a\sqrt[3]{a} = 5a\sqrt[3]{a}$

26. $5H\sqrt[3]{H^2}$ **27.** $\sqrt[5]{x^2y^5y} + \sqrt[5]{x^5x^2y} = y\sqrt[5]{x^2y} + x\sqrt[5]{x^2y} = (y + x)\sqrt[5]{x^2y}$

28. $(b + a)\sqrt[5]{a^3b^3}$

29. $\frac{3}{1}\sqrt{\frac{1}{6} \cdot \frac{6}{6}} + \sqrt{4 \cdot 3} - \frac{5}{1}\sqrt{\frac{3}{2} \cdot \frac{2}{2}} = \frac{3}{6}\sqrt{6} + 2\sqrt{3} - \frac{5}{2}\sqrt{6} = \left(\frac{1}{2} - \frac{5}{2}\right)\sqrt{6} + 2\sqrt{3} = -2\sqrt{6} + 2\sqrt{3}$

30. $\sqrt{10} + 2\sqrt{5}$

31. $\frac{10}{2}\sqrt{5b} - \frac{3b}{2}\sqrt{\frac{4}{5b} \cdot \frac{5b}{5b}} = 5\sqrt{5b} - \frac{3b}{2} \cdot \frac{2}{5b}\sqrt{5b} = 5\sqrt{5b} - \frac{3}{5}\sqrt{5b} = \left(5 - \frac{3}{5}\right)\sqrt{5b} = \left(\frac{25}{5} - \frac{3}{5}\right)\sqrt{5b} = \frac{22}{5}\sqrt{5b}$

32. $\frac{7x}{2}\sqrt[3]{4}$

33. $\frac{2k}{1}\sqrt[4]{\frac{3}{2^3k} \cdot \frac{2k^3}{2k^3}} - \frac{1}{k}\sqrt[4]{\frac{2k^3}{3^3} \cdot \frac{3}{3}} + \frac{5k^2}{1}\sqrt[4]{\frac{6}{k^2} \cdot \frac{k^2}{k^2}}$
$= \frac{2k}{1} \cdot \frac{1}{2k}\sqrt[4]{6k^3} - \frac{1}{k} \cdot \frac{1}{3}\sqrt[4]{6k^3} + \frac{5k^2}{1} \cdot \frac{1}{k}\sqrt[4]{6k^2}$
$= \sqrt[4]{6k^3} - \frac{1}{3k}\sqrt[4]{6k^3} + 5k\sqrt[4]{6k^2} = \left(1 - \frac{1}{3k}\right)\sqrt[4]{6k^3} + 5k\sqrt[4]{6k^2}$
$= \left(\frac{3k}{3k} - \frac{1}{3k}\right)\sqrt[4]{6k^3} + 5k\sqrt[4]{6k^2} = \frac{3k - 1}{3k}\sqrt[4]{6k^3} + 5k\sqrt[4]{6k^2}$

34. $2a\sqrt[3]{4a}$ **35.** $\sqrt{(2x + 1)^2} = 2x + 1$ **36.** $4a + 1$

37. $\frac{-10 - \sqrt{52}}{6} = \frac{-10 - 2\sqrt{13}}{6} = \frac{2(-5 - \sqrt{13})}{6} = \frac{-5 - \sqrt{13}}{3}$

38. $\frac{13 + 5\sqrt{5}}{2}$ **39.** $\frac{5 - \sqrt{13}}{6}$ **40.** $\frac{-9 + \sqrt{17}}{16}$

41. $\frac{7 + \sqrt{57}}{4}$ **42.** $\frac{11 + \sqrt{149}}{14}$

Exercises 7.7 (page 346)

1. $\sqrt{3 \cdot 3} = 3$ **2.** 7 **3.** $\sqrt[3]{3 \cdot 9} = \sqrt[3]{3^3} = 3$ **4.** 4

5. $\sqrt[4]{9 \cdot 9} = \sqrt[4]{3^4} = 3$ **6.** 5

7. $\sqrt{100a^2b^3} = \sqrt{10^2a^2b^2b} = 10ab\sqrt{b}$ **8.** $9xy\sqrt{x}$

9. $3 \cdot 2\sqrt[5]{2a^3b \cdot 2^4a^2b} = 6\sqrt[5]{2^5a^5b^2} = 6 \cdot 2a\sqrt[5]{b^2} = 12a\sqrt[5]{b^2}$

10. $40b\sqrt[5]{c^3}$ **11.** $5\sqrt{7}(5\sqrt{7}) = 25\sqrt{7 \cdot 7} = 25 \cdot 7 = 175$

12. 96 **13.** $\sqrt{2}(\sqrt{2}) + \sqrt{2}(1) = 2 + \sqrt{2}$ **14.** $3 + \sqrt{3}$

15. $\sqrt{x}(\sqrt{x}) - \sqrt{x}(3) = x - 3\sqrt{x}$ **16.** $4\sqrt{y} - y$

17. $\sqrt{3}(2\sqrt{3}) + \sqrt{3}(1) = 2 \cdot 3 + \sqrt{3} = 6 + \sqrt{3}$

18. $15 + \sqrt{5}$

19. $\sqrt{3x}(\sqrt{3x}) - \sqrt{3x}(4\sqrt{12}) = \sqrt{3x \cdot 3x} - 4\sqrt{6^2x} = 3x - 4 \cdot 6\sqrt{x} = 3x - 24\sqrt{x}$

20. $5\sqrt{2a} + 15a$ **21.** $7 + 3\sqrt{7} + 2\sqrt{7} + 6 = 13 + 5\sqrt{7}$

22. $11 + 6\sqrt{3}$ **23.** $5^2 - (\sqrt{3})^2 = 25 - 3 = 22$ **24.** 2

25. $(2\sqrt{3})^2 + 2(2\sqrt{3})(-5) + (-5)^2 = 12 - 20\sqrt{3} + 25 = 37 - 20\sqrt{3}$

26. $59 - 30\sqrt{2}$

27. $2 \cdot 5 \cdot 2\sqrt{3 \cdot 7^2x^6xy^4y} = 20 \cdot 7x^3y^2\sqrt{3xy} = 140x^3y^2\sqrt{3xy}$

28. $300x^4y^4\sqrt{2xy}$ **29.** $3^2(\sqrt{2x + 5})^2 = 9(2x + 5) = 18x + 45$

30. $|48x - 32|$ **31.** $(\sqrt{2x})^2 - (3)^2 = 2x - 9$ **32.** $5x - 49$

33. $(\sqrt{xy})^2 - 2(\sqrt{xy})(6\sqrt{y}) + (6\sqrt{y})^2 = xy - 12y\sqrt{x} + 36y$

34. $ab + 4a\sqrt{b} + 4a$

We use the binomial theorem in Exercises 35–40.

35. $(\sqrt[5]{x})^5 + 5(\sqrt[5]{x})^4 + 10(\sqrt[5]{x})^3 + 10(\sqrt[5]{x})^2 + 5\sqrt[5]{x} + 1 = x + 5\sqrt[5]{x^4} + 10\sqrt[5]{x^3} + 10\sqrt[5]{x^2} + 5\sqrt[5]{x} + 1$

36. $1 + 3\sqrt[3]{y} + 3\sqrt[3]{y^2} + y$

37. $(x^{1/4})^4 + 4(x^{1/4})^3 + 6(x^{1/4})^2 + 4(x^{1/4})^1 + 1 = x + 4x^{3/4} + 6x^{1/2} + 4x^{1/4} + 1$

38. $x + 2x^{1/2}y^{1/2} + y$, or $x + 2\sqrt{xy} + y$

39. $[1 - 4\sqrt{2} + 6(\sqrt{2})^2 - 4(\sqrt{2})^3 + (\sqrt{2})^4] - 1 = 1 - 4\sqrt{2} + 6(2) - 4(2)\sqrt{2} + 4 - 1 = -4\sqrt{2} + 12 - 8\sqrt{2} + 4 = 16 - 12\sqrt{2}$

40. $8 + 5\sqrt{2}$

Exercises 7.8 (page 350)

1. $\sqrt{\frac{32}{2}} = \sqrt{16} = 4$ **2.** 7

3. $\sqrt[3]{\frac{5}{2^2} \cdot \frac{2}{2}} = \sqrt[3]{\frac{10}{2^3}} = \frac{\sqrt[3]{10}}{2}$, or $\frac{1}{2}\sqrt[3]{10}$ **4.** $\frac{\sqrt[3]{28}}{2}$, or $\frac{1}{2}\sqrt[3]{28}$

5. $\frac{3}{1}\sqrt[4]{\frac{15x}{5x}} = 3\sqrt[4]{3}$ **6.** $3\sqrt[4]{6}$

7. $\sqrt[5]{\frac{128z^7}{2z}} = \sqrt[5]{64z^6} = \sqrt[5]{2^6z^6} = 2z\sqrt[5]{2z}$ **8.** $3b\sqrt[5]{3b}$

9. $\sqrt{\frac{72x^3y^2}{2xy^2}} = \sqrt{36x^2} = 6x$ **10.** $3y$

11. $\frac{\sqrt{20}}{\sqrt{5}} + \frac{5\sqrt{10}}{\sqrt{5}} = \sqrt{\frac{20}{5}} + 5\sqrt{\frac{10}{5}} = \sqrt{4} + 5\sqrt{2} = 2 + 5\sqrt{2}$

12. $2 + \sqrt{3}$ **13.** $\frac{\sqrt{75}}{\sqrt{3}} - \frac{\sqrt{6}}{\sqrt{3}} = \sqrt{\frac{75}{3}} - \sqrt{\frac{6}{3}} = \sqrt{25} - \sqrt{2} = 5 - \sqrt{2}$

14. $\sqrt{14} + \sqrt{3}$ **15.** $\frac{2}{\sqrt{7}} - \frac{\sqrt{7}}{\sqrt{7}} = \frac{2}{\sqrt{7}} \cdot \frac{\sqrt{7}}{\sqrt{7}} - 1 = \frac{2\sqrt{7}}{7} - 1$, or $\frac{2\sqrt{7} - 7}{7}$

16. $1 - \frac{4\sqrt{13}}{13}$, or $\frac{13 - 4\sqrt{13}}{13}$

17. $\frac{8\sqrt[4]{16x}}{2\sqrt[4]{2x}} + \frac{6\sqrt[4]{8x^4}}{2\sqrt[4]{2x}} = 4\sqrt[4]{\frac{16x}{2x}} + 3\sqrt[4]{\frac{8x^4}{2x}} = 4\sqrt[4]{8} + 3\sqrt[4]{4x^3}$

18. $2\sqrt[5]{27x} + 3\sqrt[5]{9x^4}$

19. $3\sqrt[4]{\frac{2^5m^2}{2m^3}} = 3\sqrt[4]{\frac{2^4}{m} \cdot \frac{m^3}{m^3}} = 3\sqrt[4]{\frac{2^4m^3}{m^4}} = 3\left(\frac{2}{m}\right)\sqrt[4]{m^3} = \frac{6}{m}\sqrt[4]{m^3}$

20. $\frac{3\sqrt[4]{H^3}}{2H}$

21. $\frac{4\sqrt[3]{8x}}{2\sqrt[3]{4x}} + \frac{6\sqrt[3]{32x^4}}{2\sqrt[3]{4x}} = 2\sqrt[3]{\frac{8x}{4x}} + 3\sqrt[3]{\frac{32x^4}{4x}} = 2\sqrt[3]{2} + 3\sqrt[3]{8x^3} = 2\sqrt[3]{2} + 3 \cdot 2x = 2\sqrt[3]{2} + 6x$

22. $6a^2 + 3\sqrt[3]{2}$

23. $\left(\frac{6}{\sqrt{3} - 1}\right)\left(\frac{\sqrt{3} + 1}{\sqrt{3} + 1}\right) = \frac{6(\sqrt{3} + 1)}{(\sqrt{3})^2 - 1^2} = \frac{6(\sqrt{3} + 1)}{3 - 1} = \frac{6(\sqrt{3} + 1)}{2} = 3(\sqrt{3} + 1) = 3\sqrt{3} + 3$

24. $5\sqrt{3} - 5$

25. $\left(\frac{\sqrt{2}}{\sqrt{3} + \sqrt{2}}\right)\left(\frac{\sqrt{3} - \sqrt{2}}{\sqrt{3} - \sqrt{2}}\right) = \frac{\sqrt{2}(\sqrt{3} - \sqrt{2})}{(\sqrt{3})^2 - (\sqrt{2})^2} = \frac{\sqrt{6} - 2}{3 - 2} = \sqrt{6} - 2$

26. $\frac{7 + \sqrt{14}}{5}$

27. $\left(\frac{\sqrt{7} + \sqrt{3}}{\sqrt{7} - \sqrt{3}}\right)\left(\frac{\sqrt{7} + \sqrt{3}}{\sqrt{7} + \sqrt{3}}\right) = \frac{(\sqrt{7})^2 + 2\sqrt{7}\sqrt{3} + (\sqrt{3})^2}{(\sqrt{7})^2 - (\sqrt{3})^2} = \frac{7 + 2\sqrt{21} + 3}{7 - 3} = \frac{10 + 2\sqrt{21}}{4} = \frac{\overset{1}{\cancel{2}}(5 + \sqrt{21})}{\underset{2}{\cancel{4}}} = \frac{5 + \sqrt{21}}{2}$

28. $\dfrac{8-\sqrt{55}}{3}$

29. $\left(\dfrac{4\sqrt{3}-\sqrt{2}}{4\sqrt{3}+\sqrt{2}}\right)\left(\dfrac{4\sqrt{3}-\sqrt{2}}{4\sqrt{3}-\sqrt{2}}\right)=\dfrac{(4\sqrt{3}-\sqrt{2})^2}{(4\sqrt{3})^2-(\sqrt{2})^2}$

$=\dfrac{48-8\sqrt{6}+2}{48-2}=\dfrac{50-8\sqrt{6}}{46}=\dfrac{2(25-4\sqrt{6})}{46}=\dfrac{25-4\sqrt{6}}{23}$

30. $2x-2\sqrt{x^2+x}+1$

31. $\sqrt{\dfrac{\cancel{(a+3)}(a-1)}{\cancel{(a+3)}(a+1)}}=\sqrt{\dfrac{a-1}{a+1}\cdot\dfrac{a+1}{a+1}}=\dfrac{\sqrt{a^2-1}}{\sqrt{(a+1)^2}}$

$=\dfrac{\sqrt{a^2-1}}{a+1}$

32. $\dfrac{\sqrt{m^2-1}}{m-1}$

Exercises 7.9 (page 352)

1. $a^{1/2}a^{1/4}=a^{1/2+1/4}=a^{3/4}=\sqrt[4]{a^3}$ 2. $\sqrt[6]{b^5}$

3. $(2^3)^{1/2}(2^4)^{1/3}=2^{3/2}2^{4/3}=2^{3/2+4/3}=2^{17/6}$
$=2^{2+5/6}=2^2\cdot 2^{5/6}=4\sqrt[6]{32}$ 4. $9\sqrt[6]{243}$

5. $x^{2/3}x^{3/4}x^{1/2}=x^{8/12+9/12+6/12}=x^{23/12}$
$=x^{1+11/12}=x^1x^{11/12}=x\sqrt[12]{x^{11}}$ 6. $y\sqrt[12]{y^7}$

7. $(-2^3z^2)^{1/3}(-z)^{1/3}(2^4z^3)^{1/4}=2^1z^{2/3}z^{1/3}(2^1z^{3/4})$
$=2^2z^{1+3/4}=4z\sqrt[4]{z^3}$ 8. $6w\sqrt[4]{w^3}$

9. $\dfrac{G^{3/4}}{G^{2/3}}=G^{3/4-2/3}=G^{9/12-8/12}=G^{1/12}=\sqrt[12]{G}$ 10. $\sqrt[10]{H^3}$

11. $\dfrac{-x^{2/3}}{x^{5/6}}=-x^{2/3-5/6}=-x^{4/6-5/6}=-x^{-1/6}$ 12. $-\dfrac{\sqrt[6]{y^5}}{y}$

$=-\dfrac{1}{x^{1/6}}=-\dfrac{1}{x^{1/6}}\cdot\dfrac{x^{5/6}}{x^{5/6}}=-\dfrac{x^{5/6}}{x^{6/6}}=-\dfrac{\sqrt[6]{x^5}}{x}$

13. $x^{1/4}y^{1/2}=x^{1/4}y^{2/4}=\sqrt[4]{x}\sqrt[4]{y^2}=\sqrt[4]{xy^2}$ 14. $\sqrt[8]{st^2}$

15. $h^{1/6}x^{3/4}=h^{2/12}x^{9/12}=\sqrt[12]{h^2}\sqrt[12]{x^9}=\sqrt[12]{h^2x^9}$ 16. $\sqrt[10]{w^4t^5}$

Sections 7.1–7.9 Review Exercises (page 354)

1. $\sqrt[4]{a^3}$ 2. $\sqrt[4]{3^3y^3}$, or $\sqrt[4]{27y^3}$ 3. $\sqrt[5]{2^2x^4}$, or $\sqrt[5]{4x^4}$

4. $b^{3/4}$ 5. $(8x^4)^{1/5}$, or $8^{1/5}x^{4/5}$ 6. $(27x^3)^{1/5}$, or $27^{1/5}x^{3/5}$

7. $(\sqrt[3]{-64})^2=(-4)^2=16$ 8. 9

9. $P^{(2/3)(2/3)}R^{(3/4)(2/3)}=P^{4/9}R^{1/2}$ 10. $\dfrac{1}{27a^{1/2}b^{5/2}}$

11. $\sqrt[3]{2^3\cdot 4}=2\sqrt[3]{4}$ 12. $-5x$ 13. $(2^4y^8)^{1/4}=2^{4/4}y^{8/4}=2y^2$

14. $m\sqrt[5]{mp}$ 15. $\sqrt{\dfrac{4x^2x}{y}\cdot\dfrac{y}{y}}=\dfrac{2|x|\sqrt{xy}}{y}$ 16. $\dfrac{5\sqrt{3y}}{y}$

17. $\sqrt[3]{8x^3x^2}=2x\sqrt[3]{x^2}$ 18. $75x$

19. $\left(\dfrac{\sqrt{3}+\sqrt{7}}{\sqrt{3}-\sqrt{7}}\right)\left(\dfrac{\sqrt{3}+\sqrt{7}}{\sqrt{3}+\sqrt{7}}\right)=\dfrac{3+\sqrt{21}+\sqrt{21}+7}{(\sqrt{3})^2-(\sqrt{7})^2}$

$=\dfrac{10+2\sqrt{21}}{-4}=-\dfrac{5+\sqrt{21}}{2}$

20. $2\sqrt{5}-2$

21. $\sqrt{2}(4\sqrt{2})+11\sqrt{2}-3=8+11\sqrt{2}-3=5+11\sqrt{2}$

22. $16+2\sqrt{39}$

23. $\sqrt[3]{8\cdot 2\cdot x^3x^2}+x\sqrt[3]{27\cdot 2x^2}=2x\sqrt[3]{2x^2}+x\cdot 3\sqrt[3]{2x^2}=5x\sqrt[3]{2x^2}$

24. 9 25. $-z^{2/3}\cdot z^{1/2}=-z^{4/6+3/6}=-z^{7/6}=-z^{1+1/6}=-z\sqrt[6]{z}$

26. $2-4x\sqrt[4]{8}+6x^2\sqrt{2}-4x^3\sqrt[4]{2}+x^4$

27. $\dfrac{G^{2/4}}{G^{1/5}}=G^{1/2-1/5}=G^{5/10-2/10}=G^{3/10}=\sqrt[10]{G^3}$

28. $\dfrac{\sqrt[4]{a^2b}}{a}$ 29. $t^{1/2}f^{2/3}=t^{3/6}f^{4/6}=\sqrt[6]{t^3}\sqrt[6]{f^4}=\sqrt[6]{t^3f^4}$

30. $3+5(2^{4/5})+10(2^{3/5})+10(2^{2/5})+5(2^{1/5})$,
or $3+5\sqrt[5]{16}+10\sqrt[5]{8}+10\sqrt[5]{4}+5\sqrt[5]{2}$

Exercises 7.10 (page 362)

We will not show finding the domains of the variables.

1. $$\begin{aligned}\sqrt{3x+1}&=5\\(\sqrt{3x+1})^2&=5^2\\3x+1&=25\\3x&=24\\x&=8\end{aligned}\qquad \textit{Check}\quad \begin{aligned}\sqrt{3x+1}&=5\\\sqrt{3(8)+1}&\stackrel{?}{=}5\\\sqrt{25}&\stackrel{?}{=}5\\5&=5\end{aligned}$$
$\{8\}$

2. $\{4\}$

3. $$\begin{aligned}\sqrt{x+1}&=\sqrt{2x-7}\\(\sqrt{x+1})^2&=(\sqrt{2x-7})^2\\x+1&=2x-7\\8&=x\\x&=8\end{aligned}\qquad \textit{Check}\quad \begin{aligned}\sqrt{x+1}&=\sqrt{2x-7}\\\sqrt{8+1}&\stackrel{?}{=}\sqrt{16-7}\\\sqrt{9}&=\sqrt{9}\end{aligned}$$
$\{8\}$

4. $\{3\}$

5. $$\begin{aligned}\sqrt[4]{4x-11}&=1\\(\sqrt[4]{4x-11})^4&=(1)^4\\4x-11&=1\\4x&=12\\x&=3\end{aligned}\qquad \textit{Check}\quad \begin{aligned}\sqrt[4]{4x-11}-1&=0\\\sqrt[4]{4(3)-11}-1&\stackrel{?}{=}0\\\sqrt[4]{1}-1&\stackrel{?}{=}0\\1-1&\stackrel{?}{=}0\\0&=0\end{aligned}$$
$\{3\}$

6. $\{5\}$

7. $$\begin{aligned}\sqrt{4x-1}&=2x\\(\sqrt{4x-1})^2&=(2x)^2\\4x-1&=4x^2\\0&=4x^2-4x+1\\0&=(2x-1)(2x-1)\\2x-1&=0\\x&=\tfrac{1}{2}\end{aligned}\qquad \textit{Check}\quad \begin{aligned}\sqrt{4x-1}&=2x\\\sqrt{4\left(\tfrac{1}{2}\right)-1}&\stackrel{?}{=}2\left(\tfrac{1}{2}\right)\\\sqrt{2-1}&\stackrel{?}{=}1\\1&=1\end{aligned}$$
$\left\{\frac{1}{2}\right\}$

8. $\left\{\frac{1}{3}\right\}$

9. $$\begin{aligned}\sqrt{x+7}&=2x-1\\(\sqrt{x+7})^2&=(2x-1)^2\\x+7&=4x^2-4x+1\\0&=4x^2-5x-6\\0&=(x-2)(4x+3)\end{aligned}$$
$x-2=0$ | $4x+3=0$
$x=2$ | $x=-\frac{3}{4}$

Check for $x=2$: $$\begin{aligned}\sqrt{x+7}&=2x-1\\\sqrt{2+7}&\stackrel{?}{=}4-1\\3&=3\end{aligned}$$

Check for $x=-\frac{3}{4}$: $$\begin{aligned}\sqrt{x+7}&=2x-1\\\sqrt{-\tfrac{3}{4}+7}&\stackrel{?}{=}2\left(-\tfrac{3}{4}\right)-1\\\sqrt{\tfrac{25}{4}}&\stackrel{?}{=}-\tfrac{3}{2}-1\\\tfrac{5}{2}&\neq-\tfrac{5}{2}\end{aligned}$$

Therefore, the solution set is $\{2\}$.

10. $\{4\}$

11. $$\begin{aligned}(\sqrt{8x}+3)^2&=(\sqrt{8x+33})^2\\8x+6\sqrt{8x}+9&=8x+33\\6\sqrt{8x}&=24\\(\sqrt{8x})^2&=4^2\\8x&=16\\x&=2\end{aligned}\qquad \textit{Check}\quad \begin{aligned}\sqrt{8\cdot 2}+3&\stackrel{?}{=}\sqrt{8\cdot 2+33}\\\sqrt{16}+3&\stackrel{?}{=}\sqrt{16+33}\\4+3&\stackrel{?}{=}\sqrt{49}\\7&=7\end{aligned}$$
$\{2\}$

12. $\{\ \}$

13. $$\begin{aligned}(\sqrt{3x+1})^2&=(\sqrt{x+4}+1)^2\\3x+1&=x+4+2\sqrt{x+4}+1\\2x-4&=2\sqrt{x+4}\\(x-2)^2&=(\sqrt{x+4})^2\\x^2-4x+4&=x+4\\x^2-5x&=0\\x(x-5)&=0\end{aligned}$$
$x=0$ | $x-5=0$
 | $x=5$

Check for $x=0$: $$\begin{aligned}\sqrt{3\cdot 0+1}-\sqrt{0+4}&\stackrel{?}{=}1\\\sqrt{1}-\sqrt{4}&\stackrel{?}{=}1\\1-2&\stackrel{?}{=}1\\-1&\neq 1\end{aligned}$$

Check for $x=5$: $$\begin{aligned}\sqrt{3\cdot 5+1}-\sqrt{5+4}&\stackrel{?}{=}1\\\sqrt{15+1}-\sqrt{9}&\stackrel{?}{=}1\\\sqrt{16}-3&\stackrel{?}{=}1\\4-3&\stackrel{?}{=}1\\1&=1\end{aligned}$$
$\{5\}$

14. $\{0, 4\}$

15.
$$\begin{aligned}
\sqrt{3x+4} &= \sqrt{2x-4} + 2 \\
(\sqrt{3x+4})^2 &= (\sqrt{2x-4} + 2)^2 \\
3x + 4 &= 2x - 4 + 4\sqrt{2x-4} + 4 \\
x + 4 &= 4\sqrt{2x-4} \\
(x+4)^2 &= (4\sqrt{2x-4})^2 \\
x^2 + 8x + 16 &= 32x - 64 \\
x^2 - 24x + 80 &= 0 \\
(x-4)(x-20) &= 0
\end{aligned}$$

$x - 4 = 0 \quad | \quad x - 20 = 0$

$x = 4 \quad | \quad x = 20$

Check for $x = 4$:
$$\sqrt{3(4)+4} - \sqrt{2(4)-4} \stackrel{?}{=} 2, \quad \sqrt{16} - \sqrt{4} \stackrel{?}{=} 2, \quad 4 - 2 = 2$$

Check for $x = 20$:
$$\sqrt{3(20)+4} - \sqrt{2(20)-4} \stackrel{?}{=} 2, \quad \sqrt{64} - \sqrt{36} \stackrel{?}{=} 2, \quad 8 - 6 \stackrel{?}{=} 2, \quad 2 = 2$$

$\{4, 20\}$

16. $\{5, 13\}$

17.
$$\begin{aligned}
\sqrt[3]{2x+3} - 2 &= 0 \\
(\sqrt[3]{2x+3})^3 &= (2)^3 \\
2x + 3 &= 8 \\
2x &= 5 \\
x &= \tfrac{5}{2}
\end{aligned}$$

Check
$$\sqrt[3]{2\left(\tfrac{5}{2}\right) + 3} - 2 \stackrel{?}{=} 0, \quad \sqrt[3]{5+3} - 2 \stackrel{?}{=} 0, \quad \sqrt[3]{8} - 2 \stackrel{?}{=} 0, \quad 2 - 2 \stackrel{?}{=} 0, \quad 0 = 0$$

$\left\{\tfrac{5}{2}\right\}$

18. $\left\{\tfrac{15}{2}\right\}$

19.
$$\begin{aligned}
\sqrt{4u+1} - \sqrt{u-2} &= \sqrt{u+3} \\
(\sqrt{4u+1} - \sqrt{u-2})^2 &= (\sqrt{u+3})^2 \\
4u + 1 - 2\sqrt{4u+1}\sqrt{u-2} + u - 2 &= u + 3 \\
4u - 4 &= 2\sqrt{(4u+1)(u-2)} \\
(2u-2)^2 &= (\sqrt{(4u+1)(u-2)})^2 \\
4u^2 - 8u + 4 &= 4u^2 - 7u - 2 \\
-u &= -6 \\
u &= 6
\end{aligned}$$

Check
$$\sqrt{4(6)+1} - \sqrt{(6)-2} \stackrel{?}{=} \sqrt{(6)+3}, \quad \sqrt{25} - \sqrt{4} \stackrel{?}{=} \sqrt{9}, \quad 5 - 2 \stackrel{?}{=} 3, \quad 3 = 3$$

$\{6\}$

20. $\{1\}$

21. $x^{1/2} = 5$, $(x^{1/2})^2 = 5^2$, $x = 25$

Check $(25)^{1/2} \stackrel{?}{=} 5$, $(5^2)^{1/2} \stackrel{?}{=} 5$, $5 = 5$

$\{25\}$

22. $\{27\}$

23.
$$\begin{aligned}
2x^{-5/3} &= 64 \\
x^{-5/3} &= 32 \\
(x^{-5/3})^{-3/5} &= (32)^{-3/5} \\
x = (2^5)^{-3/5} &= 2^{-3} = \frac{1}{2^3}, \text{ or } \frac{1}{8}
\end{aligned}$$

$\left\{\tfrac{1}{8}\right\}$ (The check will not be shown.)

24. $\left\{\tfrac{1}{4}\right\}$

25. $R = 16, I = 5$
$$5 = \sqrt{\frac{P}{16}}, \quad 25 = \frac{P}{16}, \quad P = 400$$

26. 576

27. $n = 50$, $\sigma = \frac{10}{3}$, $q = \frac{2}{3}$
$$\tfrac{10}{3} = \sqrt{50\left(\tfrac{2}{3}\right)p}, \quad \tfrac{10}{3} = \sqrt{\tfrac{100}{3}p}, \quad \tfrac{100}{9} = \tfrac{100}{3}p, \quad p = \tfrac{1}{3}$$

28. $\frac{2}{5}$

29. $T = 1.32$, $g \approx 9.8$, $\pi \approx 3.14$
$$\begin{aligned}
1.32 &\approx 2(3.14)\sqrt{\frac{L}{9.8}} \\
1.32 &\approx 6.28\sqrt{\frac{L}{9.8}} \\
(0.210\,191\,1)^2 &\approx \left(\sqrt{\frac{L}{9.8}}\right)^2 \\
0.044\,180\,3 &\approx \frac{L}{9.8} \\
L &\approx 0.432\,966\,9 \approx 0.4
\end{aligned}$$

The length of the pendulum is about 0.4 m.

30. The length of the pendulum is about 0.3 m.

31.
$$\begin{aligned}
21.50 &= 30 - \sqrt{0.01x + 2} \\
-8.5 &= -\sqrt{0.01x + 2} \\
(8.5)^2 &= (\sqrt{0.01x + 2})^2 \\
72.25 &= 0.01x + 2 \\
70.25 &= 0.01x \\
x &= 7{,}025
\end{aligned}$$

7,025 units per day are demanded.

32. 2,825 units per day are demanded.

Exercises 7.11 (page 367)

1.
$$\begin{aligned}
x^2 &= (\sqrt{6})^2 + (\sqrt{3})^2 \\
x^2 &= 6 + 3 \\
x^2 &= 9 \\
x &= \pm\sqrt{9} = \pm 3 \\
x &= 3 \quad (-3 \text{ is not in the domain.})
\end{aligned}$$

2. 4

3.
$$\begin{aligned}
10^2 + 6^2 &= x^2 \\
100 + 36 &= x^2 \\
136 &= x^2 \\
x &= \pm\sqrt{136} = \pm\sqrt{4 \cdot 34} = \pm 2\sqrt{34} \\
x &= 2\sqrt{34} \quad (-2\sqrt{34} \text{ is not in the domain.})
\end{aligned}$$

4. $4\sqrt{13}$

5.
$$\begin{aligned}
(x+1)^2 + (\sqrt{20})^2 &= (x+3)^2 \\
x^2 + 2x + 1 + 20 &= x^2 + 6x + 9 \\
12 &= 4x \\
x &= 3
\end{aligned}$$

6. 7

The checks will not be shown for Exercises 7–16.

7. Let x = length of diagonal (in in.)
$$\begin{aligned}
x^2 &= 4^2 + 4^2 \\
x^2 &= 16 + 16 \\
x^2 &= 32 \\
x &= \pm\sqrt{32} = \pm\sqrt{16 \cdot 2} = \pm 4\sqrt{2} \\
x &= 4\sqrt{2} \quad (-4\sqrt{2} \text{ is not in the domain.})
\end{aligned}$$

The diagonal is $4\sqrt{2}$ in. long.

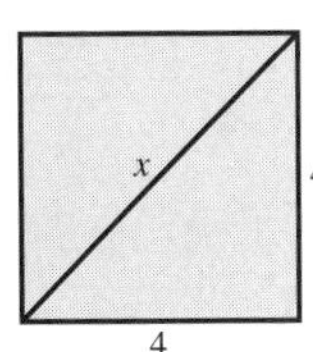

8. $3\sqrt{2}$ m

9. Let w = width (in cm)
$$\begin{aligned}
(24)^2 + w^2 &= (25)^2 \\
576 + w^2 &= 625 \\
w^2 &= 49 \\
w &= \pm\sqrt{49} = \pm 7 \\
w &= 7 \quad (-7 \text{ is not in the domain.})
\end{aligned}$$

The width is 7 cm.

10. 9 ft

11. Let x = length of one leg (in cm)

Then $2x - 4$ = length of other leg (in cm)
$$\begin{aligned}
(10)^2 &= (2x-4)^2 + x^2 \\
100 &= 4x^2 - 16x + 16 + x^2 \\
0 &= 5x^2 - 16x - 84 \\
0 &= (5x+14)(x-6)
\end{aligned}$$

$5x + 14 = 0 \quad | \quad x - 6 = 0$

$x = -\frac{14}{5} \quad | \quad x = 6$

Not in the domain $\quad | \quad 2x - 4 = 12 - 4 = 8$

One leg is 6 cm long, and the other is 8 cm long.

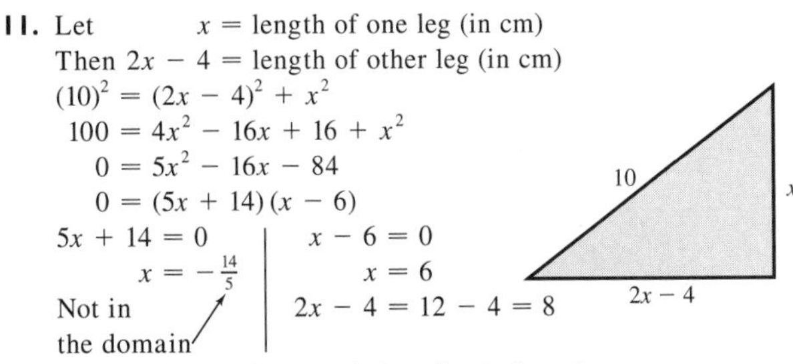

12. $\frac{9}{5}$ m, $\frac{38}{5}$ m

13. Let x = Hy's rate in mph
Then $x + 7$ = Jaime's rate in mph
$\left(x \dfrac{\text{mi}}{\cancel{\text{hr}}}\right)(1\ \cancel{\text{hr}})$, or x mi, is Hy's distance.
$\left((x+7)\dfrac{\text{mi}}{\cancel{\text{hr}}}\right)(1\ \cancel{\text{hr}})$, or $(x + 7)$ mi, is Jaime's distance.

$$x^2 + (x+7)^2 = 13^2$$
$$x^2 + x^2 + 14x + 49 = 169$$
$$2x^2 + 14x - 120 = 0$$
$$x^2 + 7x - 60 = 0$$
$$(x-5)(x+12) = 0$$
$x - 5 = 0 \mid x + 12 = 0$
$x = 5 \mid x = -12$ Reject
Hy's rate is 5 mph.

14. 4 mph

15. Let x = length of side of square (in ft)
$$x^2 + x^2 = (3.87)^2$$
$$2x^2 = 14.9769$$
$$x^2 = 7.48845$$
$$x \approx 2.7365032 \approx 2.74$$
The length of each side of the square is about 2.74 ft.

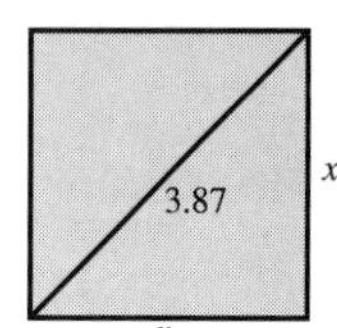

16. The length of each side of the square is about 3.00 m.

7

Exercises 7.12 (page 371)

1. $3 + \sqrt{16}\sqrt{-1} = 3 + 4i$ **2.** $4 - 5i$
3. $0 + \sqrt{64}\sqrt{-1} = 0 + 8i$ **4.** $0 + 10i$
5. $5 + \sqrt{16}\sqrt{-1}\sqrt{2} = 5 + 4i\sqrt{2}$ **6.** $6 + 3i\sqrt{2}$
7. $\sqrt{36}\sqrt{-1} + 2 = 6i + 2 = 2 + 6i$ **8.** $3 - 5i$
9. $2i - 3 = -3 + 2i$ **10.** $-4 + 3i$ **11.** $14 + 0i$
12. $-7 + 0i$ **13.** $3 - 4i = x + 2yi$
$3 = x \mid -4 = 2y$
$x = 3 \mid y = -2$
14. $x = 2, y = 5$

15. $5x - 3i = 6 - 7yi$
$5x = 6 \mid -3 = -7y$
$x = \frac{6}{5} \mid y = \dfrac{-3}{-7} = \dfrac{3}{7}$
16. $x = -\frac{3}{2}, y = -3$

17. $x\sqrt{3} - yi = 2 + i\sqrt{2}$
$x\sqrt{3} = 2 \mid -y = \sqrt{2}$
$x = \dfrac{2}{\sqrt{3}} = \dfrac{2\sqrt{3}}{3} \mid y = -\sqrt{2}$

18. $x = \dfrac{3\sqrt{2}}{4}, y = -\dfrac{\sqrt{5}}{5}$

19. $\frac{3}{4}x - \frac{1}{3}yi = \frac{3}{5}x + \frac{1}{2}yi$
$\frac{3}{4}x = \frac{3}{5}x \mid -\frac{1}{3}y = \frac{1}{2}y$
$15x = 12x \mid -2y = 3y$
$3x = 0 \mid -5y = 0$
$x = 0 \mid y = 0$
20. $x = 0, y = 0$

Exercises 7.13 (page 372)

1. $4 + 3i + 5 - i = 9 + 2i$ **2.** $3 + 3i$
3. $7 - 4i - 5 - 2i = 2 - 6i$ **4.** $4 - 4i$
5. $2 + i + 3i - 2 + 4i = 8i = 0 + 8i$ **6.** $6 - 4i$
7. $2 + 3i - x - yi = (2 - x) + (3 - y)i$
8. $(x - 7) + (-1 - y)i$
9. $9 + \sqrt{16}\sqrt{-1} + 2 + \sqrt{25}\sqrt{-1} + 6 - \sqrt{64}\sqrt{-1}$
$= 9 + 4i + 2 + 5i + 6 - 8i$
$= (9 + 2 + 6) + (4 + 5 - 8)i$
$= 17 + i$

10. $11 + 3i$ **11.** $4 + 3i - 5 + i = 3x + 2yi + 2x - 3yi$
$-1 + 4i = 5x - yi$
Therefore, $5x = -1 \mid -y = 4$
$x = -\frac{1}{5} \mid y = -4$

12. $x = \frac{3}{2}, y = 1$

13. $2 - 5i - 5 - 3i = 3x + 2yi - 5x - 3yi$
$-3 - 8i = -2x - yi$
$-3 = -2x \mid -8 = -y$
$x = \frac{3}{2} \mid y = 8$
14. $x = -\frac{4}{3}, y = \frac{6}{5}$

Exercises 7.14 (page 376)

1. $1^2 - i^2 = 1 - (-1) = 1 + 1 = 2 = 2 + 0i$ **2.** $13 + 0i$
3. $12 + 5i - 2i^2 = 12 + 5i + 2 = 14 + 5i$ **4.** $16 - 11i$
5. $12 - 22i + 6i^2 = 12 - 22i - 6 = 6 - 22i$ **6.** $-2 + 29i$
7. $(\sqrt{5})^2 - (2i)^2 = 5 - (-4) = 5 + 4 = 9 = 9 + 0i$
8. $16 + 0i$ **9.** $5i^2 - 10i = -5 - 10i$ **10.** $-12 - 6i$
11. $(2)^2 + 2(2)(5i) + (5i)^2 = 4 + 20i + 25i^2$
$= 4 + 20i - 25 = -21 + 20i$
12. $-7 - 24i$ **13.** $i^8 \cdot i^2 = 1 \cdot (-1) = -1 = -1 + 0i$
14. $0 - i$ **15.** $(i^4)^{21}i^3 = (1)(-i) = -i = 0 - i$ **16.** $0 + i$
17. $3^3i^2i = 27(-1)i = -27i = 0 - 27i$ **18.** $0 - 8i$
19. $2^4i^4 = 16(1) = 16 + 0i$ **20.** $81 + 0i$
21. $(3 - 2i)(4 + 5i) = 12 + 7i - 10i^2 = 12 + 7i + 10 = 22 + 7i$
22. $58 - 14i$ **23.** $(2 - i)^2 = 4 - 4i + (i)^2 = 4 - 4i - 1 = 3 - 4i$
24. $8 + 6i$ **25.** $[3 + i^6]^2 = [3 - 1]^2 = 2^2 = 4 + 0i$
26. $25 + 0i$ **27.** $i^{10+23} = i^{33} = (i^4)^8 \cdot i = 1 \cdot i = 0 + i$
28. $-1 + 0i$ **29.** $i^{15} + i^7 = -i + (-i) = 0 - 2i$
30. $-1 - i$
31. $[2 + (-i)^{11}]^2 = [2 + i]^2 = 4 + 4i + i^2 = 4 + 4i - 1 = 3 + 4i$
32. $8 + 6i$
33. $1 - 5i + 10i^2 - 10i^3 + 5i^4 - i^5 = 1 - 5i - 10 + 10i + 5 - i$
$= -4 + 4i$
(We used the binomial theorem.)
34. $-4 + 0i$ **35.** $2^4 - 4(2^3)i + 6(2^2)i^2 - 4(2)i^3 + i^4$
$= 16 - 32i - 24 + 8i + 1$
$= -7 - 24i$
36. $52 - 47i$

37. $(2i)^2 \stackrel{?}{=} -4$
$2^2i^2 \stackrel{?}{=} -4$
$4(-1) \stackrel{?}{=} -4$
$-4 = -4$ True

38. $(-3i)^2 \stackrel{?}{=} -9$
$(-3)^2i^2 \stackrel{?}{=} -9$
$9(-1) \stackrel{?}{=} -9$
$-9 = -9$ True

39.
$$(2 + i\sqrt{3})^2 - 4(2 + i\sqrt{3}) + 7 \stackrel{?}{=} 0$$
$$4 + 4i\sqrt{3} + i^2(\sqrt{3})^2 - 8 - 4i\sqrt{3} + 7 \stackrel{?}{=} 0$$
$$4 + 4i\sqrt{3} - 3 - 8 - 4i\sqrt{3} + 7 \stackrel{?}{=} 0$$
$0 = 0$ True

40.
$$(3 - i\sqrt{5})^2 - 6(3 - i\sqrt{5}) + 14 \stackrel{?}{=} 0$$
$$9 - 6i\sqrt{5} + i^2(\sqrt{5})^2 - 18 + 6i\sqrt{5} + 14 \stackrel{?}{=} 0$$
$$9 - 6i\sqrt{5} - 5 - 18 + 6i\sqrt{5} + 14 \stackrel{?}{=} 0$$
$0 = 0$ True

Exercises 7.15 (page 378)

1. $3 + 2i$ **2.** $5 - 4i$ **3.** $0 - 5i$ **4.** $0 + 7i$
5. $10 + 0i$ **6.** $-8 + 0i$
7. $\left(\dfrac{10}{1 + 3i}\right)\left(\dfrac{1 - 3i}{1 - 3i}\right) = \dfrac{10(1 - 3i)}{1 - 9i^2} = \dfrac{10(1 - 3i)}{1 + 9} = \dfrac{10(1 - 3i)}{10} = 1 - 3i$
8. $1 - 2i$
9. $\left(\dfrac{1 + i}{1 - i}\right)\left(\dfrac{1 + i}{1 + i}\right) = \dfrac{1 + 2i + i^2}{1 - i^2} = \dfrac{1 + 2i - 1}{1 - (-1)} = \dfrac{2i}{2} = i = 0 + i$

10. $0 - i$ **11.** $\left(\frac{8+i}{i}\right)\left(\frac{-i}{-i}\right) = \frac{-8i - i^2}{-i^2} = \frac{-8i-(-1)}{-(-1)} = 1 - 8i$

12. $-1 - 4i$ **13.** $\frac{3}{2i}\cdot\frac{i}{i} = \frac{3i}{2i^2} = \frac{3i}{-2} = 0 - \frac{3}{2}i$ **14.** $0 - \frac{4}{5}i$

15. $\left(\frac{15i}{1-2i}\right)\left(\frac{1+2i}{1+2i}\right) = \frac{15i + 30i^2}{1 - 4i^2} = \frac{15i - 30}{1+4} = \frac{15i - 30}{5}$
$= \frac{5(3i-6)}{5} = -6 + 3i$

16. $-6 + 2i$

17. $\left(\frac{4+3i}{2-i}\right)\left(\frac{2+i}{2+i}\right) = \frac{8 + 10i + 3i^2}{4 - i^2} = \frac{8 + 10i - 3}{4-(-1)}$
$= \frac{5+10i}{5} = \frac{5(1+2i)}{5} = 1 + 2i$

18. $\frac{4}{5} + \frac{1}{10}i$

Sections 7.10–7.15 Review Exercises (page 380)

1. $\sqrt{x-5} = \sqrt{3x+8}$
Domain: $\{x \mid x - 5 \geq 0\} \cap \{x \mid 3x + 8 \geq 0\} = \{x \mid x \geq 5\}$
$(\sqrt{x-5})^2 = (\sqrt{3x+8})^2$
$x - 5 = 3x + 8$
$-13 = 2x$
$x = -\frac{13}{2}$ Not in the domain
Therefore, there is no solution; { }

2. {7}

3. $\sqrt{5x-4} = \sqrt{2x+1} + 1$
$(\sqrt{5x-4})^2 = (\sqrt{2x+1} + 1)^2$
$5x - 4 = 2x + 1 + 2\sqrt{2x+1} + 1$
$3x - 6 = 2\sqrt{2x+1}$
$9x^2 - 36x + 36 = 4(2x+1)$
$9x^2 - 44x + 32 = 0$
$(x-4)(9x-8) = 0$
$x - 4 = 0 \mid 9x - 8 = 0$
$x = 4 \mid x = \frac{8}{9}$

Check for $x = 4$
$\sqrt{5x-4} - \sqrt{2x+1} = 1$
$\sqrt{5(4)-4} - \sqrt{2(4)+1} \stackrel{?}{=} 1$
$4 - 3 \stackrel{?}{=} 1$
$1 = 1$
4 is a solution.

Check for $x = \frac{8}{9}$
$\sqrt{5x-4} - \sqrt{2x+1} = 1$
$\sqrt{5\left(\frac{8}{9}\right) - 4} - \sqrt{2\left(\frac{8}{9}\right)+1} \stackrel{?}{=} 1$
$\sqrt{\frac{40-36}{9}} - \sqrt{\frac{16+9}{9}} \stackrel{?}{=} 1$
$\frac{2}{3} - \frac{5}{3} \stackrel{?}{=} 1$
$-1 \neq 1$
$\frac{8}{9}$ is not a solution.

{4}

4. $\left\{-\frac{36}{5}\right\}$ **5.** $x^{5/6} = 32$
$(x^{5/6})^{6/5} = 32^{6/5}$
$x = (2^5)^{6/5}$
$= 2^6$, or 64
{64} (The check will not be shown.)

6. {81}

7. $5 - yi = x + 6i$
$5 = x \mid -y = 6$
$x = 5 \mid y = -6$

8. $x = -1, y = -2$

9. $(3i + 2)(4 - 2i) = 12i + 8 - 6i^2 - 4i$
$= \underline{12i} + \underline{8} + \underline{6} - \underline{4i} = 14 + 8i$

10. $28 - 96i$

11. $\left(\frac{2+i}{1+3i}\right)\left(\frac{1-3i}{1-3i}\right) = \frac{2 - 3i^2 - 5i}{1 - 9i^2} = \frac{5-5i}{10} = \frac{1}{2} - \frac{1}{2}i$

12. $\frac{1}{5} + \frac{7}{5}i$ **13.** $i^{82} = (i^4)^{20}\cdot i^2 = 1^{20}(-1) = -1 = -1 + 0i$

14. $26 - 28i$

15. $(4 + \sqrt{9\cdot 3(-1)}) + (2 - \sqrt{4\cdot 3(-1)}) - (1 - \sqrt{3(-1)})$
$= (4 + 3i\sqrt{3}) + (2 - 2i\sqrt{3}) - (1 - i\sqrt{3})$
$= 4 + 3i\sqrt{3} + 2 - 2i\sqrt{3} - 1 + i\sqrt{3}$
$= 5 + 2i\sqrt{3}$

16. Width = 3 ft; length = 5 ft

17. Let x = width (in m)
Then $x + 3$ = length (in m)
$x^2 + (x+3)^2 = (\sqrt{45})^2$
$x^2 + x^2 + 6x + 9 = 45$
$2x^2 + 6x - 36 = 0$
$2(x+6)(x-3) = 0$
$2 \neq 0 \mid x + 6 = 0 \mid x - 3 = 0$
$x = -6 \mid x = 3$
Not in the domain $\mid x + 3 = 6$

Check $3^2 + 6^2 \stackrel{?}{=} (\sqrt{45})^2$
$9 + 36 \stackrel{?}{=} 45$
$45 = 45$
The width is 3 m, and the length is 6 m.

18. $\frac{3}{5}$

19. Let x = rate of southbound car
Then $x - 15$ = rate of eastbound car
$\left(x\frac{\text{mi}}{\text{hr}}\right)\left(\frac{1}{3}\text{ hr}\right)$, or $\frac{x}{3}$ mi, is the distance of the southbound car.
$\left([x-15]\frac{\text{mi}}{\text{hr}}\right)\left(\frac{1}{3}\text{ hr}\right)$, or $\frac{x-15}{3}$ mi, is the distance of the eastbound car.
$\left(\frac{x}{3}\right)^2 + \left(\frac{x-15}{3}\right)^2 = 25^2$
$\frac{x^2}{9} + \frac{x^2 - 30x + 225}{9} = 625$
$x^2 + x^2 - 30x + 225 = 5{,}625$
$2x^2 - 30x - 5{,}400 = 0$
$x^2 - 15x - 2{,}700 = 0$
$(x - 60)(x + 45) = 0$
$x - 60 = 0 \mid x + 45 = 0$
$x = 60 \mid x = -45$ Reject
(The check will not be shown.) The rate of the faster car is 60 mph.

20. The width of the box is 8 cm.

Chapter 7 Diagnostic Test (page 383)

Following each problem number is the number (in parentheses) of the textbook section where that kind of problem is discussed.

1. (7.2) $x^{1/2}x^{-1/4} = x^{1/2-1/4} = x^{1/4}$

2. (7.2) $(R^{-4/3})^3 = R^{(-4/3)(3)} = R^{-4} = \frac{1}{R^4}$

3. (7.2) $\frac{a^{5/6}}{a^{1/3}} = a^{5/6-1/3} = a^{5/6-2/6} = a^{3/6} = a^{1/2}$

4. (7.2) $\left(\frac{x^{-2/3}y^{3/5}}{x^{1/3}y}\right)^{-5/2} = (x^{-2/3-1/3}y^{3/5-1})^{-5/2} = (x^{-1}y^{-2/5})^{-5/2}$
$= x^{(-1)(-5/2)}y^{(-2/5)(-5/2)} = x^{5/2}y$

5. (7.2) $\frac{b^{2/3}}{b^{-1/5}} = b^{2/3-(-1/5)} = b^{2/3+1/5} = b^{13/15}$

6. (7.3) $\sqrt[3]{54x^6y^7} = \sqrt[3]{2(27)x^6y^6y} = 3x^2y^2\sqrt[3]{2y}$

7. (7.4) $\frac{4xy}{\sqrt{2x}} = \frac{4xy}{\sqrt{2x}}\cdot\frac{\sqrt{2x}}{\sqrt{2x}} = \frac{4xy\sqrt{2x}}{2x} = 2y\sqrt{2x}$

8. (7.5) $\sqrt[6]{a^3} = a^{3/6} = a^{1/2} = \sqrt{a}$

9. (7.6) $\sqrt{40} + \sqrt{9} = 2\sqrt{10} + 3$

7

10. (7.9) $\sqrt{x}\sqrt[3]{x} = x^{1/2}x^{1/3} = x^{1/2+1/3} = x^{5/6} = \sqrt[6]{x^5}$

11. (7.1) $(-27)^{2/3} = [(-3)^3]^{2/3} = (-3)^{3(2/3)} = (-3)^2 = 9$

12. (7.6)
$$4\sqrt{8y} + 3\sqrt{32y} = 4\sqrt{4 \cdot 2y} + 3\sqrt{16 \cdot 2y}$$
$$= 4(2)\sqrt{2y} + 3(4)\sqrt{2y} = 20\sqrt{2y}$$

13. (7.6)
$$3\sqrt{\frac{5x^2}{2}} - 5\sqrt{\frac{x^2}{10}} = 3\sqrt{\frac{5x^2}{2} \cdot \frac{2}{2}} - 5\sqrt{\frac{x^2}{10} \cdot \frac{10}{10}}$$
$$= 3\sqrt{\frac{10x^2}{4}} - 5\sqrt{\frac{10x^2}{100}}$$
$$= \frac{3x\sqrt{10}}{2} - \frac{5x\sqrt{10}}{10}$$
$$= \tfrac{3}{2}x\sqrt{10} - \tfrac{1}{2}x\sqrt{10} = x\sqrt{10}$$

14. (7.7) $\sqrt{2x^4}\sqrt{8x^3} = \sqrt{16x^6x} = 4x^3\sqrt{x}$

15. (7.7)
$$\sqrt{2x}(\sqrt{8x} - 5\sqrt{2}) = \sqrt{2x}\sqrt{8x} + \sqrt{2x}(-5\sqrt{2})$$
$$= \sqrt{16x^2} - 5\sqrt{4x} = 4x - 10\sqrt{x}$$

16. (7.8)
$$\frac{\sqrt{10x} + \sqrt{5x}}{\sqrt{5x}} = \frac{\sqrt{10x}}{\sqrt{5x}} + \frac{\sqrt{5x}}{\sqrt{5x}} = \sqrt{\frac{10x}{5x}} + 1 = \sqrt{2} + 1$$

17. (7.8)
$$\frac{5}{\sqrt{7} + \sqrt{2}} = \left(\frac{5}{\sqrt{7} + \sqrt{2}}\right)\left(\frac{\sqrt{7} - \sqrt{2}}{\sqrt{7} - \sqrt{2}}\right)$$
$$= \frac{5(\sqrt{7} - \sqrt{2})}{7 - 2} = \sqrt{7} - \sqrt{2}$$

18. (7.7)
$$(1 - \sqrt[3]{x})^3 = (1)^3 + 3(1)^2(-\sqrt[3]{x}) + 3(1)(-\sqrt[3]{x})^2 + (-\sqrt[3]{x})^3$$
$$= 1 - 3\sqrt[3]{x} + 3\sqrt[3]{x^2} - x$$

19. (7.13)
$$(5 - \sqrt{-8}) - (3 - \sqrt{-18}) = 5 - \sqrt{4 \cdot 2(-1)} - 3 + \sqrt{9 \cdot 2(-1)}$$
$$= 5 - 2i\sqrt{2} - 3 + 3i\sqrt{2} = 2 + i\sqrt{2}$$

20. (7.14)
$$(3 + i)(2 - 5i) = 6 - 15i + 2i - 5i^2$$
$$= 6 - 13i - 5(-1) = 6 - 13i + 5$$
$$= 11 - 13i$$

21. (7.15)
$$\frac{10}{1 - 3i} = \left(\frac{10}{1 - 3i}\right)\left(\frac{1 + 3i}{1 + 3i}\right) = \frac{10(1 + 3i)}{1 - 9i^2}$$
$$= \frac{10(1 + 3i)}{10} = 1 + 3i$$

22. (7.11)
$$(2 - i)^3 = 2^3 - 3(2^2)i + 3(2)i^2 - i^3$$
$$= 8 - 12i + 6(-1) - (-i)$$
$$= 8 - 6 - 12i + i$$
$$= 2 - 11i$$

23. (7.10)
$$x^{3/2} = 8$$
$$(x^{3/2})^{2/3} = 8^{2/3} = (2^3)^{2/3}$$
$$x = 2^2$$
$$x = 4$$

Check
$$x^{3/2} = 8$$
$$(4)^{3/2} \stackrel{?}{=} 2^3$$
$$(2^2)^{3/2} \stackrel{?}{=} 2^3$$
$$2^3 = 2^3$$
$$\{4\}$$

24. (7.10)
$$\sqrt{x - 3} = x - 5$$
$$(\sqrt{x - 3})^2 = (x - 5)^2$$
$$x - 3 = x^2 - 10x + 25$$
$$0 = x^2 - 11x + 28$$
$$0 = (x - 4)(x - 7)$$

$x - 4 = 0$ | $x - 7 = 0$

$x = 4$ | $x = 7$

Check for $x = 4$
$$\sqrt{x - 3} + 5 = x$$
$$\sqrt{4 - 3} + 5 \stackrel{?}{=} 4$$
$$\sqrt{1} + 5 \stackrel{?}{=} 4$$
$$1 + 5 \neq 4$$
Therefore, 4 is not a solution.

Check for $x = 7$
$$\sqrt{x - 3} + 5 = x$$
$$\sqrt{7 - 3} + 5 \stackrel{?}{=} 7$$
$$\sqrt{4} + 5 \stackrel{?}{=} 7$$
$$2 + 5 \stackrel{?}{=} 7$$
$$7 = 7$$
Therefore, 7 is a solution.

The solution set is $\{7\}$.

25. (7.11) Let x = number of cm in shorter leg

Then $x + 2$ = number of cm in hypotenuse
$$(x + 2)^2 = x^2 + (\sqrt{12})^2$$
$$x^2 + 4x + 4 = x^2 + 12$$
$$4x = 8$$
$$x = 2$$

Check
$$(2 + 2)^2 \stackrel{?}{=} 2^2 + (\sqrt{12})^2$$
$$4^2 \stackrel{?}{=} 4 + 12$$
$$16 = 16$$

The shorter leg of the right triangle is 2 cm long.

Chapters 1–7 Cumulative Review Exercises (page 383)

1. LCD $= (a + 3)(a - 3)(a - 1)$
$$\left(\frac{6}{(a + 3)(a - 3)}\right)\left(\frac{a - 1}{a - 1}\right) - \left(\frac{2}{(a - 3)(a - 1)}\right)\left(\frac{a + 3}{a + 3}\right)$$
$$= \frac{6a - 6}{(a + 3)(a - 3)(a - 1)} - \frac{2a + 6}{(a + 3)(a - 3)(a - 1)}$$
$$= \frac{6a - 6 - 2a - 6}{(a + 3)(a - 3)(a - 1)} = \frac{4a - 12}{(a + 3)(a - 3)(a - 1)}$$
$$= \frac{4(a - 3)}{(a + 3)(a - 3)(a - 1)} = \frac{4}{(a + 3)(a - 1)}$$

2. $\dfrac{x^2 + 2x + 4}{x^2}$

3.
$$\frac{\frac{4}{x} \cdot \frac{x^2}{1} - \frac{8}{x^2} \cdot \frac{x^2}{1}}{\frac{1}{x} \cdot \frac{x^2}{1} - \frac{2}{x^2} \cdot \frac{x^2}{1}} = \frac{4x - 8}{x - 2} = \frac{4(x - 2)}{x - 2} = 4$$

4. $5x + 2\sqrt{5x}$

5.
$$(\sqrt{26})^2 - 2\sqrt{26}\sqrt{10} + (\sqrt{10})^2 = 26 - 2\sqrt{2^2 \cdot 13 \cdot 5} + 10$$
$$= 36 - 4\sqrt{65}$$

6. $-12 - 6\sqrt{5}$

7.
$$12x - 3 - 3 + 2x = 8x - 12$$
$$14x - 6 = 8x - 12$$
$$6x = -6$$
$$x = -1$$
$$\{-1\}$$

8. $\{x \mid x \leq 5\}$

9.
$$\sqrt{2x + 2} = 1 + \sqrt{3x - 12}$$
$$(\sqrt{2x + 2})^2 = (1 + \sqrt{3x - 12})^2$$
$$2x + 2 = 1 + 2(1)(\sqrt{3x - 12}) + 3x - 12$$
$$2x + 2 = 3x - 11 + 2\sqrt{3x - 12}$$
$$-x + 13 = 2\sqrt{3x - 12}$$
$$(-x + 13)^2 = (2\sqrt{3x - 12})^2$$
$$x^2 - 26x + 169 = 4(3x - 12)$$
$$x^2 - 26x + 169 = 12x - 48$$
$$x^2 - 38x + 217 = 0$$
$$(x - 7)(x - 31) = 0$$

$x - 7 = 0$ | $x - 31 = 0$

$x = 7$ | $x = 31$

Check for $x = 7$
$$\sqrt{2x + 2} = 1 + \sqrt{3x - 12}$$
$$\sqrt{2(7) + 2} \stackrel{?}{=} 1 + \sqrt{3(7) - 12}$$
$$\sqrt{16} \stackrel{?}{=} 1 + \sqrt{9}$$
$$4 \stackrel{?}{=} 1 + 3$$
$$4 = 4$$

Check for $x = 31$
$$\sqrt{2x + 2} = 1 + \sqrt{3x - 12}$$
$$\sqrt{2(31) + 2} \stackrel{?}{=} 1 + \sqrt{3(31) - 12}$$
$$\sqrt{64} \stackrel{?}{=} 1 + \sqrt{81}$$
$$8 \neq 1 + 9$$

Therefore, the only solution is 7; $\{7\}$

10. $\{-3\}$

11.
$$6x^2 + 11x = 10$$
$$6x^2 + 11x - 10 = 0$$
$$(3x - 2)(2x + 5) = 0$$

$3x - 2 = 0$ | $2x + 5 = 0$

$3x = 2$ | $2x = -5$

$x = \frac{2}{3}$ | $x = -\frac{5}{2}$
$$\left\{\tfrac{2}{3}, -\tfrac{5}{2}\right\}$$

12. $a = \dfrac{bc}{b - c}$

13. If $|2x - 5| > 3$,

$2x - 5 > 3$ or $2x - 5 < -3$

$2x > 8$ or $2x < 2$

$x > 4$ or $x < 1$

$\{x \mid x > 4\} \cup \{x \mid x < 1\}$

14. $\frac{30}{11}$ hr $= 2\frac{8}{11}$ hr

15. Let $2x$ = number of cm in width
Then $3x$ = number of cm in length

$$(2x)(3x) = 54$$
$$6x^2 = 54$$
$$x^2 = 9$$
$$x = \pm 3$$

-3 is not in the domain; $x = 3$

$$2x = 6$$
$$3x = 9$$

Check (6 cm)(9 cm) = 54 sq cm, and the ratio of 6 cm to 9 cm is 2 to 3. The width is 6 cm, and the length is 9 cm.

16. \$14,200 at 7.6%, \$12,300 at 7.3%

17. Let x = speed of current (in mph)
Then $27 - x$ = speed of boat going upstream (in mph)
and $27 + x$ = speed of boat going downstream (in mph)

$$\text{Time upstream} = \frac{120 \text{ mi}}{(27 - x)\frac{\text{mi}}{\text{hr}}} = \frac{120}{27 - x}\text{ hr}$$

$$\text{Time downstream} = \frac{150 \text{ mi}}{(27 + x)\frac{\text{mi}}{\text{hr}}} = \frac{150}{27 + x}\text{ hr}$$

$$\frac{120}{27 - x} = \frac{150}{27 + x} \quad \text{The times are equal.}$$
$$120(27 + x) = 150(27 - x)$$
$$4(27 + x) = 5(27 - x) \quad \text{(Dividing both sides by 30)}$$
$$108 + 4x = 135 - 5x$$
$$9x = 27$$
$$x = 3$$

(The check will not be shown.) The speed of the current is 3 mph.

18. 9 yd by 9 yd

Exercises 8.1 (page 390)

1. a. (3, 0) **b.** (0, 5) **c.** (−5, 2) **d.** (−4, −3)

2. a. (4, 3) **b.** (−6, 0) **c.** (3, −4) **d.** (0, −3)

3.

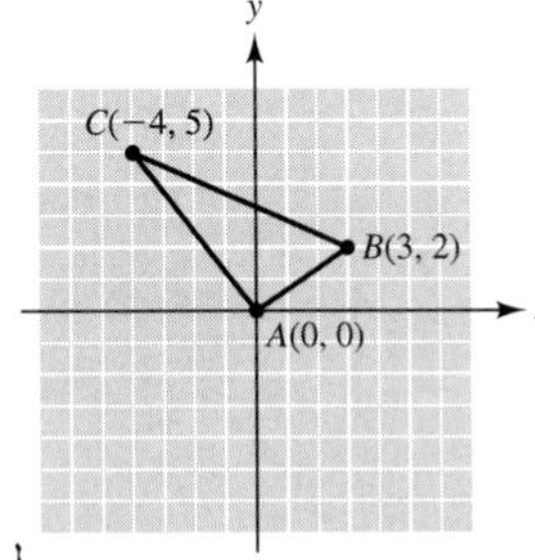

4. B(−2, 4), C(3, 5), A(−2, −3)

5. a. $d = \sqrt{[2 - (-2)]^2 + [1 - (-2)]^2}$
$= \sqrt{4^2 + 3^2} = \sqrt{16 + 9} = \sqrt{25} = 5$

b. $d = \sqrt{[3 - (-3)]^2 + [-1 - 3]^2}$
$= \sqrt{36 + 16} = \sqrt{52} = \sqrt{4 \cdot 13} = 2\sqrt{13}$

c. $|-2 - 5| = |-7| = 7$

d. $|-5 - (-2)| = |-5 + 2| = |-3| = 3$

e. $d = \sqrt{(4 - 0)^2 + (6 - 0)^2} = \sqrt{16 + 36} = \sqrt{52}$
$= \sqrt{4 \cdot 13} = 2\sqrt{13}$

6. a. 13 **b.** $\sqrt{74}$ **c.** 6 **d.** 6 **e.** $3\sqrt{13}$

7. $|AB| = \sqrt{[4 - (-2)]^2 + (2 - 2)^2} = \sqrt{6^2} = 6$
$|BC| = \sqrt{(6 - 4)^2 + (8 - 2)^2} = \sqrt{4 + 36} = \sqrt{40}$
$= \sqrt{4 \cdot 10} = 2\sqrt{10}$
$|AC| = \sqrt{[6 - (-2)]^2 + (8 - 2)^2} = \sqrt{64 + 36} = \sqrt{100} = 10$
Perimeter $= 6 + 2\sqrt{10} + 10 = 16 + 2\sqrt{10}$

8. $16 + 4\sqrt{5}$

9. $|AB| = \sqrt{[5 - (-3)]^2 + [-1 - (-2)]^2} = \sqrt{64 + 1} = \sqrt{65}$
$|BC| = \sqrt{(3 - 5)^2 + [2 - (-1)]^2} = \sqrt{4 + 9} = \sqrt{13}$
$|AC| = \sqrt{[3 - (-3)]^2 + [2 - (-2)]^2} = \sqrt{36 + 16} = \sqrt{52}$
$(|AB|)^2 \stackrel{?}{=} (|BC|)^2 + (|AC|)^2$
$(\sqrt{65})^2 \stackrel{?}{=} (\sqrt{13})^2 + (\sqrt{52})^2$
$65 = 13 + 52$
Therefore, the triangle is a right triangle.

10. The triangle is a right triangle.

11. Domain = {2, 3, 0, −3}
Range = {−1, 4, 2, −2}

12. Domain = {−4, 0, 3, 1, −3}
Range = {0, −2, 5, −3}

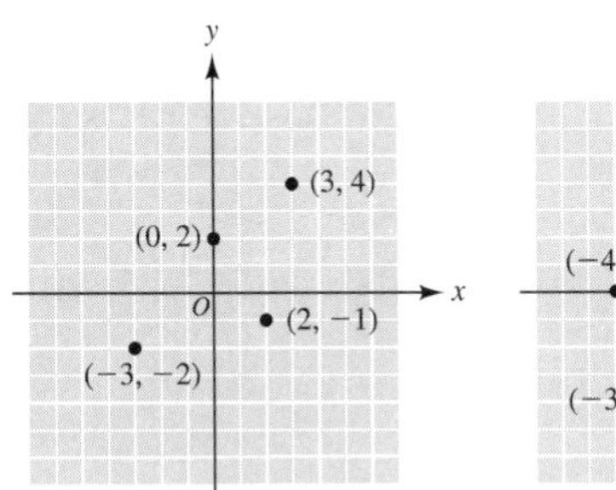

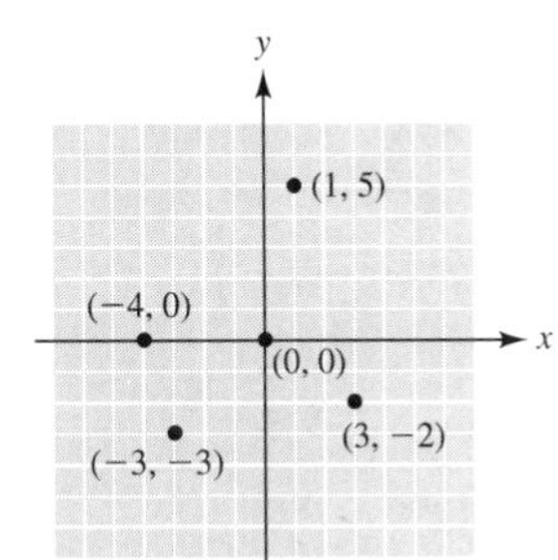

Exercises 8.2 (page 400)

1.

x	y
2	0
0	3
4	−3

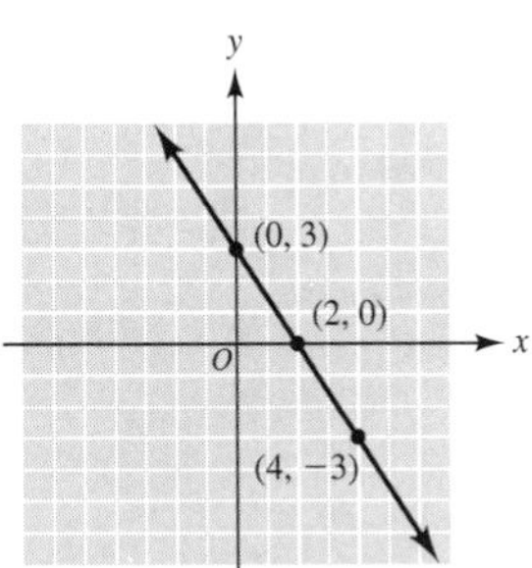

2.

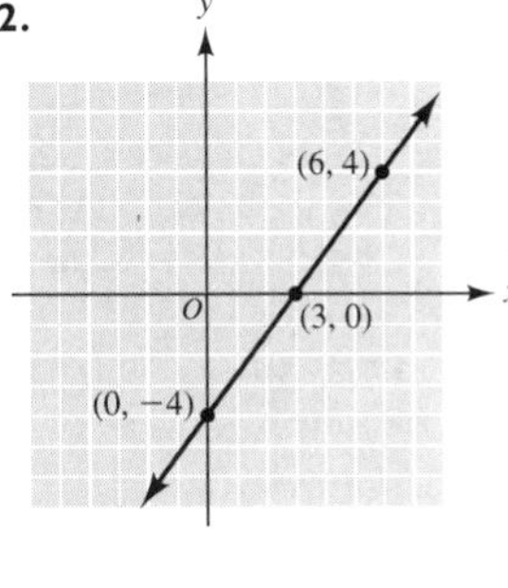

3.

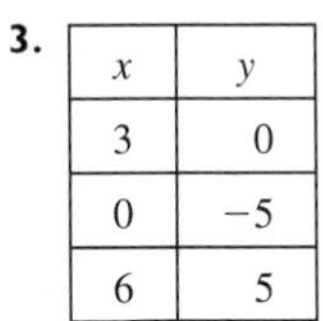

x	y
3	0
0	−5
6	5

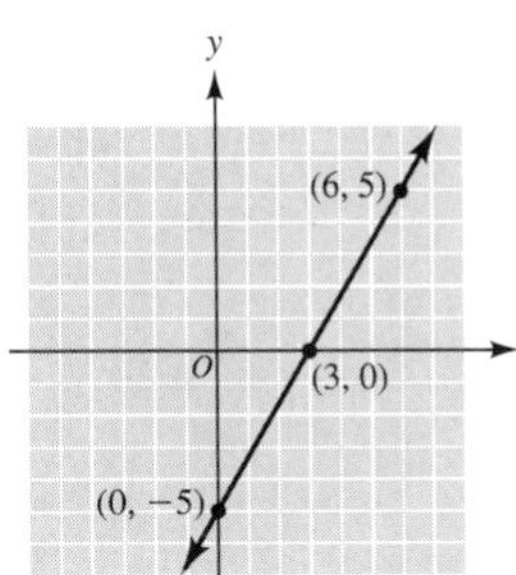

4.

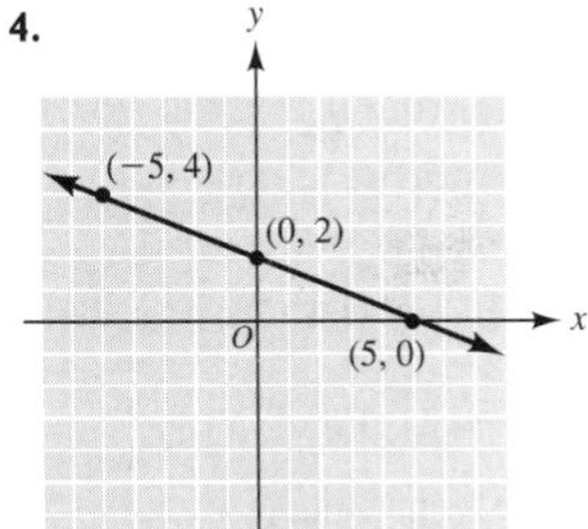

5.

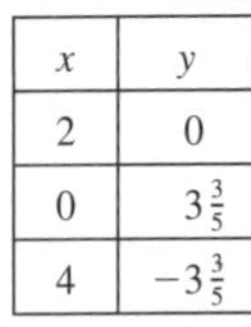

x	y
2	0
0	$3\frac{3}{5}$
4	$-3\frac{3}{5}$

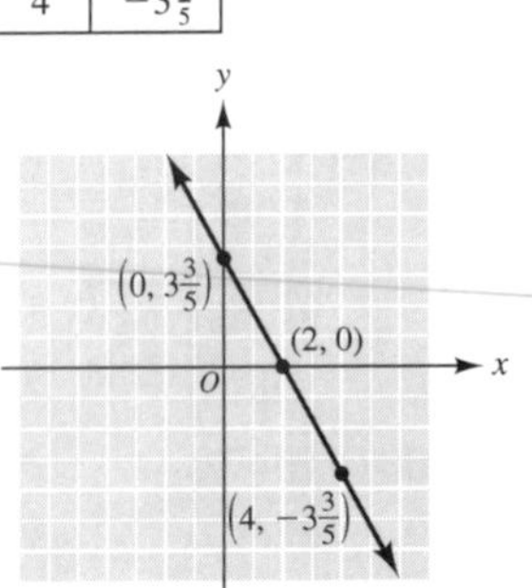

6.

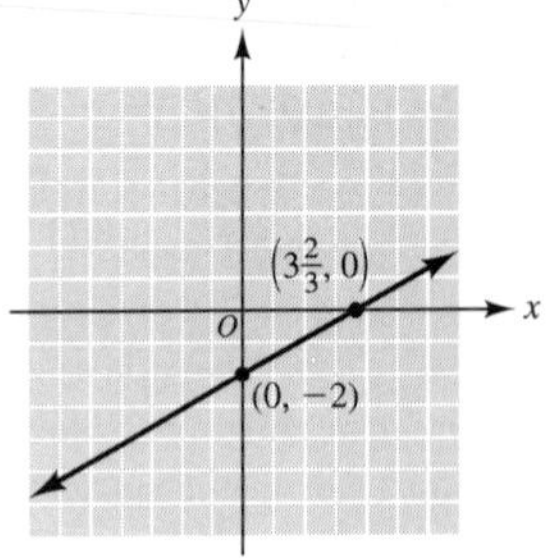
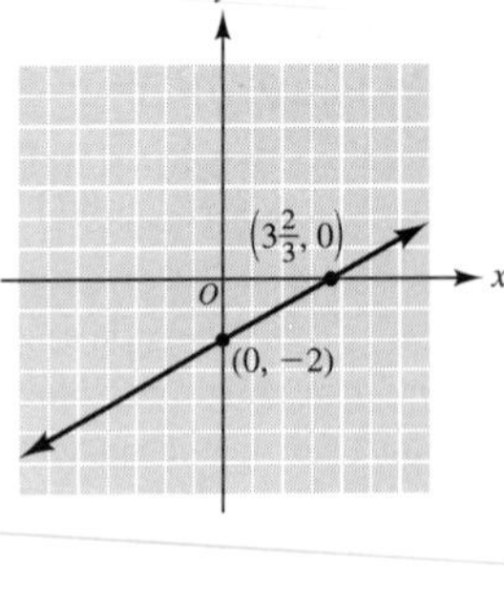

7.

x	y
$2\frac{1}{10}$	0
0	-3
4	$2\frac{5}{7}$

y
$\left(4, 2\frac{5}{7}\right)$
x
O
$\left(2\frac{1}{10}, 0\right)$
(0, −3)

8.

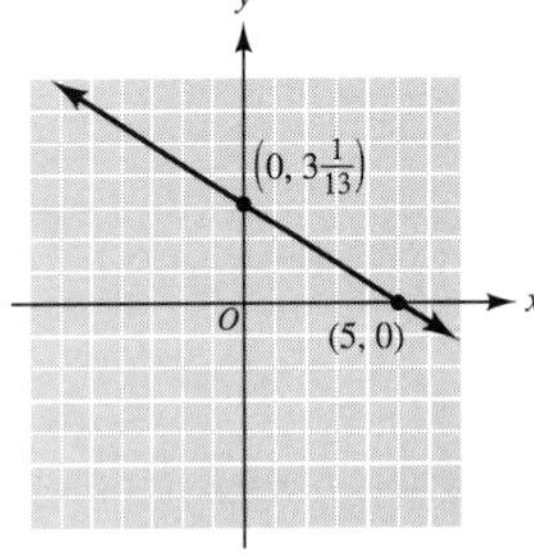

9.

x	y
$3\frac{4}{7}$	0
0	$2\frac{7}{9}$
2	$1\frac{2}{9}$

y
$\left(0, 2\frac{7}{9}\right)$
$\left(2, 1\frac{2}{9}\right)$
x
O
$\left(3\frac{4}{7}, 0\right)$

10.

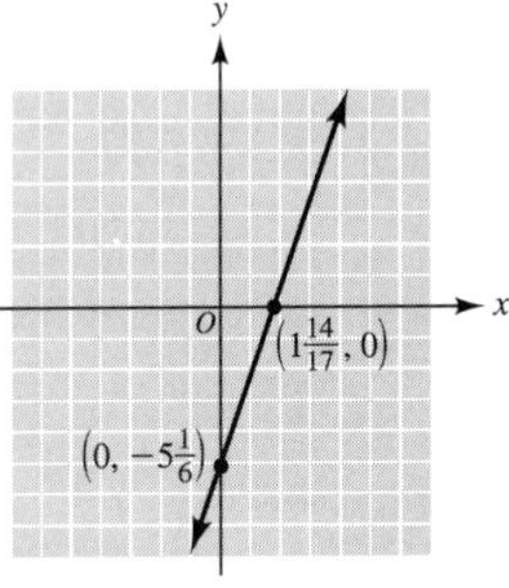

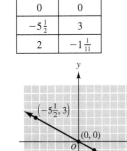

11.

x	y
$5\frac{1}{8}$	0
0	$-2\frac{13}{14}$
3	$-1\frac{3}{14}$

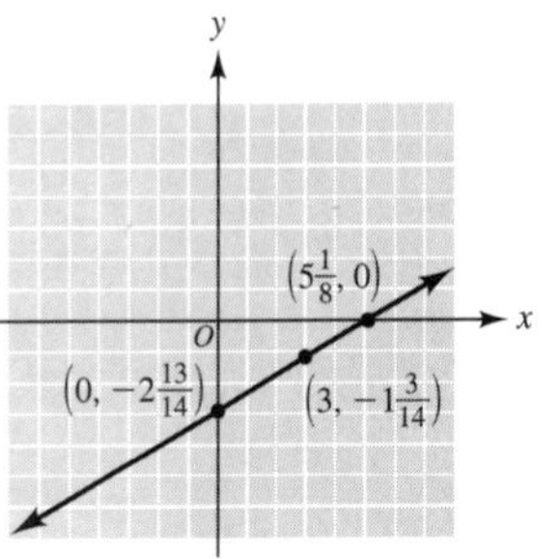

12.

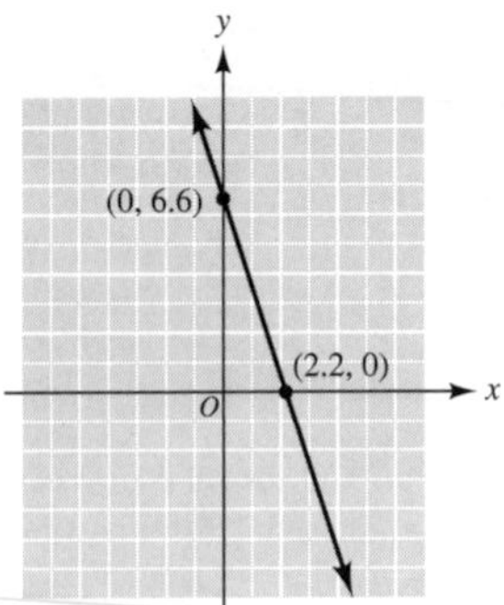

13.

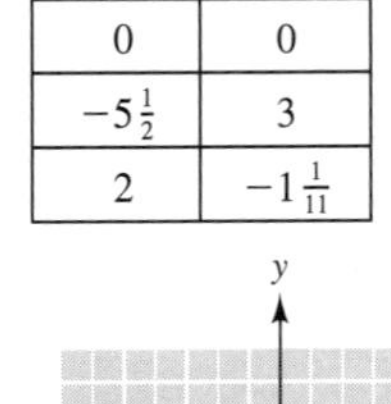

x	y
0	0
$-5\frac{1}{2}$	3
2	$-1\frac{1}{11}$

y
$\left(-5\frac{1}{2}, 3\right)$
(0, 0)
x
O
$\left(2, -1\frac{1}{11}\right)$

14.

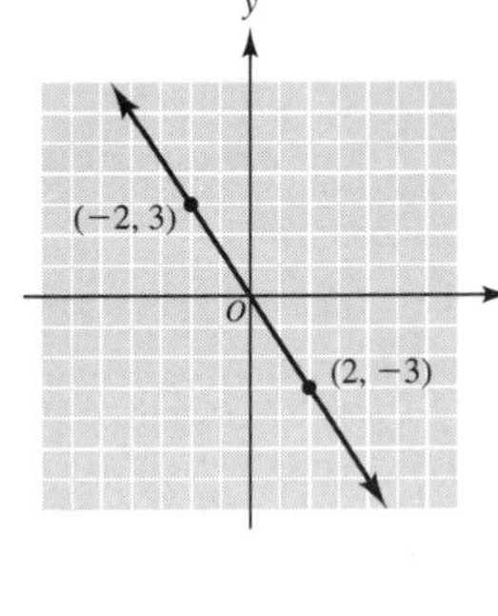

15.

x	y
0	0
1	-2
-1	2

y
(−1, 2)
x
O
(1, −2)

16.

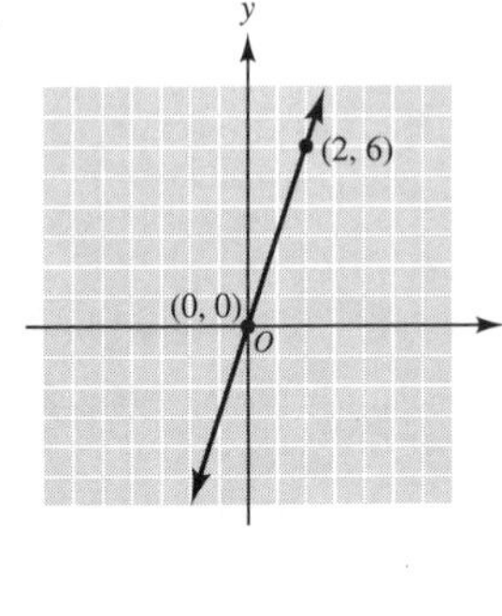

17.

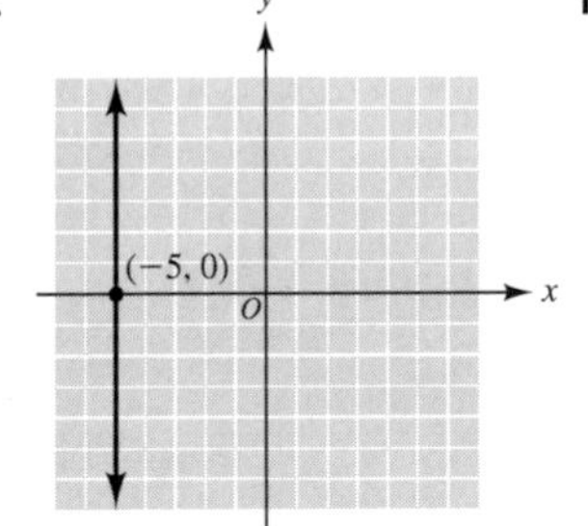

18.

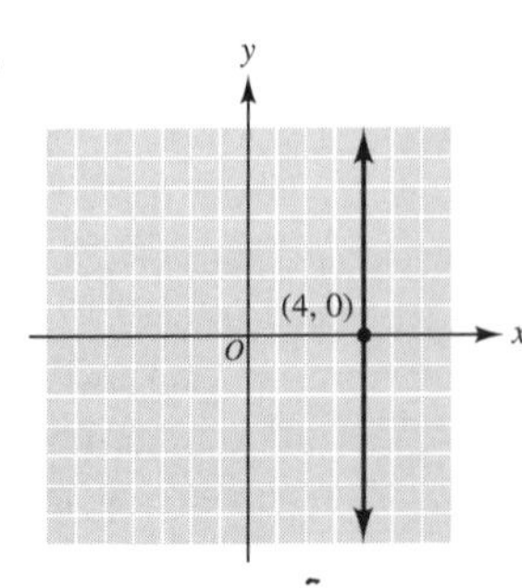

8

19. **20.**

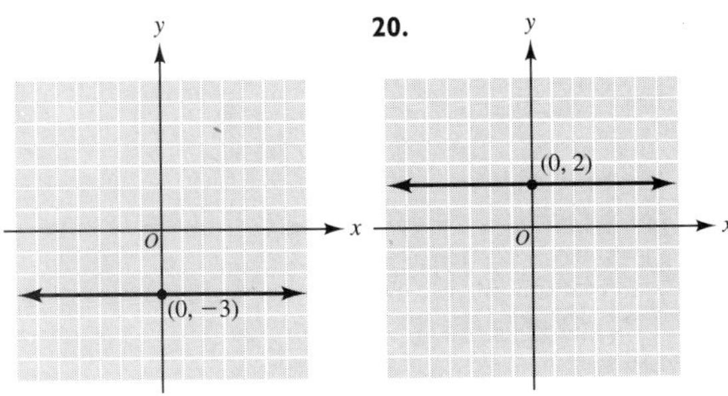

21.

x	y
$\frac{2}{7}$	0
0	$\frac{2}{5}$
3	$-3\frac{4}{5}$

22.

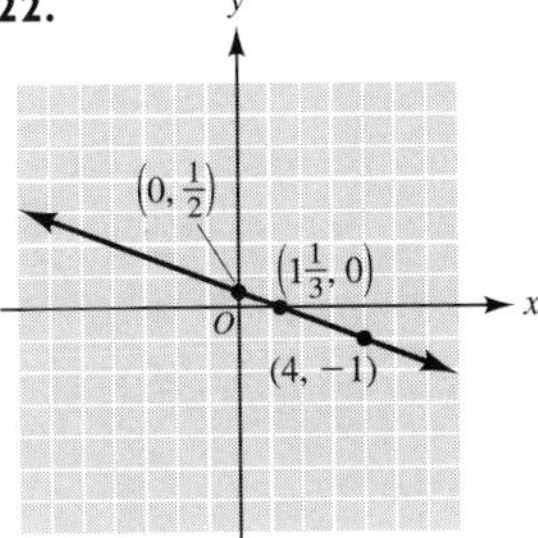

23.

x	y
2	0
0	−1
−2	−2

24.

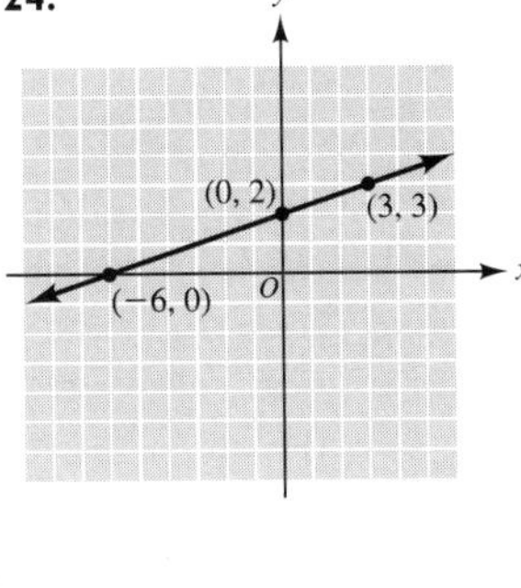

25.

x	y
5	0
0	$-2\frac{1}{7}$
3	$-\frac{6}{7}$

26.

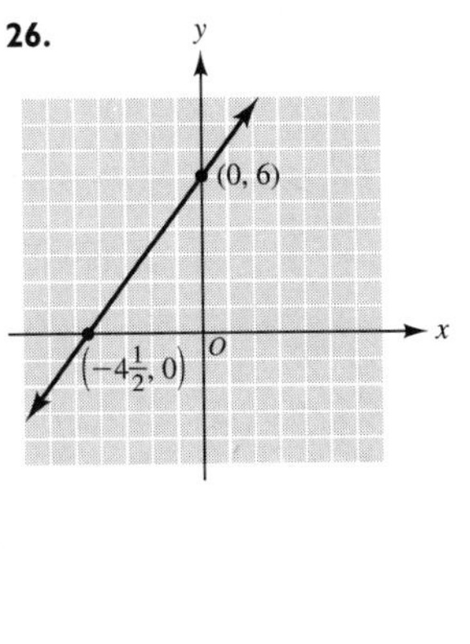

In Exercises 27–30, it is best if the units on the two axes are not the same.

27.

x	y
0	−100
−2	0
−1	−50

28.

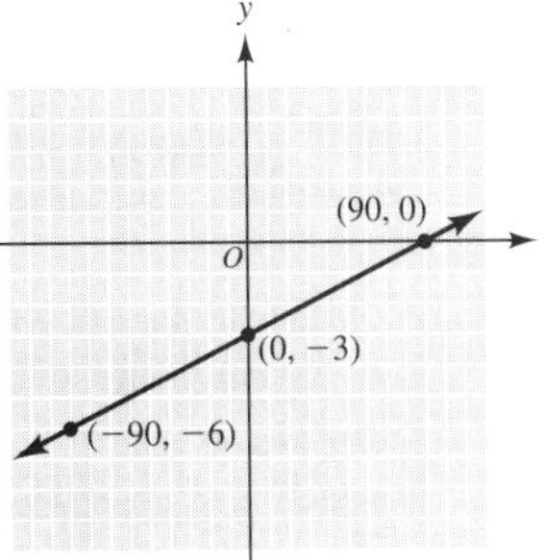

29.

x	y
0	0
50	1
−50	−1

30.

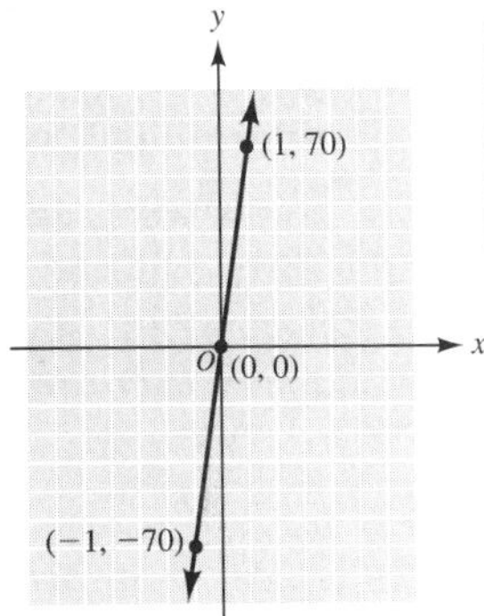

31.

x	y
1	−2
3	−6
5	−10

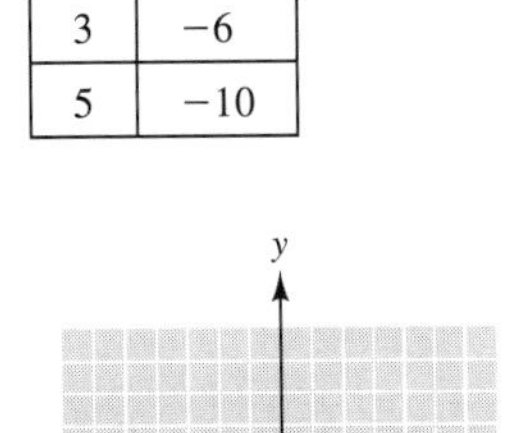

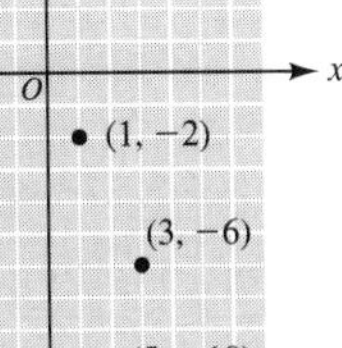

32.

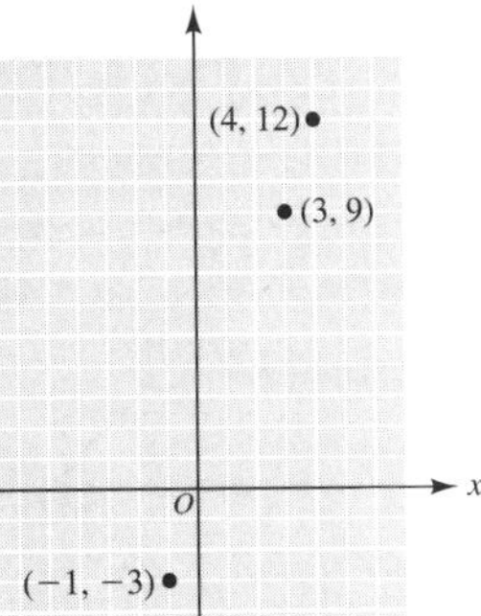

8

33. **34.**

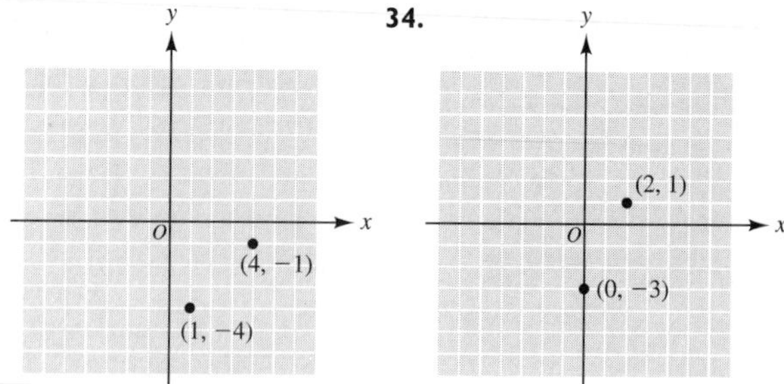

35. a. $5x - 7y = 18$

x	y
$3\frac{3}{5}$	0
0	$-2\frac{4}{7}$

$2x + 3y = -16$

x	y
−8	0
0	$-5\frac{1}{3}$

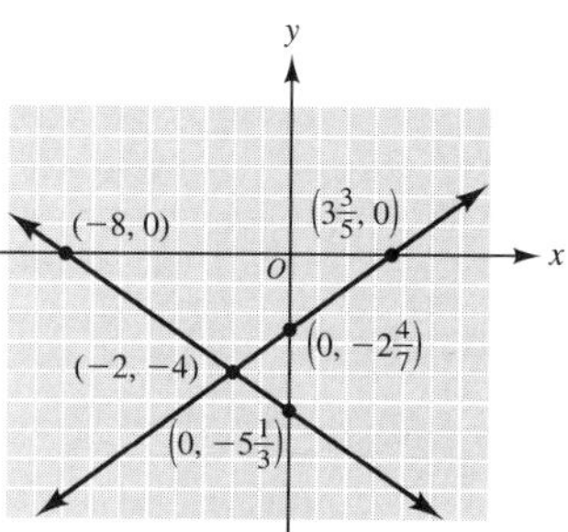

b. $(-2, -4)$

36. a.

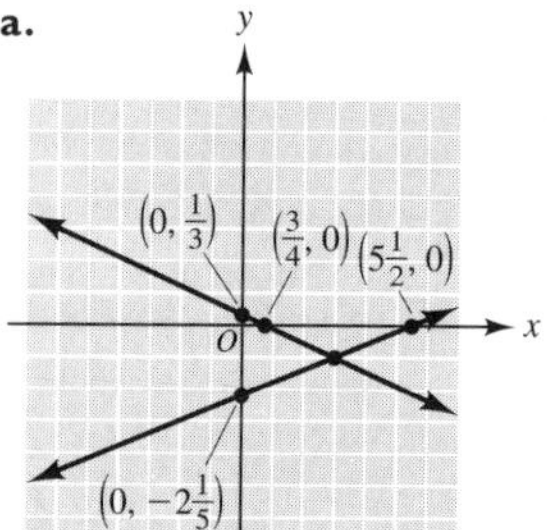

b. $(3, -1)$

Exercises 8.3 (page 407)

1. $m = \dfrac{6-4}{10-1} = \dfrac{2}{9}$ **2.** $\dfrac{5}{6}$

3. $m = \dfrac{(-7)-(-5)}{1-(-5)} = \dfrac{-7+5}{1+5} = \dfrac{-2}{6} = \dfrac{-1}{3}$ **4.** $-\frac{1}{2}$

5. $m = \dfrac{(-5)-(-5)}{2-(-7)} = \dfrac{0}{9} = 0$ **6.** 0

7. $m = \dfrac{-2-3}{-4-(-4)} = \dfrac{-5}{0}$ Not defined; m does not exist.

8. Does not exist

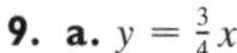

9. a. $y = \frac{3}{4}x$

x	y
0	0
4	3
−4	−3

$m = \dfrac{3-0}{4-0} = \dfrac{3}{4}$

b. $y = \frac{3}{4}x + 5$

x	y
0	5
4	8
−4	2

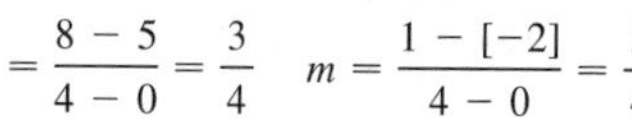

$m = \dfrac{8-5}{4-0} = \dfrac{3}{4}$

c. $y = \frac{3}{4}x - 2$

x	y
0	−2
4	1
8	4

$m = \dfrac{1-[-2]}{4-0} = \dfrac{3}{4}$

y
b. $y = \frac{3}{4}x + 5$
a. $y = \frac{3}{4}x$
c. $y = \frac{3}{4}x - 2$
(0, 5) (−4, 2) (4, 3) (4, 1) O x (−4, −3) (0, −2)

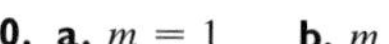
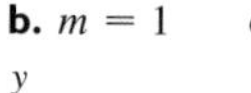

10. a. $m = 1$ **b.** $m = 1$ **c.** $m = 1$

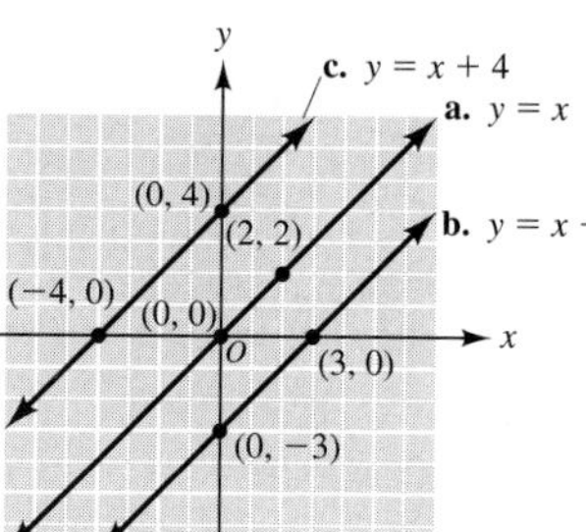

11.

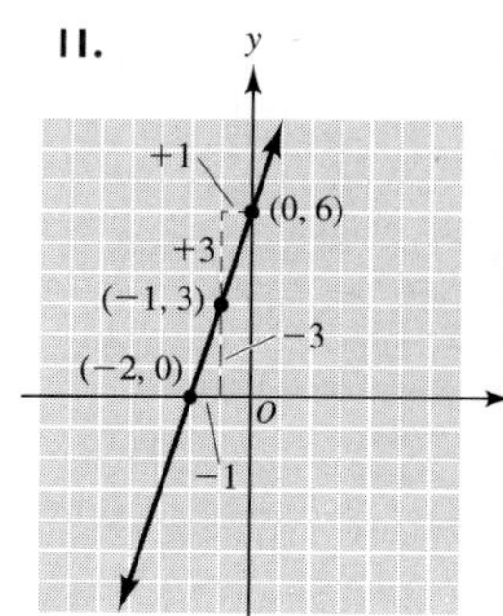

12.

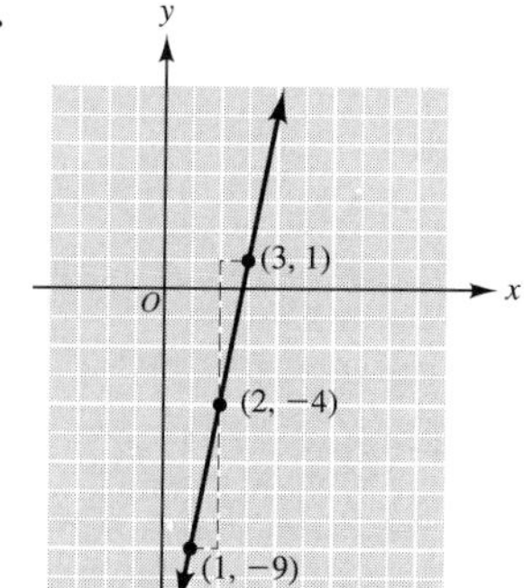

13.

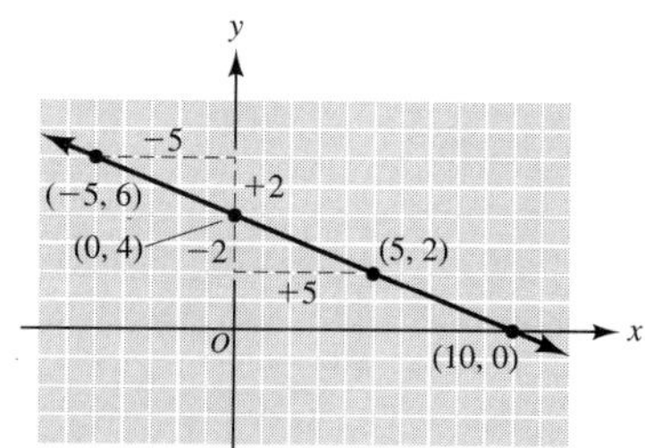

14.

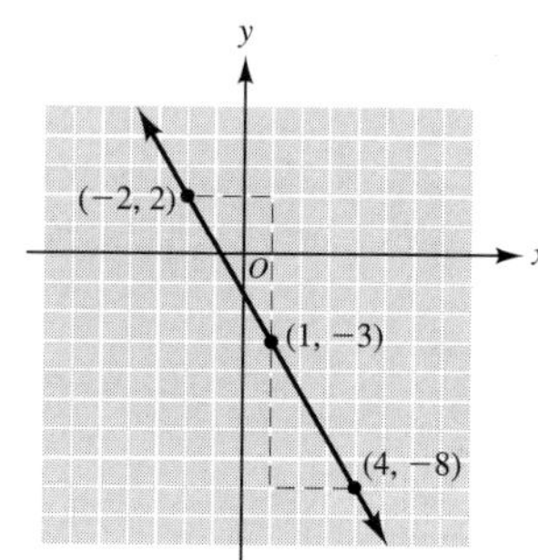

Exercises 8.4 (page 415)

1. $5x - 3y + 7 = 0$ **2.** $4x + 9y - 3 = 0$

3. $\frac{\overset{5}{\cancel{10}}}{1}\left(\frac{x}{\underset{1}{\cancel{2}}}\right) - \frac{\overset{2}{\cancel{10}}}{1}\left(\frac{y}{\underset{1}{\cancel{5}}}\right) = \left(\frac{10}{1}\right)\left(\frac{1}{1}\right)$

$5x - 2y = 10$

$5x - 2y - 10 = 0$

4. $6x - 7y + 42 = 0$

5. $\frac{3}{1}\left(\frac{y}{1}\right) = \frac{\overset{1}{\cancel{3}}}{1}\left(-\frac{5}{\underset{1}{\cancel{3}}}\right)x + \left(\frac{3}{1}\right)\left(\frac{4}{1}\right)$

$3y = -5x + 12$

$5x + 3y - 12 = 0$

6. $3x + 8y + 40 = 0$ **7.** $4x - 4y = 11 - 2x - 6y$

$6x + 2y - 11 = 0$

8. $25x - 8y - 15 = 0$ **9.** $y - (-3) = \frac{1}{5}(x - 4)$

$5y + 15 = x - 4$

$x - 5y - 19 = 0$

10. $5x + 6y + 16 = 0$ **11.** $y - 5 = \frac{1}{4}[x - (-6)]$

$4y - 20 = x + 6$

$x - 4y + 26 = 0$

12. $4x + 5y + 22 = 0$ **13.** $y - 3 = -4(x - [-1])$

$y - 3 = -4x - 4$

$4x + y + 1 = 0$

14. $2x - y - 6 = 0$ **15.** $y = \frac{5}{7}x - 3$

$7y = 5x - 21$

$5x - 7y - 21 = 0$

16. $x + 4y + 8 = 0$ **17.** $y = -\frac{4}{3}x + \frac{1}{2}$

$6y = -8x + 3$

$8x + 6y - 3 = 0$

18. $12x + 20y - 15 = 0$ **19.** $y = 0x + 5$

$y - 5 = 0$

20. $y - 7 = 0$

21. The line is horizontal; the equation is $y = 3$, or $y - 3 = 0$.

22. $y + 5 = 0$

23. The line is vertical; the equation is $x = 7$, or $x - 7 = 0$.

24. $x + 6 = 0$ **25. a.** $-5y = -4x - 20$

$y = \frac{-4x - 20}{-5}$

$y = \frac{4}{5}x + 4$

b. $m = \frac{4}{5}$

c. y-intercept $= 4$

26. a. $y = -\frac{8}{3}x + 8$ **b.** $-\frac{8}{3}$ **c.** 8

27. $2x + 9y + 15 = 0$

$9y = -2x - 15$

a. $y = -\frac{2}{9}x - \frac{5}{3}$

b. $m = -\frac{2}{9}$

c. y-intercept $= -\frac{5}{3}$

28. a. $y = \frac{18}{5}x - \frac{12}{5}$ **b.** $\frac{18}{5}$ **c.** $-\frac{12}{5}$

29. $m = \frac{4 - (-1)}{6 - 8} = \frac{5}{-2}$

$y - 4 = \frac{-5}{2}(x - 6)$

$2y - 8 = -5x + 30$

$5x + 2y - 38 = 0$

30. $3x + 2y - 17 = 0$

31. $m = \frac{4 - 0}{7 - 10} = \frac{4}{-3}$

$y - 0 = \frac{-4}{3}(x - 10)$

$3y = -4x + 40$

$4x + 3y - 40 = 0$

32. $5x - 2y + 20 = 0$

33. $m = \frac{(-1) - 3}{(-3) - (-9)} = \frac{-4}{6} = \frac{-2}{3}$

$y - (-1) = \frac{-2}{3}[x - (-3)]$

$3y + 3 = -2x - 6$

$2x + 3y + 9 = 0$

34. $3x + 4y + 17 = 0$

35. The line must have the same slope as $3x - 5y = 6$.

$-5y = -3x + 6$

$y = \frac{3}{5}x - \frac{6}{5}$

$m = \frac{3}{5}$

Slope of required line is $\frac{3}{5}$.

$y - 7 = \frac{3}{5}[x - (-4)]$

$5y - 35 = 3x + 12$

$0 = 3x - 5y + 47$

36. $7x + 4y - 36 = 0$

37. Slope of $2x + 4y = 3$:

$4y = -2x + 3$

$y = -\frac{1}{2}x + \frac{3}{4}$

$m = -\frac{1}{2}$

Slope of required line is 2.

$y - 2 = 2(x - 6)$

$y - 2 = 2x - 12$

$2x - y - 10 = 0$

38. $2x + y - 9 = 0$

39. The line must have the same slope as $3x + 5y - 12 = 0$.

$5y = -3x + 12$

$y = \frac{-3}{5}x + \frac{12}{5}$

$m = \frac{-3}{5}$

Slope of required line is $\frac{-3}{5}$.

x-intercept $= (4, 0)$

$y - 0 = \frac{-3}{5}(x - 4)$

$5y = -3x + 12$

$3x + 5y - 12 = 0$

40. $9x - 14y + 27 = 0$

41. x-intercept $= (-6, 0)$

y-intercept $= (0, 4)$

$m = \frac{4 - 0}{0 - (-6)} = \frac{4}{6} = \frac{2}{3}$

Using $(0, 4)$:

$y - 4 = \frac{2}{3}(x - 0)$

$3y - 12 = 2x$

$2x - 3y + 12 = 0$

42. $4x - 5y - 60 = 0$

43.

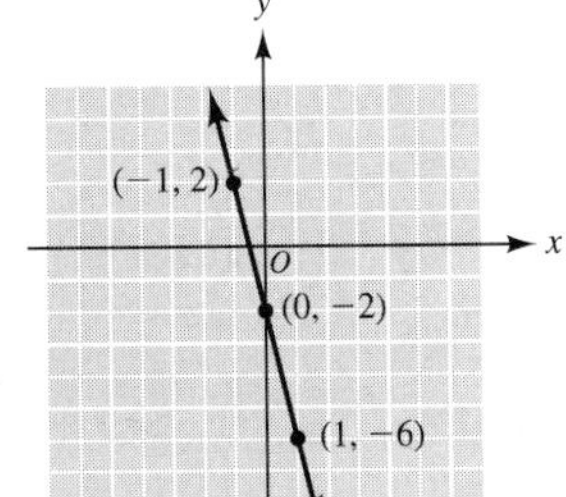

44.

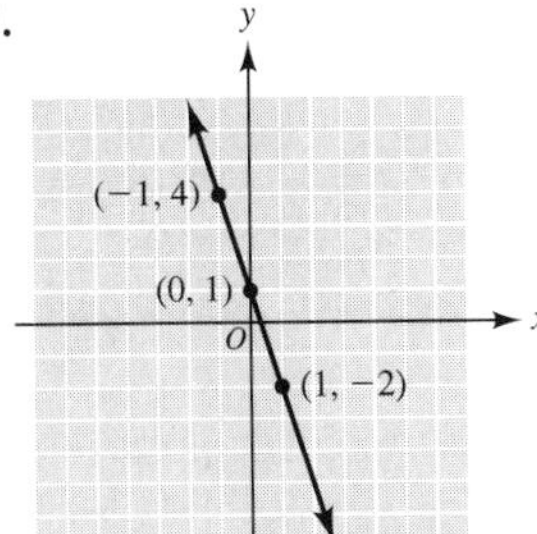

8

45.

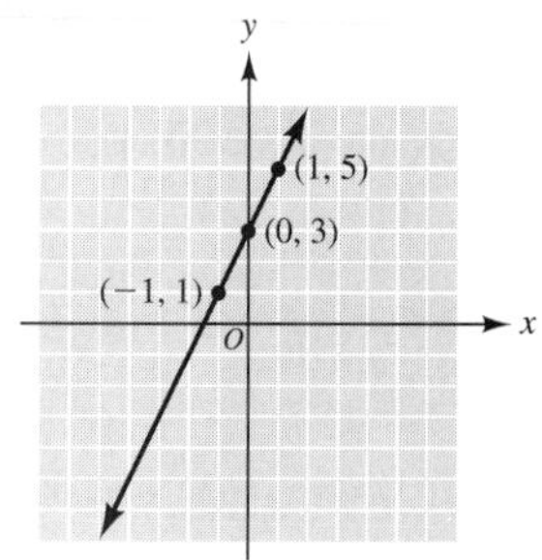

46.

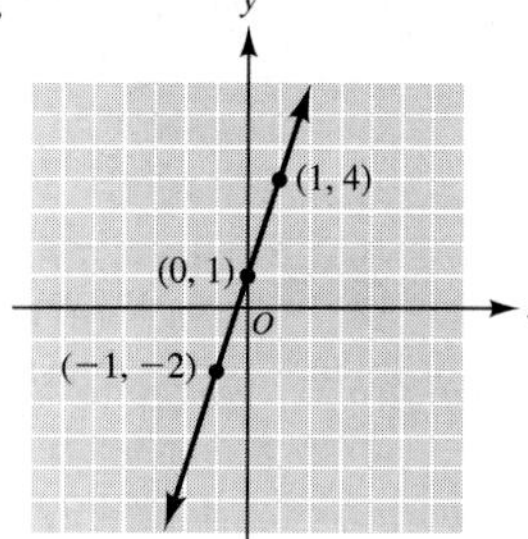

47.

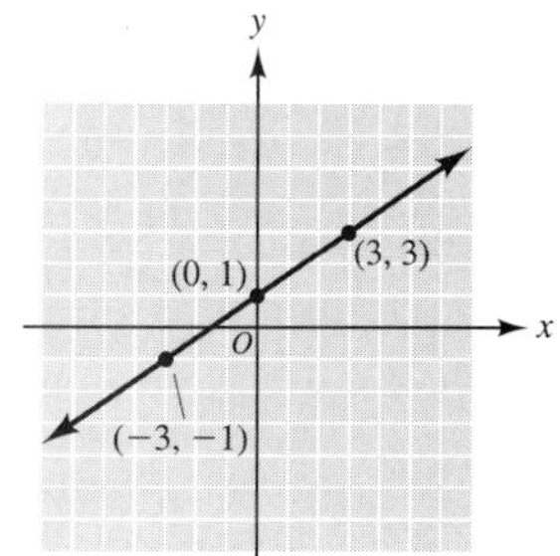

48.

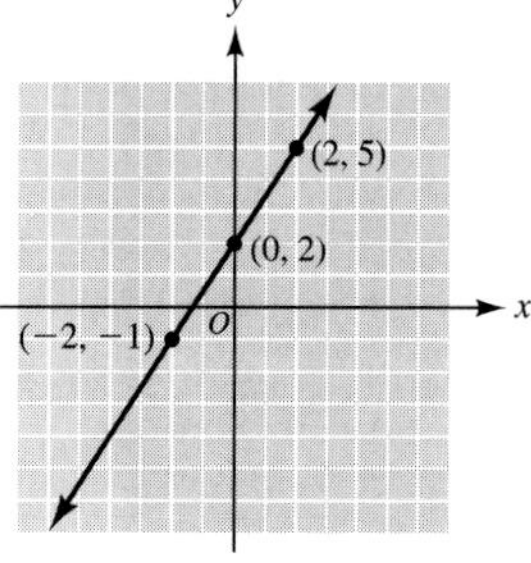

Exercises 8.5 (page 422)

1. Boundary line: $4x + 5y = 20$
Intercepts are (5, 0) and (0, 4).
Dashed line because the inequality is $<$.
Half-plane includes (0, 0) because
$4(0) + 5(0) < 20$
$0 < 20$ True

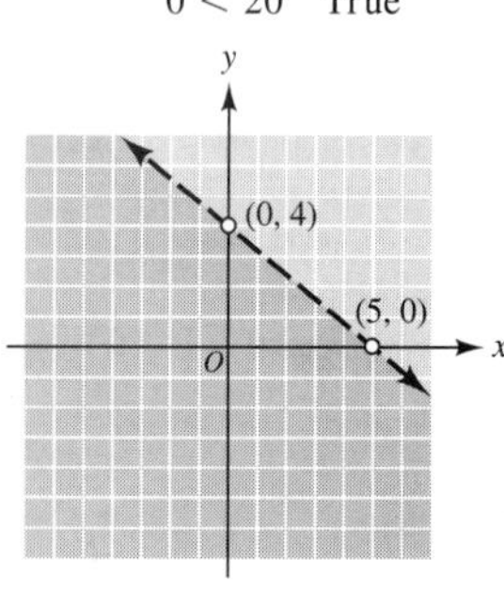

2.

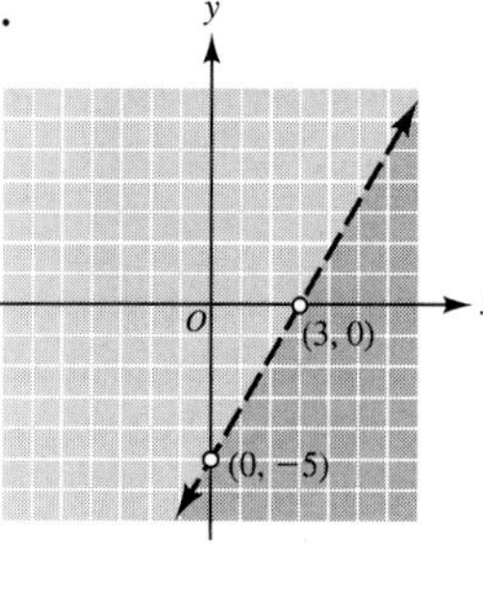

3. Boundary line: $3x - 8y = -16$
Intercepts are $\left(-5\frac{1}{3}, 0\right)$ and (0, 2).
Dashed line because the inequality is $>$.
Half-plane includes (0, 0) because
$3(0) - 8(0) > -16$
$0 > -16$ True

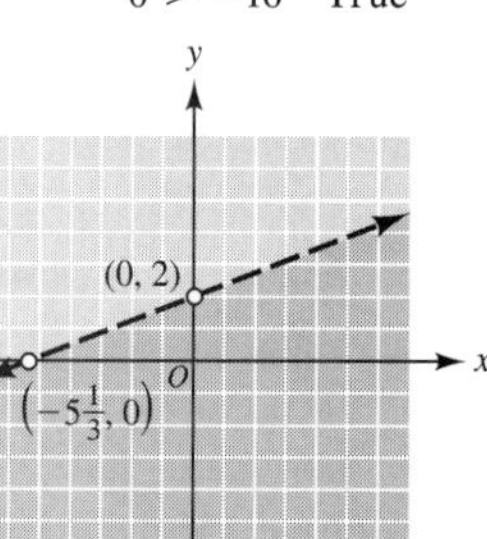

4.

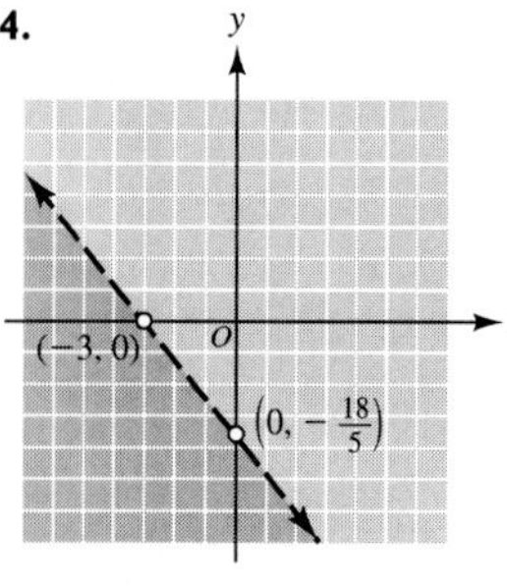

5. Boundary line: $9x + 7y = -27$
Intercepts are (−3, 0) and $\left(0, -3\frac{6}{7}\right)$.
Solid line because the inequality is $\leq$.
Half-plane does not include (0, 0) because
$9(0) + 7(0) \leq -27$
$0 \leq -27$ False

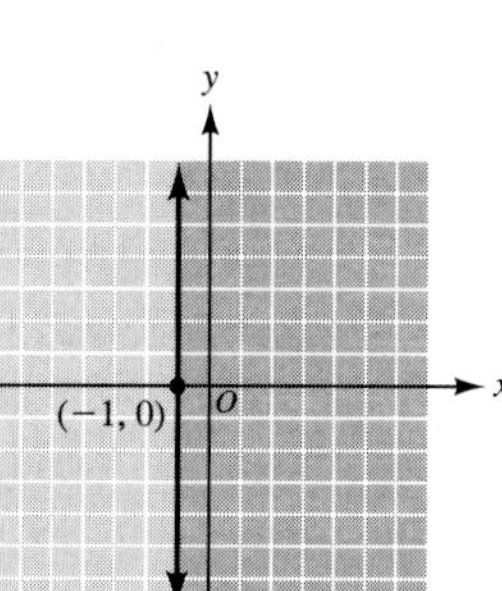

6.

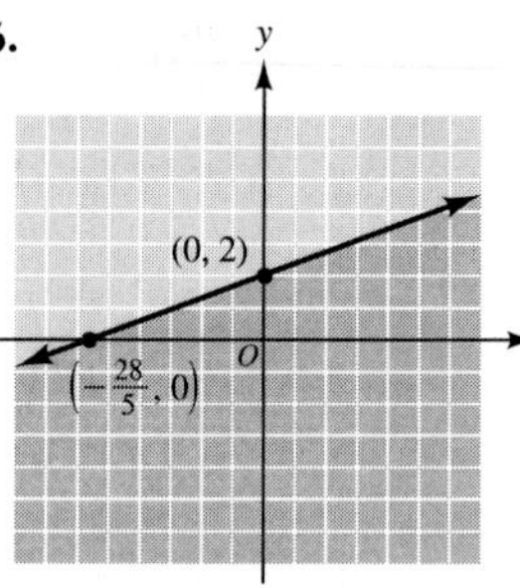

7. Boundary line: $x = -1$
[a vertical line through (−1, 0)]
Solid line because the inequality is $\geq$.
Half-plane includes (0, 0) because $0 \geq -1$ True

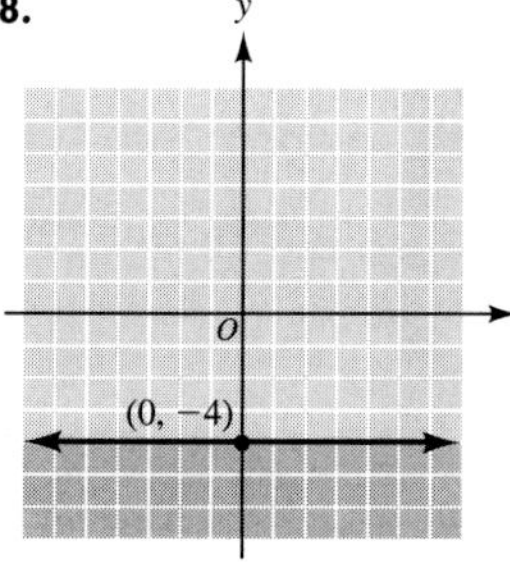

8.

9. Boundary line: $6x - 13y = 0$
Intercepts are (0, 0);
another point is $\left(6, 2\frac{10}{13}\right)$.
Dashed line because the inequality is $>$.
Boundary line passes through the origin.
Half-plane does not include (1, 1) because
$6(1) - 13(1) > 0$
$-7 > 0$ False

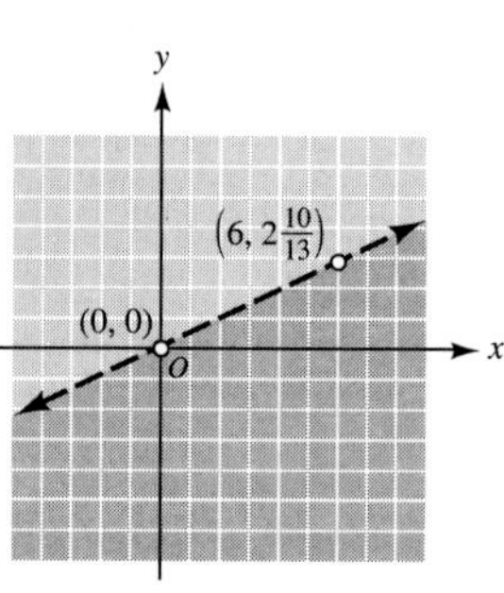

10.

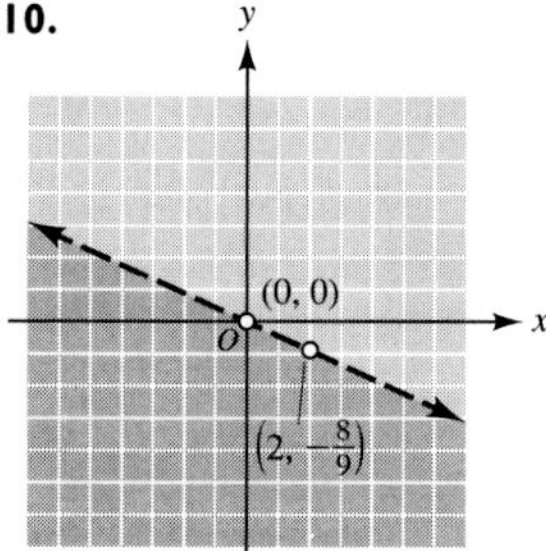

11. Boundary line: $14x + 3y = 17$
Intercepts are $\left(1\frac{3}{14}, 0\right)$ and $\left(0, 5\frac{2}{3}\right)$;
checkpoint is (1, 1).
Solid line because the inequality is $\leq$.
Half-plane includes (0, 0) because
$14(0) + 3(0) \leq 17$
$0 \leq 17$ True

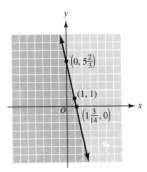

12.

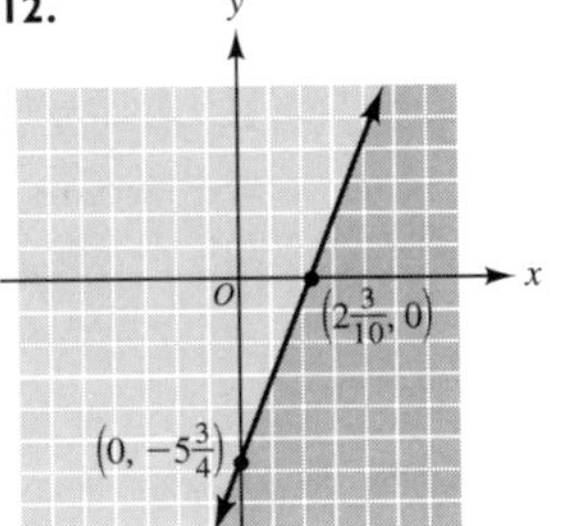

13. Boundary line: $y = -4x - 2$
y-intercept is -2; slope is -4.
Solid line because the inequality is $\geq$.
Half-plane includes (0, 0) because
$0 \geq -4(0) - 2$
$0 \geq -2$ True

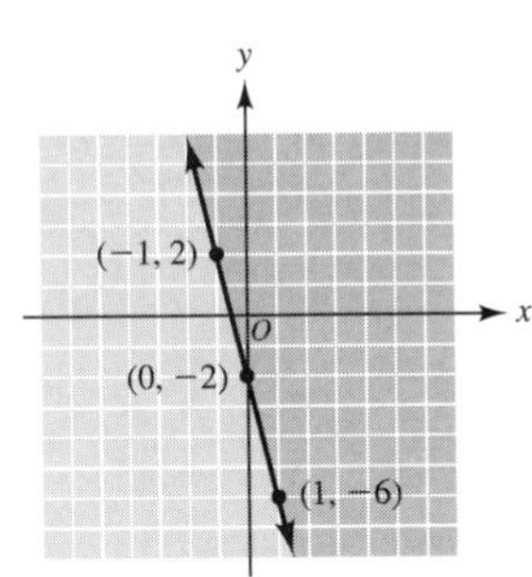

14.

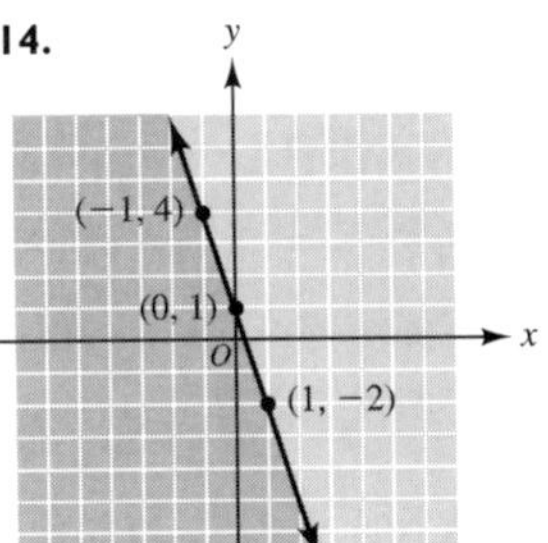

15. Boundary line: $y = 2x + 3$
y-intercept is 3; slope is 2.
Dashed line because the inequality is $>$.
Half-plane does not include (0, 0) because
$0 > 2(0) + 3$
$0 > 3$ False

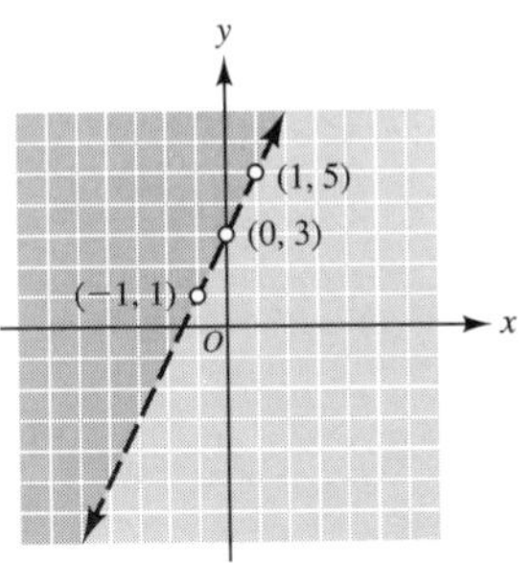

16.

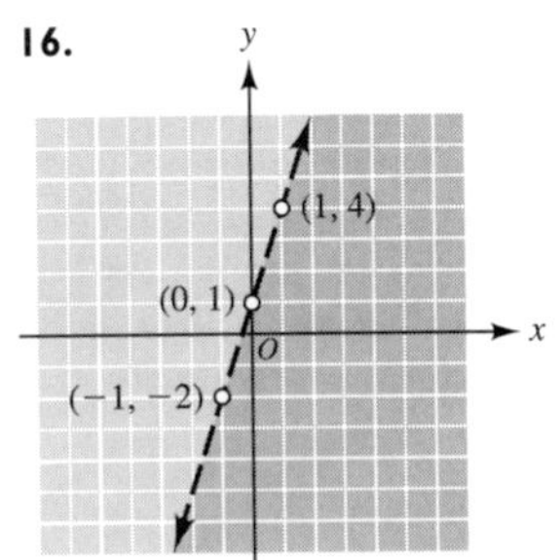
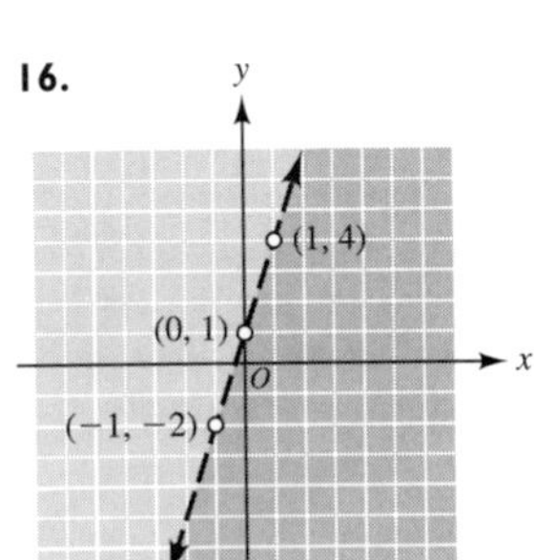

17. $\overset{1}{\cancel{4}}\left(\frac{x}{\underset{1}{\cancel{4}}}\right) - \overset{2}{\cancel{4}}\left(\frac{y}{\underset{1}{\cancel{2}}}\right) > 4(1)$

$x - 2y > 4$
Boundary line: $x - 2y = 4$
Intercepts are (4, 0) and (0, −2).
Dashed line because the inequality is $>$. Half-plane does not include (0, 0) because
$(0) - 2(0) > 4$
$0 > 4$ False

18.

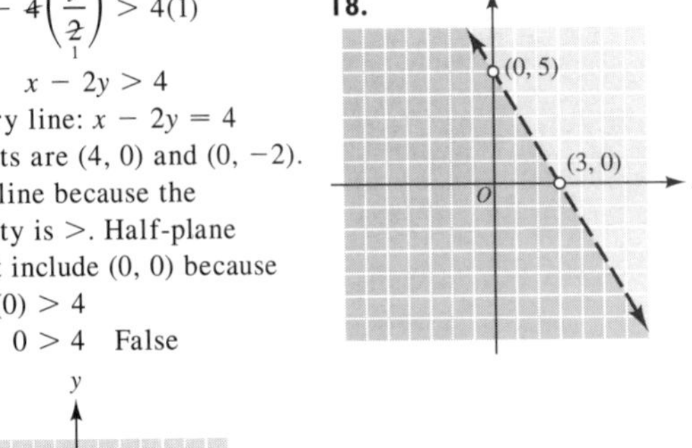

19. $4x + 8 + 7 \leq 15 - 6x$
$10x \leq 0$
$x \leq 0$
Boundary line: $x = 0$
[a vertical line through (0, 0)]
Solid line because the inequality is $\leq$. Half-plane does not include (1, 0) because
$1 \leq 0$ False

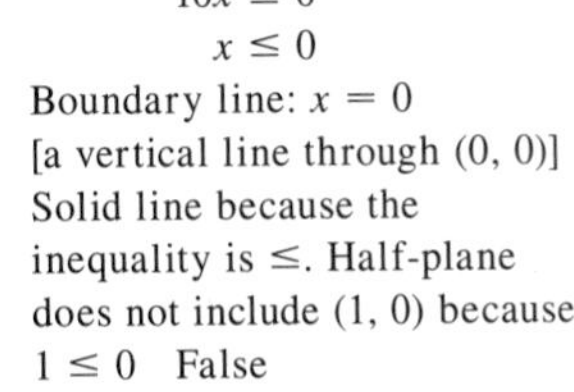

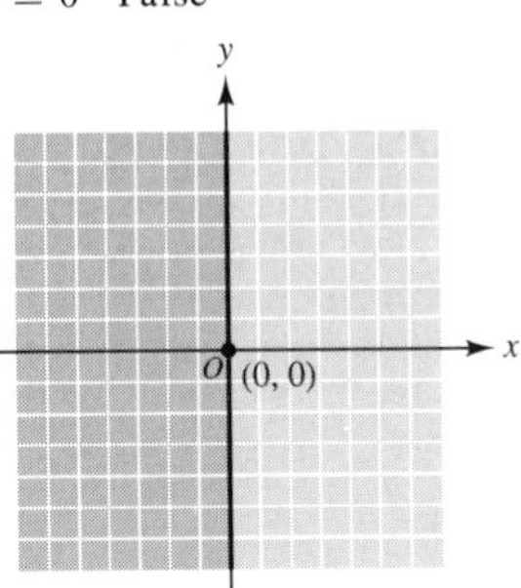

20.

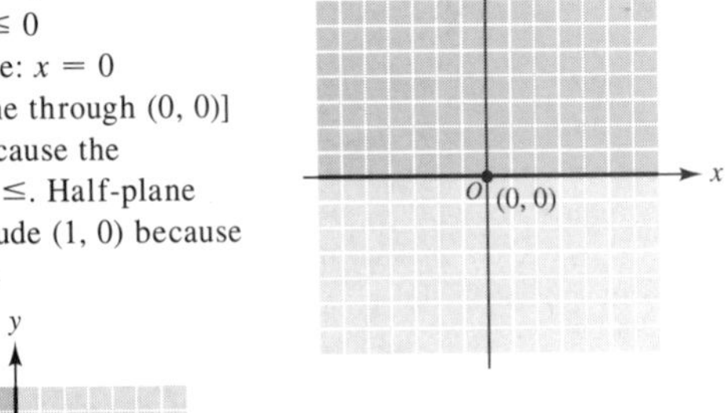

21. $\overset{2}{\cancel{6}}\left(\frac{2x + y}{\underset{1}{\cancel{3}}}\right) + \overset{3}{\cancel{6}}\left(\frac{-(x - y)}{\underset{1}{\cancel{2}}}\right) \geq \overset{1}{\cancel{6}}\left(\frac{5}{\underset{1}{\cancel{6}}}\right)$

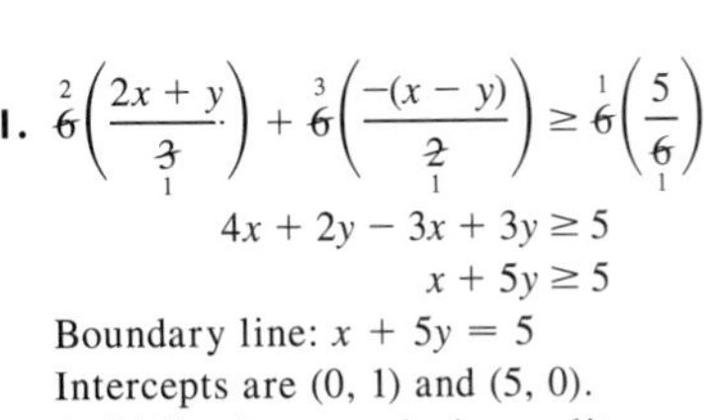

$4x + 2y - 3x + 3y \geq 5$
$x + 5y \geq 5$
Boundary line: $x + 5y = 5$
Intercepts are (0, 1) and (5, 0).
Solid line because the inequality is $\geq$. Half-plane does not include (0, 0) because
$(0) + 5(0) \geq 5$
$0 \geq 5$ False

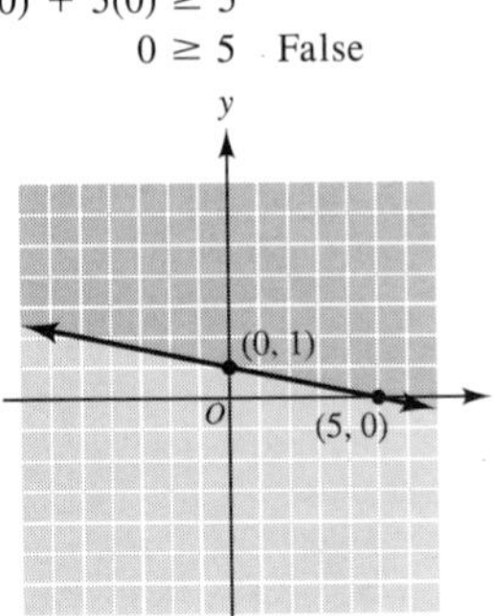

22.

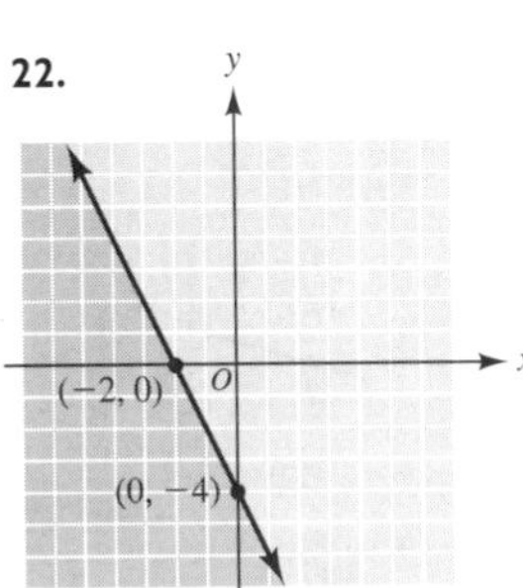
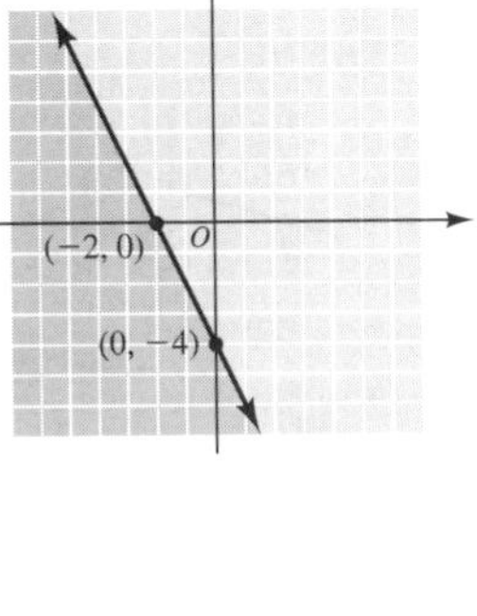

Exercises 8.6 (page 424)

The checks will not be shown.

1. The graph is a straight line. Let a *general* point on the line be (C, F). Water freezes at 0°C and at 32°F, so one point on the line is (0, 32). Water boils at 100°C and at 212°F, so another point on the line is (100, 212). The slope, in general, is $m = \dfrac{F_2 - F_1}{C_2 - C_1}$. For the given points, $m = \dfrac{212 - 32}{100 - 0} = \dfrac{180}{100} = \dfrac{9}{5}$.
Using the point (0, 32) as the fixed point, we have $F - 32 = \frac{9}{5}(C - 0)$, or $F = \frac{9}{5}C + 32$.
If $C = 37$, $F = \frac{9}{5}(37) + 32 = 66.6 + 32 = 98.6$. Therefore, the Fahrenheit temperature that corresponds to a Celsius reading of 37° is 98.6°.

2. $y = 9x + 900$; $3,600

3. Let x = units digit
Then $13 - x$ = tens digit
Original number: $10(13 - x) + x$ *Number with digits reversed*: $10x + (13 - x)$

$$
\begin{aligned}
10(13 - x) + x + 27 &= 10x + (13 - x)\\
130 - 10x + x + 27 &= 10x + 13 - x\\
-9x + 157 &= 9x + 13\\
144 &= 18x\\
x &= 8 \quad \text{Units digit}\\
13 - x = 13 - 8 &= 5 \quad \text{Tens digit}
\end{aligned}
$$

Therefore, the original number is 58.

4. 100 mi

5. Let x = volume of the tank

$$
\begin{aligned}
\tfrac{1}{3}x + 3{,}000 &= \tfrac{3}{4}x\\
4x + 36{,}000 &= 9x\\
36{,}000 &= 5x\\
x &= 7{,}200
\end{aligned}
$$

Therefore, the volume of the tank is 7,200 cu. in.

6. 52

7. Let x = hundreds digit
Then $2x$ = units digit
and $6 - (x + 2x)$ = tens digit (Sum of the digits is 6)
Original number: $100x + 10(6 - 3x) + 2x$ *Number with digits reversed*: $100(2x) + 10(6 - 3x) + x$

$$
\begin{aligned}
100(2x) + 10(6 - 3x) + x &= 198 + 100x + 10(6 - 3x) + 2x\\
200x + 60 - 30x + x &= 198 + 100x + 60 - 30x + 2x\\
171x + 60 &= 72x + 258\\
99x &= 198\\
x &= 2 \quad \text{Hundreds digit}\\
2x &= 4 \quad \text{Units digit}\\
6 - 3x &= 0 \quad \text{Tens digit}
\end{aligned}
$$

The original number is 204.

8. 45 min

9. In the original 20 lb of alloy:
40% of 20 lb is (0.40) (20 lb) = 8 lb (amount of nickel)
60% of 20 lb is (0.60) (20 lb) = 12 lb (amount of copper)
Let x = number of lb of pure copper to be added

$$\frac{x + 12}{8} = \frac{3}{1} \quad \text{Required ratio is 3 to 1.}$$

$$
\begin{aligned}
x + 12 &= 24\\
x &= 12
\end{aligned}
$$

12 lb of pure copper should be used.

10. 42 lb

11. Let x = amount invested at 5.75%
Then $13{,}000 - x$ = amount invested at 5.25%

$$
\begin{aligned}
0.0575x + 0.0525(13{,}000 - x) &= 702.50\\
575x + 525(13{,}000 - x) &= 7{,}025{,}000\\
575x + 6{,}825{,}000 - 525x &= 7{,}025{,}000\\
50x &= 200{,}000\\
x &= 4{,}000\\
13{,}000 - x &= 9{,}000
\end{aligned}
$$

$4,000 was invested at 5.75%, and $9,000 was invested at 5.25%.

12. 326

13. Let x = price of cheaper camera
Then $x + 48$ = price of expensive camera
Also, $18x$ = total price paid for cheaper cameras
and $10(x + 48)$ = total price paid for expensive cameras

$$
\begin{aligned}
18x &= 10(x + 48)\\
18x &= 10x + 480\\
8x &= 480\\
x &= 60\\
x + 48 &= 108
\end{aligned}
$$

The cheaper cameras were $60 each, and the more expensive ones were $108 each.

14. 300 children

15. The graph is a straight line. Let a *general* point on the line be (x, y). One point on the line is (8, 28), and another is (5, 17.5).
The slope, in general, is $m = \dfrac{y_2 - y_1}{x_2 - x_1}$. For the given points, $m = \dfrac{28 - 17.5}{8 - 5} = \dfrac{10.5}{3} = 3.5$. Using the point (8, 28) as the fixed point, we have $y - 28 = 3.5(x - 8)$, or $y = 3.5x$.
If $x = 10$, $y = 3.5(10) = 35$. Therefore, the force necessary to stretch the spring 10 in. is 35 lb.

Sections 8.1–8.6 Review Exercises (page 427)

1. a. $d = \sqrt{[2 - (-3)]^2 + [-4 - (-4)]^2} = \sqrt{5^2 + 0^2} = 5$
b. $d = \sqrt{[4 - (-2)]^2 + [1 - (-3)]^2} = \sqrt{6^2 + 4^2} = \sqrt{36 + 16} = \sqrt{52} = \sqrt{4 \cdot 13} = 2\sqrt{13}$

2. Domain is $\{0, -2, 3\}$; range is $\{5, 3, -4, 0\}$.

3. Domain is $\{-2, 0, 6, 8\}$; range is $\{-6, -3, 6, 9\}$.

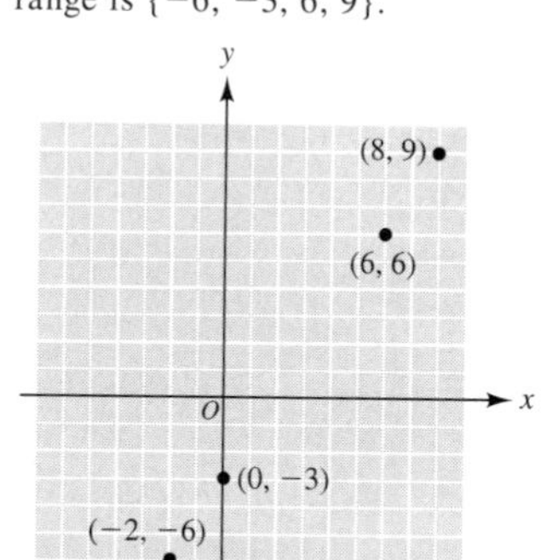

4.

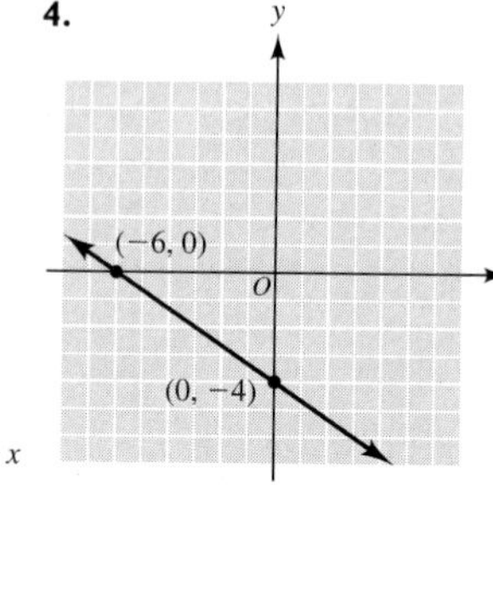

5.

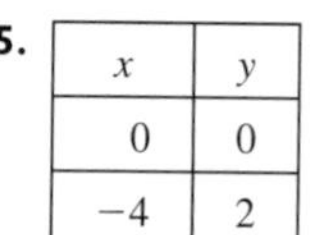

x	y
0	0
−4	2

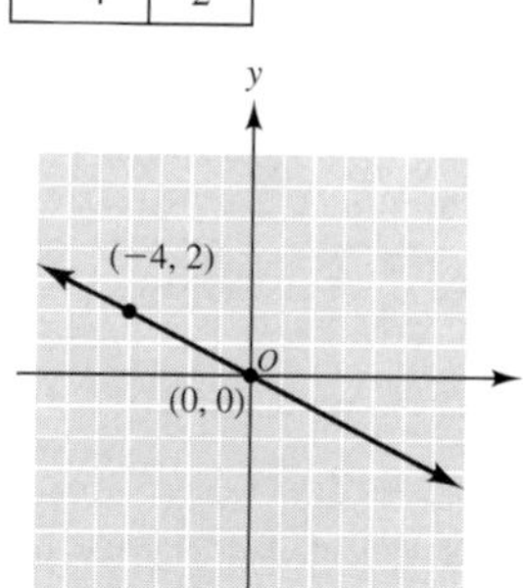

6.

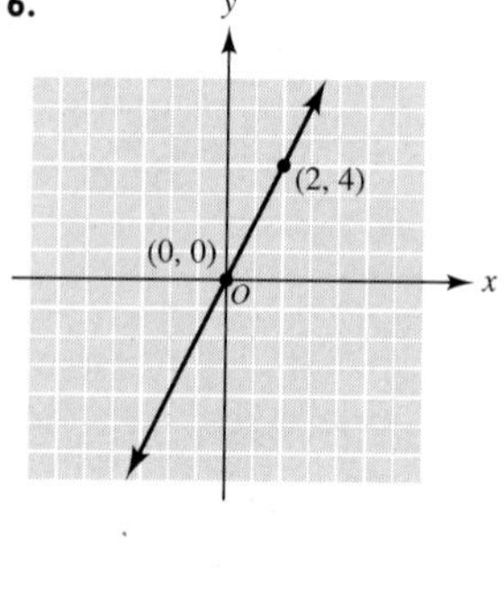

8

7. $\overset{4}{\cancel{20}}\left(\dfrac{2x + 3y}{\underset{1}{\cancel{5}}}\right) + \overset{5}{\cancel{20}}\left(\dfrac{-(x - 3y)}{\underset{1}{\cancel{4}}}\right) = \overset{2}{\cancel{20}}\left(\dfrac{9}{\underset{1}{\cancel{10}}}\right)$

$$4(2x + 3y) + 5(-x + 3y) = 2(9)$$
$$8x + 12y - 5x + 15y = 18$$
$$3x + 27y = 18$$
$$x + 9y = 6$$

x	y
0	$\frac{2}{3}$
6	0
−3	1

(−3, 1) $(0, \frac{2}{3})$ (6, 0)

8. $-\frac{9}{5}$ **9.** $m = \dfrac{2 - 2}{3 - (-6)} = \dfrac{0}{9} = 0$ **10.** $3x + 4y + 8 = 0$

11.
$$y = mx + b$$
$$y = -\tfrac{1}{2}x + 6$$
$$2y = -x + 12$$
$$x + 2y - 12 = 0$$

12. $7x - 3y + 2 = 0$

13. $10\left(\dfrac{2x}{5} - \dfrac{3}{2}y\right) = 3(10)$
$$4x - 15y = 30$$
$$\frac{15y}{15} = \frac{4x - 30}{15}$$
$$y = \tfrac{4}{15}x - 2$$
$m = \frac{4}{15}$, y-intercept $= -2$

14.

(0, 3)

15. Boundary line:
$2x - 5y = 10$

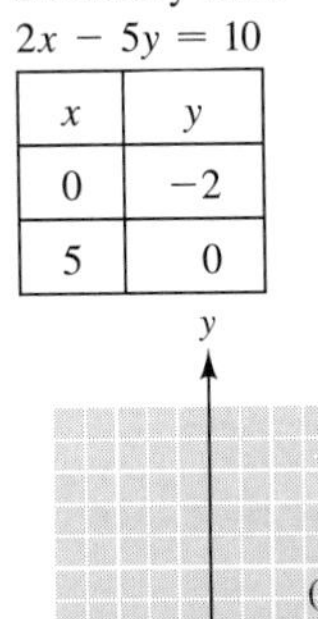

x	y
0	−2
5	0

(5, 0) (0, −2)

16.

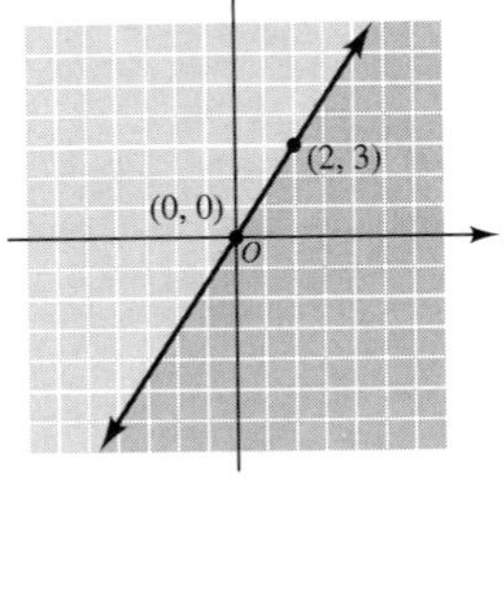

17. The graph is a straight line. Let a *general* point on the line be (x, y). One point on the line is (6, 18), and another is (8, 24). The slope, in general, is $m = \dfrac{y_2 - y_1}{x_2 - x_1}$. For the given points, $m = \dfrac{24 - 18}{8 - 6} = \dfrac{6}{2} = 3$. Using the point (6, 18) as the fixed point, we have $y - 18 = 3(x - 6)$, or $y = 3x$.

If $x = 12$, $y = 3(12) = 36$. Therefore, the force necessary to stretch the spring 12 in. is 36 lb.

18. 83 **19.** Let x = amount invested at 12%
Then $27{,}000 - x$ = amount invested at 8%
$$0.12x + 0.08(27{,}000 - x) = 2{,}780$$
$$12x + 8(27{,}000 - x) = 278{,}000$$
$$12x + 216{,}000 - 8x = 278{,}000$$
$$4x = 62{,}000$$
$$x = 15{,}500$$
$$27{,}000 - x = 11{,}500$$
(The check will not be shown.)
\$15,500 was invested at 12% and \$11,500 at 8%.

20. The distance between Ms. Sontag's home and her work is 15 mi.

Chapter 8 Diagnostic Test (page 433)

Following each problem number is the number (in parentheses) of the textbook section where that kind of problem is discussed.

1. (8.1)

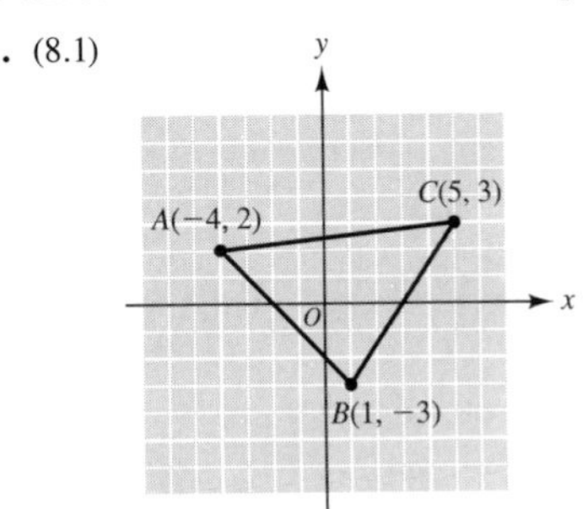

2. (8.3, 8.4) $x_1 = -4$, $y_1 = 2$, $x_2 = 1$, and $y_2 = -3$
$$m = \frac{-3 - 2}{1 - (-4)} = \frac{-5}{5} = -1$$
Using (1, −3) as the fixed point, we have
$$y - (-3) = -1(x - 1)$$
$$y + 3 = -x + 1$$
$$x + y + 2 = 0$$

3. (8.2) If $y = 0$, $x = -2$; the x-intercept is $(-2, 0)$.
If $x = 0$, $y = -2$; the y-intercept is $(0, -2)$.

4. (8.4)
$$y = mx + b$$
$$y = \tfrac{6}{5}x - 4$$
$$5y = 6x - 20$$
$$6x - 5y - 20 = 0$$

5. (8.4) $3x - 5y = 15$
$$-5y = -3x + 15$$
$$y = \tfrac{3}{5}x - 3$$
The slope is $\frac{3}{5}$; the y-intercept is −3.

6. (8.4) The slope must be $\frac{3}{5}$.
$$y - 6 = \tfrac{3}{5}(x + 1) \quad \text{Point-slope form}$$
$$5(y - 6) = 3(x + 1)$$
$$5y - 30 = 3x + 3$$
$$3x - 5y + 33 = 0 \quad \text{General form}$$

7. (8.2) $x - 2y = 6$
If $x = 0$, $y = -3$.
If $y = 0$, $x = 6$.

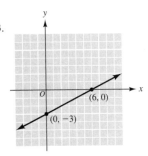

(6, 0) (0, −3)

8

8. (8.5) $4x - 3y \le -12$
Boundary line:
$4x - 3y = -12$
If $x = 0$, $y = 4$.
If $y = 0$, $x = -3$.
Boundary line is solid because the inequality is $\le$.
Test point: (0, 0)
$4x - 3y \le -12$
$4(0) - 3(0) \le -12$
$0 \le -12$ False

9. (8.1) **a.** $\{(-4, -5), (2, 4), (4, -2), (-2, 3), (2, -1)\}$

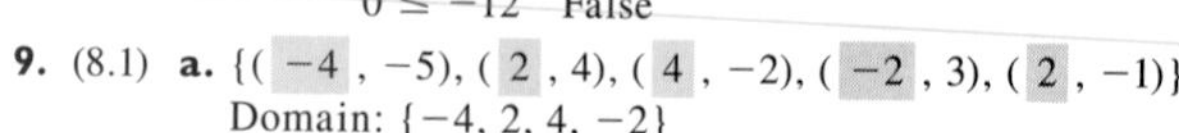

Domain: $\{-4, 2, 4, -2\}$

b. $\{(-4, -5), (2, 4), (4, -2), (-2, 3), (2, -1)\}$
Range: $\{-5, 4, -2, 3, -1\}$

c.

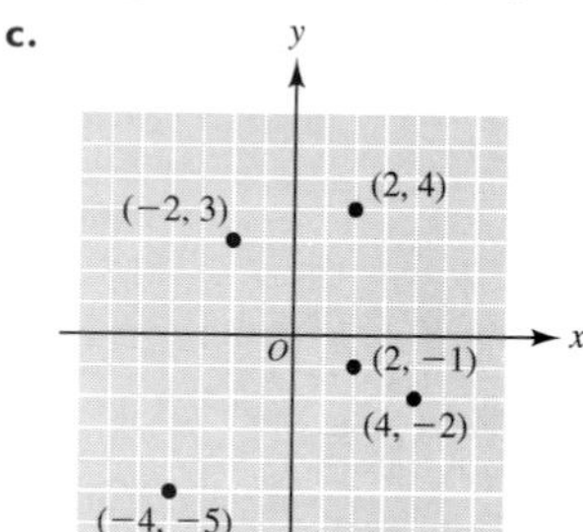

10. (8.6) The graph is a straight line. Let a *general* point on the line be (x, y). One point on the line is (8, 20), and another is (10, 25). The slope, in general, is $m = \dfrac{y_2 - y_1}{x_2 - x_1}$. For the given points, $m = \dfrac{25 - 20}{10 - 8} = \dfrac{5}{2} = 2.5$. Using the point (8, 20) as the fixed point, we have $y - 20 = 2.5(x - 8)$, or $y = 2.5x$. If $x = 11$, $y = 2.5(11) = 27.5$. Therefore, the force necessary to stretch the spring 11 in. is 27.5 lb.

Chapters 1–8 Cumulative Review Exercises (page 433)

1. $2x^4 - 24x^2 + 54 = 2(x^4 - 12x^2 + 27)$
$= 2(x^2 - 3)(x^2 - 9)$
$= 2(x^2 - 3)(x + 3)(x - 3)$

2. $ab(2a - 3b)(4a^2 + 6ab + 9b^2)$

3. $\sqrt[3]{-8a^5} = -\sqrt[3]{2^3a^3a^2} = -2a\sqrt[3]{a^2}$ **4.** $7x^2\sqrt{2} + 4x^2\sqrt{2x}$

5. $\left(\dfrac{3 + 2i}{2 - i}\right)\left(\dfrac{2 + i}{2 + i}\right) = \dfrac{6 + 7i + 2i^2}{4 - i^2} = \dfrac{6 + 7i - 2}{4 - (-1)} = \dfrac{4 + 7i}{5}$
$= \dfrac{4}{5} + \dfrac{7}{5}i$

6. $2 + 2i$ **7.** $(3 + i)^4 = 3^4 + 4(3^3)i + 6(3^2)i^2 + 4(3)i^3 + i^4$
$= 81 + 108i + 54(-1) + 12(-i) + 1$
$= 81 - 54 + 1 + (108 - 12)i$
$= 28 + 96i$

8. $r = \dfrac{S - a}{S}$ **9.** $x^{-1/3} = 3$
$(x^{-1/3})^{-3} = (3)^{-3}$
$x = \dfrac{1}{3^3} = \dfrac{1}{27}$
$\left\{\frac{1}{27}\right\}$

10. $3x^2 - 2x - 3$

11. $7x + 2y = 14$
$2y = -7x + 14$
$y = -\frac{7}{2}x + 7$
The slope is $m = -\frac{7}{2}$; the y-intercept is 7.

12.

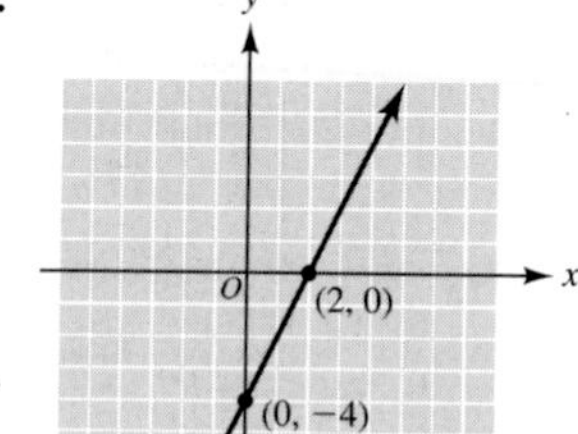

13. $m = \dfrac{4 - (-2)}{-3 - 1} = \dfrac{6}{-4} = -\dfrac{3}{2}$
$y - 4 = -\frac{3}{2}(x - [-3])$
$2(y - 4) = -3(x + 3)$
$2y - 8 = -3x - 9$
$3x + 2y + 1 = 0$

14. The cheaper stereos were \$70 each, and the more expensive ones were \$125 each.

15. Let x = number of lb of cashews
Then $60 - x$ = number of lb of peanuts
$7.40x + 2.80(60 - x) = 4.18(60)$
$740x + 280(60 - x) = 418(60)$
$74x + 28(60 - x) = 418(6)$
$74x + 1{,}680 - 28x = 2{,}508$
$46x = 828$
$x = 18$
$60 - x = 42$
Check 18 lb @ \$7.40 = \$133.20
42 lb @ \$2.80 = \$117.60
\$250.80
60 lb @ \$4.18 = \$250.80
The grocer should use 18 lb of cashews and 42 lb of peanuts.

16. 88 mi

17. Let x = length of shorter leg
Then $x + 7$ = length of longer leg
and $x + 9$ = length of hypotenuse
$x^2 + (x + 7)^2 = (x + 9)^2$
$x^2 + x^2 + 14x + 49 = x^2 + 18x + 81$
$x^2 - 4x - 32 = 0$
$(x - 8)(x + 4) = 0$
$x - 8 = 0$ | $x + 4 = 0$
$x = 8$ | $x = -4$ Reject
$x + 7 = 15$
$x + 9 = 17$

(The check will not be shown.) The shorter leg is 8 cm, the longer leg is 15 cm, and the hypotenuse is 17 cm.

18. Gwen has 5 quarters, 8 dimes, and 3 nickels.

Exercises 9.1 (page 441)

1. (b) and (c) **2.** (b), (c), and (d)

3. a. Yes **b.** No **c.** No **d.** Yes

4. a. No **b.** Yes **c.** Yes **d.** No

5. a. The domain is the set of all real numbers. (The curve extends infinitely far to the left and to the right.) The range is the set of all real numbers. (The curve extends infinitely far up and down.)

b. The domain is the set of all real numbers, and so is the range. (The equation is the equation of a straight line that is not a horizontal or vertical line.)

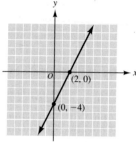

c. For the domain: $x - 5 \geq 0 \Rightarrow x \geq 5$. Therefore, the domain is $\{x \mid x \geq 5\}$. For the range, we know that the square root sign has an understood plus sign in front of it. Therefore, y cannot be negative. As x gets larger and larger, so does $x - 5$, and so does $\sqrt{x-5}$. Therefore, the range is $\{y \mid y \geq 0\}$.

d. The domain is $\{1, 4, 7\}$. If $x = 1$, $y = 0$. If $x = 4$, $y = -3$. If $x = 7$, $y = -6$. Therefore, the range is $\{0, -3, -6\}$.

e. The domain is $\{3, 4, 2\}$, or $\{2, 3, 4\}$; the range is $\{-6, 2, 4\}$.

6. **a.** Domain: $\{x \mid -3 \leq x \leq 3\}$; range: $\{y \mid -2 \leq y \leq 0\}$

b. Domain: $\{x \mid x \geq -3\}$; range: $\{y \mid y \geq 0\}$

c. Domain: $\{x \mid x \in R\}$; range: $\{y \mid y \in R\}$

d. Domain: $\{0, 1\}$; range: $\{-4, -1\}$

e. Domain: $\{3, 5, 6, 8\}$; range: $\{1\}$

7. w is a function of n because exactly one value of w corresponds to each value of n.
The domain is $\{1, 2, 3, 4, 5, 6, 7, 8\}$.
The range is $\{4, 8, 12, 16, 20, 24, 28, 32\}$.

8. y is a function of x because exactly one value of y corresponds to each value of x.
The domain is $\{1, 2, 3, 4, 5, 6, 7, 8\}$.
The range is $\{1, 4, 8, 16, 32, 64, 128, 256\}$.

9. If $x = -5$, $y = 4$
If $x = -1$, $y = \frac{12}{5}$
If $x = 0$, $y = 2$
If $x = 2$, $y = \frac{6}{5}$
Range: $\left\{4, \frac{12}{5}, 2, \frac{6}{5}\right\}$

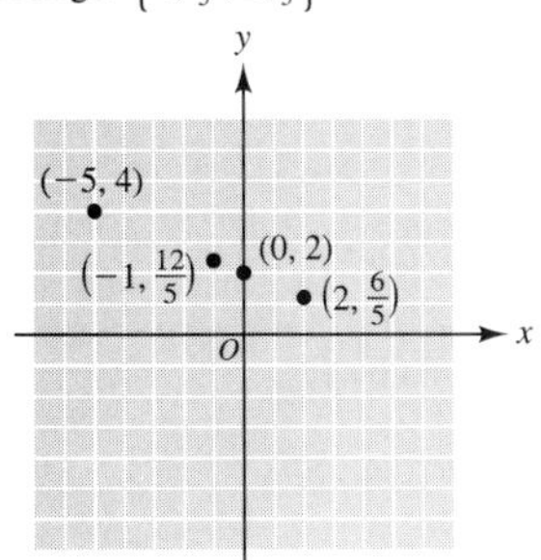

10. Range: $\{-6, -3, 0, 3\}$

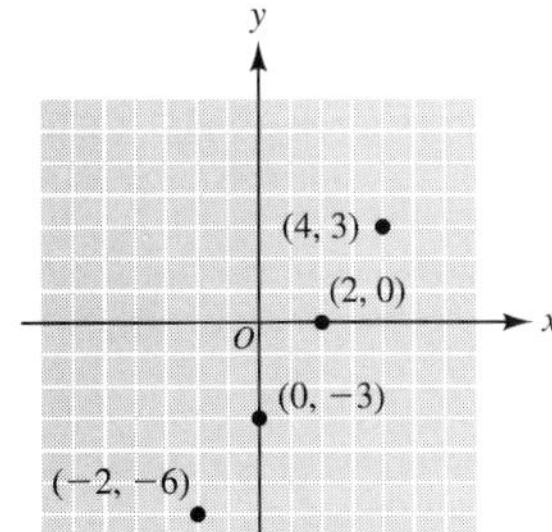

11. $y = 2x - 3$

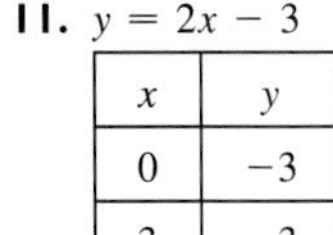

x	y
0	−3
3	3

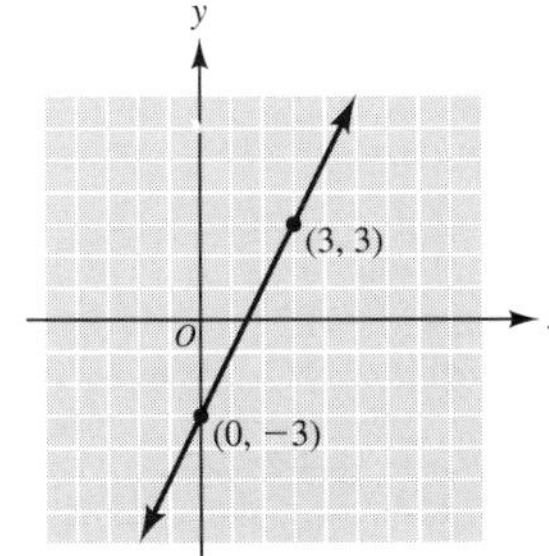

12.

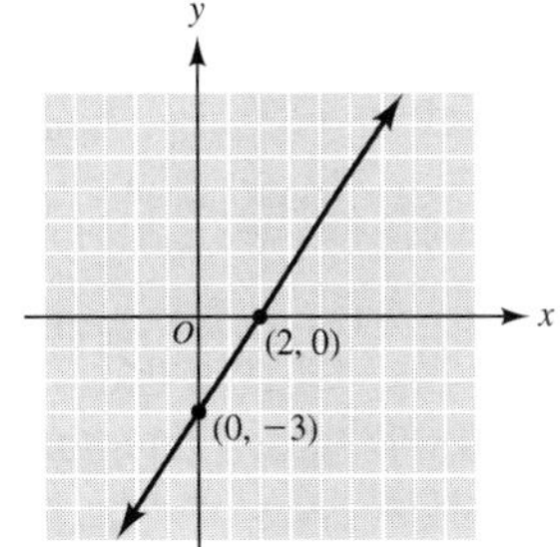

Exercises 9.2 (page 448)

1. **a.** $f(2) = 3(2) - 1 = 6 - 1 = 5$

b. $f(0) = 3(0) - 1 = -1$

c. $f(a - 2) = 3(a - 2) - 1 = 3a - 6 - 1 = 3a - 7$

d. $f(x + 2) = 3(x + 2) - 1 = 3x + 6 - 1 = 3x + 5$

2. **a.** $f(3) = 13$ **b.** $f(-5) = -19$ **c.** $f(0) = 1$

d. $f(x - 2) = 4x - 7$

3. $f(5) = 2(5)^2 - 3 = 50 - 3 = 47$
$f(2) = 2(2)^2 - 3 = 5$
Therefore, $\dfrac{f(5) - f(2)}{6} = \dfrac{47 - 5}{6} = \dfrac{42}{6} = 7.$

4. 8

5. $f(3) = 3(3)^2 - 2(3) + 4 = 27 - 6 + 4 = 25$
$f(1) = 3(1)^2 - 2(1) + 4 = 3 - 2 + 4 = 5$
$f(0) = 3(0)^2 - 2(0) + 4 = 4$
Therefore, $2f(3) + 4f(1) - 3f(0) = 2(25) + 4(5) - 3(4)$
$= 50 + 20 - 12 = 58$

6. 46

7. $f(-3) = (-3)^3 = -27$
$g(2) = \frac{1}{2}$
$6g(2) = 6\left(\frac{1}{2}\right) = 3$
Therefore, $f(-3) - 6g(2) = -27 - 3 = -30.$

8. 29

9. $H(2) = 3(2)^2 - 2(2) + 4 = 12 - 4 + 4 = 12$
$K(3) = (3) - (3)^2 = 3 - 9 = -6$
Therefore, $2H(2) - 3K(3) = 2(12) - 3(-6) = 24 + 18 = 42.$

10. -4

11.
$$\begin{aligned}
f(x) &= x^2 - x \\
f(x + h) &= (x + h)^2 - (x + h) \\
&= x^2 + 2xh + h^2 - x - h \\
\frac{f(x + h) - f(x)}{h} &= \frac{x^2 + 2xh + h^2 - x - h - x^2 + x}{h} \\
&= \frac{2xh + h^2 - h}{h} = \frac{2xh}{h} + \frac{h^2}{h} + \frac{-h}{h} \\
&= 2x + h - 1
\end{aligned}$$

12. $6x + 3h$

13. $f(x) = x^4$ ← We will use the binomial theorem
$$\begin{aligned}
\frac{f(x + h) - f(x)}{h} &= \frac{(x + h)^4 - x^4}{h} \\
&= \frac{\cancel{x^4} + 4x^3h + 6x^2h^2 + 4xh^3 + h^4 - \cancel{x^4}}{h} \\
&= \frac{4x^3h + 6x^2h^2 + 4xh^3 + h^4}{h} \\
&= \frac{\overset{1}{\cancel{h}}(4x^3 + 6x^2h + 4xh^2 + h^3)}{\underset{1}{\cancel{h}}} \\
&= 4x^3 + 6x^2h + 4xh^2 + h^3
\end{aligned}$$

14. $3x^2 + 3xh + h^2$

9

15. $A(r) = \pi r^2$
$C(r) = 2\pi r$
Then $\dfrac{3A(r) - 2C(r)}{\pi r} = \dfrac{3\pi r^2 - 2(2\pi r)}{\pi r}$
$= \dfrac{3\pi r^2}{\pi r} + \dfrac{-4\pi r}{\pi r} = 3r - 4$

16. $5r + 6$

17. $h(2) = 64(2) - 16(2)^2 = 128 - 64 = 64$
The arrow is 64 ft high.

18. 112 ft

19. $C(100) = 500 + 20(100) - 0.1(100^2)$
$= 500 + 2{,}000 - 0.1(10{,}000)$
$= 2{,}500 - 1{,}000 = 1{,}500$
It costs $1,500 to manufacture 100 bushings per day.

20. $400

21. $g(3, -4) = 5(3)^2 - 2(-4)^2 + 7(3) - 4(-4)$
$= 45 - 32 + 21 + 16 = 50$

22. 10

23. $f(100, 0.08, 12) = 100(1 + 0.08)^{12} = 100(1.08)^{12}$
≈ 251.82

24. 2,979,659,632

25. $D_f = \{x \mid x \geq 4\}$ because $\sqrt{x - 4}$ is not real when $x - 4 < 0$ or $x < 4$.
$R_f = \{y \mid y \geq 0\}$ because $y = \sqrt{x - 4}$, and the principal square root cannot be negative.

26. $D_f = \left\{x \mid x \geq \frac{16}{9}\right\}$; $R_f = \{y \mid y \leq 0\}$

Exercises 9.3 (page 452)

1. a.

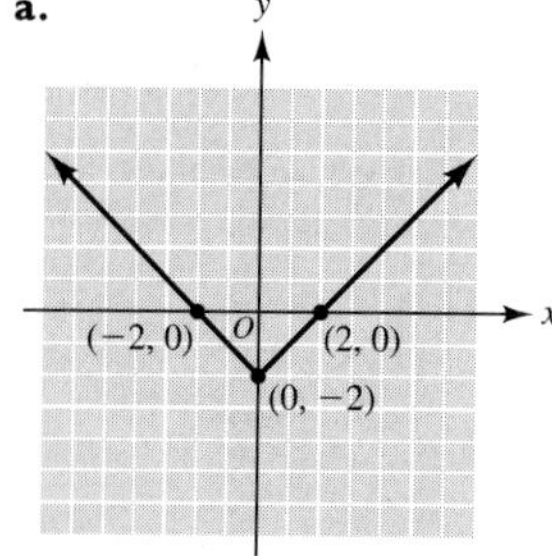

2. a.

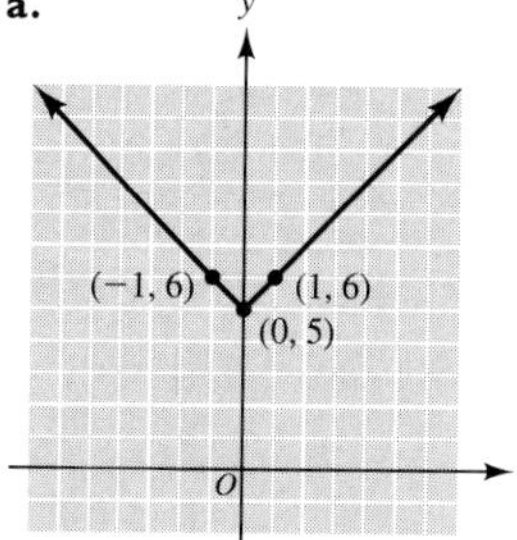

b.

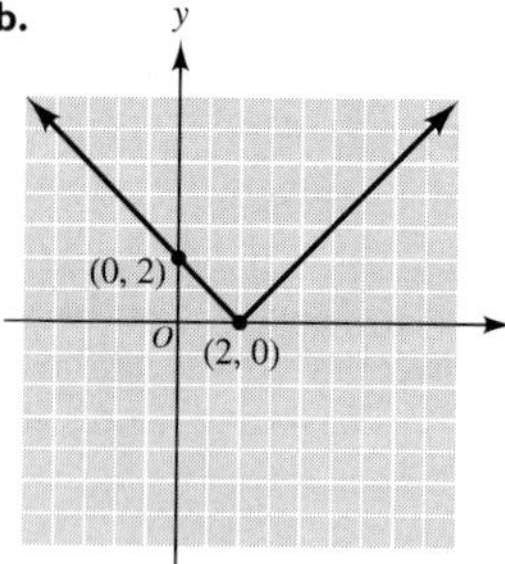

b.

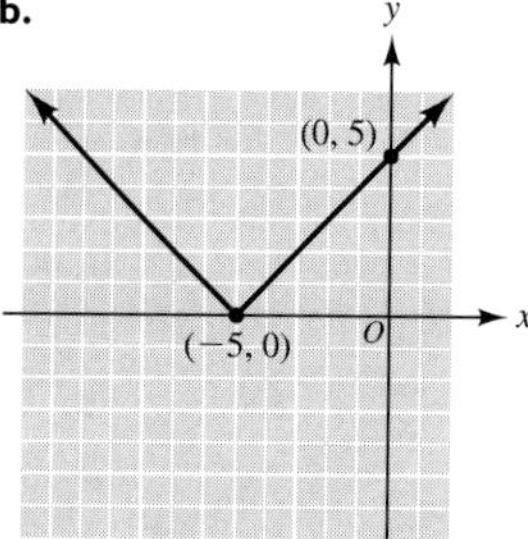

c.

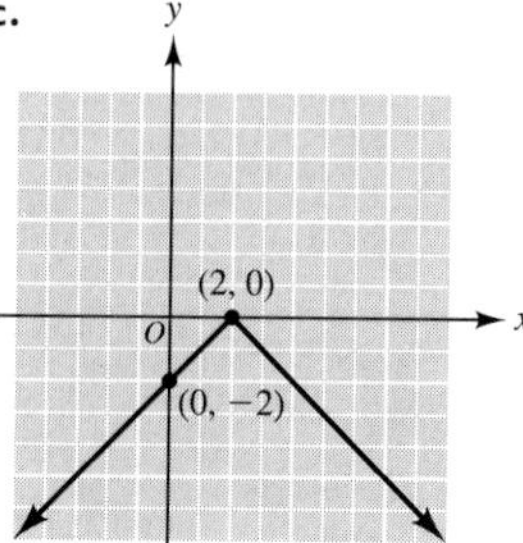

c.

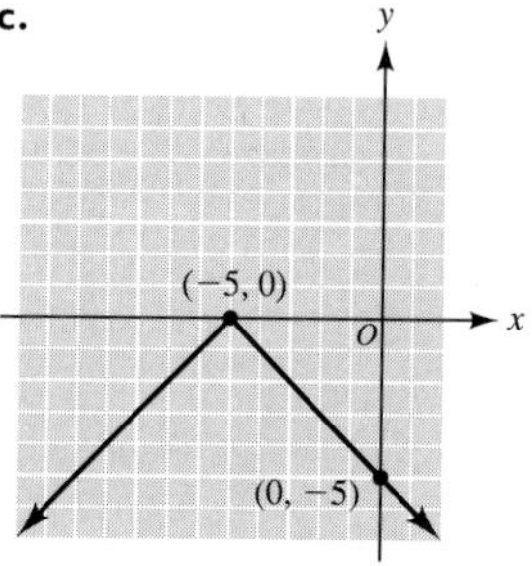

3. a.

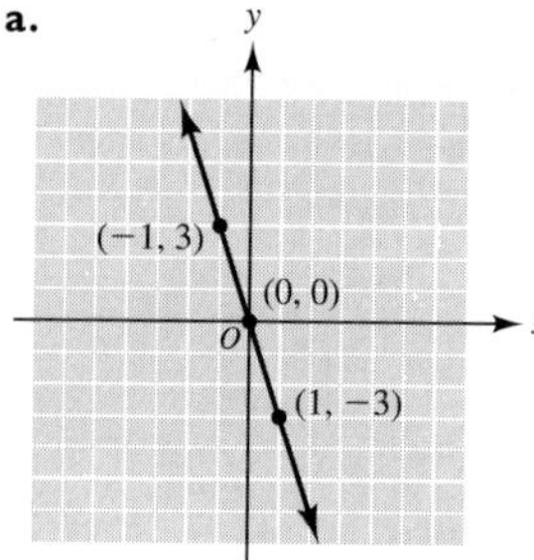

4. a.

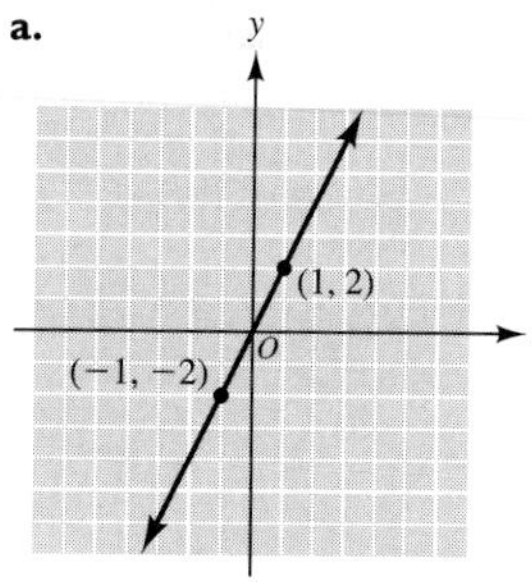

b.

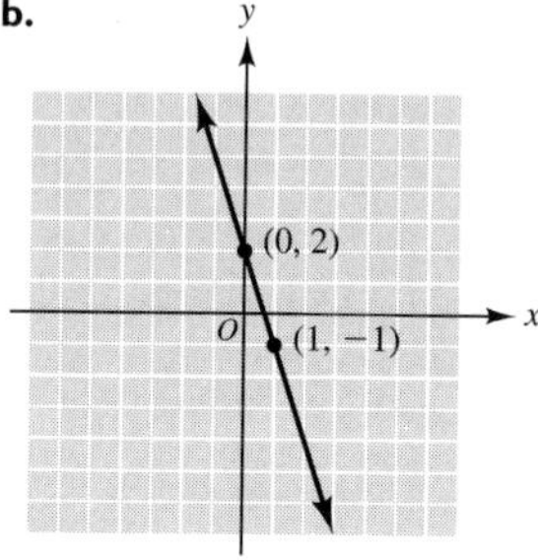

b.

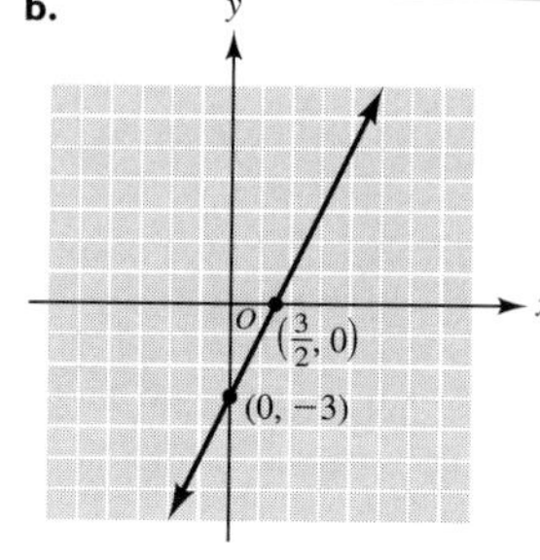

c.

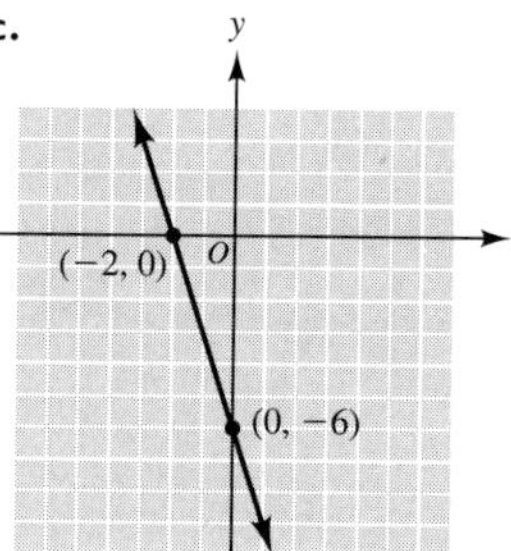

c.

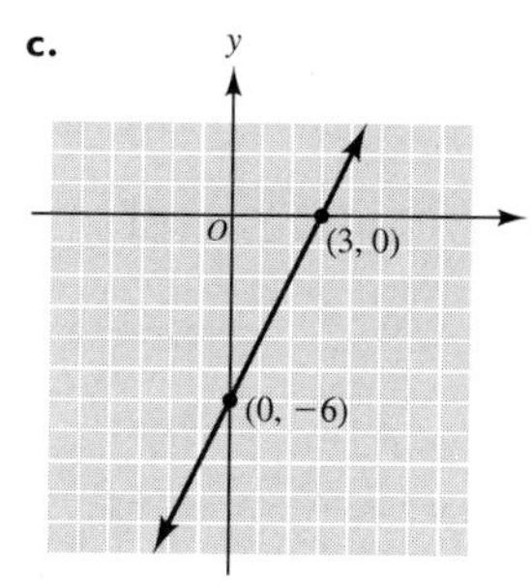

Exercises 9.4 (page 455)

1. The domain of f is R; the domain of g is R; the intersection of those domains is R.
$(f + g)(x) = 7x - 1 + 2 - 5x = 2x + 1$
$(f - g)(x) = 7x - 1 - (2 - 5x) = 7x - 1 - 2 + 5x$
$= 12x - 3$
$(f \cdot g)(x) = (7x - 1)(-5x + 2) = -35x^2 + 19x - 2$
$\left(\dfrac{f}{g}\right)(x) = \dfrac{7x - 1}{2 - 5x}$, with domain the set of all real numbers except $\frac{2}{5}$

2. The domain of f is R; the domain of g is R; the intersection of those domains is R.
$(f + g)(x) = -6x - 2$
$(f - g)(x) = -6$
$(f \cdot g)(x) = 9x^2 + 6x - 8$
$\left(\dfrac{f}{g}\right)(x) = \dfrac{-3x - 4}{2 - 3x}$, with domain the set of all real numbers except $\frac{2}{3}$

3. The domain of f is R; the domain of g is R; the intersection of those domains is R.
$(f + g)(x) = 2 - x + x^2 + 3 = x^2 - x + 5$
$(f - g)(x) = 2 - x - (x^2 + 3) = 2 - x - x^2 - 3$
$= -x^2 - x - 1$
$(f \cdot g)(x) = (2 - x)(3 + x^2) = -x^3 + 2x^2 - 3x + 6$
$\left(\dfrac{f}{g}\right)(x) = \dfrac{2 - x}{x^2 + 3}$, with domain R (no value of x can make $x^2 + 3$ equal 0)

4. The domain of f is R; the domain of g is R; the intersection of those domains is R.
$(f + g)(x) = -x^2 + x + 12$
$(f - g)(x) = x^2 + x$
$(f \cdot g)(x) = -x^3 - 6x^2 + 6x + 36$
$\left(\frac{f}{g}\right)(x) = \frac{6 + x}{6 - x^2}$, with domain the set of all real numbers except $\pm\sqrt{6}$

5. The domain of f is R; the domain of g is R; the intersection of those domains is R.
$(f + g)(x) = 7x - 1 + 7x - 1 = 14x - 2$
$(f - g)(x) = 7x - 1 - (7x - 1) = 7x - 1 - 7x + 1 = 0$
$(f \cdot g)(x) = (7x - 1)(7x - 1) = 49x^2 - 14x + 1$
$\left(\frac{f}{g}\right)(x) = \frac{7x - 1}{7x - 1} = 1$, with domain the set of all real numbers except $\frac{1}{7}$

6. The domain of f is R; the domain of g is R; the intersection of those domains is R.
$(f + g)(x) = \frac{145x}{12}$
$(f - g)(x) = \frac{143x}{12}$
$(f \cdot g)(x) = x^2$
$\left(\frac{f}{g}\right)(x) = 144$, with domain the set of all real numbers except 0

7. The domain of f is $\{x \mid x \le 7\}$; the domain of g is R.
$(f \circ g)(x) = f(g(x)) = \sqrt{7 - (x^2 + 3)} = \sqrt{4 - x^2}$
$(g \circ f)(x) = g(f(x)) = (\sqrt{7 - x})^2 + 3 = 7 - x + 3 = 10 - x$ for $x \le 7$; the domain of $g \circ f$ is $\{x \mid x \le 7\}$. (The domain can't be larger than the domain of f.)

8. The domain of f is R; the domain of g is R.
$(f \circ g)(x) = 2 - 3\sqrt[3]{x} + 3\sqrt[3]{x^2} - x$, with domain R
$(g \circ f)(x) = 1 - \sqrt[3]{x^3 + 1}$, with domain R

9. The domain of f is R; the domain of g is R.
$(f \circ g)(x) = f(g(x)) = 5\left(\frac{x + 1}{5}\right) - 1 = x + 1 - 1 = x$, with domain R
$(g \circ f)(x) = g(f(x)) = \frac{5x - 1 + 1}{5} = \frac{5x}{5} = x$, with domain R

10. The domain of f is R; the domain of g is R.
$(f \circ g)(x) = x$, with domain R
$(g \circ f)(x) = x$, with domain R

11. The domain of f is R; the domain of g is the set of all real numbers except $-\frac{2}{3}$.
$(f \circ g)(x) = f(g(x)) = 3\left(\frac{1}{3x + 2}\right) + 2 = \frac{3}{3x + 2} + \frac{6x + 4}{3x + 2}$
$= \frac{6x + 7}{3x + 2}$, with domain the set of all real numbers except $-\frac{2}{3}$
$(g \circ f)(x) = g(f(x)) = \frac{1}{3(3x + 2) + 2} = \frac{1}{9x + 6 + 2}$
$= \frac{1}{9x + 8}$, with domain the set of all real numbers except $-\frac{8}{9}$

12. The domain of f is R; the domain of g is the set of all real numbers except $\frac{7}{2}$.
$(f \circ g)(x) = \frac{47 - 14x}{7 - 2x}$, with domain the set of all real numbers except $\frac{7}{2}$
$(g \circ f)(x) = \frac{1}{4x - 7}$, with domain the set of all real numbers except $\frac{7}{4}$

Exercises 9.5 (page 460)

1. The domains of f and g are both R.
$(f \circ g)(x) = f(g(x)) = 5\left(\frac{x + 1}{5}\right) - 1 = x + 1 - 1 = x$
$(g \circ f)(x) = g(f(x)) = \frac{5x - 1 + 1}{5} = \frac{5x}{5} = x$
Therefore, f and g are inverses of each other.

2. The domains of f and g are both R.
$(f \circ g)(x) = f(g(x)) = \frac{2x - 8 + 8}{2} = \frac{2x}{2} = x$
$(g \circ f)(x) = g(f(x)) = 2\left(\frac{x + 8}{2}\right) - 8 = x + 8 - 8 = x$
Therefore, f and g are inverses of each other.

3. $(f \circ g)(x) = f(g(x)) = 3\left(\frac{1}{3x + 2}\right) + 2 = \frac{3}{3x + 2} + \frac{6x + 4}{3x + 2}$
$= \frac{6x + 7}{3x + 2}$
Since $(f \circ g)(x) \ne x$, f and g are *not* inverses of each other.

4. $(f \circ g)(x) = f(g(x)) = 7 - 2\left(\frac{1}{7 - 2x}\right)$
$= \frac{49 - 14x}{7 - 2x} + \frac{-2}{7 - 2x} = \frac{47 - 14x}{7 - 2x}$
Since $(f \circ g)(x) \ne x$, f and g are *not* inverses of each other.

5. f is one-to-one, since no ordered pairs with the same first coordinates have different second coordinates, *and* no ordered pairs with the same second coordinates have different first coordinates.
$f^{-1} = \{(5, -2), (-3, 3), (-4, -5), (-2, 4)\}$
The domain of f is $\{-2, 3, -5, 4\}$, or $\{-5, -2, 3, 4\}$, and the range of f is $\{5, -3, -4, -2\}$, or $\{-4, -3, -2, 5\}$.
The domain of f^{-1} is $\{-4, -3, -2, 5\}$, and the range of f^{-1} is $\{-5, -2, 3, 4\}$.

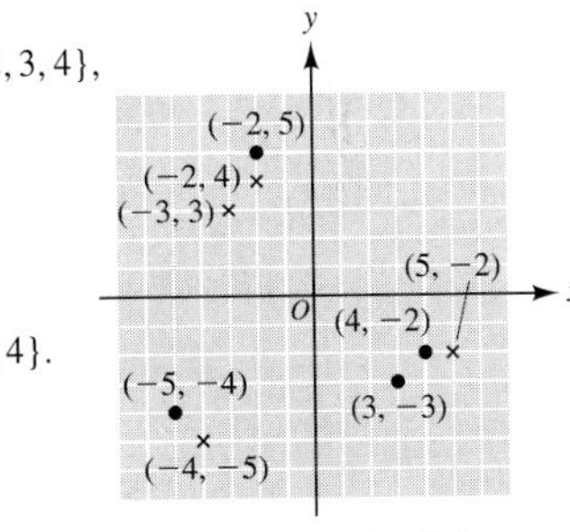

• indicates a point of f
× indicates a point of f^{-1}

6. f is one-to-one, since no ordered pairs with the same first coordinates have different second coordinates, *and* no ordered pairs with the same second coordinates have different first coordinates.
$f^{-1} = \{(-4, 4), (3, 0), (4, 3), (1, -2), (-2, 5)\}$
The domain of f is $\{-2, 0, 3, 4, 5\}$, and the range of f is $\{-4, -2, 1, 3, 4\}$.
The domain of f^{-1} is $\{-4, -2, 1, 3, 4\}$, and the range of f^{-1} is $\{-2, 0, 3, 4, 5\}$.

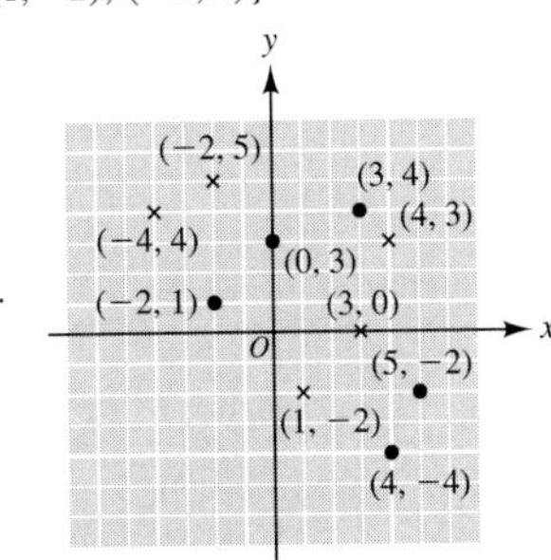

• indicates a point of f
× indicates a point of f^{-1}

7. $x = 5 - 2y \Rightarrow y = \dfrac{5 - x}{2} = f^{-1}(x)$

$f(x) = 5 - 2x \quad f^{-1}(x) = \dfrac{5 - x}{2}$

x	y
$2\frac{1}{2}$	0
0	5

x	y
5	0
0	$2\frac{1}{2}$

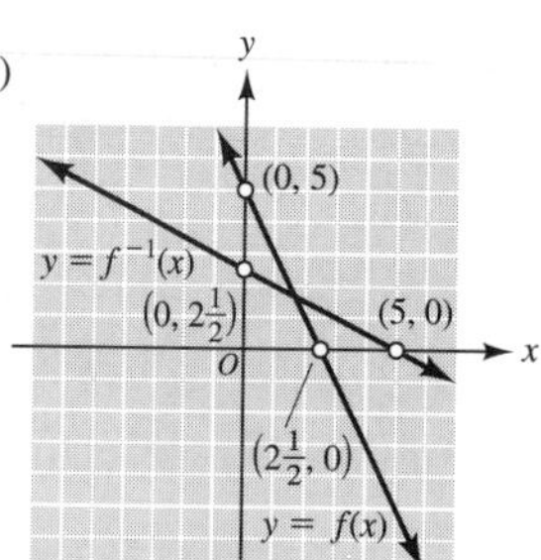

8. $f^{-1}(x) = \dfrac{x + 10}{3}$

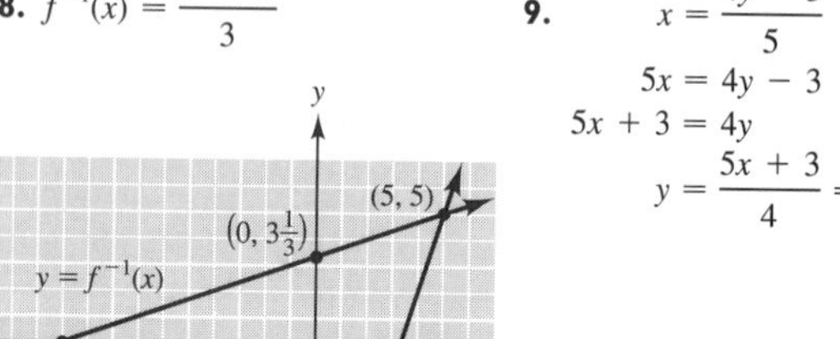

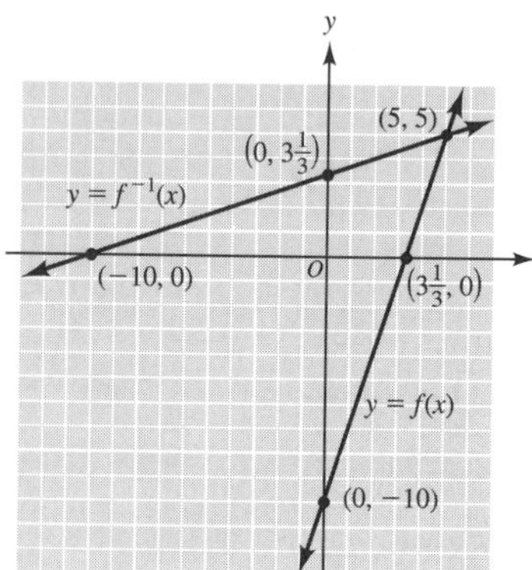

9.
$$x = \frac{4y - 3}{5}$$
$$5x = 4y - 3$$
$$5x + 3 = 4y$$
$$y = \frac{5x + 3}{4} = f^{-1}(x)$$

10. $f^{-1}(x) = \dfrac{3x + 7}{2}$

11.
$$x = \frac{5}{y + 2}$$
$$x(y + 2) = 5$$
$$xy + 2x = 5$$
$$xy = 5 - 2x$$
$$y = \frac{5 - 2x}{x} = f^{-1}(x)$$

12. $f^{-1}(x) = \dfrac{x + 10}{2x}$

9

Exercises 9.6 (page 468)

1.

x	y
−3	12
−2	7
−1	4
0	3
1	4
2	7
3	12

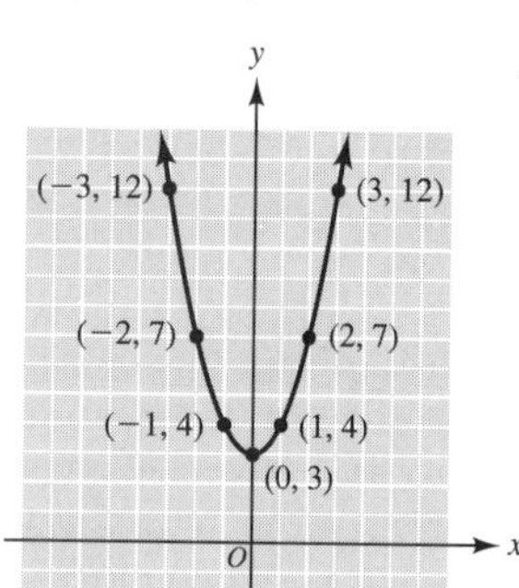

2.

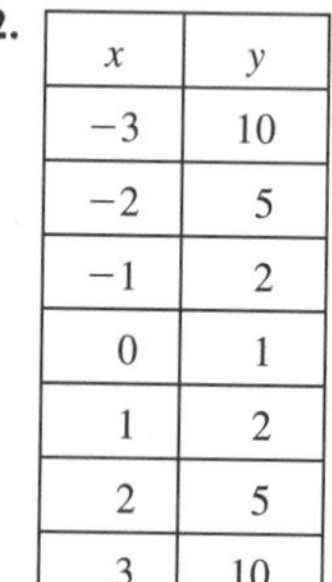

x	y
−3	10
−2	5
−1	2
0	1
1	2
2	5
3	10

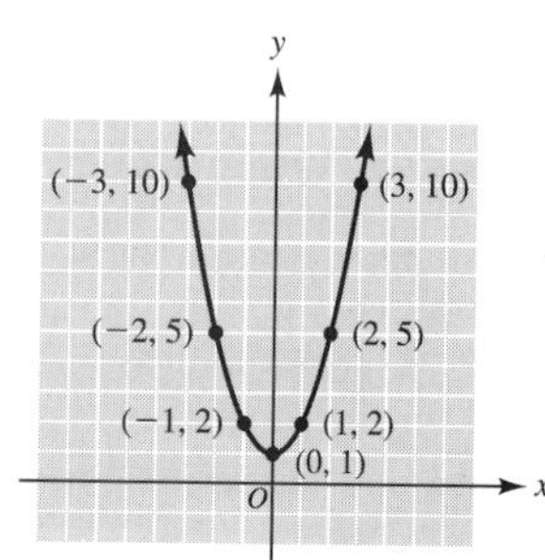

3.

x	y
−4	9
−3	4
−2	1
−1	0
0	1
1	4
2	9

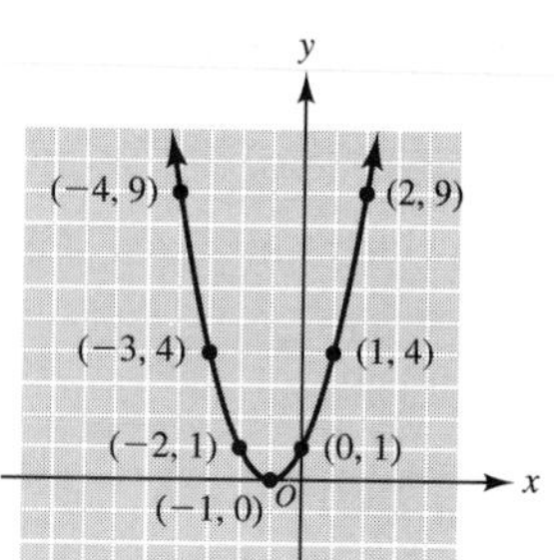

4.

x	y
−6	9
−5	4
−4	1
−3	0
−2	1
−1	4
0	9

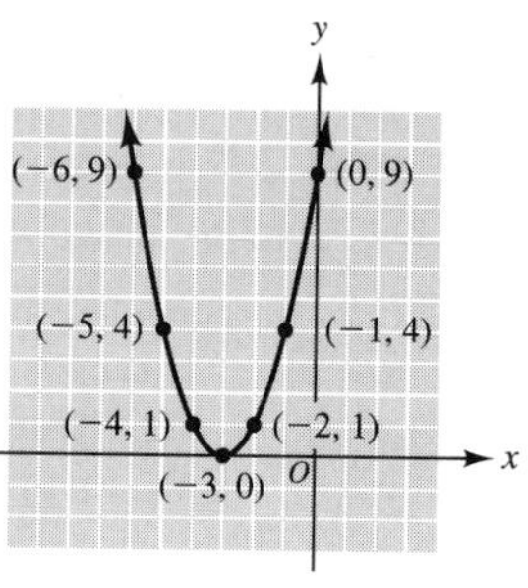

5.

x	y
−2	8
−1	3
0	0
1	−1
2	0
3	3
4	8

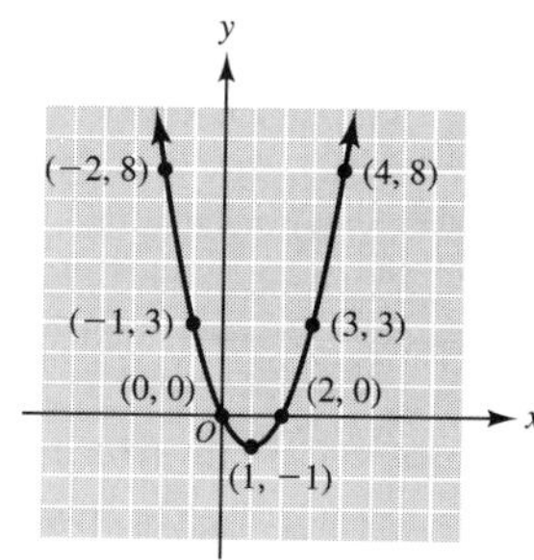

6.

x	y
−2	−10
−1	−4
0	0
1	2
2	2
3	0
4	−4

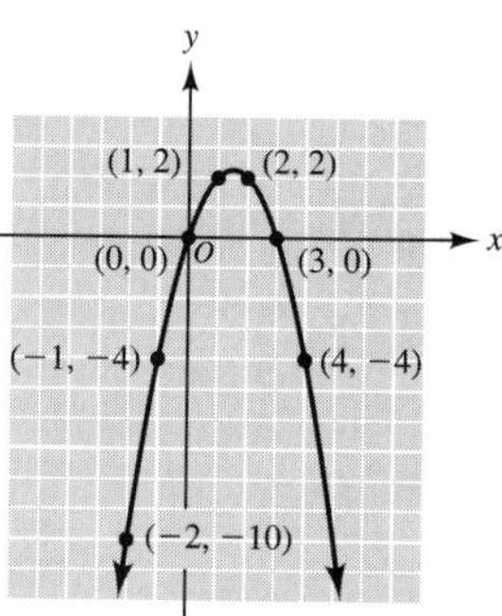

7.

x	y
−2	−8
−1	−3
0	0
1	1
2	0
3	−3
4	−8

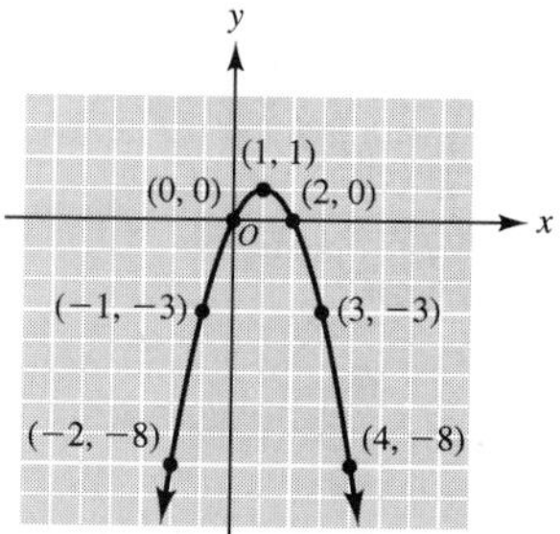

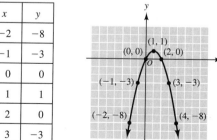

8.

x	y
−3	3
−2	0
−1	−1
0	0
1	3

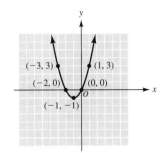

9.

x	y
−2	−8
−1	−1
0	0
1	1
2	8

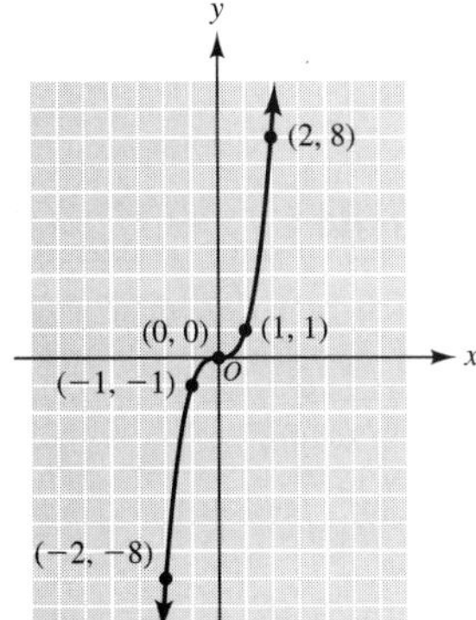

10.

x	y
−2	−6
−1	1
0	2
1	3
2	10

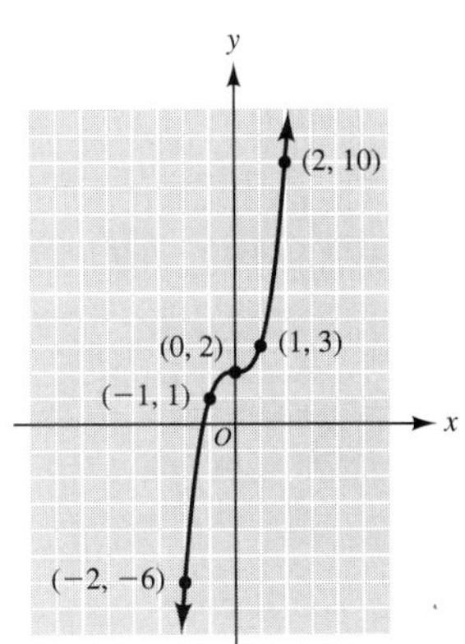

11.

x	y
−3	−8
−2	−1
−1	0
0	1
1	8

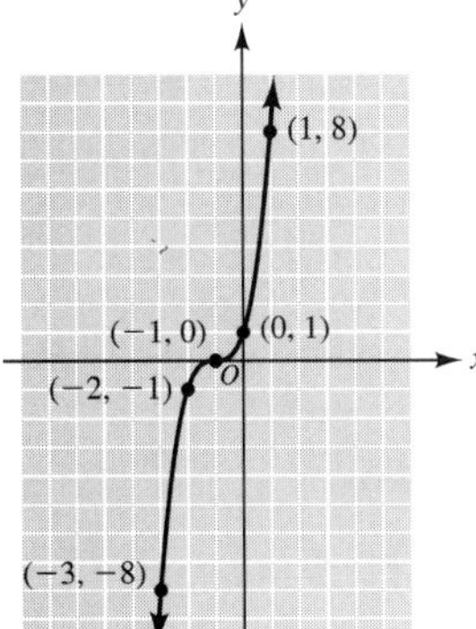

12.

x	y
−1	−8
0	−1
1	0
2	1
3	8

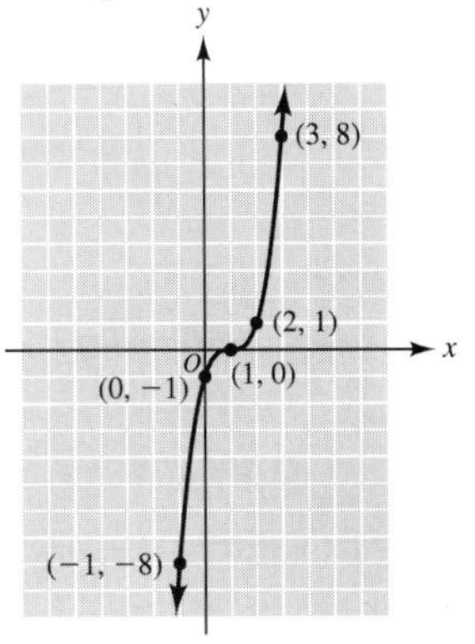

13. We find the range here by using synthetic division. Substitution can be used, instead.

x	1	−2	−13	20 $f(x)$
−4	1	−6	11	−24
−3	1	−5	2	14
−2	1	−4	−5	30
0	1	−2	−13	20
2	1	0	−13	−6
3	1	1	−10	−10
4	1	2	−5	0
5	1	3	2	30

Range: $\{-24, 14, 30, 20, -6, -10, 0\}$

14. $\{18, 32, 12, -32, -48, -52, -38, 0\}$

Exercises 9.7 (page 473)

1. Condition 1: $x_1 = 3$, $y_1 = 12$
Condition 2: $x_2 = 5$, $y_2 = ?$

$y = kx$
$12 = k(3)$
$k = 4$
$y_2 = 4(5)$
$y_2 = 20$

Alternative method:
$$\frac{y_1}{y_2} = \frac{x_1}{x_2}$$
$$\frac{12}{y_2} = \frac{3}{5}$$
$$3y_2 = (5)(12)$$
$$y_2 = \frac{(5)(12)}{3} = 20$$

2. $y = kx$; $k = 2.5$; $y_2 = -15$

3. $A = ks^2$
$-27 = k(-3)^2$
$-27 = 9k$
$k = -3$
$A_2 = -3(5)^2 = -75$

4. $R = kx^2$; $k = 2$; $R_2 = 200$

5. $l = kF$
$4 = k(5)$
$k = \frac{4}{5}$
$l_2 = \left(\frac{4}{5}\right)(2) = \frac{8}{5}$
The change in length is $\frac{8}{5}$ in. when a 2-lb force is applied.

6. $m = kg$; 391 mi

7. $s = kt^2$
$64 = k(2)^2$
$k = 16$
$s_2 = (16)(5)^2 = 16(25) = 400$
The object will fall 400 ft in 5 sec.

8. $R = kv^2$; the air resistance is $\frac{6{,}400}{9}$ at 80 mph.

9. Let M = number of mi on 50 gal of gasoline
$M = kG$
$161 = k(23)$
$k = 7$
$M_2 = 7(50) = 350$

Alternative method:
$$\frac{161 \text{ mi}}{23 \text{ gal}} = \frac{M \text{ mi}}{50 \text{ gal}}$$
$$23\,M = (161)(50)$$
$$M = 350$$
Leon can expect to drive 350 mi on 50 gal of gasoline.

10. $k = 5$; $m = 375$

11. Let x = height of tree
$$\frac{5 \text{ ft}}{3 \text{ ft}} = \frac{x \text{ ft}}{27 \text{ ft}}$$
$$3x = 5(27)$$
$$x = 45$$
The tree is 45 ft tall.

12. 33 ft

9

13.
$$C = kr$$
$$47.1 = k(7.5)$$
$$k = 6.28$$
$$C_2 = 6.28(4.5)$$
$$C_2 = 28.26$$

Alternative method:
$$\frac{47.1}{C} = \frac{7.5}{4.5}$$
$$7.5C = (47.1)(4.5)$$
$$7.5C = 211.95$$
$$C = 28.26$$

14. $P = kd$; $k = 0.433$; $P = 7.794$

15.
$$A = kr^2$$
$$28.26 = k(3)^2$$
$$k = \frac{28.26}{9} = 3.14$$
$$A_2 = 3.14(6)^2 = 113.04$$

Alternative method:
$$\frac{28.26}{A} = \frac{3^2}{6^2}$$
$$A = \frac{(28.26)(36)}{9} = 113.04$$

16. $S = kr^2$; $k = 12.56$; $S = 200.96$

17. Let S = amount of sediment carried by current and s = speed of current
Condition 1: $S_1 = 1$, $s_1 = 2$
Condition 2: $S_2 = ?$, $s_2 = 4$
$$\frac{S_1}{S_2} = \frac{s_1^6}{s_2^6}$$
$$\frac{1}{S_2} = \frac{2^6}{4^6}$$
$$S_2 = \frac{4^6}{2^6} = \frac{(2^2)^6}{2^6} = \frac{2^{12}}{2^6} = 2^{12-6} = 2^6 = 64$$
The stream carries 64 units of sediment. This shows that when the speed of the current is doubled, its destructive power becomes 64 times as great.

18. \$213.25

Exercises 9.8 (page 477)

1.
$$y = \frac{k}{x}$$
$$7 = \frac{k}{2}$$
$$k = 14$$
$$y_2 = \frac{14}{-7} = -2$$

2. $z = \frac{k}{w}$; $k = -48$; $z = -6$

3.
$$P = \frac{k}{V}$$
$$18 = \frac{k}{15}$$
$$k = 270$$
$$P_2 = \frac{270}{10} = 27$$

4. $s = \frac{k}{t}$; $k = 40$; $s = 10$

5. Let t = time for machine B to complete an order
$$t = \frac{k}{r}$$
$$3 = \frac{k}{375}$$
$$k = 1{,}125$$
$$t_2 = \frac{1{,}125}{225} = 5$$
It would take machine B 5 hr to complete an order.

6. $k = 2{,}475$; 7.5 hr

7.
$$y = \frac{k}{x^2}$$
$$9 = \frac{k}{4^2}$$
$$k = 9(16) = 144$$
$$y_2 = \frac{144}{3^2} = \frac{144}{9} = 16$$

8. $C = \frac{k}{v^2}$; $k = 72$; $C = 2$

9.
$$F = \frac{k}{d^2}$$
$$3 = \frac{k}{4^2}$$
$$k = 48$$
$$F_2 = \frac{48}{2^2} = 12$$

10. $6\frac{2}{3}$ ft-candles

11. a. If the length is l and the width is w, $l = \frac{k}{w}$.

b.
$$15 = \frac{k}{3}$$
$$k = 45$$
$$l = \frac{45}{5} = 9$$
The length is 9 cm.

12. a. If the altitude is h and the base is b, $h = \frac{k}{b}$.

b. $k = 60$; $h = 6$; the altitude is 6 cm.

13.
$$I = \frac{k}{d^2}$$
$$75 = \frac{k}{50^2}$$
$$k = 187{,}500$$
$$I_2 = \frac{187{,}500}{125^2} = \frac{187{,}500}{15{,}625} = 12$$
The sound intensity is 12 db at 125 ft.

14. $t = \frac{k}{r}$; it will take $\frac{35}{3}$ hr to get the job done.

Exercises 9.9 (page 481)

1.
$$z = kxy$$
$$-36 = k(-3)(2)$$
$$k = 6$$
$$z_2 = (6)(4)(3) = 72$$

2. $A = kLW$; $k = 4$; $A = 84$

3.
$$I = kPr$$
$$115.50 = k(880)(0.0875)$$
$$k = 1.5$$
$$I_2 = (1.5)(760)(0.0925) = 105.45$$
The interest earned would be \$105.45.

4. $k = 3$; $I = \$356.25$

5.
$$F = kAV^2$$
$$1.8 = k(1)(20^2)$$
$$k = 0.0045$$
$$F_2 = 0.0045(2)(60^2) = 32.4$$
The force is 32.4 lb.

6. $P = 2{,}474.7$

7.
$$z = \frac{kx}{y}$$
$$12 = \frac{k(6)}{2}$$
$$k = 4$$
$$z_2 = \frac{(4)(-8)}{-4} = 8$$

8. $k = 4$; $R = 1.6$

9.
$$e = \frac{kPL}{A}$$
$$3 = \frac{k(2.4)(45)}{0.9}$$
$$k = 0.025$$
$$e_2 = \frac{0.025(1.5)(40)}{0.75} = 2$$

10. 300 lb

Sections 9.1–9.9 Review Exercises (page 483)

1. (a) and (c) **2.** (b) and (c)

3. **a.** $\mathcal{R}$ is a function (no two ordered pairs have the same first coordinate but different second coordinates).
   b. $\mathcal{R}$ is not a one-to-one function [the ordered pairs $(-1, -2)$ and $(4, -2)$ have the same second coordinate but different first coordinates].
   c. Does not apply ($\mathcal{R}$ isn't a one-to-one function.)

4. Yes

5. **a.** $f(2) = 3(2)^2 - 5(2) + 4 = 12 - 10 + 4 = 6$
   b. $f(0) = 3(0)^2 - 5(0) + 4 = 0 - 0 + 4 = 4$
   c. $f(x-1) = 3(x-1)^2 - 5(x-1) + 4 = 3(x^2 - 2x + 1) - 5x + 5 + 4 = 3x^2 - 11x + 12$

6. -22

7.
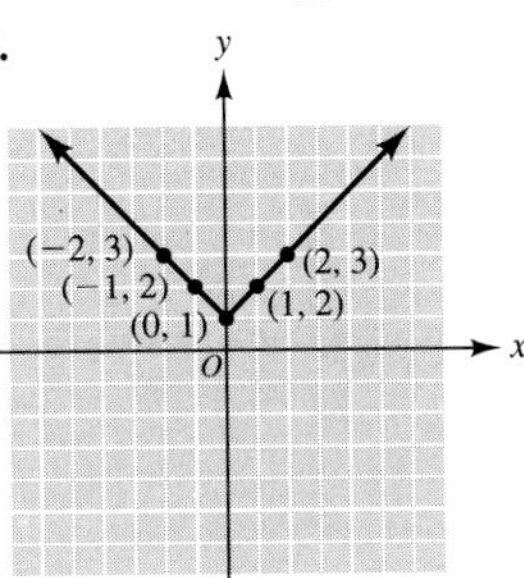

8.
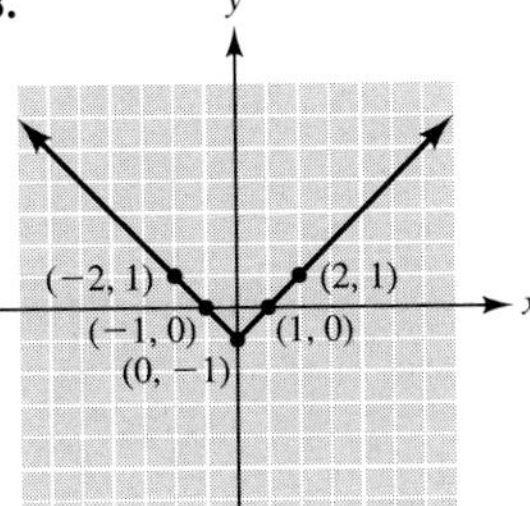

9.
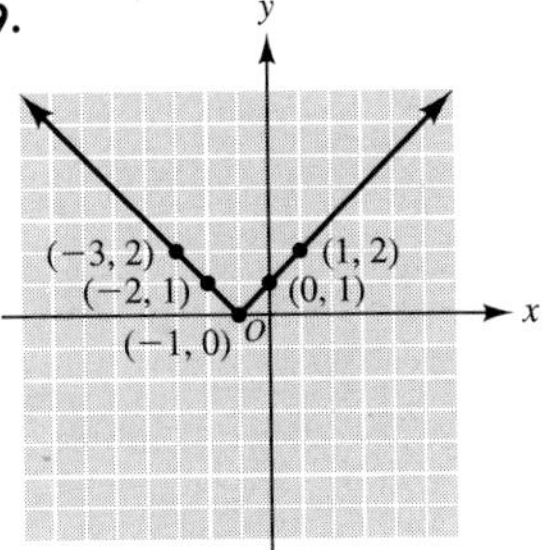

10. $(f+g)(x) = 4x^2 - 3x + 11$
    $(f-g)(x) = 4x^2 + 3x + 1$
    $(f \cdot g)(x) = -12x^3 + 20x^2 - 18x + 30$
    $\left(\dfrac{f}{g}\right)(x) = \dfrac{4x^2 + 6}{-3x + 5},\ x \neq \dfrac{5}{3}$

11. $(f \circ g)(x) = f(g(x)) = 4(-3x + 5)^2 + 6 = 4(9x^2 - 30x + 25) + 6 = 36x^2 - 120x + 100 + 6 = 36x^2 - 120x + 106$
    $(g \circ f)(x) = g(f(x)) = -3(4x^2 + 6) + 5 = -12x^2 - 18 + 5 = -12x^2 - 13$

12. The domains of f and g are both R.
    $(f \circ g)(x) = f(g(x)) = 7\left(\dfrac{x - 9}{7}\right) + 9 = x - 9 + 9 = x$
    $(g \circ f)(x) = g(f(x)) = \dfrac{(7x + 9) - 9}{7} = \dfrac{7x}{7} = x$
    Since $(f \circ g)(x)$ and $(g \circ f)(x)$ both equal x, f and g are inverses of each other.

13. $3x = y + 6$
    $y = 3x - 6 = f^{-1}(x)$
    $f(x) = \dfrac{x + 6}{3}$ $\quad f^{-1}(x) = 3x - 6$

x	y
-6	0
0	2

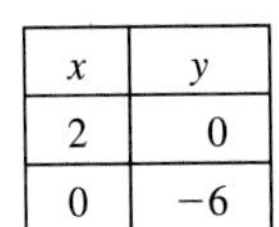

x	y
2	0
0	-6

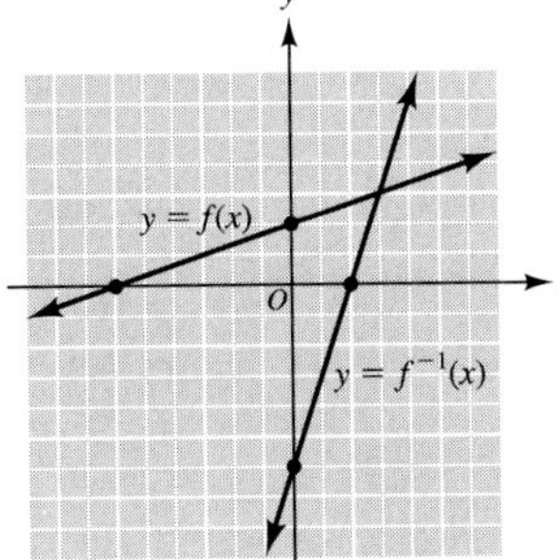

The domain is the set of all real numbers for $y = f(x)$ and for $y = f^{-1}(x)$. The range is the set of all real numbers for $y = f(x)$ and for $y = f^{-1}(x)$.

14. $f^{-1}(x) = \dfrac{2 - 2x}{5}$

The domain is the set of all real numbers for $y = f(x)$ and for $y = f^{-1}(x)$. The range is the set of all real numbers for $y = f(x)$ and for $y = f^{-1}(x)$.

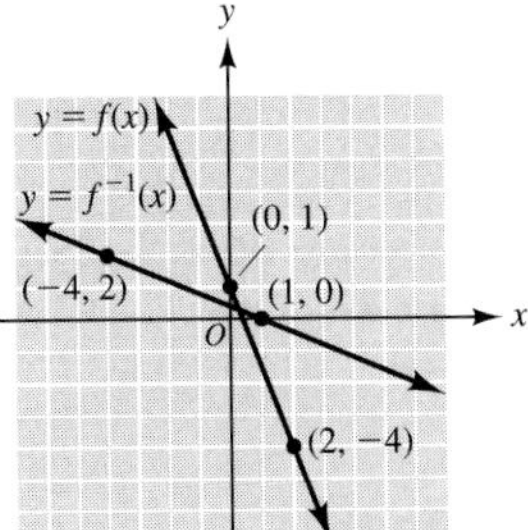

15. $x = \dfrac{9}{2 - 7y}$
    $x(2 - 7y) = 9$
    $2x - 7xy = 9$
    $2x - 9 = 7xy$
    $y = \dfrac{2x - 9}{7x} = f^{-1}(x)$

16. $f^{-1}(x) = \dfrac{10x + 11}{8x}$

17.

x	y
-4	8
-2	2
-1	$\frac{1}{2}$
0	0
1	$\frac{1}{2}$
2	2
4	8

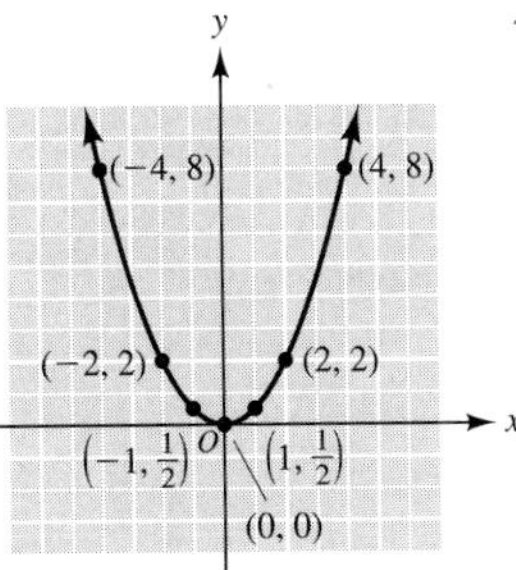

18.

x	y
-4	9
-2	3
-1	$\frac{3}{2}$
0	1
1	$\frac{3}{2}$
2	3
4	9

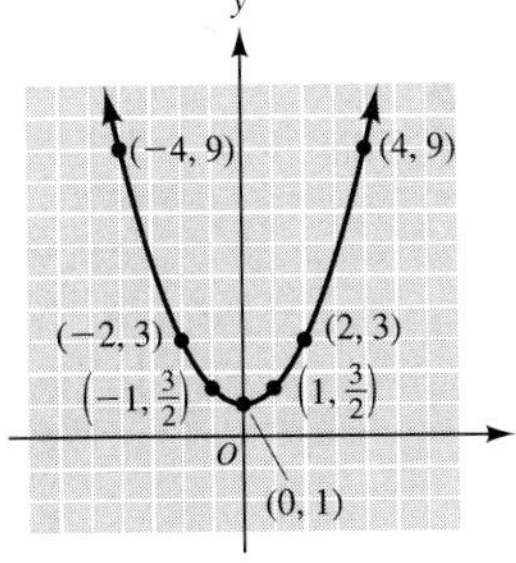

19.

x	y
-3	-18
-2	-2
-1	2
0	0
1	-2
2	2
3	18

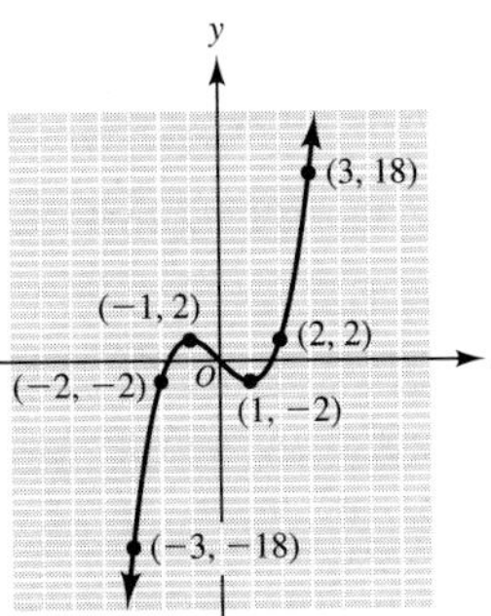

20. $k = 32$; $c = 728$

21. $S = kr^2$
    $615.44 = k(7^2)$
    $k = \dfrac{615.44}{49} = 12.56$
    $S_2 = (12.56)(10^2) = 1{,}256$

22. $k = 500$; $t_2 = 6\frac{2}{3}$
    It would take machine B $6\frac{2}{3}$ hr to complete the order.

Chapter 9 Diagnostic Test (page 489)

Following each problem number is the number (in parentheses) of the textbook section where that kind of problem is discussed.

1. (9.1) (b) (No vertical line can intersect the graph in more than one point.)

2. (9.1, 9.5) **a.** $\mathcal{R}$ is a function (no two ordered pairs have the same first coordinate but different second coordinates).

(9.5) **b.** $\mathcal{R}$ is a one-to-one function (it *is* a function, and no two ordered pairs have the same second coordinate but different first coordinates).

(9.5) **c.** $\mathcal{R}^{-1} = \{(5, 1), (7, 3), (-2, 5), (0, 0)\}$

3. (9.1) Yes, $y = -4x + 9$ is a function. Its domain is R.

4. (9.2) $f(x) = 3(x)^2 - 5$

a. $f(-2) = 3(-2)^2 - 5 = 12 - 5 = 7$

b. $f(4) = 3(4)^2 - 5 = 48 - 5 = 43$

c. $\dfrac{f(4) - f(-2)}{6} = \dfrac{43 - 7}{6} = \dfrac{36}{6} = 6$

5. (9.3)

a.

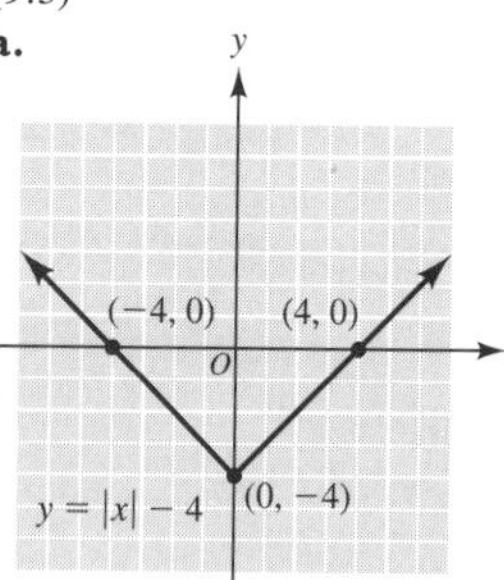

b.

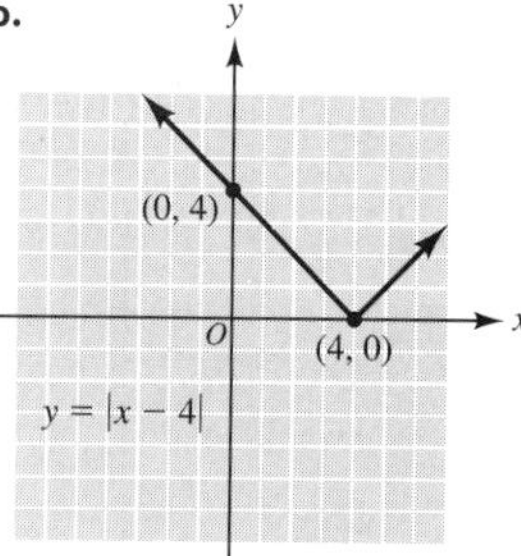

6. (9.4) **a.** $(f + g)(x) = 7x - 3 + 5x^2 + 2 = 5x^2 + 7x - 1$

b. $(f - g)(x) = 7x - 3 - (5x^2 + 2)$
$= 7x - 3 - 5x^2 - 2 = -5x^2 + 7x - 5$

c. $(f \cdot g)(x) = (7x - 3)(5x^2 + 2)$
$= 35x^3 - 15x^2 + 14x - 6$

d. $\left(\dfrac{f}{g}\right)(x) = \dfrac{7x - 3}{5x^2 + 2}$

7. (9.5) The domains of f and g are both R.

a. $(f \circ g)(x) = f(g(x)) = 8\left(\dfrac{x + 1}{8}\right) - 1 = x + 1 - 1 = x$

b. $(g \circ f)(x) = g(f(x)) = \dfrac{(8x - 1) + 1}{8} = \dfrac{8x}{8} = x$

c. f and g are inverses of each other, because the domain of f equals the range of g and vice versa, and $(f \circ g)(x)$ and $(g \circ f)(x)$ both equal x.

8. (9.6) $y = 1 + x - x^2$

$y = 1 + (-2) - (-2)^2 = -5$
$y = 1 + (-1) - (-1)^2 = -1$
$y = 1 + (0) - (0)^2 = 1$
$y = 1 + (1) - (1)^2 = 1$
$y = 1 + (2) - (2)^2 = -1$
$y = 1 + (3) - (3)^2 = -5$

x	y
−2	−5
−1	−1
0	1
1	1
2	−1
3	−5

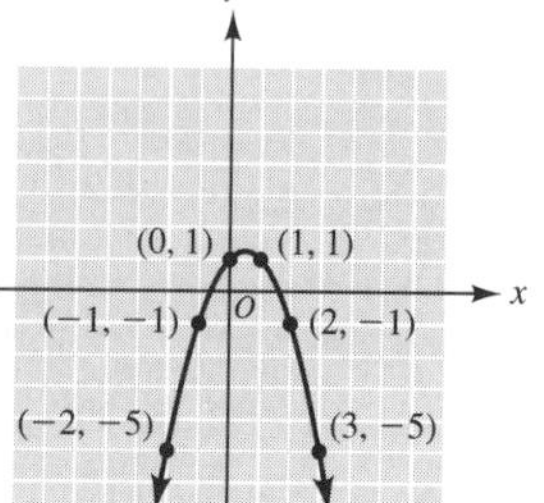

9. a. (9.5) $y = f(x) = -\frac{3}{2}x + 1$

$x = -\frac{3}{2}y + 1$
$2x = -3y + 2$
$3y = 2 - 2x$
$y = \dfrac{2 - 2x}{3} = f^{-1}(x)$

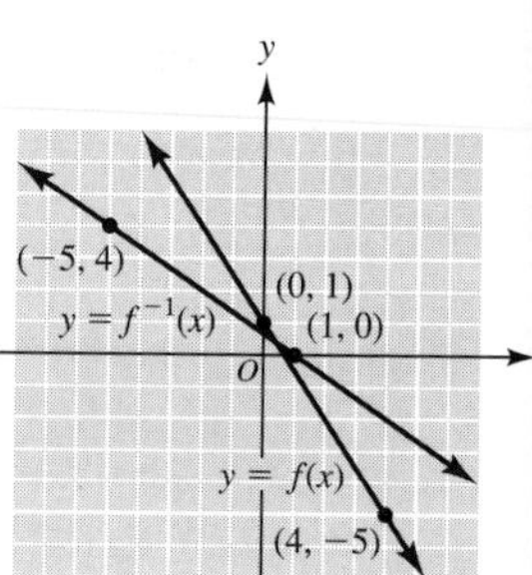

$f(x) = -\dfrac{3}{2}x + 1$

x	y
$\frac{2}{3}$	0
0	1
4	−5

$f^{-1}(x) = \dfrac{2 - 2x}{3}$

x	y
1	0
0	$\frac{2}{3}$
−5	4

b. (9.5) $y = f(x) = \dfrac{15}{4(2 - 5x)}$

$D_f =$ all real numbers except $\frac{2}{5} = R_{f^{-1}}$

$x = \dfrac{15}{4(2 - 5y)}$
$4x(2 - 5y) = 15$
$8x - 20xy = 15$
$y = \dfrac{8x - 15}{20x} = f^{-1}(x)$

$D_{f^{-1}} =$ all real numbers except $0 = R_f$

10. (9.8, 9.9) $w = \dfrac{kxy}{z^2}$

$20 = \dfrac{k(8)(6)}{(12)^2}$
$k = 60$
$w_2 = \dfrac{(60)(6)(10)}{(5)^2} = 144$

Chapters 1–9 Cumulative Review Exercises (page 489)

1. a. True **b.** False **c.** False **d.** False **e.** False **f.** False **g.** False **h.** False **i.** False **j.** False

2. $x + y - 3 = 0$

3. $s = \frac{1}{2}gt^2$
$s = \frac{1}{2}(32)(2)^2 = \frac{1}{2}(32)(4) = 64$

4. $2x^2y^2 - 10x^2y + 7xy$

5. $6 - \{4 - [3x - 10 + 6x]\} = 6 - \{4 - [9x - 10]\}$
$= 6 - \{4 - 9x + 10\}$
$= 6 - \{14 - 9x\}$
$= 6 - 14 + 9x$
$= 9x - 8$

6. $\{-1\}$

7. $12 - 12x + 15 \geq 18 - 9x - 6$
$27 - 12x \geq 12 - 9x$
$-3x \geq -15$
$\dfrac{-3x}{-3} \leq \dfrac{-15}{-3}$
$x \leq 5$
$\{x \mid x \leq 5\}$

8. $\{x \mid -2 < x < 2\}$

9. If $\left|\dfrac{2x - 6}{4}\right| = 1$,

$\dfrac{2x - 6}{4} = 1$ or $\dfrac{2x - 6}{4} = -1$

$2x - 6 = 4$ $\quad$ $2x - 6 = -4$
$2x = 10$ $\quad$ $2x = 2$
$x = 5$ $\quad$ $x = 1$

$\{5, 1\}$

10. $\{x \mid -2 < x < 3\}$

The checks will not be shown for Exercises 11–14.

11. Let $7x$ = length of rectangle (in in.)
Let $4x$ = width of rectangle (in in.)

$2(7x) + 2(4x) = 66$
$14x + 8x = 66$
$22x = 66$
$x = 3$
$7x = 21$
$4x = 12$

The length is 21 in., and the width is 12 in.

12. 485

13. Let x = number of dollars for each of four people to invest
Then $x - 4{,}250$ = number of dollars for each of five people to invest

$4x = 5(x - 4{,}250)$ The amounts invested are equal.
$4x = 5x - 21{,}250$
$21{,}250 = x$

The cost of the property is 4($21,250), or $85,000.

14. The speed of the river is 3 mph.

Exercises 10.1 (page 498)

1. 20.0855 **2.** 1.6487 **3.** 0.0498 **4.** 0.6065

5. $y = 4^x$
If $x = -2$, $y = 4^{-2} = \frac{1}{16}$
If $x = -1$, $y = 4^{-1} = \frac{1}{4}$
If $x = 0$, $y = 4^0 = 1$
If $x = 1$, $y = 4^1 = 4$
If $x = 2$, $y = 4^2 = 16$

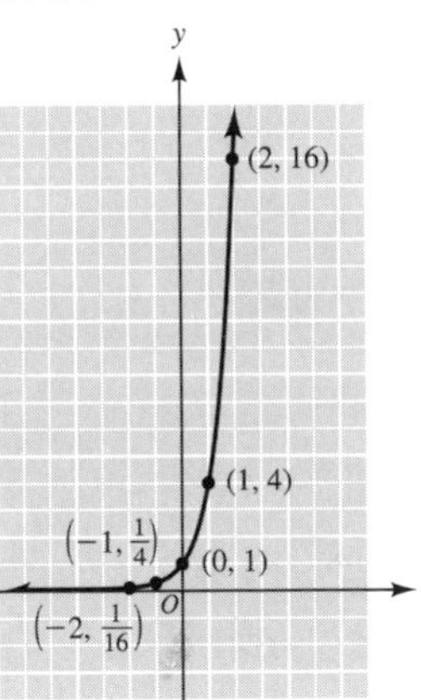

6.

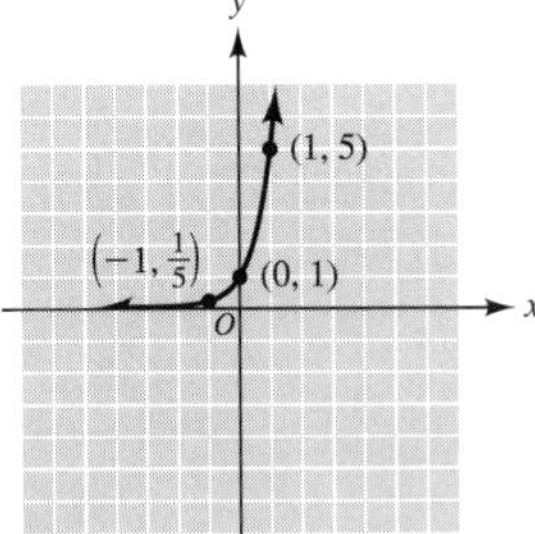

7. $y = e^x$
If $x = -2$, $y = e^{-2} = 0.1$
If $x = -1$, $y = e^{-1} = 0.4$
If $x = 0$, $y = e^0 = 1$
If $x = 1$, $y = e^1 = 2.7$
If $x = 2$, $y = e^2 = 7.4$

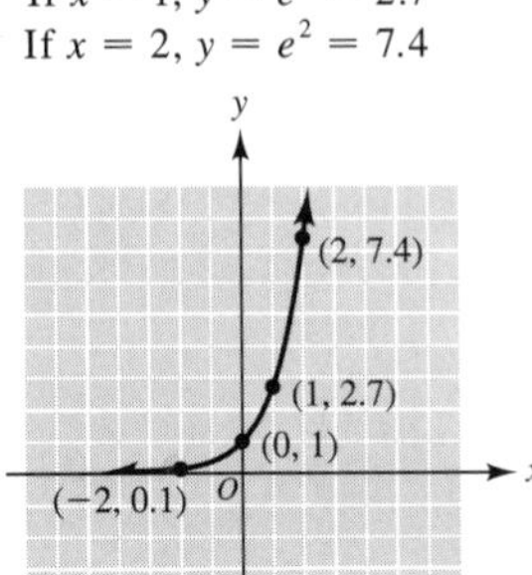

8.

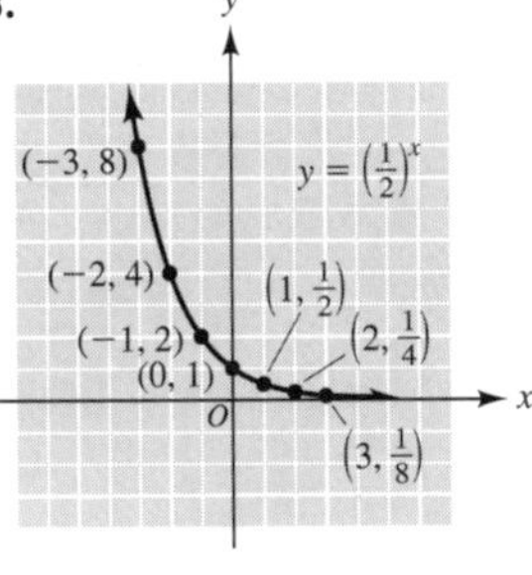

9.

x	y
−3	$\frac{8}{27}$
−2	$\frac{4}{9}$
−1	$\frac{2}{3}$
0	1
1	$\frac{3}{2}$
2	$\frac{9}{4}$
3	$\frac{27}{8}$
4	$\frac{81}{16}$
5	$\frac{243}{32}$

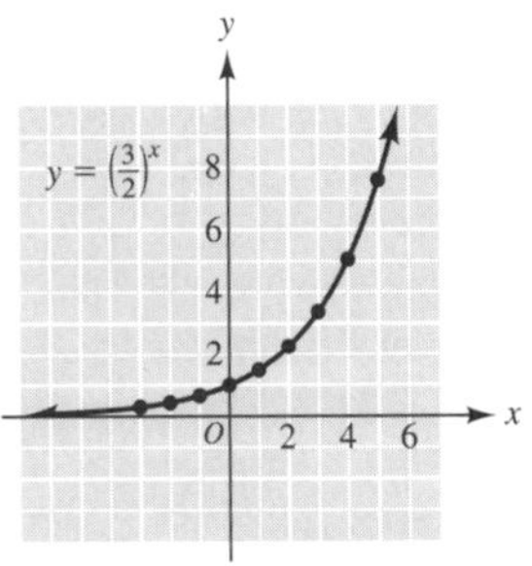

10.

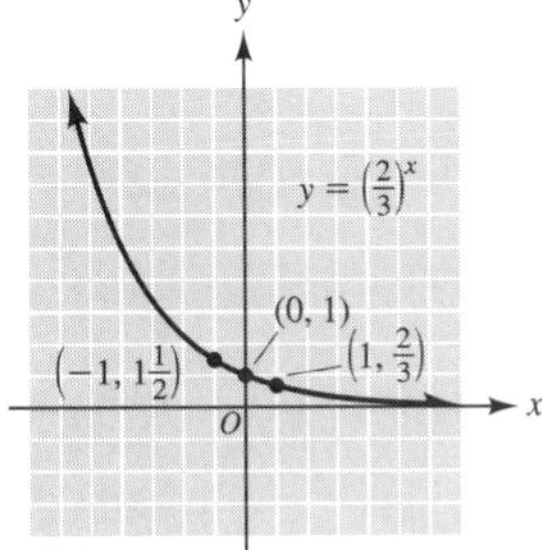

11. The exponential form for $y = \log_2 x$ is $x = 2^y$.
The exponential form for $y = \log_{10} x$ is $x = 10^y$.

$x = 2^y$

x	y
$\frac{1}{4}$	−2
$\frac{1}{2}$	−1
1	0
2	1
4	2
8	3

$x = 10^y$

x	y
0.01	−2
0.1	−1
1	0
10	1

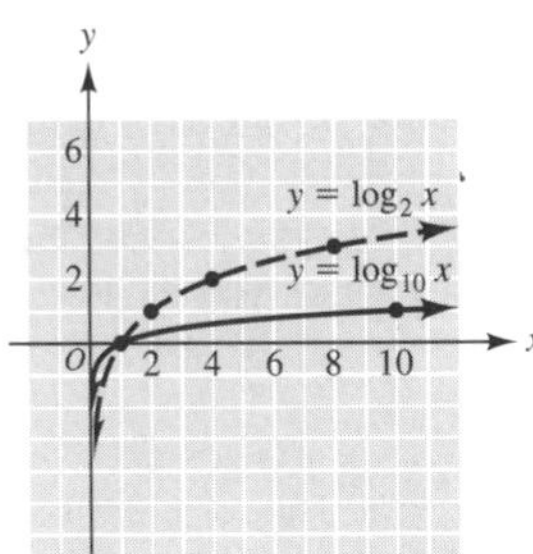

12.

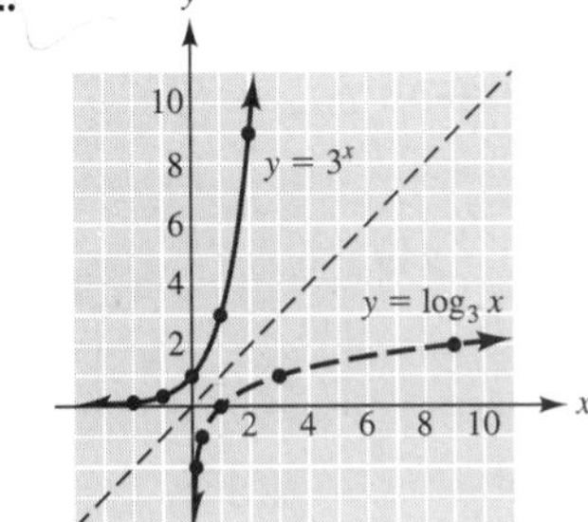

13. $y = e^x$

x	y
−1	0.4
0	1
1	2.7
2	7.4

$y = \log_e x$

x	y
0.4	−1
1	0
2.7	1
7.4	2

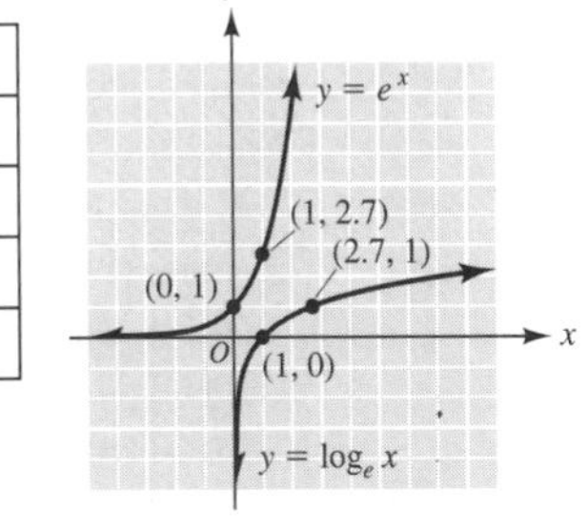

Exercises 10.2 (page 499)

1. $\log_3 9 = 2$ **2.** $\log_4 64 = 3$ **3.** $\log_{10} 1{,}000 = 3$
4. $\log_{10} 100{,}000 = 5$ **5.** $\log_2 16 = 4$ **6.** $\log_4 16 = 2$
7. $\log_3 \frac{1}{9} = -2$ **8.** $\log_2 \frac{1}{8} = -3$ **9.** $\log_{12} 1 = 0$
10. $\log_8 1 = 0$ **11.** $\log_{16} 4 = \frac{1}{2}$ **12.** $\log_8 2 = \frac{1}{3}$
13. $8^2 = 64$ **14.** $2^5 = 32$ **15.** $7^2 = 49$ **16.** $4^3 = 64$
17. $5^0 = 1$ **18.** $6^0 = 1$ **19.** $9^{1/2} = 3$ **20.** $8^{1/3} = 2$
21. $10^2 = 100$ **22.** $10^3 = 1{,}000$
23. $10^{-3} = \dfrac{1}{10^3} = \dfrac{1}{1{,}000} = 0.001$ **24.** $10^{-2} = 0.01$

Exercises 10.3 (page 502)

1. $N = 5^2 = 25$ **2.** 32 **3.** $3^x = 9 = 3^2$, $x = 2$ **4.** 4
5. $b^3 = 27 = 3^3$, $b = 3$ **6.** 3 **7.** $5^x = 125 = 5^3$, $x = 3$ **8.** 4
9. $10^x = 10^{-4}$, $x = -4$ **10.** -3 **11.** $\left(\frac{3}{2}\right)^2 = N$, $N = \frac{9}{4}$ **12.** $\frac{125}{27}$

13.
$$9^x = \tfrac{1}{3}$$
$$(3^2)^x = 3^{-1}$$
$$3^{2x} = 3^{-1}$$
$$2x = -1$$
$$x = -\tfrac{1}{2}$$

14. -2

15.
$$b^{1.5} = 8 = 2^3$$
$$b^{3/2} = 2^3$$
$$(b^{3/2})^{2/3} = (2^3)^{2/3}$$
$$b = 2^2 = 4$$

16. 25 **17.** $2^{-2} = N$, $N = \dfrac{1}{2^2} = \dfrac{1}{4}$ **18.** 0.01

19. Let $x = \log_5 25 \Leftrightarrow 5^x = 25 = 5^2$
$$x = 2$$
$$\log_5 25 = 2$$
20. 4

21. Let $x = \log_{10} 10{,}000 \Leftrightarrow 10^x = 10{,}000 = 10^4$
$$x = 4$$
$$\log_{10} 10{,}000 = 4$$
22. 5

23. Let $x = \log_4 8 \Leftrightarrow 4^x = 8$
$$(2^2)^x = 2^3$$
$$2^{2x} = 2^3$$
$$2x = 3$$
$$x = \tfrac{3}{2}$$
$$\log_4 8 = \tfrac{3}{2}$$
24. $\frac{2}{3}$

25. Let $x = \log_3 3^4 \Leftrightarrow 3^x = 3^4$
$$x = 4$$
$$\log_3 3^4 = 4$$
26. 5

27. Let $x = \log_{16} 16 \Leftrightarrow 16^x = 16$
$$x = 1$$
$$\log_{16} 16 = 1$$
28. 1

29. Let $x = \log_8 1 \Leftrightarrow 8^x = 1 = 8^0$
$$x = 0$$
$$\log_8 1 = 0$$
30. 0

Exercises 10.4 (page 506)

1. $\log_{10} 31 + \log_{10} 7$ **2.** $\log_{10} 17 + \log_{10} 29$
3. $\log_{10} 41 - \log_{10} 13$ **4.** $\log_{10} 19 - \log_{10} 23$ **5.** $3 \log_{10} 19$
6. $4 \log_{10} 7$ **7.** $\log_{10} 75^{1/5} = \frac{1}{5} \log_{10} 75$ **8.** $\frac{1}{4} \log_{10} 38$
9. $\log_{10} 35\sqrt{73} - \log_{10} (1.06)^8 = \log_{10} 35 + \frac{1}{2} \log_{10} 73 - 8 \log_{10} 1.06$
10. $\log_{10} 27 + \frac{1}{2} \log_{10} 31 - 10 \log_{10} 1.03$
11. $\log_{10} (2)(7) = \log_{10} 2 + \log_{10} 7$
$$\approx 0.301 + 0.845 = 1.146$$
12. 1.322

13.
$$\log_{10} 9 - \log_{10} 7 = \log_{10} 3^2 - \log_{10} 7$$
$$= 2 \log_{10} 3 - \log_{10} 7$$
$$\approx 2(0.477) - 0.845$$
$$\approx 0.954 - 0.845 = 0.109$$
14. 0.243

15.
$$\log_{10} (27)^{1/2} = \log_{10} (3^3)^{1/2}$$
$$= \log_{10} 3^{3/2} = \tfrac{3}{2} \log_{10} 3$$
$$\approx (1.5)(0.477) = 0.7155$$
16. 0.4515

17.
$$\log_{10} (6^2)^2 = \log_{10} 6^4 = 4 \log_{10} 6$$
$$= 4 \log_{10} (2)(3)$$
$$= 4[\log_{10} 2 + \log_{10} 3]$$
$$\approx 4[0.301 + 0.477]$$
$$= 4(0.778) = 3.112$$
18. 3.982

19.
$$\log_{10} (2)(3)(10^3) = \log_{10} 2 + \log_{10} 3 + 3 \log_{10} 10$$
$$\approx 0.301 + 0.477 + 3 = 3.778$$

20. 3.146 **21.** $\log_b xy$ **22.** $\log_b x^4 y^2$
23. $\log_b x^2 - \log_b y^3 = \log_b \dfrac{x^2}{y^3}$ **24.** $2 \log_b x$
25.
$$3 \log_b x - 6 \log_b y = \log_b x^3 - \log_b y^6$$
$$= \log_b \frac{x^3}{y^6}$$
26. $\log_b y$

27.
$$\log_b (x^2 - y^2) - \log_b (x + y)^3 = \log_b \frac{x^2 - y^2}{(x + y)^3}$$
$$= \log_b \frac{(x + y)(x - y)}{(x + y)^3}$$
$$= \log_b \frac{(x - y)}{(x + y)^2}$$

28. $\log_b (x + z)$
29.
$$\log_b (2xy)^2 - \log_b 3xy^2 + \log_b 3x$$
$$= \log_b 4x^2y^2 - \log_b 3xy^2 + \log_b 3x$$
$$= \log_b \frac{4x^2y^2(3x)}{3xy^2} = \log_b 4x^2 = \log_b (2x)^2 = 2 \log_b 2x$$
30. $\log_b 3y^2$

Exercises 10.5 (page 513)

1. **a.** 0.4771 **b.** 3.4771 **c.** 1.4771 **d.** 1.0986 **e.** 8.0064 **f.** 3.4012
2. **a.** 1.0000 **b.** 2.0000 **c.** 5.0000 **d.** 2.3026 **e.** 4.6052 **f.** 11.5129
3. **a.** -1.0000 **b.** -2.0000 **c.** -3.0000 **d.** -2.3026 **e.** -4.6052 **f.** -6.9078
4. **a.** -0.5229 **b.** -3.5229 **c.** -5.5229 **d.** -1.2040 **e.** -8.1117 **f.** -12.7169
5. 2.35 **6.** 4.06 **7.** 0.851 **8.** 0.755 **9.** 2,560
10. 65.1 **11.** 0.003 06 **12.** 0.000 409 **13.** 15.2
14. 230 **15.** 0.0754 **16.** 0.005 78

Exercises 10.6 (page 518)

1.
$$(3^3)^x = \frac{1}{3^2}$$
$$3^{3x} = 3^{-2}$$
$$3x = -2$$
$$x = -\tfrac{2}{3}$$
2. $-\frac{4}{3}$

3.
$$(2^2)^x = \frac{1}{2^3}$$
$$2^{2x} = 2^{-3}$$
$$2x = -3$$
$$x = -\tfrac{3}{2}$$
4. $-\frac{2}{3}$

5.
$$(5^2)^{2x+3} = 5^{x-1}$$
$$5^{4x+6} = 5^{x-1}$$
$$4x + 6 = x - 1$$
$$3x = -7$$
$$x = -\tfrac{7}{3}$$
6. 1

7.
$$\begin{aligned} 2^x &= 3 \\ \log 2^x &= \log 3 \\ x\log 2 &= \log 3 \\ x &= \frac{\log 3}{\log 2} \approx \frac{0.4771}{0.3010} \approx 1.59 \end{aligned}$$
(1.58 if log 3 and log 2 are not rounded off)

8. 0.861

9.
$$\begin{aligned} e^x &= 8 \\ \ln e^x &= \ln 8 \\ x\ln e &= \ln 8 \\ x &\approx 2.0794 \\ x &\approx 2.08 \end{aligned}$$

10. 3.00

11.
$$\begin{aligned} (7.43)^{x+1} &= 9.55 \\ \log(7.43)^{x+1} &= \log 9.55 \\ (x+1)\log 7.43 &= \log 9.55 \\ x+1 &= \frac{\log 9.55}{\log 7.43} \approx \frac{0.9800}{0.8710} \\ x+1 &\approx 1.125 \\ x &\approx 0.125 \end{aligned}$$

12. 2.20

13.
$$\begin{aligned} (8.71)^{2x+1} &= 8.57 \\ \log(8.71)^{2x+1} &= \log 8.57 \\ (2x+1)\log 8.71 &= \log 8.57 \\ 2x+1 &= \frac{\log 8.57}{\log 8.71} \approx \frac{0.932\,98}{0.940\,02} \\ 2x+1 &\approx 0.992\,51 \\ x &\approx -0.003\,74 \end{aligned}$$
($-0.003\,72$ if tables are used)

14. 0.500

15.
$$\begin{aligned} e^{3x+4} &= 5 \\ \ln e^{3x+4} &= \ln 5 \\ (3x+4)\ln e &\approx 1.609 \\ 3x+4 &\approx 1.609 \\ 3x &\approx -2.391 \\ x &\approx -0.797 \end{aligned}$$

16. 1.93

17.
$$\begin{aligned} \log(3x-1)(4) &= \log(9x+2) \\ (3x-1)(4) &= 9x+2 \\ 12x-4 &= 9x+2 \\ 3x &= 6 \\ x &= 2 \end{aligned}$$
Check
$$\begin{aligned} \log[3(2)-1] + \log 4 &\stackrel{?}{=} \log[9(2)+2] \\ \log 5 + \log 4 &\stackrel{?}{=} \log 20 \\ \log(5)(4) &\stackrel{?}{=} \log 20 \\ \log 20 &= \log 20 \end{aligned}$$
2 is a solution.

18. 2

19.
$$\begin{aligned} \ln\frac{x+4}{3} &= \ln(x-2) \\ \frac{x+4}{3} &= x-2 \\ x+4 &= 3(x-2) \\ x+4 &= 3x-6 \\ 10 &= 2x \\ x &= 5 \end{aligned}$$
Check
$$\begin{aligned} \ln(5+4) - \ln 3 &\stackrel{?}{=} \ln(5-2) \\ \ln\tfrac{9}{3} &\stackrel{?}{=} \ln 3 \\ \ln 3 &= \ln 3 \end{aligned}$$
5 is a solution.

20. 2

21.
$$\begin{aligned} \log\frac{5x+2}{x-1} &= 0.7782 \\ \frac{5x+2}{x-1} &= 10^{0.7782} \\ \frac{5x+2}{x-1} &\approx 6 \\ 5x+2 &\approx 6x-6 \\ 8 &\approx x \end{aligned}$$
Check for $x \approx 8.00$
$$\begin{aligned} \log[5(8)+2] - \log(8-1) &\stackrel{?}{=} 0.7782 \\ \log 42 - \log 7 &\stackrel{?}{=} 0.7782 \\ \log\tfrac{42}{7} &\stackrel{?}{=} 0.7782 \\ \log 6 &\stackrel{?}{=} 0.7782 \\ 0.77815 &\approx 0.7782 \end{aligned}$$
8.00 is a solution.

22. 2.00

23.
$$\begin{aligned} \log x(7-x) &= \log 10 \\ x(7-x) &= 10 \\ 7x - x^2 &= 10 \\ x^2 - 7x + 10 &= 0 \\ (x-2)(x-5) &= 0 \end{aligned}$$
$x - 2 = 0 \quad | \quad x - 5 = 0$
$x = 0 \quad | \quad x = 5$

Check for $x = 2$
$$\begin{aligned} \log 2 + \log(7-2) &\stackrel{?}{=} \log 10 \\ \log(2)(5) &\stackrel{?}{=} \log 10 \\ \log 10 &= \log 10 \end{aligned}$$
Check for $x = 5$
$$\begin{aligned} \log 5 + \log(7-5) &\stackrel{?}{=} \log 10 \\ \log(5)(2) &\stackrel{?}{=} \log 10 \\ \log 10 &= \log 10 \end{aligned}$$
Both 2 and 5 are solutions.

24. 1 and 10

25.
$$\begin{aligned} \ln x(x-3) &= \ln 4 \\ x(x-3) &= 4 \\ x^2 - 3x &= 4 \\ x^2 - 3x - 4 &= 0 \\ (x-4)(x+1) &= 0 \end{aligned}$$
$x - 4 = 0 \quad | \quad x + 1 = 0$
$x = 4 \quad | \quad x = -1$

Check for $x = 4$
$$\begin{aligned} \ln 4 + \ln(4-3) &\stackrel{?}{=} \ln 4 \\ \ln 4 + \ln 1 &\stackrel{?}{=} \ln 4 \\ \ln 4 + 0 &\stackrel{?}{=} \ln 4 \\ \ln 4 &= \ln 4 \end{aligned}$$
Check for $x = -1$
$$\ln(-1) + \ln(-1-3) \stackrel{?}{=} \ln 4$$
Not real numbers

-1 *is not* a solution; 4 is the only solution.

26. 2

27.
$$\begin{aligned} \log(x+1)(x-2) &= \log 10 \\ x^2 - x - 2 &= 10 \\ x^2 - x - 12 &= 0 \\ (x-4)(x+3) &= 0 \end{aligned}$$
$x - 4 = 0 \quad | \quad x + 3 = 0$
$x = 4 \quad | \quad x = -3$

Check for $x = 4$
$$\begin{aligned} \log(4+1) + \log(4-2) &\stackrel{?}{=} 1 \\ \log 5 + \log 2 &\stackrel{?}{=} 1 \\ \log(5)(2) = \log 10 &= 1 \end{aligned}$$
Check for $x = -3$
$$\begin{aligned} \log(-3+1) + \log(-3-2) &\stackrel{?}{=} 1 \\ \log(-2) + \log(-5) &\stackrel{?}{=} 1 \end{aligned}$$
Not real numbers

-3 is *not* a solution; 4 is the only solution.

28. 4

29.
$$\begin{aligned} \log\frac{10x}{x-450} &= 2 \\ \frac{10x}{x-450} &= 10^2 \\ 10x &= 100(x-450) \\ 10x &= 100x - 45{,}000 \\ 45{,}000 &= 90x \\ x &= 500 \end{aligned}$$
Check
$$\begin{aligned} \log 10(500) - \log(500-450) &\stackrel{?}{=} 2 \\ \log(5{,}000) - \log(50) &\stackrel{?}{=} 2 \\ \log\tfrac{5{,}000}{50} &\stackrel{?}{=} 2 \\ \log 100 &= 2 \end{aligned}$$
500 is a solution.

30. 12

Exercises 10.7 (page 522)

Note: Your calculator displays may vary slightly from those shown below.

1. $P = \$1{,}250;\ r = 0.055;\ t = 20$

a. $A = P(1 + r)^t$
$A = \$1{,}250(1 + 0.055)^{20}$
$= \$1{,}250(1.055)^{20}$
$\approx \$1{,}250(2.917757)$
$\approx \$3{,}647.20$

b. $A = P\left(1 + \dfrac{r}{k}\right)^{kt};\ k = 12$
$A = \$1{,}250\left(1 + \frac{0.055}{12}\right)^{12(20)}$
$= \$1{,}250(1 + 0.00458333)^{240}$
$= \$1{,}250(1.00458333)^{240}$
$\approx \$1{,}250(2.9966255)$
$\approx \$3{,}745.78$

c. $A = P\left(1 + \dfrac{r}{k}\right)^{kt};\ k = 365$
$A = \$1{,}250\left(1 + \frac{0.055}{365}\right)^{365(20)}$
$= \$1{,}250(1 + 0.000150685)^{7300}$
$= \$1{,}250(1.000150685)^{7300}$
$\approx \$1{,}250(3.0039164)$
$\approx \$3{,}754.90$

d. $A = Pe^{rt}$
$A = \$1{,}250e^{0.055(20)}$
$= \$1{,}250e^{1.1}$
$\approx \$1{,}250(3.0041660)$
$\approx \$3{,}755.21$

2. a. \$4,372.64 **b.** \$4,436.73 **c.** \$4,442.62 **d.** \$4,442.83

3. $P = \$1{,}500;\ r = 0.0575;\ A = \$2{,}000$
$A = Pe^{rt}$
$\$2{,}000 = \$1{,}500e^{0.0575t}$
$1.333\,333\,3 \approx e^{0.0575t}$
$\log_e 1.333\,333\,3 \approx 0.0575t$
$0.287\,682\,1 \approx 0.0575t$
$t \approx 5.003$ yr

4. 8.546 yr

5. The formula is $y = Ce^{0.035t}$.

a. $C = 500;\ t = 3$
$y = 500e^{0.035(3)}$
$= 500e^{0.105}$
$\approx 500(1.110\,711)$
≈ 555.4
There will be about 555.4 bacteria present.

b. $y = 800;\ C = 500$
$800 = 500e^{0.035t}$
$1.6 = e^{0.035t}$
$\ln 1.6 = \ln e^{0.035t}$
$0.470\,003\,6 \approx 0.035t \ln e \quad (\ln e = 1)$
$0.035t \approx 0.470\,003\,6$
$t \approx 13.43$
It will take about 13.43 days.

6. a. 946.1 bacteria **b.** 23.01 days

7. The formula is $y = Ce^{-0.3t}$; $y = 80$; $C = 100$
$80 = 100e^{-0.3t}$
$0.8 = e^{-0.3t}$
$\ln 0.8 = \ln e^{-0.3t}$
$-0.223\,144 \approx -0.3t \ln e \quad (\ln e = 1)$
$-0.3t \approx -0.223\,144$
$t \approx 0.7438$ yr

8. 0.7754 yr

Exercises 10.8 (page 524)

1. $\log_2 156 = \dfrac{\log_{10} 156}{\log_{10} 2} \approx \dfrac{2.193\,12}{0.301\,03} \approx 7.285$ **2.** 4.954

3. $\log_{12} 7.54 = \dfrac{\log_{10} 7.54}{\log_{10} 12} \approx \dfrac{0.877\,37}{1.079\,18} \approx 0.8130$ **4.** 0.7602

5. $\log_e 3.04 \approx \dfrac{\log_{10} 3.04}{\log_{10} 2.718} \approx \dfrac{0.482\,87}{0.434\,25} \approx 1.112$ **6.** 1.406

7. $\log_{6.8} 0.507 = \dfrac{\log 0.507}{\log 6.8} \approx \dfrac{-0.294\,99}{0.832\,51} \approx -0.3543$

8. -1.651

Sections 10.1–10.8 Review Exercises (page 526)

1. $\log_3 81 = 4$ **2.** $4^{-2} = 0.0625$ **3.** $10^x = 1{,}000$
$10^x = 10^3$
$x = 3$

4. -2 **5.** $9^{3/2} = N$
$N = (3^2)^{3/2} = 3^3 = 27$

6. 2

7. $10^x = 145.6$
$\log 10^x = \log 145.6$
$x \log 10 = \log 145.6 \quad (\log 10 = 1)$
$x \approx 2.1632$

8. $13 \log x$

9. $\log\frac{3}{5} + \log\frac{5}{3} = \log\left(\frac{3}{5}\right)\left(\frac{5}{3}\right) = \log 1 = 0$ **10.** $-\log 2x$

11. 1.4062 **12.** -3.0966 **13.** -3.3755 **14.** 8.5755

15. 2,554 **16.** 12.00 **17.** $(3^4)^{x-1} = \dfrac{1}{3^2}$
$3^{4x-4} = 3^{-2}$
$4x - 4 = -2$
$4x = 2$
$x = \frac{1}{2}$

18. $\frac{3}{2}$

19. $\log_4 75 = \dfrac{\log_{10} 75}{\log_{10} 4} \approx \dfrac{1.8751}{0.6021} \approx 3.114$

20. The inverse function of $y = 6^x$ is $y = \log_6 x$.

$y = 6^x$

x	y
-2	$\frac{1}{36}$
-1	$\frac{1}{6}$
0	1
1	6
2	36

$y = \log_6 x$

x	y
$\frac{1}{36}$	-2
$\frac{1}{6}$	-1
1	0
6	1
36	2

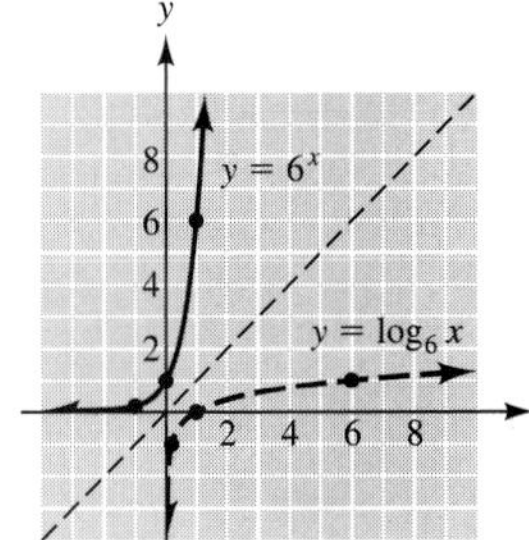

21. We use $A = Pe^{rt}$, where $A = \$2{,}500$, $P = \$2{,}000$, $r = 0.058$.
$2{,}500 = 2{,}000e^{0.058t}$
$1.25 = e^{0.058t}$
$\ln 1.25 = \ln e^{0.058t}$
$0.223\,144 \approx 0.058t \ln e \quad (\ln e = 1)$
$0.058t \approx 0.223\,144$
$t \approx 3.8$
It will be about 3.8 yr before there is \$2,500 in the account.

Chapter 10 Diagnostic Test (page 529)

Following each problem number is the number (in parentheses) of the textbook section where that kind of problem is discussed.

1. (10.2) **a.** *Exponential form* *Logarithmic form*
$2^4 = 16 \iff \log_2 16 = 4$

b. *Logarithmic form* *Exponential form*
$\log_{2.5} 6.25 = 2 \iff (2.5)^2 = 6.25$

2. (10.2) **a.** $\log_4 N = 3 \iff 4^3 = N$
$N = 64$

b. $\log_{10} 10^{-2} = x \iff 10^x = 10^{-2}$
$x = -2$

c. $\log_b 6 = 1 \iff b^1 = 6$
$b = 6$

d. $\log_5 1 = x \iff 5^x = 1 = 5^0$
$x = 0$

e. $\log_{0.5} N = -2 \iff (0.5)^{-2} = N$
$\left(\frac{1}{2}\right)^{-2} = N$
$2^2 = N$
$N = 4$

3. (10.3) **a.** $\log x + \log y - \log z = \log \frac{xy}{z}$

b. $\frac{1}{2}\log x^4 + 2\log x = \log(x^4)^{1/2} + 2\log x$
$= \log x^2 + 2\log x$
$= 2\log x + 2\log x = 4\log x$

4. (10.3) $\log(x^2 - 9) - \log(x - 3)$
$= \log(x + 3)(x - 3) - \log(x - 3)$
$= \log(x + 3) + \log(x - 3) - \log(x - 3) = \log(x + 3)$

5. (10.5) $\log(3x + 5) - \log 7 = \log(x - 1)$

$$\log\frac{(3x+5)}{7} = \log(x-1)$$
$$\frac{3x+5}{7} = \frac{x-1}{1}$$
$$3x + 5 = 7x - 7$$
$$12 = 4x$$
$$x = 3$$

Check
$\log(3x + 5) - \log 7 = \log(x - 1)$
$\log[3(3) + 5] - \log 7 \overset{?}{=} \log(3 - 1)$
$\log 14 - \log 7 \overset{?}{=} \log 2$
$\log \frac{14}{7} \overset{?}{=} \log 2$
$\log 2 = \log 2$

3 is a solution.

6. (10.5) $\log(x + 8) + \log(x - 2) = \log 11$
$\log(x + 8)(x - 2) = \log 11$
$x^2 + 6x - 16 = 11$
$x^2 + 6x - 27 = 0$
$(x - 3)(x + 9) = 0$

$x - 3 = 0$ | $x + 9 = 0$
$x = 3$ | $x = -9$

Check for $x = 3$
$\log(3 + 8) + \log(3 - 2) \overset{?}{=} \log 11$
$\log 11 + \log 1 \overset{?}{=} \log 11$
$\log 11 = \log 11$

Check for $x = -9$
$\log(-9 + 8) = \log(-1)$, which is not real.

Therefore, 3 is the only solution.

7. (10.5) $e^{3x-4} = 8$
$\ln e^{3x-4} = \ln 8$
$(3x - 4)\ln e \approx 2.079\,441\,5$
$3x - 4 \approx 2.079\,441\,5$
$3x \approx 6.079\,441\,5$
$x \approx 2.03$

8. (10.8) Find $\log_2 718$.

$$\log_2 N = \frac{\log_{10} N}{\log_{10} 2}$$
$$\log_2 718 = \frac{\log 718}{\log 2} \approx \frac{2.8561}{0.3010} \approx 9.49$$

9. (10.1) The inverse function of $y = 7^x$ is $y = \log_7 x$.

$y = 7^x$
If $x = -1$, $y = 7^{-1} = \frac{1}{7}$
If $x = 0$, $y = 7^0 = 1$
If $x = 1$, $y = 7^1 = 7$

$y = \log_7 x$
$x = 7^y$
If $y = -1$, $x = 7^{-1} = \frac{1}{7}$
If $y = 0$, $x = 7^0 = 1$
If $y = 1$, $x = 7^1 = 7$

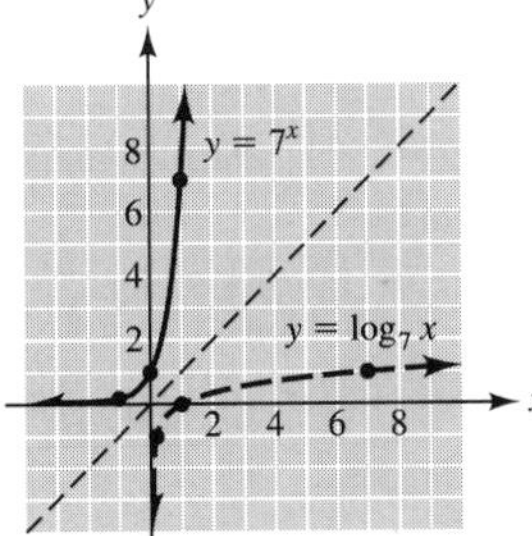

10. (10.6) The formula is $y = Ce^{0.05t}$; $C = 1{,}500$

a. $t = 2$
$y = 1{,}500e^{0.05(2)}$
$y = 1{,}500e^{0.1}$
$y \approx 1{,}500(1.105\,170\,9)$
$y \approx 1{,}658$
There will be about 1,658 bacteria present.

b. $C = 1{,}500$; $y = 4{,}500$
$4{,}500 = 1{,}500e^{0.05t}$
$3 = e^{0.05t}$
$\ln 3 = \ln e^{0.05t}$ Taking ln of both sides
$1.098\,612\,3 \approx 0.05t \ln e$ ($\ln e = 1$)
$0.05t \approx 1.098\,612\,3$
$t \approx 22$
It will take about 22 hr for the bacteria to triple.

Chapters 1–10 Cumulative Review Exercises (page 529)

1. $\frac{7}{(x-4)(x+3)}\frac{(x-1)}{(x-1)} - \frac{3}{(x-4)(x-1)}\frac{(x+3)}{(x+3)}$
$= \frac{7x - 7 - 3x - 9}{(x-4)(x+3)(x-1)} = \frac{4x - 16}{(x-4)(x+3)(x-1)}$
$= \frac{4(x-4)}{(x-4)(x+3)(x-1)} = \frac{4}{(x+3)(x-1)}$

2. $\frac{2}{y + x}$

3.

x	y
$3\frac{1}{2}$	0
0	−1

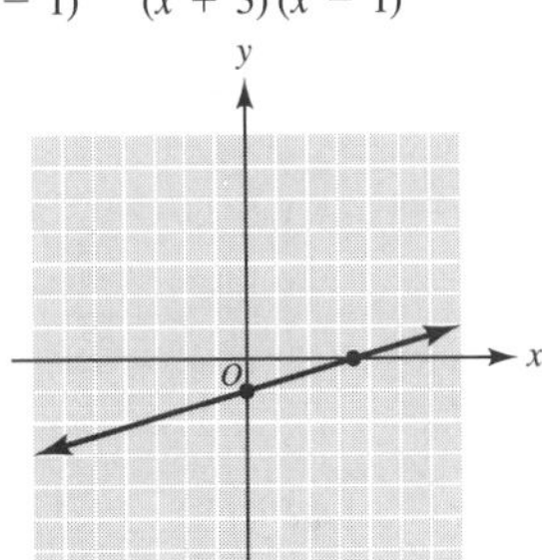

4.

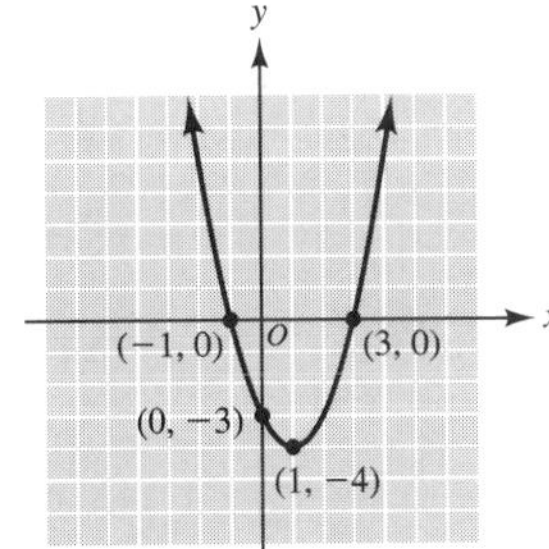

5. LCD = 15
$\frac{15}{1}\cdot\frac{2x}{3} - \frac{15}{1}\cdot\frac{1}{1} \le \frac{15}{1}\cdot\frac{x+2}{5}$
$10x - 15 \le 3x + 6$
$7x \le 21$
$x \le 3$
$\{x \mid x \le 3\}$

6. $\{-2\}$

7. $\log(5x + 2)(3) = \log(12x + 15)$
$(5x + 2)(3) = (12x + 15)$
$15x + 6 = 12x + 15$
$3x = 9$
$x = 3$

Check
$\log(5x + 2) + \log 3 = \log(12x + 15)$
$\log[5(3) + 2] + \log 3 \overset{?}{=} \log[12(3) + 15]$
$\log 17 + \log 3 \overset{?}{=} \log 51$
$\log(17)(3) \overset{?}{=} \log 51$
$\log 51 = \log 51$ True

$\{3\}$

8. $\{0.536\}$

9. The prime numbers between 7 and 31 are 11, 13, 17, 19, 23, and 29; the required prime is 11.

10. $\sqrt{5}$ m **11.** $d = \sqrt{[1-3]^2 + [2-(-4)]^2}$
$= \sqrt{4+36} = \sqrt{40} = 2\sqrt{10}$

12. $3x + 7y + 42 = 0$

13. $10x - 6y + 15 = 0 \Leftrightarrow y = \frac{5}{3}x + \frac{5}{2}$. We must write the equation of the line that passes through $(9, -13)$ and that has a slope of $\frac{5}{3}$.

$$y - (-13) = \tfrac{5}{3}(x-9)$$
$$3(y+13) = 5(x-9)$$
$$3y + 39 = 5x - 45$$
$$5x - 3y - 84 = 0$$

14. 1 **15.** $-\frac{1}{3}$ **16.** $\log_2 128 = 7$ **17.** -1.5058

18. 72.0

19. We use $A = Pe^{rt}$, where $A = \$5{,}000$, $P = \$3{,}000$, $r = 0.056$.

$$\$5{,}000 = \$3{,}000e^{0.056t}$$
$$1.666\,666\,7 \approx e^{0.056t}$$
$$\ln 1.666\,666\,7 \approx \ln e^{0.056t} \quad \text{Taking ln of both sides}$$
$$0.510\,825\,6 \approx 0.056t \ln e$$
$$0.056t \approx 0.510\,825\,6$$
$$t \approx 9.12$$

It will be about 9.12 yr before there is \$5,000 in the account.

20. $2\frac{1}{2}$ L

21. Let x = number of hr for Maria to do the job
Then $x + 2$ = number of hr for Claudio to do the job

$\dfrac{1 \text{ job}}{x \text{ hr}}$ = Maria's rate Maria works for $\left(2 + 3\frac{1}{3}\right)$ hr, or $\frac{16}{3}$ hr.

$\dfrac{1 \text{ job}}{(x+2) \text{ hr}}$ = Claudio's rate Claudio works for $3\frac{1}{3}$ hr, or $\frac{10}{3}$ hr.

$$\left(\frac{1 \text{ job}}{x \text{ hr}}\right)\left(5\frac{1}{3} \text{ hr}\right) + \left(\frac{1 \text{ job}}{(x+2) \text{ hr}}\right)\left(3\frac{1}{3} \text{ hr}\right) = 1 \text{ job}$$
$$\left(\frac{1}{x}\right)\left(\frac{16}{3}\right) + \left(\frac{1}{x+2}\right)\left(\frac{10}{3}\right) = 1$$
$$\frac{3x(x+2)}{1}\left(\frac{1}{x}\right)\left(\frac{16}{3}\right) + \frac{3x(x+2)}{1}\left(\frac{1}{x+2}\right)\left(\frac{10}{3}\right) = \frac{3x(x+2)}{1}(1)$$
$$(x+2)(16) + 10x = 3x(x+2)$$
$$16x + 32 + 10x = 3x^2 + 6x$$
$$0 = 3x^2 - 20x - 32$$
$$0 = (3x+4)(x-8)$$

$3x + 4 = 0$	$x - 8 = 0$
$3x = -4$	$x = 8$
$x = -\frac{4}{3}$ ↑ Reject	$x + 2 = 10$

It would take Maria 8 hr and Claudio 10 hr to do the job.

Exercises 11.1 (page 536)

1. $3x^2 + 5x - 2 = 0$; $a = 3$, $b = 5$, $c = -2$

2. $2x^2 - 3x - 5 = 0$; $a = 2$, $b = -3$, $c = -5$

3. $3x^2 + 0x - 4 = 0$; $a = 3$; $b = 0$, $c = -4$

4. $x^2 + 0x - 16 = 0$; $a = 1$, $b = 0$, $c = -16$

5. $4x = 12 + 3x^2$
$0 = 3x^2 - 4x + 12$; $a = 3$, $b = -4$, $c = 12$

6. $2x^2 - 3x + 10 = 0$; $a = 2$, $b = -3$, $c = 10$

7. $x^2 + 2x - 4 = 0$; $a = 1$, $b = 2$, $c = -4$

8. $x^2 - 4x + 20 = 0$; $a = 1$, $b = -4$, $c = 20$

9. $3x^2 - 6x = x^2 - 4x - 5$
$2x^2 - 2x + 5 = 0$; $a = 2$, $b = -2$, $c = 5$

10. $13x^2 + 20x + 12 = 0$; $a = 13$, $b = 20$, $c = 12$

11. $x^2 = 27$
$x = \pm\sqrt{27} = \pm 3\sqrt{3}$
$\{3\sqrt{3}, -3\sqrt{3}\}$

12. $\{2\sqrt{2}, -2\sqrt{2}\}$

13. $x^2 = -16$
$x = \pm\sqrt{-16}$
$x = \pm 4i$
$\{4i, -4i\}$

14. $\{9i, -9i\}$

15. $x^2 = 7$
$x = \pm\sqrt{7}$
$\{\sqrt{7}, -\sqrt{7}\}$

16. $\{\sqrt{13}, -\sqrt{13}\}$

17. $x^2 = -12$
$x = \pm\sqrt{-12}$
$x = \pm 2i\sqrt{3}$
$\{2i\sqrt{3}, -2i\sqrt{3}\}$

18. $\{5i\sqrt{3}, -5i\sqrt{3}\}$

19. $0 = 8x^3 - 12x$
$0 = 4x(2x^2 - 3)$

$4x = 0$	$2x^2 - 3 = 0$
$x = 0$	$2x^2 = 3$
	$x^2 = \frac{3}{2}$
	$x = \pm\sqrt{\frac{3}{2}}$
	$x = \pm\sqrt{\frac{3}{2} \cdot \frac{2}{2}} = \pm\frac{\sqrt{6}}{2}$

$\left\{0, \frac{\sqrt{6}}{2}, -\frac{\sqrt{6}}{2}\right\}$

20. $\left\{0, \frac{\sqrt{3}}{2}, -\frac{\sqrt{3}}{2}\right\}$

21. $x^2 = -\frac{4}{5}$
$x = \pm\sqrt{-\frac{4}{5}}$
$x = \pm\frac{2i}{\sqrt{5}} = \pm\frac{2i}{\sqrt{5}} \cdot \frac{\sqrt{5}}{\sqrt{5}}$
$x = \pm\frac{2\sqrt{5}}{5}i$
$\left\{\frac{2\sqrt{5}}{5}i, -\frac{2\sqrt{5}}{5}i\right\}$

22. $\left\{\frac{5\sqrt{3}}{3}i, -\frac{5\sqrt{3}}{3}i\right\}$

23. $2x^2 = 12x$
$2x^2 - 12x = 0$
$2x(x - 6) = 0$

$2x = 0$	$x - 6 = 0$
$x = 0$	$x = 6$

$\{0, 6\}$

24. $\left\{0, \frac{1}{10}\right\}$

25. $x^2 - 2x = 2x^2 + 3x$
$x^2 + 5x = 0$
$x(x + 5) = 0$
$x = 0 \mid x = -5$
$\{0, -5\}$

26. $\{0, 5\}$

27. $3x(x + 1) = x(x + 2)$
$3x^2 + 3x = x^2 + 2x$
$2x^2 + x = 0$
$x(2x + 1) = 0$

$x = 0$ Not in the domain	$2x + 1 = 0$
	$x = -\frac{1}{2}$

$\left\{-\frac{1}{2}\right\}$

28. $\{2\}$

29. $x^{-2/3} = 16$
$(x^{-2/3})^{-3/2} = \pm 16^{-3/2}$
$x = \pm\frac{1}{16^{3/2}}$
$= \pm\frac{1}{(\sqrt{16})^3} = \pm\frac{1}{4^3} = \pm\frac{1}{64}$
$\left\{\frac{1}{64}, -\frac{1}{64}\right\}$ (Both check.)

30. $\{32, -32\}$

31. $(x-3)^2 = 35$

$x - 3 = \pm\sqrt{35}$

$x = 3 \pm \sqrt{35}$

$x \approx 3 \pm 5.92$

$\{3+\sqrt{35}, 3-\sqrt{35}\}$, or $\{8.92, -2.92\}$

32. $\{7+\sqrt{3}, 7-\sqrt{3}\}$, or $\{8.73, 5.27\}$

33. $(x+2)^2 = 5$

$x + 2 = \pm\sqrt{5}$

$x = -2 \pm \sqrt{5}$

$x \approx -2 \pm 2.24$

$\{-2+\sqrt{5}, -2-\sqrt{5}\}$, or $\{0.24, -4.24\}$

34. $\{-11+\sqrt{13}, -11-\sqrt{13}\}$, or $\{-7.39, -14.61\}$

35. $(x+6)^2 = 18$

$x + 6 = \pm\sqrt{18}$

$x = -6 \pm 3\sqrt{2}$

$x \approx -6 \pm 4.24$

$\{-6+3\sqrt{2}, -6-3\sqrt{2}\}$, or $\{-1.76, -10.24\}$

36. $\{-1+5\sqrt{2}, -1-5\sqrt{2}\}$, or $\{6.07, -8.07\}$

37. $(x-5)^2 = -15$

$x - 5 = \pm\sqrt{-15}$

$x = 5 \pm i\sqrt{15}$

$x \approx 5 \pm 3.87i$

$\{5+i\sqrt{15}, 5-i\sqrt{15}\}$, or $\{5+3.87i, 5-3.87i\}$

38. $\{12+i\sqrt{17}, 12-i\sqrt{17}\}$, or $\{12+4.12i, 12-4.12i\}$

39. Let x = length of one side (in cm)

$x^2 + x^2 = (\sqrt{32})^2$

$2x^2 = 32$

$x^2 = 16$

$x = 4 \mid x = -4$ Not in the domain

The length of a side is 4 cm.

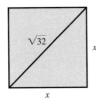

$\sqrt{32}$ x x

40. $9\sqrt{2}$ in.

41. Let x = length of diagonal (in cm)

$7^2 + 10^2 = x^2$

$49 + 100 = x^2$

$\sqrt{149} = x$ ($x = -\sqrt{149}$ is not in the domain.)

The diagonal is $\sqrt{149}$ cm long.

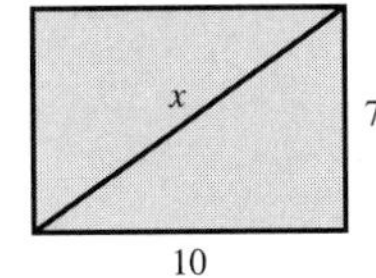

42. $4\sqrt{13}$ cm

Exercises 11.2 (page 540)

1. Let $z = x^2$: $z^2 - 37z + 36 = 0$

$(z-36)(z-1) = 0$

$z - 36 = 0$	$z - 1 = 0$
$z = 36$	$z = 1$
$x^2 = 36$	$x^2 = 1$
$x = \pm 6$	$x = \pm 1$

$\{-6, -1, 1, 6\}$

2. $\{-3, -2, 2, 3\}$

3. Let $y = z^{-2}$: $y^2 - 10y + 9 = 0$

$(y-9)(y-1) = 0$

$y - 9 = 0$	$y - 1 = 0$
$y = 9$	$y = 1$
$z^{-2} = 9$	$z^{-2} = 1$
$\frac{1}{z^2} = 9$	$\frac{1}{z^2} = 1$
$9z^2 = 1$	$z^2 = 1$
$z^2 = \frac{1}{9}$	$z = \pm 1$
$z = \pm\frac{1}{3}$	

$\left\{1, -1, \frac{1}{3}, -\frac{1}{3}\right\}$

4. $\left\{1, -1, \frac{1}{2}, -\frac{1}{2}\right\}$

5. Let $a = y^{1/3}$: $a^2 - 5a + 4 = 0$

$(a-4)(a-1) = 0$

$a - 4 = 0$	$a - 1 = 0$
$a = 4$	$a = 1$
$y^{1/3} = 4$	$y^{1/3} = 1$
$(y^{1/3})^3 = 4^3$	$(y^{1/3})^3 = 1^3$
$y = 64$	$y = 1$

$\{1, 64\}$

6. $\{1, 729\}$

7. Let $a = z^{-2}$: $a^2 - 4a = 0$

$a(a-4) = 0$

$a = 0$	$a = 4$
$z^{-2} = 0$	$z^{-2} = 4$
$\frac{1}{z^2} = 0$	$\frac{1}{z^2} = \frac{4}{1}$
$1 = 0$ False	$4z^2 = 1$
	$z^2 = \frac{1}{4}$
	$z = \pm\frac{1}{2}$

$\left\{\frac{1}{2}, -\frac{1}{2}\right\}$

8. $\{\frac{1}{3}, -\frac{1}{3}\}$

9. Let $a = K^{-1/3}$

$a^2 = K^{-2/3}$

$a^2 + 2a + 1 = 0$

$(a+1)(a+1) = 0$

$a + 1 = 0$ (Same answer for both factors.)

$a = -1$

$K^{-1/3} = -1$

$(K^{-1/3})^{-3} = (-1)^{-3} = \frac{1}{(-1)^3} = \frac{1}{-1} = -1$

$K = -1$

$\{-1\}$

10. $\{1\}$

11. Let $a = (x^2 - 4x)$

$a^2 = (x^2 - 4x)^2$

$a^2 - a - 20 = 0$

$(a+4)(a-5) = 0$

$a + 4 = 0$	$a - 5 = 0$	
$a = -4$	$a = 5$	
$x^2 - 4x = -4$	$x^2 - 4x = 5$	
$x^2 - 4x + 4 = 0$	$x^2 - 4x - 5 = 0$	
$(x-2)(x-2) = 0$	$(x-5)(x+1) = 0$	
$x - 2 = 0$	$x - 5 = 0$	$x + 1 = 0$
$x = 2$	$x = 5$	$x = -1$

$\{-1, 2, 5\}$

12. $\{-1, 1, 3\}$

13. Let w = width

Then $2w$ = length

Area $= w(2w) = 2w^2$

Perimeter $= 2w + 2(2w) = 6w$

w $2w$

$2w^2 + 6w = 80$

$2w^2 + 6w - 80 = 0$

$w^2 + 3w - 40 = 0$

$(w+8)(w-5) = 0$

$w + 8 = 0$	$w - 5 = 0$
$w = -8$	$w = 5$
Not in the domain	$2w = 10$

The width is 5, and the length is 10.

14. Width = 4; length = 12

15. Let r = speed from Los Angeles to Mexico (in mph)

Then $r + 20$ = speed returning (in mph)

$\text{Time} = \frac{\text{distance}}{\text{rate}}$

Time to go + time to return = 5 hr

$$\frac{120}{r} + \frac{120}{r+20} = 5$$

$$\frac{24}{r} + \frac{24}{r+20} = 1 \quad \text{Dividing both sides by 5}$$

LCD = $r(r + 20)$

$$\frac{r(r+20)}{1}\cdot\frac{24}{r} + \frac{r(r+20)}{1}\cdot\frac{24}{r+20} = \frac{r(r+20)}{1}\cdot 1$$

$$24r + 480 + 24r = r^2 + 20r$$

$$r^2 - 28r - 480 = 0$$

$$(r + 12)(r - 40) = 0$$

$r + 12 = 0$, $r = -12$ Not in the domain | $r - 40 = 0$, $r = 40$

Bruce's speed going to Mexico was 40 mph.

16. 30 mph

Exercises 11.3 (page 548)

1.
$$x^2 - 6x = 11$$
$$x^2 - 6x + 9 = 9 + 11$$
$$(x - 3)^2 = 20$$
$$x - 3 = \pm\sqrt{20}$$
$$x - 3 = \pm\sqrt{4 \cdot 5}$$
$$x = 3 \pm 2\sqrt{5}$$
$\{3 + 2\sqrt{5}, 3 - 2\sqrt{5}\}$

2. $\{5 + 2\sqrt{3}, 5 - 2\sqrt{3}\}$

3.
$$x^2 - 4x = 13$$
$$x^2 - 4x + 4 = 13 + 4$$
$$(x - 2)^2 = 17$$
$$x - 2 = \pm\sqrt{17}$$
$$x = 2 \pm \sqrt{17}$$
$\{2 + \sqrt{17}, 2 - \sqrt{17}\}$

4. $\{4 + 2i, 4 - 2i\}$

5.
$$x^2 - 2x = 2$$
$$x^2 - 2x + 1 = 2 + 1$$
$$(x - 1)^2 = 3$$
$$x - 1 = \pm\sqrt{3}$$
$$x = 1 \pm \sqrt{3}$$
$\{1 + \sqrt{3}, 1 - \sqrt{3}\}$

6. $\left\{\frac{2 + \sqrt{3}}{2}, \frac{2 - \sqrt{3}}{2}\right\}$

7.
$$x^2 + 24x = 2{,}881$$
$$x^2 + 24x + 144 = 2{,}881 + 144$$
$$(x + 12)^2 = 3{,}025$$
$$x + 12 = \pm\sqrt{3{,}025}$$
$$x = -12 \pm 55$$
$\{43, -67\}$

8. $\{39, -67\}$

9.
$$x^2 + 6x = 8{,}091$$
$$x^2 + 6x + 9 = 8{,}091 + 9$$
$$(x + 3)^2 = 8{,}100$$
$$x + 3 = \pm\sqrt{8{,}100}$$
$$x = -3 \pm 90$$
$\{87, -93\}$

10. $\{72, -64\}$

11. $3x^2 - x - 2 = 0$; $a = 3, b = -1, c = -2$

$$x = \frac{-(-1) \pm \sqrt{(-1)^2 - 4(3)(-2)}}{2(3)} = \frac{1 \pm \sqrt{1 + 24}}{6} = \frac{1 \pm \sqrt{25}}{6}$$

$$= \frac{1 \pm 5}{6} = \begin{cases} \frac{1+5}{6} = \frac{6}{6} = 1 \\ \frac{1-5}{6} = \frac{-4}{6} = -\frac{2}{3} \end{cases}$$

$\{1, -\frac{2}{3}\}$

12. $\{\frac{1}{2}, -2\}$

13. $x^2 - 4x + 1 = 0$; $a = 1, b = -4, c = 1$

$$x = \frac{-(-4) \pm \sqrt{(-4)^2 - 4(1)(1)}}{2(1)} = \frac{4 \pm \sqrt{16 - 4}}{2(1)} = \frac{4 \pm \sqrt{12}}{2} = \frac{4 \pm 2\sqrt{3}}{2} = 2 \pm \sqrt{3}$$

$\{2 + \sqrt{3}, 2 - \sqrt{3}\}$

14. $\{2 + \sqrt{5}, 2 - \sqrt{5}\}$

15. $x^2 - 4x + 2 = 0$; $a = 1, b = -4, c = 2$

$$x = \frac{-(-4) \pm \sqrt{(-4)^2 - 4(1)(2)}}{2(1)} = \frac{4 \pm \sqrt{16 - 8}}{2} = \frac{4 \pm \sqrt{8}}{2} = \frac{4 \pm 2\sqrt{2}}{2} = 2 \pm \sqrt{2}$$

$\{2 + \sqrt{2}, 2 - \sqrt{2}\}$

16. $\{1 + \sqrt{3}, 1 - \sqrt{3}\}$

17. $x^2 + x + 5 = 0$; $a = 1, b = 1, c = 5$

$$x = \frac{-(1) \pm \sqrt{(1)^2 - 4(1)(5)}}{2(1)} = \frac{-1 \pm \sqrt{1 - 20}}{2} = \frac{-(1) \pm \sqrt{-19}}{2} = \frac{-1 \pm i\sqrt{19}}{2}$$

$\left\{\frac{-1 + i\sqrt{19}}{2}, \frac{-1 - i\sqrt{19}}{2}\right\}$

18. $\left\{\frac{-1 + 3i\sqrt{3}}{2}, \frac{-1 - 3i\sqrt{3}}{2}\right\}$

19. $3x^2 + 2x + 1 = 0$; $a = 3, b = 2, c = 1$

$$x = \frac{-2 \pm \sqrt{(2)^2 - 4(3)(1)}}{2(3)} = \frac{-2 \pm \sqrt{4 - 12}}{6} = \frac{-2 \pm \sqrt{-8}}{6} = \frac{-2 \pm 2i\sqrt{2}}{6} = \frac{-1 \pm i\sqrt{2}}{3}$$

$\left\{\frac{-1 + i\sqrt{2}}{3}, \frac{-1 - i\sqrt{2}}{3}\right\}$

20. $\left\{\frac{-3 + i\sqrt{23}}{8}, \frac{-3 - i\sqrt{23}}{8}\right\}$

21. $2x^2 - 8x + 9 = 0$; $a = 2, b = -8, c = 9$

$$x = \frac{-(-8) \pm \sqrt{(-8)^2 - 4(2)(9)}}{2(2)} = \frac{8 \pm \sqrt{64 - 72}}{4} = \frac{8 \pm \sqrt{-8}}{4} = \frac{8 \pm 2i\sqrt{2}}{4}$$

$$= \frac{\overset{1}{\cancel{2}}(4 \pm i\sqrt{2})}{\underset{2}{\cancel{4}}} = \frac{4 \pm i\sqrt{2}}{2}$$

$\left\{\frac{4 + i\sqrt{2}}{2}, \frac{4 - i\sqrt{2}}{2}\right\}$

22. $\left\{\frac{3 + i\sqrt{3}}{3}, \frac{3 - i\sqrt{3}}{3}\right\}$

23. $3x\left(x + \frac{1}{3}\right) = 3x\left(\frac{-1}{3x}\right)$

$$3x^2 + x = -1$$

$3x^2 + x + 1 = 0$; $a = 3, b = 1, c = 1$

$$x = \frac{-1 \pm \sqrt{1^2 - 4(3)(1)}}{2(3)} = \frac{-1 \pm \sqrt{1 - 12}}{6} = \frac{-1 \pm \sqrt{-11}}{6} = \frac{-1 \pm i\sqrt{11}}{6}$$

$\left\{\frac{-1 + i\sqrt{11}}{6}, \frac{-1 - i\sqrt{11}}{6}\right\}$

24. $\left\{\dfrac{-1+i\sqrt{15}}{8}, \dfrac{-1-i\sqrt{15}}{8}\right\}$

25. $2x^2 - 5x + 7 = 0;\ a = 2,\ b = -5,\ c = 7$

$$x = \frac{-(-5) \pm \sqrt{(-5)^2 - 4(2)(7)}}{2(2)} = \frac{5 \pm \sqrt{25 - 56}}{4} = \frac{5 \pm \sqrt{-31}}{4} = \frac{5 \pm i\sqrt{31}}{4}$$

$\left\{\dfrac{5+i\sqrt{31}}{4}, \dfrac{5-i\sqrt{31}}{4}\right\}$

26. $\left\{\dfrac{5+i\sqrt{47}}{6}, \dfrac{5-i\sqrt{47}}{6}\right\}$

27. $5x^2 - 102x - 891 = 0;\ a = 5,\ b = -102,\ c = -891$

$$x = \frac{-(-102) \pm \sqrt{(-102)^2 - 4(5)(-891)}}{2(5)} = \frac{102 \pm \sqrt{10{,}404 + 17{,}820}}{10} = \frac{102 \pm \sqrt{28{,}224}}{10}$$

$$= \frac{102 \pm 168}{10} = \begin{cases} \dfrac{102 + 168}{10} = \dfrac{270}{10} = 27 \\ \dfrac{102 - 168}{10} = \dfrac{-66}{10} = -6.6 \end{cases}$$

$\{27, -6.6\}$, or $\left\{27, -6\frac{3}{5}\right\}$

28. $\{32, -7.75\}$, or $\left\{32, -7\frac{3}{4}\right\}$

29. $x^2 - 4x + 5 = 0;\ a = 1,\ b = -4,\ c = 5$

$$x = \frac{-(-4) \pm \sqrt{(-4)^2 - 4(1)(5)}}{2(1)} = \frac{4 \pm \sqrt{16 - 20}}{2} = \frac{4 \pm \sqrt{-4}}{2} = \frac{4 \pm 2i}{2} = 2 \pm i$$

$\{2 + i, 2 - i\}$

30. $\{3 + i, 3 - i\}$

The checks will not be shown for Exercises 31–38.

31. Let x = width of rectangle in cm
Then $x + 30$ = length of rectangle in cm
and $(x + 30)x$ = area in sq. cm

$$x^2 + 30x = 6{,}664$$
$$x^2 + 30x + 225 = 6{,}664 + 225 \quad \text{Completing the square}$$
$$(x + 15)^2 = 6{,}889$$
$$x + 15 = \pm 83$$
$$x = -15 \pm 83$$

$x = -15 + 83$, $x = 68$, $x + 30 = 98$ | $x = -15 - 83 = -98$ Reject

The width of the rectangle is 68 cm (and the length is 98 cm).

32. The length of the rectangle is 61 in. (and the width is 29 in.).

33. Let x = Bill's actual rate (in words per min)
Then $x + 20$ = the faster rate (in words per min)

Solving $rt = w$ for t, we have $t = \dfrac{w}{r}$.

$$\text{Actual time} = \frac{825}{x}\text{ min}$$
$$\text{Time if Bill typed faster} = \frac{825}{x + 20}\text{ min}$$
$$\frac{825}{x} - 4 = \frac{825}{x + 20}$$
$$x(x + 20)\left(\frac{825}{x} - 4\right) = x(x + 20)\left(\frac{825}{x + 20}\right)$$
$$825(x + 20) - 4x(x + 20) = 825x$$
$$825x + 16{,}500 - 4x^2 - 80x = 825x$$
$$-4x^2 - 80x + 16{,}500 = 0$$
$$x^2 + 20x = 4{,}125 \quad \text{Dividing both sides by } -4$$
$$x^2 + 20x + 100 = 4{,}125 + 100 \quad \text{Completing the square}$$
$$(x + 10)^2 = 4{,}225$$
$$x + 10 = \pm\sqrt{4{,}225}$$
$$x = -10 \pm 65$$

$x = -10 + 65$, $x = 55$ | $x = -10 - 65 = -75$ Reject

Bill's actual rate of typing is 55 words per min.

34. The pump works at the rate of 29 gal per min.

35. Let w = width (in yd)
Then $w + 2$ = length (in yd)
and $w(w + 2)$ = area (in sq. yd)

$$w(w + 2) = 2$$
$$w^2 + 2w - 2 = 0;\ a = 1,\ b = 2,\ c = -2$$
$$w = \frac{-(2) \pm \sqrt{(2)^2 - 4(1)(-2)}}{2(1)} = \frac{-2 \pm \sqrt{4 + 8}}{2} = \frac{-2 \pm \sqrt{12}}{2} = \frac{-2 \pm 2\sqrt{3}}{2}$$
$$= -1 \pm \sqrt{3} = \begin{cases} -1 + \sqrt{3} \approx -1 + 1.732 = 0.732 \approx 0.73 \\ -1 - \sqrt{3} \approx -1 - 1.732 = -2.73 \quad \text{Not in the domain} \end{cases}$$
$$w + 2 = -1 + \sqrt{3} + 2 = 1 + \sqrt{3} \approx 2.73$$

The width of the rectangle is $(-1 + \sqrt{3})$ yd, or about 0.73 yd, and the length is $(1 + \sqrt{3})$ yd, or about 2.73 yd.

36. The width is $(-2 + \sqrt{10})$ cm, or about 1.16 cm, and the length is $(2 + \sqrt{10})$ cm, or about 5.16 cm.

37. Let x = length of a side
Perimeter = $4x$
Area = x^2

$$4x = 4 + x^2$$
$$x^2 - 4x + 4 = 0$$
$$(x - 2)(x - 2) = 0$$
$$x = 2$$

The length of a side is 2 units.

38. The length of a side is $2 + \sqrt{6}$, or about 4.45, units.

Exercises 11.4 (page 553)

1. $x^2 - x - 12 = 0;\ a = 1,\ b = -1,\ c = -12$
$b^2 - 4ac = (-1)^2 - 4(1)(-12) = 1 + 48 = 49$
Since 49 is positive and a perfect square, there are two distinct real, rational roots.

2. There are two distinct real, rational roots.

3. $6x^2 - 7x - 2 = 0;\ a = 6,\ b = -7,\ c = -2$
$b^2 - 4ac = (-7)^2 - 4(6)(-2) = 49 + 48 = 97$
Since 97 is positive but not a perfect square, there are two distinct real, irrational roots (and they are conjugates).

4. There are two distinct real, irrational, conjugate roots.

5. $x^2 - 4x + 4 = 0;\ a = 1,\ b = -4,\ c = 4$
$b^2 - 4ac = (-4)^2 - 4(1)(4) = 16 - 16 = 0$
Since the discriminant is 0, there is one real, rational root of multiplicity two.

6. There is one real, rational root of multiplicity two.

7. $9x^2 - 6x + 2 = 0;\ a = 9,\ b = -6,\ c = 2$
$b^2 - 4ac = (-6)^2 - 4(9)(2) = 36 - 72 = -36$
Since the discriminant is negative, there are two distinct, complex conjugate roots.

8. There are two distinct, complex conjugate roots.

9. $x = 4$, $x - 4 = 0$ | $x = -2$, $x + 2 = 0$
$$(x - 4)(x + 2) = 0$$
$$x^2 - 2x - 8 = 0$$

11

10. $x^2 + x - 6 = 0$

11. $x = 0 \quad\Big|\quad x = 5$

$x = 0 \quad\Big|\quad x - 5 = 0$

$x(x-5) = 0$

$x^2 - 5x = 0$

12. $x^2 - 6x = 0$

13. $x = 2 + \sqrt{3} \quad\Big|\quad x = 2 - \sqrt{3}$

$x - 2 - \sqrt{3} = 0 \quad\Big|\quad x - 2 + \sqrt{3} = 0$

$$\begin{aligned}[(x-2) - \sqrt{3}][(x-2) + \sqrt{3}] &= 0\\ (x-2)^2 - (\sqrt{3})^2 &= 0\\ x^2 - 4x + 4 - 3 &= 0\\ x^2 - 4x + 1 &= 0\end{aligned}$$

14. $x^2 - 6x + 4 = 0$

15. $x = \frac{1}{2} \quad\Big|\quad x = \frac{2}{3}$

$2x = 1 \quad\Big|\quad 3x = 2$

$2x - 1 = 0 \quad\Big|\quad 3x - 2 = 0$

$$\begin{aligned}(2x-1)(3x-2) &= 0\\ 6x^2 - 7x + 2 &= 0\end{aligned}$$

16. $15x^2 - 19x + 6 = 0$

17. $x = \dfrac{1 + i\sqrt{3}}{2} \quad\Big|\quad x = \dfrac{1 - i\sqrt{3}}{2}$

$2x = 1 + i\sqrt{3} \quad\Big|\quad 2x = 1 - i\sqrt{3}$

$2x - 1 - i\sqrt{3} = 0 \quad\Big|\quad 2x - 1 + i\sqrt{3} = 0$

$$\begin{aligned}[(2x-1) - i\sqrt{3}][(2x-1) + i\sqrt{3}] &= 0\\ (2x-1)^2 - (i\sqrt{3})^2 &= 0\\ 4x^2 - 4x + 1 + 3 &= 0\\ 4x^2 - 4x + 4 &= 0\\ x^2 - x + 1 &= 0\end{aligned}$$

18. $3x^2 - 2x + 1 = 0$

19. $x = 1 \quad\Big|\quad x = 3 \quad\Big|\quad x = 4$

$x - 1 = 0 \quad\Big|\quad x - 3 = 0 \quad\Big|\quad x - 4 = 0$

$$\begin{aligned}(x-1)(x-3)(x-4) &= 0\\ (x^2 - 4x + 3)(x-4) &= 0\\ x^3 - 8x^2 + 19x - 12 &= 0\end{aligned}$$

20. $x^3 - 8x^2 + 17x - 10 = 0$

21. $x = 3 \quad\Big|\quad x = -2i \quad\Big|\quad x = 2i$

$x - 3 = 0 \quad\Big|\quad x + 2i = 0 \quad\Big|\quad x - 2i = 0$

$$\begin{aligned}(x-3)(x+2i)(x-2i) &= 0\\ (x-3)(x^2+4) &= 0\\ x^3 - 3x^2 + 4x - 12 &= 0\end{aligned}$$

22. $x^3 - 5x^2 + 9x - 45 = 0$

23. If $5 - i$ is a root, $5 + i$ is a root.

$x = 5 - i \quad\Big|\quad x = 5 + i$

$x - 5 + i = 0 \quad\Big|\quad x - 5 - i = 0$

$$\begin{aligned}([x-5] + i)([x-5] - i) &= 0\\ [x-5]^2 - i^2 &= 0\\ x^2 - 10x + 25 - (-1) &= 0\\ x^2 - 10x + 26 &= 0\end{aligned}$$

24. $x^2 - 4x + 13 = 0$

25. If $1 - 2\sqrt{5}$ is a root, $1 + 2\sqrt{5}$ is a root.

$x = 1 - 2\sqrt{5} \quad\Big|\quad x = 1 + 2\sqrt{5}$

$x - 1 + 2\sqrt{5} = 0 \quad\Big|\quad x - 1 - 2\sqrt{5} = 0$

$$\begin{aligned}([x-1] + 2\sqrt{5})([x-1] - 2\sqrt{5}) &= 0\\ [x-1]^2 - (2\sqrt{5})^2 &= 0\\ x^2 - 2x + 1 - 20 &= 0\\ x^2 - 2x - 19 &= 0\end{aligned}$$

26. $x^2 - 6x - 41 = 0$

27. If $-3i$ is a root, $+3i$ is a root.

$x = 2 \quad\Big|\quad x = -3i \quad\Big|\quad x = 3i$

$x - 2 = 0 \quad\Big|\quad x + 3i = 0 \quad\Big|\quad x - 3i = 0$

$$\begin{aligned}(x-2)(x+3i)(x-3i) &= 0\\ (x-2)(x^2+9) &= 0\\ x^3 - 2x^2 + 9x - 18 &= 0\end{aligned}$$

28. $x^3 + 3x^2 + 36x + 108 = 0$

Sections 11.1–11.4 Review Exercises (page 556)

1.
$$\begin{aligned}x^2 + x - 6 &= 0\\ (x+3)(x-2) &= 0\end{aligned}$$

$x + 3 = 0 \quad\Big|\quad x - 2 = 0$

$x = -3 \quad\Big|\quad x = 2$

$\{-3, 2\}$

2. $\{5, -2\}$

3. $x^2 - 2x - 4 = 0;\ a = 1,\ b = -2,\ c = -4$

$$x = \frac{-(-2) \pm \sqrt{(-2)^2 - 4(1)(-4)}}{2(1)}$$

$$= \frac{2 \pm \sqrt{4 + 16}}{2} = \frac{2 \pm \sqrt{20}}{2} = \frac{2 \pm 2\sqrt{5}}{2} = 1 \pm \sqrt{5}$$

$\{1 + \sqrt{5}, 1 - \sqrt{5}\}$

4. $\{2 + \sqrt{2}, 2 - \sqrt{2}\}$

5. $x^2 - 2x + 5 = 0;\ a = 1,\ b = -2,\ c = 5$

$$x = \frac{-(-2) \pm \sqrt{(-2)^2 - 4(1)(5)}}{2(1)}$$

$$= \frac{2 \pm \sqrt{-16}}{2} = \frac{2 \pm 4i}{2} = 1 \pm 2i$$

$\{1 + 2i, 1 - 2i\}$

6. $\{17, -25\}$

7.
$$\begin{aligned}16x^2 &= 9\\ x^2 &= \tfrac{9}{16}\\ x &= \pm\tfrac{3}{4}\end{aligned}$$

$\left\{\frac{3}{4}, -\frac{3}{4}\right\}$

8. $\left\{\frac{5}{6}, -\frac{5}{6}\right\}$

9. LCD $= 3(x-2)$

$$\frac{\overset{1}{\cancel{3}}(x-2)}{1}\cdot\frac{(x+2)}{\underset{1}{\cancel{3}}} = \frac{3\overset{1}{\cancel{(x-2)}}}{1}\cdot\frac{1}{\underset{1}{\cancel{(x-2)}}} + \frac{\overset{1}{\cancel{3}}(x-2)}{1}\cdot\frac{2}{\underset{1}{\cancel{3}}}$$

$$\begin{aligned}(x-2)(x+2) &= 3 + 2(x-2)\\ x^2 - 4 &= 3 + 2x - 4\\ x^2 - 2x - 3 &= 0\\ (x-3)(x+1) &= 0\end{aligned}$$

$x - 3 = 0 \quad\Big|\quad x + 1 = 0$

$x = 3 \quad\Big|\quad x = -1$

$\{3, -1\}$

10. $\{-4, 2\}$

11.
$$\begin{aligned}x^2 + 3x - 10 &= 3x - 2x^2 + 2\\ 3x^2 &= 12\\ x^2 &= 4\\ x &= \pm 2\end{aligned}$$

$\{2, -2\}$

12. $\left\{\dfrac{3\sqrt{5}}{5}, -\dfrac{3\sqrt{5}}{5}\right\}$

13. Let $z = (x^2 - 4x)$

Then $z^2 = (x^2 - 4x)^2$

$$\begin{aligned}z^2 + 5z + 4 &= 0\\ (z+4)(z+1) &= 0\end{aligned}$$

$z + 4 = 0 \quad\Big|\quad z + 1 = 0$

$z = -4 \quad\Big|\quad z = -1$

$x^2 - 4x = -4 \quad\Big|\quad x^2 - 4x = -1$

$x^2 - 4x + 4 = 0 \quad\Big|\quad x^2 - 4x + 1 = 0$

$(x-2)^2 = 0 \quad\Big|\quad x = \dfrac{-(-4) \pm \sqrt{(-4)^2 - 4(1)(1)}}{2(1)}$

$x - 2 = 0 \quad\Big|\quad = \dfrac{4 \pm \sqrt{16 - 4}}{2}$

$x = 2 \quad\Big|\quad = \dfrac{4 \pm \sqrt{12}}{2} = \dfrac{4}{2} \pm \dfrac{2\sqrt{3}}{2} = 2 \pm \sqrt{3}$

$\{2, 2 + \sqrt{3}, 2 - \sqrt{3}\}$

14. $\{-3, -3 + \sqrt{5}, -3 - \sqrt{5}\}$

15.
$$\begin{aligned}(x^2 - 1)(x^2 - 64) &= 0\\ (x-1)(x+1)(x+8)(x-8) &= 0\end{aligned}$$

$x - 1 = 0$ | $x + 1 = 0$ | $x + 8 = 0$ | $x - 8 = 0$
$x = 1$ | $x = -1$ | $x = -8$ | $x = 8$

$\{1, -1, 8, -8\}$

16. $\{1, -1, 9, -9\}$

17. $x^2 - 6x + 7 = 0$; $a = 1, b = -6, c = 7$
$b^2 - 4ac = (-6)^2 - 4(1)(7) = 36 - 28 = 8 > 0$
Because 8 is positive but is not a perfect square, there are two distinct real roots, and they are irrational.

18. There are two distinct real roots, and they are irrational.

19. Let x = first odd integer
Then $x + 2$ = second odd integer
$(x)(x + 2) - 14 = 85$
$x^2 + 2x - 99 = 0$
$(x + 11)(x - 9) = 0$
$x + 11 = 0$ | $x - 9 = 0$
$x = -11$ | $x = 9$
$x + 2 = -9$ | $x + 2 = 11$
There are two answers: The integers are -11 and -9, or the integers are 9 and 11.

20. 7 and 8, or -8 and -7

21. Let x = width (in in.)
Then $x + 3$ = length (in in.)
$x(x + 3) = 8$
$x^2 + 3x - 8 = 0$

$$x = \frac{-3 \pm \sqrt{9 + 32}}{2} = \frac{-3 \pm \sqrt{41}}{2}$$

$$\approx \frac{-3 \pm 6.403}{2} = \begin{cases} \dfrac{-3 - 6.403}{2} \approx -4.70 & \text{Not in the domain} \\ \dfrac{-3 + 6.403}{2} \approx 1.70 \end{cases}$$

$$x + 3 = \frac{-3 + \sqrt{41}}{2} + 3 = \frac{3 + \sqrt{41}}{2} \approx \frac{3 + 6.403}{2} \approx 4.70$$

The width is $\dfrac{-3 + \sqrt{41}}{2}$ in. (about 1.70 in.), and the length is $\dfrac{3 + \sqrt{41}}{2}$ in. (about 4.70 in.).

22. One leg is $\dfrac{-2 + \sqrt{14}}{2}$ cm ≈ 0.87 cm.
The other leg is $\dfrac{2 + \sqrt{14}}{2}$ cm ≈ 2.87 cm.

23. $x = 1 - \sqrt{7}$ | $x = 1 + \sqrt{7}$
$x - 1 + \sqrt{7} = 0$ | $x - 1 - \sqrt{7} = 0$
$([x - 1] + \sqrt{7})([x - 1] - \sqrt{7}) = 0$
$[x - 1]^2 - [\sqrt{7}]^2 = 0$
$x^2 - 2x + 1 - 7 = 0$
$x^2 - 2x - 6 = 0$

24. $x^3 - 2x^2 + 6x - 12 = 0$

Exercises 11.5 (page 568)

1. Let $y = 0$
$x^2 - 2x - 3 = 0$
$(x - 3)(x + 1) = 0$
$x - 3 = 0$ | $x + 1 = 0$
$x = 3$ | $x = -1$
x-intercepts: -1 and 3
y-intercept: -3
Axis of symmetry:
$$x = \frac{-(-2)}{2} = 1$$
$f(1) = 1^2 - 2(1) - 3$
$= 1 - 2 - 3 = -4$
Vertex at $(1, -4)$; opens upward

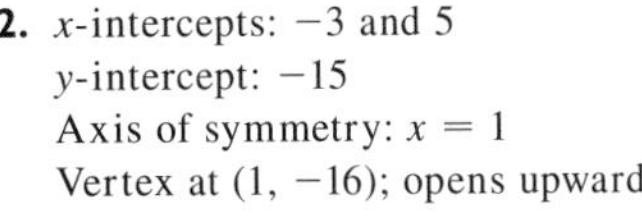

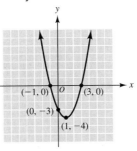

2. x-intercepts: -3 and 5
y-intercept: -15
Axis of symmetry: $x = 1$
Vertex at $(1, -16)$; opens upward

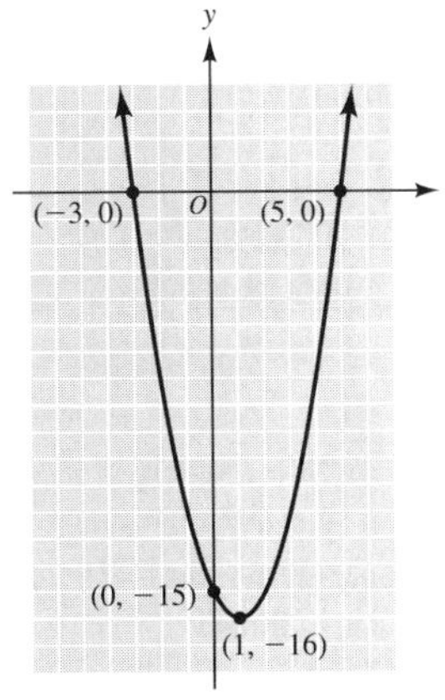

3. Let $y = 0$; then $x^2 - 2x - 13 = 0$

$$x = \frac{-(-2) \pm \sqrt{(-2)^2 - 4(1)(-13)}}{2(1)}$$
$$= \frac{2 \pm \sqrt{4 + 52}}{2} = \frac{2 \pm \sqrt{56}}{2}$$
$$= \frac{2 \pm \sqrt{4 \cdot 14}}{2} = \frac{2 \pm 2\sqrt{14}}{2}$$
$$= 1 \pm \sqrt{14}$$

x-intercepts: $1 + \sqrt{14}$ and $1 - \sqrt{14}$
y-intercept: -13
Axis of symmetry: $x = \dfrac{-(-2)}{2} = 1$
$f(1) = 1^2 - 2(1) - 13 = -14$
Vertex at $(1, -14)$; opens upward

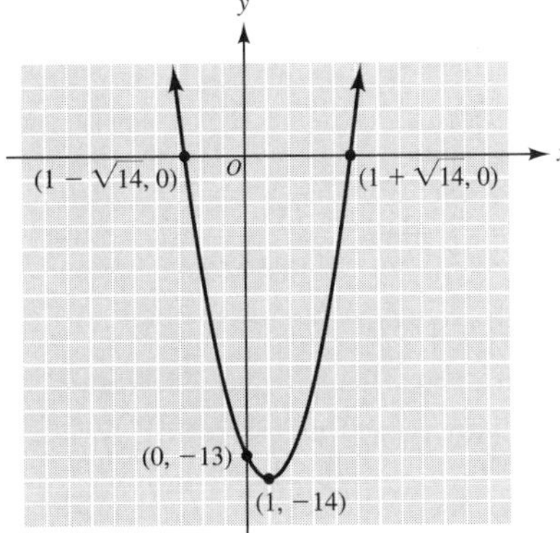

4. x-intercepts: $2 + 2\sqrt{3}$ and $2 - 2\sqrt{3}$
y-intercept: -8
Axis of symmetry: $x = 2$
Vertex at $(2, -12)$; opens upward

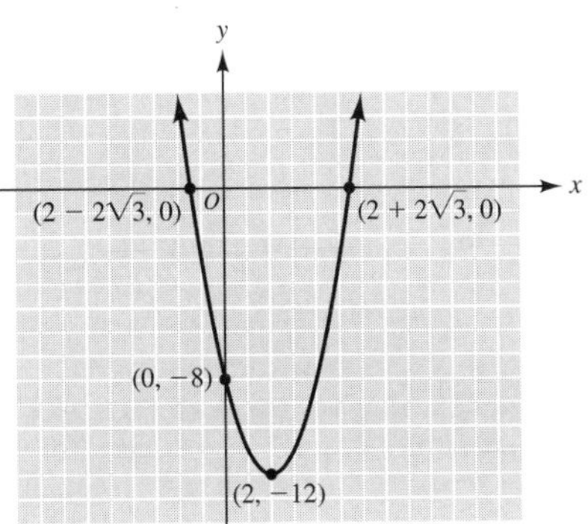

5. Let $y = 0$
$x^2 - 4x + 3 = 0$
$(x - 1)(x - 3) = 0$
$x - 1 = 0 \mid x - 3 = 0$
$x = 1 \mid x = 3$
x-intercepts: 1 and 3
y-intercept: 3
Axis of symmetry:
$x = \dfrac{-(-4)}{2} = 2$
$f(2) = 2^2 - 4(2) + 3 = -1$
Vertex at $(2, -1)$; opens upward

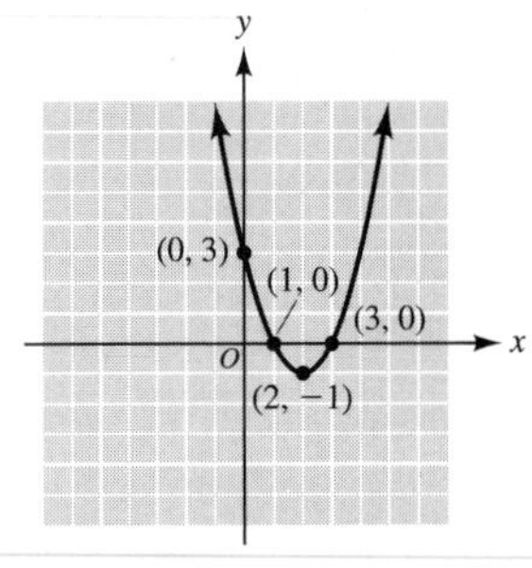

6. x-intercepts: 2 and -4
y-intercept: -8
Axis of symmetry: $x = -1$
Vertex at $(-1, -9)$;
opens upward

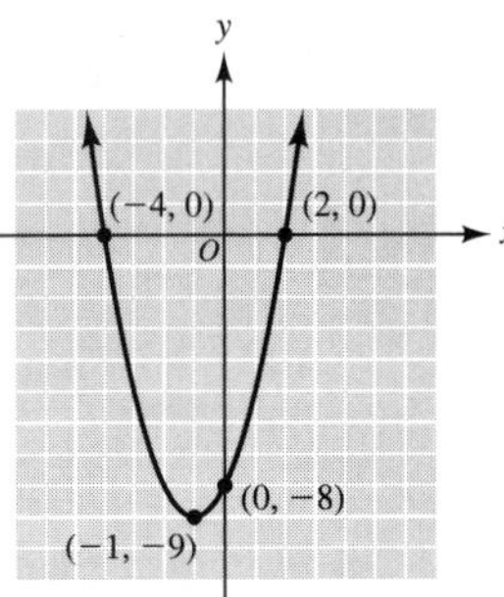

7. Let $y = 0$
$x^2 + 3x - 10 = 0$
$(x - 2)(x + 5) = 0$
$x - 2 = 0 \mid x + 5 = 0$
$x = 2 \mid x = -5$
x-intercepts: 2 and -5
y-intercept: -10
Axis of symmetry: $x = -\frac{3}{2}$
$f\left(-\frac{3}{2}\right) = \left(-\frac{3}{2}\right)^2 + 3\left(-\frac{3}{2}\right) - 10$
$= -\frac{49}{4}$
Vertex at $\left(-\frac{3}{2}, -\frac{49}{4}\right)$;
opens upward

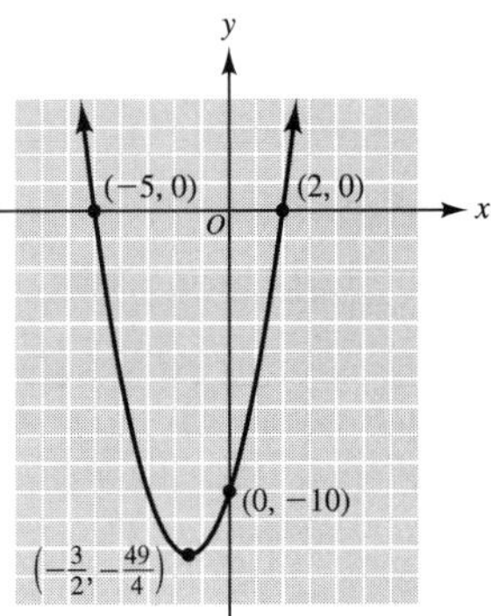

8. x-intercepts: 4 and $-\frac{1}{2}$
y-intercept: -4
Axis of symmetry: $x = \frac{7}{4}$
Vertex at $\left(\frac{7}{4}, -\frac{81}{8}\right)$; opens upward

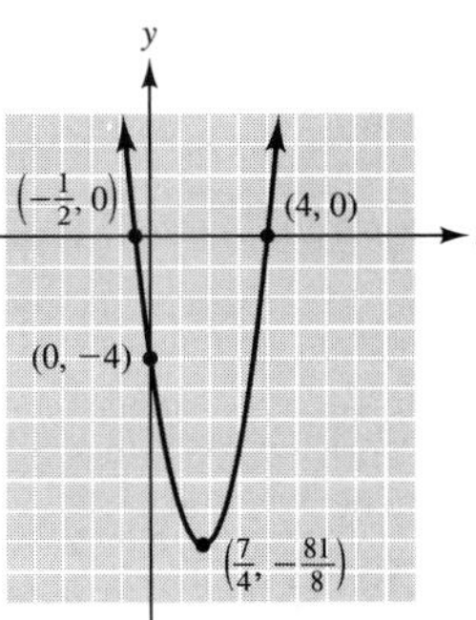

9. $y = f(x) = x^2 - 8x + 12$
Let $y = 0$
$0 = (x - 2)(x - 6)$
$x - 2 = 0 \mid x - 6 = 0$
$x = 2 \mid x = 6$
x-intercepts: 2 and 6
y-intercept: 12
Axis of symmetry:
$x = -\dfrac{-8}{2} = 4$
$f(4) = 4^2 - 8(4) + 12$
$= 16 - 32 + 12$
$= -4$
Vertex at $(4, -4)$; opens upward

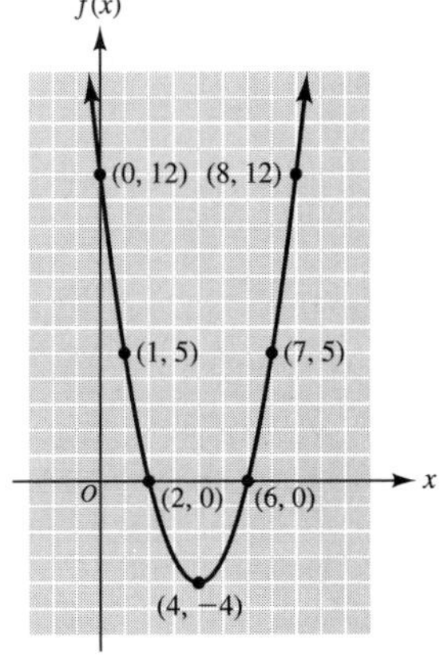

10. x-intercepts: $2 \pm \sqrt{10}$
y-intercept: -6
Axis of symmetry: $x = 2$
Vertex at $(2, -10)$;
opens upward

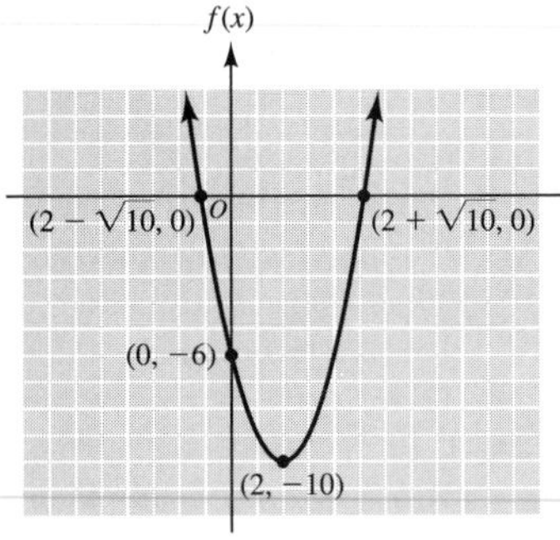

11. $y = f(x) = 3 + 2x - x^2$
Let $y = 0$
$0 = 3 + 2x - x^2$
$0 = (3 - x)(1 + x)$
$3 - x = 0 \mid 1 + x = 0$
$x = 3 \mid x = -1$
x-intercepts: 3 and -1
y-intercept: 3
Axis of symmetry:
$x = -\dfrac{2}{2(-1)} = 1$
$f(1) = 2(1) - 1^2 + 3 = 4$
Vertex at $(1, 4)$; opens downward

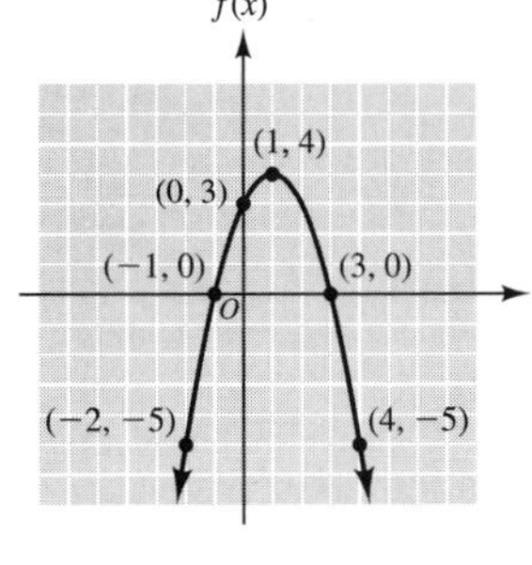

12. x-intercepts: -5 and 1
y-intercept: 5
Axis of symmetry: $x = -2$
Vertex at $(-2, 9)$; opens downward

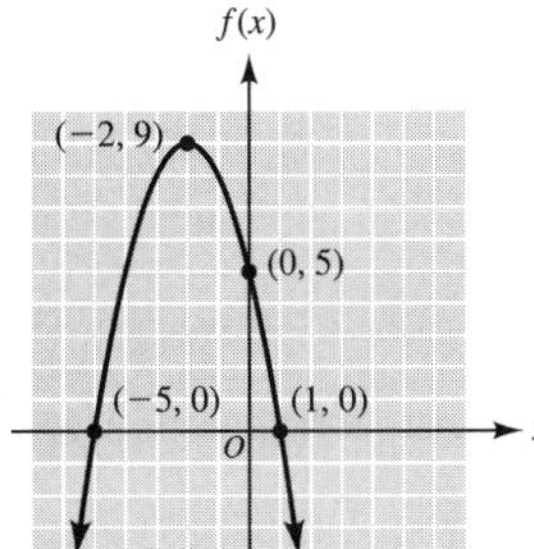

13. $y = f(x) = x^2 - 6x + 10$
Let $y = 0$; then $x^2 - 6x + 10 = 0$
$x = \dfrac{-(-6) \pm \sqrt{(-6)^2 - 4(1)(10)}}{2(1)}$
$= \dfrac{6 \pm \sqrt{36 - 40}}{2} = \dfrac{6 \pm \sqrt{-4}}{2}$
$= \dfrac{6 \pm 2i}{2} = 3 \pm i$
This means that the curve does not cross the x-axis.
y-intercept: 10
Axis of symmetry: $x = -\dfrac{-6}{2} = 3$
Vertex at $(3, 1)$; opens upward

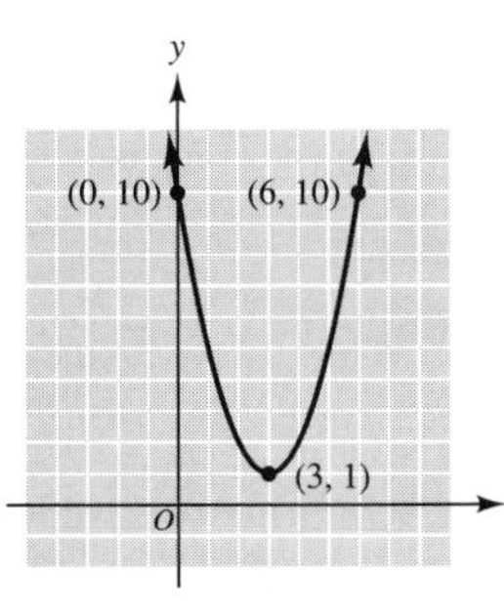

14. No x-intercepts
y-intercept: 11
Axis of symmetry: $x = 3$
Vertex at $(3, 2)$; opens upward

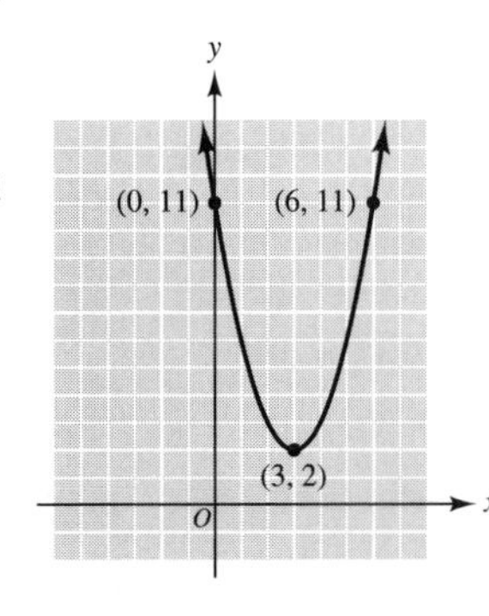

15. Let $y = 0$
$0 = 4x^2 - 9$
$9 = 4x^2$
$x^2 = \frac{9}{4}$
$x = \pm\sqrt{\frac{9}{4}} = \pm\frac{3}{2}$
x-intercepts: $\pm\frac{3}{2}$
y-intercept: -9
Axis of symmetry: $x = \dfrac{-0}{2(4)} = 0$
$f(0) = -9$
Vertex at $(0, -9)$; opens upward

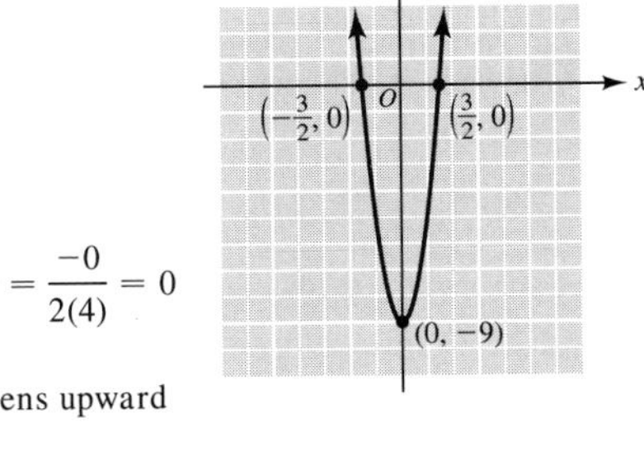

16. x-intercepts: $\pm\frac{2}{3}$
y-intercept: -4
Axis of symmetry: $x = 0$
Vertex at $(0, -4)$; opens upward

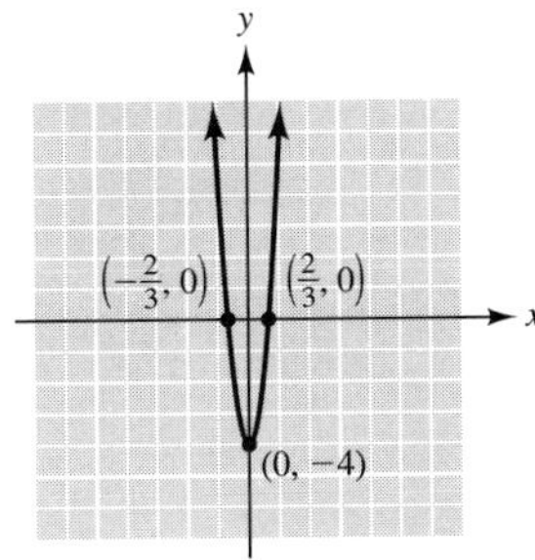

17. $y = f(x) = x^2 - 4x + 6$
If we let $y = 0$, there are no real solutions (no x-intercepts).
y-intercept: 6
Axis of symmetry:
$x = -\dfrac{-4}{2} = 2$
Vertex at $(2, 2)$; opens upward

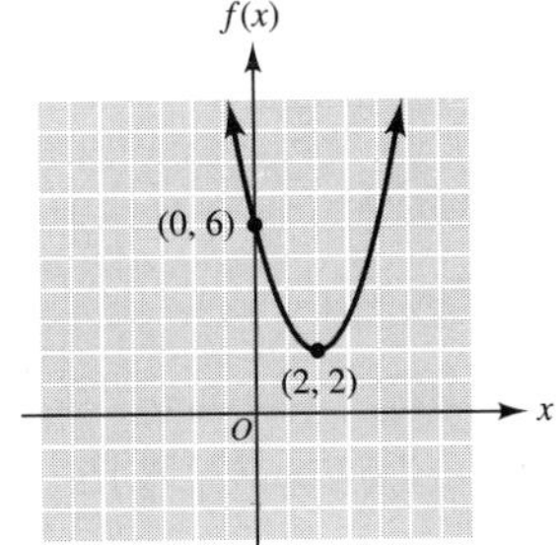

18. No x-intercepts
y-intercept: -2
Axis of symmetry: $x = 1$
Vertex at $(1, -1)$;
opens downward

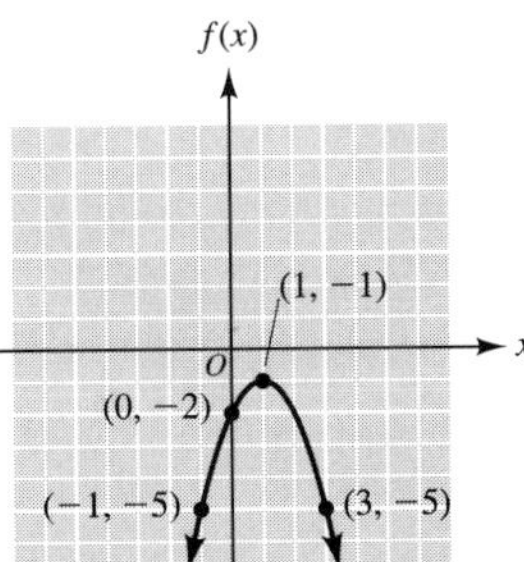

19. $y = x^2 - 6$ is equivalent to $y + 6 = x^2$; the graph has the same size and shape as $y = x^2$. The vertex is at $(0, -6)$.

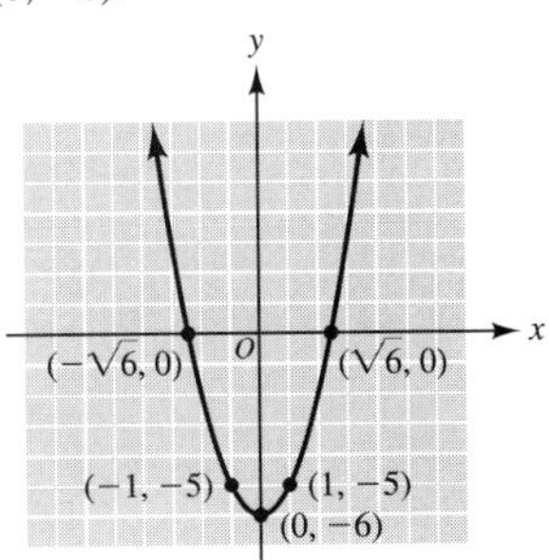

20.

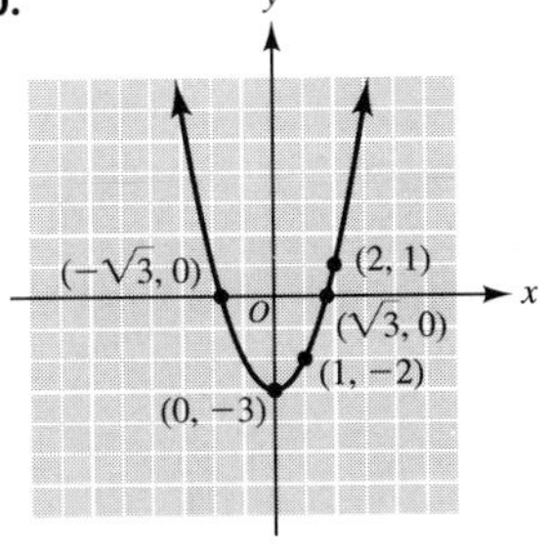

21. The graph has the same size and shape as the graph of $y = x^2$. The vertex is at $(-2, 3)$.

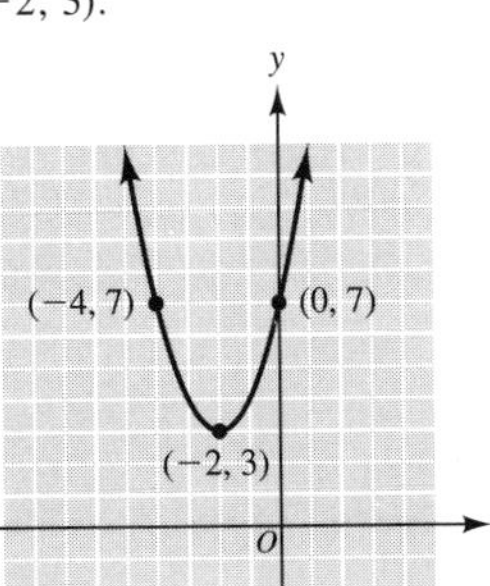

22.

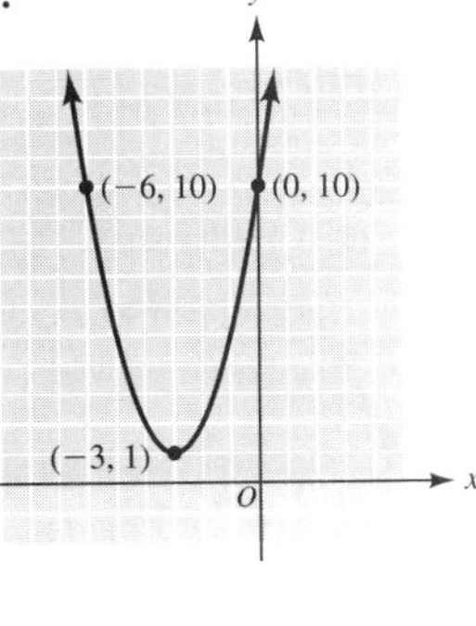

23. The graph has the same size and shape as the graph of $y = x^2$. The vertex is at $(1, -4)$.

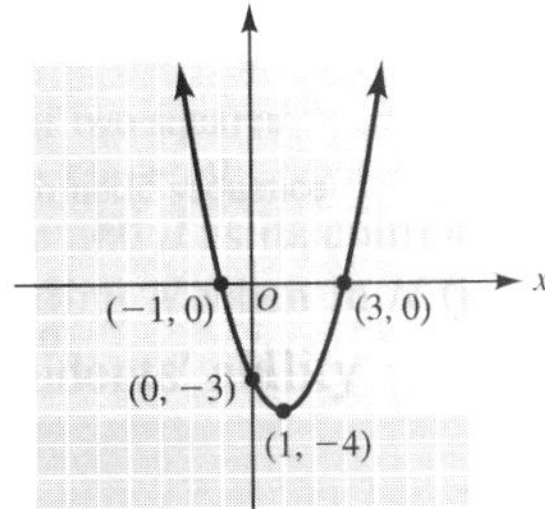

24.

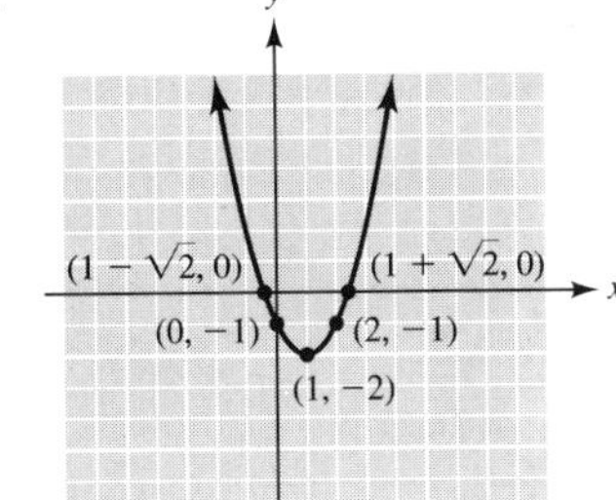

25. $y = (x - 1)^2 + 3$ is equivalent to $y - 3 = (x - 1)^2$; the graph has the same size and shape as the graph of $y = x^2$. The vertex is at $(1, 3)$.

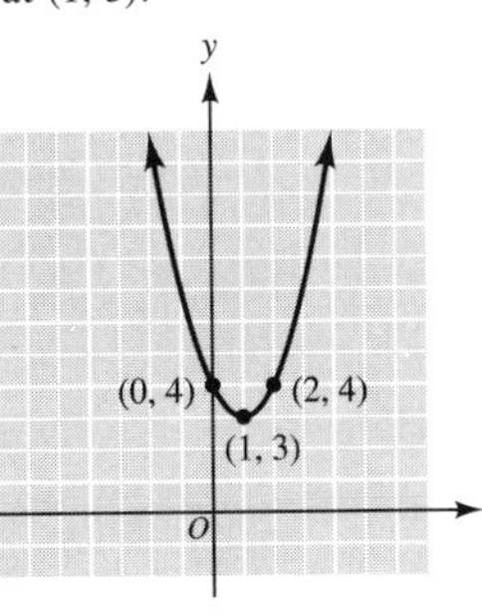

26.

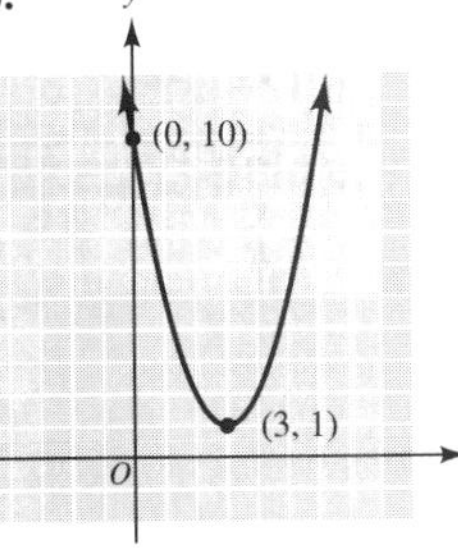

II

27. $f(x) = x^2 - 6x + 7$
$a > 0$
$f(x)$ has a minimum.
$-\dfrac{b}{2a} = -\left(\dfrac{-6}{2(1)}\right) = 3$
$f(3) = (3)^2 - 6(3) + 7$
$= 9 - 18 + 7$
$= -2$ Minimum
Vertex: $(3, -2)$

28. Minimum = 1
Vertex: (2, 1)

29. $y = -2x^2 + 8x - 3$
$a < 0$
$f(x)$ has a maximum.
$-\dfrac{b}{2a} = -\dfrac{8}{2(-2)} = 2$
$f(2) = -2(4) + 8(2) - 3 = 5$ Maximum
Vertex: (2, 5)

30. Maximum = 7
Vertex: (1, 7)

31. $f(x) = -\frac{1}{2}x^2 + x + \frac{3}{2}$
$a < 0$
$f(x)$ has a maximum.
$-\dfrac{b}{2a} = -\dfrac{1}{2\left(-\frac{1}{2}\right)} = 1$
$f(1) = -\frac{1}{2}(1)^2 + (1) + \frac{3}{2} = 2$ Maximum
Vertex: (1, 2)

32. Maximum = 3
Vertex: (−2, 3)

Exercises 11.6 (page 577)

1. $x^2 + y^2 = 9$
Circle; center at (0, 0);
radius is 3.

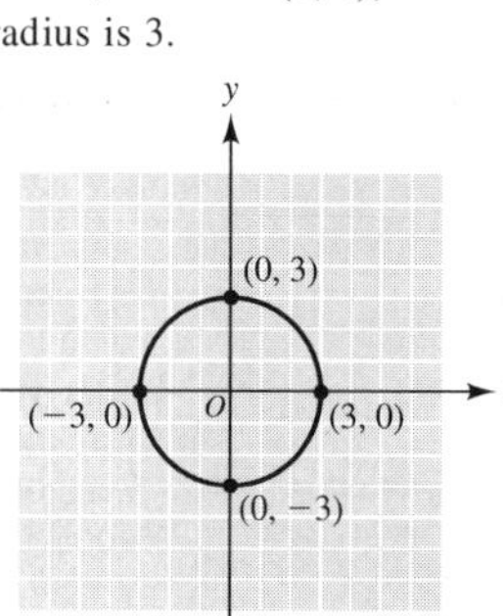

2. Circle

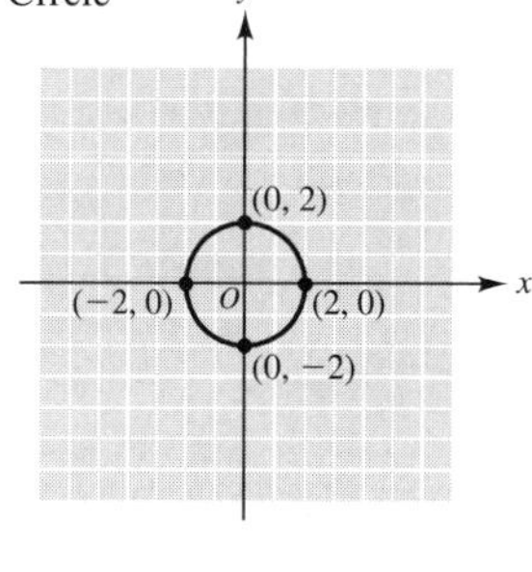

3. $y = x^2 + 2x - 3$
Parabola
x-intercepts: $0 = x^2 + 2x - 3$
$0 = (x + 3)(x - 1)$
$x = -3$ or $x = 1$
y-intercept: $y = 0 + 0 - 3$
$= -3$
Axis of symmetry:
$x = -\dfrac{b}{2a} = -\dfrac{2}{2} = -1$
$f(-1) = (-1)^2 + 2(-1) - 3 = -4$
Vertex at (−1, −4)

4. Parabola

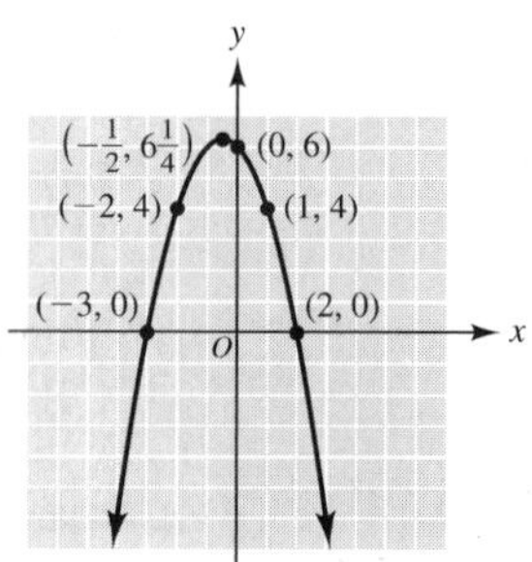

5. $3x^2 + 3y^2 = 21$
$x^2 + y^2 = 7$
Circle; center at (0, 0);
radius is $\sqrt{7} \approx 2.6$.

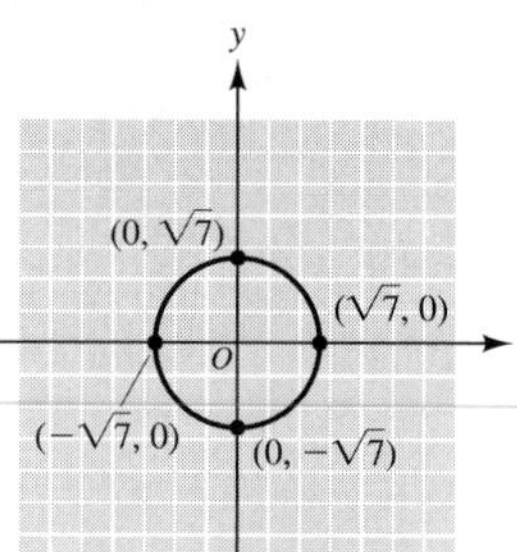

6. Ellipse

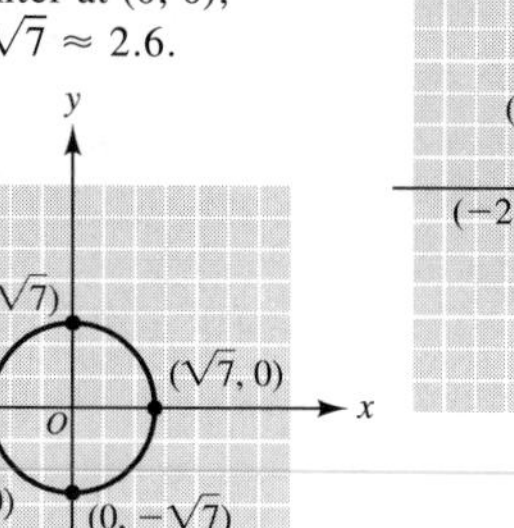

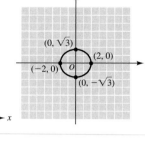

7. $4x^2 + 9y^2 = 36$
Ellipse
x-intercepts: $4x^2 + 9(0)^2 = 36$
$4x^2 = 36$
$x^2 = 9$
$x = \pm 3$
y-intercepts: $4(0)^2 + 9y^2 = 36$
$9y^2 = 36$
$y^2 = 4$
$y = \pm 2$

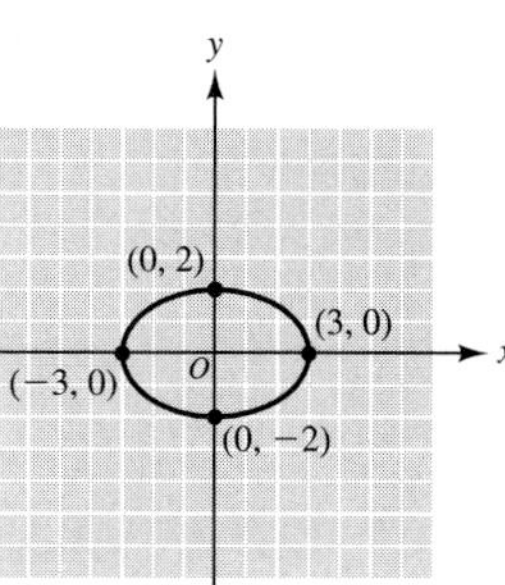

8. Ellipse

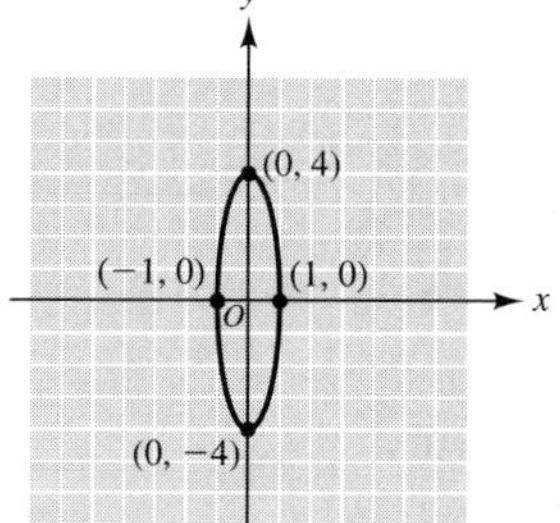

9. $9x^2 - 4y^2 = 36$
Hyperbola
x-intercepts: $9x^2 - 4(0)^2 = 36$
$9x^2 = 36$
$x^2 = \pm 4$
$x = \pm 2$
(Graph does not intersect y-axis.)
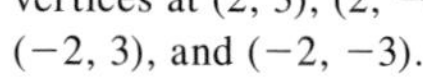
The rectangle of reference has vertices at (2, 3), (2, −3), (−2, 3), and (−2, −3).

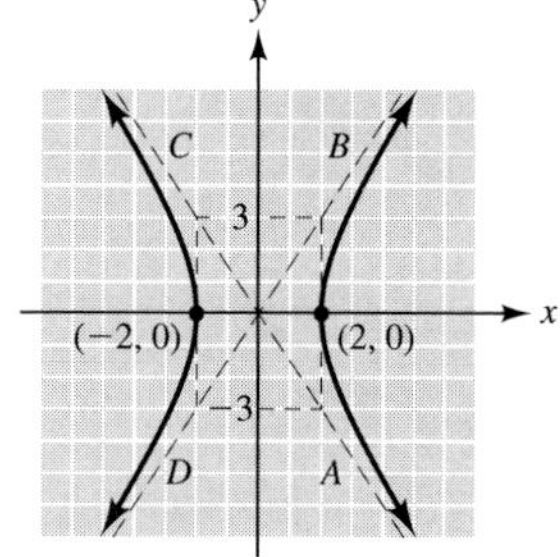

10. Hyperbola

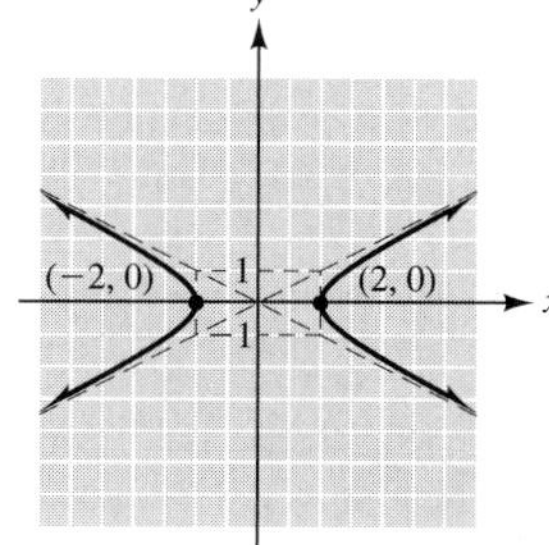

11. $y^2 = 8x$
Parabola
When $x = 1$, $y^2 = 8$
$y = \pm\sqrt{8} \approx \pm 2.8$
When $x = 2$, $y^2 = 16$
$y = \pm 4$
When $x = 0$, $y^2 = 0$
$y = 0$
Vertex at (0, 0)

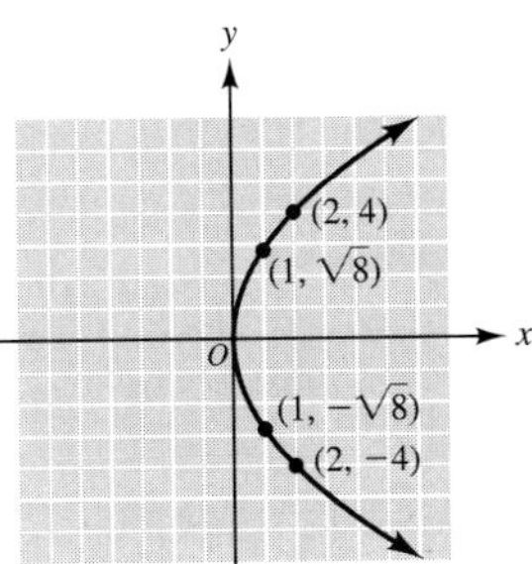

12. Parabola

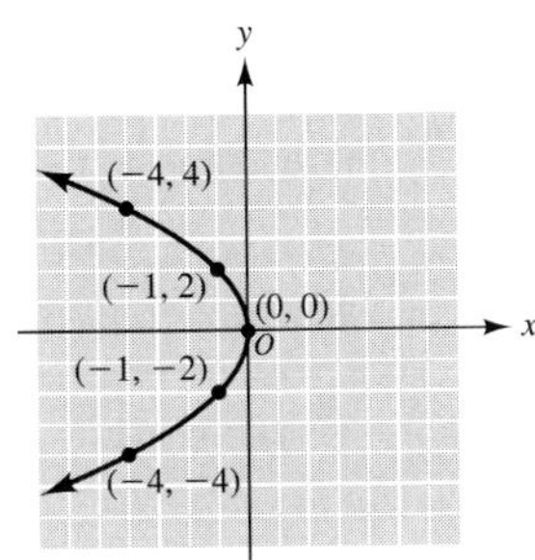

13. $(x^2 + 2x + ?) + (y^2 - 4y + ?) = -1$
$(x^2 + 2x + 1) + (y^2 - 4y + 4) = -1 + 1 + 4$
$(x + 1)^2 + (y - 2)^2 = 4$ is in the form
$(x - h)^2 + (y - k)^2 = r^2$, where $h = -1$, $k = 2$, and $r = 2$.
The conic is a circle with its center at $(-1, 2)$ and with a radius of 2.

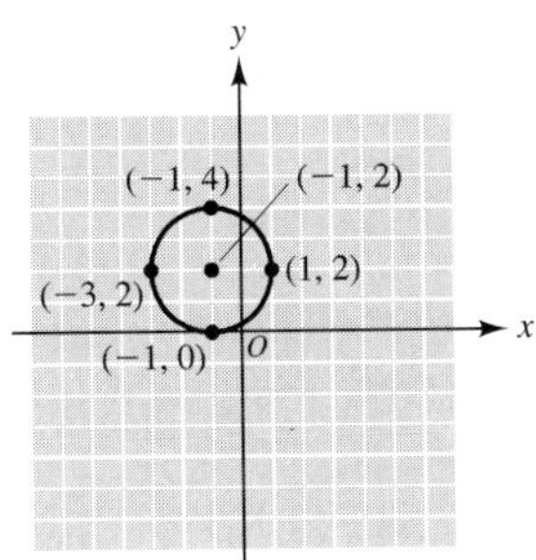

14. Circle; center at $(2, -3)$, radius is 1.

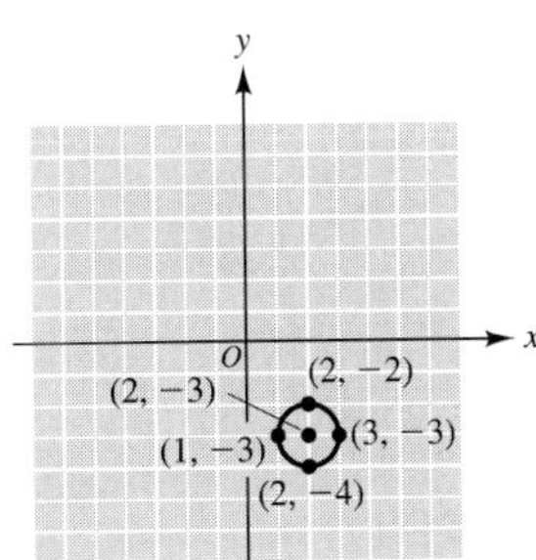

15. $x = -y^2 - 2y$
$x = -(y^2 + 2y + ?)$
$x - 1 = -(y^2 + 2y + 1)$
$x - 1 = -(y + 1)^2$
The conic is a parabola that opens to the left; the vertex is at $(1, -1)$. The line of symmetry is the line
$y = \dfrac{-(-2)}{2(-1)} = -1.$

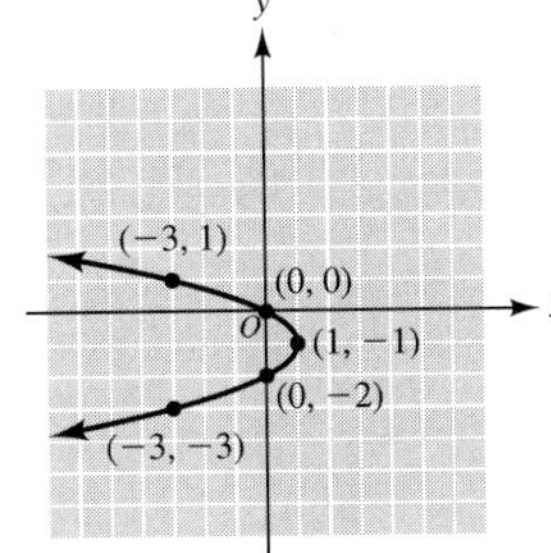

16. The conic is a parabola that opens to the right; the vertex is at $(-2, 1)$. The line of symmetry is the line $y = 1$.

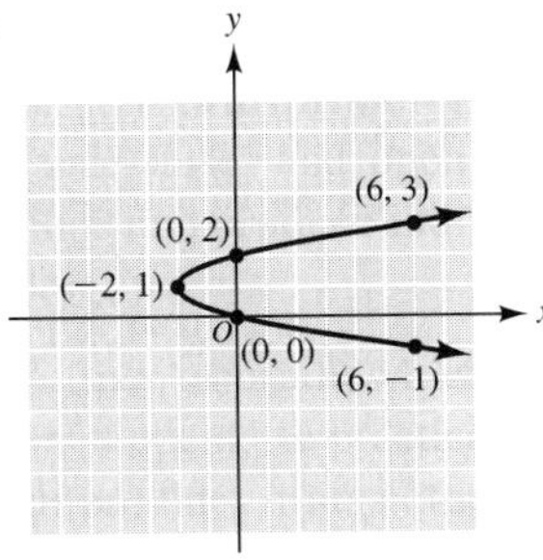

Exercises 11.7 (page 586)

Note: Some exercises are shown using method 1 and some using method 2.

1. $(x + 1)(x - 2) < 0$
Let $f(x) = (x + 1)(x - 2) = 0$
Then $x + 1 = 0$ | $x - 2 = 0$
$x = -1$ | $x = 2$
x-intercepts: -1 and 2
$f(x) = (x + 1)(x - 2)$
$f(0) = (1)(-2) = -2$
The curve goes through $(0, -2)$.
Solution set: $\{x \mid -1 < x < 2\}$, or $(-1, 2)$

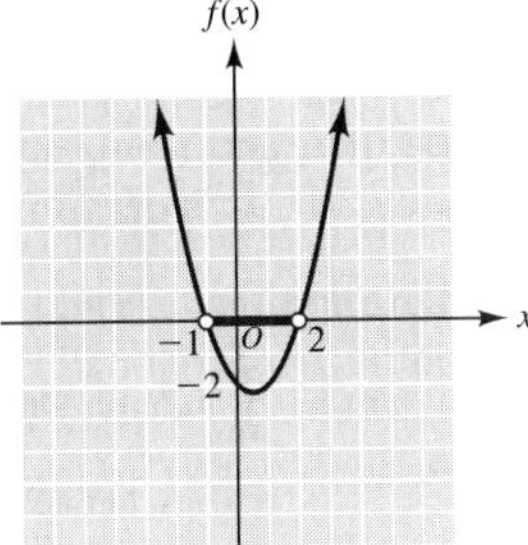

2. $\{x \mid x < -1\} \cup \{x \mid x > 2\}$, or $(-\infty, -1) \cup (2, +\infty)$

3. $3 - 2x - x^2 \geq 0$
Let $f(x) = (3 + x)(1 - x) = 0$
Then $3 + x = 0$ | $1 - x = 0$
$x = -3$ | $x = 1$
Critical points: -3 and 1

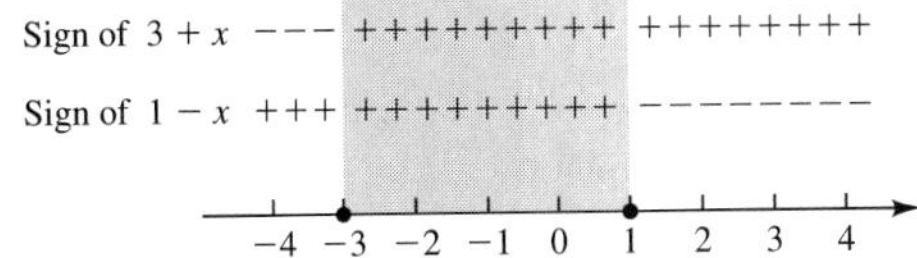

The inequality was $\geq$. We select the interval on which both signs are the same. The graph is

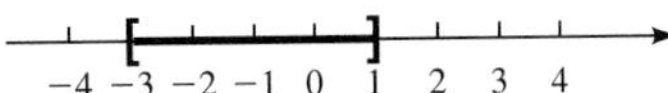

Solution set: $\{x \mid -3 \leq x \leq 1\}$, or $[-3, 1]$

4. $\{x \mid x \leq -2\} \cup \{x \mid x \geq 1\}$, or $(-\infty, -2] \cup [1, +\infty)$

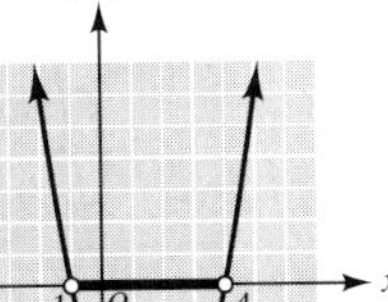

5. $x^2 - 3x - 4 < 0$
Let $f(x) = x^2 - 3x - 4 = 0$
Then $(x - 4)(x + 1) = 0$
x-intercepts: 4 and -1
$f(0) = 0 - 0 - 4 = -4$
The curve goes through $(0, -4)$.
Solution set: $\{x \mid -1 < x < 4\}$, or $(-1, 4)$

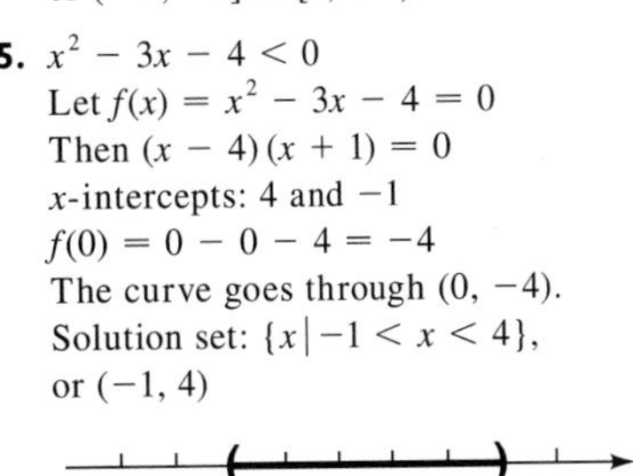

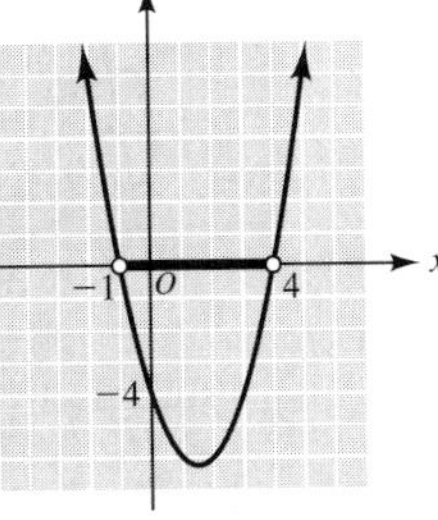

11

6. $\{x \mid x < -5\} \cup \{x \mid x > 1\}$, or $(-\infty, -5) \cup (1, +\infty)$

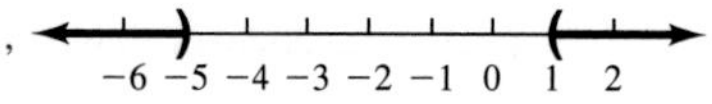

7. $x^2 - 6x + 7 > 0$

Let $f(x) = x^2 - 6x + 7 = 0$

$$x = \frac{-(-6) \pm \sqrt{36 - 28}}{2} = \frac{6 \pm \sqrt{8}}{2}$$

$$= \frac{6 \pm 2\sqrt{2}}{2} = 3 \pm \sqrt{2}$$

x-intercepts: $3 - \sqrt{2} \approx 1.6$ and $3 + \sqrt{2} \approx 4.4$

$f(0) = 0 - 0 + 7 = 7$

The graph goes through $(0, 7)$.

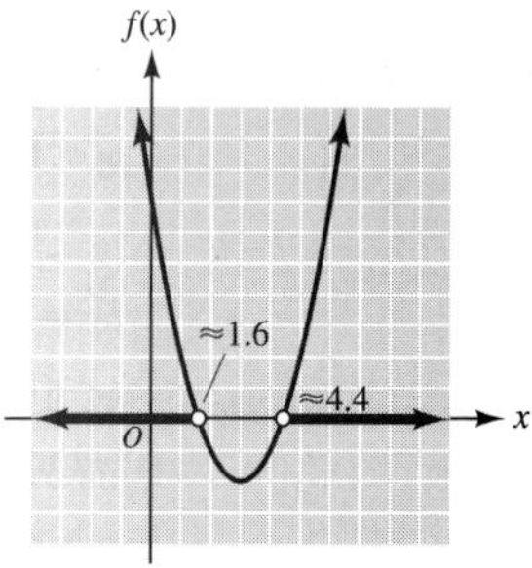

Solution set: $\{x \mid x < 3 - \sqrt{2}\} \cup \{x \mid x > 3 + \sqrt{2}\}$, or $(-\infty, 3 - \sqrt{2}) \cup (3 + \sqrt{2}, +\infty)$

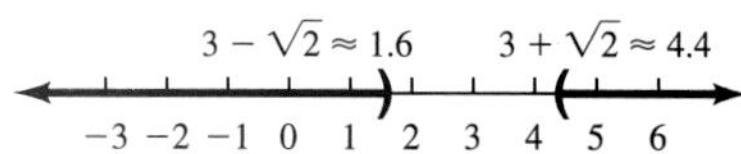

8. $\{x \mid 4 - \sqrt{3} < x < 4 + \sqrt{3}\}$, or $(4 - \sqrt{3}, 4 + \sqrt{3})$

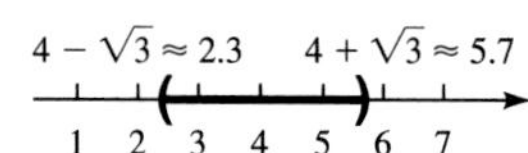

9. $x^2 \le 5x$

Let $f(x) = x^2 - 5x = 0$

Then $x(x - 5) = 0$

x-intercepts: 0 and 5

$f(1) = 1 - 5 = -4$

The graph goes through $(1, -4)$.

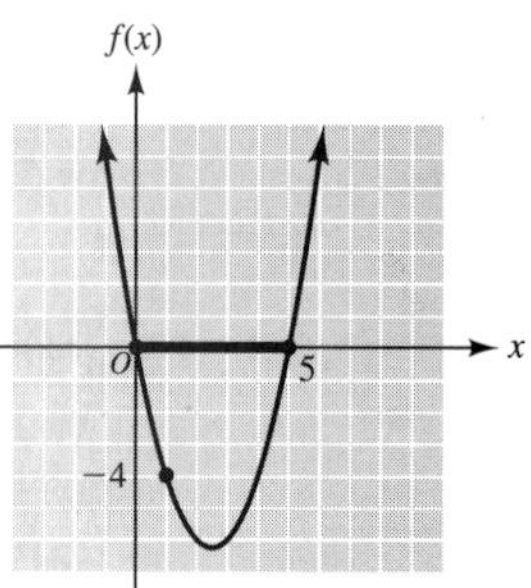

Solution set: $\{x \mid 0 \le x \le 5\}$, or $[0, 5]$

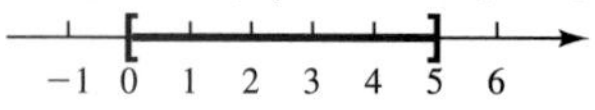

10. $\{x \mid x \le 0\} \cup \{x \mid x \ge 3\}$, or $(-\infty, 0] \cup [3, +\infty)$

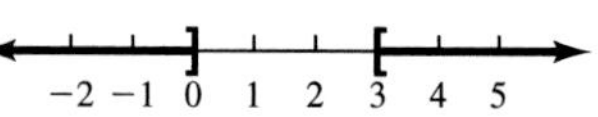

11. $3x - x^2 > 0$

Let $f(x) = x(3 - x) = 0$

Critical points: 0 and 3

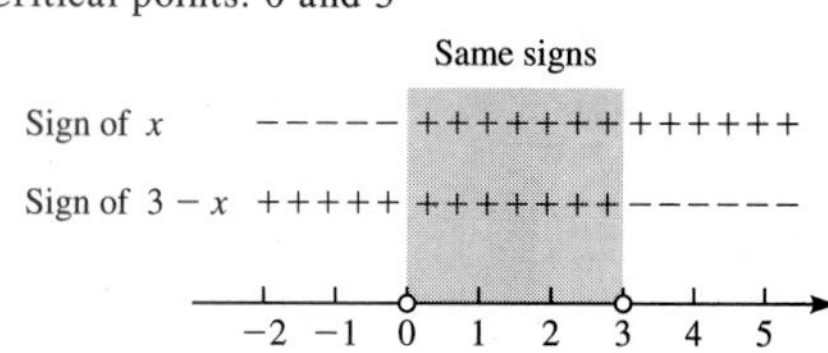

We select the interval on which both signs are the same. The graph is

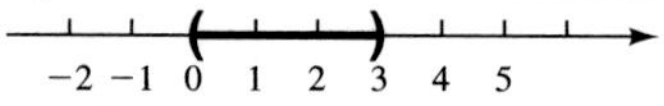

Solution set: $\{x \mid 0 < x < 3\}$, or $(0, 3)$

12. $\{x \mid x < 0\} \cup \{x \mid x > 7\}$, or $(-\infty, 0) \cup (7, +\infty)$

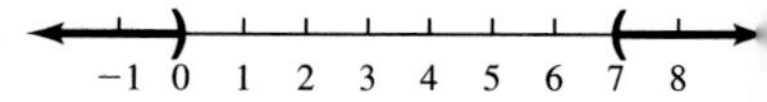

13. $x^3 - 3x^2 - x + 3 > 0$

Let $f(x) = x^3 - 3x^2 - x + 3 = 0$

Factoring by grouping, we have

$x^3 - 3x^2 - x + 3 = (x - 3)(x + 1)(x - 1)$.

Critical points: -1, 1, and 3

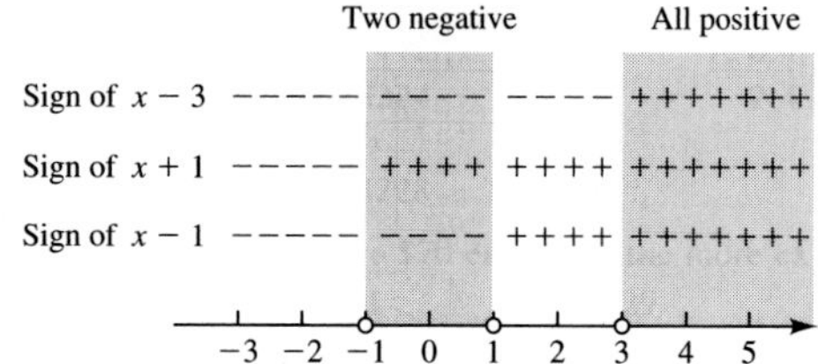

We select the intervals on which there are an *even* number of negative signs or where all the signs are positive. The graph is

Solution set: $\{x \mid -1 < x < 1\} \cup \{x \mid x > 3\}$, or $(-1, 1) \cup (3, +\infty)$

14. $\{x \mid -2 < x < -1\} \cup \{x \mid x > 1\}$, or $(-2, -1) \cup (1, +\infty)$

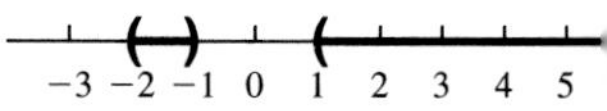

15. Let $f(x) = (x + 1)(x - 2)^2 = 0$

Critical points: -1 and 2

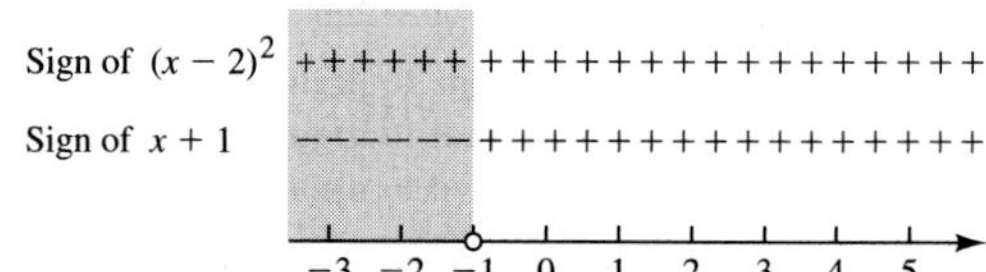

We select the interval on which the signs are different. The graph is

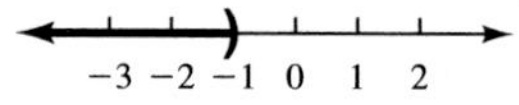

Solution set: $\{x \mid x < -1\}$, or $(-\infty, -1)$

16. $\{x \mid x < -2\} \cup \{x \mid -2 < x < 1\}$, or $(-\infty, -2) \cup (-2, 1)$

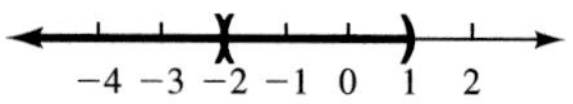

17. $\dfrac{x - 2}{x + 3} \ge 0$

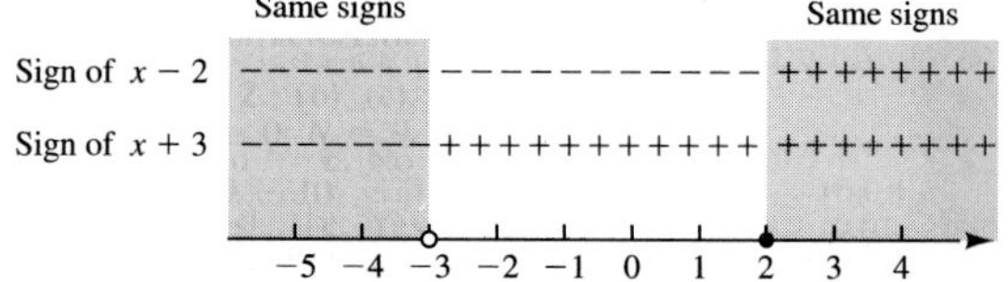

We select the interval on which the signs are the same. (Remember that x cannot equal -3.) The graph is

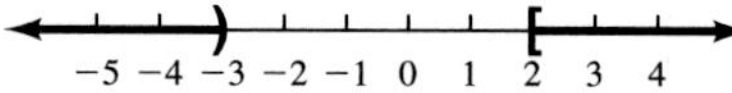

Solution set: $\{x \mid x < -3\} \cup \{x \mid x \ge 2\}$, or $(-\infty, -3) \cup [2, +\infty)$

18. $\{x \mid -4 \le x < 2\}$, or $[-4, 2)$

−5 −4 −3 −2 −1 0 1 2 3 4

19. $\dfrac{2x+5}{3x-1} \le 0$

	Same signs	Different signs	Same signs
Sign of $2x+5$	− − − − − − −	+ + + + + +	+ + + + + + + + + +
Sign of $3x-1$	− − − − − − −	− − − − − −	+ + + + + + + + + +

−4 −3 −2 −1 0 1 2 3 4

We select the interval on which the signs are different. The graph is

−4 −3 −2 −1 0 1 2 3 4

Solution set: $\left\{x \middle| -2\frac{1}{2} \le x < \frac{1}{3}\right\}$, or $\left[-2\frac{1}{2}, \frac{1}{3}\right)$

20. $\left\{x \middle| x < -2\frac{1}{2} \text{ or } x \ge 1\frac{1}{3}\right\}$, or $\left(-\infty, -2\frac{1}{2}\right) \cup \left[1\frac{1}{3}, +\infty\right)$

−4 −3 −2 −1 0 1 2 3 4

21. We multiply by $(x+1)^2$, which is positive. This clears fractions and does not change the sense of the inequality.

$$\frac{(x+1)^2}{1} \cdot \frac{x}{1} < \frac{(x+1)^2}{1} \cdot \frac{2}{x+1}$$
$$(x+1)^2 x < 2(x+1)$$
$$(x+1)^2 x - 2(x+1) < 0$$
$$(x+1)[(x+1)x - 2] < 0$$
$$(x+1)(x^2 + x - 2) < 0$$
$$(x+1)(x+2)(x-1) < 0$$

Let $(x+1)(x+2)(x-1) = 0$

Critical points: -2, -1, and 1

	Three negative	Two negative	One negative	
Sign of $x+1$	− − − − − − −	− −	+ + + +	+ + + + + + +
Sign of $x+2$	− − − − − − −	+ +	+ + + +	+ + + + + + +
Sign of $x-1$	− − − − − − −	− −	− − − −	+ + + + + + +

−4 −3 −2 −1 0 1 2 3

We select the intervals that have an *odd* number of negative signs. The graph is

−4 −3 −2 −1 0 1 2 3

Solution set: $\{x \mid x < -2\} \cup \{x \mid -1 < x < 1\}$, or $(-\infty, -2) \cup (-1, 1)$

22. $\{x \mid -4 < x < -3\} \cup \{x \mid x > 1\}$, or $(-4, -3) \cup (1, +\infty)$

−5 −4 −3 −2 −1 0 1 2 3

23.
$$x - \frac{8}{x+2} \ge 0$$
$$\frac{x(x+2) - 8}{x+2} \ge 0$$
$$\frac{x^2 + 2x - 8}{x+2} \ge 0$$
$$\frac{(x+4)(x-2)}{x+2} \ge 0$$

	Three negative	Two negative	One negative	No negative
Sign of $x+4$	− − − − −	+ + + +	+ + + + + + + + +	+ + + + + + +
Sign of $x-2$	− − − − −	− − − −	− − − − − − − − −	+ + + + + + +
Sign of $x+2$	− − − − −	− − − −	+ + + + + + + + +	+ + + + + + +

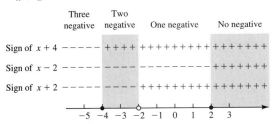

−5 −4 −3 −2 −1 0 1 2 3

We select the intervals that have an even number of negative signs or no negative signs. The graph is

−5 −4 −3 −2 −1 0 1 2 3

Solution set: $\{x \mid -4 \le x < -2\} \cup \{x \mid x \ge 2\}$, or $[-4, -2) \cup [2, +\infty)$

24. $\{x \mid x \le -1 \text{ or } 3 < x \le 4\}$, or $(-\infty, -1] \cup (3, 4]$

−3 −2 −1 0 1 2 3 4 5

25. We must solve $x^2 - 4x - 12 \ge 0$, or $(x+2)(x-6) \ge 0$.
Critical points: -2 and 6

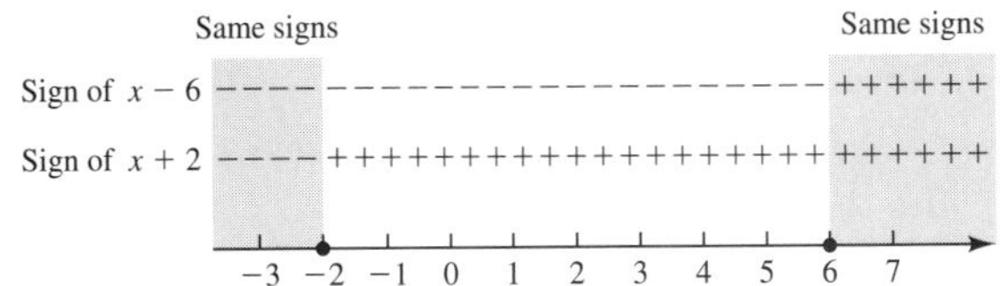

We select the intervals on which the signs are the same. The graph is

−3 −2 −1 0 1 2 3 4 5 6 7

Solution set: $\{x \mid x \le -2\} \cup \{x \mid x \ge 6\}$, or $(-\infty, -2] \cup [6, +\infty)$

26. $\{x \mid x \le -5\} \cup \{x \mid x \ge 2\}$, or $(-\infty, -5] \cup [2, +\infty)$

27. We must solve $4 + 3x - x^2 \ge 0$, or $(4 - x)(1 + x) \ge 0$.
Critical points: 4 and -1

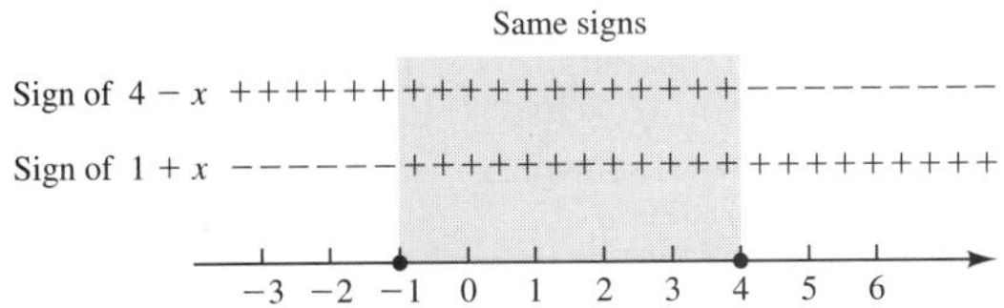

We select the interval on which the signs are the same. The graph is

−3 −2 −1 0 1 2 3 4 5 6

Solution set: $\{x \mid -1 \le x \le 4\}$, or $[-1, 4]$

28. $\{x \mid -5 \le x \le 1\}$, or $[-5, 1]$

Exercises 11.8 (page 589)

1. $y \le x^2 + 5x + 6$. We first graph $y = x^2 + 5x + 6$; x-intercepts are -2 and -3; y-intercept is 6. We try $(0, 0)$ in the inequality; $0 \le 0^2 + 0 + 6$ is true. Therefore, we shade the region that contains $(0, 0)$.

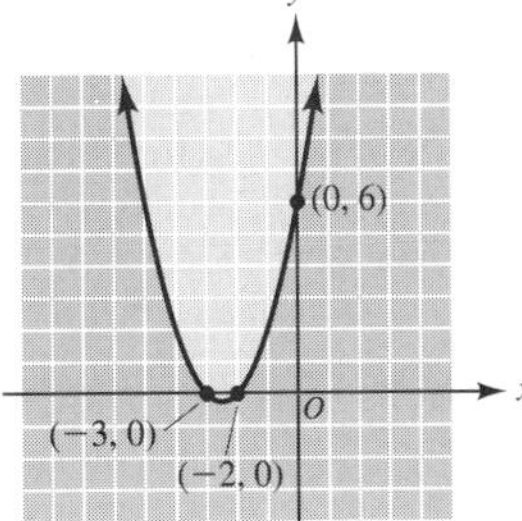

2.

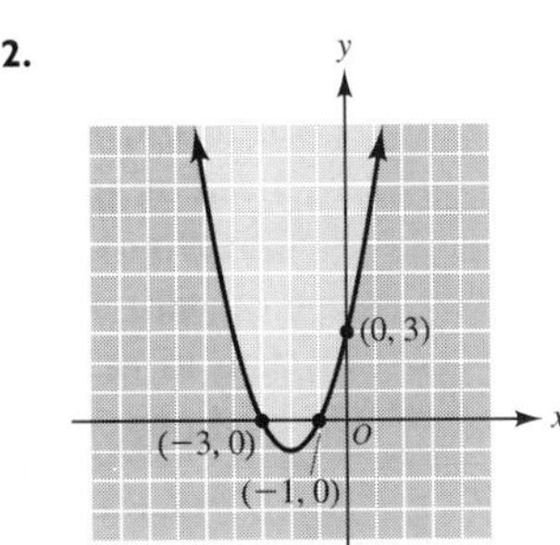

3. $x^2 + 4y^2 > 4$. We first graph $\frac{x^2}{4} + \frac{y^2}{1} = 1$, using a dotted line. It is an ellipse with x-intercepts 2 and -2 and y-intercepts 1 and -1. Then we try $(0, 0)$ in the inequality: $0^2 + 4(0^2) > 4$ is a false statement. Therefore, we shade the region that does not contain $(0, 0)$.

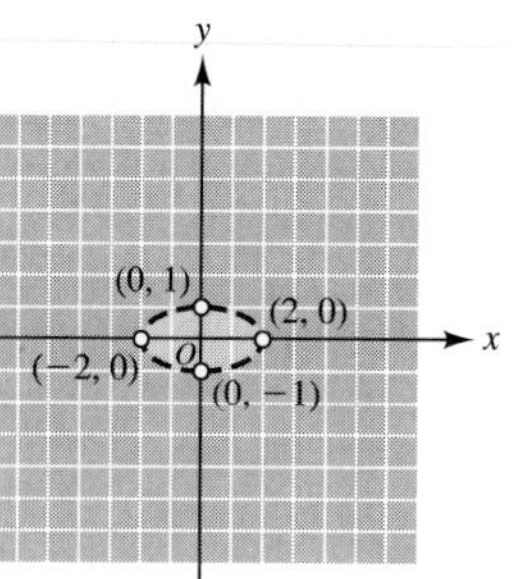

4.

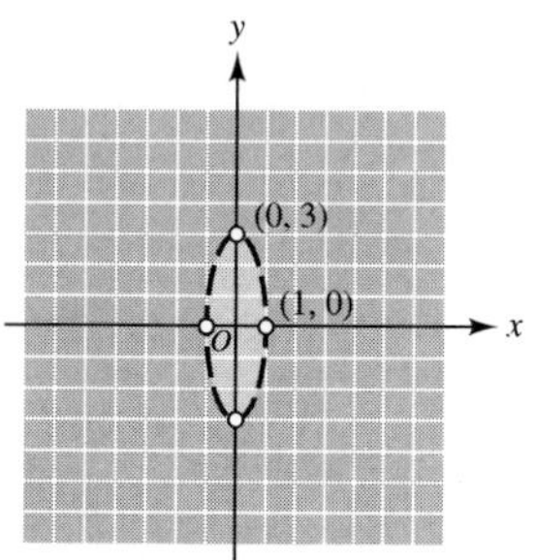

5. We first graph $(x + 2)^2 + (y - 3)^2 = 9$; this is a circle with its center at $(-2, 3)$ and with a radius of 3. Then we try $(0, 0)$ in the inequality: $(0 + 2)^2 + (0 - 3)^2 \leq 9$ is a false statement. Therefore, we shade the region that does not contain $(0, 0)$.

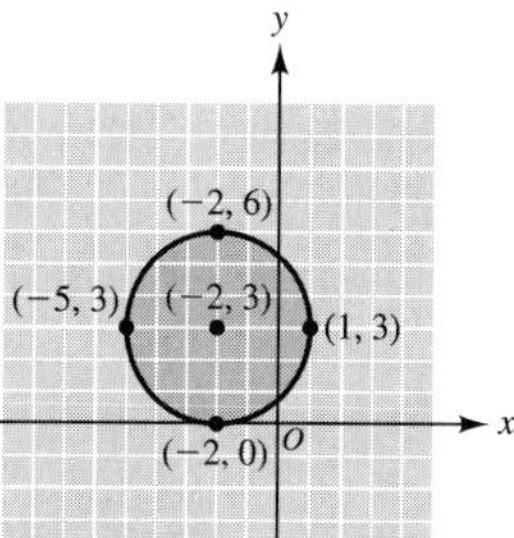

6.

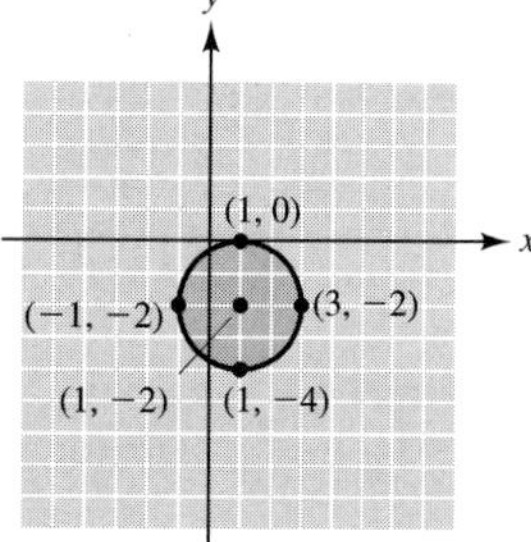

7. $4x^2 - y^2 < 4$. We first graph $4x^2 - y^2 = 4$, using a dotted line. It is a hyperbola with x-intercepts 1 and -1. We try $(0, 0)$ in the inequality: $4(0^2) - 0^2 < 4$ is a true statement. We shade the region that contains $(0, 0)$.

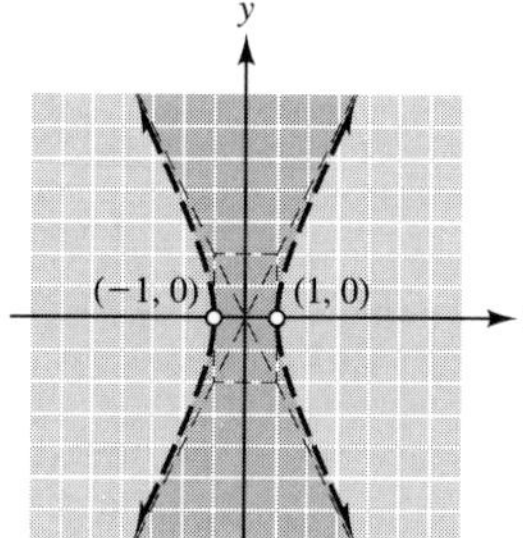

8.

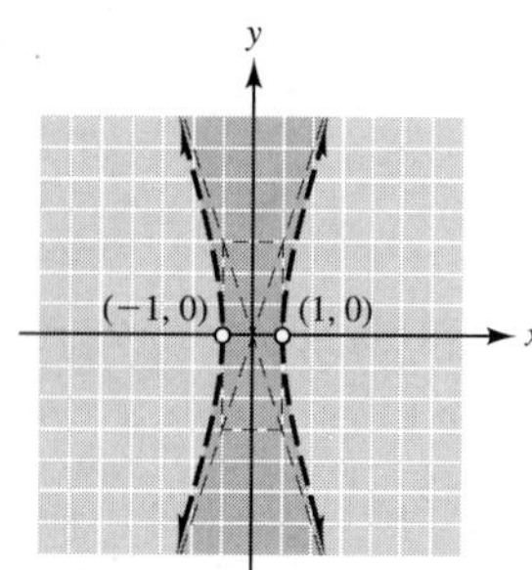

Sections 11.5–11.8 Review Exercises (page 592)

1. a. Let $f(x) = 0$

$x^2 - 2x - 8 = 0$

$(x - 4)(x + 2) = 0$

$x - 4 = 0 \quad | \quad x + 2 = 0$

$x = 4 \quad | \quad x = -2$

x-intercepts: 4 and -2

b. $f(0) = 0^2 - 2(0) - 8 = -8$

y-intercept: -8

c. Axis of symmetry: $x = -\frac{-2}{2} = 1$

$f(1) = 1^2 - 2(1) - 8 = -9$

Vertex at $(1, -9)$

d.

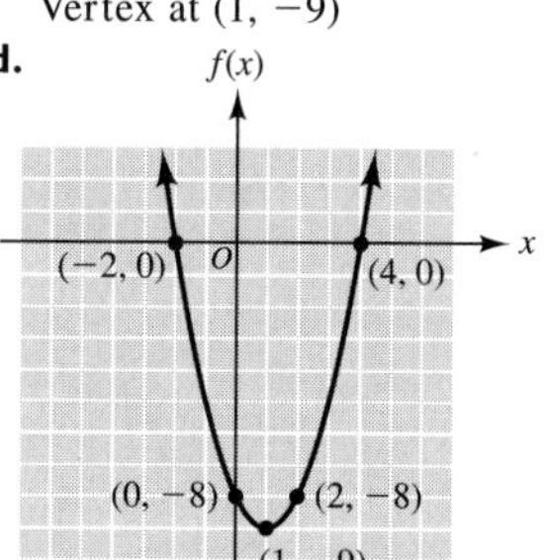

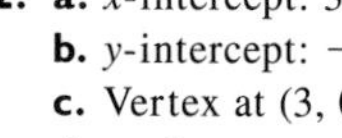

2. a. x-intercept: 3

b. y-intercept: -9

c. Vertex at $(3, 0)$

d.

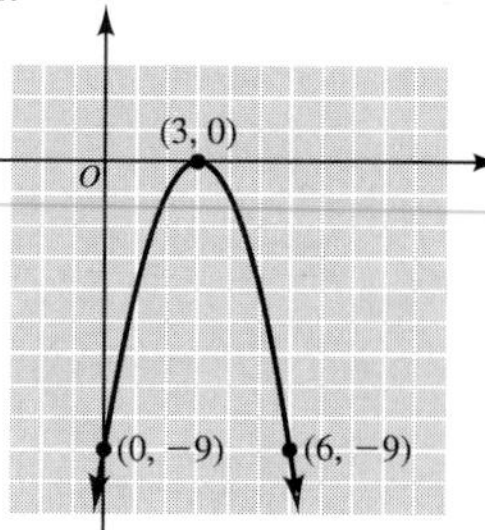

3. $(4 - x)(2 + x) > 0$

Let $f(x) = (4 - x)(2 + x) = 0$

Then $4 - x = 0 \quad | \quad 2 + x = 0$

$x = 4 \quad | \quad x = -2$

x-intercepts: -2 and 4

$f(0) = (4)(2) = 8$

The curve goes through $(0, 8)$.

Solution set: $\{x \mid -2 < x < 4\}$, or $(-2, 4)$

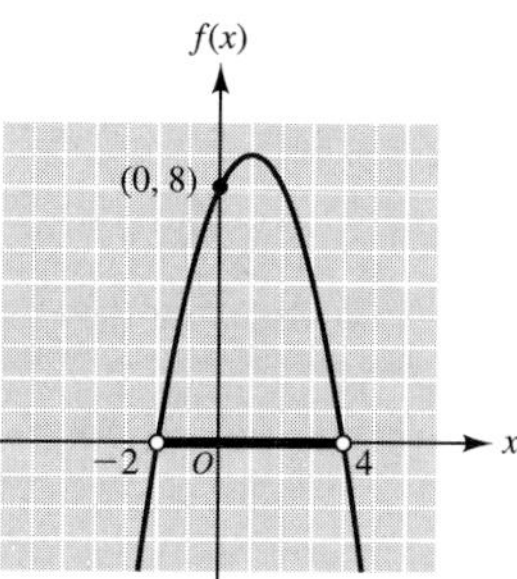

4. $\{x \mid x < 1\} \cup \{x \mid 2 < x < 3\}$, or $(-\infty, 1) \cup (2, 3)$

5. $x - \frac{3}{x + 2} > 0$

$\frac{x^2 + 2x - 3}{x + 2} > 0$

$\frac{(x + 3)(x - 1)}{x + 2} > 0$

Critical points: -3, 1, and -2

	Two negatives		No negatives
Sign of $x + 3$	– – – – – –	+ + + + + + + + +	+ + + + + +
Sign of $x - 1$	– – – – – –	– – – – – – – – –	+ + + + + +
Sign of $x + 2$	– – – – – –	– – + + + + + + +	+ + + + + +

$-5 \quad -4 \quad -3 \quad -2 \quad -1 \quad 0 \quad 1 \quad 2$

We select the intervals that have an even number of negative signs or no negative signs. The graph is

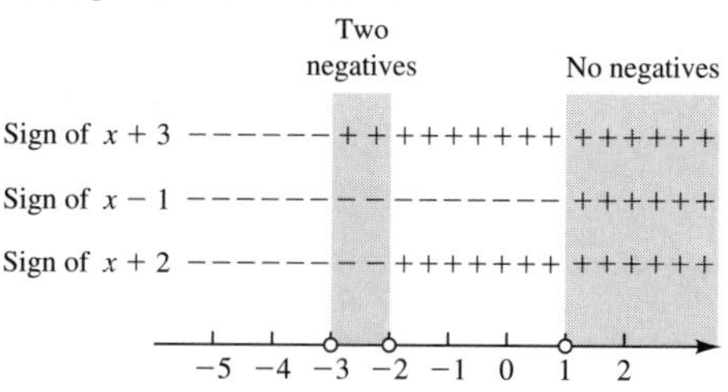

Solution set: $\{x \mid -3 < x < -2\} \cup \{x \mid x > 1\}$, or $(-3, -2) \cup (1, +\infty)$

6. $\{x \mid -1 \leq x \leq 6\}$

7. $x^2 + y^2 = 9$
Circle; center at the origin; radius is 3.

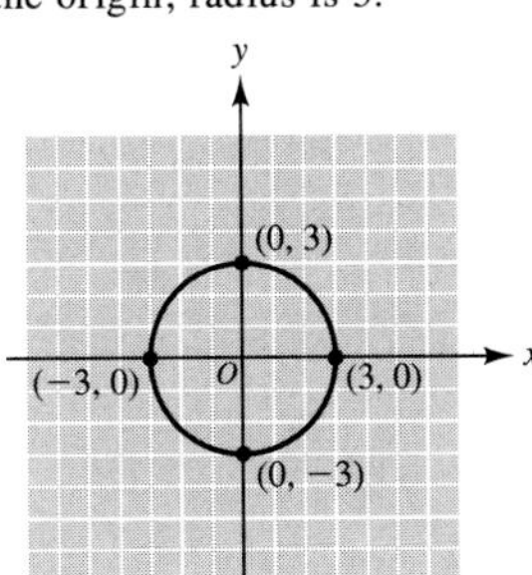

8. Hyperbola

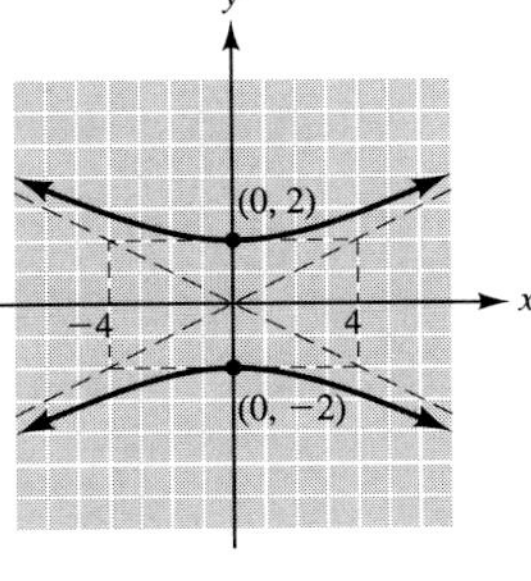

9. Ellipse

x-intercepts: $9x^2 + 4(0)^2 = 36$
$9x^2 = 36$
$x = \pm 2$

y-intercepts: $9(0)^2 + 4y^2 = 36$
$4y^2 = 36$
$y = \pm 3$

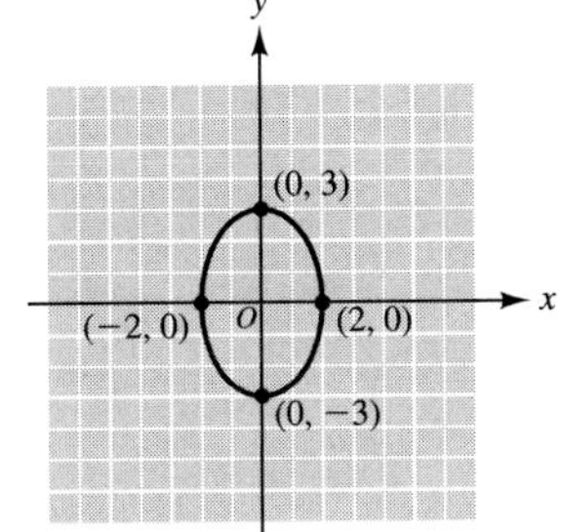

10.

Chapter 11 Diagnostic Test (page 595)

Following each problem number is the number (in parentheses) of the textbook section where that kind of problem is discussed. (The checks are not always shown.)

1. (11.1) **a.**
$2x^2 = 6x$
$2x^2 - 6x = 0$
$2x(x-3) = 0$
$2x = 0 \quad | \quad x - 3 = 0$
$x = 0 \quad | \quad x = 3$
$\{0, 3\}$

b.
$2x^2 = 18$
$2x^2 - 18 = 0$
$2(x^2 - 9) = 0$
$2(x+3)(x-3) = 0$
$x + 3 = 0 \quad | \quad x - 3 = 0$
$x = -3 \quad | \quad x = 3$
$\{-3, 3\}$

2. (11.2) $\dfrac{x-1}{2} + \dfrac{4}{x+1} = 2 \quad \text{LCD} = 2(x+1)$

$$\frac{2(x+1)}{1} \cdot \frac{(x-1)}{2} + \frac{2(x+1)}{1} \cdot \frac{4}{x+1} = \frac{2(x+1)}{1} \cdot \frac{2}{1}$$
$$(x+1)(x-1) + 2 \cdot 4 = 4(x+1)$$
$$x^2 - 1 + 8 = 4x + 4$$
$$x^2 - 4x + 3 = 0$$
$$(x-1)(x-3) = 0$$

$x - 1 = 0 \quad | \quad x - 3 = 0$
$x = 1 \quad | \quad x = 3$

Check for $x = 1$
$\dfrac{x-1}{2} + \dfrac{4}{x+1} = 2$
$\dfrac{(1)-1}{2} + \dfrac{4}{(1)+1} \stackrel{?}{=} 2$
$0 + 2 = 2$

Check for $x = 3$
$\dfrac{x-1}{2} + \dfrac{4}{x+1} = 2$
$\dfrac{(3)-1}{2} + \dfrac{4}{(3)+1} \stackrel{?}{=} 2$
$1 + 1 = 2$

$\{1, 3\}$

3. (11.2) $2x^{2/3} + 3x^{1/3} = 2$
Let $z = x^{1/3}$
Then $z^2 = x^{2/3}$
Equation becomes $2z^2 + 3z - 2 = 0$
$(z+2)(2z-1) = 0$

$z + 2 = 0 \quad | \quad 2z - 1 = 0$
$z = -2 \quad | \quad z = \frac{1}{2}$
$x^{1/3} = -2 \quad | \quad x^{1/3} = \frac{1}{2}$
$(x^{1/3})^3 = (-2)^3 \quad | \quad (x^{1/3})^3 = \left(\frac{1}{2}\right)^3$
$x = -8 \quad | \quad x = \frac{1}{8}$

Check for $x = -8$
$2(-8)^{2/3} + 3(-8)^{1/3} \stackrel{?}{=} 2$
$2(4) + 3(-2) \stackrel{?}{=} 2$
True $8 - 6 = 2$

Check for $x = \frac{1}{8}$
$2\left(\frac{1}{8}\right)^{2/3} + 3\left(\frac{1}{8}\right)^{1/3} \stackrel{?}{=} 2$
$2\left(\frac{1}{4}\right) + 3\left(\frac{1}{2}\right) \stackrel{?}{=} 2$
True $\frac{1}{2} + \frac{3}{2} = 2$

$\left\{-8, \frac{1}{8}\right\}$

4. (11.3) **a.** $x^2 = 6x - 7$
$x^2 - 6x + 7 = 0$
The left side will not factor; we use the quadratic formula:
$$x = \frac{-(-6) \pm \sqrt{(-6)^2 - 4(7)}}{2(1)} = \frac{6 \pm \sqrt{36 - 28}}{2}$$
$$= \frac{6 \pm \sqrt{8}}{2} = \frac{6 \pm 2\sqrt{2}}{2} = 3 \pm \sqrt{2}$$
$\{3 + \sqrt{2}, 3 - \sqrt{2}\}$

b. $3x^2 + 7x - 1$ will not factor; we use the quadratic formula, with $a = 3$, $b = 7$, $c = -1$:
$$x = \frac{-7 \pm \sqrt{7^2 - 4(3)(-1)}}{2(3)} = \frac{-7 \pm \sqrt{49 + 12}}{6}$$
$$= \frac{-7 \pm \sqrt{61}}{6}$$
$$\left\{\frac{-7 + \sqrt{61}}{6}, \frac{-7 - \sqrt{61}}{6}\right\}$$

5. (11.3) $x^2 + 6x + 10$ will not factor; we use the quadratic formula, with $a = 1$, $b = 6$, $c = 10$:
$$x = \frac{-6 \pm \sqrt{6^2 - 4(1)(10)}}{2(1)} = \frac{-6 \pm \sqrt{36 - 40}}{2}$$
$$= \frac{-6 \pm \sqrt{-4}}{2} = \frac{-6 + 2i}{2} = \frac{-6}{2} \pm \frac{2i}{2} = -3 \pm i$$
$\{-3 + i, -3 - i\}$

6. (11.4) **a.** $25x^2 - 20x + 7 = 0$; $a = 25$, $b = -20$, $c = 7$
$b^2 - 4ac = (-20)^2 - 4(25)(7) = 400 - 700 = -300$
The roots are complex conjugates.

b. If $2 + i\sqrt{3}$ is a root, $2 - i\sqrt{3}$ is also a root.
$x = 2 + i\sqrt{3}$, $x = 2 - i\sqrt{3}$, and $x = 3$
$x - 2 - i\sqrt{3} = 0$, $x - 2 + i\sqrt{3} = 0$, and
$x - 3 = 0$ and so
$$([x-2] - i\sqrt{3})([x-2] + i\sqrt{3})(x-3) = 0$$
$$([x-2]^2 - [i\sqrt{3}]^2)(x-3) = 0$$
$$(x^2 - 4x + 4 - 3i^2)(x-3) = 0$$
$$(x^2 - 4x + 4 + 3)(x-3) = 0$$
$$(x^2 - 4x + 7)(x-3) = 0$$
$$x^3 - 7x^2 + 19x - 21 = 0$$

11

7. (11.6) $25x^2 + 16y^2 = 400$
Ellipse
x-intercepts: $25x^2 + 16(0)^2 = 400$
$25x^2 = 400$
$x^2 = 16$
$x = \pm 4$
y-intercepts: $25(0)^2 + 16y^2 = 400$
$16y^2 = 400$
$y^2 = 25$
$y = \pm 5$

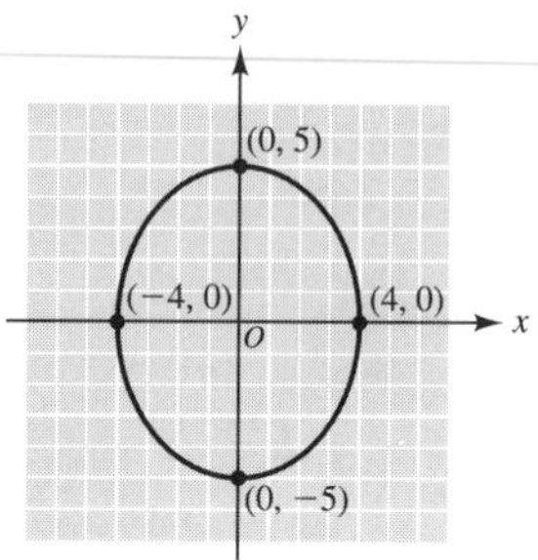

8. (11.7) **a.** $x^2 - 6x + 5 < 0$
$(x - 1)(x - 5) < 0$
Let $(x - 1)(x - 5) = 0$
Then $x - 1 = 0 \quad | \quad x - 5 = 0$
$x = 1 \quad | \quad x = 5$
x-intercepts: 1 and 5
$f(x) = (x - 1)(x - 5)$
$f(0) = (-1)(-5) = 5$
The curve goes through (0, 5).

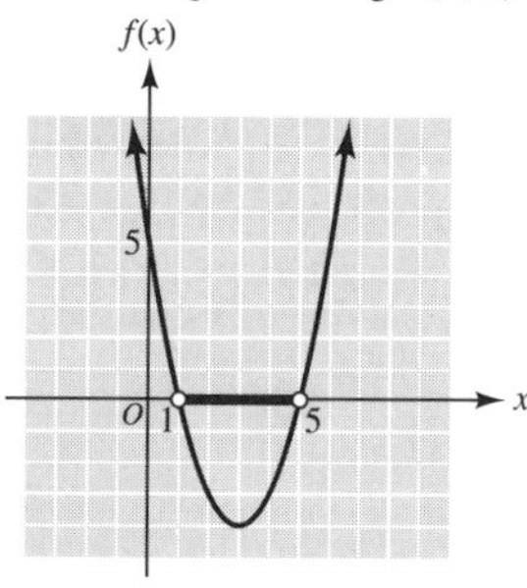

Solution set: $\{x \mid 1 < x < 5\}$, or $(1, 5)$

b. $x^2 + 2x - 8 \ge 0$
$(x + 4)(x - 2) \ge 0$
Let $f(x) = (x + 4)(x - 2) = 0$
Then $x + 4 = 0 \quad | \quad x - 2 = 0$
$x = -4 \quad | \quad x = 2$
x-intercepts: -4 and 2
$f(x) = (x + 4)(x - 2)$
$f(0) = (4)(-2) = -8$
The curve goes through $(0, -8)$.

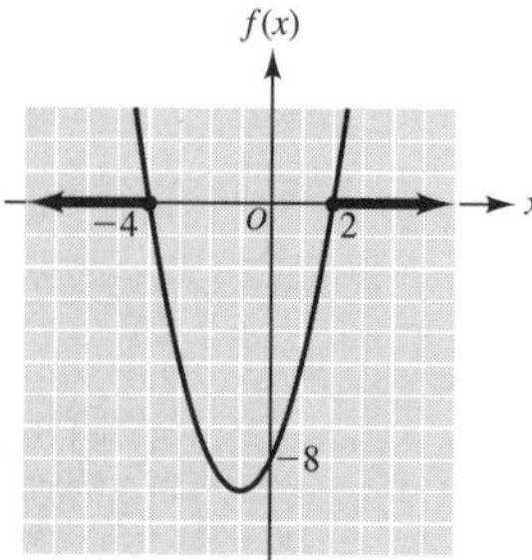

Solution set: $\{x \mid x \le -4\} \cup \{x \mid x \ge 2\}$, or $(-\infty, -4] \cup [2, +\infty)$

9. (11.5) $f(x) = x^2 - 4x$

a. Axis of symmetry: $x = -\dfrac{b}{2a} = -\dfrac{-4}{2(1)} = 2$

b. $f(2) = (2)^2 - 4(2)$
$= 4 - 8 = -4$
Vertex at $(2, -4)$

c. $f(x) = x(x - 4) = 0$
$x = 0 \quad | \quad x - 4 = 0$
$x = 4$
x-intercepts: 0 and 4

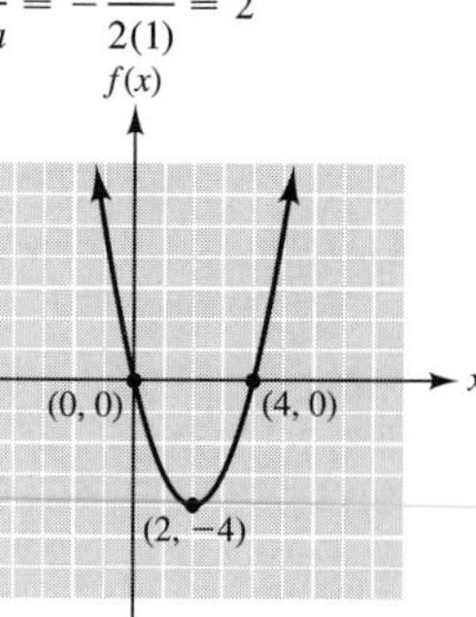

10. (11.3) Let x = number of cm in width
Then $x + 18$ = number of cm in length
and $x(x + 18)$ = area (in sq. cm)
Area is 8,944
$x(x + 18) = 8{,}944$
$x^2 + 18x = 8{,}944$
$x^2 + 18x + 81 = 8{,}944 + 81$ Completing the square
$(x + 9)^2 = 9{,}025$
$x + 9 = \pm 95$ Using Property 1
$x = -9 \pm 95 = \begin{cases} -9 + 95 \\ -9 - 95 \quad \text{Reject} \end{cases}$
$x = 86$
The width of the rectangle is 86 cm.

Chapters 1–11 Cumulative Review Exercises (page 595)

1. $m = \dfrac{-1 - 5}{4 - 2} = \dfrac{-6}{2} = -3$

2. $x + 2y - 4 = 0$

3. If $|4 - 3x| \ge 10$,
$4 - 3x \ge 10$ or $4 - 3x \le -10$
$-3x \ge 6$ or $-3x \le -14$
$\dfrac{-3x}{-3} \le \dfrac{6}{-3}$ or $\dfrac{-3x}{-3} \ge \dfrac{-14}{-3}$
$x \le -2$ or $x \ge \frac{14}{3}$
$\{x \mid x \le -2\} \cup \left\{x \mid x \ge \frac{14}{3}\right\}$

4. $\{-2\}$

5. $4x - 2y < 8$
Boundary line: $4x - 2y = 8$
Boundary line is dashed because = is not included in <. Half-plane includes (0, 0) because
$4(0) - 2(0) < 8$
$0 < 8$ True

x	y
2	0
0	−4

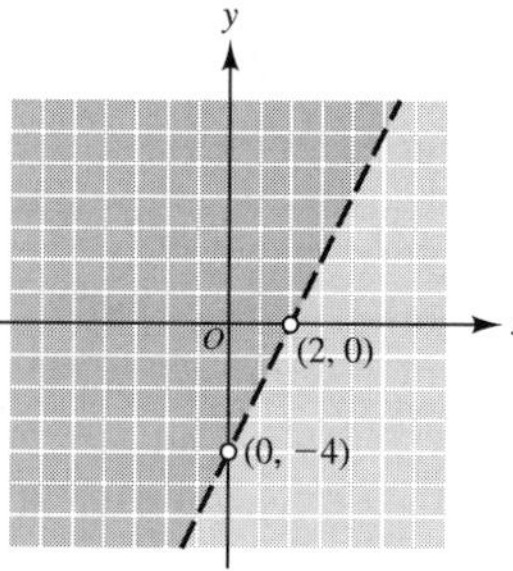

6. $f^{-1}(x) = \dfrac{x + 3}{2}$

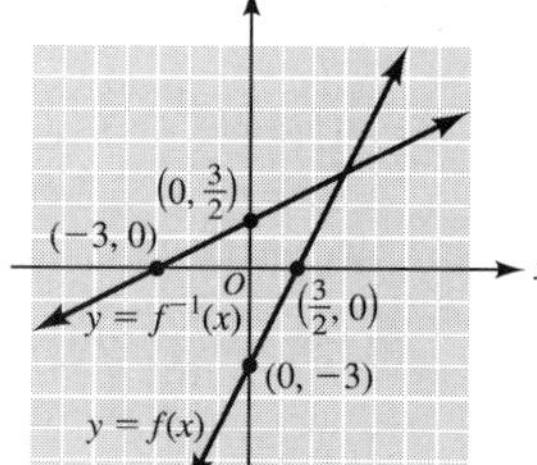

7. $\left(\frac{x^{-3}y^4}{x^2y}\right)^{-2} = (x^{-3-2}y^{4-1})^{-2} = (x^{-5}y^3)^{-2} = x^{10}y^{-6} = \frac{x^{10}}{y^6}$

8. $\frac{4}{x^{2/5}}$ **9.** $a^{1/3} \cdot a^{-1/2} = a^{1/3+(-1/2)} = a^{2/6+(-3/6)} = a^{-1/6} = \frac{1}{a^{1/6}}$

10. $\left\{\frac{1+3i\sqrt{3}}{2}, \frac{1-3i\sqrt{3}}{2}\right\}$ **11.** $3x^2 + x - 3 = 0;$

$a = 3, b = 1, c = -3$

$$x = \frac{-1 \pm \sqrt{1^2 - 4(3)(-3)}}{2(3)} = \frac{-1 \pm \sqrt{1+36}}{6} = \frac{-1 \pm \sqrt{37}}{6}$$

$\left\{\frac{-1+\sqrt{37}}{6}, \frac{-1-\sqrt{37}}{6}\right\}$

12. 1.956 **13.** $\log(x + 11) - \log(x + 1) = \log 6$ **14.** $b = 16$

$$\log\frac{x+11}{x+1} = \log 6$$
$$\frac{x+11}{x+1} = 6$$
$$6(x+1) = x + 11$$
$$6x + 6 = x + 11$$
$$5x = 5$$
$$x = 1$$

15. $\log_7 \frac{1}{7} = x$

$7^x = \frac{1}{7} = 7^{-1}$

$x = -1$

16.

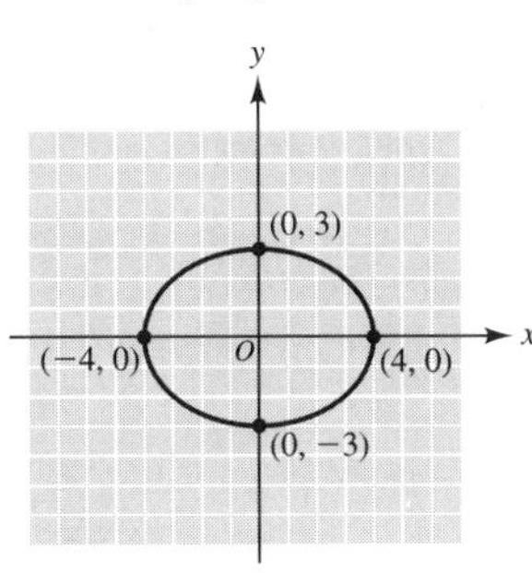

17. We must solve $4 - 3x - x^2 \geq 0$, or $(4 + x)(1 - x) \geq 0$.
Critical points: -4 and 1

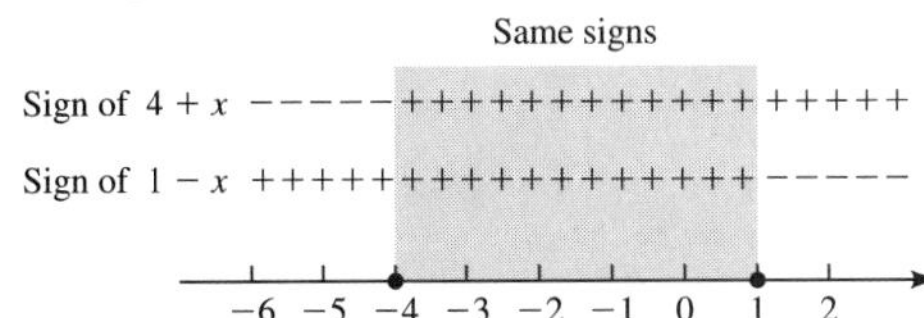

The domain of $y = f(x) = \sqrt{4 - 3x - x^2}$ is $\{x \mid -4 \leq x \leq 1\}$.

18. The length is 8 yd, and the width is 6 yd.

19. Let x = number of hr for Merwin to do the job
Then $x + 3$ = number of hr for Mina to do the same job

Merwin's rate: $\frac{1 \text{ job}}{x \text{ hr}}$; Mina's rate: $\frac{1 \text{ job}}{(x+3) \text{ hr}}$

Mina works a total of 8 hr, and Merwin works 3 hr.

$$\begin{pmatrix}\text{Work done}\\ \text{by Mina}\end{pmatrix} + \begin{pmatrix}\text{work done}\\ \text{by Merwin}\end{pmatrix} = \begin{pmatrix}\text{total work done}\\ \text{to complete job}\end{pmatrix}$$

$$8\left(\frac{1}{x+3}\right) + 3\left(\frac{1}{x}\right) = 1; \quad \text{LCD} = x(x+3)$$

$$\frac{x(x+3)}{1}\cdot\frac{8}{x+3} + \frac{x(x+3)}{1}\cdot\frac{3}{x} = \frac{x(x+3)}{1}\cdot 1$$
$$8x + 3x + 9 = x^2 + 3x$$
$$x^2 - 8x - 9 = 0$$
$$(x-9)(x+1) = 0$$

$x + 1 = 0$, $x = -1$ — Not in the domain | $x - 9 = 0$, $x = 9$, $x + 3 = 12$

It takes 9 hr for Merwin to do the job and 12 hr for Mina to do the job.

20. Colin was walking $\frac{-3+\sqrt{41}}{2}$ mph, or about 1.7 mph.

21. Let t = number of yr for money to double

$$A = Pe^{rt}$$
$$2{,}000 = 1{,}000e^{0.0625t}$$
$$2 = e^{0.0625t}$$
$$\ln 2 = \ln e^{0.0625t}$$
$$0.6931 \approx 0.0625t$$
$$t \approx 11$$

It will take about 11 yr for the money to double.

22. The lengths of the sides are 16 cm, 32 cm, and 40 cm.

Exercises 12.1 (page 599)

1. a. In $x + y = 8$: $1 + 7 = 8$ True
in $x - y = 2$: $1 - 7 = 2$ False; no
b. In $x + y = 8$: $0 + 8 = 8$ True
in $x - y = 2$: $0 - 8 = 2$ False; no
c. In $x + y = 8$: $5 + 3 = 8$ True
in $x - y = 2$: $5 - 3 = 2$ True; yes

2. a. No **b.** Yes **c.** No

3. a. In $2x - 3y + 7z = 14$: $10 - 3 + 7 = 14$ True
in $4x + 3y - 5z = 0$: $20 + 3 - 5 = 0$ False; no
b. In $2x - 3y + 7z = 14$: $4 + 3 + 7 = 14$ True
in $4x + 3y - 5z = 0$: $8 - 3 - 5 = 0$ True
in $3x - 6y - 2z = 10$: $6 + 6 - 2 = 10$ True; yes
c. In $2x - 3y + 7z = 14$: $0 - 0 + 14 = 14$ True
in $4x + 3y - 5z = 0$: $0 + 0 - 10 = 0$ False; no

4. a. No **b.** No **c.** Yes

Exercises 12.2 (page 604)

1. (1) $2x + y = 6$
Intercepts: (3, 0), (0, 6)
(2) $x - y = 0$
Intercepts: (0, 0)
Additional point: (5, 5)

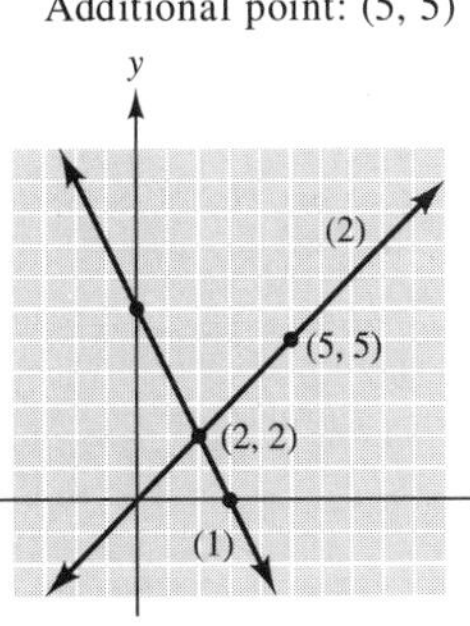

$\{(2, 2)\}$

2.

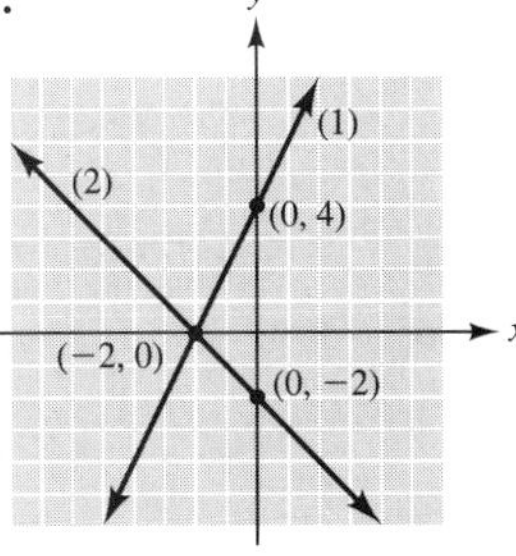

$\{(-2, 0)\}$

3. (1) $x + 2y = 3$
Intercepts: (3, 0), $\left(0, \frac{3}{2}\right)$
(2) $3x - y = -5$
Intercepts: $\left(-\frac{5}{3}, 0\right)$, (0, 5)

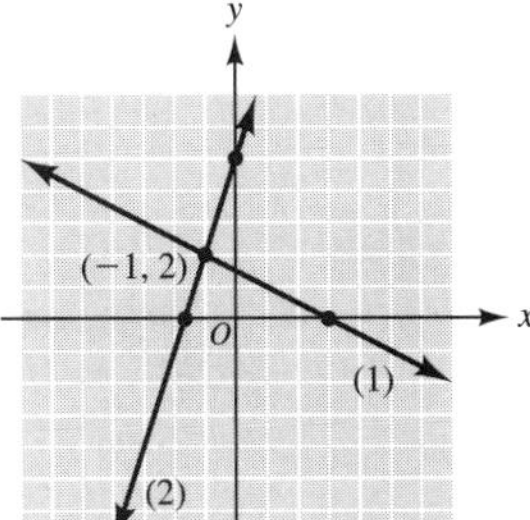

$\{(-1, 2)\}$

4.

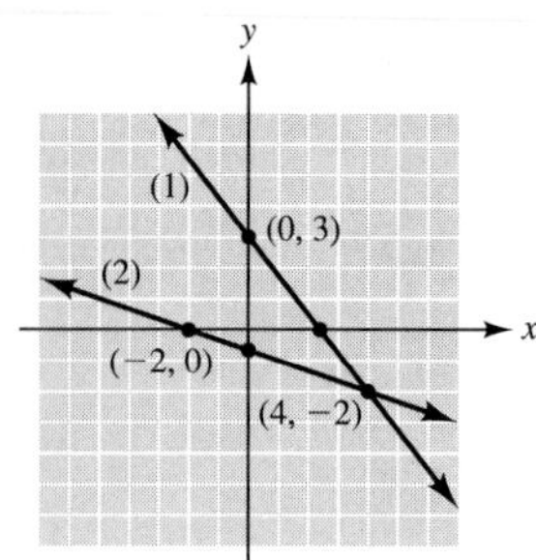

$\{(4, -2)\}$

5. (1) $x + 2y = 0$
Intercept: $(0, 0)$
Additional point: $(4, -2)$
(2) $2x - y = 0$
Intercept: $(0, 0)$
Additional point: $(3, 6)$

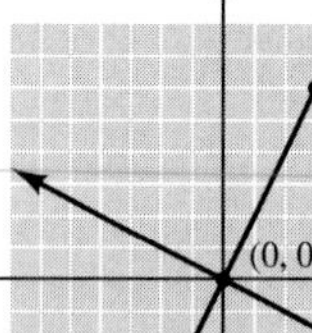

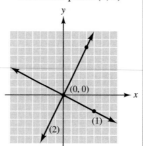

$\{(0, 0)\}$

6.

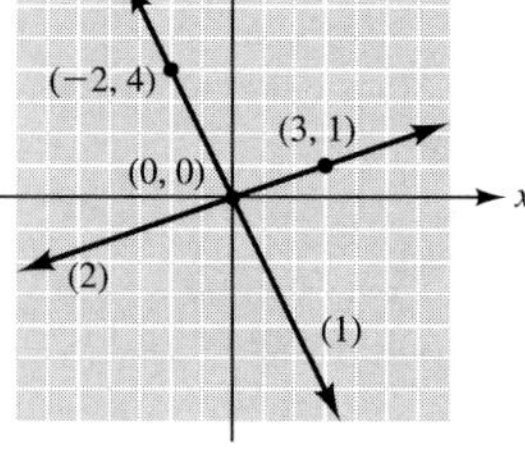

$\{(0, 0)\}$

7. (1) $8x - 5y = 15$
Intercepts: $\left(1\frac{7}{8}, 0\right)$, $(0, -3)$
(2) $10y - 16x = 16$
Intercepts: $(-1, 0)$, $\left(0, 1\frac{3}{5}\right)$

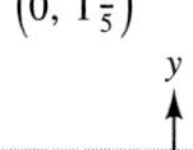

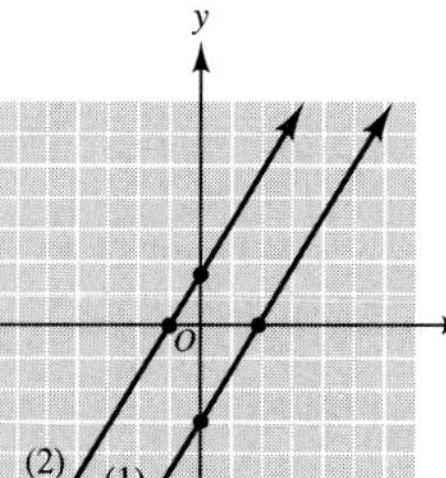

$\{\ \}$

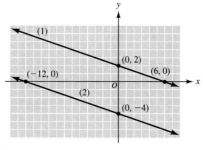

8.

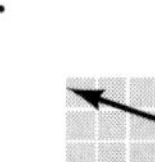

$\{\ \}$

9. (1) $8x - 10y = 16$
Intercepts: $(2, 0)$, $\left(0, -1\frac{3}{5}\right)$
(2) $15y - 12x = -24$
Intercepts: $(2, 0)$, $\left(0, -1\frac{3}{5}\right)$
Since both lines have the same intercepts, they are the same line.
Infinite number of solutions (dependent); $\left\{\left(t, \frac{4t - 8}{5}\right), t \in R\right\}$

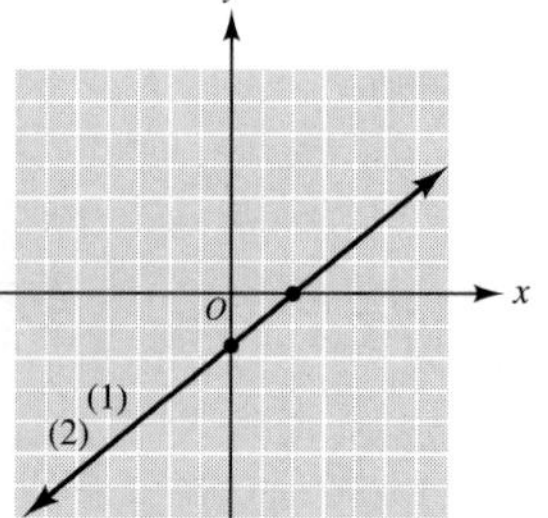

10. Infinite number of solutions (dependent); $\left\{\left(t, \frac{-35 - 7t}{15}\right), t \in R\right\}$

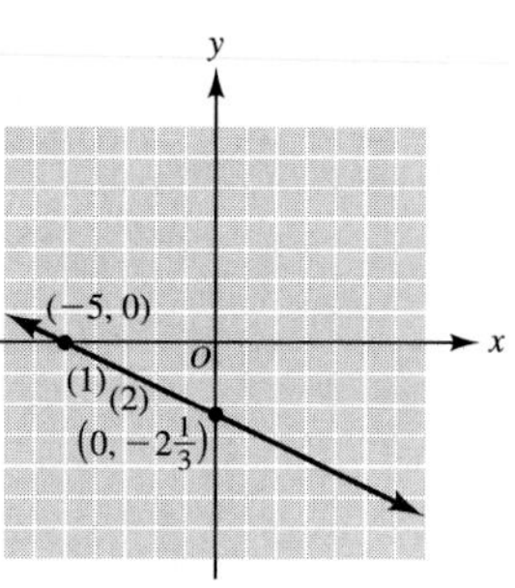

Exercises 12.3 (page 610)

1.
$$\begin{array}{rrcl} -1] & 3x - y = 11 & \Rightarrow & -3x + y = -11 \\ & 3x + 2y = -4 & \Rightarrow & 3x + 2y = -4 \\ \hline & & & 3y = -15 \\ & & & y = -5 \end{array}$$

Substituting -5 for y in (1):
(1) $3x - y = 11$
$3x - (-5) = 11$
$3x + 5 = 11$
$3x = 6$
$x = 2$
$\{(2, -5)\}$

2. $\{(-3, 4)\}$

3.
$$\begin{array}{rrrcl} \backslash 4 & 1] & 8x + 15y = 11 & \Rightarrow & 8x + 15y = 11 \\ \not{-8} & -2] & 4x - y = 31 & \Rightarrow & -8x + 2y = -62 \\ \hline & & & & 17y = -51 \\ & & & & y = -3 \end{array}$$

Substituting -3 for y in (2):
(2) $4x - y = 31$
$4x - (-3) = 31$
$4x + 3 = 31$
$4x = 28$
$x = 7$
$\{(7, -3)\}$

4. $\{(6, 3)\}$

5.
$$\begin{array}{rrrcl} \backslash 9 & 3] & 7x - 3y = 3 & \Rightarrow & 21x - 9y = 9 \\ \not{-3} & -1] & 20x - 9y = 12 & \Rightarrow & -20x + 9y = -12 \\ \hline & & & & x = -3 \end{array}$$

Substituting -3 for x in (1):
(1) $7x - 3y = 3$
$7(-3) - 3y = 3$
$-21 - 3y = 3$
$-3y = 24$
$y = -8$
$\{(-3, -8)\}$

6. $\{(9, -13)\}$

7.
$$\begin{array}{rrrcl} \backslash 4 & 2] & 6x + 5y = 0 & \Rightarrow & 12x + 10y = 0 \\ \not{-6} & -3] & 4x - 3y = 38 & \Rightarrow & -12x + 9y = -114 \\ \hline & & & & 19y = -114 \\ & & & & y = -6 \end{array}$$

Substituting -6 for y in (1):
(1) $6x + 5y = 0$
$6x + 5(-6) = 0$
$6x - 30 = 0$
$6x = 30$
$x = 5$
$\{(5, -6)\}$

8. $\{(-4, 10)\}$

9.
$$\begin{array}{rrcl} -2] & 4x + 6y = 5 & \Rightarrow & -8x - 12y = -10 \\ & 8x + 12y = 7 & \Rightarrow & 8x + 12y = 7 \\ \hline & & & 0 = -3 \end{array}$$
A false statement

$\{\ \}$

12

10. { }

11. $-2]$
$$\begin{aligned} 7x - 3y &= 5 \Rightarrow -14x + 6y = -10 \\ 14x - 6y &= 10 \Rightarrow \underline{\ \ 14x - 6y = \ \ 10} \\ & \qquad\qquad\qquad 0 = \ \ 0 \quad \text{A true statement} \end{aligned}$$

$\left\{\left(t, \dfrac{7t - 5}{3}\right), t \in R\right\}$

12. $\left\{\left(t, \dfrac{2t - 4}{3}\right), t \in R\right\}$

13. 6] 3] / 4] 2]
$$\begin{aligned} 9x + 4y &= -4 \Rightarrow 27x + 12y = -12 \\ 15x - 6y &= 25 \Rightarrow \underline{30x - 12y = \ \ 50} \\ & \qquad\qquad 57x \qquad\ = \ \ 38 \\ & \qquad\qquad\qquad x = \tfrac{38}{57} = \tfrac{2}{3} \end{aligned}$$

Substituting $\frac{2}{3}$ for x in (1):

(1) $9x + 4y = -4$

$$\begin{aligned} \overset{3}{\cancel{9}}\left(\frac{2}{\underset{1}{\cancel{3}}}\right) + 4y &= -4 \\ 6 + 4y &= -4 \\ 4y &= -10 \\ y &= -\tfrac{10}{4} = -\tfrac{5}{2} \end{aligned}$$

$\left\{\left(\frac{2}{3}, -\frac{5}{2}\right)\right\}$

14. $\left\{\left(-\frac{7}{4}, \frac{2}{5}\right)\right\}$

15. Add $15x$ to both sides of equation (2) so that like terms are in the same column:

15 5] / 9 $-3]$
$$\begin{aligned} 9x + 10y &= -3 \Rightarrow \ \ 45x + 50y = -15 \\ 15x + 14y &= \ \ 7 \Rightarrow \underline{-45x - 42y = -21} \\ & \qquad\qquad\qquad\quad 8y = -36 \\ & \qquad\qquad\qquad\quad y = -\tfrac{36}{8} = -\tfrac{9}{2} \end{aligned}$$

Substituting $-\frac{9}{2}$ for y in (1):

(1) $9x + 10y = -3$

$$\begin{aligned} 9x + \overset{5}{\cancel{10}}\left(-\frac{9}{\underset{1}{\cancel{2}}}\right) &= -3 \\ 9x - 45 &= -3 \\ 9x &= 42 \\ x &= \tfrac{42}{9} = \tfrac{14}{3} \end{aligned}$$

$\left\{\left(\frac{14}{3}, -\frac{9}{2}\right)\right\}$

16. $\left\{\left(\frac{3}{7}, \frac{5}{6}\right)\right\}$

Exercises 12.4 (page 615)

1. (1) $7x + 4y = 4$

(2) $y = 6 - 3x$

Substituting $6 - 3x$ for y in (1):

(1)
$$\begin{aligned} 7x + 4y &= 4 \\ 7x + 4(6 - 3x) &= 4 \\ 7x + 24 - 12x &= 4 \\ -5x &= -20 \\ x &= 4 \end{aligned}$$

Substituting 4 for x in $y = 6 - 3x$:

$y = 6 - 3(4) = 6 - 12 = -6$

$\{(4, -6)\}$

2. $\{(-7, 3)\}$

3. (1) $5x - 4y = -1$

(2) $3x + y = -38 \Rightarrow y = -3x - 38$

Substituting $-3x - 38$ for y in (1):

(1)
$$\begin{aligned} 5x - 4y &= -1 \\ 5x - 4(-3x - 38) &= -1 \\ 5x + 12x + 152 &= -1 \\ 17x &= -153 \\ x &= -9 \end{aligned}$$

Substituting -9 for x in $y = -3x - 38$:

$y = -3(-9) - 38 = 27 - 38 = -11$

$\{(-9, -11)\}$

4. $\{(-5, -4)\}$

5. (1) $8x - 5y = 4$

(2) $x - 2y = -16 \Rightarrow x = 2y - 16$

Substituting $2y - 16$ for x in (1):

$$\begin{aligned} (1)\quad 8(2y - 16) - 5y &= 4 \\ 16y - 128 - 5y &= 4 \\ 11y &= 132 \\ y &= 12 \end{aligned}$$

Substituting 12 for y in $x = 2y - 16$:

$x = 2(12) - 16 = 24 - 16 = 8$

$\{(8, 12)\}$

6. $\{(10, 16)\}$

7. (1) $15x + 5y = 8$

(2) $6x + 2y = -10 \Rightarrow y = -3x - 5$

Substituting $-3x - 5$ for y in (1):

$$\begin{aligned} (1)\quad 15x + 5(-3x - 5) &= 8 \\ 15x - 15x - 25 &= 8 \\ -25 &= 8 \quad \text{A false statement} \end{aligned}$$

{ }

8. { }

9. (1) $20x - 10y = 70$

(2) $6x - 3y = 21 \Rightarrow y = 2x - 7$

Substituting $2x - 7$ for y in (1):

$$\begin{aligned} (1)\quad 20x - 10(2x - 7) &= 70 \\ 20x - 20x + 70 &= 70 \\ 70 &= 70 \quad \text{A true statement} \end{aligned}$$

$\{(t, 2t - 7), t \in R\}$

10. $\{(5s + 9, s), s \in R\}$

11. (1) $8x + 4y = 7$

(2) $3x + 6y = 6 \Rightarrow 3x = 6 - 6y$

$x = \dfrac{6 - 6y}{3} = 2 - 2y$

Substituting $2 - 2y$ for x in (1):

(1)
$$\begin{aligned} 8x + 4y &= 7 \\ 8(2 - 2y) + 4y &= 7 \\ 16 - 16y + 4y &= 7 \\ -12y &= -9 \\ y &= \tfrac{3}{4} \end{aligned}$$

Substituting $\frac{3}{4}$ for y in $x = 2 - 2y$:

$x = 2 - 2\left(\frac{3}{4}\right) = \frac{1}{2}$

$\left\{\left(\frac{1}{2}, \frac{3}{4}\right)\right\}$

12. $\left\{\left(\frac{3}{5}, \frac{1}{4}\right)\right\}$

13. (1) $4x + 4y = 3 \Rightarrow 4x = 3 - 4y$

$x = \dfrac{3 - 4y}{4}$

(2) $6x + 12y = -6$

Substituting $\dfrac{3 - 4y}{4}$ for x in (2):

(2)
$$\begin{aligned} 6x + 12y &= -6 \\ \overset{3}{\cancel{6}}\left(\frac{3 - 4y}{\underset{2}{\cancel{4}}}\right) + 12y &= -6 \quad \text{LCD} = 2 \\ \frac{\overset{1}{\cancel{2}}}{1} \cdot 3\left(\frac{3 - 4y}{\underset{1}{\cancel{2}}}\right) + 2(12y) &= 2(-6) \\ 9 - 12y + 24y &= -12 \\ 12y &= -21 \\ y &= -\tfrac{21}{12} = -\tfrac{7}{4} = -1\tfrac{3}{4} \end{aligned}$$

Substituting $-\frac{7}{4}$ for y in $x = \dfrac{3 - 4y}{4}$:

$x = \dfrac{3 - 4\left(-\frac{7}{4}\right)}{4} = \dfrac{3 + 7}{4} = \dfrac{10}{4} = 2\frac{1}{2}$

$\left\{\left(2\frac{1}{2}, -1\frac{3}{4}\right)\right\}$

14. $\left\{\left(2\frac{1}{2}, -2\frac{1}{3}\right)\right\}$

Exercises 12.5 (page 618)

1.

$$\begin{aligned}
(1)\quad 2x + y + z &= 4\\
(2)\quad x - y + 3z &= -2\\
(3)\quad x + y + 2z &= 1
\end{aligned}$$

$$\begin{aligned}
(1) + (2):\quad 3x \qquad + 4z &= 2\\
(2) + (3):\quad 2x \qquad + 5z &= -1
\end{aligned}$$

Next, eliminate x:

$$\begin{aligned}
2]\quad 3x + 4z = 2 \Rightarrow 6x + 8z &= 4\\
-3]\quad 2x + 5z = -1 \Rightarrow -6x - 15z &= 3\\
-7z &= 7\\
z &= -1
\end{aligned}$$

Substituting -1 for z in $3x + 4z = 2$:

$$\begin{aligned}
3x - 4 &= 2\\
3x &= 6\\
x &= 2
\end{aligned}$$

Substituting 2 for x and -1 for z in (3):

$$\begin{aligned}
(3)\quad x + y + 2z &= 1\\
2 + y - 2 &= 1\\
y &= 1
\end{aligned}$$

$\{(2, 1, -1)\}$

2. $\{(1, -2, 2)\}$

3.

$$\begin{aligned}
(1)\quad x + 2y + 2z &= 0\\
(2)\quad 2x - y + z &= -3\\
(3)\quad 4x + 2y + 3z &= 2
\end{aligned}$$

$$\begin{aligned}
(1) + 2(2):\quad 5x \qquad + 4z &= -6 \quad (4)\\
2(2) + (3):\quad 8x \qquad + 5z &= -4\\
-5]\quad 5x + 4z = -6 \Rightarrow -25x - 20z &= 30\\
4]\quad 8x + 5z = -4 \Rightarrow 32x + 20z &= -16\\
7x &= 14\\
x &= 2
\end{aligned}$$

Substituting 2 for x in (4):

$$\begin{aligned}
(4)\quad 5x + 4z &= -6\\
5(2) + 4z &= -6\\
4z &= -16\\
z &= -4
\end{aligned}$$

Substituting 2 for x and -4 for z in (1):

$$\begin{aligned}
(1)\quad x + 2y + 2z &= 0\\
2 + 2y + 2(-4) &= 0\\
2y &= 6\\
y &= 3
\end{aligned}$$

$\{(2, 3, -4)\}$

4. $\{(3, -4, 2)\}$

5.

$$\begin{aligned}
(1)\quad x \qquad + 2z &= 7\\
(2)\quad 2x - y \qquad &= 5\\
(3)\quad 2y + z &= 4
\end{aligned}$$

$$\begin{aligned}
2(2) + (3):\quad 4x \qquad + z &= 14 \quad (4)\\
(1)\quad x \qquad + 2z &= 7\\
2(4):\quad 8x \qquad + 2z &= 28\\
-(1):\quad -x \qquad - 2z &= -7\\
7x &= 21\\
x &= 3
\end{aligned}$$

Substituting 3 for x in (4):

$$\begin{aligned}
(4)\quad 4x + z &= 14\\
4(3) + z &= 14\\
z &= 2
\end{aligned}$$

Substituting 2 for z in (3):

$$\begin{aligned}
(3)\quad 2y + z &= 4\\
2y + 2 &= 4\\
2y &= 2\\
y &= 1
\end{aligned}$$

$\{(3, 1, 2)\}$

6. $\{(2, -1, 3)\}$

7.

$$\begin{aligned}
(1)\quad 2x + 3y + z &= 7\\
(2)\quad 4x \qquad - 2z &= -6\\
(3)\quad 6y - z &= 0
\end{aligned}$$

$$\begin{aligned}
2(3):\quad 12y - 2z &= 0\\
-(2):\quad -4x \qquad + 2z &= +6\\
2(3) - (2):\quad -4x + 12y &= 6 \quad (4)\\
(1) + (3):\quad 2x + 9y &= 7 \quad (5)\\
\tfrac{1}{2}(4):\quad -2x + 6y &= 3\\
(5) + \tfrac{1}{2}(4):\quad 15y &= 10\\
y &= \tfrac{2}{3}
\end{aligned}$$

Substituting $\frac{2}{3}$ for y in (3):

$$\begin{aligned}
(3)\quad 6y - z &= 0\\
6\left(\tfrac{2}{3}\right) - z &= 0\\
4 - z &= 0\\
z &= 4
\end{aligned}$$

Substituting $\frac{2}{3}$ for y and 4 for z in (1):

$$\begin{aligned}
(1)\quad 2x + 3y + z &= 7\\
2x + 3\left(\tfrac{2}{3}\right) + 4 &= 7\\
2x + 2 + 4 &= 7\\
2x &= 1\\
x &= \tfrac{1}{2}
\end{aligned}$$

$\left\{\left(\frac{1}{2}, \frac{2}{3}, 4\right)\right\}$

8. $\left\{\left(\frac{3}{4}, \frac{2}{5}, -1\right)\right\}$

9.

$$\begin{aligned}
(1)\quad x + y + z + w &= 5\\
(2)\quad 2x - y + 2z - w &= -2\\
(3)\quad x + 2y - z - 2w &= -1\\
(4)\quad -x + 3y + 3z + w &= 1
\end{aligned}$$

$$\begin{aligned}
(1) + (2):\quad 3x \qquad + 3z &= 3 \quad (5)\\
2(1) + (3):\quad 3x + 4y + z &= 9 \quad (6)\\
(2) + (4):\quad x + 2y + 5z &= -1 \quad (7)\\
(6)\quad 3x + 4y + z &= 9\\
-2(7):\quad -2x - 4y - 10z &= 2\\
(6) - 2(7):\quad x \qquad - 9z &= 11 \quad (8)\\
-\tfrac{1}{3}(5):\quad -x \qquad - z &= -1 \quad (9)\\
(8) + (9):\quad -10z &= 10\\
z &= -1
\end{aligned}$$

Substituting -1 for z in $-1(9)$:

$$\begin{aligned}
-1(9)\quad x + z &= 1\\
x + (-1) &= 1\\
x &= 2
\end{aligned}$$

Substituting 2 for x and -1 for z in (7):

$$\begin{aligned}
(7)\quad x + 2y + 5z &= -1\\
2 + 2y - 5 &= -1\\
2y &= 2\\
y &= 1
\end{aligned}$$

Substituting 2 for x, 1 for y, and -1 for z in (1):

$$\begin{aligned}
(1)\quad x + y + z + w &= 5\\
2 + 1 - 1 + w &= 5\\
w &= 3
\end{aligned}$$

$\{(2, 1, -1, 3)\}$

10. $\{(1, 2, 3, -2)\}$

11.

$$\begin{aligned}
(1)\quad 6x + 4y + 9z + 5w &= -3\\
(2)\quad 2x + 8y - 6z + 15w &= 8\\
(3)\quad 4x - 4y + 3z - 10w &= -3\\
(4)\quad 2x - 4y + 3z - 5w &= -1
\end{aligned}$$

$$\begin{aligned}
(1) + (4):\quad 8x \qquad + 12z &= -4 \quad (5)\\
2(1) + (3):\quad 16x + 4y + 21z &= -9 \quad (6)\\
(2) + 3(4):\quad 8x - 4y + 3z &= 5 \quad (7)\\
(5)\quad 8x \qquad + 12z &= -4\\
(6) + (7):\quad 24x \qquad + 24z &= -4 \quad (8)\\
3(5) - (8):\quad 12z &= -8\\
z &= -\tfrac{2}{3}
\end{aligned}$$

Substituting $-\frac{2}{3}$ for z in (5):

$$\begin{aligned} (5)\quad 8x + 12z &= -4 \\ 8x + 12\left(-\tfrac{2}{3}\right) &= -4 \\ 8x - 8 &= -4 \\ 8x &= 4 \\ x &= \tfrac{4}{8} = \tfrac{1}{2} \end{aligned}$$

Substituting $\frac{1}{2}$ for x and $-\frac{2}{3}$ for z in (7):

$$\begin{aligned} (7)\quad 8x - 4y + 3z &= 5 \\ 8\left(\tfrac{1}{2}\right) - 4y + 3\left(-\tfrac{2}{3}\right) &= 5 \\ 4 - 4y - 2 &= 5 \\ -4y &= 3 \\ y &= -\tfrac{3}{4} \end{aligned}$$

Substituting $\frac{1}{2}$ for x, $-\frac{3}{4}$ for y, and $-\frac{2}{3}$ for z in (4):

$$\begin{aligned} (4)\quad 2x - 4y + 3z - 5w &= -1 \\ 2\left(\tfrac{1}{2}\right) - 4\left(-\tfrac{3}{4}\right) + 3\left(-\tfrac{2}{3}\right) - 5w &= -1 \\ 1 + 3 - 2 - 5w &= -1 \\ -5w &= -3 \\ w &= \tfrac{3}{5} \end{aligned}$$

$\left\{\left(\frac{1}{2}, -\frac{3}{4}, -\frac{2}{3}, \frac{3}{5}\right)\right\}$

12. $\left\{\left(-\frac{5}{6}, \frac{1}{3}, \frac{3}{2}, -\frac{1}{4}\right)\right\}$

Sections 12.1–12.5 Review Exercises (page 619)

1. (1) $4x + 5y = 22$
(2) $3x - 2y = 5$
(1) Intercepts: $\left(5\frac{1}{2}, 0\right)$ and $\left(0, 4\frac{2}{5}\right)$
(2) Intercepts: $\left(1\frac{2}{3}, 0\right)$ and $\left(0, -2\frac{1}{2}\right)$

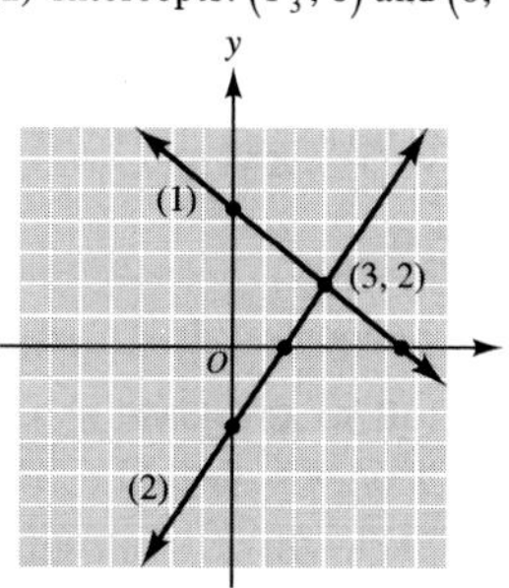

$\{(3, 2)\}$

2.

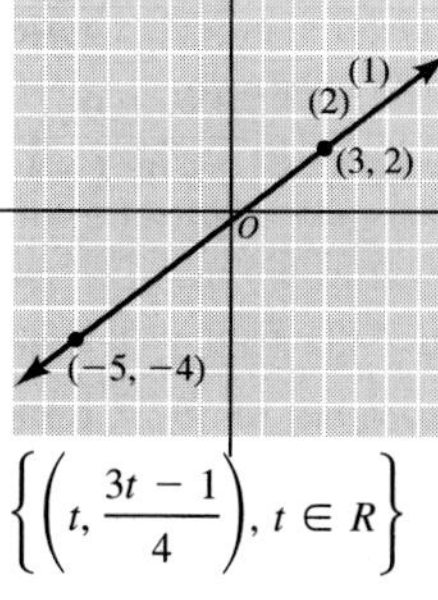

$\left\{\left(t, \frac{3t - 1}{4}\right), t \in R\right\}$

3. (1) $2x - 3y = 3$
(2) $3y - 2x = 6$
(1) Intercepts: $\left(1\frac{1}{2}, 0\right)$ and $(0, -1)$
(2) Intercepts: $(-3, 0)$ and $(0, 2)$

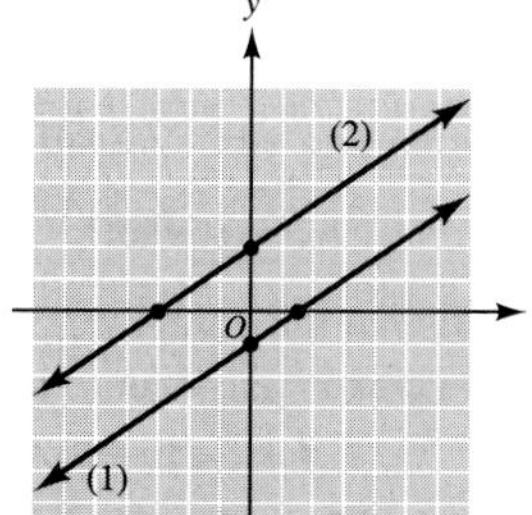

$\{\ \}$

4. $\left\{\left(4\frac{1}{2}, -3\frac{1}{2}\right)\right\}$

5.
$$\begin{aligned} -3]\ \ 4x - 8y = 4 &\Rightarrow -12x + 24y = -12 \\ 4]\ \ 3x - 6y = 3 &\Rightarrow \underline{\ \ 12x - 24y = \ \ 12} \\ & \qquad\qquad\quad 0 = \ \ 0 \end{aligned}$$

Many solutions: $\left\{\left(t, \frac{t - 1}{2}\right), t \in R\right\}$

6. $\{\ \}$

7. (1) $x = y + 2$
(2) $4x - 5y = 3$

Substituting $y + 2$ for x in (2):

$$\begin{aligned} (2)\quad 4x - 5y &= 3 \\ 4(y + 2) - 5y &= 3 \\ 4y + 8 - 5y &= 3 \\ -y &= -5 \\ y &= 5 \end{aligned}$$

Substituting 5 for y in $x = y + 2$:
$x = 5 + 2 = 7$

$\{(7, 5)\}$

8. $\{(4, 9)\}$

9.
$$\begin{aligned} (1)\quad 2x + y - z &= 1 \\ (2)\quad 3x - y + 2z &= 3 \\ (3)\quad x + 2y + 3z &= -6 \end{aligned}$$

$$\begin{aligned} (1) + (2){:}\quad 5x + z &= 4 \quad (4) \\ 2(2) + (3){:}\quad 7x + 7z &= 0 \quad (5) \\ \tfrac{1}{7}(5){:}\quad x + z &= 0 \quad (6) \\ (4)\quad 5x + z &= 4 \quad (4) \\ (4) - (6){:}\quad 4x &= 4 \\ x &= 1 \end{aligned}$$

Substituting 1 for x in (6):

$$\begin{aligned} (6)\quad x + z &= 0 \\ 1 + z &= 0 \\ z &= -1 \end{aligned}$$

Substituting 1 for x and -1 for z in (1):

$$\begin{aligned} (1)\quad 2x + y - z &= 1 \\ 2 + y + 1 &= 1 \\ y &= -2 \end{aligned}$$

$\{(1, -2, -1)\}$

10. $\{(2, 1, -3)\}$

Exercises 12.6 (page 630)

1.
$$\begin{bmatrix} 2 & 1 & 1 & 4 \\ 1 & -1 & 3 & -2 \\ 1 & 1 & 2 & 1 \end{bmatrix} \sim \begin{matrix} -2] \\ \\ \\ \end{matrix} \begin{bmatrix} 1 & -1 & 3 & -2 \\ 2 & 1 & 1 & 4 \\ 1 & 1 & 2 & 1 \end{bmatrix}$$

$$\sim \begin{matrix} -1] \\ \\ \\ \end{matrix} \begin{bmatrix} 1 & -1 & 3 & -2 \\ 0 & 3 & -5 & 8 \\ 1 & 1 & 2 & 1 \end{bmatrix} \sim \begin{matrix} \\ \frac{1}{3}] \\ \\ \end{matrix} \begin{bmatrix} 1 & -1 & 3 & -2 \\ 0 & 3 & -5 & 8 \\ 0 & 2 & -1 & 3 \end{bmatrix}$$

$$\sim \begin{matrix} \\ -2] \\ \\ \end{matrix} \begin{bmatrix} 1 & -1 & 3 & -2 \\ 0 & 1 & -\frac{5}{3} & \frac{8}{3} \\ 0 & 2 & -1 & 3 \end{bmatrix} \sim \begin{matrix} \\ \\ \frac{3}{7}] \end{matrix} \begin{bmatrix} 1 & -1 & 3 & -2 \\ 0 & 1 & -\frac{5}{3} & \frac{8}{3} \\ 0 & 0 & \frac{7}{3} & -\frac{7}{3} \end{bmatrix}$$

$$\sim \begin{bmatrix} 1 & -1 & 3 & -2 \\ 0 & 1 & -\frac{5}{3} & \frac{8}{3} \\ 0 & 0 & 1 & -1 \end{bmatrix}$$

$z = -1$
$y - \frac{5}{3}(-1) = \frac{8}{3}$
$y = 1$
$x - 1 + 3(-1) = -2$
$x = 2$

$\{(2, 1, -1)\}$

2. $\{(1, -2, 2)\}$

3.
$$\begin{matrix} \frac{1}{2}] \\ \\ \\ \end{matrix} \begin{bmatrix} 2 & 3 & 1 & 7 \\ 4 & 0 & -2 & -6 \\ 0 & 6 & -1 & 0 \end{bmatrix} \sim \begin{matrix} -4] \\ \\ \\ \end{matrix} \begin{bmatrix} 1 & \frac{3}{2} & \frac{1}{2} & \frac{7}{2} \\ 4 & 0 & -2 & -6 \\ 0 & 6 & -1 & 0 \end{bmatrix}$$

$$\sim \begin{matrix} \\ -\frac{1}{6}] \\ \\ \end{matrix} \begin{bmatrix} 1 & \frac{3}{2} & \frac{1}{2} & \frac{7}{2} \\ 0 & -6 & -4 & -20 \\ 0 & 6 & -1 & 0 \end{bmatrix} \sim \begin{matrix} \\ -6] \\ \\ \end{matrix} \begin{bmatrix} 1 & \frac{3}{2} & \frac{1}{2} & \frac{7}{2} \\ 0 & 1 & \frac{2}{3} & \frac{10}{3} \\ 0 & 6 & -1 & 0 \end{bmatrix}$$

$$\sim \begin{matrix} \\ \\ -\frac{1}{5}] \end{matrix} \begin{bmatrix} 1 & \frac{3}{2} & \frac{1}{2} & \frac{7}{2} \\ 0 & 1 & \frac{2}{3} & \frac{10}{3} \\ 0 & 0 & -5 & -20 \end{bmatrix} \sim \begin{bmatrix} 1 & \frac{3}{2} & \frac{1}{2} & \frac{7}{2} \\ 0 & 1 & \frac{2}{3} & \frac{10}{3} \\ 0 & 0 & 1 & 4 \end{bmatrix}$$

$z = 4$
$y + \frac{2}{3}(4) = \frac{10}{3}$
$y = \frac{2}{3}$
$x + \frac{3}{2}\left(\frac{2}{3}\right) + \frac{1}{2}(4) = \frac{7}{2}$
$x = \frac{1}{2}$

$\left\{\left(\frac{1}{2}, \frac{2}{3}, 4\right)\right\}$

4. $\{(2, -1, 3)\}$

5. $-2]\ \begin{bmatrix} 1 & 1 & 1 & 1 & 5 \\ 2 & -1 & 2 & -1 & -2 \\ 1 & 2 & -1 & -2 & -1 \\ -1 & 3 & 3 & 1 & 1 \end{bmatrix} \sim -1]\ \begin{bmatrix} 1 & 1 & 1 & 1 & 5 \\ 0 & -3 & 0 & -3 & -12 \\ 1 & 2 & -1 & -2 & -1 \\ -1 & 3 & 3 & 1 & 1 \end{bmatrix}$

$\sim 1]\ \begin{bmatrix} 1 & 1 & 1 & 1 & 5 \\ 0 & -3 & 0 & -3 & -12 \\ 0 & 1 & -2 & -3 & -6 \\ -1 & 3 & 3 & 1 & 1 \end{bmatrix} \sim -\tfrac{1}{3}]\ \begin{bmatrix} 1 & 1 & 1 & 1 & 5 \\ 0 & -3 & 0 & -3 & -12 \\ 0 & 1 & -2 & -3 & -6 \\ 0 & 4 & 4 & 2 & 6 \end{bmatrix}$

$\sim -1]\ \begin{bmatrix} 1 & 1 & 1 & 1 & 5 \\ 0 & 1 & 0 & 1 & 4 \\ 0 & 1 & -2 & -3 & -6 \\ 0 & 4 & 4 & 2 & 6 \end{bmatrix} \sim -4]\ \begin{bmatrix} 1 & 1 & 1 & 1 & 5 \\ 0 & 1 & 0 & 1 & 4 \\ 0 & 0 & -2 & -4 & -10 \\ 0 & 4 & 4 & 2 & 6 \end{bmatrix}$

$\sim -\tfrac{1}{2}]\ \begin{bmatrix} 1 & 1 & 1 & 1 & 5 \\ 0 & 1 & 0 & 1 & 4 \\ 0 & 0 & -2 & -4 & -10 \\ 0 & 0 & 4 & -2 & -10 \end{bmatrix} \sim -4]\ \begin{bmatrix} 1 & 1 & 1 & 1 & 5 \\ 0 & 1 & 0 & 1 & 4 \\ 0 & 0 & 1 & 2 & 5 \\ 0 & 0 & 4 & -2 & -10 \end{bmatrix}$

$\sim -\tfrac{1}{10}]\ \begin{bmatrix} 1 & 1 & 1 & 1 & 5 \\ 0 & 1 & 0 & 1 & 4 \\ 0 & 0 & 1 & 2 & 5 \\ 0 & 0 & 0 & -10 & -30 \end{bmatrix} \sim \begin{bmatrix} 1 & 1 & 1 & 1 & 5 \\ 0 & 1 & 0 & 1 & 4 \\ 0 & 0 & 1 & 2 & 5 \\ 0 & 0 & 0 & 1 & 3 \end{bmatrix}$

$w = 3$
$z + 2(3) = 5$
$z = -1$
$y + 0(-1) + 1(3) = 4$
$y = 1$
$x + 1 + (-1) + 3 = 5$
$x = 2$
$\{(2, 1, -1, 3)\}$

6. $\{(1, 2, 3, -2)\}$

Exercises 12.7 (page 634)

1. $\begin{vmatrix} 3 & 4 \\ 2 & 5 \end{vmatrix} = (3)(5) - (2)(4) = 15 - 8 = 7$ **2.** 22

3. $\begin{vmatrix} 2 & -4 \\ 5 & -3 \end{vmatrix} = 2(-3) - 5(-4) = -6 + 20 = 14$ **4.** 40

5. $\begin{vmatrix} -7 & -3 \\ 5 & 8 \end{vmatrix} = (-7)(8) - (5)(-3) = -56 - (-15) = -41$

6. 0 **7.** $\begin{vmatrix} 2 & -4 \\ 3 & x \end{vmatrix} = 20$ **8.** 3

$2x - (3)(-4) = 20$
$2x + 12 = 20$
$2x = 8$
$x = 4$

9. a. $\begin{vmatrix} 4 & -1 \\ -3 & 0 \end{vmatrix}$ **b.** $-\begin{vmatrix} 4 & -1 \\ -3 & 0 \end{vmatrix}$ **c.** $\begin{vmatrix} 1 & 3 \\ 4 & -1 \end{vmatrix}$

d. $-\begin{vmatrix} 1 & 3 \\ 4 & -1 \end{vmatrix}$ **e.** $\begin{vmatrix} 2 & 3 \\ 5 & -1 \end{vmatrix}$ **f.** $\begin{vmatrix} 2 & 3 \\ 5 & -1 \end{vmatrix}$

10. a. $\begin{vmatrix} 0 & 1 \\ 1 & 5 \end{vmatrix}$ **b.** $\begin{vmatrix} 0 & 1 \\ 1 & 5 \end{vmatrix}$ **c.** $\begin{vmatrix} 2 & 0 \\ -3 & -2 \end{vmatrix}$ **d.** $-\begin{vmatrix} 2 & 0 \\ -3 & -2 \end{vmatrix}$

e. $\begin{vmatrix} 2 & 1 \\ 4 & 5 \end{vmatrix}$ **f.** $-\begin{vmatrix} 2 & 1 \\ 4 & 5 \end{vmatrix}$

11. $\begin{vmatrix} 1 & 2 & 1 \\ 3 & 1 & 2 \\ 4 & 2 & 0 \end{vmatrix} = (1)\begin{vmatrix} 3 & 1 \\ 4 & 2 \end{vmatrix} - (2)\begin{vmatrix} 1 & 2 \\ 4 & 2 \end{vmatrix} + (0)\begin{vmatrix} 1 & 2 \\ 3 & 1 \end{vmatrix}$

$= 1(6 - 4) - 2(2 - 8) + 0 = 2 + 12 = 14$

12. 18

13. $\begin{vmatrix} 1 & 3 & -2 \\ -1 & 2 & -3 \\ 0 & 4 & 1 \end{vmatrix} = 0\begin{vmatrix} 3 & -2 \\ 2 & -3 \end{vmatrix} - 4\begin{vmatrix} 1 & -2 \\ -1 & -3 \end{vmatrix} + 1\begin{vmatrix} 1 & 3 \\ -1 & 2 \end{vmatrix}$

$= 0 - 4(-3 - 2) + 1(2 + 3)$
$= 0 + 20 + 5 = 25$

14. -10

15. $\begin{vmatrix} 1 & -2 & 3 \\ -3 & 4 & 0 \\ 2 & 6 & 5 \end{vmatrix} = (3)\begin{vmatrix} -3 & 4 \\ 2 & 6 \end{vmatrix} - (0) + (5)\begin{vmatrix} 1 & -2 \\ -3 & 4 \end{vmatrix}$

$= 3(-18 - 8) - 0 + 5(4 - 6)$
$= 3(-26) + 5(-2) = -78 - 10 = -88$

16. -84 **17.** $\begin{vmatrix} 6 & 7 & 8 \\ -6 & 7 & -9 \\ 0 & 0 & -2 \end{vmatrix} = 0 - 0 - 2\begin{vmatrix} 6 & 7 \\ -6 & 7 \end{vmatrix}$

$= 0 - 0 - 2(42 + 42) = -168$

18. -180 **19.** $\begin{vmatrix} x & 0 & 1 \\ 0 & 2 & 3 \\ 4 & -1 & -2 \end{vmatrix} = 6$ **20.** -2

$x\begin{vmatrix} 2 & 3 \\ -1 & -2 \end{vmatrix} + 1\begin{vmatrix} 0 & 2 \\ 4 & -1 \end{vmatrix} = 6$
$x(-1) + 1(-8) = 6$
$-x - 8 = 6$
$x = -14$

Exercises 12.8 (page 640)

1. $x = \dfrac{\begin{vmatrix} 11 & -1 \\ -4 & 2 \end{vmatrix}}{\begin{vmatrix} 3 & -1 \\ 3 & 2 \end{vmatrix}} = \dfrac{22 - 4}{6 - (-3)} = \dfrac{18}{9} = 2$ **2.** $\{(-3, 4)\}$

$y = \dfrac{\begin{vmatrix} 3 & 11 \\ 3 & -4 \end{vmatrix}}{\begin{vmatrix} 3 & -1 \\ 3 & 2 \end{vmatrix}} = \dfrac{-12 - 33}{6 - (-3)} = \dfrac{-45}{9} = -5$

$\{(2, -5)\}$

3. $x = \dfrac{\begin{vmatrix} 11 & 15 \\ 31 & -1 \end{vmatrix}}{\begin{vmatrix} 8 & 15 \\ 4 & -1 \end{vmatrix}} = \dfrac{-11 - 465}{-8 - 60} = \dfrac{-476}{-68} = 7$ **4.** $\{(6, 3)\}$

$y = \dfrac{\begin{vmatrix} 8 & 11 \\ 4 & 31 \end{vmatrix}}{\begin{vmatrix} 8 & 15 \\ 4 & -1 \end{vmatrix}} = \dfrac{248 - 44}{-8 - 60} = \dfrac{204}{-68} = -3$

$\{(7, -3)\}$

5. $x = \dfrac{\begin{vmatrix} 3 & -3 \\ 12 & -9 \end{vmatrix}}{\begin{vmatrix} 7 & -3 \\ 20 & -9 \end{vmatrix}} = \dfrac{-27 - (-36)}{-63 - (-60)} = \dfrac{9}{-3} = -3$ **6.** $\{(9, -13)\}$

$y = \dfrac{\begin{vmatrix} 7 & 3 \\ 20 & 12 \end{vmatrix}}{\begin{vmatrix} 7 & -3 \\ 20 & -9 \end{vmatrix}} = \dfrac{84 - 60}{-63 - (-60)} = \dfrac{24}{-3} = -8$

$\{(-3, -8)\}$

7. $x = \dfrac{\begin{vmatrix} 0 & 5 \\ 38 & -3 \end{vmatrix}}{\begin{vmatrix} 6 & 5 \\ 4 & -3 \end{vmatrix}} = \dfrac{0 - 190}{-18 - 20} = \dfrac{-190}{-38} = 5$

$y = \dfrac{\begin{vmatrix} 6 & 0 \\ 4 & 38 \end{vmatrix}}{\begin{vmatrix} 6 & 5 \\ 4 & -3 \end{vmatrix}} = \dfrac{228 - 0}{-18 - 20} = \dfrac{228}{-38} = -6$

$\{(5, -6)\}$

8. $\{(-4, 10)\}$

9. $x = \dfrac{\begin{vmatrix} 5 & 6 \\ 7 & 12 \end{vmatrix}}{\begin{vmatrix} 4 & 6 \\ 8 & 12 \end{vmatrix}} = \dfrac{60 - 42}{48 - 48} = \dfrac{18}{0}$ Not defined

The denominator is zero and the numerator is not zero; { }

10. { }

11. $x = \dfrac{\begin{vmatrix} 5 & -3 \\ 10 & -6 \end{vmatrix}}{\begin{vmatrix} 7 & -3 \\ 14 & -6 \end{vmatrix}} = \dfrac{-30 - (-30)}{-42 - (-42)} = \dfrac{0}{0}$ Not defined

The numerator and denominator are both zero; the system is dependent.

12. Dependent

13. $D = \begin{vmatrix} 2 & 1 & 1 \\ 1 & -1 & 3 \\ 1 & 1 & 2 \end{vmatrix} = (2)\begin{vmatrix} -1 & 3 \\ 1 & 2 \end{vmatrix} - (1)\begin{vmatrix} 1 & 3 \\ 1 & 2 \end{vmatrix} + (1)\begin{vmatrix} 1 & -1 \\ 1 & 1 \end{vmatrix}$

$= 2(-5) - 1(-1) + 1(2) = -10 + 1 + 2 = -7$

$D_x = \begin{vmatrix} 4 & 1 & 1 \\ -2 & -1 & 3 \\ 1 & 1 & 2 \end{vmatrix}$

$= (4)\begin{vmatrix} -1 & 3 \\ 1 & 2 \end{vmatrix} - (-2)\begin{vmatrix} 1 & 1 \\ 1 & 2 \end{vmatrix} + (1)\begin{vmatrix} 1 & 1 \\ -1 & 3 \end{vmatrix}$

$= 4(-5) + 2(1) + 1(4) = -20 + 2 + 4 = -14$

$D_y = \begin{vmatrix} 2 & 4 & 1 \\ 1 & -2 & 3 \\ 1 & 1 & 2 \end{vmatrix}$

$= (2)\begin{vmatrix} -2 & 3 \\ 1 & 2 \end{vmatrix} - (4)\begin{vmatrix} 1 & 3 \\ 1 & 2 \end{vmatrix} + (1)\begin{vmatrix} 1 & -2 \\ 1 & 1 \end{vmatrix}$

$= 2(-7) - 4(-1) + 1(3) = -14 + 4 + 3 = -7$

$D_z = \begin{vmatrix} 2 & 1 & 4 \\ 1 & -1 & -2 \\ 1 & 1 & 1 \end{vmatrix}$

$= (2)\begin{vmatrix} -1 & -2 \\ 1 & 1 \end{vmatrix} - (1)\begin{vmatrix} 1 & -2 \\ 1 & 1 \end{vmatrix} + (4)\begin{vmatrix} 1 & -1 \\ 1 & 1 \end{vmatrix}$

$= 2(1) - 1(3) + 4(2) = 2 - 3 + 8 = 7$

$x = \dfrac{D_x}{D} = \dfrac{-14}{-7} = 2,\ y = \dfrac{D_y}{D} = \dfrac{-7}{-7} = 1,\ z = \dfrac{D_z}{D} = \dfrac{7}{-7} = -1$

$\{(2, 1, -1)\}$

14. $\{(1, -2, 2)\}$

15. $D = \begin{vmatrix} 2 & 3 & 1 \\ 4 & 0 & -2 \\ 0 & 6 & -1 \end{vmatrix} = (0) - (6)\begin{vmatrix} 2 & 1 \\ 4 & -2 \end{vmatrix} + (-1)\begin{vmatrix} 2 & 3 \\ 4 & 0 \end{vmatrix}$

$= -6(-8) - 1(-12) = 48 + 12 = 60$

$D_x = \begin{vmatrix} 7 & 3 & 1 \\ -6 & 0 & -2 \\ 0 & 6 & -1 \end{vmatrix} = (0) - (6)\begin{vmatrix} 7 & 1 \\ -6 & -2 \end{vmatrix} + (-1)\begin{vmatrix} 7 & 3 \\ -6 & 0 \end{vmatrix}$

$= 0 - 6(-8) - 1(18) = 48 - 18 = 30$

$D_y = \begin{vmatrix} 2 & 7 & 1 \\ 4 & -6 & -2 \\ 0 & 0 & -1 \end{vmatrix} = +(-1)\begin{vmatrix} 2 & 7 \\ 4 & -6 \end{vmatrix} = -1(-40) = 40$

$D_z = \begin{vmatrix} 2 & 3 & 7 \\ 4 & 0 & -6 \\ 0 & 6 & 0 \end{vmatrix} = -(6)\begin{vmatrix} 2 & 7 \\ 4 & -6 \end{vmatrix} = -6(-40) = 240$

$x = \dfrac{D_x}{D} = \dfrac{30}{60} = \dfrac{1}{2},\ y = \dfrac{D_y}{D} = \dfrac{40}{60} = \dfrac{2}{3},\ z = \dfrac{D_z}{D} = \dfrac{240}{60} = 4$

$\{(\frac{1}{2}, \frac{2}{3}, 4)\}$

16. $\{(2, -1, 3)\}$

Exercises 12.9 (page 643)

The checks for these exercises will not be shown.

1. Let x = one number
Let y = the other number

$$\begin{aligned} (1)\quad & x + y = 30 \\ (2)\quad & x - y = 12 \\ (1) + (2):\ & 2x = 42 \\ & x = 21 \end{aligned}$$

Substituting 21 for x in (1):

$$\begin{aligned} (1)\quad x + y &= 30 \\ 21 + y &= 30 \\ y &= 9 \end{aligned}$$

The numbers are 21 and 9.

2. 125° and 55°

3. Let x = number of nickels
Let y = number of quarters

(1) $x + y = 15$
(2) $5x + 25y = 175$

Solving (1) for x:

(1) $x + y = 15$
$x = 15 - y$

Substituting $15 - y$ for x in (2):

$$\begin{aligned} (2)\quad 5x + 25y &= 175 \\ 5(15 - y) + 25y &= 175 \\ 75 - 5y + 25y &= 175 \\ 20y &= 100 \\ y &= 5 \end{aligned}$$

Substituting 5 for y in $x = 15 - y$:

$x = 15 - 5 = 10$

Beatrice has 5 quarters and 10 nickels.

4. 15 dimes, 7 half-dollars

5. Let x = numerator
Let y = denominator

(1) $\dfrac{x}{y} = \dfrac{2}{3} \Rightarrow 3x = 2y$

(2) $\dfrac{x + 10}{y - 5} = 1 \Rightarrow x + 10 = y - 5$

$x = y - 15$

Substituting $y - 15$ for x in $3x = 2y$:

$$\begin{aligned} 3(y - 15) &= 2y \\ 3y - 45 &= 2y \\ y &= 45 \end{aligned}$$

Substituting 45 for y in $x = y - 15$:

$x = 45 - 15 = 30$

The original fraction was $\frac{30}{45}$.

6. $\frac{12}{24}$

7. Let u = units digit
Let t = tens digit
Let h = hundreds digit

$$\begin{array}{lrcl} (1) & h + t + u & = & 20 \\ (2) & t - u & = & 3 \\ (3) & h + t & = & 15 \\ \hline (1)+(2): & h + 2t & = & 23 \quad (4) \\ (3): & h + t & = & 15 \\ \hline (4)-(3): & t & = & 8 \end{array}$$

Substituting 8 for t in (3) $\Rightarrow h = 7$
Substituting 8 for t in (2) $\Rightarrow u = 5$
The number is 785.

8. 498

9. Let a = number of hr for Albert to do the job
Let b = number of hr for Bill to do the job
Let c = number of hr for Carlos to do the job
Then Albert's rate is $\frac{1 \text{ job}}{a \text{ hr}}$, Bill's rate is $\frac{1 \text{ job}}{b \text{ hr}}$, and Carlos's rate is $\frac{1 \text{ job}}{c \text{ hr}}$.

$$\begin{array}{lrcl} (1) & \frac{2}{a} + \frac{2}{b} + \frac{2}{c} & = & 1 \\ (2) & \frac{3}{b} + \frac{3}{c} & = & 1 \\ (3) & \frac{4}{a} + \frac{4}{b} & = & 1 \\ \hline (1) & \frac{2}{a} + \frac{2}{b} + \frac{2}{c} & = & 1 \\ \frac{1}{2}(3): & \frac{2}{a} + \frac{2}{b} & = & \frac{1}{2} \\ \hline (1) - \frac{1}{2}(3): & \frac{2}{c} & = & \frac{1}{2} \\ & c & = & 4 \end{array}$$

Substituting 4 for c in (2) $\Rightarrow b = 12$
Substituting 12 for b in (3) $\Rightarrow a = 6$
It would take Albert 6 hr, Bill 12 hr, and Carlos 4 hr to do the job alone.

10. A: 4 hr; B: 8 hr; C: 12 hr

11. Let x = average speed of plane in still air (in mph)
Let y = average speed of wind (in mph)
Then $x - y$ = average speed flying against the wind (in mph)
and $x + y$ = average speed flying with the wind (in mph)
Formula to use: $rt = d$

(1) $(x - y)\frac{11}{2} = 2{,}750 \Rightarrow 11(x - y) = 2(2{,}750)$
(2) $(x + y)(5) = 2{,}750 \Rightarrow 5(x + y) = 2{,}750$

$$\begin{array}{lrcl} \frac{1}{11}] & 11x - 11y = 5{,}500 \Rightarrow x - y & = & 500 \\ \frac{1}{5}] & 5x + 5y = 2{,}750 \Rightarrow x + y & = & 550 \\ \hline & 2x & = & 1{,}050 \\ & x & = & 525 \\ & 525 + y & = & 550 \\ & y & = & 25 \end{array}$$

The speed of the plane in still air is 525 mph, and the speed of the wind is 25 mph.

12. 540 mph = speed of plane; 60 mph = speed of wind

13. Let x = number of lb of grade A coffee
Let y = number of lb of grade B coffee

$$\begin{array}{lrcl} (1) & x + y & = & 90 \\ (2) & 3.85x + 3.65y & = & 338.90 \\ -365(1): & -365x - 365y & = & -32{,}850 \\ 100(2): & 385x + 365y & = & 33{,}890 \\ \hline & 20x & = & 1{,}040 \\ & x & = & 52 \\ & 52 + y & = & 90 \\ & y & = & 38 \end{array}$$

The mixture should contain 52 lb of grade A coffee and 38 lb of grade B coffee.

14. 37 lb of grade A coffee; 43 lb of grade B coffee

15. Let u = units digit
Let t = tens digit

$$\begin{array}{lrcl} (1) & t + u & = & 12 \\ (2) & (10t + u) - (10u + t) = 54 \Rightarrow 9t - 9u & = & 54 \\ \frac{1}{9}(2): & t - u & = & 6 \\ (1) & t + u & = & 12 \\ \hline (1) + \frac{1}{9}(2): & 2t & = & 18 \\ & t & = & 9 \end{array}$$

Substituting 9 for t in (1) $\Rightarrow u = 3$
The number is 93.

16. 86

17. Let x = number of 18¢ stamps
Let y = number of 22¢ stamps
Let z = number of 45¢ stamps
(1) $x + y + z = 29$
(2) $18x + 22y + 45z = 721$
(3) $y = 2x$
Substituting (3) into (1) and into (2):

$$\begin{array}{lrcl} (1) & x + y + z & = & 29 \\ & x + 2x + z & = & 29 \\ & 3x + z & = & 29 \quad (4) \\ (2) & 18x + 22y + 45z & = & 721 \\ & 18x + 44x + 45z & = & 721 \\ & 62x + 45z & = & 721 \quad (5) \\ -45(4): & -135x - 45z & = & -1{,}305 \\ (5) & 62x + 45z & = & 721 \\ \hline -45(4) - (5): & -73x & = & -584 \\ & x & = & 8 \end{array}$$

Substituting 8 for x in (3) $\Rightarrow y = 16$
Substituting 8 for x and 16 for y in (1) $\Rightarrow z = 5$
Tom bought eight 18¢ stamps, sixteen 22¢ stamps, and five 45¢ stamps.

18. Six 18¢ stamps; twelve 22¢ stamps; three 45¢ stamps

19. Let t = cost of tie (in cents)
Let p = cost of pin (in cents)
(1) $t + p = 110$
(2) $t = 100 + p$
Substituting $100 + p$ for t in (1):

$$\begin{array}{lrcl} (1) & t + p & = & 110 \\ & 100 + p + p & = & 110 \\ & 2p & = & 10 \\ & p & = & 5 \\ & t & = & 100 + p = 100 + 5 = 105 \end{array}$$

The pin costs 5¢, and the tie costs \$1.05.

20. 7 on the upper branch; 5 on the lower branch

Exercises 12.10 (page 650)

1. (1) $x^2 = 2y$
(2) $x - y = -4$
Solving (2) for y:
(2) $x - y = -4$
$y = x + 4$
Substituting $x + 4$ for y in (1):

$$\begin{array}{lrcl} (1) & x^2 & = & 2y \\ & x^2 & = & 2(x + 4) \\ & x^2 & = & 2x + 8 \\ & x^2 - 2x - 8 & = & 0 \\ & (x - 4)(x + 2) & = & 0 \end{array}$$

2. $\{(2, 1), (-4, 4)\}$

$x - 4 = 0 \quad | \quad x + 2 = 0$

$x = 4 \quad | \quad x = -2$

Substituting 4 for x in $y = x + 4$:

$y = 8$

Substituting -2 for x in $y = x + 4$:

$y = 2$

$\{(4, 8), (-2, 2)\}$

3. (1) $x^2 = 4y$

(2) $x - y = 1$

Solving (2) for y:

(2) $x - y = 1$

$y = x - 1$

Substituting $x - 1$ for y in (1):

(1) $x^2 = 4y$

$x^2 = 4(x - 1)$

$x^2 = 4x - 4$

$x^2 - 4x + 4 = 0$

$(x - 2)(x - 2) = 0$

Therefore, $x = 2$.

Substituting 2 for x in (2):

(2) $x - y = 1$

$2 - y = 1$

$y = 1$

$\{(2, 1)\}$

4. $\{(-5, 0), (4, 3)\}$

5. (1) $xy = 4$

(2) $x - 2y = 2$

Solving (2) for x:

(2) $x - 2y = 2$

$x = 2y + 2$

Substituting $2y + 2$ for x in (1):

(1) $xy = 4$

$(2y + 2)y = 4$

$2y^2 + 2y - 4 = 0$

$y^2 + y - 2 = 0$

$(y + 2)(y - 1) = 0$

$y + 2 = 0 \quad | \quad y - 1 = 0$

$y = -2 \quad | \quad y = 1$

Substituting -2 for y in $xy = 4$:

$x(-2) = 4$

$x = -2$

Substituting 1 for y in $xy = 4$:

$x(1) = 4$

$x = 4$

$\{(-2, -2), (4, 1)\}$

6. $\{(-i\sqrt{3}, i\sqrt{3}), (i\sqrt{3}, -i\sqrt{3})\}$ is the algebraic solution set. The curves do not intersect.

7. (1) $x^2 + y^2 = 61$

(2) $x^2 - y^2 = 11$

(1) + (2): $2x^2 = 72$

$x^2 = 36$

$x = \pm 6$

Substituting ± 6 for x in (1):

(1) $x^2 + y^2 = 61$

$(\pm 6)^2 + y^2 = 61$

$36 + y^2 = 61$

$y^2 = 25$

$y = \pm 5$

$\{(6, 5), (6, -5), (-6, 5), (-6, -5)\}$

8. $\{(2, 0), (-2, 0), (0, -1)\}$

9. (1) $2x^2 + 3y^2 = 21$

(2) $x^2 + 2y^2 = 12$

$-1]\ 2x^2 + 3y^2 = 21 \Rightarrow -2x^2 - 3y^2 = -21$

$2]\ 1x^2 + 2y^2 = 12 \Rightarrow 2x^2 + 4y^2 = 24$

$y^2 = 3$

$y = \pm\sqrt{3}$

Substituting $\pm\sqrt{3}$ for y in (2):

(2) $x^2 + 2y^2 = 12$

$x^2 + 2(\pm\sqrt{3})^2 = 12$

$x^2 + 6 = 12$

$x^2 = 6$

$x = \pm\sqrt{6}$

$\{(\sqrt{6}, \sqrt{3}), (\sqrt{6}, -\sqrt{3}), (-\sqrt{6}, \sqrt{3}), (-\sqrt{6}, -\sqrt{3})\}$

10. $\{(3\sqrt{2}, \sqrt{2}), (3\sqrt{2}, -\sqrt{2}), (-3\sqrt{2}, \sqrt{2}), (-3\sqrt{2}, -\sqrt{2})\}$

Exercises 12.11 (page 653)

1. (1) $4x - 3y > -12$

(2) $y > 2$

Boundary line for (1): $4x - 3y = -12$

Intercepts: (0, 4) and (−3, 0)

Boundary line for (2): $y = 2$, a horizontal line that passes through (0, 2)

Both lines are dashed because equality is not included.

The origin is in the correct half-plane for (1) because $4(0) - 3(0) > -12$ is true. The origin is not in the correct half-plane for (2) because $0 > 2$ is false.

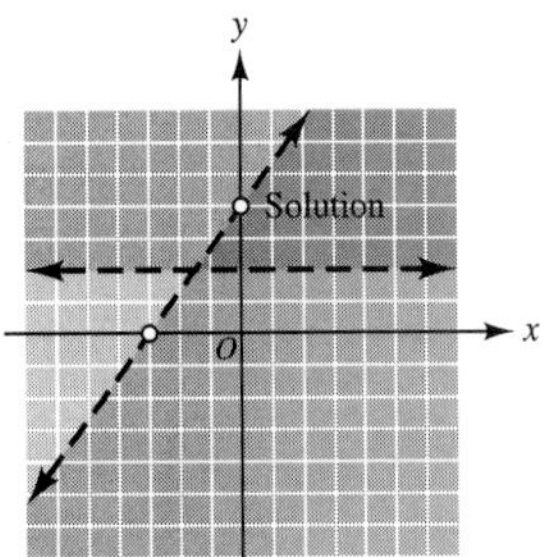

2.

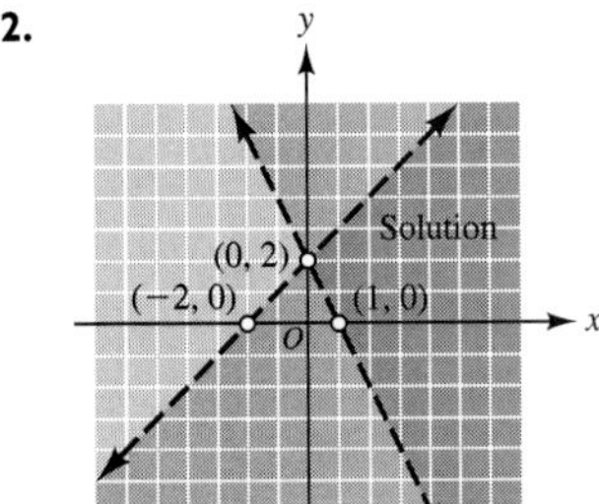

3. (1) $2x - y \le 2$

(2) $x + y \ge 5$

Boundary line for (1): $2x - y = 2$

Intercepts: (0, −2) and (1, 0)

Boundary line for (2): $x + y = 5$

Intercepts: (0, 5) and (5, 0)

Both lines are solid because equality is included.

The origin is in the correct half-plane for (1), but not for (2).

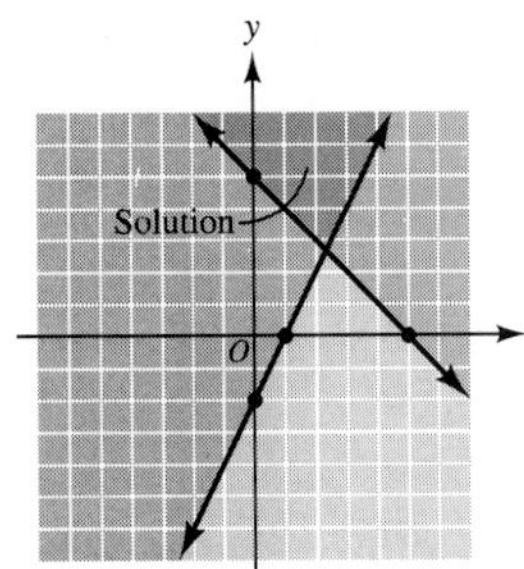

4.

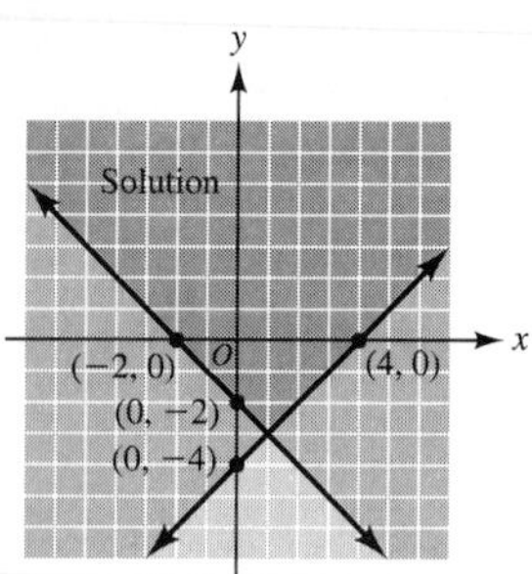

5. (1) $2x + y < 0$
(2) $x - y \geq -3$
Boundary line for (1): $2x + y = 0$
Points on line: (0, 0) and (−1, 2)
The point (−1, 0) is in the correct half-plane for (1).
Boundary line for (2): $x - y = -3$
Points on line: (0, 3) and (−3, 0)
The origin is in the correct half-plane for (2).

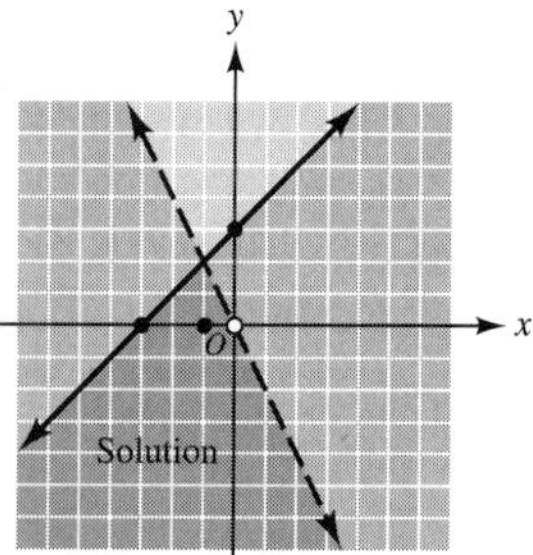

The boundary line for (1) is dashed because equality is not included; the boundary line for (2) is solid because equality is included.

6.

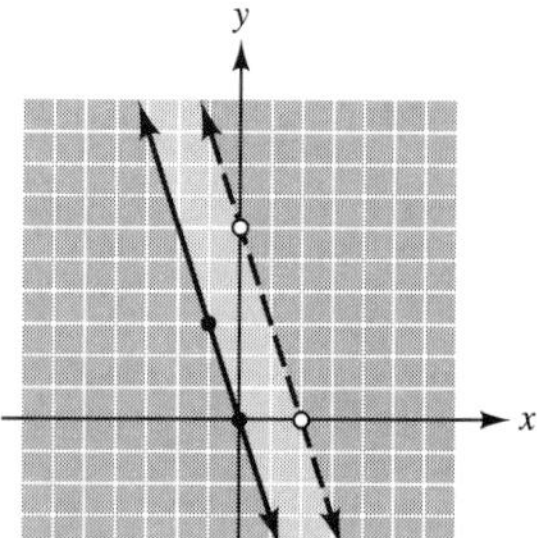

No solution

7. (1) $3x - 2y < 6$
(2) $x + 2y \leq 4$
(3) $6x + y > -6$
Boundary line for (1):
$3x - 2y = 6$
Intercepts: (2, 0) and (0, −3)
The half-plane for (1) includes (0, 0) because $3(0) - 2(0) < 6$.
Boundary line for (2):
$x + 2y = 4$
Intercepts: (4, 0) and (0, 2)
The half-plane for (2) includes (0, 0) because $(0) + 2(0) \leq 4$.
Boundary line for (3): $6x + y = -6$
Intercepts: (−1, 0) and (0, −6)
The half-plane for (3) includes (0, 0) because $6(0) + (0) > -6$.
Shaded area is solution.
The boundary lines for (1) and (3) are dashed because equality is not included; the boundary line for (2) is solid because equality is included.

8.

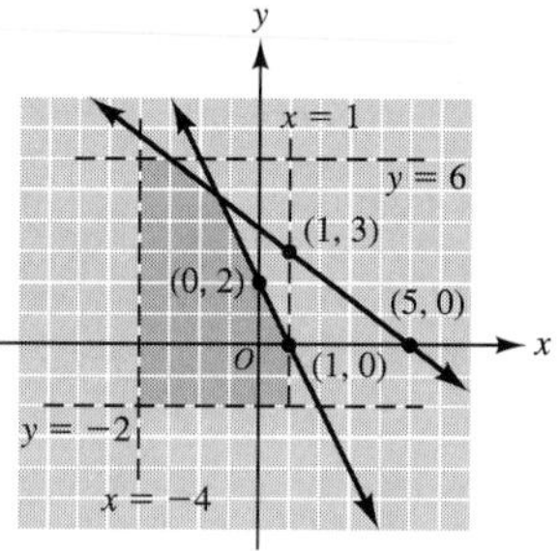

Shaded area is solution.

9. Boundary for $\frac{x^2}{9} + \frac{y^2}{4} < 1$ is $\frac{x^2}{9} + \frac{y^2}{4} = 1$, which is an ellipse with intercepts (±3, 0) and (0, ±2). Boundary for $\frac{x^2}{4} + \frac{y^2}{9} < 1$ is $\frac{x^2}{4} + \frac{y^2}{9} = 1$, which is an ellipse with intercepts (±2, 0) and (0, ±3). Both boundaries must be graphed as dashed curves. Substituting (0, 0) in both inequalities gives true statements.

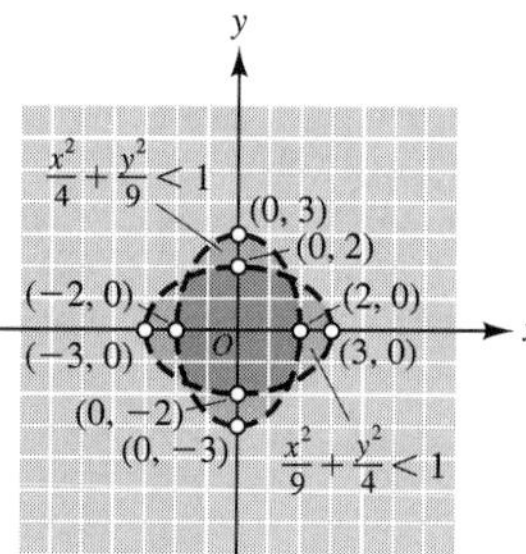

The final answer is heavily shaded.

10.

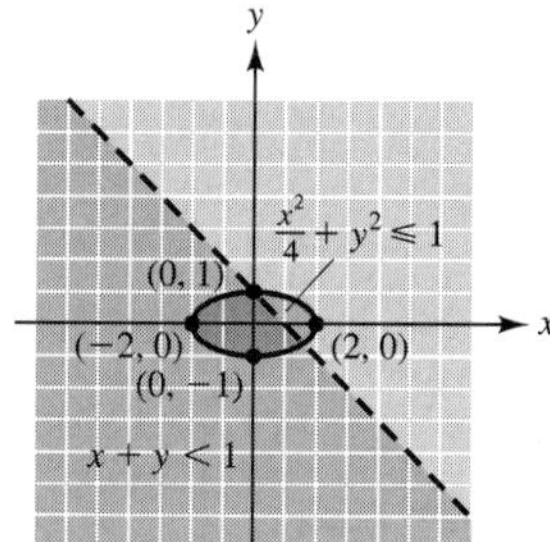

The final answer is heavily shaded.

11. Boundary for $y > 1 - x^2$ is $y = 1 - x^2$, which is a parabola with intercepts (±1, 0) and (0, 1). Boundary for $x^2 + y^2 < 4$ is $x^2 + y^2 = 4$, which is a circle with intercepts (±2, 0) and (0, ±2). Both curves must be graphed as dashed curves. Substituting (0, 0) in $y > 1 - x^2$ gives a false statement; we must shade the region that does not contain the origin. Substituting (0, 0) in $x^2 + y^2 < 4$ gives a true statement; we must shade the region that contains the origin. The final answer is heavily shaded.

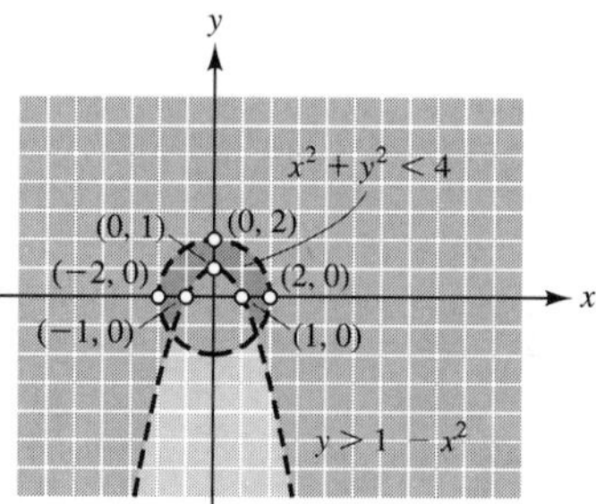

12

12.

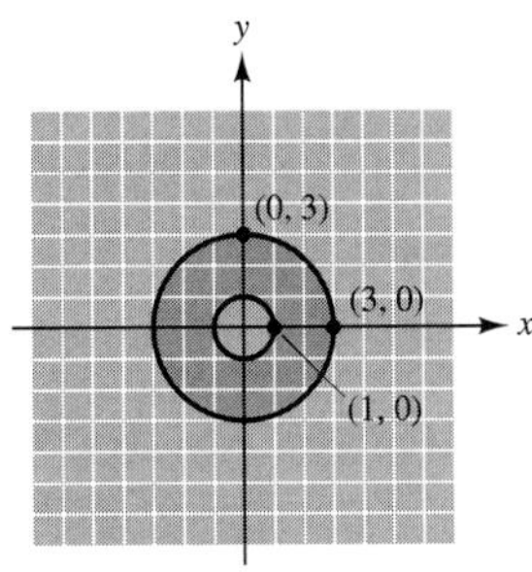

13. Boundary for $x^2 + y^2 \le 9$ is $x^2 + y^2 = 9$, which is a circle of radius 3 with its center at the origin; it must be graphed with a solid line. Boundary for $y \ge x - 2$ is a straight line with intercepts (2, 0) and (0, −2); it must be graphed with a solid line. Substituting (0, 0) in $x^2 + y^2 \le 9$ gives a true statement; we shade the region that contains the origin (the region *inside* the circle). Substituting (0, 0) in $y \ge x - 2$ gives a true statement; we shade the half-plane that contains the origin. The final answer is heavily shaded.

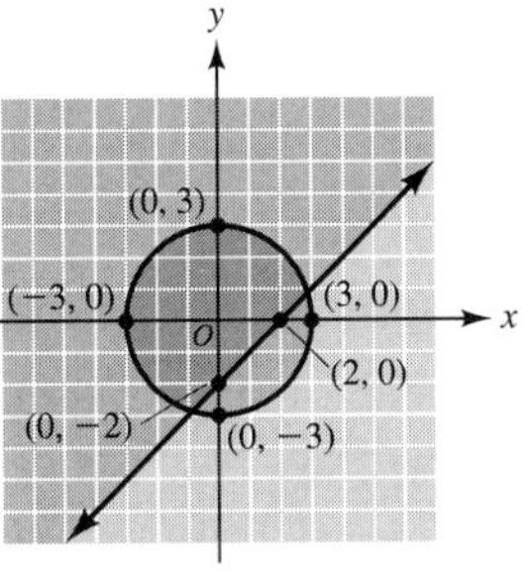

14.

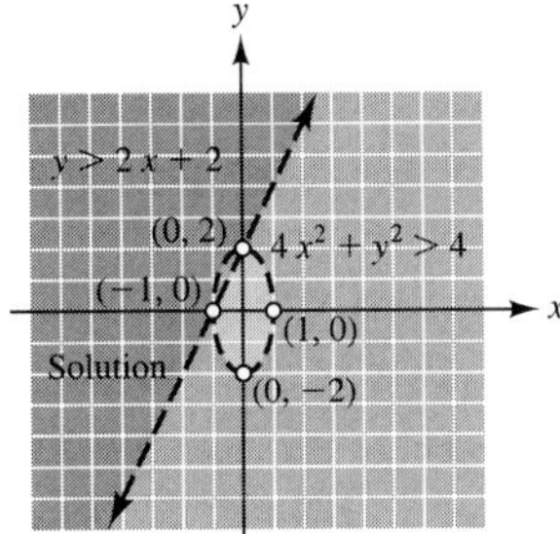

Sections 12.6–12.11 Review Exercises (page 655)

1. Using the matrix method:

$$\begin{bmatrix} 3 & -2 & 8 \\ 2 & 5 & -1 \end{bmatrix} \sim \begin{bmatrix} 1 & -\frac{2}{3} & \frac{8}{3} \\ 2 & 5 & -1 \end{bmatrix}$$

$$\sim \begin{bmatrix} 1 & -\frac{2}{3} & \frac{8}{3} \\ 0 & \frac{19}{3} & -\frac{19}{3} \end{bmatrix} \sim \begin{bmatrix} 1 & -\frac{2}{3} & \frac{8}{3} \\ 0 & 1 & -1 \end{bmatrix}$$

$y = -1$

$x - \frac{2}{3}(-1) = \frac{8}{3}$

$x + \frac{2}{3} = \frac{8}{3}$

$x = \frac{6}{3} = 2$

Using Cramer's rule:

$$x = \frac{D_x}{D} = \frac{\begin{vmatrix} 8 & -2 \\ -1 & 5 \end{vmatrix}}{\begin{vmatrix} 3 & -2 \\ 2 & 5 \end{vmatrix}} = \frac{40 - 2}{15 + 4} = \frac{38}{19} = 2$$

$$y = \frac{D_y}{D} = \frac{\begin{vmatrix} 3 & 8 \\ 2 & -1 \end{vmatrix}}{\begin{vmatrix} 3 & -2 \\ 2 & 5 \end{vmatrix}} = \frac{-3 - 16}{15 + 4} = \frac{-19}{19} = -1$$

$\{(2, -1)\}$

2. $\{(-1, 3, 2)\}$

3. Using the matrix method:

$$\begin{bmatrix} 5 & 2 & 1 \\ 7 & -6 & 8 \end{bmatrix} \sim \begin{bmatrix} 1 & \frac{2}{5} & \frac{1}{5} \\ 7 & -6 & 8 \end{bmatrix} \sim \begin{bmatrix} 1 & \frac{2}{5} & \frac{1}{5} \\ 0 & -\frac{44}{5} & \frac{33}{5} \end{bmatrix} \sim \begin{bmatrix} 1 & \frac{2}{5} & \frac{1}{5} \\ 0 & 1 & -\frac{3}{4} \end{bmatrix}$$

$y = -\frac{3}{4}$

$x + \frac{2}{5}\left(-\frac{3}{4}\right) = \frac{1}{5}$

$x - \frac{3}{10} = \frac{1}{5}$

$x = \frac{5}{10} = \frac{1}{2}$

Using Cramer's rule:

$$x = \frac{D_x}{D} = \frac{\begin{vmatrix} 1 & 2 \\ 8 & -6 \end{vmatrix}}{\begin{vmatrix} 5 & 2 \\ 7 & -6 \end{vmatrix}} = \frac{-6 - 16}{-30 - 14} = \frac{-22}{-44} = \frac{1}{2}$$

$$y = \frac{D_y}{D} = \frac{\begin{vmatrix} 5 & 1 \\ 7 & 8 \end{vmatrix}}{\begin{vmatrix} 5 & 2 \\ 7 & -6 \end{vmatrix}} = \frac{40 - 7}{-30 - 14} = \frac{33}{-44} = -\frac{3}{4}$$

$\left\{\left(\frac{1}{2}, -\frac{3}{4}\right)\right\}$

4. $\{(-3, 2, -1)\}$

5. (1) $x - 2y = -1$

(2) $2x^2 - 3y^2 = 6$

Solving (1) for x:

(1) $x - 2y = -1$

$x = 2y - 1$

Substituting $2y - 1$ for x in (2):

(2) $2x^2 - 3y^2 = 6$

$2(2y - 1)^2 - 3y^2 = 6$

$2(4y^2 - 4y + 1) - 3y^2 = 6$

$8y^2 - 8y + 2 - 3y^2 = 6$

$5y^2 - 8y - 4 = 0$

$(y - 2)(5y + 2) = 0$

$y - 2 = 0$	$5y + 2 = 0$
$y = 2$	$5y = -2$
	$y = -\frac{2}{5}$

Substituting 2 for y in $x = 2y - 1$:

$x = 2(2) - 1 = 3$

(3, 2) is a solution.

Substituting $-\frac{2}{5}$ for y in $x = 2y - 1$:

$x = 2\left(-\frac{2}{5}\right) - 1 = -\frac{4}{5} - 1 = -\frac{9}{5}$

$\left(-\frac{9}{5}, -\frac{2}{5}\right)$ is a solution.

$\left\{(3, 2), \left(-\frac{9}{5}, -\frac{2}{5}\right)\right\}$

6. $\left\{(4, -2), \left(-\frac{20}{7}, \frac{2}{7}\right)\right\}$

7. (1) $x + 4y \le 4$

(2) $3x + 2y > 2$

Boundary line for (1): $x + 4y = 4$

Points on line: (0, 1) and (4, 0)

The origin is in the correct half-plane for (1).

Boundary line for (2): $3x + 2y = 2$

Points on line: (0, 1) and (2, −2)

The origin is not in the correct half-plane for (2).

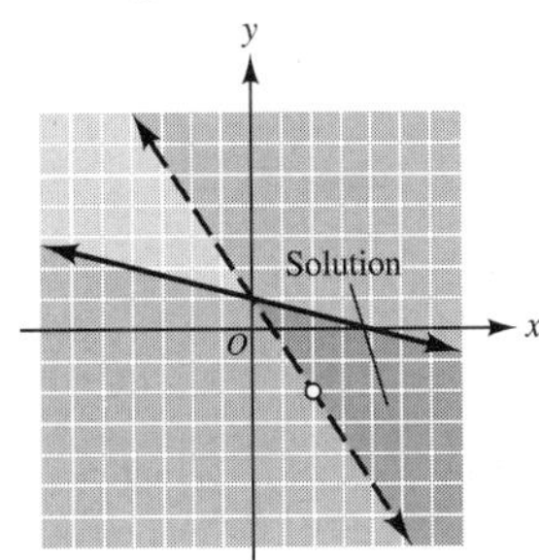

8.

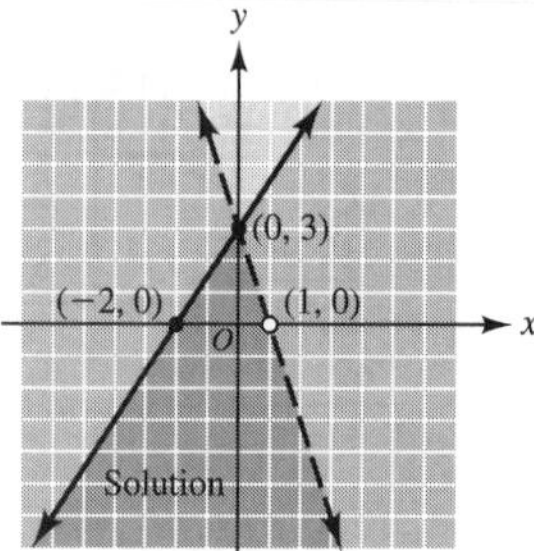

9. Boundary for $y \le 12 + 4x - x^2$ is $y = 12 + 4x - x^2$, which is a parabola. Its intercepts are $(-2, 0)$, $(6, 0)$, and $(0, 12)$, and its vertex is at $(2, 16)$; it is graphed with a solid line. Boundary for $x^2 + y^2 \le 16$ is $x^2 + y^2 = 16$, which is a circle of radius 4 with its center at the origin; it must be graphed with a solid line. Substituting $(0, 0)$ in $y \le 12 + 4x - x^2$ gives a true statement; we shade the region that contains the origin (the region "inside" the parabola). Substituting $(0, 0)$ in $x^2 + y^2 \le 16$ gives a true statement; we shade the region that contains the origin (the region *inside* the circle). The final answer is the heavily shaded region.

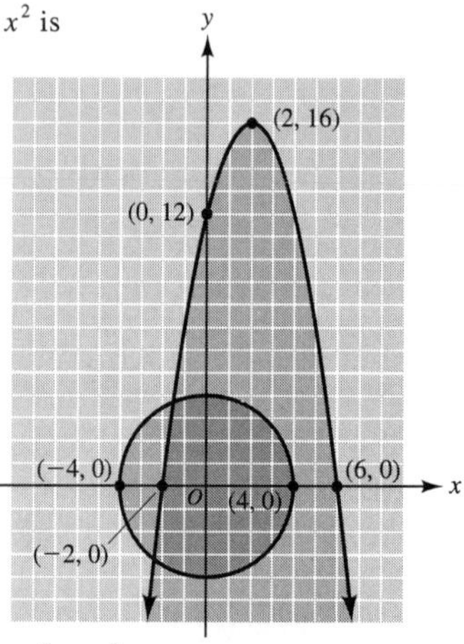

10. One pair is -4, 3, and another is -3, 4.

11. Let x = number of \$10 rolls
Let y = number of \$8 rolls
(1) $x + y = 80 \Rightarrow x = 80 - y$
(2) $10x + 8y = 730$
Substituting $80 - y$ for x in (2):

$$\begin{aligned} (2)\quad 10x + 8y &= 730 \\ 10(80 - y) + 8y &= 730 \\ 800 - 10y + 8y &= 730 \\ -2y &= -70 \\ y &= 35 \end{aligned}$$

Substituting 35 for y in $x = 80 - y$:
$$x = 80 - 35 = 45$$
The office bought 35 \$8 rolls and 45 \$10 rolls.

12. 24 mph is the average speed of Jennifer's boat; 8 mph is the average speed of the river.

Chapter 12 Diagnostic Test (page 659)

Following each problem number is the number (in parentheses) of the textbook section where that kind of problem is discussed.

1. (12.2) (1) $3x + 2y = 4$
If $x = 0$,
$3(0) + 2y = 4 \Rightarrow y = 2$
If $y = 0$,
$3x + 2(0) = 4 \Rightarrow x = \frac{4}{3}$
(2) $x - y = 3$
If $x = 0$,
$0 - y = 3 \Rightarrow y = -3$
If $y = 0$,
$x - 0 = 3 \Rightarrow x = 3$
$\{(2, -1)\}$

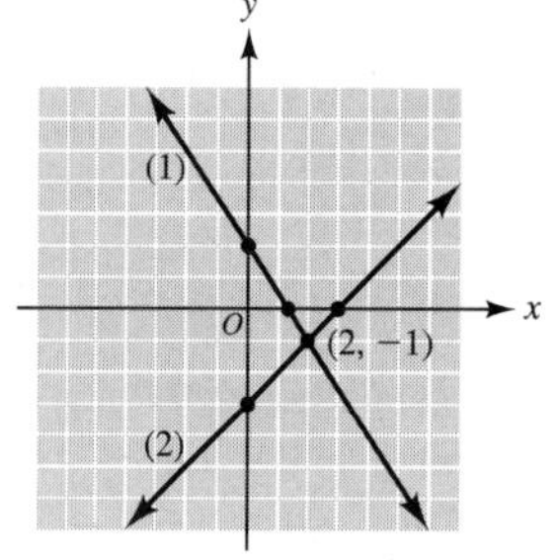

2. (12.3)

$$\begin{aligned} -2]\quad 4x - 3y = 13 &\Rightarrow -8x + 6y = -26 \\ 3]\quad 5x - 2y = 4 &\Rightarrow 15x - 6y = 12 \\ &\quad\ 7x \qquad\ \ = -14 \\ &\qquad\qquad\ \ x = -2 \end{aligned}$$

Substituting -2 for x in (2):

$$\begin{aligned} (2)\quad 5x - 2y &= 4 \\ 5(-2) - 2y &= 4 \\ -10 - 2y &= 4 \\ -2y &= 14 \\ y &= -7 \end{aligned}$$

$\{(-2, -7)\}$

3. (12.3 or 12.4) (1) $5x + 4y = 23$
(2) $3x + 2y = 9 \Rightarrow 2y = 9 - 3x$
$$y = \frac{9 - 3x}{2}$$
Substituting $\frac{9 - 3x}{2}$ for y in (1):

$$\begin{aligned} (1)\quad 5x + 4y &= 23 \\ 5x + \frac{\overset{2}{\cancel{4}}}{1}\left(\frac{9 - 3x}{\underset{1}{\cancel{2}}}\right) &= 23 \\ 5x + 2(9 - 3x) &= 23 \\ 5x + 18 - 6x &= 23 \\ -x &= 5 \\ x &= -5 \end{aligned}$$

Substituting -5 for x in $y = \frac{9 - 3x}{2}$:

$$\begin{aligned} y &= \frac{9 - 3(-5)}{2} \\ &= \frac{9 + 15}{2} = \frac{24}{2} = 12 \end{aligned}$$

$\{(-5, 12)\}$

4. (12.3)

$$\begin{aligned} 16 \quad 2]\quad 15x + 8y = -18 &\Rightarrow 30x + 16y = -36 \\ 8 \quad -1]\quad 9x + 16y = -8 &\Rightarrow -9x - 16y = +8 \\ &\quad\ 21x \qquad\ \ = -28 \\ &\qquad\qquad\ \ x = \frac{-28}{21} \\ &\qquad\qquad\quad\ = \frac{-4}{3} \end{aligned}$$

Substituting $\frac{-4}{3}$ for x in (1):

$$\begin{aligned} (1)\quad 15x + 8y &= -18 \\ \frac{\overset{5}{\cancel{15}}}{1}\left(\frac{-4}{\underset{1}{\cancel{3}}}\right) + 8y &= -18 \\ -20 + 8y &= -18 \\ 8y &= 2 \\ y &= \tfrac{2}{8} = \tfrac{1}{4} \end{aligned}$$

$\{(-\frac{4}{3}, \frac{1}{4})\}$

5. (12.3)

$$\begin{aligned} 4 \quad 2]\quad -10x + 35y = -18 &\Rightarrow -20x + 70y = -36 \\ 10 \quad 5]\quad 4x - 14y = 8 &\Rightarrow 20x - 70y = 40 \\ &\qquad\qquad\quad 0 = 4 \end{aligned}$$

False

$\{\ \}$

6. (12.5)

$$\begin{aligned} (1)\quad x + y + z &= 0 \\ (2)\quad 2x \qquad - 3z &= 5 \\ (3)\quad 3y + 4z &= 3 \\ 2(1) - (2):\quad 2y + 5z &= -5 \quad (4) \\ (3)\quad 3y + 4z &= 3 \\ 3(4) - 2(3):\quad 7z &= -21 \\ z &= -3 \end{aligned}$$

Substituting -3 for z in (2) $\Rightarrow x = -2$
Substituting -3 for z in (3) $\Rightarrow y = 5$
$\{(-2, 5, -3)\}$

7. (12.10) (1) $y^2 = 8x$

(2) $3x + y = 2 \Rightarrow y = 2 - 3x$

Substituting $2 - 3x$ for y in (1):

$$\begin{aligned}
(1) \qquad y^2 &= 8x \\
(2 - 3x)^2 &= 8x \\
4 - 12x + 9x^2 &= 8x \\
9x^2 - 20x + 4 &= 0 \\
(x - 2)(9x - 2) &= 0
\end{aligned}$$

$x - 2 = 0$ | $9x - 2 = 0$

$x = 2$ | $x = \frac{2}{9}$

If $x = 2$, $y = 2 - 3x = 2 - 3(2) = -4$

If $x = \frac{2}{9}$, $y = 2 - 3\left(\frac{2}{9}\right) = \frac{4}{3}$

$\left\{(2, -4), \left(\frac{2}{9}, \frac{4}{3}\right)\right\}$

8. (12.11) (1) $2x + 3y \le 6$

Boundary line $2x + 3y = 6$ is solid because equality is included.

If $y = 0$, $2x + 3(0) = 6 \Rightarrow x = 3$

If $x = 0$, $2(0) + 3y = 6 \Rightarrow y = 2$

Substituting (0, 0) in (1):

$2(0) + 3(0) \le 6$

$0 \le 6$ True

The half-plane containing (0, 0) is the solution of (1).

(2) $y - 2x < 2$

Boundary line $y - 2x = 2$ is dashed because equality is *not* included.

If $y = 0$, $0 - 2x = 2 \Rightarrow x = -1$

If $x = 0$, $y - 2(0) = 2 \Rightarrow y = 2$

Substituting (0, 0) in (2):

$0 - 2(0) < 2$

$0 < 2$ True

The half-plane containing (0, 0) is the solution of (2).

The final answer is the heavily shaded region.

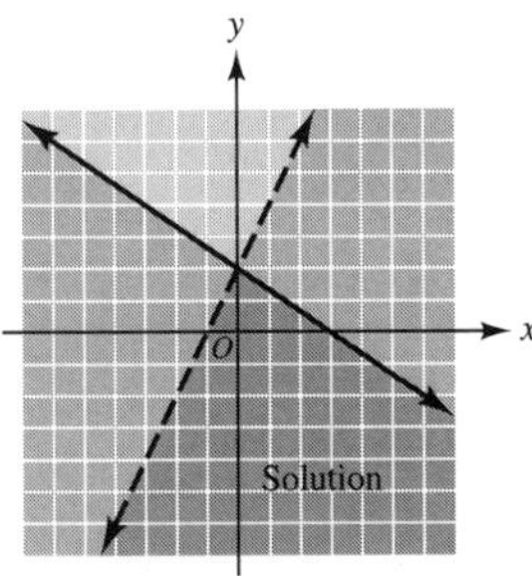

9. (12.6) Using the matrix method:

$$\begin{matrix} -1] \\ \llcorner\!\to \\ \end{matrix}\begin{bmatrix} 1 & -1 & -1 & 0 \\ 1 & 3 & 1 & 4 \\ 7 & -2 & -5 & 2 \end{bmatrix} \sim \begin{matrix} -7] \\ \\ \llcorner\!\to \end{matrix}\begin{bmatrix} 1 & -1 & -1 & 0 \\ 0 & 4 & 2 & 4 \\ 7 & -2 & -5 & 2 \end{bmatrix}$$

$$\sim \begin{matrix} \\ \frac{1}{4}] \\ \\ \end{matrix}\begin{bmatrix} 1 & -1 & -1 & 0 \\ 0 & 4 & 2 & 4 \\ 0 & 5 & 2 & 2 \end{bmatrix} \sim \begin{matrix} \\ -5] \\ \llcorner\!\to \end{matrix}\begin{bmatrix} 1 & -1 & -1 & 0 \\ 0 & 1 & \frac{1}{2} & 1 \\ 0 & 5 & 2 & 2 \end{bmatrix}$$

$$\sim \begin{matrix} \\ \\ -2] \end{matrix}\begin{bmatrix} 1 & -1 & -1 & 0 \\ 0 & 1 & \frac{1}{2} & 1 \\ 0 & 0 & -\frac{1}{2} & -3 \end{bmatrix} \sim \begin{bmatrix} 1 & -1 & -1 & 0 \\ 0 & 1 & \frac{1}{2} & 1 \\ 0 & 0 & 1 & 6 \end{bmatrix}$$

$z = 6$

$y + \frac{1}{2}(6) = 1$

$y = -2$

$x - 1(-2) - 1(6) = 0$

$x = 4$

(12.8) Using Cramer's rule:

$$D = \begin{vmatrix} 1 & -1 & -1 \\ 1 & 3 & 1 \\ 7 & -2 & -5 \end{vmatrix} = -2; \qquad D_x = \begin{vmatrix} 0 & -1 & -1 \\ 4 & 3 & 1 \\ 2 & -2 & -5 \end{vmatrix} = -8;$$

$$D_y = \begin{vmatrix} 1 & 0 & -1 \\ 1 & 4 & 1 \\ 7 & 2 & -5 \end{vmatrix} = 4; \qquad D_z = \begin{vmatrix} 1 & -1 & 0 \\ 1 & 3 & 4 \\ 7 & -2 & 2 \end{vmatrix} = -12$$

$$x = \frac{D_x}{D} = \frac{-8}{-2} = 4; \qquad y = \frac{D_y}{D} = \frac{4}{-2} = -2;$$

$$z = \frac{D_z}{D} = \frac{-12}{-2} = 6$$

$\{(4, -2, 6)\}$

10. (12.9) Let b = speed of boat (in mph)

Let c = speed of stream (in mph)

Then $b - c$ = speed upstream (in mph)

and $b + c$ = speed downstream (in mph)

$(b - c)\dfrac{\text{mi}}{\cancel{\text{hr}}}(2\ \cancel{\text{hr}})$ = distance upstream (in mi)

$(b + c)\dfrac{\text{mi}}{\cancel{\text{hr}}}(1\ \cancel{\text{hr}})$ = distance downstream (in mi)

(1) $2(b - c) = 30 \Rightarrow b - c = 15$

(2) $1(b + c) = 30 \Rightarrow \underline{b + c = 30}$

$2b = 45$

$b = \frac{45}{2}$, or $22\frac{1}{2}$

Substituting $\frac{45}{2}$ for b in (2):

(2) $b + c = 30$

$\frac{45}{2} + c = 30$

$c = \frac{60}{2} - \frac{45}{2} = \frac{15}{2}$, or $7\frac{1}{2}$

Check Speed downstream is $\left(22\frac{1}{2} + 7\frac{1}{2}\right)$ mph, or 30 mph; speed upstream is $\left(22\frac{1}{2} - 7\frac{1}{2}\right)$ mph, or 15 mph. Distance upstream is $15\dfrac{\text{mi}}{\text{hr}}(2\text{ hr}) = 30$ mi. Distance downstream is $30\dfrac{\text{mi}}{\text{hr}}(1\text{ hr}) = 30$ mi. Distances are equal. Therefore, the speed of the boat in still water is $22\frac{1}{2}$ mph, and the speed of the river is $7\frac{1}{2}$ mph.

Chapters 1–12 Cumulative Review Exercises (page 659)

1. $$\frac{x}{y(x-y)} + \frac{y}{x(y-x)} = \frac{x}{y(x-y)} \cdot \frac{x}{x} + \frac{-y}{x(x-y)} \cdot \frac{y}{y} = \frac{x^2 - y^2}{xy(x-y)} = \frac{(x+y)(x-y)}{xy(x-y)} = \frac{x+y}{xy}$$

2. $x^2 - 2x + 3 + \dfrac{29}{x-3}$

3. $(2^2x^2)^{1/6} = 2^{2/6}x^{2/6} = 2^{1/3}x^{1/3} = \sqrt[3]{2x}$

4. $\dfrac{|b|}{4|a|}\sqrt{6b}$

5. $m = \dfrac{4-2}{2-(-1)} = \dfrac{2}{3}$

$$\begin{aligned}
y - y_1 &= m(x - x_1) \\
y - 2 &= \tfrac{2}{3}[x - (-1)] \\
3(y - 2) &= 2(x + 1) \\
3y - 6 &= 2x + 2 \\
0 &= 2x - 3y + 8
\end{aligned}$$

6. $\{x \mid 2 < x < 6\}$

7. $4x^2 = 25$

$x^2 = \frac{25}{4}$

$x = \pm\sqrt{\frac{25}{4}}$

$x = \pm\frac{5}{2}$

$\left\{\frac{5}{2}, -\frac{5}{2}\right\}$

8. $\{5, 7\}$

9. $x^2 - 4x + 5 = 0;\ a = 1,\ b = -4,\ c = 5$

$$x = \frac{-(-4) \pm \sqrt{(-4)^2 - 4(1)(5)}}{2(1)} = \frac{4 \pm \sqrt{-4}}{2} = \frac{4 \pm 2i}{2} = 2 \pm i$$

$\{2 + i, 2 - i\}$

10. No **11.** $\frac{3}{5}$ **12.** $-\frac{5}{3}$

13. $-2]\ \ 4x + 3y = 8 \Rightarrow -8x - 6y = -16$
$\quad 8x + 7y = 12 \Rightarrow \ \ 8x + 7y = 12$
$\quad y = -4$

$4x + 3(-4) = 8$
$4x - 12 = 8$
$4x = 20$
$x = 5$
$\{(5, -4)\}$

14. { }

15.
(1) $2x + 3y + z = 4$
(2) $x + 4y - z = 0$
(3) $3x + y - z = -5$
(1) + (2): $3x + 7y = 4$ (4)
(1) + (3): $5x + 4y = -1$ (5)
$5]\ 3x + 7y = 4 \Rightarrow 15x + 35y = 20$ (6)
$-3]\ 5x + 4y = -1 \Rightarrow -15x - 12y = 3$ (7)
(6) + (7): $23y = 23$
$y = 1$

Substituting 1 for y in (4):
(4) $3x + 7y = 4$
$3x + 7(1) = 4$
$3x = -3$
$x = -1$

Substituting 1 for y and -1 for x in (1):
(1) $2x + 3y + z = 4$
$2(-1) + 3(1) + z = 4$
$-2 + 3 + z = 4$
$z = 3$
$\{(-1, 1, 3)\}$

16.

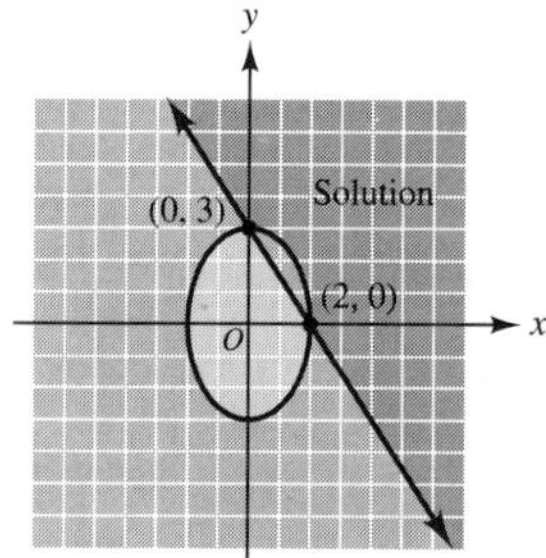

17. We must solve the inequality $12 - 4x - x^2 \geq 0$. Let $y = g(x) = (6 + x)(2 - x) = 0$. Critical points are -6 and 2.

Same signs
Sign of $6 + x$ – – – – – – +
Sign of $2 - x$ + – – – – – – – –
$-8\ \ -7\ \ -6\ \ -5\ \ -4\ \ -3\ \ -2\ \ -1\ \ 0\ \ 1\ \ 2\ \ 3\ \ 4$

We select the interval on which both signs are the same. The graph is
$-8\ \ -7\ \ -6\ \ -5\ \ -4\ \ -3\ \ -2\ \ -1\ \ 0\ \ 1\ \ 2\ \ 3\ \ 4$

The domain is $\{x \mid -6 \leq x \leq 2\}$.

The checks will not be shown for Exercises 18–21.

18. Length = 12 cm; width = 5 cm

19. Let x = number of hr for Jeannie to do the job alone
Then $x + 2$ = number of hr for Darryl to do the job alone

$\frac{1}{x}$ = Jeannie's rate (in jobs per hr)

$\frac{1}{x + 2}$ = Darryl's rate (in jobs per hr)

$$\left(\frac{1}{x}\right)(3) + \left(\frac{1}{x+2}\right)(4) = 1$$
$$\frac{3}{x} + \frac{4}{x + 2} = 1$$
$$x(x + 2)\left(\frac{3}{x} + \frac{4}{x + 2}\right) = 1(x)(x + 2)$$
$$3(x + 2) + 4x = x^2 + 2x$$
$$3x + 6 + 4x = x^2 + 2x$$
$$7x + 6 = x^2 + 2x$$
$$0 = x^2 - 5x - 6$$
$$0 = (x - 6)(x + 1)$$

$x - 6 = 0 \mid x + 1 = 0$
$x = 6 \mid x = -1$ Not in the domain
$x + 2 = 8$

It would take 8 hr for Darryl and 6 hr for Jeannie to do the job alone.

20. Barbara's average speed: 9 mph; Pat's average speed: 12 mph

21. Let x = number of L of pure alcohol to be added
There will be $(x + 3)$ L of the mixture.
$(1.00)x + 0.50(3) = 0.70(x + 3)$
$10x + 5(3) = 7(x + 3)$
$10x + 15 = 7x + 21$
$3x = 6$
$x = 2$
(The check will not be shown.) 2 L of pure alcohol should be added.

Exercises 13.1 (page 665)

1. 25, 30, 35 (Add 5 to preceding term.) **2.** 17, 20, 23

3. 9, 7, 5 (Add -2 to preceding term.) **4.** $2, \frac{7}{3}, \frac{8}{3}$

5. $a_n = n + 4$
$a_1 = 1 + 4 = 5$
$a_2 = 2 + 4 = 6$
$a_3 = 3 + 4 = 7$
The first three terms are 5, 6, 7.

6. 3, 5, 7, 9

7. $a_n = \frac{1 - n}{n}$
$a_1 = \frac{1 - (1)}{(1)} = 0$
$a_2 = \frac{1 - (2)}{(2)} = -\frac{1}{2}$
$a_3 = \frac{1 - (3)}{(3)} = -\frac{2}{3}$
$a_4 = \frac{1 - (4)}{(4)} = -\frac{3}{4}$
The first four terms are $0, -\frac{1}{2}, -\frac{2}{3}, -\frac{3}{4}$.

8. 0, 1, 3

9. $a_n = n^2 - 1$
$a_1 = 1^2 - 1 = 0$
$a_2 = 2^2 - 1 = 3$
$a_3 = 3^2 - 1 = 8$
The first three terms are 0, 3, 8.

10. 2, 9, 28, 65

11. $a_n = 2n - 3$
$a_1 = 2(1) - 3 = -1$
$a_2 = 2(2) - 3 = 1$
$a_3 = 2(3) - 3 = 3$
$a_4 = 2(4) - 3 = 5$
$S_4 = a_1 + a_2 + a_3 + a_4$
$= -1 + 1 + 3 + 5 = 8$

12. $10\frac{3}{4}$

13. $a_n = \dfrac{n-1}{n+1}$

$a_1 = \dfrac{1-1}{1+1} = 0$

$a_2 = \dfrac{2-1}{2+1} = \dfrac{1}{3}$

$a_3 = \dfrac{3-1}{3+1} = \dfrac{1}{2}$

$S_3 = 0 + \frac{1}{3} + \frac{1}{2} = \frac{5}{6}$

14. 45 **15.** $\frac{1}{7} + \frac{2}{7} + \frac{3}{7} + \frac{4}{7} + \frac{5}{7}$

16. $\frac{1}{11} + \frac{2}{11} + \frac{3}{11} + \frac{4}{11} + \frac{5}{11}$

17. $(2 \cdot 1 + 1)^2 + (2 \cdot 2 + 1)^2 + (2 \cdot 3 + 1)^2 + \cdots = 3^2 + 5^2 + 7^2 + \cdots$

18. $3^3 + 4^3 + 5^3 + \cdots$

Exercises 13.2 (page 670)

1. 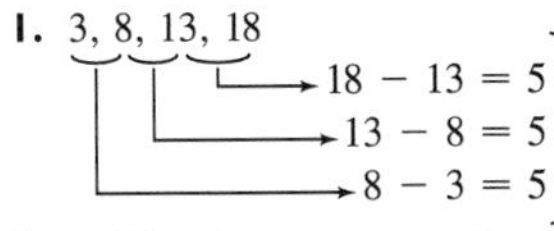

3, 8, 13, 18

$18 - 13 = 5$, $13 - 8 = 5$, $8 - 3 = 5$ — Arithmetic sequence because all the differences are the same; $d = 5$

2. Arithmetic sequence; $d = 4$

3. 7, 4, 1, −2

$-2 - 1 = -3$, $1 - 4 = -3$, $4 - 7 = -3$ — Arithmetic sequence because all the differences are the same; $d = -3$

4. Arithmetic sequence; $d = -5$

5. 4, $5\frac{1}{2}$, 7, 9

$9 - 7 = 2$, $7 - 5\frac{1}{2} = 1\frac{1}{2}$ — *Not* an arithmetic sequence because the differences are not all the same

6. Not an arithmetic sequence

7. $2x - 1, x, 1, -x + 2$

$-x + 2 - 1 = -x + 1$, $1 - x = -x + 1$, $x - (2x - 1) = -x + 1$ — Arithmetic sequence because all the differences are the same; $d = -x + 1$

8. Arithmetic sequence; $d = x - 1$

9. $a_1, a_1 + d, a_1 + 2d, a_1 + 3d$
$= 5, 5 + (-7), 5 + 2(-7), 5 + 3(-7)$
$= 5, -2, -9, -16$

10. 4, −1, −6, −11, −16

11. $-8, -2, 4, \ldots$

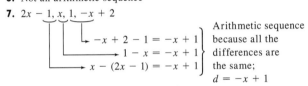

$4 - (-2) = 6 = d$
$-2 - (-8) = 6$

$a_{31} = a_1 + (31 - 1)d$
$a_{31} = -8 + 30(6) = 172$

12. 155

13. $x, 2x + 1, 3x + 2, \ldots$
$d = (2x + 1) - x = x + 1$
$a_{11} = a_1 + (11 - 1)d = x + 10(x + 1)$
$a_{11} = 11x + 10$

14. $10z - 7$

15. $a_5 = a_1 + (5 - 1)d$
$31 = 7 + 4d$
$24 = 4d$
$d = 6$
Arithmetic sequence: 7, 7 + 6, 7 + 2(6), 7 + 3(6), 7 + 4(6), or 7, 13, 19, 25, 31

16. 6, 15, 24, 33, 42, 51 **17.** $2 + 4 + \cdots + 100$
$S_{50} = \frac{50}{2}(2 + 100) = 2{,}550$

18. 2,500 **19.** $a_1 = 2$, $a_{100} = 200$, and $n = 100$;

$$S_{100} = \frac{\overset{50}{\cancel{100}}(2 + 200)}{\underset{1}{\cancel{2}}} = 10{,}100$$

20. 2,500

21. $a_1 = 3(1) - 4 = -1$, $a_{10} = 3(10) - 4 = 26$, and $n = 10$;

$$S_{10} = \frac{10(-1 + 26)}{2} = 5(25) = 125$$

22. −530

23. $a_1 = 5(3) = 15$, $a_8 = 5(8) = 40$, and $n = 8 - 3 + 1 = 6$;

$$S_6 = \frac{\overset{3}{\cancel{6}}(15 + 40)}{\underset{1}{\cancel{2}}} = 165$$

24. 270

25.
$a_6 = a_1 + (6 - 1)d \Rightarrow a_1 + 5d = 15 \quad (1)$
$a_{12} = a_1 + (12 - 1)d \Rightarrow a_1 + 11d = 39 \quad (2)$
(2) − (1): $6d = 24$
$d = 4$

Substituting 4 for d in (1):
$a_1 + 5(4) = 15$
$a_1 + 20 = 15$
$a_1 = -5$

26. $d = 5$
$a_1 = -8$

27. $16 = -5 + (n - 1)(3)$
$21 = 3n - 3$
$3n = 24$
$n = 8$
$S_8 = \frac{8}{2}(-5 + 16) = 4(11) = 44$

28. $n = 9$
$S_9 = 81$

29. $S_n = \dfrac{n}{2}(a_1 + a_n)$
$44 = \dfrac{n}{2}(5 + 17)$
$44 = \dfrac{n}{2}(22)$
$n = 4$
$a_4 = 5 + 3d$
$17 = 5 + 3d$
$3d = 12$
$d = 4$

30. $n = 8$
$d = \frac{39}{7}$

31. $a_n = a_1 + 8\left(\frac{3}{2}\right)$
$a_n = a_1 + 12$

$S_n = \dfrac{n}{2}(a_1 + a_n)$
$-\frac{9}{4} = \frac{9}{2}(a_1 + a_n)$
Substituting $a_1 + 12$ for a_n:
$-\frac{9}{4} = \frac{9}{2}(a_1 + a_1 + 12)$
$-\frac{9}{4} = \frac{9}{2}(2a_1 + 12)$
$-\frac{9}{4} = 9a_1 + 54$
$-9 = 36a_1 + 216$
$a_1 = -\frac{25}{4}$
Substituting $-\frac{25}{4}$ for a_1 in $a_n = a_1 + 12$:
$a_n = -\frac{25}{4} + \frac{48}{4} = \frac{23}{4}$

32. $a_1 = -\frac{3}{2}$
$a_n = a_7 = 3$

33. $a_1 = 16$, $a_2 = 48$ $\rangle$ $d = 48 - 16 = 32$
$a_3 = 80$
⋮
$a_{10} = a_1 + 9d = 16 + 9(32) = 16 + 288 = 304$
$a_{12} = 304 + 64 = 368$
$S_{12} = \frac{12}{2}(16 + 368) = 6(384) = 2{,}304$
The rock falls 304 ft during the tenth second and 2,304 ft during the first 12 sec.

34. \$12.40

Exercises 13.3 (page 676)

1. 4, 12, 36, 108

$\frac{108}{36} = 3$, $\frac{36}{12} = 3$, $\frac{12}{4} = 3$ Geometric sequence because the ratios are all equal; $r = 3$

2. Geometric sequence; $r = 2$

3. $-5, 15, -45, 135, \ldots$

$\frac{135}{-45} = -3$, $\frac{-45}{15} = -3$, $\frac{15}{-5} = -3$ Geometric sequence because the ratios are all equal; $r = -3$

4. Geometric sequence; $r = -4$

5. $2, \frac{1}{2}, \frac{1}{8}, \frac{1}{16}$

$\frac{1}{16} \div \frac{1}{8} = \frac{1}{16} \cdot \frac{8}{1} = \frac{1}{2}$, $\frac{1}{8} \div \frac{1}{2} = \frac{1}{8} \cdot \frac{2}{1} = \frac{1}{4}$ *Not* a geometric sequence because all ratios are not equal

6. Not a geometric sequence

7. $5x, 10xy, 20xy^2, 40xy^3, \ldots$

$\frac{40xy^3}{20xy^2} = 2y$, $\frac{20xy^2}{10xy} = 2y$, $\frac{10xy}{5x} = 2y$ Geometric sequence because the ratios are all equal; $r = 2y$

8. Geometric sequence; $r = 3z$

9. $12 - 36 = -24$; $4 - 12 = -8$, the sequence is not arithmetic. $12 \div 36 = \frac{1}{3}$; $4 \div 12 = \frac{1}{3}$; $\frac{4}{3} \div 4 = \frac{4}{3} \cdot \frac{1}{4} = \frac{1}{3}$; the sequence is geometric.

10. Neither

11. $7 - 5 = 2$; $10 - 7 = 3$; the sequence is not arithmetic. $7 \div 5 = \frac{7}{5}$; $10 \div 7 = \frac{10}{7}$; the sequence is not geometric. Neither

12. Geometric

13. $a_1 = 12$
$a_2 = 12\left(\frac{1}{3}\right) = 4$
$a_3 = 4\left(\frac{1}{3}\right) = \frac{4}{3}$
$a_4 = \frac{4}{3}\left(\frac{1}{3}\right) = \frac{4}{9}$
$a_5 = \frac{4}{9}\left(\frac{1}{3}\right) = \frac{4}{27}$
$12, 4, \frac{4}{3}, \frac{4}{9}, \frac{4}{27}$

14. $8, 12, 18, 27, 40\frac{1}{2}$

15. $-9, -6, -4, \ldots$

$r = \frac{-4}{-6} = \frac{2}{3}$

$a_7 = a_1r^6 = -9\left(\frac{2}{3}\right)^6 = -\frac{64}{81}$

16. $-\frac{6{,}561}{32}$

17. $16x, 8xy, 4xy^2, \ldots$

$\frac{8xy}{16x} = \frac{y}{2} = r$

$a_8 = a_1r^7 = 16x\left(\frac{y}{2}\right)^7 = \frac{xy^7}{8}$

18. $\frac{x^{12}y}{27}$

19. $a_5 = a_1r^4$
$54 = \frac{2}{3}r^4$
$81 = r^4$
$r = \pm 3$
One solution: $r = 3$
$\frac{2}{3}, 2, 6, 18, 54$
The other solution: $r = -3$
$\frac{2}{3}, -2, 6, -18, 54$

20. One solution: $r = 5$
$\frac{3}{25}, \frac{3}{5}, 3, 15, 75$
The other solution: $r = -5$
$\frac{3}{25}, -\frac{3}{5}, 3, -15, 75$

21. $a_1 = 3$, $r = \frac{1}{3}$, and $n = 5$;

$$S_5 = \frac{3\left[1 - \left(\frac{1}{3}\right)^5\right]}{1 - \frac{1}{3}} = \frac{3\left(1 - \frac{1}{243}\right)}{\frac{2}{3}} = \frac{3\left(\frac{242}{243}\right)}{\frac{2}{3}} = \frac{242}{81} \cdot \frac{3}{2} = \frac{121}{27}$$

22. $\frac{635}{64}$

23. $a_1 = \frac{3}{4}$, $r = \frac{3}{4}$, and $n = 6$;

$$S_6 = \frac{\frac{3}{4}\left[1 - \left(\frac{3}{4}\right)^6\right]}{1 - \frac{3}{4}} = \frac{\frac{3}{4}\left[1 - \frac{729}{4{,}096}\right]}{\frac{1}{4}} = \frac{3}{4}\left[\frac{4{,}096 - 729}{4{,}096}\right] \cdot \frac{4}{1}$$
$$= \frac{3(3{,}367)}{4{,}096} = \frac{10{,}101}{4{,}096}$$

24. $\frac{21{,}845}{524{,}288}$

25. $a_5 = a_1r^4$
$80 = a_1\left(\frac{2}{3}\right)^4$
$80 = \frac{16}{81}a_1$
$a_1 = 405$
$$S_5 = \frac{a_1(1 - r^5)}{1 - r}$$
$$= \frac{405\left[1 - \left(\frac{2}{3}\right)^5\right]}{1 - \frac{2}{3}} = \frac{405\left[1 - \frac{32}{243}\right]}{\frac{1}{3}} = \frac{405}{1}\left(\frac{211}{243}\right) \cdot \frac{3}{1} = 1{,}055$$

26. $a_1 = 5$
$S_5 = 155$

27. $a_3 = a_1r^2 = 28$ (1)
$a_5 = a_1r^4 = \frac{112}{9}$ (2)
Dividing (2) by (1):
$r^2 = \frac{112}{9} \div 28 = \frac{112}{9} \cdot \frac{1}{28} = \frac{4}{9}$
$r = \pm\frac{2}{3}$
One solution: Substituting $\frac{2}{3}$ for r in (1):
$a_1\left(\frac{2}{3}\right)^2 = 28$
$\frac{4}{9}a_1 = 28$
$a_1 = 63$
$$S_5 = \frac{63\left[1 - \left(\frac{2}{3}\right)^5\right]}{1 - \frac{2}{3}} = \frac{63 - \frac{224}{27}}{\frac{1}{3}} = \frac{1{,}477}{27} \cdot \frac{3}{1} = \frac{1{,}477}{9}$$
The other solution: Substituting $-\frac{2}{3}$ for r in (1):
$a_1\left(-\frac{2}{3}\right)^2 = 28$
$a_1 = 63$
$$S_5 = \frac{63\left[1 - \left(-\frac{2}{3}\right)^5\right]}{1 - \left(-\frac{2}{3}\right)} = \frac{63 - \left(-\frac{224}{27}\right)}{\frac{5}{3}} = \frac{1{,}925}{27} \cdot \frac{3}{5} = \frac{385}{9}$$

28. One solution: $r_1 = \frac{1}{4}$
$a_1 = 1{,}536$
$S_4 = 2{,}040$

The other solution: $r_2 = -\frac{1}{4}$
$a_1 = -1{,}536$
$S_4 = -1{,}224$

29. $a_n = a_1r^{n-1}$
$3 = a_1\left(\frac{1}{2}\right)^{n-1}$
(1) $\frac{3}{2} = a_1\left(\frac{1}{2}\right)^n$
$$S_n = \frac{a_1(1 - r^n)}{1 - r}$$
$$189 = \frac{a_1\left[1 - \left(\frac{1}{2}\right)^n\right]}{1 - \frac{1}{2}} = \frac{a_1 - a_1\left(\frac{1}{2}\right)^n}{\frac{1}{2}}$$
From (1)
$$189 = \frac{a_1 - \frac{3}{2}}{\frac{1}{2}} = 2a_1 - 3$$
$192 = 2a_1$
$a_1 = 96$
Substituting 96 for a_1 in (1):
$\frac{3}{2} = 96\left(\frac{1}{2}\right)^n$
$\left(\frac{1}{2}\right)^6 = \left(\frac{1}{2}\right)^n$ $\frac{3}{2} \div 96 = \frac{1}{64} = \left(\frac{1}{2}\right)^6$
$6 = n$

30. $a_1 = 108$
$n = 4$

31. $S_3 = \dfrac{a_1\left(1 - \left[\frac{6}{5}\right]^3\right)}{1 - \frac{6}{5}}$

$= a_1\left(-\frac{91}{125}\right)\left(-\frac{5}{1}\right) = \frac{91}{25}a_1$

$\frac{91}{25}a_1 = 22{,}750$

$a_1 = \$6{,}250$

$a_5 = a_1 r^4$

$= 6{,}250\left(\frac{6}{5}\right)^4 = 6{,}250\left(\frac{1{,}296}{625}\right)$

$= \$12{,}960$

32. $6,250

33. $a_n = a_1 r^{n-1}$

a. $a_{10} = 1(2)^9 = 512¢ = \$5.12$

b. $a_{31} = 1(2)^{30} = 1{,}073{,}741{,}824¢ = \$10{,}737{,}418.24$

c. $S_{31} = \dfrac{a_1(1 - r^n)}{1 - r} = \dfrac{1(1 - 2^{31})}{1 - 2} = 2^{31} - 1$

$= 2{,}147{,}483{,}646¢ = \$21{,}474{,}836.46$

Exercises 13.4 (page 680)

1. $3 + 1 + \frac{1}{3} + \cdots$; $\frac{1}{3} = r$

$S_\infty = \dfrac{a_1}{1 - r} = \dfrac{3}{\frac{2}{3}} = \dfrac{9}{2}$

2. $\frac{81}{10}$

3. $\frac{4}{3} + 1 + \frac{3}{4} + \cdots$; $\dfrac{\frac{3}{4}}{1} = \dfrac{3}{4} = r$

$S_\infty = \dfrac{a_1}{1 - r} = \dfrac{\frac{4}{3}}{1 - \frac{3}{4}} = \dfrac{\frac{4}{3}}{\frac{1}{4}} = \dfrac{16}{3}$

4. $\frac{1}{9}$

5. $-6 - 4 - \frac{8}{3} - \cdots$; $\dfrac{-4}{-6} = \dfrac{2}{3} = r$

$S_\infty = \dfrac{a_1}{1 - r} = \dfrac{-6}{1 - \frac{2}{3}} = \dfrac{-6}{\frac{1}{3}} = -18$

6. $-\frac{343}{2}$

7. $0.\overline{2} = 0.222\ldots$

$= 0.2 + 0.02 + 0.002 + 0.0002 + \cdots$; $\dfrac{0.02}{0.2} = 0.1 = r$

$S_\infty = \dfrac{a_1}{1 - r} = \dfrac{0.2}{1 - 0.1} = \dfrac{0.2}{0.9} = \dfrac{2}{9}$

8. $\frac{7}{33}$

9. $0.0\overline{54} = 0.05454\ldots$

$= 0.054 + 0.00054 + \cdots$; $\dfrac{0.00054}{0.054} = 0.01 = r$

$S_\infty = \dfrac{a_1}{1 - r} = \dfrac{0.054}{1 - 0.01} = \dfrac{0.054}{0.99} = \dfrac{54}{990} = \dfrac{3}{55}$

10. $\frac{13}{330}$

11. $8.6\overline{4} = 8.6444\ldots$

$= 8.6 + 0.04 + 0.004 + 0.0004 + \cdots$; $r = 0.1$

$S_\infty = \dfrac{0.04}{1 - 0.1} = \dfrac{0.04}{0.9} = \dfrac{4}{90} = \dfrac{2}{45}$

$8.6444\ldots = \frac{8}{1} + \frac{6}{10} + \frac{2}{45}$

$= \frac{720}{90} + \frac{54}{90} + \frac{4}{90} = \frac{778}{90} = \frac{389}{45}$

12. $\frac{79}{15}$

13. $a_1 = 5$, $r = -\frac{3}{4}$, and $\left|-\frac{3}{4}\right| < 1$

$S_\infty = \dfrac{5}{1 - \left(-\frac{3}{4}\right)} = \dfrac{5}{\frac{7}{4}} = \dfrac{20}{7}$

14. $\frac{5}{3}$

15. The geometric series of the *heavy* lines is $6 + 4 + \frac{8}{3} + \cdots$.

$S_\infty = \dfrac{a_1}{1 - r} = \dfrac{6}{1 - \frac{2}{3}} = \dfrac{6}{\frac{1}{3}} = 18$

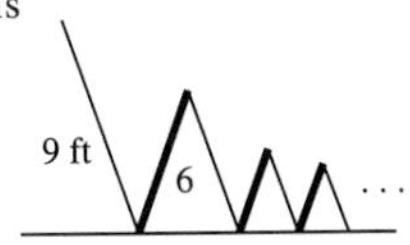

This distance is doubled to include all but the first drop: $2(18) = 36$.
The total distance traveled = 9 ft + 36 ft = 45 ft.

16. 40 ft

17.

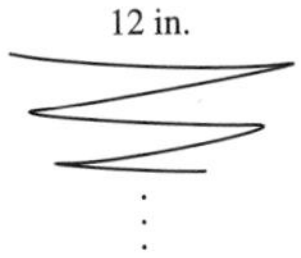

From the sketch, we see that we have a different situation from the bouncing ball described in Exercise 15. This is a single geometric series.

$S_\infty = \dfrac{12}{1 - \frac{9}{10}} = \dfrac{12}{\frac{1}{10}} = 120$

The total distance is 120 in.

Sections 13.1–13.4 Review Exercises (page 681)

1. $7, 5, 3, \ldots$

$3 - 5 = -2$
$5 - 7 = -2$ } Arithmetic sequence; $d = -2$

2. Geometric sequence; $r = 3$

3. $\frac{1}{2}, \frac{1}{4}, \frac{1}{6}, \cdots$

$\frac{1}{6} - \frac{1}{4} = -\frac{1}{12}$
$\frac{1}{4} - \frac{1}{2} = -\frac{1}{4}$ } *Not* an arithmetic sequence

$\dfrac{\frac{1}{6}}{\frac{1}{4}} \neq \dfrac{\frac{1}{4}}{\frac{1}{2}}$ } *Not* a geometric sequence

Neither

4. Geometric sequence; $r = -\frac{1}{3}$

5. $3x - 2, 2x - 1, x$

$x - (2x - 1) = -x + 1$
$(2x - 1) - (3x - 2) = -x + 1$ } Arithmetic sequence; $d = -x + 1$

6. $-4, -2, 0, 2$
$d = 2$
$a_{30} = 54$
$S_{30} = 750$

7. $a_n = \left(\frac{1}{2}\right)^n$

$a_1 = \left(\frac{1}{2}\right)^1 = \frac{1}{2}$

$a_2 = \left(\frac{1}{2}\right)^2 = \frac{1}{4}$

$a_3 = \left(\frac{1}{2}\right)^3 = \frac{1}{8}$

$\frac{1}{2}, \frac{1}{4}, \frac{1}{8}$

$S_n = \dfrac{a_1(1 - r^n)}{1 - r}$

$S_5 = \dfrac{\frac{1}{2}\left(1 - \frac{1}{32}\right)}{\frac{1}{2}} = \dfrac{31}{32}$

$S_\infty = \dfrac{a_1}{1 - r} = \dfrac{\frac{1}{2}}{1 - \frac{1}{2}} = 1$

8. $n = 16$
$d = \frac{4}{5}$

9. $a_n = a_1 + 6\left(\frac{3}{2}\right) = a_1 + 9$

$S_n = \dfrac{n(a_1 + a_n)}{2}$

$\dfrac{7}{2} = \dfrac{7(2a_1 + 9)}{2}$

$1 = 2a_1 + 9$

$-8 = 2a_1$

$-4 = a_1$

$a_n = a_7 = -4 + 9 = 5$

10. $a_1 = \frac{81}{16}$
$S_6 = \frac{133}{48}$

11. $a_5 = \frac{32}{9} = a_1 r^4$ (1)

$a_3 = 8 = a_1 r^2$ (2)

Dividing (1) by (2):

$\frac{4}{9} = r^2$

$\pm\frac{2}{3} = r$

One solution: Substituting $\frac{2}{3}$ for r in (2):

$8 = a_1\left(\frac{2}{3}\right)^2$

$8 = a_1 \cdot \frac{4}{9}$

$18 = a_1$

$S_5 = \dfrac{a_1(1 - r^5)}{1 - r} = \dfrac{18\left[1 - \left(\frac{2}{3}\right)^5\right]}{1 - \frac{2}{3}} = \dfrac{18\left(1 - \frac{32}{243}\right)}{\frac{1}{3}} = \dfrac{18}{1}\left(\dfrac{211}{243}\right) \cdot \dfrac{3}{1}$

$= \frac{422}{9}$

The other solution: Substituting $-\frac{2}{3}$ for r in (2):

$$8 = a_1\left(-\tfrac{2}{3}\right)^2$$
$$8 = a_1 \cdot \tfrac{4}{9}$$
$$18 = a_1$$
$$S_5 = \frac{18\left[1 - \left(-\frac{2}{3}\right)^5\right]}{1 - \left(-\frac{2}{3}\right)} = \frac{18\left(1 + \frac{32}{243}\right)}{\frac{5}{3}} = \frac{18}{1}\left(\frac{275}{243}\right)\cdot\frac{3}{5} = \frac{110}{9}$$

12. $\dfrac{32{,}813}{9{,}990}$

13. Heavy lines: $6 + \frac{9}{2} + \cdots$

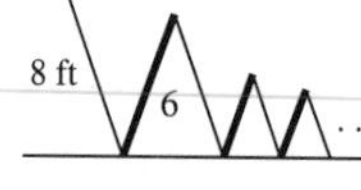

$$S_\infty = \frac{6}{1 - \frac{3}{4}} = \frac{6}{\frac{1}{4}} = 24$$

Total distance $= 8 \text{ ft} + 2(24 \text{ ft})$
$= 56 \text{ ft}$

14. $\frac{9}{11} + \frac{12}{11} + \frac{15}{11} + \frac{18}{11} + \frac{21}{11}$

15. The series is geometric; $a_1 = 2$, $r = \frac{1}{4}$, and $n = 6$;

$$S_6 = \frac{2\left[1 - \left(\frac{1}{4}\right)^6\right]}{1 - \frac{1}{4}} = \frac{2\left(1 - \frac{1}{4{,}096}\right)}{\frac{3}{4}} = \frac{2\left(\frac{4{,}095}{4{,}096}\right)}{\frac{3}{4}} = \frac{4{,}095}{2{,}048}\cdot\frac{4}{3} = \frac{1{,}365}{512}$$

Chapter 13 Diagnostic Test (page 685)

Following each problem number is the number (in parentheses) of the textbook section where that kind of problem is discussed.

1. (13.1) $a_n = \dfrac{2n-1}{n}$ (The sequence is not arithmetic and it is not geometric.)

$$S_4 = \frac{2(1)-1}{1} + \frac{2(2)-1}{2} + \frac{2(3)-1}{3} + \frac{2(4)-1}{4}$$
$$= 1 + \tfrac{3}{2} + \tfrac{5}{3} + \tfrac{7}{4} = \tfrac{71}{12}$$

2. (13.2, 13.3) **a.** $8, -20, 50, \ldots$

$\dfrac{50}{-20} = -\dfrac{5}{2}$, $\dfrac{-20}{8} = -\dfrac{5}{2}$ Geometric sequence; $r = -\frac{5}{2}$

b. $\frac{1}{2}, \frac{3}{4}, 1, \frac{5}{4}, \frac{3}{2}, \ldots$

$1 - \frac{3}{4} = \frac{1}{4}$, $\frac{3}{4} - \frac{1}{2} = \frac{1}{4}$ Arithmetic sequence; $d = \frac{1}{4}$

(Remaining differences are also $\frac{1}{4}$.)

c. $2x - 1, 3x, 4x + 2, \ldots$

$4x + 2 - 3x = x + 2$
$3x - (2x - 1) = x + 1$ *Not* an arithmetic sequence

$\dfrac{4x+2}{3x} \neq \dfrac{3x}{2x-1}$ *Not* a geometric sequence

Neither

d. $\dfrac{c^4}{16}, -\dfrac{c^3}{8}, \dfrac{c^2}{4}, \ldots$

$\dfrac{c^2}{4} \div \left(-\dfrac{c^3}{8}\right) = \dfrac{c^2}{4}\cdot\left(-\dfrac{8}{c^3}\right) = -\dfrac{2}{c}$

$-\dfrac{c^3}{8} \div \dfrac{c^4}{16} = -\dfrac{c^3}{8}\cdot\dfrac{16}{c^4} = -\dfrac{2}{c}$ Geometric sequence; $r = -\dfrac{2}{c}$

3. (13.2)

a. $x + 1, (x + 1) + (x - 1), (x + 1) + 2(x - 1),$
$(x + 1) + 3(x - 1), (x + 1) + 4(x - 1)$
$= x + 1, 2x, 3x - 1, 4x - 2, 5x - 3$

b. $a_5 = a_1 + 4d$
$-2 = 2 + 4d$
$-4 = 4d$
$-1 = d$
$2, 2 + (-1), 2 + 2(-1), 2 + 3(-1), 2 + 4(-1)$
$= 2, 1, 0, -1, -2$

c. $1 - 6h, 2 - 4h, 3 - 2h, \ldots$
$(2 - 4h) - (1 - 6h) = 1 + 2h = d$
$a_{15} = a_1 + 14d = (1 - 6h) + 14(1 + 2h) = 15 + 22h$

4. (13.2) **a.**
$a_7 = a_1 + 6d \Rightarrow 2 = a_1 + 6d$ (1)
$a_3 = a_1 + 2d \Rightarrow 1 = a_1 + 2d$ (2)
(1) − (2): $1 = 4d$
$\frac{1}{4} = d$

Substituting $\frac{1}{4}$ for d in (2):
(2) $1 = a_1 + 2\left(\frac{1}{4}\right)$
$1 = a_1 + \frac{1}{2}$
$\frac{1}{2} = a_1$

b. $S_n = \dfrac{n}{2}(a_1 + a_n)$
$7 = \dfrac{n}{2}(10 - 8)$
$7 = n$
$a_n = a_1 + (n - 1)d$
$-8 = 10 + 6d$
$-18 = 6d$
$-3 = d$

5. (13.3) $\dfrac{c^4}{16}, \dfrac{c^4}{16}\left(-\dfrac{2}{c}\right), \dfrac{c^4}{16}\left(-\dfrac{2}{c}\right)^2, \dfrac{c^4}{16}\left(-\dfrac{2}{c}\right)^3, \dfrac{c^4}{16}\left(-\dfrac{2}{c}\right)^4$
$= \dfrac{c^4}{16}, -\dfrac{c^3}{8}, \dfrac{c^2}{4}, -\dfrac{c}{2}, 1$

6. (13.3) $a_4 = a_1r^3 \Rightarrow -8 = a_1r^3$ (1)
$a_2 = a_1r \Rightarrow -18 = a_1r$ (2)
Dividing (1) by (2):
$\dfrac{-8}{-18} = r^2$
$r^2 = \frac{4}{9}$
$r = \pm\frac{2}{3}$

One solution: Substituting $\frac{2}{3}$ for r in (2):
(2) $-18 = a_1r$
$-18 = a_1\left(\frac{2}{3}\right)$
$a_1 = -27$
$$S_4 = \frac{a_1(1 - r^4)}{1 - r} = \frac{-27\left(1 - \frac{16}{81}\right)}{1 - \left(\frac{2}{3}\right)}$$
$$S_4 = \frac{-27\left(\frac{65}{81}\right)}{\frac{1}{3}} = -65$$

The other solution: Substituting $-\frac{2}{3}$ for r in (2):
(2) $-18 = a_1r$
$-18 = a_1\left(-\frac{2}{3}\right)$
$a_1 = 27$
$$S_4 = \frac{a_1(1 - r^4)}{1 - r} = \frac{27\left(1 - \frac{16}{81}\right)}{1 - \left(-\frac{2}{3}\right)}$$
$$S_4 = \frac{27\left(\frac{65}{81}\right)}{\frac{5}{3}} = \frac{65}{5} = 13$$

7. (13.3)
$a_n = a_1r^{n-1}$
$a_nr = a_1r^n$
$-16\left(-\frac{2}{3}\right) = a_1\left(-\frac{2}{3}\right)^n$
(1) $\frac{32}{3} = a_1\left(-\frac{2}{3}\right)^n$
$$S_n = \frac{a_1(1 - r^n)}{1 - r}$$
$$26 = \frac{a_1\left[1 - \left(-\frac{2}{3}\right)^n\right]}{1 - \left(-\frac{2}{3}\right)}$$
From (1)
$$26 = \frac{a_1 - a_1\left(-\frac{2}{3}\right)^n}{\frac{5}{3}} = \frac{a_1 - \frac{32}{3}}{\frac{5}{3}}$$
$$26 = \frac{3a_1 - 32}{5}$$
$130 = 3a_1 - 32$
$54 = a_1$

Substituting 54 for a_1 in (1):

$$\frac{32}{3} = 54\left(-\frac{2}{3}\right)^n$$
$$\frac{16}{81} = \left(-\frac{2}{3}\right)^n$$
$$\left(\frac{2}{3}\right)^4 = \left(-\frac{2}{3}\right)^n \qquad \left(\frac{2}{3}\right)^4 = \left(-\frac{2}{3}\right)^4$$
$$4 = n$$

8. (13.4) $8, -4, 2, -1, \ldots$

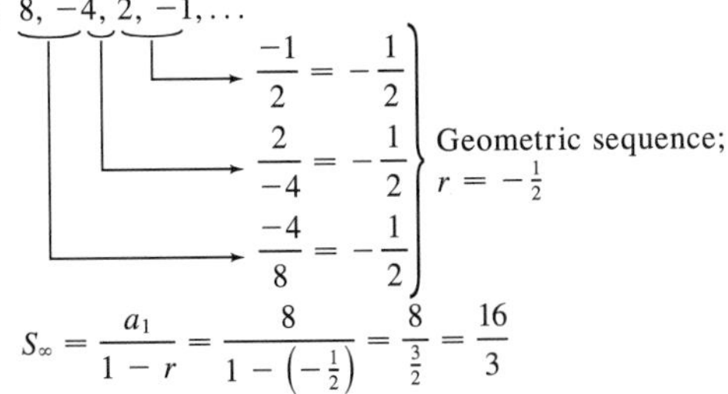

$$S_\infty = \frac{a_1}{1 - r} = \frac{8}{1 - \left(-\frac{1}{2}\right)} = \frac{8}{\frac{3}{2}} = \frac{16}{3}$$

9. (13.4) $3.0\overline{3} = 3.0333\ldots = 3 + 0.0333\ldots$

$$0.0333\ldots = 0.03 + 0.003 + 0.0003 + \cdots$$
$$\frac{0.0003}{0.003} = \frac{1}{10} = r$$
$$0.0333\ldots = \frac{0.03}{1 - \frac{1}{10}} = \frac{0.03}{0.9} = \frac{3}{90} = \frac{1}{30}$$
$$3.0333\ldots = 3 + \frac{1}{30} = \frac{91}{30}$$

10. (13.4) The series of heavy lines:

$$S_\infty = \frac{4}{1 - \frac{2}{3}} = \frac{4}{\frac{1}{3}} = 12$$

Total distance $= 6 \text{ ft} + 2(12 \text{ ft})$
$= 6 \text{ ft} + 24 \text{ ft} = 30 \text{ ft}$

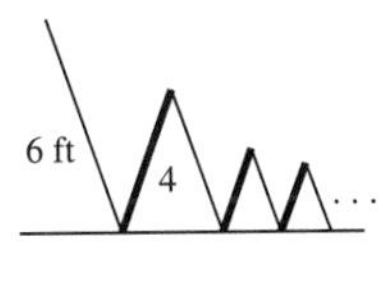

Chapters 1–13 Cumulative Review Exercises (page 685)

1. $\left(\frac{x^5z^2}{y^4}\right)^{-2} = \left(\frac{y^4}{x^5z^2}\right)^2 = \frac{y^8}{x^{10}z^4}$ **2.** $\frac{1}{64}a^{1/5}$

3. $\sqrt{27x^3y^4} + 3xy^2\sqrt{12x} - x^2y^2\sqrt{\frac{3 \cdot x}{x \cdot x}}$

$$= 3xy^2\sqrt{3x} + 3xy^2(2)\sqrt{3x} - \frac{x^2y^2}{x}\sqrt{3x}$$
$$= 3xy^2\sqrt{3x} + 6xy^2\sqrt{3x} - xy^2\sqrt{3x} = 8xy^2\sqrt{3x}$$

4. $3(\sqrt{7} + \sqrt{2})$, or $3\sqrt{7} + 3\sqrt{2}$

5. $\frac{(6 - 5i)(6 - 5i)}{(6 + 5i)(6 - 5i)} = \frac{36 - 60i + 25i^2}{36 + 25}$
$= \frac{(36 - 25) - 60i}{61} = \frac{11}{61} - \frac{60}{61}i$

6. $\frac{(x + 1)(x - 1)}{x^2 - 2x + 4}$

7. $\frac{x(2)(x + 3) - 9(2)(x - 3) - 1}{2(x - 1)(x - 3)}$
$= \frac{2x^2 + 6x - 18x + 54 - 1}{2(x - 3)(x + 3)}$
$= \frac{2x^2 - 12x + 53}{2x^2 - 18}$

8. $x^2 + 4x - 3 + \frac{-9x + 4}{x^2 - x + 1}$

9. $\frac{x + 1}{(3x - 1)(x + 5)} \cdot \frac{(3x - 1)(2x - 3)}{(2x - 3)(x + 1)} = \frac{1}{x + 5}$

10. $x^5 + 10x^4 + 40x^3 + 80x^2 + 80x + 32$

11. $(x^2 - 4)\left(\frac{6 - x}{x^2 - 4} - \frac{x}{x + 2}\right) = 2(x^2 - 4)$

$$6 - x - x(x - 2) = 2x^2 - 8$$
$$6 - x - x^2 + 2x = 2x^2 - 8$$
$$0 = 3x^2 - x - 14$$
$$0 = (3x - 7)(x + 2)$$

$3x - 7 = 0 \;\big|\; x + 2 = 0$
$3x = 7 \;\big|\; x = -2$ Not in the domain
$x = \frac{7}{3}$

(The check for $\frac{7}{3}$ will not be shown.) The solution set is $\left\{\frac{7}{3}\right\}$.

12. $\{3\}$

13.
$$x - 3 = \sqrt{x + 9}$$
$$(x - 3)^2 = (\sqrt{x + 9})^2$$
$$x^2 - 6x + 9 = x + 9$$
$$x^2 - 7x = 0$$
$$x(x - 7) = 0$$

$x = 0 \;\big|\; x - 7 = 0$
$ \;\big|\; x = 7$

Check for $x = 0$
$0 \stackrel{?}{=} \sqrt{0 + 9} + 3$
$0 \stackrel{?}{=} \sqrt{9} + 3$
$0 \neq 3 + 3$

Check for $x = 7$
$7 \stackrel{?}{=} \sqrt{7 + 9} + 3$
$7 \stackrel{?}{=} \sqrt{16} + 3$
$7 = 4 + 3$

The solution set is $\{7\}$.

14. $\{x \mid x \le -5 \text{ or } x \ge 3\}$, or, in interval notation, $(-\infty, -5] \cup [3, +\infty)$

15. a. $\log_5 \frac{1}{25} = x$
$5^x = \frac{1}{25} = 5^{-2}$
$x = -2$
$\{-2\}$

b. $\log_b 4 = \frac{1}{2}$
$b^{1/2} = 4$
$(b^{1/2})^2 = 4^2$
$b = 16$
$\{16\}$

(The checks will not be shown.)

16. $\{2 + 2\sqrt{3}, 2 - 2\sqrt{3}\}$

17.
3] $2x + 4y = 0 \Rightarrow 6x + 12y = 0$
4] $5x - 3y = 13 \Rightarrow 20x - 12y = 52$

$$26x = 52$$
$$x = 2$$

$$2x + 4y = 0$$
$$2(2) + 4y = 0$$
$$4 + 4y = 0$$
$$4y = -4$$
$$y = -1$$

$\{(2, -1)\}$

18. $\{(1, -2, 2)\}$

19. If $\left|\frac{x}{2} + 3\right| < 4$,

$$-4 < \frac{x}{2} + 3 < 4$$
$$-4 - 3 < \frac{x}{2} + 3 - 3 < 4 - 3$$
$$-7 < \frac{x}{2} < 1$$
$$-14 < x < 2$$

$\{x \mid -14 < x < 2\}$, or $(-14, 2)$

20.

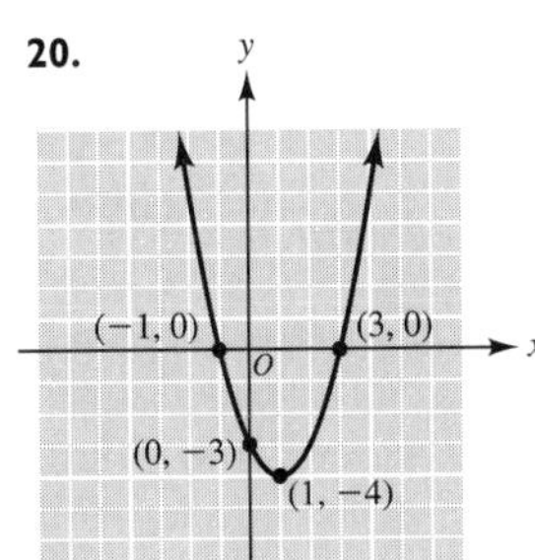

21. (1) $2x - 5y \ge 10$
(2) $5x + y < 5$
Boundary line for (1):
$2x - 5y = 10$
Intercepts: $(0, -2)$ and $(5, 0)$
Boundary line for (2):
$5x + y = 5$
Intercepts: $(0, 5)$ and $(1, 0)$
Origin is in the correct half-plane for (2), but not for (1).

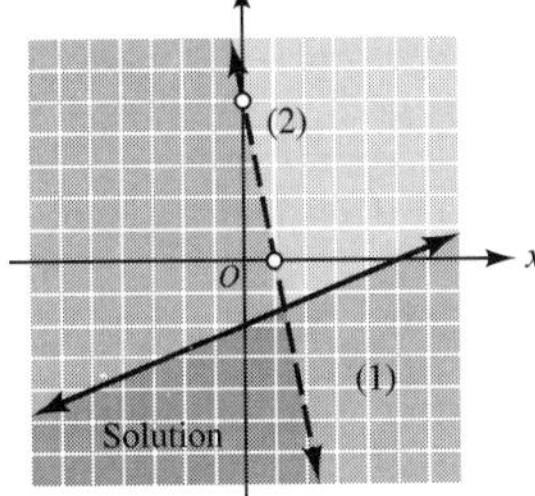

22. a. All real numbers except $\frac{5}{3}$

b. $f(x + h) = \frac{2}{3(x + h) - 5}$, or $\frac{2}{3x + 3h - 5}$

23. $y = f(x) = \dfrac{2}{3x - 5}$

$x = \dfrac{2}{3y - 5}$

$x(3y - 5) = 2$

$3xy - 5x = 2$

$3xy = 5x + 2$

$y = f^{-1}(x) = \dfrac{5x + 2}{3x}$

24. a. $2\sqrt{26}$ **b.** $m = -5$

25. $y - (-4) = -5(x - 3)$

$y + 4 = -5x + 15$

$5x + y - 11 = 0$

26. $d = \frac{9}{2}$

$n = 5$

27. $a_4 = a_1r^3$

$-1 = a_1\left(-\frac{2}{7}\right)^3$

$a_1 = \frac{343}{8}$

$S_5 = \dfrac{a_1(1 - r^5)}{1 - r} = \dfrac{\frac{343}{8}\left[1 - \left(-\frac{2}{7}\right)^5\right]}{1 - \left(-\frac{2}{7}\right)} = \dfrac{\frac{343}{8}\left[1 + \frac{32}{16{,}807}\right]}{\frac{9}{7}}$

$= \frac{7}{9}\left(\frac{343}{8}\right)\left(\frac{16{,}839}{16{,}807}\right) = \frac{1{,}871}{56}$

28. Two answers: 6, 7, and 8 or −3, −2, and −1

29. Let x = number of lb of cheaper coffee
Let y = number of lb of more expensive coffee

$$\begin{array}{rrl}
(1) & x + y &= 30 \\
(2) & 3.50x + 4.25y &= 4.00(30) \\
\hline
-350(1): & -350x - 350y &= -10{,}500 \\
100(2): & 350x + 425y &= 12{,}000 \\
\hline
-350(1) + 100(2): & 75y &= 1{,}500 \\
 & y &= 20 \\
 & x &= 30 - y \\
 & &= 30 - 20 = 10
\end{array}$$

(The check will not be shown.) 10 lb of the cheaper coffee and 20 lb of the more expensive coffee should be used.

30. John's rate is 9 mph, and Roberta's rate is 12 mph.

Exercises B.1 (page 693)

1. 3‸86. Characteristic is 2. **2.** 1

3. 5‸67 Characteristic is 0. **4.** 1

5. 0.5‸16 Characteristic is −1, written as 9 − 10.

6. −2, written as 8 − 10 **7.** 9‸3000000. Characteristic is 7.

8. 5 **9.** 0.00008‸06 Characteristic is −5, written as 5 − 10.

10. −4, written as 6 − 10 **11.** 7‸8000. Characteristic is 4.

12. 3 **13.** 5 (same as exponent of 10) **14.** 4

15. −3 (same as exponent of 10), written as 7 − 10

16. −5, written as 5 − 10

Exercises B.2 (page 695)

1. log 7‸54. ≈ 2.8774 **2.** 2.2695 **3.** log 1‸7. ≈ 1.2304

4. 1.4624 **5.** log 3‸350. ≈ 3.5250 **6.** 3.6637

7. log 7‸000. ≈ 3.8451 **8.** 2.3010

9. log 0.06‸04 ≈ 8.7810 − 10 **10.** 8.2695 − 10

11. log(5.64×10^3) ≈ 3.7513 **12.** 6.3304 − 10

Exercises B.3 (page 696)

1. 1.2809 **2.** 1.6487 **3.** 2.0919 **4.** 2.0149

5. ln 8.3 + ln 10 ≈ 2.1163 + 2.3026 = 4.4189 **6.** 4.1271

7. ln 0.2 − ln 100 ≈ (8.3906 − 10) − 4.6052 = −6.2146

8. −5.1160

Exercises B.4 (page 698)

1.
log 23.30 ≈ 1.3674
log 23.35 ≈ 1.3683
log 23.40 ≈ 1.3692
(5, 10; 9, 18)
18 × 0.5 = 9.0

2. 1.4448

3.
log 3.060 ≈ 0.4857
log 3.062 ≈ 0.4860
log 3.070 ≈ 0.4871
(2, 10; 3, 14)
14 × 0.2 = 2.8 ≈ 3

4. 0.6126

5.
log 0.06‸640 ≈ 8.8222 − 10
log 0.06644 ≈ 8.8224 − 10
log 0.06650 ≈ 8.8228 − 10
(4, 10; 2, 6)
6 × 0.4 = 2.4 ≈ 2

6. 9.7663 − 10

7.
log 1‸50.0 ≈ 2.1761
log 150.7 ≈ 2.1781
log 151.0 ≈ 2.1790
(7, 10; 20, 29)
29 × 0.7 = 20.3 ≈ 20

8. 1.3197

9.

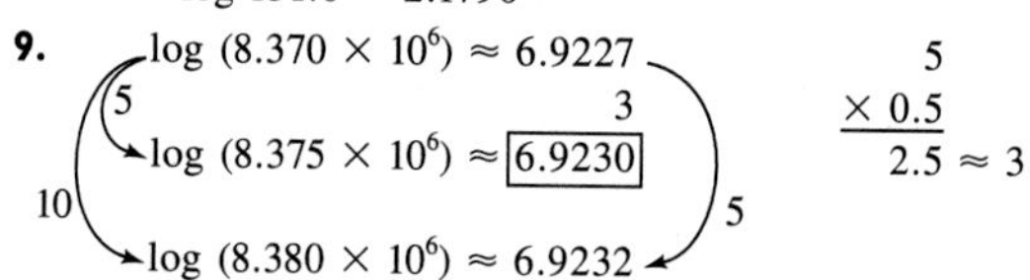

log (8.370×10^6) ≈ 6.9227
log (8.375×10^6) ≈ 6.9230
log (8.380×10^6) ≈ 6.9232
(5, 10; 3, 5)
5 × 0.5 = 2.5 ≈ 3

10. 5.5883 − 10

11. Because we are using a four-place table, we first round off 324.38 to 324.4.

log 3‸24.0 ≈ 2.5105
log 324.4 ≈ 2.5111
log 325.0 ≈ 2.5119
(4, 10; 6, 14)
14 × 0.4 = 5.6 ≈ 6

12. 1.8754

Exercises B.5 (page 701)

1. log N = 3 . 5478
(3 — Characteristic; 5478 — Mantissa)

Locate mantissa 0.5478 in body of Table II and read digits for N:
3‸53
N ≈ 3‸530. = 3,530
(3 — Characteristic)

2. 276

3. log N = 0.9605
(0 — Characteristic; 9605 — Mantissa)
9‸13
Characteristic is 0; N ≈ 9.13

4. 7.67

5. log N = 9.2529 − 10
1‸79
Characteristic is 9 − 10 = −1; N ≈ 0.1‸79 = 0.179

6. 0.0134

7. log N = 3.5051
3‸20
Characteristic is 3; N ≈ 3‸200. = 3,200

8. 430

9. $\log N = 4.0588$

$\log 1_{\wedge}140 \approx 0.0569$

$\frac{19}{38} \times 10 = 5 \to 5$ (10; 19; 38)

$\log \boxed{1_{\wedge}145} \approx 0.0588$

$\log 1_{\wedge}150 \approx 0.0607$

Characteristic is 4; $N \approx 1_{\wedge}1450. = 11{,}450$ (decimal point moved 4)

10. 1,215

11. $\log N = 6.9900 - 10$

$\log 9_{\wedge}770 \approx 0.9899$

$\frac{1}{4} \times 10 = 2.5 \approx 3 \to 3$ (10; 1; 4)

$\log \boxed{9_{\wedge}773} \approx 0.9900$

$\log 9_{\wedge}780 \approx 0.9903$

Characteristic is -4; $N \approx 0.0009_{\wedge}773 = 0.000\,977\,3$ (decimal point moved -4)

12. 0.009 112

13. $\log N = 7.7168 - 10$

$5_{\wedge}21$

Characteristic is $7 - 10 = -3$; $N \approx 0.005_{\wedge}21 = 0.005\,21$ (decimal point moved -3)

14. 0.000 008 73

15. $\log N = 1.7120$

$\log 5_{\wedge}150 \approx 0.7118$

$\frac{2}{8} \times 10 = 2.5 \approx 3 \to 3$ (10; 2; 8)

$\log \boxed{5_{\wedge}153} \approx 0.7120$

$\log 5_{\wedge}160 \approx 0.7126$

Characteristic is 1; $N \approx 5_{\wedge}1.53 = 51.53$ (decimal point moved 1)

16. 97,120

Exercises B.6 (page 704)

1. Let $N = (74.3)(0.618)$

$$\begin{aligned} \log 74.3 &\approx 1.8710 \\ \log 0.618 &\approx 9.7910 - 10 \quad (+) \\ \hline \log N &\approx 11.6620 - 10 \\ &\approx 1.6620 \\ N &\approx 45.9 \end{aligned}$$

2. 4.68

3. Let $N = \dfrac{562}{21.4}$

$$\begin{aligned} \log 562 &\approx 2.7497 \\ \log 21.4 &\approx 1.3304 \quad (-) \\ \hline \log N &\approx 1.4193 \\ N &\approx 26.3 \end{aligned}$$

4. 21.3

5. Let $N = (1.09)^5$

$$\begin{aligned} \log N &= 5 \log 1.09 \\ \log 1.09 &\approx 0.0374 \\ 5 \log 1.09 &\approx 5(0.0374) \\ \log N &\approx 0.1870 \\ N &\approx 1.54 \end{aligned}$$

6. 134

7. Let $N = \sqrt[3]{0.444} = (0.444)^{1/3}$

Then $\log N = \frac{1}{3} \log 0.444$

$$\begin{aligned} \log 0.444 &\approx 9.6474 - 10 \\ &\approx 29.6474 - 30 \\ \tfrac{1}{3} \log 0.444 &\approx \tfrac{1}{3}(29.6474 - 30) \\ \log N &\approx 9.8825 - 10 \\ N &\approx 0.7_{\wedge}630 = 0.7630 \end{aligned}$$

8. 0.973

9. $2.863 + \log 38.46 \approx 2.863 + 1.5850 = 4.448 \approx 4.45$

10. 0.565

11. Let $N = \sqrt[5]{\dfrac{(5.86)(17.4)}{\sqrt{450}}}$

$$\begin{aligned} \log 5.86 &\approx 0.7679 \\ \log 17.4 &\approx 1.2405 \quad (+) \\ \hline & 2.0084 \\ \log 450 &\approx 2.6532 \\ \tfrac{1}{2} \log 450 &\approx 1.3266 \quad (-) \\ \hline & 0.6818 \\ \log N &\approx \tfrac{1}{5}(0.6818) \\ &\approx 0.1364 \\ N &\approx 1.37 \end{aligned}$$

12. 0.403

INDEX

A

B

C

D

E

F

G

H

I

J

K

L

M

Q

R

S

PROPERTIES OF THE REAL NUMBER SYSTEM

Page

The closure property of addition: If a and b represent any real numbers, then their sum, $a + b$, is a real number. 22

The additive identity element is 0. If a represents any real number, then $a + 0 = a$ and $0 + a = a$. 23

Additive inverses: For every real number a, there exists a real number $-a$, called the additive inverse of a, such that $a + (-a) = 0$ and $-a + a = 0$. 23

The closure property of multiplication: If a and b represent any real numbers, then their product, ab, is a real number. 24

The multiplicative identity element is 1. If a represents any real number, then $a \cdot 1 = a$ and $1 \cdot a = a$. 24

Multiplicative inverses: For every *nonzero* real number a, there exists a real number $\frac{1}{a}$, called the multiplicative inverse, or reciprocal, of a, such that $a\left(\frac{1}{a}\right) = 1$ and $\left(\frac{1}{a}\right)a = 1$. 25

The commutative property of addition: If a and b represent any real numbers, then $a + b = b + a$. 30

The associative property of addition: If a, b and c represent any real numbers, then $(a + b) + c = a + (b + c)$. 31

The commutative property of multiplication: If a and b represent any real numbers, then $ab = ba$. 30

The associative property of multiplication: If a, b and c represent any real numbers, then $(ab)c = a(bc)$. 31

Multiplication is distributive over addition: If a, b and c represent any real numbers, then $a(b + c) = ab + ac$. 32

The multiplication property of zero: If a represents any real number, then $a \cdot 0 = 0$ and $0 \cdot a = 0$. 25

Division involving zero: 26

$\frac{0}{a} = 0$ $\quad \frac{a}{0}$ is not possible $\quad \frac{0}{0}$ cannot be determined

Order of operations 43

1. If operations are indicated inside grouping symbols, those operations within grouping symbols should be performed first. If grouping symbols appear within other grouping symbols, remove the *innermost* grouping symbols first.
2. The evaluation then proceeds *in this order*:

   First: Powers and roots are done.

   Next: Multiplication and division are done *in order, from left to right.*

   Last: Addition and subtraction are done *in order, from left to right.*

The zero-factor property If $a, b \in R$ and if $ab = 0$, then $a = 0$ or $b = 0$. 238

PROPERTIES AND FORMULAS

The properties of equality and of inequalities

		Page
$a = a$	The reflexive property	2
If $a = b$, then $b = a$	The symmetric property	2
If $a = b$ and $b = c$, then $a = c$	The transitive property	2
If $a = b$, then b can be substituted for a	The substitution principle	2
If $a < b$ and $b < c$, then $a < c$	The transitive property of *less than*	18
If $a > b$ and $b > c$, then $a > c$	The transitive property of *greater than*	18
If $a = b$, then $a + c = b + c$	The addition property of equality	101
If $a = b$, then $a - c = b - c$	The subtraction property of equality	101
If $a = b$, then $ac = bc$	The multiplication property of equality	101
If $a = b$, and if $c \neq 0$, then $\frac{a}{c} = \frac{b}{c}$	The division property of equality	102
If $a \leq b$, then $a + c \leq b + c$	The addition property of inequalities	111
If $a \leq b$, then $a - c \leq b - c$	The subtraction property of inequalities	111
If $a \leq b$, and if $c > 0$, then $ac \leq bc$ If $a \leq b$, and if $c < 0$, then $ac \geq bc$	The multiplication properties of inequalities	111
If $a \leq b$, and if $c > 0$, then $\frac{a}{c} \leq \frac{b}{c}$ If $a \leq b$, and if $c < 0$, then $\frac{a}{c} \geq \frac{b}{c}$	The division properties of inequalities	111

The properties of exponents

$x^m \cdot x^n = x^{m+n}$	The product rule	62
$(x^m)^n = x^{mn}$	The power rule	62
$(xy)^n = x^n y^n$	The rule for raising a product to a power	62
$\frac{x^m}{x^n} = x^{m-n}$, if $x \neq 0$	The quotient rule	62
$\left(\frac{x}{y}\right)^n = \frac{x^n}{y^n}$, if $y \neq 0$	The rule for raising a quotient to a power	62
$x^0 = 1$, if $x \neq 0$	The zero exponent rule	65
$x^{-n} = \frac{1}{x^n}$, if $x \neq 0$	The negative exponent rule, form 1	66
$\frac{1}{x^{-n}} = x^n$, if $x \neq 0$	The negative exponent rule, form 2	67
$\left(\frac{x}{y}\right)^{-n} = \left(\frac{y}{x}\right)^n$, if $x \neq 0$ and $y \neq 0$	The negative exponent rule, form 3	67
$\left(\frac{x^a y^b}{z^c}\right)^n = \frac{x^{an} y^{bn}}{z^{cn}}$	The general property of exponents	71

Properties used in solving absolute value equations and inequalities

If $\|X\| = k$, then $X = k$ or $X = -k$	Property 1 for absolute values	123
If $\|X\| \leq k$, then $-k \leq X \leq k$	Property 2 for absolute values	125
If $\|X\| \geq k$, then $X \geq k$ or $X \leq -k$	Property 3 for absolute values	128